# BIOMATERIALS FOR 3D TUMOR MODELING

# BIOMATERIALS FOR 3D TUMOR MODELING

*Edited by*


SUBHAS C. KUNDU

*3B's Research Group, I3Bs – Research Institute on Biomaterials, Biodegradables and Biomimetics, University of Minho, Headquarters of the European Institute of Excellence on Tissue Engineering and Regenerative Medicine, Guimarães, Portugal*

*ICVS/3B's – PT Government Associate Laboratory, Braga/Guimarães, Portugal*

RUI L. REIS

*3B's Research Group, I3Bs – Research Institute on Biomaterials, Biodegradables and Biomimetics, University of Minho, Headquarters of the European Institute of Excellence on Tissue Engineering and Regenerative Medicine, Guimarães, Portugal*

*ICVS/3B's – PT Government Associate Laboratory, Braga/Guimarães, Portugal*



ELSEVIER

Elsevier
Radarweg 29, PO Box 211, 1000 AE Amsterdam, Netherlands
The Boulevard, Langford Lane, Kidlington, Oxford OX5 1GB, United Kingdom
50 Hampshire Street, 5th Floor, Cambridge, MA 02139, United States



**Notices**

Knowledge and best practice in this field are constantly changing. As new research and experience broaden our understanding, changes in research methods, professional practices, or medical treatment may become necessary.

Practitioners and researchers must always rely on their own experience and knowledge in evaluating and using any information, methods, compounds, or experiments described herein. In using such information or methods they should be mindful of their own safety and the safety of others, including parties for whom they have a professional responsibility.

To the fullest extent of the law, neither the Publisher nor the authors, contributors, or editors, assume any liability for any injury and/or damage to persons or property as a matter of products liability, negligence or otherwise, or from any use or operation of any methods, products, instructions, or ideas contained in the material herein.

**British Library Cataloguing-in-Publication Data**
A catalogue record for this book is available from the British Library

**Library of Congress Cataloging-in-Publication Data**
A catalog record for this book is available from the Library of Congress

ISBN: 978-0-12-818128-7

For Information on all Elsevier publications
visit our website at https://www.elsevier.com/books-and-journals

*Publisher:* Matthew Deans
*Acquisitions Editor:* Sabrina Webber
*Editorial Project Manager:* Joshua Mearns
*Production Project Manager:* Anitha Sivaraj
*Cover Designer:* Christian J. Bilbow

Typeset by MPS Limited, Chennai, India

# Contents

# List of Contributors

**Enrico Almici** Nanobioengineering Group, Institute for Bioengineering of Catalonia (IBEC), The Barcelona Institute of Science and Technology (BIST), Barcelona, Spain; Department of Electronics and Biomedical Engineering, University of Barcelona, Barcelona, Spain; Networking Biomedical Investigation Center for Bioengineering, Biomaterials and Nanomedicine (CIBER-BBN), Madrid, Spain

**M. Alonso** Nano-Oncology and Translational Therapeutics Unit, Health Research Institute of Santiago de Compostela (IDIS), SERGAS, Santiago de Compostela, Spain; Cancer Research Network (CIBERONC), Spain

**Sara Amorim** 3B's Research Group, I3Bs – Research Institute on Biomaterials, Biodegradables and Biomimetics, University of Minho, Headquarters of the European Institute of Excellence on Tissue Engineering and Regenerative Medicine, Guimarães, Portugal; ICVS/3B's - PT Government Associate Laboratory, Braga/Guimarães, Portugal

**Esteban A. Astudillo-Ortiz** 3B's Research Group, I3Bs – Research Institute on Biomaterials, Biodegradables and Biomimetics, University of Minho, Guimarães, Portugal; ICVS/3B's–PT Government Associate Laboratory, Braga/Guimarães, Portugal; GIRO Research Group in Oral Rehabilitation, University of Cuenca, Cuenca, Ecuado

**Pedro S. Babo** 3B's Research Group, I3Bs – Research Institute on Biomaterials, Biodegradables and Biomimetics, University of Minho, Headquarters of the European Institute of Excellence on Tissue Engineering and Regenerative Medicine, Guimarães, Portugal; ICVS/3B's–PT Government Associate Laboratory, Braga/Guimarães, Portugal

**Fátima Baltazar** Life and Health Science Research Institute (ICVS), School of Medicine, University of Minho, Braga, Portugal; ICVS/3B's–PT Government Associate Laboratory, Braga/Guimarães, Portugal

**Ana I. Barbosa** 3B's Research Group, I3Bs – Research Institute on Biomaterials, Biodegradables and Biomimetics, University of Minho, Headquarters of the European Institute of Excellence on Tissue Engineering and Regenerative Medicine, Guimarães, Portugal; ICVS/3B's – PT Government Associate Laboratory, Braga/Guimarães, Portugal

**Virginia Brancato** 3B's Research Group, I3Bs – Research Institute on Biomaterials, Biodegradables and Biomimetics, University of Minho, Headquarters of the European Institute of Excellence on Tissue Engineering and Regenerative Medicine, Guimarães, Portugal; ICVS/3B's–PT Government Associate Laboratory, Braga/Guimarães, Portugal

**David Caballero** 3B's Research Group, I3Bs – Research Institute on Biomaterials, Biodegradables and Biomimetics, University of Minho, Headquarters of the European Institute of Excellence on Tissue Engineering and Regenerative Medicine, Guimarães, Portugal; ICVS/3B's – PT Government Associate Laboratory, Braga/Guimarães, Portugal

**Paolo Caccavale** University of Basilicata, College of Engineering, Modeling and Prototyping Laboratory—ModProLab, Potenza, Italy

**Bo Cai** Research Center for Tissue Engineering and Regenerative Medicine, Union Hospital, Tongji Medical College, Huazhong University of Science and Technology, Wuhan, P.R. China

**S. Calabuig-Fariñas** Molecular Oncology Laboratory, General University Hospital of Valencia Research Foundation, Valencia, Spain; Cancer Research Network (CIBERONC), Spain; Department of Pathology, Universitat de València, Valencia, Spain

**Ana C. Carvalho** 3B's Research Group, I3Bs – Research Institute on Biomaterials, Biodegradables and Biomimetics, University of Minho, Headquarters of the European Institute of Excellence on Tissue Engineering and Regenerative Medicine, Guimarães, Portugal; ICVS/3B's, PT Government Associate Laboratory, Braga/Guimarães, Portugal

**Sónia Pires Celeiro** Life and Health Science Research Institute (ICVS), School of Medicine, University of Minho, Braga, Portugal; ICVS/3B's–PT Government Associate Laboratory, Braga/Guimarães, Portugal

**Ileana L. Co** Department of Chemical Engineering and Applied Chemistry, Institute for Biomaterials and Biomedical Engineering, University of Toronto, Toronto, ON, Canada

**Vitor M. Correlo** 3B's Research Group, I3Bs – Research Institute on Biomaterials, Biodegradables and Biomimetics, University of Minho, Headquarters of the European Institute of Excellence on Tissue Engineering and Regenerative Medicine, Guimarães, Portugal; ICVS/3B's – PT Government Associate Laboratory, Braga/Guimarães, Portugal

**Lara S. Costard** Tissue Engineering Research Group (TERG), Department of Anatomy and Regenerative Medicine, Royal College of Surgeons in Ireland (RCSI), Dublin, Ireland

**Caroline M. Curtin** Tissue Engineering Research Group (TERG), Department of Anatomy and Regenerative Medicine, Royal College of Surgeons in Ireland (RCSI), Dublin, Ireland; Trinity Centre of Bioengineering (TCBE), Trinity College Dublin (TCD), Dublin, Ireland; Advanced Materials and Bioengineering Research Centre (AMBER), RCSI and TCD, Dublin, Ireland.

**Serena Danti** Department of Civil and Industrial Engineering, University of Pisa, Pisa, Italy

**Subhayan Das** School of Medical Science and Technology, Indian Institute of Technology Kharagpur, Kharagpur, India

**Rupesh Dash** Institute of Life Sciences, Bhubaneswar, India

**Maria Valeria De Bonis** University of Basilicata, College of Engineering, Modeling and Prototyping Laboratory—ModProLab, Potenza, Italy

**Maria de la Fuente** Nano-Oncology and Translational Therapeutics Unit, Health Research Institute of Santiago de Compostela (IDIS), SERGAS, Santiago de Compostela, Spain; Cancer Research Network (CIBERONC), Spain

**Nandini Dhiman** Department of Biomedical Engineering, Indian Institute of Technology Hyderabad, Kandi, India; Department of Chemistry and Biotechnology, Faculty of Science, Engineering, and Technology, Swinburne University of Technology, Hawthorn, VIC, Australia

**Laura di Blasio** Department of Oncology, University of Torino, Turin, Italy; Candiolo Cancer Institute FPO-IRCCS, Candiolo, Italy

**Eleonora Dondossola** David H. Koch Center for Applied Research of Genitourinary Cancers, The University of Texas MD Anderson Cancer Center, Houston, TX, United States

**Daniela F. Duarte Campos** Department of Materials Science & Engineering, Stanford University, CA, United States

**Mohammad Eltohamy** Glass Research Department, National Research Centre, Giza, Egypt

**Vítor M. Gaspar** Department of Chemistry, CICECO—Aveiro Institute of Materials, University of Aveiro, Campus Universitário de Santiago, Aveiro, Portugal

**Roya Ghods** Oncopathology Research Center, Iran University of Medical Sciences, Tehran, Iran; Department of Molecular Medicine, Faculty of Advanced Technologies in Medicine, Iran University of Medicine Sciences, Tehran, Iran

**Mazaher Gholipourmalekabadi** Department of Molecular Medicine, Faculty of Advanced Technologies in Medicine, Iran University of Medicine Sciences, Tehran, Iran; Cellular and Molecular Research Center, Iran University of Medical Sciences, Tehran, Iran; Department of Tissue Engineering & Regenerative Medicine, Faculty of Advanced Technologies in Medicine, Iran University of Medical Sciences, Tehran, Iran

**Manuela E. Gomes** 3B's Research Group, I3Bs – Research Institute on Biomaterials, Biodegradables and Biomimetics, University of Minho, Headquarters of the European Institute of Excellence on Tissue Engineering and Regenerative Medicine, Guimarães, Portugal; ICVS/3B's–PT Government Associate Laboratory, Braga/Guimarães, Portugal

**Michael M. Gottesman** Laboratory of Cell Biology (LCB), Center for Cancer Research, National Cancer Institute (NCI), National Institutes of Health, Bethesda, MD, United States

**M.A. Grimaudo** Nano-Oncology and Translational Therapeutics Unit, Health Research Institute of Santiago de Compostela (IDIS), SERGAS, Santiago de Compostela, Spain

**David Grosh** Trans-NIH Shared Resource on Biomedical Engineering and Physical Science (BEPS), National Institute of Biomedical Imaging and Bioengineering (NIBIB), National Institutes of Health, Bethesda, MD, United States

**Priyanka Gupta** Bioprocess and Biochemical Engineering Group (BioProChem), Department of Chemical and Process Engineering, University of Surrey, Surrey, United Kingdom

**A. Herreros-Pomares** Molecular Oncology Laboratory, General University Hospital of Valencia Research Foundation, Valencia, Spain; Cancer Research Network (CIBERONC), Spain

**E. Jantus-Lewintre** Molecular Oncology Laboratory, General University Hospital of Valencia Research Foundation, Valencia, Spain; Cancer Research Network (CIBERONC), Spain; Department of Biotechnology, Universitat Politècnica de València, Valencia, Spain

**Bikash Chandra Jena** School of Medical Science and Technology, Indian Institute of Technology Kharagpur, Kharagpur, India

**Brad A. Krajina** Department of Materials Science & Engineering, Stanford University, CA, United States

**Banani Kundu** 3B's Research Group, I3Bs—Research Institute on Biomaterials, Biodegradables and Biomimetics, University of Minho, Headquarters of the European Institute of Excellence on Tissue Engineering and Regenerative Medicine, Guimarães, Portugal; ICVS/3B's—PT Government Associate Laboratory, Braga/Guimarães, Portugal

**Moumita Kundu** School of Medical Science and Technology, Indian Institute of Technology Kharagpur, Kharagpur, India

**Subhas C. Kundu** 3B's Research Group, I3Bs—Research Institute on Biomaterials, Biodegradables and Biomimetics, University of Minho, Headquarters of the European Institute of Excellence on Tissue Engineering and Regenerative Medicine, Guimarães, Portugal; ICVS/3B's—PT Government Associate Laboratory, Braga/Guimarães, Portugal

**Natalie Landon-Brace** Department of Chemical Engineering and Applied Chemistry, Institute for Biomaterials and Biomedical Engineering, University of Toronto, Toronto, ON, Canada

**Simon Latour** Department of Chemical Engineering and Applied Chemistry, Institute for Biomaterials and Biomedical Engineering, University of Toronto, Toronto, ON, Canada

**Nancy T. Li** Department of Chemical Engineering and Applied Chemistry, Institute for Biomaterials and Biomedical Engineering, University of Toronto, Toronto, ON, Canada

**Maria Angélica Luque-González** 3B's Research Group, I3Bs – Research Institute on Biomaterials, Biodegradables and Biomimetics, University of Minho, Headquarters of the European Institute of Excellence on Tissue Engineering and Regenerative Medicine, Guimarães, Portugal; ICVS/3B's – PT Government Associate Laboratory, Braga/Guimarães, Portugal

**Qiying Lv** Research Center for Tissue Engineering and Regenerative Medicine, Union Hospital, Tongji Medical College, Huazhong University of Science and Technology, Wuhan, P.R. China

**Tanushree Mahapatra** Institute of Life Sciences, Bhubaneswar, India

**Mahitosh Mandal** School of Medical Science and Technology, Indian Institute of Technology Kharagpur, Kharagpur, India

**João F. Mano** Department of Chemistry, CICECO—Aveiro Institute of Materials, University of Aveiro, Campus Universitário de Santiago, Aveiro, Portugal

**Alexandra P. Marques** 3B's Research Group, I3Bs—Research Institute on Biomaterials, Biodegradables and Biomimetics, University of Minho, Headquarters of the European Institute of Excellence on Tissue Engineering and Regenerative Medicine, Braga/Guimarães, Portugal; ICVS/3B's—PT Government Associate Laboratory, Braga/Guimarães, Portugal

**Albino Martins** 3B's Research Group, I3Bs – Research Institute on Biomaterials, Biodegradables and Biomimetics, University of Minho, Headquarters of the European Institute of Excellence on Tissue Engineering and Regenerative Medicine, Guimarães, Portugal; ICVS/3B's, PT Government Associate Laboratory, Braga/Guimarães, Portugal

**Alison P. McGuigan** Department of Chemical Engineering and Applied Chemistry, Institute for Biomaterials and Biomedical Engineering, University of Toronto, Toronto, ON, Canada

**Sibasish Mohanty** Institute of Life Sciences, Bhubaneswar, India; Regional Center for Biotechnology, Faridabad, India

**Pallavi Mohapatra** Institute of Life Sciences, Bhubaneswar, India; Regional Center for Biotechnology, Faridabad, India

**Maria V. Monteiro** Department of Chemistry, CICECO—Aveiro Institute of Materials, University of Aveiro, Campus Universitário de Santiago, Aveiro, Portugal

**Joan Montero** Nanobioengineering Group, Institute for Bioengineering of Catalonia (IBEC), The Barcelona Institute of Science and Technology (BIST), Barcelona, Spain

**Nicole Y. Morgan** Trans-NIH Shared Resource on Biomedical Engineering and Physical Science (BEPS), National Institute of Biomedical Imaging and Bioengineering (NIBIB), National Institutes of Health, Bethesda, MD, United States

**Antonella Motta** BIOtech Research Center, Department of Industrial Engineering and European Institute of Excellence on Tissue Engineering and Regenerative Medicine, University of Trento, Trento, Italy

**Nuno N. Neves** 3B's Research Group, I3Bs – Research Institute on Biomaterials, Biodegradables and Biomimetics, University of Minho, Headquarters of the European Institute of Excellence on Tissue Engineering and Regenerative Medicine, Guimarães, Portugal; ICVS/3B's, PT Government Associate Laboratory, Braga/Guimarães, Portugal

**Catarina Oliveira** 3B's Research Group, I3Bs – Research Institute on Biomaterials, Biodegradables and Biomimetics, University of Minho, Headquarters of the European Institute of Excellence on Tissue Engineering and Regenerative Medicine, Guimarães, Portugal; ICVS/3B's, PT Government Associate Laboratory, Braga/Guimarães, Portugal

**Joaquim M. Oliveira** 3B's Research Group, I3Bs—Research Institute on Biomaterials, Biodegradables and Biomimetics, University of Minho, Headquarters of the European Institute of Excellence on Tissue Engineering and Regenerative Medicine, Guimarães, Portugal; ICVS/3B's—PT Government Associated Laboratory, Braga/Guimarães, Portugal

**Claudia Paindelli** David H. Koch Center for Applied Research of Genitourinary Cancers, The University of Texas MD Anderson Cancer Center, Houston, TX, United States; Department of Cell Biology, Radboud Institute for Molecular Life Sciences, Radboud University Medical Center, Nijmegen, The Netherlands

**Lara Pierantoni** 3B's Research Group, I3Bs—Research Institute on Biomaterials, Biodegradables and Biomimetics, University of Minho, Headquarters of the European Institute of Excellence on Tissue Engineering and Regenerative Medicine, Guimarães, Portugal; ICVS/3B's—PT Government Associated Laboratory, Braga/Guimarães, Portugal

**Ricardo A. Pires** 3B's Research Group, I3Bs – Research Institute on Biomaterials, Biodegradables and Biomimetics, University of Minho, Headquarters of the European Institute of Excellence on Tissue Engineering and Regenerative Medicine, Guimarães, Portugal; ICVS/3B's - PT Government Associate Laboratory, Braga/Guimarães, Portugal

**Thomas J. Pohida** Signal Processing and Instrumentation Section (SPIS), Computational Bioscience and Engineering Laboratory, Center for Information Technology, National Institutes of Health, Bethesda, MD, United States

**Luca Primo** Department of Oncology, University of Torino, Turin, Italy; Candiolo Cancer Institute FPO-IRCCS, Candiolo, Italy

**Manashi Priyadarshini** Institute of Life Sciences, Bhubaneswar, India; KIIT School of Biotechnology, KIIT University, Bhubaneswar, India

**Harumi Ramanayake** Tissue Engineering Research Group (TERG), Department of Anatomy and Regenerative Medicine, Royal College of Surgeons in Ireland (RCSI), Dublin, Ireland

**Subha Narayan Rath** Department of Biomedical Engineering, Indian Institute of Technology Hyderabad, Kandi, India

**Rita Rebelo** 3B's Research Group, I3Bs – Research Institute on Biomaterials, Biodegradables and Biomimetics, University of Minho, Headquarters of the European Institute of Excellence on Tissue Engineering and Regenerative Medicine, Guimarães, Portugal; ICVS/3B's – PT Government Associate Laboratory, Braga/Guimarães, Portugal

**Rui L. Reis** 3B's Research Group, I3Bs—Research Institute on Biomaterials, Biodegradables and Biomimetics, University of Minho, Headquarters of the European Institute of Excellence on Tissue Engineering and Regenerative Medicine, Guimarães, Portugal; ICVS/3B's—PT Government Associated Laboratory, Braga/Guimarães, Portugal

**J.P. Ribeiro** 3B's Research Group, I3Bs – Research Institute on Biomaterials, Biodegradables and Biomimetics, University of Minho, Headquarters of the European Institute of Excellence on Tissue Engineering and Regenerative Medicine, Guimarães, Portugal; ICVS/3B's–PT Government Associate Laboratory, Braga/Guimarães, Portugal

**Claudio Ricci** National Interuniversity Consortium of Materials Science and Technology (INSTM), Florence, Italy; Department of Civil and Industrial Engineering, University of Pisa, Pisa, Italy

**Robert W. Robey** Laboratory of Cell Biology (LCB), Center for Cancer Research, National Cancer Institute (NCI), National Institutes of Health, Bethesda, MD, United States

**Gianpaolo Ruocco** University of Basilicata, College of Engineering, Modeling and Prototyping Laboratory—ModProLab, Potenza, Italy

**Josep Samitier** Nanobioengineering Group, Institute for Bioengineering of Catalonia (IBEC), The Barcelona Institute of Science and Technology (BIST), Barcelona, Spain; Department of Electronics and Biomedical Engineering, University of Barcelona, Barcelona, Spain; Networking Biomedical Investigation Center for Bioengineering, Biomaterials and Nanomedicine (CIBER-BBN), Madrid, Spain

**Roghayeh Sheervalilou** Cellular and Molecular Research Center, Resistant Tuberculosis Institute, Zahedan University of Medical Sciences, Zahedan, Iran

**Tiago H. Silva** 3B's Research Group, I3Bs – Research Institute on Biomaterials, Biodegradables and Biomimetics, University of Minho, Headquarters of the European Institute of Excellence on Tissue Engineering and Regenerative Medicine, Guimarães, Portugal; ICVS/3B's, PT Government Associate Laboratory, Braga/Guimarães, Portugal

**Joana Silva-Correia** 3B's Research Group, I3Bs—Research Institute on Biomaterials, Biodegradables and Biomimetics of University of Minho, Headquarters of the European Institute of Excellence on Tissue Engineering and Regenerative Medicine, Guimarães, Portugal; ICVS/3B's—PT Government Associated Laboratory, Braga/Guimarães, Portugal

**Ana I. Soares** 3B's Research Group, I3Bs—Research Institute on Biomaterials, Biodegradables and Biomimetics, University of Minho, Headquarters of the European Institute of Excellence on Tissue Engineering and Regenerative Medicine, Braga/Guimarães, Portugal; ICVS/3B's—PT Government Associate Laboratory, Braga/Guimarães, Portugal

**Marianela Vara-Messler** Department of Oncology, University of Torino, Turin, Italy; Candiolo Cancer Institute FPO-IRCCS, Candiolo, Italy

**Eirini Velliou** Bioprocess and Biochemical Engineering Group (BioProChem), Department of Chemical and Process Engineering, University of Surrey, Surrey, United Kingdom

**Marta Viana-Pereira** Life and Health Science Research Institute (ICVS), School of Medicine, University of Minho, Braga, Portugal; ICVS/3B's—PT Government Associate Laboratory, Braga/Guimarães, Portugal

**Lin Wang** Research Center for Tissue Engineering and Regenerative Medicine, Union Hospital, Tongji Medical College, Huazhong University of Science and Technology, Wuhan, P.R. China; Department of Clinical Laboratory, Union Hospital, Tongji Medical College, Huazhong University of Science and Technology, Wuhan, P.R. China

**Zheng Wang** Research Center for Tissue Engineering and Regenerative Medicine, Union Hospital, Tongji Medical College, Huazhong University of Science and Technology, Wuhan, P.R. China; Department of Gastrointestinal Surgery, Union Hospital, Tongji Medical College, Huazhong University of Science and Technology, Wuhan, P.R. China.

**William J. Wulftange** Trans-NIH Shared Resource on Biomedical Engineering and Physical Science (BEPS), National Institute of Biomedical Imaging and Bioengineering (NIBIB), National Institutes of Health, Bethesda, MD, United States; Laboratory of Cell Biology (LCB), Center for Cancer Research, National Cancer Institute (NCI), National Institutes of Health, Bethesda, MD, United States

**Amir Zarebkohan** Department of Medical Nanotechnology, Faculty of Advanced Medical Sciences, Tabriz University of Medical Sciences, Tabriz, Iran; Drug Applied Research Center, Tabriz University of Medical Sciences, Tabriz, Iran

# Preface

Traditionally, in vitro approaches employed to model tumor progression involve the use of tissue culture flasks and a single cell population. This oversimplified scenario is in contrast with the dynamic tumor microenvironment that is observed in vivo, which is three-dimensional (3D) and contains multiple cell types. During the recent years, a new generation of predictive tumor models has been emerging. This new paradigm in tumor models combines the advantages of 3D biomaterials with the most recent advancements in the fields of tissue engineering, nanotechnology, cell biology, biochemistry, 3D bioprinting, and computer simulation. As a result, more realistic and predictive platform(s) have been created, defining the roadmap for their future use as preclinical drug/therapy screening platforms in cancer research.

The editors feel very proud to have collected a diverse variety of relevant chapters from recognized international leaders, which include their personal and forward-looking vision in the fields. The chapters report further on the latest trends of biomaterials for 3D cancer modeling, and in particular, about their uses as physiologically relevant platforms for drug discovery and screening as well as for mechanistic investigations. The editors believe that this book will be an additional support and stimuli for designing new experiments in different research areas, such as biomaterials, tissue engineering, regenerative medicine, biomedical devices, pharmacology, and preclinical experimentations. Furthermore, this book will be in general helpful to the material science, biology, and medical scientific communities working in the field of cancer-related research. Overall, this new paradigm in 3D in vitro tumor modeling will univocally improve the field of cancer diagnosis and treatment.

Finally, we are grateful to all the authors and co-authors for their contributions in this book. We are also thankful to the entire Elsevier team for their continuous support and help during the edition of this book. We sincerely acknowledge the financial support from the project Forefront Research in 3D disease cancer models as in vitro screening technologies (FoReCaST), which is supported by the European Union Framework Program for Research and Innovation Horizon 2020 under grant agreement no. 668983. This ERA-Chairs widening spreading of excellence project aims to develop a new generation of 3D in vitro tumor models for drug screening applications and has, therefore, motivated us to edit this unique book. We also sincerely thank all the FoReCaST collaborators and team-members, namely Drs. Joaquim M. Oliveira, Vitor M. Correlo, David Caballero, Banani Kundu, Virginia Brancato, and Ricardo Pires for their encouragement during the entire course of action for editing this book.

**The Editors,**

*Subhas C. Kundu*

*Rui L. Reis*

August 1st
3B's Research Group, I3Bs-Research Institute for Biomaterials, Biodegradables, and Biomimetics, University of Minho, Guimarães, Portugal

# SECTION I

# Engineering biomaterials for 3D cancer modelling

CHAPTER

# 1

# Trends in biomaterials for three-dimensional cancer modeling

*David Caballero*[1,2], *Rui L. Reis*[1,2] *and Subhas C. Kundu*[1,2]

[1]3B's Research Group, I3Bs – Research Institute on Biomaterials, Biodegradables and Biomimetics, University of Minho, Headquarters of the European Institute of Excellence on Tissue Engineering and Regenerative Medicine, Guimarães, Portugal [2]ICVS/3B's – PT Government Associate Laboratory, Braga/Guimarães, Portugal

## Abbreviations

| | |
|---|---|
| *2D* | Two-dimensions |
| *3D* | Three-dimensions |
| *4D* | Four-dimensions |
| *ADMET* | Adsorption, distribution, metabolism, excretion, and toxicity |
| *CAFs* | Cancer-associated fibroblasts |
| *CDMs* | Cell-derived matrices |
| *CTCs* | Circulating tumor cells |
| *ECM* | Extracellular matrix |
| *EMEA* | European Medicines Agency |
| *EMT* | Epithelial-to-mesenchymal transition |
| *EPR* | Enhanced permeability and retention |
| *FDA* | Federal drug agency |
| *HA* | Hyaluronic acid |
| *HUVEC* | Human umbilical vein endothelial cells |
| *MMPs* | Matrix metalloproteases |
| *PDMS* | Polydimethylsiloxane |
| *PEG* | Poly(ethylene glycol) |
| *TAMs* | Tumor-associated macrophages |
| *TME* | Tumor microenvironment |
| *ToC* | Tumor-on-a-chip |
| *VEGF* | Vascular endothelial growth factor |

*Biomaterials for 3D Tumor Modeling*
DOI: https://doi.org/10.1016/B978-0-12-818128-7.00001-0

## 1.1 A historical introduction

### 1.1.1 In vitro and in vivo models: an overview

During the last decade, the type of biological assays that are used for extracting information about the efficiency of drugs (including cancer-related compounds) have dramatically changed. The reason is that a large amount of these drugs fail when they are tested in pre-clinical assays. This is because most pre-clinical drug evaluations rely on simplified in vitro assays based on flat two-dimensional (2D) surfaces. This type of assay poorly correlates with the human disease state. Therein, the cells display artificial phenotypes and perturbed gene expressions. In general, the drugs respond differently than in vivo. Ex vivo (e.g., biopsies) and in vivo (e.g., animal) models are also employed for drug evaluation. In cancer research, these models display certain advantages over 2D surfaces, such as a greater biological complexity. This makes the drugs to produce native-like responses. However, ex vivo models typically lack perfusion and are not representative of the heterogeneity of the tumor. In contrast, in vivo (animal) models are highly dynamic systems, but they are very costly, lack the human immune system, and are ethically controversial. In addition, regardless of the type of animal model, it is extremely difficult to investigate cellular and physiological interactions on this type of models. More advanced tumor models are patient-derived xenografts, where a surgically resected tumor sample of a patient is engrafted into an immunodeficient mice. However, these models are extremely expensive and time-consuming, they are associated with ethical concerns, and individual parameters cannot be isolated [1].

### 1.1.2 A paradigm shift

To resolve the abovementioned issues, a paradigm shift has occurred since the late 1990s about how the cells are studied [2,3]. Indeed, there is a huge difference between cells cultured on flat surfaces and/or on three-dimensional (3D) environments. Indeed, seminal works using 3D in vitro models have demonstrated the important differences in the phenotypes and activities between cells grown in 2D surfaces and 3D cultures. The latter recapitulate the important interactions between—healthy and malignant—cells and the surrounding 3D environment [4,5]. As a result, the cancer research community, including academic researchers, pharma/biotech industries, and clinicians, have started to move toward the third dimension. In this regard, 2D assays in the form of traditional tissue culture flasks or Petri dishes, are being replaced by more sophisticated 3D tissue-engineered cell culture microenvironments. Therein, the cells display several similarities with the native scenario, and therefore, are considered as physiologically relevant environments that bridge the gap between obsolete flat materials and the native scenario. In particular, the early events of cancer progression can closely be reproduced in 3D culture platforms, where the cells display phenotypes, morphodynamics, and gene expression patterns much closer to the in vivo physiological microenvironment. In addition, as previously mentioned, this is of upmost importance during the development of drugs, where the cells respond to compounds as they do in vivo.

Finally, the combination of cutting-edge nanotechnologies with advances in materials science and tissue engineering tools, have created a powerful toolbox in the field of biomedical and health sciences capable to create a new generation of 3D biomaterials with unprecedented possibilities in cancer research. These biomaterials are capable (1) to copycat the tremendous complexity (physical, biological, structural, biochemical, and rheological) of the natural tumor microenvironment (TME); (2) to investigate in a physiologically relevant environment the pathogenesis of tumors; and (3) to screen the efficiency and pharmacokinetics/pharmacodynamics of drugs in a reliable manner. All these advanced capabilities make this new generation of 3D biomaterials ideal candidates for 3D tumor modeling.

### 1.1.3 Three-dimensional biomaterials for cancer modeling

The pioneering work by the Bissell lab in this field changed completely the vision and paradigm of how cancer cells need to be cultured in vitro [6]. The third dimension reproduces the physiological behavior of cells in the human body. As an example, the inhibition of β1 integrin in human breast cancer cells seeded in a 3D model reversed the malignant phenotype of cells resulting into normal morphologies and functional phenotypes; cells lost their abnormal shapes and patterns of growth. This work shows clearly that the context (i.e., environment) where cells are cultured influences the phenotype and response of cells, and in particular, cancerous cells. Similarly, the way how tumors are cultured in vitro also influences the migration capacity of cells and their motility mechanism [7]. In this regard, new migration modes, which are not observed in 2D, are revealed when tumor cells are seeded in 3D. In this case, the cells switch to an amoeba-like form of migration. This new type of migration mechanism enhances the capacity of cancer cells to squeeze to physical constrictions of the extracellular matrix (ECM) and invade the tumor stroma.

Cancer researchers have focused in developing biomimetic 3D platforms with controlled composition, bioactive functional groups, architecture, mechanical properties, and degradation rates, capable to reproduce the mechanochemical and hydrodynamic properties of the TME, overcoming the limitations of standard biomaterials. These biomaterials may also display "smart" characteristics and respond to the presence of an external stimulus, such as pH variations or temperature, or to self-assemble following specific configurations [8].

Typical biomaterials for 3D cancer modeling include the use of natural-based hydrogels or scaffolds that are rich in proteins of the ECM, such as collagen, gelatin, Matrigel™, or blends of them. These biomaterials display interesting mechanochemical properties, which can be modulated to reproduce those of the native scenario. This is of upmost importance because solid evidences have shown that both mechanical and (bio)chemical cues regulate multiple cellular functions and fate in a process known as *mechanotransduction*. However, these biomaterials also display certain drawbacks, such as the batch-to-batch variation that impedes the direct comparison of the results obtained by different laboratories. Other type of biomaterials with more controlled composition and more reproducible for cancer modeling include silk fibroin, gellan gum hydrogels/scaffolds, or blends of them [9–12]. Finally, more advanced biomaterials also include 3D topographical substrates, 3D bioprinted scaffolds, or microfluidic platforms.

### 1.1.4 From the lab to the clinic

Three-dimensional biomaterial-based cancer models have attract the interest of pharmaceutical and biotechnological companies who have realized their potential in developing better drugs in a cheaper way. Plenty of companies are selling these types of Federal Drug Agency (FDA)- and European Medicines Agency (EMEA)-approved 3D biomaterial platforms for cancer research applications. However, and despite the generous funds that have been provided by several national funding agencies to boost their development and their initial expectations, most of the current research is still based on flat surfaces. This may also explain the decrease in the development of innovative anticancer drugs and therapies during the last decade. New approaches and characteristics are indeed needed to promote their adoption by the market. In particular, the integration of materials science with nanotechnology, tissue engineering, and biosensor technology is expected to boost and solidify this emerging field of research.

Overall, 3D biomaterials provide an improved in vitro model for cancer research, which reproduce in a better manner, what happens in the living tissues of the human body. These advanced models provide multiple advantages when compared with flat, labor-intensive, and expensive animal models. Therefore, they may bridge the gap between both type of models, even though 3D biomaterials and related technologies are not expected to completely replace animal experimentation. However, they will definitely contribute to improve (1) our knowledge in tumor biology by providing new insights about tumor pathogenesis; (2) the efficacy of drug development and testing, and (3) to reduce animal experimentation. In this chapter, we describe the current trends in the field of 3D biomaterials for cancer modeling, their main applications, limitations, and discuss about their present and future challenges.

## 1.2 The three-dimensional tumor microenvironment

The TME is of extreme biochemical, structural (mechanical and architectural), cellular, and rheological complexity. As mentioned above, it is recognized as a key player in the mechanochemical mechanism of tumor progression and metastasis. During this process, there is a synergistic interaction between the tumor cells and the surrounding 3D stroma that guides tumor cell invasion. This malignant interaction can be reproduced by engineering in vitro tumor models made of biomimetic and biodegradable 3D biomaterials. In the following, we briefly discuss about this synergistic interaction, as well as how 3D biomaterials can be employed as a model to copycat the complexity of the TME.

### 1.2.1 The tumor and its three-dimensional environment: a synergistic interaction

The composition of the TME is highly complex [13]. It includes the primary tumor mass made of a heterogeneous population of cancer cells, stroma cells [immune cells—T lymphocytes, B lymphocytes, NK/NKT cells, tumor-associated macrophages (TAMs), myeloid-derived suppressor cells dendritic cells, tumor-associated neutrophils—, tumor vasculature and lymphatic cells, cancer-associated fibroblasts (CAFs), pericytes, and/or stem cells], the ECM, and a diverse variety of secreted factors. An excellent summary of the function of these cells in the TME is provided in Ref. [13]. There is a strong interplay between the tumor and stromal cells, which have a dynamic and tumor-promoting function during all stages of tumorigenesis. Indeed, the contribution of the TME in tumor progression is widely recognized [14–16]. This is exemplified during the initial steps of tumor progression, when cells escape from the primary tumor and migrate along the ECM fibers (mainly made of collagen and elastin) that have been reorganized and cross-linked into parallel bundles by CAFs. These cells eventually intravasate into the blood/lymphatic vessels, which display an enhanced permeability. After intravasation, cancer cells circulate along the vasculature. These cells are coined as *circulating tumor cells* (CTCs), which eventually extravasate and invade at a distant tissue or organ. Finally, these cells proliferate and establish a secondary tumor site.

As described above, the ECM of the TME has a critical role in tumor progression [17–19]. It not only provides a physical environment for the cells to adhere but also is fundamental for the directed migration of cancer cells. It also contains key compounds that interact with cells and provides the tissue with its stiffness. The TME also contains other types of biochemical, structural, and rheological cues that contribute to carcinogenesis. Biochemical cues include a diverse variety of cytokines, chemokines, growth factors, and inflammatory/matrix remodeling enzymes. The presence of matrix metalloproteases (MMPs) is of particular interest. They have the function of degrading and remodeling the ECM, and are secreted by cancer cells, TAMs, and CAFs. Because of this degradation, chemokines, growth factors, and angiogenic factors [e.g., vascular endothelial growth factor (VEGF)] are released into the TME. Structural cues include the rigidity of the ECM as well as its architecture. The tumor is typically stiffer than the surrounding healthy tissue as a consequence of an increased ECM deposition by CAFs, even though cancer cells are softer than their healthy counterparts [20]. Finally, rheological cues include the hydrodynamic properties of the TME, such as an enhanced interstitial fluid flow, which plays a fundamental role in triggering the invasion of cells into the vasculature. Overall, all these cues synergize with the cells and regulate all the events occurring in the cascade of tumor growth and metastasis.

Finally, it is worth noting that the importance of the TME in tumor dissemination is so high that therapies targeting the tumor ECM have become very popular to improve the response of the tumor to chemotherapy and the access of drugs [21]. Indeed, recent evidences have shown that the TME can modulate the therapeutic responses and resistance to therapies [22]. This confirms the importance of modeling the TME for the design of more efficient anticancer drugs.

### 1.2.2 Biomaterials as a model of the tumor niche

Despite the accepted importance of the TME in the growth and progression of a tumor, its exact role is not well understood yet. This is mainly a consequence of the high complexity of the TME and its constituents, which makes difficult to identify the mechanistic determinants of disease progression. For this reason, large efforts have been invested in

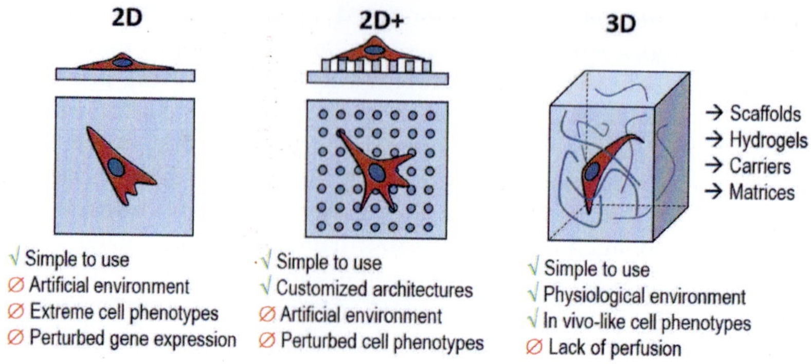


**FIGURE 1.1** Schematics of the different type of models typically employed in cancer research and other biomedical-related applications.

modeling with precision the high cellular, biochemical, and structural heterogeneity of the tumor niche. Standard tumor models based on flat surfaces coated with purified proteins of the ECM cannot reproduce this high heterogeneity, and in particular, the 3D architecture of the tumor niche (see Fig. 1.1). This results in a limited information about tumor etiology, disease mechanism, and drug efficiency. In contrast, engineered 3D biomaterials can reproduce better the intrinsic properties of the native TME. For this, the integration of innovative TE strategies, nanotechnology tools, biosensing technologies, and material sciences, has permitted the design and development of biomimetic 3D biomaterials for cancer research with unprecedented properties. These 3D biomaterials can be engineered to provide precise control on the (1) composition, (2) mechanical properties, (3) self-assembling properties, and (4) degradability of the biomaterial that could not be achieved with standard 2D cultures.

During the last decade, 3D in vitro cancer models have widely been employed to understand better the tumor biology and drug efficiency. Indeed, there is a large body of literature describing 3D-based biomaterials as models of the tumor niche for different applications. Typically, 3D biomaterials for tumor modeling can be classified into (1) *scaffold-based* or (2) *scaffold-free*. The formers include stiff biomaterials acting as supporting structures and hydrogels, whereas the latter include multicellular tumor spheroids [23] or organoids [24]. Other existing 3D-based methods typically employed in cancer research include matrix-based models, microcarriers, or topographically structured surfaces [25] (see Fig. 1.1). All these methods have in common that they provide to cancer and stromal cells a biomimetic 3D multicellular microenvironment imitating the native scenario. Overall, these systems may bridge the gap between the conventional 2D in vitro and animal testing models [26].

#### 1.2.2.1 *Scaffold-based biomaterials*

In vivo, tumor cells are supported by a complex 3D ECM, which influences their response to external stimuli. This native physical microenvironment can be mimicked using scaffold-based biomaterials, which provide 3D architectural support to tumor cells [10]. In this condition, tumor cells display phenotypes reminiscent of in vivo condition, and therefore, these are ideal platforms for drug discovery, screening, and testing applications [11,27–29]. As an example, MCF7 breast cancer cells seeded into 3D chitosan scaffolds and exposed to *tamoxifen*, an antiestrogen drug typically employed in the treatment of this type of tumor [30], produced lactate levels similar to those observed in vivo. Interestingly, the $IC_{50}$ was 10-fold higher for the 3D model than that of the flat surface.

Scaffold-based biomaterials are also employed as models to investigate the interaction between tumor cells and other cell types (malignant and healthy), such as immune cells, endothelial cells, stem cells, and others [12]. Typically, the material used for the fabrication of such scaffold-based 3D models is made of a single component or a blend of different materials [9]. Indeed, the selection of this material is crucial to copycat the mechanochemical properties of the native ECM. A specific type of scaffold are *hydrogels*, which are cross-linked networks of polymeric chains containing high amount of water; this facilitates the exchange of nutrients and cellular metabolites. Hydrogels have excellent mechanical properties, which can be modulated typically by changing polymer concentrations or cross-linker densities to match those of the target tissues or ECMs [31]. Typically, the mechanical properties of hydrogels are, in general, static; that is, they do not evolve over time. Others display reversible properties by incorporating reversible chemistries, which allow them to reproduce the dynamic properties of the native ECM [32]. One example of such a dynamic behavior is tumor formation, where there is a continuous stiffening of the ECM [33]. Importantly, specific bioactive moieties can be incorporated within the hydrogel, such as adhesive sites, to influence the behavior of the encapsulated (cancer) cells or matrix metalloproteinases (MMPs). The latter make the biomaterial susceptible to degradation by cell-secreted proteases mimicking the naturally occurring remodeling of the ECM by cancer cells.

Scaffold-based materials can be classified as *natural*, *synthetic*, or hybrid:

- *Natural-derived*: This type of biomaterials display better biocompatibility, biodegradability, and bioavailability. They also have a higher similarity to the native ECM, and therefore, cells interact better with this type of biomaterials. In addition, the mechanical and physicochemical properties of natural-derived hydrogels can be modulated, and their morphology adapted. Typical examples include natural ECM-derived proteins, such as collagen, laminin, hyaluronic acid (HA), fibrinogen, or reconstituted basement membrane, such as Matrigel™ (Table 1.1). However, these biomaterials present certain limitations, such as batch-to-batch variability, complex molecular composition, or uncontrolled degradation. In addition, they show a limited ability to alter their stiffness and the density of cell adhesion peptides. This makes challenging to control the influence of specific parameters of the ECM on cancer cells. Other type of natural-derived biomaterials include other types of proteins, such as silk fibroin [29], and polysaccharides, such as chitosan, gellan gum, or alginate (Table 1.1). These biomaterials have a broader range of chemical and mechanical properties. However, they lack cell adhesion sites and often require chemical modification for cross-linking.
- *Synthetic-derived*: This type of materials typically display better mechanical properties compared to natural ones. They are, in general, easier to synthesize than their natural counterparts and provide better experimental control over biochemical and mechanical

**TABLE 1.1** Examples of natural and synthetic origin biomaterials used in cancer research.

| Material | Characteristics and applications |
|---|---|
| ***Natural origin*** | |
| Matrigel™ | Derived from mouse tumor; contains angiogenic and tumor promoting factors; cytocompatibility; cell adhesion sites; tunable physical properties; possible presence of undefined substances. |
| Collagen I | Most prominent ECM protein of the TME; multiple cross-linking methods; biocompatibility; biodegradability; angiogenesis potential. |
| Fibrinogen | Blood clotting; cellular and matrix interactions; neoplasia; architecture mimics native ECM; 3D microenvironment for cancer growth. |
| HA | Glycosaminoglycan found in extracellular tissue in many parts of body; major component of native brain ECM; used for studying tumor migration processes. |
| Chitosan | Analysis of interaction of prostate cancer tumor cells with immune cells; formation of tumor spheroids. |
| Alginate | Properties for cell transplantation, drug delivery, and TE; suitable for hydrogel microspheres; promotes conversion of cultured cancer cells to a more malignant in vivo-like phenotype; nonadhesive to cells. |
| Cell-derived matrices (CDMs) | Distinctly different cell morphology, aggregation pattern, proliferation profile, and invasive potential; however, these matrices do not fully represent the composition and structure of the TME. |
| Silk fibroin | Unique mechanical properties; good biocompatibility; well-controlled degradability; versatile processability. |
| Agarose | Amenable mechanical and biological properties; more stable than traditional natural hydrogels. |
| ***Synthetic origin*** | |
| Polyethylene glycol | Biocompatibility; high-water content; multitunable properties; specific biological functionalities can be covalently incorporated. |
| Poly(lactic-*co*-glycolic) acid | Highly porous scaffolds; convenient to handle; amenable to large-scale use. |
| Poly ε-caprolactone | Biologically inert synthetic polymer; high porosity; large surface area:volume ratio for cellular attachment; tunable fiber diameter; low cost. |
| Synthetic peptides | Controlled amino acid composition for easy incorporation of specific biological relevant ligands. |

*Adapted from Carvalho MR, et al. Evaluating biomaterial- and microfluidic-based 3D tumor models. Trends Biotechnol 2015;33(11):667–78 [1] with permission of Elsevier.*

properties of the TME. They also display low batch-to-batch variability and are more reproducible. Examples of synthetic hydrogels used for biomedical applications include poly(acrylamide), poly(ethylene glycol) (PEG), poly(vinyl alcohol), poly(methacrylic acid), or poly(acrylic acid), and their derivatives (Table 1.1). However, synthetic-derived biomaterials lack natural cell adhesion sites and, in general, are not remodeled by cells. However, they can easily be functionalized with well-controlled density of cell adhesion ligands, like RGD or MMP-sensitive peptides.

- *Hybrid (bio)materials*: This type of material includes the combination of two or more biomaterials from natural and/or synthetic origin. Therefore, the resulting biomaterial individually or synergistically displays all the advantages of the constituent parts or provide an added value, such as the presence of specific moieties. In some cases, one of the materials is used as a sacrificial template to engineer microstructures, such as channels, for cells to adhere or for flowing a fluid. Hybrid biomaterials include PEG-conjugated proteins, biofunctionalized alginate, or other type of blends, such as gellan gum and silk fibroin [9], which has recently shown its performance to model osteosarcoma. Overall, hybrid biomaterials provide combinatorial advantages of both natural and synthetic materials for the development of 3D bioengineered scaffolds for cancer research applications.

Altogether, scaffold-based biomaterials can be engineered mechanically, chemically, and structurally to modulate the morphodynamics of cells as well as to mimic the 3D architecture of the native TME.

#### 1.2.2.2 Matrix-based

Matrix-based biomaterials try to reproduce the fibrillar structure and composition of the native ECM network. This category of 3D biomaterial can again be of natural- or synthetic origin and include a diverse subset of fabrication methods. The main methods though are based on (1) cell-derived matrices (CDMs), (2) decellularized tissues, and (3) electrospun meshwork.

- *CDMs*: These are natural ECMs that are deposited by cells, typically fibroblasts, on tissue culture dishes, which closely mimic the random stromal fiber organization and molecular content of the native ECMs [34]. CDMs provide a rich molecular content, which is similar to the in vivo scenario, with thicknesses that are in the order of 10–20 μm. These cues stimulate signaling pathways that regulate adhesion, survival, proliferation, polarity, and cell migration. A derivation of this approach is the use of a guiding template to guide the growth of 3D matrices with specific morphologies. This is of special relevance when trying to mimic the TME, where the collagen fibers are aligned into parallel bundles. This approach was recently followed by combining the confluent culture of NIH3T3 fibroblasts with microfabricated guiding templates to direct the 3D growth of well-defined extracellular networks [35]. The resulting ECM model recapitulated the structural and biomolecular complexity of features typically found in vivo. Overall, CDMs can be employed as a 3D model to reproduce the architecture and mechanochemical properties of the TME. Importantly, by manipulating the culture conditions (e.g., culture media composition and additives, morphology of the culture substrates, cell types, culture periods, passage number, and others), a significant control over CDM composition and microstructure can be achieved. For this, CAFs can be used to obtain a CDM with tumor-like properties.
- *Tissue/organ-derived decellularized matrices*: Decellularized tissues and organs are extensively used in tissue engineering and regenerative medicine applications. This method is based on the removal of the cellular content of a tissue or organ resulting with only the matrix component. This contains a similar composition and 3D architecture of the native ECM. For this, different decellularization protocols have been

reported, which vary depending on the tissue or organ of interest [36]. Typically, physical and chemical treatments are carried out. The formers include agitation, mechanical massage or pressure, or snap freezing, that disrupt the cell membrane. The latter includes the use of detergents, ionic solutions, or enzymatic treatments, such as trypsin (or equivalent) to disrupt cell membranes and cell–cell/ECM connections (see Refs. [36,37] for a detailed description about the different decellularization methods). Depending on the method employed, this may affect the resulting biochemical composition, mechanical properties, and 3D architecture of the remaining matrix. Importantly, this approach may also be utilized for developing physiologically relevant 3D models of cancer. However, the composition and morphology of cancerous ECM is very heterogeneous and differ among patients with the same type of tumor.

- *Electrospun meshwork*: This method uses electrospinning to deposit a 3D layer of micro- and nanosized fibers of synthetic or natural origin that support the growth of a tumor in 3D mats [38]. The fibers can be randomly oriented or oriented into parallel bundles. This method allows a precise control over certain matrix parameters, such as porosity, fiber diameter, elasticity, and surface roughness. Similar to CDMs, cells seeded on aligned electrospun fibers have the ability to polarize, orient, and migrate along fiber main axis, thus reproducing the initial events of cancer cell progression. However, and in general, electrospun fibers are incapable to mimic the biomolecular and mechanical properties of the native scenario, and therefore are far from being physiologically relevant [39].

#### 1.2.2.3 Microcarrier-based

Microcarriers are small porous or nonporous beads, which are typically employed in 3D cell culture to provide a 3D physical support to cells [40]. Microcarriers are typically in the range of 100–500 μm in diameter and are made of biocompatible materials, including synthetic (e.g., dextran, plastic, glass) and natural (e.g., gelatin, collagen, Matrigel$^{TM}$, glycosaminoglycans, cellulose) origin. In addition, the outer surface can also be coated with different biomaterials (e.g., chitosan, hyaluroan, and others). Microcarrier beads are typically added inside a bioreactor or spinner flasks to promote cell adhesion. Therein, the cells adhere and proliferate generating their own ECM and forming 3D structures reminiscent of the native TME. This system is very convenient for those cells that do not form spontaneously cell aggregates, or for including different cell types, thus facilitating cell–cell interaction. Microcarriers also allow a more efficient nutrient and gas transfer between the media and the attached cells compared with 2D surfaces. Large number of cell-laden microcarriers can be suspended in cell culture media and cultured within a bioreactor with well-controlled environmental conditions, or even be loaded and 3D (bio) printed in hydrogels [41]. Interestingly, a diverse variety of commercial microcarriers displaying different porosites are available (e.g., microsized pores: Cytodex, GE Healthcare; SoloHill, Pall; macrosized pores: Cytopore, Cytoline, FibraCell) [42]. The different porosity will, at the end, determine the final application of the microcarriers. For macrosized pore carriers, the cells can grow both outside and inside the pores. In contrast, on smaller pore carriers, the cells can only grow at the surface and the pores may be used to modulate cellular responses to biological and molecular gradient changes.

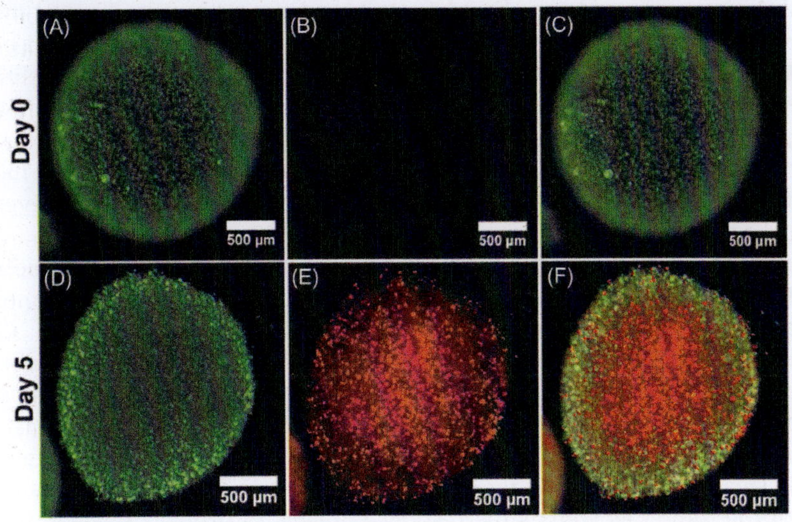


**FIGURE 1.2 Tumor cells seeded on 3D spherical microcarriers display in vivo-like features.** Fluorescence images showing the formation of a necrotic core on cancer cells seeded in a 3D microenvironment after 5 days of culture. The dead cells are stained in red, whereas green represents live cells. This behavior is characteristic of tumors in vivo. Source: *Reproduced with permission by the American Chemical Society from Pradhan S, Chaudhury CS, Lipke EA. Dual-phase, surface tension-based fabrication method for generation of tumor millibeads. Langmuir 2014;30 (13):3817–825 [44].*

This type of system is very attractive for 3D cancer modeling because it supports the aggregation of cancer and stromal cells, preserves cell phenotype, allows coculture and cell–cell/ECM interaction studies, and different bead coatings can be produced. In addition, they can reproduce important behavior of native tumors, such as the formation of a necrotic core (see Fig. 1.2). Another advantage of microcarriers is their high surface area-to-volume ratio and their compatibility with other culturing methods, thus resulting into advanced tumor models. For example, microcarriers can be integrated within microfluidic devices to facilitate the harvesting of cancerous cells [43]. This is of high interest to collect the cells after drug screening on in vivo like conditions. However, this type of model also displays certain limitations, including lack of perfusion, shear stress, and in general, vasculature.

### *1.2.2.4 Scaffold-free: tumor spheroids*

In this case, no material is employed for the culture of cells. Instead, the capability of cells to self-assemble is used for the formation of 3D self-standing structures. This category includes the use of 3D multicellular *tumor spheroids*, which are compact aggregates of tumor cells with an extensive 3D network of cell–ECM and cell–cell interactions [45]. This is one of the most used 3D cancer models, because they exhibit a 3D architecture and a rich cellular heterogeneity close to that of tumors in vivo [46]. The cells on the periphery of the spheroid actively proliferate, whereas those located in the inner core remain quiescent or necrotic when the spheroids grow beyond 400–600 μm.

Tumor microsized spheroids can be fabricated by different means. Typical methodology uses nonadhesive culture methods, such as nonadherent plates (flat or round bottom; e.g., agarose plates), magnetic levitation, or rotational culture. Other methods rely on forcing the aggregation of cells in a confined space, such as the hanging drop method or by centrifuging the cells in suspension. Finally, the encapsulation of cells within a 3D biomaterial (e.g., hydrogel) can also be employed to form tumor spheroids. These methods rely on the ability of cells to establish cell–cell contacts, and may take from several days to weeks to form aggregates of the desired size.

Tumor spheroids can be utilized to investigate the effect of drugs in a biomimetic 3D environment. Indeed, it has widely been reported that cancer cells are much more resistant to chemotherapeutic drugs when grown as spheroids compared with flat cultures [47]; therefore, they provide a much more realistic prediction of in vivo drug responses. As an example, a pioneering work on this field showed that the $IC_{50}$ of *paclitaxel* and *cisplatin*, two typical anticancerous drugs, was 100-fold higher in both MCF7 human breast carcinoma and DLD-1 human colon carcinoma spheroids compared with the same cells seeded on flat surfaces [48]. This shows that cell spheroids mimic better in vitro the phenotype and genotypic levels of the tumor, as well as the response of cells to drugs. However, in vivo, other cell types, such as stromal, endothelial, and immune cells have demonstrated to contribute significantly to tumor progression. In this regard, the formation of heterotypic multicellular 3D tumor spheroids and accompanying cells can significantly improve the biological relevance of the obtained data. This approach was recently followed to build a triple coculture pancreatic tumor model [49]. The model included pancreatic cancer cells (PANC-1), fibroblasts (MRC-5), and endothelial cells (HUVEC), which self-assembled to form 3D spheroids. Importantly, the influence of the TME was evaluated by exposing the triple coculture spheroids to *doxorubicin* and *gemcitabine*, and compared compared with monoculture spheroids. The obtained results showed a higher resistance of the triple coculture spheroids to chemotherapy, confirming the critical role of the multicellular microenvironment on the proliferation of the tumor, and importantly, similar to in vivo scenario.

#### 1.2.2.5 Microstructured surfaces

The combination of biomaterials with nanotechnology tools, such as microfabrication techniques and replica molding, have permitted the development of advanced surfaces containing 3D topographic micro- and nanostructures for cancer-related studies [50]. These surfaces with well-defined 3D morphologies recapitulate topographical features typically encountered in vivo, with length scales in the order of microns and nanometers that match the intrinsic scale of cells and their organelles. As an example, an array of topographically engineered parallel microgrooves can mimic the alignment of the ECM that occurs during tumor progression. Similarly, microstructured channels or pillars can be employed to investigate the effect of cellular confinement and the migration mode of cancer cells into the surrounding TME. Next, micropatterned substrates functionalized with specific molecules can be utilized to study the interactions of these molecules with cancer cells with a high-spatial control. This allows the precise isolation and assessment of the effects of biochemical or biophysical cues in disease progression with high precision. For instance, the interaction of CD44 membrane, a membrane receptor of cancer cells, with HA, a glycosaminoglycan component of the ECM highly expressed in several tumors, was investigated in the mechanism of tumor progression [51]. A microcontact printed array of HA was employed to investigate its interaction with colon and breast cancer cells finding that tumor cells adhered and elongated protrusions preferentially on top of HA-containing motifs.

Typical materials employed for the fabrication of microstructured surfaces include natural- and synthetic-derived hydrogels described above, and other types of polymers and elastomers, such as polysterene, polycarbonate, polymethylmethacrylate, and polydimethylsiloxane (PDMS), among others. Depending on the selected material, the microstructures can be manufactured by means of a diverse set of technologies, which are mainly based on printing, molding, and embossing (see below).

## 1.3 Engineering the native tumor microenvironment using custom-designed three-dimensional biomaterials

As described above, the TME plays a fundamental role in tumorigenesis, regulating the formation, growth, and dissemination of the cancer cells. This regulatory role has attracted the interest of the cancer biology community in developing new strategies to model the TME to reproduce in vitro the structural, biological, and mechanochemical complexity of the native scenario. These new strategies are based on (1) tissue engineering, or (2) nanotechnology approaches, or a combination of both. In both cases, they use custom-designed 3D biomaterials to study cell–cell/ECM interactions. Taken together, these strategies may find a diverse variety of applications including fundamental mechanistic studies (e.g., new insights into the cells–TME interaction) and/or, therapeutic applications (e.g., drug discovery and screening). In the following, we describe the most typical approaches employed to engineer 3D TME models using custom-designed biomaterials.

### 1.3.1 Tissue engineering approaches

During the last decade, a diverse set of tissue engineering strategies have been developed for the fabrication of advanced 3D models of cancer [1,27,52]. These strategies combine—cancer and healthy—cells with biochemical/mechanical signaling and customized biomaterials. The aim is to copycat the complexity of the native TME with respect to its 3D architecture, structural organization, mechanical properties, and biocompatibility. In the following, we briefly describe some of the most typical TE-based approaches to fabricate 3D models of the TME, and comment on their main applications and limitations.

#### 1.3.1.1 *Freeze-drying*

Freeze-drying, or lyophilization, is one of the most common tissue engineering-based methods employed for the fabrication of porous 3D scaffolds [53–55]. Typically, a natural- or synthetic-derived hydrogel is frozen trapping the polymeric fibers between the growing ice crystals. The ice crystals are next sublimated during the drying phase resulting into the formation of a highly porous solid scaffold with a network of interconnected pores (Fig. 1.3). This pore network structure is heterogeneous in shape and distribution. The

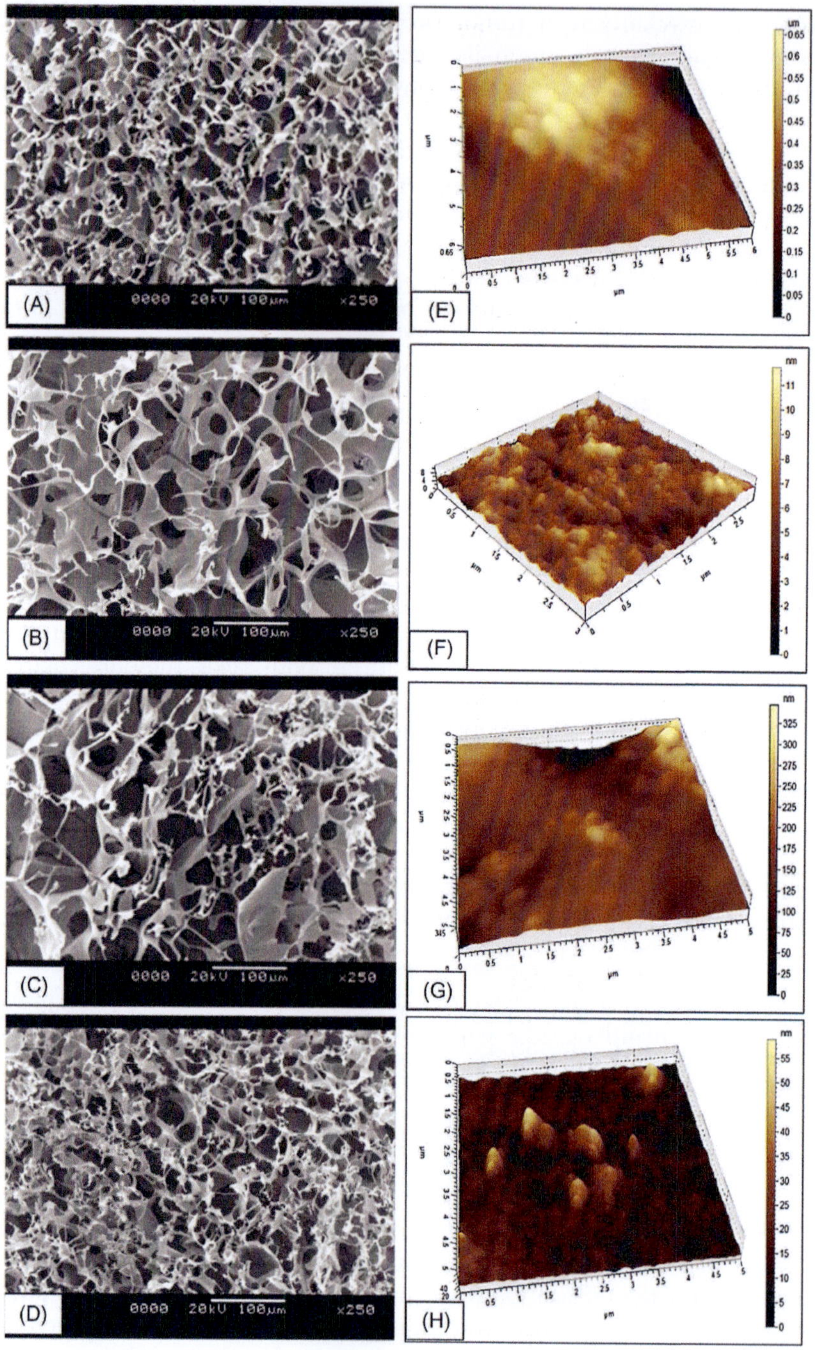


FIGURE 1.3 **Freeze-dried 3D scaffolds.** (A–D) Scanning electron microscopy and (E–H) atomic force microscopy images of freeze-dried scaffolds. (A and E) Silk fibroin; (B and F) thelebolan; (C and G) silk fibroin–thelebolan blend; and (D and H) genipin-cross linked fibroin–thelebolan blend. Source: *Reproduced with permission by Elsevier from Mukhopadhyay SK, et al. Silk fibroin-thelebolan matrix: a promising chemopreventive scaffold for soft tissue cancer. Colloids Surf B Biointerfaces 2017;155:379–89.*

pore size can be controlled by altering the freezing process, such as the cooling rate [56]. This is of extreme importance since the pore size is observed to influence the adhesion, growth, and phenotype of a wide variety of cell types. Importantly, besides the adhesion and transport of cells, freeze-dried scaffolds also permit the flow of nutrients and gases.

Even though freeze-drying is widely employed for the fabrication of 3D scaffolds, it also has certain limitations. Among all of them, the fabrication of scaffolds with this technique may be very time consuming. It may also generate small and irregular pores with heterogeneous sizes [57]. In addition, the fabrication of hierarchical-structured scaffolds is challenging, thus limiting the applicability of this method in complex and ordered environments, such as in the TME.

#### *1.3.1.2 Photopolymerization*

Photopolymerization is a method that employs visible or UV light to form in situ cross-linked polymer structures at physiological pH and at room temperature. Typically, this technique uses hydrogels and free radical photopolymerization due to their numerous advantages, including biocompatibility, high-water content, high versatility, injectability, or in situ gelation capability, and others. Importantly, hydrogels can be UV-gelated by introducing nonreversible covalent bonds [58]. Indeed, it is possible to localize spatially the irradiation offering both physical and chemical control over the polymer network.

The capability of in situ polymerization in a minimally invasive manner (e.g., injection) makes this type of materials very attractive for tissue engineering applications, including cancer therapeutics. Indeed, photopolymerization is a very promising technique with high-clinical impact due to its capacity to entrap cells and drugs, such as specific anticancerous compounds. Then, the drug-loaded hydrogel can be injected in the tumor region and photopolymerized for localized and sustained delivery of the drug. As an example, a photopolymerizable PEG-DMA-based hydrogel was loaded with *temozolomide*, an anticancerous drug, and injected in a nude mice on a human glioblastoma model [59]. Results showed that the size of the tumor from mice treated with the drug-loaded hydrogel significantly decreased compared with nontreated animals. This demonstrates the potential of this approach in cancer therapeutics.

The number of photopolymerizable hydrogels is vast, and includes PEG derivatives, polyvinyl alcohol, alginate, chitosan, gelatin, or HA, among others. It is worth mentioning that photopolymerizable hydrogels have wide industrial and clinical applications. They provide certain advantages compared with hydrogels that are cross-linked by other means (temperature, chemically, or others). For a detailed description about this approach, the readers may refer to a specialized review on the topic [60].

#### *1.3.1.3 Three-dimensional bioprinting*

Three-dimensional printing is defined as an additive manufacturing process by which materials are directly patterned layer-by-layer to form complex 3D architectures [61]. When the employed materials are from biological origin, that is bioinks, such as living cells, biological macromolecules (e.g., proteins from the ECM), or biomaterials (e.g., gelatin), this method is coined *3D bioprinting*. During the last decade, 3D bioprinting has been gaining huge popularity due to its versatility, performance, and compatibility with a wide range of biomaterials. It offers several advantages in comparison with other (bio)fabrication techniques, such as photolithography or soft lithography (see below). Among others, 3D bioprinting has the ability to (1) create scaffolds displaying complex 3D architectures with high accuracy; (2) directly print living cells; (3) print different cells types and biomaterials; (4) cross-link photopolymerizable hydrogels *in situ;* (5) produce 3D scaffolds fast and at low cost; and (6) be high throughput and reproducible.

Three-dimensional bioprinting is not just a single technique but a collection of different methods. Herein, we will focus in those that use biological material, including cells, in particular by inkjet printing. For details on the other methodologies, readers are referred to Section 1.3.2.2.2: Three-dimensional printing, and references therein.

Inkjet 3D bioprinting method involves the printing of microsized droplets of a bioink at the print head (or nozzle), typically assisted by a piezoelectric actuator or a heater. The former produces a mechanical pulse to force the bioink from the printhead as droplets, whereas the latter generates air–pressure pulses to generate the droplets [61]. Usually, two different methods can be followed to deposit cells, namely one- (or direct) and two-step (or indirect) fabrication methods. One-step bioprinting is based on the deposition of a mixture of hydrogel and cells, whereas in two-step bioprinting, the cells are seeded after the 3D printing of the scaffold. Finally, the printing of the bioink can be combined with other methods, such as UV-photopolymerization, where each hydrogel layer can be cross-linked before the printing of the other ones. Importantly, different bioinks can be combined during the printing process, resulting into the fabrication of complex 3D structures. This may be very appealing for cancer modeling.

The 3D bioprinting represents a significant improvement in cancer research compared with standard 2D flat surfaces. In particular, they can reproduce better the 3D architecture and cellular content of the TME, thus facilitating physiologically relevant cell–cell/ECM interactions [61]. Among other features characteristic of the native TME, 3D bioprinted models allow to study tumor heterogeneity or angiogenesis, or the development of 3D tumor models for anticancer drug discovery, screening, and testing.

## 1.3.2 Nanotechnology approaches

Advances in nanotechnology have permitted the development of a set of innovative microfabrication techniques for 3D cell-based studies. These technologies are derived from the microelectronics industry and are capable to create micro- and nanosized structures, which are of the same scale of cells and their organelles. Typically, microelectronic chips are created using solid materials, such as silicon (or silicon-based) and glass, combining photolithography approaches and chemical etching, or by means of more sophisticated methods, such as electron-beam lithography. As an alternative to these approaches, nanotechnologists have developed a collection of techniques based on *molding, printing,* and *templating,* which are compatible with—soft—biomaterials, such as hydrogels and other polymeric materials [62]. This collection of techniques are coined *"soft lithography"* since they are based on using "soft" elastomeric materials. In the following, we briefly describe this collection of techniques and their main applications in the field of 3D cancer modeling.

#### 1.3.2.1 Molding

This method includes a subset of different techniques used for the fabrication and replication of 3D structures. These are (1) *replica molding*, (2) *hot embossing*, (3) *microthermoforming*, and (4) *injection molding*, which are briefly described next. For more details, the readers may refer to the specialized literature on this topic.

- *Replica molding*: It uses a microengineered silicon-based mold, typically fabricated by UV-photolithography and wet etching methods, to replicate the microstructures of interest. Usually, these are replicated using an elastomer, such as PDMS, mixing the prepolymer and the cross-linker in a 10:1 ratio (w/w or v/v), or in hydrogel, such as collagen. Typically, the replicated microstructures are coated with a protein of the ECM, such as fibronectin, before seeding the cells. This methodology allows the fabrication of structures in the range of about 1–5 μm in lateral resolution and up to 200 μm in thickness. However, more complex structures can be fabricated by aligning, stacking, and bonding different layers.
- *Hot embossing*: It uses a mold with micro- or nanosized features to replicate them on a thermoplastic film. This film is first warmed to reach its glass transition temperature, which depends on each polymer, to soften. Next, the flat film is structured under the application of high pressure against the mold during a specific time. Then, the temperature is decreased (maintaining the pressure) and the mold is withdrawn from the solidified film. This process is relatively fast for simple structures. Longer cycle time may be required to emboss components that are more complex. Finally, hot embossing is compatible with different thermoplastics, such as polymethylmethacrylate, polysterene, or polycarbonate.
- *Microthermoforming*: This method shares some similarities with hot embossing. In this case, a thin thermoplastic film is heated up to its crystallization temperature, which is then thermoformed into thin-walled microstructures by a 3D stretching using vacuum or compressed air [63]. This procedure displays certain advantages compared to hot embossing, including the possibility to fabricate hollow microstructures or the fabrication of microstructures with spherical bottoms. However, it also displays some limitations, such as the limit of resolution and the possibility to fabricate high-aspect ratio structures.
- *Injection molding*: It uses a rotating forward screw into which a granular polymer is introduced and heated above its glass transition temperature. This continues until the polymer is melted when the nozzle is reached. Then, it is injected into a closed microstructured mold and is cooled down prior demolding the polymer. Due to its simplicity, and despite its intrinsic limitations (e.g., possibility of blocking of the injection channel), this methodology is still the favorite of the industry for the fabrication of disposable devices for biomedical applications.

#### 1.3.2.2 Printing

Nanotechnology-based printing technology includes a very different subset of methodologies, such as (1) *(2D) microcontact printing*, (2) *3D printing*, or (3) *4D printing*.

#### 1.3.2.2.1 (Two-dimensional) microcontact printing

It uses a topographically patterned mold, coined "stamp" and typically made of PDMS, to transfer by physical contact an array of a "bioink" of interest (biomolecules, proteins, cells, and similar) to the surface of a substrate [64]. Different combinations of micro- and nanosized features can be created (typically >100 nm). Even though this is considered as a 2D microfabrication technique, some 3D topographies can be achieved by patterning molecules of different dimensions or by 3D structuring the surface [65].

#### 1.3.2.2.2 Three-dimensional printing

It uses a thin nozzle (or extruder) to deposit on a surface a "bioink" of interest in a layer-by-layer manner. Structures ranging from several μm to cm dimensions can be created in short-time duration (minutes), even though more complex samples containing nm-sized features may take longer (several hours). This methodology includes different printing types, which differ on their working principle and (bio)material employed. In particular, nanotechnology-based 3D printing methods includes (1) *extrusion*, (2) *stereolithography*, (3) *two-photon polymerization*, and (4) *inkjet deposition* printing (Fig. 1.4), which are summarized here.

- *Inkjet deposition printing* is considered as a tissue engineering approach for the deposition of a biomaterial on a dissolvable matrix to print layer-by-layer 3D structures (Fig. 1.4A, and details in Section 1.3.1.3). This method can be performed within a support bath, which can be removed at the end (Fig. 1.4B).

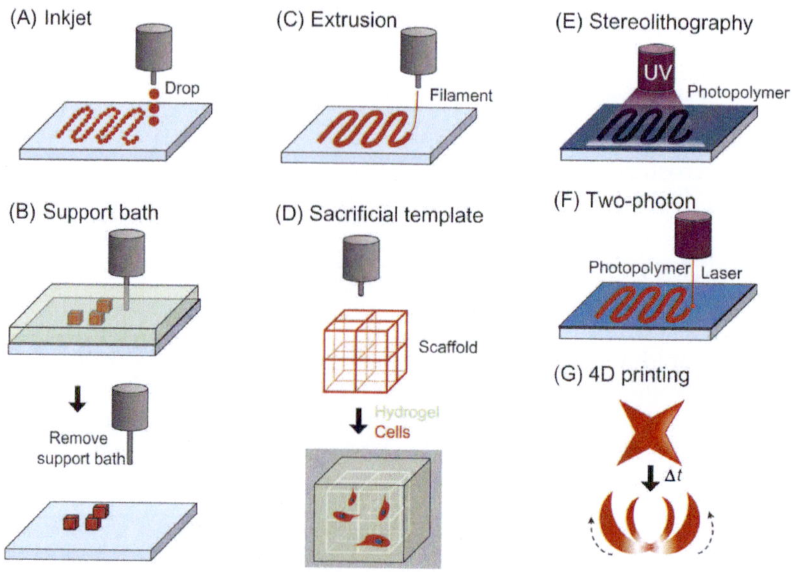


**FIGURE 1.4 Bioprinting methods.** Schematic representation of the most typical bioprinting methods. (A) Inkjet-based; (B) Support bath-based; (C) Extrusion-based; (D) Sacrificial templating; (E) Stereolithography-based; (F) Two-photon (laser)-based; and (G) 4D printing.

- *Extrusion printing* uses a thermoresistive polymer to dispense through an extruder a thin and melted filament. This is typically employed to engineer 3D supporting or sacrificial structures of variable geometries to allow the adhesion of cells (Fig. 1.4C). This method can be utilized for the fabrication of biodegradable (sacrificial) templates for the deposition of cell-laden hydrogels (Fig. 1.4D).
- *Stereolithography printing* utilizes projected UV light (or laser) to generate 3D structures by consecutively photopolymerizing a thin layer of UV curable material layer-by-layer onto a substrate submerged in a tank that contains liquid photosensitive material (Fig. 1.4E).
- *Two-photon printing* deploys a near-infrared femtosecond laser to polymerize a photocurable epoxy into 3D structures (Fig. 1.4F).

#### 1.3.2.2.3 Four-dimensional printing

This technique is a variation of 3D printing where the fourth dimension (4D) refers to a time-dependent evolution of the 3D printed objects. The 4D printing improves the static behavior of a 3D printed object, which limits structural deformation and functional motion [66]. In 4D printing, there is a response of the 3D printed object over time upon internal or external stimuli, including changes in pH, magnetic field application, temperature/humidity variation, or light stimulation. This leads to a change in object shape, size, and/or functions with time (Fig. 1.4G). The exciting capabilities of this new technology leads to the development of new "programmable" (bio)materials for a diverse variety of applications. As an example, complex flower architectures were 4D printed using an hydrogel bioink that mimicked the composition of plant cell walls showing a localized and anisotropic swelling behavior [67]. This behavior was controlled by the alignment of cellulose fibrils along prescribed 4D printing pathways. Similarly, by combining soft lithography, molding, and 3D printing methods, an autonomous *soft* robotic device composed solely of soft biomaterials was fabricated [68]. The robot was powered by a chemical reaction controlled by integrated microfluidic channels. The catalytic decomposition of hydrogen peroxide using platinum produced large amounts of gas, which flowed along the integrated microfluidic channels and was employed to generate actuation on the extremities of the robot. Interestingly, this strategy can be utilized to fabricate other type of "smart" biorobots, with specific functions, paving the way toward a new generation of completely soft, biocompatible, and autonomous machines.

This new paradigm in 4D printing is associated with the development of novel "smart" biomaterials. The most common materials employed in 4D printing are hydrogels (natural- and synthetic-origin hydrogels) and polymers (e.g., PDMS). Other biomaterials, such as lipids and cells are also employed (see below). These stimuli-responsive biomaterials are able to actuate through shape deformations in response to an external stimulus present in the surrounding microenvironment. The combination of 4D printing with biological materials and cells leads to the development of 4D bioprinting, an extension of 4D printing method [69]. In this case, the 3D bioprinted cells or biocompatible materials are considered as the "smart" stimuli-responsive materials. Here, a smart 4D bioprinted material can be programmed to detect cancer cells and proceed to the precise release of a chemotherapeutic compound locally. As an example, 4D printed soft robots made of a hydrogel bilayer structure containing iron oxide particles were fabricated for cancer drug delivery applications [70]. The robots were sensitive to changes in pH and magnetic field responsive, which was utilized to drive them toward the region of interest and release an encapsulated anticancer drug (*docetaxel*) in the mammary carcinoma tissue region upon the detection of acidic pH (2.6).

Overall, 4D (bio)printing offers multiple advantages compared with static 3D printed systems, in particular for cancer therapeutics and drug delivery applications. However, there are still challenges, which require to overcome. This includes the development of a larger library of biocompatible responsive materials, which can be printed and stimulated. For more information about 4D (bio)printing of smart biomaterials, the readers may consult the recent reviews on the topic [69,71].

## 1.4 Advanced models of the three-dimensional tumor microenvironment

As aforementioned, 3D tumor models have several advantages compared with those models based on either flat surfaces or nonrepresentative animal models. However, 3D tumor models also display certain limitations. This is because tumor growth and metastasis include a very complex collection of events that standard 3D tumor models cannot completely recapitulate. These events include the uncontrolled growth of the primary tumor, followed by the invasion of cancer cells into the surrounding stroma, and the formation of new microvasculature and lymphatic vessels (angiogenesis and lymphangiogenesis, respectively), Next, cancer cells (and ctDNA/RNA, exosomes, and antibodies) intravasate into the microcirculation system resulting into CTCs. These cells transit through the vasculature and eventually extravasate to colonize a distant tissue, forming a micrometastatic niche. Traditional 3D tumor models cannot reproduce some of these events, which incorporate dynamic phenomena or the presence of precisely controlled 3D configurations. Advanced models capable to reproduce such complex features would provide a more realistic model of the in vivo TME improving our understanding of the tumor biology and the development of more efficient therapies. In this regard, microfluidic-based models of tumors, or *tumor-on-a-chip* (ToC) devices, and 3D bioprinted tumor models can provide an advanced biomimetic microenvironment capable to reproduce the complex morphodynamic signature of tumors, and therefore, to elucidate better the behavior of drugs [25,72]. In the following, we describe this new generation of advanced tumor models.

### 1.4.1 Microfluidics-based models

Microfluidic models are dynamic 3D cell culture devices that contain engineered microsized channels populated with one or several cell types. These channels are typically perfused with cell culture media or body fluids for extended periods of time (hours to days, or longer). Microfluidics-based devices are ideal candidates to copycat the biological complexity, tissue–tissue interfaces, hierarchical multicellular architecture, biochemical gradients, and the hydrodynamics of the native environment in the human body. When combined with specific cell types, they can emulate the functional unit of an organ or tissue within a custom-made 3D environment. This results into the development of the so-called *Organs-on-a-Chip* models [73]. This type of models offers a large plethora of advantages compared with their static 2D and 3D counterparts, and have already attracted the attention of the pharmaceutical and biotechnology industry as well as regulatory agencies, such as the FDA. Indeed, different organs-on-a-chip models have been developed so far, including the lung, heart, brain, spleen, bone marrow, kidney, pancreas, gut, and many others [73]. Importantly, these individual models can be interconnected following a hierarchical and physiologically relevant order, resulting into *Multiorgan-on-a-Chip* or *Body-on-a-Chip* models [74]. This type of microfluidic models can better assess the response of the different organs and tissues to drugs and its metabolites [75], pathogenic bacteria [76], virus [77], as well as other type of internal and external insults, including cancer metastasis. Indeed, (multi-)organs-on-a-chip models of cancer can be employed to investigate the mechanistic determinants of tumor progression and organ specificity in metastasis. This has the potential to significantly improve our understanding in tumor etiology or to assess the efficiency of new therapeutic compounds [25,72,78]. In the following, we briefly describe how microfluidic-based tumor models can be used in cancer research.

#### 1.4.1.1 *Microfluidic-based models of tumors: tumor-on-a-chip*

The combination of organ-on-a-chip technology with cancer—and other—cells results into the development of microfluidic-based *Cancer-* or *ToC* devices [25,72]. ToC models can satisfactorily reproduce the important dynamic events that occur in the metastatic cascade. During the last decade, numerous ToC models have been reported in the literature reproducing one or several of these events, including (1) *tumor growth*, (2) *invasion*, (3) *angiogenesis/lymphangiogenesis*, (4) *intravasation*, (5) *transit of CTCs along the vasculature*, (6) *extravasation*, and (7) *metastatic niche formation*, among others [25] (Fig. 1.5). These models have provided fundamental insights into the mechanistic determinants of tumor

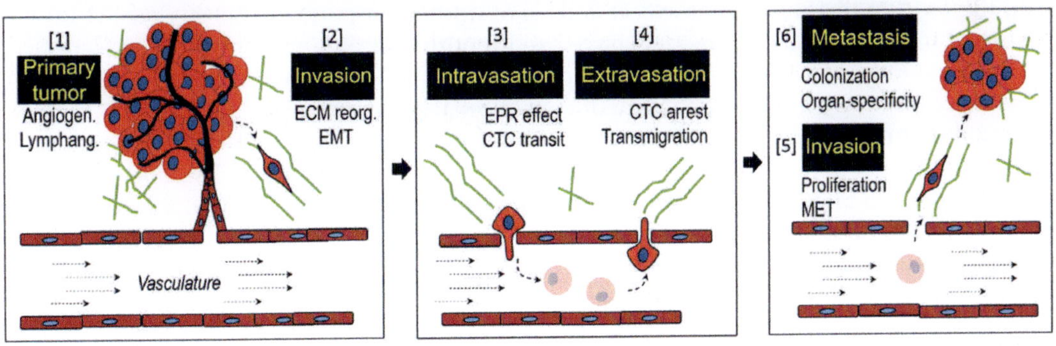


FIGURE 1.5 Scheme showing the different steps occurring in the cascade of cancer metastasis together with the characteristic events. [1]: Growth of the primary tumor and new blood/lymphatic vessels; [2]: Cancer cells undergo EMT and migrate directionally along the reorganized ECM; [3]: Cancer cells intravasate into the vasculature favored by its higher permeability and retention effect; [4]: The cancer cells transit along the vasculature as CTCs, eventually arrest on the vasculature wall, and extravasate; [5]: Cells invade the surrounding stroma, proliferate, and undergo mesenchymal-to-epithelial transition (MET); and [6]: Cancer cells colonize and form a secondary metastatic niche.

dissemination, and importantly, have proven instrumental in drug discovery. Herein, we briefly describe several applications of ToC models.

1. *Tumor growth-on-a-chip*: The growth of the primary tumor is the first event in the multistep process of cancer metastasis. During this event, the tumor progressively increases its capacity of proliferation and invasion. Tumors are characterized by a heterogeneous population of cells that interact with the surrounding stroma (ECM, stromal cells, and endothelial cells). Indeed, this synergistic interaction plays a fundamental role in the growth and progression of the tumor. This has attracted intense research interest to create such models. In one model, a ToC device containing a heterotypic triple coculture of breast cancer cells, stromal cells, and monocytes, was employed to recapitulate in vitro the growing TME [55]. This model allowed assessing how different cell types interacted through paracrine signaling via production of TGF-β by breast cancer cells and the expression of the corresponding receptors by stromal cells. This work provided key insights into how tumor and stromal cells responded to microenvironments of varying complexities.

   During tumor growth, there is also a continuous remodeling and changes in the composition of the ECM. This remodeling, and the effects of ECM and stromal composition in tumor growth, was investigated using a ToC device containing a 3D microenvironment with different combinations of ECM molecules (collagen, fibronectin, and laminin) [79]. A coculture of T47D breast carcinoma cells with human mammary fibroblasts showed that the presence of fibroblasts promoted an increased growth of the tumor, in particular, for fibronectin- or laminin-rich ECMs.
2. *Invasion-on-a-chip*: The invasion of tumor cells into the surrounding stroma is a crucial step in the metastatic cascade. During this event, the ECM is reorganized into bundles of aligned fibers promoted by the stromal cells. Then, after undergoing epithelial-to-mesenchymal transition (EMT), cancer cells migrate directionally toward the vasculature, initiating metastasis. Identifying the contribution of the individual factors in the TME to the invasion of cancer cells is of upmost importance. In this regard, several invasion-on-a-chip models have been reported aiming to unravel the effect of structural (e.g., ECM stiffness and/or architecture) and biochemical cues (e.g., cytokine gradients) in the invasion capability of cancer cells. As an example, a novel ToC model was recently employed to investigate the effect of 3D matrices with controlled architecture on the invasion of MBA-MD-231 breast cancer cells [80]. The ToC contained two microchannels separated by a porous electrospun 3D PCL matrix with different fiber diameters. Interestingly, the different diameters did not affect the invasion capability of cancer cells, which were stimulated with a chemotactic gradient. Only differences in protrusion morphology were observed. Similarly, a ToC device was employed to investigate the different invasion capabilities of the heterogeneous population of SGC-7901 human gastric cancer cells [81]. To this aim, a fetal bovine serum gradient was generated on-chip to stimulate the migration of cells. The difference in the invasive potential of the distinct subpopulation of cancer cells permitted their separation between high- and low-metastatic cells. Importantly, these results were validated in vivo using a mouse model, confirming the relevance of ToC models.

3. *Angiogenesis/lymphangiogenesis-on-a-chip*: During tumor growth, new vascular and lymphatic vessels are generated. These are two fundamental and highly dynamic phenomena in cancer progression, which are coined *angiogenesis* and *lymphangiogenesis*, respectively. Different ToC models of the human microcirculation system have been reported. These models differ in their complexities and cellular contents, ranging from simple and individual vasculature or lymphatic system-on-a-chip, to more complex and complete microcirculation-on-a-chip models [82]. In particular, the number of ToC models reproducing vascular sprouting and angiogenesis formation is vast and recent excellent reviews on the topic are available [72,83,84]. In a seminal paper, the angiogenic sprouting during morphogenesis was replicated in a ToC model, which contained two channels embedded within a 3D collagen I matrix [85]. One of the channels was lined with endothelial cells, whereas the other one was perfused with proangiogenic factors to induce sprouting. Capillary sprouts were generated, which eventually interconnected both channels to establish new functional capillary tubes. Importantly, this model was used to evaluate the effect of angiogenic inhibitors and stimulators, therefore, validating its clinical impact.

   The contribution of the lymphatic system in tumor metastasis attracts less attention in comparison to its vascular counterpart, despite the solid evidences about its crucial role during cancer progression. Indeed, a few ToC models of the lymphatic system and lymphangiogenesis have been reported. In one of these models, the mechanism of lymphatic vessel sprouting was investigated [86]. Interestingly, it was found that lymphangiogenesis was stimulated by a synergistic interaction between mechanical stimuli (interstitial flow and endothelial–stromal interaction) and biochemical cues (series of prolymphangiogenic factors). Finally, a complete microcirculation-on-a-chip device was reported for investigating vascular permeability [82]. The ToC model contained both the blood and lymphatic vessels. For this, blood vascular and lymphatic endothelial cells were seeded back-to-back on a porous membrane defining both channels. Both cell types formed a confluent layer demonstrated by the expression of tight and adherent junction markers, thus showing a reduced endothelial permeability. When histamine, an inflammation-promoting substance, was injected, an increase in vascular permeability was observed. Importantly, this mimicked the phenomena occurring during the dissemination of cancer cells, highlighting the utility of this type of device in cancer research.

4. *Intravasation-on-a-chip*: Intravasation is the invasion of tumor cells into the microcirculation system. This is one of the most critical points during metastasis, since cancer cells can transit along the vasculature and invade a distant organ. In order to shed light into the mechanochemical mechanism of cancer cell intravasation, several ToC models have been reported, mainly to investigate the interaction between cancer cells and the endothelium in a 3D environment. In one of these models, the tumor–endothelium interface was recreated using a three-channel microfluidic chip containing endothelial and tumoral cells, which were interconnected via a 3D ECM hydrogel [87]. This pioneering ToC was employed to investigate the dynamic interplay between vessel permeability, cancer cell intravasation capability, and the paracrine signaling loops involving macrophages and tumor cells. In particular, it was found that signaling with macrophages via secretion of tumor necrosis factor $\alpha$ resulted in

enhanced endothelial barrier impairment that favored cancer cell invasion. More recently, a similar ToC was used to study tumor progression and intravasation of different cancer cell lines and patient-derived breast cancer organoids [88]. The model comprised one central chamber containing endothelial cells extracted from cord blood and normal human lung fibroblasts, which after 1 week, self-assembled forming a perfused quiescent microvascular 3D network in fibrin gel. This chamber was interconnected to two adjacent ones that encapsulated the tumor cells in Matrigel$^{TM}$. The intravasation of cancer cells into the microvasculature was boosted by adding TGF-β, an EMT promoter. Indeed, the results showed an enhancement of mesenchymal markers and the invasion of cells into the microvascular chamber. Importantly, intravasation of breast cancer cells into the vasculature was observed, highlighting the utility of this ToC device in recapitulating key hallmarks of cancer metastasis.

5. *CTC transit-on-a-chip*: After intravasation, cancer cells transit along the vasculature to become CTCs. Several ToC devices have been developed to capture, isolate, and analyze in peripheral blood CTCs. In addition, other tumor-associated biomarkers, such as ctDNA/RNA, antibodies, and exosomes, also present in peripheral blood representative of cancer metastasis, can be captured and analyzed. This type of *liquid biopsy* approach provides very valuable information about the origin of the tumor and its etiology, and therefore, it is clinically relevant. Different methods based on ToC devices have been reported for the isolation of this type of biomarkers from different tumor origins. For CTCs, different methods based on hydrodynamic separation, physical or chemical entrapment, or magnetic detection/isolation, have been described [25,89]. Recently, a ToC model made of carbon nanotubes was engineered to specifically capture CTCs from peripheral blood of breast cancer patients with high purity and with extremely high sensitivity [90]. This device exploited the preferential adherence of CTCs on the nanotube surface. It was first validated by spiking GFP-labeled MDA-MB-231 breast cancer cells in the blood of wild mice, and next, with samples from breast cancer patients from different stages. These features made this ToC device very promising for clinical applications, since it could potentially detect the appearance of cancer at very early stages, thus improving the diagnosis of cancer patients. Recently, inertial forces were employed to separate CTCs in a microfluidic device taking advantage of the difference in size with respect the other cell types (white blood cells) [91]. Interestingly, the device was tested using clinical samples from nonsmall-cell-lung-cancer patients exhibiting a high-separation efficiency. Overall, this method demonstrated to be very efficient without the need of sophisticated or biological equipment for the specific detection of CTCs.
6. *Extravasation-on-a-chip*: The extravasation of cancer cells out of the vasculature is the most essential step in the metastatic cascade. During this event, the CTCs arrest in the wall of the vasculature and escape invading the surrounding stroma forming a secondary metastatic niche. This event has been widely investigated and several ToC models have been outlined. In one ToC model, a 3D microvascular network was engineered within a fibrin hydrogel matrix and perfused with cancer cells via lumen openings [92,93]. The device contained three 3D hydrogel chambers separated by media channels, which were employed for the delivery and exchange of soluble factors and small particles. In the central chamber, HUVEC cells were encapsulated, whereas the

adjacent chambers contained human lung fibroblasts. After 4–5 days of culture, and due to fibroblasts paracrine signaling, HUVEC cells elongated forming vacuoles and connecting with neighboring cells to form stable lumen structures that connected with the media channels. Next, MDA-MB-231 breast cancer cells were perfused and eventually invaded the microvascular network. Interestingly, the cancer cells transmigrated across the endothelium invading the surrounding hydrogel region.

Other extravasation ToC models have also been observed. Most of these models share some similarities, but mostly focus in just one heterotypic cell–cell interaction [94,95]. Recently, a ToC device was used to investigate both CTC–endothelium and CTC–secondary site interactions during cancer cell extravasation [96]. The model integrated cancer and endothelial cells, and a 3D bone-like microenvironment to assess how organ tropism affected the extravasation capability of cancer cells from different tissues. The results showed that extravasation was associated with the upregulation of late-stage metastatic biomarkers along with specific proteases. Organ-specificity was investigated, finding that the bone microenvironment enhanced the invasion of breast, bladder and ovarian cancer cells.

7. *Metastasis-on-a-chip*: These are considered the most complete and complex ToC models, since they include most of the events occurring in the cascade of cancer metastasis. In particular, these models deal with organ specificity, hallmark of cancer metastasis. Due to the technical and biological complexity to model all these events, very few ToC models capable to reproduce the multiple components and events occurring in metastasis, have been reported. Recently, a multisite metastasis-on-a-chip system was described to assess the metastatic preference of cancer cells [97]. In particular, the device integrated different interconnected chambers. The first chamber contained the primary tumor site made of HCT-116 colorectal cancer cells. This was connected through microchannels, which mimicked the vasculature, to several downstream chambers that contained secondary liver, lung, and endothelial constructs embedded within a 3D HA/gelatin matrix. The results indicated that HCT-116 cells invaded the vasculature-like channel and preferentially metastasized to the liver and lung constructs, similar to in vivo scenario. Similarly, a multiorgan-on-a-chip approach was used to investigate organ-specificity in lung metastasis. Besides the primary tumor cells, the ToC model included the microvascular endothelium, immune cells, fibroblasts, and three distant organs to mimic lung metastasis to the brain (astrocytes), bone (osteocytes), and liver (hepatocytes). Interestingly, the cells from the respective metastatic sites overexpressed specific proteins characteristic of the native scenario. Importantly, these results were validated in vivo using a mice model.

Overall, these works demonstrate the clinical relevance of ToC models for investigating organ-specificity in cancer metastasis. Other works are currently working in this direction to increase the physiological relevance of the system. As an example, the mechanism of metastasis from breast to bone has recently been investigated using a human bone perivascular niche-on-a-chip [98]. The obtained results have exhibited the critical role of interstitial flow in the formation of stable vasculature that mediates cancer cell colonization.

#### 1.4.1.2 Drug discovery and screening on-chip

The compatibility of ToC models with advanced microscopy and biosensing techniques permits the investigation of perturbed biochemical, metabolic, and genetic profiles. This may provide key insights into the mechanistic determinants of tumor growth and progression as well as in the response of the tumor to treatment. Indeed, ToC models have significantly contributed to the understanding of the pharmacokinetics and pharmacodynamics of drugs in microcirculation, for assessing the efficiency of drug-loaded carriers in healthy and cancerous tissues, and to investigate the transport of nutrients, cytokines, and oxygen in the TME [78,83]. In one of these models, the efficiency of drug-loaded dendrimer nanoparticles was recently investigated using a ToC platform for the precise and specific targeting of HCT-116 colorectal cancer cells embedded within Matrigel™ [99]. Interestingly, this ToC model permitted to assess on-chip the response of cells to *gemcitabine* in a dose-dependent and radial manner. Overall, this work highlights the potential of ToC systems for the efficient screening of anticancerous drugs.

#### 1.4.1.3 Reproducing dynamic events on-chip

An essential characteristic of ToC models is their abilities to recreate dynamic events on-chip. This includes physiological fluid flow (vasculature and interstitial), mechanical deformation, biochemical cues, and multicellular architectures, among other characteristics. This has allowed to reproduce with high precision the mechanochemical and dynamic microenvironment of tumors. These dynamic events are well exemplified in the seminal paper reporting a *Lung-on-a-Chip* model, where pulmonary cells were exposed to cyclic stretching to mimic physiological breathing [100]. Interestingly, in a follow-up work, this model was lined with tumor cells to create in vitro a human orthotopic model of nonsmall-cell lung cancer [75]. This ToC model reproduced similar behavior and responses to therapies that were observed in human patients, thus confirming the clinical relevance of the model. Surprisingly, results showed that breathing motion suppressed tumor growth and invasion, which may explain the high level of resistance to therapy in patients with minimal residual cancer in regions of the lung. Overall, this work demonstrates the capability of ToC models to reproduce physiologically-relevant dynamic events on-chip. This makes ToC devices very promising for clinical applications.

#### 1.4.1.4 Personalized tumor-on-a-chip models

ToC models are well-suited to be combined with primary cells or different tissue types (i.e., tumor biopsy) from patients to reflect individuals physiology. This results into personalized ToC models of a cancer patient with high clinical relevance [101]. These personalized ToC models allow assessing the effect of specific chemotherapeutic strategies on individual patients. In addition, these models can be perfused with blood from the same patient, and therefore, tumor biomarkers, such as CTCs, can be captured and analyzed. Several examples of personalized ToC have been reported. As an example, patient-derived breast cancer organoids were seeded for longer time periods in a ToC model to investigate the physiological delivery of drugs to the tumor [88]. The organoids formed robust sprouting within the chip, reproducing in vivo features of the tumor. Next, *paclitaxel* was injected, which reduced tumor size in about 50%. Overall, this work demonstrated the capability of personalized ToC models to assess the response of patient tumors to chemotherapy. Finally, this high degree of personalization provides new opportunities for boosting patient-specific assessment of drug efficacy and safety. This paves the way toward a personalized medicine, where a treatment can be first evaluated on-chip, and then may be applied on the patient.

#### 1.4.1.5 *Manufacturing methods of a tumor-on-a-chip*

The basic components of a ToC model include a microfluidic chip, tumor cells, other cell types (stromal, endothelial, and others), 3D ECM material for cells encapsulation/adhesion or surface coating (e.g., collagen, Matrigel™, or a mixture of them), microfluidic tubing for interconnection, and external/internal pumping system (e.g., syringe pump). Typically, the microfluidic chip is first designed using computer-assisted software. Its design depends on a multitude of factors, including the event aiming to reproduce, the architecture of the TME to mimic, the amount and type of cells to populate, or the type(s) of fluid(s) to perfuse, among other parameters. The final components can differ from one chip to another, but they typically share common components (e.g., tumor reservoir, channels, porous membrane/slits, and others) and read-outs (e.g., invasion of cells, gradient sensing, quantification of gene expressions, and others). The microfluidic chip is typically fabricated using the same technology developed for microelectronics industry. As aforementioned, standard UV-photolithography and soft lithography methods are usually employed, which make the fabrication of microfluidic-based devices simple, fast, and at reduced cost. Finally, ToC models are usually made of biocompatible and optically transparent polymeric material (e.g., PDMS), plastic (e.g., PC, PS), hydrogel (e.g., collagen, Matrigel™), or glass, or a combination of them, replicated from the fabricated master. Then, tubing and external microfluidic pumps are connected to perfuse the device.

### 1.4.2 Three-dimensional bioprinted models

The extreme complexity of the TME has boosted the need for a new generation of biofabrication techniques capable to reproduce more accurately all the spatial, cellular, and biochemical characteristics of the tumor and surrounding microenvironment. This higher accuracy is expected to contribute in solving one of the major challenges in cancer research: the translation of potential drug candidates from the lab to the clinic by developing personalized 3D tumor models. Similar to microfluidic-based models, 3D bioprinting has rapidly evolved to precisely design and fabricate in vitro complex living tissues with a well-defined architecture [102]. This technique allows for the printing of different cell types and other critical components, such as the supporting matrix (typically, hydrogel-based biomaterials), with precise spatial control, following a hierarchical order, and with well-controlled combinations. This allows recapitulating in vitro fundamental cell–cell/ECM physiological interactions of the native scenario. Finally, 3D bioprinting also permits the printing of functional biomaterials, such as programmable capsules, to precisely (spatially and temporally) control the release of compounds, by extrinsic or intrinsic (e.g., degradation) means, thus allowing a prolonged and controlled therapeutic dose [103,104].

The advanced performance of 3D bioprinting has attracted the interest of cancer biologists to model the complex structure of the TME and to investigate tumor–stroma interactions [105–107]. This is because 3D bioprinting can reproduce the heterogeneous content and cues (physical and biochemical) of the native TME. As mentioned above, the cells are deposited following a hierarchic order in an engineered matrix, where they can mature, self-organize, and deposit their own matrix proteins. Recently, this approach was used to bioprint mm-sized 3D tumor tissues from different tumor subtypes of breast and pancreatic cancer. These tissues were surrounded by different stromal cell types (primary human mammary fibroblasts, HUVECs, adipocytes, and primary human bone-marrow-derived mesenchymal stem cells) [106]. The cancerous and stromal cells were encapsulated into a hydrogel-based bioink during printing, which eventually was replaced by the matrix generated by the printed cells. This resulted into a purely cellular structure reminiscent of in vivo neoplastic tissues. The obtained results showed that the bioprinted tumoral tissues displayed in vivo-like features. This was demonstrated by the observed interactions between the tumor cells and the TME and by the response of cancer cells to chemotherapeutic treatments. Interestingly, this method showed the promising capabilities to model patient-specific tumors by using primary patient tissue.

One of the main disadvantages of 3D bioprinting is the thickness of the generated tissues. They are typically thin (~1–2 mm) and only survive for relatively short periods of time (~2 weeks). This is in most cases due to the inability to perfuse satisfactorily the system. Embedding endothelial cells within the 3D bioprinted tissues may result in the generation of thick tissues containing functional vasculature that may allow their cultures for longer durations [108,109]. In this regard, a new method was reported to bioprint thick (~1 cm) 3D cell-laden, vascularized tissue that could be perfused on-chip for a long time period (6 weeks) [108]. HUVEC, human neonatal dermal fibroblasts, and human bone-marrow-derived cells were encapsulated within a gelatin and fibrinogen blend 3D matrix, which was cross-linked by a dual enzymatic, thrombin and transglutaminase strategy. Such thick tissues may be very advantageous for tissue engineering and regenerative medicine applications. In this regard, and as a proof of concept, the 3D bioprinted tissue was perfused with growth factors to differentiate the mesenchymal stem cells toward an osteogenic lineage. Even though this work was not directly related to cancer research, it demonstrated its strength for building physiologically relevant cancer models.

Overall, 3D bioprinted models of tumors offer the possibility to reproduce accurately the native spatial, biochemical, and cellular environment of tumors. This is of upmost importance to understand better how complex tumor–stromal interactions contribute to tumor growth, progression, and therapeutic response. This procedure/technique may eventually improve the clinical translation of potential anticancer drugs and therapeutics.

## 1.5 Applications of three-dimensional tumor models in cancer therapeutics

Besides the use of 3D biomaterials-based tumor models for investigating the mechanistic determinants of the disease, they also provide a powerful platform for cancer therapeutics. This includes the development and screening of new anticancerous drugs to assess the effect of drug metabolites in organs and tissues, or to investigate organ specificity of drug-loaded carriers, among other applications. Herein, we briefly describe some of these applications. For details, the readers may refer to specialized reviews on the topic and references therein.

### 1.5.1 Drug discovery, development, and screening

In the drug discovery and development pipeline, the cost of producing new cancer drugs is estimated to be approximately $1 billion dollars for each compound that reaches the market [110]. Indeed, only 1 out of 10 drug candidates ultimately succeeds. These poor results are the direct consequence of the low performance of the prescreening and testing models used to predict the effect of the drugs. The higher performance demonstrated by 3D biomaterials-based tumor models compared to their flat counterparts have triggered a paradigm shift in this field. They can recapitulate tumor–stromal interactions, which are of upmost importance in therapy response. This permits to reproduce with higher fidelity the in vivo scenario and the native therapeutic response of drugs in the human body [111]. This has attracted the interest of big pharmaceutical and biotechnology companies who are now interested in the development of a new generation of tumor models for cancer drug testing.

This new paradigm in cancer research is associated with the integration of patient-derived samples. This will univocally improve the clinical relevance of the obtained data helping physicians in selecting the best therapeutic approach for each patient. For this, 3D biomaterials-based models may also be combined with other technologies, such as microfluidics (i.e., organ-on-chip) or 3D bioprinting to also improve the throughput characteristics. Importantly, this will allow to efficiently analyze the adsorption, distribution, metabolism, excretion, and toxicity (ADMET) of anticancerous drugs and their metabolites within a realistic model. Indeed, there is a vast amount of literature about the use of 3D tumor models for ADMET applications [112–114]. Some of them include high-throughput characteristics, which are very attractive for the pharmaceutical and biotechnological industry [115]. In particular, the potential toxic effect of drug metabolites is one of the most critical steps in the pipeline of drug discovery. 3D biomaterials-based tumor models can be useful to assess cytotoxicity but cannot capture the effects of drug metabolites that become apparent at the organ systems level. Again, microfluidic devices provide an advanced platform to investigate the toxicity of drug metabolites by the interconnection of different organ-on-a-chip models (i.e., multiorgan-on-a-chip). This approach has recently been followed to evaluate the efficacy and off-target toxicity of anticancer drugs using multiple human cancer and healthy cells from different organs and recirculating serum-free medium containing different anticancer compounds [74]. The obtained results indicated that some of these compounds were dependent on drug metabolism by liver cells. Overall, this work showed how the combination of 3D biomaterials and ToC models provide an efficient preclinical drug-testing platform to evaluate on- and off-target toxicity of drugs and their metabolites. These multiorgan-on-a-chip models, when combined with cells from patients, may result in the development of personalized *body-* or *patient-on-a-chip* platforms capable to model systemic metastasis on-chip (Fig. 1.6).

Finally, 3D tumor models may be employed to assess the acquisition of drug resistance, a hallmark in cancer research. In one of these models, a 3D silk fibroin scaffold was used

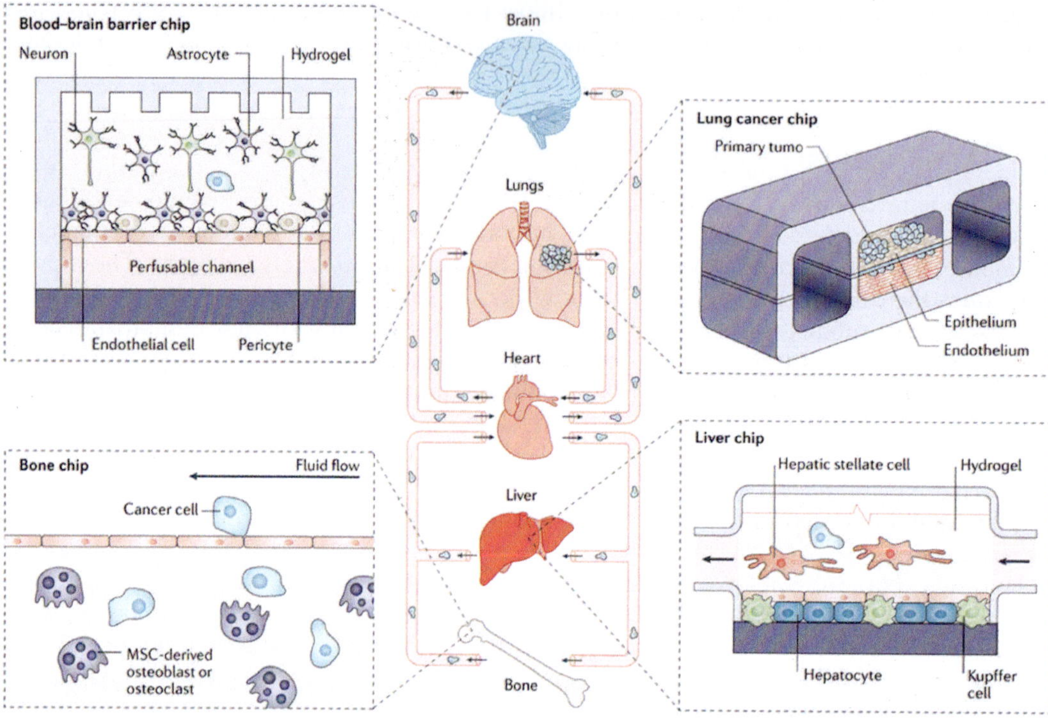


**FIGURE 1.6 Multiorgan or body-on-a-chip for 3D cancer metastasis studies.** Schematic representation on how the interconnection of individual organs-on-a-chip can be employed to study organ-specificity in cancer metastasis. The represented example shows the interconnection of individual organ models (lung, liver, bone, and brain). Here, lung cancer cells located in the lung-on-a-chip model invade the vascular channel and migrate to the other organ models. The formation of metastasis in the distant organs (liver, bone, and brain; the most common sites for lung cancer metastasis) may eventually be detected by labeling the lung cancer cells with a fluorophore. Source: *Reproduced with permission from Nature Publishing Group from Sontheimer-Phelps A, Hassell BA, and Ingber DE. Modelling cancer in microfluidic human organs-on-chips. Nat Rev Cancer 2019;19(2):65–81.*

as a metastasis model in a coculture of human breast adenocarcinoma, mesenchymal stem cells, and osteoblast-like cells [29]. Upon the addition of *paclitaxel* (25 μM), an anticancerous compounds, a significantly higher cell viability was observed for a coculture of breast cancer cells with osteoblast-like cells compared with human breast cancer cells alone. This suggested that the multicellular 3D environment supplied by the osteoblasts-like cells increased drug resistance to the breast cancer cells. Overall, this work demonstrated that the screening of new chemotherapeutic compounds in a native-like multicellular environment is of upmost importance to evaluate the actual effects of drugs and their resistance.

## 1.5.2 Transport and delivery of drugs

The specific and selective targeting of drugs into the tumor are one of the most challenging issues in the field of cancer therapeutics. In order to improve the selectivity and specificity of the drug, different targeting strategies, such as nanoparticles, have been proposed. As an example, porous silk fibroin nanoparticles were recently developed for the specific delivery and targeting of drugs to cancer cells [116]. In this work, *doxorubicin* was encapsulated within the nanoparticles, which were also decorated with folic acid. The latter was employed to specifically target the folate receptor of cancer cells. The obtained results showed that at low pH, high-ionic strength and high-enzyme concentration, there was a specific targeting of the nanoparticles and drug release toward the cancerous cells. Overall, this work shows the great potential of nanocarriers for targeted drug delivery, which may revolutionize cancer treatment due to their enhanced selectivity.

However, the encapsulation of drugs within nanoparticles or other type of carriers is associated with some systemic effects, such as corona formation or body clearance, which hamper their efficiency. In order to investigate this and other effects, several 3D high-throughput models of the tumor vasculature with improved resemblance to in vivo systems have been proposed to investigate the specificity of the drug transport and delivery [83]. These models include ToC-based platforms of human tumor blood vessels or 3D (bio) printed models of the human vasculature. These advanced models can better reproduce several of the key steps of the systemic drug delivery process, such as extravasation, immune response, and endothelial targeting. In addition, these models provide key insights to understand the behavior of nanocarriers in the vasculature, allowing the study of crucial phenomena involved in drug delivery, such as vascular targeting, the EPR effect, protein corona formation, or immune clearance.

Finally, and even though static 3D tumor models are widely employed to assess the effect of drugs, most of them are based on organ-on-a-chip technology, since the combination of microfluidics with 3D biomaterials can control better the cellular, structural, and hydrodynamic properties of the native TME, allowing also the real-time readout of drug testing data. In particular, they allow to investigate the role of the vascular and lymphatic system in the mechanism of drug delivery [83]. As a result, several ToC models have been developed to investigate the efficient delivery and specific targeting of therapeutic agents into the tumors and predict drug-induced responses on-chip [99].

## 1.6 Limitations of biomaterials-based three-dimensional tumor models

Even though biomaterials-based 3D tumor models display a higher performance compared with o standard 2D cell culture plates, the latter are still widely used for the early stages of drug discovery and screening. The main reasons behind the low adoption of 3D tumor models can be listed as follows:

1. *Physiological relevance*: Despite the high complexity of 3D models compared with flat cell culture system, they are still incapable to recapitulate fully all the structural, cellular, and biochemical compositions of the in vivo tissue and disease pathology. Similarly, the selection of the biomaterial is crucial. The appropriate biomaterial should be physiologically relevant but versatile enough to be simple to use and fine-tuned to meet the tissue-specific characteristics of an in vivo-like microenvironment.

2. *Simplicity, utility, and reproducibility*: The 3D tumor models should be easy to fabricate, store, and manipulate by nonexperienced users. At the end, this may determine whether the model is useful or not. In this regard, the models should address the issue of labor intensiveness and material cost as they are intended for the market. For this, there is a need for a "democratization" in the type and configuration of the model employed, which may differ from lab to lab. In this regard, clear guidelines are to be framed for choices of biomaterials, model design, and cellular content. This is in particular evident in the field of ToC models, where the proposed designs significantly differ from one and the other. This has an associated impact/concern on the reproducibility of the results. Several companies have proposed the use of commercial chips with generic and versatile designs in order to democratize the field of tumor models.
3. *Integration of biosensing and screening capabilities*: The 3D tumor models should be capable to integrate biosensing and screening capabilities in order to monitor fundamental parameters of tumor physiology, such as the expression of key cancer biomarkers or the release of compounds. For this, the improvement of the high-throughput and high-content screening capabilities (e.g., >384 multiwell plates) of this type of tumor models will univocally boost their adoption by the pharmaceutical and biotechnological market [117].
4. *Compatibility with existing technologies*: The 3D tumor models must be compatible with current technologies (assay and detection methods), such as advanced optical microscopy for imaging or automated screening setup for analysis. The former is of upmost importance, since 3D structures integrated within the models need to be visualized at high resolution avoiding light scattering, absorption, and poor light penetration with prolonged imaging acquisition times [114].
5. *Use of relevant cells and clinical/ethical regulations*: The route toward precision and personalized medicine requires the use of cells from the individual patients into the models. This is associated with some clinical and ethical concerns, including patient consent for getting the sample. Further, strict guidelines must be followed for the collection of cells.

If the aforementioned limitations are addressed, this may eventually rule out the eternal question of physiological relevance of all tumor models. If successful, the field of tumor modeling and 3D drug discovery and screening may move into the era of precision medicine, which may finally facilitate the development of personalized cancer therapies.

## 1.7 Future of three-dimensional biomaterials for cancer research

The emergence of 3D biomaterials-based platforms in the field of cancer research has already attracted the attention of big pharmaceutical companies interested in a new generation of tumor models for cancer drug discovery, screening, and testing. However, as mentioned above, there are still some limitations, which are threatening their adoption. This new paradigm in cancer research is associated with the use of patient-derived samples to help physicians in selecting the best therapeutic approach for every individual patient. For this, 3D biomaterials-based models may be combined with other technologies, such as microfluidics or 3D bioprinting to improve the throughput characteristics and physiological relevance of the tumor models. This will allow to analyze efficiently the ADMET of drugs and their metabolites.

Future challenges will also be related to the use of 4D biomaterials-based tumor models, where the morphodynamics of the 3D matrix can be controlled over time. This is of extreme relevance because the majority of the 3D biomaterials typically used for tumor modeling do not evolve over time (i.e., their physical properties do not change after biomaterial formation), in contrast to what happens in the native scenario, where extra- and intracellular matrix mechanics are far from constant. The use of novel biomaterials with reversible characteristics capable to evolve over time to capture the dynamic nature of the cellular microenvironment has been reported [32]. In some cases, this new type of 3D biomaterials can precisely expand or contract upon, for example, changes in temperature in a well-controlled and reversible manner [118]. These advanced and dynamic 4D biomaterials can provide greater temporal control on cellular behavior, which may allow for more advanced tumor models. This will provide unprecedented information about how tumor cells evolve over time and contribute to develop therapies that are more efficient.

Finally, 3D biomaterials are currently enabling a prolific field of cancer research dedicated to developing novel biomaterials for vaccine-based cancer immunotherapy. The use of cancer vaccines as a method to generate immune reactivity against tumors has already been explored since the discovery of tumor-specific antigens [119,120]. Despite some exceptions with the development of vaccines against hepatitis B and human papillomavirus, which are the causes of liver and cervical cancer, respectively [120], cancer vaccines have traditionally suffered from weak immunogenicity, systemic toxicity, and off-target effects; this hampers their broad clinical translation [121]. The combination of 3D implantable (or injectable) biomaterials with cancer vaccine components (e.g., DNA, mRNA, peptide/protein, antigens, or immunomodulators) enable an enhanced efficacy and reduced toxicity of cancer vaccines, thus overcoming their traditional limitations, such as limited duration of action and specificity. This is because specially-designed biomaterials can control how these components are distributed within the material and locally released to activate endogenous immune cells inside the body to attack cancer cells. Typically, this type of biomaterial employs porous scaffolds made from a biodegradable polymer (e.g., poly(lactic-*co*-glycolic acid) (PLGA), alginate-based hydrogels, mesoporous silica microrods [122]), which are injected with inactivated antigens from the patient's tumor cells. However, other types of biomaterials, such as liposomes or polymeric nanoparticles can be employed. Therein, immunostimulatory molecules can be infused to attract and activate dendritic cells to stimulate a host response. This new 3D biomaterial-based strategy holds very promising for improved cancer immunotherapy. For further details about the use of 3D biomaterials as cancer vaccines, the readers may refer to the recent reviews on the topic [121–123].

## 1.8 Final remarks and conclusions

Even though 2D in vitro tumor models have historically provided key insights into the mechanistic determinants of physiopathological phenomena, and have helped in the discovery and screening of new drugs, studies in flat environments are becoming less informative, and therefore, models that are more efficient are required. Biomaterials provide a unique opportunity to build a new generation of 3D in vitro models that reproduce the structural and biochemical complexity of tumors and surrounding ECM. These 3D in vitro models provide a more realistic environment to investigate cell–cell, cell–ECM interactions, and the response of cancer cells to drugs. Under these circumstances, the cells display phenotypes and gene expressions similar to those in vivo (living tissues in the body) thus enabling the identification of predictive biomarkers associated with tumor origin or resistance. Indeed, the cells grown on 3D microenvironments exhibit higher resistance to cancer drugs [124]. For this reason, it is of upmost importance to engineer biomaterials-based tumor models to study the response of cancer cells to a specific therapy before testing them in patients. In this chapter, we have described the most typical approaches for the development of 3D biomaterials-based methods to better model tumors in vitro. We have also highlighted their importance as well as their limitations. Finally, we have outlined the new generation of advanced 3D tumor models, which will contribute to develop efficient anticancerous drugs. In this regard, the introduction of microfluidic systems as perfusable vascularized models and 3D bioprinted systems have provided valuable insights into the key steps of cancer metastasis, including angiogenesis, intravasation, or extravasation, among other phenomena. In particular, ToC models have contributed to understand better how the local microenvironmental cues and biochemical gradients regulate tumor progression or the impact of dynamic events in the invasion capability of cancer cells. Finally, the introduction of different cell types and tissues have permitted to recapitulate native-like cancer behaviors and organ specificity in cancer metastasis. Overall, 3D biomaterials-based tumor models provide a unique 3D native-like microenvironment to investigate molecular and cellular activities responsible of tumor growth and progression. This type of assays may be adopted by the clinics and pharmaceutical/biotechnological industries due to their capacities to mimic human-specific disease states and to their efficiencies to assess the drug responses. Finally, 3D biomaterials-based tumor models offer unprecedented possibilities to investigate how environmental factors contribute to tumor appearance and progression, which may contribute in a better understanding of the mechanism of tumor occurrence.

## Acknowledgments

This work is financially supported by the European Union Framework Programme for Research and Innovation Horizon 2020 on Forefront Research in 3D Disease Cancer Models as in vitro Screening Technologies (FoReCaST) under grant agreement no. 668983; the Portuguese Foundation for Science and Technology (FCT) under the program CEEC Individual 2017 (CEECIND/00352/2017) to D.C.; and the FCT under the scope of the project Modeling Cancer Metastasis into the Human Microcirculation System using a Multiorgan-on-a-Chip Approach (2MATCH) (02/SAICT/2017 – nº 028070) funded by the Programa Operacional Regional do Norte supported by FEDER to D.C. and S.C.K.

## References

[1] Carvalho MR, et al. Evaluating biomaterial- and microfluidic-based 3D tumor models. Trends Biotechnol 2015;33(11):667–78.

[2] Jacks T, Weinberg RA. Taking the study of cancer cell survival to a new dimension. Cell 2002;111(7):923–5.

[3] Abbott A. Biology's new dimension. Nature 2003;424(6951):870–2.

[4] Bissell MJ, Radisky D. Putting tumours in context. Nat Rev Cancer 2001;1(1):46–54.

[5] Cukierman E, et al. Taking cell-matrix adhesions to the third dimension. Science 2001;294(5547):1708–12.

[6] Weaver VM, et al. Reversion of the malignant phenotype of human breast cells in three-dimensional culture and in vivo by integrin blocking antibodies. J Cell Biol 1997;137(1):231–45.

[7] Wolf K, et al. Compensation mechanism in tumor cell migration. J Cell Biol 2003;160(2):267–77.

[8] Rajangam K, et al. Peptide amphiphile nanostructure–heparin interactions and their relationship to bioactivity. Biomaterials 2008;29(23):3298–305.

[9] Kundu B, et al. Mechanical property of hydrogel and the presence of adipose stem cells in tumor stroma affect spheroid formation in 3D osteosarcoma model. ACS Appl Mater Interfaces 2019;16(16):14548–59.

[10] Mukhopadhyay SK, et al. Silk fibroin-thelebolan matrix: a promising chemopreventive scaffold for soft tissue cancer. Colloids Surf B Biointerfaces 2017;155:379–89.

[11] Subia B, et al. Target specific delivery of anticancer drug in silk fibroin based 3D distribution model of bone–breast cancer cells. ACS Appl Mater Interfaces 2015;7(4):2269–79.

[12] Talukdar S, Kundu SC. Engineered 3D silk-based metastasis models: interactions between human breast adenocarcinoma, mesenchymal stem cells and osteoblast-like cells. Adv Funct Mater 2013;23(42):5249–60.

[13] Balkwill FR, Capasso M, Hagemann T. The tumor microenvironment at a glance. J Cell Sci 2012;125 (23):5591–6.

[14] Bissell MJ, Hines WC. Why don't we get more cancer? A proposed role of the microenvironment in restraining cancer progression. Nat Med 2011;17:320.

[15] Lim B, et al. Inflammatory breast cancer biology: the tumour microenvironment is key. Nat Rev Cancer 2018;18(8):485–99.

[16] Pickup MW, Mouw JK, Weaver VM. The extracellular matrix modulates the hallmarks of cancer. EMBO Rep 2014;15(12):1243–53.

[17] Salvatore V, et al. The tumor microenvironment promotes cancer progression and cell migration. Oncotarget 2016;8(6):9608–16.

[18] Binnewies M, et al. Understanding the tumor immune microenvironment (TIME) for effective therapy. Nat Med 2018;24(5):541–50.

[19] Gouirand V, Guillaumond F, Vasseur S. Influence of the tumor microenvironment on cancer cells metabolic reprogramming. Front Oncol 2018;8(117).

[20] Plodinec M, et al. The nanomechanical signature of breast cancer. Nat Nanotech 2012;7:757.

[21] Provenzano Paolo P, et al. Enzymatic targeting of the stroma ablates physical barriers to treatment of pancreatic ductal adenocarcinoma. Cancer Cell 2012;21(3):418–29.

[22] Son B, et al. The role of tumor microenvironment in therapeutic resistance. Oncotarget 2016;8(3):3933–45.

[23] Weiswald L-B, Bellet D, Dangles-Marie V. Spherical cancer models in tumor biology. Neoplasia (NY) 2015;17 (1):1–15.

[24] Drost J, Clevers H. Organoids in cancer research. Nat Rev Cancer 2018;18(7):407–18.

[25] Caballero D, et al. Organ-on-chip models of cancer metastasis for future personalized medicine: from chip to the patient. Biomaterials 2017;149:98–115.

[26] Katt ME, et al. In vitro tumor models: advantages, disadvantages, variables, and selecting the right platform. Front Bioeng Biotechnol 2016;4(12):12.

[27] Rijal G, Li W. A versatile 3D tissue matrix scaffold system for tumor modeling and drug screening. Sci Adv 2017;3(9):e1700764.

[28] Dey T, et al. Cytotoxicity and sustained release of modified divinylsulfone from silk based 3D construct. J Mater Sci Mater Med 2015;26(11):263.

[29] Talukdar S, Kundu SC. A non-mulberry silk fibroin protein based 3D in vitro tumor model for evaluation of anticancer drug activity. Adv Funct Mater 2012;22(22):4778–88.

[30] Dhiman HK, Ray AR, Panda AK. Three-dimensional chitosan scaffold-based MCF-7 cell culture for the determination of the cytotoxicity of tamoxifen. Biomaterials 2005;26(9):979–86.

[31] Discher DE, Janmey P, Wang Y-L. Tissue cells feel and respond to the stiffness of their substrate. Science 2005;310(5751):1139–43.

[32] Rosales AM, Anseth KS. The design of reversible hydrogels to capture extracellular matrix dynamics. Nat Rev Mater 2016;1:15012.

[33] Levental KR, et al. Matrix crosslinking forces tumor progression by enhancing integrin signaling. Cell 2009;139(5):891–906.

[34] Franco-Barraza J, et al. Preparation of extracellular matrices produced by cultured and primary fibroblasts. Curr Protoc Cell Biol 2016;71:10.9.1–10.9.34.

[35] Caballero D, Samitier J. Topological control of extracellular matrix growth: a native-like model for cell morphodynamics studies. ACS Appl Mater Interfaces 2017;9(4):4159–70.

[36] Gilbert TW, Sellaro TL, Badylak SF. Decellularization of tissues and organs. Biomaterials 2006;27 (19):3675–83.

[37] Keane TJ, Swinehart IT, Badylak SF. Methods of tissue decellularization used for preparation of biologic scaffolds and in vivo relevance. Methods 2015;84:25–34.

[38] Bhardwaj N, Kundu SC. Electrospinning: a fascinating fiber fabrication technique. Biotechnol Adv 2010;28 (3):325–47.

[39] Poláková L, et al. Electrospun nanofibers for local anticancer therapy: review of in vivo activity. Int J Pharm 2019;558:268–83.

[40] Brancato V, et al. Bioengineered tumoral microtissues recapitulate desmoplastic reaction of pancreatic cancer. Acta Biomater 2017;49:152–66.

[41] Turner AEB, Flynn LE. Design and characterization of tissue-specific extracellular matrix-derived microcarriers. Tissue Eng Part C Methods 2012;18(3):186–97.

[42] Clapp KP, Castan A, Lindskog EK. Chapter 24 - Upstream processing equipment. In: Jagschies G, et al., editors. Biopharmaceutical processing. Amsterdam: Elsevier; 2018. p. 457–76.

[43] Abeille F, et al. Continuous microcarrier-based cell culture in a benchtop microfluidic bioreactor. Lab Chip 2014;14(18):3510–18.

[44] Pradhan S, Chaudhury CS, Lipke EA. Dual-phase, surface tension-based fabrication method for generation of tumor millibeads. Langmuir 2014;30(13):3817–25.

[45] Ong S-M, et al. Engineering a scaffold-free 3D tumor model for in vitro drug penetration studies. Biomaterials 2010;31(6):1180–90.

[46] Achilli T-M, Meyer J, Morgan JR. Advances in the formation, use and understanding of multi-cellular spheroids. Expert Opin Biol Ther 2012;12(10):1347–60.

[47] Elliott NT, Yuan F. A review of three-dimensional in vitro tissue models for drug discovery and transport studies. J Pharm Sci 2011;100(1):59–74.

[48] Nicholson KM, Bibby MC, Phillips RM. Influence of drug exposure parameters on the activity of paclitaxel in multicellular spheroids. Eur J Cancer 1997;33(8):1291–8.

[49] Lazzari G, et al. Multicellular spheroid based on a triple co-culture: a novel 3D model to mimic pancreatic tumor complexity. Acta Biomater 2018;78:296–307.

[50] Peela N, et al. A three dimensional micropatterned tumor model for breast cancer cell migration studies. Biomaterials 2016;81:72–83.

[51] Dickinson LE, et al. Functional surfaces for high-resolution analysis of cancer cell interactions on exogenous hyaluronic acid. Biomaterials 2010;31(20):5472–8.

[52] Villasante A, Vunjak-Novakovic G. Tissue-engineered models of human tumors for cancer research. Expert Opin Drug Discov 2015;10(3):257–68.

[53] Kundu B, et al. Silk fibroin biomaterials for tissue regenerations. Adv Drug Deliv Rev 2013;65(4):457–70.

[54] Mandal BB, Kundu SC. Non-bioengineered silk fibroin protein 3D scaffolds for potential biotechnological and tissue engineering applications. Macromol Biosci 2008;8(9):807–18.

[55] Regier MC, et al. Transitions from mono- to co- to tri-culture uniquely affect gene expression in breast cancer, stromal, and immune compartments. Biomed Microdevices 2016;18(4):70.

[56] O'Brien FJ, et al. Influence of freezing rate on pore structure in freeze-dried collagen-GAG scaffolds. Biomaterials 2004;25(6):1077–86.

[57] Tomihata K, Ikada Y. In vitro and in vivo degradation of films of chitin and its deacetylated derivatives. Biomaterials 1997;18(7):567–75.

[58] Van Tomme SR, Storm G, Hennink WE. In situ gelling hydrogels for pharmaceutical and biomedical applications. Int J Pharm 2008;355(1):1–18.

[59] Fourniols T, et al. Temozolomide-loaded photopolymerizable PEG-DMA-based hydrogel for the treatment of glioblastoma. J Control Release 2015;210:95–104.

[60] Eaton DF. Chapter 20: Photopolymers: an enabler of industrial innovation. In: Dessauer R, editor. Photochemistry, history and commercial applications of hexaarylbiimidazoles. Amsterdam: Elsevier Science B.V.; 2006. p. 207–14.

[61] Knowlton S, et al. Bioprinting for cancer research. Trends Biotechnol 2015;33(9):504–13.

[62] Weibel DB, DiLuzio WR, Whitesides GM. Microfabrication meets microbiology. Nat Rev Microbiol 2007;5:209.

[63] Truckenmüller R, et al. Thermoforming of film-based biomedical microdevices. Adv Mater 2011;23 (11):1311–29.

[64] Caballero D, et al. Direct patterning of anti-human serum albumin antibodies on aldehyde-terminated silicon nitride surfaces for HSA protein detection. Small 2009;5(13):1531–4.

[65] Borowiec J, et al. 3D microcontact printing for combined chemical and topographical patterning on porous cell culture membrane. ACS Appl Mater Interfaces 2018;10(26):22857–65.

[66] Jadhav RG, Das AK. Chapter 10 - Four dimensional printing in healthcare. In: Kalaskar DM, editor. 3D printing in medicine. Cambridge: Woodhead Publishing; 2017. p. 207–18.

[67] Sydney Gladman A, et al. Biomimetic 4D printing. Nat Mater 2016;15:413.

[68] Wehner M, et al. An integrated design and fabrication strategy for entirely soft, autonomous robots. Nature 2016;536:451.

[69] Gao B, et al. 4D bioprinting for biomedical applications. Trends Biotechnol 2016;34(9):746–56.

[70] Li H, et al. Magnetic actuated pH-responsive hydrogel-based soft micro-robot for targeted drug delivery. Smart Mater Struct 2016;25(2):027001.

[71] Lui YS, et al. 4D printing and stimuli-responsive materials in biomedical aspects. Acta Biomater 2019;92:19–36.

[72] Sontheimer-Phelps A, Hassell BA, Ingber DE. Modelling cancer in microfluidic human organs-on-chips. Nat Rev Cancer 2019;19(2):65–81.

[73] Bhatia SN, Ingber DE. Microfluidic organs-on-chips. Nat Biotechnol 2014;32:760.

[74] McAleer CW, et al. Multi-organ system for the evaluation of efficacy and off-target toxicity of anticancer therapeutics. Sci Transl Med 2019;11(497):eaav1386.

[75] Hassell BA, et al. Human organ chip models recapitulate orthotopic lung cancer growth, therapeutic responses, and tumor dormancy in vitro. Cell Rep 2017;21(2):508–16.

[76] Kim HJ, et al. Contributions of microbiome and mechanical deformation to intestinal bacterial overgrowth and inflammation in a human gut-on-a-chip. Proc Natl Acad Sci U S A 2016;113(1):E7–15.

[77] Villenave R, et al. Human gut-on-a-chip supports polarized infection of coxsackie B1 virus in vitro. PLoS One 2017;12(2):e0169412.

[78] Esch EW, Bahinski A, Huh D. Organs-on-chips at the frontiers of drug discovery. Nat Rev Drug Discov 2015;14(4):248–60.

[79] Montanez-Sauri SI, et al. Enabling screening in 3D microenvironments: probing matrix and stromal effects on the morphology and proliferation of T47D breast carcinoma cells. Integr Biol 2013;5(3):631–40.

[80] Eslami Amirabadi H, et al. A novel method to understand tumor cell invasion: integrating extracellular matrix mimicking layers in microfluidic chips by "selective curing". Biomed Microdevices 2017;19(4):92.

[81] Chen Z-z, et al. Establishment of a gastric cancer subline with high metastatic potential using a novel microfluidic system. Sci Rep 2016;6:38376.

[82] Sato M, et al. Microcirculation-on-a-chip: a microfluidic platform for assaying blood- and lymphatic-vessel permeability. PLoS One 2015;10(9):e0137301.

[83] Caballero D, et al. Tumour-vessel-on-a-chip models for drug delivery. Lab Chip 2017;17(22):3760–71.

[84] Chen L-J, Kaji H. Modeling angiogenesis with micro- and nanotechnology. Lab Chip 2017;17(24):4186–219.

[85] Nguyen D-HT, et al. Biomimetic model to reconstitute angiogenic sprouting morphogenesis in vitro. Proc Natl Acad Sci U S A 2013;110(17):6712–17.

[86] Kim S, Chung M, Jeon NL. Three-dimensional biomimetic model to reconstitute sprouting lymphangiogenesis in vitro. Biomaterials 2016;78:115–28.

[87] Zervantonakis IK, et al. Three-dimensional microfluidic model for tumor cell intravasation and endothelial barrier function. Proc Natl Acad Sci U S A 2012;109(34):13515–20.

[88] Shirure VS, et al. Tumor-on-a-chip platform to investigate progression and drug sensitivity in cell lines and patient-derived organoids. Lab Chip 2018;18(23):3687–702.

[89] Vaidyanathan R, et al. Cancer diagnosis: from tumor to liquid biopsy and beyond. Lab Chip 2019;19(1):11–34.
[90] Loeian MS, et al. Liquid biopsy using the nanotube-CTC-chip: capture of invasive CTCs with high purity using preferential adherence in breast cancer patients. Lab Chip 2019;19(11):1899–915.
[91] Zhou J, et al. Isolation of circulating tumor cells in non-small-cell-lung-cancer patients using a multi-flow microfluidic channel. Microsyst Nanoeng 2019;5(1):8.
[92] Chen MB, et al. On-chip human microvasculature assay for visualization and quantification of tumor cell extravasation dynamics. Nat Protoc 2017;12:865.
[93] Chen MB, et al. Mechanisms of tumor cell extravasation in an in vitro microvascular network platform. Integr Biol 2013;5(10):1262–71.
[94] Jeon JS, et al. In vitro model of tumor cell extravasation. PLoS One 2013;8(2):e56910.
[95] Jeon JS, et al. Human 3D vascularized organotypic microfluidic assays to study breast cancer cell extravasation. Proc Natl Acad Sci U S A 2015;112(1):214–19.
[96] Bersini S, et al. A combined microfluidic-transcriptomic approach to characterize the extravasation potential of cancer cells. Oncotarget 2018;9(90):36110–25.
[97] Aleman J, Skardal A. A multi-site metastasis-on-a-chip microphysiological system for assessing metastatic preference of cancer cells. Biotechnol Bioeng 2019;116(4):936–44.
[98] Marturano-Kruik A, et al. Human bone perivascular niche-on-a-chip for studying metastatic colonization. Proc Natl Acad Sci U S A 2018;115(6):1256–61.
[99] Carvalho MR, et al. Colorectal tumor-on-a-chip system: a 3D tool for precision onco-nanomedicine. Sci Adv 2019;5(5):eaaw1317.
[100] Huh D, et al. Reconstituting organ-level lung functions on a chip. Science 2010;328(5986):1662–8.
[101] van den Berg A, et al. Personalised organs-on-chips: functional testing for precision medicine. Lab Chip 2019;19(2):198–205.
[102] Murphy SV, Atala A. 3D bioprinting of tissues and organs. Nat Biotechnol 2014;32:773.
[103] Gupta MK, et al. 3D printed programmable release capsules. Nano Lett 2015;15(8):5321–9.
[104] Yi H-G, et al. A 3D-printed local drug delivery patch for pancreatic cancer growth suppression. J Control Release 2016;238:231–41.
[105] Albritton JL, Miller JS. 3D bioprinting: improving in vitro models of metastasis with heterogeneous tumor microenvironments. Dis Model Mech 2017;10(1):3–14.
[106] Langer EM, et al. Modeling tumor phenotypes in vitro with three-dimensional bioprinting. Cell Rep 2019;26(3):608–23.
[107] Meng F, et al. 3D bioprinted in vitro metastatic models via reconstruction of tumor microenvironments. Adv Mater 2019;31(10):1806899.
[108] Kolesky DB, et al. Three-dimensional bioprinting of thick vascularized tissues. Proc Natl Acad Sci U S A 2016;113(12):3179–84.
[109] Pi Q, et al. Digitally tunable microfluidic bioprinting of multilayered cannular tissues. Adv Mater 2018;30(43):1706913.
[110] DiMasi JA, Hansen RW, Grabowski HG. The price of innovation: new estimates of drug development costs. J Health Econ 2003;22(2):151–85.
[111] Cox MC, et al. Toward the broad adoption of 3D tumor models in the cancer drug pipeline. ACS Biomater Sci Eng 2015;1(10):877–94.
[112] Li W, et al. A novel tissue-engineered 3D tumor model for anti-cancer drug discovery. Biofabrication 2018;11(1):015004.
[113] Puls TJ, et al. Development of a novel 3D tumor-tissue invasion model for high-throughput, high-content phenotypic drug screening. Sci Rep 2018;8(1):13039.
[114] Langhans SA. Three-dimensional in vitro cell culture models in drug discovery and drug repositioning. Front Pharmacol 2018;9:6.
[115] Ma W-Y, et al. A novel 96well-formatted micro-gap plate enabling drug response profiling on primary tumour samples. Sci Rep 2015;5:9656.
[116] Cai Y, et al. Fabricated porous silk fibroin particles for pH-responsive drug delivery and targeting of tumor cells. J Mater Sci 2018;54:3319.
[117] Montanez-Sauri SI, Beebe DJ, Sung KE. Microscale screening systems for 3D cellular microenvironments: platforms, advances, and challenges. Cell Mol Life Sci 2015;72(2):237–49.

[118] de Almeida P, et al. Cytoskeletal stiffening in synthetic hydrogels. Nat Commun 2019;10(1):609.
[119] Rousseau RF, et al. Cancer vaccines. Hematol Oncol Clin North Am 2001;15(4):741–73.
[120] Hollingsworth RE, Jansen K. Turning the corner on therapeutic cancer vaccines. NPJ Vaccines 2019;4(1):7.
[121] Zhang R, Billingsley MM, Mitchell MJ. Biomaterials for vaccine-based cancer immunotherapy. J Control Release 2018;292:256–76.
[122] Wang H, Mooney DJ. Biomaterial-assisted targeted modulation of immune cells in cancer treatment. Nat Mater 2018;17(9):761–72.
[123] Weiden J, Tel J, Figdor CG. Synthetic immune niches for cancer immunotherapy. Nat Rev Immunol 2017;18:212.
[124] Breslin S, O'Driscoll L. The relevance of using 3D cell cultures, in addition to 2D monolayer cultures, when evaluating breast cancer drug sensitivity and resistance. Oncotarget 2016;7(29):45745–56.

CHAPTER

# 2

# Bioinspired biomaterials to develop cell-rich spherical microtissues for 3D in vitro tumor modeling

*Maria V. Monteiro, Vítor M. Gaspar and João F. Mano*

**Department of Chemistry, CICECO—Aveiro Institute of Materials, University of Aveiro, Campus Universitário de Santiago, Aveiro, Portugal**

## 2.1 Introduction

Cancer is the second leading cause of death worldwide, accounting for around 9.6 million deaths in 2018 (World Health Organization). Human solid tumors are characterized by a high cellular and structural heterogeneity characterized by proliferative, hypoxic, and necrotic regions [1]. In vivo animal models and two- and three-dimensional (2D/3D) in vitro models have been employed during preclinical developmental stages to support fundamental cancer research and to screen more effective therapies for solid cancers [2]. Among these, monolayer 2D cell cultures are the gold standard for anticancer drug screening. However, such platforms are neither representative of in vivo tumors 3D architecture, cellular heterogeneity, nor include cell–extracellular matrix (ECM) interactions [3,4]. On the other hand, the use of animal models entails high costs and ethical issues, and has demonstrated a very low correlation with the data obtained in clinical trials [5]. Aiming to more closely recapitulate the architecture and tumor microenvironment (TME) components of in vivo tumors, 3D in vitro models have arisen as a robust, and cost-effective testing model for anticancer drug discovery and screening [6].

The 3D multicellular tumor spheroids (MCTS) constitute a particularly valuable type of in vitro models that were initially established by Sutherland et al. in the early 1970s. In more recent decades, considerable developments on 3D MCTS in vitro bioengineering have led to a renewed interest and to an improved potential for drug screening and fundamental tumor biology research [7–9]. In essence, 3D MCTS represent an intermediate complexity level between monolayer cultures and in vivo solid tumors generated via


DOI: https://doi.org/10.1016/B978-0-12-818128-7.00002-2

laboratory animal models [1]. Although 3D tumor spheroids recapitulate better the in vivo situation with regard to key cellular components (e.g., malignant and stromal cells), gene expression profiles and physiology, this system possesses some limitations, including the open pattern where culture media contacts the outermost layer of cells directly without the barrier provided by the tumor ECM as in the in vivo setting [1,10]. To overcome this limitation, the inclusion of ECM-mimetic biomaterials processed in the form of spherical scaffolds, namely microparticles (MPs) or microcapsules have been exploited in order to provide 3D MCTS with the biochemical and mechanical cues provided by native tumor ECM [11,12]. The inclusion of matrix-mimetic scaffolds allows the establishment of physiomimetic 3D models that mimic more closely in vivo solid tumors hallmarks, while providing a simplistic platform that is compatible with high-throughput studies including the screening of therapeutic compounds. This complexity level enables researchers to study cell–ECM interactions, to evaluate the influence of ECM components in invasion/ metastasis processes, and to screen new anticancer therapeutics.

## 2.2 Human Tumor microenvironment—key hallmarks to mimic in vitro

Malignant cells reside in a complex TME (Fig. 2.1) in which they are exposed to biomechanical stimuli and dynamic interactions with the surrounding stroma, affecting

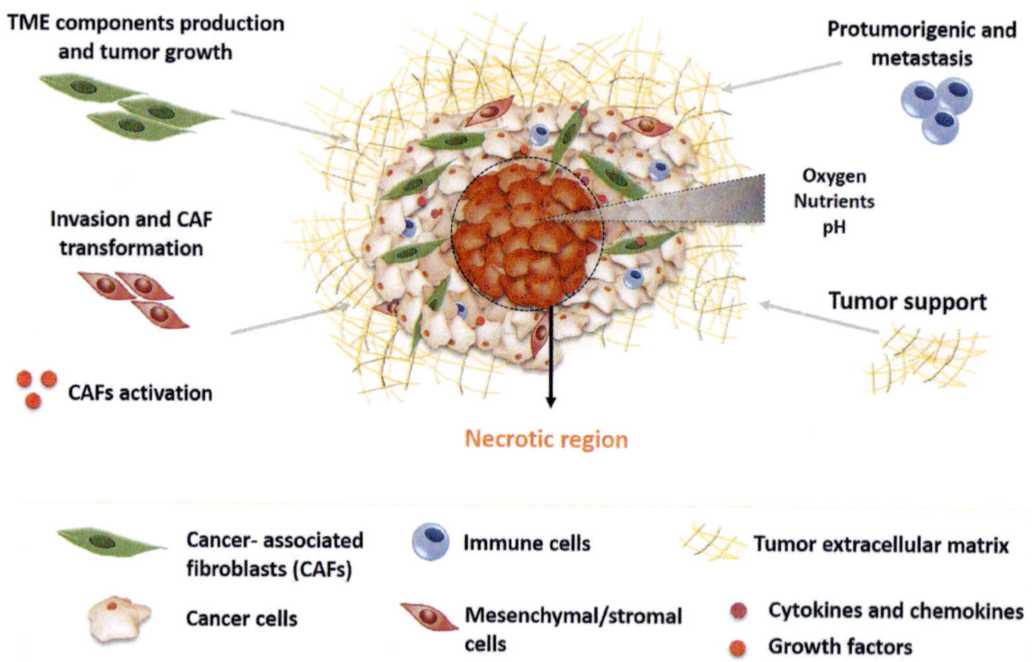


FIGURE 2.1 Schematic representation of the complex tumor microenvironment and its major components.

cancer cell behavior, tumor progression, maintenance, and drug resistance [13,14]. The native TME generally comprises cancer cells, stromal cells [e.g., cancer-associated fibroblasts (CAFs), mesenchymal, endothelial, and immune cells], and a supporting ECM. All these elements have an influential role in the physiology, structure, and progression of the tumor [15]. The ECM consists of an acellular 3D dynamic network generally secreted by cells for their structural and biochemical support [16]. In general, it is mainly comprised of water, growth factors, cytokines, fibronectin, laminin, collagen (types I–V), elastin, entactin, fibrillin, vitronectin, and hyaluronic acid [17]. This supporting matrix regulates cellular function and morphogenesis and influences nutrients/metabolites diffusion and growth factors [18]. Particularly, ECM mechanical/viscoelastic properties, permeability, composition, spatial organization, and topography affect cancer cells proliferation, epithelial–mesenchymal transition (EMT), invasion, metastasis, and the bioperformance of anticancer treatments [19,20]. During tumorigenesis, cancer cells can remodel the surrounding ECM via enzymatic degradation (e.g., matrix metalloproteinase mediated) followed by abnormal secretion and de novo deposition of ECM proteins [16]. As the tumor progresses, an increased tumor size is accompanied by proliferation and recruitment of stromal cells which secrete various growth factors [e.g., vascular endothelial growth factor (VEGF) and hypoxia-inducible factor (HIF)], which promote angiogenesis triggering new blood vessel formation to support proliferating malignant cells [16]. The excessive proliferation of cancer cells promotes the invasion of local tissue and migration to distant organs (metastatic sites) [21].

This complexity and diversity of in vivo tumors motivates the development of 3D in vitro models comprising biomimetic environments, therefore contributing to improving novel therapies preclinical validation data in comparison to that obtained from standard 2D models as it will be discussed.

## 2.3 3D In vitro tumor models—bridging the gap from 2D flat cultures to in vivo

A deeper knowledge of anticancer therapeutics response is limited by currently available disease models which fail to provide truly predictive data that can corroborate that obtained in human clinical trials [22].

One approach to study human neoplasia relies on the use of laboratory animal models aiming to replicate the biological, physiological, and dynamic microenvironment of human tumors [5]. The most routinely used animal model in drug screening and efficacy studies is the patient-derived xenograft (PDX) model which consists of inoculating human cancer cells or small fragments of tumor tissue either subcutaneously or in specific organs/tissues of immunodeficient mice [23]. This personalized medicine methodology possesses the advantage of providing the native 3D microenvironment in which tumors reside, allowing to transplant both malignant tissues and surrounding stroma to mimic the complexity of the human TME [24]. Although animal models have contributed to some key advances of cancer biology such models are low-throughput, time-consuming, expensive, difficult to scale up, and generally lack a completely functional immune system which may provide insufficient information for translating therapeutics to the clinic setting [25]. Additionally, the use of animal models also encompasses several ethical issues [2,23].

Alternatively, 2D in vitro flat cell culture models are recommended by regulatory agencies as the gold standard testing platform to understand the biomolecular mechanisms underlying tumorigenesis and to evaluate therapeutics performance [3,26]. In these models, monotypic adherent cells are cultured in a flat monolayer and form tight focal adhesions with the stiff culture plate surfaces. Such experimental setup influences cell signaling, gene expression, and drug sensitivity [4,27]. The ease of handling and straightforward analysis often amenable with high-content/high-throughput systems, the cost-effectiveness, good reproducibility, and ability to growth a myriad of different cell types render 2D cultures one of the most employed preclinical in vitro methodologies for therapeutics screening in cancer research [28,29]. However, 2D cultured cells fail to recapitulate key features of solid tumors including their complex 3D architecture, cell migration, hypoxic areas, pH gradients, and the presence of a supporting ECM, a fundamental component that influences cancer progression [22,26]. Particularly, monolayer cell cultures are unable to mimic cell–ECM crosstalk, nor the 3D organization, or the biomechanical and biochemical cues provided by the surrounding TME. Therefore, significant differences in the cellular morphology, gene expression, intracellular signaling, and drug resistance occur between monolayer cultured cells and those in in vivo tumors what decreases the realistic validity of 2D in vitro research [30].

Consequently, the search for alternative preclinical tumor models that could more accurately recapitulate the complexity of this disease is in high demand.

In this context, 3D cell cultures have emerged as a promising platform for tumor modeling and screening for anticancer therapies as they are able to closely mimic the 3D architecture of malignant tissue as well as TME [31]. The 3D in vitro cell cultures provide more physiological responses and cellular function, differentiation, drugs metabolism, gene expression patterns and protein synthesis, morphology, proliferation, and viability [32]. Adding to this tridimensional aspect, bioengineered tumor models must integrate both cellular and acellular components of the TME in order to recapitulate major tumor hallmarks [33]. In fact, the crosstalk between TME-related cells and tumor-specific ECM is well recognized to also contribute to disease progression depending on the designed platform [23]. ECM composition and setup architecture play a key role in phenotypic behavior and, consequently, the results of experiments. The design principles of 3D in vitro tissue models must therefore include cellular components ratio and phenotype, ECM-mimetic materials, the establishment of biomolecular gradients and cell–cell, cell–ECM signaling. Holistically an ideal 3D in vitro cancer model should fulfill the following criteria: (1) be representative of all the cellular classes and tumor–ECM presented in in vivo TME, (2) be capable of mimicking the interactions between cellular and supporting matrix components, (3) recapitulate biomechanical and biochemical properties, and spatial distribution TME components, and (4) reflect the invasive/metastatic behavior and drug resistance [23].

Currently available 3D in vitro tumor models range from simple multicellular spheroids comprising cell–cell aggregates (*scaffold-free 3D MCTS*), up to microtumors consisting of multiple cellular components and preexisting ECM-mimetic components (*scaffold-based 3D tumor models*) [4,22]. The following sections will report different classes of 3D in vitro tumor models with an emphasis being provided on cell-rich spherical models incorporating ECM-mimetic biomaterials.

## 2.4 Classes of 3D multicellular tumor models

### 2.4.1 Scaffold-free cell-rich 3D multicellular tumor spheroids

The scaffold-free approaches enable the establishment of 3D MCTS by culturing a cell suspension in nonadherent substrates [8,34]. MCTS models emerged as suitable in vitro testing platforms that reproduce key features found in in vivo solid tumors, including: (1) cellular heterogeneity, (2) cell–cell signaling, (3) pH/nutrient/oxygen gradients, (4) 3D ECM deposition during culture, (5) tumor growth kinetics, (6) gene expression, and (7) drug resistance [35]. The 3D MCTSs architecture comprises actively proliferating cells on the spheroid periphery with quiescent cells being located in the inner, nutrient-deprived zone [22]. Typically the size of reported spheroids ranges from 400 to 800 μm [9]. These models are easy to assemble and are amenable to high-throughput assays with good adaptability to multiwell plates for screening a large array of drug concentrations or drug–drug combinations simultaneously [17].

Currently, different technologies are available for spheroids fabrication, including: (1) hanging-drop or forced floating (static cultures), and (2) stirring-based systems such as stirring tank bioreactors or rotational bioreactors (reviewed in detail by [36,37]). Static cultures employ superhydrophobic or nonadherent surfaces [e.g., poly(2-hydroxyethyl methacrylate)], or agarose coated multiwell plates, and allow the assembly of 3D MCTSs with reproducible sizes and spheroidal morphology. This methodology prevents cellular adhesion to culture plate surfaces and promotes cell-to-cell interactions which lead to aggregates formation [17]. By using the liquid overlay methodology and culturing cells in ultralow adhesion U-bottom plates researchers generated different types of prostate cancer multicellular 3D spheroids with uniform morphology and tunable sizes depending on the initial cell density (Fig. 2.2). These highly reproducible models allowed the evaluation of different chemotherapeutics anticancer performance (e.g., MLN 4924, cisplatin, and docetaxel) via high-content screening of caspases activation, LysoTracker penetration and morphometric features [38].

Moreover, due to the versatility of these microtissue fabrication technologies, 3D MCTSs that generally incorporate cancer cell lines can also be cocultured with key stromal cells such as fibroblasts, endothelial, and/or immune system cells [39]. Since the tumor–stromal interactions in heterotypic 3D coculture spheroids provide important cues, MCTS formed by cancer cells and different stromal cells mimic more closely native tumors physiology and physical barriers in comparison to monoculture models. In fact, during culture, 3D MCTS also secrete de novo ECM, which increases microtumors density, yielding compact solid tumors similar to those obtained in vivo. Biomimetic heterotypic models are highly valuable both for fundamental research and for evaluating advanced anticancer therapeutics such as nanomedicines. Stromal–malignant coculture 3D spheroids were recently used as a testing platform for evaluating the influence of tumor–stromal components and in vitro synthesized ECM in nanoparticles penetration [40]. Tumor models were established via the liquid overlay technique in customized microwell arrays (108 wells of 200 μm depth, 400 μm diameter) (Fig. 2.3A). This platform allowed the generation of fibroblasts (3T3) and breast cancer cells (4T1) microtumors with spheroidal morphology, defined size, and with tunable cell-to-cell ratios. The heterotypic models were then used to

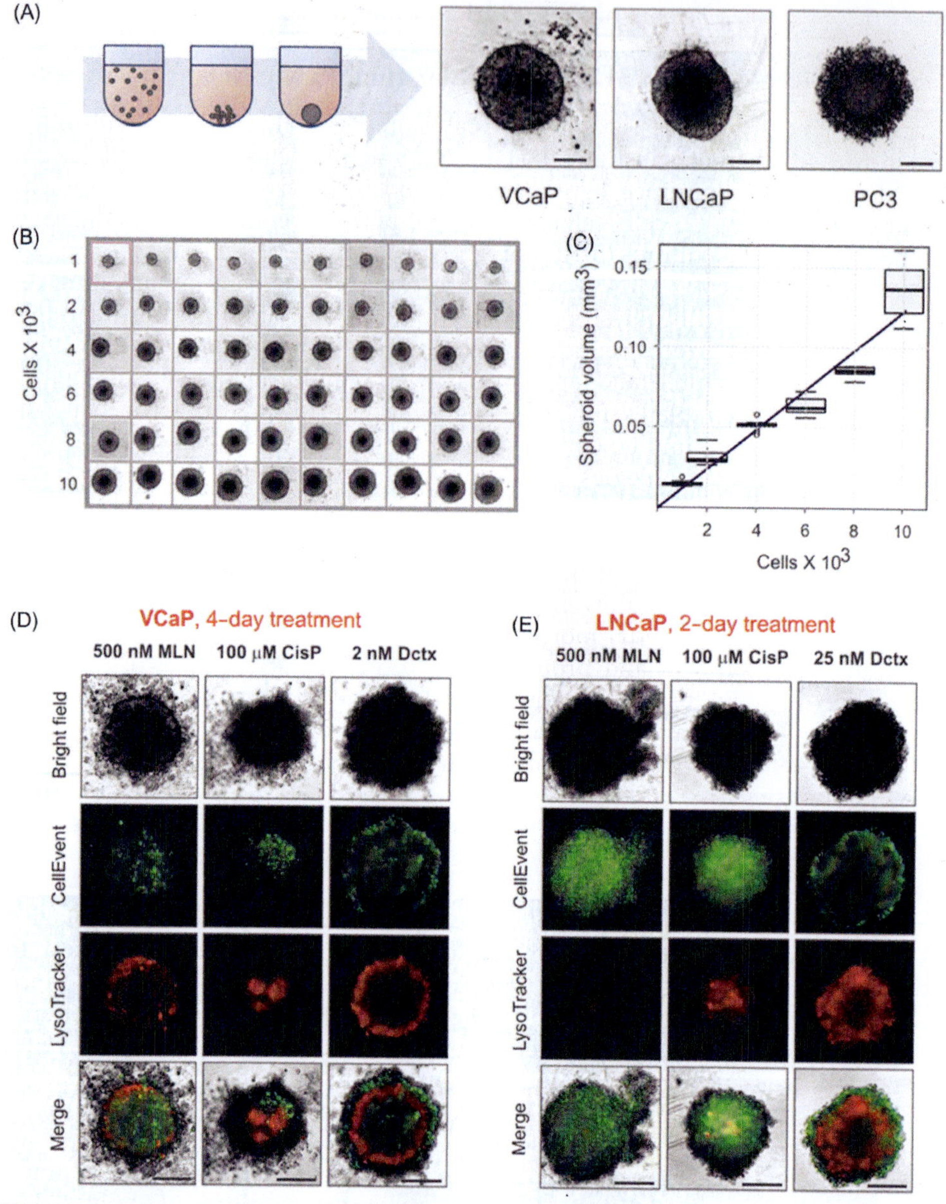


FIGURE 2.2 (A) Schematics of prostate cancer 3D spheroids assembly in ultralow adhesion U-bottom plates. Different cell lines were evaluated: *VCaP*, vertebral cancer of the prostate; *LNCaP*, lymph node carcinoma of the prostate; and *PC-3*, human Caucasian prostate adenocarcinoma. (B) Reproducibility of VCaP 3D spheroids assembled at different cell densities, during 7 days. (C) Box-plots of VCaP 3D spheroids volume according to initial cell density. (D and E) High-content analysis of VCaP and LNCaP cytotoxic phenotypes upon administration of different chemotherapeutics: MLN4924, Cisplatin, and Docetaxel. Source*: Reproduced from Mittler F, Obeid P, Rulina AV, Haguet V, Gidrol X, Balakirev MY. High-content monitoring of drug effects in a 3D spheroid model. Front Oncol 2017;7:1–19, with permission from Frontiers under a Creative Commons Attribution License (CC-BY 4.0). Additional permission from the authors have also been obtained.*

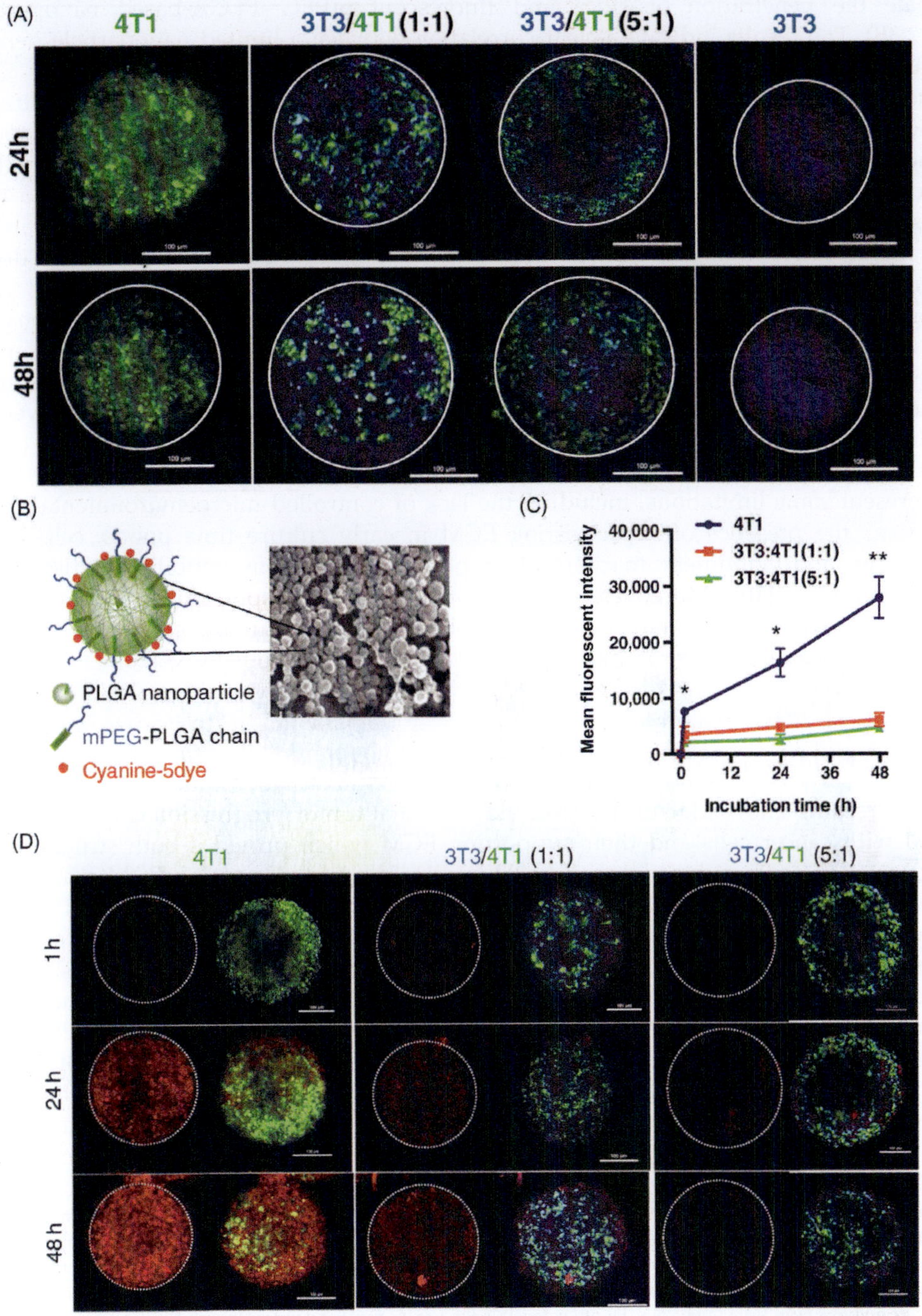


FIGURE 2.3 (A) Laser scanning confocal microscopy micrographs of monotypic and heterotypic 3D spheroid models of breast cancer at 48 h. 3T3 cells—calcein red (blue as pseudo-color) and 4T1—calcein green, respectively.

evaluate the penetration of silica and fluorescent mPEG–PLGA-based nanoparticles (Fig. 2.3B). The results indicate a clear correlation between a limited nanoparticle penetration and the presence of stromal cells in heterotypic models (Fig. 2.3C and D).

These studies evidence the importance of better recapitulating the malignant, stromal, and ECM deposition in vitro as it can significantly influence the overall bioperformance of chemotherapeutics or nanomedicines. Building on this knowledge, recent studies have aimed at following the intricate processes of de novo ECM deposition in vitro in a more precise mode. Using metabolic labeling and advanced imaging techniques researchers were recently able to evaluate nascent proteins deposition, thus contributing for a deeper understanding and modulation of the cell-secreted TME during culture [41].

Regarding dynamic cultures, stirring tank bioreactors that take advantage of mechanical forces to internally to impel culture media have been used to maintain cells in continuous suspension and lead to the formation of 3D cellular agglomerates [17]. Other types of bioreactors such as those promoting microgravity conditions have also been employed for tumor models fabrication.

Despite the remarkable advances in preclinical drug screening provided by 3D MCTS they present some limitations, including the lack of controlled microenvironmental conditions (e.g., the presence of a preexisting ECM in early culture time points, cell–matrix interactions, and biomimetic mechanical properties) and the susceptibility to dissociation during handling [10]. Moreover, ECM-mediated mass transport limitations regarding metabolites exchange and anticancer drugs delivery to native tumors cannot be accurately replicated in scaffold-free strategies due to the lack of endogenous ECM [10].

### 2.4.2 Scaffold-based 3D multicellular tumor models

An increasing knowledge of the TME indicates that tumor progression is intimately correlated with cancer cells and their supporting ECM which provides both structural and biochemical cues. Thus, bioengineering 3D in vitro models comprising ECM-mimetic biomaterials is a valuable approach for investigating TME influence on cancer cells in different scenarios, including after therapeutics administration. In in vitro 3D scaffold-based models, cancer cells proliferate in close contact with the stromal components [23]. By employing ECM-biomimetic biomaterials researchers can also investigate the fundamental interactions occurring between different cell types and the establishment of highly resistant subpopulations (e.g., cancer stem cells). Different classes of scaffold-based 3D models have been designed to suitably recapitulate tumor ECM and also accommodate the heterogeneous cellular components of the tumor. The 3D scaffold-based models also enable the establishment of drug resistant phenotypes providing an improved in vitro/in vivo

(B) Schematics of PLGA nanoparticles loaded with Cyanine-5 dye (red fluorescence). (C) Semiquantitative analysis of nanoparticles fluorescence intensity in upon administration in homotypic and heterotypic spheroids. (D) Confocal laser scanning microscopy images of monotypic and heterotypic spheroids after incubation with PLGA/Cy-5 nanoparticles. 4T1 (calcein green), 3T3 fibroblasts (calcein red-orange, pseudo-color blue). Scale bar: 100 μm. Source: *Reproduced from Priwitaningrum DL, Blondé J-BG, Sridhar A, van Baarlen J, Hennink WE, Storm G, et al. Tumor stroma-containing 3D spheroid arrays: a tool to study nanoparticle penetration. J Control Release 2016;244:257–68, with permission from Elsevier.*

correlation and exhibit a more predictive potential. Researchers have employed these concepts in the development of a 3D scaffold-based pancreatic cancer model consisting of triple cocultured pancreatic cancer cells, fibroblasts, and endothelial cells aiming to form a heterotypic MCTS combining tumor and multiple stromal components, as well as fibronectin [42]. The triple coculture enabled to replicate the in vitro cellular environment and exhibited reduced sensitivity to chemotherapy, thus closely mimicking the therapy resistance to treatments observed in vivo; however, the complex desmoplastic stroma was not fully recapitulated.

### 2.4.2.1 *Biomaterials for establishing physiomimetic 3D tumor microenvironments*

Holistically, ECM-mimetic biomaterials employed for establishing in vitro disease models can be classified as natural (e.g., collagen, fibronectin, laminin, and hyaluronic acid), naturally derived (e.g., chitosan—derived from the natural biopolymer chitin via thermochemical *N*-deacetylation and gelatin—derived from alkaline/acid hydrolysis of naturally occurring collagen), or synthetic [e.g., polycaprolactone, poly(lactic acid)] biomaterials (Table 2.1). To better model human tissues bioarchitecture and biomechanical cues, these biomaterials are generally processed into diverse classes of scaffolds including nano/microfibers, MPs/microcapsules, sponges, or hydrogels. One of the main challenges in designing biomaterial-based scaffolds is to chemically modify their building blocks to mimic the ECM of native tissues. Owing to their high water content and biochemical/biophysical similarity to the native ECM, numerous types of hydrogels have been developed for in vitro disease modeling [58,59]. The 3D hydrogel-based tumor models can recapitulate specific TME components, including: (1) key ECM proteins/glycosaminoglycans (e.g., collagen, laminin, fibronectin, entactin, heparin sulfate, hyaluronan), (2) mechanical properties, and (3) biodynamic degradability (e.g., mediated by MMPs) [60]. Hydrogels processing into highly complex anatomic size structures via 3D bioprinting technologies also make them an ideal platform for disease modeling and preclinical drug screening [61,62].

#### 2.4.2.1.1 Natural and nature-derived biomaterials for 3D tumor modeling

***Protein-based biomaterials*** Protein-based biomaterials including collagen, fibronectin, fibrin, or complex protein mixtures such as Matrigel$^{TM}$ are commonly used for 3D cancer cell culture owing to their specific biophysical and cell-adhesive properties [4]. Due to their natural origin these biomaterials exhibit ECM specific signaling molecules, cell-adhesion domains (e.g., RGD tripeptide—arginine–glycine–aspartate) and matrix metalloproteinase (MMP) cleavage sites. Among these, Matrigel is recognized as the "gold standard" scaffold for microtumors modeling in vitro [45]. This decellularized ECM (dECM) is extracted from Engelbreth–Holm–Swarm mouse sarcoma tumors and comprised primarily of laminin, fibronectin, type IV collagen, and entactin, proteoglycans, and mouse growth factors. Matrigel$^{TM}$ hydrogels have been widely employed to study cell migration, angiogenesis, and tumor progression [43].

Collagen is another protein-based biomaterial commonly used for 3D in vitro tumor modeling [44]. Collagen type I is the most abundant fibrous protein component of mammals ECM, and plays a crucial role in tumor progression, invasion, and metastasis by promoting cell adhesion and migration [16]. Collagen scaffolds can be assembled through physical or chemical crosslinking, leading to the establishment of a fibrillary scaffold.

**TABLE 2.1** Advantages and disadvantages of different bioinspired biomaterials used to produce scaffold-based 3D in vitro tumor models.

| Origin | Biomaterial | Advantages | Disadvantages | Ref. |
|---|---|---|---|---|
| Natural | Collagen | Contains cell-adhesion domains<br>Recapitulates in vivo cell–ECM interactions<br>Enzymatically degradable<br>Biomimetic properties<br>Cytocompatibility<br>Native viscoelastic environment | Batch-to-batch variability<br>Limited long-term stability<br>Difficult handling | [23,43,44] |
| | Matrigel | Tumorigenic origin<br>Enriched in ECM proteins and growth factors<br>Cytocompatibility | Batch-to-batch variability in terms of mechanical and biochemical properties<br>Exact composition is unknown | [45] |
| | Fibrin | Cytocompatibility<br>Cell-adhesive properties | Extreme susceptibility to protease-mediated degradation which limits the use for long-term 3D cell culture | [43] |
| | Fibronectin | Cell-adhesive properties<br>Interaction with growth factors (GFs)<br>Component of the TME of various tumors | High cost<br>Low mechanical properties | [46,47] |
| | Decellularized matrix | Provides native ECM components | Labor-intensive production<br>Lack of structural and architectural control<br>Variability in decellularization processes<br>Batch-to-batch variability | [48] |
| | Alginate | Low immunogenicity<br>Cytocompatibility<br>Allows cell recovery after culture | Must be modified with adhesive ligands to enable cell attachment<br>Cells cannot degrade the matrix | [2,43,49] |
| | Hyaluronic acid | Biological relevance<br>Chemical tunability<br>Versatility of crosslinking chemistries<br>Interacts with cell surface receptors (e.g., CD44 and RHAMM)<br>Component of the TME of various tumors | Lack of integrin adhesive domains (e.g., RGD)<br>Lack of MMP-cleavable sites | [43,44] |

(*Continued*)

TABLE 2.1 (Continued)

| Origin | Biomaterial | Advantages | Disadvantages | Ref. |
|---|---|---|---|---|
| Naturally derived | Chitosan (obtained from chitin deacetylation) | Cell-adhesive properties<br>Cytocompatibility<br>Ease of chemical modification<br>Low immunogenicity | Nonpresent in the TME<br>Nondegradable via matrix metalloproteinases<br>Low solubility in physiological conditions | [50,51] |
| | Gelatin (obtained from collagen hydrolysis) | Intrinsically available cell-adhesion domains<br>Enzymatically degradable<br>Biomimetic properties<br>Ease of chemical modification<br>Tunable stiffness<br>Cytocompatibility<br>Cost | Batch-to-batch variability<br>Low matrix stiffness | [52,53] |
| Synthetic | Polyacrylamide | User-defined control over interactions with cell surface receptors<br>High control over hydrogel stiffness being useful for mechanobiology studies | Must be conjugated with proteins to enable cell attachment<br>Cannot be used as 3D matrix to encapsulate cells due to toxicity of the hydrogel precursors | [2,54] |
| | Polyethylene glycol | Hydrophilic<br>Cytocompatibility<br>Inert material<br>Amenable to user-defined crosslinking chemistry | Requires combination with proteins and/or cell-adhesive biopolymers | [2,54] |
| Hybrid | Natural–synthetic conjugated polymers or Synthetic polymers–bioinstructive molecules conjugated | Defined control over the mechanical and biochemical properties<br>Biological relevance<br>Presence of ECM-like domains | Fabrication methods can be cytotoxic Fabrication and optimization can involve higher costs in comparison to other approaches | [10,27,44,55–57] |

The 3D collagen scaffolds are extensively used to mimic native ECM due to their biomimetic properties, cytocompatibility, and cell-adhesive properties without requiring further chemical modification. Adding to this, collagen platforms also provide a native viscoelastic environment where cells can reside and proliferate [23,43]. Gelatin, the product resulting from collagen hydrolysis and exhibiting the same amino acid sequence as its pristine, is also widely used to fabricate bioinstructive 3D in vitro models [52].

Although they are highly suitable for in vitro cell culture and tumor modeling, these biomaterials present some drawbacks including their complex handling/processing and animal origin, which results in batch-to-batch variations [26,43]. Furthermore, nature-derived biomaterials generally present poor mechanic properties which results in some difficulty in providing reproducible tissue specific ECM-like environments, often requiring further chemical functionalization [63].

***Polysaccharide-based biomaterials*** Nature-derived polysaccharides such as hyaluronic acid, agarose, or alginate have useful biophysical properties for disease models establishment. However, they generally require additional chemical modification with cell-adhesion motifs such as RGD peptides to promote cell attachment and proliferation in vitro [64].

Hyaluronic acid (HA) is a nonsulfated glycosaminoglycan that is distributed throughout normal tissues and tumor ECM and has been associated with tumor progression and poor patient outcome [43,44,65]. HA possess several advantages including biological relevance and chemical tunability when compared with other polysaccharides. Although unmodified HA does not support integrin-mediated cell adhesion, it interacts with numerous cell surface markers including CD44 and RHAMM (CD168). This biopolymer has the advantage of be easily modified with functional groups enabling a range of crosslinking chemistries [43]. Most commonly, HA carboxylic groups are modified with acrylate or methacrylate moieties to produce photocrosslinkable hydrogels with tunable mechanical and degradation properties [44]. Due to its role in the TME, several studies have employed HA scaffolds to assemble 3D in vitro tumor models [11,66,67]. In line with this, Liu et al. investigated the influence of matrix composition and dynamic stiffening on pancreatic tumor cells fate using a dynamic 3D gelatin/hyaluronic acid scaffold [68]. They found that HA-containing matrix inhibited tumor cell growth and induced cell phenotypic changes. In addition, mRNA expression analyses revealed unique changes in the presence of HA, a rigid microenvironment or a combination of both.

Chitosan is another polysaccharide that has been used to produce 3D scaffolds for cell culture. Its cationic nature promotes the conjugation with anionic molecules or reaction with biomaterials such as heparin, HA, or alginate [50,51]. Zhang et al. engineered a 3D scaffold composed of chitosan and alginate to promote the conversion of glioma cells to a malignant phenotype like in vivo. The authors concluded that alginate–chitosan scaffolds provided a reliable 3D microenvironment for glioma cells and thus could serve as a more effective platform to be applied into xenograft models establishment [51].

Alginate is an anionic polysaccharide extracted from brown algae species composed by β-D-mannuronic acid units and α-L-guluronic acid units. Alginate hydrogels are generally manufactured via rapid ionic crosslinking with divalent cations (e.g., $Ca^{2+}$, $Mg^{2+}$), under physiological conditions making them suitable for cell culture and at the same time preserving their high plasticity and mechanical strength [43,49]. Additionally, alginate hydrogels allow the recovery of cultured cells through hydrogel dissolution by using a chelating isotonic solution (e.g., EDTA). Such unique feature renders it highly attractive for establishing dynamic 3D cell-rich TMEs in vitro [43,69].

In contrast with Matrigel or collagen, polysaccharide-based hydrogels provide controllable and stable biophysical conditions for tumor growth, because their degradation is generally independent of cell-secreted proteolytic enzymes [64]. Polysaccharide-based hydrogels drawbacks, shared also by protein-based hydrogels, include low stiffness, limited long-term stability, and batch-to-batch variability [43].

#### 2.4.2.1.2 Synthetic biomaterials for 3D tumor modeling

Biocompatible and biodegradable synthetic biomaterials are commonly manufactured from polyethylene glycol (PEG), poly(glycolic acid) (PGA), polylactic acid (PLA), or their derivatives including poly(lactic-*co*-glycolic acid) (PLGA), poly(D,L-lactide-*co*-glycolide) (PLG), and polycaprolactone (PCL) [54]. Although their native form lacks correct representation of several biomolecular cues of the TME such as the presence of growth factors or key structural/bioactive proteins, they are easily modified with appropriate cell-adhesion ligands and biodegradable cross-linkers to overcome such limitations and thus provide control over hydrogel composition and properties [2]. Due to the absence of specific cell-adhesion domains, growth factors, or cytokines, cells cultured on these platforms lack the gene expression patterns of human tumors, are unable to metastasize or exhibit multidrug resistance phenotypes [4,70]. Furthermore, the filamentous nature of natural ECMs which controls the spatial organization of cell-adhesion ligands and mechanical signal transduction are not often recapitulated by synthetic hydrogels [2]. To overcome these issues, synthetic materials can be either modified with specific bioactive molecules. In this context, PEG-based hydrogels modified with integrin-binding RGD peptides and MMP-cleavable sites have been employed for 3D spheroids generation of different types of microtumors including brain, lung, and ovarian [2].

#### 2.4.2.1.3 Hybrid biomaterials for 3D tumor modeling

Hybrid biomaterials combine the advantages of natural proteins/polysaccharides cell-adhesive properties with synthetic polymers reproducibility and precise control over biochemical and mechanical properties, thus facilitating disease modeling. Natural and synthetic polymers can be functionalized with chemical groups to enable further control over crosslinking and consequently over the display of bioactive domains (e.g., RGD or MMP-cleavable sites) [44,55,56]. Synthetic polymers such as PEG have been combined so far with numerous natural derived polymers including chitosan, HA, and collagen [10,27,44,57].

These classes of biomaterials and hybrid biomaterial combinations have been employed to generate advanced 3D tumor models comprised by dense cellular aggregates, preexisting ECM components, and controlled morphological features as discussed in the following sections.

## 2.4.3 Generation of spherically structured cell-rich 3D tumor models

Spherically structured ECM-containing multicellular tumor models represent the biochemical/mechanical properties of solid tumors and offer the possibility to precisely control microtissues morphology/size in a reproducible manner, thus making it a useful tool for 3D modeling, compatible with drug-screening techniques and imaging assays. This provides major advantages in comparison to cell-laden ECM-hydrogel, fiber, or sponge testing platforms in which cellular agglomerates are randomly dispersed and present variable morphologies and sizes. These are major parameters which are known to influence drug performance evaluation at preclinical drug-screening stages.

Spherically structured in vitro tumor models enriched with ECM-mimetic biomaterials can be fabricated via different methodologies based on cancer cells bioencapsulation inside hydrogel microcapsules or by MP cell-supportive surfaces containing ECM-mimetic components (Fig. 2.4). Both strategies provide innovative platforms for the study of tumor biology and drug-screening assays [17].

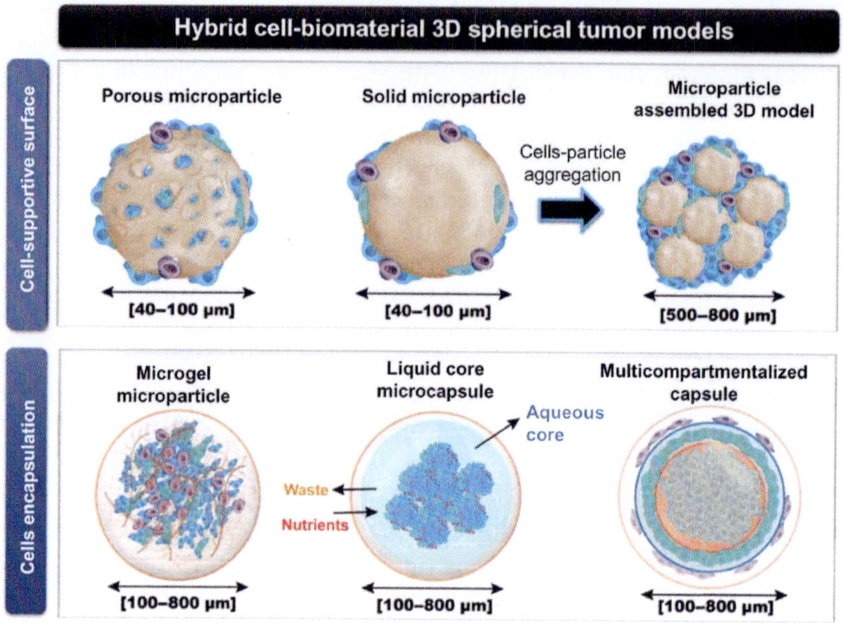


FIGURE 2.4 Schematic representation of different technologies used for assembly of spherically structured 3D in vitro tumor models. Source: *Reproduced from Ferreira LP, Gaspar VM, Mano JF. Design of spherically structured 3D in vitro tumor models - advances and prospects. Acta Biomater 2018;75:11–34, with permission from Elsevier.*

#### 2.4.3.1 Microparticles for spherically structured 3D tumor models assembly

Over the past few decades, MPs have been explored as spherical drug delivery vehicles and large-scale cell expansion carriers [71]. MP-based scaffolds can also be employed for tumor modeling as platforms to introduce previously lacking ECM-mimetic components in 3D cell agglomerates. MPs may assume a compact structure, where cells can adhere, or provide a porous scaffold allowing cellular infiltration and facilitating nutrients diffusion [71–73]. Besides, MPs can be loaded with different bioactive molecules to mimic TME signaling.

These platforms provide the necessary initial cues for characteristic cancer phenotypes establishment and a bioinstructive structure able to modulate biochemical and physical properties of tumor microtissues, thus recapitulating several aspects of TME.

MP-based scaffolds for generating in vitro tumor models can comprise natural, synthetic, or hybrid polymers [74]. Particularly, biodegradable synthetic polymers such as PLA, PGA, and their copolymers, PLGA have been extensively used to formulate MPs due to their biocompatibility, processability, chemical versatility, and mechanical properties. Synthetic polymers such as PLA and PLGA have improved mechanical strength in comparison to their natural polymer counterparts, but the later exhibit inherent cell-adhesive properties [73,74]. To surpass this limitation, MPs surface can be chemically modified to incorporate cell-adhesion motifs, or functionalized with ECM-mimetic components (e.g., collagen, gelatin, and fibronectin). To this end researchers recently fabricated PLA–PVA/chitosan MPs that provided a suitable substrate for cancer cells adhesion, resulting in the formation of 3D tumor microtissues in vitro [73]. Similarly, Ferreira and colleagues designed a multicellular 3D in vitro lung tumor model established via the incorporation of poly(ε-caprolactone) MPs coated with bioinstructive HA multilayer produced via the layer-by-layer deposition technique. The inclusion of bioinstructive MPs allowed the addition of HA, a key component of tumor ECM that is recognized to contribute to cells response to chemotherapeutics [11].

Cell-adhesion on the surface of these ECM-mimetic MP scaffolds can be achieved by suspending cells in prepolymer solution or by suspension of 3D cultures in spinner flasks containing MPs. Interestingly, the use of microgravity-simulating rotating-wall vessel (RWV) bioreactor allowed Skardal et al. to produce liver-tumor hybrid organoid by cell-culturing onto hyaluronic acid and gelatin-coated microgels, under dynamic conditions, allowing the generation of 3D microtissues, although with a limited control over their size/morphology [75].

Alternatively, MPs can also be produced using lithography, electrohydrodynamic spraying, and mechanical fragmentation methods; however these approaches are relatively low throughput and the low control over the formation of individual particles and of morphology/size-controlled microtissues [72].

#### 2.4.3.2 *Microgels for spherically structured 3D tumor models assembly*

Cell-laden microgels have originally emerged as a promising platform for cell transplantation to treat a wide range of diseases [1]. Recently, cancer cells or spheroids encapsulation in these systems has received considerable attention for tumor in vitro modeling [31]. This approach allows cancer and stromal cells to proliferate in a confined environment, establishing both cell–cell and cell–ECM interactions which dictate cell fate. The encapsulation of cancer cells in spherically shaped and size-controlled microgels has the potential to mimic avascular solid tumors in terms of growth kinetics with proliferating and necrotic cell populations and recapitulate the heterogenic microenvironment such as oxygen, glucose, and pH gradients that exist in vivo solid. Microencapsulated 3D models can be generated by ionic/thermal gelation mechanisms, emulsion, superhydrophobic, and superamphiphobic droplet generation, or electrospray by using, for example, alginate, chitosan, dextran, and PLGA.

Microgel-based models were established by Antunes et al. in Ref. [12], to study the dynamic communication between prostate cancer cells and human osteoblasts. By generating methacrylated gelatin/methacrylated hyaluronic acid microgels in superhydrophobic surfaces, the authors demonstrated that the interaction between cancer cells and osteoblasts supported the bioactivity and viability of bone-forming cells (Fig. 2.5A–D). Moreover, the in vitro generated microtumors exhibited tumor-associated characteristics, such as calcium matrix deposit. This result emphasizes the interaction between PC-3 and osteoblasts to promote the bioactivity of bone-forming cells [12]. As an alternative to coculture in the same scaffold compartment, such combinations of different cell populations can also be designed on a hierarchically structured scaffold, as demonstrated by the multilayer models produced by Fang et al. (Fig. 2.5E and F) [76].

In a different study researchers encapsulated cancer cells within PEGDA–fibrinogen hydrogel microspheres with ECM components (Fig. 2.6) [10]. This fabrication technique allowed a high-throughput generation of PEG–fibrinogen hydrogel microspheres and the

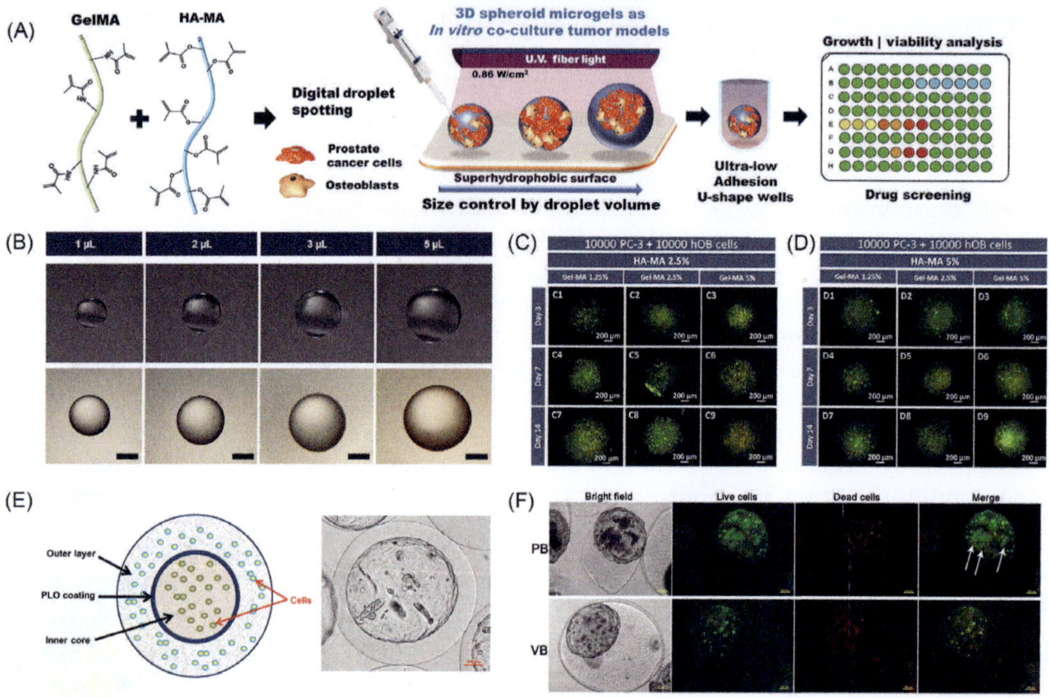


FIGURE 2.5 (A) Schematic establishment of prostate-to-bone ECM-mimetic microgel 3D tumor models using superhydrophobic (SH) surface. (B) Modulation of 3D microgel droplets size in SH surfaces by varying the dispensed volume. Scale bar: 500 μm. (C and D) Live/dead analysis of prostate cancer (PC-3) and bone cells (human osteoblasts—hOB) cocultured in ECM-biomimetic microgels comprising methacrylated gelatin/methacrylated hyaluronic acid. (E) Schematic representation of the double-layered alginate microsphere for generation of compartmentalized 3D spherical tumors; optical contrast microscopy analysis of the core–shell structure (F) Live/dead analysis of C4-2 cells stably transfected and expressing protein kinase D1 - PKD1 (protein that inhibits cel proliferation through MMP-2), after 28 days of encapsulation (PB—microcapsules with C4-2 PKD1 bioencapsulated cells at inner core and blank shell. VB—microcapsules with bioencapsulated C4-2 pcDNA3.1 control plasmid vector transfected cells at inner core and blank shell). CFDA—green channel, live cells. Propidium iodide (PI)—red channel, dead cells. Arrows represent cell colonies. Scale bar: 100 μm. Source: *(A–D) Reproduced from Antunes J, Gaspar VM, Ferreira L, Monteiro M, Henrique R, Jerónimo C, et al. In-air production of 3D co-culture tumor spheroid hydrogels for expedited drug screening. Acta Biomater 2019;94:392–409, with permission from Elsevier. (E and F) Reproduced from Fang X, Sittadjody S, Gyabaah K, Opara EC, Balaji KC. Novel 3D co-culture model for epithelial-stromal cells interaction in prostate cancer. PLoS One 2013;8:1–10, under Creative Commons License.*

ability to maintain a wide range of cancer cell types within the developed 3D models for long culture periods.

#### 2.4.3.3 *Microcapsules for spherically structured 3D tumor models assembly*

Compartmentalized hydrogel microcapsules containing ECM-mimetic scaffolds can be produced through the combination of multifluidic electrospray of hydrogel capsules with

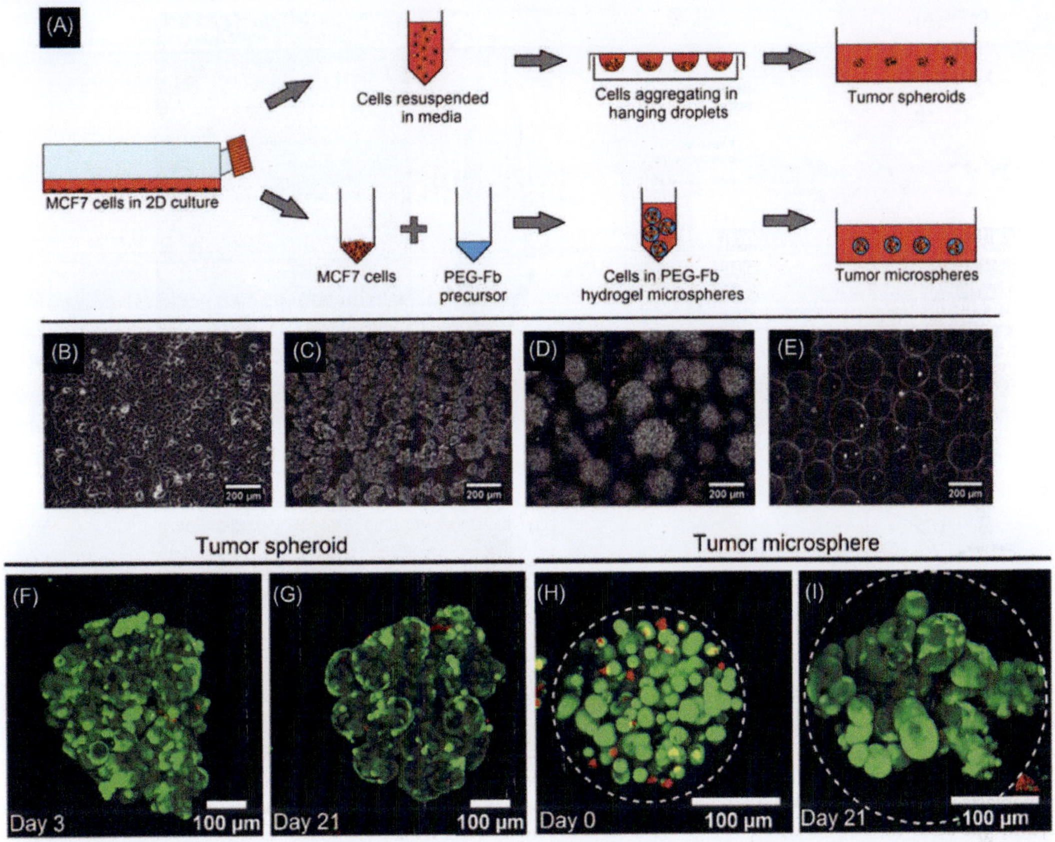


FIGURE 2.6 (A) Comparative fabrication of tumor spheroids and tumor PEG–fibrinogen-based tumor microspheres. MCF-7 cells (B) in 2D culture, (C) cultured as tumor spheroid or (D) in tumor microsphere. (E) PEG–fibrinogen microspheres formed without cells in tumor spheroids. Confocal z-projections of (F,G) tumor spheroids and (H,I) tumor microspheres labeled with live/dead probes (green: Calcein AM; red: Ethidium homodimer). White circles represent microsphere edges. Source: *Reproduced from Pradhan S, Clary JM, Seliktar D, Lipke EA. A three-dimensional spheroidal cancer model based on PEG-fibrinogen hydrogel microspheres. Biomaterials 2017;115:141–54, with permission from Elsevier.*

ionic/thermal gelation mechanisms. Through this system, cells could communicate either by direct contact or via paracrine interactions, according to the compartmentalization settings. The capacity of recapitulating direct and indirect interactions, the ability to represent both physical (hypoxic gradients) and biochemical (ECM composition) properties combined with the ability to sequentially retrieve cultured cells, makes this an ideal platform for further studies on gene expression, cell–cell signaling, and drug screening [17].

Recently, liquefied-core and multilayer capsules have emerged as an alternative to solve mass transportation concerns [77]. Besides solving the limited diffusion of hydrogels, liquified capsules allow to obtain spherically structured models with higher cellular densities, allowing to reproduce in vivo tumor mass densities. Alessandri et al. developed a microfluidic technique to produce MCTS-laden microcapsules through a microfluidic

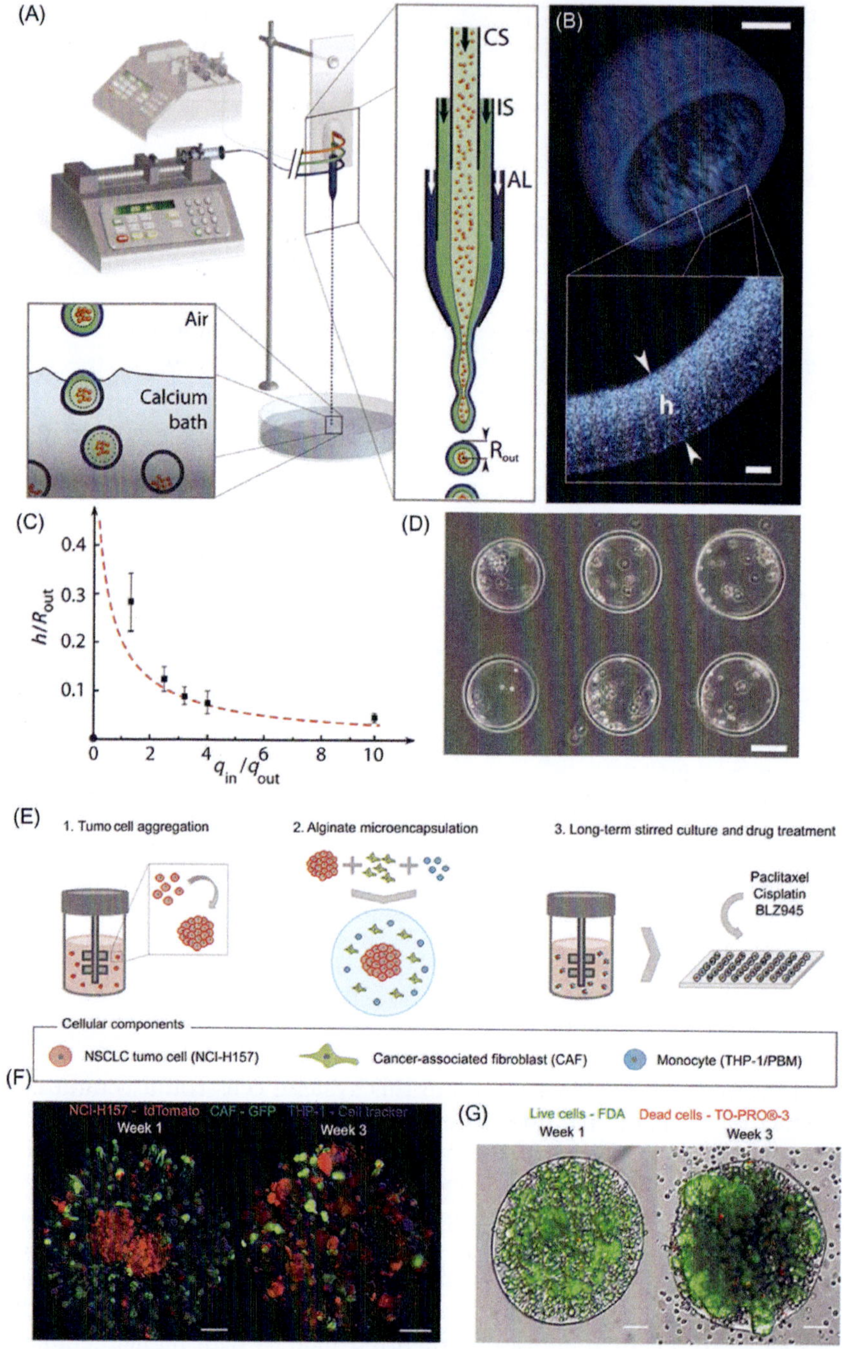


FIGURE 2.7 (A) Schematic of the microfluidic platform, comprising an external fluidic injection system, coextrusion microdevice, and an off-chip gelation bath that enables the production of alginate microcapsules.

coextrusion device (Fig. 2.7) [78]. In this study, researchers described the preparation of a simple and reproducible method to produce encapsulated spheroids in elastic alginate microcapsules with a liquid core. This system allows free flow of nutrients into the capsule and cell proliferation in a scaffold-free environment. Microencapsulated spheroid models are a suitable platform to obtain a higher rate of microtissues production, higher pluripotency, and higher degree of specific surface receptor markers.

Using a similar concept, researchers established 3D-triple coculture heterotypic tumor models comprising an ECM-mimetic matrix to recapitulate the crosstalk between cancer/stromal and immune system cells in the TME [79]. To materialize this multicomponent in vitro model, researchers employed alginate microcapsules for bioencapsulation of CAFs, monocytes, and nonsmall lung cancer cells. Such design allowed to evaluate the infiltration of monocytic and peripheral blood cells in the 3D models in vitro and to investigate the establishment of an immunosuppressive microenvironment, a key hallmark of lung cancer TME. Moreover, due to their ease of processing and handling microcapsule platforms were suitable for evaluating both chemotherapeutic and immunomodulatory treatments performance.

Overall, microgels comprised by different ECM-mimetic biomaterials exhibit a significant potential fundamental tumor modeling, invasion studies, drug-screening assays, as well as fundamental research in tumor biology and resistance, since they present inherent ECM deposition which creates a cascade effect culminating in an increased chemotherapeutics resistance when compared with conventional monolayer models. Modeling these key hallmarks of solid tumors disease is crucial for obtaining a better in vitro/in vivo correlation and reduces drugs attrition rates during development.

## 2.5 Conclusions

The absence of preclinical models capable of reproducing in vitro the hallmarks of in vivo tumors leads to a low correlation between therapies efficacy at preclinical evaluation stages and that obtained in clinical trials. Bridging this gap is crucial to expedite the development of more effective and less toxic cancer treatments. Recent evidences

Calcium-mediated gelation of the alginate shell freezes the structure of the capsule, and cells remain encapsulated. (B) Confocal image of an alginate capsule stained with high-molecular-weight fluorescent dextran. (C) Plot of the aspect ratio of produced capsules ($h/R_{out}$) as function of the ratio between: ( i) inner flow rate $q_{in} = q_{CS} + q_{IS}$ and the (ii) outer flow rate $q_{out} = q_{AL}$. Black dots represent the experimental data and the red dashed line the theoretical curve derived from volume conservation equation ($h/R_{out} = 1 - (q_{in}/q_{out}t \,/\, 1 + q_{in}/q_{out})^{1/3}$. $q$ — flow rate; $h$ — shell thickness; $R_{out}$ — outer capsule radious; AL — alginate solution; CS — cell suspension; IS — intermediate sorbitol solution, free of calcium. (D) Representative phase-contrast micrographs of individual capsules encapsulating cells. (E) Schematics of microencapsulated 3D microtumor models' fabrication. (F) Alginate microcapsules laden with triple cocultures of malignant, stromal and immune cells. (G) Live/dead analysis of 3D-triple cultures at different time points. Source: *(A–D) Reproduced from Alessandri K, Sarangi BR, Gurchenkov VV, Sinha B, Kiessling TR, Fetler L, et al. Cellular capsules as a tool for multicellular spheroid production and for investigating the mechanics of tumor progression in vitro. Proc Natl Acad Sci 2013;110:14843–14848, with permission from Proceedings of the National Academy of Sciences of the United States of America (PNAS). (E–G) Reproduced from Rebelo SP, Pinto C, Martins TR, Harrer N, Estrada MF, Loza-Alvarez P, et al. 3D-3-culture: a tool to unveil macrophage plasticity in the tumour microenvironment. Biomaterials 2018;163:185–97, with permission from Elsevier.*

demonstrate that in preclinical stages, state-of-the art in vitro 3D models show potential for a more realistic recapitulation of human tumors than their 2D counterparts and overcome several issues related to animal laboratorial models.

The discussed advances on the development of scaffold-based 3D tumors comprising ECM-mimetic biomaterials allow researchers to more faithfully recapitulate key cell–cell and cell–ECM interactions, as well as mechanical and biophysiological characteristics present in in vivo solid tumors. Combining these hallmarks with the fabrication of spherically structured 3D microtumors leverages the potential of these models to be included in high-content/high-throughput platforms and to eliminate morphology-associated variability. This allows researchers to generate more robust data regarding therapeutics anticancer activity. In the future, we envisioned that advances regarding ECM-biomimetic biomaterials mechanical properties tuning according to patient-specific tumors will further contribute to better recreate the biophysical cues present in vivo. This, allied to a deconstruction of key tumor building blocks and their bottom-up bioengineering via advanced technologies [80], will contribute for developing evermore representative models of tumor heterogeneity. Finally, engineering novel strategies for a streamlined inclusion of such models in organ-on-a-chip platforms may open a new paradigm for evaluating therapeutics activity in dynamic conditions that better simulate those found human tumors.

## References

[1] Zhang X, Wang W, Yu W, Xie Y, Zhang X, Zhang Y, et al. Development of an in vitro multicellular tumor spheroid model using microencapsulation and its application in anticancer drug screening and testing. Biotechnol Prog 2005;21:1289–96.

[2] Li Y, Kumacheva E. Hydrogel microenvironments for cancer spheroid growth and drug screening. Sci Adv 2018;4:1–11.

[3] Cox MC, Reese LM, Bickford LR, Verbridge SS. Toward the broad adoption of 3D tumor models in the cancer drug pipeline. ACS Biomater Sci Eng 2015;1:877–94.

[4] Alemany-ribes M, Semino CE. Bioengineering 3D environments for cancer models. Adv Drug Deliv Rev 2014;79:40–9.

[5] Ruggeri BA, Camp F, Miknyoczki S. Animal models of disease: pre-clinical animal models of cancer and their applications and utility in drug discovery. Biochem Pharmacol 2014;87:150–61.

[6] Stock K, Estrada MF, Vidic S, Gjerde K, Rudisch A, Santo VE, et al. Capturing tumor complexity in vitro: comparative analysis of 2D and 3D tumor models for drug discovery. Sci Rep 2016;6:1–16.

[7] Weiswald LB, Bellet D, Dangles-Marie V. Spherical cancer models in tumor biology. Neoplasia 2015;17:1–15.

[8] Nunes AS, Correia IJ, Costa EC. 3D tumor spheroids as in vitro models to mimic in vivo human solid tumors resistance to therapeutic drugs. Biotechnol Bioeng 2018;116:206–26.

[9] Katt ME, Placone AL, Wong AD, Xu ZS, Searson PC. In vitro tumor models: advantages, disadvantages, variables, and selecting the right platform. Front Bioeng Biotechnol 2016;4:1–14.

[10] Pradhan S, Clary JM, Seliktar D, Lipke EA. A three-dimensional spheroidal cancer model based on PEG-fibrinogen hydrogel microspheres. Biomaterials 2017;115:141–54.

[11] Ferreira LP, Gaspar VM, Mano JF. Bioinstructive microparticles for self-assembly of mesenchymal stem cell-3D tumor spheroids. Biomaterials 2018;185:155–73.

[12] Antunes J, Gaspar VM, Ferreira L, Monteiro M, Henrique R, Jerónimo C, et al. In-air production of 3D co-culture tumor spheroid hydrogels for expedited drug screening. Acta Biomater 2019;94:392–409.

[13] Bregenzer ME, Horst EN, Mehta P, Novak CM, Raghavan S, Snyder CS, et al. Integrated cancer tissue engineering models for precision medicine. PLoS One 2019;14:1–40.

[14] Monteiro CF, Custódio CA, Mano JF. Three-dimensional osteosarcoma models for advancing drug discovery and development. Adv Ther 2018;2:1800108.

[15] Walker C, Mojares E, Del Rio Hernández A. Role of extracellular matrix in development and cancer progression. Int J Mol Sci 2018;19:3028–59.
[16] Mohan V, Das A, Sagi I. Emerging roles of ECM remodeling processes in cancer. Semin Cancer Biol 2020;62:192–200.
[17] Ferreira LP, Gaspar VM, Mano JF. Design of spherically structured 3D in vitro tumor models -advances and prospects. Acta Biomater 2018;75:11–34.
[18] Rozario T, DeSimone DW. The extracellular matrix in development and morphogenesis: a dynamic view. Dev Biol 2010;341:126–40.
[19] Chaudhuri O, Koshy ST, Branco Da Cunha C, Shin JW, Verbeke CS, Allison KH, et al. Extracellular matrix stiffness and composition jointly regulate the induction of malignant phenotypes in mammary epithelium. Nat Mater 2014;13:970–8.
[20] Correia AL, Bissell MJ. The tumor microenvironment is a dominant force in multidrug resistance. Drug Resist Updat 2012;15:39–49.
[21] Anderson RL, Balasas T, Callaghan J, Coombes RC, Evans J, Hall JA, et al. A framework for the development of effective anti-metastatic agents. Nat Rev Clin Oncol 2019;16:185–204.
[22] Nyga A, Cheema U, Loizidou M. 3D tumour models: novel in vitro approaches to cancer studies. J Cell Commun Signal 2011;5:239–48.
[23] Wang C, Tang Z, Zhao Y, Yao R, Li L, Sun W. Three-dimensional in vitro cancer models: a short review. Biofabrication. 2014;6:022001.
[24] Asghar W, Assal R El, Shafiee H, Pitteri S, Paulmurugan R, Demirci U. Engineering cancer microenvironments for in vitro 3-D tumor models. Biochem Pharmacol 2015;18:539–53.
[25] Aparicio S, Hidalgo M, Kung AL. Examining the utility of patient-derived xenograft mouse models. Nat Rev Cancer 2015;15:311–16.
[26] Pampaloni F, Reynaud EG, Stelzer EHK. The third dimension bridges the gap between cell culture and live tissue. Nat Rev Mol Cell Biol 2007;8:839–45.
[27] Lee J, Cuddihy MJ, Kotov NA. Three-dimensional cell culture matrices: State of the art. Tissue Eng - Part B Rev 2008;14:61–86.
[28] Duval K, Grover H, Han LH, Mou Y, Pegoraro AF, Fredberg J, et al. Modeling physiological events in 2D vs. 3D cell culture. Physiology. 2017;32:266–77.
[29] Saglam-metiner P, Gulce-iz S, Biray-avci C. Bioengineering-inspired three-dimensional culture systems: organoids to create tumor microenvironment. Gene 2018;686:203–12.
[30] Smalley KSM, Lioni M, Herlyn M. Life isn't flat: taking cancer biology to the next dimension. Vitr Cell Dev Biol - Anim 2006;42:242–7.
[31] Fong ELS, Harrington DA, Farach-Carson MC, Yu H. Heralding a new paradigm in 3D tumor modeling. Biomaterials. 2016;108:197–213.
[32] Friedrich J, Seidel C, Ebner R, Kunz-Schughart LA. Spheroid-based drug screen: Considerations and practical approach. Nat Protoc 2009;4:309–24.
[33] Langhans SA. Three-dimensional in vitro cell culture models in drug discovery and drug repositioning. Front Pharmacol 2018;9:1–14.
[34] Nath S, Devi GR. Three-dimensional culture systems in cancer research: focus on tumor spheroid model. Pharmacol Ther 2016;163:94–108.
[35] Costa EC, Moreira AF, de Melo-Diogo D, Gaspar VM, Carvalho MP, Correia IJ. 3D tumor spheroids: an overview on the tools and techniques used for their analysis. Biotechnol Adv 2016;34:1427–41.
[36] Breslin S, O'Driscoll L. Three-dimensional cell culture: the missing link in drug discovery. Drug Discov Today 2013;18:240–9.
[37] Picollet-D'hahan N, Dolega ME, Liguori L, Marquette C, Le Gac S, Gidrol X, et al. A 3D toolbox to enhance physiological relevance of human tissue models. Trends Biotechnol 2016;34:757–69.
[38] Mittler F, Obeïd P, Rulina AV, Haguet V, Gidrol X, Balakirev MY. High-content monitoring of drug effects in a 3D spheroid model. Front Oncol 2017;7:1–19.
[39] Laschke MW, Menger MD. Life is 3D: boosting spheroid function for tissue engineering. Trends Biotechnol 2017;35:133–44.
[40] Priwitaningrum DL, Blondé J-BG, Sridhar A, van Baarlen J, Hennink WE, Storm G, et al. Tumor stroma-containing 3D spheroid arrays: a tool to study nanoparticle penetration. J Control Release 2016;244:257–68.

[41] Loebel C, Mauck RL, Burdick JA. Local nascent protein deposition and remodelling guide mesenchymal stromal cell mechanosensing and fate in three-dimensional hydrogels. Nat Mater 2019;18:883–91.
[42] Lazzari G, Nicolas V, Matsusaki M, Akashi M, Couvreur P, Mura S. Multicellular spheroid based on a triple co-culture: a novel 3D model to mimic pancreatic tumor complexity. Acta Biomater 2018;78:296–307.
[43] Caliari S, Burdick JA. A practical guide to hydrogels for cell culture. Nat Methods 2016;13:405–14.
[44] Liu Z, Vunjak-Novakovic G. Modeling tumor microenvironments using custom-designed biomaterial scaffolds. Curr Opin Chem Eng 2016;11:94–105.
[45] Benton G, Arnaoutova I, George J, Kleinman HK, Koblinski J. Matrigel: from discovery and ECM mimicry to assays and models for cancer research. Adv Drug Deliv Rev 2014;79–80:3–18.
[46] Llopis-Hernández V, Cantini M, González-García C, Salmerón-Sánchez M. Material-based strategies to engineer fibronectin matrices for regenerative medicine. Int Mater Rev 2015;60:245–63.
[47] Zhu J, Clark RAF. Fibronectin at select sites binds multiple growth factors and enhances their activity: expansion of the collaborative ECM-GF paradigm. J Invest Dermatol 2014;134:895–901.
[48] Dzobo K, Motaung KSCM, Adesida A. Recent trends in decellularized extracellular matrix bioinks for 3D printing: an updated review. Int J Mol Sci 2019;20:4628.
[49] Estrada MF, Rebelo P, Davies EJ, Pinto MT, Pereira H, Santo VE, et al. Biomaterials modelling the tumour microenvironment in long-term microencapsulated 3D co-cultures recapitulates phenotypic features of disease progression. Biomaterials 2016;78:50–61.
[50] Kim IY, Seo SJ, Moon HS, Yoo MK, Park IY, Kim BC, et al. Chitosan and its derivatives for tissue engineering applications. Biotechnol Adv 2008;26:1–21.
[51] Kievit FM, Florczyk SJ, Leung MC, Veiseh O, Park JO, Disis ML, et al. Chitosan-alginate 3D scaffolds as a mimic of the glioma tumor microenvironment. Biomaterials 2010;31:5903–10.
[52] Yin J, Yan M, Wang Y, Fu J, Suo H. 3D bioprinting of low-concentration cell-laden gelatin methacrylate (GelMA) bioinks with a two-step cross-linking strategy. ACS Appl Mater Interfaces 2018;10:6849–57.
[53] Nichol JW, Koshy S, Bae H, Hwang CM, Khademhosseini A. Cell-laden microengineered gelatin methacrylate hydrogels. Biomaterials. 2011;31:5536–44.
[54] Pradhan S, Hassani I, Clary JM, Lipke EA. Polymeric biomaterials for in vitro cancer tissue engineering and drug testing applications. Tissue Eng Part B Rev 2011;22:1–52.
[55] Weiss MS, Bernabé BP, Shikanov A, Bluver DA, Mui MD, Shin S, et al. The impact of adhesion peptides within hydrogels on the phenotype and signaling of normal and cancerous mammary epithelial cells. Biomaterials 2012;33:3548–59.
[56] Zhu J. Bioactive modification of poly(ethylene glycol) hydrogels for tissue engineering. Biomaterials 2010;31:4639–56.
[57] Ouasti S, Donno R, Cellesi F, Sherratt MJ, Terenghi G, Tirelli N. Biomaterials Network connectivity, mechanical properties and cell adhesion for hyaluronic acid/PEG hydrogels. Biomaterials. 2011;32:6456–70.
[58] Dave PN, Gor A. Natural polysaccharide-based hydrogels and nanomaterials. In: Hussain CM, editor. Handbook of nanomaterials for industrial applications. Newark, NJ: Elsevier; 2018. p. 36–66.
[59] Liaw CY, Ji S, Guvendiren M. Engineering 3D hydrogels for personalized in vitro human tissue models. Adv Healthc Mater 2018;7:1701165.
[60] Zhang YS, Khademhosseini A. Advances in engineering hydrogels. Science (80-) 2017;356:1–10.
[61] Heinrich MA, Bansal R, Lammers T, Zhang YS, Michel Schiffelers R, Prakash J. 3D-bioprinted mini-brain: a glioblastoma model to study cellular interactions and therapeutics. Adv Mater 2019;31:1806590.
[62] Cao X, Ashfaq R, Cheng F, Maharjan S, Li J, Ying G, et al. A tumor-on-a-chip system with bioprinted blood and lymphatic vessel pair. Adv Funct Mater 2019;29:1807173.
[63] Yao H, Wang J, Mi S. Photo processing for biomedical hydrogels design and functionality: a review. Polymers 2017;10:1–27.
[64] Zanoni M, Pignatta S, Arienti C, Bonafè M, Zanoni M, Pignatta S, et al. Anticancer drug discovery using multicellular tumor spheroid models. Expert Opin Drug Discov 2019;14:289–301.
[65] Stern R. Hyaluronan metabolism: a major paradox in cancer biology. Pathol Biol 2005;53:372–82.
[66] Shen YI, Abaci HE, Krupski Y, Weng LC, Burdick JA, Gerecht S. Hyaluronic acid hydrogel stiffness and oxygen tension affect cancer cell fate and endothelial sprouting. Biomater Sci 2014;2:655–65.
[67] Gurski LA, Jha AK, Zhang C, Jia X, Farach-Carson MC. Hyaluronic acid-based hydrogels as 3D matrices for in vitro evaluation of chemotherapeutic drugs using poorly adherent prostate cancer cells. Biomaterials 2009;30:6076–85.

[68] Liu HY, Korc M, Lin CC. Biomimetic and enzyme-responsive dynamic hydrogels for studying cell-matrix interactions in pancreatic ductal adenocarcinoma. Biomaterials 2018;160:24–36.

[69] Torres AL, Bidarra SJ, Pinto MT, Aguiar PC, Silva EA, Barrias CC. Guiding morphogenesis in cell-instructive microgels for therapeutic angiogenesis. Biomaterials 2018;154:34–47.

[70] Ravi M, Paramesh V, Kaviya SR, Anuradha E, Paul Solomon FD. 3D cell culture systems: advantages and applications. J Cell Physiol 2015;230:16–26.

[71] Neto MD, Oliveira MB, Mano JF. Microparticles in contact with cells: from carriers to multifunctional tissue modulators. Trends Biotechnol 2019;37:1011–28.

[72] Daly AC, Riley L, Segura T, Burdick JA. Hydrogel microparticles for biomedical applications. Nat Rev Mater 2020;5:20–43.

[73] Horning JL, Sahoo SK, Vijayaraghavalu S, Dimitrijevic S, Vasir JK, Jain TK, et al. 3-D tumor model for in vitro evaluation of anticancer drugs. Mol Pharm 2008;5:849–62.

[74] Sahoo SK, Panda AK, Labhasetwar V. Characterization of porous PLGA/PLA microparticles as a scaffold for three dimensional growth of breast cancer cells. Biomacromolecules 2005;6:1132–9.

[75] Skardal A, Devarasetty M, Rodman C, Atala A, Soker S. Liver-tumor hybrid organoids for modeling tumor growth and drug response in vitro. Ann Biomed Eng 2015;43:2361–73.

[76] Fang X, Sittadjody S, Gyabaah K, Opara EC, Balaji KC. Novel 3D co-culture model for epithelial-stromal cells interaction in prostate cancer. PLoS One 2013;8:1–10.

[77] Correia CR, Bjørge IM, Zeng J, Matsusaki M, Mano JF. Liquefied microcapsules as dual-microcarriers for 3D + 3D bottom-up tissue engineering. Adv Healthc Mater 2019;8:1901221.

[78] Alessandri K, Sarangi BR, Gurchenkov VV, Sinha B, Kiessling TR, Fetler L, et al. Cellular capsules as a tool for multicellular spheroid production and for investigating the mechanics of tumor progression in vitro. Proc Natl Acad Sci 2013;110:14843–8.

[79] Rebelo SP, Pinto C, Martins TR, Harrer N, Estrada MF, Loza-Alvarez P, et al. 3D-3-culture: a tool to unveil macrophage plasticity in the tumour microenvironment. Biomaterials. 2018;163:185–97.

[80] Gaspar VM, Lavrador P, Borges J, Oliveira MB, Mano JF. Advanced bottom-up engineering of living architectures. Adv Mater 2019;32:1903975.

CHAPTER

# 3

# Biofabrication of 3D tumor models in cancer research

M.A. Grimaudo[1,*], A. Herreros-Pomares[2,3,*], M. Alonso[1,3], S. Calabuig-Fariñas[2,3,4], E. Jantus-Lewintre[2,3,5] and Maria de la Fuente[1,3]

[1]Nano-Oncology and Translational Therapeutics Unit, Health Research Institute of Santiago de Compostela (IDIS), SERGAS, Santiago de Compostela, Spain [2]Molecular Oncology Laboratory, General University Hospital of Valencia Research Foundation, Valencia, Spain [3]Cancer Research Network (CIBERONC), Spain [4]Department of Pathology, Universitat de València, Valencia, Spain [5]Department of Biotechnology, Universitat Politècnica de València, Valencia, Spain

## 3.1 Current challenges in oncology

According to the World Health Organization (WHO), cancer is a generic term that defines a large group of diseases characterized by the growth of abnormal cells beyond their usual boundaries, which can affect almost any part of the body. Indeed, there are more than 100 types of cancers, and many subtypes of tumors can be found in each specific organ. In 2018, 18.1 million new cases of cancer were diagnosed and 9.6 million people died due to cancer. Lung cancer (11.6%), breast cancer (11.6%), prostate cancer (7.1%), and colorectal cancer (6.1%) were the most commonly diagnosed types of cancer, whereas lung cancer (18.4%), colorectal cancer (9.2%), stomach cancer (8.2%), and liver cancer (8.2) were the leading causes of cancer death worldwide (Fig. 3.1) [1].

Research has shown that cancer has the most devastating economic impact of any cause of death in the world. The health care cost of cancer has been estimated at $53.9 billion within the European Union and $80.2 billion in the United States per year. This is only

* Both authors contributed equally to this work.

* Both authors contributed equally to this work.

Biomaterials for 3D Tumor Modeling
DOI: https://doi.org/10.1016/B978-0-12-818128-7.00003-4

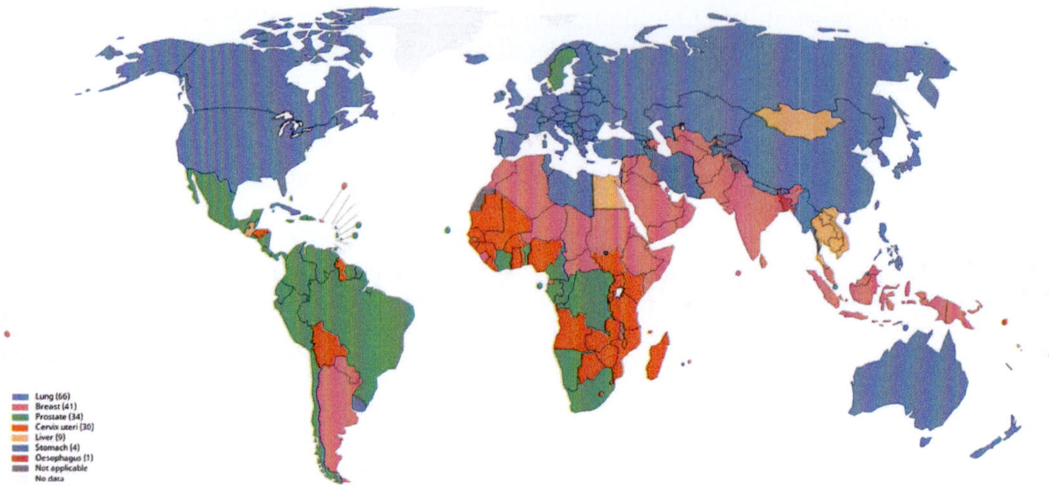


**FIGURE 3.1** Top cancer according to mortality (estimated age-standardized) per country for both sexes in 2018. *Source: GLOBOCAN 2018, graph production: IARC.*

around 40% of the total cost of cancer to society, which takes into account productivity loss among other factors. One of the major costs of cancer is the treatment, which achieves around a 45% 10-year survival rate. As such, the development of highly effective therapies is an essential research field, and drug discovery and efficacy assays require new methods of producing proper in vitro recreations of solid tumors [2].

Furthermore, cancers exist in an extraordinary variety of types and subtypes with significant differences in frequency and mortality, and it is the genetic and phenotypic variation that makes each tumor individually unique. In cancer, abnormal cells have defects in their regulatory mechanisms, making them able to grow uncontrolledly and spread into other tissues/organs. In addition to this reductionist view of cancer, there are more dimensions of complexity. In the past decade, Hanahan and Weinberg suggested that all cancers share some characteristics which are gained during multistep tumorigenesis and commonly known as hallmarks of cancer [3]. The manner of acquisition of these features as well as the order in which they are acquired varies significantly across different neoplasms. Accordingly, tumor cells are characterized by generating autocrine and paracrine growth signals and being insensitive to growth-inhibitory signals. In addition, neoplastic cells are able to evade programmed cell death, that is, apoptosis and autophagy, reprogramming energy metabolism to maintain their growth, and enabling replicative immortality by inducing the activity of the telomerase enzyme which protects telomeres length over time.

The majority of cancers have in common to be caused by somatic mutations at the beginning, although differences can be found in populations of tumor cells. The genetic diversity is an unavoidable consequence of the genome instability, which is caused by defects affecting the DNA damage and repair mechanisms in combination with a large number of cell divisions [4]. Our understanding of the biological processes generating these mutations is still limited, but next generation sequencing (NGS) has allowed the integration of the advances in cancer genomics into clinical care. This has facilitated several major mass-sequencing genome projects—such as the International Cancer Genome Consortium (ICGC) and The Cancer Genome Atlas (TCGA)—for almost every type of tumor and has identified tens of thousands of tumor mutations [5,6]. These approaches have led to a broader understanding of the cancer nature, allowed to routinely detect driver mutations that are critical for precision oncology, and promoted the off-label use of targeted therapies for patients with similar mutation profile regardless of the tumor origin. In this scenario, when molecularly defined subsets of patients have been identified and personalized therapies can be applied, the development of proper tumor models to test innovative anticancer therapeutics is of great relevancy.

## 3.2 The tumor microenvironment

Tumors are complex tissues in which cancer cells interact with the surrounding environment and can recruit different cell types to perform as active collaborators, creating tumor microenvironments (TME) in which proliferation and invasion are favored. Cancer cells are interwoven with fibroblasts, endothelial cells, pericytes, and inflammatory cells, constituting these cell types, together with the extracellular matrix (ECM), the so called TME (Fig. 3.2).

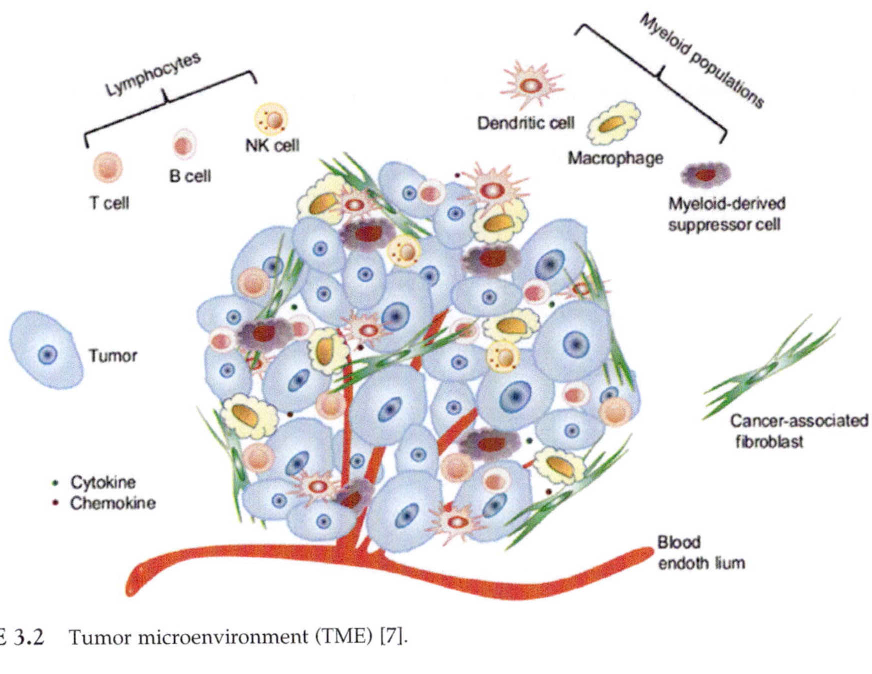


FIGURE 3.2 Tumor microenvironment (TME) [7].

Tumor cells significantly impact the local TME, causing an expansion and activation of different stromal cell types, and modifying their phenotypes and functions, overall converting them into tumor-supporting cells [8,9]. In order to contribute to the tumor growth and maintenance, cancer cells have the capacity to induce angiogenesis by producing proangiogenic factors, and it has become clear that inflammation can contribute to tumorigenesis by supplying different molecules to the TME such as growth factors and extracellular matrix-modifying enzymes. Additionally, it has been proven that neoplastic cells have developed different strategies to avoid being detected by the immune system and favoring an immunosuppressive contexture. The TME affects as well the behavior of tumor cells, by secreting tumor-promoting growth factors and cytokines, supporting their growth and capacity to invade neighboring or distant tissues, to proliferate, migrate, and become resistant to treatment [10–12]. It has also become clear that the contribution of stromal cells to cancer development is not limited to the primary tumor growth, but also involves the metastatic process [12,13]. Recent reports have indicated emerging roles for stromal cells, such as cancer associated fibroblasts (CAFs), in the modulation of cancer stem cells (CSCs) traits and in the formation of metastases [14].

In addition, immune cells are recruited to the primary tumor and produce multiple signals, promoting tumor growth, cell invasion and metastasis, genomic instability, and inflammation [15]. Among all, lymphocytes and myeloid cells, such as macrophages and dendritic cells, have been shown to contribute extensively to tumor-induced immunosuppression and to create a chronic inflammatory microenvironment that promote the growth and facilitate the survival of cancer cells [12,15]. Among the different immune cells found in the TME, tumor-associated macrophages (TAMs) represent around 50% of the tumor mass, and constitute a driving force in the initiation, proliferation, metastasis, angiogenesis, matrix remodeling, and suppression of adaptive immunity in some tumor types [16,17]. In vitro studies have shown that other tumor-infiltrating cells such as lymphocytes also produce a myriad of signaling molecules and cytokines able to promote tumor development and progression through angiogenesis, matrix remodeling, and suppression of the adaptive immunity [18]. Even in cases in which the immune cells are actively driving the initial responses to targeted therapies, a wide array of immune suppressive mechanisms may eventually evolve, ultimately leading to tumor progression [11]. Both innate and adaptive responses have also been suggested to impact on the outcome of conventional treatments with chemo- and radiotherapy [10,19].

The ECM, a complex and dynamic biopolymer network, provides tumor cells with integrated biophysical and biochemical cues able to influence cell survival and differentiation signaling, cell–cell interactions, cell mechanics, and tissue homeostasis as well [8]. Indeed, the ECM is known to impact drug sensitivity in human carcinomas in vivo [20] and in vitro [21]. Any perturbation of ECM synthesis, degradation, density, and rigidity can considerably influence the capacity of the TME to promote cancer cell proliferation, migration, and invasion, as well as modulate inflammatory responses and lymphangiogenesis [22].

Desmoplasia, that is, the fibrotic stromal reaction associated with most carcinomas, is characterized by the local deposition of different fibrillary collagen types and often correlates with adverse prognosis [12]. The collagen fiber network serves as both a transducer of exogenous forces and a source of endogenous microscale mechanical properties, and there is a growing body of experimental and theoretical work investigating the relationship among microarchitecture-derived cell–scale matrix mechanics, cellular biomechanics, and bulk mechanical properties [8]. As a matter of fact, the stiffness of a matrix and its susceptibility to remodeling by cellular contractile processes, matrix secretion, and enzymatic degradation can affect the distribution of cell surface integrin receptor, the types of cell adhesion and cytoskeletal structures formed [23], as well as the paracrine signaling in the TME [22].

Therefore, models allowing an improved understanding of TME–tumor interactions, and to investigate new therapeutic options, would have a profound effect on the development of innovative therapeutic strategies.

## 3.3 Development of the cancer therapeutics field

Cancer treatment for solid tumors generally consists of a combined approach using surgical, radiation, and drug treatment, and essentially depends on the tumor localization and stage, the anatomopathological classification, and the performance status (PS) of the patient. For most tumor types, surgery is the standard of care for patients who are in early stages and have a good PS. However, the great problem for cancer patients is the relapse and the diagnosis at advanced stages, when surgery is no longer an option. In these cases, drug treatment and radiotherapy have been for many years the standard therapies and palliative intention their main use.

Anticancer drugs can be classified into chemotherapy, hormone, immunotherapy, and targeted therapy. Untargeted chemotherapeutic agents include a variety of molecules and target broad pathways, being responsible for an extensive list of side effects, including cardiotoxicity, peripheral neuropathy, alopecia, inflammation of the intestines, anemia, or immune suppression. Fortunately, the management of cancer has evolved substantially over the last years. Specific antitarget therapies have emerged, which have increased patients' survival and decreased the toxicity produced by conventional treatments. Good examples are tyrosine-kinase inhibitors (TKIs), such as osimertinib, sunitinib, or lapatinib, which are specific treatments for patients bearing particular molecular alterations. For instance, patients harboring metastatic lung adenocarcinoma with mutations or deletions in EGFR or rearrangements in ALK or ROS1 can be offered several TKIs lines of treatment with negligible toxicity profiles and large increases in overall survival [24,25]. However, these drugs have a limited effect over time because of the frequent emergence of drug resistance. This resistance is frequently related to the appearance of specific clones, emphasizing the role of intratumor heterogeneity as one way of explaining resistance mechanisms to target agents [26].

Recently, cancer immunotherapy has also set a standard for the treatment of certain types of cancer. It refers to a number of approaches intended to activate the immune system in order to induce objective responses and disease stabilization, including antibodies that block suppressive immune-checkpoint pathways, cellular therapies based on dendritic and engineered T cells, lymphocyte-activating cytokines, and vaccines that trigger antigen-specific immune responses in tumors. Among them, blocking antibodies specific for the immune-checkpoint molecules cytotoxic T lymphocyte-associated antigen-4 (CTLA-4), programmed cell death receptor-1 (PD-1), or its ligand (PD-L1) have become game-changers in clinical cancer therapy, being approved by the American and the European regulatory agencies for the treatment of many cancers, including melanoma, lung cancer, urothelial cancer, bladder cancer, head and neck cancer, hepatocellular cancer, gastric cancer, and kidney cancer [27]. Moreover, the first engineered T-cell therapies have been developed and approved for clinical use in certain hematological malignancies [28]. Despite these major advances, the clinical use of immunotherapies faces several challenges in terms of efficacy and safety. Regarding efficacy, only 10%–30% of patients respond to immunotherapies and robust biomarkers to predict patient responses are limited. In order to improve response rates, combination therapies seem attractive, but grade-3 or grade-4 immune-related toxicities, including hepatic, gastrointestinal, and renal disorders, have been reported after treatment with anti-CTLA-4 and anti-PD-1/PD-L1 antibodies [29]. Besides, immunotherapy can induce autoimmune adverse effects, triggering attacks on healthy tissues. As happened with IL-2 therapy, immunotherapies can cause cytokine release syndrome and vascular leak syndrome, which lead to severe hypotension, fever, renal dysfunction, and other potentially lethal side effects [21].

With the development of more and more therapeutic approaches designed to target cancer, it is imperative to assess the various models to test them before they enter into clinical trials. Drug development is a very inefficient, tedious and expensive process, facts that are limiting the access of new drugs to clinic. Indeed about 95% of all of new anticancer compounds fail in Phase 1 to Phase 3 clinical testing [30]. The poor efficacy of many experimental anticancer drugs in clinical trials is partly justified by the lack of predictive potential of current preclinical models [31]. As a matter of fact, drugs are optimized to be highly effective based mainly in results generated in conventional two-dimensional (2D) cell culture models and in animals in vivo [32,33]. The 2D cultures present simplicity, low cost, and are well-established methods for investigating cancer cell behavior, but they fail to resemble the tumor environment by loss of biological, chemical, and mechanical cues since they are not able to replicate cell–cell and cell–extracellular interactions [34] and physical and chemical gradients present in TME [35]. In contrast, animal models offer an analogous natural three-dimensional (3D) microenvironment, but they have strong limitations. Mouse xenograft tumor models are frequently used to determine the response to experimental anticancer drugs in vivo, however there is growing evidence that the data obtained from subcutaneous xenograft models poorly translate into positive clinical responses [36]. PDX models, much more representative of human tumors and with a better predictive capacity, are technically demanding, time-consuming, and cost intense, thus strongly limiting their application in the drug discovery process [31]. An interesting alternative model system is the use of 3D models to bridge the gap between 2D cell culture and complex animal models. These models, as detailed in the next section, present soluble gradients, adhesions distributed in all three dimensions, and no prescribed polarity, allowing testing drug delivering technologies, studying the interaction between tumors and immune system or identifying predictive biomarkers as possible tools for the future management of cancer [37,38].

To summarize, the usage of reliable tumor models to determine if a treatment is likely to be effective, would improve the clinical management of cancer patients, and pave the way toward the implementation of personalized medicine. Current tumor models show limitations that need to be overcome. The development of cancer therapeutics could undoubtedly benefit from the development of novel 3D tumor models able to represent some features of real tumors and in the meantime reduce the number of animals in preclinical studies. In the next section the principal methodology for fabricating 3D models will be discussed.

## 3.4 3D tumor models in cancer research

As stated before, 3D tumor models, capable of recapitulating some key features of real tumors, are promising tools for allowing a more accurate preclinical screening of cancer therapeutics, prior to their evaluation in vivo. Indeed, 3D tumor models represent valuable tools for (1) a more accurate preclinical screening of drugs and drug delivery systems and (2) the identification of candidates with the highest chances of success for further evaluation in vivo [39–41]. Cells grown in 3D matrices or as 3D aggregates (multicellular spheroids) much better recapitulate the in vivo architecture of tissues and tumors as well as the genetic patterns [37,42–44]. Additionally, in vitro 3D models have facilitated progress in the understanding of the different stages of cancer progression [45].

Considering that the TME is playing an active role in cancer, this must therefore be taking into consideration in the search of innovative anticancer therapeutics. The 3D in vitro model systems of the TME have raised an increasing interest in the study of cancer and in the drug discovery and drug delivery fields [31]. Different organotypic models based on cell cocultivation have been reported, demonstrating the functional interplay between tumor and stromal fibroblasts [46] or tumor and immune cells [19,22,47]. Moreover, additional cell types, like fibroblast, endothelial, and immune cells can be introduced to the model, thus raising the complexity of the system [23,46,48–50]. Indeed, numerous studies in the last years suggest that tumor-immune system interactions may be investigated by using 3D culture models [19,51].

The 3D culture models, ranging from simple cancer cell spheroids to models comprising multiple cell lines [33] such as fibroblasts, endothelial, and immune cells [19,46,48,49,52] with the purpose of reproducing architectural characteristics of cancer and the TME, can be broadly classified as nonscaffold- and scaffold-based, and prepared following different methodologies, as described next and summarized in Fig. 3.3.

### 3.4.1 Nonscaffold-based 3D cell culture methods

Multicellular tumor spheroids (MCST) are scaffold-free spherical self-assembled aggregates of cancer cells displaying an intermediate complexity between 2D in vitro cell cultures (i.e., cell monolayers) and in vivo solid tumors with which they share important similarities [33]. Importantly, as previously mentioned, they can be established from a single cell type or can be multicellular mixtures of tumor, stromal, and immune cells [23,39].

Different methodologies have been reported for the preparation of MCTS. One of the most common procedures is the use of bioreactors, as for example, spinner flasks as well as rotational culture systems, in which cell aggregation and spheroid formation are mediated by a continuous spinning that keeps in motion a cell suspension, supporting cell-to-cell interactions, and avoiding their attachment to the container wall [41,56]. A different

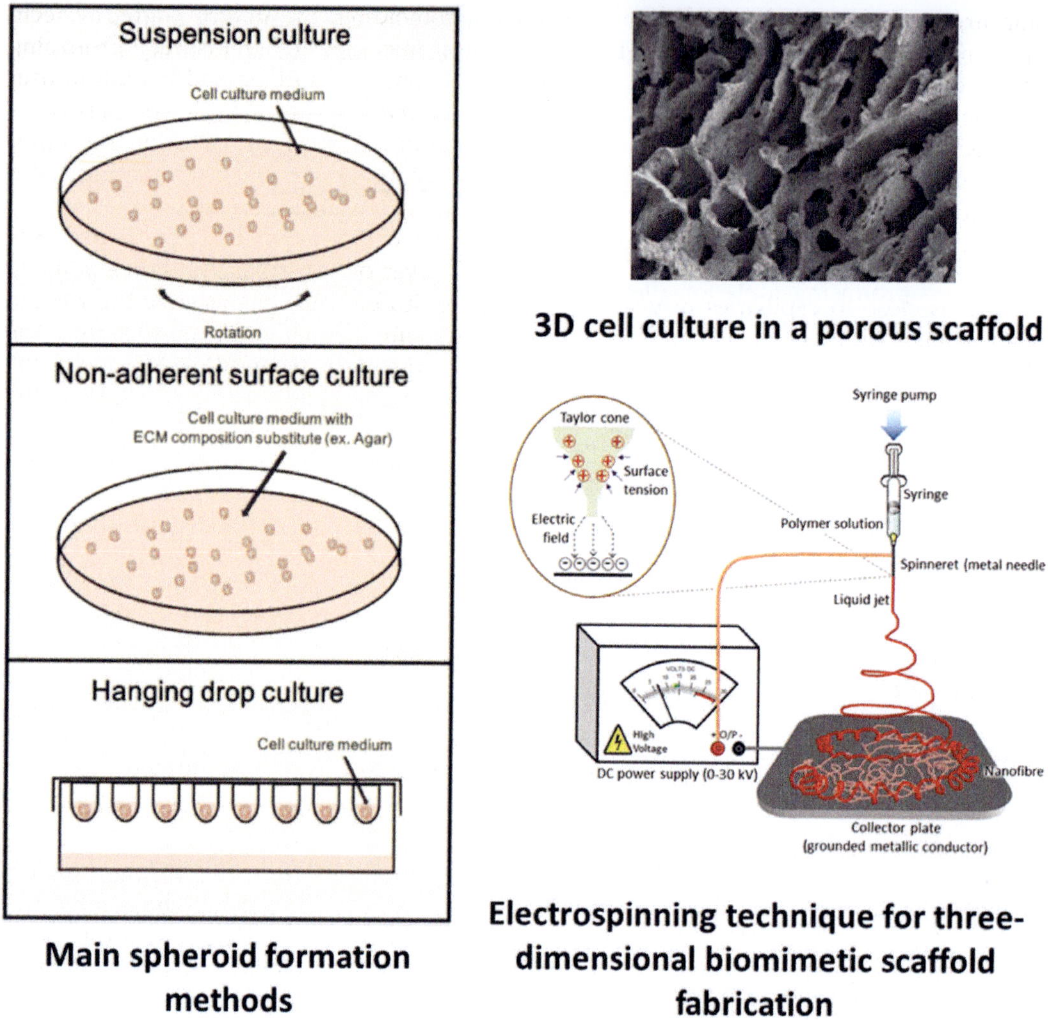


FIGURE 3.3 Principal methods for spheroids preparation and scaffold-based 3D culture methods. Source: *Modified from Huang B-W, Gao JQ. Application of 3D cultured multicellular spheroid tumor models in tumor-targeted drug delivery system research. J Control Release 2018;270:246–59 [53], Ghosal K, Chandra A, Praveen, G, Snigdha S, Roy C, Agatemor C, et al. Electrospinning over solvent casting: tuning of mechanical properties of membranes. Sci Rep 2018;8 (1):5058 [54], and O'Brien FJ. Biomaterials & scaffolds for tissue engineering. Mater Today 2011;14(3):88–95 [55] with permission.*

technique is the liquid overlay technique, which relies on the use of nonadherent 96-well plates, in which a coating with poly-2-hydroxytehyl methacrylate [57,58] (poly-HEMA) or agarose prevents cell attachment [59,60]. This method allows the preparation of both mono and heterotype spheroids, whose size and morphology can be tuned by varying the number of cells seeded in each well. However, a major drawback is the time-consuming plate-coating step. Additionally, spheroids with a defined size and composition have been generated using arrays of microwells fabricated with the micromolding, the photolithography technique, and microfluidics [61]. Among these technics, microfluidics seems very promising because of its capability to combine a high number of spheroids formed and minimum user handling [62]. Additionally, the viability of cells can be maintained longer in comparison to other techniques thanks to the continuous flowing culture medium [63]. Finally, 3D tumor models made of multicellular spheroids surrounded by a polymer scaffold have been also proposed as an alternative to simple spheroids suspended in cell culture medium [39,64].

Importantly, MCTS can mimic the avascular region of the in vivo tumor tissues [39]. Indeed, highly proliferating cells are in the external layer of spheroids and correspond to tumor cells close to capillaries in vivo. Conversely, quiescent cells characterize the middle layer, because cell metabolism decreases progressively with the increasing distance from the spheroid periphery. Finally, in the spheroid core hypoxia, nutrients shortage and metabolic waste accumulation result in cell necrosis [65]. Another important feature of MCTS is the presence of a network of structural (i.e., collagen and elastin) and adhesive (i.e., fibronectin and laminin) ECM proteins embedded via gel of glycosaminoglycans and proteoglycans [39].

Importantly, the 3D structure of MCTs creates a penetration path which needs to be overcome by any agent to be tested. For this reason, they are widely used for testing chemotherapeutics [33] (Fig. 3.4). Indeed, coculture of immunocompetent cells might provide important information related to chemoattraction and activation of specific cell subsets and establish a relevant tool for the test of innovative antiinflammatory compounds or immunostimulating agents for potential antitumoral impact [19].

Spheroids characterization in terms of cytocompatibility is usually carried out using different assays. Among all, the quantifications of the acid phosphatase (APH) activity and the intracellular adenosine triphosphate (ATP) levels do not require preliminary dissociation of spheroids into single-cell suspensions. Moreover, well-established markers for determination of mitotic indices (by quantifying phospho-H3 staining), proliferation indices (Ki67 or PCNA), or apoptosis (cleaved caspase 3 or TUNEL staining) are also used [31].

For the visual assessment of morphological changes, optical microscopy is the most widely used technique (mean diameter, minimum diameter, maximum diameter, volume, area, and circularity) in spheroids. Moreover, qualitative assessment of proliferating or dead/necrotic cells can be performed by confocal laser scanning microscopy (CLSM) imaging following spheroid incubation with various live/dead reagents [39]. Expression of constitutively active luciferase in tumor cells can be used as a marker for cell number/cell viability to detect the tumor specific drug response in coculture with unlabeled stromal cells [66].

Monitoring of fluorescent drugs and dye-labeled nanocarriers by CLSM has been largely applied for evaluating their uptake and penetration into individual spheroids [39]. Compared with traditional CLSM, significant improvement in the evaluation of the penetration into living tumor spheroids can be obtained with two-photon (TPM) or multiphoton microscopy (MPM) in which the energies of two (or more) photons are combined to promote the transition of a fluorescent marker to an excited state [67].

### 3.4.2 Scaffold-based 3D cell culture methods

Scaffolds have been widely used in tissue engineering, for the purpose of wound healing and bone regeneration. However, they have been adapted as matrices for cell

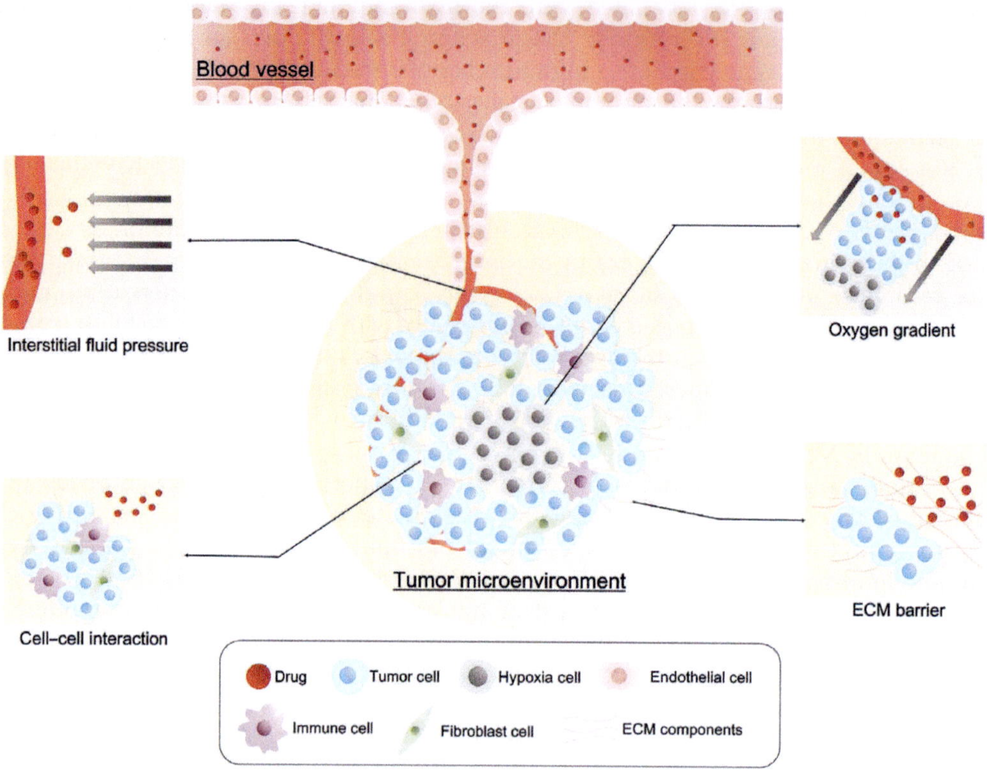


FIGURE 3.4 Transport barriers for drug delivery in tumors that can be mimicked with MCST. Source: *From Huang B-W, Gao JQ. Application of 3D cultured multicellular spheroid tumor models in tumor-targeted drug delivery system research. J Control Release 2018;270:246–59 [53] with permission.*

culture and for the investigation of proliferation, growth, and migration of cancer cells [33]. Indeed, hydrogels can be designed to mimic stiffness and other mechanical properties of tumors in order to understand their impact on tumor invasiveness and metastasis [68,69]. Both natural [64,68,70,71] and synthetic materials [37,72,73] have been used as the scaffolding materials for the engineering of 3D tumor models. Natural scaffolds hydrogels are reported to be mechanically weak, but provide the cells with a biological environment to proliferate in. Synthetic scaffolds are mechanically stronger than their natural counterparts, but cell adhesion to these polymers is much poorer [74]. Cell behavior will also depend on various properties of these biomaterials including their chemistry, material surface properties (hydrophobicity, hydrophilicity, and topography), and structure (porosity, surface area, and interconnectivity) [75].

As commented above, different tumor models are based on the use of scaffold technology, that is, porous natural or synthetic materials with tailored biological, chemical and mechanical properties providing a 3D structure to cells for growth and tissue formation [19]. Due to their ability to simulate the nature of most soft tissues, hydrogels are a highly attractive material for developing synthetic ECM analogs. These reticulated structures of crosslinked polymer chains possess high water contents, facile transport of oxygen, nutrients and waste, as well as realistic transport of soluble factors. Furthermore, many hydrogels can be formed under mild, cytocompatible conditions and are easily modified to possess cell adhesion ligands, desired viscoelasticity, and degradability [76,77].

In the last decades, various fabrication methods for construction of 3D biomimetic scaffolds have been developed. The traditional methods include phase-separation, freeze drying, and self-assembly, while recently electrospinning and bioprinting are gaining much attention [78,79]. Electrospinning is a technique used widely for fabricating fibers with diameters in the nano- or microscale range. In the electrospinning process, a very high voltage is applied to a capillary tube filled with polymer solution or melt, which is held at the tip of the capillary via surface tension. When an electrical field is applied, the charge repulsion within the polymer solution or melt will overcome the surface tension of the polymer solution and a jet is formed. At the end, the ejected polymer solution and melt repel each other and the solvent evaporation form fibers as the jet travels to the collector [78]. An example of electrospun scaffold has been recently developed by Pal et al. [80]. Authors highlighted that the designed 3D hybrid scaffold enhances cell proliferation and induces cancer cells (breast cancer cells, MDA-MB-231 and gastric cancer cells, MKN74) to undergo epithelial–mesenchymal transition (EMT), as demonstrated by significant upregulation of EMT associated transcriptional factors (Snail1, Zeb1, and Twist2) and mesenchymal markers and downregulation of E-cadherin.

The 3D bioprinting allows for the generation of tissues incorporating a variety of cell types in a complex and defined spatial architecture, and rapid advances in its use have been shown toward modeling both normal tissues and disease states [81]. Among all, this technique may provide an additional way for obtaining desired 3D cell structures for implementation in cancer models in vitro [81–83]. Huang et al. [82] designed a 3D in vitro microchip in hydrogel by using 3D projection printing aimed at mimicking 3D vascular morphology to analyze differences in the behavior of cancer cells. Zhao et al. [83] reported a method of 3D printing for Hela cells and gelatin/alginate/fibrinogen hydrogels to construct in vitro cervical tumor models, showing that Hela cells in 3D printed models displayed higher MMP protein expression and higher chemoresistance than those in 2D culture. Langer et al. [81] designed a tumor tissue model similar to solid tumor architecture in which a core tumor cell bioink was surrounded on all sides by a normal stromal cell bioink, finding that cellular proliferation, extracellular matrix deposition, and cellular migration are altered in response to extrinsic signals or therapies.

Several methodologies could be used for studying hydrogels microstructure. Among all, scanning electron microscopy (SEM) and transmission electron microscopy (TEM) are widely used [84]. Additionally, pore size distribution could be studied as it is reported to highly affect cells viability, proliferation, and differentiation on the scaffolds [85]. The assessment of scaffolds properties to guide cell proliferation and differentiation could be performed using the same methodologies used for multicellular spheroids (see Section 3.4.1). The relationship between tissue elasticity and cell signaling and response has significant implications for many types of disease, especially cancer. In fact, increased ECM stiffness, along with elevated collagen concentration, is a hallmark of many tumors. For these reasons, mechanical properties should be well defined by mechanical compression tests [84].

Successful studies have been already reported in literature. For example, Szot et al. [68] bioengineered collagen I hydrogel culturing the scaffolds with human breast cancer MDA-MB-231 cells, obtaining a significant upregulation of vascular endothelial growth factor (VEGF)-A gene expression. Xu et al. [64] designed a hyaluronic acid (HA)-based bilayer hydrogel system for supporting the tumoroid formation from LNCaP prostate cancer cells (PCa), but also for simulating their reciprocal interactions with the tumor-associated stroma. Interestingly, compared with cells cultured on 2D, the engineered tumoroids significantly increased the expression of two proangiogenic factors, vascular endothelial growth factor-165 (VEGF165), and interleukin-8 (IL-8).

## 3.5 Evaluation of anticancer therapeutics in 3D tumor models

### 3.5.1 Drug screening/drug resistance

Several studies have been reported for the use of spheroids for drug screening and drug resistance mechanisms investigation. Wenzel et al. [86] incubated T47D (human breast cancer cells) spheroids for 3 days either with cisplatin or paclitaxel, two routinely used chemotherapeutic agents. Similar to their toxic effect in 2D cell cultures, these compounds induced wide spread cell death in T47D MCTS after 3 days of incubation. However, authors observed that cell death only occurred in outer spheroid regions. In contrast, treatment with staurosporine, a general cytotoxic agent, led to complete dissolution of the tumor spheroid and wide spread cell death (Fig. 3.5). Authors concluded that dormant inner MCTS cells can resist cisplatin and paclitaxel therapy as these drugs act on certain component of the respiratory chain. The same research group designed a high-content based setup for the identification of substances specifically targeting the respiratory chain and inducing cell death in the spheroid core showing that floating spheroids of breast, colon, and prostate cancer cells are valuable tools for anticancer drug testing.

A spheroid model of liver carcinoma showed that the drug response of the tumor cells declined with increasing complexity, that is, addition of fibroblasts during spheroid aggregation and/or addition of a collagen matrix [48]. In a similar way, Vorsmann et al. [87] demonstrated that skin equivalent-embedded melanoma spheroids were only affected by

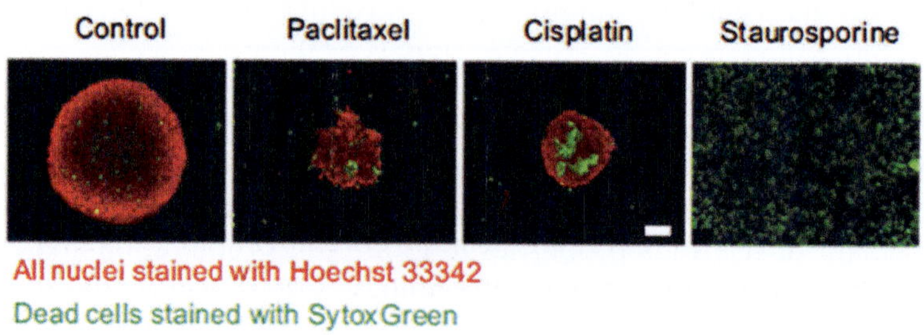


**FIGURE 3.5** Cytotoxic effects of paclitaxel (100 μM), cisplatin (100 μM), and staurosporine (10 μM) in human breast cancer cells spheroids. Source: *With permission from Wenzel C, Riefke B, Grundemann S, Krebs A, Christian S, Prinz F, et al. 3D high-content screening for the identification of compounds that target cells in dormant tumor spheroid regions. Exp Cell Res 2014;323(1):131–43.*

the TRAIL/cisplatin combination, while TRIAL/UV light and TRAIL/cisplatin killed the melanoma cell lines grown in conventional culture plates.

McMillin et al. [66] used a cocultured model of malignant tumor cells stably expressing luciferase (luc) with nonmalignant luc negative stromal cells for selective quantification of tumor cell viability, in presence versus absence of stromal cells or drug treatment. Indeed, bioluminescent signal, proportional to the number of viable luc-expressing tumor cells, allowed for specific quantification of tumor cell viability in cocultures. A coculture model has been also developed by Alonso-Nocelo et al. [40] who studied tumor-immune cells interactions in the TME by coculturing nonsmall cell lung cancer (NSCLC) cells and T lymphocytes. The study of secretome revealed a subset of different cancer-related inflammation proteins pointing out an intercellular communication. Additionally, authors pointed out differences in the response to chemotherapy.

The use of alginate-based scaffolds for cell encapsulation in cancer drug testing has been also described for nonsmall cell lung cancer (NSCLC) [88] and brain cancer [89], respectively. In both cases, embedded cells in these 3D cultures revealed higher resistance to chemotherapeutic as compared with 2D ones. In accordance to the previous work, Fischbach et al. [37] studied a 3D model of human oral cancer by culturing oral squamous carcinoma cells (OSCC-3) within highly porous scaffolds fabricated from synthetic poly (lactide-*co*-glycolide) (PLG). Authors treated both 2D monolayer cultures and 3D PLG tumor models with the PI3-kinase inhibitor LY294002, a model drug earlier used to study the efficacy of anti-OSCC-3 tumor therapy. Tumor cells cultured under 2D conditions were sensitive to the drug, whereas those present within the engineered 3D PLG TME were not.

Finally, Gurski et al. [90] investigated a biomimetic and 3D culture system for poorly adherent bone metastatic prostate cancer cells (C4-2B) as an in vitro platform for anticancer drug screening. The HA hydrogel system was used to test the efficacy of anticancer drugs including camptothecin, docetaxel, and rapamycin, alone and in combination, including specificity, dose, and time responses. Authors found that responses of cells to antineoplastics differed between the 3D HA hydrogel and 2D monolayer systems.

To summarize, spheroids have largely shown a high versatility for the possibility of coculturing different cancer and TME cells and testing novel therapeutics, showing the potentiality to be a fundamental step before in vivo testing.

### 3.5.2 Anticancer nanomedicines

In the last years, nanotechnology has gained a large interest in cancer therapy thanks to several advantages. Indeed, nanoparticles (NP) can enhance the solubility and stability of several therapeutic agents. Moreover, NP characteristics can be properly engineered for controlling the pharmacokinetics of the loaded drug and obtain a proper targeting and selectivity to cancer cells [41].

Currently, initial evaluations of NP delivery formulations are usually completed in monolayer, or 2D cell culture systems. However, successful NP delivery observed in 2D cell culture studies does not translate to similar results in vivo. Whereas NPs delivered in a bulk solution to a 2D cell culture typically reach and bind to cells relatively unimpeded, the same NPs delivered in vivo will be hindered by extracellular factors such as the relatively small pore sizes within the extracellular matrix and the tortuosity of the interstitium. Indeed, the 3D organization of a tumor mass fundamentally alters the diffusion profile for drugs, both through the cell–cell contacts and cell–matrix interactions [41,64].

For these reasons, 3D platforms, as spheroids and hydrogels, have been proposed for the evaluation of NPs diffusion properties into tumors and interaction with cancer cells [64,91].

The 3D MCTS, thanks to their similarity in morphology and biological microenvironment to solid tumors, have been used as a robust tool for easy polymer NP screening [92,93] and to accurately predict the in vivo behavior of the nanocarriers as a function of their specific physicochemical properties [94]. Ma et al. [93] generated HeLa cell-derived spheroids using the liquid overlay method for comparing the penetration of doxorubicin (DOX) and DOX-loaded synthetic micelles into 3D HeLa spheroid versus HeLa cells grown in a traditional 2D system. When HeLa cells were grown in a monolayer, DOX was absorbed rapidly, whereas the drug entered spheroids much more slowly in HeLa cells spheroids, demonstrating that HeLa cells in 3D structures provide a multicellular resistance model that mimics the chemotherapy resistance often found in solid tumors in vivo. Then, authors studied the penetration of three types of DOX-loaded micelles, methoxy polyethylene glycol-poly(lactic-*co*-glycolic)acid (mPEG-PLGA), gelatin-polylactic acid (Gel-PLA), and poly(L-aspartic acid-*co*-L-lactide)-1,2-dipalmitoyl-sn-glycero-3-phosphoethanolamine (PAL-DPPE) in the 2D and 3D cultured HeLa cells. In the monolayer cells, the intracellular fluorescence intensity of mPEG-PLGA- and PAL-DPPE-treated cells was almost the same, whereas the Gel-PLA-treated cells had weaker intracellular fluorescence. In the 3D culture system, mPEG-PLGA-treated cells had moderate fluorescence, whereas PAL-DPPE-treated cells had the strongest fluorescence, which indicated deeper penetration, and Gel-PLA-treated cells had the weakest fluorescence, presenting the shortest translocation.

Curcumin-loaded biodegradable and nonbiodegradable NPs prepared by self-assembly of bovine serum albumin (BSA) and conjugated with poly(ε-caprolactone) (PCL) or poly (methyl methacrylate) (PMMA) were investigated in LNCaP prostate cancer MCTS in comparison with conventional 2D cultures [95]. The latter did not reveal appreciable differences between the two types of NPs, which were both rapidly taken up by cells and induced similar cytotoxicity. In contrast, the MTCS allowed the detection of the faster and deeper accumulation of BSA-PMMA NPs, which occurred via a rapid sequence of endo- and exocytosis events.

Godugu et al. [88] obtained multicellular spheroids (composed by H460, A549, and H1650 NSCLC cells) in the size range of 100–300 μm using 6- and 96-well AlgiMatrix scaffolds and results in terms of uptake studies for drug (DOX) and NP were compared with the conventional 2D monolayer cultures. Increased chemoresistance for docetaxel was found in spheroid cultures compared with 2D culture systems. Indeed, IC50 values for the 3D spheroid cultures ranged from 10- to 60-fold higher than their 2D counterparts. Authors suggested that the poor penetration of the chemotherapeutic drugs into the in vitro tumor spheroid might be the reason for the observed resistance, in a similar way as happens in vivo. Fluorescence analysis of fluorescent uptake with free DOX and DIO oil-loaded NPs indicated that compared with NPs, free DOX produced 5.56-folds higher fluorescent intensities. The poor penetrating nature exhibited by spheroids demonstrate very limited NP uptake. Similarly, in vivo tumors have also been shown to exhibit poor NP penetration due to the presence of their highly fibrous nature.

On their behalf, applications of coculture models are really useful for investigating the specificity of NP carriers. Meng et al. [96] developed a brain tumor coculture model consisting of cancerous cell aggregates and organotypic brain slices. Evaluation of the uptake of poly(glycerol-adipate) NPs revealed that in 3D cocultures, NPs were predominantly taken up by cancerous cells rather than normal cells. In contrast, in monolayer cultures, NPs were taken up more readily by normal cells. Results suggested that the NPs may be useful as anticancer formulations with reduced toxicity to normal cells.

Alonso-Nocelo et al. [40] evaluated the potential of PEGylated polyglutamic acid nanocapsules for the treatment of advanced metastatic cancer in a 3D model obtaining coculturing A549 lung tumor cells and Jurkat lymphocytes. Authors demonstrated that nanocapsules are able to penetrate into the tumor mass targeting cancer cells with respect to immune cells (Fig. 3.6).

Currently, MCTS have also been widely used to evaluate the tumoral behavior of targeted NPs. Teijeiro-Valiño et al. [97] reported nanocarriers consisting of hyaluronic acid (HA) nanocapsules (NC) conjugated with the tumor homing peptide (tLyp1), for the selective targeting of tumor cells. tLyp1 functionalized HA NCs were loaded with docetaxel and tested in 3D cocultures of the lung adenocarcinoma cell line A549 and Jurkat E6.1 lymphocytes. Successfully, authors demonstrated nanocarriers specifically targeted cancer cells and did not interact with lymphocytes. Indeed, DiD fluorescence was detected around the cell nuclei or the cytoplasm of cancer cells, while no fluorescence was observed within lymphocytes.

Perche et al. [98] described the evaluation of DOX-loaded micelles targeting using an ovarian cancer cell spheroid model. DOX was encapsulated in polyethylene glycol–phosphatidyl ethanolamine (PEG–PE) conjugated micelles. The DOX-loaded PEG–PE

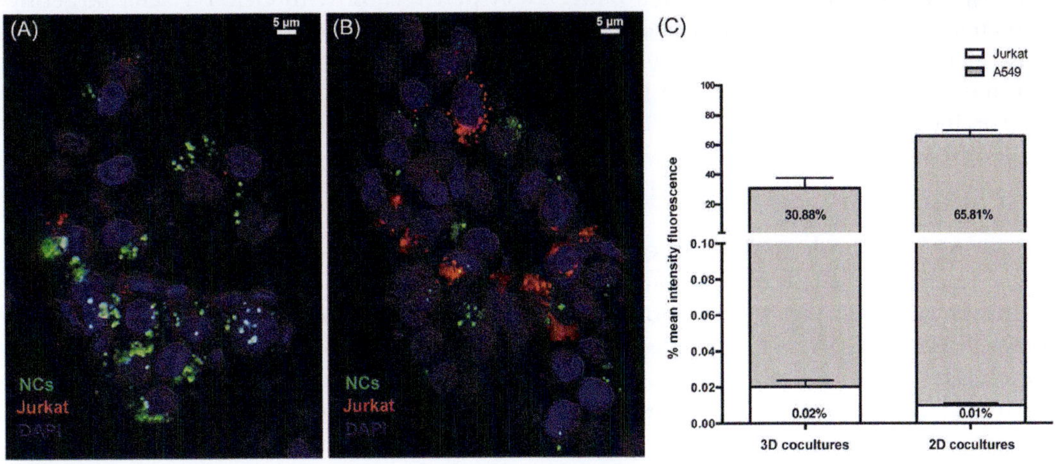


FIGURE 3.6 Interaction of PEGylated polyglutamic acid nanocapsules into lung cancer cells and lymphocytes [40]. a–b) confocal microscopy z-stacks showing nanocapsules (NCs) accumulation into cells; c) graphical representation of the uptake of DiD-loaded NCs in 3D and 2D co-cultures [40].

micelles (MDOX) were further decorated with a cancer cell-specific monoclonal 2C5 antibody to obtain DOX-loaded immunomicelles (2C5-MDOX). Superior accumulation of 2C5-MDOX compared with free DOX or untargeted MDOX in spheroids was evidenced by flow cytometry, fluorescence, and confocal microscopy.

Oishi et al. [99] demonstrated that PEGylated polyplexes (lac-PEGylated polyplexes) composed of poly(L-lysine) and lactosylated poly(ethylene glycol)-small interfering RNA conjugate inhibit the growth of multicellular HuH-7 (human hepatocarcinoma cell lines) for up to 21 days. The lactose moieties on their periphery were selected as targeting ligands for the asialoglycoprotein-receptor-expressing HuH-7 cells. OligofectAMINE/siRNA showed no growth-inhibitory effect against HuH-7 spheroids, even though the lipoplexes showed in 2D cultures a stronger growth-inhibitory effect than the lac-PEGylated polyplexes. Additionally, the FITC-tagged conjugate in the lac-PEGylated polyplexes showed smooth penetration into the HuH-7 spheroids compared with that in the lipoplexes, as observed by confocal fluorescence scanning microscopy.

Roy et al. [60] investigated the anticancer activity of alginate coated chitosan NPs encapsulating cell-permeable dominant negative surviving (SR9) with locked nucleic acid (LNA) aptamers targeting EpCAM and nucleolin (termed as nanobullets). Authors observed a higher intensity of nanobullets in both the periphery and the core of the multicellular tumor spheroids compared with untargeted NPs. Indeed, the nanobullets were found to be highly effective as they led to a 2.26-fold reduction at 24 h and a 4.95-fold reduction in the spheroid size at 72 h in comparison to untargeted NPs.

Sarisozen et al. [59] decorated PEG–PE-based polymeric micelles with glucose transporter-1 (GLUT-1) antibody single chain fragment variable (scFv) as the ligand to promote blood–brain barrier transport and glioblastoma targeting. When distribution of the rhodamine-labeled micelles was followed in the spheroids of U87MG cells, the profile clearly indicates that the GLUT-1 scFv targeting allows micelles not only to penetrate deeper layers of the spheroids but also at a higher extent, compared with nontargeted micelles. The data clearly indicate a better DOX penetration with GLUT-1 scFv targeting compared with nontargeted micelles encapsulated DOX or its free-base.

In conclusion, the large amount of reported characterization studies of NPs in 3D culture models have highlighted the importance of these systems for assessing the efficacy and/or the selective targeting of the NPs prior to in vivo testing and clinical practice implementation.

## 3.6 Implementation of 3D tumor models in a clinical setting

The advances in the understanding of the tumorigenesis and the interaction between tumor cells and TME have provided valuable information which, undoubtedly, is impacting on the design of new therapeutic strategies in different tumor types. Proof of this is the increasing number of patient-derived 3D tumor models being studied for their potential use as improved platforms for drug discovery and personalized medicine, which is the ultimate objective, as each cancer is unique. With the substantial advances in biomedical research and the improvements in diagnosis and therapies of the past decades, personalized medicine has become a more realistic goal, although clinical implementation of 3D tumor models requires a high quality of methodological and clinical validation comparable to other biomarkers. For a 3D modeling to be universally adopted, it must be cheap and efficient, providing reliable results that can be translated to clinical practice. For this to happen, tumor culturing methods need to be rationalized and standardized, and high-throughput methods of analyzing the chemosensitivity must be implemented.

Despite all the technological advancements and improvements that are still required, researchers are putting a great deal of effort in order to directly convert results obtained in 3D tumor models into feasible treatment options for patients, as can be noticed by the number of clinical trials ongoing and already finishing recruitment (Table 3.1). The most advanced application of these models is their usage to predict patients' response to cancer therapies. In metastatic colorectal cancer, a prospective, multicenter trial (NCT02849106) in patients with indication for treatment with a combination including FOLFOX/FOLFIRI associated to an anti-VEGF drug has recently completed recruitment. In this study, patient's metastatic or primitive tumor biopsies before treatment were collected and were used to establish 3D ex vivo cultures, where chemograms were made to estimate the predictive value of this technique for treatment response. Patient response was evaluated by the RECIST assessment and is currently being compared with the results obtained by the chemogram. Subsequently, a phase II study (NCT03251612) is recruiting metastatic colorectal cancer patients to assess the predictive value of in vitro testing anticancer therapy sensitivity on tumorspheres. In this study, patient-derived tumorspheres are treated with standard anticancer treatments, including oxaliplatin, bevacizumab, pembrolizumab, or sorafenib, among others, and, according to the result of the sensitivity analysis, patients are treated with one drug, another, or a combination. Similarly, a new clinical trial (NCT03997617) started this summer based on a novel and centralized approach named Integrated Personalized Functional Profiling for patients with metastatic gastrointestinal cancer or recurrent glioblastoma. Based on recent improvements in cancer biopsy-derived 3D culture technologies, this approach will screen primary cultures with approved cancer drugs to generate personalized treatment recommendation to the clinician. Furthermore, the use of 3D cancer organoids has particularly increased in recent years with greater focus on the aspect of personalized medicine than on early stage of drug discovery. Using this strategy, colorectal, breast, ovarian, gastric, bladder, and liver cancer organoids have been used to relatively low throughput screen target-known inhibitors and chemotherapy drugs [100]. Importantly, patient-derived organoids exhibit interpatient and intratumor heterogeneity in their drug responses, representing proof of the concept that cancer organoids are a useful platform for drug screening. In addition, they found that some drugs are highly heterogeneous in the responses among organoids, while others are uniformly toxic in all cases. Based on this approach, observational studies are recruiting lung (NCT03979170) and breast (NCT03925233) cancer patients to establish a feasible drug screening method on organoid models which provide drug sensitivity information applicable in clinical practice.

Another application that has particular interest is the use of 3D tumor models for high-throughput screening for drug discovery, but the amount of available cells is a critical factor. To overcome this issue, some investigators have included the generation of xenograft tumors, and the obtaining of 3D models from these xenografts. In fact, a clinical trial (NCT02910895) is recruiting sarcoma patients to study the passage possibility of

**TABLE 3.1** Tumor model approaches in clinical development.

| Study | Type | Phase | Enrollment | Purpose | Status | Sponsor | Identifier |
|---|---|---|---|---|---|---|---|
| Predictive response to chemotherapy (FOLFOX or FOLFIRI) by ex vivo culture 3D technique in metastatic colorectal cancer: a preliminary study | Interventional | – | 22 | Diagnostic | Recruitment completed | Institut Curie | NCT02849106 |
| Predictive value of drug sensitivity testing tumorspheres from patients with metastatic colorectal cancer | Interventional | Phase II | 90 | Treatment | Recruiting | Vejle Hospital | NCT03251612 |
| Personalized functional profiling in metastatic gastrointestinal cancer or recurrent glioblastoma patients | Interventional | – | 20 | Diagnostic | Recruiting | Luxembourg Institute of Health | NCT03997617 |
| Patient-derived organoids of lung cancer to test drug response | Observational | – | 50 | Basic Science | Recruiting | University Hospital, Geneva | NCT03979170 |
| Sensitivity detection and drug resistance mechanism of breast cancer therapeutic drugs based on organ-like culture | Observational | – | 300 | Basic Science | Recruiting | Xijing Hospital | NCT03925233 |
| Development of a platform of patient-derived xenografts (PDX) and 2D/3D cell cultures of soft tissue sarcomas (STS): protocol to obtain tumor material from patients with STS | Interventional | – | 40 | Basic Science | Recruiting | The Netherlands Cancer Institute | NCT02910895 |
| Dendritic cell immunotherapy against cancer stem cells in glioblastoma patients receiving standard therapy (DEN-STEM) | Interventional | Phase II/III | 60 | Treatment | Recruiting | Oslo University Hospital | NCT03548571 |
| Identification of novel genes involved in cancer cell extravasation by transcriptional profiling of primary organ-specific endothelial cells and in vitro 3D models | Observational | – | 31 | Basic Science | Recruitment completed | Istituto Ortopedico Galeazzi | NCT04047459 |
| Identification and characterization of predictive factors of onset of bone metastases in cancer patients | Observational | – | 25 | Basic Science | Recruiting | Istituto Ortopedico Rizzoli | NCT03895216 |

xenografts, 2D, and 3D cultures. Additionally, important efforts have been done in colorectal cancer to evaluate thousands of drugs with automated systems. Using organoid handlers and reagent dispensers, Kondo et al. screened a library of 2427 drugs, including approved drugs and small molecules with known targets, in colorectal cancer spheroids, finding drugs which potentially suppressed their viability [101].

The 3D tumor models have also become an attractive tool to analyze the interaction between the immune system and tumor cells. Based on this idea, 3D tumor cultures have been implemented to stimulate dendritic cells used for dendritic cell-based cancer vaccines. This approach is specially advanced in glioblastoma patients, where a randomized Phase II/III trial (NCT03548571) of dendritic cell immunotherapy against CSCs is ongoing. Tumor biopsies are cultured under sphere-forming conditions and RNA is extracted and purified. At specified intervals, patients randomized to the vaccine group receive intradermal injections of dendritic cells previously transfected with mRNA from autologous CSCs, surviving, and hTERT. Microfluidic devices are meaning a revolution as well, letting to measure the recognition capacity of tumor cells by immune cells after immunotherapy treatments or by engineered T cells preinfusion [63].

Finally, 3D cultures are potential models to analyze cell invasion, migration, and metastatic potential under well-controlled conditions. An observational study (NCT04047459) has recently completed recruitment focused on breast cancer metastatic. Using microfluidic in vitro organ-specific 3D models to find differences between bone and skeletal muscle

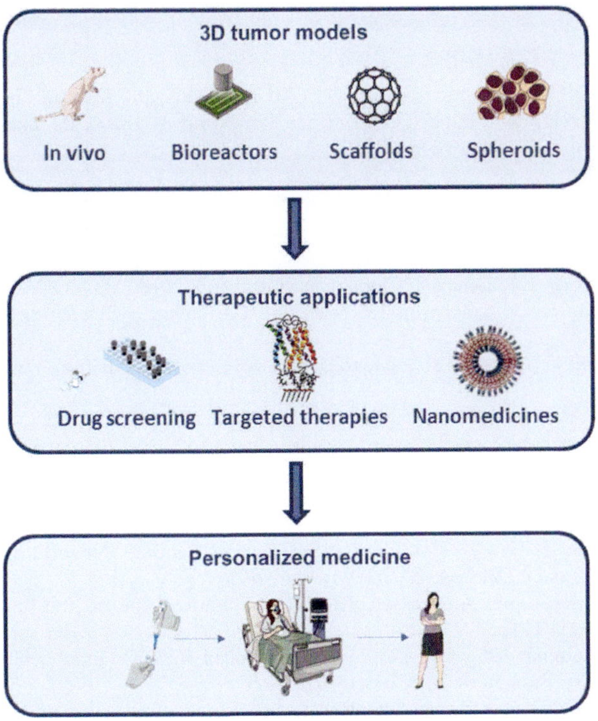


FIGURE 3.7 Diagram of the therapeutic applications of 3D tumor models for their implementation in clinical practice to achieve a personalized medicine.

microenvironments, which are, respectively, preferential and nonpreferential metastasisation sites for breast cancer cells; and the aim of the study is to identify specific pathways driving breast cancer cell extravasation. Similarly, a second observational study (NCT03895216) is now ongoing trying to analyze the cross-talk between tumor exosomes and bone tissue cells within the tumor niche in 3D models from metastatic bone patients.

## 3.7 Final remarks

Major advances in 3D tumor models have enabled rapid development of these technologies, starting their way from bench to bedside (Fig. 3.7). The drive toward precision medicine has resulted in increased interest in adapting in vitro tumor models for patient-specific therapies, clinical management, and assessment of metastatic potential. The ideal model would incorporate both treatment efficacy and toxicity in a representative 3D patient-derived tumor model.

However, much more work needs to be carried out in order to design approaches that can feasibly be translated to patients. Some essential design criteria for clinical translation include stability of the technique, scalability, low production cost, and complexity. The next generation of 3D tumor models will likely include combinations of existing models and the incorporation of new technologies to overcome these limitations and narrow the gap between tumors in patients and personalized in vitro models.

## Acknowledgments

The authors acknowledge funding from the ERA-NET Euronanomed III project METASTARG, the Instituto de Salud Carlos III (ISCIII) and European Regional Development Fund (FEDER) (PI18/00176 and AC18/00045), and Fundación Científica de la Asociación Española Contra el Cancer (IDEAS18153).

## References

[1] Bray F, Ferlay J, Soerjomataram I, Siegel RL, Torre LA, Jemal A. Global cancer statistics 2018: GLOBOCAN estimates of incidence and mortality worldwide for 36 cancers in 185 countries. CA Cancer J Clin 2018;68 (6):394–424.

[2] Bartlett R, Everett W, Lim S, Loizidou NG,M, Jell G, Tan A, et al. Personalized in vitro cancer modeling - fantasy or reality? Transl Oncol 2014;7(6):657–64.

[3] Hanahan D, Weinberg RA. Hallmarks of cancer: the next generation. Cell 2011;144(5):646–74.

[4] Marusyk A, Almendro V, Polyak K. Intra-tumour heterogeneity: a looking glass for cancer? Nat Rev Cancer 2012;12(5):323–34.

[5] Weinstein JN, Collisson EA, Mills GB, Shaw KR, Ozenberger BA, Ellrott K, et al. The cancer genome atlas pan-cancer analysis project. Nat Genet 2013;45(10):1113–20.

[6] Zhang J, Baran J, Cros A, Guberman JM, Haider S, Hsu J, et al. International Cancer Genome Consortium Data Portal--a one-stop shop for cancer genomics data. Database (Oxf) 2011;2011:bar026.

[7] Cui Y, Guo G. Immunomodulatory function of the tumor suppressor p53 in host immune response and the tumor microenvironment. Int J Mol Sci 2016;17(11):E1942.

[8] Carey SP, Kraning-Rush CM, Williams RM, Reinhart-King CA. Biophysical control of invasive tumor cell behavior by extracellular matrix microarchitecture. Biomaterials 2012;33(16):4157–65.

[9] Bissell MJ, Radisky D. Putting tumours in context. Nat Rev Cancer 2001;1(1):46–54.

[10] Klemm F, Joyce JA. Microenvironmental regulation of therapeutic response in cancer. Trends Cell Biol 2015;25(4):198–213.

[11] Junttila MR, de Sauvage FJ. Influence of tumour micro-environment heterogeneity on therapeutic response. Nature 2013;501(7467):346–54.

[12] Sounni NE, Noel A. Targeting the tumor microenvironment for cancer therapy. Clin Chem 2013;59(1):85–93.

[13] Noël A, Gutiérrez-Fernández A, Sounni NE, Behrendt N, Maquoi E, Lund IK, et al. New and paradoxical roles of matrix metalloproteinases in the tumor microenvironment. Front Pharmacol 2012;3:140.

[14] Togo S, Polanska UM, Horimoto Y, Orimo A. Carcinoma-associated fibroblasts are a promising therapeutic target. Cancers (Basel) 2013;5(1):149–69.

[15] Burkholder B, Huang RY, Burgess R, Luo S, Jones VS, Zhang W, et al. Tumor-induced perturbations of cytokines and immune cell networks. Biochim Biophys Acta 2014;1845(2):182–201.

[16] Coffelt SB, Hughes R, Lewis CE. Tumor-associated macrophages: effectors of angiogenesis and tumor progression. Biochim Biophys Acta 2009;1796(1):11–18.

[17] Pollard JW. Tumour-educated macrophages promote tumour progression and metastasis. Nat Rev Cancer 2004;4(1):71–8.

[18] Croci DO, Zacarias Fluck MF, Rico MJ, Matar P, Rabinovich GA, Scharovsky OG. Dynamic cross-talk between tumor and immune cells in orchestrating the immunosuppressive network at the tumor microenvironment. Cancer Immunol Immunother 2007;56(11):1687–700.

[19] Hirt C, Papadimitropoulos A, Mele V, Muraro MG, Mengus C, Iezzi G, et al. "In vitro" 3D models of tumor-immune system interaction. Adv Drug Deliv Rev 2014;79–80:145–54.

[20] Rintoul RC, Sethi T. Extracellular matrix regulation of drug resistance in small-cell lung cancer. Clin Sci (Lond) 2002;102(4):417–24.

[21] Lee JM, Mhawech-Fauceglia P, Lee N, Parsanian LC, Lin YG, Gayther SA, et al. A three-dimensional microenvironment alters protein expression and chemosensitivity of epithelial ovarian cancer cells in vitro. Lab Invest 2013;93(5):528–42.

[22] Alonso-Nocelo M, Raimondo TM, Vining KH, Lopez-Lopez R, de la Fuente M, Mooney DJ. Matrix stiffness and tumor-associated macrophages modulate epithelial to mesenchymal transition of human adenocarcinoma cells. Biofabrication 2018;10(3):035004.

[23] Yamada KM, Cukierman E. Modeling tissue morphogenesis and cancer in 3D. Cell 2007;130(4):601–10.

[24] Lynch TJ, Bell DW, Sordella R, Gurubhagavatula S, Okimoto RA, Brannigan BW, et al. Activating mutations in the epidermal growth factor receptor underlying responsiveness of non-small-cell lung cancer to gefitinib. N Engl J Med 2004;350(21):2129–39.

[25] Shaw AT, Kim DW, Nakagawa K, Seto T, Crino L, Ahn MJ, et al. Crizotinib versus chemotherapy in advanced ALK-positive lung cancer. N Engl J Med 2013;368(25):2385–94.

[26] Aparicio S, Caldas C. The implications of clonal genome evolution for cancer medicine. N Engl J Med 2013;368(9):842–51.

[27] Riley RS, June CH, Langer R, Mitchell MJ. Delivery technologies for cancer immunotherapy. Nat Rev Drug Discov 2019;18(3):175–96.

[28] Schuster SJ, Bishop MR, Tam CS, Waller EK, Borchmann P, McGuirk JP, et al. Tisagenlecleucel in adult relapsed or refractory diffuse large B-cell lymphoma. N Engl J Med 2019;380(1):45–56.

[29] Wolchok JD, Kluger H, Callahan MK, Postow MA, Rizvi NA, Lesokhin AM, et al. Nivolumab plus ipilimumab in advanced melanoma. N Engl J Med 2013;369(2):122–33.

[30] Kola I, Landis J. Can the pharmaceutical industry reduce attrition rates? Nat Rev Drug Discov 2004;3(8):711–16.

[31] Unger C, Kramer N, Walzl A, Scherzer M, Hengstschlager M, Dolznig H. Modeling human carcinomas: physiologically relevant 3D models to improve anti-cancer drug development. Adv Drug Deliv Rev 2014;79–80:50–67.

[32] Kamb A. What's wrong with our cancer models? Nat Rev Drug Discov 2005;4(2):161–5.

[33] Nyga A, Cheema U, Loizidou M. 3D tumour models: novel in vitro approaches to cancer studies. J Cell Commun Signal 2011;5(3):239–48.

[34] Cukierman E, Pankov R, Stevens DR, Yamada KM. Taking cell-matrix adhesions to the third dimension. Science 2001;294(5547):1708–12.

[35] Heldin C-H, Rubin K, Pietras K, Ostman A, Heldin CH, Rubin K, et al. High interstitial fluid pressure - an obstacle in cancer therapy. Nat Rev Cancer 4: 806–813. Nat Rev Cancer 2004;4:806–13.

[36] Voskoglou-Nomikos T, Pater JL, Seymour L. Clinical predictive value of the in vitro cell line, human xenograft, and mouse allograft preclinical cancer models. Clin Cancer Res 2003;9(11):4227–39.

[37] Fischbach C, Chen R, Matsumoto T, Schmelzle T, Brugge JS, Polverini PJ, et al. Engineering tumors with 3D scaffolds. Nat Methods 2007;4(10):855–60.

[38] Benien P, Swami A. 3D tumor models: history, advances and future perspectives. Future Oncol 2014;10 (7):1311–27.

[39] Lazzari G, Nicolas V, Matsusaki M, Akashi M, Couvreur P, Mura S. Multicellular spheroid based on a triple co-culture: a novel 3D model to mimic pancreatic tumor complexity. Acta Biomater 2018;78:296–307.

[40] Alonso-Nocelo M, Abellan-Pose R, Vidal A, Abal M, Csaba N, Alonso MJ, et al. Selective interaction of PEGylated polyglutamic acid nanocapsules with cancer cells in a 3D model of a metastatic lymph node. J Nanobiotechnol 2016;14:51. Available from: https://doi.org/10.1186/s12951-016-0207-8.

[41] Goodman TT, Ng CP, Pun SH. 3-D tissue culture systems for the evaluation and optimization of nanoparticle-based drug carriers. Bioconjug Chem 2008;19(10):1951–9.

[42] Kunz-Schughart LA, Heyder P, Schroeder J, Knuechel R. A heterologous 3-D coculture model of breast tumor cells and fibroblasts to study tumor-associated fibroblast differentiation. Exp Cell Res 2001;266(1):74–86.

[43] Takagi A, Watanabe M, Ishii Y, Morita J, Hirokawa Y, Matsuzaki T, et al. Three-dimensional cellular spheroid formation provides human prostate tumor cells with tissue-like features. Anticancer Res 2007;27 (1A):45–53.

[44] Pickl M, Ries CH. Comparison of 3D and 2D tumor models reveals enhanced HER2 activation in 3D associated with an increased response to trastuzumab. Oncogene 2009;28(3):461–8.

[45] Adjei IM, Blanka S. Modulation of the tumor microenvironment for cancer treatment: a biomaterials approach. J Funct Biomater 2015;6(1):81–103.

[46] Dolznig H, Rupp C, Puri C, Haslinger C, Schweifer N, Wieser E, et al. Modeling colon adenocarcinomas in vitro a 3D co-culture system induces cancer-relevant pathways upon tumor cell and stromal fibroblast interaction. Am J Pathol 2011;179(1):487–501.

[47] Florczyk SJ, Kievit FM, Wang K, Erickson AE, Ellenbogen RG, Zhang M. 3D porous chitosan-alginate scaffolds promote proliferation and enrichment of cancer stem-like cells. J Mater Chem B 2016;4(38):6326–34.

[48] Yip D, Cho CH. A multicellular 3D heterospheroid model of liver tumor and stromal cells in collagen gel for anti-cancer drug testing. Biochem Biophys Res Commun 2013;433(3):327–32.

[49] Seano G, Chiaverina G, Gagliardi PA, di Blasio L, Sessa R, Bussolino F, et al. Modeling human tumor angiogenesis in a three-dimensional culture system. Blood 2013;121(21):e129–37.

[50] Linde N, Gutschalk CM, Hoffmann C, Yilmaz D, Mueller MM. Integrating macrophages into organotypic co-cultures: a 3D in vitro model to study tumor-associated macrophages. PLoS One 2012;7(7):e40058.

[51] Alonso-Nocelo M, Abuin C, Lopez-Lopez R, de la Fuente M. Development and characterization of a three-dimensional co-culture model of tumor T cell infiltration. Biofabrication 2016;8(2):025002.

[52] Kundu B, Bastos A, Brancato V, et al. Mechanical property of hydrogels and the presence of adipose stem cells in tumor stroma affect spheroid formation in the 3D osteosarcoma model. ACS Appl Mater Interfaces 2019;11(6):14548–59.

[53] Huang B-W, Gao J-Q. Application of 3D cultured multicellular spheroid tumor models in tumor-targeted drug delivery system research. J Control Release 2018;270:246–59.

[54] Ghosal K, Chandra A, Praveen G, Snigdha S, Roy C, Agatemor, et al. Electrospinning over solvent casting: tuning of mechanical properties of membranes. Sci Rep 2018;8(1):5058.

[55] O'Brien FJ. Biomaterials & scaffolds for tissue engineering. Mater Today 2011;14(3):88–95.

[56] Breslin S, O'Driscoll L. Three-dimensional cell culture: the missing link in drug discovery. Drug Discov Today 2013;18(5–6):240–9.

[57] Grainger SJ, Serna JV, Sunny S, Zhou Y, Deng CX, El-Sayed MEH. Pulsed ultrasound enhances nanoparticle penetration into breast cancer spheroids. Mol Pharm 2010;7(6):2006–19.

[58] Wei X, Senanayake TH, Warren G, Vinogradov SV. Hyaluronic acid-based nanogel-drug conjugates with enhanced anticancer activity designed for the targeting of CD44-positive and drug-resistant tumors. Bioconjug Chem 2013;24(4):658–68.

[59] Sarisozen C, Dhokai S, Tsikudo EG, Luther E, Rachman IM, Torchilin VP. Nanomedicine based curcumin and doxorubicin combination treatment of glioblastoma with scFv-targeted micelles: in vitro evaluation on 2D and 3D tumor models. Eur J Pharm Biopharm 2016;108:54–67.

[60] Roy K, Kanwar RK, Antonio Cheung CH, Lee Fleming C, Veedu RN, Krishnakumar S, et al. Locked nucleic acid modified bi-specific aptamer-targeted nanoparticles carrying survivin antagonist towards effective colon cancer therapy. RSC Adv 2015;5(37):29008–16.

[61] Rodrigues T, Kundu B, Silva-Correia J, Kundu SC, Oliveira JM, Reis RL, et al. Emerging tumor spheroids technologies for 3D in vitro cancer modeling. Pharmacol Therap 2018;184:201–11.

[62] Marimuthu M, Rousset N, St-Georges-Robillard A, Lateef MA, Ferland M, Mes-Masson AM, et al. Multi-size spheroid formation using microfluidic funnels. Lab Chip 2018;18(2):304–14.

[63] Sontheimer-Phelps A, Hassell BA, Ingber DE. Modelling cancer in microfluidic human organs-on-chips. Nat Rev Cancer 2019;19(2):65–81.

[64] Xu X, Gurski LA, Zhang C, Harrington DA, Farach-Carson MC, Jia X. Recreating the tumor microenvironment in a bilayer, hyaluronic acid hydrogel construct for the growth of prostate cancer spheroids. Biomaterials 2012;33(35):9049–60.

[65] Lin R-Z, Chang H-Y. Recent advances in three-dimensional multicellular spheroid culture for biomedical research. Biotechnol J 2008;3(9-10):1172–84.

[66] McMillin DW, Delmore J, Weisberg E, Negri JM, Geer DC, Klippel S, et al. Tumor cell-specific bioluminescence platform to identify stroma-induced changes to anticancer drug activity. Nat Med 2010;16 (4):483–9.

[67] Helmchen F, Denk W. Deep tissue two-photon microscopy. Nat Methods 2005;2(12):932–40.

[68] Szot CS, Buchanan CF, Freeman JW, Rylander MN. 3D in vitro bioengineered tumors based on collagen I hydrogels. Biomaterials 2011;32(31):7905–12.

[69] Song HH, Park KM, Gerecht S. Hydrogels to model 3D in vitro microenvironment of tumor vascularization. Adv Drug Deliv Rev 2014;79–80:19–29.

[70] Chen L, Xiao Z, Meng Y, Zhao Y, Han J, Su G, et al. The enhancement of cancer stem cell properties of MCF-7 cells in 3D collagen scaffolds for modeling of cancer and anti-cancer drugs. Biomaterials 2012;33 (5):1437–44.

[71] Cross VL, Zheng Y, Won Choi N, Verbridge SS, Sutermaster BA, Bonassar LJ, et al. Dense type I collagen matrices that support cellular remodeling and microfabrication for studies of tumor angiogenesis and vasculogenesis in vitro. Biomaterials 2010;31(33):8596–607.

[72] Fong EL, Lamhamedi-Cherradi SE, Burdett E, Ramamoorthy V, Lazar AJ, Kasper FK, et al. Modeling Ewing sarcoma tumors in vitro with 3D scaffolds. Proc Natl Acad Sci U S A 2013;110(16):6500–5.

[73] Loessner D, Stok KS, Lutolf MP, Hutmacher DW, Clements JA, Rizzi SC. Bioengineered 3D platform to explore cell-ECM interactions and drug resistance of epithelial ovarian cancer cells. Biomaterials 2010;31 (32):8494–506.

[74] Prestwich GD. Evaluating drug efficacy and toxicology in three dimensions: using synthetic extracellular matrices in drug discovery. Acc Chem Res 2008;41(1):139–48.

[75] Zhu J, Marchant RE. Design properties of hydrogel tissue-engineering scaffolds. Expert Rev Med Devices 2011;8(5):607–26.

[76] Tibbitt MW, Anseth KS. Hydrogels as extracellular matrix mimics for 3D cell culture. Biotechnol Bioeng 2009;103(4):655–63.

[77] Geckil H, Xu F, Zhang X, Moon S, Demirci U. Engineering hydrogels as extracellular matrix mimics. Nanomed (Lond) 2010;5(3):469–84.

[78] Lu T, Li Y, Chen T. Techniques for fabrication and construction of three-dimensional scaffolds for tissue engineering. Int J Nanomed 2013;8:337–50.

[79] Jakab K, Damon B, Neagu A, Kachurin A, Forgacs G. Three-dimensional tissue constructs built by bioprinting. Biorheology 2006;43(3,4):509–13.

[80] Pal M, Chen H, Lee BH, Lee JYH, Yip YS, Tan NS, et al. Epithelial-mesenchymal transition of cancer cells using bioengineered hybrid scaffold composed of hydrogel/3D-fibrous framework. Sci Rep 2019;9(1):8997.

[81] Langer EM, Allen-Petersen BL, King SM, Kendsersky ND, Turnidge MA, Kuziel GM, et al. Modeling tumor phenotypes in vitro with three-dimensional bioprinting. Cell Rep 2019;26(3):608–23.

[82] Huang TQ, Qu X, Liu J, Chen S. 3D printing of biomimetic microstructures for cancer cell migration. Biomed Microdevices 2014;16(1):127–32.

[83] Zhao Y, Yao R, Ouyang L, Ding H, Zhang T, Zhang K, et al. Three-dimensional printing of Hela cells for cervical tumor model in vitro. Biofabrication 2014;6(3):035001.

[84] Antoine EE, Vlachos PP, Rylander MN. Review of collagen I hydrogels for bioengineered tissue microenvironments: characterization of mechanics, structure, and transport. Tissue Eng Part B Rev 2014;20(6):683–96.
[85] Loh QL, Choong C. Three-dimensional scaffolds for tissue engineering applications: role of porosity and pore size. Tissue Eng Part B Rev 2013;19(6):485–502.
[86] Wenzel C, Riefke B, Grundemann S, Krebs A, Christian S, Prinz F, et al. 3D high-content screening for the identification of compounds that target cells in dormant tumor spheroid regions. Exp Cell Res 2014;323(1):131–43.
[87] Vorsmann H, Groeber F, Walles H, Busch S, Beissert S, Walczak H, et al. Development of a human three-dimensional organotypic skin-melanoma spheroid model for in vitro drug testing. Cell Death Dis 2013;4:e719.
[88] Godugu C, Patel AR, Desai U, Andey T, Sams A, Singh M. AlgiMatrix based 3D cell culture system as an in-vitro tumor model for anticancer studies. PLoS One 2013;8(1):e53708.
[89] Lee DW, Choi YS, Seo YJ, Lee MY, Jeon SY, Ku B, et al. High-throughput screening (HTS) of anticancer drug efficacy on a micropillar/microwell chip platform. Anal Chem 2014;86(1):535–42.
[90] Gurski LA, Jha AK, Zhang C, Jia X, Farach-Carson MC. Hyaluronic acid-based hydrogels as 3D matrices for in vitro evaluation of chemotherapeutic drugs using poorly adherent prostate cancer cells. Biomaterials 2009;30(30):6076–85.
[91] Lee J, Lilly GD, Doty RC, Podsiadlo P, Kotov NA. In vitro toxicity testing of nanoparticles in 3D cell culture. Small 2009;5(10):1213–21.
[92] Goodman TT, Olive PL, Pun SH. Increased nanoparticle penetration in collagenase-treated multicellular spheroids. Int J Nanomed 2007;2(2):265–74.
[93] Ma H-l, Jiang Q, Han S, Wu Y, Tomshine JC, Wang D, et al. Multicellular tumor spheroids as an in vivo–like tumor model for three-dimensional imaging of chemotherapeutic and nano material cellular penetration. Mol Imaging 2012;11(6):487–98.
[94] Mikhail AS, Eetezadi S, Ekdawi SN, Stewart J, Allen C. Image-based analysis of the size- and time-dependent penetration of polymeric micelles in multicellular tumor spheroids and tumor xenografts. Int J Pharm 2014;464(1):168–77.
[95] Jiang Y, Lu H, Dag A, Hart-Smith G, Stenzel MH. Albumin–polymer conjugate nanoparticles and their interactions with prostate cancer cells in 2D and 3D culture: comparison between PMMA and PCL. J Mater Chem B 2016;4(11):2017–27.
[96] Meng W, Kallinteri P, Walker DA, Parker TL, Garnett MC. Evaluation of poly (glycerol-adipate) nanoparticle uptake in an in vitro 3-D brain tumor co-culture model. Exp Biol Med 2007;232(8):1100–8.
[97] Teijeiro-Valiño C, Yebra-Pimentel E, Guerra-Varela J, Csaba N, Alonso MJ, Sánchez L. Assessment of the permeability and toxicity of polymeric nanocapsules using the zebrafish model. Nanomedicine (London, Engl) 2017;12(17):2069–82.
[98] Perche F, Patel NR, Torchilin VP. Accumulation and toxicity of antibody-targeted doxorubicin-loaded PEG-PE micelles in ovarian cancer cell spheroid model. J Control Release 2012;164(1):95–102.
[99] Oishi M, Nagasaki Y, Nishiyama N, Itaka K, Takagi M, Shimamoto A, et al. Enhanced growth inhibition of hepatic multicellular tumor spheroids by lactosylated poly(ethylene glycol)-siRNA conjugate formulated in PEGylated polyplexes. ChemMedChem 2007;2(9):1290–7.
[100] Kondo J, Inoue M. Application of cancer organoid model for drug screening and personalized therapy. Cells 2019;8(5):470.
[101] Kondo J, Ekawa T, Endo H, Yamazaki K, Tanaka N, Kukita Y, et al. High-throughput screening in colorectal cancer tissue-originated spheroids. Cancer Sci 2019;110(1):345–55.

CHAPTER

# 4

# Biomatrices that mimic the cancer extracellular environment

*Sara Amorim*[1,2], *Rui L. Reis*[1,2] *and Ricardo A. Pires*[1,2]

[1]3B's Research Group,I3Bs—Research Institute on Biomaterials, Biodegradables and Biomimetics, University of Minho, Headquarters of the European Institute of Excellence on Tissue Engineering and Regenerative Medicine, Guimarães, Portugal [2]ICVS/3B's - PT Government Associate Laboratory, Braga/Guimarães, Portugal

## 4.1 Introduction

The tumor microenvironment (TME) is a complex matrix of blood vessels, malignant and nonmalignant cells, proteins, glycosaminoglycans, signaling molecules, among other components of the extracellular matrix (ECM) [1,2]. The ECM is the major component of the TME and has been increasingly linked to cancer progression and metastasis [3–5]. The dynamic character of the ECM involves a series of mechanisms related with its production, degradation, and remodeling. In turn, these mechanisms regulate the normal functioning of the cells and the composition of their extracellular environment. In pathological frameworks, such as the TME, these key regulatory features are disturbed altering the composition of the ECM through the uncontrolled synthesis and degradation of proteins and glycans that accumulate on the basement membrane and the extracellular/pericellular space. In addition, the reorganization of ECM components, alters the cross-linking extension of the interstitial matrix, affecting the homeostasis and normal functioning of tissues [6–8] (Fig. 4.1). In this view, the complexity of the ECM in healthy tissues is significantly amplified in the TMEs.

During the last few years, the development of three-dimensional (3D) TME mimics emerged as model platforms for the in vitro study of cancer, as well as the behavior of its cells as a function of the composition of the TME. These models have been used for the screening of the role of each TME component on tumor progression and invasiveness. These 3D models provide architectural support for cell–cell interactions, offering the biochemical and mechanical cues of the 3D TME. Moreover, the third dimension of the models, confers the ability to tune key parameters, such as porosity and water content, which, in turn, regulates the transport of oxygen and nutrients, as well as the two-way exchange of soluble biochemical factors [10]. The controlled

*Biomaterials for 3D Tumor Modeling*
DOI: https://doi.org/10.1016/B978-0-12-818128-7.00004-6

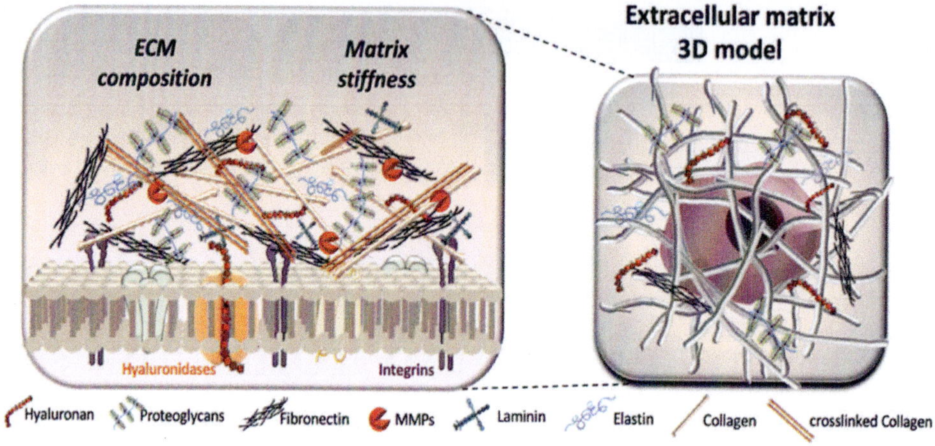


FIGURE 4.1 Schematic representation of the dysregulated ECM in the TME. In summary, there is: an accumulation of ECM, through the overexpression of HA, and an increment of the ECM stiffness through the cross-linking of collagen, leading to an upregulation of matrix metalloproteinases (MMPs), accelerating ECM degradation and promoting the extravasation of tumor cells [9].

diffusion and distribution of these elements throughout the 3D network are fundamental to recapitulate the cancer cell behavior and to control the homeostasis of the TME. These properties can be tuned through the chemical and/or physical manipulation of the 3D models.

In addition to its spatial organization, the TME is composed of different types of collagens (types I–V), elastin, fibronectin (FN), laminin, as well as different glycoproteins, glycosaminoglycans, proteoglycans, and ECM regulators [11]. The recent developments to recreate the TME network includes the natural- and synthetic-based materials. Using specific ECM constituents, such collagen [12], the glycosaminoglycan hyaluronic acid (HA) [13], or natural nonmammalian materials, such as alginate (Alg) [14], it is possible to recreate in vitro cell–cell and cell–ECM interactions that recapitulate the TME. Advances on the development of 3D TME models, led to the use of synthetic polymers with bioactive, biocompatible, and biodegradable properties. The recreation of the cellular surroundings using synthetic materials allowed to: (1) manipulate and functionalize the materials, with specific bioactive motifs able to induce cellular adhesion, spreading, and/or migration [15]; (2) easily tune the mechanical properties of the 3D networks through the manipulation of the concentration of the polymer or the type and extension of cross-linking [16].

In the following sections we will review the recently reported methodologies and materials used to develop 3D TME mimics, based on the abovementioned strategies. We will also make an overview of the advantages and disadvantages of each one, in an attempt to elucidate the reader about which ones are more suitable to recapitulate the TME of different tumors.

## 4.2 The three-dimensional in vitro models

### 4.2.1 Natural-based models

Natural-based materials have been exploited in the development of 3D models that mimic the TME. They are regarded as mimics of the ECM due to their close resemblance to the chemical and morphological presentation of its components. In fact, the chemical similarity between the proposed natural materials, for example, proteins and polysaccharides, with the ones that are present in the ECM that constitutes the TME allows the presentation of different bioactive motifs (e.g., amino acid sequences, and monosaccharides and their sequences) that are recognized by the cells and able to modulate their behavior [17]. In the following sections, we will give an overview on the different biomaterials that have been used to mimic the 3D TME and in particular its' ECM. These include the use of proteins, glycoproteins, and polysaccharides extracted from natural sources. The advantages and disadvantages of each one will be explained in detail and an overview of the most relevant biochemical features from the in vivo TME will be provided that they are able to recapitulate in these in vitro 3D models.

#### 4.2.1.1 *Protein-based systems*

Collagens are the most abundant type of proteins present in the ECM and are the most tested for the development of 3D tissue models. They are present in nature (and in the human body) under different conformations and aminoacid sequences. The most common one is type I but there are more than 20 different types that have been already identified [18]. One of the main advantages of using collagen type I to produce 3D matrices is associated with the mild cross-linking conditions (i.e., pH and temperature) [19], which is compatible with cell encapsulation. The self-assembly capacity of collagens is another advantage, as they readily organize into fibers [20,21], generating meshes of different pore sizes and densities [21,22]. The generation of 3D collagen-based models have been used to screen its role in the TME. In fact, it is known that the increased expression and deposition of collagens on the ECM, is correlated with cancer progression and poor patient prognosis on different cancers, for example, breast cancer [23]. MDA-MB-231 and MCF-7 cells (human breast cancer cell lines) cultured on 3D collagen matrices showed to be more elongated with higher invasive behavior, which was correlated with fibril thickness, intrafibrillar cross-linking, as well as the fibril's stiffness (Fig. 4.2) [24]. On the other hand, Plou et al. demonstrated that the nonsmall cell lung cancer (NSCLC) H1299 cells adhere and proliferate in highly concentrated collagen matrices. However, they present reduced migration and tend to spontaneously form cell clusters while inhibiting single cell invasion [25]. The variation of the thickness of collagen I-based hydrogel (between 1.5 and 3.0 mm) was also tested to study the impact on cell–cell and cell–matrix interactions using MDA-MB-231 cells [26]. The expression of hypoxia-inducible factor (HIF)-1$\alpha$, associated to the formation of tumor spheroids, and the angiogenic potential through the expression of vascular endothelial growth factor (VEGF) was evaluated. The results showed that in the higher thickness of collagen hydrogel, 3.0 mm, HIF-1$\alpha$ and VEGF-A gene expression was upregulated, as a result of lack of oxygen and nutrient-induced cell death after day 5.

Similar to collagen, the glycoprotein FN forms fibrillar structures in the ECM [5]. FN occurs in a soluble form in the plasma; however, it can also self-assemble into linear and branched supramolecular fibers, that surrounds the cells, allowing the interplay between ECM components [27]. In contrast to collagen, FN is secreted as a soluble dimer and its cell–surface interaction is necessary for its assembly into longer fibrils [28]. To our knowledge, FN on its own has not been used for the formulation of 3D matrices; however, due to its overexpression in the interstitial ECM and its association with cancer cell migration

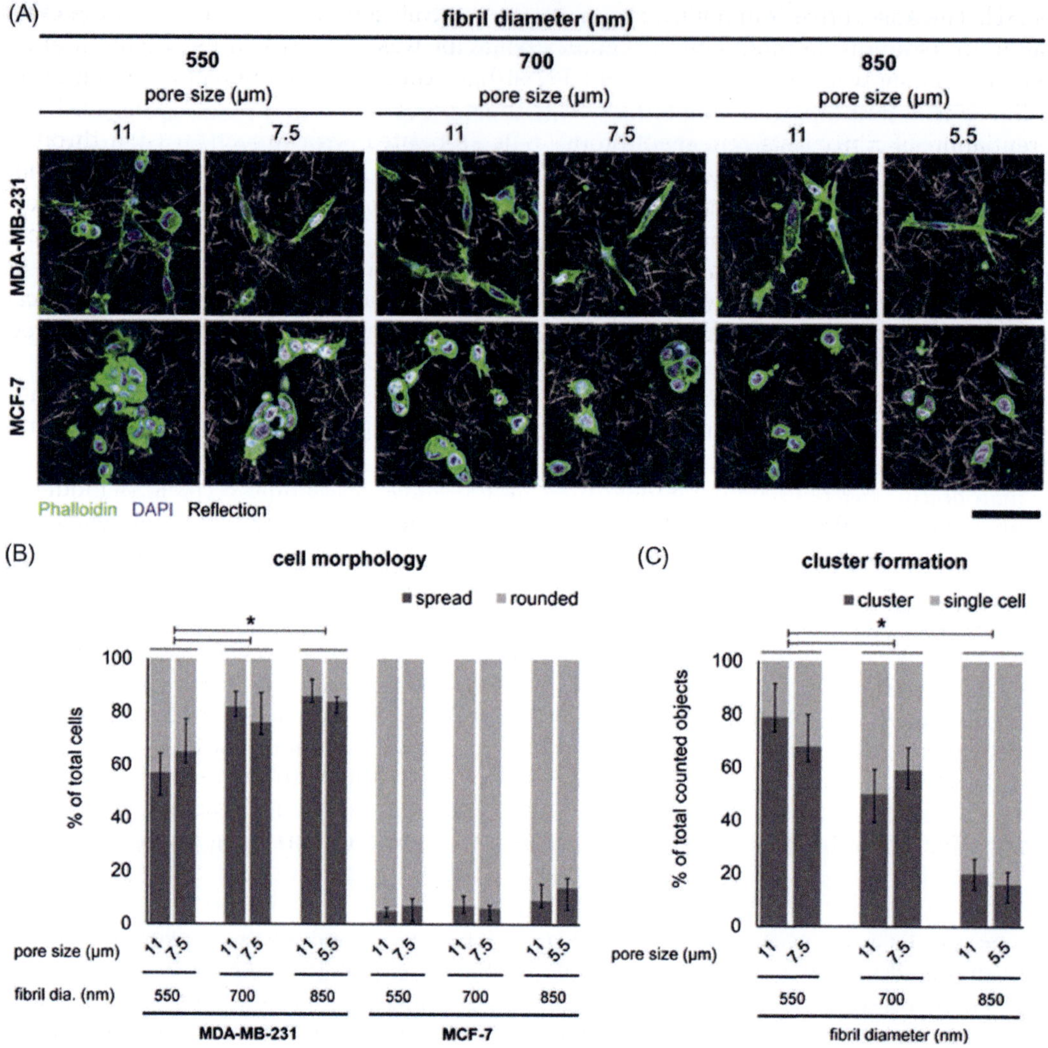


FIGURE 4.2 Influence of collagen fibril diameter and hydrogel pore size on cluster formation and cell morphology of breast cancer cells, MDA-MB-231 and MCF-7. (A) Confocal laser scanning microscopy images of breast cancer cells seeded on the 3D collagen I matrices; (B) percentage of spread and rounded cells; and (C) quantification of single cells and clusters formation cultivated in the different collagen matrices. Source: *Reprinted from Sapudom J, et al. Fibril bending stiffness of 3D collagen matrices instructs spreading and clustering of invasive and non-invasive breast cancer cells. Biomaterials 2019;193:47–57 Copyright (2019), with permission from Elsevier.*

and tumor growth [29], it has been used in the formulation of different 3D cancer models, as a cellular adhesive component. The combination of collagen and FN are reported as a close mimic of the ECM. The presence of increasing concentrations of FN within a collagen matrix has been studied and it has been reported that it alters the collagen fibrillation kinetics, where a faster fibril formation results in an altered microstructure network

[30,31]. The adsorptive immobilization of FN on the collagen network, as well as the evaluation of its ability to modulate the cellular behavior was investigated by Sapudom et al. The authors showed that the presence of FN at high concentrations induces the migration of fibroblasts, accompanied by a reduction of their proliferation rate [30]. The induced differentiation of fibroblasts into myofibroblasts is associated with cancer invasion through the production of proinvasive proteinases [32]; however, Sapudom et al. [30] did not observe any cellular differentiation in the developed collagen–FN 3D matrices. An important feature to be considered, and that is relevant to be recapitulated in the 3D cancer models, is the fact that the remodeling and activation of fibroblasts induces a stiffening of the ECM matrix and tissue desmoplasia [33]. These observations show that the mechanical properties of the ECM present in the TME is an important parameter to be considered when developing cancer models, a topic that will be discussed further in a later section.

Laminin, is another glycoprotein present at the cellular basement membrane that has been used to model the 3D laminin-rich ECM. It is secreted by healthy myofibroblasts and plays a crucial role in maintaining epithelial polarity. In breast tumors, this polarity is altered due to the reduced expression of laminin [34]. A 3D laminin-rich ECM model is commercially available (3D IrECM). It has been used to study the cancer cells behaviors, mimicking physiologic in vivo growth conditions [35]. A panel of cancer cells was already evaluated under this 3D culture condition, by screening their morphologies, proliferation, and expression of signaling proteins. Cells seeded on this 3D model exhibited different morphologies (round, mass, grape-like, and stellate), which was dependent on the cell type, and was associated with gene and protein expression profiles, as well as correlated with invasiveness characteristics of the tested tumor cells [36]. The 3D IrECM has also been applied to evaluate the response of human breast cancer cells to doxorubicin (a common anticancer drug). Under the laminin-rich microenvironment, MCF-7 cells presented morphological alterations, a decreased proliferation rate, and an elevated sensitivity to doxorubicin [37], showing the suitability of the model for drug-screening purposes.

#### 4.2.1.2 Polysaccharide-based systems

The polysaccharides that are present in the ECM [i.e., the glycosaminoglycans (GAGs)], play an important role in the biological milieu. They usually regulate the cell–cell communication and cell–ECM interactions [38]. They can also initiate intracellular pathways, through the interaction with cell membrane receptors, that modulate gene expression and protein secretion, as well as cell proliferation and differentiation [39]. The GAGs are composed of repeating disaccharides units, one is an amino sugar (D-glucosamine or D-galactosamine), and the other is a uronic acid residue (D-glucuronic acid or iduronic acid) [40]. One or both of this units can be additionally sulfated, generating a range of five different GAGs, namely, HA, heparan sulfate (HS), Heparin (Hep), chondroitin sulfate (CS), and dermatan sulfate (DS). These different types of GAGs are distinguished by the degree of sulfation and the distribution of their sulfation motifs throughout the repeating unit and the polymeric chain [41]. These structural features present in the GAGs, namely the changes on their sulfation degree, have been linked to cancer cell migration and tumor aggressiveness [42–45].

The most studied and the only nonsulfated GAG, associated to cancer pathology, is HA. The deposition of HA in the TME has been related to the proliferation and migration of cancer cells. In particular, the molecular weight ($M_w$) of HA have been reported to generate different biological outcomes. While, in some studies, low $M_w$ HA (i.e., oligomers) are associated with tumor invasiveness, others suggest that HAs of high $M_w$ are the cause of invasive cellular character [46–48]. Moreover, the CD44 transmembrane protein (a major HA receptor) is known to be overexpressed in some cancers (e.g., colorectal, pancreatic, and gastric), in particular the isoforms between CD44v6 and CD44v15 have been related to cancer migration and invasiveness [49].

Different approaches have been used to fabricate 3D HA-based culture systems to study its influence on cancer progression [13]. Usually HA-based systems require cross-linking to generate stable 3D structures that resemble the in vivo HA network. The most common procedure to generate 3D HA hydrogels is to chemically modify HA into its methacrylated form (Me-HA), that can be cross-linked by UV light [50–52]. The 3D glioblastoma (GMB) models were developed using these Me-HA-based hydrogels [53], where cellular morphology and motility were dependent on the hydrogel stiffness. Pedron et al. also used Me-HA but in combination with methacrylated-gelatin or PEG; and they found that the addition of Me-HA (0.3–0.5 wt.%) modulated the glioma proliferation and morphology, as well as the expression of invasive-associated genes (VEGF, HIF-1, MMP-9, and FN) [54]. Modification of HA with aldehyde (HAALD) and hydrazide (HAADH) groups have also been described. In this case, the hydrogel can be cross-linked via the hydrazine bonds, allowing the encapsulation of prostate cancer cells. This system was proposed as a platform to screen anticancer drugs, namely, camptothecin, docetaxel, and rapamycin [55,56]. The mixture of HA with Alg is another approach to immobilize and mimic the presentation of HA under a 3D environment. In this case, the physical dispersion of HA into an Alg solution [57] or the cross-linking of HA–Alg through the oxidation of Alg has been described. In both cases, the generated systems were used to study the role of the HA's $M_w$s on breast cancer cells, namely on their proliferation and invasiveness [13]. The presence of HA of low $M_w$ (35 kDa) significantly enhances the migration ability of SKBR3 (breast cancer) cells, as well as 4T1 (pancreatic cancer cells); and the behavior that was suppressed by the HA of higher $M_w$ (117 kDa). Moreover, it was observed that the HA of 35 kDa potentiates the epithelial–mesenchymal transition (EMT) phenotype, associated with cancer metastasis.

Among the sulfated GAGs, CS, a linear polysaccharide, in particular its sulfation degree has been reported to be involved in cancer cell proliferation, migration, and adhesion, being one relevant factor that induces cancer progression [58,59]. CS is present on the cell surface and found in extracellular/pericellular space covalently attached to different core proteins [60]. In the TME, CS is detected at high concentration within the collagen network, altering its structure and viscoelastic properties [61]. The 3D models generated by the combination of collagen and CS (Coll–CS) have been reported. The pore sizes of the Coll–CS structures have been shown to influence cell adhesion, where larger pores reduce cell viability, and smaller pores (with the correspondent increase in surface area) lead to higher viability of MC3T3 cells [62]. Coll–CS hydrogels have also been used as a neuroblastoma model using two different cell lines: KellyLuc and KellyCis83Luc. It was observed cellular penetration and proliferation into the Coll–CS matrix, with increased resistance to cisplatin (platin-containing drugs are used in neuroblastoma treatment) when compared with 2D cell cultures [63], a result that resembles better the in vivo environment. Other researchers have reported the modification of Alg hydrogels with CS, for the encapsulation of hepatocellular carcinoma cells (HCCLM3), as well as head and neck squamous carcinoma cells (HNSCC). They found an upregulation on the gene expression of matrix metalloproteinase (MMPs), namely MMP-2, MMP-9, and MMP-14, on cells cultured in Alg–CS with higher invasion ability [64].

#### *4.2.1.3 Other natural occurring materials*

In the previous sections we outlined a series of materials present in the human biological tissues, in both healthy and pathological states, and that have been used to develop 3D models that mimic the ECM present in the TME. However, there are a vast range of other natural materials (that are not present in the human body) but, due to its chemical similarity, has been tested as 3D TME mimics.

Another widely studied material for the development of 3D cancer model is Alg, a polysaccharide obtained from brown seaweeds that has been widely used in the biomedical field, in particular as a biomaterial for 3D cell culture [65]. Its main advantages are related with the easy handling, the possibility to cross-link and gel under mild conditions (e.g., in the presence of divalent cations, $Ca^{2+}$), the ability to tune the physical properties of the Alg hydrogels through the manipulation of the concentration of Alg and its cross-linking agent (e.g., $Ca^{2+}$) [66]. However, and despite these important advantages, the main drawback of using Alg-based systems for cell culture studies is the limited cell adhesion. To overcome this weakness, Alg has been modified with cell adhesive peptide sequences, such as Arg–Gly–Asp (RGD) [67], or has been formulated through its combination with other biomolecules present in the ECM, such as collagen or hyaluronic acid, as detailed below [65,68]. The blending of Alg with collagen, an ECM-occurring protein, has been tested in the development of 3D models suitable for the encapsulation of human breast cancer cells [69,70], as well as a combination of MDA-MB-231 cells and human mammary fibroblast (HMF) spheroids. Under this coculture conditions it was observed the secretion of CXCL12–CXCR4 signaling molecules as inductors of directional cell migration into the 3D hydrogel [70]. The incorporation of HA on Alg matrices has been also reported to mimic the cancer stem cells (CSCs) microenvironment and used to isolate and enrich CSCs from 4T1 breast cancer cells. The incorporation of HA and growth factors, EGF and bFGF, on the Alg matrices, allowed the efficient isolation of CSCs that highly expresses CD44 and SCa-1, with low expression of CD24, usually referred as cancer stem cells markers [71]. The encapsulation of mono- and coculture of human dermal fibroblasts (HDF) and human breast cancer cells (MCF-7) have also been attempted [72]: in this case, the 3D culture system provided a favorable microenvironment to study the crosstalk between tumor cells and fibroblasts. It was detected that the synthesis and deposition of ECM and the accumulation of soluble factors promoted tumor progression.

CHI is another widely used material in the biomedical field. It is a polysaccharide derived from the partially deacetylation of chitin, that occurs in nature in the shells of crab and shrimp (among other sources). Its partial deacetylation under alkaline conditions yields CHI [73], a cationic polymer, presenting cell adhesive properties due to its positive charge. CHI is also biocompatible, biodegradable, and nontoxic [74,75]. CHI scaffolds with different degrees of deacetylation has been used as a support for the culture of human breast cancer cells (MCF-7). Increased proliferation rate and formation of cells' clusters have been correlated with a higher degree of deacetylation and higher swelling ability.

This CHI-based system revealed to be a 3D model to consider for drug-screening purposes, namely the evaluation of anticancer drugs [76]. The 3D culture systems based on CHI have also been described to mimic the glioma TME: the combination of CHI and Alg has been proposed to support the viability of hepatocellular carcinoma cells (HCC) [77], prostate cancer cells [78], U-87 MG [79,80], and U-118 MG [80] human glioma cells. In these latter cases, glioma cells showed round morphology with interconnectivity characteristic of solid tumor cells. Moreover, the upregulation of growth factors upon culture on these 3D systems, rather than on 2D surfaces, indicates that cells are able to modify its surroundings into a more invasive environment [80]. CHI has also been formulated with HA for the generation of 3D systems that are able to support the encapsulation of glioblastoma cells (GBM). This blending of CHI and HA was used as a close mimic of the native TME, through the incorporation of HA, a TME-occurring GAG. The incorporation of HA induced the expression of CD44, a stem cell surface marker, on GBM cells. The stemness of GBM cells cultured in the CHI/HA 3D system was confirmed by the genetic upregulation of stem cells markers, namely, Nestin, Musashi-1, and GFAP. Moreover, under this culture conditions, genes associated with tumor invasiveness, metastasis, and EMT, such as SNAIL1, SNAIL2, and TWIST were altered in the GBM cells [81,82]. The EMT process was also observed on pancreatic cancer cells encapsulated on the HA-rich CHI model, through the upregulation of different markers, such as E-cadherin, N-cadherin, Vimentin, TWIST1, and SNAIL [83].

### 4.2.2 Synthetic and other biobased models

Synthetic materials, such as polyethylene glycol (PEG), polyglycolide (PGA) polylactide-*co*-glycolide (PLGA), or biobased ones, such as poly(lactic acid) (PLA) have been also described for the formulation of 3D systems with tuneable biochemical and mechanical properties [84]. The main drawback of these materials is the limited cell adhesion, which hinders the cell's cytoskeleton organization, necessary for cell proliferation and migration. To overcome this hurdle, these 3D matrices can be functionalized with: (1) the RGD peptide sequence, a known cell-adhesion motif present in FN but also identified in many other ECM proteins such laminin and collagen [85–87]; (2) peptide cross-linkers that are degradable by MMPs, allowing the cells to degrade the matrix, creating space for cell proliferation and migration [88,89]. PEG is widely used as a scaffold material to mimic the ECM part of the TME [16,90]. PEG conjugated with fibrinogen (PEG-Fb) enables integrin-mediated signaling mechanisms required for cell attachment, spreading, and proliferation. This system has been proposed as 3D photocross-linkable hydrogels able to encapsulate breast cancer cells [91,92]. It has been shown that PEG-Fb of varying composition influences the morphology of MCF-7 and SK-BR-3 cells with epithelial phenotype, while inducing the formation of colonies. In addition, MDA-MB-231 cells presenting mesenchymal character, exhibited invasive behavior under the same experimental conditions. The physical properties of these hydrogels (namely their cross-linking density and higher pore size) promoted the formation of a high number of cellular protrusions and an invasive character [91]. Microspheres of PLA-based hydrogels have been used to mimic the tumor shape and as a support that facilitates the formation of tumor spheroids [92]. The advantage of using PLA to generate 3D tumor models is based on its processing flexibility using different methodologies. As an example, it is possible to produce PLA-based electrospun scaffolds, using different PLA concentration (12% and 15%) [93], that supported the viability of the breast cancer cells MDA-MB-231. In this case, an increase in the STAT3 phosphorylation levels, a transcription factor that regulates the expression of different genes crucial for tumor progression was observed. In addition, the physical characteristics of PLA has been tuned in the development of 3D TME models, including variations on the pore sizes, filament density, infill pattern, or layer height [94]. In this case, the MDA-MB-231 cells were tested and, upon 3 days of cell culture, an upregulation of ALDH, which is associated to the stemness character of cancer cells was detected. Variations on the intensity of this upregulation was dependent on the pore size of the model. PLGA has also been used as a 3D TME mimic. In this case, the U251 glioma cells were evaluated, in particular their resistance to hypoxia conditions [95]. PLGA presents low stability under physiological conditions due to its fast degradation rates [96]. In order to improve the stability of the TME mimic, PCL has been tested as an alternative to PLGA. Its suitability to mimic the TME has been evaluated using bone cancer cells [97]. In particular, the Ewing sarcoma cells cultured on electrospun PCL 3D system presented enhanced resistance to doxorubicin, showing favorable results on the application of these materials for the screening of the activity of anticancer therapeutic agents. Moreover, PCL-based models are also suitable to mimic the TME of breast cancers being closer to the physiological environment of the subpopulation of breast cancer stem cells (BCSCs) mimicking its native behavior [98].

### 4.2.3 Mimicking the tumor microenvironment mechanical features

The TME, is a complex environment with different cell types and ECM molecules. Importantly, it provides mechanical support for cell adhesion and motility and, consequently, to tumor growth [8,99]. The TME stiffness influences the ability of cancer cells to spread and invade the neighboring tissues. In fact, invasive cells tend to alter the TME mechanical properties, through the deposition of different ECM proteins that changes its biomechanical characteristics [100]. In general, the cancer stroma is usually stiffer than the healthy one, increasing from 150 to 400 Pa, and in premalignant breast tissues an increased ECM remodeling and stiffness is observed ($\sim$350 Pa) contributing to the development of malignancy [101,102]. One of the major contributors for the increased stiffness of the TME is the deposition of fibrillar collagen. The cross-linking and reorganization of collagen into thick and organized fibers, induces the generation of intracellular tension, inducing an increment of cellular traction forces, while enhancing ECM stiffness [103]. In this context, collagen has been explored to generate 3D cancer models to study the intracellular tension in response to collagen concentration (Fig. 4.3).

Collagen concentrations between 1 and 4 $mg\ mL^{-1}$, increased the Young's modulus of the hydrogels from approximately 370 to 1200 Pa, a range that is comparable to the physiological stiffness of healthy soft tissues, for example, lung, breast, or endothelial tissues [105]. The breast adenocarcinoma cells MDA-MB-231, were cultured on these collagen hydrogels of varying stiffness values and a direct correlation was observed between the

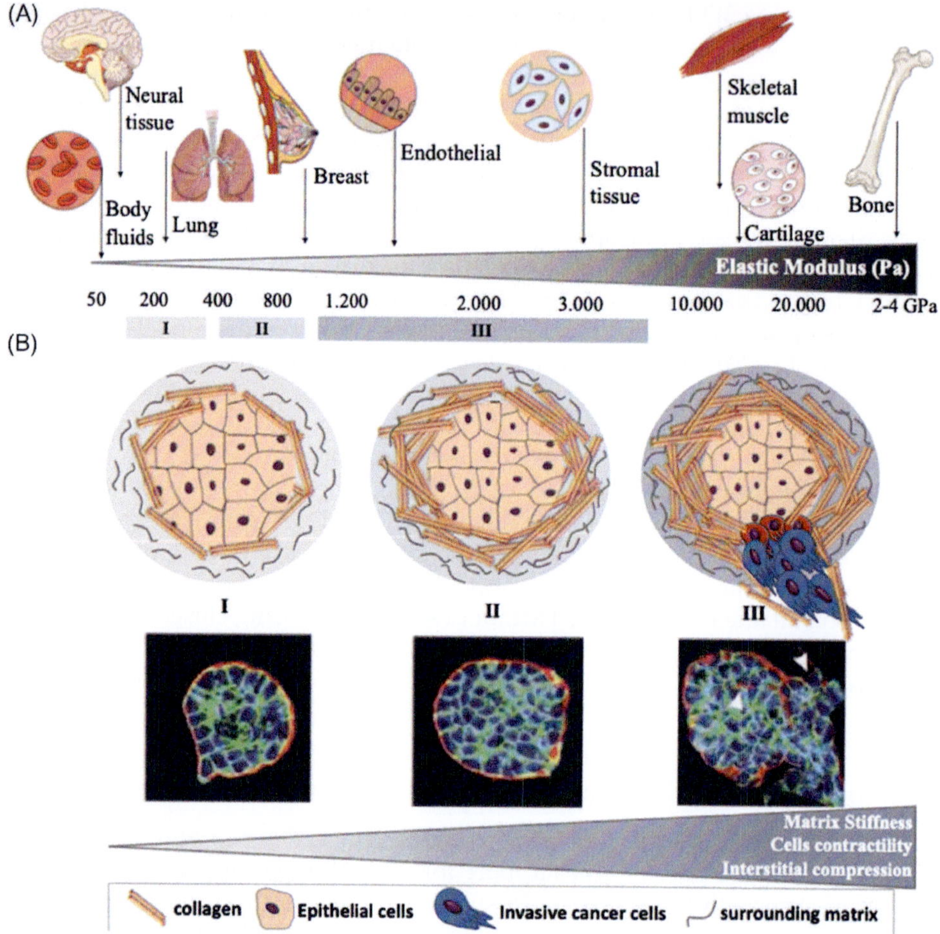


FIGURE 4.3 (A) Biomechanical properties of human tissues and their elastic modulus values and (B) increased accumulation and alignment of collagen fibers with high stiffness on the extracellular and pericellular matrix. Disruption of the basement membrane promoting invasion and tumor migration. Source: *Adapted from Paszek MJ, et al. Tensional homeostasis and the malignant phenotype. Cancer Cell 2005;8(3):241–54 [104] and Cox TR, Erler JT. Remodeling and homeostasis of the extracellular matrix: implications for fibrotic diseases and cancer. Dis Model Mech 2011;4(2):165–78 [101]. Copyright (2005), with permission from Elsevier.*

cellular and hydrogel stiffness [106]. In more detail, the modulus of the TME of breast cancer tissues can reach a value of 1500 Pa, close to the one for the stiffer collagen hydrogels. Wullkopf et al. used the same concentration of collagen to screen the invasive potential of cells from different sources. Breast (4T1, 67NR, MDA-MB-231, and MCF-7), colorectal (SW620 and SW480), and pancreatic ($KP^{R172H}C$ and KPflC) cancer cells were evaluated in the presence of 1 and 4 $mg\,mL^{-1}$ of collagen hydrogels. In general, more invasive cells (MDA-MD-231, 4T1, $KP^{R172H}C$, and SW620) increased their viscoelastic properties with the increment of the hydrogel stiffness. However, this type of change is dependent of the type of cells, for example, the MDA-MB-231 and $KP^{R172H}C$ changed from a viscous cytoplasm to a more elastic phenotype when exposed to the denser collagen networks, while the 4T1 and SW620 cells, that were more elastic on soft matrices, had an opposite response, increasing their viscosity on dense collagen matrices [105]. Moreover, pediatric osteosarcoma cell line U2OS and the breast adenocarcinoma cells MDA-MB-231, behave differently when culture on 3D collagen hydrogels with modulus from 10 to 200 kPa. [107] Larger osteosarcoma spheroids (335–395) were formed on the stiffest hydrogels, whereas bigger adenocarcinoma spheroids (550–800 μm) are present in softer gel. Actually, the formation of cancer cells spheroids is also sensitive to the surrounding matrix stiffness. MCF-7 breast cancer cells were encapsulated into 3D PEG–heparin hydrogels and the formation of tumor spheroids were evaluated in response to the mechanical properties of the surrounding 3D environment [108]. The PEG–heparin matrix presenting a Young's modulus between 2000 and 20,000 Pa showed a relevant impact on the size of cellular spheroids. The mechanical stress induced by the stiffer hydrogels lead to an increased cellular packing density in the spheroids, without a relevant impact on cell proliferation or apoptosis. MCF-7 cells isolated from stiffer hydrogels showed increase CK8/18 expression that are reported to play a critical role in protecting tissues and cells from mechanical stress [109]. In fact, the mechanical properties of PEG-based hydrogels modulate the proliferation of prostate cancer cells (LNCaP). Higher inhibition of cellular proliferation was observed in hydrogels prepared with 2.5% of PEG, when compared with the hydrogels prepared with 1.5% or 2% of PEG. This was described as a consequence of the depletion of nutrients in the cellular environment due to their limited diffusion through the TME [90]. The chemical modification of PEG with thiol groups, allows the cross-linking of the hydrogels through the formation of PEG–dithiol linkages [110,111]. This strategy is also a feasible approach to tune the mechanical properties of the hydrogels, by changing the concentrations of linker and initiator, without changing the PEG concentration. The creation of patterned regions with alternated stiffness and concentrations of RGD sequences has been also tested. Results with HT-1080 human fibrosarcoma cells, showed that, well cultured on softer (100 Pa) regions, cells invert their direction or form aggregates at the interface with stiffer domains (i.e., 360 Pa). The same type of cells seeded on stiffer regions were less motile, while they tend to migrate to the softer domains [88,112]. 3D Alg–Collagen blends have been also reported has a platform to study the impact of mechanical properties of the extracellular space in the morphology and phenotype of breast cancer-associated fibroblasts (bCAFs). It was found a linear relationship between the hydrogel's stiffness, cell morphology, and expression of mechanosensitive proteins. In particular, bCAFs assume a spindle-like shape on soft matrices ($\approx$100 Pa), in contrast, cells encapsulated on stiffer networks ($\approx$900 Pa) presented a round shape. The expression and nuclear translocation of YAP/TAZ mechanosignaling protein was also observed in softer 3D matrices [69]. Alg beads of varying stiffness, from 21 to 105 kPa were used as 3D models to encapsulate human head and neck squamous cell carcinoma (HNSCC) tumor initiating cells (TICs) [113]. In this case, CD44 and ABCG2 were used as TICs markers and an upregulation was detected on cells seeded on a matrix of moderate stiffness (i.e., 70 kPa). The same matrix induced high-cellular proliferation, however, this Alg-based system does not provide any enriched adhesive sites for cellular attachment. Despite this drawback, cell–cell contacts leading to the formation of spheroids and to the secretion of ECM, maintaining cellular viability was detected. Fibrin-based gels with stiffness of 90 Pa [114] were also used to evaluate its capacity to support the proliferation of murine B16-F1 melanoma cells. This system was able to induce the formation of larger tumor spheroids, when compared with the gels of higher stiffness (i.e., 420 and 1050 Pa). Moreover, when different cell types were compared, namely, murine H22 hepatocarcinoma cells, murine EL4 lymphoma cells (suspension growth cell types), murine P815 lymphoblast-like mastocytoma cells, human A2780 ovarian cancer cells, and human HepG2 liver carcinoma cells (all adherent-dependent cells), the softer 90 Pa fibrin hydrogels were preferable for the formation of spheroids. In general, the impact of the mechanical properties in cancer cell behavior is dependent on the specific cell type and their original TME mechanical features.

## 4.3 Conclusions and future remarks

The development of matrices that mimic the characteristics of the TME is of great interest and is a reliable approach for the in vitro study of cancer behavior under different experimental conditions. The inability of the two-dimensional (2D) systems to completely recapitulate the TME and the in vivo cellular behavior, led to the recent developments that tries to mimic the TME under a 3D configuration. The present state of the proposed 3D models mimic some important biochemical and biomechanical features, closer to the ones that are present in the in vivo environment. The use of natural materials to generate this 3D environment has the major advantage of being a close mimic of components present in the native ECM; however, its main disadvantage is the presence of a high batch to batch variability. The use of materials that present a high level of control over their chemical composition is desirable to fully mimic the TME. Analyzing the presently proposed systems, it is clear that they can recapitulate some of the TME biochemical or biomechanical features (e.g., porosity, stiffness, topography, and investigate single components involved in the tumor progression by the incorporation of single ECM cues, such as growth factors, cytokines, and enzymes). However, a complete mimic of the TME is still to be developed. Anyway, the present systems are able to study the TME dynamics, as well as to evaluate the efficacy of anticancer compounds under a drug-screening approach. The further increase in complexity of these 3D systems (being closer mimics of the TME) is expected to lead to the generation of models that better recapitulate the in vivo scenario. These 3D matrices expected to be proposed in the near future can be a valid alternative to the use of animal models, where relevant conclusions can be drawn on the mechanisms of tumor progression and the suitability of these systems to evaluate the efficacy of different anticancer drugs to treat cancer or to control its progression.

## References

[1] Hui LL, Chen Y. Tumor microenvironment: sanctuary of the devil. Cancer Lett 2015;368(1):7–13.
[2] Quail DF, Joyce JA. Microenvironmental regulation of tumor progression and metastasis. Nat Med 2013;19 (11):1423–37.
[3] Filipe EC, Chitty JL, Cox TR. Charting the unexplored extracellular matrix in cancer. Int J Exp Pathol 2018;99 (2):58–76.
[4] Xiong G-F, Xu R. Function of cancer cell-derived extracellular matrix in tumor progression. JCMT 2016;2(9).

[5] Walker C, Mojares E, Del Rio Hernandez A. Role of extracellular matrix in development and cancer progression. Int J Mol Sci 2018;19(10).

[6] Bonnans C, Chou J, Werb Z. Remodelling the extracellular matrix in development and disease. Nat Rev Mol Cell Biol 2014;15(12):786–801.

[7] Lu PF, Weaver VM, Werb Z. The extracellular matrix: a dynamic niche in cancer progression. J Cell Biol 2012;196(4):395–406.

[8] Malandrino A, et al. Complex mechanics of the heterogeneous extracellular matrix in cancer. Extreme Mech Lett 2018;21:25–34.

[9] Kaushik N, et al. Proinvasive extracellular matrix remodeling for tumor progression. Arch Pharm Res 2019;42(1):40–7.

[10] Tibbitt MW, Anseth KS. Hydrogels as extracellular matrix mimics for 3D cell culture. Biotechnol Bioeng 2009;103(4):655–63.

[11] Hynes RO, Naba A. Overview of the matrisome-an inventory of extracellular matrix constituents and functions. Cold Spring Harb Perspect Biol 2012;4(1):a004903.

[12] Sapudom J, et al. Molecular weight specific impact of soluble and immobilized hyaluronan on CD44 expressing melanoma cells in 3D collagen matrices. Acta Biomater 2017;50:259–70.

[13] Zhao YF, et al. Modulating three-dimensional microenvironment with hyaluronan of different molecular weights alters breast cancer cell invasion behavior. ACS Appl Mater Interfaces 2017;9(11):9327–38.

[14] Canibano-Hernandez A, et al. Alginate microcapsules incorporating hyaluronic acid recreate closer in vivo environment for mesenchymal stem cells. Mol Pharm 2017;14(7):2390–9.

[15] Weiss MS, et al. The impact of adhesion peptides within hydrogels on the phenotype and signaling of normal and cancerous mammary epithelial cells. Biomaterials 2012;33(13):3548–59.

[16] Pranav S, et al. Cancer cell migration within 3D layer-by-layer microfabricated photocrosslinked PEG scaffolds with tunable stiffness. Biomaterials 2012;33(29):7064–70.

[17] Turner NJ, Badylak SF. Biologic scaffolds for musculotendinous tissue repair. Eur Cell Mater 2013;25:130–43.

[18] Kular JK, Basu S, Sharma RI. The extracellular matrix: structure, composition, age-related differences, tools for analysis and applications for tissue engineering. J Tissue Eng 2014;5 2041731414557112.

[19] Chuang CH, et al. Comparison of covalently and physically cross-linked collagen hydrogels on mediating vascular network formation for engineering adipose tissue. Artif Cell Nanomed Biotechnol 2018;46(Suppl. 3):S434–47.

[20] Chevallay B, Herbage D. Collagen-based biomaterials as 3D scaffold for cell cultures: applications for tissue engineering and gene therapy. Med Biol Eng Comput 2000;38(2):211–18.

[21] Sapudom J, et al. The phenotype of cancer cell invasion controlled by fibril diameter and pore size of 3D collagen networks. Biomaterials 2015;52:367–75.

[22] Wolf K, et al. Physical limits of cell migration: control by ECM space and nuclear deformation and tuning by proteolysis and traction force. J Cell Biol 2013;201(7):1069–84.

[23] Zhang H, et al. Membrane associated collagen XIII promotes cancer metastasis and enhances anoikis resistance. Breast Cancer Res 2018;20(1):116.

[24] Sapudom J, et al. Fibril bending stiffness of 3D collagen matrices instructs spreading and clustering of invasive and non-invasive breast cancer cells. Biomaterials 2019;193:47–57.

[25] Plou J, et al. From individual to collective 3D cancer dissemination: roles of collagen concentration and TGF-beta. Sci Rep 2018;8(1):12723.

[26] Szot CS, et al. 3D in vitro bioengineered tumors based on collagen I hydrogels. Biomaterials 2011;32(31):7905–12.

[27] Theocharis AD, Manou D, Karamanos NK. The extracellular matrix as a multitasking player in disease. FEBS J 286(15):2830–69.

[28] Mao Y, Schwarzbauer JE. Fibronectin fibrillogenesis, a cell-mediated matrix assembly process. Matrix Biol 2005;24(6):389–99.

[29] Wang JP, Hielscher A. Fibronectin: how its aberrant expression in tumors may improve therapeutic targeting. J Cancer 2017;8(4):674–82.

[30] Sapudom J, et al. The interplay of fibronectin functionalization and TGF-beta1 presence on fibroblast proliferation, differentiation and migration in 3D matrices. Biomater Sci 2015;3(9):1291–301.

[31] Sapudom J, Pompe T. Biomimetic tumor microenvironments based on collagen matrices. Biomater Sci 2018;6 (8):2009–24.

[32] De Wever O, et al. Stromal myofibroblasts are drivers of invasive cancer growth. Int J Cancer 2008;123 (10):2229–38.
[33] Frantz C, Stewart KM, Weaver VM. The extracellular matrix at a glance. J Cell Sci 2010;123(24):4195–200.
[34] Insua-Rodriguez J, Oskarsson T. The extracellular matrix in breast cancer. Adv Drug Deliv Rev 2016;97:41–55.
[35] Eke I, et al. 3D matrix-based cell cultures: automated analysis of tumor cell survival and proliferation. Int J Oncol 2016;48(1):313–21.
[36] Kenny PA, et al. The morphologies of breast cancer cell lines in three-dimensional assays correlate with their profiles of gene expression. Mol Oncol 2007;1(1):84–96.
[37] Gomes LR, Vessoni AT, Menck CF. Three-dimensional microenvironment confers enhanced sensitivity to doxorubicin by reducing p53-dependent induction of autophagy. Oncogene 2015;34(42):5329–40.
[38] Stavolone L, Lionetti V. Extracellular matrix in plants and animals: hooks and locks for viruses. Front Microbiol 2017;8:1760.
[39] Kowitsch A, Chhalotre A, Groth T. Effect of thiolated glycosaminoglycans on the behavior of breast cancer cells: toward the development of in vitro models of cancer. Int J Artif Organs 2017;40(1):31–9.
[40] Wang M, et al. Glycosaminoglycans (GAGs) and GAG mimetics regulate the behavior of stem cell differentiation. Colloids Surf B: Biointerfaces 2017;150:175–82.
[41] Richter RP, et al. Glycosaminoglycans in extracellular matrix organisation: are concepts from soft matter physics key to understanding the formation of perineuronal nets? Curr Opin Struct Biol 2018;50:65–74.
[42] Soares da Costa D, Reis RL, Pashkuleva I. Sulfation of glycosaminoglycans and its implications in human health and disorders. Annu Rev Biomed Eng 2017;19:1–26.
[43] Fernandez-Vega I, et al. Specific genes involved in synthesis and editing of heparan sulfate proteoglycans show altered expression patterns in breast cancer. BMC Cancer 2013;13.
[44] Kalathas D, et al. Alterations of glycosaminoglycan disaccharide content and composition in colorectal cancer: Structural and expressional studies. Oncol Rep 2009;22(2):369–75.
[45] Vallen MJE, et al. Highly sulfated chondroitin sulfates, a novel class of prognostic biomarkers in ovarian cancer tissue. Gynecol Oncol 2012;127(1):202–9.
[46] Kultti A, et al. Accumulation of extracellular hyaluronan by hyaluronan synthase 3 promotes tumor growth and modulates the pancreatic cancer microenvironment. Biomed Res Int 2014;2014:817613.
[47] Toole BP. Hyaluronan: from extracellular glue to pericellular cue. Nat Rev Cancer 2004;4(7):528–39.
[48] Price ZK, Lokman NA, Ricciardelli C. Differing roles of hyaluronan molecular weight on cancer cell behavior and chemotherapy resistance. Cancers (Basel) 2018;10(12):E482.
[49] Senbanjo LT, Chellaiah MA. CD44: a multifunctional cell surface adhesion receptor is a regulator of progression and metastasis of cancer cells. Front Cell Dev Biol 2017;5:18.
[50] Freudenberg U, et al. Glycosaminoglycan-based biohybrid hydrogels: a sweet and smart choice for multifunctional biomaterials. Adv Mater 2016;28(40):8861–91.
[51] Gil MS, et al. Injectable hydrogel-incorporated cancer cell-specific cisplatin releasing nanogels for targeted drug delivery. J Mater Chem B 2017;5(34):7140–52.
[52] Wang Y, et al. 3D hydrogel breast cancer models for studying the effects of hypoxia on epithelial to mesenchymal transition. Oncotarget 2018;9(63):32191–203.
[53] Ananthanarayanan B, Kim Y, Kumar S. Elucidating the mechanobiology of malignant brain tumors using a brain matrix-mimetic hyaluronic acid hydrogel platform. Biomaterials 2011;32(31):7913–23.
[54] Pedron S, Becka E, Harley BAC. Regulation of glioma cell phenotype in 3D matrices by hyaluronic acid. Biomaterials 2013;34(30):7408–17.
[55] Gurski LA, et al. Hyaluronic acid-based hydrogels as 3D matrices for in vitro evaluation of chemotherapeutic drugs using poorly adherent prostate cancer cells. Biomaterials 2009;30(30):6076–85.
[56] Gurski LA, et al. Hyaluronan (HA) interacting proteins RHAMM and hyaluronidase impact prostate cancer cell behavior and invadopodia formation in 3D HA-based hydrogels. PLoS One 2012;7(11):e50075.
[57] Tang Y, et al. Three-dimensional prostate tumor model based on a hyaluronic acid-alginate hydrogel for evaluation of anti-cancer drug efficacy. J Biomater Sci Polym Ed 2017;28(14):1603–16.
[58] Mikami T, Kitagawa H. Biosynthesis and function of chondroitin sulfate. Biochim Biophys Acta (BBA) 2013;1830(10):4719–33.
[59] Theocharis AD, et al. Chondroitin sulfate as a key molecule in the development of atherosclerosis and cancer progression. Adv Pharmacol 2006;53:281–95.

[60] Sugahara K, Kitagawa H. Recent advances in the study of the biosynthesis and functions of sulfated glycosaminoglycans. Curr Opin Struct Biol 2000;10(5):518–27.

[61] Brauchle E, et al. Biomechanical and biomolecular characterization of extracellular matrix structures in human colon carcinomas. Matrix Biol 2018;68-69:180–93.

[62] O'Brien FJ, et al. The effect of pore size on cell adhesion in collagen-GAG scaffolds. Biomaterials 2005;26 (4):433–41.

[63] Curtin C, et al. A physiologically relevant 3D collagen-based scaffold-neuroblastoma cell system exhibits chemosensitivity similar to orthotopic xenograft models. Acta Biomater 2018;70:84–97.

[64] Liu Y, et al. Development of a biomimetic chondroitin sulfate-modified hydrogel to enhance the metastasis of tumor cells. Sci Rep 2016;6:29858.

[65] Andersen T, Auk-Emblem P, Dornish M. 3D cell culture in alginate hydrogels. Microarrays (Basel) 2015;4 (2):133–61.

[66] Wisdom K, Chaudhuri O. 3D cell culture in interpenetrating networks of alginate and rBM matrix. Methods Mol Biol 2017;1612:29–37.

[67] Kang SW, et al. The effect of conjugating RGD into 3D alginate hydrogels on adipogenic differentiation of human adipose-derived stromal cells. Macromol Biosci 2011;11(5):673–9.

[68] Little CJ, Kulyk WM, Chen X. The effect of chondroitin sulphate and hyaluronic acid on chondrocytes cultured within a fibrin-alginate hydrogel. J Funct Biomater 2014;5(3):197–210.

[69] Cao H, et al. Mechanoregulation of cancer-associated fibroblast phenotype in three-dimensional interpenetrating hydrogel networks. Langmuir 2019;35(23):7487–95.

[70] Liu C, et al. Hybrid collagen alginate hydrogel as a platform for 3D tumor spheroid invasion. Acta Biomater 2018;75:213–25.

[71] Qiao SP, et al. An alginate-based platform for cancer stem cell research. Acta Biomater 2016;37:83–92.

[72] Estrada MF, et al. Modelling the tumour microenvironment in long-term microencapsulated 3D co-cultures recapitulates phenotypic features of disease progression. Biomaterials 2016;78:50–61.

[73] Rinaudo M. Main properties and current applications of some polysaccharides as biomaterials. Polym Int 2008;57(3):397–430.

[74] Jayakumar R, et al. Biomedical applications of chitin and chitosan based nanomaterials-a short review. Carbohydr Polym 2010;82(2):227–32.

[75] Ahmadi F, et al. Chitosan based hydrogels: characteristics and pharmaceutical applications. Res Pharm Sci 2015;10(1):1–16.

[76] Dhiman HK, Ray AR, Panda AK. Characterization and evaluation of chitosan matrix for in vitro growth of MCF-7 breast cancer cell lines. Biomaterials 2004;25(21):5147–54.

[77] Leung M, et al. Chitosan-alginate scaffold culture system for hepatocellular carcinoma increases malignancy and drug resistance. Pharm Res 2010;27(9):1939–48.

[78] Florczyk SJ, et al. 3D porous chitosan-alginate scaffolds: a new matrix for studying prostate cancer cell-lymphocyte interactions in vitro. Adv Healthc Mater 2012;1(5):590–9.

[79] Kievit FM, et al. Modeling the tumor microenvironment using chitosan-alginate scaffolds to control the stem-like state of glioblastoma cells. Biomater Sci 2016;4(4):610–13.

[80] Kievit FM, et al. Chitosan-alginate 3D scaffolds as a mimic of the glioma tumor microenvironment. Biomaterials 2010;31(22):5903–10.

[81] Wang K, et al. Culture on 3D chitosan-hyaluronic acid scaffolds enhances stem cell marker expression and drug resistance in human glioblastoma cancer stem cells. Adv Healthc Mater 2016;5 (24):3173–81.

[82] Florczyk SJ, et al. Porous chitosan-hyaluronic acid scaffolds as a mimic of glioblastoma microenvironment ECM. Biomaterials 2013;34(38):10143–50.

[83] Wong CW, et al. Biomaterial substrate-derived compact cellular spheroids mimicking the behavior of pancreatic cancer and microenvironment. Biomaterials 2019;213:119202.

[84] Gu L, Mooney DJ. Biomaterials and emerging anticancer therapeutics: engineering the microenvironment. Nat Rev Cancer 2016;16(1):56–66.

[85] Cruz-Acuna R, Garcia AJ. Synthetic hydrogels mimicking basement membrane matrices to promote cell-matrix interactions. Matrix Biol 2017;57-58:324–33.

[86] Enemchukwu NO, et al. Synthetic matrices reveal contributions of ECM biophysical and biochemical properties to epithelial morphogenesis. J Cell Biol 2016;212(1):113–24.

[87] Bellis SL. Advantages of RGD peptides for directing cell association with biomaterials. Biomaterials 2011;32 (18):4205–10.
[88] Singh SP, et al. A peptide functionalized poly(ethylene glycol) (PEG) hydrogel for investigating the influence of biochemical and biophysical matrix properties on tumor cell migration. Biomater Sci 2014;2(7):1024–34.
[89] Sridhar BV, et al. Development of a cellularly degradable PEG hydrogel to promote articular cartilage extracellular matrix deposition. Adv Healthc Mater 2015;4(5):702–13.
[90] Sieh S, et al. Phenotypic characterization of prostate cancer LNCaP cells cultured within a bioengineered microenvironment. PLoS One 2012;7(9):e40217.
[91] Pradhan S, et al. PEG-fibrinogen hydrogels for three-dimensional breast cancer cell culture. J Biomed Mater Res A 2017;105(1):236–52.
[92] Pradhan S, et al. A three-dimensional spheroidal cancer model based on PEG-fibrinogen hydrogel microspheres. Biomaterials 2017;115:141–54.
[93] Polonio-Alcala E, et al. PLA electrospun scaffolds for three-dimensional triple-negative breast cancer cell culture. Polymers (Basel) 2019;11(5):916.
[94] Polonio-Alcala E, et al. Screening of additive manufactured scaffolds designs for triple negative breast cancer 3D cell culture and stem-like expansion. Int J Mol Sci 2018;19(10):3148.
[95] Kim JW, Ho WJ, Wu BM. The role of the 3D environment in hypoxia-induced drug and apoptosis resistance. Anticancer Res 2011;31(10):3237–45.
[96] Rezvantalab S, et al. PLGA-based nanoparticles in cancer treatment. Front Pharmacol 2018;9:1260.
[97] Fong EL, et al. Modeling Ewing sarcoma tumors in vitro with 3D scaffolds. Proc Natl Acad Sci U S A 2013;110(16):6500–5.
[98] Rabionet M, et al. Electrospinning PCL scaffolds manufacture for three-dimensional breast cancer cell culture. Polymers (Basel) 2017;9(8):328.
[99] Yingzi L, et al. Effects of mechanical properties on tumor invasion: insights from a cellular model. Conf Proc IEEE Eng Med Biol Soc 2014;2014:6818–21.
[100] Yu H, Mouw JK, Weaver VM. Forcing form and function: biomechanical regulation of tumor evolution. Trends Cell Biol 2011;21(1):47–56.
[101] Cox TR, Erler JT. Remodeling and homeostasis of the extracellular matrix: implications for fibrotic diseases and cancer. Dis Model Mech 2011;4(2):165–78.
[102] Butcher DT, Alliston T, Weaver VM. A tense situation: forcing tumour progression. Nat Rev Cancer 2009;9 (2):108–22.
[103] Northcott JM, et al. Feeling stress: the mechanics of cancer progression and aggression. Front Cell Dev Biol 2018;6:17.
[104] Paszek MJ, et al. Tensional homeostasis and the malignant phenotype. Cancer Cell 2005;8(3):241–54.
[105] Wullkopf L, et al. Cancer cells' ability to mechanically adjust to extracellular matrix stiffness correlates with their invasive potential. Mol Biol Cell 2018;29(20):2378–85.
[106] Kim JE, et al. Characterization of the mechanical properties of cancer cells in 3D matrices in response to collagen concentration and cytoskeletal inhibitors. Integr Biol (Camb) 2018;10(4):232–41.
[107] Charoen KM, et al. Embedded multicellular spheroids as a biomimetic 3D cancer model for evaluating drug and drug-device combinations. Biomaterials 2014;35(7):2264–71.
[108] Taubenberger AV, et al. 3D microenvironment stiffness regulates tumor spheroid growth and mechanics via p21 and ROCK. bioRxiv 2019;586784.
[109] Pastuszak M, et al. Cytokeratins in gastroenterology. Syst Rev Prz Gastroenterol 2015;10(2):61–70.
[110] Lin CC, Raza A, Shih H. PEG hydrogels formed by thiol-ene photo-click chemistry and their effect on the formation and recovery of insulin-secreting cell spheroids. Biomaterials 2011;32(36):9685–95.
[111] Xin S, Wyman OM, Alge DL. Assembly of PEG microgels into porous cell-instructive 3D scaffolds via thiol-ene click chemistry. Adv Healthc Mater 2018;7(11):e1800160.
[112] Hoffmann JC, West JL. Three-dimensional photolithographic micropatterning: a novel tool to probe the complexities of cell migration. Integr Biol (Camb) 2013;5(5):817–27.
[113] Liu C, et al. Potential effect of matrix stiffness on the enrichment of tumor initiating cells under three-dimensional culture conditions. Exp Cell Res 2015;330(1):123–34.
[114] Liu J, et al. Soft fibrin gels promote selection and growth of tumorigenic cells. Nat Mater 2012;11(8):734–41.

CHAPTER

# 5

# 3D neuroblastoma in vitro models using engineered cell-derived matrices

*Enrico Almici*[1,2,3], *David Caballero*[4], *Joan Montero*[1] *and Josep Samitier*[1,2,3]

[1]Nanobioengineering Group, Institute for Bioengineering of Catalonia (IBEC), The Barcelona Institute of Science and Technology (BIST), Barcelona, Spain [2]Department of Electronics and Biomedical Engineering, University of Barcelona, Barcelona, Spain [3]Networking Biomedical Investigation Center for Bioengineering, Biomaterials and Nanomedicine (CIBER-BBN), Madrid, Spain [4]3B's Research Group, I3Bs Research Institute on Biomaterials, Biodegradables and Biomimetics, University of Minho, Headquarters of the European Institute of Excellence on Tissue Engineering and Regenerative Medicine, Guimarães, Portugal

## 5.1 Introduction

Neuroblastoma (NB) is a tumor of the sympathetic lineage that affects mostly infants. The clinical presentation is very heterogeneous; a portion of patients experience uncontrolled resorption of the mass, while the majority present an aggressive, metastatic, relapsing disease [1]. Histologic features, age, and genetic abnormalities correlate with the risk of relapse and are exploited to stratify patients and tailor treatment. A significant proportion of patients experience relapse and only few survive, since no standard therapy is available [2]. Tumor-associated collagen signatures are currently applied as a prognostic marker for many adult tumors [3]. Indeed, recent publications on the genetic signature and histopathological presentation of NB in subsets of high-risk NB patients (HR-NB) highlight the presence of stromal remodeling [4].

Nonetheless, the role of the tumor microenvironment (TME) in disease progression and treatment response is not fully understood. The development of innovative three-dimensional (3D) in vitro culture models may foster the study of how tumors interact with the TME [5,6]. This physiological model aims to bridge the gap between simpler in vitro—two-dimensional (2D)—platforms (e.g., Petri dishes) and model animals (e.g., mice). In particular, decellularized ex vivo tissues are the most accurate in vitro systems


DOI: https://doi.org/10.1016/B978-0-12-818128-7.00005-8

capable of recapitulating the complex composition and the microarchitecture of tissue niches. Chemical and enzymatic treatments are usually applied to eliminate cells and retrieve a structured, functional scaffold [7]. Extracellular matrices (ECMs) derived from ex vivo decellularized tissues best retain and resemble the native tissue ultrastructure. Nonetheless, sample availability and denaturation of the biomaterial due to a long decellularization period limit their application [7,8].

Moreover, scaffold reproducibility is hindered by the natural heterogeneity of the biological system, limiting their application throughput. Overall, there is a need for a model capable of overcoming the aforementioned problems. Cell-derived matrices (CDMs) employ long-term cell culture to obtain a structured and functional ECM construct in vitro [7]. Due to their limited thickness, decellularization conditions are milder; therefore, CDMs preserve ECM molecule functionality more accurately. Moreover, CDM deposition can be designed to model specific properties of the native tissue. Herein, we discuss the potential of CDMs in studying the interplay between NB cells and the ECM, and its active remodeling. We focus on the fabrication of CDMs with controlled composition and architecture for NB research applications. We conclude by discussing computational advances and imaging techniques that enable the system to be exploited to study cancer cell–ECM interactions.

## 5.2 Neuroblastoma

NB is a malignant tumor of the peripheral nervous system. It accounts for 6% of childhood cancers and is the most common and fatal type of cancer in infants [9]. It is responsible for 15% of cancer-related deaths in children, and in 50% of cases, with metastasis and resistance to chemotherapy observed at the time of detection [10]. The median age at diagnosis is 18 months, and the majority of patients diagnosed younger than 10 years. The prevalence and prognosis of NB are affected by age. The 5-year survival rate is 90% for patients under 12 months of age, which decreases to 66% in patients aged between 10 and 14 years [10].

NB develops as a result of the impaired functionality of neuroblast cells that arise from the neural crest (NC) [2]. During development, NC cells undergo an epithelial-to-mesenchymal transition (EMT) by which they develop into multiple tissues and cell types, migrating throughout the body [11]. NB usually starts in the adrenal glands, but can originate anywhere along the sympathetic nervous system, such as the paraspinal ganglia in the chest and the abdomen [2]. Migratory and invasive behaviors are iterated in highly malignant NBs, which typically present with metastasis. The most common secondary sites are the bone marrow, followed by the lymph nodes and the liver [12]. In histological sections, the nervous tissue appears altered, lacking a functional relationship between neural and supportive cells (i.e., Schwann cells). The proportion of neuroblastic and Schwann-like cells that populate the microenvironment defines the level of tissue differentiation, which is inversely correlated with tumor malignancy [1,2,13]. The clinical presentation of NB is diverse resulting in a widely variable prognosis. Some patients experience spontaneous regression or differentiation of the mass, while the majority present with treatment-resistant and highly metastatic tumors [1,14].

To aid in determining patient treatment, the classification system of the International Neuroblastoma Risk Group (INRG) defines the risk of relapse into the categories low, intermediate, and high, correlating prognosis with patient age and stage, tissue histology, amplification of specific genes, such as MYCN, and chromosomal aberrations [1,13,15]. HR-NB is associated with advanced stages, older age (>12–18 months), unfavorable histology, and MYCN amplification. This subgroup accounts for 30%–50% of NB patients and presents the most enigmatic progression and therapeutic response [2].

Low- and intermediate-risk NBs are usually observed or treated with surgery alone or surgery combined with chemotherapy. They generally have a favorable outcome, with an overall survival rate of over 80% [2]. HR-NB patients receive intensive multiagent therapy; however, prognosis remains poor with long-term survival of less than 50% [10,14]. Usually, HR-NB treatment relies on surgery, multiple chemotherapeutic drugs, bone marrow transplantation, radiotherapy, immunotherapy, and maintenance therapies (tretinoin) [16]. Nonetheless, almost 60% of treated patients experience recurrence. Relapsed and refractory NB patients are challenging to treat and there is no standard cure [12]. Overall, there is an urgent need for new and more effective treatments.

Current evidence correlates the most aggressive phenotype of HR-NB, termed ultra HR-NB, with alterations in ECM architecture, composition, and vascularity, as well as with the expression of biomarkers of stemness and vasculogenic mimicry [4,5,17–19]. Further deepening our knowledge on the relationship between the malignancy and the microenvironment is important, particularly to disclose the limiting (e.g., Schwann stroma) and potentiating effects [e.g., tumor-derived endothelial cells; tumor-associated fibroblasts (TAFs)] that appear to characterize several aspects of the tumor [20]. A better understanding of fundamental molecular and biological processes correlated with ultra HR-NB, including the regulatory pathways controlling metastasis and drug resistance, may lead to improved stratification of high-risk patients and could form the basis for new therapeutic strategies to enhance survival [21,22].

### 5.2.1 Evidence of cell–extracellular matrix interaction in neuroblastoma

Tumor–stroma interaction is extremely complex and variable, depending on cell phenotype and matrix organization [23]. Solid evidence suggests that it has numerous implications in the regulation and modulation of cancer cell invasion and migration, proliferation, and the metabolic and secretory activity of various tumors [24].

Overexpression of fibrillary and reticular collagen is observed in many carcinomas, resulting in stiffer ECMs [3,25]. Collagens, such as type I and III, as well as collagen-modifying enzymes [metalloproteinase and lysyl oxidase (LOX)] are abnormally expressed in tumors compared with healthy connective tissue [23]. Matrix deposition and crosslinking cause changes in fiber organization and alignment that also result in an altered microenvironment and cancer response [26]. Tumor stiffening is associated with increased malignancy and metastatic behaviors in breast, prostate, and liver cancers [27–29]. Other components of the ECM, such as glycosaminoglycans (GAGs; large hydrophilic aggregates of disaccharides present in all mammalian tissue), concur to this physicochemical network. GAGs have biofiltering, scaffolding, and cell anchoring properties implicated in the modulation of biomechanical interactions and cytokine storage in the TME [30]. Densification and remodeling of the matrix cause biomolecule release from the ECM reservoir, which furthers regulate cancer malignancy and acts on inflammatory and immune responses [19]. ECM physical properties and remodeling can influence further aspects of tumor progression, such as vasculogenesis, chronic fibrotic inflammation, migration, and immune cell activation [11,23]. These processes synergize to regulate tumoral progression, tissue damage, and drug response, ultimately affecting patient symptomology and histopathology [31]. Therefore, morphometric analysis has been proposed to characterize ECM deposition and turnover to stratify prognostic classification in several tumors [11].

Noguera et al. characterized the presentation of ECM structural elements in the biopsies of NB patients (Fig. 5.1). They analyzed the correlation between these features, current parameters employed to predict the risk of relapse, and prognosis [4,17,18,32]. Loss of tissue architecture and altered composition promote malignant progression, and characteristic ECM elements correlate with prognosis (relapse or overall survival) [18]. As an example, GAGs, as well as collagen I and III fibers, presented an altered organization and crosslinking in high-risk primary and relapsed NB patients, with the presence of treatment-damaged tissue [4,18]. Moreover, different collagen networks were found to further characterize ultra HR-NB [18].

A cross-talk between the Schwannian stroma (Fig. 5.1A) and neuroblasts has been reported to impair NB growth and inhibit angiogenesis, according to positive prognostic features [33,34]. An aligned collagen matrix is considered to enhance the differentiation of human NC stem cells towards Schwann cells, while a poorly porous ECM correlates with malignant tissue degeneration [35,36]. Further insights on NB dependence on the microenvironment correspond with the presence of tumor-associated fibroblasts (peculiar populations associated with chronic fibrosis in inflamed tissues, such as tumors) with increased vascularity and stiffness [37]. NB cells interact with the ECM through the expression of several integrins and surface receptors, whose effect depends on the degree of expression and ECM properties, generating a potential feedback loop of malignant induction [36,38]. At the in vitro level, the effect of tissue composition and architecture on signaling pathways that influence cancer progression and vice versa can be investigated [5]. Topography, composition, and mechanical properties of 3D cellular microenvironments are linked to gene expression and cell morphology changes of NB cultures. For example, Kumar et al. showed in vitro that stiffness impacts NB cell morphology, cell proliferation, and MYCN expression [39]. Interestingly, CDMs modified to present graded ratios of GAGs showed an important regulatory effect on ECM composition to differentiation treatments currently applied as maintenance therapy in high-risk patients (retinoic acid) [40]. Further aspects of tumor progression and therapeutic response could be disclosed to understand how biomechanical forces and changes affect cell–cell and cell–matrix cross-talk, altering tumor cell fate and morphology.

In conclusion, physiologic, orthotopic in vitro models are a promising road to identify potential therapeutic targets, test novel agents, and develop personalized in vitro models. Biomimicry of native ECM is fundamental to assess in vitro the effect of the entire array of cell receptors as well as changes in cell morphology and migration strategies present in real tumors. In this chapter, we will focus on the use of CDMs to engineer ECM with controlled composition and architecture to closely reproduce the properties of NB tumors in different stages and anatomic sites.

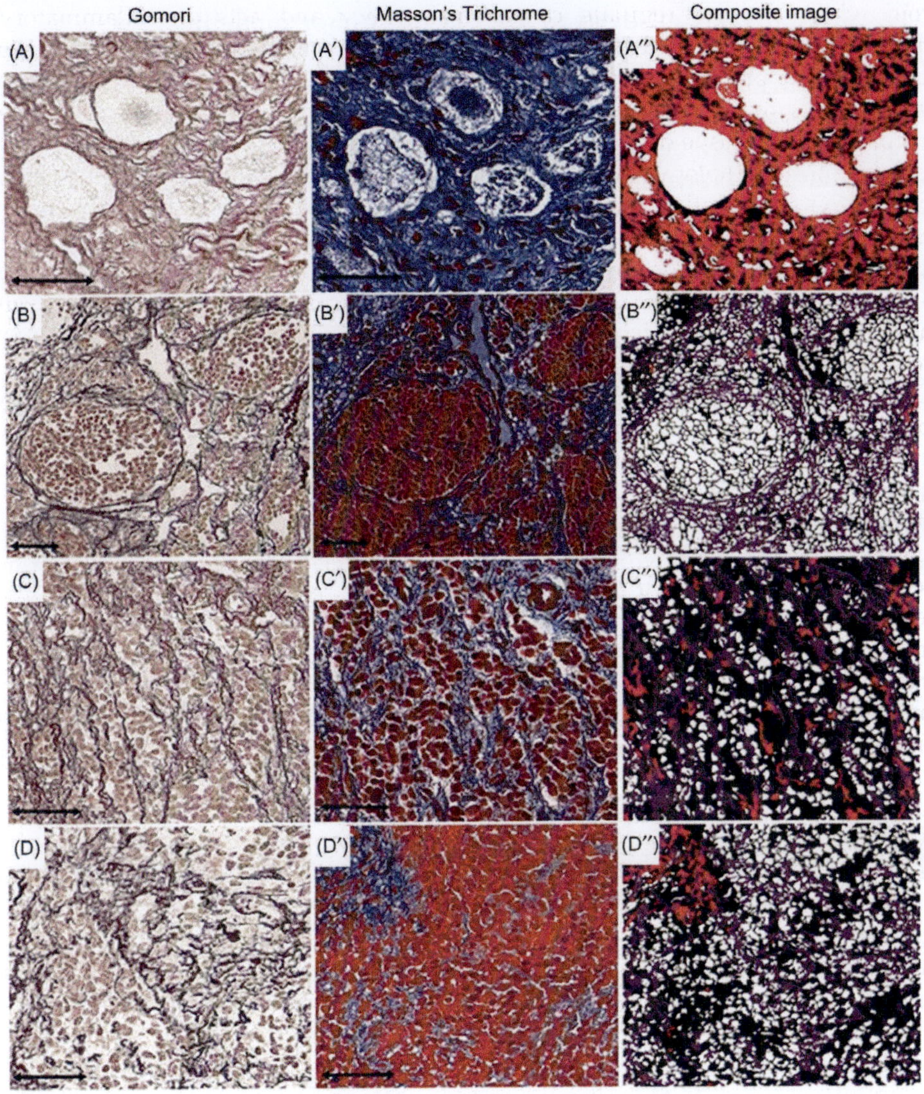


**FIGURE 5.1** Histological presentation of ECM components in NB patients of increasing malignancy. Gomori and Masson's Trichrome are two conventional histological staining techniques employed to highlight the collagenous architecture in patient biopsies. (A–D) In black, the reticulin fiber networks (Col III) and brown, neuroblastic cells. (A′–D′) In intense blue, collagen type I bundles and in red, neuroblastic cells. (A″–D″) Composite images with reticulin fibers highlighted in purple, collagen type I fibers highlighted in pink, and cells represented in white. (A, A′, and A″) Individual ganglionic cells of a ganglioneuroma are separated from one another by a fibrous ECM composed mostly of collagen type I bundles. (B, B′, and B″) Neuroblasts form aggregates of various sizes surrounded by reticulin fibers and scant collagen type I fibers. (C, C′, and C″) Neuroblasts arranged in rows accompanied by reticulin fibers. (D, D′, and D″) Neuroblasts are intermingled in reticulin fibers networks. Scale bars correspond to 50 μm. Source: *Reproduced with permission from Tadeo I, Berbegall AP, Navarro S, Castel V, Noguera R. A stiff extracellular matrix is associated with malignancy in peripheral neuroblastic tumors. Pediatr Blood Cancer 2017;64(9):e26449.*

## 5.3 Cell-derived matrices in tumor modeling

CDMs are native-like ECMs produced by culturing in vitro cells in well-controlled media conditions for a specific period before decellularization. Furthermore, CDMs are usually employed to investigate cell–ECM interactions in in vivo-like conditions (Fig. 5.2). Due to their intrinsic properties, cell morphology, migration, and secretion can be easily accessed and characterized, which is of particular interest in a CDM tumor model. The production process of CDMs is simple, but not trivial to optimize using tumoral or primary cells, due to their complex regulation and heterogeneous response. In addition, the fabrication of such matrices shows lower scalability and reproducibility compared with standard hydrogels and bioengineered scaffolds. Nonetheless, CDMs holds great promise in tumor modeling due to the complex biomimetic ultrastructure of the biological material deposited [24]. In addition, CDMs can be easily modified with regard to architecture and composition, thus mimicking functional and pathologic niches. Multiple protocols for CDM fabrication have been reported for the production of native-like microenvironments for a vast plethora of applications [41–43]. Herein, we present the most relevant applications of CDMs in cancer research.

It has been shown with hydrogel-based cancer models that the 3D microenvironment and its biochemical properties are crucial for reproducing the interactions between cells

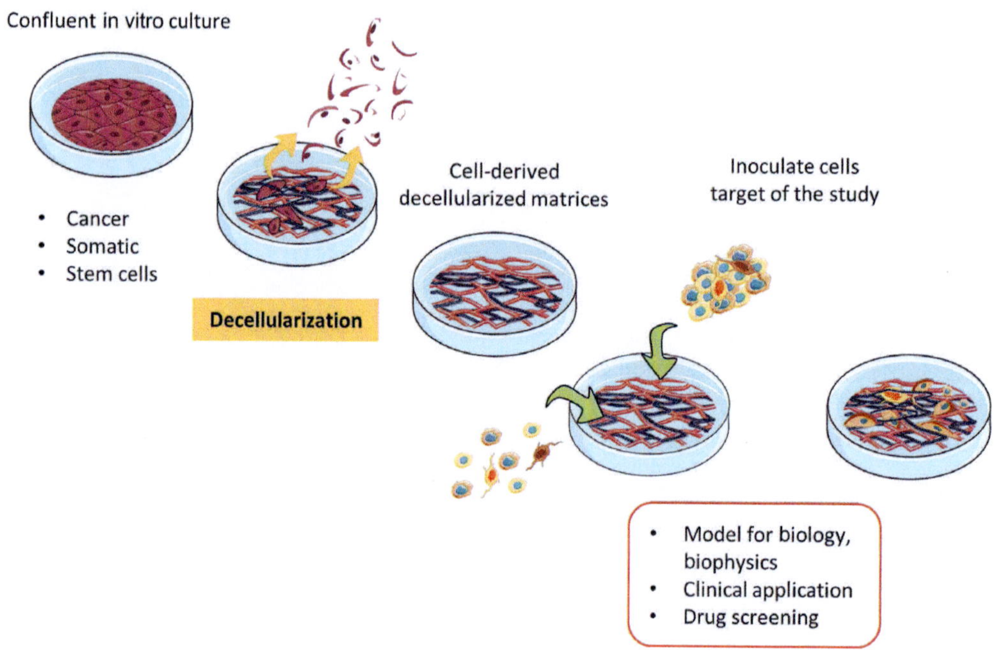


FIGURE 5.2 Production and application of cell-derived matrices. The selected cell source is cultured at confluence for several days in well-controlled media conditions prior to decellularization. Finally, the target cell culture is seeded on the CDM to study cell behavior.

and the ECM [6,19]. However, the conventional hydrogels poorly mimic the complex molecular composition of the native ECM and its associated interactions with cells.

Several groups exploited CDMs to evaluate the effect of cell–ECM interaction in cancer progression and drug response. Table 5.1 shows the most relevant studies describing the use of CDMs for cancer research. These studies define some of the major trends in the application of CDMs in this field. Composition and architecture can be modulated, employing cells of different origin and malignancy. Primary and secondary sites of invasion can be modeled controlling CDM properties by using different cell sources. Similarly, CDMs can be exploited to study cell and tumor biology, as well as pathways involved in chemotherapy resistance [7].

Cukierman et al. investigated the effect of CDMs derived from NIH-3T3 fibroblasts on different epithelial tumors [43,44]. In NIH-3T3 CDMs, the response of inoculated cell lines varied independently upon cancer type. For example, metastatic ovarian carcinoma cells showed strong morphological changes but almost no effect on proliferation, while lung carcinoma cells proliferated less with moderate morphological changes (Fig. 5.3). Colorectal cancer cell lines did not show proliferation nor morphological changes [44].

**TABLE 5.1** Cell-derived matrices in cancer modeling.

| CDM | Cell source | Target cell | Findings | Ref |
|---|---|---|---|---|
| Normal | Fibroblasts (NIH-3T3) | Various (HCT116, NCI-H460, PA-1, COLO 205, PANC-1, MCF-7, SW620, HCT116/p53, HS 578T, PA-1/E6, and MCF-10A) | • Increased chemoresistance via integrin β1 survival signal | [44] |
| Normal and cancer | Fibroblasts (NIH-3T3 and CAFs) | Breast cancer MDA-MB-231, MCF-7, and breast endothelial MCF-10A | • Activated PI3K-Akt signaling via integrin β1<br>• Perturbed morphology and cell migration | [45] |
| Benign tumor and cancer | Breast (MDA-MB-231, MCF-7, and MCF-10A) | Breast cancer MDA-MB-231, MCF-7, and breast endothelial MCF-10A | • Promoted proliferation on invasive MDA-MB-231 CDMs<br>• Suppressed proliferation on healthy MCF-10A CDMs<br>• Increased chemoresistance on invasive MDA-MB-231 CDMs | [46] |
| Normal | Rat smooth muscle cells (R22CIF), human fibroblasts (T-1) | Neuroblastoma (HTLA23) | • Enhanced responsiveness to retinoic acid<br>• Low ratio of chondroitin sulfate/hyaluronic acid inhibits effect of differentiation agents | [47] |
| Normal and cancer | Colon (HT-29, SW480, and CCD-841-CoN) | Colon cancer HT-29 and SW480 | • Increased chemoresistance on invasive HT-29 CDMs via Akt activation and ABCB1 upregulation<br>• Promoted EMT on invasive HT-29 CDMs via TGF-β signaling | [48] |
| Cancer | Tongue (oral carcinoma HN12) | Oral carcinoma HN12 | • Increased chemoresistance via talin- and focal adhesion kinase-mediated signals | [49] |

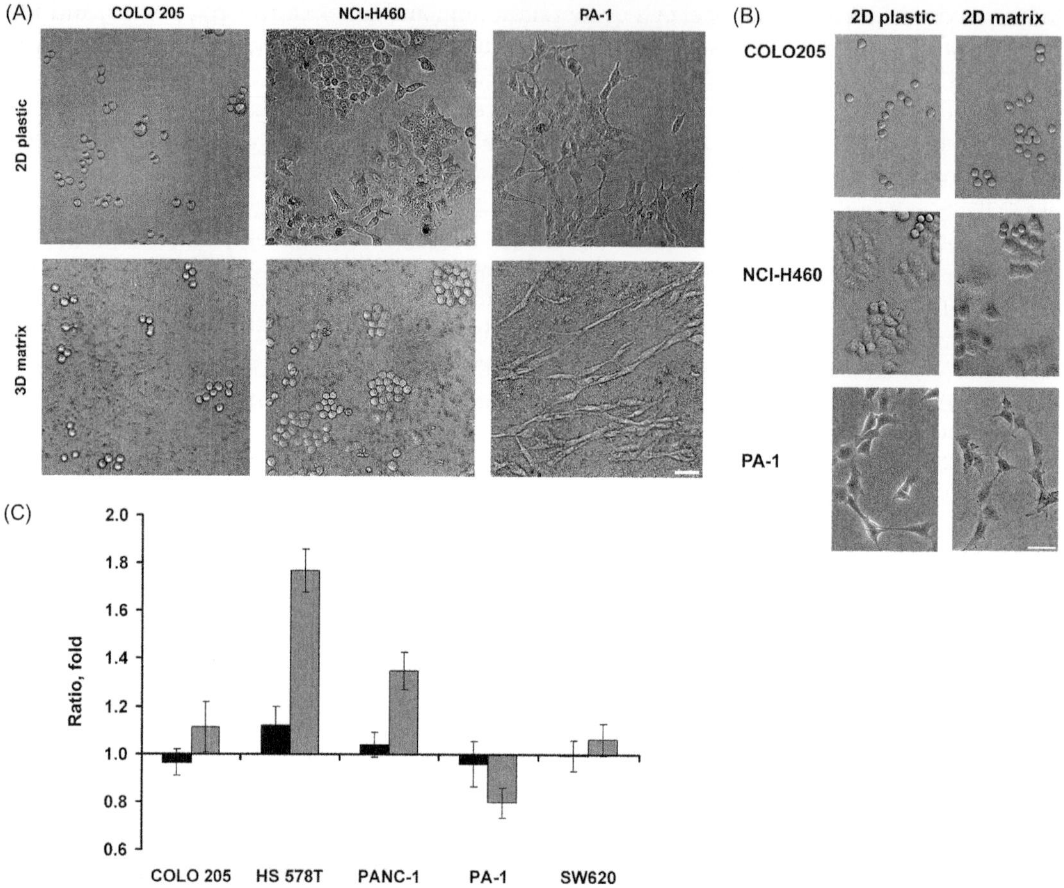


FIGURE 5.3 (A) Carcinomas of different origin show variable modification of cell morphology when cultured in fibroblast CDMs. COLO 205, NCI-H460, and PA-1 present distinct characteristic responses to CDM culturing. Round, scant cells insensitive to cell body polarization; spread cells gaining amoeboid, spindle-like, or spread morphologies. (B) No morphological alteration is observed when CDMs are solubilized and used as a 2D substrate. (C) Ratios of 2D matrix:plastic (in black) or 3D matrix:plastic (in gray). $IC_{50}$ values for different carcinoma cell lines 3 days after treatment with Taxol (10–100 nM). Cells demonstrating an altered morphology or proliferation when cultured on CDMs show a matrix-dependent drug response (positive, e.g., PANC-1, either negative, e.g., PA-1). NCI-H460, human large cell lung cancer; COLO 205, human colorectal adenocarcinoma; HS 578T, human breast carcinoma; PANC-1, human pancreatic carcinoma; PA-1, human ovarian teratocarcinoma; SW620, human colorectal adenocarcinoma. Source: *Adapted with permission from Serebriiskii I, Castelló-Cros R, Lamb A, Golemis EA, Cukierman E. Fibroblast-derived 3D matrix differentially regulates the growth and drug-responsiveness of human cancer cells. Matrix Biol 2008;27(6):573–85.*

CDMs significantly increased the resistance of human pancreatic carcinoma, breast carcinoma, and colorectal adenocarcinoma cell lines to Taxol (a mitosis inhibitor that acts via microtubule stabilization) through integrin β1-dependent pathways [44]. Furthermore, CDMs derived from TAFs comparing breast cancer cell lines of increasing malignancy were investigated [45]. TAF CDMs were responsible for activating PI3K/Akt pathways (associated with drug resistance, cell cycle regulation, and other cancer pathways), and for inducing motile morphologies (mesenchymal or amoeboid-like), especially in highly invasive malignant cells [45]. In addition, invasion was observed to be dependent on β1 integrin. Although its inhibition did not affect cell velocity, cellular morphology, and directionality were affected [45]. Altogether, the works mentioned above indicate that cells seeded on native-like CDMs show an altered morphology and/or proliferation profile, a 3D matrix-dependent drug response, and, importantly, similarity to an in vivo scenario.

Matsushima and Bogenmann [47] employed rat smooth cells to investigate the effect of the CDM on human NB cell differentiation. The impact of retinoic acid in inducing cell differentiation was investigated, which is a maintenance therapy for high-risk NB. Trypsin digestion and/or high-salt extraction modulated the ratio of chondroitin sulfate to hyaluronic acid (HA) present in the CDM, which was crucial in the capacity of the matrix to promote NB differentiation. Altering this ratio did not result in cell differentiation induction by retinoic acid or nerve growth factor.

Sansing et al. employed metastatic oral carcinoma CDMs to study the effect of integrin on cisplatin resistance. The obtained results showed that the $IC_{50}$ increased threefold for cells cultured on CDMs compared with 2D culture [49]. Hoshiba et al. studied invasive, noninvasive, and benign breast cancer cells within staged tumorigenesis-mimicking CDMs produced by the same cell lines [46,48,50]. Malignant cells attached and proliferated according to the degree of malignancy of the underlying ECM, while benign cells followed an opposite trend [46]. Increased fluorouracil (5-FU) resistance in invasive cancer cells grown on malignant CDMs was observed. Chemoresistance in general increased, especially when cells were cultured on CDMs from the same cell type or of a higher malignant phenotype [48]. Culture substrates for CDM deposition were shown to affect chemoresistance induction in replated cells.

In conclusion, the cellular microenvironment provided by fibroblast CDMs induce native-like phenotypes in several tumor cells and affect their response to chemotherapeutic drugs compared with 2D culture. This effect is enhanced by the matrices produced by cells associated with malignant tissue remodulation. Similarly, cancer CDMs influenced drug resistance with increasing malignancy in cells of identical lineage. Finally, because of technological and engineering advances, more specialized CDM environments for cancer research can be designed. In the next sections, we summarize the most relevant approaches employed to engineer constructs with controlled characteristics, as well as the techniques that enable a connection with information on CDM composition and remodeling to cell processes characteristic of tumor progression.

## 5.4 Engineering cell-derived matrix deposition

CDM variability can be controlled via different biochemical and physical means. This enables CDMs to be engineered with specific properties that resemble patient histology [51]. The main parameters requiring control to engineer specific CDM are: (1) cell source, (2) culture medium composition, (3) culture substrate and conditions, (4) decellularization agents, and (5) chemical and physical modifications (Fig. 5.4).

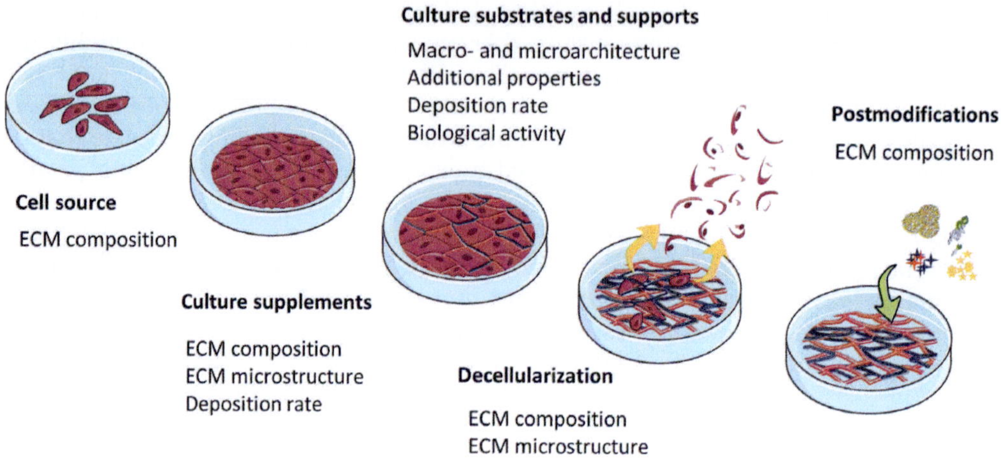


FIGURE 5.4 Schematic overview of the fabrication steps of engineered CDMs. (i) Cell source defines CDM composition, (ii) culture medium composition affects CDM composition, microstructure, and deposition rate. (iii) Culture substrates and conditions modulate CDM macro- and microarchitecture, deposition rate, and CDM biochemical and mechanical properties. (iv) Decellularization agents modify CDM composition and microstructure, while (v) chemical and physical modifications can be employed to control CDM composition and bioactive properties.

## 5.4.1 Cell source

The secretory profile of the cell source impacts the composition and architecture of the CDM, thus determining its functionality and application. Somatic, stem, and cancer cells are the major sources of ECM deposition and remodeling during tissue morphogenesis, healing, and disease [48]. Indeed, most CDM applications reported in the literature exploit these cell types. Cultured cells maintain the expression profile of the source tissue; therefore, under proper culture conditions, it is possible to derive niches of different anatomical sites to model tissue targets of metastasis [51].

CDMs derived from fibroblasts cell lines are selected for the robust deposition of fibrillary structures that model mesenchymal and stroma enriched microenvironments [51]. Primary fibroblasts secrete the highest amount of ECM in vitro, while immortalized cell lines, such as NIH-3T3, secrete matrices with more consistent properties. Fibroblasts generate a matrix rich in FN and collagen I and III, but generally lack more specific molecules, unless stimulated accordingly. Cukierman et al. focus on the effect of the phenotype of source cells on CDM properties, describing the use of fibroblasts, immortalized and from primary sources [52]. Fibroblasts from malignant and activated phenotypes were demonstrated to result in CDMs with distinct composition and microstructure. Fibroblast activation protein (FAP), a serine protease selectively produced by TAFs in over 90% of epithelial tumors, induces increased deposition of FN and collagens, in conjunction with thicker fibers, with higher degrees of alignment and crosslinking (Fig. 5.5) [53]. FAP enzymatic activity locally modifies stromal ECM components, thus facilitating the formation of a permissive microenvironment. Finally, the authors show that increased levels of FN and collagen fiber organization in FAP CDMs promote enhanced migration directionality and velocity of pancreatic cancer cells [53].

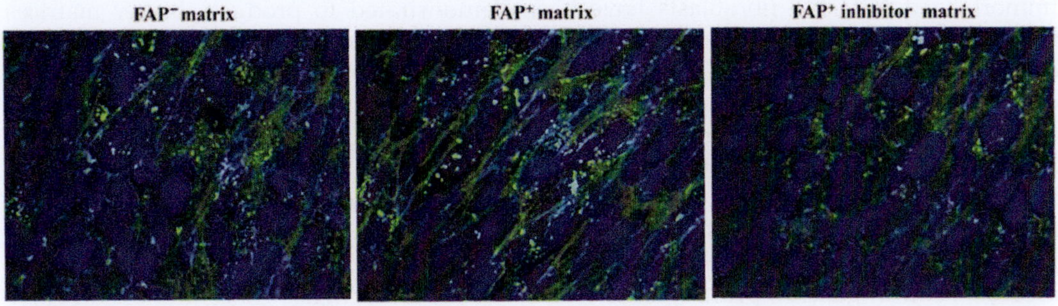


**FIGURE 5.5** Fibroblast-activated protein (FAP) expression in NIH-3T3 cells induces matrix fiber alignment (FN in green, DAPI in blue). Source: *Reprinted with permission from Lee H-O, Mullins SR, Franco-Barraza J, Valianou M, Cukierman E, Cheng JD. FAP-overexpressing fibroblasts produce an extracellular matrix that enhances invasive velocity and directionality of pancreatic cancer cells. BMC Cancer 2011;11:245.*

Smooth muscle cells deposit a highly stable matrix rich in elastin and GAG. Matsushima and Bogenmann [47] exploit these cells to compare the chondroitin sulfate: HA ratio in CDMs. Careful selection of the cell source and decellularization technique enabled an evaluation of the differences in CDM composition on NB differentiation, as previously mentioned. In another study, LOX, an enzyme that initiates covalent crosslinking of elastin precursors, increased elastin synthesis, and crosslinking in vitro [54].

Pathological niches can be obtained by exploiting TAFs and cancer cells. Depending on the cell secretome and malignancy, the matrix recapitulates the structural properties and composition of the tumoral ECM [46]. Cancer cells of increasing malignant phenotypes can produce in vitro CDMs that retain the tumor stromagenesis characteristic of the particular stage [51]. Hoshiba et al. reported a method to engineer matrices mimicking different stages of tumorigenesis by culturing cells of increasing breast and colon malignant phenotypes [46,48,50].

Finally, gene expression can be tuned to induce the deposition of exogenous molecules. For example, Higuchi et al. used transfected human embryonic kidney cells to deposit a CDM enriched in laminin [55]. Acting on the FAP protein, immortalized fibroblasts were shown to be transfected to secrete a matrix expressing comparable properties and functionalities to those obtained from fibroblasts of the innate active phenotype (i.e., TAFs) [53]. Furthermore, genetic engineering can also be exploited to introduce functional features from the researchers' perspective. For example, Erikson et al. developed a fibroblast cell line that secretes green fluorescein protein (GFP)-tagged FN, which facilitates the evaluation of ECM properties and cell–ECM interaction [56].

Overall, CDMs produced with different cell types can be used to mimic specific stages of tumor progression and cell–ECM interaction at primary or secondary sites of tumor invasion.

### 5.4.2 Culture medium composition

The conditions under which cells are kept in culture also influence the resulting CDM. Medium additives play crucial roles in determining the composition and quality of the deposited matrix. Few studies discuss the importance of serum origin.

Immortalized NIH-3T3 fibroblasts have been demonstrated to produce thicker matrices with calf serum. However, the use of fetal bovine serum resulted in the fibroblasts producing thinner matrices with higher stability [57]. Further modifications to the media can be exploited to stabilize and accelerate collagen deposition [51]. Soucy et al. report that ascorbate treatment of lung fibroblasts produce a twofold increase in CDM collagen I content [58]. The antioxidant and antiscorbutic activity of ascorbic acid is fundamental in stabilizing and stimulating collagen deposition in vitro [59]. Further studies confirmed that the ascorbic acid in variable concentrations is necessary to obtain matrices enriched in collagen I [60]. NIH-3T3 fibroblasts are commonly supplemented with $50\,\mu g\,mL^{-1}$ of ascorbic acid. In general, primary cells would benefit from higher concentrations at early time points; however, the optimal regime needs to be determined for each fibroblastic cell type [52,57]. Interestingly, Davidson et al. show that ascorbate treatment inhibited elastin deposition by 70% in skin fibroblasts [61]. Nonetheless, ascorbic acid displays low stability and oxidizes quickly under culture conditions; therefore, cells are exposed to alternated regimes of antioxidant activity over time, which can influence the ECM deposition rate and increase intra- and interbatch variabilities. More stable forms of ascorbic acid can be identified from food industry applications, where the vitamin is employed primarily in the form of sodium, calcium, and magnesium salts, as well as other vitamers encountered in nature. They can still act as cofactors of prolyl hydroxylase, contributing to protocollagen assembly, but present lower antiscurvy activity. Transforming growth factor β1 and other cytokines can foster collagen deposition [62]. Similarly, specific enzymes, such as LOX, are shown to foster elastin synthesis, crosslinking, and matrix deposition in adult rat aortic smooth muscle cells [54]. Lastly, a novel approach proposes the modification of culture media by the addition of inert macromolecules (10–500 kDa), such as dextran and commercial hyperbranched saccharides (e.g., Ficoll). As a result, cells sensed a denser microenvironment and secreted a more consistent CDM (Fig. 5.6) [63,64].

### 5.4.3 Culture substrates and conditions

Substantial evidence demonstrates the potential to control, design, and monitor cell response in CDMs with pre-defined characteristics to efficiently assay cell–ECM interactions. Moreover, independently modulating CDM morphology and composition is also possible [65]. A critical parameter for CDM production is the supporting substrate that holds and guides matrix deposition. FN and gelatin coatings are generally employed to ensure CDM attachment to the substrate, enhancing CDM quality and stability [57]. Other compounds can also be immobilized, which may depend on the cell used for matrix growth. For example, hepatocyte CDMs were grown on PVLA, a polymer containing galactose groups that enhance hepatocyte adhesion by specific surface receptors, and on poly (L-lysine) (PLL), which enhanced laminin and FN deposition, respectively [66]. Similarly, poly(octadecene-alt-maleic anhydride) (POMA) can be deposited in thin layers to functionalize glass surfaces. POMA favors the binding of FN and provides a stable attachment for cells for CDM growth [67]. Interestingly, specific coatings can be employed, which, under certain conditions, facilitate the CDM detachment. Some of these

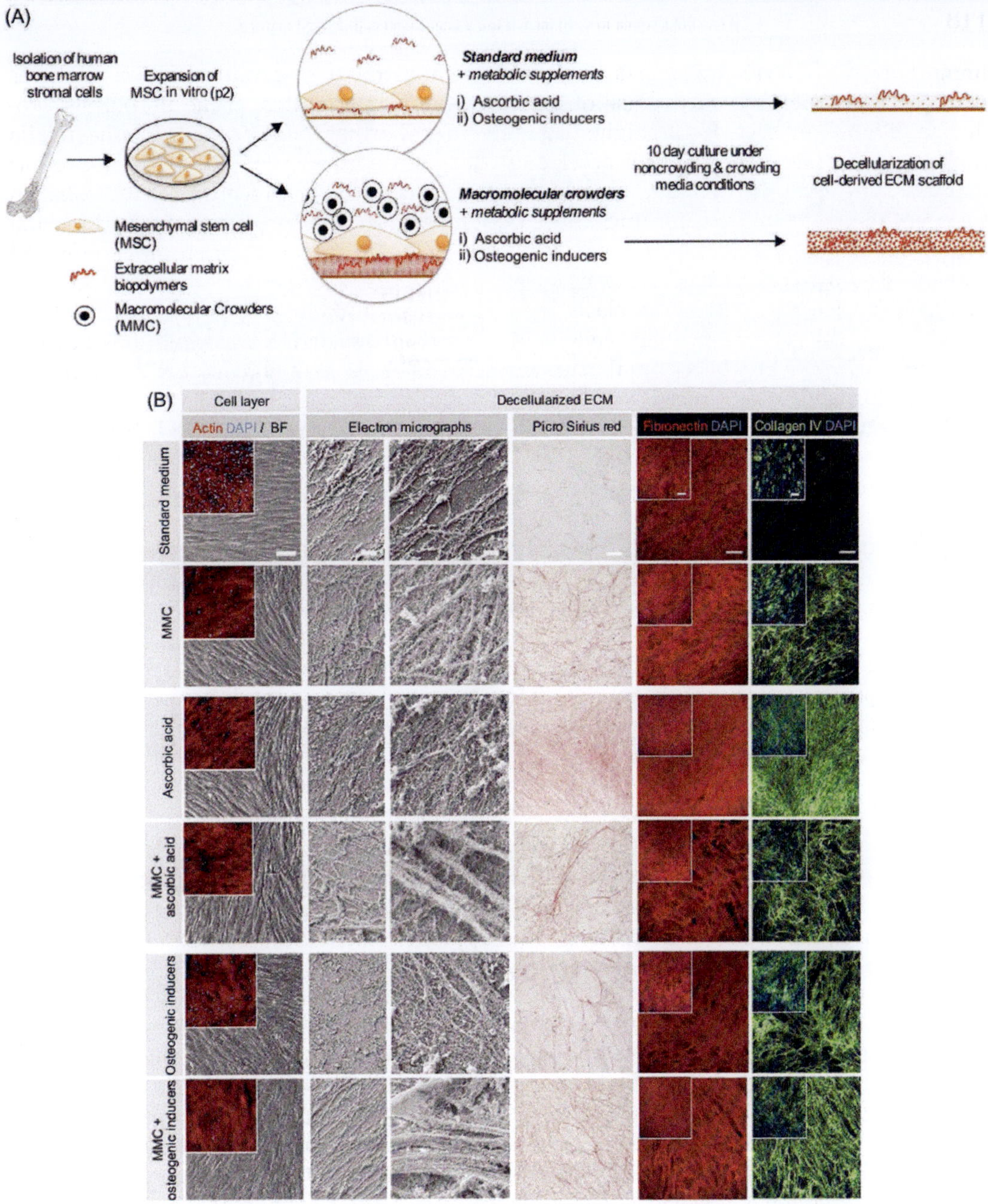


**FIGURE 5.6** Macromolecular crowding can enhance CDM deposition yield. (A) Mesenchymal stem cells (MSC) were cultured for 10 days under standard and macromolecular crowding culture conditions with additional media supplementation of ascorbic acid or osteogenic inducers. After decellularization, cell-free CDMs were investigated for their physicochemical characteristics. (B) CDMs exhibit characteristic differences dependent on culture conditions. (Left) Bright field images of the MSC cell layer at day 10 of culture prior to decellularization. Scale bar: 20 μm; actin and DAPI staining in small insets. (Right) Electron micrographs, Picro Sirius red (conventional histological staining for collagen structure), fibronectin, and collagen IV staining (intact cell layer prior to decellularization in small insets) of decellularized CDMs. Scale bars: 10 μm, 2 μm, 100 μm, respectively. Source: *Reprinted with permission from Prewitz MC, et al. Extracellular matrix deposition of bone marrow stroma enhanced by macromolecular crowding. Biomaterials 2015;73:60–9.*

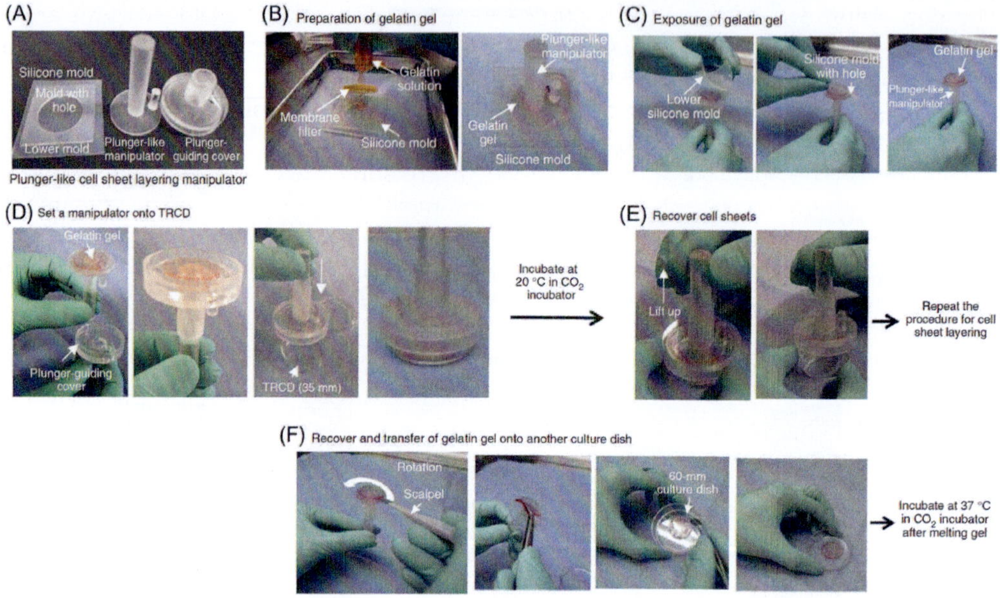


**FIGURE 5.7** Sequence showing the procedure for multiple stacking of detached cell sheets. A plunger-like manipulator is layered with a gelatin layer that temporarily binds to the cell sheet. Incubation at 20°C alters the hydrophylic properties of the culture substrate (hydrophobic–hydrophilic threshold temperature of 32°C). Therefore, the sheet can be recovered, easily manipulated, and stacked. Source: *Reprinted with permission from Haraguchi Y, et al. Fabrication of functional three-dimensional tissues by stacking cell sheets in vitro. Nat Protoc 2012;7 (5):850–8.*

coatings are indeed commercially available. As an example, UpCell employs a temperature-responsive polymer (PIPAAm) immobilized on the surface of culture ware [68]. Below 32°C, the polymer becomes hydrophilic, enabling the detachment of cell sheets in the form of thin films. Interestingly, this approach can be modified to stack several CDMs, resulting in thick 3D CDM sheets (Fig. 5.7) [69].

CDMs can also be deposited on inorganic materials as well as on synthetic- and natural-derived polymer templates, resulting in hybrid CDMs supported by 3D porous scaffolds [51]. This hybrid support provides enhanced mechanical strength, bioactive cues, and larger (3D) dimensionality of tens of micrometers [57]. By using supports with specific 3D architectures, CDMs of desirable shapes, such as patches, tubes, and capsules, can be engineered. The 3D printing further enhances the design capabilities of CDM-hybrid materials. Hoshiba and Gong employ polylactic acid (PLA) to print 3D scaffolds covered by a fibrosarcoma-derived CDM [70]. In some cases, the support material can be removed, resulting in CDMs with complex architectures. For example, polylactic-*co*-glycolic acid (PLGA) templates can be solubilized selectively with trisodium phosphate after CDM deposition, therefore obtaining pure CDMs of high dimensionality and porosity [71]. The techniques described above are particularly suitable for embedding CDMs in more complex culture systems, such as microfluidic systems, which can sustain long-term experiments.

The deposition of CDMs on 3D scaffolds generates the highest yield per unit area. It also enables the comparison between CDMs with synthetic scaffolds and ex vivo tissues that can be produced in much higher dimensionality. Nonetheless, culturing cells in sheets enables substrates to be exploited that are more reproducible and cost-effective. Microfabrication and soft lithography techniques can be readily applied in the laboratory to obtain specific patterns to be employed as culture substrates. Alignment, interspacing, length, and geometry of the substrate patterns influence the morphology and dynamics of the cultured cells [72]. Indeed, several studies show that cells polarize along the direction set by the underlying pattern, secreting ECM proteins following the cue [73]. This strategy was followed by Caballero et al. using micropatterned PDMS guiding templates (Fig. 5.8). These patterns induce NIH-3T3 fibroblasts to increase collagen packing and anisotropy, as well as FN bundle deposition. The resulting matrices mimic the architecture encountered in pathological ECM remodeling [65]. Similarly, aligned matrices were also obtained using human mesenchymal stem cells (hMSCs) cultured on parallel nanogrooves. The obtained results showed that the expression pattern of ECM-related genes by hMSCs on the obtained CDMs was altered [73]. Interestingly, we demonstrated this technique to be suitable in obtaining CDMs of different microarchitectures to study cell morphology and migration in 3D [74]. Finally, human fetal lung fibroblasts are employed to produce a fibrous 3D matrix capable of inducing tubulogenesis [75].

### 5.4.4 Decellularization agents

Decellularization is the most crucial aspect of CDM production since it defines the final micro/nanostructure and composition of the matrices [76]. CDMs can be stripped of cells via chemical agents, or enzymatic and physical treatments. Optimization of a protocol comprising subsequent or combined passages of these agents is fundamental to ensure CDMs with low DNA contamination and well-defined properties. Sodium dodecyl sulfate (SDS), sodium deoxycholate (SDC), and Triton X-100 are the most widely applied detergents [77]. Surfactants solubilize cytoplasmic and nuclear lipid membranes. Compared with ex vivo tissues, milder and shorter decellularization times can be employed; however, the protein structure is affected unless tightly controlled. SDS and SDC achieve higher extraction efficiencies, yet nonionic detergents (i.e., Triton X-100) are preferred as they disturb the ECM to a lesser degree. In addition to detergents, alkaline and acid solvents solubilize cytoplasmic components and disrupt nucleic acids. However, these reagents risk extracting GAGs from the matrices [7]. Alternatively, chelating agents, such as EDTA sequester metal ions, detach cells from the ECM, whereas hypotonic or hypertonic treatments can effectively lyse cells by osmotic shock but cannot remove the residues. Freeze–thaw cycles are also proposed to lyse cells. Nonetheless, ice crystal formation can disrupt the ECM structure. Each of these methods presents both advantages and disadvantages; thus, combinations of these agents are usually employed to ensure complete cell depletion and preserve ECM function. The most common buffers employed in CDM production include 0.1%–1% SDS and 0.1%–1% Triton X-100 alone or in combination with 20–50 mM $NH_4OH$ [77]. EDTA and deionized water can be added at the beginning of the treatment to ensure a better efficacy of the subsequent steps in removing nuclear and

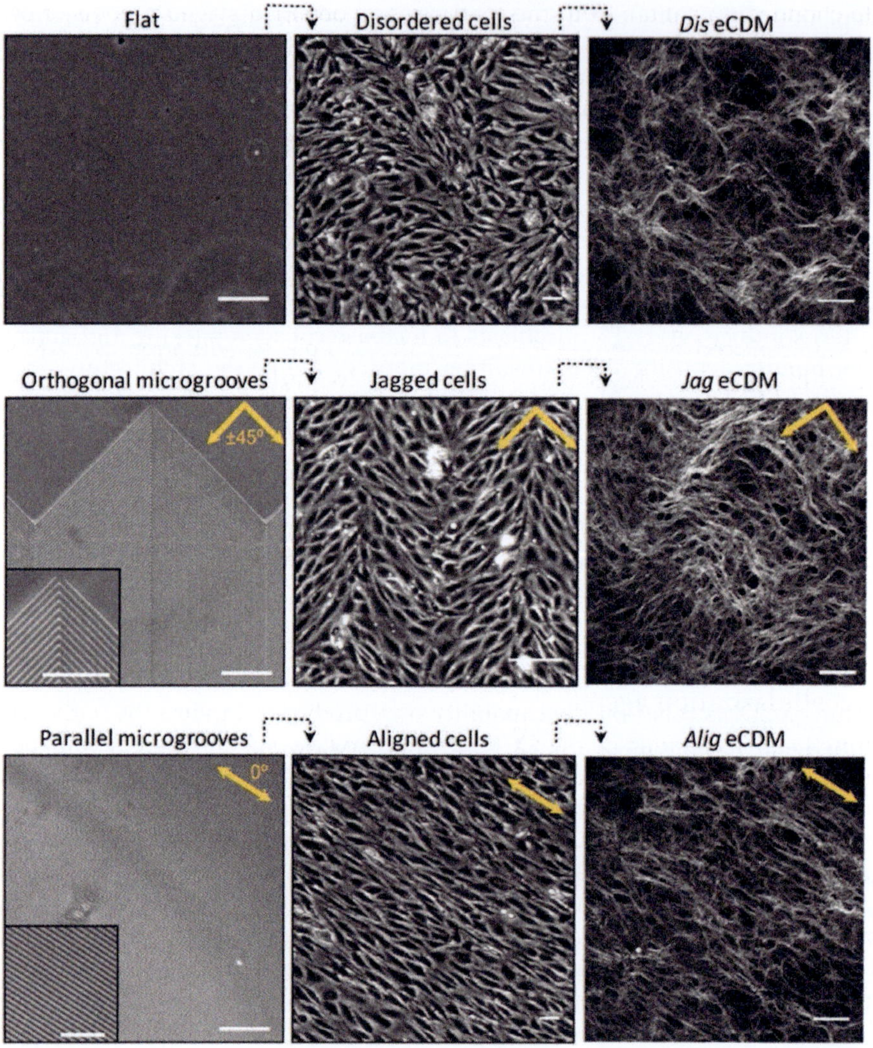


FIGURE 5.8 (Left) Phase-contrast microscopy images of PDMS microfabricated substrates. (Middle) NIH-3T3 fibroblasts seeded on top of the PDMS microfabricated substrates. (Right) SHG images of the extracted eCDMs showing different architectures. (Upper) Disordered, (middle) jagged-orthogonal, and (lower) aligned eCDMs, respectively. Scale bars: 100 μm. Source: *Reprinted with permission from Caballero D, Samitier J. Topological control of extracellular matrix growth: a native-like model for cell morphodynamics studies. ACS Appl Mater Interfaces 2017;9 (4):4159–70.*

cytosolic material [57]. Nucleases can be employed to remove nucleic acids efficiently without major drawbacks but cannot remove cytosolic components. Therefore, DNase and RNase are employed as the last step to further deplete nucleic contamination of the CDMs. Other enzymes can also be applied to cleave specific components, such as dispase, which disrupts FN and collagen IV selectively [78]. Chen et al. employed chondroitinase to deplete chondroitin sulfate from the matrices and obtained staged models of osteogenesis [79]. Furthermore, trypsin digestion and/or high-salt extraction have been used to modulate HA and chondroitin sulfate content in CDMs [47]. In conclusion, decellularization conditions require optimization to ensure that the CDM retains proper micro- and nanoarchitecture, functionality, and lacks contaminants [76].

### 5.4.5 Chemical and physical modifications

Modulatory proteins and soluble factors that physically absorb to the ECM are depleted by the decellularization process. Williams and Schwarzbauer show fibulin-1 and tenascin-C can be reintroduced to reproduce in vitro the complex modulation of FAK signaling [80]. Interestingly, click chemistry is employed to enable further CDM enrichment. Azide-modified monosaccharides (1,3,4,6-tetra-*O*-acetyl-*N*-azidoacetylgalactosamine: $Ac_4GalNAz$) are added to the culture media and integrated within the matrices without affecting their functionality [81]. An azide–alkyne reaction is then exploited to tune the bioactive properties of CDMs to provide a route to include additional fluorescent or bioactive moieties.

## 5.5 Cell-derived matrices and cell morphodynamic characterization

The value of CDMs relies on the capability of reproducing in vitro the native conformation of functional molecules of the ECM, while providing a system that can be assayed with high spatiotemporal resolution. Engineered CDMs can contribute to the understanding of cell–matrix signaling, cell morphology, and gene expression within the context of 3D ECMs and in vivo microenvironments [41]. Recent advances in imaging techniques and biophysical tools permit the processing and analysis of a broad range of assays that investigate the link between ECM architecture and composition, as well as tumoral processes, such as metastasis and tissue remodeling.

CDM characterization is crucial for cell morphodynamic studies. Preliminary experiments must be conducted to characterize CDM decellularization efficiency [7]. Nucleic material (i.e., DNA) is usually detected via cell nuclei staining (e.g., hematoxylin and Hoechst 33258/33343 staining; no signal upon visual observation) or by measuring DNA content (<200 bp DNA fragment length). Additional intracellular components are often monitored using fluorescent-labeled phalloidin for fibrous actin detection and immunocytochemistry of cytosolic proteins [77]. CDM composition can be assessed using histological techniques. Nonnucleic components can be detected through eosin staining, whereas alcian blue and toluidine blue staining are used for GAGs. Collagens are detected using sirius red and azan staining [70]. Similarly, immunochemistry is employed to detect specific ECM proteins [43]. Ultimately, mass spectrometry can provide exhaustive compositional analysis of the CDM [82].

Additionally, characterization of the mechanical properties of CDMs is important, since substrate stiffness affects cell growth and differentiation [83]. CDM bulk properties can be estimated with commercial microcompression machines and rheometers [84]. However, both techniques are insensitive to spatial heterogeneities and are destructive. Atomic force microscopy (AFM) is used to locally and quantitatively measure CDM elasticity at the micrometer length scale, exploiting micro- and nanoindentation probes [58]. Furthermore, magnetic and optical tweezers are utilized for microrheology experiments, whereby matrices are deformed at a constant force and reference bead displacement measured. Young's modulus can be easily calculated assuming that the restoring force is being exerted by a linear elastic medium, acceptable in the range of forces exerted by cells [85]. This approach provides much more realistic measures of inner CDM properties since AFM probes engage primarily with the fibers on the surface [86].

Real-time matrix visualization is central to detect morphodynamic cell processes and CDM remodeling. Collagen fibers can be visualized in live samples without fluorescent tagging through confocal reflectance microscopy or second harmonic generation (SHG). Confocal reflectance microscopy collects the backward scatter of light from the specimen and can be obtained using a standard confocal microscope [87]. Collagen fibers are noncentrosymmetric; therefore, they produce a strong and specific SHG signal upon excitation with a multiphoton laser beam in the infrared spectrum. Fibers can be detected in confocal setups at half of the incident wavelength exciting the signal [88]. Lastly, coherent anti-Stokes Raman scattering (CARS) imaging makes use of molecular vibrations and has been proposed to visualize collagen and elastin fibers and discern them from cellular structures [89].

Enrichment of cell culture media with fluorescence tagged proteins (e.g., GFP-FN) during CDM growth is proposed to obtain CDMs that incorporate contrasting agents [86]. Fluorescent CDMs can also be produced from cells that secrete fluorescence tagged matrix proteins [56]. Interestingly, real-time monitoring of fluorescent-tagged proteins incorporated in the CDM is reported for ECM remodeling [90]. Multichannel images can be integrated into co-localization and particle tracking analysis to extract information on matrix deformation and deposition, cell migration, attachment, and secretion [91]. Finally, image processing is fundamental for application reliability and reproducibility. Instruments employed in clinical histology laboratories usually provide incorporated software that enables complex visualization graphics (e.g., tomography-like, from slide reconstruction), integrated image processing, and analysis tools. Similarly, expensive imaging setups commonly provide high-quality analysis and post-processing software incorporated in the microscope main package. Many groups work on customizing the application of these tools to study novel morphometric markers [17]. Commercial and nonprofit systems are available to process images looking for quantification and localization of cell processes. Image Pro-plus software, CellProfiler, and ImageScope are diffused suites with multiple algorithms for image segmentation, morphometric quantification and plotting tools already incorporated. ImageJ is the most widely used software due to its open-source availability and broad applicability [92]. Regarding the quantification of CDM fiber network properties, specific approaches are employed. Fast Fourier transform analysis highlights fiber alignment as a bulk [93]. Furthermore, the CT-FIRE Matlab plugin combines transform filtering and automated tracking algorithms to extract fiber length, alignment, and interspacing distributions [94].

## 5.6 Cell-derived matrix capture relevant processes involved in neuroblastoma malignancy

The presence and functional interconnection of ECM constituents in CDMs make it possible to study the cohort effect of their morphological presentation on cell migration and mechanotransduction in NB regulation. As presented through this section, careful design of all the production steps is fundamental to engineer CDMs as reliable and reproducible in vitro models for tumor niches.

Experimental design driven by clinical observations can better define the properties and goals of the system. NB CDM models should provide a composition that resembles patient histological hallmarks, critical components of the nerve tissue, and most common metastatic sites. Thus, it could be feasible to assess the link between regulatory pathways (i.e., MYCN activity, GD2, integrin β1 and β3) with NB histopathological presentation and drug response. Following novel publications on NB stromal hallmarks, such as by Noguera et al., GAGs and collagenous structure (Fig. 5.1), together with ECM physicochemical properties, require further investigation [32]. Indeed, Kumar et al. demonstrate that substrate mechanical properties influence cell morphology, proliferation, and expression of MYCN in vitro [39]. In CDMs, we can highlight the effect of composition and architecture alterations encountered in NB histological sections on pathways that influence cancer progression and vice versa [5]. As initially explored by Matsushima and Bogenmann, neuroblasts are sensitive to GAG distribution in CDMs [47]. Interestingly, the authors explored different cell sources and decellularization techniques to prepare CDMs with graded GAGs ratios, which further affected cell behavior. Moreover, CDM generation with variable fiber anisotropy, density, and porosity can highlight ECM regulation of cell migration and survival (Fig. 5.8) [74]. Engineered NIH-3T3 or TAFs CDMs can be readily applied to manipulate matrix architecture (Figs. 5.8 and 5.5, respectively). In fact, in preliminary experiments, HR-NB models responded to the CDM 3D matrix, presenting a heterogeneous set of cell morphologies and migration patterns (data not published) [53,65]. Moreover, CDMs can contribute to determining which ECM elements regulate tumor progression [45]. Finally, CDMs can be generated to model healthy tissues commonly affected by NB progression, such as nerve tissue during tumor invasion and bone metastasis. In Section 5.4, specific examples show a set of techniques to define CDM composition, generating tissue-specific models of variable composition (i.e., laminin, collagen, and elastin enriched/deprived). Therefore, investigation of cell adaptation and metastasis initiation is possible, as well as the leading causes for NB-related death.

CDMs are employed to investigate how different stages of ECM remodulation alter cancer response to treatment, such as general antimetabolic, antimitotic chemotherapy, and NB maintenance therapies (Table 5.1) [44,47,48]. Due to NB heterogeneity and the relatively low number of cases, optimal therapeutic regimes could be identified upon investigating patient subgroups in vitro [22]. In this respect, CDMs provide multiple biomimetic ECM models that can mimic patient histopathological complexity. Furthermore, primary cells can be exploited to generate patient-specific models. Therefore, by recapitulating the complex and heterogeneous set of biochemical and biomechanical forces that the ECM exerts on malignant cells, we can better address drug efficacy and avoid resistance in NB.

## 5.7 Conclusions

In the last few decades, decellularization techniques have been studied extensively. Tissue and CDMs have gained immense interest due to the unique architecture and composition of the ECM that can be preserved and repurposed as a scaffold or in vitro model. Currently, CDMs are gaining interest for applications in cancer biology and drug discovery. This system tests cell–ECM interactions due to the presence and functional interconnection of ECM constituents and the possibility to study the cohort effect of their morphological presentation on cell migration and mechanotransduction. Engineering enables us to modulate CDM properties to mimic NB patient tissue and ease the collection of quantitative studies to boost our understanding of the role of ECM in NB progression, which can lead to the identification of novel therapeutic regimes and drug candidates.

## Acknowledgments

Figs. 5.2 and 5.4 were composed by the authors, modifying the sources from Servier Medical Art by Servier (licensed under a Creative Commons Attribution 3.0 Unported License CC BY 3.0).

This work was partially supported by Project (RTI2018-097038-B-C21), funded by the Spanish Ministry of Economy and Competitiveness under the National Program of R&D. E.A. acknowledges the support of "La Caixa" Banking Foundation (LCF/BQ/IN18/11660055) and European Union's Horizon 2020 research and innovation program under the Marie Skłodowska-Curie grant agreement No. 713673. J.M. acknowledges the Ramon y Cajal Program, Ministerio de Economia y Competitividad (RYC-2015-18357). D.C acknowledges the support of the Portuguese Foundation for Science and Technology (CEECIND/00352/2017 and 02/SAICT/2017-nº 028070-2MATCH) and H2020-EU (FoReCaST-668983). The Nanobioengineering group is supported by the Commission for Universities and Research of the Department of Innovation, Universities, and Enterprise of the Generalitat de Catalunya (2017 SGR 1079) and is part of the CERCA Program/Generalitat de Catalunya. CIBER-BBN is an initiative funded by the VI National R&D&I Plan 2008–11, Iniciativa Ingenio 2010, Consolider Program, CIBER Actions and financed by the Instituto de Salud Carlos III with assistance from the European Regional Development Fund.

## References

[1] Cohn SL, et al. The international neuroblastoma risk group (INRG) classification system: an INRG task force report. J Clin Oncol 2009;27(2):289–97.
[2] Matthay KK, et al. Neuroblastoma. Nat Rev Dis Primers 2016;2(1):16078.
[3] Eliceiri K, Conklin M, Bredfeldt J, Mackie T, Liu Y, Keely P. Automated quantification of aligned collagen for human breast carcinoma prognosis. J Pathol Inf 2014;5(1):28.
[4] Tadeo I, Berbegall AP, Navarro S, Castel V, Noguera R. A stiff extracellular matrix is associated with malignancy in peripheral neuroblastic tumors. Pediatr Blood Cancer 2017;64(9):e26449.
[5] Villasante A, et al. Vascularized tissue-engineered model for studying drug resistance in neuroblastoma. Theranostics 2017;7(17):4099–117.
[6] Curtin C, et al. A physiologically relevant 3D collagen-based scaffold–neuroblastoma cell system exhibits chemosensitivity similar to orthotopic xenograft models. Acta Biomater 2018;70:84–97.
[7] Hoshiba T. Decellularized extracellular matrix for cancer research. Materials (Basel) 2019;12(8):1311.
[8] White LJ, et al. The impact of detergents on the tissue decellularization process: a ToF-SIMS study. Acta Biomater 2017;50:207.
[9] Stiller CA, Parkin DM. International variations in the incidence of neuroblastoma. Int J Cancer 1992;52 (4):538–43.
[10] PDQ Pediatric Treatment Editorial Board. Neuroblastoma treatment (PDQ®): health professional version. Bethesda: National Cancer Institute; 2002.

[11] Johnsen JI, Dyberg C, Wickström M. Neuroblastoma—A neural crest derived embryonal malignancy. Front Mol Neurosci 2019;12:9.
[12] Simon T, Berthold F, Borkhardt A, Kremens B, De Carolis B, Hero B. Treatment and outcomes of patients with relapsed, high-risk neuroblastoma: results of German trials. Pediatr Blood Cancer 2011;56(4):578–83.
[13] Monclair T, et al. The International Neuroblastoma Risk Group (INRG) staging system: an INRG task force report. J Clin Oncol 2009;27(2):298–303.
[14] Speleman F, Park JR, Henderson TO. Neuroblastoma: a tough nut to crack. Am Soc Clin Oncol Educ B 2016;(36):e548–57.
[15] Brisse HJ, et al. Guidelines for imaging and staging of neuroblastic tumors: consensus report from the International Neuroblastoma Risk Group project. Radiology 2011;261(1):243–57.
[16] Smith V, Foster J. High-risk neuroblastoma treatment review. Child (Basel, Switzerland) 2018;5(9).
[17] Tadeo I, et al. Quantitative modeling of clinical, cellular, and extracellular matrix variables suggest prognostic indicators in cancer: a model in neuroblastoma. Pediatr Res 2014;75(2):302–14.
[18] Tadeo I, Berbegall AP, Escudero LM, Álvaro T, Noguera R. Biotensegrity of the extracellular matrix: physiology, dynamic mechanical balance, and implications in oncology and mechanotherapy. Front Oncol 2014;4:1–10.
[19] Pandian V, Ramraj S, Khan FH, Azim T, Aravindan N. Metastatic neuroblastoma cancer stem cells exhibit flexible plasticity and adaptive stemness signaling. Stem Cell Res Ther 2015;6(1):400.
[20] Borriello L, Seeger RC, Asgharzadeh S, DeClerck YA. More than the genes, the tumor microenvironment in neuroblastoma. Cancer Lett 2016;380(1):304–14.
[21] Cole KA, Maris JM. New strategies in refractory and recurrent neuroblastoma: translational opportunities to impact patient outcome. Clin Cancer Res 2012;18(9):2423–8.
[22] Fletcher JI, Ziegler DS, Trahair TN, Marshall GM, Haber M, Norris MD. Too many targets, not enough patients: rethinking neuroblastoma clinical trials. Nat Rev Cancer 2018;18(6):389–400.
[23] Burgos-Panadero R, Lucantoni F, Gamero-Sandemetrio E, de la Cruz-Merino L, Álvaro T, Noguera R. The tumour microenvironment as an integrated framework to understand cancer biology. Cancer Lett 2019;461:112–22.
[24] Yamada KM, Cukierman E. Modeling tissue morphogenesis and cancer in 3D. Cell 2007;130(4):601–10.
[25] Veta M, Pluim JPW, van Diest PJ, Viergever MA. Breast cancer histopathology image analysis: a review. IEEE Trans Biomed Eng 2014;61(5):1400–11.
[26] Levental KR, et al. Matrix crosslinking forces tumor progression by enhancing integrin signaling. Cell 2009;139(5):891–906.
[27] Nahon P, et al. Liver stiffness measurement in patients with cirrhosis and hepatocellular carcinoma: a case-control study. Eur J Gastroenterol Hepatol 2009;21(2):214–19.
[28] Huang S, Ingber DE. A non-genetic basis for cancer progression and metastasis: self-organizing attractors in cell regulatory networks. Breast Dis 2006;26:27–54.
[29] Hoyt K, et al. Tissue elasticity properties as biomarkers for prostate cancer. Cancer Biomarkers 2008;4(4–5):213–25.
[30] Oxford JT, Reeck JC, Hardy MJ. Extracellular matrix in development and disease. Int J Mol Sci 2019;20(1):1–24.
[31] Ingber DE. Cancer as a disease of epithelial-mesenchymal interactions and extracellular matrix regulation. Differentiation 2002;70(9–10):547–60.
[32] Noguera R, Nieto OA, Tadeo I, Fariñas F, Alvaro T. Extracellular matrix, biotensegrity and tumor microenvironment. An update and overview. Histol Histopathol 2012;27(6):693–705.
[33] Liu S, Tian Y, Chlenski A, Yang Q, Salwen HR, Cohn SL. 'Cross-talk' between Schwannian stroma and neuroblasts promotes neuroblastoma tumor differentiation and inhibits angiogenesis. Cancer Lett 2005;228(1–2):125–31.
[34] Chlenski A, et al. SPARC is a key Schwannian-derived inhibitor controlling neuroblastoma tumor angiogenesis. Cancer Res 2002;62(24):7357–63.
[35] Ren Y-J, et al. Enhanced differentiation of human neural crest stem cells towards the Schwann cell lineage by aligned electrospun fiber matrix. Acta Biomater 2013;9(8):7727–36.
[36] Li GN, Livi LL, Gourd CM, Deweerd ES, Hoffman-Kim D. Genomic and morphological changes of neuroblastoma cells in response to three-dimensional matrices. Tissue Eng 2007;13(5):1035–47.

[37] Zeine R, et al. Presence of cancer-associated fibroblasts inversely correlates with Schwannian stroma in neuroblastoma tumors. Mod Pathol 2009;22(7):950–8.
[38] Favrot MC, et al. Expression of integrin receptors on 45 clinical neuroblastoma specimens. Int J Cancer 1991;49(3):347–55.
[39] Lam WA, Cao L, Umesh V, Keung AJ, Sen S, Kumar S. Extracellular matrix rigidity modulates neuroblastoma cell differentiation and N-myc expression. Mol Cancer 2010;9(1):35.
[40] Amoroso L, Haupt R, Garaventa A, Ponzoni M. Investigational drugs in phase II clinical trials for the treatment of neuroblastoma. Expert Opin Investig Drugs 2017;26(11):1281–93.
[41] Kaukonen R, Jacquemet G, Hamidi H, Ivaska J. Cell-derived matrices for studying cell proliferation and directional migration in a complex 3D microenvironment. Nat Protoc 2017;12(11):2376–90.
[42] Scherzer MT, et al. Fibroblast-derived extracellular matrices: an alternative cell culture system that increases metastatic cellular properties. PLoS One 2015;10(9):1–17.
[43] Franco-Barraza J, Beacham DA, Amatangelo MD, Cukierman E. Preparation of extracellular matrices produced by cultured and primary fibroblasts. Curr Protoc Cell Biol 2016;71:10.9.1–10.9.34.
[44] Serebriiskii I, Castelló-Cros R, Lamb A, Golemis EA, Cukierman E. Fibroblast-derived 3D matrix differentially regulates the growth and drug-responsiveness of human cancer cells. Matrix Biol 2008;27 (6):573–85.
[45] Castelló-Cros R, Khan DR, Simons J, Valianou M, Cukierman E. Staged stromal extracellular 3D matrices differentially regulate breast cancer cell responses through PI3K and beta1-integrins. BMC Cancer 2009;9:94.
[46] Hoshiba T, Tanaka M. Breast cancer cell behaviors on staged tumorigenesis-mimicking matrices derived from tumor cells at various malignant stages. Biochem Biophys Res Commun 2013;439(2):291–6.
[47] Matsushima H, Bogenmann E. Modulation of neuroblastoma cell differentiation by the extracellular matrix. Int J Cancer 1992;51(5):727–32.
[48] Hoshiba T, Tanaka M. Optimization of the tissue source, malignancy, and initial substrate of tumor cell-derived matrices to increase cancer cell chemoresistance against 5-fluorouracil. Biochem Biophys Res Commun 2015;457(3):353–7.
[49] Sansing HA, et al. Integrin αβ1, αvβ, α6β effectors p130Cas, Src and talin regulate carcinoma invasion and chemoresistance. Biochem Biophys Res Commun 2011;406(2):171–6.
[50] Hoshiba T, et al. Decellularized extracellular matrix as an *in vitro* model to study the comprehensive roles of the ECM in stem cell differentiation. Stem Cell Int 2016;2016:1–10.
[51] Hoshiba T. Cultured cell-derived decellularized matrices: a review towards the next decade. J Mater Chem B 2017;5(23):4322–31.
[52] Beacham DA, Amatangelo MD, Cukierman E. Preparation of extracellular matrices produced by cultured and primary fibroblasts. Curr Protoc Cell Biol 2006;33(1):10.9.1–10.9.21.
[53] Lee H-O, Mullins SR, Franco-Barraza J, Valianou M, Cukierman E, Cheng JD. FAP-overexpressing fibroblasts produce an extracellular matrix that enhances invasive velocity and directionality of pancreatic cancer cells. BMC Cancer 2011;11:245.
[54] Kothapalli CR, Ramamurthi A. Lysyl oxidase enhances elastin synthesis and matrix formation by vascular smooth muscle cells. J Tissue Eng Regen Med 2009;3(8):655–61.
[55] Higuchi Y, et al. Synthesized basement membranes direct the differentiation of mouse embryonic stem cells into pancreatic lineages. J Cell Sci 2010;123(Pt 16):2733–42.
[56] Ohashi T, Kiehart DP, Erickson HP. Dynamics and elasticity of the fibronectin matrix in living cell culture visualized by fibronectin-green fluorescent protein. Proc Natl Acad Sci U S A 1999;99:5139–43.
[57] Castelló-Cros R, Cukierman E. Stromagenesis during tumorigenesis: characterization of tumor-associated fibroblasts and stroma-derived 3D matrices. Methods Mol Biol 2009;522:275.
[58] Soucy PA, Werbin J, Heinz W, Hoh JH, Romer LH. Microelastic properties of lung cell-derived extracellular matrix. Acta Biomater 2011;7(1):96–105.
[59] Chan D, Lamande SR, Cole WG, Bateman JF. Regulation of procollagen synthesis and processing during ascorbate-induced extracellular matrix accumulation *in vitro*. Biochem J 1990;269(1):175–81.
[60] Grinnell F, Fukamizu H, Pawelek P, Nakagawa S. Collagen processing, crosslinking, and fibril bundle assembly in matrix produced by fibroblasts in long-term cultures supplemented with ascorbic acid. Exp Cell Res 1989;181(2):483–91.

[61] Davidson JM, LuValle PA, Zoia O, Quaglino D, Giro M. Ascorbate differentially regulates elastin and collagen biosynthesis in vascular smooth muscle cells and skin fibroblasts by pretranslational mechanisms. J Biol Chem 1997;272(1):345–52.

[62] Qureshi OS, et al. An immunofluorescence assay for extracellular matrix components highlights the role of epithelial cells in producing a stable, fibrillar extracellular matrix. Biol Open 2017;6(10):1423–33.

[63] Prewitz MC, et al. Extracellular matrix deposition of bone marrow stroma enhanced by macromolecular crowding. Biomaterials 2015;73:60–9.

[64] Benny P, Raghunath M. Making microenvironments: A look into incorporating macromolecular crowding into in vitro experiments, to generate biomimetic microenvironments which are capable of directing cell function for tissue engineering applications. J Tissue Eng 2017;8. p. 204173141773046.

[65] Caballero D, Samitier J. Topological control of extracellular matrix growth: a native-like model for cell morphodynamics studies. ACS Appl Mater Interfaces 2017;9(4):4159–70.

[66] Hoshiba T, Cho CS, Murakawa A, Okahata Y, Akaike T. The effect of natural extracellular matrix deposited on synthetic polymers on cultured primary hepatocytes. Biomaterials 2006;27(26):4519–28.

[67] Prewitz MC, et al. Tightly anchored tissue-mimetic matrices as instructive stem cell microenvironments. Nat Methods 2013;10(8):788–94.

[68] Kobayashi J, Okano T. Fabrication of a thermoresponsive cell culture dish: a key technology for cell sheet tissue engineering. Sci Technol Adv Mater 2010;11(1):014111.

[69] Haraguchi Y, et al. Fabrication of functional three-dimensional tissues by stacking cell sheets in vitro. Nat Protoc 2012;7(5):850–8.

[70] Hoshiba T, Gong J. Fabrication of cell-derived decellularized matrices on three-dimensional (3D)-printed biodegradable polymer scaffolds. Microsyst Technol 2018;24(1):613–17.

[71] Lu H, Hoshiba T, Kawazoe N, Koda I, Song M, Chen G. Cultured cell-derived extracellular matrix scaffolds for tissue engineering. Biomaterials 2011;32(36):9658–66.

[72] Kim DH, Provenzano PP, Smith CL, Levchenko A. Matrix nanotopography as a regulator of cell function. J Cell Biol 2012;197(3):351–60.

[73] Ozguldez HO, Cha J, Hong Y, Koh I, Kim P. Nanoengineered, cell-derived extracellular matrix influences ECM-related gene expression of mesenchymal stem cells. Biomater Res 2018;22(1):1–9.

[74] Caballero D, Palacios L, Freitas PP, Samitier J. An interplay between matrix anisotropy and actomyosin contractility regulates 3D-directed cell migration. Adv Funct Mater 2017;27(35):1702322.

[75] Soucy PA, Hoh M, Heinz W, Hoh J, Romer L. Oriented matrix promotes directional tubulogenesis. Acta Biomater 2015;11(1):264–73.

[76] Crapo PM, Gilbert TW, Badylak SF. An overview of tissue and whole organ decellularization processes. Biomaterials 2011;32(12):3233–43.

[77] Lu H, Hoshiba T, Kawazoe N, Chen G. Comparison of decellularization techniques for preparation of extracellular matrix scaffolds derived from three-dimensional cell culture. J Biomed Mater Res Part A 2012;100A (9):2507–16.

[78] Hopkinson A, et al. Optimization of amniotic membrane (AM) denuding for tissue engineering. Tissue Eng Part C Methods 2008;14(4):371–81.

[79] Hoshiba T, Kawazoe N, Chen G. Mechanism of regulation of PPARG expression of mesenchymal stem cells by osteogenesis-mimicking extracellular matrices. Biosci Biotechnol Biochem 2011;75(11):2099–104.

[80] Williams SA, Schwarzbauer JE. A shared mechanism of adhesion modulation for tenascin-C and fibulin-1. Mol Biol Cell 2009;20(4):1141–9.

[81] Ruff SM, et al. clickECM: development of a cell-derived extracellular matrix with azide functionalities. Acta Biomater 2017;52:159–70.

[82] Aguado BA, et al. Extracellular matrix mediators of metastatic cell colonization characterized using scaffold mimics of the pre-metastatic niche. Acta Biomater 2016;33:13–24.

[83] Discher DE, Janmey P, Wang YL. Tissue cells feel and respond to the stiffness of their substrate. Science 2005;310(5751):1139–43.

[84] Zaman MH, et al. Migration of tumor cells in 3D matrices is governed by matrix stiffness along with cell-matrix adhesion and proteolysis. Proc Natl Acad Sci U S A 2006;103(29):10889–94.

[85] Laurent VM, et al. Assessment of mechanical properties of adherent living cells by bead micromanipulation: comparison of magnetic twisting cytometry vs optical tweezers. J Biomech Eng 2002;124(4):408–21.

[86] Godeau A. Cyclic contractions contribute to 3D cell motility. Biological Physics [physics.bio-ph]. Université de Strasbourg; 2016 [English].

[87] Friedl P, Maaser K, Klein CE, Niggemann B, Krohne G, Zänker KS. Migration of highly aggressive MV3 melanoma cells in 3-dimensional collagen lattices results in local matrix reorganization and shedding of alpha2 and beta1 integrins and CD44. Cancer Res. 1997;57(10):2061–70.

[88] Pfeffer CP, Olsen BR, Ganikhanov F, Légaré F. Multimodal nonlinear optical imaging of collagen arrays. J Struct Biol 2008;164(1):140–5.

[89] Wang H-W, Le TT, Cheng J-X. Label-free imaging of arterial cells and extracellular matrix using a multimodal CARS microscope. Opt Commun 2008;281(7):1813–22.

[90] Sivakumar P, Czirok A, Rongish BJ, Divakara VP, Wang YP, Dallas SL. New insights into extracellular matrix assembly and reorganization from dynamic imaging of extracellular matrix proteins in living osteoblasts. J Cell Sci 2006;119(7):1350–60.

[91] Keely P, Nain A. Capturing relevant extracellular matrices for investigating cell migration. F1000Res 2015;4:1408.

[92] Schneider CA, Rasband WS, Eliceiri KW. NIH Image to ImageJ: 25 years of image analysis. Nat Methods 2012;9(7):671–5.

[93] Sun M, Bloom AB, Zaman MH. Rapid quantification of 3D collagen fiber alignment and fiber intersection correlations with high sensitivity. PLoS One 2015;10(7):e0131814.

[94] Bredfeldt JS, et al. Computational segmentation of collagen fibers from second-harmonic generation images of breast cancer. J Biomed Opt 2014;19(1):016007.

CHAPTER

# 6

# 3D culture systems as models for solid tumors and cancer metabolism

*Sónia Pires Celeiro*[1,2], *Fátima Baltazar*[1,2] *and Marta Viana-Pereira*[1,2]

[1]Life and Health Science Research Institute (ICVS), School of Medicine, University of Minho, Braga, Portugal [2]ICVS/3B's—PT Government Associate Laboratory, Braga/Guimarães, Portugal

## Abbreviations

| | |
|---|---|
| **ADP** | Adenosine diphosphate |
| **AMP** | Adenosine monophosphate |
| **AMPK** | AMP kinase |
| **ATP** | Adenosine triphosphate |
| **CAF** | Cancer-associated fibroblasts |
| **CSC** | Cancer stem cell |
| **CTLA-4** | Cytotoxic T-lymphocyte-associated protein 4 |
| **DNA** | Deoxyribonucleic acid |
| **ECM** | Extracellular matrix |
| **EGF** | Epidermal growth factor |
| **EMA** | European Medicines Agency |
| **ETC** | Electron transport chain |
| **FDA** | Food and Drug Administration |
| **GLUT-1** | Glucose transporter-1 |
| **HER-2** | Human epidermal growth factor receptor-2 |
| **HGF** | Hepatocyte growth factor |
| **HIF** | Hypoxia-inducible factors |
| **HIF-1α** | Hypoxia-inducible factor-1α |
| **IL-6** | Interleukin-6 |
| **iNOS** | Inducible-nitric oxide synthase |
| **LDH** | Lactate dehydrogenase |
| **LDHA** | Lactate dehydrogenase-A |

*Biomaterials for 3D Tumor Modeling*
DOI: https://doi.org/10.1016/B978-0-12-818128-7.00006-X

| | |
|---|---|
| **MCT** | Monocarboxylate transporter |
| **MMPs** | Matrix metalloproteinases |
| **NK** | Natural killer |
| **NO** | Nitric oxide |
| **OXPHOS** | Oxidative phosphorylation |
| **PBS** | Phosphate-buffered saline |
| **PD-1** | Programmed cell death 1 |
| **PD-L1** | Programmed cell death-ligand 1 |
| **ROS** | Reactive oxygen species |
| **TAM** | Tumor-associated macrophage |
| **TCA** | Tricarboxylic acid |
| **TGF-β** | Transforming growth factor-β |
| **TILs** | Tumor-infiltrating lymphocytes |
| **TME** | Tumor microenvironment |
| **ULA** | Ultra-low-attachment |
| **VEGF** | Vascular endothelial growth factor |
| **2D** | Two-dimensional |
| **3D** | Three-dimensional |

## 6.1 Introduction

Cancer is the second leading cause of death worldwide and, in 2030, the incidence and mortality numbers are expected to be approximately 26 and 17 million, respectively [1,2]. The rise of these numbers and the great absence of curative therapies, for this disease, confirm the urgent need to seek for more efficient therapeutic strategies.

Despite the innumerous studies and rapid therapeutic advances in the oncobiology field, the development of treatment resistance is a major concern and, in several cancer types, the mechanisms that lead to tumor recurrence and progression are not completely understood. Classically, anticancer treatments are based on surgery, radiotherapy, chemotherapy, or their combinations, and those are still the most commonly used clinical approaches [3]. Nevertheless, both radiotherapy and chemotherapy, have low specificity and severe side effects and their efficacy is limited, by cancer resistance and disease progression [4–6]. Thus, innumerous studies aim to evaluate and develop more efficient therapeutic strategies to overcome solid tumors resistance [6].

Several mechanisms of therapeutic resistance in cancer have been described. Hypoxia, acidity, cell-cycle arrest, DNA mutations, cell–cell, and cell–extracellular matrix (ECM) interactions as well as cancer cells metabolic alterations, are some of the resistance mechanisms that promote drug inhibition or degradation [7,8]. The tumor microenvironment (TME) of a cancer has a heterogenic cellular constitution, including malignant cells, immune cells, endothelial cells, and cancer-associated fibroblasts (CAF) [9]. This heterogenic cellular constitution and their signaling cross-talk contribute to tumor angiogenesis, proliferation, invasion, metastasis, and therapeutic resistance [7,10]. In order to survive and sustain their malignant features, cancer cells reprogram their metabolism, increasing the metabolic fluxes to generate energy, redox equivalents, and macromolecules [11].

Cancer metabolic profiles are heterogeneous and different for each solid tumor depending on the TME, cellular origin, genetic landscape, among other factors [12,13].

Glucose metabolism is the main deregulated metabolic pathway in cancer. The cancer cells of the inner region of a solid tumor, under a selective pressure of low levels of oxygen, have to perform an anaerobic metabolism of glucose, in order to obtain energy [7]. Interestingly, Otto Warburg also verified that even under aerobic conditions, cancer cells favor a glycolytic metabolism (Warburg effect) [14]. Glycolytic metabolism influences the tumor local inflammatory status to support procarcinogenic processes [15]. Moreover, increasing evidence confirms that the metabolic reprogramming of cancer cells can induce chemo-, radio-, and immunotherapy resistance [16,17]. Altogether these demonstrate the potential for targeting dysregulated metabolic pathways, and the importance of including cancer metabolism evaluation in the development of effective therapies to fight solid tumors resistance.

Anticancer compounds, like all other medicines, need to respect the requirements established by the regulatory agencies like Food and Drug Administration (FDA) and European Medicines Agency (EMA). The drug development process involves in vitro and in vivo efficacy and toxicity assays and other preclinical studies, before clinical trials evaluation [18]. Researchers commonly use two-dimensional (2D) cell cultures as the in vitro model for drug screening. However, these cell culture systems are unable to replicate different characteristics of cancer in vivo and, therefore, numerous ineffective drugs proceed to in vivo testing, which implies an overuse of animals and a higher cost of the drug development process [19]. In order to reduce these costs and the number of animals used, in vitro models capable of more accurately reproducing an in vivo system are needed [20].

Three-dimensional (3D) in vitro cell culture models, that have been increasingly explored in recent decades, have allowed scientists to better resemble the in vivo tumor. These culture models can better reproduce the TME in terms of cellular heterogeneity, nutrient gradient, cell–cell interaction, and oxygen gradient [21]. Therefore, 3D culture systems seem to be a bridge between the 2D cell culture models and in vivo studies, allowing researchers to more accurately evaluate cancer biology and cancer metabolism and consequently obtain better outcomes in terms of cancer behavior and therapeutic strategies.

This chapter describes the TME and the metabolic heterogeneity of solid tumors, exploring the 2D in vitro model limitations and the potential of the 3D cell culture systems.

## 6.2 Solid tumors: tumor microenvironment and tumorigenesis

Tumorigenesis is a complex process, that leads to the deregulation of a variety of mechanisms for tumor advantage. Hanahan and Weinberg, in 2011, suggested the 10 hallmarks of cancer, distinct mechanisms essential for tumor growth, and metastatic spreading [22]. The 10 hallmarks of cancer are promotion of tumor inflammation, evasion to immune destruction, reprogramming of energy metabolism, genome instability and mutation, evasion from growth suppressors, activation of invasion and metastasis, angiogenesis induction, replicative immortality, resistance to cell death, and sustained proliferative signaling [22]. Altogether, these hallmarks demonstrate the complexity of this disease. Moreover, cancer stem cells (CSCs) have been identified in different solid tumors including brain, breast, lung, gastric, prostate, and pancreatic cancer, increasing cancer heterogeneity and complexity [23,24]. CSCs are characterized by distinct properties like self-renewal and differentiation capacity. It has also been demonstrated that CSC can rebuild the tumor after chemotherapy, associating these subpopulation of cancer cells with resistance to therapy [25].

The TME of a solid tumor has a heterogeneous cellular constitution of malignant (tumor cells) and nonmalignant cells [9]. Nonmalignant cells found in the TME are: immune cells [T lymphocytes, B lymphocytes, and tumor-associated macrophages (TAMs)]; CAF; endothelial cells; and pericytes (Fig. 6.1) [9,26]. Despite being nonmalignant, the role of these cells in the TME is associated to a protumorigenic phenotype [10]. The different cell types of the TME interact through the secretion of cytokines and growth factors to promote cell proliferation, invasion, therapeutic resistance, and other malignant features (Fig. 6.1). For example, the secretion of transforming growth factor-β (TGF-β) and hepatocyte growth factor (HGF) by CAF cells in the TME prompt cancer cell invasion [27]. Moreover, within the immune cells several studies show that TAM are associated to angiogenesis and the degradation of the ECM through vascular endothelial growth factor (VEGF) and matrix metalloproteinases (MMPs) secretion, respectively [28]. VEGF acts on endothelial cells to form new blood vessels for tumor supply of oxygen and nutrients and, therefore, enhance tumor growth [29]. Interleukin-6 (IL-6), a proinflammatory cytokine, affects several tumorigenesis processes and treatment efficacy [30]. This cytokine is released in TME by different cell types, including cancer cells, CAFs, and immune cells. In gastric cancer, IL-6 secretion by CAFs is associated with chemotherapeutic resistance through JAK1-STAT3 pathway activation [31]. Thus, this TME interaction deregulates a variety of mechanisms for tumorigenesis. The presence of nonmalignant cells also influences the TME conditions in terms of oxygen and nutrients delivery to tumor cells. Alterations of the microenvironment conditions, like pH level [32], nutrient supply [33], reactive oxygen

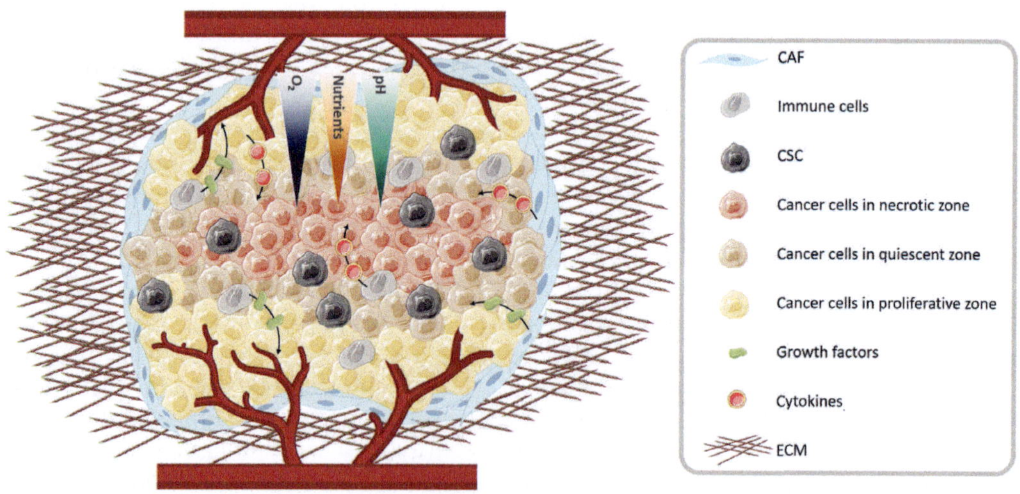


**FIGURE 6.1** Cell heterogeneity, signaling, and biological zones of a solid tumor. Different cell types are found on TME (cancer cells, CSC, CAF, and immune cells) that interact through the secretion of cytokines and growth factors to promote tumor progression. Tumor development leads to oxygen, nutrient, and pH gradients responsible for the distinct biological zones (proliferative, quiescent, and necrotic) verified in a tumor mass.

species (ROS) [34], and hypoxia [35] can occur, and consequently, malignant cells need to adapt to these microenvironmental stresses for cancer survival and growth.

At the starting point of solid tumors development, the vasculature of the host organ supplies the tumor with oxygen and nutrients and is responsible for the elimination of metabolic wastes [36]. However, with tumor expansion in size the host vasculature is insufficient and consequently, the inner tumoral regions have lower levels of oxygen and nutrients and higher levels of metabolic wastes. These gradients are responsible for proliferative, quiescent and necrotic zones verified in a tumor (Fig. 6.1). At this stage, tumor cells gain the capacity to enhance angiogenesis to build new tumoral vasculature [36]. However, it remains inadequate for the solid tumor expansion demands, leading to an oxygen gradient and a hypoxic environment of the inner tumor cells [37]. To adapt to this microenvironmental stresses tumoral cells alter their metabolic profile [14]. This metabolic reprogramming capacity, a hallmark of cancer, is related to tumor progression and anticancer therapeutic resistance [38].

## 6.3 Cancer metabolism: influence in tumor microenvironment

Cancer cells acquire different capabilities that favor malignant growth, progression, and survival. To better understand these tumorigenic capabilities 10 hallmarks of cancer were established, among which figures metabolic reprogramming [22].

As described before, cancer cells can modify their metabolic programming to fulfill the demands of a growing tumor and adapt to the TME [39]. Some of these metabolic alterations are present across most types of cancer and their study allows scientists to better understand how metabolic reprogramming is related to cancer development and progression [40]. Moreover, the study of cancer metabolism allows the identification of cancer-specific metabolic changes and the development of targeted therapeutic strategies for patients. Cancer metabolism is an attracting field among oncobiological research, with great potential for cancer therapy.

In cancer, glucose metabolism is the main deregulated metabolic pathway [15]. Physiologically, normal cells produce their energy through oxidative phosphorylation (OXPHOS), in the presence of oxygen, and perform glycolysis, by uptaking glucose and producing lactate, in response to hypoxia. However, under pathological circumstances, cancer cells favor the uptake of glucose and production of lactate both in normoxic and hypoxic conditions, the well-recognized aerobic glycolysis or Warburg effect [14,40,41]. Warburg effect is the most common example of a reprogrammed metabolic pathway in cancer. Most of the tumor cells produce their energy through high-rate glycolysis followed by lactic acid fermentation. Glycolysis is a less efficient metabolic pathway, in the sense that it only produces two adenosine triphosphate (ATP) molecules per molecule of glucose; however it allows tumor cells to produce energy in a rapid way [42,43]. This metabolic switch to a glycolytic profile is very important in the development of solid tumors.

As described before, the expansion of the solid tumor mass promotes lower levels of oxygen and nutrients in their central regions and, consequently, tumor cells reprogram their metabolic activity to sustain this tumor anabolic growth [34,39]. The hypoxic environment upregulates the expression of hypoxia-inducible factors (HIF) and these molecules can influence malignant features, such as angiogenesis and cell proliferation capacity [44]. Moreover, in most of the solid tumors, the metabolic shift from OXPHOS to glycolysis seems to be regulated by hypoxia-inducible factor-1α (HIF-1α) that upregulates the expression of glucose transporter-1 (GLUT-1) and glycolytic enzymes to increase glycolysis rate [44,45]. Higher glycolytic rates, in turn, provide glycolytic intermediates that undergo into subsidiary metabolic pathways and generate nucleotides, lipids, amino acids, and NADPH de novo that benefit cell proliferation [40,46]. This metabolic pathway converts glucose into pyruvate and the fate of this metabolite depends on the supply of oxygen in the TME. In the presence of oxygen, pyruvate is oxidized to produce acetyl-coenzyme A for the tricarboxylic acid (TCA) cycle [45]. On the other hand, under low-oxygen levels, pyruvate is reduced to lactate [45], through lactate dehydrogenase-A (LDHA) [40]. The hypoxic environment and HIF-1α upregulation in solid tumors, leads to LDHA expression, enhancing the production of lactate [47], and thus, pyruvate oxidation is compromised and TCA cycle downregulated [48].

The acid overproduction leads to necrosis or apoptosis. Thus, to maintain the glycolytic phenotype and resist the acidic environment (high-lactate levels), cancer cells upregulate monocarboxylate transporters (MCTs). MCTs are responsible not only for pH regulation, but also for the co-transport of lactate with a proton ($H^+$), which consequently promotes glycolysis and continuous conversion of pyruvate to lactate [49,50]. MCT1 and MCT4 isoforms are overexpressed in many cancers, being associated with cancer aggressiveness [50]. Moreover, HIF-1α expression also contributes to continuous glycolytic metabolism and resistance to the acidic environment by MCT4 upregulation [49]. Lactate accumulation in TME through a higher glycolysis rate promotes an acidic environment (lactic acidosis) that enhances cancer aggressiveness, more specifically, malignant cells proliferation and invasive capacity, and influences therapeutic efficacy (Fig. 6.2) [51,52].

Some drugs, like doxorubicin, are protonated in this environment, and consequently, present higher difficulties in transposing the cellular membrane [7,52,53]. Moreover, lactic acidosis is important for rescuing cancer cells from glucose deprivation by inducing a dormant, nonproliferative state where cells are maintained in the G0/G1 cell-cycle phase, which consumes lower levels of glucose [54]. Lactic acidosis also promotes tumor cells autophagy, an antiapoptotic adaptive tumor mechanism to nutrient starvation. Moreover, to inhibit the glucose deprivation induced apoptosis, lactic acidosis is also related to AKT activation that inhibits proapoptotic activity [54]. Despite this nonproliferative condition, cancer cells still express cytokines, chemokines, and growth factors important for tumorigenesis and anticancer therapies resistance [7].

Besides this glycolytic profile, both metabolic processes, OXPHOS and glycolysis, are able to coexist in cancer cells. This high-metabolic plasticity allows cancer cells to adapt to distinct environmental conditions and, consequently, turns the identification of effective metabolic targets more difficult [52,55]. Under nutrient depletion conditions, to sustain the mitochondrial function for cell survival and growth, cancer cells use glutamine and other amino acids to undergo TCA cycle for oxidative ATP production [40,56]. Additionally, when the TME becomes nutrient deprived the ATP demand of tumor cells decreases. In this situation, two adenosine diphosphate (ADP) molecules can be transformed into one adenosine monophosphate (AMP) and one ATP [57]. Increased AMP levels result in the activation of AMP kinase (AMPK) which consequently activates other metabolic pathways, including fatty acid oxidation, for ATP production [58].

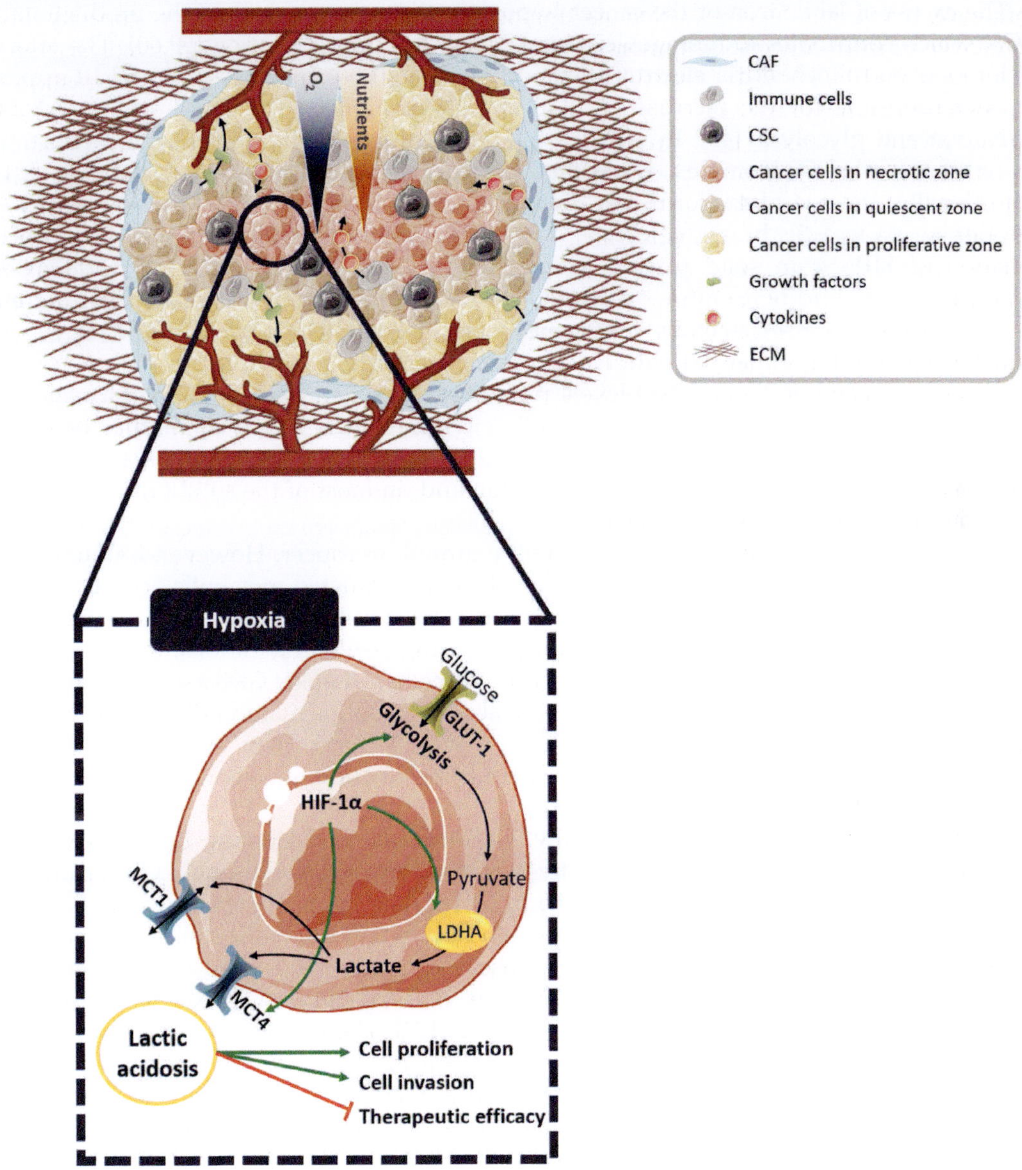


FIGURE 6.2 The influence of hypoxic environment in tumor metabolism and TME conditions. Upregulation of HIF-1α expression leads to a metabolic shift from OXPHOS to glycolysis. The high glycolysis rate, under hypoxic conditions, promote lactic acidosis that is related to cell proliferation, invasion, and therapeutic resistance in tumor.

In addition to glucose metabolism, iron metabolism seems to play a role in cancer biology; however this connection is currently poorly explored. Nevertheless, it is known that iron interferes with malignant transformation, tumor progression (once transformed, malignant cells require high amounts of iron for proliferation), and TME regulation [59].

The excess of labile iron or the cancer hypoxic environment enhances the production of ROS, which contributes to mutagenicity and malignant transformation [59,60]. The effects of hypoxia on mitochondria electron transport chain (ETC) complexes I, II, and III appear to be responsible for the increase of ROS and these higher ROS levels sustain AMPK activation and glycolysis [45]. In fact, the treatment of pancreatic cancer with metformin, an inhibitor of ETC complex I, resulted in the increase of patients' survival [61]. Additionally, the use of metformin in diabetic patients is associated with a lower risk of developing pancreatic, breast, colorectal, and liver cancers [62–65]. Similar to ROS, upregulation of HIF-1α in solid tumors promotes the expression of inducible-nitric oxide synthase (iNOS) leading to nitric oxide (NO) production [66,67]. NO is related to different tumorigenesis processes, such as angiogenesis and cellular invasion. VEGF, MMP-2, and MMP-9 expression is enhanced with NO expression in solid tumors [68].

Increased lactate, ROS, and NO levels promote changes in local inflammatory state to facilitate tumor development [15]. This acidic environment is linked to another hallmark of cancer, the capacity of cancer cells to evade the immune system surveillance [15,22]. Tumors have multiple means of immune evasion and, in most of the solid tumors, effector antitumor immune cells are suppressed.

The immune system can prevent or slow the growth of cancer. However, immune cells often suffer from metabolic stress as a result of the unregulated metabolic activity of cancer cells, leading to a suppression of antitumor immunity [69]. The activation of effector antitumor immune cells has diverse stages, promoting several alterations, including their cellular metabolism. For example, helper CD4 + T cells, cytotoxic CD8 + T cells, and natural killer (NK) cells perform a glycolytic metabolism upon activation [70]. However, the high-glycolytic rates of tumor cells impose an altered metabolism in tumor-infiltrating lymphocytes (TILs) and consequently an impaired production of activated effector antitumor cells [69,71].

Similar to TILs, TAMs are influenced by TME conditions. TAMs are highly present in solid tumors and interact with TME in diverse ways. In response to the TME conditions, these immune cells may change their profile from a proinflammatory (M1) to an antiinflammatory state (M2). M1 state can prompt the efficacy of anticancer therapies. On the other hand, the M2 profile is related to a poorer disease outcome, resistance to treatment, and immune evasion capacity [72]. In hypoxic areas of solid tumors, TAMs demonstrate a proangiogenic and immunosuppressive profile—M2 [73]. The tumor supporting function of TAM in hypoxia can be displayed by two mechanisms. First, the hypoxic environment polarizes infiltrating TAM to an M2 phenotype that releases iron into TME and promotes cell proliferation (macrophages regulate iron homeostasis, M1 is responsible for iron sequestration, and M2 for releasing it) [74,75]. Second, hypoxic TAMs promote a shift to an oxidative metabolism that leads to increased TME glucose availability and, as a consequence, endothelial cell activation for angiogenesis and metastasis [76]. In this context, a potential therapeutic strategy can be linked to metabolic interventions that change TME conditions and manipulate TAM differentiation in combination with immunotherapeutic agents.

Immunotherapeutic agents, in particular, monoclonal antibodies directed to the antiprogrammed cell death 1 (PD-1) receptor (e.g., nivolumab, pembrolizumab) or its ligand PD-L1 (e.g., atezolizumab, durvalumab, and avelumab) and cytotoxic T-lymphocyte-associated protein (CTLA-4) (e.g., ipilimumab), have shown promising responses. Some of these drugs have also been approved for the treatment of patients with advanced melanoma, lung cancer, urothelial cancer, head and neck squamous cell carcinoma, and renal cell cancer, among others [77]. In melanoma, LDHA and lactate levels are negatively correlated to T cells and NK cells and overall survival [78]. On the other hand, also in melanoma, low-serum levels of lactate dehydrogenase (LDH) are associated with better therapeutic outcomes to the immunotherapeutic agent pembrolizumab, a monoclonal antibody directed to the PD-1 receptor [79].

Altogether this demonstrates that glucose metabolism, leading to lactate production and secretion can influence the tumor immune state and the immunotherapy efficacy. This favors cancer metabolic pathways as potential therapeutic targets. Under a hypoxic environment, the inhibition of lactate production prompted by LDHA could also increase immunotherapy efficacy [69].

Overall, metabolic interventions in combination with immunotherapeutic agents seem to have a therapeutic potential in cancer. To evaluate and develop these therapeutic strategies in vitro models are used. As demonstrated, cancer metabolism and the TME in solid tumors are extremely important and influence the efficacy of therapeutic strategies. Thus, it is crucial to have an in vitro model that can reproduce a real solid tumor [80].

## 6.4 Solid tumors in vitro models

Experimental in vitro models are widely used in biological and medical research for innumerous research fields, including cancer research [81–84].

In cancer research in vitro models are crucial, allowing researchers to evaluate cancer biology, the efficacy of anticancer therapies, and to perform drug screening. In vitro models have contributed to the oncobiology field with different findings on the metastatic process, migration, proliferation, and cancer metabolism, as others [84]. In vitro models for solid tumors vary from 2D cell culture to more complex 3D systems. Different in vitro models were developed in the last decades to improve the cancer research outcomes, being essential to evaluate the potential of therapeutic candidates before the application in vivo or in the clinics [18,84].

### 6.4.1 2D cell culture systems in cancer research

The 2D cell monolayers are the most traditional in vitro models used to study malignant cell behavior and cancer response to therapy. Many researchers and the pharmaceutical industry commonly use these cell culture systems, since the experimental conditions can be controlled and the results are highly reproducible. In drug development, 2D models remain the most commonly used in vitro method for drug screening [20]. The 2D cell culture models have largely contributed to the understanding of malignant cell behavior; however evidence has shown that cell behaviors in a 2D monolayer diverge from those found in vivo [85,86].

As described before, cancer is one of the most complex diseases characterized by cell heterogeneity and the capacity to adapt to the different selective pressures of the TME.

Thus, it is crucial that the in vitro models used for oncobiology research do not highly deviate from what happens in vivo. One of the main differences between 2D cell culture systems and solid tumors is that, in this culture model, cells are distributed into a monolayer form that is not the natural architecture of tumors. Inner cells of a tumor mass have limited access to nutrients, oxygen, metabolites, and other limitations that are not reproduced in a 2D cell culture model. Also, cell–cell interactions in a monolayer or a tumor mass are performed in distinct ways [86,87]. Besides that, 2D cell culture systems may alter the morphology of the malignant cells, which will consequently affect cell function, division, secretion, and signaling [86]. Also, the composition and arrangement of the ECM is affected and the extracellular signals do not mimic a normal ECM [85,87,88]. Additionally, differences in the interactions of the tumor adherent cells with the external environment can lead to loss of cell polarity which contributes to changes in the cellular responses to various mechanisms [86,89,90].

Finally, 2D cell culture systems, usually, are monocultures allowing the study of one cell type only, which is not representative of the tumor in vivo development. As described above, the TME is very important in solid tumors to acquire and sustain the malignant phenotype. Moreover, interactions between tumor cells and TME affect different cellular features, such as viability, proliferation, differentiation, gene and protein expression, sensitivity to stimuli, drug metabolism, and others [86].

Altogether, this evidence demonstrates the need for culture models that better replicate what happens in vivo. The 3D cell culture systems have become an important tool in cell biology understanding and cancer research, by improving in vitro studies and better representing what happens in vivo [91].

### 6.4.2 3D cell culture systems

The 3D cell culture systems development improved in vitro experiments, as these models reproduce more accurately in vivo conditions. 3D systems portrait a microenvironment where cells can grow and interact with its extracellular surroundings in a 3D form, mimicking microtumors [92]. The capacity to replicate better the in vivo environment improved cell and cancer biology research, for example, in the prediction of drug efficacy and toxicity [93,94]. The 3D models allow researchers to form a microtumor that simulates the microenvironment conditions of an in vivo solid tumor and the hypoxic tumor core [92].

The 3D culture models most commonly used are: Tumor-tissue explants, tumor-*on-a-chip* and multicellular tumor spheroids [95].

Tumor-tissue explants are mostly used for in vitro drug discovery. In this model, human tumor tissues collected from biopsies are cultured in tissue culture plates and exposed to the drugs to evaluate treatment efficacy [95,96].

Tumor-*on-a-chip* is a 3D culture model that allows the co-culturing of malignant cells with other cell types on a microfluidic device. This microfluidic system allows researchers to replicate cellular vasculature and the biological fluidics that are present in the TME. Building a functional unit of a tumor in a microfluidic device improves the studies on cancer cell behavior and interactions with the TME, metastatic processes, detection of circulating tumor cells, and drug screening [95,97–99].

Multicellular tumor spheroids can be obtained only with cancer cells, or in aggregates with other cell types, respectively mono- or co-cultures. The morphology of the multicellular tumor spheroids can be spherical or have other irregular shape and depends on the nature of the cell, cell culture conditions, and cell interactions. This model replicates in vivo tumor characteristics including cellular heterogeneity, nutrient, oxygen, and metabolic wastes gradient, proliferative and necrotic zones, as well as gene and protein expression profiles [95]. Due to these advantageous features, this 3D cell culture model is the most well characterized for cancer research.

Different techniques are available for obtaining tumor spheroids and many of them are being optimized to produce large-scale tumor spheroids under reproducible conditions. These methodologies can be divided into two main classes, the scaffold 3D cell cultures and the nonscaffold 3D cell cultures [100]. Besides these two main classes, specialized 3D culture platforms have been developed to allow the establishment of multicellular tumor spheroids, for example, micropatterned plates [101]. All the main 3D cell culture systems discussed above have advantages and disadvantages that are summarized in Table 6.1.

Scaffold-based 3D cell cultures are obtained through a 3D matrix (biological or synthetic hydrogels), that is used to support the 3D organization of the tumor and to mimic the ECM architecture. Cells are embedded in this matrix and the chemical and mechanical properties of this platform influence the tumor cells, allowing the researcher to select a 3D cell culture scaffold, that better adjusts to the objectives of the work [101]. One of the limitations of these methodologies is the incapacity to replicate the mass transport gradient of an in vivo tumor environment [102,103].

Hydrogels are used to encapsulate the cancer cells and form a microporous scaffold resembling a native ECM. This matrix contains proteins and other ECM components and allows the cells to adhere, proliferate, and migrate in a 3D manner. The interconnected

**TABLE 6.1** Advantages and disadvantages of 3D cell culture model used in cancer field [95,97–99].

| 3D cell culture model | Advantages | Disadvantages |
|---|---|---|
| Tissue explant | Preservation of the original architecture of tumor tissue (tumor cells and components of the TME) | Low reproducibility;<br>Difficult to apply distinct assays;<br>Tissue degeneration |
| Tumor-*on-a-chip* | Able to produce co-cultures;<br>Real time recording data;<br>Reproducibility of the TME (cellular vasculature and biological fluids movement) | Complex device;<br>Size limitation |
| Scaffold-based tumor spheroids | Able to produce co-cultures;<br>Reproducibility of the TME and ECM architecture;<br>High-throughput | Fail to reproduce mass transport gradient;<br>Low reproducibility;<br>Expensive |
| Nonscaffold based tumor spheroids | Simple;<br>Able to produce co-cultures;<br>Reproducibility of the TME;<br>High-throughput;<br>Less expensive model | Spheroids are not uniform;<br>Fail to reproduce the ECM architecture |

*3D*, three-dimensional; *TME*, tumor microenvironment; *ECM*, extracellular matrix.

microporous of this hydrophilic structure facilitates water retention and consequently the transport of soluble factors as nutrients, metabolic wastes, and oxygen [95,101]. Hydrogels can have a synthetic or biological origin, in the first case, ECM properties as stiffness or porosity can be manipulated with external stimuli, like temperature and pH [104]. Moreover, some synthetic scaffolds allow the encapsulation of specific molecules, such as growth factors or hormones, to enhance specific cellular features, like proliferation [101]. Thus, with this kind of scaffold, chemical, physical, and mechanical properties are well defined. The ability to control the properties of the synthetic scaffolds allows researchers to obtain reproducible results. Polyacrylic acid, polyethylene glycol, and polyglycolic acid are examples of the main synthetic materials used in synthetic hydrogels [93]. The disadvantages of the synthetic hydrogels are mostly the cost of preparation and the fact that their morphology does not entirely mimic the filamentous nature of an in vivo ECM [105].

Hydrogel scaffolds from biological sources, like matrigel, fibrin, and collagen, sustain different physiopathological functions similar to the ECM observed in vivo [101]. These protein-based hydrogels are used in this 3D scaffolds due to their cell-adhesive properties [105]. Multicellular tumor spheroids with these biological matrixes are capable to respond to TME changes through the presence of integrins in the matrix that allow signal transduction. However, this kind of scaffold tends to easily dissociate and have endogenous soluble factors that add variability to the culture conditions and consequently hampers reproducibility [95]. Matrigel is the main biological hydrogel scaffold used for multicellular tumor spheroids, being used in 3D cell culture models of different malignancies, like breast, colorectal, and prostate cancer cell lines [105–108].

Some cell lines can form cell aggregates without the use of a scaffold. Nonscaffold culture models form heterogeneous cellular aggregates that are commonly named "spheroids." This type of system is usually less expensive, simpler, and highly reproducible. Spheroids with a diameter greater than 500 μm are characterized by inner hypoxic and necrotic regions featuring in vivo tumors [21]. Additionally, cells from a 3D spheroid, during the culture formation produce their own ECM components [93,100]. The need for scaffolds in these cellular aggregates becomes apparent with spheroids that have a thickness greater than 1–2 mm. In these cases, problems with cell death arise principally due to the lack of nutrients and waste metabolites exchange [92]. Different methodologies for the construction of spheroids are available: spontaneous aggregation [109,110], spinner flasks [111], rotatory cell culture systems [112], liquid overlay cultures [102], hanging drop cultures [113,114], microcarrier beads [102], ultra-low-attachment (ULA) plates [21], and magnetic levitation [115,116].

In liquid overlay cultures, cell suspension grow on a nonadhesive substrate, like agar/agarose or Matrigel, to form the spheroid. This is a simple, easy to handle, and economical method, that allows the generation of many spheroids [117]. Limitations related to this technique are the variations obtained in the size and shape of the spheroids [102,117].

Hanging drop technique is based on droplets of the cell suspension. This technique can be performed on specialized plates [101] or, for example, distributed on the underside of a petri dish lid, and gravity promotes the formation of the spheroids. Surface tension allows the attachment of the droplet to the dish and the phosphate-buffered saline (PBS) solution on the petri dish prevents its dehydration [118]. With this method, the coating of plates with a substrate is not required and spheroids are obtained in a rapid manner and with a controlled size [102]. Additionally, various cell types can be co-cultured with this low-cost method, which is another advantage of the technique [93]. A big limitation of this methodology is cell viability, one of the main cell features affected due to the osmolarity caused by the evaporation of the media, and also the limited size of the spheroid obtained [95].

ULA plates are specialized well plates that allow the reproducible production of spheroids [21]. These plates are coated with polystyrene that reduces the attachment of the cells and promotes cell aggregation [101]. When compared with the hanging drop method, this technique allows the use of higher volumes and is, consequently, easier to obtain larger spheroids [101]. It also allows for large-scale production of spheroids, which is less time consuming than most other techniques and is reproducible [21]. The main disadvantages of ULA plates are the fact that not all cancer cell lines are capable of forming tight spheroids with this method, and are more expensive than regular culture plates used in liquid overlay and hanging drop culture systems [21].

Different methods to obtain 3D cell culture models are now available and scientists have the opportunity to select a 3D cell culture method that better suits to the objective of their work. The development of these in vitro models improved different scientific areas, including the cancer research field.

## 6.5 3D cell culture systems in cancer research

The 2D cell culture systems are still the most widely used in vitro models in cancer research, while 3D models are increasing on popularity. This is because 3D culture systems are time consuming, more difficult to replicate and more expensive, while the 2D results are easier to analyze and most literature includes 2D models. The 2D cell culture models differ from the 3D system in many technical characteristics described in Table 6.2 [85–87,119].

The 3D tumor spheroids, as explained, are able to reflect the in vivo TME in terms of: cell heterogeneity; nutrient, oxygen, and metabolic waste gradients; gene and protein

**TABLE 6.2** Main culture characteristics of 2D and 3D cell culture models [21,86,93,119].

| Culture characteristics | 2D cell culture | 3D cell culture |
|---|---|---|
| Time | Cell culture is obtained within hours | Culture formation is achieved in a few hours to a few days |
| Reproducibility | Simple cell culture with high reproducibility | Lower reproducibility |
| Cost | Less expensive system | Some of the materials and techniques to obtain 3D cell culture models are more expensive |
| Results | Easier to analyze; Higher amount of comparative literature | Some cell lines cannot form 3D cell culture models. Analysis of 3D cell culture results can be more difficult |

*2D*, two-dimensional; *3D*, three-dimensional.

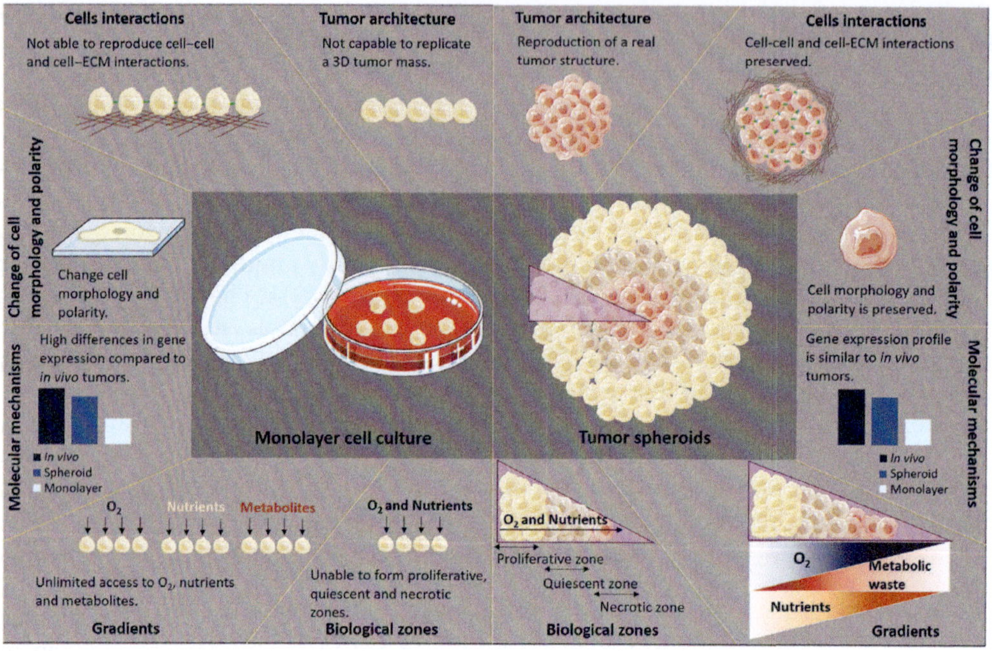


**FIGURE 6.3** Comparison of the main in vivo tumor features in a 2D monolayer cell culture and a 3D tumor spheroid.

expression profiles; cell–cell and cell–ECM interactions; tumor and ECM architecture; biological zones (proliferative, quiescent, and necrotic zones) and, cell morphology and polarity [21,87,117,120,121]. Thus, 3D cell culture systems are more similar to the mass of in vivo solid tumors compared with 2D systems (Fig. 6.3) [86], being a bridge between 2D in vitro studies and in vivo animal models [93]. The fact that 3D models can more accurately replicate what happens in TME in vivo allows researchers to study and understand with greater certainty tumor behavior and their related mechanisms, such as cell proliferation, invasion, metastasis, and survival [85,122].

Studies have demonstrated not only the limitations of the 2D models, but also the supremacy of the 3D systems in different cellular characteristics and capacities. 2D cultures grow on solid, flat support that forces cells to polarize and, consequently, excessive nutrition and oxygenation without hypoxic areas, in contrast to the gradient found in a real solid tumor mass (Fig. 6.3) [85,122,123]. Various studies using different cancer cell lines have demonstrated high differences in these two cell culture models in the activation of cell signaling pathways [124,125]. For example, in some breast and ovarian cancer cell lines, the human epidermal growth factor receptor-2 (HER-2), related to cancer aggressiveness and poor prognosis, was activated in 3D models, as opposed to 2D results [126,127].

The interactions between cancer and other malignant and nonmalignant cells of the TME are important for tumorigenesis. It is well described, that the TME of a solid tumor has a heterogeneous cellular constitution that influences tumor progression [9]. Prostate tumor spheroids were used to evaluate tumor induced angiogenesis. For that, a co-culture of the spheroids with embryonic stem cells was performed to understand the angiogenic responses induced by the cancer cells. This study allowed the researchers to understand that tumor vascularization promotes growth stimulation, higher oxygen consumption, and an increase in proangiogenic genes expression in the tumor spheroid [128]. In another study, spheroids co-cultures of different cancer cells and fibroblasts demonstrated an increase in cell proliferation. This increase was related to the higher secretion of epidermal growth factor (EGF), HGF, and IL-6 observed in these 3D co-cultures [129]. Interestingly, the pancreatic spheroids in co-culture demonstrated a higher survival percentage to cetuximab, an anti-EGFR monoclonal antibody, when compared with the monoculture [129]. Also, the co-cultivation of spheroid cancer cells and CAF, resulted in increased expression of proliferation and invasive markers, TGF-β, N-cadherin, and vimentin. Moreover, cancer and cancer-CAF spheroids demonstrated a higher resistance to sorafenib and cisplatin treatment compared with the 2D cultures [130].

The presence of the CSC, a cancer cell subpopulation, is also known to increase tumor heterogeneity and complexity. As previously described, CSC subpopulation in tumors is associated with the recurrence of the tumor after therapy [25]. Cancer cells obtained from biopsies allow researchers to obtain spheroids composed of CSC or with stem cell properties as confirmed using the CD133 marker. In spheroids obtained from human colon adenocarcinoma tissue, the positive population for this marker was related to treatment resistance to the chemotherapeutic agents, oxaliplatin and fluorouracil [131]. Moreover, a higher CD133-positive cell population could be obtained in spheroids models when compared with the 2D culture of the U87 and U373 glioma cell lines [132]. Studies in glioblastoma and breast cancer showed that the presence of CSC in tumor spheroids was related to higher resistance to different therapies than 2D cultures [133–135]. Head and neck squamous cell carcinoma cell lines were used to evaluate cell viability after drug treatment. Results from this study demonstrated that the increase of CSC associated proteins (namely, CDH1, NANOG, and SOX2) observed in 3D models was related to lower sensitivity to the therapeutic agents when compared with 2D cultures [136].

In addition to the cellular constitution of a TME, the ECM, which provides structural support to cells, also influences tumor development. In cancer, the cross-talk between ECM and cells is deregulated and leads to higher deposition of ECM proteins, like collagen and fibronectin. This ECM deregulation influences cell proliferation and survival [137]. In pancreatic cancer cells, deregulation of the cell–ECM interactions by integrins appears to enhance invasion capacity [138]. As referred previously, 2D models are not able to replicate the ECM architecture and the cell–ECM interactions present in human tumors. For example, in glioma spheroids, the expression of collagen was enhanced, while in 2D cultures no significant expression was identified [7]. In soft sarcoma and osteosarcoma, 3D systems presented chemoresistance to different drugs, due to collagen and fibronectin upregulation, when compared with 2D cultures [139]. In solid tumors, ECM protein deposition through cell–ECM and cell–cell interactions leads to increased tissue density. Consequently, this represents a physical barrier that reduces the penetration of therapeutic compounds and their delivery to cells, increasing tumor resistance to treatment (Fig. 6.4) [140,141].

Overall, in 3D systems, cells usually exhibit reduced sensitivity to therapeutic agents compared with 2D models (Fig. 6.4). Many drugs that seem to be effective in vitro have little or no efficacy in vivo or in clinical trials. Part of this problem has been attributed to the

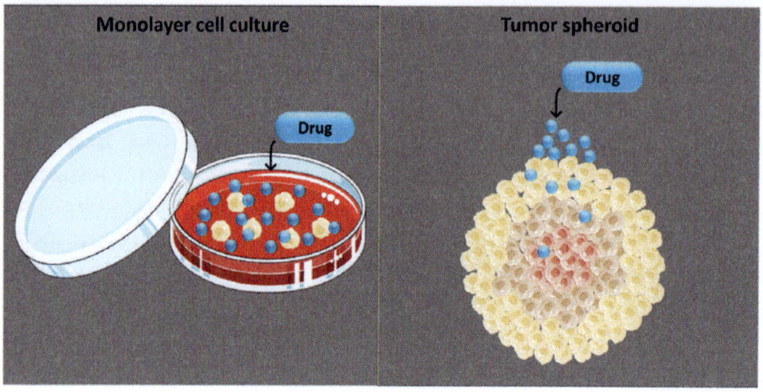


**FIGURE 6.4** Therapeutic agents in a 2D monolayer cell culture and 3D tumor spheroid.

fact that in vitro experiments were mostly done in 2D culture models. Although 2D culture models had improved the knowledge about cancer biology, they represent a poor drug development model, with a drug failure rate of 95% [92,142]. As such, novel drugs must undergo in vivo experiments before human clinical trials. Nevertheless, there are many issues related to animal models, including higher costs, species differences, and ethical issues and animal models should be replaced by other models whenever possible to reduce the number of animals that are used in cancer research [91,143,144]. Novel culture systems like 3D cell culture models can be an answer to minimize the number of in vivo experiments [91,144]. For example, a novel 3D tumor-tissue invasion model for high-throughput and reproducible studies was developed by Puls et al. [145]. This model allowed the authors to control the tumor and the surrounding tissue compartments in a 96-well fabrication system and address tumor cells invasion of pancreatic ductal adenocarcinoma. Moreover, a co-culture of CAFs with a pancreatic cell line demonstrated significant differences in invasion capacity with or without CAFs. Importantly, the Puls et al. model also allowed to confirm the effects of different compounds on proliferation and invasion in a single assay [145].

Methodologies to form 3D tumor spheroids as well as the methods for assessing different characteristics, such as cell proliferation, migration, and invasion, have been developed to improve and make this model more commonly used in cancer research [122]. Besides cancer biology, 3D cell culture models can also be used to study TME interactions, through co-cultures with different cell types, CSC and ECM influence, among others [146]. All of these factors influence the therapeutic response of a tumor. Thus, 3D cell culture models are a potential accurate tool for drug efficacy and toxicity studies.

## 6.6 3D cell culture systems for study cancer metabolism

Reprogramming energy metabolism is considered one of the hallmarks of cancer proposed by Hanahan and Weinberg, as referred previously. The alterations of metabolic activity in cancer cells allow them to sustain cell growth and division [22].

In summary of what was previously described, cancer cells are characterized by metabolic plasticity, allowing OXPHOS and glycolysis metabolic modes to coexist in a tumor. Still, cancer cells even in the presence of oxygen reprogram their metabolism to perform glycolysis, preferring to convert glucose into lactate independently of the oxygen levels [43].

The majority of the solid tumors are characterized by a central hypoxic core that triggers some changes in gene expression, lactate production, and is responsible for TME acidification [95]. The adaptation of cancer cells to the microenvironment promotes the modulation of cancer metabolism. This metabolic reprogramming hampers the identification of metabolic targets and is one of the factors responsible for anticancer therapeutic resistance. The development of culture systems that ease researchers in the process to reproduce cancer metabolic plasticity and the respective TME influence is important. This will enhance the identification of potential metabolic markers and therapeutic processes of drug evaluation [52,55,147]. The 3D models allow researchers to approximate their studies to what happens in vivo, taking into consideration the TME. Moreover, nutrient and oxygen gradients that lead to a hypoxic tumor core and influence cancer metabolism are reproduced in these models, whereas 2D culture models fail to reproduce cancer metabolic activity due to lack of TME interactions and gradients of nutrient and oxygen [80,86,148,149].

With spheroids growth, a gradient of oxygen is established, and the inner cells are in hypoxia. This oxygen gradient was demonstrated in various studies that measured oxygen content by oxygen-sensitive microelectrodes or fiber-optic oxygen sensors [150]. Oxygen consumption can be evaluated in spheroids through oxoplates. A plate that is coated with an oxygen-sensing indicator and a reference dye. The fluorescence intensity is measured along the time to obtain oxygen consumption [151]. The levels of ROS can also be measured through a similar process [152].

Hypoxia in cancer leads to the accumulation of HIF-1$\alpha$, which consequently promotes a shift to glycolytic metabolism [153]. The studies of glycolysis in 3D cell culture models had demonstrated an increase in glucose uptake and lactate production when tumor spheroids are larger [154,155]. This demonstrates that spheroids are capable to mimic the metabolic shift that can occur in vivo. Glucose uptake in solid tumors is mediated by GLUT-1 and others. The expression of GLUT-1 is higher in breast cancer spheroids in contrast to the monolayer culture [156]. Similarly, glucose consumption and lactate production are higher in glioma spheroids than in monolayer cell cultures [157]. For glucose uptake quantification "imaging bioluminescence" can be used, which allows mapping and quantification of ATP, glucose, and lactate in different areas of the spheroid [158].

Metabolic switch promoted by HIF-1$\alpha$ leads to the increase of lactate levels, acidification of the TME and secretion of proangiogenic factors [95,159]. While HIF-1$\alpha$ expression was observed in HeLa spheroids this protein was not expressed, in 2D models [160]. In different cancer models, like esophageal cancer and melanoma, 3D spheroids presented HIF-1$\alpha$ upregulation which significantly enhanced VEGF expression levels, when compared with 2D systems [161,162]. VEGF expression enhances angiogenesis and cancer development, as described previously [28,29]. Besides the secretion of proangiogenic factors, TME acidification of solid tumors was shown to influence drug uptake. In colon cancer spheroids, doxorubicin uptake decreased proportionally with the pH value [163]. As previously described, TME acidification may induce a nonproliferative state in solid tumors cells [54] and this subpopulation of nonproliferative cells was responsible for increased therapeutic resistance to doxorubicin in lung and breast cancer spheroids compared with the 2D model [164,165]. Doxorubicin and other drugs are more effective in proliferative cells, which may explain these results.

As stated above, tumor cells are more sensitive to drugs in 2D culture systems than in 3D models. This can be explained by reduced access to compounds in the medium or due to some physiological differences found in 3D systems such as hypoxia [86]. The adaptation of tumor cells in the inner layer to hypoxic conditions leads to cell selection and, consequently, to more aggressive and resistant tumor cells [44,117,166].

In this context, 3D cell culture systems seem to have high potential as tumor models for studying cancer metabolism. The development of simple, rapid, and reproducible systems that allow researchers to obtain important information about cancer metabolism is critical to drug discovery. Moreover, the inclusion of automated imaging and analysis facilitates high-throughput screening. For example, Russel et al. have developed a microchamber system to obtain the metabolic profile of 2D cultures, 3D spheroids (with minimal dispersions for oxygen consumption rate within spheroids replicas), microtissue from tumors and normal organs. In this system, spheroids or a microtissue in a single well of 96-well plates could be measured for oxygen uptake and extracellular acidification rate. Moreover, the response to different drugs confirmed that some drugs effective in 2D models did not have the same result in 3D systems [80].

The novel 3D tumor-tissue invasion model of Puls et al. and this microchamber system of Russel et al. are promising ways to evaluate the efficacy and toxicity of drugs, diminishing in vivo studies and having higher efficient drugs under clinical trials [80,145].

## 6.7 Conclusions

Cancer is one of the main health problems, so understanding the pathological processes of this disease is crucial. Innovative methodologies are always being produced to offer better outcomes in the cancer research field and ultimately to oncological patients.

The 2D models have increased the knowledge about cancer complexity in the past years. However, they do not reproduce an actual solid tumor mass and fail to mimic various aspects of the tumors, including their metabolic complexity. During the last decades, 3D cell culture models have demonstrated their advantages not only in increasing knowledge about cancer cell behavior and metabolism, but also in enabling the identification of powerful targets and better assessing the effectiveness of anticancer drugs.

Throughout this chapter, the complexity of TME and cancer metabolism has been explored. Cancer cells are able to adapt their metabolic processes to the microenvironment and this plasticity turns the identification of metabolic targets more difficult. In this sense, it is essential to have 3D models that can better resemble the in vivo TME.

Overall, it is thought that the 3D cell culture systems can fulfill the gap between 2D and in vivo models to diminish the use of animal models. Moreover, these systems can improve the outcomes in cancer biology and drug development fields.

## Acknowledgments

The authors acknowledge Dr. Julieta Afonso (Life and Health Sciences Research Institute (ICVS)/School of Medicine/University of Minho, Braga, Portugal) for her useful suggestions. Financial support was provided by grants from the Bolsa LPCC/Fundação PT 2018 to SPC; FCT—CTTI-27/18- ICVS (14)-perfil 11 to MVP; Northern Portugal Regional Operational Programme (NORTE 2020), under the Portugal 2020 Partnership Agreement, through the European Regional Development Fund (FEDER)—NORTE-01–0145-FEDER-000013 and NORTE-01–0145-FEDER-000023; FEDER, through the Competitiveness Factors Operational Programme (COMPETE), and FCT—POCI-01–0145-FEDER-007038.


## Conflict of interest

Nothing to declare.

## References


[1] WHO. WHO cancer fact sheet 2018, <https://www.who.int/news-room/fact-sheets/detail/cancer>; 2018 [acessed 12.05.2019].

[2] Thun MJ, DeLancey JO, Center MM, Jemal A, Ward EM. The global burden of cancer: priorities for prevention. Carcinogenesis. 2010;31(1):100–10.

[3] Chabner BA, Roberts TG. Chemotherapy and the war on cancer. Nat Rev Cancer 2005;5(1):65–72.

[4] Yadav P, Shankar BS. Radio resistance in breast cancer cells is mediated through TGF-β signalling, hybrid epithelial-mesenchymal phenotype and cancer stem cells. Biomed Pharmacother 2019;111:119–30.

[5] Moreira AF, Dias DR, Correia IJ. Stimuli-responsive mesoporous silica nanoparticles for cancer therapy: a review. Microporous Mesoporous Mater 2016;236:141–57.

[6] Barnes TA, Amir E, Templeton AJ, Gomez-Garcia S, Navarro B, Seruga B, et al. Efficacy, safety, tolerability and price of newly approved drugs in solid tumors. Cancer Treat Rev 2017;56:1–7.

[7] Nunes AS, Barros AS, Costa EC, Moreira AF, Correia IJ. 3D tumor spheroids as in vitro models to mimic in vivo human solid tumors resistance to therapeutic drugs. Biotechnol Bioeng 2019;116(1):206–26.

[8] Burrell RA, Swanton C. Tumour heterogeneity and the evolution of polyclonal drug resistance. Mol Oncol 2014;8(6):1095–111.

[9] Balkwill FR, Capasso M, Hagemann T. The tumor microenvironment at a glance. J Cell Sci 2012;125(23):5591–6.

[10] Sun Y. Tumor microenvironment and cancer therapy resistance. Cancer Lett 2016;380(1):205–15.

[11] Boroughs LK, Deberardinis RJ. Metabolic pathways promoting cancer cell survival and growth. Nat Cell Biol 2015;17(4):351–9.

[12] Elia I, Schmieder R, Christen S, Fendt SM. Organ-specific cancer metabolism and its potential for therapy. Handb Exp Pharmacol 2016;233:321–53.

[13] Muir A, Danai LV, Vander Heiden MG. Microenvironmental regulation of cancer cell metabolism: implications for experimental design and translational studies. Dis Model Mech 2018;11(8):dmm035758.

[14] Warburg O. On the origin of cancer cells. Science 1956;123(3191):309–14.

[15] Neagu M, Constantin C, Popescu ID, Zipeto D, Tzanakakis G, Nikitovic D, et al. Inflammation and metabolism in cancer cell—mitochondria key player. Front Oncol 2019;9:348.

[16] Ramapriyan R, Caetano MS, Barsoumian HB, Mafra ACP, Zambalde EP, Menon H, et al. Altered cancer metabolism in mechanisms of immunotherapy resistance. Pharmacol Therap 2019;195:162–71.

[17] Lin J, Xia L, Liang J, Han Y, Wang H, Oyang L, et al. The roles of glucose metabolic reprogramming in chemo- and radio-resistance. J Exp Clin Cancer Res 2019;38:218.

[18] Singh S, Sharma B, Kanwar SS, Kumar A. Lead phytochemicals for anticancer drug development. Front Plant Sci 2016;7:1667.

[19] Menshykau D. Emerging technologies for prediction of drug candidate efficacy in the preclinical pipeline. Drug Discov Today 2017;22(11):1598–603.

[20] Chatzinikolaidou M. Cell spheroids: the new frontiers in vitro models for cancer drug validation. Drug Discov Today 2016;21(9):1553–60.
[21] Vinci M, Gowan S, Boxall F, Patterson L, Zimmermann M, Court W, et al. Advances in establishment and analysis of three-dimensional tumor spheroid-based functional assays for target validation and drug evaluation. BMC Biol 2012;10:29.
[22] Hanahan D, Weinberg RA. Hallmarks of cancer: the next generation. Cell 2011;144(5):646–74.
[23] Chang JC. Cancer stem cells: role in tumor growth, recurrence, metastasis, and treatment resistance. Medicine (Baltim) 2016;95(1):S20–5.
[24] Nimmakayala RK, Batra SK, Ponnusamy MP. Unraveling the journey of cancer stem cells from origin to metastasis. Biochim Biophys Acta - Rev Cancer 2019;1871(1):50–63.
[25] Vidal SJ, Rodriguez-Bravo V, Galsky M, Cordon-Cardo C, Domingo-Domenech J. Targeting cancer stem cells to suppress acquired chemotherapy resistance. Oncogene 2014;33(36):4451–63.
[26] Ham SL, Joshi R, Thakuri PS, Tavana H. Liquid-based three-dimensional tumor models for cancer research and drug discovery. Exp Biol Med 2016;241(9):939–54.
[27] Erdogan B, Webb DJ. Cancer-associated fibroblasts modulate growth factor signaling and extracellular matrix remodeling to regulate tumor metastasis. Biochem Soc Trans 2017;45(1):229–36.
[28] Albini A, Bruno A, Noonan DM, Mortara L. Contribution to tumor angiogenesis from innate immune cells within the tumor microenvironment: implications for immunotherapy. Front Immunol 2018;9:527.
[29] Hida K, Maishi N, Annan DA, Hida Y. Contribution of tumor endothelial cells in cancer progression. Int J Mol Sci 2018;19(5):1–12.
[30] Masjedi A, Hashemi V, Hojjat-Farsangi M, Ghalamfarsa G, Azizi G, Yousefi M, et al. The significant role of interleukin-6 and its signaling pathway in the immunopathogenesis and treatment of breast cancer. Biomed Pharmacother 2018;108:1415–24.
[31] Ham IH, Oh HJ, Jin H, Bae CA, Jeon SM, Choi KS, et al. Targeting interleukin-6 as a strategy to overcome stroma-induced resistance to chemotherapy in gastric cancer. Mol Cancer 2019;18(1):68.
[32] Thews O, Riemann A. Tumor pH and metastasis: a malignant process beyond hypoxia. Cancer Metastasis Rev 2019;38(1–2):113–29.
[33] Sun L, Song L, Wan Q, Wu G, Li X, Wang Y, et al. CMyc-mediated activation of serine biosynthesis pathway is critical for cancer progression under nutrient deprivation conditions. Cell Res 2015;25(4):429–44.
[34] Kumari S, Badana AK, Murali Mohan G, Shailender G, Malla RR. Reactive oxygen species: a key constituent in cancer survival. Biomarker Insights 2018;13 117727191875539.
[35] Miranda-Gonçalves V, Granja S, Martinho O, Honavar M, Pojo M, Costa BM, et al. Hypoxia-mediated upregulation of MCT1 expression supports the glycolytic phenotype of glioblastomas. Oncotarget 2016;7(29):46335–53.
[36] Horsman MR, Vaupel P. Pathophysiological basis for the formation of the tumor microenvironment. Front Oncol 2016;6:66.
[37] Graham K, Unger E. Overcoming tumor hypoxia as a barrier to radiotherapy, chemotherapy and immunotherapy in cancer treatment. Int J Nanomed 2018;13:6049–58.
[38] Ganapathy-Kanniappan S, Geschwind JFH. Tumor glycolysis as a target for cancer therapy: progress and prospects. Mol Cancer 2013;12:152.
[39] Linchong S, Caixia S, Shi-ting L, Huafeng Z, Ping G. Metabolic reprogramming for cancer cells and their microenvironment: beyond the Warburg effect. Biochim Biophys Acta - Rev Cancer 2019;8(5):55.
[40] DeBerardinis RJ, Chandel NS. Fundamentals of cancer metabolism. Sci Adv 2016;2(5).
[41] Koppenol WH, Bounds PL, Dang CV. Otto Warburg's contributions to current concepts of cancer metabolism. Nat Rev Cancer 2011;11(5):325–37.
[42] DeBerardinis RJ, Sayed N, Ditsworth D, Thompson CB. Brick by brick: metabolism and tumor cell growth. Curr Opin Genet Dev 2008;18(1):54–61.
[43] Vander HMG, Cantley LC, Thompson CB. Understanding the warburg effect: the metabolic requirements of cell proliferation. Science 2009;324(5930):1029–33.
[44] Vaupel P. The role of hypoxia-induced factors in tumor progression. Oncologist 2004;9:10–17.
[45] Yu L, Chen X, Sun X, Wang L, Chen S. The glycolytic switch in tumors: how many players are involved? J Cancer 2017;8(17):3430–40.
[46] Li Z, Zhang H. Reprogramming of glucose, fatty acid and amino acid metabolism for cancer progression. Cell Mol Life Sci 2016;73(2):377–92.

[47] Eales KL, Hollinshead KER, Tennant DA. Hypoxia and metabolic adaptation of cancer cells. Oncogenesis 2016;5:e190.
[48] Gentric G, Mieulet V, Mechta-Grigoriou F. Heterogeneity in cancer metabolism: new concepts in an old field. Antioxid Redox Signal 2017;26(9):462–85.
[49] Granja S, Tavares-Valente D, Queirós O, Baltazar F. Value of pH regulators in the diagnosis, prognosis and treatment of cancer. Semin Cancer Biol 2017;43:17–34.
[50] Pinheiro C, Longatto-Filho A, Azevedo-Silva J, Casal M, Schmitt FC, Baltazar F. Role of monocarboxylate transporters in human cancers: state of the art. J Bioenerg Biomembr 2012;44(1):127–39.
[51] Gillies RJ, Gatenby RA. Adaptive landscapes and emergent phenotypes: why do cancers have high glycolysis? J Bioenerg Biomembr 2007;39(3):251–7.
[52] Granja S, Pinheiro C, Reis R, Martinho O, Baltazar F. Glucose addiction in cancer therapy: advances and drawbacks. Curr Drug Metab 2015;16(3):221–42.
[53] Tavares-Valente D, Baltazar F, Moreira R, Queirós O. Cancer cell bioenergetics and pH regulation influence breast cancer cell resistance to paclitaxel and doxorubicin. J Bioenerg Biomembr 2013;45(5):467–75.
[54] Hu X, Chao M, Wu H. Central role of lactate and proton in cancer cell resistance to glucose deprivation and its clinical translation. Signal Transduct Target Ther 2017;2:16047.
[55] Yu L, Lu M, Jia D, Ma J, Ben-Jacob E, Levine H, et al. Modeling the genetic regulation of cancer metabolism: interplay between glycolysis and oxidative phosphorylation. Cancer Res 2017;77(7):1564–74.
[56] Metallo CM, Gameiro PA, Bell EL, Mattaini KR, Yang J, Hiller K, et al. Reductive glutamine metabolism by IDH1 mediates lipogenesis under hypoxia. Nature 2012;481(7381):380–4.
[57] Lanning NJ, Looyenga BD, Kauffman AL, Niemi NM, Sudderth J, DeBerardinis RJ, et al. A mitochondrial RNAi screen defines cellular bioenergetic determinants and identifies an adenylate kinase as a key regulator of ATP levels. Cell Rep 2014;7(3):907–17.
[58] Hardie DG, Schaffer BE, Brunet A. AMPK: an energy-sensing pathway with multiple inputs and outputs. Trends Cell Biol 2016;26(3):190–201.
[59] Pfeifhofer-Obermair C, Tymoszuk P, Petzer V, Weiss G, Nairz M. Iron in the tumor microenvironment-connecting the dots. Front Oncol 2018;8:549.
[60] Tafani M, Sansone L, Limana F, Arcangeli T, De Santis E, Polese M, et al. The interplay of reactive oxygen species, hypoxia, inflammation, and sirtuins in cancer initiation and progression. Oxid Med Cell Longev 2016;2016:3907147.
[61] Kordes S, Pollak MN, Zwinderman AH, Mathôt RA, Weterman MJ, Beeker A, et al. Metformin in patients with advanced pancreatic cancer: a double-blind, randomised, placebo-controlled phase 2 trial. Lancet Oncol 2015;16(7):839–47.
[62] Wang Z, Lai ST, Xie L, Zhao JD, Ma NY, Zhu J, et al. Metformin is associated with reduced risk of pancreatic cancer in patients with type 2 diabetes mellitus: a systematic review and meta-analysis. Diabetes Res Clin Pract 2014;106(1):19–26.
[63] Col NF, Ochs L, Springmann V, Aragaki AK, Chlebowski RT. Metformin and breast cancer risk: a meta-analysis and critical literature review. Breast Cancer Res Treat 2012;135(3):639–46.
[64] Nie Z, Zhu H, Gu M. Reduced colorectal cancer incidence in type 2 diabetic patients treated with metformin: a meta-analysis. Pharm Biol 2016;54(11):2636–42.
[65] Zhang H, Gao C, Fang L, Zhao HC, Yao SK. Metformin and reduced risk of hepatocellular carcinoma in diabetic patients: a meta-analysis. Scand J Gastroenterol 2013;48(1):78–87.
[66] Umbrello M, Dyson A, Feelisch M, Singer M. The key role of nitric oxide in hypoxia: hypoxic vasodilation and energy supply-demand matching. Antioxid Redox Signal 2013;19(14):1690–710.
[67] Burrows N, Cane G, Robson M, Gaude E, Howat WJ, Szlosarek PW, et al. Hypoxia-induced nitric oxide production and tumour perfusion is inhibited by pegylated arginine deiminase (ADI-PEG20). Sci Rep 2016;6:22950.
[68] Choudhari SK, Chaudhary M, Bagde S, Gadbail AR, Joshi V. Nitric oxide and cancer: a review. World J Surg Oncol 2013;11:118.
[69] Li X, Wenes M, Romero P, Huang SCC, Fendt SM, Ho PC. Navigating metabolic pathways to enhance antitumour immunity and immunotherapy. Nat Rev Clin Oncol 2019;16(7):425–41.
[70] Bantug GR, Galluzzi L, Kroemer G, Hess C. The spectrum of T cell metabolism in health and disease. Nat Rev Immunol 2018;18(1):19–34.
[71] Chang CH, Qiu J, O'Sullivan D, Buck MD, Noguchi T, Curtis JD, et al. Metabolic competition in the tumor microenvironment is a driver of cancer progression. Cell 2015;162(6):1229–41.

[72] Vitale I, Manic G, Coussens LM, Kroemer G, Galluzzi L. Macrophages and metabolism in the tumor microenvironment. Cell Metab 2019;30(1):36–50.
[73] Laoui D, Van Overmeire E, Conza GD, Aldeni C, Keirsse J, Morias Y, et al. Tumor hypoxia does not drive differentiation of tumor-associated macrophages but rather fine-tunes the M2-like macrophage population. Cancer Res 2014;74(1):24–30.
[74] Mertens C, Akam EA, Rehwald C, Brüne B, Tomat E, Jung M. Intracellular iron chelation modulates the macrophage iron phenotype with consequences on tumor progression. PLoS One 2016;11(11):e0166164.
[75] Mertens C, Mora J, Ören B, Grein S, Winslow S, Scholich K, et al. Macrophage-derived lipocalin-2 transports iron in the tumor microenvironment. Oncoimmunology 2018;7(3):e1408751.
[76] Wenes M, Shang M, Di Matteo M, Goveia J, Martín-Pérez R, Serneels J, et al. Macrophage metabolism controls tumor blood vessel morphogenesis and metastasis. Cell Metab 2016;24(5):701–15.
[77] Kruger S, Ilmer M, Kobold S, Cadilha BL, Endres S, Ormanns S, et al. Advances in cancer immunotherapy 2019 - latest trends. J Exp Clin Cancer Res 2019;38(1):268.
[78] Brand A, Singer K, Koehl GE, Kolitzus M, Schoenhammer G, Thiel A, et al. LDHA-associated lactic acid production blunts tumor immunosurveillance by T and NK cells. Cell Metab 2016;24(5):657–71.
[79] Weide B, Martens A, Hassel JC, Berking C, Postow MA, Bisschop K, et al. Baseline biomarkers for outcome of melanoma patients treated with pembrolizumab. Clin Cancer Res 2016;22(22):5487–96.
[80] Russell S, Wojtkowiak J, Neilson A, Gillies RJ. Metabolic profiling of healthy and cancerous tissues in 2D and 3D. Sci Rep 2017;7(1):15285.
[81] Stoppelkamp S, Bell HS, Palacios-Filardo J, Shewan DA, Riedel G, Platt B. In vitro modelling of Alzheimer's disease: degeneration and cell death induced by viral delivery of amyloid and tau. Exp Neurol 2011;229(2):226–37.
[82] Zhang YS, Oklu R, Albadawi H. Bioengineered in vitro models of thrombosis: methods and techniques. Cardiovasc Diagn Ther 2017;7(3):S329–35.
[83] Ahluwalia A, Misto A, Vozzi F, Magliaro C, Mattei G, Marescotti MC, et al. Systemic and vascular inflammation in an in-vitro model of central obesity. PLoS One 2018;13(2):e0192824.
[84] Katt ME, Placone AL, Wong AD, Xu ZS, Searson PC. In vitro tumor models: advantages, disadvantages, variables, and selecting the right platform. Front Bioeng Biotechnol 2016;4:12.
[85] Betriu N, Semino C. Development of a 3D co-culture system as a cancer model using a self-assembling peptide scaffold. Gels 2018;4(3):65.
[86] Kapałczyńska M, Kolenda T, Przybyła W, Zajączkowska M, Teresiak A, Filas V, et al. 2D and 3D cell cultures – a comparison of different types of cancer cell cultures. Arch Med Sci 2018;14(4):910–19.
[87] Pampaloni F, Reynaud EG, Stelzer EHK. The third dimension bridges the gap between cell culture and live tissue. Nat Rev Mol Cell Biol 2007;8(10):839–45.
[88] Griffith LG, Swartz MA. Capturing complex 3D tissue physiology in vitro. Nat Rev Mol Cell Biol 2006;7(3):211–24.
[89] Mseka T, Bamburg JR, Cramer LP. ADF/cofilin family proteins control formation of oriented actin-filament bundles in the cell body to trigger fibroblast polarization. J Cell Sci 2007;120(24):4332–44.
[90] Weaver VM, Lelièvre S, Lakins JN, Chrenek MA, Jones JCR, Giancotti F, et al. β4 integrin-dependent formation of polarized three-dimensional architecture confers resistance to apoptosis in normal and malignant mammary epithelium. Cancer Cell 2002;2(3):205–16.
[91] Lv D, Hu Z, Lu L, Lu H, Xu X. Three-dimensional cell culture: a powerful tool in tumor research and drug discovery. Oncol Lett 2017;14(6):6999–7010.
[92] Antoni D, Burckel H, Josset E, Noel G. Three-dimensional cell culture: a breakthrough in vivo. Int J Mol Sci 2015;16(3):5517–27.
[93] Joseph JS, Malindisa ST, Ntwasa M. Two-dimensional (2D) and three-dimensional (3D) cell culturing in drug discovery. IntechOpen 2018;1:13.
[94] Koledova Z. 3D cell culture: an introduction. Methods Mol Biol 2017;1612:1–11.
[95] Nath S, Devi GR. Three-dimensional culture systems in cancer research: focus on tumor spheroid model. Pharmacol Ther 2016;163:94–108.
[96] Ritter CA, Perez-Torres M, Rinehart C, Guix M, Dugger T, Engelman JA, et al. Human breast cancer cells selected for resistance to trastuzumab in vivo overexpress epidermal growth factor receptor and ErbB ligands and remain dependent on the ErbB receptor network. Clin Cancer Res 2007;13(16):4909–19.

[97] Albanese A, Lam AK, Sykes EA, Rocheleau JV, Chan WCW. Tumour-on-a-chip provides an optical window into nanoparticle tissue transport. Nat Commun 2013;4:2718.

[98] Esch EW, Bahinski A, Huh D. Organs-on-chips at the frontiers of drug discovery. Nat Rev Drug Discov 2015;14(4):248–60.

[99] Roma-Rodrigues C, Mendes R, Baptista PV, Fernandes AR. Targeting tumor microenvironment for cancer therapy. Int J Mol Sci 2019;20(10):840.

[100] Costa EC, Moreira AF, de Melo-Diogo D, Gaspar VM, Carvalho MP, Correia IJ. 3D tumor spheroids: an overview on the tools and techniques used for their analysis. Biotechnol Adv 2016;34(8):1427–41.

[101] Langhans SA. Three-dimensional in vitro cell culture models in drug discovery and drug repositioning. Front Pharmacol 2018;9:6.

[102] Benien P, Swami A. 3D tumor models: history, advances and future perspectives. Future Oncol 2014;10 (7):1311–27.

[103] Burdett E, Kasper FK, Mikos AG, Ludwig JA. Engineering tumors: a tissue engineering perspective in cancer biology. Tissue Eng Part B Rev 2010;16(3):351–9.

[104] Zhang YS, Khademhosseini A. Advances in engineering hydrogels. Science 2017;356(6337):eaaf3627.

[105] Li Y, Kumacheva E. Hydrogel microenvironments for cancer spheroid growth and drug screening. Sci Adv 2018;4(4):eaas8998.

[106] Lee GY, Kenny PA, Lee EH, Bissell MJ. Three-dimensional culture models of normal and malignant breast epithelial cells. Nat Meth 2007;4(4):359–65.

[107] Yeung TM, Gandhi SC, Wilding JL, Muschel R, Bodmer WF. Cancer stem cells from colorectal cancer-derived cell lines. Proc Natl Acad Sci U S A 2010;107(8):3722–7.

[108] Lang SH, Sharrard RM, Stark M, Villette JM, Maitland NJ. Prostate epithelial cell lines form spheroids with evidence of glandular differentiation in three-dimensional Matrigel cultures. Br J Cancer 2001;85 (4):590–9.

[109] Singh SK, Clarke ID, Terasaki M, Bonn VE, Hawkins C, Squire J, et al. Identification of a cancer stem cell in human brain tumors. Cancer Res 2003;63(18):5821–8.

[110] Deisboeck TS, Berens ME, Kansal AR, Torquato S, Stemmer-Rachamimov AO, Chiocca EA. Pattern of self-organization in tumour systems: complex growth dynamics in a novel brain tumour spheroid model. Cell Prolif 2001;34(2):115–34.

[111] Lin RZ, Chang HY. Recent advances in three-dimensional multicellular spheroid culture for biomedical research. Biotechnol J 2008;3(9–10):1172–84.

[112] Mazzoleni G, Di Lorenzo D, Steimberg N. Modelling tissues in 3D: the next future of pharmaco-toxicology and food research? Genes Nutr 2009;4(1):13–22.

[113] Foty R. A simple hanging drop cell culture protocol for generation of 3D spheroids. J Vis Exp 2011;51:2720.

[114] Nagelkerke A, Bussink J, Sweep FCGJ, Span PN. Generation of multicellular tumor spheroids of breast cancer cells: how to go three-dimensional. Anal Biochem 2013;437(1):17–19.

[115] Haisler WL, Timm DM, Gage JA, Tseng H, Killian TC, Souza GR. Three-dimensional cell culturing by magnetic levitation. Nat Protoc 2013;8(10):1940–9.

[116] Türker E, Demirçak N, Arslan-Yildiz A. Scaffold-free three-dimensional cell culturing using magnetic levitation. Biomater Sci 2018;6(7):1745–53.

[117] Lazzari G, Couvreur P, Mura S. Multicellular tumor spheroids: a relevant 3D model for the: in vitro preclinical investigation of polymer nanomedicines. Polym Chem 2017;8(34):4947–69.

[118] Jørgensen A, Young J, Nielsen JE, Joensen UN, Toft BG, Rajpert-De Meyts E, et al. Hanging drop cultures of human testis and testis cancer samples: a model used to investigate activin treatment effects in a preserved niche. Br J Cancer 2014;110(10):2604–14.

[119] Hoarau-Véchot J, Rafii A, Touboul C, Pasquier J. Halfway between 2D and animal models: are 3D cultures the ideal tool to study cancer-microenvironment interactions? Int J Mol Sci 2018;19(1):E181.

[120] Ghosh S, Spagnoli GC, Martin I, Ploegert S, Demougin P, Heberer M, et al. Three-dimensional culture of melanoma cells profoundly affects gene expression profile: a high density oligonucleotide array study. J Cell Physiol 2005;204(2):522–31.

[121] De Witt Hamer PC, Van Tilborg AAG, Eijk PP, Sminia P, Troost D, Van Noorden CJF, et al. The genomic profile of human malignant glioma is altered early in primary cell culture and preserved in spheroids. Oncogene 2008;27(14):2091–6.

[122] Duval K, Grover H, Han L-H, Mou Y, Pegoraro AF, Fredberg J, et al. Modeling physiological events in 2D vs. 3D cell culture. Physiology 2017;32(4):266–77.
[123] Imamura Y, Mukohara T, Shimono Y, Funakoshi Y, Chayahara N, Toyoda M, et al. Comparison of 2D- and 3D-culture models as drug-testing platforms in breast cancer. Oncol Rep 2015;33(4):1837–43.
[124] Riedl A, Schlederer M, Pudelko K, Stadler M, Walter S, Unterleuthner D, et al. Comparison of cancer cells in 2D vs 3D culture reveals differences in AKT–mTOR–S6K signaling and drug responses. J Cell Sci 2017;130(1):203–18.
[125] Ekert JE, Johnson K, Strake B, Pardinas J, Jarantow S, Perkinson R, et al. Three-dimensional lung tumor microenvironment modulates therapeutic compound responsiveness in vitro - implication for drug development. PLoS One 2014;9(3):e92248.
[126] Pickl M, Ries CH. Comparison of 3D and 2D tumor models reveals enhanced HER2 activation in 3D associated with an increased response to trastuzumab. Oncogene 2009;28(3):461–8.
[127] Weigelt B, Lo AT, Park CC, Gray JW, Bissell MJ. HER2 signaling pathway activation and response of breast cancer cells to HER2-targeting agents is dependent strongly on the 3D microenvironment. Breast Cancer Res Treat 2010;122(1):35–43.
[128] Wartenberg M, Dönmez F, Ling FC, Acker H, Hescheler J, Sauer H. Tumor-induced angiogenesis studied in confrontation cultures of multicellular tumor spheroids and embryoid bodies grown from pluripotent embryonic stem cells. FASEB J 2002;15(6):995–1005.
[129] Majety M, Pradel LP, Gies M, Ries CH. Fibroblasts influence survival and therapeutic response in a 3D co-culture model. PLoS One 2015;10(6):e0127948.
[130] Lee J, Shin D, Roh JL. Development of an in vitro cell-sheet cancer model for chemotherapeutic screening. Theranostics 2018;8(14):3964–73.
[131] Todaro M, Alea MP, Di Stefano AB, Cammareri P, Vermeulen L, Iovino F, et al. Colon cancer stem cells dictate tumor growth and resist cell death by production of interleukin-4. Cell Stem Cell 2007;1(4):389–402.
[132] Bhuvanalakshmi G, Arfuso F, Millward M, Dharmarajan A, Warrier S. Secreted frizzled-related protein 4 inhibits glioma stem-like cells by reversing epithelial to mesenchymal transition, inducing apoptosis and decreasing cancer stem cell properties. PLoS One 2015;10(6):e0127517.
[133] Reynolds DS, Tevis KM, Blessing WA, Colson YL, Zaman MH, Grinstaff MW. Breast cancer spheroids reveal a differential cancer stem cell response to chemotherapeutic treatment. Sci Rep 2017;7(1):10382.
[134] Rodríguez CE, Berardi DE, Abrigo M, Todaro LB, Bal de Kier Joffé ED, Fiszman GL. Breast cancer stem cells are involved in Trastuzumab resistance through the HER2 modulation in 3D culture. J Cell Biochem 2018;119(2):1381–91.
[135] Wang K, Kievit FM, Erickson AE, Silber JR, Ellenbogen RG, Zhang M. Culture on 3D chitosan-hyaluronic acid scaffolds enhances stem cell marker expression and drug resistance in human glioblastoma cancer stem cells. Adv Healthc Mater 2016;5(24):3173–81.
[136] Melissaridou S, Wiechec E, Magan M, Jain MV, Chung MK, Farnebo L, et al. The effect of 2D and 3D cell cultures on treatment response, EMT profile and stem cell features in head and neck cancer. Cancer Cell Int 2019;19:16.
[137] Xiong GF, Xu R. Function of cancer cell-derived extracellular matrix in tumor progression. J Cancer Metastasis Treat 2016;2:357–64.
[138] Timpson P, McGhee EJ, Morton JP, Von Kriegsheim A, Schwarz JP, Karim SA, et al. Spatial regulation of RhoA activity during pancreatic cancer cell invasion driven by mutant p53. Cancer Res 2011;71(3):747–57.
[139] Bai C, Yang M, Fan Z, Li S, Gao T, Fang Z. Associations of chemo- and radio-resistant phenotypes with the gap junction, adhesion and extracellular matrix in a three-dimensional culture model of soft sarcoma. J Exp Clin Cancer Res 2015;34:58.
[140] Longati P, Jia X, Eimer J, Wagman A, Witt MR, Rehnmark S, et al. 3D pancreatic carcinoma spheroids induce a matrix-rich, chemoresistant phenotype offering a better model for drug testing. BMC Cancer 2013;13:95.
[141] Wang X, Zhen X, Wang J, Zhang J, Wu W, Jiang X. Doxorubicin delivery to 3D multicellular spheroids and tumors based on boronic acid-rich chitosan nanoparticles. Biomaterials 2013;34(19):4667–79.
[142] Hutchinson L, Kirk R. High drug attrition rates—where are we going wrong? Nat Rev Clin Oncol 2011;8 (4):189–90.
[143] Russell WMS, Burch RL. The principles of humane experimental technique. London: Methuen Co, Ltd; 1959.

[144] Wästfelt M, Fadeel B, Henter JI. A journey of hope: lessons learned from studies on rare diseases and orphan drugs. J Intern Med 2006;260(1):1–10.
[145] Puls TJ, Tan X, Husain M, Whittington CF, Fishel ML, Voytik-Harbin SL. Development of a novel 3D tumor-tissue invasion model for high-throughput, high-content phenotypic drug screening. Sci Rep 2018;8 (1):13039.
[146] Yamada KM, Cukierman E. Modeling tissue morphogenesis and cancer in 3D. Cell 2007;130(4):601–10.
[147] Trédan O, Galmarini CM, Patel K, Tannock IF. Drug resistance and the solid tumor microenvironment. J Natl Cancer Inst 2007;99(19):1441–54.
[148] Marushima H, Shibata SI, Asakura T, Matsuura T, Maehashi H, Ishii Y, et al. Three-dimensional culture promotes reconstitution of the tumor-specific hypoxic microenvironment under TGFβ stimulation. Int J Oncol 2011;39(5):1327–36.
[149] Sant S, Johnston PA. The production of 3D tumor spheroids for cancer drug discovery. Drug Discov Today Technol 2017;23:27–36.
[150] Grimes DR, Kelly C, Bloch K, Partridge M. A method for estimating the oxygen consumption rate in multicellular tumour spheroids. J R Soc Interface 2014;11(92):20131124.
[151] Cook CC, Kim A, Terao S, Gotoh A, Higuchi M. Consumption of oxygen: a mitochondrial-generated progression signal of advanced cancer. Cell Death Dis 2012;3:e258.
[152] Wartenberg M, Schallenberg M, Hescheler J, Sauer H. Reactive oxygen species-mediated regulation of eNOS and iNOS expression in multicellular prostate tumor spheroids. Int J Cancer 2003;104(3):274–82.
[153] Sulkowska M, Wincewicz A, Sulkowski S, Koda M, Kanczuga-Koda L. Relations of TGF-β1 with HIF-1α, GLUT-1 and longer survival of colorectal cancer patients. Pathology 2009;41(3):254–60.
[154] Khaitan D, Chandna S, Arya MB, Dwarakanath BS. Differential mechanisms of radiosensitization by 2-deoxy-D-glucose in the monolayers and multicellular spheroids of a human glioma cell line. Cancer Biol Ther 2006;5(9):1142–51.
[155] Liao J, Qian F, Tchabo N, Mhawech-Fauceglia P, Beck A, Qian Z, et al. Ovarian cancer spheroid cells with stem cell-like properties contribute to tumor generation, metastasis and chemotherapy resistance through hypoxia-resistant metabolism. PLoS One 2014;9(1):e84941.
[156] Pereira PMR, Berisha N, Bhupathiraju NVSDK, Fernandes R, Tomé JPC, Drain CM. Cancer cell spheroids are a better screen for the photodynamic efficiency of glycosylated photosensitizers. PLoS One 2017;12(5): e0177737.
[157] Khaitan D, Chandna S, Arya MB, Dwarakanath BS. Establishment and characterization of multicellular spheroids from a human glioma cell line; implications for tumor therapy. J Transl Med 2006;4:12.
[158] Walenta S, Doetsch J, Mueller-Klieser W, Kunz-Schughart LA. Metabolic imaging in multicellular spheroids of oncogene-transfected fibroblasts. J Histochem Cytochem 2000;48(4):509–22.
[159] Gottfried E, Kunz-Schughart LA, Ebner S, Mueller-Klieser W, Hoves S, Andreesen R, et al. Tumor-derived lactic acid modulates dendritic cell activation and antigen expression. Blood 2006;107(5):2013–21.
[160] Tian X, Wang W, Zhang Q, Zhao L, Wei J, Xing H, et al. Hypoxia-inducible factor-1α enhances the malignant phenotype of multicellular spheroid HeLa cells in vitro. Oncol Lett 2010;1(5):893–7.
[161] Terashima J, Sampei S, Iidzuka M, Ohsakama A, Tachikawa C, Satoh J, et al. VEGF expression is regulated by HIF-1α and ARNT in 3D KYSE-70, esophageal cancer cell spheroids. Cell Biol Int 2016;40(11):1187–94.
[162] Qin Y, Roszik J, Chattopadhyay C, Hashimoto Y, Liu C, Cooper ZA, et al. Hypoxia-driven mechanism of vemurafenib resistance in melanoma. Mol Cancer Ther 2016;15(10):2442–54.
[163] Swietach P, Hulikova A, Patiar S, Vaughan-Jones RD, Harris AL. Importance of intracellular pH in determining the uptake and efficacy of the weakly basic chemotherapeutic drug, doxorubicin. PLoS One 2012;7 (4):e35949.
[164] Barrera-Rodríguez R, Fuentes JM. Multidrug resistance characterization in multicellular tumour spheroids from two human lung cancer cell lines. Cancer Cell Int 2015;15:47.
[165] Gong X, Lin C, Cheng J, Su J, Zhao H, Liu T, et al. Generation of multicellular tumor spheroids with microwell-based agarose scaffolds for drug testing. PLoS One 2015;10(6):e0130348.
[166] Rohwer N, Cramer T. Hypoxia-mediated drug resistance: novel insights on the functional interaction of HIFs and cell death pathways. Drug Resist Updat 2011;14(3):191–201.

CHAPTER

# 7

# Biomaterials as ECM-like matrices for 3D in vitro tumor models

*Lara Pierantoni*[1,2], *Joana Silva-Correia*[1,2], *Antonella Motta*[3], *Rui L. Reis*[1,2] and *Joaquim M. Oliveira*[1,2]

[1]3B's Research Group, I3Bs—Research Institute on Biomaterials, Biodegradables and Biomimetics of University of Minho, Headquarters of the European Institute of Excellence on Tissue Engineering and Regenerative Medicine, Guimarães, Portugal [2]ICVS/3B's—PT Government Associated Laboratory, Braga/Guimarães, Portugal [3]BIOtech Research Center, Department of Industrial Engineering and European Institute of Excellence on Tissue Engineering and Regenerative Medicine, University of Trento, Trento, Italy

## Abbreviations

**3D** three-dimensional
**bFGF** basic fibroblast growth factor
**ECM** extracellular matrix
**GG** gellan gum
**HA** hyaluronic acid
**IL-8** interleukin 8
**MSCs** mesenchymal stem cells
**MW** molecular weight
**PCL** poly(ε-caprolactone)
**PEG** poly(ethylene glycol)
**PGA** poly(vinyl alcohol)
**PLA** poly(lactic acid)
**PLGA** poly(lactic-*co*-glycolic acid)
**TE** tissue engineering
**TME** tumor microenvironment
**VEGF** vascular endothelial growth factor


DOI: https://doi.org/10.1016/B978-0-12-818128-7.00007-1

## 7.1 Introduction

It is unfortunately quite common to see on the news people deceased due to cancer disease, with Steve Jobs one of the sadly known cases happened in the recent years. In fact, cancer is the second leading cause of mortality worldwide, responsible for approximately 9.6 million deaths in 2018 [1,2]. The initiation and development of a tumor mass is caused by a concurrent set of changes occurring in cell components, such as genomic, transcriptomic, and proteomic alterations [3–5]. At the DNA level, mutations usually can affect specific genes, namely oncogenes and tumor suppressor genes, which usually have a key role in controlling molecular pathways involved in cell survival, growth, and proliferation [6–9]. The genetic alterations impair the tight and controlled balance of these molecular pathways, thus leading to an uncontrolled cells proliferation and consequently to tumor formation. In the past years, the key phenomena driving tumor initiation and progression have been firstly identified by Hanahan and Weinberg, and collectively defined as hallmarks of cancer [10,11]. Namely sustaining proliferative signaling, evading growth suppressors, avoiding immune destruction, enabling replicative immortality, tumor-promoting inflammation, activating invasion and metastasis, inducing angiogenesis, genome instability and mutation, resisting cell death, and deregulating cellular energetics, make cancer a complex disease of altered cell molecular pathways.

Cells constituting a tissue or organ are surrounded by the extracellular matrix (ECM), a complex system comprising different structural components, which helps to maintain the physical architecture and biological homeostasis of the tissue [12,13]. The ECM is mainly composed by two different macromolecule types, namely proteoglycans and fibrous proteins, such as collagen, elastin, fibronectin, and laminin [12]. The interaction between cells–ECM is particularly important in tumor development. Cancer cells in fact release molecules that degrade the surrounding ECM and alter the behavior of the adjacent stromal cells. In turn, they will release growth factors used by cancer cells to increase their tumorigenicity, migration, and invasion potential, by establishing the so-called tumor microenvironment (TME) (Fig. 7.1) [14,15]. Cancer research studies have been usually carried out relying on the use of two-dimensional (2D) cell cultures and animal models, but these systems lack the three-dimensional (3D) molecular and mechanical environment provided by the ECM and clinical translation, respectively [16,17]. In recent years, the continuous technological progresses in the TE field are paving the way for the development of 3D tissue and disease models [18–20]. Specifically, an increased interest is being given on the development of in vitro 3D tumor models as a benchtop tool to better study the molecular development of a specific type of cancer of interest, in order to find novel possible targets for treatment [21–23].

## 7.2 Biomaterials as ECM-like matrices for cancer 3D in vitro models

Nowadays, there are many different biomaterials available for the generation of in vitro TE models. Depending on the originating source, they are divided into two main categories, synthetic- and natural-based. Table 7.1 summarizes the different materials types,

FIGURE 7.1 Schematic representation of the TME. The TME is formed by a heterogeneous complex cellular and ECM network. It is constituted of actively proliferating cancer cells, stromal cells embedded in the surrounding ECM, blood vessels (BV), and different types of immune infiltrating cells. The interplay between all these components can generate a multitude of different molecules and growth factors, leading to an increased tumorigenicity and invasion potential of the cancer cells.

elucidating their main advantages and disadvantages, and their application as ECM in different in vitro tumor models.

Biomaterials can be processed with many different techniques to produce constructs with diverse composition and architecture, which will ultimately influence the response and behavior of the incorporated cancer cell type. In the following sections, we will elucidate the characteristics of different materials types, and their main applications in the development of 3D in vitro tumor models.

### 7.2.1 Synthetic

Synthetic biomaterials, as the name suggests, do not derive from natural biological structures, but they are generated through chemical synthesis [50]. They possess the advantages of a relatively simple processing with high reproducibility, easy modification, and tunable degradation [50]. There is a wide variety of synthetic biomaterials available for TE applications, with PEG, PCL, PGA, and PLA as examples of the most commonly used ones [51]. PLA is composed of lactic acid (LA) monomers, which provides the D or L chiral form to the generated PLA polyester [52]. To improve its hydrophilicity and bioactivity, LA has been coupled with glycolic acid (GA), generating PLGA [52]. This polymer, in addition, has a fast and controllable degradation depending on the LA:GA ratios [53]. PEG is formed by the repetition of many ethylene glycol $[-(CH_2CH_2O)_n]$

TABLE 7.1 Summary of the main types of biomaterials used for the generation of 3D in vitro tumor models.

| Origin | Biomaterial type | | Advantages | Disadvantages | Cancer model | Reference(s) |
|---|---|---|---|---|---|---|
| Synthetic | Polyesters [poly (ε-caprolactone), PCL; poly (lactic acid), PLA; poly(lactic-*co*-glycolic acid), PLGA] | | Highly tunable degradability and mechanical properties; biocompatible | Biologically inert; poor protein adsorption; lack of cell adhesion and activation motifs | Breast cancer | [24] |
| | Poly(ethylene glycol) (PEG) | | Capable of absorbing great amounts of water; possibility of multiple functionalization with active biomolecules; biocompatible | Biologically inert; poor protein adsorption; lack of cell adhesion and activation motifs | Breast cancer | [25] |
| | Self-assembling peptide hydrogel (RAD16-I) matrix (BD PuraMatrix) | | Self-assembling; tunable fibrils assembling and degradation; easy functionalization | Expensive, unfeasible for large-scale cell cultures; short stability; poor mechanical properties | Lung adenocarcinoma; ovarian cancer; pancreatic ductal adenocarcinoma (PDAC) | [26–28] |
| Natural | Matrigel | | "Gold standard" for 3D cultures; great cell biocompatibility | Batch-to-batch variability; undefined composition; poor mechanical properties | Lung adenocarcinoma; melanoma; colorectal carcinoma | [26,29–31] |
| | Proteins | Silk fibroin (SF) and derivatives | Easily available; tunable processing; excellent mechanical properties; biocompatible | Instability of the derived liquid solution; some silk types lack cell adhesion moieties; batch-to-batch variability | Colorectal cancer (CRC); osteosarcoma; glioblastoma and ependymoma | [32–34] |
| | | Collagen (COL) | Major component of the ECM; Easily available; tunable architectural assembling; biocompatible | Batch-to-batch variability | Clear cell renal cell carcinoma; breast cancer; Ewing sarcoma (ES) | [35–37] |
| | | Gelatin | Low cost production; easy processability; biodegradable; biocompatible | Batch-to-batch variability; poor mechanical properties | Breast cancer; bladder cancer; glioblastoma | [38–40] |

*(Continued)*

**TABLE 7.1** (Continued)

| Origin | Biomaterial type | | Advantages | Disadvantages | Cancer model | Reference(s) |
|---|---|---|---|---|---|---|
| | Polysaccharides | Hyaluronic acid (HA) | Highly present in the ECM; great water absorbing capacity; biocompatible | Batch-to-batch variability | Breast cancer; prostate cancer | [41–43] |
| | | Gellan gum (GG) | Low cost production; Easy processability and tunability; high water uptake; biocompatible | Lack of cell recognition moieties; batch-to-batch variability | Osteosarcoma; lung; ovarian; hepatic cancer; nonsmall cell lung cancer (NSCLC) | [33,44,45] |
| | Decellularized matrices (dECM) | | Fully comprise the tissue architecture complexity; biocompatible | Batch-to-batch variability; decellularization method can remove tissue's biomolecules | CRC; breast cancer; hepatic cancer | [46–48] |
| | Cell-derived matrices | | Cell types specific; tunable production; biocompatible | Batch-to-batch variability | CRC | [49] |

monomeric parts [54]. This polyether can be produced with various degrees of polymerization and structures, giving the possibility to functionalize it with different biological groups [54]. Due to the lack of intrinsic cellular recognition moieties, synthetic biomaterials are minorly used for 3D tumor models generation. Nevertheless, research groups are working to improve their relevance in the field of cancer TE. For example, Rijal et al. [24] fabricated PLGA and PCL scaffolds that were able to support the viability and proliferation of different breast cancer cells. In addition, when implanted in mice, the in vitro tumor model showed cancer cells proliferation and levels of genes expression consistent with the ones observed in human. In another study, Pradhan and Slater [25] used PEG-based hydrogels (Fig. 7.2) to study the influence of matrix composition and stiffness on MDA-MB-231 breast cancer cells. They observed that RGDS (integrin ligating peptide) ligand density and hydrogels' degradability can influence the breast cancer cells behavior, passing from a dormant and inactive state to a high proliferative and invasiveness-prone one. These systems could be used to study the mechanisms underlying cancer cells dormancy, thus leading to novel therapeutic approaches against metastasis formation.

Another category of synthetic materials used for 3D cell cultures is the one of self-assembling peptides. Namely PuraMatrix is an amphiphilic peptide formed by 16 repeating arginine–alanine–aspartate–alanine sequences (RADARADARADARADA) [55]. In the presence of even small amounts of monovalent cations, the peptide self-assemble in

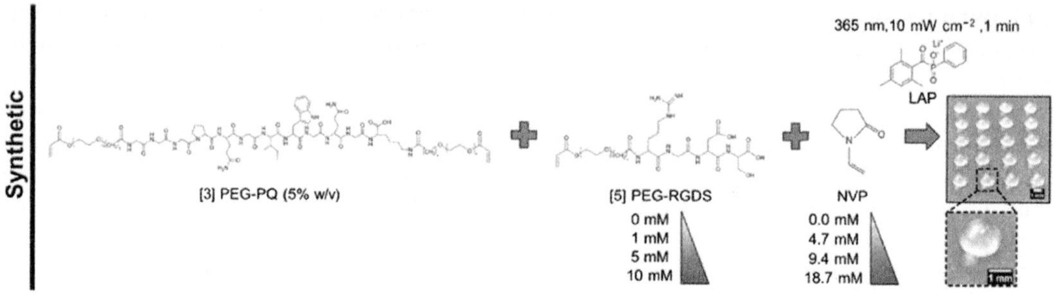


FIGURE 7.2 Schematic representation of the reaction mechanism for the generation of PEG-RGDS hydrogels reported by Pradhan and Slater [25]. Source: *Adapted and reproduced with permission from Pradhan S, Slater JH. Tunable hydrogels for controlling phenotypic cancer cell states to model breast cancer dormancy and reactivation. Biomaterials 2019;215:119177 [25].*

a β-sheet configuration, forming a stable nanofibrillar hydrogel, with a structure similar to the one found physiologically in the ECM [55]. Thus, as being able to mimic the ECM nanomesh architecture, PuraMatrix has been used to in vitro mimic pathological disease mechanisms and 3D tumor formation [26–28].

## 7.2.2 Natural-based

Biopolymers of natural origin, as the name suggests, are derived from animal or plant sources, thus having the advantage of being biocompatible. Since they are derived from intrinsic components of the ECM, or possess a similar physicochemical structure, they are extensively used to recapitulate matrix architecture and composition for the design of 3D in vitro tumor models. Within this category, Matrigel has been considered for years as the "gold standard" for 3D cancer cells culture [56]. Matrigel is derived from the basement membrane of tumor cells from Engelbreth–Holm–Swarm mouse, which contains a multitude of different proteins and growth factors [57]. Thus, by mimicking the molecular composition of the TME, Matrigel has been widely used for in vitro 3D tumor cells culture and studies [29–31]. Despite an undefined molecular composition, Matrigel also presents poor mechanical properties. As a strategy toward the improvement of its stiffness, other types of biomaterials (e.g., alginate) have been blended with Matrigel [58]. This highlights the need to exploit other types of material substrates, in order to generate 3D in vitro tumor models with specifically established composition and biomechanical cues. Given the multitude of natural-based biopolymers availability, the next subsections focus on the description of the recently used natural-based biomaterials for in vitro tumor models generation.

### *7.2.2.1 Proteins*

Silk fibers have been widely used for centuries in the textile industry for fabrics production [59]. Silk can be produced by different types of arthropods (e.g., spiders and silkworms) [59]. Depending on the originating species, the inner structure and physical properties of the silk protein will change [60–62]. Specifically, *Bombyx mori* silk fibroin

(SF), obtained from the degumming process of silkworm cocoons, has been widely used in biomedical applications due to its easy production, tunable degradation, and the capacity to undergo different processing techniques for TE disease model generation [63–65]. In particular, SF has been lately used for the 3D culture and the generation of in vitro cancer models, thanks to its excellent biocompatibility and mechanical properties, thus providing biomechanical cues to the cultured tumor cells [66,67]. Carvalho et al. [32] used enzymatically crosslinked silk fibroin (eSF) hydrogels in different concentrations in a dynamic, microfluidic chip to study the extravasation potential of HCT-116 CRC cells (Fig. 7.3). As compared with the control material, that is, Matrigel, the developed 2 wt.% eSF hydrogels showed an increased cell migration in the 3D microfluidic device, which was even higher in the presence of hVCAM-1. This molecule is involved in the adhesion of invasive cancer cells to the endothelial compartment, and its levels are found increased in CRC patients. This model on-a-chip could help to further understand the underlying mechanisms of invasiveness and metastasis of this type of cancer.

Kundu et al. [33] instead combined SF together with GG to obtain different blended hydrogels to generate an osteosarcoma 3D cancer model, by coculturing adipose-derived stem cells (ASCs) with Saos-2 osteosarcoma cells. They observed that the presence of ASCs induces the formation of spheroids of Saos-2 cancer cells, but only when cultured in specific GG–SF hydrogels ratio (i.e., GG75:S25). In addition, as compared with GG, cancer cells cultured in GG75:S25 hydrogels showed an increased alkaline phosphatase expression and upregulation of genes considered biomarkers of osteosarcoma, such as osteopontin, osteocalcin, and RUNX2. Altogether these results highlight the importance of the interplay between stromal and tumor cells in cancer development, which is related to specific TME stiffness. To study the interactions between brain cancer cells and the surrounding ECM, Sood et al. [34] used patient-derived pediatric anaplastic ependymoma and adult glioblastoma cancer cells, included in previously developed compartmentalized SF scaffolds. To better mimic the complex ECM environment, SF scaffolds were infused either with collagen type I or HA hydrogels combined with decellularized, porcine-derived fetal or adult brain ECM. Researchers observed that these two types of cancer exhibited different metabolic activity and behavior even when seeded in the same brain ECM

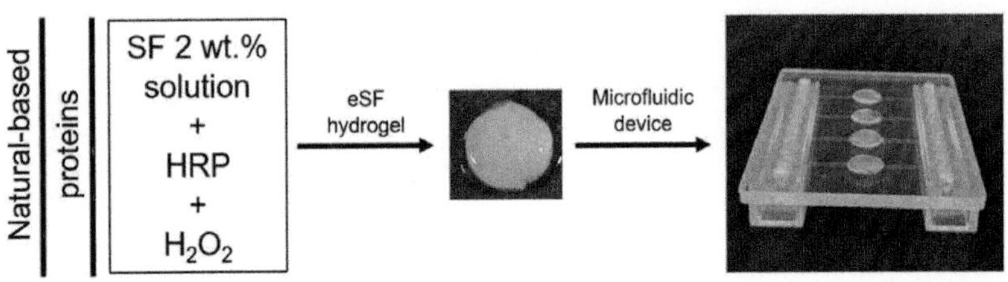


**FIGURE 7.3** Schematic representation of eSF hydrogels incorporation into a microfluidic platform presented by Carvalho et al. [32]. For eSF hydrogels preparation, SF 2 wt.% solution was used as a substrate for the enzymatic crosslinking, mediated by horseradish peroxidase (HRP) and hydrogen peroxide ($H_2O_2$) as oxidizer. eSF hydrogels were then incorporated into a microfluidic chip to study CRC cells extravasation potential. Source: *Adapted and reproduced with permission from Carvalho MR, Maia FR, Vieira S, Reis RL, Oliveira JM. Microfluidics: tuning enzymatically crosslinked silk fibroin hydrogel properties for the development of a colorectal cancer extravasation 3D model on a chip (Global Challenges 5-6/2018). Glob Chall 2018;2:1870164 [32].*

environment, highlighted by a distinct transcriptomic and molecular release profile. The overall results of this study further prove that cancer cells interaction with the surrounding ECM is tumor-type specific, and that this 3D bioengineered model could be used to study novel therapeutics against diverse types of brain cancers.

Collagen is one of the major structural components of tissues' ECM. There are 28 different types of collagens present in our body, with tissues and organs specific abundance and distribution, even though collagen type I is the most known and used [68]. Collagen I is first synthesized by cells as procollagen subunits, which are subsequently assembled in the α-helix (secondary structure), triple helix (tertiary structure), fibrils (quaternary structure), and supramolecular aggregates, following this order, respectively [68]. The collagen fibers network, also by interacting with several other components of the ECM, is necessary to maintain the elasticity and mechanical properties of the tissues. In fact, alterations in collagen amounts or structures give rise to a great number of hereditary and pathological disorders [69]. Specifically in the case of cancer, alterations in collagen fibers formation, with a consequent increase in the tumor stroma stiffness, is linked to a higher tumor cells' carcinogenic potential [70]. So given its natural presence in the tissue microenvironment, and its established role in influencing cancer cells behavior, collagen has been widely used as biomaterial for 3D tumor models generation [71]. For example, Miller et al. [35] used collagen I gels to generate a clear cell renal cell carcinoma (ccRCC) on-a-chip, using patient-derived cancer cells and endothelial cells. Under the influence of the flow and the surrounding matrix, ccRCC tumors induced a strong angiogenesis activation in the surrounding endothelial cells, with sprouting formation. This system highlights the importance of cancer and endothelial cells interplay, making the ccRCC-on-a-chip a suitable model to better study the molecular mechanisms underlying their interactions, and thus possible novel drug targets. Hume et al. [36] were able to develop a 3D engineered breast adipose tissue model, based on anisotropic collagen I scaffold and breast-derived human MSCs, which were subsequently immortalized. They demonstrated that these cells, in the presence of the collagen fibers, were able to differentiate into mature adipocytes resembling the adipose tissue architecture found in breast tumors. In addition, the presence of adipose cells enhanced the migration of metastatic MDA-MB-231 breast cancer cells, as compared with empty scaffold controls. The developed 3D model highlights the important role of adipose cells on the migration capability of breast cancer cells, which can provide the basis for possible novel therapeutic strategies. Marturano-Kruik et al. [37] combined collagen I together with HA to produce 3D porous scaffolds for the development of an ES model (Fig. 7.4). Researchers incorporated ES patient-derived cells and ES cancer cell lines into the models and cultured them in a bioreactor under physiological mechanical stimulation. As compared with control models cultured in static conditions, they observed that the ones cultured in the bioreactor activate ERK1/2 kinases, leading to the downstream activation of the transcription factor RUNX2, which is involved in tumor cells survival and development of drug resistance. This study further highlights the importance of biophysical and mechanical cues of bioengineered 3D tumor models to be used for drug screening.

The hydrolyzation of collagen fibrils and meshes originates another commonly used type of biomaterial, namely gelatin [72]. It can be obtained from different animal sources, such as porcine, bovine, and fish, which will lead to distinct gelatin's structure and

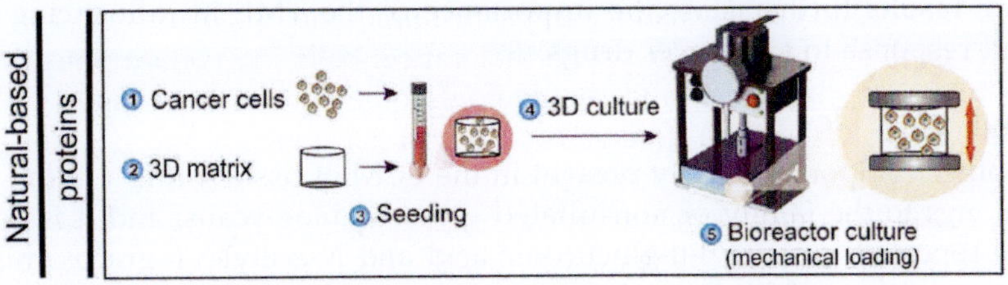


**FIGURE 7.4** Experimental workflow of the study undertaken by Marturano-Kruik et al. [37]. ES cancer cells (1) were seeded onto a 3D collagen I-HA scaffold (2, 3), and then 3D cultured in a bioreactor under mechanical stimuli (4, 5). Source: *Adapted and reproduced with permission from Marturano-Kruik A, et al. Biomechanical regulation of drug sensitivity in an engineered model of human tumor. Biomaterials 2018;150:150–61 [37].*

aminoacidic and protein composition [72]. Gelatin, possessing all the cell biological cues derived from collagen, has proven to be a great biomaterial for 3D cells culture, but with poor mechanical properties. For this reason, gelatin is usually modified (e.g., by methacrylation) or blended with different other natural-based materials, to improve its mechanostability and further mimic the ECM structure and composition [73,74]. The similar biological cues to the native ECM make gelatin a good candidate for 3D in vitro tumor models generation. For example, Brancato et al. [38] developed a 3D breast cancer microtissue based on porous gelatin microbeads to study the efficacy of doxorubicin, using MCF-7 breast cancer cells cocultured with cancer-associated fibroblasts (CAF) in dynamic conditions. As compared with the standard spheroid model, the CAF present in the 3D microtissue were able to produce and assemble their own ECM, simulating the one observed in the TME in vivo. Even though the developed 3D models were more subjected to doxorubicin diffusion and cell death, during time the presence of CAF generated drug resistance, by inducing the expression of the multidrug resistance P-gp glycoprotein. These findings highlight the importance of cancer cells and TME stromal cells interactions in the development of chemotherapeutics resistance, and the capacity of this model to recapitulate in vitro the mechanisms of drug resistance that are observed in the native tumor, proving its suitability as a drug-screening platform. Kim et al. [39] used gelatin methacryloyl as a bioink to bioprint a 3D scaffold containing the 5637 bladder cancer cells line, resembling the native diseased tissue. As compared with 2D-cultured T24 bladder tumor cells, 5637 cells cultured in the 3D bioprinted scaffold showed higher proliferation and expression of E- and N-cadherin, indicating higher cells interaction. In addition, 3D cultured cells exhibited lower sensitivity to rapamycin and Bacillus Calmette–Guérin (BCG), anticancer drugs typically used in this specific type of cancer. Overall, these results proved the suitability of the 3D bioprinting technique to generate a bladder tissue-like TME, which can be used for screening of novel therapeutics. Ngo and Harley [40] instead studied the interaction of the perivascular niche (PVN) with the related glioblastoma (GBM) tumor formation, by culturing endothelial and stromal cells together with GBM tumor cells on methacrylated-gelatin hydrogels. As compared with GBM and perivascular niche cells monoculture, the integrated PVN–GBM model showed an upregulation of genes involved in angiogenesis, ECM remodeling, and in molecular pathways related to GBM switch to a malignant phenotype. Moreover, the PVN model showed less responsiveness to temozolomide, one of the major therapeutics used to treat this type of cancer.

Overall these results further state the importance of the TME in influencing tumor cells behavior and response to anticancer drugs.

#### *7.2.2.2 Polysaccharides*

HA is another component highly present in the ECM of tissues. It is a linear polysaccharide that belongs to the family of nonsulfated glycosaminoglycans, and it is composed by disaccharide repeated units of β-D-glucuronic acid and *N*-acetyl-β-D-glucosamine [75]. HA helps to maintain the architecture of the stroma, capable of absorbing great amounts of water, thus preserving the moisture of the tissue. HA chains can be divided in low molecular weight ($10^4$ Da $\leq$ MW $<$ $10^6$ Da) and high molecular weight (MW $\geq$ $10^6$ Da), which can induce different biological responses [75,76]. Recently, studies have shown the role of HA in mediating cancer cell–TME interaction [77,78]. In fact, accumulation of HA has been correlated with higher cancer cells malignancy and poor prognosis in patients [79,80]. For these reasons, HA has been widely used as a biomaterial for 3D cancer cells culture and for tumor models generation. For example, Suo et al. [41] used two HA derivatives to generate HA hydrogels for the mimicry of the breast cancer ECM (Fig. 7.5). As compared with standard 2D culture, MCF-7 breast cancer cells seeded on HA hydrogels showed an increased cell growth and proliferation, upregulation of pro-tumor growth factors (e.g., VEGF, bFGF, IL-8), and an enhanced migration, invasion, and tumorigenic potential. These results demonstrate the suitability of HA hydrogels to mimic the physiological cues of the ECM of breast cancer.

In another experimental setup, the interaction between breast cancer cells and endothelial cells was studied by Kassim et al. [42] using MDA-MB-231 and HMEC cells seeded on HA-crosslinked hydrogels. They observed that in the 3D environment provided by HA, endothelial cells assemble in spheroids, with the cancer cells surrounding them. In addition, levels of VEGF, MMP-2, and MMP-9 change during the coculture time, decreasing until 6 days and then increasing at day 12. This could be possibly related to a period of quiescence of the cancer cells, induced by their restored polarity. The overall results of this study further highlight the important role of the 3D tissue structure in the tumor cells malignant behavior. Instead of using it singularly, Tang et al. [43] combined HA and alginate to develop an in vitro 3D prostate tumor model, by encapsulating PCa cells into HA-alginate hybrid hydrogels. They observed that in the 3D matrix, cancer cells were forming spheroids, and presented an increased expression of EMT-associated factors and higher drug tolerance as compared with standard 2D cultures. Overall, these findings further highlight the importance of a detailed reconstruction of the 3D TME to closely recapitulate cancer development as in the native human compartment.

Another widely used biomaterial in the TE field is GG. It is an anionic, linear polysaccharide produced by the *Sphingomonas elodea* bacteria species [81]. GG can present different degrees of acetylation, which will influence its rheological behavior. In fact, a higher degree of acetylation will lead to the formation of thermoreversible and soft gels, whereas a low degree of acetylation will generate highly thermally stable but brittle gels [82]. Even though its lack of cells' recognition and adhesion moieties, GG has been modified and combined with many different molecules to improve its biological relevance [81]. In this way, GG has been widely used for drug and cell therapies delivery applications and in the TE cancer field, as previously shown by the work of Kundu et al. [33]. In another example,

Natural-based polysaccharides

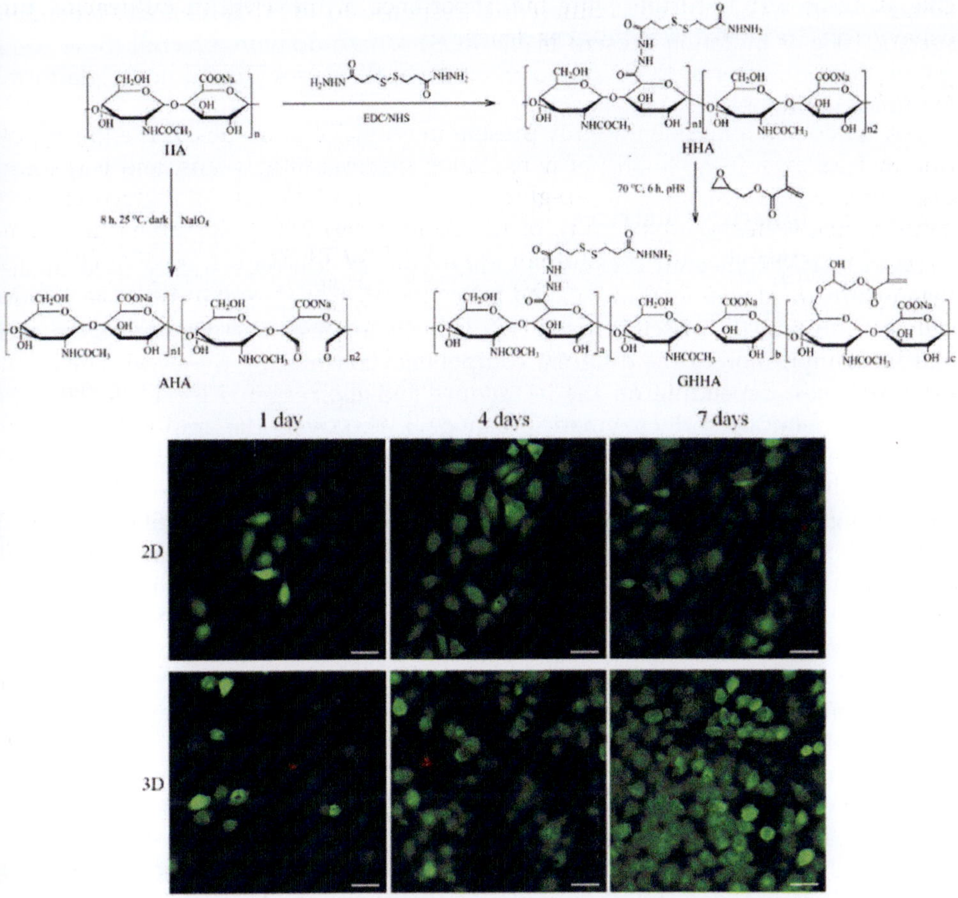


**FIGURE 7.5** Composition and properties of HA hydrogels proposed by Suo et al. [41]. The top part of the image shows the chemical composition of HA derivatives used for hydrogels preparation. In the bottom, MCF-7 breast cancer cells behavior is represented, upon culture on 2D surfaces and on the 3D developed HA hydrogels. Source: *Adapted and reproduced with permission from Suo A, et al. Dual-degradable and injectable hyaluronic acid hydrogel mimicking extracellular matrix for 3D culture of breast cancer MCF-7 cells. Carbohydr Polym 2019;211:336–48 [41].*

Aihara et al. [44] added GG to specific cell culture media in order to generate a homogeneous 3D culture of A549, SKOV3, and HepG2 cancer cells-derived spheroids. They observed that the addition of GG, by generating a low attachment environment, improved cancer cells proliferation and sensitivity to epidermal growth factor (EGF) inhibitors, which are commonly used as targeted therapy. Therefore, this 3D culture technique could be used for the screening of novel anticancer therapeutics in a high-throughput manner. In another experimental setup, Gong et al. [45] instead used GG to study the mechanisms of ROS1 fusion gene's inhibitors resistance in NSCLC. The authors observed that HCC78 cells, cultured with GG, presented an upregulation of the ROS1 gene and a down-regulation of the EGFR pathway activity. In addition, subclones of HCC78 cancer cells, cultured with GG, showed a differential response to ROS1 inhibitors, dependent on the specific type of mutation present in the ROS1 kinase domain. Overall these results could mimic the mutations of ROS1 found in patients, thus providing a novel platform for testing novel inhibitors.

### 7.2.3 Decellularized matrices

Besides self-engineered constructs, other types of scaffolds widely used in the TE and regenerative medicine field are dECM [83]. They derive from the complete removal of the cells constituting the specific tissue or organ of interest, but preserving the architecture and the composition of the different components constituting the ECM. The decellularization processes, depending on the techniques and the reagents used, can be divided into physical, chemical, and enzymatic methods [84]. Given the importance of the cancer cells–ECM interaction, also mediated by its structural and mechanical cues, dECM have started to be used for the generation of cancer models. The research group of Piccoli et al. [46], using a detergent–enzymatic process, were able to decellularize CRC matrices derived from tumor biopsies, maintaining the physiological structural architecture of the tissue. In addition, when HT-29 adenocarcinoma cells were used to repopulate the acellular matrices for 5 days of culture, they started to overexpress IL-8, essential for cancer development. This highlights the promising use of these 3D CRC matrices for preclinical screenings and drug development. With a similar experimental setup, instead, Jin et al. [47] were able to decellularize normal tissue and breast cancer biopsies using sodium lauryl ether sulfate. In addition, they observed that after 30 days of repopulation with MCF-7 adenocarcinoma cells, breast cancer-derived dECM induced MCF-7s′ epithelial to mesenchymal transition (EMT) and proliferation; whereas normal dECM inhibited EMT and induced cancer cells apoptosis. This study highlights the use of 3D cancer models based on dECM to study novel molecular pathways to target breast cancer development. In another work, Lu et al. [48] demonstrated that the combination of decellularized liver matrix components and gelatin-methacryloyl scaffold, into a microfluidic chip, enhance hepatic HepG2 cancer cells viability and behavior under flow (Fig. 7.6). This fact is due to the biochemical and biophysical cues provided by the decellularized liver matrix factors, which contribute to the reestablishment of the 3D physiological TME of the liver, highlighting the possible use of this liver tumor on-a-chip for pharmacological studies.

Besides complete decellularized tissues, latest trends are focused on the development of cell-derived matrices (CDM) to be further explored for TE applications [85]. CDM can be originated either from primary cells or cell lines, which are cultured for a prolonged period of time to allow their ECM deposition [85]. Subsequently, the cell compartment is carefully removed, in order to preserve the ECM structure. Depending on the cell's types of origin, the consequent CDM will present specific microenvironmental cues. For this reason, CDM generated from cancer cells are currently being investigated for finding application as 3D substrates, in order to study in depth tumor development in specifically derived stroma architecture. As an example, Hoshiba and Tanaka [49] studied the mechanisms of CRC cells resistance to 5-fluorouracil (5-FU), by culturing them on CDM with

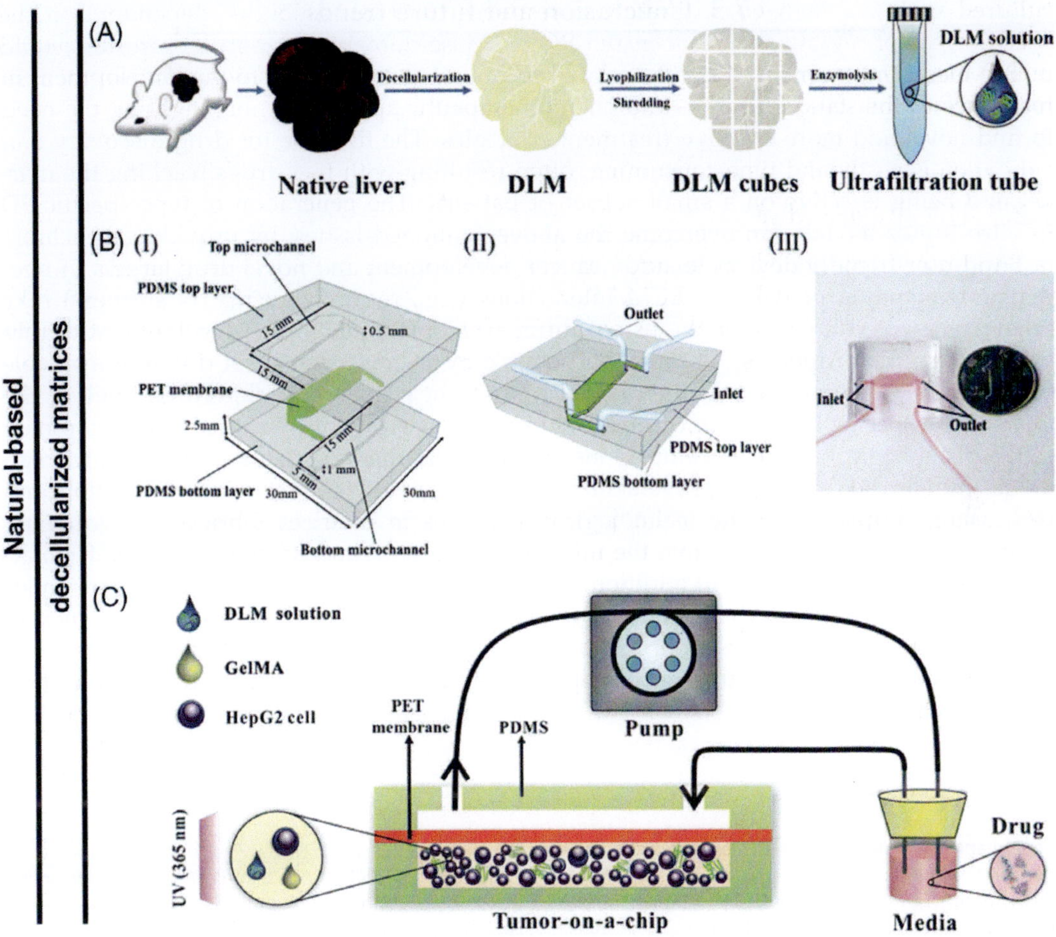


FIGURE 7.6 Schematic representation of the experimental setup of the work performed by Lu et al. [48]. From the top to the bottom part, the image shows the process of liver decellularization, the microfluidic platform fabrication, and the final assembly of the tumor-on-a-chip. Source: *Reproduced with permission from Lu S, et al. Development of a biomimetic liver tumor-on-a-chip model based on decellularized liver matrix for toxicity testing. Lab Chip 2018;18:3379–92 [48].*

different levels of malignancy. To do so, they utilized different CRC cell lines to obtain the related tumor ECM: HT-29 (high malignancy), SW480 (low malignancy), and normal CCD-841-CoN cells (as a control). As compared with the low malignancy state and the control, 5-FU resistance increased in CRC cells cultured on highly malignant HT-29 CDM, through the activation of Akt and the upregulation of multidrug resistance proteins. The reported results further highlight the important role of the ECM in the acquired resistance to anticancer drugs.

## 7.3 Conclusion and future trends

The increasing number of deaths related to cancer is mostly due to the development of resistance to the standard radio- and chemotherapeutic approaches, highlighting the need to find novel and more effective treatment strategies. The timeline for drug discovery and validation is costly and time-consuming, often resulting with few drugs reaching the market and being effective on a small subset of patients. The generation of type-specific 3D in vitro tumor models can overcome the abovementioned issues, by providing benchtop, fast and user-friendly devices to study cancer development and novel drug targets. A deeper understanding of the cell–ECM interactions (e.g., response to matrix stiffness) may open up new possibilities for the use of different biomaterials in the development of reliable in vitro tumor models. This will be possible by the exploitation of different available biomaterials and tune them to mimic the physicochemical and architectural cues of the ECM of the specific type of tumor under study, thus recapitulating its native development as the physiological one (e.g., cell response to host immune modulatory effect) as observed in the human body. The improvement in biomaterials' synthesis, functionalization and processing, coupled with the technological advances in matrices fabrication/processing techniques, will further ameliorate the mimicry of the TME and its behavior, thus facilitating their clinical translation. In addition, the further implementation of these tumor models with patient-derived cells, and combination with dynamic culturing platforms such as microfluidics and bioreactors in order to modulate cellular response to fluid shear stress, will pave the way for high-throughput approaches and efficient drug screenings and personalized therapies.

## Acknowledgments

This chapter is being part of the project FROnTHERA (NORTE-01-0145-FEDER-000023), funded by the Norte Portugal Regional Operational Programme (NORTE 2020), under the PORTUGAL 2020 Partnership Agreement, through the European Regional Development Fund (ERDF). Lara Pierantoni thanks the Portuguese Science Foundation (FCT) for her PhD scholarship (PD/BD/150520/2019). J. Silva-Correia and J. M. Oliveira also thanks the FCT for the funds provided under the program Investigador FCT 2015 (IF/00115/2015 and IF/01285/2015, respectively). This research has also been supported by REMIX project (H2020-MSCA-RISE-2017), funded by the European Union's Horizon 2020 research and innovation programme, under the Marie Sklodowska-Curie grant agreement n. 778078.

## References

[1] American Cancer Society. Cancer facts and figures 2017. Genes Dev 2017;21:2525–38.
[2] World Health Organization (WHO). GLOBOCAN 2018, International Agency for Research on Cancer (IARC). <http://globocan.iarc.fr/Pages/fact_sheets_cancer.aspx>; 2018.
[3] Jeggo PA, Pearl LH, Carr AM. DNA repair, genome stability and cancer: a historical perspective. Nat Rev Cancer 2016;16:35–42.
[4] Tubbs A, Nussenzweig A. Endogenous DNA damage as a source of genomic instability in cancer. Cell 2017;168:644–56.
[5] Ben-Aroya S, Levanon EY. A-to-I RNA editing: an overlooked source of cancer mutations. Cancer Cell 2018;33:789–90.
[6] Morris LGT, Chan TA. Therapeutic targeting of tumor suppressor genes. Cancer 2015;121:1357–68.

[7] Pierotti MA, Frattini M, Molinari F, Sozzi G, Croce CM. Oncogenes. In: Pierotti MA, Sozzi G, Croce CM, editors. Holland-Frei cancer medicine. Boston, MA: American Cancer Society; 2017. p. 1–22.
[8] Bailey MH, et al. Comprehensive characterization of cancer driver genes and mutations. Cell 2018;173:371–85.
[9] Sanchez-Vega F, et al. Oncogenic signaling pathways in the cancer genome atlas. Cell 2018;173:321–37.
[10] Hanahan D, Weinberg RA. The hallmarks of cancer. Cell 2000;100:57–70.
[11] Hanahan D, Weinberg RA. Hallmarks of cancer: the next generation. Cell 2011;144:646–74.
[12] Theocharis AD, Skandalis SS, Gialeli C, Karamanos NK. Extracellular matrix structure. Adv Drug Deliv Rev 2016;97:4–27.
[13] Vogel V. Unraveling the mechanobiology of extracellular matrix. Annu Rev Physiol 2018;80:353–87.
[14] Kai FB, Drain AP, Weaver VM. The extracellular matrix modulates the metastatic journey. Dev Cell 2019;49:332–46.
[15] Erdogan B, Webb DJ. Cancer-associated fibroblasts modulate growth factor signaling and extracellular matrix remodeling to regulate tumor metastasis. Biochem Soc Trans 2017;45:229–36.
[16] Duval K, et al. Modeling physiological events in 2D vs. 3D cell culture. Physiology 2017;32:266–77.
[17] Gengenbacher N, Singhal M, Augustin HG. Preclinical mouse solid tumour models: status quo, challenges and perspectives. Nat Rev Cancer 2017;17:751–65.
[18] Khademhosseini A, Langer R. A decade of progress in tissue engineering. Nat Protoc 2016;11:1775–81.
[19] Gomes ME, Rodrigues MT, Domingues RMA, Reis RL. Tissue engineering and regenerative medicine: new trends and directions - a year in review. Tissue Eng - Part B Rev 2017;23:211–24.
[20] Dutta D, Heo I, Clevers H. Disease modeling in stem cell-derived 3D organoid systems. Trends Mol Med 2017;23:393–410.
[21] Bielecka ZF, Maliszewska-Olejniczak K, Safir IJ, Szczylik C, Czarnecka AM. Three-dimensional cell culture model utilization in cancer stem cell research. Biol Rev 2017;92:1505–20.
[22] Ferreira LP, Gaspar VM, Mano JF. Design of spherically structured 3D in vitro tumor models: advances and prospects. Acta Biomater 2018;75:11–34.
[23] Santo VE, et al. Drug screening in 3D in vitro tumor models: overcoming current pitfalls of efficacy read--outs. Biotechnol J 2017;12:1–18.
[24] Rijal G, Bathula C, Li W. Application of synthetic polymeric scaffolds in breast cancer 3D tissue cultures and animal tumor models. Int J Biomater 2017;2017:8074890.
[25] Pradhan S, Slater JH. Tunable hydrogels for controlling phenotypic cancer cell states to model breast cancer dormancy and reactivation. Biomaterials 2019;215:119177.
[26] Prina-Mello A, et al. Culturing substrates influence the morphological, mechanical and biochemical features of lung adenocarcinoma cells cultured in 2D or 3D. Tissue Cell 2018;50:15–30.
[27] Yang Z, et al. A miniature cell pattern formation of ovarian cancer cell lines on self-assembling peptide nanofiber-coated coverslip and in vitro chemosensitivity assay. J Nanosci Nanotechnol 2018;18:2370–8.
[28] Betriu N, Semino C. Development of a 3D co-culture system as a cancer model using a self-assembling peptide scaffold. Gels 2018;4:65.
[29] Wessels D, et al. Melanoma cells undergo aggressive coalescence in a 3D Matrigel model that is repressed byanti-CD44. PLoS One 2017;12.
[30] Zoetemelk M, Rausch M, Colin DJ, Dormond O, Nowak-Sliwinska P. Short-term 3D culture systems of various complexity for treatment optimization of colorectal carcinoma. Sci Rep 2019;9:1–14.
[31] Schnalzger TE, et al. 3D model for CAR-mediated cytotoxicity using patient-derived colorectal cancer organoids. EMBO J 2019;38:1–15.
[32] Carvalho MR, Maia FR, Vieira S, Reis RL, Oliveira JM. Microfluidics: tuning enzymatically crosslinked silk fibroin hydrogel properties for the development of a colorectal cancer extravasation 3D model on a chip (Global Challenges 5-6/2018). Glob Chall 2018;2:1870164.
[33] Kundu B, et al. Mechanical property of hydrogels and the presence of adipose stem cells in tumor stroma affect spheroid formation in the 3D osteosarcoma model. ACS Appl Mater Interfaces 2019;11:14548–59.
[34] Sood D, et al. 3D extracellular matrix microenvironment in bioengineered tissue models of primary pediatric and adult brain tumors. Nat Commun 2019;10:1–14.
[35] Miller CP, Tsuchida C, Zheng Y, Himmelfarb J, Akilesh S. A 3D human renal cell carcinoma-on-a-chip for the study of tumor angiogenesis. Neoplasia (US) 2018;20:610–20.
[36] Hume RD, et al. An engineered human adipose/collagen model for in vitro breast cancer cell migration studies. Tissue Eng - Part A 2018;24:1309–19.

[37] Marturano-Kruik A, et al. Biomechanical regulation of drug sensitivity in an engineered model of human tumor. Biomaterials 2018;150:150–61.
[38] Brancato V, et al. 3D breast cancer microtissue reveals the role of tumor microenvironment on the transport and efficacy of free-doxorubicin in vitro. Acta Biomater 2018;75:200–12.
[39] Kim MJ, et al. Structure establishment of three-dimensional (3D) cell culture printing model for bladder cancer. PLoS One 2019;14:1–15.
[40] Ngo MT, Harley BAC. Perivascular signals alter global gene expression profile of glioblastoma and response to temozolomide in a gelatin hydrogel. Biomaterials 2019;198:122–34.
[41] Suo A, et al. Dual-degradable and injectable hyaluronic acid hydrogel mimicking extracellular matrix for 3D culture of breast cancer MCF-7 cells. Carbohydr Polym 2019;211:336–48.
[42] Kassim YL, Al Tawil E, Buquet C, Le Cerf D, PierreVannier J. Three dimensional tumor engineering by co-culture of breast tumor and endothelial cells using a hyaluronic acid hydrogel model. J Clin Exp Oncol 2017;06:6–13.
[43] Tang Y, et al. Three-dimensional prostate tumor model based on a hyaluronic acid-alginate hydrogel for evaluation of anti-cancer drug efficacy. J Biomater Sci Polym Ed 2017;28:1603–16.
[44] Aihara A, Abe N, Saruhashi K, Kanaki T, Nishino T. Novel 3-D cell culture system for in vitro evaluation of anticancer drugs under anchorage-independent conditions. Cancer Sci 2016;107:1858–66.
[45] Gong B, Oh-hara T, Fujita N, Katayama R. 3D culture system containing gellan gum restores oncogene dependence in ROS1 rearrangements non-small cell lung cancer. Biochem Biophys Res Commun 2018;501:527–33.
[46] Piccoli M, et al. Decellularized colorectal cancer matrix as bioactive microenvironment for in vitro 3D cancer research. J Cell Physiol 2018;233:5937–48.
[47] Jin Q, et al. Decellularized breast matrix as bioactive microenvironment for in vitro three-dimensional cancer culture. J Cell Physiol 2019;234:3425–35.
[48] Lu S, et al. Development of a biomimetic liver tumor-on-a-chip model based on decellularized liver matrix for toxicity testing. Lab Chip 2018;18:3379–92.
[49] Hoshiba T, Tanaka M. Decellularized matrices as in vitro models of extracellular matrix in tumor tissues at different malignant levels: mechanism of 5-fluorouracil resistance in colorectal tumor cells. Biochim Biophys Acta - Mol Cell Res 2016;1863:2749–57.
[50] Hacker MC, Krieghoff J., Mikos AG. Synthetic polymers. In: Atala A, Lanza RP, Nerem RM, Mikos T, editors. Principles of regenerative medicine. London: Elsevier; 2019. p. 559–590.
[51] Pradhan S, Hassani I, Clary JM, Lipke EA. Polymeric biomaterials for in vitro cancer tissue engineering and drug testing applications. Tissue Eng Part B Rev 2016;22:470–84.
[52] Manavitehrani I, et al. Biomedical applications of biodegradable polyesters. Polymers (Basel) 2016;8:E20.
[53] Kapoor DN, et al. PLGA: a unique polymer for drug delivery. Ther Deliv 2015;6:41–58.
[54] D'Souza AA, Shegokar R. Polyethylene glycol (PEG): a versatile polymer for pharmaceutical applications. Expert Opin Drug Deliv 2016;13:1257–75.
[55] Caliari SR, Burdick JA. A practical guide to hydrogels for cell culture. Nat Methods 2016;13:405–14.
[56] Li Y, Kumacheva E. Hydrogel microenvironments for cancer spheroid growth and drug screening. Sci Adv 2018;4:1–11.
[57] Benton G, Arnaoutova I, George J, Kleinman HK, Koblinski J. Matrigel: from discovery and ECM mimicry to assays and models for cancer research. Adv Drug Deliv Rev 2014;79:3–18.
[58] Cavo M, et al. A new cell-laden 3D Alginate-Matrigel hydrogel resembles human breast cancer cell malignant morphology, spread and invasion capability observed 'in vivo'. Sci Rep 2018;8:1–12.
[59] Holland C, Numata K, Rnjak-Kovacina J, Seib FP. The biomedical use of silk: past, present, future. Adv Healthc Mater 2019;8.
[60] Kundu B, Rajkhowa R, Kundu SC, Wang X. Silk fibroin biomaterials for tissue regenerations. Adv Drug Deliv Rev 2013;65:457–70.
[61] Koh LD, et al. Structures, mechanical properties and applications of silk fibroin materials. Prog Polym Sci 2015;46:86–110.
[62] Yarger JL, Cherry BR, Van Der Vaart A. Uncovering the structure-function relationship in spider silk. Nat Rev Mater 2018;3:18008.
[63] Thurber AE, Omenetto FG, Kaplan DL. In vivo bioresponses to silk proteins. Biomaterials 2015;71:145–57.

[64] Qi Y, et al. A review of structure construction of silk fibroin biomaterials from single structures to multi-level structures. Int J Mol Sci 2017;18:237.

[65] Zhang L, Liu X, Li G, Wang P, Yang Y. Tailoring degradation rates of silk fibroin scaffolds for tissue engineering. J Biomed Mater Res - Part A 2019;107:104–13.

[66] Jastrzebska K, et al. Silk as an innovative biomaterial for cancer therapy. Rep Pract Oncol Radiother 2015;20:87–98.

[67] Dondajewska E, Juzwa W, Mackiewicz A, Dams-Kozlowska H. Heterotypic breast cancer model based on a silk fibroin scaffold to study the tumor microenvironment. Oncotarget 2018;9:4935–50.

[68] Liu X, Zheng C, Luo X, Wang X, Jiang H. Recent advances of collagen-based biomaterials: multi-hierarchical structure, modification and biomedical applications. Mater Sci Eng C 2019;99:1509–22.

[69] Arseni L, Lombardi A, Orioli D. From structure to phenotype: impact of collagen alterations on human health. Int J Mol Sci 2018;19:E1407.

[70] Morris BA, et al. Collagen matrix density drives the metabolic shift in breast cancer cells. EBioMedicine 2016;13:146–56.

[71] Sapudom J, Pompe T. Biomimetic tumor microenvironments based on collagen matrices. Biomater Sci 2018;6:2009–24.

[72] Duconseille A, Astruc T, Quintana N, Meersman F, Sante-Lhoutellier V. Gelatin structure and composition linked to hard capsule dissolution: a review. Food Hydrocoll 2015;43:360–76.

[73] Klotz BJ, Gawlitta D, Rosenberg AJWP, Malda J, Melchels FPW. Gelatin-methacryloyl hydrogels: towards biofabrication-based tissue repair. Trends Biotechnol 2016;34:394–407.

[74] Afewerki S, Sheikhi A, Kannan S, Ahadian S, Khademhosseini A. Gelatin-polysaccharide composite scaffolds for 3D cell culture and tissue engineering: towards natural therapeutics. Bioeng Transl Med 2019;4:96–115.

[75] Gallo N, et al. Hyaluronic acid for advanced therapies: promises and challenges. Eur Polym J 2019;117:134–47.

[76] Cyphert JM, Trempus CS, Garantziotis S. Size matters: molecular weight specificity of hyaluronan effects in cell biology. Int J Cell Biol 2015;2015:563818.

[77] Nikitovic D, Tzardi M, Berdiaki A, Tsatsakis A, Tzanakakis GN. Cancer microenvironment and inflammation: role of hyaluronan. Front Immunol 2015;6:1–7.

[78] Senbanjo LT, Chellaiah MA. CD44: a multifunctional cell surface adhesion receptor is a regulator of progression and metastasis of cancer cells. Front Cell Dev Biol 2017;5:18.

[79] Li X, et al. Parallel accumulation of tumor hyaluronan, collagen, and other drivers of tumor progression. Clin Cancer Res 2018;24:4798–807.

[80] Seufferlein T, Ducreux M, Hidalgo M, Prager G, Van Cutsem E. More than a gel - hyaluronic acid, a central component in the microenvironment of pancreatic cancer. Eur Oncol Haematol 2018;14:40–4.

[81] Palumbo FS, Federico S, Pitarresi G, Fiorica C, Giammona G. Gellan gum-based delivery systems of therapeutic agents and cells. Carbohydr Polym 2019;229:115430. Available from: https://doi.org/10.1016/j.carbpol.2019.115430.

[82] Kirchmajer DM, Steinhoff B, Warren H, Clark R, In Het Panhuis M. Enhanced gelation properties of purified gellan gum. Carbohydr Res 2014;388:125–9.

[83] Taylor DA, Sampaio LC, Ferdous Z, Gobin AS, Taite LJ. Decellularized matrices in regenerative medicine. Acta Biomater 2018;74:74–89.

[84] Yao Q, et al. Recent development and biomedical applications of decellularized extracellular matrix biomaterials. Mater Sci Eng C 2019;104:109942.

[85] Fitzpatrick LE, McDevitt TC. Cell-derived matrices for tissue engineering and regenerative medicine applications. Biomater Sci 2015;3:12–24.

CHAPTER

# 8

# Three-dimensional in vitro models of angiogenesis

*Laura di Blasio*[1,2], *Marianela Vara-Messler*[1,2] *and Luca Primo*[1,2]

[1]Department of Oncology, University of Torino, Turin, Italy [2]Candiolo Cancer Institute FPO-IRCCS, Candiolo, Italy

## 8.1 Vessels formation and tumor angiogenesis

Blood vessels supply oxygen, nutrients and provide gateways for immune surveillance. Endothelial cells (ECs) line the inner surface of vessels to support tissue growth and repair. As this network nourishes all tissues, it is not surprising that structural or functional vessel abnormalities contribute to many diseases. Angiogenesis is the process that leads to the formation of new blood vessels both during development and in pathological conditions. This process of remodeling and expansion of a preexisting vessel network occurs mainly by endothelial sprouting and intussusceptive microvascular growth [1].

Physiologically, the process of angiogenesis begins at implantation, and during placental and embryonic development. In later life, the process is prominent in the healing of wounds, osteogenesis, as well as in ovulatory-related and nonovulatory-related reproductive processes. Furthermore, pathological angiogenesis is related to inflammatory disorders, pulmonary hypertension, blinding eye disease, and the development of both benign and malignant tumors [2].

The angiogenic process is regulated by many factors like local composition of the extracellular matrix (ECM) and physical contact between ECs and pericytes, but definitely vascular endothelial growth factor (VEGF)-A is the major regulator of angiogenesis. VEGF-A is part of a large family of potent angiogenic regulators including placental growth factor (PlGF), VEGF-B, VEGF-C, and VEGF-D [3]. The cellular response to VEGF-A is finally tuned by a receptor system that includes different family members: binding of VEGF-A to the receptor tyrosine kinase VEGFR2 (also known as KDR or FLK1) promotes EC differentiation, proliferation, and sprouting [1], in contrast VEGFR1 signaling seems to

*Biomaterials for 3D Tumor Modeling*
DOI: https://doi.org/10.1016/B978-0-12-818128-7.00008-3

be dispensable for endothelial cell function, but it is essential for the migration of hematopoietic cells [4].

The angiogenic sprouting starts with the selection of some ECs within the capillary vessel wall, so-called tip cells, which guide the growing vessels. The selection mechanism of the tip cells by the Delta-like-4 (Dll4) expression and Notch signaling between adjacent EC, has been described in detail: when a potential tip cell is selected by a strong VEGF stimulation, Dll4 expression increases activating Notch in the neighboring ECs, so-called stalk cells, which are inhibited to form new tip cells reducing VEGFR2 expression [1,5]. Then the tip sprouting ECs acquire an invasive and motile phenotype, activating proteases to degrade the surrounding basement membrane (BM). Alterations in Dll4/Notch signaling pathway determine several defects, such as enhanced formation of tip cells, resulting in dramatically increased sprouting, branching, and fusion of endothelial tubes [5,6].

The growth and the elongation of the newly formed blood vessel need to be polarized and directional. Indeed, VEGF-A drives these mechanisms through the prominent VEGFR2 expression in the tip cell: a gradient of a matrix-anchored isoform of VEGF-A acts as chemoattractive signal that promotes the polarized extension of tip cell filopodia. Another concentration gradient is generated since the newly formed blood vessel is highly unstable, which is platelet-derived growth factor B (PDGFB). This promotes the recruitment of pericytes that express the PDGF receptor β (PDGFRβ), ensuring a stabilizing effect on the endothelium of growing vessels by supporting mural cells [7].

In the setting of tumor angiogenesis, in the early 1970s, Judah Folkman hypothesized that tumor growth is dependent on angiogenesis and subsequently experimental work demonstrated that for tumor grow beyond a size of 1–2 $mm^3$ a substantial new blood supply must develop to support the increasing metabolic requirements [8]. Besides growing, tumor vessels serve also as routes for tumor cells to metastasize [9]. Following seminal observations by Folkman and other investigators, angiogenesis became a vivid area in cancer research and stimulated the search for conceptually new treatments [2,10].

The increased release of proangiogenic factors is one of the most common hallmarks of cancer and it is usually due to metabolic stress, hypoxia, low pH, hypoglycemia, inflammation, and cell stress [11]. This is the reason why the main therapeutic strategies followed to develop antiangiogenic agents involve the inhibition of proangiogenic factors. Angiogenic inhibitors such as Bevacizumab, a humanized anti-VEGF antibody, have been used for the past 15 years. However, the clinical benefits of antiangiogenic drugs have been limited, resulting in negligible improvements in prognosis [12].

## 8.2 Vascular extracellular matrix

The vascular ECM is a critical aspect for the angiogenenic process, as it provides both support for vascular endothelium and confers molecular signals essential for all stages of blood vessel formation, including sprouting, lumen formation, maturation, and ultimately vessel stabilization [13]. Several evidences support a model in which the vascular BM, a matrix layer rich of laminin, serves to maintain the integrity of the mature endothelium but also to insulate ECs from interstitial or provisional matrix [13]. Local degradation of vascular BM by endothelial podosomes proteases exposes ECs to fibrillar interstitial matrix and activates signaling pathways that drive cytoskeletal reorganization and sprouting morphogenesis [14]. Many of these interstitial and provisional components are capable of supporting chemotactic migration. Therefore, sprouting angiogenesis assays are frequently performed in hydrogel made of fibrillar collagens, which promote EC migration. Conversely, gel of basement membrane extract (BME), commercially known as Matrigel or Cultrex, is exploited for vessels differentiation assays.

### 8.2.1 Vascular basement membrane composition

Collagen type IV and laminins are the two major BM components that self-assemble to form independent networks that confer structural stability and biological activity, respectively. These two networks are connected by perlecans and nidogen [15].

The Collagen type IV is a nonfibril forming collagen and is derived from six genetically distinct α-chain polypeptides, with 56 possible combinations of homotrimers. Collagen type IV self-assembles into organized networks and thereby it is crucial for BM stability and assembly [16].

Laminins are composed of an α, β, and γ chain that assemble to form an approximately crossed-shaped or Y-shaped molecule. Up to 18 laminin isoforms have been identified by five α, four β, and three γ chains composition, by which is determined also laminins name. The assembly into the BM network is usually mediated by the N-terminal laminin domains, while the C-terminal domains contain the majority of receptor binding sites and are entirely constituted of α chain sequences [13].

EC express laminin α4 and α5 chains that combine with laminin β1 and γ1 chains to form laminins 411 and 511 (or isotypes 8 and 10), respectively [17]. Usually the α4 chain is ubiquitously found along different vessel types independent of their stage of development, while the α5 chain has a more varied distribution according to vessel type and it appears postnatally, approximately at the time of pericyte recruitment. Laminins are considered to be responsible for the biological functions of BMs and beta integrins constitute the major class of receptor for laminins of vBM. Integrins α6β1 and α3β1 have been reported to bind vascular laminin; however, there is no data on the signals transduced by these receptors when engaged by laminin.

Since vBM is difficult to obtain in sufficient amount to generate three-dimensional (3D) gel, it is generally replaced by hydrogel of BME purified from Engelbreth–Holm–Swarm mouse sarcoma, which contains laminin 111. This nonvascular laminin could affect in vitro EC biological response, although specific biochemical and cellular effects for each laminin type have not been completely characterized.

### 8.2.2 Interstitial matrix

During angiogenesis, the quiescent endothelial layer becomes "activated" and ECs breach the BM and migrate into surrounding tissue containing different complements of ECM proteins, which can include collagens and fibronectin in interstitial ECM or fibrinogen and fibronectin in provisional matrices generated after vascular injury and during wound healing.

The interstitial ECM is mainly composed of elastic fibers and fibrillar collagens, collagen I being the most abundant protein. It is crucial for establishing and maintaining the mechanical stability of the vascular system, as demonstrated in mice deficient for the $\alpha 1$ chain of collagen I, which die in embryo due to rupture of major blood vessels [18].

The "provisional ECM" is formed by plasma proteins, such as fibrin, fibrinogen, and fibronectin, seeping into the wound site along with hyaluronan, either from the plasma or released from migrating mesenchymal cells. These components interact to form a cross-linked gel which acts as a temporary scaffold for cellular events needed to repair the wound. Thus, the provisional matrix serves as a scaffold and substrate for other cells entering into the wound bed, such as fibroblasts and smooth muscle cells.

Interstitial matrix proteins and matricryptic sites, derived from proteolytic degradation of these proteins, interact with integrins, which are key receptors in stimulating EC activation as well as concurrent sprouting and tubular morphogenesis [19–21]. The major integrins responsible for these interactions are $\alpha_2\beta_1$, $\alpha_1\beta_1$, and $\alpha_v\beta_3$, $\alpha_5\beta_1$, which are collagen and fibrin/fibronectin receptors, respectively [22,23].

## 8.3 Endothelial cells-based 3D angiogenesis models

Vasculogenesis is the in situ primary differentiation of ECs from an undifferentiated precursor cell, while angiogenesis is the subsequent "sprouting" and organization of ECs into blood vessels [24,25]. During vasculogenesis a putative precursor cell differentiates to form an angioblast, which then will give rise to ECs, and a stem cell for the hematopoietic lineages. The analysis of early vascular development is possible by using embryoid bodies (EBs) obtained in vitro from embryonic stem cells (ES), derived from the inner cell mass of mouse blastocysts [26].

The maturation of the primitive vascular network through the process of angiogenesis is a very complex process, which comprises several steps: (1) EC proliferation, (2) EC migration, (3) building up of connections between EC and ECM, (4) lumen formation, and (5) remodeling and pruning [27]. These individual steps of the angiogenic process can be mimicked by in vitro assays using human primary EC. EC can be isolated from different human tissues and maintain typical endothelial markers and the ability to form capillary-like structures.

The most common assays to evaluate vascular morphogenesis in vitro are described below.

### 8.3.1 Vascular differentiation in embryoid body

Murine ES cell lines possess the potential to generate all embryonic cell lineages when they undergo differentiation [28]. When cultured in the presence of leukemia inhibitory factor (LIF), which inhibits their differentiation, ES cells maintain their pluripotential state. With the removal of LIF, ES cells spontaneously differentiate into cyst-like structures, termed as EBs, which contain derivatives of the three primitive germ layers. The appearance of blood island-like structures that consist of immature hematopoietic cells surrounded by ECs and the formation of vascular-like channels have been reported on the surface of cystic EBs (Fig. 8.1A), suggesting that ES cells produce all the factors necessary for the induction of vasculogenesis [29,30].

Differentiation of ES cells into EBs can be done in different ways [31]. The simplest method involves the removal of ES from self-renewing culture conditions and culturing

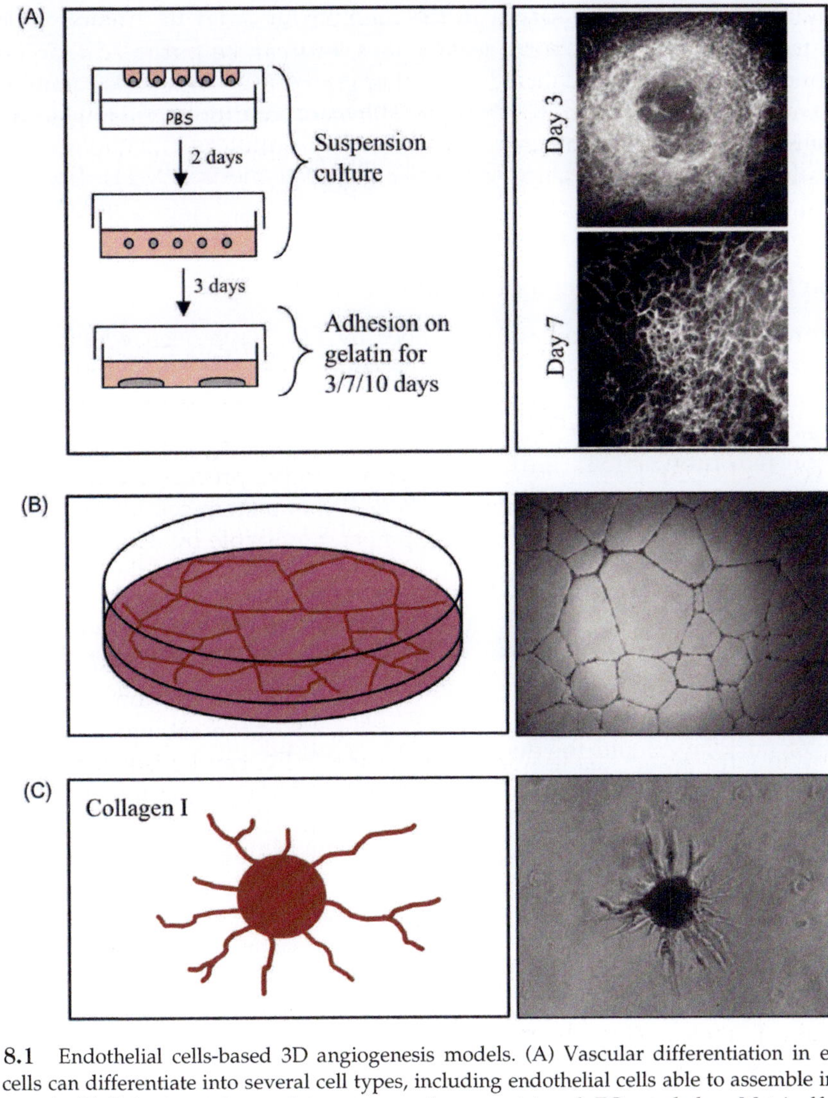


FIGURE 8.1 Endothelial cells-based 3D angiogenesis models. (A) Vascular differentiation in embryoidbody model: ES cells can differentiate into several cell types, including endothelial cells able to assemble into a complex vascular network. (B) Tube formation on basement membrane matrix gel: ECs seeded on Matrigelform a branching network of capillary-like structures after 6 h of culture. (C) Sprouting from EC spheroids in collagen gel: when spheroids of ECs are embedded in collagen type I gel and stimulated with angiogenic factors,ECs are able to invade the gel to form capillary-like structures.

them in suspension, either directly in liquid media or in methylcellulose supplemented media in bacterial grade petri dishes. The main disadvantage of this approach is the formation of very heterogeneous EBs, varying in both size and shape. A second method of initiating EB development is to culture the ES cells in hanging drops. In this way, it is possible to control the size of the generated EBs by varying the number of cells placed in each drop (typically 400 cells in 30 L drop). Once formed, such EBs are transferred to standard suspension culture in bacterial dishes to grow and complete their development. This method requires the addition of serum in the medium in order to promote efficient and uniform EB formation. Usually after at least 3 days, EBs can be put in adhesion culture to be easily observed and analyzed by immunostaining of ECs with CD31 pan-endothelial marker. Interestingly, several groups have established conditions leading to angiogenic sprouting of EBs in type I 3D collagen gels [32,33]. Sprouting of CD31 positive cells was specifically stimulated in the presence of VEGF and FGF2.

### 8.3.2 Tube formation on basement membrane matrix gel

The in vitro formation of capillary-like tubes by ECs have been widely used to test either the role of different genes in vascular morphogenesis or the antiangiogenic properties of several compounds. In this assay, EC are typically cultured on a layer of matrices, such as fibrin, collagen, or BME, which induce their attachment, migration, and differentiation into tubules [34]. BME is a mixture of extracellular and BM proteins, as described in the precious paragraph. Fig. 8.1B shows the typical formation of a branching network of capillary-like structures after 6 h of culture of EC on BME, in the presence of VEGF-A.

However, although BME pushes EC toward their differentiation very rapidly, it is still debating whether the tubules formed on BME possess a proper lumen [35]. Moreover, various non-EC types, such as fibroblasts or cancer cells, are able to form tubules on BME, suggesting that tubulogenesis of ECs may not represent true differentiation process [36]. Another issue to consider during the setup of this assay is the optimal amount of BME matrix (thin or thick layer) and the density of EC, because both of these may influence the formation of capillary-like structures [37].

In recent years, it has been developed a computer-assisted platform to test antiangiogenic properties of different agents by scaling down the Matrigel assay for use in 384-well and 1536-well plates [38]. In these conditions ECs assemble into highly reproducible tubules in each well. These modifications allow circumventing some of the problems associated with the assay, in particular the large quantity of expensive BME used per test and the inability to quantify a whole well. Indeed, through a computer-assisted integrated platform for capturing and processing images of complete wells, it is possible to measure the total number of nodes, the length of tubes, and their connections in each well. Tubulogenesis assay can also be performed by co-culturing EC with stromal cells (such as fibroblasts or smooth muscle cells), with or without the supply of an ECM [39,40]. Stromal cells, for example, fibroblasts, secrete the necessary matrix components that support the tubule formation. Tubules formed in this condition have lumen and are more heterogeneous in length compared with those assembled in BME assay [36,39]. However, this assay is more time consuming than the BME assay (12–14 days compared with 4–24 h, respectively). Furthermore, other issues have to be considered: first, the exact composition of the matrix secreted by the fibroblasts cannot be defined a priori; second, undefined interactions between ECs and the fibroblasts may occur, making hard to study the specific effect (s) of drugs or other compounds on ECs.

### 8.3.3 Sprouting from endothelial cell spheroids in collagen gel

This assay is based on the sprouting of capillary-like structures originating from gel-embedded spherical aggregates of EC (spheroids). This assay was originally developed by Korff and Augustin [41] who observed that while EC in suspension culture undergo anoikis, cells allowed to aggregate become responsive to VEGF and FGF2. Indeed, when spheroids of EC are embedded in collagen type I gel and stimulated with angiogenic factors, EC are able to invade the gel to form complex network of capillary-like structures (Fig. 8.1C). This assay is particularly suited to study angiogenic regulators because of the very low level of spontaneous angiogenesis, even in presence of serum [42,43].

Spheroids of EC can be fabricated using different techniques, including spontaneous formation using (ultra)low binding plates, hanging drop culture, spinner-flask culture, liquid overlay, rotating wall vessel culture, external force, cell sheets, microfluidics, and micromodel nonadhesive hydrogels [44]. The driving force for the formation of cell aggregates is cell–cell adhesion mediated by cadherin. Spheroid size can be controlled by changing the cell number and should result in structures 100–500 μm in diameter, which permits the free diffusion of oxygen [45]. This assay has several advantages: spheroid fabrication requires very low cell number, is inexpensive, easy, permits variable cell-type composition, and is applicable to high-throughput screening [46].

Surely, the hanging drop method is the most cost-effective and likely take the least amount of specialized skill set. The spheroid structure is obtained by pipetting a single-cell suspension of 800–1000 tumor cells in a 20-μL droplet on the inside of the lid of a cell culture plate. Gravity enforced self-aggregation is facilitated by inverting the lid. After 48–72 h, spheroids can be collected and embedded into collagen gel, with or without angiogenic factors [47].

Quantitative and qualitative evaluations, including imaging techniques such as bright-field and fluorescence microscopy, molecular marker assessment by immunohistochemistry or immunofluorescence, or functional assays such as cell viability and survival assessment, can be performed on spheroids.

Recently, 3D tumor organoids from different tissues have become fundamental for cancer research studies [48]. Although these cultures are very representative of tumor behavior, they miss the various components of tumor microenvironment, such as fibroblasts, immune cells, and vessels. Thus, it is very interesting the possibility to coculture tumor organoids with spheroids of EC, to simulate the complexity of tumor microenvironment [49]. This kind of coculture allows studying different aspect of tumor biology: for example, the interaction of capillary-like structures with tumor organoids, but also the impact of therapies on both tumor and ECs.

## 8.4 Vascular explant-based 3D angiogenesis models

To bridge the gap between endothelial cultured cells and in vivo animal models, it has been functional the use of ex vivo assays, in which explants of vascular tissues are embedded in ECM gels producing new vascular sprouts that differentiate in capillary-like structure. Rat and mouse aortic ring (AR) assays are frequently used to evaluate the efficacy of pro- and antiangiogenic molecules and to exploit the many available genetically modified mouse lines [50]. Adaptations of this assay allow to directly monitor the angiogenic process and to follow the dynamics of endothelial tip cells [7,42,51].

### 8.4.1 Rat aortic ring sprouting assay

The property of the rat aorta wall to sprout and form microvessels when excised and embedded in gels was first described in the early 1980s by Nicosia and led to the development of the aortic ring assay [52,53]. This assay is widely used to test the efficacy of angiogenesis inhibitors and to study mechanisms of vessels sprouting. Angiogenesis in this system is mainly driven by endogenous growth factors released by the aorta and its outgrowth in response to the injury of the dissection procedure. While this is convenient for studying angiogenesis inhibitors it could be confounding when the response of angiogenic factors is tested. However, the ability to sprout in absence of serum is a very useful feature. Probably the main advantage of the AR assay is the presence of pericytes, macrophages, and fibroblasts during the EC spouting mimicking all stages of vessel development, as well as the mechanisms involved in the morphogenetic process. Moreover, since many assays can be prepared from the thoracic aorta of a single animal, the angiogenic response can be precisely quantitated, even in real time. The method, in brief, required the thoracic aorta excision from the animal, then, under a dissecting microscope, the aorta is cleaned of fibroadipose tissue, rinsed with PBS or culture medium, and cut into 0.5–1-mm-long rings. Aortic rings are then embedded individually into collagen or fibrin gel and incubated in a humidified $CO_2$ incubator at 37°C for 7–21 days. The angiogenic sprouting can be visualized by standard inverted bright-field microscope and quantified by image analysis considering the total sprouts length or the projected sprouting area. Alternatively, the angiogenic response can be scored by counting microvessel sprouts and branches [54].

### 8.4.2 Mouse aortic ring sprouting assay

The original rat aortic ring method has been further translated to mice. The increasing availability of transgenic mouse models has indeed force to apply this approach to mice and develop this assay. Genetically altered mice have revealed unexpected roles for previously characterized proteins involved in angiogenesis, therefore it is natural to exploit mouse AR for studying mechanism of sprouting process. As a result, protocols for mouse aortic ring assays have been refined over several years to provide a versatile ex vivo assay (Fig. 8.2A). The general protocol for mouse AR culture reproduces that of rat AR and involves dissection of the aorta, sectioning of rings, embedding in gel and imaging and

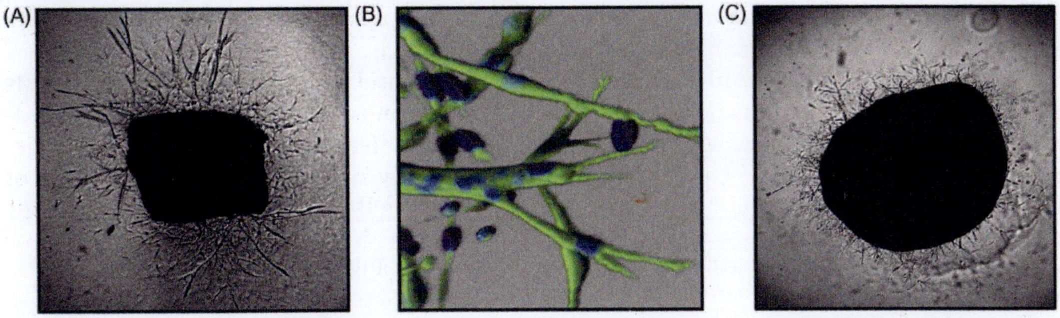


FIGURE 8.2 **Vascular explant-based 3D angiogenesis models.** (A) Tubular structures sprouted by mouse aortic ring embedded in collagen type I gel. The network of vessels originating from endothelial internal layer of aorta is clearly visible by bright-field microscopy. (B) 3D rendering of mAR tubular structure observed by high-magnification confocal microscopy. mAR have been transduced with lentivirus expressing GFP. In blue are nuclei stained with DAPI, in green GFP. (C) Human arterial ring embedded in Matrigel gel and stimulated with VEGF-A. Sprouted EC form an evident network.

quantifying microvessel sprouts [54,55]. Differently form rat AR, mouse AR requires serum in the culture medium while the growth factors response is preferably observed after overnight starvation before the embedding in gel [55,56].

Mouse AR have been also genetically modified by lentiviral or adenoviral transduction to knockdown genes for the duration of the experiment using small hairpin RNA [42,56]. In addition, this protocol has been adapted for gene overexpression by inserting an expression clone in place of the small hairpin RNA sequence [57]. Interesting experiments of cell dynamics during sprouting angiogenesis have been performed taking advantage of transgenic mouse models expressing fluorescent proteins, both as expression reporters and fusion proteins (Fig. 8.2B). Endothelial EGFP-positive cells have revealed the "cell mixing" during sprouting and elongation of new vessels, while NG2 fluorescent pericytes allowed to follow pericytes replication and migration dynamics directly on endothelial structures [58,59]. Similarly, live-imaging experiment with AR from Lifeact-GFP mice, which express fluorescent actin, showed how endothelial podosomes control the vessel sprouting and branching [14].

### 8.4.3 Human arterial ring angiogenesis assay

In the human arterial ring angiogenesis (hAR) assay, human umbilical artery rings are embedded in BME, producing capillary-like structures, which recapitulate different steps of angiogenesis, including EC sprouting, migration, and differentiation into capillaries (Fig. 8.2C) [60]. Contrary to previously described similar assays (rat and mouse ARs, which autonomously form capillary-like structures when matrix-embedded), vessel outgrowth of hARs is completely dependent on the addition of angiogenic factors in the culture medium. The hAR assay offers the possibility for the screening of antiangiogenic drugs, preventing failures caused by a lack of interspecies cross-reactivity. This assay is relatively cheap, using biological material that would otherwise be discarded, and it requires fewer resources than in vivo assays in mice or other small animal models. Moreover, compared with cell cultures, which are commonly constituted a unique cell type, this method has the advantage that it takes into account that blood vessels are a 3D network composed by human ECs associated with human stromal cells. Human arterial rings can also be cultured with tumor spheroids giving rise to branched structures, which mimic tumor vessels [60].

## 8.5 Microvessels on a chip

Recent advances in microfabrication and biological integration have propelled the implementation and improvement of microfluidic systems, which model the microvascular function. Microfluidic approaches have enabled unprecedented levels of control of chemical gradients, fluid flow, matrix composition, and cell–cell interactions, all of which can be integrated to provide a physiologically relevant context for studying angiogenesis [61]. Within the past 15 years these microfluidic have progressed significantly from primarily two-dimensional (2D) planar microchannel device to 3D organ on a chip [62,63].

In vitro microvessel models can be categorized based on several aspects: fabrication approach, material, complexity, origin of EC, self-organization, and presence of fluid flow. Every of these features is relevant and can be combined. Here, we briefly described microfluidics systems with architectural features reminiscent of vasculature, based on main microfabrication technologies used for their fabrication.

### 8.5.1 Microfluidics-based devices

The majority of microfluidic devices are constructed by soft lithography or replica molding of an elastomeric and biocompatible polymer, which frequently is PDMS. The major advantage of these methods is that microvasculature geometry and dimensions can be easily controlled based on different microfabrication techniques with high flexibility. These devices enable to manipulate tiny amounts of fluids, using microstructures with characteristic dimensions on the order of tens to hundreds of micrometers. Within this size scale, microfluidic devices are especially suitable for biological applications because of the scale of microchannels correspond well with the native cellular microenvironment [64]. Several microfluidic-based models of blood vessels have been developed. They differ in their design and complexity, ranging from simple hollow channels to 3D cavities filled with an ECM gel, organized in different geometries. Basic systems are designed with one or two channels separated by microchannels or micropillars, and coated with ECs (Fig. 8.3A). These microfluidic tools are useful to study EC responses to mechanical stress, oxygen tensions and chemotactic stimuli. Moreover, they can assess the influence of microenvironment and heterotypic cell–cell interactions effect by co-culturing EC with other cell types such as cancer cells and stromal cells (Fig. 8.3B) [65]. Other applications of microfluidics devices are shear stress and laminar flow studies. Different types of pumping devices are typically used to allow fluids to flow through microfluidic chips in a controlled manner, namely pneumatic pressure-driven, syringe, or peristaltic pumps [62]. Notably, thanks to these systems several studies have demonstrated that interstitial flow potently induces angiogenesis [66,67]. Frequently, the microchannels of these devices are filled with a collagen or fibrin mixture to form hydrogel, which contain or can be

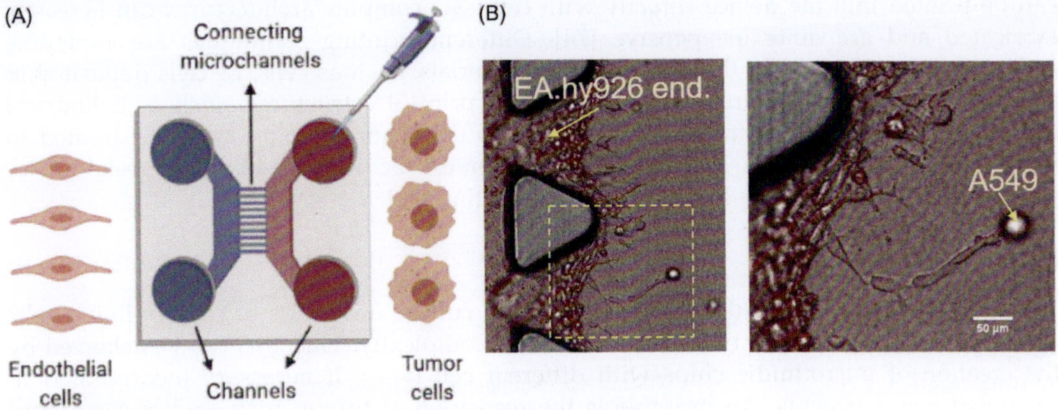


**FIGURE 8.3 Microfluidic devices used in the study of angiogenesis.** (A) Illustration of a double channel design with connecting microchannels. Channel systems can be double or triple, and are arrayable. Vascularized tumor conditions can be achieved by included mixed cell populations using both channels. (B) Directed migration of endothelial cells toward tumor cells on-chip (Left). Optical microscopy image showing the migration of EA. hy926 endothelial cells toward A549 lung tumor microspheroids seeded in a 3D collagen I matrix in a microfluidic device. The triangular structures separate the endothelial and tumor cell chambers. (Right) Magnified view of the dashed region highlighting the migration of endothelial cells toward the tumor cells. Source: *Image provided by David Caballero (3B's Research Group—University of Minho, Portugal).*

colonized by EC. These confined ECM-derived gels permit studies in a real 3D-like microenvironment, together with vascularization studies and heterotypic cell–cell interactions, providing evidence for the interplay between endothelial barrier function and tumor cell intravasation [65,68–70]. Generally biological hydrogels are confined in a central channel between two parallel microfluidic channels. This design has enable to develop microfluidic device for cell motility analysis in a 3D environment under spatially and temporally controlled chemical stimuli [71].

## 8.5.2 3D bioprinting and sacrificial templating

One limitation to PDMS replica molding is that this process produces microchannels with rectangular cross sections. To address this limitation, alternative microfabrication techniques have been used to form engineered microvessels with circular cross sections.

One example is sacrificial templating that involves the use of a sacrificial structure to cast a synthetic or ECM-based material. Then hollow channels can be filled with ECs and cell culture media [72]. Within these microfluidic devices, angiogenesis can be induced from either EC-lined microvessel or vasculogenesis-derived microvascular network. It was reported that these vascularized microtissues could be cultured and remained viable for up to 40 days and after applying the transendothelial flow, the sprouting angiogenesis could be induced from an EC monolayer [70] or by VEGF gradient [73].

Recent microfabrication of vessels and tubular networks by 3D printing and bioprinting overcomes many of the limitations of the soft-lithography approaches. In this printing process different types of biomaterials, including polymers, natural and synthetic hydrogels, can be printed into the device directly with cells. 3D complex architectures can be easily fabricated and are quite inexpensive [74]. Different printing techniques are exploited, including inkjet deposition that dispenses biomaterials, such as cells, or cells deposited in a dissolvable matrix precursor layer-by-layer to print 3D structures, such as cylindrical blood vessels [75]. The restrain of 3D printing is the spatial resolution that is limited to millimeter-sized capillaries. This limit of resolution can be overcome by using related bio-printing techniques, such as laser-assisted bioprinting [76].

### 8.5.3 Organ-on-a-chip

A main development of these microfluidic cell culture devices is to contain living cells arranged to simulate organ tissue with increased complexity. This goal can be achieved by the creation of microfluidic chips with different cell types, if necessary incorporated in separated compartments. An example is the recreation of tumor microenvironment, containing ECs, tumor-associate microphages, fibroblasts, and tumor cells. The 3D microtumor models supported by perfused human microvasculature for anticancer drug screening have shown that the microvessels are sensitive to the effects of antiangiogenic drugs such as pazopanib, linifanib, cabozantinib, axitinib, as well as vascular disrupting agents such as vincristine and taxol [77–79]. Alternatively different devices can be connected in series to mimic the different tissues (e.g., liver, spleen, tumor) that play a role in targeted drug delivery [80,81].

## 8.6 Future perspectives

Overall, 3D angiogenesis models are closer to the in vivo environment than the 2D ones because they take into account more angiogenesis steps allowing to study the dynamic of single biological events as well as the molecular manipulation. The 3D angiogenesis assays are based on the ability of activated ECs to invade 3D substrates, proliferate, migrate, or differentiate in the 3D configuration. Physiological vascularized tumor conditions can be achieved by using special devices with or without scaffolds and including mixed cell populations such as tumor, stroma, immune cells as well as ECs that line the vessels. Although significant progress has been made toward vascularized tumor models in the past decade, numerous challenges remain to be addressed to develop more novel micro-physiological systems that can closely mimic organ-specific physiology in vivo. The recent establishment of organoid technology is dramatically improving the efficiency to generate human tissue analog in vitro. However, organoids frequently lack the vascular network. The development of technologies allowing the generation of vascularized and perfused organoids could significantly advance our ability to grow small organ in vitro and to generate drug response predictive models.

## Acknowledgments

The authors acknowledge the contributions to this work by Giorgio Seano and Giulia Chiaverina.

This work was supported by Associazione Italiana per la Ricerca sul Cancro (AIRC AIG 18675 to L.P.), Fondazione Piemontese per la Ricerca sul Cancro, Ministero della Salute (FPRC 5x1000 2014 to L.P.), Fondazione CRT-2017 (to L.P.) and AIOM 2017 (to L.P.).

## References

[1] Adams RH, Alitalo K. Molecular regulation of angiogenesis and lymphangiogenesis. Nat Rev Mol Cell Biol 2007;8(6):464–78.
[2] Potente M, Gerhardt H, Carmeliet P. Basic and therapeutic aspects of angiogenesis. Cell 2011;146(6):873–87.
[3] Ferrara N, Gerber HP, LeCouter J. The biology of VEGF and its receptors. Nat Med 2003;9(6):669–76.
[4] Olsson AK, et al. VEGF receptor signalling - in control of vascular function. Nat Rev Mol Cell Biol 2006;7(5):359–71.
[5] Hellstrom M, et al. Dll4 signalling through Notch1 regulates formation of tip cells during angiogenesis. Nature 2007;445(7129):776–80.
[6] Sainson RC, et al. Cell-autonomous notch signaling regulates endothelial cell branching and proliferation during vascular tubulogenesis. FASEB J 2005;19(8):1027–9.
[7] Gerhardt H, Betsholtz C. Endothelial-pericyte interactions in angiogenesis. Cell Tissue Res 2003;314(1):15–23.
[8] Folkman J. Tumor angiogenesis: therapeutic implications. N Engl J Med 1971;285(21):1182–6.
[9] Folkman J. Angiogenesis in cancer, vascular, rheumatoid and other disease. Nat Med 1995;1(1):27–31.
[10] Quintieri L, Selmy M, Indraccolo S. Metabolic effects of antiangiogenic drugs in tumors: therapeutic implications. Biochem Pharmacol 2014;89(2):162–70.
[11] Hanahan D, Weinberg RA. Hallmarks of cancer: the next generation. Cell 2011;144(5):646–74.
[12] Singh M, Ferrara N. Modeling and predicting clinical efficacy for drugs targeting the tumor milieu. Nat Biotechnol 2012;30(7):648–57.
[13] Davis GE, Senger DR. Endothelial extracellular matrix: biosynthesis, remodeling, and functions during vascular morphogenesis and neovessel stabilization. Circ Res 2005;97(11):1093–107.
[14] Seano G, et al. Endothelial podosome rosettes regulate vascular branching in tumour angiogenesis. Nat Cell Biol 2014;16(10):931–41 1-8.
[15] Marchand M, et al. Extracellular matrix scaffolding in angiogenesis and capillary homeostasis. Semin Cell Dev Biol 2019;89:147–56.
[16] Kalluri R. Basement membranes: structure, assembly and role in tumour angiogenesis. Nat Rev Cancer 2003;3(6):422–33.
[17] Yousif LF, Di Russo J, Sorokin L. Laminin isoforms in endothelial and perivascular basement membranes. Cell Adh Migr 2013;7(1):101–10.
[18] Lohler J, Timpl R, Jaenisch R. Embryonic lethal mutation in mouse collagen I gene causes rupture of blood vessels and is associated with erythropoietic and mesenchymal cell death. Cell 1984;38(2):597–607.
[19] Arroyo AG, Iruela-Arispe ML. Extracellular matrix, inflammation, and the angiogenic response. Cardiovasc Res 2010;86(2):226–35.
[20] Taverna D, Hynes RO. Reduced blood vessel formation and tumor growth in alpha5-integrin-negative teratocarcinomas and embryoid bodies. Cancer Res 2001;61(13):5255–61.
[21] Whelan MC, Senger DR. Collagen I initiates endothelial cell morphogenesis by inducing actin polymerization through suppression of cyclic AMP and protein kinase A. J Biol Chem 2003;278(1):327–34.
[22] Senger DR, et al. The alpha(1)beta(1) and alpha(2)beta(1) integrins provide critical support for vascular endothelial growth factor signaling, endothelial cell migration, and tumor angiogenesis. Am J Pathol 2002;160(1):195–204.
[23] Alghisi GC, Ruegg C. Vascular integrins in tumor angiogenesis: mediators and therapeutic targets. Endothelium 2006;13(2):113–35.
[24] Houser JW, Ackerman GA, Knouff RA. Vasculogenesis and erythropoiesis in the living yolk sac of the chick embryo. A phase microscopic study. Anat Rec 1961;140:29–43.
[25] Risau W, Lemmon V. Changes in the vascular extracellular matrix during embryonic vasculogenesis and angiogenesis. Dev Biol 1988;125(2):441–50.
[26] Risau W, et al. Vasculogenesis and angiogenesis in embryonic-stem-cell-derived embryoid bodies. Development 1988;102(3):471–8.
[27] Risau W. Mechanisms of angiogenesis. Nature 1997;386(6626):671–4.
[28] Smith AG. Mouse embryo stem cells: their identification, propagation and manipulation. Semin Cell Biol 1992;3(6):385–99.
[29] Doetschman TC, et al. The in vitro development of blastocyst-derived embryonic stem cell lines: formation of visceral yolk sac, blood islands and myocardium. J Embryol Exp Morphol 1985;87:27–45.

[30] Wang R, Clark R, Bautch VL. Embryonic stem cell-derived cystic embryoid bodies form vascular channels: an in vitro model of blood vessel development. Development 1992;114(2):303–16.
[31] Brickman JM, Serup P. Properties of embryoid bodies. Wiley Interdiscip Rev Dev Biol 2017;6(2):e259.
[32] Feraud O, Cao Y, Vittet D. Embryonic stem cell-derived embryoid bodies development in collagen gels recapitulates sprouting angiogenesis. Lab Invest 2001;81(12):1669–81.
[33] Hermant B, et al. Development of a one-step embryonic stem cell-based assay for the screening of sprouting angiogenesis. BMC Biotechnol 2007;7:20.
[34] Lawley TJ, Kubota Y. Induction of morphologic differentiation of endothelial cells in culture. J Invest Dermatol 1989;93(2 Suppl.):59S–61S.
[35] Auerbach R, et al. Angiogenesis assays: a critical overview. Clin Chem 2003;49(1):32–40.
[36] Donovan D, et al. Comparison of three in vitro human 'angiogenesis' assays with capillaries formed in vivo. Angiogenesis 2001;4(2):113–21.
[37] Staton CA, Reed MW, Brown NJ. A critical analysis of current in vitro and in vivo angiogenesis assays. Int J Exp Pathol 2009;90(3):195–221.
[38] Sanz L, et al. Development of a computer-assisted high-throughput screening platform for anti-angiogenic testing. Microvasc Res 2002;63(3):335–9.
[39] Bishop ET, et al. An in vitro model of angiogenesis: basic features. Angiogenesis 1999;3(4):335–44.
[40] Montesano R, Pepper MS, Orci L. Paracrine induction of angiogenesis in vitro by Swiss 3T3 fibroblasts. J Cell Sci 1993;105(Pt 4):1013–24.
[41] Korff T, Augustin HG. Integration of endothelial cells in multicellular spheroids prevents apoptosis and induces differentiation. J Cell Biol 1998;143(5):1341–52.
[42] Primo L, et al. Increased expression of alpha6 integrin in endothelial cells unveils a proangiogenic role for basement membrane. Cancer Res 2010;70(14):5759–69.
[43] Sessa R, et al. The miR-126 regulates angiopoietin-1 signaling and vessel maturation by targeting p85beta. Biochim Biophys Acta 2012;1823(10):1925–35.
[44] Achilli TM, Meyer J, Morgan JR. Advances in the formation, use and understanding of multi-cellular spheroids. Expert Opin Biol Ther 2012;12(10):1347–60.
[45] Mueller-Klieser W. Method for the determination of oxygen consumption rates and diffusion coefficients in multicellular spheroids. Biophys J 1984;46(3):343–8.
[46] Friedrich J, et al. Spheroid-based drug screen: considerations and practical approach. Nat Protoc 2009;4 (3):309–24.
[47] di Blasio L, Bussolino F, Primo L. Three-dimensional in vitro assay of endothelial cell invasion and capillary tube morphogenesis. Methods Mol Biol 2015;1214:41–7.
[48] Drost J, Clevers H. Organoids in cancer research. Nat Rev Cancer 2018;18(7):407–18.
[49] Tuveson D, Clevers H. Cancer modeling meets human organoid technology. Science 2019;364(6444):952–5.
[50] Aplin AC, Nicosia RF. The rat aortic ring model of angiogenesis. Methods Mol Biol 2015;1214:255–64.
[51] Schmidt M, et al. EGFL7 regulates the collective migration of endothelial cells by restricting their spatial distribution. Development 2007;134(16):2913–23.
[52] Nicosia RF. The aortic ring model of angiogenesis: a quarter century of search and discovery. J Cell Mol Med 2009;13(10):4113–36.
[53] Nicosia RF, Tchao R, Leighton J. Histotypic angiogenesis in vitro: light microscopic, ultrastructural, and radioautographic studies. In Vitro 1982;18(6):538–49.
[54] Blacher S, et al. Improved quantification of angiogenesis in the rat aortic ring assay. Angiogenesis 2001;4 (2):133–42.
[55] Masson VV, et al. Mouse aortic ring assay: a new approach of the molecular genetics of angiogenesis. Biol Proced Online 2002;4:24–31.
[56] Baker M, et al. Use of the mouse aortic ring assay to study angiogenesis. Nat Protoc 2011;7(1):89–104.
[57] Hajitou A, et al. The antitumoral effect of endostatin and angiostatin is associated with a down-regulation of vascular endothelial growth factor expression in tumor cells. FASEB J 2002;16(13):1802–4.
[58] Arima S, et al. Angiogenic morphogenesis driven by dynamic and heterogeneous collective endothelial cell movement. Development 2011;138(21):4763–76.
[59] Chiaverina G, et al. Dynamic interplay between pericytes and endothelial cells during sprouting angiogenesis. Cells 2019;8(9):1109.

[60] Seano G, et al. Modeling human tumor angiogenesis in a three-dimensional culture system. Blood 2013;121 (21):e129–37.

[61] Nowak-Sliwinska P, et al. Consensus guidelines for the use and interpretation of angiogenesis assays. Angiogenesis 2018;21(3):425–532.

[62] Caballero D, et al. Tumour-vessel-on-a-chip models for drug delivery. Lab Chip 2017;17(22):3760–71.

[63] Huh D, Hamilton GA, Ingber DE. From 3D cell culture to organs-on-chips. Trends Cell Biol 2011;21 (12):745–54.

[64] Whitesides GM. The origins and the future of microfluidics. Nature 2006;442(7101):368–73.

[65] Zervantonakis IK, et al. Microfluidic devices for studying heterotypic cell-cell interactions and tissue specimen cultures under controlled microenvironments. Biomicrofluidics 2011;5(1):13406.

[66] Galie PA, et al. Fluid shear stress threshold regulates angiogenic sprouting. Proc Natl Acad Sci U S A 2014;111(22):7968–73.

[67] Song JW, Munn LL. Fluid forces control endothelial sprouting. Proc Natl Acad Sci U S A 2011;108 (37):15342–7.

[68] Yeon JH, et al. In vitro formation and characterization of a perfusable three-dimensional tubular capillary network in microfluidic devices. Lab Chip 2012;12(16):2815–22.

[69] Buchanan CF, et al. Flow shear stress regulates endothelial barrier function and expression of angiogenic factors in a 3D microfluidic tumor vascular model. Cell Adh Migr 2014;8(5):517–24.

[70] Vickerman V, Kamm RD. Mechanism of a flow-gated angiogenesis switch: early signaling events at cell-matrix and cell-cell junctions. Integr Biol (Camb) 2012;4(8):863–74.

[71] Marasso SL, et al. Optimized design and fabrication of a microfluidic platform to study single cells and multicellular aggregates in 3D. Microfluid Nanofluid 2017;21(2):29.

[72] Zheng Y, et al. In vitro microvessels for the study of angiogenesis and thrombosis. Proc Natl Acad Sci U S A 2012;109(24):9342–7.

[73] Jeong GS, et al. Sprouting angiogenesis under a chemical gradient regulated by interactions with an endothelial monolayer in a microfluidic platform. Anal Chem 2011;83(22):8454–9.

[74] Miri AK, et al. Multiscale bioprinting of vascularized models. Biomaterials 2019;198:204–16.

[75] Miller JS, et al. Rapid casting of patterned vascular networks for perfusable engineered three-dimensional tissues. Nat Mater 2012;11(9):768–74.

[76] Devillard R, et al. Cell patterning by laser-assisted bioprinting. Methods Cell Biol 2014;119:159–74.

[77] Sobrino A, et al. 3D microtumors in vitro supported by perfused vascular networks. Sci Rep 2016;6:31589.

[78] Xu Z, et al. Application of a microfluidic chip-based 3D co-culture to test drug sensitivity for individualized treatment of lung cancer. Biomaterials 2013;34(16):4109–17.

[79] Phan DTT, et al. A vascularized and perfused organ-on-a-chip platform for large-scale drug screening applications. Lab Chip 2017;17(3):511–20.

[80] Sung JH, et al. Recent advances in body-on-a-chip systems. Anal Chem 2019;91(1):330–51.

[81] Carvalho MR, et al. Evaluating biomaterial- and microfluidic-based 3D tumor models. Trends Biotechnol 2015;33(11):667–78.

CHAPTER

# 9

# Metastasis in three-dimensional biomaterials

*Banani Kundu*[1,2], *Rui L. Reis*[1,2] *and Subhas C. Kundu*[1,2]

[1]3B's Research Group, I3Bs—Research Institute on Biomaterials, Biodegradables and Biomimetics, University of Minho, Headquarters of the European Institute of Excellence on Tissue Engineering and Regenerative Medicine, Guimarães, Portugal [2]ICVS/3B's—PT Government Associate Laboratory, Braga/Guimarães, Portugal

## 9.1 Why biomaterial is needed in cancer modeling?

In 2002, the National Cancer Institute announced the beginning of a new US$ 40 million per year program entitled, "Signatures of Cancer Cell and its Microenvironment," which pioneers the three-dimensional (3D) culture work in cancer [1]. Further, in the commentary article, Hutmacher (2010) commented, "Biomaterials offer cancer research the third dimension" [2]. Contrary to the conventional two-dimensional cell culture and 3D spheroid culture system, encapsulation of cancer cells within the 3D biomaterials mimics the spatial organization of cancer cells more closely to physiological disease niche with relevant to *in vivo*-like cell mechanobiology. In order to model the stages of metastasis, the understanding of metastasis cascade is needed.

Cancer metastasis is a complex dynamic multistep process involving tumor growth, angiogenesis, stromal penetration, intravasation, extravasation, and colonization at secondary sites (illustrated in Fig. 9.1). Understanding the process of metastasis necessitates the study of not only the tumor cells but also the dynamic and reciprocal cross-talk between tumor cells and the adjacent tumor stroma, namely tumor microenvironment (or niche). Tumor microenvironment is heterogenous containing extracellular matrix (ECM), which embeds diverse types of cells [e.g. cancer-associated fibroblasts (CAFs), endothelial cells (ECs), stem cells, etc.], proteins and signaling molecules. The properties of tumor ECM modulate as tumor grows and metastasize. For example, CAFs and tumor cells alike alter the fibrillar ECM of tumor tissue into more cross-linked, aligned, denser, stiffer, and less

*Biomaterials for 3D Tumor Modeling*
DOI: https://doi.org/10.1016/B978-0-12-818128-7.00009-5

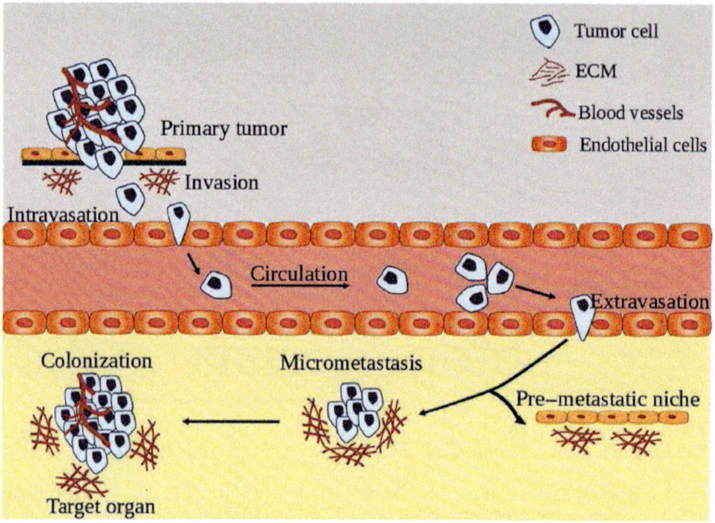


FIGURE 9.1 Schematic illustration of stages of cancer metastasis. (1) Primary tumor: The mutation of a single-cell results uncontrolled cell division leading to a mass of abnormal cells, which further acquire additional mutation and form primary tumor. This heterogenous mass starts remodeling the ECM. (2) Invasion: Cancer cells within the primary tumor lose their cell–cell attachment, break down the basement membrane and invade the surrounding stroma, characterized by change in cell shape, the epithelial to mesenchymal transition. (3) The migration of cancer cells into blood or lymphatic vessel from tumor adjacent stroma. (4) Extravasation: The exit of cancer cells from blood or lymphatic vessel into secondary site through the endothelial lining. (5) Colonization: Malignant tumor grows at secondary site far distant site from primary tumor.

porous stroma [3]. In case of breast tumors, the patients can actually experience the stiffened disease stroma.

The journey of biomaterials in modeling tumor biology has started in early 1980s, when Bissell *et al.*, demonstrated the changes in extracellular milieu affecting the expression of mammary glands genes, introducing the phenomenon of "dynamic reciprocity" between cells and its microenvironment [4]. This pioneer work indicates that the ECM components (like collagen and/or fibronectin) interact with plasma membrane of cells and ultimately connect to cell cytoskeleton through specific adhesion, namely focal adhesion. The discovery of "dynamic reciprocity" phenomenon in cancer is possible due to the use of tissue-derived biomaterials, that is, Matrigel [5]. The model thus developed demonstrates easy discrimination between normal and malignant human breast epithelial cells by growth and differentiation pattern in 3D [6]. The distinctive role of ECM in cancer is further acknowledged by "The seed and soil theory," proposed by Stephen Paget (1889) [7]. In this theory, the cancer cells are represented as seed, while their interaction with the ECM is termed as soil—together, attempted to explain the tissue-specific spreading pattern of cancer cells (tropism) as per cancer types. Paget has illustrated that the unique characteristics of the soil (ECM) lead the cancer cells to metastasize to certain locations.

The factors of ECM such as pore size, stiffness, viscoelasticity, ligand accessibility, etc., manipulate the expression of ECM molecules that influence the ECM-modifying enzymes, which in turn directs cell–ECM interactions. Hence, while the traditional cancer research focuses on biochemical mechanisms, the recent most approaches involve in investigating the regulatory role of ECM in advancement of cancer by employing diverse biomaterials.

This chapter summarizes the use of biomaterials to model tumor microenvironment. First, unique characteristics of different biomaterials of natural and synthetic origin that are used so far in tumor modeling is reviewed. This is followed by the different regulatory cues of ECM that affect disease progression and metastasis. Then the current status of the state-of-the-art of stepwise modeling of cancer metastasis using biomaterials and microengineering technologies is examined. The chapter concludes by noting the most recent trends in using biomaterial to model metastatic recurrences, namely tumor dormancy and reactivation.

## 9.2 Biomaterials employed in tumor ECM modeling

The discovery of "dynamic reciprocity" is made through the use of Matrigel [5]. Matrigel is a solubilized, gelatinous protein mixture extracted from Engelbreth–Hotm–Swarm mouse sarcoma, and consists of ECM proteins like laminin, type IV collagen, heparin sulfate proteoglycans, entactin, and cocktail of growth factors. Hence, Matrigel closely mimics the *in vivo* cancer stroma [8].

### 9.2.1 Naturally derived biomaterials

Apart from Matrigel, decellularized matrices such as human adipose tissue, small intestinal submucosal tissue [9] and rat lung [10] have also been used to model 3D human tumors. A compelling variation of decellularized ECM is the neo-synthesized matrix obtained by prolong culturing of patient-derived primary human osteoblasts in high density *in vitro* to mimic the bone stroma [11] or culturing the highly invasive metastatic human cancer cell lines [12].

#### *9.2.1.1 Collagen*

Collagen is the most abundant ECM protein in human, which makes it a popular choice for tumor modeling [13]. Taken into consideration the signature role of collagen in cancer biology, numerous collagen-based models are developed to investigate the different events of cancer metastasis such as EMT of cancer cells, tumor–stromal interaction, cancer cell migration and invasion, metabolic adaptations to hypoxia, and tumor vasculature. For instance, one of the pioneer work of tumor angiogenesis modeling using collagen hydrogel reveals the supporting role of fibroblasts in tumor angiogenesis [14]. The collagen used for modeling the tumor niche are derived from rat tail, bovine skin, and human placenta [15]. The major limitation of collagen-based models is fast degradation rate of collagen, resulting in the shrinking of the hydrogel [16]. Collagen has also been found as chief matrix component in desmoplastic tumors of breast, pancreas, lung, colorectal, and skin [17]. Cancer cells bind collagen through integrin-binding motif Arg–Gly–Asp (commonly known as RGD) [18].

#### 9.2.1.2 *Gelatin*

Gelatin is denatured collagen type I. The presence of integrin-binding motifs provides it the biocompatibility, while the existence of proteolytically degradable domains imparts it the biodegradability [19]. The polymer backbone of gelatin must be cross-linked in order to obtain a stable hydrophilic network at body temperature. Functionalization of gelatin with methacryloyl groups results a photocross-linkable GelMA matrix, which facilitates the formation of ovarian cancer spheroids [20]. These spheroids lead to metastasis when implanted.

#### 9.2.1.3 *Laminin-rich extracellular matrix*

Laminin-rich ECM (lrECM) is a solubilized protein extract obtained from Engelbreth–Holm–Swarm mouse sarcoma cells. Laminin is an integral component of BM, responsible for polarized epithelial morphogenesis [21]. Interestingly, the malignant subtypes of cancer cells are able to distinguish the distinct morphological variation in 3D lrECM from their nonmalignant counterpart. lrECM supports the interaction of cancer cells with ECM through integrin-mediated signaling. It also promotes the production of endogenous fibronectin and together elevate the survival, proliferation, and invasion of cancer cells [15].

#### 9.2.1.4 *Alginate*

Alginate is naturally occurring polysaccharides obtained from brown seaweeds. It lacks integrin-binding sites. The negative charge of alginate induces easy crosslinking in presence of diverse divalent cations, resulting in uniform reproducible alginate beads [22]. Size is one of the most critical parameters in cancer radiation therapies in correlation of depth and dose, or investigating the effectiveness of radiosensitizing or radioprotective drugs. Alginate salts, such as solution of sodium alginate is administered orally to the patient of local advanced nonsmall cell lung cancer undergoing concurrent chemotherapy [23] in order to minimize the therapeutic side effects, namely esophagitis. Sodium alginate is viscous enough to remain in esophageal mucosa of the patient, providing preventive effect against radiation-induced esophagitis. In *in vitro* models, alginate is popularly used as core–shell structure to facilitate the spontaneous cancer spheroid formation within a microcapsule of 200–300 μm diameter [24]. These microcapsules then can be used to prolong cancer cell culture using bioreactors.

#### 9.2.1.5 *Chitosan*

Chitosan is a linear polysaccharide procured from crustaceous chitins after partial deacetylation. Chitin structurally resembles the glycosaminoglycans of native ECM. The functional groups and surface of chitin promote the attachment of cells [25]. Culturing cancer cells (colon and hepatocarcinoma) on chitosan biomaterials provide spongy effects and upregulate the expression of cancer stem cell (CSC) genes like OCT4, CD133, CD44, NANOG, and epithelial marker EpCAM [26]. Apart from modeling, chitosan is also used as targeting moiety to eliminate $CD44^+$ breast cancer cells [27].

#### 9.2.1.6 *Hyaluronic acid*

Hyaluronic acid (HA) is anionic natural polymer and has been reported to be overexpressed in tumors [15]. HA interacts with cells through cell surface receptors like CD44 and receptor for hyaluronan mediated motility (RHAMM). HA also acts as an attractive substrate for binding cytokines. The accumulation of HA around tumor correlates with greater invasiveness of cancer cells, enhanced malignancy, and poor patient outcomes [28]. The ease in production, ready chemical modification, and regulatory role of HA in cancer make it a popular cancer biomaterial. HA hydrogels successfully support the prostate cancer PDX cells with continuous expression of androgen [29]. However, HA lacks cell-adhesive RGD peptides, therefore, fails to support integrin-mediated cell engagement. Hence, to obtain biomimetic hydrogel, HA is chemically modified with cell-adhesive cysteine containing peptides (PolyRGD) [30] and GRGDS (Gly–Arg–Gly–Asp–Ser) [31], MMP-degradable peptides [31], or grafting with chitosan [32].

#### *9.2.1.7 Silk*

Silk is natural proteinaceous biopolymer obtained from silkworms. Silk as suture is FDA-approved, exhibiting well tolerating response to platelets [33]. Similar to col I, silk protein fibroin is fibrous in nature, supporting prolong culture of cancer cells including multiple myeloma, breast, prostate [34], liver [33], and bone [16]. Topographical cues of silk are able to obtain neuronal phenotype of neuroprogenitor cells in absence of any neuronal growth factor [35]. In addition, silk fibroin of nonmulberry silkworm possess the integrin-binding RGD motifs, imparting superior cellular response over other silks [36,37].

Other naturally derived material worth mentioning is the porous scaffold of chitosan-alginate, which preferentially binds and supports the growth of $CD133^+$ CSC-like cells [38]. CSCs are considered as playing detrimental role in multi-drug resistance and available in very low concentration in human body. The isolation of CSCs is very laborious, time consuming, and costly. The porous scaffold of chitosan-alginate, therefore, act as alternative isolation tool for CSCs.

### 9.2.2 Synthetic biomaterials

In natural biomaterials, it is very difficult to decouple the effect of chemical parameters (like presence of RGD motifs, enzyme degrading site, etc.) from physical ones (such as pore size, stiffness, etc.), therefore, the individual ECM properties vary relative to each other. Thus, it is impossible to specifically predict the impact of individual ECM biomaterials on cell migration. For instance, with increase in concentration of ECM, there is an alteration in ECM stiffness as well as the amount of cell-adhesive cues. Furthermore, natural biomaterials suffer from batch-to-batch variations, which influences the experimental reproducibility even in case of use of commercial products like Matrigel. Therefore, synthetic materials are developed. Synthetic materials serve as "blank slate," while allow the engineering of unique ECM characteristics based on tumor types with precise control.

#### *9.2.2.1 Polyethylene glycol and its derivatives*

The rapid sol-to-gel transition of polyethylene glycol (PEG) makes it a popular alternative to natural biomaterials. In addition, the relative ease in functionalization of PEG with cell-binding sites (e.g., RGD peptide), enzymatic degrading sites {MMP-degradable sequences like GGGPQGIWGQGK (PQ) peptide [39]; dithiothreitol (DTT, KCGPQGIWGQCK) [40], epidermal growth factor receptor (EGFR) [40], and peptide inhibitor (NYQQNC) [40]} facilitate the cellular adhesion and remodeling of the ECM by cancer cells. PEG is also chemically modified with natural materials like HA [41], fibrinogen (PEG-DA) [42] to augment its bioactivity. The metastatic cancer cells including patient-derived cancer cells, cultured on these semisynthetic materials form spheroids, the hallmarks of malignant cancer. Aqueous solution of PEG when mixes with aqueous solution of dextran (DEX) over certain concentration, results into an aqueous two-phase immiscible system (ATPS) [43]. The denser aqueous DEX-phase partition the cancer cells from aqueous immersion PEG phase, while no interphase barrier for nutrients and drugs. The PEG-DEX APTS serves as high-throughput drug screening platform for single or dual-agent testing.

#### 9.2.2.2 *Poly(lactic-co-glycolic) acid*

Poly(lactic-*co*-glycolic) acid (PLGA), the copolymer of polylactic acid and polyglycolic acid, is an FDA-approved biocompatible, biodegradable polymer facilitating the adherence and proliferation of cancer cells [15]. The $\alpha_5$-integrin receptors are upregulated in cancer cells [oral squamous carcinoma cells (OSCC-3)] cultured in 3D PLGA—characteristic phenomenon of invasive cancer [44].

#### 9.2.2.3 *Polycaprolactone*

Polycaprolactone (PCL) is obtained through polycaprolactone of a hydroxycarboxylic acid (6-hydroxyhexanoic acid). The viscoelasticity, malleable properties, low-melting point ($\sim 60^\circ C$), solubility in diverse organic solvents like acetone, chloroform, benzene, dichloromethane, dimethylformamide, etc., absence of isomers and low cost make it a popular synthetic biomaterial [45]. In addition, the slow biodegradability, bioresorbability, and well biocompatibility makes it a potential candidate for biomedical engineering and cancer modeling.

#### 9.2.2.4 *Polyacrylamide*

Polyacrylamide (PA) hydrogels are produced by crosslinking the reacting acrylamide monomer with bis-acrylamide cross-linker in presence of ammonium persulfate and tetramethylenediamine. PA lacks any cell-adhesive site. The conjugation or blending of protein improves the cytocompatibility of PA hydrogels [46]. PA exhibits metastatic cell memory [47] and widely used as elastic support material to investigate the traction force in cancer cells [48].

#### 9.2.2.5 *Polydimethylsiloxane*

Polydimethylsiloxane (PDMS) is the most commonly used polymer in microfluidic application, both industrial and academic purposes due to ease in processing and flexibility to create patterned surfaces. PDMS exhibits unique characteristics such as gas permeability, tunable elasticity, biological inertness, optical transparency, and inexpensiveness [49]. The byproducts of PDMS degradation is not harmful to living organism. PDMS surfaces take part in nonspecific absorption of small-molecules, including certain drugs, which is currently circumvented using different experimental setup and computational modeling [50]. To overcome this, alternative thermoplastics such as polyurethanes, ethylene butylene styrene, cyclic olefin polymers, and copolymers are exploited. However, to obtain a mechanically actuatable organ chip, a flexible biodegradable porous membrane with minimal absorption of nutrients and drugs is needed to be developed.

#### 9.2.2.6 *Thermoresponsive polymers*

Thermoresponsive polymers allow easy encapsulation of cancer cells at 37°C and re-collection of spheroids by liquefying the gels at low temperature. For example, the poly (N-isopropylacrylamide-*co*-acrylic acid) (PNIPAM) microgels able to construct and release of human hepatocarcinoma cells spheroids (HepG2). Galactosylated PNIPAM-AA microgels exhibit greater liver-specific functionality including enhanced secretion of albumin and urea [51].

The synthetic materials are also modified with cancer cell-secreted exosomes or haptoglobin to create bioengineered niche [52]. Together, the different biomaterials (natural, synthetic, or semisynthetic) incorporate different aspects of *in vitro* cancer microenvironment helping the cancer biologist to create reductionist systems recapitulating stepwise process of metastasis within a system.

## 9.3 Properties of cell surrounding matrix/niche contribute to tumor cell migration

### 9.3.1 Pore size

The pore sizes impact directly the cellular migration by imposing the resistance. If the cross-section of elongated migrating cells is similar or slightly less than the pore size of the surrounding network, the cells experience little or no resistance or steric hindrance [53]. When the pore size is large, the cell migration speed declines due to the loss of contact of cells with surrounding fibers, which either pull or push the cell body onward [54]. Large pore size allows the ameboid movement of round shaped cell through the network. High density of fibers results decrease in pore size, leading to the entrapment of cells within the network and loss of cellular migration [54]. Cells at this point, either force the network fibers apart leading to matrix remodeling or reshaping itself to pass through the small pores, resulting in cellular deformity. When the pore size is too small, the degradation of surrounding matrix is triggered, which in turn alters the topography of matrix (discussed in next section) [55]. Large scale cellular deformity is regulated by nucleus through nuclear lamins A/C [56]. The mechanical resistance associated with high-matrix density facilitates protease-dependent invasion of cancer cells by degradation of matrix fibers through matrix metalloproteinases (MMPs) [57]. Interestingly, the inhibition of MMPs results the switch to an MMP-independent mode of migration by cancer cells [58].

### 9.3.2 Topography or contact guidance

Aligned matrix fibers are typical features of many healthy normal tissues [59]. Cells respond to this matrix orientation by polarizing their cytoskeleton and migrate along the aligned fibers, a phenomenon known as "contact guidance" [60]. The nano- or microscale surface topography of matrix fibrils and hierarchical assembly patterns have been known to guide cancer cell migration [61]. For example, a shift from native mesh-like topography of collagen into rapidly aligned collagen fibers is reported for breast cancer in association with cancer-associated stromal cell activity and termed as "highway" [62]. The outwardly radiating pattern of collagen fibers nurture the invasive cell survival pathways, such as PI3K/Akt, p53, MAPK, ERK/MAPK, and Rho/ROCK [63–66]. The arrays of matrix fibers of same density gradient usually interpreted as a cue demonstrating in opposite direction by the cells and the cancer cells navigate either down or up the same gradient based on their genetic status [67]. For example, melanoma cells with different degree of aggressiveness exhibit opposite topotactic guidance toward same topographic gradient, a phenomenon which can be useful in diagnostic and prognostic tool designing.

### 9.3.3 Stiffness

Stiffness plays influential role in cancer metastasis. In context of cancer, both the stiffness of cells (i.e., the elastic modulus of each cell) and surrounding matrix are interwind and the difference between them is the determinant of the type of interaction takes place between them. Metastatic cancer cells appeared to be softer than their benign counterpart [68]. In context to the cell, the tumor-associated niche appears stiffer than the surrounding healthy tissue [69]. Stiff tumor niche is caused by fibrosis and increased deposition of ECM components by the rapidly growing cells that in turn enhance the interstitial pressure and chaotic microvasculature [70]. Additionally, the tight alignment of collagen fiber and dense crosslinking due to the upregulation of lysyl oxidase enzyme (LOX) in case of breast cancer, increases the stiffness of the disease niche. This facilitates Rho A-mediated actomyosin contraction, reinforcement of focal adhesions—leading to cell protrusions, high-traction forces, and elongated cell shape [71]. As a result, the migration of cells is directed toward the chemotactic gradient and in this case toward the regions of higher stiffness due to high-collagen content. This leads to increased confinement that consequences a switch from single-cell migration to collective one. In collective migration, the cells may follow the leading cells or self-organize; this is an ongoing research effort [72].

### 9.3.4 Matrix rheology

The force, which is required to move apart the matrix fiber is a characteristic mechanical phenomenon, known as "Matrix rheology." It includes compressibility, the strain-dependent plasticity, strain–rate dependent viscoelastic shear modulus, and frequency. During the progression of cancer, highly proliferative cancer cells and stromal cells generate growth-induced stress that transmits through niche. The niche components, precisely the collagen takes active part to transmit the stress across the tumor and to the surrounding tissues. Volumetric compaction of collagen gel is commonly used to mimic the elevated interstitial fluid pressure in tumor microenvironment to investigate the signaling pathways [73]. The interstitial flow enforces a physical drag on cancer cells that activates the mechanotransducive machinery to direct the migration of cells upstream of the follow [74]. Cyclic forces imposed by the circulatory system also contribute to mechanosensitive process of cancer cells by Rho-mediated mechanosensitive pathway [75]. The comparison between decellularized matrix of both healthy and cancerous colon tissue reveals high storage and loss modulus of diseased tissue compared with the healthy one, correlating differential rheological property of disease niche with disease progression [76].

### 9.3.5 Ligand accessibility

The ligands present in the matrix regulate the adhesion, viability, proliferation, migration, and prosurvival pathways that help cells escaping the drug treatment [77]. The upregulation of matrix proteins, such as collagen VI is commonly associated with advanced stage ovarian cancer [78] and drug resistance. The investigation of cross-talk between RGD receptor (e.g., arginyl-glycyl-aspartic acid) of cell surface with epidermal growth factor (EGF) ligands indicates that at a proper proportion of these two motifs, EGF controls the focal adhesions of cancer cells [79]. A less controlled ligand presentation can also regulate cancer progression. For example, the 3D presentation of laminin, an ECM protein, increases DNA methylation that imparts more invasive phenotype of cancer cells [79]. Laminin is reported to convey resistance to extracellular-signal-regulated kinase (ERK) inhibition [77]. Fibronectin in association with HA evokes the resistance to proto-oncogene B-rapidly accelerated fibrosarcoma (BRAF) inhibition. The availability of fibronectin and collagen also alters the stiffness of cancer cells *in vitro* [80]. Besides chemical ligands, nanostructured flexible materials obtained via block copolymer micelle nanolithography and their controlled spacing in cancer microenvironment through nanolithography is ongoing research focus to unreveal the relationship between ligand composition and spacing with cancer progression and chemoresistance [81].

These characteristics of cancer matrix/niche has been considered during the *in vitro* modeling of cancer in pursuit of closely mimic the *in vivo* cancer microenvironment (Fig. 9.2).

## 9.4 Biomaterial-based stepwise modeling of cancer metastasis *in vitro*

Advancement in tissue engineering, microfabrication techniques, and biomaterials enable the development of complex *in vitro* models by incorporating multiple cell types, three dimension (by biomaterials), and spatial and temporal presentation of soluble factors [82]. For convenience, these models of metastasis are discussed emphasizing on the type of biomaterials used and the phenomenon of stage of the disease studied.

### 9.4.1 Tumor initiation and progression

The reductionist approach to model tumor initiation involves the encapsulation of tumor cells within bulk hydrogel [83,84]. For example, the encapsulation of breast cancer cells within the hybrid hydrogel of collagen–laminin–fibronectin–HA demonstrates the cellular organization similar to that of human breast [85]. Ultraviolet (UV)-based photocross-linkable flexible platform of gelatin methacrylamide (GelMA) is another popular hydrogel choice to investigate the effects of biophysical cues such as pore sizes and

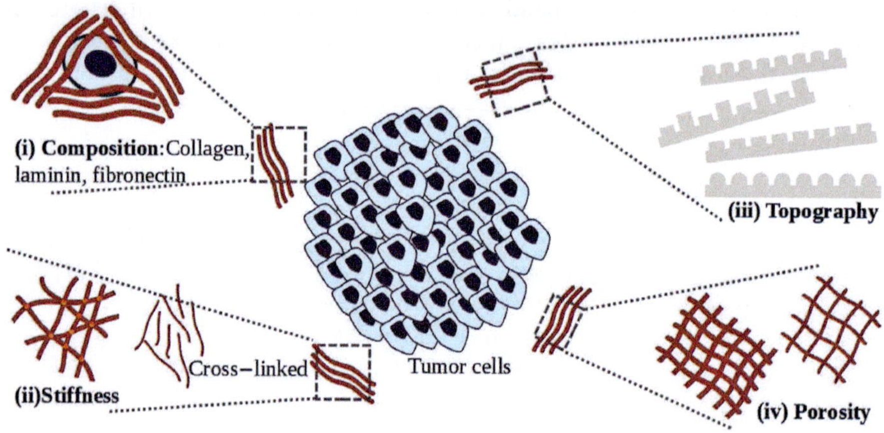


FIGURE 9.2 Designing cancer niche (Adapted from Beri P, Matte BF, Fattet L, Kim D, Yang J, Engler AJ, Biomaterials to model and measure epithelial cancers. Nat Rev Mater 2018;3(11):418–30 [3]). Cancer niche reveals distinct dynamic ECM. Characteristics critical for modeling the ECM are—(1) Composition: Both the type (protein like collagen or polysaccharide like HA) and concentration of ligand affect the behavior of cells and prompt the epithelial to mesenchymal transition (EMT). (2) Stiffness: Stiffness of ECM also impacts cell behavior. It can be regulated by the entanglement of polymer chain or crosslinking. (3) Topography: The topographical features of niche regulate cell adhesion. Materials can be imprinted with specific micronanographical features like pits that induce cell transformation. (4) Porosity: The pore size and porosity can be modulated by the density of bulk polymer and size of the droplet in emulsion. Small pore size inhibits cell migration, while large pore size reduces the cell speed.

stiffness on disease progression [20]. Loose aggregates of ovarian cancer cells are observed within low concentration of GelMA (2.5% w/v) compared with nonuniform spheroid at higher concentration hydrogel (7% w/v). Interestingly, uniform spheroids with tight intercellular junctions are formed in GelMA at 5% (w/v) concentration. These approaches allow to study single-cell response toward the external stimuli but exhibits relatively low throughput and greater dependency on self-organization of cancer cells into the architecture [86].

To overcome the limitations, microcapsule (or microsphere), microfabrication techniques such as soft-lithography and micromolding technique are used. The most exploited capsule biomaterial is alginate, which allows the encapsulation of the tumor cells. Cell encapsulated capsule with mechanical sensors allow the monitoring of the dynamic pressure change exerted by the growing tumor cells on the surrounding microenvironment while undergo structural reorganization [87]. During post-confluency, the cells at the periphery of spheroids demonstrate more motile characteristics and elongated morphology, suggesting the role of microenvironmental stress in cancer invasion. Alternatively, collagen is also reported to be used as a capsule material [88].

Microfabrication techniques are used to design microwells of diverse materials such as PDMS [89], polyethylene glycol dimethacrylate (PEGDMA) [90], and agarose [91] to enable the formation of 3D tumor spheroids within a well-characterized controlled biophysical architecture. The concave microwells of collagen [92] or PDMS facilitate spheroid formation [89,92]. Cues such as tensional stress can be imitated by using photolithography to fabricate the 3D hollow ducts of collagen—a structure similar to that of lobular architecture of breast with higher mechanical stress at the proximal ends of the ducts compared with center. The co-culture of mammary epithelial cells with breast cancer cells within the duct reveals high-migration rate of invasive cells situated at the end, indicating the detrimental role of biophysical cue such as tensional stress in cancer motility [93]. To unwind the role of ECM on transformation of tumor cells into invasive cancer cells, composite hydrogel of collagen, Matrigel, is injected into the Y-shaped microfluidic platform [94]. The hydrogels are then co-seeded with cancer cells and mammary fibroblasts. At the ECM interface between fibroblast and cancer cells, cancer cells are transformed from circular morphology into protrusion developing phenotype by downregulating E-cadherin and break down the basement membrane (BME)—hence, remodel the ECM.

Ongoing efforts include the use of shear-thinning composite hydrogel "ink" of gelatin–alginate–fibronectin to develop a layer-by-layer 3D lattice structure with interconnected channels [95]. This approach employs the use of shear-thinning properties of biomaterial to protect cells during extrusion (Box 9.1).

### 9.4.2 Tumor angiogenesis

Tumors beyond 1–2 $mm^3$ size require to connect with vascular system to fulfill their requirement of oxygen and nutrients [97]. Tumors turn to dormant state in absence of the tumor angiogenesis. Larger tumors exhibit a typical characteristic, known as "angiogenic switch" involving the secretion of proangiogenic growth factors like vascular endothelial growth factors (VEGF-A, VEGF-B, VEGF-C) and fibroblast growth factors (FGF-1, FGF-2), which attract ECs toward tumor site [98]. In body, microvasculature exists in contact with a continual ECM, namely BME, which is composed of collagen I, collagen IV, fibronectin, different forms of laminin, perlecan, and several proteoglycans [99]. Hence, biologically

**BOX 9.1**

**Outlook**

- The heterogeneity at tumor niche suggests the inability of a single biomaterial to recapitulate *in vivo*-like characteristics. Instead it pushes the development of a multicomponent hybrid system.
- Decellularized ECM suitably represents the mechanical ECM of the tumor. However, the method of ECM extraction induces alterations in the final composition of biomaterials.
- Choice of 3D model is critical and specific, based on the aim of the study. For example, the synthesized ECM components such as collagen, fibronectin, and HA by healthy fibroblast and cancer-associated fibroblasts are significantly different in case of a microbeads-based model in contrast to no difference in 3D spheroid-based model [96].

derived ECM components are popularly adapted to induce endothelial network and angiogenesis. Collagen, Matrigel, and fibrin facilitate proangiogenic behavior of ECs. In addition, the density, pore sizes, anisotropy, stiffness, and degradation of the matrix as well as the presentation of integrin-binding cues affect the angiogenesis. High density of collagen leading to the development of shorter, less branched poorly connected microvessels compared with low-density collagen [100]. On the other hand, too low density of collagen (e.g., $0.7\ \mathrm{mg\ mL^{-1}}$) is unable to provide adequate structural support, resulting in unstable, broken sprouts [101]. Intermediate collagen density (1.7 and $1.2\ \mathrm{mg\ mL^{-1}}$) promotes the maximum number of endothelial sprouts. It is hypothesized that matrix density is closely linked with matrix anisotropy. When the matrix density is moderate, the random collagen fibers start aligning themselves near the sprouting vessels. Cells, which encounter less anisotropic disturbance, are thought to induce local branching [102]. During angiogenesis, the BME is proteolytically degraded under physiological condition to enable sprouting from main vessel. MMPs, specifically MT1-MMP, MMP-2, and MMP-9 are necessitated for sprouting angiogenesis [103]. Peptide such as RGD (Arg–Gly–Asp) is known to promote angiogenesis as reported by 89% published biomaterial-based investigations between 1970 and 2018 [104]. A substitution of aspartic acid (D) with glutamic acid (E) in RGD sequence results complete abrogation of sprouting [105]. Another short-adhesive peptide is Tyr–Ile–Gly–Ser–Arg (YIGSR), which is known to regulate the morphology of ECs and maximize the vasculogenesis, when applied in combination with RGD peptides [106]. Other angiogenic peptides reported so far are IKVAV (isoleucine–lysine–valine–alanine–valine, laminin-111 motif) or GFOGER (glycine–phenylalanine–hydroxyproline–glycine–glutamic acid–arginine, collagen I motif) [107].

One of the earliest reports of tumor angiogenesis demonstrates a sandwich model of collagen hydrogel (containing human foreskin fibroblasts) clung with a fibrin hydrogel [embedded with prostate cancer cells (PC-3) and human umbilical vein endothelial cells (HUVECs)] [14]. The formation of tube by HUVECs is observed in presence of fibroblasts in tricellular culture, indicating the significance of supporting fibroblasts in tumor angiogenesis. However, the position of fibroblast in a co-construct model is crucial for angiogenic outcome. For instance, colon fibroblasts (CF56) and HUVECs-laden collagen hydrogel when surround a collagen core containing human colon adenocarcinoma (HT29), HUVECs are failed to form a tubule [108]. The nature of cancer cells is also a determinant in tumor angiogenic model. For example, in the co-construct culture of telomerase-immortalized human microvascular ECs containing collagen I hydrogel with MDA-MB-231 or MCF7 laden collagen I construct [109], the invasion of ECs and formation of tubular network are observed only in presence of MDA-MB-231 cells, suggesting invasive phenotype of cancer cells induces the metastasis. Several of these approaches are adapted in pioneer microengineered techniques such as MDA-MB-231 breast cancer cell laden collagen hydrogels casted into PDMS mold and covered with ECs. The angiogenic sprouting is evaluated under normoxic (normal oxygen concentration, typically 20%–21%) and hypoxic (deprive of adequate oxygen, oxygen tension ranges in 1%–5%) conditions. The hypoxia and 3D culture synergistically upregulate the rate of neovascularization and the release of VEGF and IL-8.

The cellular heterogeneity in cancerous niche also influences the angiogenesis. To confirm this hypothesis, a microfluidic platform with three channels separated from each other by two acellular collagen components is fabricated [110]. The central channel is coated with collagen and topped with human microvascular endothelial cells (HMVECs). Different cells such as human glioblastoma, mouse smooth muscle, and rat mammary adenocarcinoma cells are seeded in other channels and the spontaneous angiogenic sprouting toward rat mammary adenocarcinoma cells is observed. No observable neovascular formation toward human glioblastoma is exhibited, while angiogenesis is inhibited in presence of mouse smooth muscle cells. There is distinct difference in vascular phenotype between cancer induced vascularization compared with fibroblasts induced one. The soluble factors from glioblastoma forms inefficient vasculature with immature lumens and random fusions resulting convolutions among angiogenic sprouts [111].

To obtain variable thick tissue engineered constructs ($>1$ cm), an intricate endothelial network is created within bulk hydrogel by using a sacrificial mold. Copolymers such as Pluronic F127, an FDA-approved triblock of PEO [poly(ethyleneoxide)]–PPO [poly (propyleneoxide)]–PEO, which undergoes sol–gel transition over a temperature and concentration range, is exploited to create hollow channels within bulk hydrogel materials [112]. Postfabrication, Pluronic F127 is removed by changing the temperature to obtain the hollow channels. The hollow channels are subsequently seeded with HUVECs and connected with a peristaltic pump to obtain a perfusing endothelial network. This maintains the viability of cells up to 3 months. Controlled degradability is also imparted to synthetic polymers by incorporating MMP-degradable sites [113] (Box 9.2).

### 9.4.3 Modeling of tumor invasion or migration

The vascularization of tumor provides it the potentiality to metastasize, leading to the breakdown of BME and migrate through it. The local invasion is initiated by loss of tissue polarity and tight intercellular junctions—a hallmark of cancer metastasis known as "epithelial–mesenchymal transition" (EMT). During EMT transition, cancer cells upregulate mesenchymal markers like N-cadherin and downregulate epithelial markers like E-cadherin and become highly motile [114]. The motile cells typically exhibit higher number of protrusions and myosin-based contractility [115]. Tumor cells may undergo

BOX 9.2

**Outlook**

- The unique characteristics of organ-specific ECs such as nonfenestrated capillaries with limited diffusion in brain, skin, lung, and heart compared with well-fenestrated liver capillaries with discontinuous endothelium actively taking part in big molecules exchange should be taken into consideration to model tissue-specific tumor angiogenesis.
- In current years micro-engineered approaches involving the use of prefabricated channels for lining ECs are replaced by bulk natural hydrogels due to their angiogenic stimulating ability.

single or collective cell migration. In multicellular migration of tumor cells, the leader cells create microtracks by proteolytically degrading the ECM and the follower cells migrate into this track [116]. On the other hand, single tumor cell may first attach to vascular endothelium, grows into small colony/spheroid and migrate [117]. This process is guided by "tip cells," which generate the migratory tracks by interacting with ECM [116]. The present understanding of the sequential stages of tumor invasion is chiefly based on *in vitro* models such as spheroids, transwell assay, tumor-microvessel models, and microfluidic models. However, the current chapter solely focus on the approaches involving biomaterials.

#### 9.4.3.1 *Spheroids*

Spheroids are defined as cellular aggregates obtained in suspension or embedded within 3D matrix [84]. The cancer spheroids are widely encapsulated within 3D matrix to assess the migration and invasion of cancer cells. For example, heterotypic tumor spheroids of highly invasive human breast cancer cells (MDA-MB-231) and non-tumorigenic human breast epithelial cells (MCF-10A) are encapsulated in collagen hydrogel to investigate the contribution of tumor heterogeneity in cell migration [118]. The fibers of collagen are aligned parallel to the boundary of homotypic spheroids of MCF-10A. The addition of MDA-MB-231 to the spheroid culture results perpendicular reorientation of collagen fibers at the spheroid boundary, facilitating the cell migration. The MDA-MB-231 cells proteolytically degrade the surrounding matrix, forming microtracks through which MCF-10A cells invade into the surrounding matrix. This model is also well-suited to elucidate the interaction between the "leader" and the "follower" cells in 3D tumor niche. Further, advancement of this model involves the use of BME extract, which acts as chemotactic gradient or chemical clue in cancer cell invasion [119]. The spheroids of MCF-10A and oncogenically transformed MCF-10A-HRas are coated with a shell of BME, followed by encapsulation within collagen I hydrogel. Interestingly, MCF-10A spheroids alone without any BME coating exhibit sheet-like architecture. In contrast, MCF-10A spheroids with BME–collagen encapsulation fail to breach BME and lose the ability of migration. But in presence of MCF-10A-HRas, the MCF-10A cells breach the BME, invade the stroma and realign the collagen fibers at the invasion front. The stiffness of the collagen layer serves as regulatory factor for fibroblasts activation. Stiffer gels [obtained by suspended collagen with poly (ethylene glycol)-di(succinic acid N-hydroxysuccinimide ester)] cause high contractility in fibroblasts and allow the dissemination of tumor cells [120]. These models can be exploited to understand the role of secretory molecules or soluble factors secreted by fibroblasts and ECs in cancer metastasis.

#### 9.4.3.2 *Transwell-based models*

Transwell-based models are popularly used to investigate the migration and invasion of cancer cells. Currently, a number of transwell inserts or assay kits are commercially marketed by Corning, Lonza, Thermo Fisher Scientific, Nunc, Abcam, etc. For migration, a porous membrane (typically of 8-μm pore diameter) is used to quantify the number of cells that migrate across the membrane within a predetermined time interval (24–72 h) [121]. To incorporate complexity, a layer of ECM protein such as collagen [122], Matrigel [121], or laminin [123] is deposited on the porous membrane in a range of thickness up to 1 mm [124]. ECs are also plated onto the porous support to mimic the tight cell–cell junctions between ECs and the underlying BME produced by ECs to model extravasation process [123]. Interestingly, by switching the position of the ECs in aforementioned model, another modeling approach is obtained, known as transendothelial assay and popularly used to study brain–capillary epithelium. Though these models getting popularity due to the simplicity and ease in incorporating with commonly used culture plates, but suffer from poor reproducibility in creating well-defined biophysical architecture.

#### 9.4.3.3 *Microfluidic models*

Avascular microfluidic platforms are primarily employed to investigate the migration of cancer cells with reference to chemotactic gradients in a high-throughput scale. Experimental measurements usually include the determination of migration speed along the channel as a function of experimental conditions such as channel dimensions, coating material, solute gradients, and the presence of obstacles [82]. Using photolithography, arrays of microtracks with unique architecture are imprinted on collagen on a single chip [125]. The cancer cells (MDA-MB-231) and fibroblasts (MCF-10A) move faster within the microtracks compared with nonpatterned collagen even in absence of MT-MMP1 activity. This observation can be a practicable explanation to comparatively low-success rate of MMP inhibitors in therapeutic clinical trial. Femtosecond-pulsed laser is used to fabricate cage-like structures (e.g., pillars, ridges) with variable pore sizes [126]. The simultaneous culture of metastatic and normal cells in this structure indicates a pore-size threshold to discriminate between metastatic cancer cells and normal epithelial cells. The shape and size of the pore pose stark difference in the speed and mechanisms of cancer cell motility [127]. Miniscule microchannels of a microfluidic platform is able to measure single-cell motility of an invasive cancer [128]. Microfluidic device-based studies largely contribute to the discovery of ECM factors that regulate ECM alteration, genes that involve in cancer cell migration and the proteins act as chemoattractant such as EGFR or CXCL12 [129] (Box 9.3).

BOX 9.3

**Outlook**

- The invasion studies are carried out using cell lines, which often lose their *in vivo*-like characteristics. The current thrives toward precision medicine need the adaptation of these models using patient-derived cells for patient-specific therapies and clinical medicine.
- Cytoskeleton dynamics play significant role in cancer cell migration irrespective of MMT activity [130]. Hence, actin contractility inhibitors like myosin II inhibitors (CT04, Y27632, blebbistatin, and M17), myosin light chain kinase, and ROCK are expected to have more success in clinical trial over MMP inhibitors

### 9.4.4 Intravasation models

Intravasation, as described by National Cancer Institute, the invasion of cancer cells through the lining of BME of blood or lymphatic vessel into the circulating system and disseminate to distant tissues. The process of cancer intravasation is broadly subdivided into two—(1) Active intravasation: the migration of cancer cells toward blood or lymphatic vessels is driven by chemokine gradient; and (2) passive intravasation: the shedding of small random piece of cellular aggregates from growing tumor mass due to high stress or poor network [131]. The intravasation of cancer cells through blood vessel is termed as hematogenous spread, while the intravasation through lymphatic nodes is known as lymphogenous spread. Due to the complexity of *in vivo* process, the complete *in vitro* modeling of intravasation is yet to achieve. Hence, the level of intravasation is currently determined by indirect methods like counting the vascular-arrested cancer cells in distal organs or circulating cancer cells in peripheral blood. The presence of endothelium is necessitated to model intravasation. The intravasation models in 3D includes prevascularized tumor spheroids, PDMS microchannels with cultured ECs, bulk biomaterials having cultured EC monolayer. Cancer cells are incorporated into this using hydrogel or magnetic force-mediated cell patterning.

#### 9.4.4.1 *Prevascularized spheroids*

Co-culture of cancer cells with ECs results a prevascularized tissue, which is further embedded within ECM hydrogel, typically in fibrin [132] or in Matrigel [133]. While intravasation is observed by cancer cells (SW620 breast cancer cells), it is worth noting that hypoxic condition has stronger effect on intravasation by upregulating Slug, an EMT marker.

#### 9.4.4.2 *Microfluidic devices*

In microfluidic devices, the microvascular network is formed within microchannels coated with layers of hydrogel, leading to an open lumen structure allowing the perfusion through the network. The ECM materials such as collagen [134] and/or Matrigel [135] are deposited between two microfluidic channels containing ECs and tumor cells separately. The device is then imaged for cancer cell intravasation. Further, inclusion of cells like tumor-associated macrophages or chemokines like TNF-$\alpha$ within the ECM layer suggest greater rate of intravasation. These devices can readily be used as high-throughput screening platform to monitor the dynamics of intravasation and cellular cross-talk.

#### 9.4.4.3 *Magnetic force-based cell patterning*

To obtain microvessel on capillary scale, ECs (HUVECs) are seeded uniformly on Matrigel [136]. Cancer spheroids (B16f1 melanoma cells) are topped on this EC layer using magnetic force. Quantitative observation reveals elongated invasive morphology of cancer cells in close proximity of vessels and higher expression of MMP-9, IL-6 molecules that are known to contribute in matrix remodeling (Box 9.4).

BOX 9.4

**Outlook**

- The intravasation through lymphatic vessel is poorly exploited. The phenomenon of lymphangiogenesis, that is, the formation of neolymphatic vessels from pre-existing lymphatic vessels and incorporation of cancer intravasation within a microengineered model has yet to explore.
- The phenomenon of intravasation has been studied through the paracellular mode of migration, damaging the intercellular endothelial junctions. But the transcellular intravasation, that is, the entry of cancer cells through EC body is majorly overlooked.

### 9.4.5 Extravasation and colonization

Extravasation is defined as the active passage of the surviving circulatory tumor cells across the endothelial barrier, puncturing the underlying BME and enter into the surrounding tissue to colonize. Extravasation, the hallmark of cancer cell dissemination, follows one of two possible mode to lodge into specific secondary sites. According to Steven Paget's "Seed and soil" hypothesis, the unique microenvironment interplays the guiding role between specific cancer cells (seed) and their catering to distant organ (soil). The second opinion is the trapping of larger cancer cells compared with small circulating cells and aggregated with platelets [137]. However, these two theories are not mutually exclusive—the tissue tropism involves mechanical trapping by vascular system as well as biochemical interaction with the respective soil. Together, the extravasation and organ-specific colonization is a dynamic process regulated by the mechanical–chemical cues, cell–cell, and cell–ECM interactions.

The conventional extravasation model includes Zebrafish embryos and mice tail-vein injections. However, these models lack human physiological relevance. *In vitro* modeling involves multi-channel microfluidic platforms, where tumor cells are loaded into one or more channels which adhere to the endothelium. The tumor cells transmigrate through endothelium into adjacent gels [138]. Alternatively, a microvascular network has been set up within ECM gel, which is perfused with media containing cancer cells. In this setup, the migrating cancer cells either attach or enter into smaller vessels, resembling the entrapment in the capillaries of target organ under physiological condition [139]. Recent investigation includes organ-specific cells into the ECM gel, leading to different extravasation rate exhibited by a particular cancer cells—recapitulate the tissue tropism *in vitro* [140]. Apart from extravasation mechanism, these organ-specific microfluidic platforms are also used to investigate the effect of proinflammatory cues such as TNF-$\alpha$ [126], chemokines like CXCL12 [141] or ligand accessibility like integrin $\beta1$ [142] during extravasation. The ECM materials used for the extravasation studies are collagen [140], fibrin [140], or Matrigel [141] (Box 9.5).

BOX 9.5

Outlook

- The small volume and cell number of microfluidics limit its coupling with many traditional biochemical or genetic analysis (e.g., real-time monitoring of signaling factors, protein analysis, RNA-sequencing, etc.) to check cellular functionality beyond imaging.
- Establishment of organ-specific microfluidic models with high-cell densities or structural hierarchy like pancreas, liver, and kidney remain a challenge.

## 9.5 Biomaterial-based *in vitro* models of cancer dormancy and reactivation

Though the current most advancement in early diagnosis of cancer and treatment improves significantly the patient outcome and survival. However, the most challenging aspect of treatment of metastatic disease is the metastatic recurrence [143]. Post-surgery, the investigation of long-term survival outcome suggests that the chance of metastatic recurrence and death typically follows two distinctive patterns—(1) at 1−2 years and (2) at 5 years post-surgery. The disseminated cancer cells undergo a phase of latency or dormancy in patient. This period of dormancy may range from few months to decades based on the types of cancer, its molecular characteristics, age, gender and life-style of patient, immune status of host, and others [144]. The surviving cancer cells at secondary niche (e.g., bone marrow, brain perivascular niche) are believed to survive as single cell [145]. The cell cycle of these cells is arrested at $G_0$ phase with transformed metabolic profile and an induction of anti-apoptotic survival pathway along with multiple factors including the altered ECM properties, signaling from secondary cell types and biochemical factors. Hence, the engineering approaches of tumor dormancy are broadly classified into—(1) ECM-induced, (2) cell-signaling-induced, (3) biochemical-induced, and (4) drug-induced models of tumor dormancy. In this section, emphasis is only given to biomaterial-based model of tumor dormancy and subsequent reactivation.

The typical approach to induce the ECM-mediated dormancy is the physical confinement of cancer cells within the dense ECM that restricts the proliferation, spreading and invasion of cancer cells while facilitate the apoptosis process, thereby regulating the balance of overall cancer cell population [143] (Fig. 9.3). Several biomaterials such as collagen, Matrigel, agarose, PEG, and PCL are reported to use to confine the single cancer cell or cancer spheroids by modulating their crosslinking density that in turns causes the change in pore size, stiffness, and rate of degradation of ECM. Interestingly, cancer cells respond differently to similar matrix conditions. For example, breast cancer cells of epithelial origin (i.e., MCF7) exhibit round shape, ameboid migration, and a decreased cluster formation in presence of collagen fibril with diameter of 850 nm [146]. In contrast, breast cancer cells of mesenchymal origin (i.e., MDA-MB-231) exhibit a morphological transition from rounded

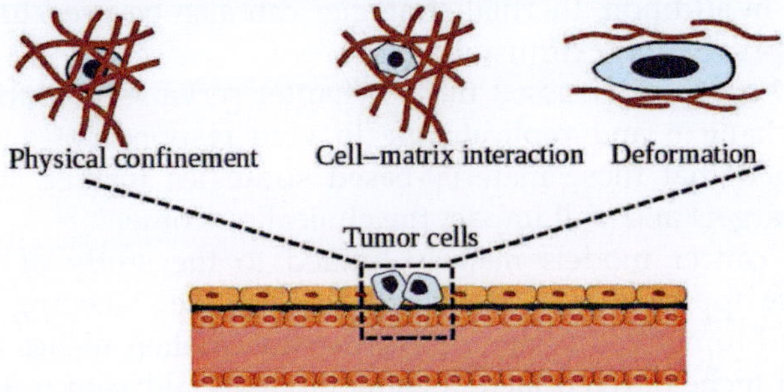


FIGURE 9.3 Modes of ECM-induced dormancy induction (Adapted from Carey SP, Rahman A, Kraning-Rush CM, Romero B, Somasegar S, Torre OM, et al. Comparative mechanisms of cancer cell migration through 3D matrix and physiological microtracks. Am J Physiol Cell Physiol 2015;308(6):C436–C47 [129]). The ECM imposes biophysical constraints such as physical confinement, which induces the dormancy. The interaction of cancer cells with ECM further reactive the disease with time.

to elongated morphology. This is suggestive that thicker fibrils promote greater cell–ECM interactions resulting in more polarization of mesenchymal MDA-MB-231 cells, which is evoked by enhanced force. While less clustering for epithelial MCF7 cells is due to the stronger cell–ECM interaction than cell–cell interaction. This also highlights the effect of cell type being used to attain the desired dormant tumor cell morphology during modeling. Given these issues, a clear consensus on the relationship of solid stress imparted by ECM and the physical restriction of cancer cells confined within ECM is established that maintain the quiescence and dormancy. The inherent ECM cues of the secondary niche also induce dormancy in different cancer cells. For example, the growth of mammary epithelial cells is arrested by the adhesion of a single integrin unit to laminin III, a BME component [147]. The presence of osteopontin in bone marrow, lung and brain perivascular confers dormancy signature to cancer cells [148]. Overall, the intelligent designing of biomaterial platforms can unravel the insight of molecular mechanism of metastatic recurrence.

## 9.6 Concluding remarks

In this chapter, the material-based approaches to model cancer metastasis are presented. Despite the advancement of biomaterial science in developing composite biomaterials with the goal specific functional modification of ECM, Matrigel is still the most recognized cancer biomaterial. Therefore, the effort of material scientists to find an alternative tunable reproducible biomaterial is still continuing.

Undoubtedly, the biomaterial-based modeling that employing the fundamentals of tissue engineering technology is a breakthrough in cancer diagnosis and treatment. Potential therapeutic avenues open up such as disruption of disease ECM physically, that is, the use of high-intensity ultrasound to improve the penetration of drug carrier [149] or chemical agents, that is, the use of proteases or cancer specific ECM-degrading enzyme such as hyaluronidase [150]. In addition, thermal strategies can also be used to determine the collagenous ECM to improve tumor diffusivity [151].

The strategies and models discussed in this chapter serve as potential tool to recapitulate the disease mechanism and replicate the in vivo response of a neo-therapeutic. In future, it is envisioned that these material-based strategies further improve our understanding of specific targets and will impact the clinical outcome.

Biomaterial-based cancer models majorly limited to the study of tumorigenesis and metastasis. The future approach may involve to sensitize the "Sleeping giants," that is, the dormant disseminated tumor cells, which are protected by their niches from current therapies. Identification of niche-derived factors using biomaterial-based modeling can serve as potential target in the development of dormancy-targeted drugs [148]. For example, the inhibition of $\alpha_5\beta_3$ integrin of metastatic niche inhibits tumor progression. Further, the striking parallels between dormant cancer cells and cancer stem-like cells [152] contribute to the dilemma if they both origin from same dormant population or of different in origin and need further understanding using bioengineered models.

The major hurdle of translation of these model-based findings into clinical practice is the patient-specificity. The ambitious concept of precision medicine is solely dependent on our ability to handle "Big data" including genomics, transcriptomics, mutation, ethnicity, sex, etc.—information from molecular to social level [153]. This requires intra- and interdisciplinary collaboration among biologist, oncologists, material scientists, physician–patients to shape the precision cancer medicine paradigm.

## Acknowledgments

The work was supported by the European Union Framework Programme for Research and Innovation Horizon 2020 under grant agreement nọ 668983—FoReCaST.

## References

[1] The Nation's Investment in Cancer Research: A Plan and Budget Proposal for Fiscal Year 2004, Prepared by the Director of the National Cancer Institute.
[2] Hutmacher DW. Biomaterials offer cancer research the third dimension. Nat Mater 2010;9(2):90–3.
[3] Beri P, Matte BF, Fattet L, Kim D, Yang J, Engler AJ. Biomaterials to model and measure epithelial cancers. Nat Rev Mater 2018;3(11):418–30.
[4] Bissell MJ, Hall HG, Parry G. How does the extracellular matrix direct gene expression. J Theor Biol 1982;99:31–68.
[5] Orkin RW, Gehron P, McGoodwin EB, Martin GR, Valentine T, Swarm R. A murine tumor producing a matrix of basement membrane. J Exp Med 1977;145:204–20.
[6] Petersen OW, Ronnovjessen L, Howlett AR, Bissell MJ. Interaction with basement membrane serves to rapidly distinguish growth and differentiation pattern of normal and malignant human breast epithelial cells. Proc Natl Acad Sci U S A 1992;89:9064–8.
[7] Paget S. The distribution of secondary growths in cancer of the breast. Lancet 1889;133:571–3.
[8] Liu Z, Vunjak-Novakovic G. Modeling tumor microenvironments using custom-designed biomaterial scaffolds. Curr Opin Chem Eng 2016;11:94–105.
[9] Mishra DK, Thrall MJ, Baird BN, Ott HC, Blackmon SH, Kurie JM, et al. Human lung cancer cells grown on acellular rat lung matrix create perfusable tumor nodules. Ann Thorac Surg 2012;93:1075–81.

[10] Stratmann AT, Fecher D, Wangorsch G, Gottlich C, Walles T, Walles H, et al. Establishment of a human 3D lung cancer model based on a biological tissue matrix combined with a boolean in silico model. Mol Oncol 2014;8:351–65.
[11] Reichert JC, Quent VM, Burke LJ, Stansfield SH, Clements JA, Hutmacher DW. Mineralized human primary osteoblast matrices as a model system to analyse interactions of prostate cancer cells with the bone microenvironment. Biomaterials. 2010;31:7928–36.
[12] Hoshiba T, Tanaka M. Breast cancer cell behaviors on staged tumorigenesis-mimicking matrices derived from tumor cells at various malignant stages. Biochem Biophys Res Commun 2013;439:291–6.
[13] Rijal G, Li W. Native-mimicking in vitro microenvironment: an elusive and seductive future for tumor modeling and tissue engineering. J Biol Eng 2018;12:20.
[14] Janvier R, Sourla A, Koutsilieris M, Doillon CJ. Stromal fibroblasts are required for PC-3 human prostate cancer cells to produce capillary-like formation of endothelial cells in a three-dimensional co-culture system. Anticancer Res 1997;17(3A):1551–7.
[15] Thakuri PS, Liu C, Luker GD, Tavana H. Biomaterials-based approaches to tumor spheroid and organoid modeling. Adv Healthc Mater 2018;7(6):e1700980.
[16] Kundu B, Bastos ARF, Brancato V, Cerqueira MT, Oliveira JM, Correlo VM, et al. Mechanical property of hydrogels and the presence of adipose stem cells in tumor stroma affect spheroid formation in the 3D osteosarcoma model. ACS Appl Mater Interfaces 2019;11(16):14548–59.
[17] Bray LJ, Werner C. Evaluation of three-dimensional in vitro models to study tumor angiogenesis. ACS Biomater Sci Eng 2018;4(2):337–46.
[18] Szot CS, Buchanan CF, Freeman JW, Rylander MN. 3D in vitro bioengineered tumors based on collagen I hydrogels. Biomaterials 2011;32(31):7905–12.
[19] Klotz BJ, Gawlitta D, Rosenberg AJ, Malda J, Melchels FP. Gelatin-methacryloyl hydrogels: towards biofabrication-based tissue repair. Trends Biotechnol 2016;34:394–407.
[20] Kaemmerer E, Melchels FP, Holzapfel BM, Meckel T, Hutmacher DW, Loessner D. Gelatine methacrylamide-based hydrogels: an alternative three-dimensional cancer cell culture system. Acta Biomater 2014;10 (6):2551–62.
[21] Bissell MJ, Kenny PA, Radisky DC. Microenvironmental regulators of tissue structure and function also regulate tumor induction and progression: the role of extracellular matrix and its degrading enzymes. Cold Spring Harb Symp Quant Biol 2005;70:343–56.
[22] Read GH, Miura N, Carter JL, Kines KT, Yamamoto K, Devasahayam N, et al. Three-dimensional alginate hydrogels for radiobiological and metabolic studies of cancer cells. Colloids Surf B Biointerfaces 2018;171:197–204.
[23] Ninomiya K, Ichihara E, Hotta K, Sone N, Murakami T, Harada D, et al. Three-arm randomized trial of sodium alginate for preventing radiation-induced esophagitis in locally advanced non-small cell lung cancer receiving concurrent chemoradiotherapy: the OLCSG1401 study protocol. Clin Lung Cancer 2017;18 (2):245–9.
[24] Wang Q, Li S, Xie Y, Yu W, Xiong Y, Ma X, et al. Cytoskeletal reorganization and repolarization of hepatocarcinoma cells in APA microcapsule to mimic native tumor characteristics. Hepatol Res 2006;35 (2):96–103.
[25] Croisier F, Jérôme C. Chitosan-based biomaterials for tissue engineering. Eur Polym J 2013;49(4):780–7892.
[26] Chang PH, Sekine K, Chao HM, Hsu SH, Chern E. Chitosan promotes cancer progression and stem cell properties in association with Wnt signaling in colon and hepatocellular carcinoma cells. Sci Rep 2017;8:45751.
[27] Rao W, Wang H, Han J, Zhao S, Dumbleton J, Agarwal P, et al. Chitosan-decorated doxorubicin-encapsulated nanoparticle targets and eliminates tumor reinitiating cancer stem-like cells. ACS Nano 2015;9 (6):5725–40.
[28] Liu N, Lapcevich RK, Underhill CB, Han Z, Gao F, Swartz G, et al. Metastatin: a hyaluronan-binding complex from cartilage that inhibits tumor growth. Cancer Res 2001;61(3):1022–8.
[29] Fong EL, Martinez M, Yang J, Mikos AG, Navone NM, Harrington DA, et al. Hydrogel-based 3D model of patient-derived prostate xenograft tumors suitable for drug screening. Mol Pharm 2014;11(7):2040–50.
[30] Hao Y, Zerdoum AB, Stuffer AJ, Rajasekaran AK, Jia X. Biomimetic hydrogels incorporating polymeric cell-adhesive peptide to promote the 3D assembly of tumoroids. Biomacromolecules 2016;17(11):3750–60.

[31] Fong EL, Wan X, Yang J, Morgado M, Mikos AG, Harrington DA, et al. A 3D in vitro model of patient-derived prostate cancer xenograft for controlled interrogation of in vivo tumor-stromal interactions. Biomaterials 2016;77:164–72.

[32] Huang YJ, Hsu SH. Acquisition of epithelial-mesenchymal transition and cancer stem-like phenotypes within chitosan-hyaluronan membrane-derived 3D tumor spheroids. Biomaterials 2014;35(38):10070–9.

[33] Kundu B, Schlimp CJ, Nürnberger S, Redl H, Kundu SC. Thromboelastometric and platelet responses to silk biomaterials. Sci Rep 2014;4:4945.

[34] Dadwal U, Falank C, Fairfield H, Linehan S, Rosen CJ, Kaplan DL, et al. Tissue-engineered 3D cancer-in-bone modeling: silk and PUR protocols. Bonekey Rep 2016;5:842.

[35] Kundu B, Eltohamy M, Yadavalli VK, Kundu SC, Kim HW. Biomimetic designing of functional silk nano-topography using self-assembly. ACS Appl Mater Interfaces 2016;8:28458–67.

[36] Sahu N, Baligar P, Midha S, Kundu B, Bhattacharjee M, Mukherjee S, et al. Nonmulberry silk fibroin scaffold shows superior osteoconductivity than mulberry silk fibroin in calvarial bone regeneration. Adv Healthc Mater 2015;4:1709–21.

[37] Wang J, Zhang S, Xing T, Kundu B, Li M, Kundu SC, et al. Ion-induced fabrication of silk fibroin nanoparticles from Chinese oak tasar *Antheraea pernyi*. Int J Biol Macromol 2015;79:316–25.

[38] Kievit FM, Florczyk SJ, Leung MC, Wang K, Wu JD, Silber JR, et al. Proliferation and enrichment of $CD133^+$ glioblastoma cancer stem cells on 3D chitosan- alginate scaffolds. Biomaterials. 2014;35:9137–43.

[39] Loessner D, Stok KS, Lutolf MP, Hutmacher DW, Clements JA, Rizzi SC. Bioengineered 3D platform to explore cell-ECM interactions and drug resistance of epithelial ovarian cancer cells. Biomaterials 2010;31:8494–506.

[40] Ki CS, Shih H, Lin C-C. Effect of 3D matrix compositions on the efficacy of EGFR inhibition in pancreatic ductal adenocarcinoma cells. Biomacromolecules 2015;14:3017–26.

[41] Engel BJ, Constantinou PE, Sablatura LK, Doty NJ, Carson DD, Farach-Carson MC, et al. Multilayered, hyaluronic acid-based hydrogel formulations suitable for automated 3D high throughput drug screening of cancer-stromal cell cocultures. Adv Healthc Mater 2015;4(11):1664–74.

[42] Almany L, Seliktar D. Biosynthetic hydrogel scaffolds made from fibrinogen and polyethylene glycol for 3D cell cultures. Biomaterials 2005;26(15):2467–77.

[43] Albertsson P. Partition of cell particles and macromolecules. 3rd ed New York: Wiley; 1986.

[44] Maschler S, Wirl G, Spring H, Bredow DV, Sordat I, Beug H, et al. Tumor cell invasiveness correlates with changes in integrin expression and localization. Oncogene 2005;24(12):2032–41.

[45] Rabionet M, Yeste M, Puig T, Ciurana J. Electrospinning PCL scaffolds manufacture for three-dimensional breast cancer cell culture. Polymers (Basel) 2017;9(8).

[46] Kundu B, Kundu SC. Silk sericin/polyacrylamide in situ forming hydrogels for dermal reconstruction. Biomaterials 2012;33:7456–67.

[47] Nasrollahi S, Walter C, Loza AJ, Schimizzi GV, Longmore GD, Pathak A. Past matrix stiffness primes epithelial cells and regulates their future collective migration through a mechanical memory. Biomaterials 2017;146:146–55.

[48] Kraning-Rush CM, Califano JP, Reinhart-King CA. Cellular traction stresses increase with increasing metastatic potential. PLoS One 2012;7:e32572.

[49] Sia SK, Whitesides GM. Microfluidic devices fabricated in poly(dimethylsiloxane) for biological studies. Electrophoresis 2003;24:3563–76.

[50] Sontheimer-Phelps A, Hassell BA, Ingber DE. Modelling cancer in microfluidic human organs-on-chips. Nat Rev Cancer 2019;19(2):65–81.

[51] Wu Y, Zhao Z, Guan Y, Zhang Y. Galactosylated reversible hydrogels as scaffold for HepG2 spheroid generation. Acta Biomater 2014;10(5):1965–74.

[52] Seib FP, Berry JE, Shiozawa Y, Taichman RS, Kaplan DL. Tissue engineering a surrogate niche for metastatic cancer cells. Biomaterials 2015;51:313–19.

[53] Brábek J, Mierke CT, Rösel D, Veselý P, Fabry B. The role of the tissue microenvironment in the regulation of cancer cell motility and invasion. Cell Commun Signal 2010;8:22.

[54] Harley BA, Kim HD, Zaman MH, Yannas IV, Lauffenburger DA, Gibson LJ. Microarchitecture of three-dimensional scaffolds influences cell migration behavior via junction interactions. Biophys J 2008;95:4013–24.

[55] Sabeh F, Shimizu-Hirota R, Weiss SJ. Protease-dependent versus -independent cancer cell invasion programs: three-dimensional amoeboid movement revisited. J Cell Biol 2009;185(1):11–19.

[56] Dahl KN, Ribeiro AJ, Lammerding J. Nuclear shape, mechanics, and mechanotransduction. Circ Res 2008;102:1307–18.
[57] Aung A, Seo YN, Lu S, Wang Y, Jamora C, del Alamo JC, et al. 3D traction stresses activate protease-dependent invasion of cancer cells. Biophys J 2014;107:2528–37.
[58] Wolf K, Mazo I, Leung H, Engelke K, Von Andrian UH, Deryugina EI, et al. Compensation mechanism in tumor cell migration: mesenchymal–amoeboid transition after blocking of pericellular proteolysis. J Cell Biol 2003;160:267–77.
[59] Liliensiek SJ, Nealey P, Murphy CJ. Characterization of endothelial basement membrane nanotopography in rhesus macaque as a guide for vessel tissue engineering. Tissue Eng Part A 2009;15:2643–51.
[60] Teixeira AI, Abrams GA, Bertics PJ, Murphy CJ, Nealey PF. Epithelial contact guidance on well-defined micro- and nanostructured substrates. J Cell Sci 2003;116:1881–92.
[61] Biela SA, Su Y, Spatz JP, Kemkemer R. Different sensitivity of human endothelial cells, smooth muscle cells and fibroblasts to topography in the nano–micro range. Acta Biomater 2009;5:2460–6.
[62] Sidani M, Wyckoff J, Xue C, Segall JE, Condeelis J. Probing the microenvironment of mammary tumors using multiphoton microscopy. J Mammary Gland Biol Neoplasia 2006;11:151–63.
[63] Carragher NO, Frame MC. Focal adhesion and actin dynamics: a place where kinases and proteases meet to promote invasion. Trends Cell Biol 2004;14:241–9.
[64] Zhao J, Guan JL. Signal transduction by focal adhesion kinase in cancer. Cancer Metastasis Rev 2009;28:35–49.
[65] Bommert K, Bargou RC, Stühmer T. Signalling and survival pathways in multiple myeloma. Eur J Cancer 2006;42:1574–80.
[66] Schlaepfer DD, Hauck CR, Sieg DJ. Signaling through focal adhesion kinase. Prog Biophys Mol Biol 1999;71:435–78.
[67] Jeon JS, Bersini S, Gilardi M, Dubini G, Charest JL, Moretti M, et al. Human 3D vascularized organotypic microfluidic assays to study breast cancer cell extravasation. Proc Natl Acad Sci U S A 2015;112:214–19.
[68] Fuhrmann A, Staunton JR, Nandakumar V, Banyai N, Davies PC, Ros R. AFM stiffness nanotomography of normal, metaplastic and dysplastic human esophageal cells. Phys Biol 2011;8:015007.
[69] Levental KR, Yu H, Kass L, Lakins JN, Egeblad M, Erler JT, et al. Matrix crosslinking forces tumor progression by enhancing integrin signaling. Cell 2009;139(5):891–906.
[70] Milosevic M, Fyles A, Hedley D, Hill R. The human tumor microenvironment: in-vasive (needle) measurement of oxygen and interstitial fluid pressure. Semin Radiat Oncol 2004;14:249–58.
[71] Peyton SR, Kim PD, Ghajar CM, Seliktar D, Putnam AJ. The effects of matrix stiffness and RhoA on the phenotypic plasticity of smooth muscle cells in a 3-D biosynthetic hydrogel system. Biomaterials 2008;29:2597–607.
[72] Sharma Y, Vargas DA, Pegoraro AF, Lepzelter D, Weitz DA, Zaman M. Collective motion of mammalian cell cohorts in 3D. Integr Biol 2015;7(12):1526–33.
[73] Berg A, Ekwall AK, Rubin K, Stjernschantz J, Reed RK. Effect of PGE1, PGI2, and PGF2$\alpha$ analogs on collagen gel compaction in vitro and interstitial pressure in vivo. Am J Physiol Heart Circ Physiol 1998;274:H663–71.
[74] Polacheck WJ, Charest JL, Kamm RD. Interstitial flow influences direction of tumor cell migration through competing mechanisms. Proc Natl Acad Sci U S A 2011;108:11115–20.
[75] Ghosh K, Thodeti CK, Dudley AC, Mammoto A, Klagsbrun M, Ingber DE. Tumor-derived endothelial cells exhibit aberrant Rho-mediated mechanosensing and abnormal angiogenesis in vitro. Proc Natl Acad Sci U S A 2008;105:11305–10.
[76] Romero-López M, Trinh AL, Sobrino A, Hatch MM, Keating MT, Fimbres C, et al. Recapitulating the human tumor microenvironment: colon tumor-derived extracellular matrix promotes angiogenesis and tumor cell growth. Biomaterials 2017;116:118–29.
[77] Blehm BH, Jiang N, Kotobuki Y, Tanner K. Deconstructing the role of the ECM mi- croenvironment on drug efficacy targeting MAPK signaling in a pre-clinical platform for cutaneous melanoma. Biomaterials 2015;56:129–39.
[78] Sherman-Baust CA, Weeraratna AT, Rangel LB, Pizer ES, Cho KR, Schwartz DR, et al. Remodeling of the extracellular matrix through overex- pression of collagen VI contributes to cisplatin resistance in ovarian cancer cells. Cancer Cell 2003;3(4):377–86.
[79] Shahal T, Geiger B, Dunlop IE, Spatz JP. Regulation of integrin adhesions by varying the density of substrate-bound epidermal growth factor. Biointerphases 2012;7(1-4):23.

[80] Teng Y, Qiu J, Zheng Y, Luo X, Zhang L. Effects of type I collagen and fibronectin on regulation of breast cancer cell biological and biomechanical characteristics. J Med Biol Eng 2014;34:62–8.
[81] Glass R, Möller M, Spatz JP. Block copolymer micelle nanolithography. Nanotechnology 2003;14:1153–60.
[82] Katt ME, Placone AL, Wong AD, Xu ZS, Searson PC. In vitro tumor models: advantages, disadvantages, variables, and selecting the right platform. Front Bioeng Biotechnol 2016;12(4):12.
[83] Nyga A, Cheema U, Loizidou M. 3D tumour models: novel in vitro approaches to cancer studies. J Cell Commun Signal 2011;5(3):239–48.
[84] Rodrigues T, Kundu B, Silva-Correia J, Kundu SC, Oliveira JM, Reis RL, et al. Emerging tumor spheroids technologies for 3D in vitro cancer modeling. Pharmacol Ther 2018;184:201–11.
[85] Sokol ES, Miller DH, Breggia A, Spencer KC, Arendt LM, Gupta PB. Growth of human breast tissues from patient cells in 3D hydrogel scaffolds. Breast Cancer Res 2016;18(1):19.
[86] Weigelt B, Ghajar CM, Bissell MJ. The need for complex 3D culture models to unravel novel pathways and identify accurate biomarkers in breast cancer. Adv Drug Deliv Rev 2014;69-70:42–51.
[87] Alessandri K, Sarangi BR, Gurchenkov VV, Sinha B, Kiessling TR, Fetler L, et al. Cellular capsules as a tool for multicellular spheroid production and for investigating the mechanics of tumor progression in vitro. Proc Natl Acad Sci U S A 2013;110(37):14843–8.
[88] Jang M, Koh I, Lee SJ, Cheong JH, Kim P. Droplet-based microtumor model to assess cell-ECM interactions and drug resistance of gastric cancer cells. Sci Rep 2017;7:41541.
[89] Liu T, Chien CC, Parkinson L, Thierry B. Advanced micromachining of concave microwells for long term on-chip culture of multicellular tumor spheroids. ACS Appl Mater Interfaces 2014;6(11):8090–7.
[90] Singh M, Close DA, Mukundan S, Johnston PA, Sant S. Production of uniform 3D microtumors in hydrogel microwell arrays for measurement of viability, morphology, and signaling pathway activation. Assay Drug Dev Technol 2015;13(9):570–83.
[91] Tang Y, Liu J, Chen Y. Agarose multi-wells for tumour spheroid formation and anti-cancer drug test. Microelectron Eng 2016;158:41–5.
[92] Ma J, Zhang X, Liu Y, Yu H, Liu L, Shi Y, et al. Patterning hypoxic multicellular spheroids in a 3D matrix - a promising method for anti-tumor drug screening. Biotechnol J 2016;11(1):127–34.
[93] Boghaert E, Gleghorn JP, Lee K, Gjorevski N, Radisky DC, Nelson CM. Host epithelial geometry regulates breast cancer cell invasiveness. Proc Natl Acad Sci U S A 2012;109(48):19632–7.
[94] Sung KE, Yang N, Pehlke C, Keely PJ, Eliceiri KW, Friedl A, et al. Transition to invasion in breast cancer: a microfluidic in vitro model enables examination of spatial and temporal effects. Integr Biol 2011;3 (4):439–50.
[95] Zhao Y, Yao R, Ouyang L, Ding H, Zhang T, Zhang K, et al. Three-dimensional printing of Hela cells for cervical tumor model in vitro. Biofabrication 2014;6(3):035001.
[96] Brancato V, Garziano A, Gioiella F, Urciuolo F, Imparato G, Panzetta V, et al. 3D is not enough: building up a cell instructive micro- environment for tumoral stroma microtissues. Acta Biomater 2017;47:1–13.
[97] Bergers G, Benjamin LE. Tumorigenesis and the angiogenic switch. Nat Rev Cancer 2003;3(6):401–10.
[98] Bray LJ, Binner M, Holzheu A, Friedrichs J, Freudenberg U, Hutmacher DW, et al. Multi-parametric hydrogels support 3D in vitro bioengineered microenvironment models of tumour angiogenesis. Biomaterials 2015;53:609–20.
[99] Nguyen EH, Murphy WL. Customizable biomaterials as tools for advanced anti-angiogenic drug discovery. Biomaterials 2018;181:53–66.
[100] Edgar LT, Underwood CJ, Guilkey JE, Hoying JB, Weiss JA. Extracellular matrix density regulates the rate of neovessel growth and branching in sprouting angiogenesis. PloS One 2014;9:1–10.
[101] Shamloo A, Mohammadaliha N, Heilshorn SC, Bauer AL. A comparative study of collagen matrix density effect on endothelial sprout formation using experimental and computational approaches. Ann Biomed Eng 2016;44:929–41.
[102] Bauer AL, Jackson TL, Jiang Y. Topography of extracellular matrix mediates vascular morphogenesis and migration speeds in angiogenesis. PLoS Comput Biol 2009;5:e1000445.
[103] Crosby CO, Zoldan J. Mimicking the physical cues of the ECM in angiogenic biomaterials. Regen Biomater 2019;6:61–73.
[104] Huettner N, Dargaville TR, Forget A. Discovering cell-adhesion peptides in tissue engineering: beyond RGD. Trends Biotechnol 2018;36:372–83.

[105] Miller JS, Shen CJ, Legant WR, et al. Bioactive hydrogels made from step-growth derived PEG-peptide macromers. Biomaterials 2010;31:3736–43.

[106] Ali S, Saik JE, Gould DJ, Dickinson ME, West JL. Immobilization of cell-adhesive laminin peptides in degradable PEGDA hydrogels influences endothelial cell tubulogenesis. BioRes Open Access 2013;2:241–9.

[107] Taubenberger AV, Bray LJ, Haller B, Shaposhnykov A, Binner M, Freudenberg U, et al. 3D extracellular matrix interactions modulate tumour cell growth, invasion and angiogenesis in engineered tumour microenvironments. Acta Biomater 2016;36:73–85.

[108] Nyga A, Loizidou M, Emberton M, Cheema U. A novel tissue engineered three-dimensional in vitro colorectal cancer model. Acta Biomater 2013;9(8):7917–26.

[109] Szot CS, Buchanan CF, Freeman JW, Rylander MN. In vitro angiogenesis induced by tumor-endothelial cell co-culture in bilayered, collagen I hydrogel bioengineered tumors. Tissue Eng Part C 2013;19(11):864–74.

[110] Chung S, Sudo R, Mack PJ, Wan CR, Vickerman V, Kamm RD. Cell migration into scaffolds under co-culture conditions in a microfluidic platform. Lab Chip 2009;9(2):269–75.

[111] Kim S, Lee H, Chung M, Jeon NL. Engineering of functional, perfusable 3D microvascular networks on a chip. Lab Chip 2013;13(8):1489–500.

[112] Kolesky DB, Homan KA, Skylar-Scott MA, Lewis JA. Three-dimensional bioprinting of thick vascularized tissues. Proc Natl Acad Sci U S A 2016;113(12):3179–84.

[113] Chwalek K, Levental KR, Tsurkan MV, Zieris A, Freudenberg U, Werner C. Two-tier hydrogel degradation to boost endothelial cell morphogenesis. Biomaterials 2011;32(36):9649–57.

[114] Kalluri R, Weinberg RA. The basics of epithelial-mesenchymal transition. J Clin Investig 2009;119(6):1420–8.

[115] Yamaguchi H, Condeelis J. Regulation of the actin cytoskeleton in cancer cell migration and invasion. Biochim Biophys Acta (BBA) - Mol Cell Res 2007;1773(5):642–52.

[116] Friedl P, Alexander S. Cancer invasion and the microenvironment: plasticity and reciprocity. Cell 2011;147 (5):992–1009.

[117] Al-Mehdi AB, Tozawa K, Fisher AB, Shientag L, Lee A, Muschel RJ. Intravascular origin of metastasis from the proliferation of endothelium-attached tumor cells: a new model for metastasis. Nat Med 2000;6(1):100–2.

[118] Carey SP, Starchenko A, McGregor AL, Reinhart-King CA. Leading malignant cells initiate collective epithelial cell invasion in a three-dimensional heterotypic tumor spheroid model. Clin Exp Metastasis 2013;30 (5):615–30.

[119] Guzman A, Sánchez Alemany V, Nguyen Y, Zhang CR, Kaufman LJ. A novel 3D in vitro metastasis model elucidates differential invasive strategies during and after breaching basement membrane. Biomaterials 2017;115:19–29.

[120] McLane JS, Ligon LA. Stiffened extracellular matrix and signaling from stromal fibroblasts via osteoprotegerin regulate tumor cell invasion in a 3-D tumor in situ model. Cancer Microenviron 2016;9(2-3):127–39.

[121] Marshall J. Transwell® invasion assays. In: Wells CM, Parsons M, editors. Cell Migration. London (UK): Springer; 2011. p. 97–110.

[122] Wang FQ, Ariztia EV, Boyd LR, Horton FR, Smicun Y, Hetherington JA, et al. Lysophosphatidic acid (LPA) effects on endometrial carcinoma in vitro proliferation, invasion, and matrix metalloproteinase activity. Gynecol Oncol 2010;117(1):88–95.

[123] Rath BH, Fair JM, Jamal M, Camphausen K, Tofilon PJ. Astrocytes enhance the invasion potential of glioblastoma stem-like cells. PLoS One 2013;8:e54752.

[124] Kramer N, Walzl A, Unger C, Rosner M, Krupitza G, Hengstschläger M, et al. In vitro cell migration and invasion assays. Mutat Res 2013;752(1):10–24.

[125] Kraning-Rush CM, Carey SP, Lampi MC, Reinhart-King CA. Microfabricated collagen tracks facilitate single cell metastatic invasion in 3D. Integr Biol (Camb) 2013;5(3):606–16.

[126] Spagnolo B, Brunetti V, Leménager G, De Luca E, Sileo L, Pellegrino T, et al. Three-dimensional cage-like microscaffolds for cell invasion studies. Sci Rep 2015;5:10531.

[127] Schneider J, Bachmann T, Franco D, Richner P, Galliker P, Tiwari MK, et al. A novel 3D integrated platform for the high- resolution study of cell migration plasticity. Macromol Biosci 2013;13(8):973–83.

[128] Stroka KM, Jiang H, Chen SH, Tong Z, Wirtz D, Sun SX, et al. Water permeation drives tumor cell migration in confined microenvironments. Cell 2014;157(3):611–23.

[129] Torisawa YS, Mosadegh B, Bersano-Begey T, Steele JM, Luker KE, Luker GD, et al. Microfluidic platform for chemotaxis in gradients formed by CXCL12 source-sink cells. Integr Biol (Camb) 2010;2(11-12):680–6.

[130] Carey SP, Rahman A, Kraning-Rush CM, Romero B, Somasegar S, Torre OM, et al. Comparative mechanisms of cancer cell migration through 3D matrix and physiological microtracks. Am J Physiol Cell Physiol 2015;308(6):C436–47.

[131] Bockhorn M, Jain RK, Munn LL. Active versus passive mechanisms in metastasis: do cancer cells crawl into vessels, or are they pushed? Lancet Oncol 2007;8(5):444–8.

[132] Ehsan SM, Welch-Reardon KM, Waterman ML, Hughes CC, George SC. A three-dimensional in vitro model of tumor cell intravasation. Integr Biol 2014;6(6):603–10.

[133] De Moor L, Merovci I, Baetens S, Verstraeten J, Kowalska P, Krysko DV, et al. High-throughput fabrication of vascularized spheroids for bioprinting. Biofabrication 2018;10(3):035009.

[134] Zervantonakis IK, Hughes-Alford SK, Charest JL, Condeelis JS, Gertler FB, Kamm RD. Three-dimensional microfluidic model for tumor cell intravasation and endothelial barrier function. Proc Natl Acad Sci 2012;109(34):13515–20.

[135] Han W, Chen S, Yuan W, Fan Q, Tian J, Wang X, et al. Oriented collagen fibers direct tumor cell intravasation. Proc Natl Acad Sci U S A 2016;113(40):11208–13.

[136] Yamamoto S, Hotta MM, Okochi M, Honda H. Effect of vascular formed endothelial cell network on the invasive capacity of melanoma using the in vitro 3D Co-culture patterning model. PLoS One 2014;9(7): e103502.

[137] Psaila B, Lyden D. The metastatic niche: adapting the foreign soil. Nat Rev Cancer 2009;9(4):285–93.

[138] Riahi R, Yang YL, Kim H, Jiang L, Wong PK, Zohar Y. A microfluidic model for organ-specific extravasation of circulating tumor cells. Biomicrofluidics 2014;8(2):024103.

[139] Chen MB, Whisler JA, Jeon JS, Kamm RD. Mechanisms of tumor cell extravasation in an in vitro microvascular network platform. Integr Biol Quant Biosci Nano Macro 2013;5:1262–71.

[140] Bersini S, Jeon JS, Dubini G, Arrigoni C, Chung S, Charest JL, et al. A microfluidic 3D invitro model for specificity of breast cancer metastasis to bone. Biomaterials 2014;35(8):2454–61.

[141] Zhang Q, Liu T, Qin J. A microfluidic-based device for study of trans- endothelial invasion of tumor aggregates in realtime. Lab Chip 2012;12(16):2837–42.

[142] Gilardi M, Bersini S, Kamm RD, Moretti M, Vanoni M. Abstract PR07: dissection of cancer cells extravasation through human vascularized 3D microfluidic model: the major role of talin-1. Cancer Res 2016;76(7).

[143] Pradhan S, Sperduto JL, Farino CJ, Slater JH. Engineered in vitro models of tumor dormancy and reactivation. J Biol Eng 2018;12:37.

[144] Uhr JW, Pantel K. Controversies in clinical cancer dormancy. Proc Natl Acad Sci U S A 2011;108 (30):12396–400.

[145] van der Toom EE, Verdone JE, Pienta KJ. Disseminated tumor cells and dormancy in prostate cancer metastasis. Curr Opin Biotechnol 2016;40:9–15.

[146] Sapudom J, Rubner S, Martin S, Kurth T, Riedel S, Mierke CT, et al. The phenotype of cancer cell invasion controlled by fibril diameter and pore size of 3D collagen networks. Biomaterials 2015;52:367–75.

[147] Boudreau N, Sympson CJ, Werb Z, Bissell MJ. Suppression of ICE and apoptosis in mammary epithelial cells by extracellular matrix. Science 1995;267:891–3.

[148] Ghajar CM, Peinado H, Mori H, Matei IR, Evason KJ, Brazier H, et al. The perivascular niche regulates breast tumour dormancy. Nat Cell Biol 2013;15(7):807–17.

[149] Lee S, Han H, Koo H, Na JH, Yoon HY, Lee KE, et al. Extracellular matrix remodeling in vivo for enhancing tumor-targeting efficiency of nanoparticle drug carriers using the pulsed high intensity focused ultrasound. J Control Release 2017;26368–78.

[150] Provenzano PP, Cuevas C, Chang AE, Goel VK, Von Hoff DD, Hingorani SR. Enzymatic targeting of the stroma ablates physical barriers to treatment of pancreatic ductal adenocarcinoma. Cancer Cell 2012;21 (3):418–29.

[151] Raeesi V, Chan WCW. Improving nanoparticle diffusion through tumor collagen matrix by photo-thermal gold nanorods. Nanoscale 2016;8(25):12524–30.

[152] He J, Xiong L, Li Q, Lin L, Miao X, Yan S, et al. 3D modeling of cancer stem cell niche. Oncotarget. 2018;9 (1):1326–45.

[153] Butte AJ. Big data opens a window onto wellness. Nat Biotechnol 2017;35(8):720–1.

CHAPTER

# 10

# 3D cancer spheroids and microtissues


*Virginia Brancato*[1,2], *Rui L. Reis*[1,2] *and Subhas C. Kundu*[1,2]

[1]3B's Research Group, I3Bs – Research Institute on Biomaterials, Biodegradables and Biomimetics, University of Minho, Headquarters of the European Institute of Excellence on Tissue Engineering and Regenerative Medicine, Guimarães, Portugal [2]ICVS/3B's–PT Government Associate Laboratory, Braga/Guimarães, Portugal


## Abbreviations

| | |
|---|---|
| **ECM** | extracellular matrix |
| **TME** | tumor microenvironment |
| **HA** | hyaluronic acid |
| **EHS** | Engelbreth–Holm–Swarm |
| **CAF** | cancer-associated fibroblasts |
| **HMF** | human mammary fibroblasts |
| **FGFR** | fibroblast growth factor receptor |
| **VEGF** | vascular endothelial growth factor |
| **FGF** | fibroblast growth factor |
| **GEM** | global eukaryotic microcarriers |
| **TAMs** | tumor-associated macrophages |
| **TGF-β** | tumor growth factor beta |
| **CIK** | cytokine-induced killer |
| **PMBC** | peripheral blood mononuclear cell |
| **TIL** | tumor-infiltrating lymphocytes |
| **NSCLC** | nonsmall cell lung cancer |
| **HTS** | high-throughput screening |
| **HCS** | high-content screening |

## 10.1 Introduction

The National Cancer Institute provides a definition of cancer as a disease in which abnormal cells divide without control and can invade the nearby tissues. This definition does not reveal the complexity of the genetic, cellular, and environmental crosstalk that trigger the cancer development. Cancer can be considered as a genetic disease since it


*Biomaterials for 3D Tumor Modeling*
DOI: https://doi.org/10.1016/B978-0-12-818128-7.00010-1

is caused by an accumulation of mutations on oncogenes or tumor suppressor genes. These genes are responsible for promoting or suppressing cell growth and their mutations give place to the neoplastic transformation [1]. Hanahan and Weinberg published an interesting and exhaustive article describing the general features of this disease, in which it defines for the first time as the "hallmarks of cancer" [2]. They state that cancer cells replicate out of control, evade growth suppressor signals, activate invasion and metastasis, enable replicative immortality, resist to cell death, and induce angiogenesis. After a decade, the progress in cancer biology widened the perspective on this disease, putting in the context the deregulation of the cellular metabolism and the involvement of immune system as double-face player in cancer progression (Fig. 10.1). Moreover, researchers described that cancer is triggered by genomic instability and mutation as well as tumor promoting inflammation. These two features are classified as enabling characteristics [2,3]. Cancer may arise from almost any part of the body tissues with genetic heterogeneities that make even harder to treat this disease. Furthermore, cancer growth proceeds through several stages that transform it from benign or more tissue-confined malignant or metastatic.

Cancer cells are able to modify the TME and invade the stromal tissue surrounding them. Through the process of metastasis, the cancer cells penetrate the blood vessel, and are transported by the bloodstream. They are finally able to extravasate in other tissues. Cancer cells also develop mechanisms that enable them to escape the immune system cells [4]. The percentage of cells that successfully reach the secondary site and start a new

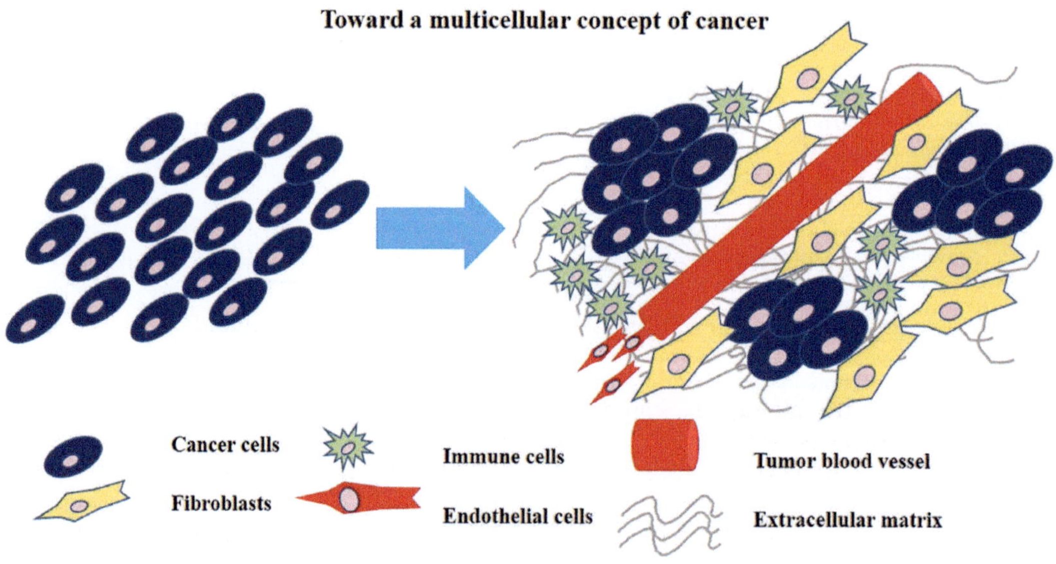


FIGURE 10.1 From a cancer cell-based to a multicellular-based concept of tumor microenvironment (TME). Knowledge of cancer disease is the results of studies on cancer cells extrapolated by their native microenvironment. The interest on TME enhances the investigation of the interaction among the several cell populations involved in tumor progression (cancer cells, fibroblasts, endothelial, and immune system cells). The multicellular concept for cancer has also broadened the spectrum of anticancer treatments.

tumor is near to 0.01%. Even if the percentage is low, from a treatment perspective, the metastasis remains the worst stage to cure and 90% of cancer deaths are caused by secondary site tumor [5]. Unraveling the tumor complexity is necessary to spot the different cell types that together with the extracellular matrix (ECM) contribute to the unique structure of the tumor tissue. In the last decades, the cancer biologists have begun to move their interest toward TME. Researchers investigated the role of stromal cells (fibroblasts, endothelial, and immune system cells) in the cancer progression and metastasis, leaving behind the reductionist image of tumor as pool of partially homogeneous cancer cells (Fig. 10.1). Alongside the genetic and biochemical cues that alter the cellular behavior, mechanical properties of TME and the ECM stiffness reveal as potential regulators of cell fate. Moreover, stroma gains attention since its potentiality as target for next generation anticancer drugs [6]. Recent progress in tissue engineering and biotechnology leads to the development of advanced three-dimensional (3D) in vitro cancer models. Biomaterials support the growth of cancer cells and are able to mimic the TME. Oncologists, cancer researchers, and pharmaceutical companies are joining the efforts to achieve a standard 3D in vitro model, which may provide most of the features to mimic the general cancer microenvironmental condition as in vivo. However, this interest leads to the myriad of 3D cancer models based on various tissue-engineering approaches. Unfortunately, due to a large variability among the 3D cancer models, there is still an open debate for their possible involvement as preclinical tools for testing drugs in a reproducible and reliable ways. Moreover, the on-model-fits-all paradigm cannot faithfully copy the complexities of the tumor tissues. In this chapter, we pay attention to the cellular components of the TME and the 3D in vitro models that aim to recapitulate the crosstalk among these cell types including the cancer cells. Particularly, the chapter focuses on the spheroids and the 3D cell-laden microcarrier models, since they are the promising tools for drug screening. The chapter highlights the differences between the two approaches and how it is possible to combine the cell types in order to better mimic the TME. The features of the models are recapitulated in the Fig. 10.2.

## 10.2 Biomaterials advances tumor cell culture to the third dimension

### 10.2.1 Biodegradable microcarriers to develop in vitro 3D heterotypic tumor models

Microcarriers are mainly exploited to amplify the cells for tissue engineering and regenerative purposes, since they allow the cells to preserve their morphologies following the good manufacturing practice [7]. In 1967, van Wezel reported for the first time the use of diethylaminoethyl cellulose (DEAE) Sephadex A50 microcarriers as support for anchorage-dependent cells expansion [8]. Microcarriers allow a close spatial contact of the cells that is relevant in case of co-cultures with different types of cells. Microcarriers are characterized by high porosity and interconnectivity that allow the growth of cells throughout the entire surface. The range of diameters of the pores of microcarriers spans from 50 to 400 μm [9]. Microcarriers are available commercially, for example, from Hillex, Cultispher, Cytodex, GELIBEAD, Global Eukaryotic Microcarrier

| | Spheroid models | Microcarriers-based models |
|---|---|---|
| | 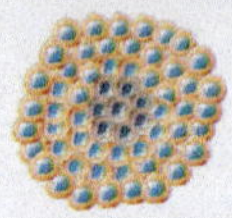 | 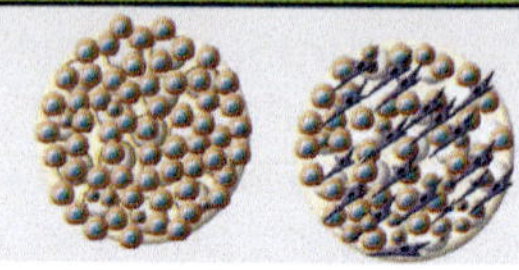 |
| Derived from cell line monoculture or co-culture | +++ | +++ |
| Derived from patient-derived cells | + | ++ |
| Nutrient and oxygen gradient | +++ | ++ |
| Hypoxic core | +++ | ++ |
| ECM formation | + | +++ |
| Suitable for HTS | +++ | + |
| Suitable for microfluidic | ++ | ++ |
| Miniaturized multiwell format and suitable for microplater reader | ++ | ++ |
| Short-term culture | +++ | + |
| Long-term culture | + | ++ |
| Use of large amount of cells per unit | + | +++ |
| Controllable size distribution | + | + |

FIGURE 10.2 Comparison of 3D spheroid models and microcarriers-based models. Spheroids are the gold standard in cancer research since they are easy to fabricate and manipulate. They can be embedded in different biomaterials to supply the lack of ECM production. Microcarriers-based microtissues have the advantages of the production of endogenous ECM and the suitability to long-term culture. However, spheroids can be incorporated into high-throughput screening (HTS) systems more easily than microtissues that require an extra manual step to collect the unit from the dynamic culture flasks.

(GEM), and Ventregel. The coating is customizable to support the growth of different cell types, and assays are performed with fluorescence, luminescence, or absorbance detection methods [10]. Cultispher microcarriers are also the substrate to grow cancer cells and test the efficacy of chemotherapeutic in vitro [11]. The results on A549 cells, show a decrease in the efficacy of the drug, when the cells are cultured in a 3D architecture on the microcarriers in comparison to standard two-dimensional (2D) culture [11]. One of the advantages of the cell culture on microcarriers is the nondestructive collection of the samples for multiple time points' analyses. Cytodex 3 microcarriers are useful to grow breast cancer patient-derived cells together with fibroblasts in a dynamic bioreactor. The model is compared with the fresh breast cancer tissue in terms of DNA content, cell cycle, and tissue morphology [12]. Alginate is widely used to prepare freeze-dried microcarriers. In particular, alginate/chitosan microcarriers enriched with fibronectin or heparin/basic fibroblasts growth factors support the proliferation of hepatocarcinoma cells. The cells in 3D spatial organization are able to express albumin, which is one of the main features of the liver cells [13]. In the next section, we further describe the use of microcarriers for the co-culture of cancer cells with fibroblasts, endothelial, and immune system cells.

### 10.2.2 Exogenous extracellular matrix as support for the growth of tumor spheroids

Cell–cell contact is an important hallmark that drives tumor progression and metastasis. Spheroids are tightly packed cells' aggregate that are able to recapitulate the cell–cell interaction. This promotes the exchange of adhesion molecules necessary to tumor progression. However, spheroids lack of ECM and may fail to recapitulate the interactions between the cells and the TME. Biomaterials may overcome this disadvantage by offering biochemical and mechanical cues that may recapitulate the in vivo condition. Scaffolds and hydrogels, made up of different biomaterials, promote 3D architecture to cancer cells. This mimics the tumor core and the gradient of nutrients and drug diffusion [14,15]. Hydrogels are cross-linked networks of polymeric biomaterials with high percentage of water. Hydrogels can be classified according to the origin of polymers as naturals [e.g., alginate, silk, gelatin, collagen, chitosan, gellan gum, hyaluronic acid (HA), and fibrin] or synthetics (e.g., polyethylene glycol, polyacrylamide, polyacrylic acid, polyvinyl alcohol, and polyvinylpyrrolidone) [16,17]. Hydrogels' mechanical and biochemical properties can be modulated by changing the polymer's network or playing with the cross-linking of the polymers. Moreover, the bioactivity of the hydrogels can be enhanced by incorporating growth factors or peptides with cells adhesion sequences. The porosity of the hydrogels allows the growth and proliferation of cells. It has been demonstrated that HA-based hydrogel supports the growth of patient-derived prostate cancer cells and promote the spheroid formations [18]. The most used biomaterial in the production of spheroid-based hydrogels is collagen I. It is demonstrated that collagen hydrogels can be obtained via enzymatic or physical cross-linking of a rat-tail or bovine skin collagen solution, recapitulating several tumor models. The fibrils density and diameter can also trigger the invasion and migration potential of the cancer cells. Guzman et al. showed how a fibrillar substrate such as collagen enhances the cell polarization and migration in comparison to the other types of hydrogels [matrigel or nonfibrillar basement membrane extract (BME)]. A blended hydrogel of BME/collagen modulate the migration and can be useful to assess the cell behavior in different environments [19]. MCF-7 cells (nonmetastatic breast cancer cells) acquire a mesenchymal phenotype when they are cultured in 3D collagen hydrogels [20] showing an upregulation of the epithelial–mesenchymal-transition markers and a downregulation of the E-cadherin proteins. This shows aggressive and invasive phenotype of MCF-7 cells when they are cultured in 3D in comparison to 2D [20]. Matrigel is commonly used as exogenous ECM for cancer cell cultures in 3D. It derives by the basement membrane secreted by the Engelbreth–Holm–Swarm (EHS) mouse sarcoma cells. Matrigel is a mixture of laminin, collagen IV, perlecan, and entactin together with other proteases and growth factors. Even though Matrigel composition is not precisely defined, this composite is the preferential matrix to grow spheroids and organoids [21]. The success of matrigel in the development of 3D cell culture is due to its interesting biochemical and mechanical cues, and the easy-to-handle protocol to perform cancer migration and invasion assays.

Alginate is made up of two repeating units (α-L-guluronic acid and β-D-mannuronic acid). Alginate is obtained from a brown seaweed and it is employed as exogenous matrix to encapsulate cancer cells. The modulation of the concentration of the two monomers affects the physico-chemical properties like pH, molecular weight, mechanical, and ionic strength [22]. A microfluidic-based approach exploits the properties of alginate to generate one- and two-shell layered hydrogel to encapsulate breast cancer cells. The authors demonstrate how the presence of the second shell enhanced the proliferation and the 3D organization of the cells [23]. HA is classified as glycosaminoglycans and it is composed of repeating units of D-glucuronic acid and D-N-acetylglucosamine. HA is a key player in cancer progression and metastasis. The principal cell surface receptor of HA is the transmembrane glycoprotein CD44. Besides the accumulation of HA in the TME, the HA molecular weight is the key factor affecting the malignant transformation of tumor cells. The equilibrium between HA synthases and hyaluronidases modulates the amount and composition of HAs in the ECM, triggering the pathological state of a tissue [24]. Recently, the importance of how the HA is exposed to the cellular receptors and the influence of the various HAs molecular weight is demonstrated. The authors show how the cell adhesion and motility, invadopodia formation, and cell cluster organization is affected by high-molecular weights of HA [25]. Three cell lines (SA87, NCI-H460, and H460M) from primary tumor and metastasis sites are encapsulated in HA hydrogels and their responses to increasing concentration of doxorubicin and 5-fluoruracil are evaluated. The results are compared with the 2D cell culture of the same cell lines, showing that 3D tumor cell cultures are more resistant to the drug effects [26]. HA hydrogels can enhance the growth and the metastatic behavior of bone metastatic prostate cancer cells. These cells also exhibit more resistant phenotypes when they are treated with camptothecin, docetaxel, and rapamycin, alone or in combination to the same cells cultured on dishes [27].

Besides hydrogels, biomaterials are used to fabricate scaffolds to support the cell growth. The scaffold surface has to enhance the cell adhesion and proliferation. Scaffolds are also highly porous structures so the cells can migrate throughout the structure. Ready-to-use off-the-shelf scaffolds with various shapes are now available. Biomaterials are also blended together to enhance the mechanical properties. For this aim, silk protein fibroin mixed to different materials promotes better 3D architecture and cell growth. Silk protein fibroins are isolated from different species of silkworms *Bombyx mori* (mulberry) and, as an example, *Antheraea mylitta* (nonmulberry). Fibroins are already used in many applications for tissue engineering and regenerative medicine. The biophysical properties of silk fibroin are modulated to enhance the cell growth and to mimic better TME [28,29]. Silk fibroin blended to chitosan is exploited to culture hepatocarcinoma cell lines (HepG2). The researchers demonstrate that lower the silk fibroin concentration in the scaffold, the bigger is the porous size. Mechanical property, water stability, and porosity are also controlled [30]. Another attempt to develop a 3D lung cancer model is made by embedding A549 cell in silk fibroin/chitosan blended scaffolds. This model is also used as drug testing platform for the anticancer activity of doxorubicin. Porous dimension and organization are parameters to take into account when a 3D tumor model is developed. The silk fibroin/chitosan scaffold is treated with n-butanol that promotes a symmetrical pore network, with a porous size of about 35 μm, suitable for nutrients diffusion and cells growth [31]. Due to its high versatility, silk fibroin is blended along with gellan gum in four ratio. Adipose stem cells and osteosarcoma cells (Saos-2) are seeded on the scaffolds and are cultured in static condition for 14 days. In the case of co-culture and only in one of the blend silk fibroin/gellan gum ratio, the cells organize themselves into spheroids. Interestingly, this work sheds light on the contribution of different mechanical cues in promoting organization of cells [32].

## 10.3 Recapitulating the tumor–stroma crosstalk in spheroid and microtissue models

### 10.3.1 The role of cancer-associated fibroblasts in promoting cancer progression

In physiological conditions, the fibroblasts are deputed for the synthesis and secretion of the ECM components. During inflammation stages, as wound healing and fibrosis, fibroblasts acquire a contractile phenotype with the expression of α-smooth muscle actin (α-SMA) and start the synthesis of ECM components [33]. When the inflammation is depleted, the activated fibroblasts or myofibroblasts undergo apoptosis. In tumor, fibroblasts, or namely cancer-associated fibroblasts (CAF), escape apoptosis, over-proliferate, and keep on altering the ECM [34]. This phenomenon occurs in many solid tumors affecting breast, prostate, colon, lung, and pancreas. Some proteins as α-SMA, platelet-derived growth factor receptors-β (PDGFr-β), and fibroblast activation protein (FAP) are identified as CAF markers because they are upregulated in tumors. A dynamic crosstalk is orchestrated between cancer cells and CAF, where cancer cells activate fibroblasts by mean of tumor growth factor beta (TGF-β) signaling. In turn, CAF seems to send oncogenic signal to the cellular environment. This initiates the tumor transformation [35].

### 10.3.2 Co-cultured spheroid models

Besides the cell types included in the 3D in vitro tumor models, it is also relevant to consider the type of biomaterials to embed the cells. Collagen is one of the main proteins in the TME and collagen hydrogels are used to promote 3D cells architecture. To increase the mechanical property of collagen, alginate is used to obtain blend hydrogel. Human mammary fibroblasts (HMF) and MDA-MB-231 cells, seeded on to the collagen/alginate hydrogels, are able to migrate inside the scaffolds. The model is also able to mimic the cytokines crosstalk that underpins the tumor signaling. Due to the tunable mechanical properties of the hydrogels it is possible to study how the environmental condition can affect cancer and stroma cells [36]. Besides spheroids, different approaches have been adopted to copy the interplay among the cells and ECM in vitro. The cell–ECM interaction is achieved using poly(ε-caprolactone) microparticles coated with a HA multilayer produced via the layer-by-layer deposition technique. A triculture model is established allowing the seeding of human dermal fibroblasts, bone marrow derived mesenchymal stem cells, and nonsmall cell lung cancer (NSCLC) cells (A549) on layered microparticles, acting as anchoring substrates. The triculture can be useful to shed light on the mechanism of drug-resistance occurring in vivo. The authors prove that the presence of mesenchymal stem cells impairs the doxorubicin-induced cytotoxicity. Moreover, the model triggers endogenous ECM deposition and high dense cellular mass, which recapitulate the main features of the cancer in vivo [37].

Recently, a novel 3D co-cultured tumor model, namely microtissues, is developed by means of a dynamic system. Here, cells are seeded on porous microbeads and are grown in spinner flasks. This demonstrates that the spheroid and microtissue display different tissue mimicking capabilities, due to the different 3D culture systems promoting differently ECM assembly [38]. Authors also propose a co-cultured model for pancreatic cancer. The heterotypic cancer model replicates normal fibroblasts activation into CAF, in response to the crosstalk with pancreatic and breast cancer cells and may be used as drug screening platform [39,40]. This 3D culture system is also used as screening platform for MMP-2 sensitive nanoparticles against breast cancer [41]. Due to the size of microtissue at a millimetric scale, it is possible to fabricate a tumor on chip, able to replicate the tumor–stroma crosstalk. This allows on-line monitoring of ECM remodeling by means of confocal and multiphoton laser scanning microscopies [42].

## 10.4 Vascularized microtumor models

### 10.4.1 Endothelial cells promote invasion and migration of cancer cells

Endothelial cells are important players in the tumor progression because they are involved in the angiogenic switching. In this case, they start the synthesis of new blood vessels. In normal condition, the endothelial network is dynamic and hierarchical. It controls the nutrients and oxygen distributions to the tissues and organs as well as it helps in the trafficking of leukocytes. In TME, the blood vessels are less robust, somehow leaky and chaotic. Tumors cannot grow more than 1–2 mm in volume if they lack of vascularization. It is clear how the new blood vessels support the tumor tissues growth [43]. Endothelial cells respond to the signals sent by tumor cells and CAF in terms of growth factors like vascular endothelial growth factor family (VEGF).

### 10.4.2 Multicellular spheroids to recapitulate the tumor angiogenesis

Spheroids are considered as a golden standard to investigate the mechanisms underpinning intravasation and extravasation in metastatic process. Many strategies are adopted to recapitulate cancer angiogenesis. The 3D architecture of spheroids mimics the different gradient of the nutrient diffusion inside the tumor. Recently, Klimkiewicz et al. demonstrated that a model of 3D melanoma spheroids that incorporate early progenitors of the endothelial cells, copy the modulation in chemokines and miRNA in normal and hypoxic condition. This work is extremely relevant because they shed light on the role of the miRNA as powerful regulators of the cellular activities in tumor angiogenesis and can be considered as drug targets [44]. New tools and more complex 3D tumor models can help the researchers to investigate the transport mechanisms and drug action in order to get more information on chemotherapeutic efficacy. These efforts may lead to the development of more efficient and precise drug-delivery systems. An interesting 3D model mimicking the colon adenocarcinoma is made by using a double layer of collagen. To resemble the tumor core, HT29 cancer cells are embedded in the inner and denser part. Fibroblasts and endothelial cells populate a less dense collagen matrix that surrounds the inner core.

The model allows studying the cancer cells migration in the outer matrix driven by the interaction with fibroblasts and endothelial cells. Moreover, HT29 cancer cells are able to secrete VEGF that is considered as a marker for the angiogenesis regulation [45]. A recent work demonstrates the inhibitory role of endothelial cells when co-cultured with colon cancer cells in a multicellular spheroid model. The model is based on manual seeding of cancer and endothelial cells in Matrigel. The 3D co-culture is challenged with Dasatinib (a BCR/ABL and Src family tyrosine kinase inhibitor) or Erdafitinib [a fibroblast growth factor receptor (FGFR) kinase inhibitor] and the cancer cells invasion capability is fully invalidated. A similar behavior is observed when endothelial cells were co-cultured with cancer cells. These results are confirmed in vivo. This shows the relevant role that spheroids have as preclinical relevant model to test chemotherapeutic agents [46]. Scaffold-free multicellular spheroids can recapitulate the microenvironment of tumors that are very difficult to treat. Among these, pancreatic adenocarcinoma shows a very dense ECM and a disorganized vasculature that hinders the drug diffusion. At this aim, multicellular spheroids are obtained using pancreatic cancer cells (Panc-1), lung fibroblast, and endothelial cells and are deeply characterized in terms of morphology, cell viability, and for the expression of genes and proteins related to the ECM. A layer-by-layer nanofilm of fibronectin and gelatin is deposited at the surface of the fibroblasts to boost the adhesion of PANC-1 and HUVEC cells in the spheroids. The spheroid co-cultures and tricultures are not affected by the treatment with doxorubicin and gemcitabine, the two standard-of-care in the fight of pancreatic cancer [47]. A triple culture made up of NSCLC cells, normal fibroblasts, and endothelial cells is developed following the hanging drop technology. The 3D triculture, developed in 5 days, shows a precise cellular organization where the endothelial cells are located nearby the fibroblasts. This organization mainly occurs when endothelial cells are added after 5 days of co-culture of the fibroblasts and cancer cells. These cells during the 5 days secretion of proangiogenic factors (VEGF and FGF), which may trigger the migration of the endothelial cells toward the fibroblasts compartment. The spheroid triculture is challenged with two antiangiogenic drugs (bevacizumab and nindetanib). They exert different effects on the endothelial cells, due to the different targets of the drugs. This model is relevant since the authors are able to recapitulate the interaction between cancer cells and stroma without adding exogenous cytokines or ECM component [48].

### 10.4.3 Tumor microtissues as 3D bioengineered architecture to study cancer vascularization

Microcarriers are used to amplify the cells for tissue engineering purpose in order to preserve morphological features of the cells. The convenience in using microcarriers is the possibility to co-culture various cell types in a closed spatial organization and to grow the cells that do not spontaneously aggregate as 3D structures [49]. The 2D cell cultures are not suitable for studies that are longer that 72 h, since it is necessary to subculture the cells and expose them again to the tested compounds. Bioreactors sustain the cell growth in dynamic condition allowing the expansion of anchorage-dependent cells by mean of microcarriers. Endothelial cell line such as EAhy 926 is used to investigate the effect of nanoparticles toxicity in long-term dynamic culture. Endothelial cells are seeded on Global

Eucaryotic Microcarrier GEM in a stirrer tank bioreactor and cytotoxicity tests were performed in 96-well plate. The GEM are coated with different ECM protein (gelatin, laminin, fibronectin, collagen type I and IV, basal membrane) to mimic the in vivo condition and promote the cell adhesion and growth. This study reveals that the nanoparticles are accumulated in the lysosome and this finding could be assessed only by means of long-term culture. This is not the proper 3D tumor model but it paves way for dynamic culture [50]. An interesting approach is the use of gelatin porous microcarriers to build triculture of breast cancer cells, fibroblasts, and endothelial cells by using a dynamic system such as spinner flasks. The strength of the model is the capability of synthesize–assemble–disassemble endogenous ECM, which provides the mechanical and biochemical cues for the study of the tumor angiogenesis in vitro. The authors carry out the dynamic triculture up to 28 days, with no signal of loss of fibroblasts or endothelial cells. This is a great advantage since the model can be exploited to study long-term drug effects in a more realistic model [51]. Currently, the study of the interaction between cancer- and endothelial cells is deeply investigated by means of tumor-on-chip approach. Tumor-on-chip allows online monitoring by means of advanced imaging techniques, which provides new insight to phenomena's such as tumor cells invasion and migration. This is not possible to study in 2D cell culture [52].

## 10.5 The contribution of immune system cells in microtumors

### 10.5.1 Macrophage: the double side of the same player

Immune system interaction with TME garners attention in the recent years. Immune system cells are now considered as a potential target for cancer therapy [53]. Large amounts of the so-called tumor associated macrophages (TAMs) are detected in cancer tissues. It is observed that macrophages play a double-edged role in cancer progression. They are classified into two categories, according to their behaviors, namely, as supporter or inhibitor of cancer [54]. Macrophages that inhibit the growth or kill cancer cells are called M1-type. They exert their function through the synthesis of nitric oxide. Macrophages, namely M2-type, promote the cancer growth or repair by means of secreting ornithine and other tumor-inducing factors such as VEGF, EGF, and TGF-β in the microenvironment. It is hypothesized that the lack of immunogenic stimuli in the TME seems to negatively influence the innate responses. In this case M1-type macrophages are involved [55]. Notwithstanding the progress in unraveling the crosstalk between immune system and cancer, the use of a chemotherapeutic approach involving macrophages is still far away. Acute side effects of some immunotherapeutic approaches slow down the translation of immunotherapy from the bench to the bedside.

### 10.5.2 Spheroids incorporating the immune system cells

Cytokine-induced killer (CIK) cells are a promising tool to affect tumor growth and progression. However, conventional 2D cell cultures could not mimic the mechanisms of CIK cells infiltration in the ECM and the consequential perturbation of the solid tumor. At this aim, a collagen-based scaffold offers gastric cancer cells the 3D architecture and promotes the spheroids growth. The 3D gastric spheroids are challenged with CIK cells and their cytotoxic effects against the solid tumor as well as the invasive capability of the CIK cells are evaluated. Their results show that CIK cells affect the 3D gastric spheroids slower than the 2D cell culture. This finding supports the hypothesis that 3D cancer cell cultures can mimic the better in in vivo condition [56]. The role of macrophage in the TME can change according to the different macrophages population that may infiltrate the hypoxic core of the tumor. This modulates the tumor ECM [57]. It becomes necessary to recapitulate the different conditions by means of 3D tumor models. In another work, a heterotypic spheroid model is made up of triple negative breast cancer cells and macrophage. This is compared with other 3D models where spheroids interact with macrophages embedded in a collagen ECM-like hydrogel. The authors investigate the paracrine interaction mediated by cytokines in the two different models and how chemotherapeutic agents can reveal the protective role of macrophages in preserving the growth capabilities of the cancer cells [58]. Patient-derived cells present some limitations since it is difficult to establish continuous tumor cell line from explanted tumor. These limitations hamper the possibility to exploit the patient-derived cells potential to study the drug response profiles of tumor specific patients. Resected tumor tissues can be maintained in culture up to 44 days together with the stromal compartment. This system recapitulates the molecular features of the primary tumor of the patients. They can be exploited either to study the drug response in vitro or to detect new biomarkers that can enhance the cancer prediction [59]. The 3D cancer models can be more complex with the addition of different cell types in the culture. This is the case of a 3D colorectal cancer spheroid model obtained by seeding colon cancer cells and fibroblasts in a hanging drop plate from InSphero. After 3 days of culture, heterotypic spheroids are formed and peripheral blood mononuclear cell (PMBC) is added to each well to induce the infiltration of monocytes and leukocytes. The 3D model is then challenged with different candidates for immunotherapy treatments such as tumor- and fibroblast-targeted IgG-IL2v (CEA-IL2v and FAP-IL2v, respectively). The results show an enhanced leukocyte infiltration and a cytotoxic effect against tumor cells. This model may closely mimic the in vivo condition and can be further improved by adding patient-derived tumor containing tumor-infiltrating lymphocytes (TIL) instead of PMBC derived from a healthy donor [60]. A multicellular spheroid model provides interesting information on the crosstalk among cancer cells, fibroblasts, and macrophage in terms of migration and growth. Algimatrix 3D culture system is used for growing the breast cancer cells, fibroblasts, and macrophages for promoting the spheroid formation. The study reveals that the presence of stroma cells as fibroblasts and macrophages affected the invasive capability of the cancer cells and this feature is strongly associated with the tumor malignant progression [61]. To demonstrate further the close link between tumor progression and TAMs, a triculture spheroids model is obtained and the TAMs infiltrating capability is investigated. The researchers demonstrate that the chemoattractants and their receptors mediate the infiltration of macrophages in the tumor by means of fibroblasts associated with the cancer cells. This model demonstrates the crucial role of the stroma in the progression of the tumor via paracrine signal pathway [62]. Another attempt of recapitulating the interaction between tumor cells and macrophages demonstrates the presence of M2-polarized macrophages when they are co-cultured with tumor cells and fibroblasts. This mimics the same crosstalk that can be achieved in vivo by means of IL-4 stimulation [63].

### 10.5.3 3D complex architecture to copycat the immune-competence in tumors

It is demonstrated that the solid tumors can recruit circulating monocytes and trigger their polarizations from an antiinflammatory to a more invasive phenotype (better known as transition from M1 to M2 phenotype). So far, the role of infiltrating macrophages is not yet understood. The main knowledge in this field derives from immunohistochemical analysis. Hence, it is necessary to boost the skill of tissue engineering to develop reliable models to recapitulate the interactions between immune system cells and tumor tissues. Cancer spheroids are not the only strategies to study the interactions between tumor and immune system cells. Spinner flasks are an interesting tool to develop cancer microtissues based on microcarriers. The cells are encapsulated in alginate microspheres together with other cell types. An appropriate example comes from a recently published research that aims to elucidate the role of tumor-associated macrophages in a subtype of lung cancer (NSCLC). They encapsulate tumor spheroids, CAFs, and monocytes in alginate microspheres, according to a well-established protocol of dynamic cell culture. They demonstrate that the monocytes recruited by the tumor tissues, are induced to polarize into a M2-like phenotype. They conclude that the polarization of monocytes toward a M2 phenotype is achieved only when both cancer cells and fibroblasts are co-cultured in the alginate beads [64]. Online monitoring of cancer cells migration is achieved by means of microfluidic chips that are compatible with advanced fluorescence microscopy and time-lapse recording. To monitor the interaction between melanoma cells and spleen cells, a microfluidic chip is developed where cell signals can diffuse through the microchannels. It mimics the body small capillary that allows the cell migration. The level of invasiveness of B12 melanoma cells stimulated by spleen cells is increased when IRF-8 KO cells were used. This highlights the role of immunocompetent cells in promoting cancer cells invasiveness [65]. The translation of biomedical devices from the bench to the market is still a challenge. Operating with patient-derived biopsy to investigate deeper how TILs can affect the tumor growth. A microfluidic model, named ex vivo immuno-oncology dynamic environment for tumor biopsies (EVIDENT), aims to recapitulate the interaction between tumor biopsy and TILs. The chip is connected with a pump that drives the flow of TILs toward the tumor biopsy. The dynamic system allows the maintaining of the biopsy for several days and the monitoring in real-time, high resolution of the interaction among the cells. Custom image analytic algorithms based on machine learning support the experimental results. This approach allows an automation in the quantification and analysis of the acquired images, pushing the proposed model closer to be used in the clinical trials [66].

## 10.6 Spheroids as screening platform for drug testing

### 10.6.1 The importance of moving 3D culture to high-throughput screening approaches

Metastasis is the main cause of death in case of cancer relapse. Invasive cells activate specific signaling pathway to remodel the TME. Tumor spheroids are an interesting tool to dissect the step of the invasion and migration in cancer progression. It is necessary to combine the study of spheroids growth and invasion capability with automated analysis. This approach could predict the therapy response and investigate the mechanisms of single-cell drug resistance. The group led by Friedl developed a 3D melanoma spheroid culture in a high-density collagen matrix. Melanoma spheroid undergoes a radiation-based treatment and the invasion capability is monitored by confocal microscopy. The images were analyzed by means of an automated workflow that was able to distinguish an intact and apoptotic nuclei. The workflow can be applied to a high-throughput analysis of the mechanisms of cell invasion and metastasis [67]. The 3D cancer culture shows a different expression of genes and proteins in comparison to 2D culture. A 3D spheroid co-culture is made by using colon cancer cell lines (GFP positive) and mouse NIH3T3 fibroblasts, which are used to identify the targets of the RNAi treatment. This interesting model permits the monitoring of the impact of the depletion of determined genes (e.g., Kif11/Eg5) in 3D and 2D by mean of fluorescence intensity measurement and phenotypic changes over time. This automated system allows a functional gene analysis that enhances the discovery of target genes in the drug development [68]. We already have discussed about the alginate-coated core magnetic GEM that are compatible with the high-throughput approach. These nanocarriers offer a high-curved surface to enhance cells growth [50]. A microwell chip is fabricated to grow human hepatoma cells in alginate. The 3D culture is used to test the toxicity of Bisphenol A, assessing the function of different metabolic enzymes by means of high-throughput assay. The system allows testing of 28 enzymes for each chip with a sample volume of 2000 times less than the one that requires for a 96-well plate. The work is a step further in finding methods for toxicity screening alternative to the animal models [69]. Kuraray, a Japanese company, commercializes micro-Space cell culture plates. The system is suitable to grow a high number of 3D organoids in 5 days recapitulating cancer tissue. In particular, hepatocytes (HepG2 cells) are cultured on the Micro-Space plates to carry out the drug testing. The plate is an example of high-throughput exploitation of the cell culture space. This is a 24-well format where each well is divided into 200 μm long × 200 μm wide × 100 μm deep square microsized compartments [70]. One of the limitations of the use of high-throughput approach is the uncontrolled size of the spheroids that do not allow the replicabilities of the results of the drug testings. The variabilities of cell density and spheroids size directly affect the pharmacokinetic and pharmacodynamic assessments. These valuable strategies need improvement to control and standardize the size of the spheroids in order to make these techniques more appealing for the pharma companies. This approach may enhance their uses as drug screening or toxicity assessment platforms.

### 10.6.2 The development of novel methodology for solving high-content imaging problem in preclinical study models

The analysis of 3D in vitro tumor models by means of high-content screening (HCS) techniques cannot fully be accomplished. Heterogeneity in spheroids size and shape, lack of tools to analyze thick structures after staining is one of the problems. The difficulty to automatize the spheroids culture according to an industrial setting is still disadvantageous that prevents the analysis of 3D spheroids with HCS approach. New microscopic techniques and the increase of tissue transparency are the powerful tools to improve the resolution and the speed of the imaging of 3D structures. Micropatterned wells are the substrate to grow spheroids without the use of agarose or matrigel. They are not suitable for automatized procedures due to the need of heating and cooling block, respectively, for their preparations. In the system, the spheroids adhere to the adhesive micropatterned wells, so changes of medium are easy to carry out. Moreover, the possibility to perform long-term culture makes the spheroids a platform to screen the effects of drugs. Finally, the adhesion of spheroids on the micropatterned well allows the monitoring of each spheroids along the time without the risk of losing the track of each single spheroid. The system is fully automatized by means of a Freedom EVO 150 workstation (Tecan, Männedorf, Switzerland) and the flat micropatterned bottom of the well permits the high-content analysis of cell viability in the spheroids. Micropatterned spheroids are the helpful platforms to perform HCS assay exploiting the 3D microenvironment [71]. A 384-array of T47D spheroids is used to study the efficacy of compounds on the dormant nonproliferating cells inside the spheroids. The screening of 1120 compounds is carried out and the cell viability after the treatment is analyzed by means of a Hoechst 33342 (for total nuclei) and SytoxGreen (DNA stain of dead cells) staining. The 3D images are acquired by means of a custom-built monolithic digital scanned laser light-sheet fluorescence microscope and are analyzed with the Imaris software. The 3D model shows higher chemoresistance to the compounds [72]. In another work, high-content imaging analysis is carried out by means of Operetta high-content imaging system (PerkinElmer) equipped with Harmony software. The authors track the expansion of T47D subpopulation defined CK5 + that indicates the stemness and mesenchymal phenotype resulting in an increased chemoresistance. The T47D spheroids are stably transfected with a CK5 promoter green fluorescent protein reporter in order to track the expansion of cancer stem cell phenotype. Among 300 compounds tested, 4 hits affect the proliferation of CK5 + cells [73]. Recently, the investigation of cell spreading pattern in multicellular-spheroids is performed by phase-contrast microscopy for 36 h, collecting the images every 5 min. The advantage provided by this approach is the definition of an automated algorithm able to identify the spheroids outlines and the spreading. The spheroids are then challenged with platin-based drug that inhibits cell migration [74]. The 3D organotypic cancer models, in particular spheroids, represent a breakthrough in the drug testing pipeline. However, high-throughput imaging systems (hardware and software) together with the development of biomarkers and gene reporter readouts require be improving and implementing at industrial scale.

## 10.7 Conclusion and future trends

The 3D cancer spheroids improve the understanding of the mechanisms underpinning tumor progression and metastasis. They can predict the efficacy and toxicity of the drugs in a more realistic way, since the third dimension takes into account of nutrient and drug gradient diffusions, cell–cell, and cell–ECM interaction. The spheroids- and microcarriers-based cancer models are not fabricated following a standardized protocol. This is a disadvantageous situation when we need to compare the results of the drug-screening assays. For this reason, most of the pharmaceutical companies still rely on drug testing performed on 2D cell culture or animal models. In this chapter, we discuss about 3D tumor models that recapitulate the interactions among the different cells to create the TME, which focuses on the spheroids- and microcarriers-based models. In the future, the huge progress in game-changing and disruptive methodologies will improve the fabrication of 3D tumor models promoting their integrations with the organ-on-chip technology. A great impact is also expected from -omics analysis (genomics, proteomics, and metabolomics) as well as from advanced microscopic techniques. This will boost the introduction of 3D in vitro tumor models in the preclinical trials as the standard of investigations. This approach can pave the way to shorter and more reliable assays to validate anticancer drugs and may guarantee a more patient oriented cancer treatment.

## Acknowledgments

The work is also financially supported by the European Union Framework Programme for Research and Innovation Horizon 2020 under grant agreement no 668983—FoReCaST (Forefront Research in 3D Disease Cancer Models as in vitro Screening Technologies) and also by FCT (Portugal) project number PTDC/BTM-ORG/28168/2017 to VB and SCK.

## References

[1] Lee JY, Chaudhuri O. Regulation of breast cancer progression by extracellular matrix mechanics: insights from 3D culture models. ACS Biomater Sci Eng 2017;4:302–13.

[2] Hanahan D, Weinberg RA. The hallmarks of cancer. Cell 2000;100(1):57–70. Available from: https://doi.org/10.1007/s00262-010-0968-0.

[3] Hanahan D, Weinberg RA. Hallmarks of cancer: the next generation. Cell 2011;144(5):646–74. Available from: https://doi.org/10.1016/j.cell.2011.02.013 Inc.

[4] Mehlen P, Puisieux A. Metastasis: a question of life or death. Nat Rev Cancer 2006;6(6):449–58. Available from: https://doi.org/10.1038/nrc1886.

[5] Klein-Goldberg A, Maman S, Witz IP. The role played by the microenvironment in site-specific metastasis. Cancer Lett 2014;352(1):54–8. Available from: https://doi.org/10.1016/j.canlet.2013.08.029.

[6] Quail DF, Joyce JA. Microenvironmental regulation of tumor progression and metastasis. Nat Med 2013;19(11):1423–37. Available from: https://doi.org/10.1038/nm.3394.

[7] Lam ATL, et al. Biodegradable poly-ε-caprolactone microcarriers for efficient production of human mesenchymal stromal cells and secreted cytokines in batch and fed-batch bioreactors. Cytotherapy 2017;19(3):419–32. Available from: https://doi.org/10.1016/j.jcyt.2016.11.009.

[8] van Wezel A. Growth of cell-strains and primary cells on micro-carriers in homogeneous culture. Nature 1967;216(5110):64–5.

[9] Sun LY, et al. Functional cells cultured on microcarriers for use in regenerative medicine research. Cell Transplant 2011;20(1):49–62. Available from: https://doi.org/10.3727/096368910X532792.

[10] Li B, et al. Past, present, and future of microcarrier-based tissue engineering. J Orthop Transl 2015;3(2):51–7. Available from: https://doi.org/10.1016/j.jot.2015.02.003.

[11] Maurer BJ, et al. Growth of human tumor cells in macroporous microcarriers results in p53-independent, decreased cisplatin sensitivity relative to monolayers. Mol Pharmacol 1999;55(5):938–47. Available from: http://www.ncbi.nlm.nih.gov/pubmed/10220573.

[12] Becker JL, Blanchard DK. Characterization of primary breast carcinomas grown in three-dimensional cultures. J Surg Res 2007;142(2):256–62. Available from: https://doi.org/10.1016/j.jss.2007.03.016.

[13] Li C, et al. Chemically crosslinked alginate porous microcarriers modified with bioactive molecule for expansion of human hepatocellular carcinoma cells. J Biomed Mater Res - Part B Appl Biomater 2014;102(8):1648–58. Available from: https://doi.org/10.1002/jbm.b.33150.

[14] Charoen KM, et al. Embedded multicellular spheroids as a biomimetic 3D cancer model for evaluating drug and drug-device combinations. Biomaterials 2014;35(7):2264–71. Available from: https://doi.org/10.1016/j.biomaterials.2013.11.038.

[15] Rodrigues T, et al. Emerging tumor spheroids technologies for 3D in vitro cancer modeling. Pharmacol Therap 2018;184:201–11. Available from: https://doi.org/10.1016/j.pharmthera.2017.10.018.

[16] Liaw C-Y, Ji S, Guvendiren M. Engineering 3D hydrogels for personalized in vitro human tissue models. Adv Healthc Mater 2018;7:1701165. Available from: https://doi.org/10.1002/adhm.201701165.

[17] Ruedinger F, et al. Hydrogels for 3D mammalian cell culture: a starting guide for laboratory practice. Appl Microbiol Biotechnol 2014;99(2):623–36. Available from: https://doi.org/10.1007/s00253-014-6253-y.

[18] Fong ELS, et al. Hydrogel-based 3D model of patient-derived prostate xenograft tumors suitable for drug screening. Mol Pharm 2014;11(7):2040–50. Available from: https://doi.org/10.1021/mp500085p.

[19] Guzman A, Ziperstein MJ, Kaufman LJ. The effect of fibrillar matrix architecture on tumor cell invasion of physically challenging environments. Biomaterials 2014;35(25):6954–63. Available from: https://doi.org/10.1016/j.biomaterials.2014.04.086.

[20] Chen L, et al. The enhancement of cancer stem cell properties of MCF-7 cells in 3D collagen scaffolds for modeling of cancer and anti-cancer drugs. Biomaterials 2012;33(5):1437–44. Available from: https://doi.org/10.1016/j.biomaterials.2011.10.056.

[21] Sachs N, et al. A living biobank of breast cancer organoids captures disease heterogeneity. Cell 2017;172:373–86. Available from: https://doi.org/10.1016/j.cell.2017.11.010.

[22] Xu F, Burg KJL. Three-dimensional polymeric systems for cancer cell studies. Cytotechnology 2007;54(3):135–43. Available from: https://doi.org/10.1007/s10616-007-9065-4.

[23] Yu L, et al. Alginate core-shell beads for simplified three-dimensional tumor spheroid culture and drug screening. Biomed Microdevices 2015;17(2):33. Available from: https://doi.org/10.1007/s10544-014-9918-5.

[24] Chanmee T, Ontong P, Itano N. Hyaluronan: a modulator of the tumor microenvironment. Cancer Lett 2016;375(1):20–30. Available from: https://doi.org/10.1016/j.canlet.2016.02.031.

[25] Amorim S, et al. Molecular weight of surface immobilized hyaluronic acid influences CD44-mediated binding of gastric cancer cells. Sci Rep 2018;8(1):1–11. Available from: https://doi.org/10.1038/s41598-018-34445-0.

[26] David L, et al. Hyaluronan hydrogel: an appropriate three-dimensional model for evaluation of anticancer drug sensitivity. Acta Biomater 2008;4(2):256–63. Available from: https://doi.org/10.1016/j.actbio.2007.08.012.

[27] Gurski LA, et al. Hyaluronic acid-based hydrogels as 3D matrices for in vitro evaluation of chemotherapeutic drugs using poorly adherent prostate cancer cells. Biomaterials 2009;30(30):6076–85. Available from: https://doi.org/10.1016/j.biomaterials.2009.07.054.

[28] Kundu B, et al. Silk fibroin biomaterials for tissue regenerations. Adv Drug Deliv Rev 2013;65(4):457–70. Available from: https://doi.org/10.1016/J.ADDR.2012.09.043.

[29] Rockwood DN, et al. Materials fabrication from Bombyx mori silk fibroin. Nat Protoc 2011;6(10):1612–31. Available from: https://doi.org/10.1038/nprot.2011.379.

[30] She Z, et al. Silk fibroin/chitosan scaffold: preparation, characterization, and culture with HepG2 cell. J Mater Sci: Mater Med 2008;19(12):3545–53. Available from: https://doi.org/10.1007/s10856-008-3526-y.

[31] Li J, et al. A novel 3D in vitro tumor model based on silk fibroin/chitosan scaffolds to mimic the tumor microenvironment. ACS Appl Mater Interfaces 2018;10(43):36641–51. Available from: https://doi.org/10.1021/acsami.8b10679.

[32] Kundu B, et al. Mechanical property of hydrogels and the presence of adipose stem cells in tumor stroma affect spheroid formation in the 3D osteosarcoma model. ACS Appl Mater Interfaces 2019;11(16):14548–59. Available from: https://doi.org/10.1021/acsami.8b22724.

[33] Cirri P, Chiarugi P. Cancer-associated-fibroblasts and tumour cells: a diabolic liaison driving cancer progression. Cancer Metastasis Rev 2012;31(1–2):195–208. Available from: https://doi.org/10.1007/s10555-011-9340-x.

[34] Beacham DA, Cukierman E. Stromagenesis: the changing face of fibroblastic microenvironments during tumor progression. Semin Cancer Biol 2005;15(5 SPEC. ISS.):329–41. Available from: https://doi.org/10.1016/j.semcancer.2005.05.003.

[35] Zhang J, Liu J. Tumor stroma as targets for cancer therapy. Pharmacol Therap 2013;137(2):200–15. Available from: https://doi.org/10.1016/j.pharmthera.2012.10.003.

[36] Liu C, et al. Hybrid collagen alginate hydrogel as a platform for 3D tumor spheroid invasion. Acta Biomater 2018;75:213–25. Available from: https://doi.org/10.1016/j.actbio.2018.06.003.

[37] Ferreira LP, Gaspar VM, Mano JF. Bioinstructive microparticles for self-assembly of mesenchymal stem Cell-3D tumor spheroids. Biomaterials 2018;185:155–73. Available from: https://doi.org/10.1016/j.biomaterials.2018.09.007.

[38] Brancato V, Garziano A, et al. 3D is not enough: building up a cell instructive microenvironment for tumoral stroma microtissues. Acta Biomater 2017;47:1–13. Available from: https://doi.org/10.1016/j.actbio.2016.10.007.

[39] Brancato V, et al. 3D breast cancer microtissue reveals the role of tumor microenvironment on the transport and efficacy of free-doxorubicin in vitro. Acta Biomater 2018;75:200–12. Available from: https://doi.org/10.1016/j.actbio.2018.05.055.

[40] Brancato V, Comunanza V, et al. Bioengineered tumoral microtissues recapitulate desmoplastic reaction of pancreatic cancer. Acta Biomater 2017;49:152–66. Available from: https://doi.org/10.1016/j.actbio.2016.11.072.

[41] Brancato V, Gioiella F, et al. 3D tumor microtissues as an in vitro testing platform for microenvironmentally-triggered drug delivery systems. Acta Biomater 2017;57:47–58. Available from: https://doi.org/10.1016/j.actbio.2017.05.004.

[42] Gioiella F, et al. An engineered breast cancer model on a chip to replicate ecm-activation in vitro during tumor progression. Adv Healthc Mater 2016;5(23):3074–84. Available from: https://doi.org/10.1002/adhm.201600772.

[43] Dudley AC. Tumor endothelial cells. Cold Spring Harb Perspect Med 2012;2(3):1–18. Available from: https://doi.org/10.1101/cshperspect.a006536.

[44] Klimkiewicz K, et al. A 3D model of tumour angiogenic microenvironment to monitor hypoxia effects on cell interactions and cancer stem cell selection. Cancer Lett 2017;396:10–20. Available from: https://doi.org/10.1016/j.canlet.2017.03.006.

[45] Nyga A, et al. A novel tissue engineered three-dimensional in vitro colorectal cancer model. Acta Biomater 2013;9(8):7917–26. Available from: https://doi.org/10.1016/j.actbio.2013.04.028.

[46] Cattin S, Ramont L, Rüegg C. Characterization and in vivo validation of a three-dimensional multi-cellular culture model to study heterotypic interactions in colorectal cancer cell growth, invasion and metastasis. Front Bioeng Biotechnol 2018;6:1–14. Available from: https://doi.org/10.3389/fbioe.2018.00097.

[47] Lazzari G, et al. Multicellular spheroid based on a triple co-culture: a novel 3D model to mimic pancreatic tumor complexity. Acta Biomater 2018;78:296–307. Available from: https://doi.org/10.1016/j.actbio.2018.08.008.

[48] Amann A, et al. Development of a 3D angiogenesis model to study tumour - endothelial cell interactions and the effects of anti-angiogenic drugs. Sci Rep 2017;7(1):1–13. Available from: https://doi.org/10.1038/s41598-017-03010-6.

[49] Benien P, Swami A. 3D tumor models: history, advances and future perspectives. Future Oncol 2014;10(7):1311–27. Available from: https://doi.org/10.2217/fon.13.274.

[50] Mrakovcic M, et al. Assessment of long-term effects of nanoparticles in a microcarrier cell culture system. PLoS One 2013;8(2):e56791. Available from: https://doi.org/10.1371/journal.pone.0056791.

[51] Mazio C, et al. Recapitulating spatiotemporal tumor heterogeneity in vitro through engineered breast cancer microtissues. Acta Biomater 2018;73:236–49. Available from: https://doi.org/10.1016/j.actbio.2018.04.028.

[52] Caballero D, et al. Organ-on-chip models of cancer metastasis for future personalized medicine: from chip to the patient. Biomaterials 2017;149:98–115. Available from: https://doi.org/10.1016/j.biomaterials.2017.10.005.

[53] Ott PA, et al. Combination immunotherapy: a road map. J Immunother Cancer 2017;5(1):1–15. Available from: https://doi.org/10.1186/s40425-017-0218-5.

[54] Aras S, Raza Zaidi M. TAMeless traitors: macrophages in cancer progression and metastasis. Br J Cancer 2017;117(11):1583–91. Available from: https://doi.org/10.1038/bjc.2017.356.

[55] Liguori M, et al. Tumor-associated macrophages as incessant builders and destroyers of the cancer stroma. Cancers 2011;3(4):3740–61. Available from: https://doi.org/10.3390/cancers3043740.

[56] Sun P, et al. An engineered three-dimensional gastric tumor culture model for evaluating the antitumor activity of immune cells in vitro. Oncol Lett 2013;5(2):489–94. Available from: https://doi.org/10.3892/ol.2012.1021.

[57] Pollard JW. Macrophages define the invasive microenvironment in breast cancer. J Leukoc Biol 2008;84 (3):623–30. Available from: https://doi.org/10.1189/jlb.1107762.
[58] Tevis KM, et al. Mimicking the tumor microenvironment to regulate macrophage phenotype and assessing chemotherapeutic efficacy in embedded cancer cell/macrophage spheroid models. Acta Biomater 2017;50:271–9. Available from: https://doi.org/10.1016/j.actbio.2016.12.037.
[59] Finnberg NK, et al. Application of 3D tumoroid systems to define immune and cytotoxic therapeutic responses based on tumoroid and tissue slice culture molecular signatures. Oncotarget 2017;8(40):66747–57. Available from: https://doi.org/10.18632/oncotarget.19965.
[60] Herter S, et al. A novel three-dimensional heterotypic spheroid model for the assessment of the activity of cancer immunotherapy agents. Cancer Immunol Immunother 2017;66(1):129–40. Available from: https://doi.org/10.1007/s00262-016-1927-1.
[61] Rama-Esendagli D, et al. Spheroid formation and invasion capacity are differentially influenced by co-cultures of fibroblast and macrophage cells in breast cancer. Mol Biol Rep 2014;41(5):2885–92. Available from: https://doi.org/10.1007/s11033-014-3144-3.
[62] Ksiazkiewicz M, et al. Importance of CCL2-CCR2A/2B signaling for monocyte migration into spheroids of breast cancer-derived fibroblasts. Immunobiology 2010;215(9–10):737–47. Available from: https://doi.org/10.1016/j.imbio.2010.05.019.
[63] Linde N, et al. Integrating macrophages into organotypic co-cultures: a 3D in vitro model to study tumor-associated macrophages. PLoS One 2012;7(7). Available from: https://doi.org/10.1371/journal.pone.0040058.
[64] Rebelo SP, et al. 3D-3-culture: a tool to unveil macrophage plasticity in the tumour microenvironment. Biomaterials 2018;163:185–97. Available from: https://doi.org/10.1016/j.biomaterials.2018.02.030.
[65] Businaro L, et al. Cross talk between cancer and immune cells: exploring complex dynamics in a microfluidic environment. Lab a Chip 2013;13(2):229–39. Available from: https://doi.org/10.1039/c2lc40887b.
[66] Moore N, et al. A multiplexed microfluidic system for evaluation of dynamics of immune-tumor interactions. Lab a Chip 2018;18(13):1844–58. Available from: https://doi.org/10.1039/c8lc00256h.
[67] Veelken C, et al. Single cell-based automated quantification of therapy responses of invasive cancer spheroids in organotypic 3D culture. Methods 2017;128:139–49. Available from: https://doi.org/10.1016/j.ymeth.2017.07.015.
[68] Thoma CR, et al. A high-throughput-compatible 3D microtissue co-culture system for phenotypic RNAi screening applications. J Biomol Screen 2013;18(10):1330–7. Available from: https://doi.org/10.1177/1087057113499071.
[69] Lee DW, et al. Estimation of bisphenol A—human toxicity by 3D cell culture arrays, high throughput alternatives to animal tests. Toxicol Lett 2016;259:87–94. Available from: https://doi.org/10.1016/j.toxlet.2016.07.711.
[70] Nakamura K, et al. Evaluation of drug toxicity with hepatocytes cultured in a micro-space cell culture system. J Biosci Bioeng 2011;111(1):78–84. Available from: https://doi.org/10.1016/j.jbiosc.2010.08.008.
[71] Monjaret F, et al. Fully automated one-step production of functional 3D tumor spheroids for high-content screening. J Lab Autom 2016;21(2):268–80. Available from: https://doi.org/10.1177/2211068215607058.
[72] Wenzel C, et al. 3D high-content screening for the identification of compounds that target cells in dormant tumor spheroid regions. Exp Cell Res 2014;323(1):131–43. Available from: https://doi.org/10.1016/j.yexcr.2014.01.017.
[73] Reid BG, et al. Live multicellular tumor spheroid models for high-content imaging and screening in cancer drug discovery. Curr Chem Genom Transl Med 2014;8:27–35. Available from: https://www-ncbi-nlm-nih-gov.gate2.inist.fr/pmc/articles/PMC3941083/pdf/CCGTM-8-27.pdf.
[74] Chen D, et al. Automated contour analysis of multi-cellular spheroids spreading through high content imaging. Phys Biol 2018;15(2):026006. Available from: https://doi.org/10.1088/1478-3975/aaa27b.

CHAPTER

# 11

# Biomaterial-based in vitro models for pancreatic cancer

*Eirini Velliou[1], Priyanka Gupta[1], Claudio Ricci[2,3] and Serena Danti[3]*

[1]Bioprocess and Biochemical Engineering Group (BioProChem), Department of Chemical and Process Engineering, University of Surrey, Surrey, United Kingdom [2]National Interuniversity Consortium of Materials Science and Technology (INSTM), Florence, Italy [3]Department of Civil and Industrial Engineering, University of Pisa, Pisa, Italy

## 11.1 Introduction

Pancreatic ductal adenocarcinoma (PDAC) is a malignant epithelial neoplasm, recently estimated to be the third leading cause of death in the United States, thus becoming second only to lung cancer for mortality [1]. Indeed, the incidence for PDAC has been increasing for the past two decades at 1% annual rate. Pancreatic cancer develops because of the accumulation of genetic alterations in oncogenes and tumor-suppressing genes; however, its complex genetic landscape has not been fully elucidated yet. Biomarkers are thus still far from, a remarkable clinical relevance and many efforts are in place to fight this tumor [2].

From a histological point of view, the neoplasm is composed of mucus-secreting glandular structures deriving from the main and peripheral ductal structures. For the reason the tumor is also known with the names of pancreatic adenocarcinoma or pancreatic ductal cell carcinoma, carcinoma of the exocrine pancreas or more simply, pancreatic cancer. PDAC is, together with its variants (i.e., noncystic mucinous carcinoma, ring cell carcinoma with bezel, adenosquamous carcinoma, and undifferentiated carcinoma), the most frequent histotype of pancreatic cancer. In fact, it represents 85%–90% of all pancreatic tumors. Because of its minimal symptoms, which lead to a late diagnosis of the tumor, 75% of patients have an inoperable cancer with a poor prognosis, and a 5-year survival rate of only 5% [3]. The remaining 25% of patients are those who can also benefit from surgical treatment. However, their 5-year survival rate also barely reaches 20% [4]. This cancer has been associated with several clinical factors responsible for such high-mortality


DOI: https://doi.org/10.1016/B978-0-12-818128-7.00011-3

rates, including (1) the lack of diagnostic markers, (2) the high degree of infiltration and metastasis, and (3) the strong resistance to traditional chemotherapy.

At the time of diagnosis, most pancreatic neoplasms exceeded the organ boundaries [5]. In fact, the PDAC is capable of attacking the lymphatic system and neighboring organs (liver and lung) very early and very quickly, thus becoming extremely metastatic from the earliest stages. The lymphatic spread of pancreatic head carcinomas affects the pancreatic–duodenal lymph nodes most frequently. Diffusion to the pleura and lung through lymphatic channels is also possible. Blood metastases affect in order of frequency: liver, lung, adrenal, kidney, bone, brain, and skin [6].

PDAC originates from the ductal system and as such, it shows epithelial characteristics, but its genesis is still unknown. However, there are experimental data on genetically engineered mouse models that suggest two possible origins for this tumor. The most accepted hypothesis to date is that the PDAC develops from a metaplastic ductal acinar cell [7]. Progression from normal tissue into infiltrating tumor can start from several precursors, specifically from pancreatic intraepithelial neoplasia (PanIN), and from papillary-mucinous and cystic-mucinous neoplasms [8]. The PDAC that has been best characterized derives from PanIN, passes through various stages, and then gives rise to the development of a tumor on site (Fig. 11.1) [9]. The presence of PanIN-3 type lesions has been shown to be frequently associated with invasive PDAC [10,11].

The tumor microenvironment (TME) of PDAC is particularly rich in stroma and very heterogeneous. It includes both extracellular matrix (ECM) and various cell types, which include fibroblasts, stellate cells of the pancreas, immune system cells, endothelial cells, cancer stem cells (CSCs), and differentiated tumor cells at various stages [12,13]. In the PDAC, the failure of an early diagnosis and the inefficiency of chemotherapy seem somehow related to the complexity of his TME [14]. In fact, also for this tumor, traditional in vitro [two-dimensional (2D)] models have proved to be largely insufficient for drug screening, and animal models have generally shown poor predictive capacity [15]. Animal

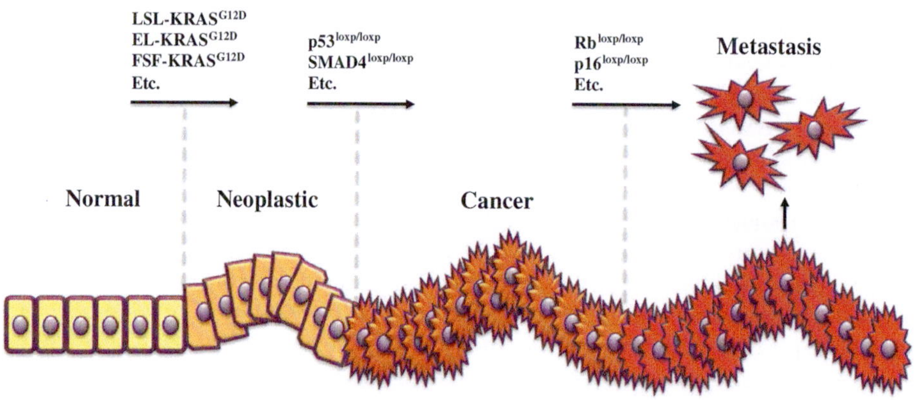


FIGURE 11.1 The diagram shows the evolution of the PDAC through the morphological changes of the epithelium in a pancreatic duct, divided by type of PanIN lesion. Some markers associated with the various stages of the tumor are also shown. Creative Common Attribution license; DeCant BT, et al. Utilizing past and present mouse systems to engineer more relevant pancreatic cancer models. Front Physiol 2014;5:464 [9].

models, even the most advanced ones, still suffer from limitations for the study of cancer. Indeed, endogenous cancers in animals can be different from their human counterparts according to biology and treatment. For example, relapse times in animals and humans occur on incomparable timescales. In xenograft models, human cancerous cells are inoculated in the animal, thus, immunodeficient hosts are needed. Consequently, the contribution of the immune system to tumor evolution and/or therapy cannot be evaluated, and new therapies, such as immunotherapy, designed to enhance antitumor immunity cannot be assessed in such models. The orthotopic site may be difficult to reach, for example, accessing the pancreas may require major surgery, or high-resolution ultrasound imaging for accurate percutaneous needle placement in the target organ. However, some advanced animal models of increased validity have been developed [16,17].

Nonetheless, an early diagnosis of the PDAC and the availability of more effect therapies are urgently needed to increase the survival of patients in the medium–long term. Achieving this remarkable goal requires a profound knowledge of pancreatic cancer biology, which is still an open challenge for scientists around the world. For all these reasons, it is necessary to develop new advanced PDAC models that simulate the tumor in more detail, genetically, morphologically, and structurally, to finally improve the diagnostic and therapeutic outcomes. The development of innovative therapies as well as the screening of existing ones can take great advantage from reliable in vitro models, which could evolve into precision and personalized medicine approaches.

## 11.2 In vitro 3D models for pancreatic cancer

Traditionally, studies of PDAC have been conducted in (1) tissue culture flasks or (2) animal models (mainly mice) [18]. Advancements in tissue engineering have enabled the construction of three-dimensional (3D) system varying from very simplistic to more complicated, which are a much more accurate representation of the human in vivo disease situation. More specifically, they offer a better/more realistic spatial cell–cell and cell–ECM interaction as well as more realistic overall structure and internal porosity [18]. Overall, such features are crucial in order to recapitulate the TME, which as described above (Section 11.1) is a complex mixture of cellular, biochemical, structural, and mechanical features which all interact in complex ways, driving the PDAC progression and resistance to treatment. There are various 3D models that have been attempted for PDAC, in the sections below the main categories of these models, that is, spheroids, hydrogels, and polymeric scaffolds are summarized.

### 11.2.1 Spheroids and organoids

Spheroid cultures are essentially cell aggregates/clusters organized in 3D (Fig. 11.2) [18–22]. In terms of complexity, they are the simplest and more widely used systems in PDAC remodeling, able to provide cell–cell and cell–ECM 3D interactions.

There are several spheroid type cultures for PDAC. For example, Longati et al. [23] developed methylcellulose-based spheroids of PDAC cell lines and observed increased ECM production, that is, fibronectin I and collagen I, lower proliferation and higher lactic

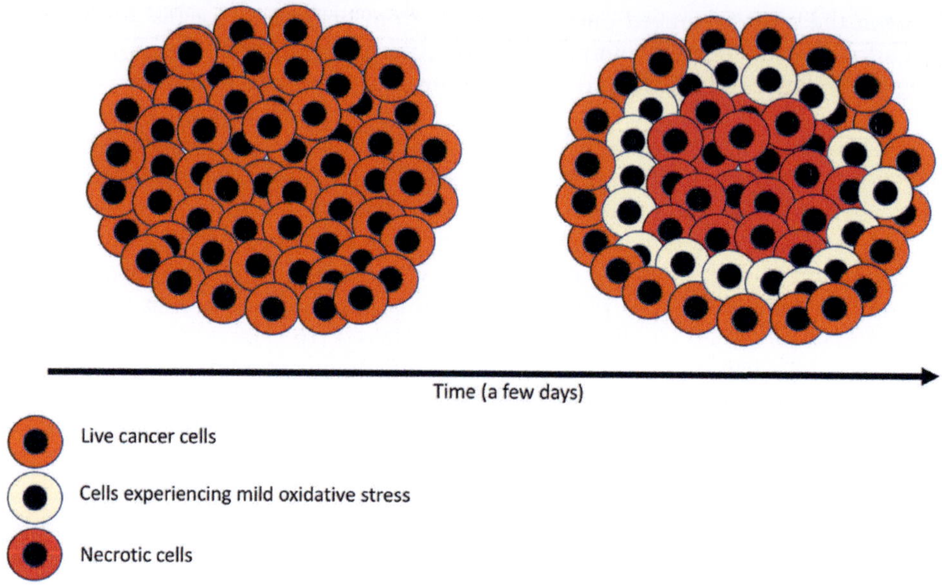


FIGURE 11.2 A spheroid cancer culture configuration. Original image from the authors.

acid accumulation in the spheroids as compared with 3D tissue culture flasks [23]. Similarly, Wen et al. [24] developed PDAC cell line-based spheroids and saw increased resistance to drug treatment (chemotherapy) in the spheroids as compared with 2D cultures [24]. Matsuda et al. [25] also developed PDAC cell line-based spheroids and observed a higher expression of cytoskeleton proteins by the cells in a spheroid configuration, as compared with 2D [25].

There are also some reported spheroid type models with increased biological complexity, that is, co-culturing in 3D cancer cells along one or more (healthy) cell types from the PDAC TME. This is an important advancement as compared with monocultures as cell–cell interactions in the TME can affect the disease progression and response to treatment (see also Section 11.1).

For example, matrigel-based or collagen-based spheroid cultures consisting of PDAC cell lines and either pancreatic stellate cells or fibroblast cell lines were developed by Froeling et al. [26] and Drifka et al. [27], respectively. Shoval et al. [28] developed a spheroid-based co-culture of PDAC cell lines and endothelial cells. Spheroid-based model consisting of PDAC cell lines, fibroblast cell lines, and endothelial cells were developed by Lazzri et al. and Beckermann et al. [29,30]. De Maggio et al. [31] developed a matrigel/collagen assisted spheroid of PDAC cell lines, stellate cells and endothelial cells. There are some studies on spheroid-based cultures of patient-derived tissues, that is, organoids. Such models have shown to maintain the patients' histological and genetic characteristics. For example, Boj et al. [33] developed matrigel-based organoids of primary PDAC cells [32,33].

Overall, the above spheroid systems have helped obtain some crucial insights on the PDAC behavior in vitro. However, the simplistic nature of the spheroids leads to several disadvantages of these structures (Table 11.1) [18,21,34–36].

**TABLE 11.1** Advantages and disadvantages of different 3D systems for pancreatic cancer remodeling.

| 3B system | Advantages + | Disadvantages − |
|---|---|---|
| Spheroids | -Provision of structure | - Variability (in size/shape) |
| | -Realistic cell–cell and cell–ECM interactions | - Diffusional limitations resulting in limited culture duration |
| | -Easy and cheap to fabricate | - Weak mechanical properties |
| Hydrogel scaffolds | -Provision of structure and porosity | - Weak mechanical properties |
| | -Realistic cell–cell and cell–ECM interactions | - Perfusion (vascularization/shear stress mimicry) can be challenging |
| | -Better mass transfer than spheroids | |
| Polymeric foam scaffolds | -Provision of structure and porosity | - Complexity in synthesis |
| | -Realistic cell–cell and cell–ECM interactions | - Expensive equipment required for fabrication |
| | -Good mechanical properties | - Postfabrication functionalization is usually required for ECM mimicry. |

More specifically, the lack of robust structure leads to abnormally high gradients of oxygen, nutrients, and metabolites, which result in necrosis within days. Therefore, spheroid-based cultures can only last a few days without requiring resuspension, the latter though is far from ideal as it disturbs the formed cell–cell and cell–ECM interactions. Furthermore, they lack mechanical strength and reproducibility in size. Finally, it is not possible to control efficiently the ECM spatial secretion within a spheroid system.

## 11.2.2 Hydrogels

Hydrogels are polymer networks, which are cross-linked and water-swollen. They are made of biocompatible polymers and have a more organized structure as compared with spheroids, enabling a better spatial organization of cells in 3D along with more realistic cell–ECM interactions (Fig. 11.3) [18,37,38]. There are some limited studies of PDAC in hydrogel-based 3D systems. For example, [39] developed poly(ethylene glycol) hydrogels cross-linked with collagen fibrils and studied the growth of PDAC within the hydrogels and observed higher proliferation and epithelial–mesenchymal transition (EMT) within the collagen rich hydrogels [39]. Raza et al. [40] used poly(ethylene glycol)–tetra-norbornene hydrogels and observed enhanced cell cluster formation of PDAC cell lined in the hydrogels. Furthermore, the cell growth was affected by the stiffness, structure characteristics, and cross-linkers used to fabricate the hydrogels [40]. Chiellini et al. developed chitosan-polyelectrolyte complex gels and monitored the growth of PDAC cell lines within them for a prolonged period, that is, 28 days [41].

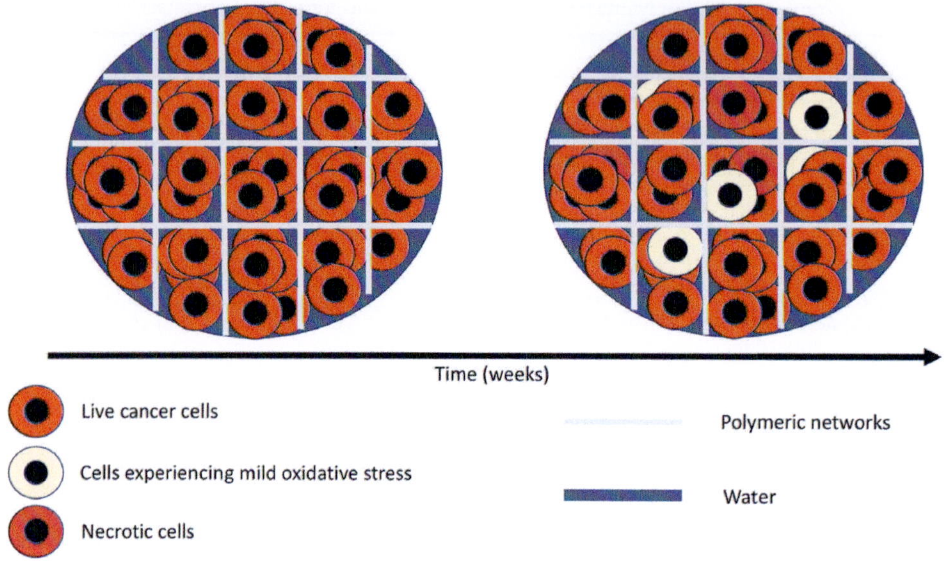


FIGURE 11.3 A hydrogel-based 3D cancer culture configuration. Original image from the authors.

Overall (Fig. 11.3, Table 11.1), hydrogels provide a better structure, porosity, biochemical, and spatial control in 3D as compared with spheroids. However, due to their generally high-water content they have relatively weak mechanical properties making the PDAC stiffness recapitulation challenging. Furthermore, it is difficult to insert perfusion for shear stress/vascularization mimicry in a hydrogel-based 3D system [18,31,35].

### 11.2.3 Polymer scaffolds

Polymeric sponge/foam type scaffolds (Fig. 11.4) are another potential 3D system for tissue remodeling in vitro [18,34,42]. These scaffolds are generally very versatile and depending on the material, the fabrication method and the operational conditions they can result in various internal structural configurations, that is, fibrous or porous structures with different levels of porosity and pore interconnectivity and mechanical properties. Such versatility is important for the recapitulation of different internal tissue structures, at the same time the potential of higher mechanical properties enables the mimicry of stiffer tissues, that is, common in PDAC, as well as the possibility of long-term perfusion, that is, such stiffer scaffolds are less affected and more stable to shear stress. However, they require specialized costly equipment for fabrication (see also Table 11.1 and Fig. 11.4) [18,34,35,42,43].

Some studies have explored the use of polymeric sponge/foam type scaffolds for pancreatic cancer remodeling. For example, Funel et al. [5,44] monitored the growth of patient-derived pancreatic cancer cells in poly(vinyl alcohol) (PVA)/gelatin sponges and observed better adhesion and higher viability in the sponges, as compared with 2D tissue culture flasks [44]. Similarly, Wang et al., observed higher proliferation of patient-derived

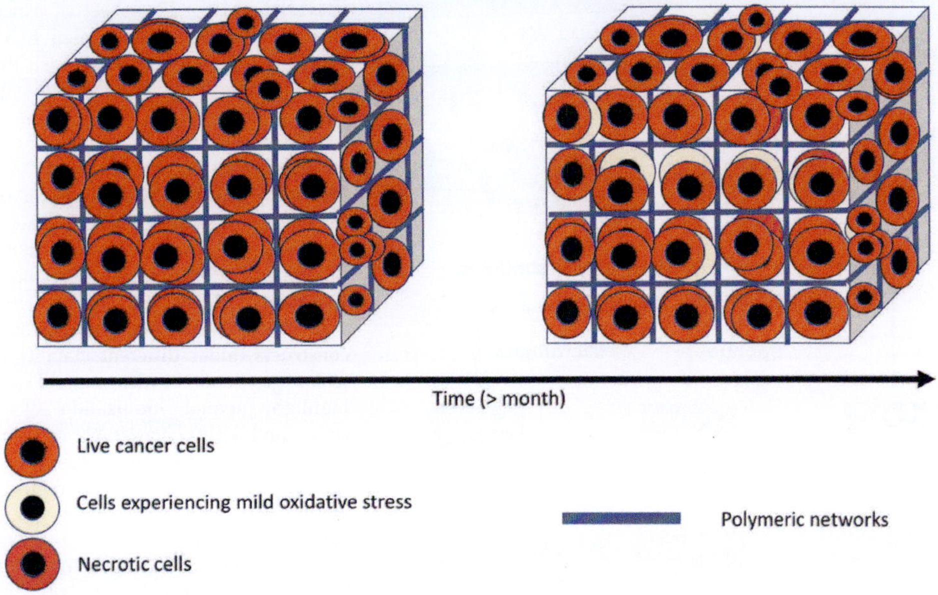


FIGURE 11.4 A polymeric scaffold-based 3D cancer culture configuration. Original image from the authors.

cells in polyglyconate/gelatin scaffolds (fabricated with electrospinning) as compared with 2D cultures [45]. He et al. [46] saw better proliferation and neoplastic formation of pancreatic cancer cells in poly(glycolide-*co*-trimethylene carbonate)–gelatin scaffolds as compared with 2D [46]. Ricci et al. [47] developed various polymer-based scaffolds of different configurations. More specifically, two different polymer configuration were used, that is, poly(ethylene oxide terephthalate)/poly(butylene terephthalate) (PEOT/PBT) and PVA/gelatin, and two structural configurations, that is, fibrous and porous scaffolds. The scaffolds supported successfully the proliferation of patient-derived pancreatic cancer cells [47]. Totti et al. [42] developed polyurethane-based highly porous scaffolds (using the Thermal Induced Phase Separation method) and surface modified them with fibronectin for ECM mimicry. Long-term (>5 weeks) proliferation of pancreatic cancer cells was observed in the scaffolds along with realistic ECM secretion, cell cluster density and spatial organization, and hypoxic gradients.

## 11.3 Using 3D models for disease understanding

Due to the growing importance of TME in understanding carcinogenic mechanisms, new biomimetic systems, such as 3D in vitro models, are currently being studied to better understand cancerous and other diseases [48,49]. As explained in detail in Section 11.2, in the simpler 3D models, the cells are agglomerated to give rise to spheroidal structures or are incorporated in gels of biological origin [49–52]. Only in recent years, an approach with biomaterials in the form of scaffolds been adopted for PDAC, which is by using the tissue engineering approach, namely cancer tissue engineering [18]. This technique in fact

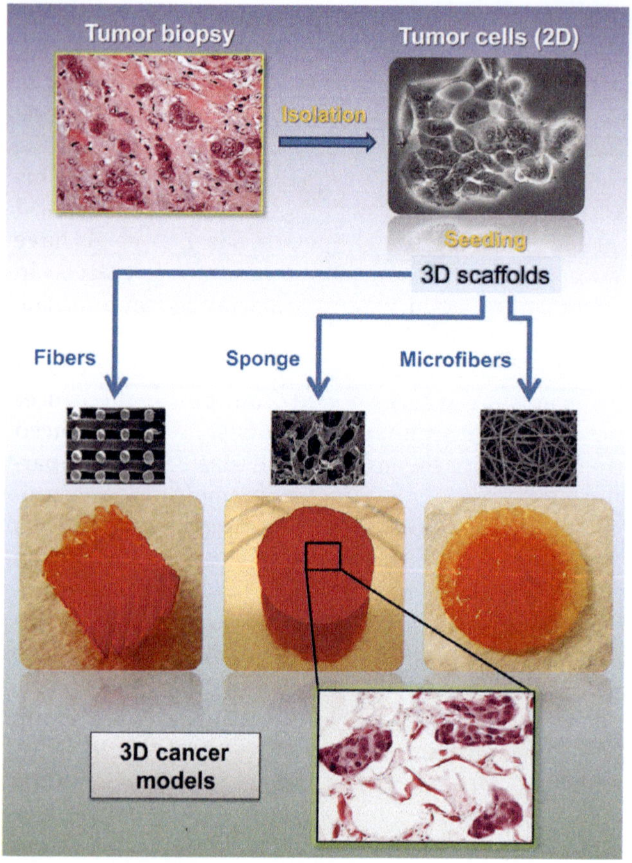


**FIGURE 11.5** Flowsheet of cancer tissue engineered models related to PDAC. Sequentially (top left corner through arrows), single images/image groups show: light micrograph of PDAC morphology (hematoxylin & eosin staining); light micrograph of isolated primary PDAC cells (PP244); scanning electron microscopy (SEM) micrograph of scaffold inner structures (i.e., 3D fiber deposition via Bioplotter, PVQA/gelatin sponge via emulsion and freeze–drying, microfibers via electrospinning); PDAC cell/scaffold constructs under different viability assays (color gradients of the scaffold surfaces highlight spatial localization of viable cells); and light micrograph of a spongy construct (hematoxylin & eosin staining) showing presence of 3D PDAC cell clusters within the scaffold pores, whose morphology mimics that of native PDAC (see the tumor biopsy micrograph). Creative Common Attribution license, [18].

offers extreme versatility, which ultimately allows complex microenvironments to be recreated in vitro, for example, biochemical–physical and mechanical phenomena, such as cell–cell and cell–ECM forces, transport phenomena, multicellular environments, which would enable the PDAC to be studied at early stages, leading to new knowledge of key factors in cancer biology [18]. The choice of the scaffold ultimately determines the most suitable environment for a cancer cell type (Fig. 11.5) [18,47].

Therefore, depending on the goal of the research, advanced 3D models can be developed and used to reproduce cancer-like tissues in vitro, which recapitulate specific phenomena of interest, useful to study the action of new drugs on selected molecular pathways.

## 11.3.1 Biomimetic role of scaffold features

A very limited number of studies make use of 3D models for the study of the PDAC. Basic 3D models use spheroids obtained from tumor-derived cell lines, such as Panc-01, Capan-1, and ASPC-1 and hydrogels as a supportive ECM, namely, polysaccharides and proteins. The role of ECM-like biomolecules is to give cells a biomimetic microenvironment for their proper growth and differentiation, which is barely evoked by purely synthetic polymers. In one of these 3D models, proposed by Longati et al. (reported also in Section 11.2.1), the spheroidal structures inserted in methylcellulose, as a polysaccharide gel, resulted in an upregulation of gene expression profiles and ECM components, while the inhibition of microRNA (miRNA) showed an increased chemoresistance [23]. This study is a typical example of the strong influence of a proper microenvironment toward cancer cells, which can be helpful to set up reliable tumor models. Hosoya et al. have reported another model, also based on an approach of trapping cells into hydrogels. In this case, protein gels based on gelatin and fibronectin were used, arranged in alternate layers together with the cells inside transwells, with the aim of reproducing some of the basic structural characteristics of the ECM [53]. This model was used to test the spread of dextran nanoparticles, using a line of murine fibroblasts derived from pancreatic cancer and normal fibroblasts as a control. The results with tumor fibroblasts showed a reduced permeability to dextran depending on the number of the layer and the size of the nanoparticles, showing analogy with the ECM of the tumor. By stimulating biomolecular recognition pathways, such biomolecules artificially added to the 3D culture generally show a catalytic effect on cancer cell behavior. Therefore, if also physical and mechanical factors are controlled, an even more powerful effect could be introduced in this scenario. In biomaterial scaffolds, specific characteristics, such as porosity, architecture, roughness, elasticity, chemical composition—including gradients, and others—can be changed and tuned to accomplish the specific application [47].

So far, a very few advanced 3D in vitro models use porous scaffolds. In 2013, the group of Prof. Hin (Xinjiang Hospital, China) focused his research on fibrous polymeric structures added with gelatin for the cultivation of pancreatic cancer cells [46]. In this study, scaffolds of glycolide–trimethylene carbonate copolymer obtained by electrospinning and cultured with CSC (selected as homogeneous subpopulation expressing the $CD24^+$ and $CD44^+$ markers) were used to implant under the skin in an animal model [46]. In addition, in this case the 3D model showed a greater neoplastic formation compared with the administration of heterogeneous cells through Matrigel, as it is more reliable and controllable.

PDAC tissue engineering landscape has availed itself of spongy scaffolds. The softness, hydrophilicity, and spongy architecture of some scaffold types seems very suitable to mimic the pancreatic tissue [44,47,54]. Macroporous sponges of PVA and gelatin (80/20 weight ratio, w/w) have first been tested in 2008 for the growth of primary cells human isolated from PDAC explants, and therefore not from commercial cell lines, reporting increased metabolic activity compared with the 2D equivalent of $1.38 \pm 0.04$ times [44]. In 2014, the same authors have carried out an explorative study on three types of porous scaffolds. PVA/gelatin sponges (70/30 w/w) have been obtained by emulsion and freeze–drying, "rigid" sponges based on the copolymer 300PEOT55PBT45, belonging to the PEOT/PBT family, obtained by salt leaching and finally electrospun PEOT/PBT fibers, always seeded with primary human PDAC cells to investigate which of these structures facilitated the morphological organization of the tumor [47]. The authors of the study found clusters with the formation of lumens, namely, ductal-like structures, only in the spongy supports, therefore within large and rounded pores. The interspaces delineated by ultrafine did not reveal the presence of these neoplasm formations, thus suggesting that the architectural aspects of the scaffold may be of primary importance for the development of tumor models through tissue engineering. The morphological features of large spherical pores resemble, better than fibrous meshes, those of glandular organs, thus resulting most appropriate to study PDAC cell biology, in particular if combined with biomolecules [54]. Preliminary studies have investigated different PVA/gelatin weight ratios (100/0 to 50/50 w/w%), to assess differences in pore size distribution and how such changes would affect cellular interactions, including mesenchymal stromal, endothelial, liver, and pancreatic cancer cells [55,56]. The higher pore size of PVA/G scaffolds, with respect to PVA/G hydrogel analogues, specifically affected cancer cell aggregation depending on the tumor type. Due to larger pore size, the sponge best sustained PDAC cell aggregation with duct-like epithelial morphostructure. Differently, hepatocellular carcinoma cells (HepG2) preferred small pore containment of hydrogels to self-assembly with a biomimetic organization [56]

By containing gelatin, a protein derived from collagen denaturation, PVA/gelatin scaffolds showed the highest expression of MMP-2 by PDAC primary cells [47]. Totti and coworkers, who employed polyurethane fibronectin-coated scaffolds to mimic ECM features of the structure of PDAC TME [42], report another example. The role of proteins and polysaccharides is fundamental to ECM, thus it provides biomimetic cues for cancerous cells. A dysregulation in the expression of MMPs, enzymes deputed to ECM remodeling, has been reported in cancerous cells to be implicated in metastatic processes [57]. However, often underestimated, the scaffold morphology does play a difference in evoking architectural features able to stimulate a more reliable behavior by cancer cells.

More studies that are systematic are necessary, also including mechanical properties, to identify the best scaffold where to study PDAC in vitro. According to our experience, soft hydrophilic polymer sponges or hydrogels incorporating ECM molecules could be a primary choice.

### 11.3.2 Tumor progression and metastasis

Considering the key role of TME on cancer progression and metastasis development, 3D culture models have become a powerful tool for investigating cancer biology and developing novel anticancer drug screening approaches. The TME is a dynamic entity, and tumor cells have learnt how to manipulate it in their favor. Tumors, such as PDAC, can recruit TME-supporting cells, such as cancer-associated fibroblasts (CAFs) and macrophages by modifying the normal tissue environment [12,13,58]. Moreover, PDAC is able to spread metastases into the lymphatic system and distant organs even at the earlier stages, which concur to a very dismal prognosis for this tumor [8,11,14]. The way PDAC cells are able to be so aggressive, much more than other cancer types, is still unknown. Therefore, additional efforts should focus on this PDAC character.

In a second study of the Chinese group, an electrophilic scaffold of polyglyconate and gelatin was tested as a 3D model of PDAC metastasis always with homogeneous CSCs, expressing $CD133^+$ and $CD184^+$ ($CXCR4^+$) [45]. In this study, it was highlighted how the scaffold has increased the incidence of neoplastic formation by rapidly inducing liver metastases. It has also been shown that folfirinox, a pharmacological cocktail, in this case has shown a superior ability to prevent such metastases compared with gemcitabine. The effectiveness of polychemotherapy must be considered related to the type of surgical treatment; therefore, the results obtained have a very specific value.

As mentioned in Section 11.2.2, Chiellini et al. have proposed a microstructured hydrogel made of chitosan or a polyelectrolyte complex with poly(g-glutamic acid) to study the behavior of BxPC-3 pancreatic cancer cell line [41]. These authors found out that the expression of several cytokines by BxPC-3 cells resulted higher in 3D than in 2D culture conditions. Among these, G-CSF and MIF are indeed involved in tumor and metastasis spread.

Owing to its asymptomatic behavior, PDAC is almost never diagnosed at an early stage and metastases often occur to the surrounding organs. The EMT mechanism has been hypothesized in recent years to underlie the migration of tumor cells to other organs [59]. Several signaling pathways regulate EMT, therefore its study is complex and multifactorial. Pal et al. [60] have recently demonstrated that 3D cultures availing of electrospun fiber scaffolds allowed EMT to be observed in several cancer cell types, namely, breast cancer cells, MDA-MB-231, and gastric cancer cells, MKN74. The 3D in vitro models of PDAC have the potential to unveil important phenomena, such as cancer cell migration, TME composition, ECM remodeling, and EMT. Froeling et al. [26] observed a more pronounced EMT transition in co-culture as compared with the monoculture, indicating that the TME plays a key role in tumor spread; thus, complex in vitro systems are needed to provide physiological relevance in disease progression [26].

The timing of expression of proteins involved in the EMT (e.g., EGFR, MMP-9, and TGF-β) in suitable 3D models, together with cell morphotype change is highly suspected for EMT occurrence. EGFR, a tyrosine kinase receptor strongly overexpressed in PDAC, is associated with the "grading" of tumor differentiation [61]. In fact, previous studies have shown a correlation between the expression of EGFR and the aggressive behavior of cancer cells, including metastatic [16]. MMPs are a family of proteases that are actively involved in the degradation of ECM. Due to their ability to reshape the TME through the destruction of the ECM, angiogenesis and recruitment of inflammatory cells, MMPs have been directly correlated with tumor development and invasion, and indirectly with the metastatic process such as those who encourage EMT process. The role of MMPs, such as MMP-2, MMP-7, and MMP-9, has been studied in the PDAC [62]. Mutation in the K-ras gene, encountered in greater than 90% of PDAC, together with the loss of Smad4 mediate tumor invasion through an increase in MMP-9 expression [62]. TGF-β has also been shown to play a role in EMT, invasion, and stroma–tumor dialog [63]. In fact, TGF-β regulates cell proliferation, differentiation, invasion, and inflammation. Recently it has emerged that a dysregulation of the TGF-β signal pathway represents one of the main areas of preclinical and clinical tumor research. The TGF-β pathway has both pro and anticancer activities, as it promotes cell cycle arrest, apoptosis, autophagy of epithelial cells, and also inhibits inflammation. On the other hand, promoting angiogenesis, cell motility, invasion, and the EMT process and tumor progression.

It is thus expected that new 3D models will aim to mimic EMT phenomena, in order to provide disclose cancerous mechanisms of PDAC and help the development of new and effective therapies for this unfavorable malignancy (Fig. 11.6).

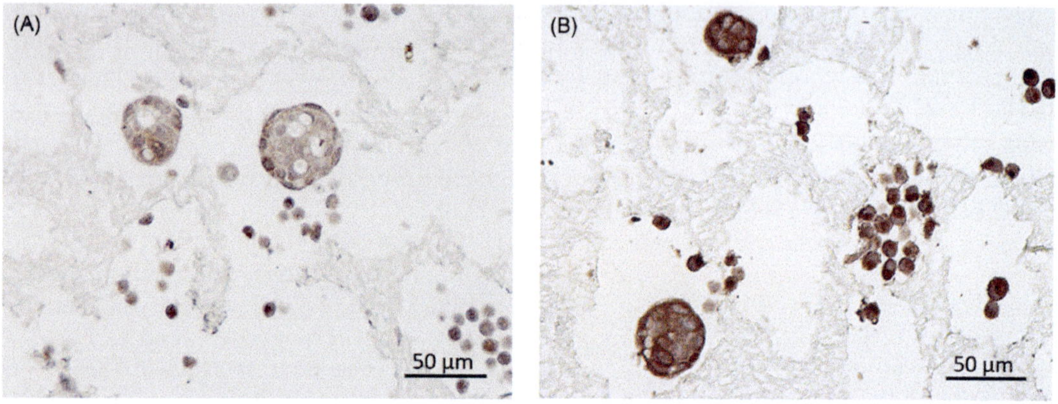


**FIGURE 11.6** Immunohistochemistry of human primary PDAC cells cultured on PVA sponges for 10 days, showing the formation of cellular aggregates with duct-like structures: (A) MMP-9 expression, (B) TGFβ overexpression. Original magnification 400 ×. Unpublished data from the authors.

## 11.4 Using 3D models for therapeutic screening

The most common treatments for PDAC are chemotherapy, radiotherapy, or a combination of the two (Cancer Research UK). Most studies on 3D models for PDAC focus on the development and characterization of the model (see also Section 11.2) and there are limited studies that focus on treatment screening with most of them reporting a higher treatment resistance in a 3D in vitro setting. The majority of 3D PDAC treatment studies in vitro have been conducted in spheroid type systems. For example, Al-Ramadan et al. conducted irradiation treatment in PDAC spheroids and observed a dose dependent cell death within the system [50]. Longati et al. observed higher resistance to the chemotherapeutic agent GEM in PDAC cell line-based spheroids, as compared with 2D and also indicated different levels of resistance for different cell lines [23]. Ware et al. developed spheroids of PDAC cell lines and stellate cells and reported less diffusion of GEM within the spheroid in presence of stellate cells, indicating the importance of the PDAC TME on treatment resistance [51]. Broekgaarden et al. reported similarly a higher tolerance of PDAC to oxaliplatin in presence of CAFs in a spheroid 3D setting [49]. Ki et al. performed PDAC treatment in hydrogels and reported higher resistance to GEM as compared with 2D cultures [39]. A very recent study performed by Gupta et al. [34] in a polymer-based highly porous scaffold reported a dose depended response of PDAC for both chemotherapy and radiotherapy. Furthermore, the porous scaffold configuration enabled the high viability of the culture for longer, allowing for treatment screening after a month and for posttreatment monitoring of 20 days. Such long timeframe is crucial to allow for a time-regime, which resembles closer the period of treatments for patients and animals [34]. Brancato et al. [64] successfully obtained human PDAC microtissues by co-culturing pancreatic cancer cells (PT45) and normal or CAFs within gelatin microcarriers in a spinner flask bioreactor. Their purpose was to recapitulate the desmoplastic reaction primed by the stroma-cancer cross-talk for a potential use of these models in drug screening [64].

Overall, as can be seen from all the above studies, the structural, biochemical and biological complexity of the 3D systems lead to high resistance to drugs and irradiation and to a generally more realistic response, indicating the importance of accounting for (1) structure, (2) biological complexity, and (3) biochemical complexity when designing 3D treatments for PDAC screening. Promising therapeutic approaches for PDAC involve (1) immunotherapy and (2) proton therapy; however, there is no reported 3D system studying those approached to date.

## 11.5 Conclusions and future trends

In the panorama of models currently present in pancreatic cancer research, tissue engineering offers the most interesting prospects. More specifically, it allows 3D cellular models to be built in which phenotypic heterogeneity and TME mimicry can be obtained. In addition, the effect of drug concentration gradients could be evaluated. Furthermore, in the porous 3D structures offered by the scaffolds, the cancer cells demonstrate realistic and physiologically relevant morphology and spatial clustering. With these models, it is possible to study phenomena such as the tumor development from early stages, the migration of malignant cells and cell–cell interactions (for both cancer–cancer cell as well as cancer–healthy cell interactions). Finally, as compared with animal models, which are more complex, costly, and difficult to reproduce, advanced 3D models are ethically sustainable, reliable, and reproducible on a large scale at a low cost, thus representing a promising approach for personalized treatment trials, as well as for the development of novel targeted and functional therapies. We are convinced that biomaterial-based 3D cancer models will open up new routes to comprehend and fight dismal tumors, such as the PDAC.

## References


[1] <https://www.med.upenn.edu/pcrc/aboutpancreaticcancer.html>.

[2] Su GH. Pancreatic cancer: methods and protocols. In: Walker JM, editor. Methods in molecular biology. 3rd ed. New York: p. v. Springer; 2019.

[3] Siegel R, et al. Cancer statistics, 2018. CA Cancer J Clin 2018;68(1):7–30.

[4] Vincent A, et al. Pancreatic cancer. Lancet. 2011;378:607–20.

[5] Funel N, et al. Laser microdissection and primary cell cultures improve pharmacogenetic analysis in pancreatic adenocarcinoma. Lab Invest 2008;88:773–84.

[6] Solcia E, et al. Tumors of the exocrine pancreas. Atlas of tumor pathology. In: Rosai J, Sobin LH, editors. Tumors of the pancreas. Washington DC: Armed Forces Institute of Pathology; 1997. p. 64–88.

[7] Lüttges J, et al. Where and when does pancreatic carcinoma start? Med Klin (Munich) 2004;99(4):191–5.

[8] Maitra A, et al. Pancreatic cancer. Annu Rev Pathol 2008;3:157–88.

[9] DeCant BT, et al. Utilizing past and present mouse systems to engineer more relevant pancreatic cancer models. Front Physiol 2014;5:464.

[10] Herreros-Villanueva M, et al. Molecular markers in pancreatic cancer diagnosis. Clin Chim Acta 2013;418:22–9.

[11] Kloppel G, Kosmahl M. Cystic lesions and neoplasms of the pancreas. The features are becoming clearer. Pancreatology 2001;1:648–55.

[12] Qian BZ, Pollard JW. Macrophage diversity enhances tumor progression and metastasis. Cell. 2010;141:39–51.

[13] Feig C, et al. The pancreas cancer microenvironment. Clin Cancer Res 2012;18:4266–76.
[14] Iovanna J, et al. Current knowledge on pancreatic cancer. Front Oncol 2012;31:2–6.
[15] Voskoglou-Nomikos T, et al. Clinical predictive value of the in vitro cell line, human xenograft, and mouse allograft preclinical cancer models. Clin Cancer Res 2003;9:4227–39.
[16] Avan A, et al. Crizotinib inhibits metabolic inactivation of gemcitabine in c-Met-driven pancreatic carcinoma. Cancer Res 2013;73:6745–56.
[17] Giovannetti E, et al. Role of CYB5A in pancreatic cancer prognosis and autophagy modulation. J Natl Cancer Inst 2014;106:djt346.
[18] Ricci C, et al. Cancer tissue engineering—New perspectives in understanding the biology of solid tumors—A critical review. OA Tissue Eng 2013;1:4.
[19] Fennema E, et al. Spheroid culture as a tool for creating 3D complex tissues. Trends Biotechnol 2013;31:108–11.
[20] Cesarz Z, Tamama K. Spheroid culture of mesenchymal stem cells. Stem Cell Int 2016;2016:9176357.
[21] Edmondson R, et al. Three-dimensional cell culture systems and their applications in drug discovery and cell-based biosensors. Assay Drug Dev Technol 2014;12:207–18.
[22] Nath S, Devi GR. Three-dimensional culture systems in cancer research: focus on tumor spheroid model. Pharmacol Ther 2016;163:94–108.
[23] Longati P, et al. 3D pancreatic carcinoma spheroids induce a matrix-rich, chemoresistant phenotype offering a better model for drug testing. BMC Cancer 2013;13:95.
[24] Wen Z, et al. A spheroid-based 3-D culture model for pancreatic cancer drug testing, using the acid phosphatase assay. Braz J Med Biol Res 2013;46:634–42.
[25] Matsuda Y, et al. Morphological and cytoskeletal changes of pancreatic cancer cells in three-dimensional spheroidal culture. Med Mol Morphol 2010;43:211–17.
[26] Froeling FE, et al. Organotypic culture model of pancreatic cancer demonstrates that stromal cells modulate E- cadherin, β-catenin, and Ezrin expression in tumor cells. Am J Pathol 2009;175:636–48.
[27] Drifka CR, et al. Human pancreatic stellate cells modulate 3D collagen alignment to promote the migration of pancreatic ductal adenocarcinoma cells. Biomed Microdevices 2016;18:105.
[28] Shoval H, et al. Tumor cells and their crosstalk with endothelial cells in 3D spheroids. Sci Rep 2017;7:10428.
[29] Lazzari G, et al. Multicellular spheroid based on a triple co-culture: a novel 3D model to mimic pancreatic tumor complexity. Acta Biomater 2018;78:296–307.
[30] Beckermann B, et al. VEGF expression by mesenchymal stem cells contributes to angiogenesis in pancreatic carcinoma. Br J Cancer 2008;99:622–31.
[31] Di Maggio F, et al. Pancreatic stellate cells regulate blood vessel density in the stroma of pancreatic ductal adenocarcinoma. Pancreatology 2016;16:995–1004.
[32] Boj SF, et al. Model organoids provide new research opportunities for ductal pancreatic cancer. Mol Cell Oncol 2015;160:324–38.
[33] Boj SF, et al. Organoid models of human and mouse ductal pancreatic cancer. Cell 2015;160:324–38.
[34] Gupta P, et al. Chemoradiotherapy screening in a novel biomimetic pancreatic cancer model. RSC Adv 2019;2019(9):41649.
[35] Burdett E, et al. Engineering tumors: a tissue engineering perspective in cancer biology. Tissue Eng Part B Rev 2010;16:351–9.
[36] Knight E, Przyborski S. Advances in 3D cell culture technologies enabling tissue-like structures to be created in vitro. J Anat 2015;227:746–56.
[37] Ahmed EM. Hydrogel: preparation, characterization, and applications: a review. J Adv Res 2015;6:105–21.
[38] Peppas NA, et al. Hydrogels in pharmaceutical formulations. Eur J Pharm Biopharm 2000;50:27–46.
[39] Ki CS, et al. Effect of 3D matrix compositions on the efficacy of EGFR inhibition in pancreatic ductal adenocarcinoma cells. Biomacromolecules. 2013;14:3017–26.
[40] Raza A, et al. The influence of matrix properties on growth and morphogenesis of human pancreatic ductal epithelial cells in 3D. Biomaterials 2013;24:5117–27.
[41] Chiellini F, et al. Modelling of pancreatic ductal adenocarcinoma in vitro with three-dimensional microstructured hydrogels. RSC Adv 2016;6:54226–35.
[42] Totti S, et al. A 3D bioinspired highly porous polymeric scaffolding system for in vitro simulation of pancreatic cancer adenocarcinoma. RSC Adv 2018;2018(8):20928.

[43] O'Brien FJ. Biomaterials and scaffolds for tissue engineering. Mater Today 2011;14:88−95.
[44] Funel N, et al. In vitro model of pancreatic ductal adenocarcinoma: new strategy to study pancreatic ductal carcinoma. J Pancreas 2008;9(6 Suppl.):810−11.
[45] Wang X, et al. A bioengineered metastatic pancreatic tumor model for mechanistic investigation of chemotherapeutic drugs. J Biotechnol 2013;166:166−73.
[46] He Q, et al. A tissue-engineered subcutaneous pancreatic cancer model for antitumor drug evaluation. Int J Nanomed 2013;8:1167−76.
[47] Ricci C, et al. Interfacing polymeric scaffolds with primary pancreatic ductal adenocarcinoma cells to develop 3D cancer models. Biomatter. 2014;4:e955386.
[48] Nyga A, et al. 3D tumour models: novel in vitro approaches to cancer studies. J Cell Commun Signal 2011;5:239−48.
[49] Broekgaarden M, et al. Modulation of redox metabolism negates cancer-associated fibroblasts-induced treatment resistance in a heterotypic 3D culture platform of pancreatic cancer. Biomaterials 2019;222:119421.
[50] Al-Ramadan A, et al. Analysis of radiation effects in two irradiated tumor spheroid models. Oncol Lett 2018;15:3008−16.
[51] Ware MJ, et al. Generation of an in vitro 3D PDAC stroma rich spheroid model. Biomaterials 2016;108:129−42.
[52] DelNero P, et al. 3D culture broadly regulates tumor cell hypoxia response and angiogenesis via pro-inflammatory pathways. Biomed Microdevices 2013;15:583−93.
[53] Hosoya H, et al. Engineering fibrotic tissue in pancreatic cancer: a novel three-dimensional model to investigate nanoparticle delivery. Biochem Biophys Res Commun 2012;419:32−7.
[54] Ricci C, Danti S. 3D models of pancreatic ductal adenocarcinoma via tissue engineering. Methods Mol Biol 2019;1882:81−95.
[55] De la Ossa JG, et al. Pore size distribution and blend composition affect in vitro prevascularized bone matrix formation on poly(vinyl alcohol)/gelatin sponges. Macromol Mater Eng 2017;302:1700300.
[56] Ricci, C., et al., (2016) Tuning scaffold pore features to accomplish biomimicry in 3D tumor models: a study on cancer cell aggregation, migration and morphotype transition. In: 10th World Biomaterials Congress, Montréal, Canada, 17−22 May, 2016. Frontiers. doi: 10.3389/conf.FBIOE.2016.01.0167.
[57] Gobin E, et al. A pan-cancer perspective of matrix metalloproteases (MMP) gene expression profile and their diagnostic/prognostic potential. BMC Cancer 2019;19:581.
[58] Dauer P, et al. Microenvironment in determining chemo-resistance in pancreatic cancer: neighborhood matters. Pancreatology 2017;17(1):7−12.
[59] Wang S, et al. Epithelial-mesenchymal transition in pancreatic cancer: a review. BioMed Res Int 2017;2017:2646148.
[60] Pal M, et al. Epithelial-mesenchymal transition of cancer cells using bioengineered hybrid scaffold composed of hydrogel/3D-fibrous framework. Sci Rep 2019;9:8997.
[61] Funel N, et al. Correlation of basal EGFR expression with pancreatic cancer grading but not with clinical outcome after gemcitabine-based treatment. Ann Oncol 2011;22(2):482−4.
[62] Bera A, et al. Oncogenic K-Ras and loss of Smad4 mediate invasion by activating an EGFR/NF-κB axis that induces expression of MMP9 and uPA in human pancreas progenitor cells. PLoS One 2013;8(12):e82282.
[63] Neuzillet C, et al. Perspectives of TGF-β inhibition in pancreatic and hepatocellular carcinomas. Oncotarget. 2014;5(1):78−94.
[64] Brancato V, et al. Bioengineered tumoral microtissues recapitulate desmoplastic reaction of pancreatic cancer. Acta Biomater 2017;49:152−66.

CHAPTER

# 12

# In vitro three-dimensional modeling for prostate cancer

*Eleonora Dondossola*[1] *and Claudia Paindelli*[1,2]

[1]David H. Koch Center for Applied Research of Genitourinary Cancers, The University of Texas MD Anderson Cancer Center, Houston, TX, United States [2]Department of Cell Biology, Radboud Institute for Molecular Life Sciences, Radboud University Medical Center, Nijmegen, The Netherlands

## 12.1 Introduction

Prostate cancer (PCa) is a leading cause of cancer-related death in males. Every year, there are an estimated 1.2 million new cases diagnosed and 359,000 deaths worldwide (7.1% of the total cancer mortality [1]). PCa incidence is higher in Australia, North America, and Western and Northern Europe, relatively high in the Caribbean, Africa, and South America, but remains low in Asia [2]. Mortality rate is generally increased in people of African descent and very low in Asian populations [2]. Family history is associated with an increased PCa incidence, suggesting a genetic predisposition, however only 9% of cases are considered truly hereditary (e.g., three or more affected relatives or at least two relatives who developed PCa before 55 years of age) [1,2]. PCa is usually suspected based on physical examination or increased circulating levels of prostate-specific antigen and confirmed by prostate biopsy. Early PCa develops as an androgen-dependent disease, which can be successfully treated by hormonal deprivation and surgery with nearly 100% survival at 5 years from diagnosis [3]. However, the disease eventually evolves toward a castration-resistant phenotype defined by cancer progression despite low-testosterone levels, which is often associated with increased incidence of metastasis, mostly in bone [4]. Despite recent approval of anticancer agents that prolong survival in advanced metastatic PCa patients (including abiraterone [5], enzalutamide [6], and radium 223 [7]), survival at 5 years drops to approximately 30%.

*Biomaterials for 3D Tumor Modeling*
DOI: https://doi.org/10.1016/B978-0-12-818128-7.00012-5

## 12.1.1 Preclinical models for addressing prostate cancer

### 12.1.1.1 *In vivo models*

Studies that apply animal models, mostly rodents, have greatly improved our understanding of the biological complexity of PCa and have supported advances in its treatment [8,9]. This includes the development of genetically modified mouse models which spontaneously develop PCa due to mutations in key regulator genes or injection of mouse or human cell lines in different sites (in the prostate, subcutaneously, in circulation, or directly in metastatic sites, such as bone) [9–11]. In vivo systems, however, display some disadvantages, including animal welfare issues, significant logistic demands, long experimental time, limited capacity for addressing multiple experimental variables, and, in xenografts, incomplete match between human PCa cells and mouse stroma [12]. As a complement to in vivo models, in vitro approaches are thus required for ease of access, better experimentally controlled conditions, diversification, and high-throughput analysis, including screening of therapeutics.

### 12.1.1.2 *In vitro models*

Traditionally, in vitro preclinical investigations of PCa biology and therapeutic response take advantage of bidimensional (2D) cultures of cell lines in plastic vessels because of their reproducibility, simple manipulation, and rapidity of growth [11,13,14]. Although they advanced the field, 2D systems display some limitations such as monocellularity, lack of 3D structures, and the absence of extracellular matrix (ECM) and stromal components, which may support PCa cell survival, growth, and resistance to therapy [15,16]. As a result, 2D monoculture models are limited in their complexity and unable to replicate tissue-specific pathophysiology.

In the past decade, a variety of three-dimensional (3D) biomimetic models have emerged that can fill a technical gap between conventional 2D approaches and preclinical in vivo models. Their features are described in the following paragraphs.

## 12.1.2 Three-dimensional in vitro models of prostate cancer

### 12.1.2.1 *Spherical cancer models*

Solid tumors growing in vivo are characterized by a 3D spatial organization within their tissue of origin; a nonhomogeneous distribution of nutrients, oxygen, and metabolites; and, often, the presence of a hypoxic core [17,18]. Spherical cancer models have been developed to better mimic these features and partially overcome the limitations of traditional 2D cell cultures. These systems include different types of sphere-like structures obtained from cancer tissues or cell lines, typically maintained as floating cultures or embedded in matrices [19] (Fig. 12.1). A broad terminology has been used to describe spherical cancer models, including spheroids, tumorspheres, and organoids [19], which have peculiar characteristics.

- *Tumor spheroids,* also identified as tumoroids, are multicellular aggregates usually obtained from immortalized cancer cell lines cultivated in standard conditions [e.g., fetal-bovine serum (FBS)-supplemented medium], either through the "hanging drop" method [20] or in nonadherent cell culture plates (Fig. 12.1A; [21]).

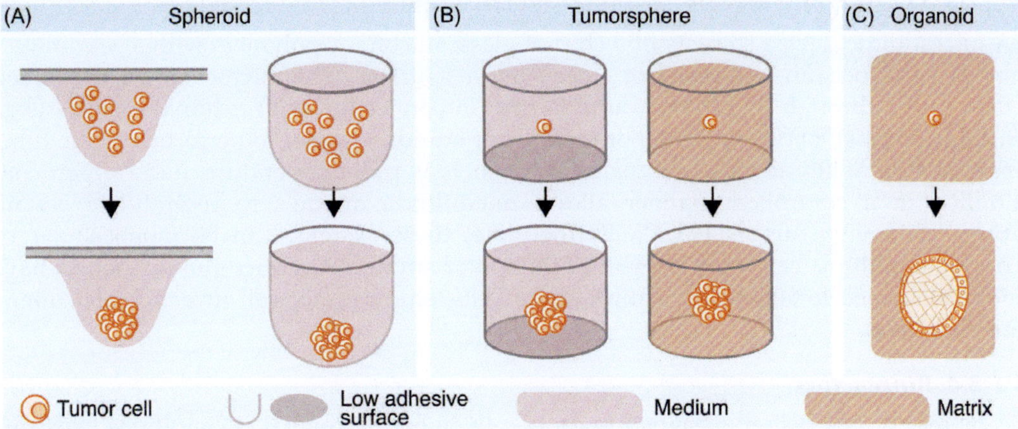


FIGURE 12.1 Spherical cancer models. The cartoon shows different modalities to generate spherical models, including (A) spheroid (hanging drop method; low-adherent culture plate); (B) tumorsphere (in medium or matrix); and (C) organoid.

- *Tumorspheres* are floating clusters derived from cancer stem cells (CSCs) isolated from cell lines or tumor tissues and processed through mechanical and enzymatic dissociation [22]. Tumorspheres are cultured at a very low density in nonadherent plates with stem cell medium (supplemented with multiple growth factors to induce stem cell proliferation) to support the formation of tumor cell clones (Fig. 12.1B; [22]). Cell aggregates similar to tumorspheres can also originate from single cells grown in matrices or on scaffolds (Fig. 12.1B; [23,24]).
- *Organoids* are 3D tissue constructs which retain a simplified, but realistic, microanatomy that mimics their corresponding organ. PCa organoids are generated from circulating tumor cells or dissociated tumor biopsies, seeded as single cells in growth factor-reduced Matrigel [25,26], and cultivated in a special supplemented medium (Fig. 12.1C). They display the histological and molecular epithelial features of their original tumors, recapitulating the diversity of PCa subsets, but lose microenvironment elements over time [25,26].

Overall, spherical cancer models, alone or in combination with bioengineered systems, have been applied to the study of primary and metastatic tumor growth, invasion, heterotypic cell–cell interactions, and therapy response.

#### 12.1.2.2 Bioengineered models

The application of recently developed tissue-engineered platforms has allowed for the establishment of multiple types of primary and metastatic PCa biomimetic models [13,14]. To this purpose, different biomaterials of natural (e.g., collagen, silk, chitosan) or polymeric (e.g., polycaprolactone, polylactide, copolymers) origin, arranged in a variety of structures, such as meshes or sponges, have been applied to model PCa, in vitro [13,27–29]. These 3D bioengineered systems can recapitulate various aspects of tumor progression, including initiation, growth, migration, and invasion, interplays with stromal cells, and mechanisms of drug response and resistance.

#### 12.1.2.3 *Microfluidic models*

A microfluidic device consists of a chip of glass, silicon, or polymers with a set of interconnected microchannels molded or etched into its surface. The microchannels are linked to the outside through inputs and outputs that, in connection with external systems (e.g., peristaltic pumps), permit injection or removal of liquids in order to perform specific functions [13]. The ability to fine-tune parameters such as pH, temperature, and oxygen concentration in a controlled manner allows microfluidic models to recapitulate certain features of in vivo tumors [13,30]. Furthermore, these systems can be monocellular or incorporate multiple cell types present in the tumor stroma [30]. Microfluidic models have mainly been used to isolate circulating tumor cells, study tumor cell invasion, and mimic bone metastasis.

#### 12.1.2.4 *Bioreactors*

A bioreactor is a system designed to grow cells in highly regulated conditions (e.g., pH, temperature, pressure, and nutrients), further enabling spatial and temporal control of multiple stimuli (e.g., growth, mechanical or electrical) to mimic physiological processes and overcome the limitations associated with conventional cultures [13,31]. As such, bioreactors support investigation of cell and tissue functions in a well-defined and controlled 3D environment and have been applied to create large aggregates of tumors cells and to model liver metastasis.

#### 12.1.2.5 *Organ explants*

This model relies on the cultivation of explanted tissues (e.g., primary tumors or bone of murine or human origin [13,32]), which can vary in thickness, ranging from thin tissue slices (200–300 μm) to bigger fragments (1–2 mm) [14,33,34]. The advantage of these systems is that they retain the complexity of the in vivo tissue, including 3D architecture and cell types [14]. Ex vivo tissue survival, usually limited to few days [14], can be extended by applying a variety of methods, such as administration of specific solutions, culture in bioreactors, or utilization of gelatin sponges [35]. Explants have been implemented to study the therapy response to antiandrogen and antiangiogenic agents.

The application of these models to address the different phases of PCa progression (Fig. 12.2) will be described in the next paragraphs.

## 12.2 Modeling primary tumors

### 12.2.1 Modeling localized prostate cancer

#### 12.2.1.1 *Monocellular models of primary tumors*

A number of 3D models of primary prostate tumors have been developed to address the pathophysiology of this disease in a more relevant fashion compared with 2D cultures to investigate the initiation of PCa, attempts have been made to recreate the features of murine prostatic tissue ex vivo. The mouse prostate is composed by four lobes (anterior, dorsal, ventral, and lateral) characterized by different shapes, histopathological properties,

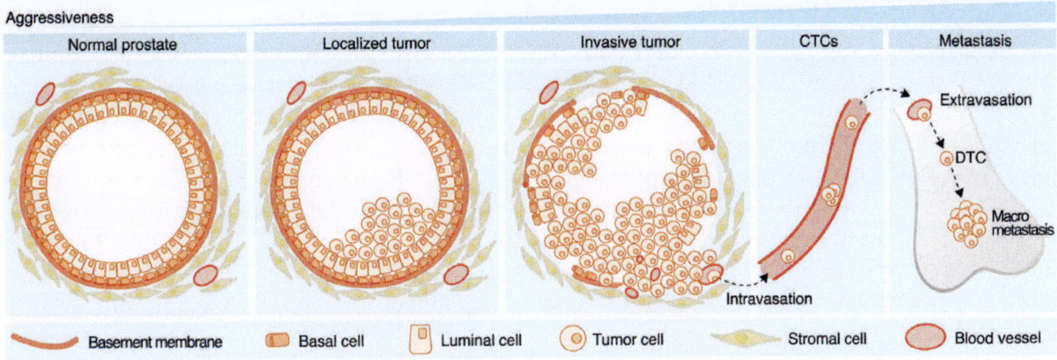


**FIGURE 12.2** Phases of PCa progression. Normal prostate undergoes an initial transformation process that generates a localized tumor. As the tumor advances, formation of new blood vessels and invasion in the surrounding tissues occur. After intravasation, tumor cells get into the systemic circulation, becoming circulating tumor cells (CTCs) and extravasate in distant sites, where they can remain dormant as disseminated tumor cells (DTCs) for years, followed by reactivation and proliferation to form macrometastasis.

and gene expression profiles [36,37]. Prostatic ducts in each lobe are lined with a layer of flat basal epithelial cells and a layer of luminal columnar secretory epithelial cells, with some interspersed neuroendocrine cells [38].

In order to enable clinically relevant investigation of prostate gland biology ex vivo, Karthaus et al. generated murine organoids by embedding dissociated epithelial cells derived from wild-type prostate in Matrigel, in the presence of standard organoid medium [25]. Prostate cells from each lobe, from both basal and luminal origin, gave rise to expanding organoids. Then, prostatic tissue from a genetically engineered mouse model carrying Pten deletion was exploited as a tool for cancer biology. These cells initially formed normal organoids endowed with a regular lumen that accumulated multiple layers over time, forming a solid spherical structure, which shared histological features with tumors developed in vivo by Pten-deficient mice (Fig. 12.3). Ex vivo silencing of *Pten* gene in wild-type prostate epithelial cell-derived organoids led to a similar outcome (Fig. 12.3) [25]. Gao et al. used an organoid system to support the long-term ex vivo culture of patient cells obtained through biopsies, which recapitulated the molecular features of diverse PCa subsets, providing the basis to generate a large inventory of patient-derived PCa cell lines [26]. These results show that both approaches can be applied to study key genes involved in PCa progression and can recapitulate in vivo phenotypes for potential use in pharmacological studies.

Besides organoids for ex vivo culture of cells directly derived from patients, which are not easily accessible and manipulable, further approaches have been developed to expand patient-derived xenograft (PDX) cells. PDX models are generated by implanting human tumors into immunodeficient mice. Neoplastic tissue can be acquired from surgical explants, biopsies, or circulating tumor cells and implanted alone or with Matrigel, usually subcutaneously [39]. As an advantage, by reproducing both intra- and intertumoral heterogeneity of clinical cancers, PDXs replicate human cancer biology with greater fidelity compared with cell lines [40]. As a limitation, however, PDX-derived cells do not survive in

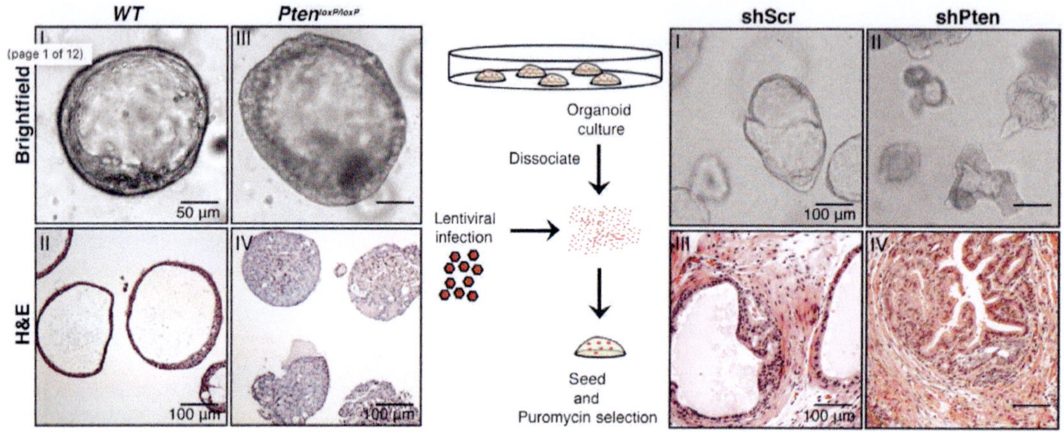


**FIGURE 12.3** Genetically engineered mouse prostate cancer models are recapitulated in organoids. Brightfield and H&E images of wild-type (WT) (I and II), PBCre PtenloxP/loxP (III and IV). In PtenloxP/loxP organoids the lumen is filled with hyperplastic cells. Schematic overview of retroviral infection. Brightfield image of shPten-RFP shScr-RFP-infected PBCre Rosa26LSL-ERG organoids (I and II) and corresponding H&E staining of shPten/shScr organoid–UGSM recombinations (III and IV) showing hyperplastic phenotype in organoids and in UGSM recombinations upon Pten knockdown. Source: *Reproduced with permission, from Karthaus WR, et al. Identification of multipotent luminal progenitor cells in human prostate organoid cultures. Cell 2014;159(1):163–75. © Elsevier (2014).*

2D cultures, and require in vivo passaging to ensure their viability and propagation. Young et al. established a spheroid culture system that supports long-term survival of PDX-derived cells, ex vivo [41]. PDXs collected from mice were minced, digested with enzymes, and processed through a cell strainer to obtain cell clusters of around 200 μm in diameter. Derived spheres were cultivated and expanded in low-attachment plates in medium supplemented with synthetic androgen R1881 and Rho kinase inhibitor Y-27632. Tumorspheres were obtained from six different PDXs belonging to the LuCaP series, expanded, and cryopreserved. Two established cell lines, LuCaP 136 and LuCaP147, were further characterized showing that after multiple passages cancer cells retained the biomarkers of their tumor of origin. Similarly, Fong et al. generated 3D PDX-derived tumoroids encapsulated in a mixture of poly(ethylene glycol)-diacrylate (PEG-DA) hydrogels and thiolated hyaluronic acid (HA) [42]. HA is a nonsulfated glycosaminoglycan expressed by both cancer and stromal cells, that is particularly enriched within the ECM and plays a role in the progression of cancer through activation of cell signaling [43]. As a result, PDX-tumoroids embedded in this matrix grew for up to two weeks and retained AR expression.

Likewise, Hao et al. have applied thiolated-HA linked to multiple molecules of a cell adhesive peptide (PolyRGD-AC) to support the formation of tumoroids from a prostate cancer cell line (LNCaP) in hydrogel [44]. These constructs can be engineered to reproduce the protein component of the ECM and allow variable tuning in terms of architecture, signaling properties, and molecular dynamics (Fig. 12.4). LNCaP cells, initially seeded in these gels as single cells, formed aggregates after 7 days in culture and proliferated for up to 28 days with minimal cell death. Tumoroids displayed E-cadherin–mediated cell–cell

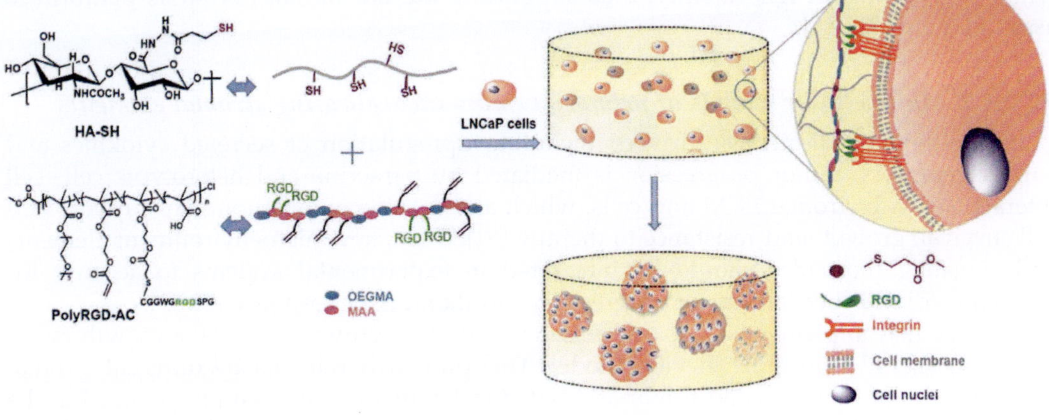


**FIGURE 12.4** Graphical representation of HA-PolyRGD hydrogels and assembly of LNCaP cell tumorspheres. *Source: Reproduced with permission, from Hao Y, et al. Biomimetic hydrogels incorporating polymeric cell-adhesive peptide to promote the 3D assembly of tumoroids. Biomacromolecules 2016;17(11):3750–60. © American Chemical Society (2016).*

interactions and cortical organization of F-actin. When compared with control gels (prepared using a scrambled protein sequence with alterations in D and G residues called PolyRDG), tumor cells grown in the presence of PolyRGD-AC peptide showed a gradual increase of cellular metabolism, which promoted the development of larger tumoroids and enhanced the expression of E-cadherin and integrin alpha 5 and 6 [44].

Besides using low-adherence culture plates or soft matrices, spheroids of PCa cells have been generated using polymeric scaffolds. Specifically, Girard et al. developed a "3P scaffold" composed of three different materials: poly(lactic-*co*-glycolic acid) (PLGA), a block copolymer of polylactic acid (PLA), and monomethoxypolyethylene glycol (mPEG) [45]. These materials formed a complex mesh structure able to support the self-assembly of tumoroids from different cancer cell lines, including PC3 cells. The resulting cell clusters were connected to the scaffold fibers via tight cell junctions that allowed anchoring and stabilization, a phenomenon not observed in the 2D culture. Furthermore, when analyzing markers typical of epithelial-to-mesenchymal transition (EMT), PC3 tumoroids cultivated on 3P scaffolds showed upregulation of the mesenchymal marker vimentin [46,47]. To determine the suitability of this system to test the response to EMT-specific inhibitors, U0126 and LY294002 (inhibiting MEK and PI3K pathways, respectively) were applied. Both molecules were shown to prevent tumoroid formation and reduced cell viability compared with untreated tumoroids, with dose-dependent cytotoxicity. This model proved also suitable for culturing tumor biopsies, administered as single-cell suspensions, showing similar results. Talukdar et al. [48] developed a model made from the silk fibroin protein matrix produced by the silkworm *Antheraea mylitta* [49] and characterized by a high porosity and elevated stability, such that it is suitable for long-term cultures. After treatment with 70% alcohol, the protein underwent a conformational change from random coil to β-sheets. These properties provided stability in water and PBS buffer, and resistance to progressive degradation in vitro [49,50]. LNCaP cells cultivated on this system for a long period of time (>50 days) grew as cell clusters and completely filled the scaffold while maintaining a high viability, thus supporting the use of silk fibroin as platform to assist PCa growth [48].

#### 12.2.1.2 Multicellular models of primary tumors incorporating stromal elements

In addition to cell-intrinsic drivers, including upregulation of secreted cytokines and surface receptors, tumor progression is mediated by paracrine and heterotypic cell–cell interactions with stromal ECM and cells, which alone or in combination may support PCa cell survival, growth, and resistance to therapy [51]. Thus, key microenvironment elements and signaling molecules should be integrated in experimental systems to account for stromal-mediated mechanisms of progression and therapeutic response [52,53].

Xu et al. [54] implemented heparin-binding epidermal growth factor-like growth factor (HB-EGF, Fig. 12.5) in a 3D in vitro model. This protein is released by different stromal cells, including fibroblasts and tumor-associated inflammatory cells, and is involved in the initiation and progression of tumorigenesis [55,56]. This model was composed of two layers of hydrogels. To mimic the paracrine interaction between stroma and cancer cells (LNCaP), the top layer contained HA-based hydrogel particles that released HB-EGF [54], while LNCaP cells were incapsulated in the bottom layer. The system supported HB-EGF dose-dependent growth of tumoroids. LNCaP cells seeded as individual cells at day 1 progressively formed clusters that increased in size over time until day 7, remaining stable up to day 15. Tumorsphere cells retained a rounded, clustered morphology with cortical actin organization and E-cadherin–mediated cell–cell junctions, whereas LNCaP cells seeded in 2D adopted a spread morphology. Furthermore, tumorspheres increased the expression of proangiogenic factors such as vascular endothelial growth factor 165 and interleukin 8. Besides investigating the role played by purified paracrine factors embedded in hydrogels, several tissue-engineered systems have been developed to address the effect of factors released by stromal cells, such as normal or cancer fibroblasts, on PCa cells [54]. Ivich et al. developed an organ-on-a-chip model featuring two compartments separated by a

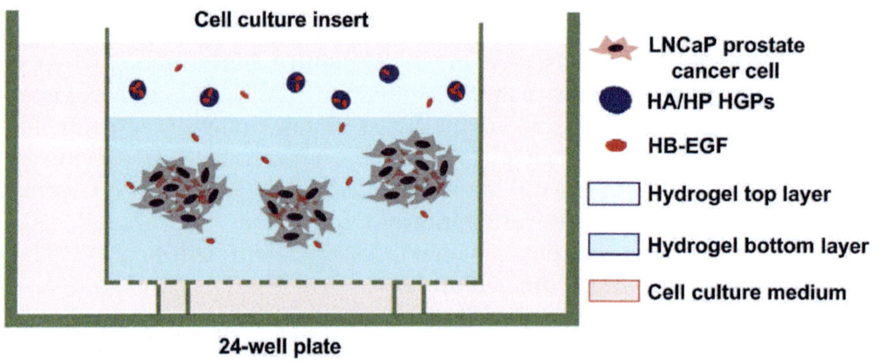


FIGURE 12.5 Graphical representation of a 3D bilayer hydrogel system (3D-HB-EGF) for the growth of tumorspheres of LNCaP cells. Source: *Reproduced with permission from Xu X, et al. Recreating the tumor microenvironment in a bilayer, hyaluronic acid hydrogel construct for the growth of prostate cancer spheroids. Biomaterials 2012;33 (35):9049–60. © Elsevier (2012).*

porous membrane, hosting PCa cells and fibroblasts, respectively [57]. Each compartment was composed of a microchannel with inlets and outlets for the independent control of medium flow and nutrients. The media flow rate determined the diffusion of nutrients and the exchange between the 2 channels. The cell types applied included human normal or cancer prostate epithelium (iPrEC or EMP, respectively) and normal human stromal fibroblasts (BHPrS1). iPrEC cells cultured in the presence of BHPrS1 cells and pretreated with R1881 synthetic androgen converted to a luminal phenotype, lost p63 and acquired AR expression. On the other hand, EMP cancerous epithelium induced the transformation of BHPrS1 stromal cells toward a cancer-associated fibroblast (CAF) phenotype, which expressed αSMA [57].

Fang et al. further addressed fibroblast–PCa cell interactions by using double-layered alginate hydrogel microspheres [58]. These microcapsules displayed an alginate inner core containing normal stromal cells (WPMY-1) and an outer layer of alginate with embedded C4-2 cancer cells (Fig. 12.6). An electrostatically linked polycationic, permselective poly-L-ornithine (PLO)-coated membrane separated the two layers, thus preventing reciprocal invasion. Each cell type survived up to 4 weeks, compartmentalized within its own layer. As proof of principle of paracrine interplays, the concentration of a soluble E-cadherin fragment (80 kDa) that correlates with an aggressive PCa phenotype was measured [59,60], revealing that coculture of epithelial and stromal cells influenced shedded E-cadherin secretion.

To investigate the tumor-promoting role of CAFs, [61,62], Pereira et al. developed a 3D model based on prostate cancer microtissues [63]. Such microtissues were made of PCL scaffolds densely seeded by normal or cancer-associated fibroblasts (collected from human prostate tissues and matched from the same patients), which extensively deposited a cell type-specific ECM to form a biphasic construct (Fig. 12.7). Fibroblasts grew over three weeks with a two- to threefold increase and maintained a high viability (>90%). In order to understand the effect of fibroblasts on prostatic cells, the shape change observed in

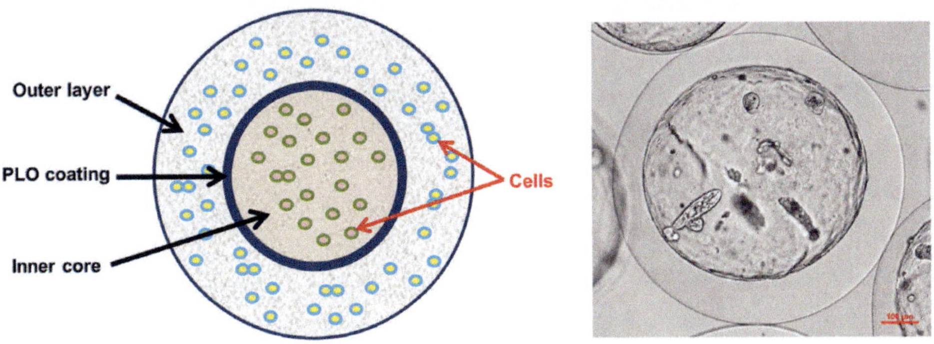


**FIGURE 12.6** Model of double-layered alginate hydrogel microsphere. Left is a cartoon model of double-layered microsphere. Inner core and outer layer were separated by PLO coating as shown. Cells growing in separate layers were indicated with red arrows. Right is an actual microsphere observed under light microscope. Stromal cells were grown in the inner core for 7 days, and outer layer is blank. Source: *Reproduced from Fang X, et al. Novel 3D co-culture model for epithelial-stromal cells interaction in prostate cancer. PLoS One 2013;8(9):e75187; use permitted under the Creative Commons Attribution License CC BY 4.0 ©.*

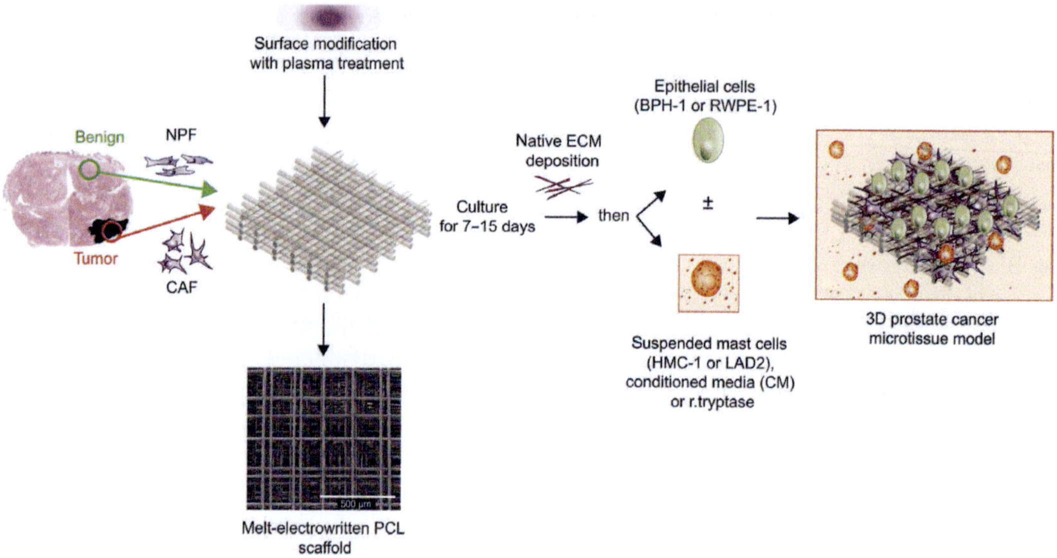


**FIGURE 12.7** Schematic of 3D prostate cancer microtissue model. Primary human prostatic fibroblasts (either NPF or CAF) were first cultured for 7–15 days in plasma-treated melt electrowritten poly(ε-caprolactone) (PCL) scaffolds. The fibroblasts deposited native ECM throughout the scaffold, forming a cell type and patient-specific 3D tissue engineered microtissue. Tagged benign epithelial cells (BPH-1 or RPWE-1) were seeded onto of the stromal microtissues to mimic the epithelial–stromal interface in prostate tissue in vivo. Mast cells (HMC-1), mast cell conditioned media (CM), or recombinant tryptase (r.tryptase) were added to the culture system to simulate different tumor microenvironment (TME) conditions. Source: *Reproduced with permission from Pereira BA, et al. Tissue engineered human prostate microtissues reveal key role of mast cell-derived tryptase in potentiating cancer-associated fibroblast (CAF)-induced morphometric transition in vitro. Biomaterials 2019;197:72–85. © Elsevier (2019).*

benign prostatic epithelia growing on the microtissues was monitored through confocal microscopy. Epithelial cell lines BPH-1 and RWPE-1 cultivated on normal fibroblast microtissues retained their epithelial conformation and a small, rounded cytoplasm with few cellular protrusions. However, the same cells growing on CAF microtissues showed a larger size, an elongated cytoplasm, and an orientation that followed the underlying stromal cell alignment. The role of mast cells, known to support tumor growth through the modulation of epithelial–stromal cross-talk, was also addressed using microtissues. Mast cells (HMC-1) further promoted malignant transformation of prostatic epithelial cells that displayed a more elongated and less-rounded cell-shape. HMC-1–conditioned medium led to similar results, suggesting that this effect was due to soluble factors. Recombinant human βI-tryptase alone recapitulated the effect of mast cells, thus suggesting a key role of mast cell-derived tryptase in potentiating CAF-induced morphometric transition in vitro [63]. Similarly, Clark et al. performed coculture experiments of patient-derived normal prostatic fibroblasts and CAFs with benign prostate epithelium followed by in vivo implantation, showing changes in the morphology, motility, and phenotype of BPH-1 cells as monitored by immunofluorescence and quantitative cell morphometric analyses [64].

### 12.2.2 Three-dimensional models to address androgen-mediated biology

Androgens [testosterone and dihydrotestosterone (DHT), its more biologically active form] play a critical role in the development of male phenotype and in the initiation and progression of PCa [65]. In order to mediate their function, androgens must bind to the androgen receptor (AR), a nuclear transcription factor that, in the absence of a ligand, is cytoplasmic and associated with heat-shock and chaperone proteins. Binding of androgens to the AR results in conformational change, dissociation of interacting proteins, and exposure of the nuclear localization signal responsible for nuclear import [65]. Upon translocation to the nucleus, the androgen/AR complex dimerizes and binds to androgen response elements, modulating transcription of downstream genes, such as prostate-specific antigen (PSA) [66]. PCa cells depend on AR activation to support their survival and growth [67]. Thus, androgen deprivation therapy represents a first-line treatment for patients with PCa, although many patients eventually develop castration-resistant prostate cancer (CRPC), usually associated with metastatic progression. "Next-generation" hormonal therapies abiraterone and enzalutamide have benefitted 45%–65% of patients with metastatic CRPC, with median responses of 9–15 months. However, resistance often arises, and limited or no benefit is experienced by the remainder of the patients [5,6].

The development of PCa cell lines expressing AR has been challenging, since the vast majority of current models are AR negative. LNCaP is the most common AR-positive cell line, and a few other AR-expressing cell lines have been developed, such as MDA PCa 2a, 2b and LNCaP sublines with higher and more stable AR expression, VCap and LuCap [11]. Zhau et al. used rotary wall vessels to coculture fibroblasts and LNCaP cell in microgravity-simulated conditions. These cells responded to DHT in an in vivo-like fashion [68].

Centenera et al. established patient-derived explants (PDE) from freshly resected prostate cancer samples [32]. PDEs cultured on gelatin sponges preserved a morphology similar to original tumor tissues, including maintenance of AR expression and stromal components, for up to 6 days (Fig. 12.8). PDEs treated with AR antagonist bicalutamide showed a heterogeneous response in terms of proliferative index (measured by Ki67 staining), with reduction, increase, or no effect in 44%, 26%, and 30% of the samples, respectively, and similar outcomes were observed upon application of enzalutamide. As a result, PDEs mimicked the heterogeneity of clinical response, highlighting the importance of using a model that displays a range of different phenotypes [32]. When PCa evolves toward a castration-resistant phenotype, tumor cells develop resistance to AR inhibition through different mechanisms, including the emergence of an AR variant defined as V7, which is constitutively active. Zadra et al. developed IPI-9119, an inhibitor that targets fatty acid synthase (FASN), an enzyme involved in de novo fatty acid synthesis overexpressed during PCa progression. IPI-9119 reduced the growth of AR-V7–driven organoids and enhanced the efficacy of enzalutamide in CRPC cells, as further confirmed in vivo. Since FASN/AR-V7 double-positive metastases were found in 77% of patients treated with enzalutamide and/or abiraterone, these results support the application of FASN inhibitors in castration-resistant PCa overexpressing AR-V7 [69]. Chua et al. generated organoids from several mouse models of PCa, which exhibited more normal or solid morphology depending on their oncogenic potential, and focused on $Nkx3.1^{CreERT2/+}$;

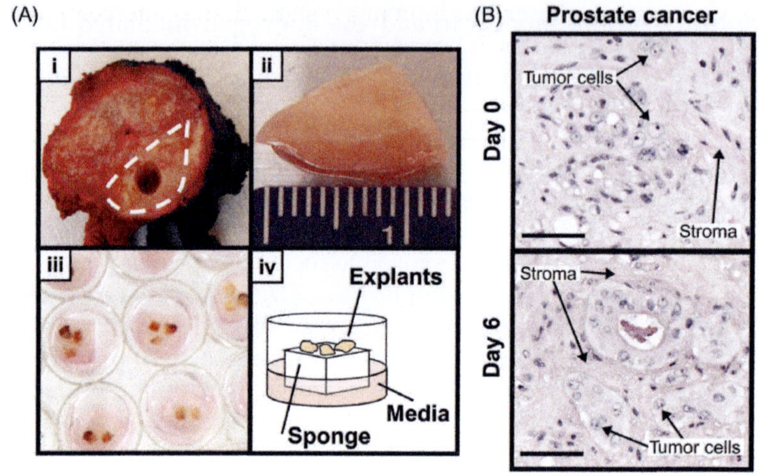


FIGURE 12.8 PDE culture sustains tissue morphology, viability, and endocrine signaling. (A) PDE tissue culture method. (i) Following surgery, a core of tumor tissue is removed by a pathologist (tumor area demarcated by broken white line), (ii) the tumor sample is dissected into 1-$mm^3$ fragments, (iii) cultured in 24-well plates on a gelatin sponge sitting in media, allowing direct comparison of treatments in matched tumor tissue, and (iv) schematic diagram of PDE setup. (B) Representative hematoxylin and eosin staining of PDEs from primary prostate tumors, showing maintenance of gross morphology following 6 days in culture. Arrows indicate examples of tumor cells and surrounding stroma. Source: *Reproduced from Centenera MM, et al. A patient-derived explant (PDE) model of hormone-dependent cancer. Mol Oncol 2018;12(9):1608–22; use permitted under the Creative Commons Attribution License CC BY 4.0 ©.*

$Pten^{flox/flox}$; R26R−YFP/ + (NP) mouse-derived organoids to test the response to enzalutamide [70]. While enzalutamide alone did not affect organoid biology, combination with MK-8669 inhibitor that reduce tumor progression in castration-resistant disease in vivo, inhibited organoid formation [70].

### 12.2.3 Three-dimensional models for prostate cancer stem cells

Stem cells are undifferentiated cells characterized by self-renewal properties and the ability to generate multiple lineages. These cells are crucial for the maintenance and repair of normal tissues and usually reside in specialized niches, which help control the balance between proliferation, differentiation, and quiescence [71]. Depending on the specific function and architecture of an organ, stem cells can remain quiescent and reactivate in specific cases (e.g., in muscle, upon injury) or support continuous tissue turnover (e.g., in the hematopoietic system, giving rise to mature blood cells) [71]. The adult prostatic stem cell population, considered quiescent, is characterized by expression of CD44, CD133, and $\alpha2\beta1$ integrin and localizes within the basal epithelial layer [72,73]. Some evidence suggests that tumor growth is linked to dysregulation of the stem cell compartment within the prostate [74].

Several 3D in vitro models have been developed in an attempt to recreate both the normal and dysfunctional self-renewing cell populations. Lang et al. modeled prostate stem cell (PSC) biology using patient-derived human prostatic tissues and compared the growth potential of stem cells in 2D versus 3D conditions [75]. Whereas monolayer cultures supported the expansion of PSCs from single cells, spheres originated in nonadherent plates maintained a quiescent phenotype. Besides culturing PSCs, Rao et al. developed a method to rapidly expand PCa stem cells in 3D microcapsules characterized by a liquid core and an alginate hydrogel shell (Fig. 12.9) [76]. Encapsulation was obtained through coaxial electrospray technology based on the injection of a solution with PCa cells and a fluid of sodium alginate into the inner and outer lumens of the coaxial needle, respectively. Under an electrostatic field, drops originating at the tip of the coaxial needle were micronized and sprayed into a 100 mM calcium chloride solution to support the polymerization of the

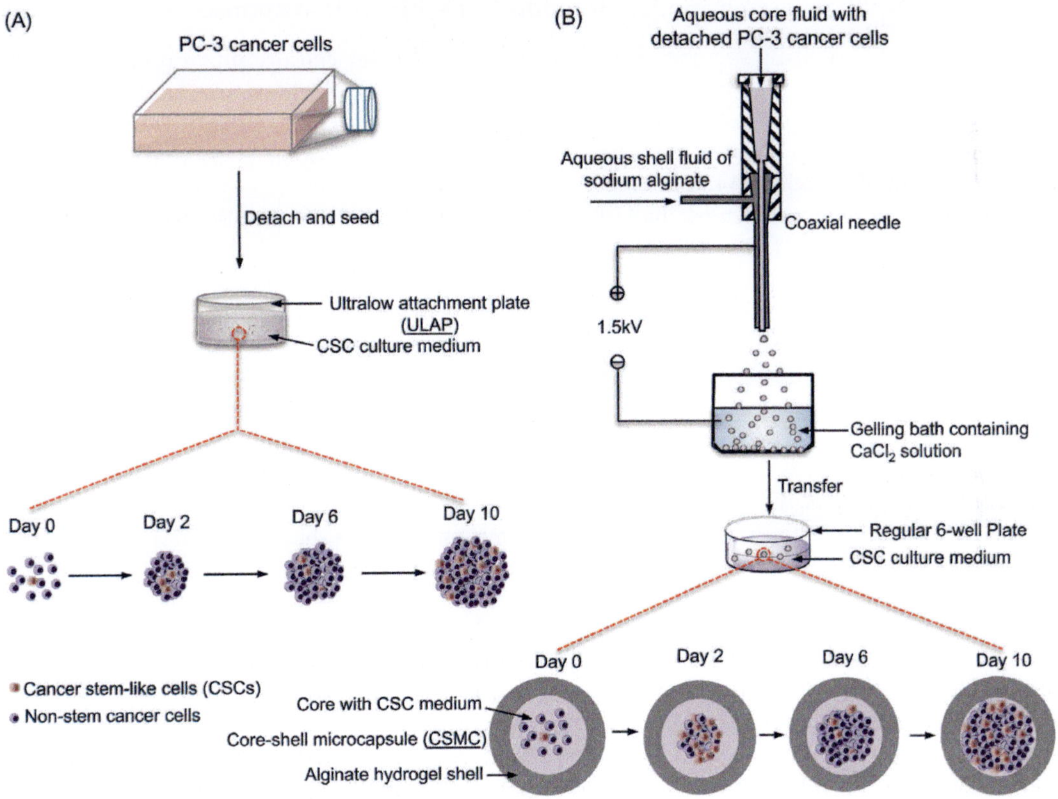


FIGURE 12.9 A schematic illustration of conventional culture in open bulk CSC medium in ultralow attachment plate (ULAP) versus the new miniaturized 3D culture in CSC medium in the core of core–shell microcapsules (CSMCs) showing the semiclosed nature of the CSMC culture as a result of the semipermeable hydrogel shell. (A) Conventional culture in ULAP. (B) Miniaturized 3D culture in CSMCs: the cell-laden CSMCs were produced by coaxial electrospray using coaxial needle in one step and transferred into regular six-well plate for culture in CSC medium. The percentage of CSCs enriched in the resultant spheroids is very much dynamic (i.e., time dependent and lowest on day 6) and it is the highest in spheroids obtained from the CSMC culture for 2 days. Source: *Reproduced with permission from Rao W, et al. Enhanced enrichment of prostate cancer stem-like cells with miniaturized 3D culture in liquid core-hydrogel shell microcapsules. Biomaterials 2014;35(27):7762–73. © Elsevier (2014).*

sodium alginate into a calcium alginate hydrogel. This system allowed for higher number of spheroids formed, in a shorter time span compared with conventional low-attachment plates, with a richer stem cell population based on the expression of surface-specific markers and gene/protein profiling. Also, spheroids obtained through the 3D culture showed an increased in vivo tumorigenicity upon injection of cells into NOD/SCID mice compared with conventional culture methods based on low-attachment plates [76]. Similar studies have been conducted by Guo et al. using rubber micromolds in conjunction with agarose to create microgrooves that engender the formation of spheroids, which expressed higher levels of CD44 and CD133 proteins than 2D cultures [77].

### 12.2.4 Three-dimensional models to address therapeutic response

The 3D culture systems support more physiological cancer cell organization and signaling response, two critical features that, besides tumor growth, regulate mechanisms of sensitivity or resistance in drug testing. 3D in vitro models have been applied to the study of multiple therapies and have highlighted important differences in therapeutic response compared with 2D models [14,78].

PDX-derived organoid systems have been applied to test different anticancer agents [79]. Watt et al. demonstrated that inhibition of CD36, a fatty acid transporter associated with aggressive disease, slowed down cancer progression in vivo [80]. Cotargeting of CD36 and C75, a fatty acid synthase inhibitor, in organoids further ablated proliferation by 95%, suggesting a new combinatorial strategy to apply in patients. Similarly, Marzi et al. developed three clinical indenoisoquinolines, noncamptothecin TOP1 inhibitors, and tested their efficacy in homologous recombination-deficient organoids (harboring BRCA2 mutations), showing hypersensitivity of these tumor cells to all drugs tested, and providing a rationale for using these agents in patients with homologous recombination defects [81]. Beshiri et al. further confirmed the suitability of BRCA2-deficient LuCaP-derived organoids to respond to PARP inhibitor olaparib, with results similar to those in patients, suggesting that PDX-derived organoids recapitulate in vivo response and provide a solid platform to test therapeutic response [82].

Xu et al. set up a 3D in vitro model to test the efficacy of nanoparticles on LNCaP cells cultured in hydrogel to generate tumorspheres [23]. The hydrogel-based model was made of HA cross-linked to acrylate groups (HA-AC) or reactive thiols (HA-SH), defined by a porosity similar to the ECM of the tumor environment. Doxorubicin was loaded in polymeric nanoparticles and administered to 3D cell cultures. LNCaP cells embedded in gels showed higher expression of multidrug resistance proteins (such as MRP1) and increased resistance to docetaxel treatment compared with cells cultured in 2D. However, greater efficacy of doxorubicin-loaded nanoparticles was shown compared with free doxorubicin, due to their ability to bypass the drug efflux function of MRP1.

Wang et al. further tested a nanoparticle-based strategy for targeted gene delivery in 3D culture conditions [83]. TrampC-2 mouse PCa cells were cultivated in a 3D scaffold made of a polyelectrolyte chitosan–alginate (CA) complex that partially reproduces the glycosaminoglycan composition of the ECM matrix (Fig. 12.10). Upon seeding on the scaffolds, TrampC-2 cells formed tumorsphere colonies, characterized by lower proliferation rate, higher malignancy, and expression of genes related to EMT, such as Twist, Snail, and

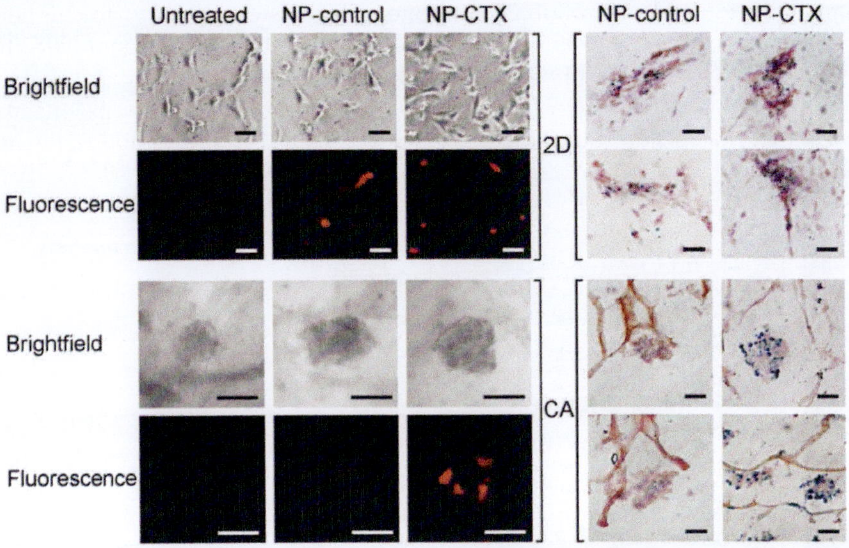


FIGURE 12.10 In vitro evaluation of targeted delivery of RFP plasmid to TrampC-2 cells in 2D cultures and CA scaffolds using control or chlorotoxin nanoparticles (NP-control; NP-CTX). Brightfield and fluorescence images of NP-control and NP-CTX transfected TrampC-2 cells growing in 2D cultures and CA scaffolds. Scale bars represent 100 μm. NP-control and NP-CTX were also detected by staining with Perls' Prussian blue. TrampC-2 cells were counterstained with nuclear fast red. Scale bars correspond to 100 μm. Source: *Reproduced with permission from Wang K, et al. 3D porous chitosan-alginate scaffolds as an in vitro model for evaluating nanoparticle-mediated tumor targeting and gene delivery to prostate cancer. Biomacromolecules 2015;16(10):3362–72. © American Chemical Society (2015).*

Sip1, compared with the 2D culture. Cells cultured on 2D plates showed no difference in gene delivery between targeted [chlorotoxin (CTX)] and nontarget control nanoparticles, whereas targeted gene delivery was observed in cells cultured in CA scaffolds [83].

Stock et al. compared different 3D culture methods to study the response of LNCaP cells and CAF to MDV-3100, a second generation anti-AR therapy, and standard chemotherapy (docetaxel) [78]. Spheroid generation occurred in low-adhesive conditions or through an alginate bioreactor (Fig. 12.11). Part of the resulting 3D aggregates were cultured in Matrigel, collagen I, or a 1:1 mixture, and all the conditions were compared with 2D cultures. Results showed a model-dependent drug response and partial resistance in MDV-3100 3D treated samples, with cultures in Matrigel less sensitive to growth inhibition [78].

Spheroids have also been used to study response to radiation therapy. Camus et al. implemented the use of surface-enhanced Raman spectroscopy (SERS) to evaluate the viability of PC3 multicellular spheroids after exposure to radiation treatment. SERS is a Raman spectroscopy technique that uses a noble metal (gold or silver) to enhance the Raman signal applied to the quantification of variables such as the pH and redox potential of living cells. Exposure to radiation treatment administered to PC3 spheroids following different schedules of radiation (single dose or fractionated) showed that PC3 cells responded with an alteration of the cellular pH (higher or lower depending on the dose) and a change in the shape of spheroids, for example, an altered, less-rounded morphology with fragmentation and erosion of the outer layer [84].

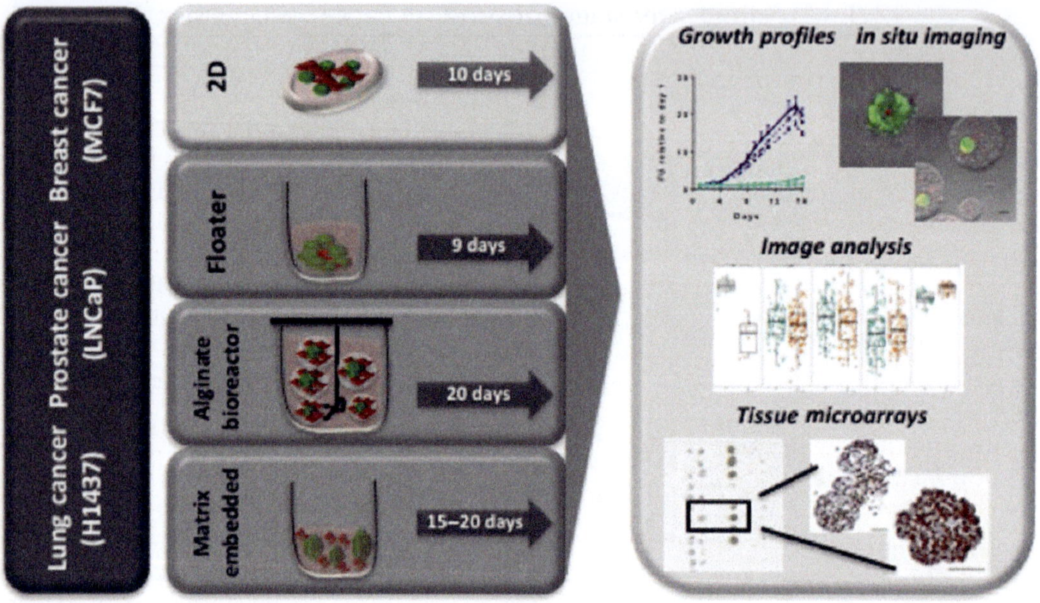


**FIGURE 12.11** The PREDECT 2D/3D model workflow. Three cell lines (MCF7, LNCaP, and H1437) belonging to three pathologies (breast, prostate, and lung) were set up as 2D or 3D models in mono- and cocultures. Growth was monitored via fluorescence, and the cultures imaged in situ by confocal microscopy and analyzed via dedicated 2D/3D imaging software, or fixed, paraffin embedded, and processed into TMAs and then analyzed via immunohistochemistry (IHC). Source: *Reproduced from Stock K, et al. Capturing tumor complexity in vitro: comparative analysis of 2D and 3D tumor models for drug discovery. Sci Rep 2016;6:28951; use permitted under the Creative Commons Attribution License CC BY 4.0 ©.*

Another potential treatment for PCa is represented by immunotherapy. Cancer cells can evade immune system surveillance using different strategies, including immune-editing [85]. The recognition of PCa cells by immune cells is efficient in a 2D setting, but when cancer cells are arranged in 3D structures, activated peripheral blood lymphocytes display reduced affinity toward cancer cells and reduced killing potential [86]. Thus, the development of 3D in vitro models would increase relevance for testing immunotherapy in vitro. Florczyk et al. created a porous 3D chitosan-alginate (CA) scaffold that mimics the structure of the ECM, mostly hyaluronan, the most abundant glycosaminoglycan of the ECM. LNCaP, C4-2, and C4-2B human PCa cell lines grown on CA scaffolds formed tumorsphere clusters characterized by a slower growth rate and higher malignancy compared with 2D and the 3D Matrigel cultures [24]. These systems were applied to analyze the interactions between peripheral blood lymphocytes and cancer cells through scanning electron microscopy. Natural killer (NK) and NK-T cells (defined as $CD3^{-}/CD56^{+}$ or $CD3^{+}/CD56^{+}$ cells, respectively) present among peripheral blood lymphocytes could interact with tumorspheres in Matrigel or on CA scaffolds, showing a positive correlation between the number of NK cells within the tumoroids and the malignancy of PCa cells, with the latter providing a more robust platform for analysis, including scanning electron microscopy, immunohistochemistry, and cell sorting [24].

# 12.3 Modeling early stages of prostate cancer progression

## 12.3.1 Modeling tumor invasion

Besides addressing the pathophysiological aspects of primary tumor growth and therapeutic response, 3D in vitro biomimetic models have been further applied to model subsequent stages of tumor progression, including mechanisms of invasion by PCa cells. During oncogenesis, dysregulated tumor cell proliferation is often coupled with increased cell motility due to altered intracellular signaling, loss of epithelial characteristics, acquisition of more mesenchymal features, and production of promigratory factors by the surrounding stromal cells [87]. As a result, single cells or groups move along tissue structures in a complex and organized fashion that involves cytoskeleton remodeling and formation of cell–cell and cell–ECM interactions [88]. This multistep process ultimately enables cancer cells to enter circulation and colonize distant organs.

To recapitulate the tumor invasion front, Yong et al. developed a microfluidic device with three parallel molded channels (defined as A, B, and C) included in a collagen gel, where PC3 and DU145 PCa cells were positioned to recreate a tumoroid of approximately 500 μM in diameter [89]. A fluid flow was pumped through a channel parallel to the tumoroid to provide nutrients and mimic the interstitial flow and pressure present in living tissues (Fig. 12.12). The invasion of cancers cells into collagen was promoted by increased fluid pressure. Furthermore, an increased collagen concentration (from 1 to 2.5 mg $mL^{-1}$) reduced the margin of invasion, but in this case fluid pressure did not affect invasion. Tumor cells at the invasive front showed a higher expression level of Ki67, a marker of cell proliferation, compared with the cells in the bulk, similar to in vivo experiments [89].

Bjork et al. recapitulated the progressive invasion of PC3 cells in Matrigel, which initially formed acinar structures organized as round cell clusters that dedifferentiated and invaded through time [90]. This system was applied to address the role of heat-shock

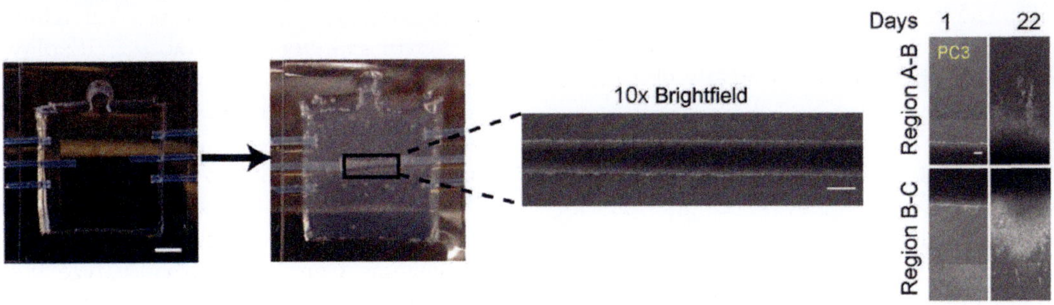


FIGURE 12.12 Long-term tumoroid culture recapitulates the invasion front from a tumor mass. Overview of tumoroid culture. Syringe needles were placed in an empty device (left) through three channel openings; scale bar = 5 mm. Fluidic channels were formed after injection and polymerization of 1 mg $mL^{-1}$ collagen gel (middle). Cancer cells were then introduced into the middle channel before both ends of the channel was sealed to form a tumoroid (right); scale bar = 500 μm. Formation of invasion front in PC3 tumoroids over 3 weeks. Scale bars = 100 μm. Source: *Reproduced from Aw Yong KM, et al. Tracking the tumor invasion front using long-term fluidic tumoroid culture. Sci Rep 2017;7(1):10784; use permitted under the Creative Commons Attribution License CC BY 4.0 ©.*

factor 2 (HSF-2), a protein involved in cell mitosis to protect cells against proteotoxicity, which is downregulated in cancer cells [91]. HSF-2 expression was higher during the formation of the acinar structures and progressively decreased during invasion, suggesting a role for this factor in maintaining the luminal phenotype and promoting epithelial maturation, as confirmed by gene expression analysis and down-regulation experiments. In line with these findings, low HSF2 expression in patient samples correlated with advanced disease stages, metastasis, and poor survival [90,92].

Since different PCa subsets could display diverse invasive functions, Harma et al. accounted for phenotypic heterogeneity by establishing a comprehensive panel of 29 transformed and non-transformed cell lines and monitored growth, invasion, and therapeutic response in 3D gels (Fig. 12.13) [93]. Normal prostate epithelial cells formed well-differentiated round spheroids

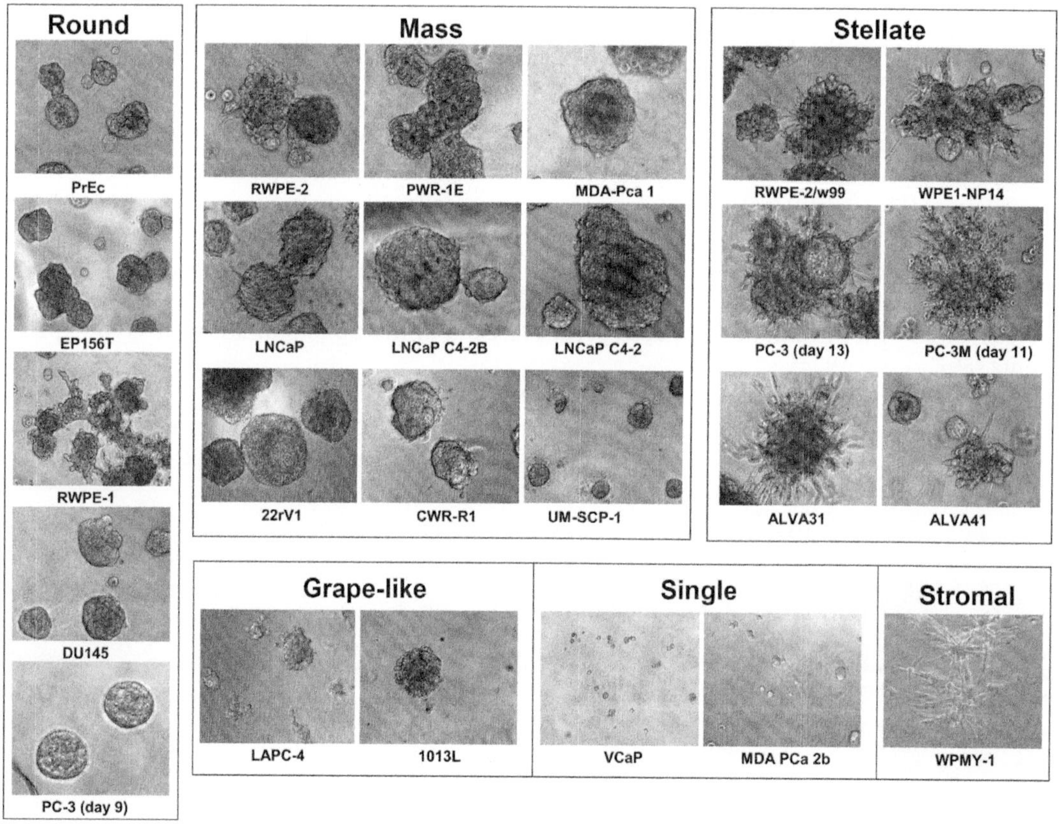


FIGURE 12.13 Representative phase contrast images of spheroids formed in 3D Matrigel culture. Most structures fell into the categories round-, mass-, stellate-, or grape-like. Some cell lines failed to form spheroids in Matrigel but persisted as single live cells for up to 2 weeks (single phenotype). Most spheroids were imaged at days 9–10 after inoculation. PC-3 spheroids, shown as round phenotype at day 9, undergo a metamorphosis to invasive/stellate phenotype at day 13. The same occurs to PC-3M at an earlier time point (days 5–8). Source: *Reproduced from Harma V, et al. A comprehensive panel of three-dimensional models for studies of prostate cancer growth, invasion and drug responses. PLoS One 2010;5(5):e10431; use permitted under the Creative Commons Attribution License CC BY 4.0.*

with epithelial polarization, lumen formation, a complete basal lamina and retained cell–cell contacts. PCa cells, instead, formed poorly differentiated spheroids that displayed invasive structures characterized by PI3-kinase, AKT, STAT/interferon, and integrin signaling pathway activation. Targeting of PI3-kinase by small-molecule inhibitors reduced invasive cell growth in PCa cells, while it was not effective in normal epithelial prostate cells or in PCa in 2D cultures [93].

In addition to gel-based systems, chip devices have also been used to study tumor cell invasion. Liu et al. addressed the migratory potential of different PCa cell lines in a 3D microfabricated landscape [94]. This chip consisted of four by four arrays, with each array constituted by 100 Tepuis, described as parallelepipeds 320 μm high and 100 × 100 μm wide (Fig. 12.14). PC3 and LNCaP cells seeded at the bottom of the chip were monitored over time for their colonization of the top of the Tepuis by confocal microscopy. As expected, the invasion potential of PC3 cells was higher than LNCaP, which failed to populate the Tepui top, suggesting that this phenomenon was due to cell contact inhibition [94].

Irimia et al. applied microfluidic devices to model invasion guided by tissue structures, such as nerves, lymphatic vessels, or other tissue-tracks [95]. Tumor cells were seeded in microchannels to generate a mechanical constraint, and their migration was measured over several hours. Cell motility occurred in the absence of external stimuli, suggesting intrinsic mechanisms mediated by mechanical confinement, with limited effects exerted by

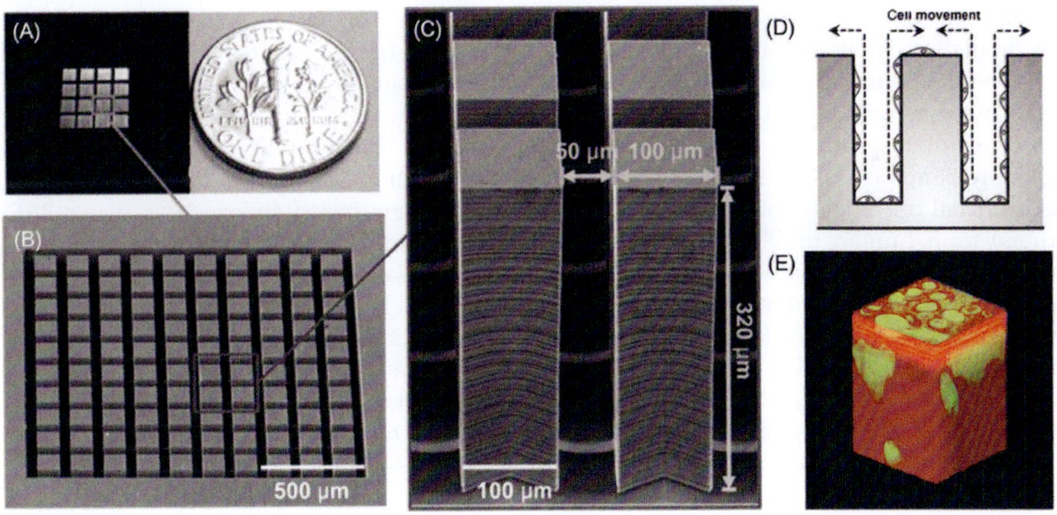


**FIGURE 12.14** The 3D microfabricated landscape. (A) Photo of the whole chip consisting of 4 × 4 arrays. A US 10-cent coin is shown on the same scale on the right. (B) SEM image corresponding to a single array in A consisting of 100 Tepuis. (C) SEM image of individual Tepuis; they are 100 × 100 × 320 μm in size and 50 μm apart (Fig. 12.2). (D) The chip is incubated, and confocal microscopy is used to observe the movement of the motile PC-3 and LNCaP cells up along the sides of the Tepuis toward the tops. (E) Three-dimensional reconstruction of an image acquired at the confocal microscope showing PC-3 cells attached to the sides and densely packed on the top. The green channel represents fluorescence of GFP expressed within the cells, and the red channel represents the direct reflection from the surface of the chip. Source: *Adapted with permission from Liu L, et al. Probing the invasiveness of prostate cancer cells in a 3D microfabricated landscape. Proc Natl Acad Sci U S A 2011;108(17):6853–56.*

agents targeting microtubules [95]. Besides employing printed devices or nontissue-specific matrices, Cazzaniga et al. used prostate tissue-derived ECM to mimic the prostate environment in invasion studies [96]. Decellularized prostate specimens from men undergoing radical prostatectomy provided the ECM support to seed primary PCa cells obtained from patients with different Gleason scores and disease stages, normal epithelial cell lines (RWPE-1, BPH-1), and PC3. As a result, epithelial cells repopulated the decellularized tissue with a morphology that resembled the normal prostate, while PCa cells invaded the matrix, with patient-derived tumors behaving accordingly to their pathological stage [96].

### 12.3.2 Modeling angiogenesis and the contribution of vessels to tumor progression

Angiogenesis, the formation of new blood vessels, is a necessary step in many physiological and pathological processes, including cancer [97,98]. Like any other tissues, tumors require oxygen and nutrient supply and waste product removal. Induction of tumor angiogenesis occurs at initial stages of development and is caused by proangiogenic stimuli, such as soluble factors or hypoxia [99]. Tumor vessels develop rapidly and irregularly, due to the overproduction of angiogenic factors, resulting in a tortuous and fenestrated vascular network, which differs from the more organized healthy counterpart [99]. Besides sustaining tumor growth, angiogenesis further supports the dissemination of tumor cells in circulation, ultimately contributing to the colonization of distant sites [98].

The establishment of 3D in vitro systems that provide insight into the role of endothelial cells in the progression of PCa has been limited by technical issues, for example, the inability to maintain primary endothelial cells in 3D coculture, the instability of natural matrices, and problematics associated with experimental monitoring [100]. In order to model PCa-endothelial cell interaction, Seano et al. developed a system based on arterial explants (AR rings) from human umbilical cord, cultivated ex vivo in a 3D basement membrane gel, under vascular endothelial growth factor (VEGF) stimulation [101]. This led to an angiogenic outgrowth with the formation of endothelial vessels surrounded by pericytes. This system was suitable to address the effect of antiangiogenic drugs, but also for studying the interaction between PCa cells and the endothelium. Indeed, the cocultivation of LNCaP and AR rings supported the development of capillary-like structures, thus confirming the hypothesis that tumor cells can release proangiogenic factors, such as VEGF [101].

Bray et al. developed a 3D bioengineered system based on cocultivation of PCa cells, human umbilical vein endothelial cells (HUVEC) and mesenchymal stem cells (MCSs) in matrix metalloproteinase (MMP)-sensitive four-arm star-shaped poly(ethylene glycol) (starPEG)-heparin hydrogels that allow local remodeling by cancer cells and favor invasion (Fig. 12.15) [100]. VEGF, FGF-2, and SDF-1 angiogenic factors were added, and RGD motifs, which were incorporated at a defined density, provided binding sites for integrins. PC3 and LNCaP cells grown in the starPEG-heparin hydrogel system displayed increased vascular recruitment and more organized vascular tubes. These models were less sensitive to chemotherapy treatment compared with 2D cultures, suggesting a role for endothelial cells in mediating chemoresistance [100].

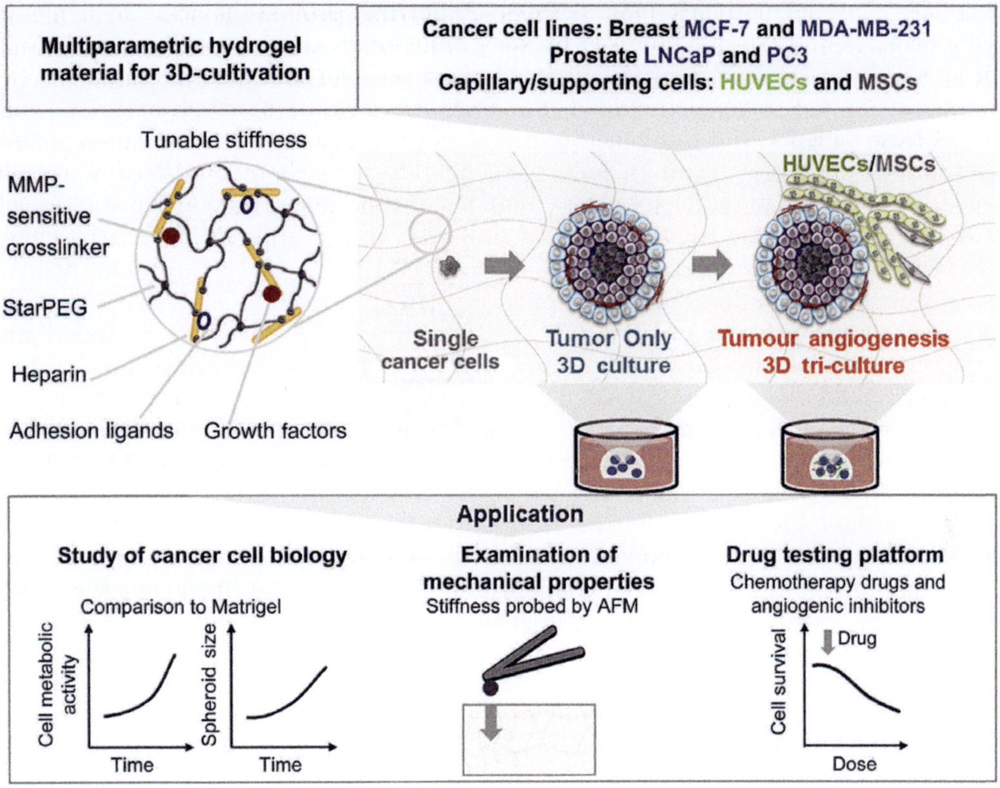


FIGURE 12.15 Multiparametric hydrogels for the development of tumor angiogenesis microenvironments. Source: *Reproduced with permission from Bray LJ, et al. Multi-parametric hydrogels support 3D in vitro bioengineered microenvironment models of tumour angiogenesis. Biomaterials 2015;53:609–20. © Elsevier (2015).*

A similar study was performed by Taubenberger et al., which applied a matrix metalloproteinase-degradable biohybrid poly(ethylene) glycol-heparin hydrogel [102]. This gel was functionalized with binding motifs for different ECM proteins (RGD, fibronectin/vitronectin; GFOGER, collagen I; or IKVAV, laminin-111) to address their effects on cocultures of cancer cells (PC3 and LNCaP) and vascular structures (MSC and HUVEC). Less aggressive LNCaP cells were not affected by the presence of specific ECM molecules, while PC3 cell growth and endothelial cell infiltration was significantly enhanced in IKVAV and GFOGER gels, suggesting a role for collagen I and laminin-111 in supporting PC3 in vitro vascularization [102].

### 12.3.3 Isolation of circulating tumor cells

Tumor cells that escape the primary tumor, invade surrounding tissues, and successfully intravasate into blood vessels enter systemic circulation as circulating tumor cells (CTCs) and are responsible for the formation of metastatic lesions [103]. Detectable levels of CTCs ($\geq$5 CTCs in 7.5 mL blood) can be identified at an advanced stage of disease and correlate with a poor prognosis [104]. The limited number of CTCs among millions of white and red blood cells in circulation represents a challenge for their detection, isolation, and subsequent analysis. Several strategies have been developed to isolate CTCs, based on immunolabeling, microfilters, affinity-based approaches, and microfluidic devices [105–107].

Augustsson et al. applied a label-free, microfluidic flow-through separation method based on acoustophoresis through forces generated by ultrasonic resonance, which can successfully discriminate between cancer and blood cells (Fig. 12.16) [105]. Recovery of PCa cells (DU145, PC3, and LNCaP) spiked in human blood ranged from 93.6% to 97.9% (97.4%–98.4% purity) for fixed cells and 72.5% to 93.9% (79.6%–99.7% purity) for nonfixed cells, with no damage induced by acoustophoresis [105]. Sarioglu et al. developed another label-free system capable of isolating CTC clusters, which are very often neglected since most of the technologies target single cells only [106]. This microchip device, called cluster-chip, could isolate tumor clusters formed by two or more cells from unprocessed blood. CTC clusters were separated through specialized bifurcating traps under low-shear stress conditions to preserve cellular integrity. Clusters of CTCs were identified in 31% ($\sim$0.28 clusters $mL^{-1}$) of metastatic PCa patients. Molecular analysis confirmed the epithelial origin of these groups of cells and further identified cluster-associated macrophages [106]. Kirby et al. developed a microfluidic device based on a geometrically enhanced differential immunocapture (GEDI) that combined an antiprostate-specific membrane antigen (PSMA) antibody with a 3D conformation suitable to capture PCa cells [107]. This system trapped a median of 54 tumor cells $mL^{-1}$ of patient blood, showing higher sensitivity compared with commercially available chips. In situ treatment of CTCs with chemotherapeutic agents that target microtubules (docetaxel and paclitaxel) revealed engagement of microtubules with the drugs in chemosensitive CTCs, while CTCs isolated from docetaxel-resistant patients did not show evidence of drug activity [107].

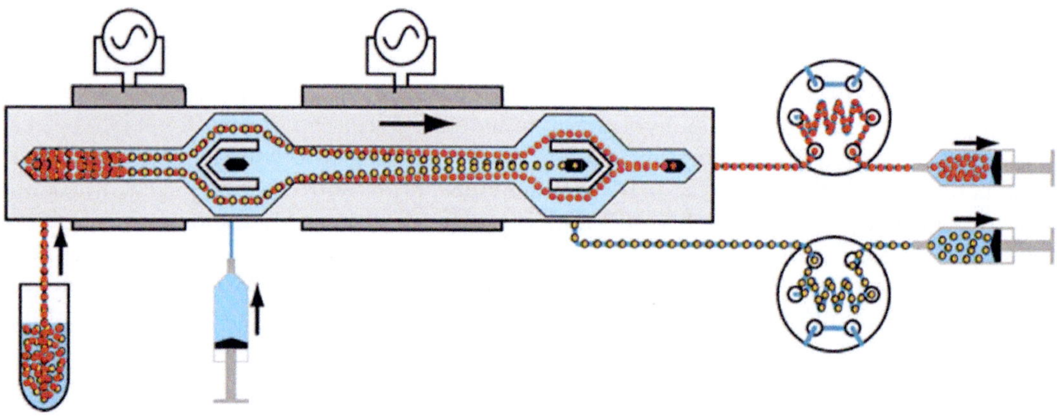

**FIGURE 12.16** Overview of the acoustophoresis microfluidic system. A schematic of the flow configuration for the acoustophoresis cell separation experiments is shown. Syringe pumps drive the flow in the outlets and in the central fluid inlet. Cell suspension is drawn from the bottom of a test tube by suction. The outlets are sampled via two sample loops, each with a volume of 100 μL. Source: *Reproduced with permission from Augustsson P, et al. Microfluidic, label-free enrichment of prostate cancer cells in blood based on acoustophoresis. Anal Chem 2012;84 (18):7954–62. © American Chemical Society (2012).*

### 12.3.4 Extravasation

After interacting with blood vessels, only a small fraction of CTCs survive into the blood stream and extravasate in distant organs to form metastases. Recently, De Menna et al. generated a microfluidic device to address PCa cell extravasation [108]. This chip-based system was patterned by standard photolithography with three compartments for cell seeding (HUVEC, hMSCs) and two microchannels for supplying medium. Vessels of 5–50 μm in diameter self-assembled in a network in 4–5 days and remained stable for up to 12 days. The lumen of vessels was perfused with PC3 cells, which extravasated over time as monitored by live imaging [108].

## 12.4 Modeling advanced stages of prostate cancer progression

Metastatic prostate cancer is a source of significant morbidity and a persistent clinical challenge, remaining lethal for 70% of patients at 5 years from the diagnosis. Bone represents the most common metastatic site (84%), followed by distant lymph nodes (10.6%) and liver (10.2%; [109]). Crucial aspects of the pathophysiology of metastatic PCa are still unclear and the response to selected therapeutic strategies, often inadequate, is followed by resistance and progressive disease relapse, which leads to detrimental consumption of the organism and death [3]. Thus, the implementation of models that account for the complexity of the niches where PCa metastases reside would help to identify the mechanisms that support disease progression and enhance the development of more effective therapeutic agents. The majority of 3D in vitro systems of metastatic PCa developed so far focus on bone. Only one mimics liver metastasis [110], and no models of other metastatic sites (e.g., lymph nodes) have been applied to the study of PCa [111]. The next paragraphs will describe the 3D systems available to model early metastatic colonization and progression.

### 12.4.1 Disseminated tumor cells

The onset of metastasis in patients with localized PCa can occur long time after prostatectomy, suggesting that tumor cells can escape prior to diagnosis and treatment. This further demonstrates that PCa cells that successfully extravasate to distant sites can survive and remain dormant for years as disseminated tumor cells (DTCs) before switching to a proliferative state and cause advanced progression [112–114]. Besides quiescence, an additional hallmark of DTCs is chemoresistance, since most of the anticancer therapy target mechanisms that are linked to cell proliferation. The molecular and physical determinants at the basis of PCa cell dormancy are still quite obscure and can be mediated by intrinsic and extrinsic factors. Grandhi et al. developed a 3D microenvironment based on Amikagel, an aminoglycoside-derived gel, and stromal cells to recapitulate, in one system, dormancy, reactivation from dormancy, and micrometastasis in different cancer types [115]. Although Amikagel was efficient in inducing arrest in the G0/G1 phase of the cell cycle followed by relapse in bladder and breast cancer cells, this system was not able to induce dormancy in PC3 cells. The mechanisms at the bases of this inefficiency are still unclear and can potentially depend on the aggressive nature of these cells. Preciado et al.

generated a platform based on stiff silica–PEG hydrogel particles, which inhibited cell proliferation and movement due to physical confinement and immobilization in a nondegradable environment. In these experiments LNCaP cells, once incorporated within the particles, remained metabolically active for 3 weeks, while other primary and cancer cells did not survive. Thus, this system can be applied to successfully induce dormancy-like status in PCa cells [116].

### 12.4.2 Three-dimensional models to address the biology of prostate cancer bone metastasis

A bone is a complex, rigid organ part of the vertebrate skeletal system that structurally supports the organism, protects soft organs, enables motility, produces blood cells and stores mineral, in particular calcium [113]. Bone is composed of a mineralized ECM and different resident cell types, which are actively involved in the establishment, growth, therapy response, and resistance of metastatic PCa cells. The expansion of tumor lesions in bone is usually associated with osteoblast and osteoclast activation which results in bone matrix remodeling and release of factors that sustain tumor growth, in the so-called "vicious cycle" of bone metastasis [113]. Other cell types that further support tumor progression in bone include endothelial cells, adipocytes, and osteocytes [117–119].

Osteoblasts, the cells responsible for bone matrix deposition, have been shown to sustain both cancer cell survival and resistance to therapy [120,121]. Sieh et al. explored the interconnection between cancer cells and osteoblasts developing a model based on polycaprolactone scaffolds seeded with human osteoblasts and further wrapped in an osteoblast-derived calcified sheet [122]. The scaffolds were fully colonized by osteoblasts that actively deposited mineralized matrix and remained viable over time. Both PC3 and LNCaP cells were cultured directly on this system. PC3 cells showed a spread and fibroblast-like morphology, while LNCaP formed more compact lesions integrated into the bone matrix. The expression of metalloproteinase MMP9 was higher in the PC3-osteoblast cocultures, while PSA was higher in LNCaP-osteoblast cocultures compared with LNCaP alone when cultivated in androgen-depleted medium. This model can thus be applied as a versatile tool for studying intercellular and cell–matrix interactions at a cellular and molecular level [122]. Bischel et al. integrated the coculture of PCa cells and osteoblasts in a microfluidic device with multiphoton detection of FAD, observing that osteotropic C4-2B cells displayed an increased protrusive phenotype and FAD fluorescent intensity compared with parental LNCaP cells, which were mediated by ROS production and decreased in the presence of antioxidants [123].

San Martin et al. applied a variety of bone mimetic models to identify the contribution of the reactive endosteal stroma to bone colonization by PCa cells [124]. A human 3D osteogenic organoid model was generated by culturing MSCs in osteogenic medium, which formed a hard, opalescent spheroid connected to the sides of the cell culture insert by fibrous tendrils. The central mass of the organoid was surrounded by a flat and compact layer of cells positive for osteocalcin, alkaline phosphatase, osteonectin, and tenascin-C, a hallmark of reactive stroma deposited during PCa progression and in sites of bone repair [124,125]. Upon seeding, VCaP and PC3 cells attached to tenascin-C high foci at the tendrils, which were degraded by PC3 osteolytic cells. Furthermore, experiments performed in Osteo Assay plates pretreated with calcium phosphate and coated with increasing concentrations of human tenascin-C showed early spreading and proliferation of VCaP cells, whereas control cultures exhibited a lag time and lower cellular density. These data supported a role for tenascin-C in mediating bone-PCa cell interactions. Immunohistochemistry analysis on patient samples revealed an association between tenascin-C and integrin α9 expression, and the inhibition of this integrin in VCaP cells ablated adhesion to tenascin-C–coated osteo-mimetic surfaces, suggesting that α9 is a cellular partner involved in interaction with tenascin-C. To further model the interplay between the reactive endosteum and PCa cells, an in ovo model was applied that consisted of a tenascin-C–coated trabecular bovine cube positioned in proximity to an organoid made of VCaP and prostate-derived MSCs on a chicken egg chorioallantoic membrane (CAM). By day 6, VCaP cells colonized the trabecular bone cubes, as identified by expression of human α9 integrin, with preferential migration to tenascin-C–coated bone fragments. These results support a role for tenascin-C in PCa cell colonization of bone [124]. Hsiao et al. fabricated a microfluidic platform for the coculture of PCa cells, osteoblasts, and endothelial cells within a niche-like structure. The device consisted of a microchannel separated into two layers by a semipermeable membrane; the upper microchannel was divided in 28 chambers for cell culture, while the lower one was continuous and contained culture medium. The microchannels and membrane were both pretreated with pluronic F108 to prevent cell adhesion. Upon injection of a mixture of fluorescent PC3 cells, HUVEC, and MC3T3 osteoblasts, multicellular spheroids formed within 1 day of culture and remained stationary during long-term cultures. These niche-like microenvironments preserved PC3 cell viability and supported a slower and more physiological growth, representing a promising platform for potential drug screening [126]. Osteocytes, terminally differentiated osteoblasts that reside in lacunae located within mineralized bone, represent another important player in the establishment of bone metastasis. Choudary et al. developed an engineered 3D in vitro model based on primary human osteocytes assembled in micron-sized microbeads stratified in a microfluidic device to mimic in vivo tissue (Fig. 12.17, [119]). Primary cells from patients with prostate lymph node metastasis integrated in the device impacted the morphology of osteocytes and affected the integrity of the underlying reconstructed bone, which showed increased apoptosis. Furthermore, PCa cells increased fibroblast growth factor 23 (FGF23) expression levels in osteocytes, as well as mineralization and alkaline phosphatase activity (two markers of bone forming lesions). These outcomes were not identified in traditional 2D cultures. As a result, this human model shows promise for investigating the role of osteocytes in metastatic PCa progression [119].

Adipocytes can also profoundly affect PCa biology in bone. These cells constitute 70% of bone volume, and growing evidence suggests that their abundance is correlated with disease progression [127]. Fat cells can stimulate PCa growth and survival through the release of growth factors and chemokines, the induction of lipid uptake, and the promotion of osteoclast activity [128]. However, few models have investigated the role of bone marrow adipocytes in the initiation and progression of PCa in bone. Herroon et al. used two systems to address the paracrine and direct effect of adipocytes on cancer cells [118]. The first 3D model was based on PC3 cells seeded on reconstituted basement membrane

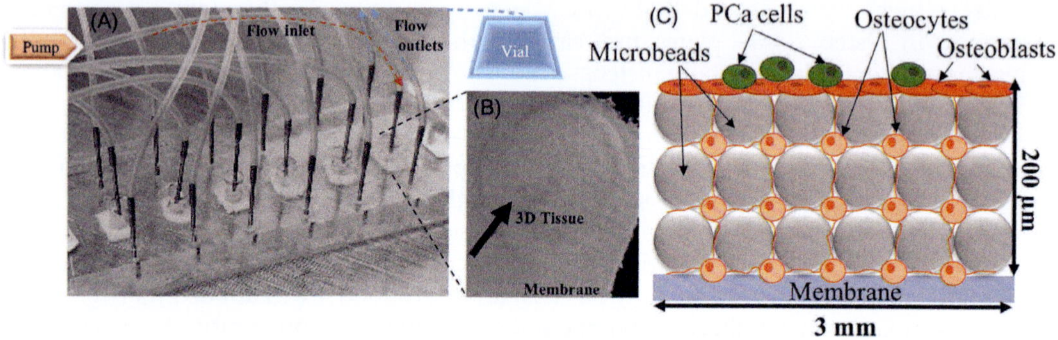


**FIGURE 12.17** Microfluidic perfusion device for engineering 3D bone tissues. (A) Actual device, containing bone tissue constructs in the central chamber with medium flowing into one inlet fed by a syringe pump and exiting through two outlets carrying effluent to a collection vial. Tissues were constructed using BCP microbeads and primary human osteocytes assembled at a 1:1 ratio. (B) Harvested 3D tissue. (C) Schematic illustration of the tissue constructs showing the location of the PCa cells, osteoblasts forming the endosteum, and the spatial distribution of osteocytes and microbeads (not drawn to scale). Source: *Reproduced from Choudhary S, et al. Human ex vivo 3D bone model recapitulates osteocyte response to metastatic prostate cancer. Sci Rep 2018;8(1):17975; use permitted under the Creative Commons Attribution License CC BY 4.0 ©.*

(BM)-coated coverslips cultured alone or in transwells with bone marrow-derived adipocytes. PC3 cells seeded as a single-cell suspension on BM gels grew as cell clusters. While cancer cells alone formed well-established and compact spheroids, cells exposed to adipocyte conditioning formed larger, more disorganized tumorspheres. In addition, the presence of paracrine signals from adipocytes stimulated PCa cell proliferation. When bone marrow macrophages were added to this model, they were able to infiltrate tumor spheroids more efficiently in the presence of adipocyte conditioning [118]. The second model was set up to examine direct interactions between PCa cells and adipocytes. Labeled PC3 cells distributed on top of a collagen I matrix invaded toward the bottom layer seeded with adipocytes, reaching them in about 96 h (Fig. 12.18). These two models provide a controllable, testable system for examination of adipocyte involvement in tumor progression in bone [118].

### 12.4.3 Three-dimensional models to address the therapeutic response of metastatic prostate cancer to bone

Besides driving disease progression, bidirectional communication between bone stroma and PCa cells is crucial in modulating therapeutic response [129]. Thus, the development of experimental systems that account for the diverse host cell partners and model the role of heterotypic cell–cell interactions has the potential to identify more clinically relevant treatment strategies for bone metastasis.

Paindelli et al. developed an in vitro 3D bone-mimetic environment (BME) by functionalizing polycaprolactone scaffolds with osteoblasts differentiated from human MSCs (Fig. 12.19) [130]. The resulting bone organotypic culture displayed 3D niche-like cellular distribution within scaffold fibers and bone matrix deposition, including calcium and

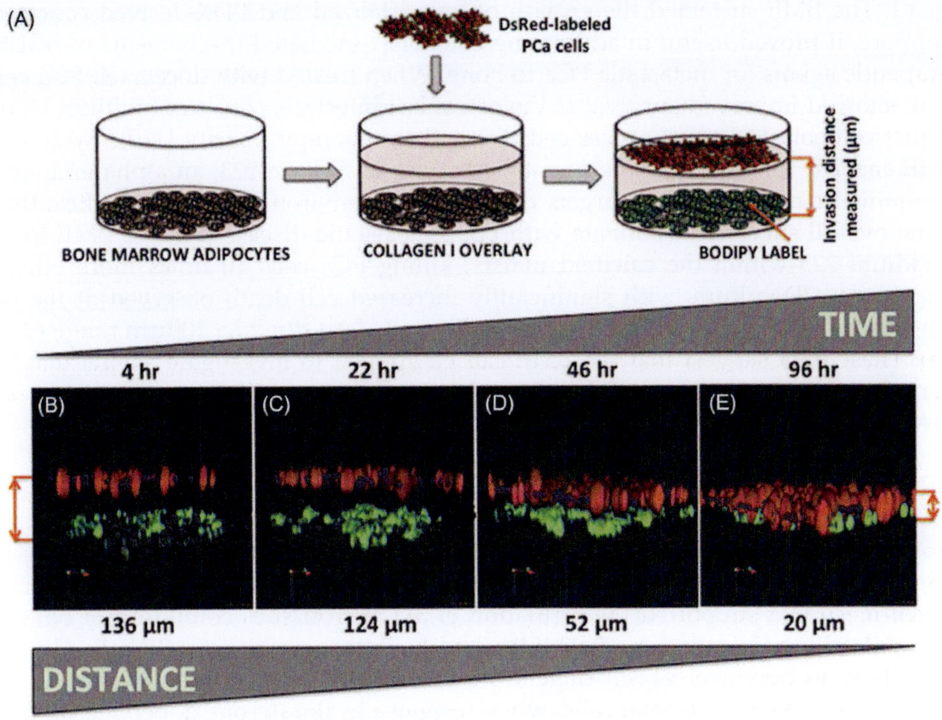


FIGURE 12.18 Three-dimensional invasion of tumor cells toward bone marrow adipocytes. (A) Diagram depicting the experimental setup. Tumor cells invaded through collagen I matrix toward bone marrow adipocytes. Adipocytes (BODIPY, green); Tumor cells (DsRed; red) for 4 h (B), 22 h (C), 46 h (D), and 96 h (E) and decreasing distance between the tumor cells and adipocytes over time is shown in micrometers. Green arrow (X), red arrow (Y), and blue arrow (Z) indicate orientation of the spheroid in 3D space. Source: *Reproduced from Herroon MK, Diedrich JD, and Podgorski I. New 3D-culture approaches to study interactions of bone marrow adipocytes with metastatic prostate cancer cells. Front Endocrinol (Lausanne) 2016;7:84; use permitted under the Creative Commons Attribution License CC BY 4.0 ©.*

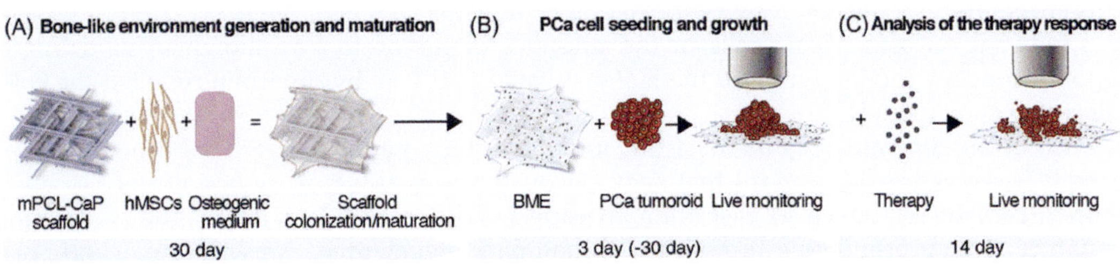


FIGURE 12.19 Schematic representation of the bone mimetic environment (BME) of metastasis. (A) Human mesenchymal stem cells (hMSCs) were seeded on a miniaturized medical grade, calcium phosphate polycaprolactone scaffold (mPCL-CaP) and maintained in osteogenic medium for 30 days that engendered osteogenic maturation. (B) PCa cells were seeded on the BME as tumoroids and monitored over time by live-cell fluorescence microscopy. (C) Established tumoroids were subjected to chemo- or radiation therapy and longitudinally monitored for therapeutic response by live-cell fluorescence microscopy. Source: *Reproduced with permission from Paindelli C, et al. Engineered bone for probing organotypic growth and therapy response of prostate cancer tumoroids in vitro. Biomaterials 2019;197:296–304. © Elsevier (2019).*

collagen I. The BME sustained the growth of immortalized and PDX-derived cancer cells. Furthermore, it proved useful in addressing osteoblast-mediated mechanisms of resistance to therapeutic agents for metastatic PCa to bone. When treated with docetaxel, PCa cells in the BME showed improved survival at the osteoblast interface, consistent with in vivo evidence that osteoblasts protect cancer cells from chemotherapy toxicity [120]. Additionally, the BME enabled testing of the therapeutic response to radium 223, an alpha emitter with calcium-mimetic properties that targets the bone microenvironment and significantly prolongs the overall survival of patients with bone metastatic disease [7]. The BME incorporated radium 223 within the calcified matrix, killing PCa cells 10 times more efficiently than in regular 2D culture, with significantly increased cell death observed at the osteoblast interface (where alpha particle exposure is maximized due to a 100 μm range of penetration). These data suggest that the BME can be applied to investigate agents that target different compartments (e.g., epithelial or bone matrix) to address treatment-specific mechanisms of response [130].

Fitzgerald et al. developed another 3D in vitro model based on coculture of PC3 or LNCaP and osteoblasts on collagen scaffolds to address delivery of small interfering RNA (siRNA) as therapeutic agents [131]. Nanoparticle-mediated gene knockdown in the 3D coculture system proved effective, suggesting a potential role for this platform in assessing gene delivery vector efficacy [131]. Mosaad et al. established a microwell platform named Microwell-mesh that supported the formation of 3D microtissues composed of bone marrow stromal cells, adipocytes or osteoblasts, and luciferase-positive C4-2B cells [132]. PCa cell sensitivity to docetaxel varied depending on culture geometry and decreased in the presence of bone marrow stromal cells, while response to abiraterone depended on geometry only and was not affected by the presence of stromal cells [132].

Fong et al. developed an advanced 3D in vitro model that allowed for the cultivation of PDXs within a bone-like environment, which could give insights about the biology and therapy resistance of PCa bone tumors (Fig. 12.20) [133]. This model is based on 3D hydrogels decorated with integrin-binding, MMP-degradable peptide sequences enriched with osteoblastic cells. PDXs proliferated within gels as multicellular aggregates, retained markers of the original tumor, and established a progressive interaction with the stromal cells. This cross-talk was mediated by fibroblast growth factor 1 (FGF1), the expression of which was induced on osteoblastic cells in the presence of tumoroids, as shown in vivo. As a proof of concept, the inhibition of the FGF-activated pathway by dovitinib reduced the cellularity of cocultures [133].

Puca et al. generated organoids from four metastatic patients affected by neuroendocrine PCa, an aggressive variant that spontaneously arises as a rare de novo tumor or as a consequence of resistance to castration therapy [134]. Organoids retained the molecular signature of their originating tumors as evaluated at the genomic, transcriptomic, and epigenomic level and were further applied to perform high-throughput drug screening using a library of 129 anticancer agents. As a result, pozotinib and vandetanib proved more effective killing of neuroendocrine PCa-derived organoids and further patient-specific sensitivities were identified, such as significant response to alisertib (an aurora kinase inhibitor) in one organoid type, consistent with similar response of the original tumor. Drug screening also identified sensitivity mechanisms predicted by genomic alterations, such as effective AKT inhibition in an organoid that harbored Pten deletion and displayed high

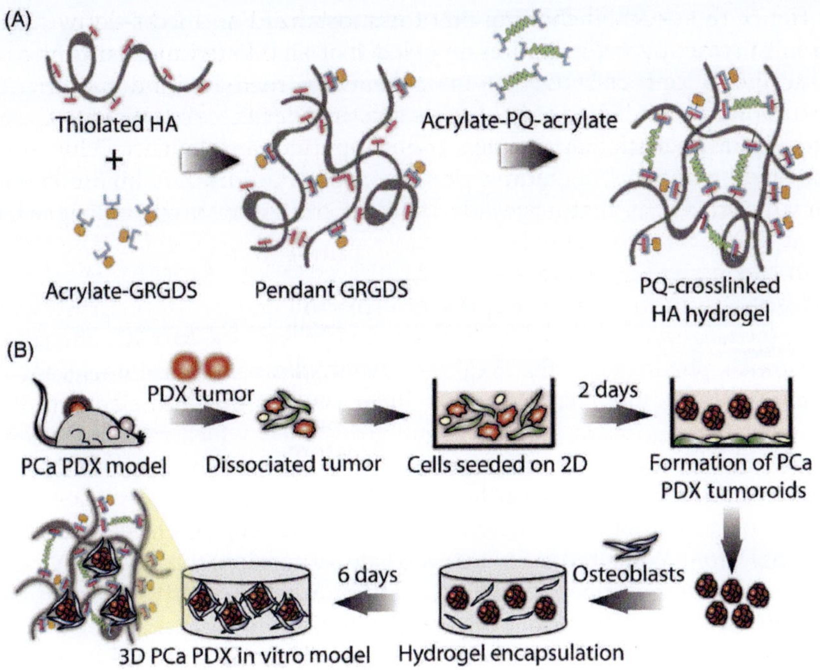


FIGURE 12.20 Schematic of 3D hydrogel system and coculture fabrication. (A) The hydrogel system comprises thiolated HA and acrylate-functionalized peptides, GRGDS (integrin-binding) and PQ (MMP-sensitive). (B) PDX tissue harvested from the animal is digested and then plated in six-well plates to allow formation of tumor aggregates. Resulting tumoroids are collected and then coencapsulated with osteoblastic cells. Source: *Reproduced with permission from Fong EL, et al. A 3D in vitro model of patient-derived prostate cancer xenograft for controlled interrogation of in vivo tumor-stromal interactions. Biomaterials 2016;77:164–72. © Elsevier (2016).*

basal levels of phospho-AKT. Combination studies with an inhibitor targeting EZH2, a mediator of progression in neuroendocrine PCa, enhanced the effects of single agents. This approach supports a potential role for organoid drug screening in identifying anticancer agents that affect cell-autonomous mechanisms of response (due to the lack of a microenvironmental component), and in predicting individual patient response to therapy [134].

## 12.4.4 Three-dimensional models of metastatic prostate cancer to the liver

In order to model liver metastasis, Yates et al. established a serum-free medium culture in a bioreactor that generated 3D liver microtissue units, constituted by hepatocytes and stromal cells, including endothelial and Kupffer cells, combined with fluorescent DU145 PCa cells [135]. Single DU145 cells attached to the liver microtissues at day 2, grew and invaded into the liver parenchyma from day 6, and expanded over 14 days. PCa cells remained viable, as determined by histology and electron microscopy, with necrosis visible only in cancer lesions with a diameter greater than 300 μm, suggesting that they received the proper amount of oxygen and nutrients due to active perfusion in the bioreactor. DU145 cells cultured in the bioreactor in the absence of multicellular hepatic units resulted in tumor cell death, indicating that their survival was mediated by liver microtissues. Electron microscopy investigation revealed that DU145 juxtaposed to the hepatic tissue and established cell–cell interactions potentially mediated by E-cadherin. Further experiments showed that E-cadherin expression in DU145 cells, which is usually low, increased in coculture with hepatocytes and supported resistance to chemotherapy via canonical survival pathways, including protein kinase B and Janus kinase [117]. This culture system supported a mechanistic understanding of PCa colonization of the liver [135].

## 12.5 Conclusion

PCa is a complex phenomenon that requires advanced, physiologically relevant experimental systems to investigate the molecular and cellular mechanisms at the basis of its establishment, early and late progression and therapeutic resistance, which remain partially obscure. In this context, the generation of a variety of 3D biomimetic platforms that recreate and address specific aspects of the disease has contributed meaningful advances to the understanding of its pathophysiological determinants. Of note, the application of human cells, obtained directly from patients or by intermediate expansion through PDX implantation, represents an effort toward more significant and accurate preclinical modeling for basic as well as translational therapeutic studies. Such cells, indeed, reproduce patient's heterogeneity and provide a more precise replica of human cancer compared with cell lines. Furthermore, their integration with human stromal components present at primary or distant sites supports the generation of fully human 3D systems that can overcome some potential shortcomings of xenograft models, for example, the incomplete match between human and mouse cells, and closely mirror patient's disease. In addition, the enrichment of in vitro models with a 3D architecture, multiple cell partners, elements of the extracellular matrices and molecular cues that account for in vivo-like mechanisms of response and resistance can lead to the development of more successful therapeutic strategies. Although 3D in vitro approaches display higher manufacturing complexity compared with conventional 2D screenings of therapeutics, they show potential for high-throughput drug testing with the advantage of improved biological and clinical relevance. Besides the experimental usefulness of 3D advanced systems, their development and enhancement align toward the principles of the 3Rs (reduction, replacement, and refinement in the use of animal models), thus representing a powerful tool not only from a scientific standpoint, but also from an ethical perspective.

Despite a comprehensive effort to model different steps of PCa development and progression, some aspects of the disease have not been properly reproduced in 3D systems yet. This includes a lack of 3D in vitro models that: tackle the role of the immune system, in both primary and metastatic tumors; investigate mechanisms of intravasation, extravasation, and dormancy; and address the biology of PCa in visceral organs or lymph nodes. Some of these approaches have been established and tested in other cancer types, such as breast or lung, and their application can be extended to PCa [30,111,136].

Overall, the integration of 3D in vitro model systems of PCa is contributing a considerable advancement in the field and shows promises for improved mechanistic understanding of PCa progression and consequent identification of more relevant therapeutic strategy.

# References

[1] Bray F, et al. Global cancer statistics 2018: GLOBOCAN estimates of incidence and mortality worldwide for 36 cancers in 185 countries. CA Cancer J Clin 2018;68(6):394–424.
[2] Ferlay J, et al. Cancer incidence and mortality worldwide: sources, methods and major patterns in GLOBOCAN 2012. Int J Cancer 2015;136(5):E359–86.
[3] Steele CB, et al. Prostate cancer survival in the United States by race and stage (2001-2009): findings from the CONCORD-2 study. Cancer 2017;123(Suppl. 24):5160–77.
[4] Scher HI, et al. Design and end points of clinical trials for patients with progressive prostate cancer and castrate levels of testosterone: recommendations of the Prostate Cancer Clinical Trials Working Group. J Clin Oncol 2008;26(7):1148–59.
[5] Faiena I, et al. Re: abiraterone for prostate cancer not previously treated with hormone therapy. Eur Urol 2018;73(6):981.
[6] Beer TM, Tombal B. Enzalutamide in metastatic prostate cancer before chemotherapy. N Engl J Med 2014;371 (18):1755–6.
[7] Parker C, et al. Alpha emitter radium-223 and survival in metastatic prostate cancer. N Engl J Med 2013;369 (3):213–23.
[8] Ellwood-Yen K, et al. Myc-driven murine prostate cancer shares molecular features with human prostate tumors. Cancer Cell 2003;4(3):223–38.
[9] Gingrich JR, et al. Androgen-independent prostate cancer progression in the TRAMP model. Cancer Res 1997;57(21):4687–91.
[10] Shirai T, et al. Experimental prostate carcinogenesis - rodent models. Mutat Res 2000;462(2-3):219–26.
[11] Cunningham D, You Z. In vitro and in vivo model systems used in prostate cancer research. J Biol Methods 2015;2(1):e17.
[12] Landgraf M, et al. Rational design of mouse models for cancer research. Trends Biotechnol 2018;36 (3):242–51.
[13] Salamanna F, et al. A systematic review on in vitro 3D bone metastases models: a new horizon to recapitulate the native clinical scenario? Oncotarget 2016;7(28):44803–20.
[14] Ellem SJ, De-Juan-Pardo EM, Risbridger GP. In vitro modeling of the prostate cancer microenvironment. Adv Drug Deliv Rev 2014;79-80:214–21.
[15] Croucher PI, McDonald MM, Martin TJ. Bone metastasis: the importance of the neighbourhood. Nat Rev Cancer 2016;16(6):373–86.
[16] McMillin DW, Negri JM, Mitsiades CS. The role of tumour-stromal interactions in modifying drug response: challenges and opportunities. Nat Rev Drug Discov 2013;12(3):217–28.
[17] Pampaloni F, Reynaud EG, Stelzer EH. The third dimension bridges the gap between cell culture and live tissue. Nat Rev Mol Cell Biol 2007;8(10):839–45.
[18] Durand RE, Raleigh JA. Identification of nonproliferating but viable hypoxic tumor cells in vivo. Cancer Res 1998;58(16):3547–50.
[19] Weiswald LB, Bellet D, Dangles-Marie V. Spherical cancer models in tumor biology. Neoplasia 2015;17 (1):1–15.
[20] Kelm JM, et al. Method for generation of homogeneous multicellular tumor spheroids applicable to a wide variety of cell types. Biotechnol Bioeng 2003;83(2):173–80.
[21] Yuhas JM, et al. A simplified method for production and growth of multicellular tumor spheroids. Cancer Res 1977;37(10):3639–43.
[22] Yu M, et al. Cancer therapy. Ex vivo culture of circulating breast tumor cells for individualized testing of drug susceptibility. Science 2014;345(6193):216–20.
[23] Xu X, et al. A hydrogel-based tumor model for the evaluation of nanoparticle-based cancer therapeutics. Biomaterials 2014;35(10):3319–30.
[24] Florczyk SJ, et al. 3D porous chitosan-alginate scaffolds: a new matrix for studying prostate cancer cell-lymphocyte interactions in vitro. Adv Healthc Mater 2012;1(5):590–9.
[25] Karthaus WR, et al. Identification of multipotent luminal progenitor cells in human prostate organoid cultures. Cell 2014;159(1):163–75.
[26] Gao D, et al. Organoid cultures derived from patients with advanced prostate cancer. Cell 2014;159 (1):176–87.

[27] Fischbach C, et al. Engineering tumors with 3D scaffolds. Nat Methods 2007;4(10):855–60.
[28] Reichert JC, et al. Mineralized human primary osteoblast matrices as a model system to analyse interactions of prostate cancer cells with the bone microenvironment. Biomaterials 2010;31(31):7928–36.
[29] Hartman O, et al. Biofunctionalization of electrospun PCL-based scaffolds with perlecan domain IV peptide to create a 3-D pharmacokinetic cancer model. Biomaterials 2010;31(21):5700–18.
[30] Bersini S, et al. A microfluidic 3D in vitro model for specificity of breast cancer metastasis to bone. Biomaterials 2014;35(8):2454–61.
[31] Asnaghi MA, et al. A double-chamber rotating bioreactor for the development of tissue-engineered hollow organs: from concept to clinical trial. Biomaterials 2009;30(29):5260–9.
[32] Centenera MM, et al. A patient-derived explant (PDE) model of hormone-dependent cancer. Mol Oncol 2018;12(9):1608–22.
[33] Curtin P, Youm H, Salih E. Three-dimensional cancer-bone metastasis model using ex-vivo co-cultures of live calvarial bones and cancer cells. Biomaterials 2012;33(4):1065–78.
[34] Nordstrand A, et al. Establishment and validation of an in vitro co-culture model to study the interactions between bone and prostate cancer cells. Clin Exp Metastasis 2009;26(8):945–53.
[35] Kreisberg JI, Brattain MG, Pretlow 2nd TG. Studies on human hyperplastic prostates maintained in organ culture. Invest Urol 1977;15(3):252–5.
[36] Berquin IM, et al. Expression signature of the mouse prostate. J Biol Chem 2005;280(43):36442–51.
[37] Marker PC, et al. Hormonal, cellular, and molecular control of prostatic development. Dev Biol 2003;253 (2):165–74.
[38] Shen MM, Abate-Shen C. Molecular genetics of prostate cancer: new prospects for old challenges. Genes Dev 2010;24(18):1967–2000.
[39] Sheth RA, et al. Patient-derived xenograft tumor models: overview and relevance to IR. J Vasc Interv Radiol 2018;29(6):880–2.
[40] Navone NM, et al. Movember GAP1 PDX project: an international collection of serially transplantable prostate cancer patient-derived xenograft (PDX) models. Prostate 2018;78(16):1262–82.
[41] Young SR, et al. Establishment and serial passage of cell cultures derived from LuCaP xenografts. Prostate 2013;73(12):1251–62.
[42] Fong EL, et al. Hydrogel-based 3D model of patient-derived prostate xenograft tumors suitable for drug screening. Mol Pharm 2014;11(7):2040–50.
[43] Jia X, et al. Hyaluronic acid-based microgels and microgel networks for vocal fold regeneration. Biomacromolecules 2006;7(12):3336–44.
[44] Hao Y, et al. Biomimetic hydrogels incorporating polymeric cell-adhesive peptide to promote the 3D assembly of tumoroids. Biomacromolecules 2016;17(11):3750–60.
[45] Girard YK, et al. A 3D fibrous scaffold inducing tumoroids: a platform for anticancer drug development. PLoS One 2013;8(10):e75345.
[46] Luo W, et al. Aberrant expression of nuclear vimentin and related epithelial-mesenchymal transition markers in nasopharyngeal carcinoma. Int J Cancer 2012;131(8):1863–73.
[47] Gravdal K, et al. A switch from E-cadherin to N-cadherin expression indicates epithelial to mesenchymal transition and is of strong and independent importance for the progress of prostate cancer. Clin Cancer Res 2007;13(23):7003–11.
[48] Talukdar S, Kundu SC. Engineered 3D silk-based metastasis models: interactions between human breast adenocarcinoma, mesenchymal stem cells and osteoblast-like cells. Adv Funct Mater 2013;23(42):5249–60.
[49] Mandal BB, Kundu SC. A novel method for dissolution and stabilization of non-mulberry silk gland protein fibroin using anionic surfactant sodium dodecyl sulfate. Biotechnol Bioeng 2008;99(6):1482–9.
[50] Mandal BB, Kundu SC. Non-bioengineered silk gland fibroin protein: characterization and evaluation of matrices for potential tissue engineering applications. Biotechnol Bioeng 2008;100(6):1237–50.
[51] Bhowmick NA, Moses HL. Tumor-stroma interactions. Curr Opin Genet Dev 2005;15(1):97–101.
[52] Joyce JA, Pollard JW. Microenvironmental regulation of metastasis. Nat Rev Cancer 2009;9(4):239–52.
[53] Verona EV, et al. Transforming growth factor-beta signaling in prostate stromal cells supports prostate carcinoma growth by up-regulating stromal genes related to tissue remodeling. Cancer Res 2007;67(12):5737–46.
[54] Xu X, et al. Recreating the tumor microenvironment in a bilayer, hyaluronic acid hydrogel construct for the growth of prostate cancer spheroids. Biomaterials 2012;33(35):9049–60.

[55] Freeman MR, et al. Heparin-binding EGF-like growth factor in the human prostate: synthesis predominantly by interstitial and vascular smooth muscle cells and action as a carcinoma cell mitogen. J Cell Biochem 1998;68(3):328–38.

[56] Iwamoto R, Mekada E. Heparin-binding EGF-like growth factor: a juxtacrine growth factor. Cytokine Growth Factor Rev 2000;11(4):335–44.

[57] Ivich F, Tran M, Tahsin S, Frank SB, Kraft A, Miranti CK, et al., Application of a microfluidic-based model of a human prostate gland for cancer research. In: Proceedings of the 12th IEEE International Conference on Nano/Molecular Medicine and Engineering, NANOMED 2018; 2018.

[58] Fang X, et al. Novel 3D co-culture model for epithelial-stromal cells interaction in prostate cancer. PLoS One 2013;8(9):e75187.

[59] Kuefer R, et al. The role of an 80 kDa fragment of E-cadherin in the metastatic progression of prostate cancer. Clin Cancer Res 2003;9(17):6447–52.

[60] Okugawa Y, et al. Clinical significance of serum soluble E-cadherin in colorectal carcinoma. J Surg Res 2012;175(2):e67–73.

[61] Olumi AF, et al. Carcinoma-associated fibroblasts direct tumor progression of initiated human prostatic epithelium. Cancer Res 1999;59(19):5002–11.

[62] Giannoni E, et al. Reciprocal activation of prostate cancer cells and cancer-associated fibroblasts stimulates epithelial-mesenchymal transition and cancer stemness. Cancer Res 2010;70(17):6945–56.

[63] Pereira BA, et al. Tissue engineered human prostate microtissues reveal key role of mast cell-derived tryptase in potentiating cancer-associated fibroblast (CAF)-induced morphometric transition in vitro. Biomaterials 2019;197:72–85.

[64] Clark AK, et al. A bioengineered microenvironment to quantitatively measure the tumorigenic properties of cancer-associated fibroblasts in human prostate cancer. Biomaterials 2013;34(20):4777–85.

[65] Heinlein CA, Chang C. Androgen receptor in prostate cancer. Endocr Rev 2004;25(2):276–308.

[66] Davey RA, Grossmann M. Androgen receptor structure, function and biology: from bench to bedside. Clin Biochem Rev 2016;37(1):3–15.

[67] Huggins C, Hodges CV. Studies on prostatic cancer: I. The effect of castration, of estrogen and of androgen injection on serum phosphatases in metastatic carcinoma of the prostate 1941. J Urol 2002;168(1):9–12.

[68] Zhau HE, et al. Establishment of a three-dimensional human prostate organoid coculture under microgravity-simulated conditions: evaluation of androgen-induced growth and PSA expression. In Vitro Cell Dev Biol Anim 1997;33(5):375–80.

[69] Zadra G, et al. Inhibition of de novo lipogenesis targets androgen receptor signaling in castration-resistant prostate cancer. Proc Natl Acad Sci U S A 2019;116(2):631–40.

[70] Chua CW, et al. Differential requirements of androgen receptor in luminal progenitors during prostate regeneration and tumor initiation. Elife 2018;7:e28768.

[71] Fuchs E, Tumbar T, Guasch G. Socializing with the neighbors: stem cells and their niche. Cell 2004;116 (6):769–78.

[72] Collins AT, et al. Identification and isolation of human prostate epithelial stem cells based on alpha(2)beta (1)-integrin expression. J Cell Sci 2001;114(Pt 21):3865–72.

[73] Richardson GD, et al. CD133, a novel marker for human prostatic epithelial stem cells. J Cell Sci 2004;117(Pt 16):3539–45.

[74] Birnie R, et al. Gene expression profiling of human prostate cancer stem cells reveals a pro-inflammatory phenotype and the importance of extracellular matrix interactions. Genome Biol 2008;9(5):R83.

[75] Lang SH, et al. Modeling the prostate stem cell niche: an evaluation of stem cell survival and expansion in vitro. Stem Cells Dev 2010;19(4):537–46.

[76] Rao W, et al. Enhanced enrichment of prostate cancer stem-like cells with miniaturized 3D culture in liquid core-hydrogel shell microcapsules. Biomaterials 2014;35(27):7762–73.

[77] Guo X, et al. Enrichment of cancer stem cells by agarose multi-well dishes and 3D spheroid culture. Cell Tissue Res 2019;375(2):397–408.

[78] Stock K, et al. Capturing tumor complexity in vitro: comparative analysis of 2D and 3D tumor models for drug discovery. Sci Rep 2016;6:28951.

[79] Nguyen HM, et al. LuCaP prostate cancer patient-derived xenografts reflect the molecular heterogeneity of advanced disease and serve as models for evaluating cancer therapeutics. Prostate 2017;77(6):654–71.

[80] Watt MJ, et al. Suppressing fatty acid uptake has therapeutic effects in preclinical models of prostate cancer. Sci Transl Med 2019;11(478):eaau5758.
[81] Marzi L, et al. The indenoisoquinoline TOP1 inhibitors selectively target homologous recombination-deficient and schlafen 11-positive cancer cells and synergize with olaparib. Clin Cancer Res 2019;25(20):6206–16.
[82] Beshiri ML, et al. A PDX/organoid biobank of advanced prostate cancers captures genomic and phenotypic heterogeneity for disease modeling and therapeutic screening. Clin Cancer Res 2018;24(17):4332–45.
[83] Wang K, et al. 3D porous chitosan-alginate scaffolds as an in vitro model for evaluating nanoparticle-mediated tumor targeting and gene delivery to prostate cancer. Biomacromolecules 2015;16(10):3362–72.
[84] Camus VL, et al. Measuring the effects of fractionated radiation therapy in a 3D prostate cancer model system using SERS nanosensors. Analyst 2016;141(17):5056–61.
[85] Schreiber RD, Old LJ, Smyth MJ. Cancer immunoediting: integrating immunity's roles in cancer suppression and promotion. Science 2011;331(6024):1565–70.
[86] Feder-Mengus C, et al. New dimensions in tumor immunology: what does 3D culture reveal? Trends Mol Med 2008;14(8):333–40.
[87] Wolf K, Friedl P. Molecular mechanisms of cancer cell invasion and plasticity. Br J Dermatol 2006;154(Suppl. 1):11–15.
[88] Friedl P, Gilmour D. Collective cell migration in morphogenesis, regeneration and cancer. Nat Rev Mol Cell Biol 2009;10(7):445–57.
[89] Aw Yong KM, et al. Tracking the tumor invasion front using long-term fluidic tumoroid culture. Sci Rep 2017;7(1):10784.
[90] Bjork JK, et al. Heat-shock factor 2 is a suppressor of prostate cancer invasion. Oncogene 2016;35(14):1770–84.
[91] Elsing AN, et al. Expression of HSF2 decreases in mitosis to enable stress-inducible transcription and cell survival. J Cell Biol 2014;206(6):735–49.
[92] Santagata S, et al. High levels of nuclear heat-shock factor 1 (HSF1) are associated with poor prognosis in breast cancer. Proc Natl Acad Sci U S A 2011;108(45):18378–83.
[93] Harma V, et al. A comprehensive panel of three-dimensional models for studies of prostate cancer growth, invasion and drug responses. PLoS One 2010;5(5):e10431.
[94] Liu L, et al. Probing the invasiveness of prostate cancer cells in a 3D microfabricated landscape. Proc Natl Acad Sci U S A 2011;108(17):6853–6.
[95] Irimia D, Toner M. Spontaneous migration of cancer cells under conditions of mechanical confinement. Integr Biol (Camb) 2009;1(8-9):506–12.
[96] Cazzaniga W, et al. Human prostate tissue-derived extracellular matrix as a model of prostate microenvironment. Eur Urol Focus 2016;2(4):400–8.
[97] Tonini T, Rossi F, Claudio PP. Molecular basis of angiogenesis and cancer. Oncogene 2003;22(42):6549–56.
[98] Zetter BR. The scientific contributions of M. Judah Folkman to cancer research. Nat Rev Cancer 2008;8(8):647–54.
[99] Bergers G, Benjamin LE. Tumorigenesis and the angiogenic switch. Nat Rev Cancer 2003;3(6):401–10.
[100] Bray LJ, et al. Multi-parametric hydrogels support 3D in vitro bioengineered microenvironment models of tumour angiogenesis. Biomaterials 2015;53:609–20.
[101] Seano G, et al. Modeling human tumor angiogenesis in a three-dimensional culture system. Blood 2013;121(21):e129–37.
[102] Taubenberger AV, et al. 3D extracellular matrix interactions modulate tumour cell growth, invasion and angiogenesis in engineered tumour microenvironments. Acta Biomater 2016;36:73–85.
[103] Fidler IJ. The pathogenesis of cancer metastasis: the 'seed and soil' hypothesis revisited. Nat Rev Cancer 2003;3(6):453–8.
[104] Danila DC, et al. Circulating tumor cell number and prognosis in progressive castration-resistant prostate cancer. Clin Cancer Res 2007;13(23):7053–8.
[105] Augustsson P, et al. Microfluidic, label-free enrichment of prostate cancer cells in blood based on acoustophoresis. Anal Chem 2012;84(18):7954–62.
[106] Sarioglu AF, et al. A microfluidic device for label-free, physical capture of circulating tumor cell clusters. Nat Methods 2015;12(7):685–91.

[107] Kirby BJ, et al. Functional characterization of circulating tumor cells with a prostate-cancer-specific microfluidic device. PLoS One 2012;7(4):e35976.
[108] De Menna M, Elisau RS, Manna F, Zeinali S, Guenat O, Kruithof-de Julio M, et al. Microvasculature on chip: cancer cells extravasation in vitro modeling. J Urol 2019;201(Suppl. 4):e1170.
[109] Gandaglia G, et al. Distribution of metastatic sites in patients with prostate cancer: a population-based analysis. Prostate 2014;74(2):210–16.
[110] Yates CC, et al. Co-culturing human prostate carcinoma cells with hepatocytes leads to increased expression of E-cadherin. Br J Cancer 2007;96(8):1246–52.
[111] Alonso-Nocelo M, et al. Selective interaction of PEGylated polyglutamic acid nanocapsules with cancer cells in a 3D model of a metastatic lymph node. J Nanobiotechnology 2016;14(1):51.
[112] van der Toom EE, Verdone JE, Pienta KJ. Disseminated tumor cells and dormancy in prostate cancer metastasis. Curr Opin Biotechnol 2016;40:9–15.
[113] Weilbaecher KN, Guise TA, McCauley LK. Cancer to bone: a fatal attraction. Nat Rev Cancer 2011;11 (6):411–25.
[114] Mishra A, et al. Homing of cancer cells to the bone. Cancer Microenviron 2011;4(3):221–35.
[115] Pavan Grandhi TS, et al. Chemomechanically engineered 3D organotypic platforms of bladder cancer dormancy and reactivation. Biomaterials 2017;142:171–85.
[116] Preciado JA, Reátegui E, Azarin SM, Lou E, Aksan A. Immobilization platform to induce quiescence in dormancy-capable cancer cells. Technology 2017;5:129.
[117] Ma B, et al. Liver protects metastatic prostate cancer from induced death by activating E-cadherin signaling. Hepatology 2016;64(5):1725–42.
[118] Herroon MK, Diedrich JD, Podgorski I. New 3D-culture approaches to study interactions of bone marrow adipocytes with metastatic prostate cancer cells. Front Endocrinol (Lausanne) 2016;7:84.
[119] Choudhary S, et al. Human ex vivo 3D bone model recapitulates osteocyte response to metastatic prostate cancer. Sci Rep 2018;8(1):17975.
[120] Zheng H, et al. Therapeutic antibody targeting tumor- and osteoblastic niche-derived jagged1 sensitizes bone metastasis to chemotherapy. Cancer Cell 2017;32(6):731–47.
[121] Lee YC, et al. Identification of bone-derived factors conferring de novo therapeutic resistance in metastatic prostate cancer. Cancer Res 2015;75(22):4949–59.
[122] Sieh S, et al. Interactions between human osteoblasts and prostate cancer cells in a novel 3D in vitro model. Organogenesis 2010;6(3):181–8.
[123] Bischel LL, et al. A microfluidic coculture and multiphoton FAD analysis assay provides insight into the influence of the bone microenvironment on prostate cancer cells. Integr Biol (Camb) 2014;6(6):627–35.
[124] San Martin R, et al. Tenascin-C and integrin alpha9 mediate interactions of prostate cancer with the bone microenvironment. Cancer Res 2017;77(21):5977–88.
[125] Tuxhorn JA, et al. Reactive stroma in human prostate cancer: induction of myofibroblast phenotype and extracellular matrix remodeling. Clin Cancer Res 2002;8(9):2912–23.
[126] Hsiao AY, et al. Microfluidic system for formation of PC-3 prostate cancer co-culture spheroids. Biomaterials 2009;30(16):3020–7.
[127] Li Z, et al. Development, regulation, metabolism and function of bone marrow adipose tissues. Bone 2018;110:134–40.
[128] Morris EV, Edwards CM. The role of bone marrow adipocytes in bone metastasis. J Bone Oncol 2016;5 (3):121–3.
[129] Klemm F, Joyce JA. Microenvironmental regulation of therapeutic response in cancer. Trends Cell Biol 2015;25(4):198–213.
[130] Paindelli C, et al. Engineered bone for probing organotypic growth and therapy response of prostate cancer tumoroids in vitro. Biomaterials 2019;197:296–304.
[131] Fitzgerald KA, et al. Nanoparticle-mediated siRNA delivery assessed in a 3D co-culture model simulating prostate cancer bone metastasis. Int J Pharm 2016;511(2):1058–69.
[132] Mosaad E, et al. Using high throughput microtissue culture to study the difference in prostate cancer cell behavior and drug response in 2D and 3D co-cultures. BMC Cancer 2018;18(1):592.
[133] Fong EL, et al. A 3D in vitro model of patient-derived prostate cancer xenograft for controlled interrogation of in vivo tumor-stromal interactions. Biomaterials 2016;77:164–72.

[134] Puca L, et al. Patient derived organoids to model rare prostate cancer phenotypes. Nat Commun 2018;9 (1):2404.

[135] Yates C, et al. Novel three-dimensional organotypic liver bioreactor to directly visualize early events in metastatic progression. Adv Cancer Res 2007;97:225–46.

[136] Aung A, et al. An engineered tumor-on-a-chip device with breast cancer-immune cell interactions for assessing T-cell recruitment. Cancer Res. 2020;80(2):263–75.

CHAPTER

# 13

# 3D in vitro cutaneous melanoma models

*Ana I. Soares*[1,2], *Rui L. Reis*[1,2] *and Alexandra P. Marques*[1,2]

[1]3B's Research Group, I3Bs—Research Institute on Biomaterials, Biodegradables and Biomimetics, University of Minho, Headquarters of the European Institute of Excellence on Tissue Engineering and Regenerative Medicine, Braga/Guimarães, Portugal [2]ICVS/3B's—PT Government Associate Laboratory, Braga/Guimarães, Portugal

## Abbreviations

| | |
|---|---|
| **2D** | two-dimensional |
| **3D** | three-dimensional |
| **3Rs** | replacement, reduction, and refinement |
| **6-4PPS** | pyrimidine (6-4) pyrimidone |
| **AJCC** | American Joint Committee on Cancer |
| **ALI** | air–liquid interface |
| **BRAFi** | BRAF inhibitor |
| **C** | cytosine |
| **CAFs** | cancer-associated fibroblasts |
| **CDK4** | cyclin-dependent kinase 4 |
| **CDK6** | cyclin-dependent kinase 6 |
| **CDNK2A** | cyclin-dependent kinase inhibitor 2a |
| **c-Met** | tyrosine-protein kinase Met |
| **CPDs** | cyclobutane pyrimidine dimers |
| **CTLA-4** | cytotoxic T-lymphocyte-associated antigen 4 |
| **DED** | de-epidermized dermis |
| **DNA** | deoxyribonucleic acid |
| **ECM** | extracellular matrix |
| **ECs** | endothelial cells |
| **FAMMM** | family atypical multiple mole melanoma |
| **FDA** | Food and Drug Administration |
| **G** | guanine |
| **HGF** | hepatocyte growth factor |
| **IL-2** | interleukin 2 |

*Biomaterials for 3D Tumor Modeling*
DOI: https://doi.org/10.1016/B978-0-12-818128-7.00013-7

| | |
|---|---|
| **LECs** | lymphatic endothelial cells |
| **M** | metastasis |
| **MAPK** | mitogen-activated protein kinase |
| **MCTs** | multicellular tumor spheroids |
| **MEKi** | MEK inhibitor |
| **MM** | metastatic melanoma |
| **N** | node |
| **NF1** | neurofibromatosis type 1 |
| **NROS** | neuroblastome RAS |
| **PD-1** | programmed cell death receptor-1 |
| **PD-L1** | programmed death ligand 1 |
| **PD-L2** | programmed death ligand 2 |
| **PTEN** | phosphatase and tensin homolog |
| **Rb** | retinoblastoma |
| **RGP** | radial growth phase |
| **ROS** | reactive oxygen species |
| **T** | thymidine |
| **T** | tumor |
| **TCGA** | cancer genome atlas |
| **TERT** | telomerase reverse transcriptase |
| **TP53** | tumor-protein p53 |
| **TRAIL** | tumor necrosis factor α-related apoptosis-inducing ligand |
| **TWT** | triple wild-type |
| **UV** | ultraviolet |
| **VGP** | vertical growth phase |

## 13.1 Introduction

Melanoma is a malignant melanocytic neoplasm arising in any anatomic location containing melanocytes, being the cutaneous melanoma the most common. Although it is the skin cancer less diagnosed (1%) it one of the most rapid to metastasize and one of the most common cancers in young adults. Despite the major advances regarding the development of new treatments, the metastatic melanoma (MM) is still responsible for more than 75% of the skin cancer deaths, in which approximately one-third in patients are under 50 years of age. The rates of cutaneous melanoma have been rising rapidly over the past few decades, doubling every 10–20 years. In 2018, were estimated more than 60,000 new cases of cutaneous melanoma worldwide, with an approximate incidence of 29,000 cases [1]. In 200 years, since the first description of melanoma published on "The Melanoses" by René Laennec, the advances to treat metastatic cutaneous melanoma is insufficient, remaining the 5-year relative survival rate to low [1].

Advanced cutaneous melanoma responds poorly to conventional treatments leading to the development of many different strategies, numerous researches, to understand the risk factors and the molecular pathogenesis, has been done to try to find better treatments. Immunotherapy and target therapy gained prominence in this area due to the notorious results in improving median overall survival [2,3]. Nevertheless, this great progress in cutaneous melanoma treatment reflects frequently the denouements of recurrent innate or acquired therapy resistance and tumor recurrence [4,5]. Due to this reality, improvements in preclinical models are necessary to recreate the features of in vivo skin structures and in this way to recreate the complex interplay of the cancerous and noncancerous cells with a more realistic microenvironment to imitate, more closely, the development and progression of the tumor in vivo.

Melanoma is a malignant melanocytic neoplasm arising in any anatomic location containing melanocytes, being the cutaneous melanoma the most common [1]. Although it is the skin cancer less diagnosed (1%) [6], it is one of the most rapid to metastasize and one of the most common cancers in young adults [6,7]. Despite the major recent regarding the development of new treatments, the MM is still responsible for more than 75% of the skin cancer deaths, in which approximately one-third in patients are under 50 years of age [8]. The rates of cutaneous melanoma have been rising rapidly over the past few decades, doubling every 10–20 years. In 2018, were estimated more than 60,000 new cases of cutaneous melanoma worldwide, with an approximate incidence of 29,000 cases [1]. In addition, the advances in the treatment of metastatic cutaneous melanoma are insufficient as demonstrated by the 5-year relative survival rate [1].

Patients with advanced cutaneous melanoma respond poorly to conventional treatments. This has led to the development of significant research to understand the risk factors and the molecular pathogenesis of this type of cancer as a way to find better outcomes. Immunotherapy and target therapy gained prominence due to the notorious results in improving median overall survival [2,3]. Nevertheless, innate or acquired resistance to the therapy and tumor recurrence are a frequent denouement [4,5]. It is then clear that there is a major need to better understand the development and progression of cutaneous melanoma model. While animal models have been extremely useful to discover what is currently known, mechanistic differences to humans impair further progresses [9]. In this context, improved preclinical models capable of recreating the complex interplay of the cancerous and noncancerous cells are a demand. In this chapter we will discuss the evolution of the in vitro cutaneous melanoma models highlighting the benefits of three-dimensional (3D) systems in relation to two-dimensional (2D) cell cultures, and their contribution so far in understanding melanoma initiation and progression, as well as in drug screening.

## 13.2 Types of melanoma

Depending on where the primary lesion occurs, melanoma can be classified into four different subtypes: [10,11] (1) cutaneous melanoma that is the most common and arises in nonglabrous skin, (2) acral melanoma that starts in the glabrous skin (palms, soles, and nail beds); (3) mucosal melanoma that develops in any mucosal epithelium and therefore is the most common in the oral and the nasal cavities, the anus/rectum, and the vulvovaginal area; and (4) uveal melanoma that originates in the uveal tract.

Over the years melanoma has been classified based on clinicopathological and histological characteristics. The first classification—Clark's classification—was based only on the depth of tumor invasion, defining the different stages of the progression of melanoma: radial growth phase (RGP), vertical growth phase (VGP), and MM [12]. Benign nevi progress into dysplastic nevi with highly proliferative atypic melanocytes that form the first tumor lesion. In the RGP, the tumor proliferates radially in sequential events confined to the epidermis with a small number of cells invading the papillary dermis. In the second stage, VGP, the melanoma cells with a more invasive and aggressive phenotype, pass through the basal membrane and progress into the dermis up to the subcutaneous tissue. The last stage, MM, concerns the systemic progression of the melanoma cells through the invasion of the blood and lymphatic systems to metastasize in distant organs. This is the most aggressive and the faster stage (Fig. 13.1).

Considering this classification, cutaneous melanoma, in particular, can be divided into different subtypes, the superficial spreading, the nodular and the lentigo maligna [13]. Superficial spreading melanoma is the most common cutaneous melanoma (70%) being more frequent in young people with intermittent sun exposure. It is characterized by a superficial spreading across the surface of the skin (radial growth). Nodular melanoma is the second most common cutaneous melanoma (15%–20%), being typically characterized by a quick and aggressive spreading into the dermis (vertical growth). It is predominant in older people with chronic sun exposure, and appears commonly in the face, head, and neck. This type of tumor has the particularity of rising as a lump frequently with black color, but which can be also red, pink, brown, or the same color as the skin. Lentigo maligna melanoma is the less common in general (10%–15%) but the most common cutaneous melanoma in older people beginning as a large freckle in an area of sun-damaged skin, such as the face, ears, and arms. It is characterized by a superficial (radial growth) and slow growth over several years before penetrating more deeply into the skin (vertical growth), changing the color from light to darker brown along the time (Table 13.1) [12].

In terms of prognosis and as the basis to define the suitable treatment, cutaneous melanoma is evaluated considering three major features: T (Tumor), N (Node), and M (Metastasis)—the TNM classification [14]. T refers to tumor thickness and the presence/absence of ulcerations, N to the spreading to lymph nodes, and M indicates the presence of metastasis. Although the depth of tumor invasion is an important prognostic indicator of the disease evolution, nowadays, it is well known, especially for the earlier stages, that

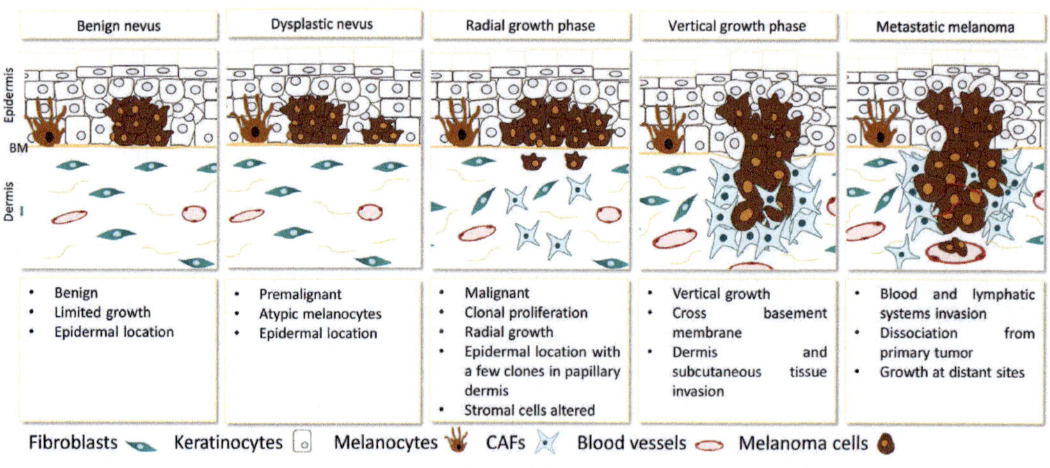


FIGURE 13.1 Cutaneous melanoma progression stages.

TABLE 13.1 Characteristics of different subtypes of cutaneous melanoma.

| Melanoma subtype | Clinicopathological characteristics |
|---|---|
| Superficial spreading melanoma | • Incidence 70%<br>• Superficial spreading (radial growth)<br>• Arises from existing nevus<br>• Commonly in young people<br>• Intermittent sun exposure |
| Nodular melanoma | • Incidence 15%–20%<br>• Quick and aggressive spreading into the dermis<br>• Arises from sun-damaged skin—commonly face head and neck<br>• Absence of radial growth<br>• Commonly in older people<br>• Chronic sun exposure |
| Lentigo maligna melanoma | • Incidence 10%–15%<br>• Superficial and slow growth<br>• Arises from sun-damaged skin—commonly face, ears, and arms<br>• The most frequent subtype in older people<br>• Chronic sun exposure |

the presence of ulcerations, as well as the involvement of the lymphatic systems, is equally important and usually associated with poor prognosis.

## 13.3 Risk factors for melanoma

Melanoma is now recognized as a multifactorial disease associated with environmental and genetic risks that are highly interconnected and directly linked to increased incidence [1]. The environmental risk, in particular the ultraviolet (UV) radiation exposure, is seen as the most critical one [15,16]. This is greatly because it is strongly correlated with pigmentation characteristics, such as light skin and eyes and red hair, being also affected by the geographic location [15,17]. In what concerns hereditable factors, genetic alterations linked to positive family and/or personal history of melanoma, and multiple atypical nevi that are extremely dependent on the environmental risks, are the main ones [18].

### 13.3.1 Ultraviolet radiation

Exposure to UV radiation is the major known environmental risk factors for cutaneous melanoma development. About 60%–70% of cutaneous melanomas are thought to be caused by UV radiation exposure from natural or artificial sunlight [19]. In contrast to most cutaneous malignancies, where the risk of development is correlated to lifetime duration of solar UV exposure [20], the risk to develop melanoma is strongly related to the number of sporadic, intermittent but intense UV exposure (i.e., sunburn), especially if started at a young age and in sites which are more susceptible to sunburn, such as the head and neck [20]. Moreover, the UV exposure risk is strongly correlated with individual phenotypic characteristics. The population with specific pigmentation characteristics, such as type I and II skin phototype (fair skin), and red hair color are more prone to be affected

[21]. This susceptibility is directly linked to the type of melanin—eumelanin (brown to black) and pheomelanin (yellowish to reddish)—produced by the melanocytes and to its protective role after being transferred to neighboring keratinocytes [21]. Pheomelanin is a redder form of melanin that is involved in the production of reactive oxygen species (ROS) upon exposure to UV radiation. If not counterbalanced by sufficient amounts of the black eumelanin, which acts as a scavenger, higher levels of pheomelanin can indirectly contribute to mutagenesis [22,23].

The effect of the UV spectrum differs according to the wavelength ranges [UVA (320–400 nm), UVB (280–320 nm), and UVC (200–280 nm)]. UVC and a large portion of UVB are blocked by the ozone stratum and do not reach the surface of the earth. The solar radiation that reaches the surface of the earth is 90%–99% comprised of UVA and only 1%–10% of UVB radiation [24]. However, UVB is approximately 1000–10,000 times more carcinogenic (per J $m^{-2}$) than UVA [25]. Interestingly, although UVA is the primary source of radiation in tanning beds, in these cases, it is 12 times more potent and potentially more harmful than the one from the sun [19].

UVA radiation promotes indirect DNA damage through the generation of ROS that are formed upon excitation of endogenous skin cellular photosensitizers such as porphyrins [26]. The ROS induce DNA oxidation damage, which can then be reflected in a loss of DNA information or deficient chromosomal rearrangement promoting erroneous DNA repair and mutagenesis. The most frequent DNA modifications are the Guanine (G) to Thymidine (T) transitions, which are known as the UVA-signature mutations [27].

In opposition, UVB causes direct excitation of DNA, inducing photoproducts such as cyclobutane pyrimidine dimers (CPDs) and pyrimidine (6-4) pyrimidone (6-4PPs) between adjacent pyrimidine bases. This leads to Cytosine (C) to T or CC to TT transitions upon DNA replication, known as the UVB-signature [28,29].

### 13.3.2 Heritable factors

Although the familial risk to develop cutaneous melanoma was described for the first time in 1820, it is not yet well understood. Only 5%–10% of cutaneous melanoma cases are considered to be linked to heritable factors and designated as familial atypical multiple mole melanoma syndrome (FAMMM) [30,31]. One of the most critical issues to identify the risk to develop FAMMM refers to the multiple and interconnected factors that contribute to it. A positive familial history in one or more first- or second-degree relatives is associated with approximately twofold increase in the risk to develop FAMMM [32]. This risk is further evidenced by a history of other cancers, especially pancreatic cancer [33] and neural systems tumors [34]. Likewise, individuals with multiple atypical nevi (>50 total nevi count) and red hair/fair skin have also associated higher disease penetrance [18].

The positive familiar history of FAMMM has been linked to the occurrence of germline mutations. However, most of these are still unknown as they do not occur in recognized susceptibility genes. The most common (30%–40%) germline mutation in FAMMM occurs in the tumor suppressor gene CDKN2A, also known as cyclin-dependent kinase inhibitor 2A which results in the inactivation of the locus [35]. Located in chromosome 9p21.3, this locus encodes two distinct proteins, p16-INK4a and p14-ARF, also tumor suppressors in genetically distinct anticancer pathways, the retinoblastoma (Rb) and the p53 pathway, respectively [36]. Mutations in the CDKN2A gene disrupt the function of proteins responsible for promoting the transcription of specific genes for the transition from G1 to S phase, causing abnormal cell-cycle arrest in the G1 phase and inappropriate mitotic division [37,38]. Loss of p14 expression results in increased MDM2-mediated degradation promoting ribosome biogenesis or inhibiting p-53-dependent cell-cycle arrest and apoptosis, which results in the incapacity to stop cell cycle and kept the proliferation of abnormal cells [39]. It is estimated that 6 in 10 Europeans with this mutation will develop melanoma until the age of 80 [40]. The germline CDK4 mutation is also found in a small percentage (<1%) of FAMMM [34]. The CDK4 oncogene is located in chromosome 12q13 and codes for the cyclin-dependent kinase 4 protein. The CDK4 mutation results in an altered protein structure rendering it insensitive to the inhibitory effects of p16, further underscoring the relevance of detective cell-cycle control for melanoma transformation [30].

## 13.4 Cutaneous melanoma development

As mentioned before, the melanocyte transformation into melanoma involves multifactorial intervenient. Nonetheless, it is the UV signature—the history (chronic or sporadic), nature (nature or artificial), and the type (UVA or UVB) of UV exposure—that determines the type of mutations found in sporadic cutaneous melanoma cells [20]. Around 50% of cutaneous melanomas are associated with BRAF mutation, commonly in codon V600, that appears earlier in life [41,42]. Cutaneous melanomas with this somatic genetic profile are associated with nevi in anatomical sites with intermittent sun exposure [20]. On the other hand, chronic sun damage is more correlated to neuroblastoma RAS (NRAS) protooncogene mutations (~30%) [10] and to the development of melanoma in anatomic sites with higher sun exposure, commonly head and neck [20]. Interestingly, this type of cutaneous melanoma is not associated with nevi and have a late age of onset [8,10]. The less frequent mutated genes induced by UV are neurofibromatosis type 1 (NF1) (10%–15%) and triple wild-type (TWT) (5%–10%) [41].

There are other types of mutations that can be found in melanomas. Mutations in the telomerase reverse transcriptase (TERT) promoter [43], for example, are found in about 30%–80% of melanomas [10]. However, this mutation per se is not sufficient to support melanoma development and the accumulation with other mutations is required. TERT-promoter mutation can lead to an intermediate lesion; however, the progression to an invasive stage requires, for example, a simultaneous mutation in cell-cycle controlling genes such as the CDKN2A. Likewise, the progression to a metastatic stage depends on the presence of a mutation in the tumor-protein p53 (TP53)- and/or phosphatase and tensin homolog (PTEN)-encoding genes [43].

## 13.5 Cutaneous melanoma treatment

Early diagnosis followed by surgical excision still is the most promising melanoma treatment. For a primary lesion in an early stage of development, surgical excision is

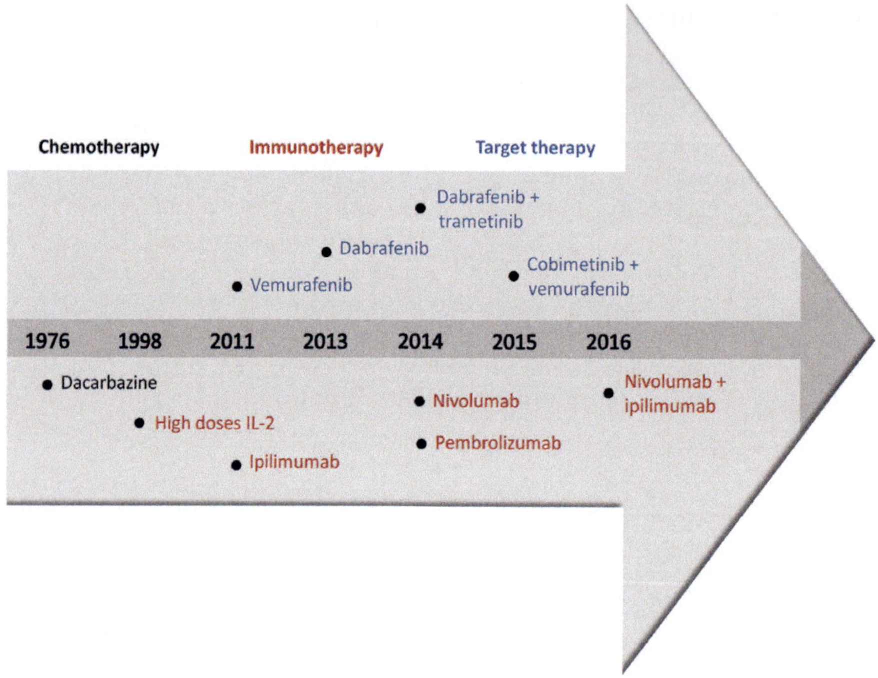


FIGURE 13.2 Timeline of the FDA treatment approved.

highly effective being associated with a survival rate of 94%–100% [2,44]. However, melanoma progression is fast and most of the patients are no longer in the early stage of melanoma development when diagnosed. Remarkable efforts have been done with the development of immunotherapies against T-cell checkpoint inhibitors, and target therapies such as selective BRAF inhibitors (BRAFi) and MEK inhibitors (MEKi) to improve the outcomes of melanoma treatments [2,3] (Fig. 13.2).

## 13.5.1 Classic approach

For many years chemotherapy with alkylating agents (dacarbazine) and high dose of interleukin 2 (IL-2) (immunotherapy) was the standard treatment for advanced cutaneous melanoma, despite the lack of clear information about successful outcomes and the severe side effects [45]. Other drugs such as temozolomide, nab-paclitaxel, paclitaxel, cisplatin treatment-resistant, and carboplatin, were then tested alone or combined with chemotherapy but, improvement in the median overall survival was not significant [46]. Treatment with interferon-α-2b was the first chemotherapy adjuvant treatment approved for MM. Just a small fraction of patients with melanoma in early premetastatic stages were successfully treated [47]. In more advanced stages, the small fraction of patients reaching durable remission were treated with high dose of IL-2, experiencing high toxicity [48]. Nowadays, chemotherapy is only considered and used when any other treatments fail [49].

### 13.5.2 Immunotherapy

The immune system plays an important role to regulate tumor growth, by identifying and eliminating malignant cells. However, these cells can manipulate the immune system and escape to its control [5]. Immunotherapies have been developed with the rationale to modulate the adaptive immune response to obtain a durable regression of the tumor and ultimately the cure [50]. One of the strategies of malignant melanoma cells to escape to the immune system control is acquiring the ability to modulate T-cell responses through the activation of the inhibitory immune checkpoints pathways to reduce the immune response [50].

The first T-cell checkpoint inhibitor (CTLA-4), ipilimumab, was approved in 2011. This drug is an antibody targeting the cytotoxic T-lymphocyte-associated antigen 4 (CTLA-4), blocking it to increase the activation and proliferation of T-cell thus boosting the immune response against the cancer cells [51]. Although ipilimumab improved median overall survival [52,53], new treatments that increase the long-term survival of patients with unresectable advanced or MM with special focus on novel checkpoints inhibitors have gradually been approved by the U.S. Food and Drug Administration (FDA) [3,14,54]. In 2014, nivolumab and pembrolizumab, targeting the programmed cell death receptor-1 (PD-1), a T-cell checkpoint inhibitor, were accepted for advanced melanoma treatment [55,56]. PD-1 constitutively expressed by T and B cells can be activated by their ligands, programmed death ligand 1 (PD-L1) and programmed death ligand 2 (PD-L2) suppressing cell activation [57]. Melanoma cells express these ligands on their surface, which results in immunosuppression [58]. Those drugs are antibodies against the immune checkpoint that block the binding PD-1−PD-L1/2 improving the immune response against the melanoma cells. Compared with the classic treatments, nivolumab has shown great improvement in the patient's response rate and in the 1-year overall survival rate in BRAF wild-type melanoma patients [49]. Due to its low toxicity, this antibody could be used as a monotherapy or combined with other treatments. A report on the 5-year outcome for the use of Nivolumab in combination with ipilimumab showed that more than 50% of patients with MM were alive and more than 70% of these remained disease-free [59]. Despite the promising results regarding the improvement of quality life and overall survival rate, there are also cases in which the patients acquire resistance to these drugs, greatly limiting increasing successes [50].

### 13.5.3 Targeted therapy

The principle behind the targeted therapies relies on directing it to specific genes, proteins, or other molecules involved in tumor growth and survival to selectively kill rapidly dividing tumor cells [60]. As described in one of the previous sections, the most frequent mutations in cutaneous melanoma are $BRAF^{V600}$ and NRAS [41], which result in the continuous activation of the mitogen-activated protein kinase (MAPK) pathway [61]. MAPK pathway consists of several kinases such as RAS and RAF, among others, which signaling in cancer cells culminates in cell proliferation, differentiation, and survival [42]. Different agents targeting the MAPK pathway, including selective BRAFi and MEKi, have been developed as therapeutic agents for melanoma [62]. Vemurafenib, the first target therapy approved by the FDA in 2011 to treat patients with advanced melanoma, is a selective inhibitor of mutant $BRAF^{V600}$ [63]. Its use resulted in an improved overall survival and progression-free survival against classic chemotherapy with dacarbazine [64]. Dabrafenib, an alternative inhibitor of mutant $BRAF^{V600}$ approved in 2013 has shown similar results [65]. However, patients treated with both drugs as monotherapies acquired resistance leading to relapse within a short period of time (<1 year) [66].

The acquired resistance from melanoma cells to these agents has been shown to result from the reactivation of the MAPK signaling pathway. Therefore, combined strategies using BRAF and MEK inhibitors have been used to improve the overall response. The combination of trametinib (MEKi) and dabrafenib (BRAFi), in comparison with the administration of dabrafenib alone, showed an improvement in the response rate and overall survival [66]. Comparable results were found with the double therapy of cobimetinib (MEKi) and vemurafenib (BRAFi) in relation to the treatment with vemurafenib alone [67].

## 13.6 In vitro models

Standard 2D cell culture has been the primary tool to understand melanoma cell biology, and to test the efficacy of drugs to treat advanced melanoma providing preliminary screening before preclinical tests [68]. While a monoculture of melanoma cells offers a more simplistic, economical, and faster solution, it lacks many of the 3D native tumor environmental factors thus affecting upstream cellular features such as morphology, division rate, and polarity. Moreover, when attempting to represent the overall pathophysiology of the melanoma, 2D systems fail to recreate the interaction with different cell types and the extracellular matrix (ECM) conditioning the assessment of heterotypic cell–cell and cell–ECM interactions. Certainly, when considering the use of standard cultures of melanoma cells for drug development, those limitations have implications in signal transduction and drug metabolism reducing accuracy and predictive power.

A wide range of 3D melanoma models has been developed and improved in the last decade [69]. Different approaches have been followed to overcome the simplicity of 2D cultures and to try to come closer to the native tumor microenvironment, being more realistic. Spheroids [70] and organotypic cutaneous melanoma models [69], respectively, simpler and more complex approaches have been used both to understand melanoma initiation and progression and to test drug response and resistance.

### 13.6.1 3D in vitro melanoma models

#### *13.6.1.1 Spheroids*

Multicellular tumor spheroids (MCTs) are 3D cell aggregates spontaneously formed when the cell–cell interactions prevail over cell–substrate ones. In relation to 2D cell cultures, MTCs have the advantage of allowing the partial recapitulation of the tumor microenvironment, namely the cell–cell and cell–ECM interactions in 3D, and the gradient of oxygen-related metabolites, nutrients, and soluble factors of a heterogeneous population, which is comparable with a tumor biopsy [71]. For example, the outer layer of a spheroid is characterized by more proliferative cells due to the proximity of oxygen and nutrients in contrast to the center part which presents a more hypoxic environment with quiescent cells mixed with the cellular debris from apoptotic and necrotic cells [72]. Importantly, these features depend on the spheroid size and are only found in spheroids larger than 500 μm in diameter [73]. In addition, MCTs also allow recreating specific molecular fingerprints such as the expression and activity of ERK in the periphery, comparable with what is observed in the tumor [74] and different from the homogeneous signature shown in 2D cultures [75]. These are the main characteristics that have supported the use of MCTs in drug testing. Spheroids of $BRAF^{V600E}$ melanoma cells treated with vemurafenib under hypoxic conditions allowed identifying the involvement of the hepatocyte growth factor (HGF)/Met signaling pathway in the resistance against vemurafenib. Additionally, this platform allowed confirming the restoration of the signaling with pharmacological inhibitors of the tyrosine-protein kinase Met (c-Met)/Akt pathways [76,77].

More complex MCTs composed of fibroblasts, human keratinocyte cell line HaCaT, and $BRAF^{V600E}$ melanoma cells (SK-MEL-28) were used to test the cytostatic drug docetaxel [78]. After treatment, the usual loss of keratinocytes differentiation observed in melanoma tumor progression was stopped and the melanoma cells in the external part of the spheroid became apoptotic. Nonetheless, the remaining melanoma cells showed an increased expression of the ABCB5, recreating a mechanism associated to tumor treatment resistance in patients with BRAF mutation. Overexpression of ABCB5 has been shown to increase the efflux of drugs from target melanoma cells [79]. Overall, this type of spheroid allows evaluating the impact of the drug not only in the cancerous but also in the noncancerous cells, being more accurate.

#### *13.6.1.2 Organotypic cutaneous melanoma models*

During melanoma progression, the whole skin microenvironment is altered. For each particular phase of progression, a very complex interplay between cancer and noncancerous neighboring cells occurs. Organotypic cutaneous melanoma models aim to reflect this complexity in order to (1) better understand the mechanisms behind melanoma progression, from the initial stage (RGP) confined to the epidermis, to more advanced stages (VGP and metastatic phase) with the invasion of the subsequent layers, (2) test the safety and efficacy of new treatments approaches in more advanced preclinical phases, and (3) study the mechanisms of therapeutic resistance.

The most common 3D in vitro cutaneous melanoma models have been built with primary keratinocytes forming a fully stratified epidermis containing melanocytes in the basal layer within the less differentiated keratinocytes, separated by a collagen IV [80,81], laminin-5 [82], and collagen VII [81] basement membrane from the dermis composed by fibroblasts embedded in bovine [80,83] or rat tail [72,81] type I collagen (Fig. 13.3A). The incorporation of different primary [82] or immortalized [72,80,81] melanoma cells with different phenotypes has allowed recapitulating specific features of the different stages of melanoma development. WM35 [80,81] and SBCL2 [72] cells (radial growth), grow as small nests radially along the basement membrane, retaining their proliferative and invasive potential. On the other hand, WM793 cells [80], WM793 [81], and WM115 [72,81] cells

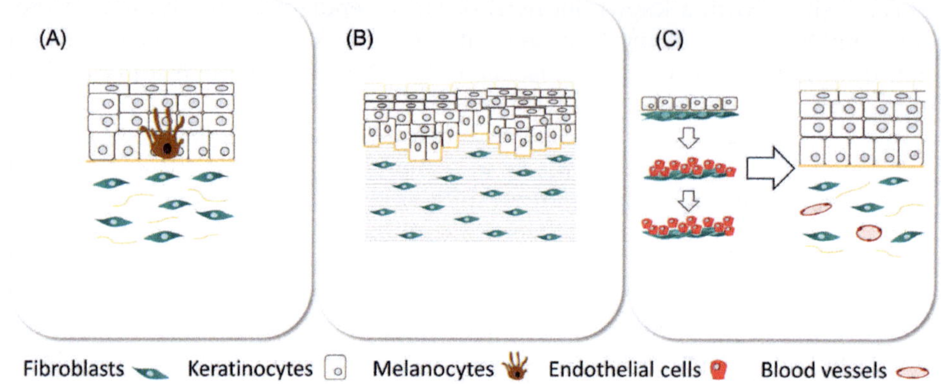


**FIGURE 13.3** Different approaches for dermal compartment recreation of 3D organotypic cutaneous melanoma models with keratinocytes stratified on top. (A) Fibroblasts embedded in type I collagen; (B) fibroblasts seeded in polystyrene—Alvetex—porous scaffold; and (C) overlapping of heterotypic cell sheets fabricated based on cell sheet technology.

(vertical growth) are capable of invading the dermal layer in a vertical manner as a big cluster of cells, while MM 1205Lu [80], WM852 [81], and 451-LU [72] cells grow in all directions colonizing the epidermis and the dermis forming small clusters, or migrate as single cells to regions distant from primary cluster as representation of a more aggressive metastatic behavior. Critically, the inclusion of these cells in the organotypic skin leads to the disruption of the basement membrane and the dermal–epidermal junction near melanoma nests during the development of the tumor. As the degradation of the basement membrane allows the invasion of the dermis layer by melanoma cells [80–82], this type of model has been used to understand the mechanisms behind melanoma initiation and progression.

In a study where WM115 and WM793-B cells (VGP) and their metastatic counterparts WM266-4 and 1205Lu were used, it was possible to confirm the correlation between the expression of CD271, a neurotrophin receptor, the aggressiveness of the melanoma cells measured by their capacity to invade the dermal part of the model [71]. A higher expression of CD271 was observed in the radial stage of melanoma which then disappeared when the dermis invasion started. The silencing and overexpression of CD271, confirmed the more and less aggressive phenotype, respectively. Melanocytes with inducted $BRAF^{V600E}$ mutation and deletion of the p53 gene, leading to apoptosis resistance, were used to create organotypic models in which the specific mechanisms underlining these mutations can be addressed [84]. The use of these $V600E^{+}/p53^{sh}$ melanocytes allowed recreating the development of pigmented lesions similar to melanoma confirming the involvement of those genes in the development of melanoma.

Despite the advantages of the organotypic cutaneous models in relation to MCTs, some organotypic cutaneous melanoma models incorporate homotypic MCTs instead of single melanoma cells [71,73,83]. Overall, the incorporation of a precise number of spheroids with a specific size and in a known localization of the epidermis, in opposition to relying on the random formation of multiple nests by the single melanoma cells, represents a more reproducible method, which can be relevant to use the model as a platform to test drugs [73]. The 3D organotypic cutaneous melanoma models containing spheroids with 500 μm in size prepared with melanoma cells from different progression stages—SBCL2 (RGP), WM-115 (VGP), and 451-LU (MM)—were shown to be more accurate in predicting therapeutic effects than 2D [85]. These melanoma cells were selected from a previous study with 18 cell lines resistant to tumor necrosis factor α-related apoptosis-inducing ligand (TRAIL) signaling. The expression of TRAIL receptor in melanoma cells was induced by a coexposure to a sublethal UVB dose or to the cytotoxic agent cisplatin. Interestingly, in the 3D model, TRAIL + cisplatin treatment killed all the melanoma cells, which are almost unaffected by the TRAIL + UVB combination. In opposition, both treatments are equally effective in the 2D cultures, confirming the importance of 3D models for drug screening in a preclinical stage [72]. This is also reinforced in another study that compared the efficacy of MAPK and AKT pharmacological inhibitors in 2D cultures and in a 3D organotypic cutaneous melanoma model containing 451Lu or SKMel28 cells (MM) [86]. Complete suppression of the invasion of the tumor was observed recreating in vivo tumor regression while in the 2D culture, only the decrease of the melanoma cell proliferation and the enhancing of the apoptosis were observed.

While the potential of the described organotypic models is unquestionable, alternatively, different approaches have been attempted using artificial scaffolds [87], and more recently a material-free strategy that relies on the cell's own capacity to assemble and produce ECM forming tissue (dermis)-like structures [88].

The biomaterial-based model has created seeding fibroblasts in polystyrene—Alvetex—porous scaffold (dermal compartment), stimulated to produce ECM before the incorporation of WM35 (RGP) or SK-mel-28 (MM) from different stages of melanoma progression and of keratinocytes (Fig. 13.3B) [87]. While enabling to recapitulate the different stages of melanoma progression, including the rupture of the continuous collagen IV- and VII-rich basement membrane, no further pathophysiological features were confirmed.

In the more recent material-free organotypic melanoma model, heterotypic cell sheets of fibroblasts, endothelial cells (ECs) and lymphatic endothelial cells (LECs) were prepared to obtain vascularized dermis-like analogs (Fig. 13.3C). Keratinocytes mixed with spheroids of melanoma cells were cultured on top before exposure to the air–liquid interface (ALI) in order to obtain a stratified epidermis [88]. In this model the stromal cells produce their own ECM, thus recreating more closely tumor microenvironment with relevant cell–matrix interactions. Moreover, the incorporation of both endothelial and lymphatic endothelial cells expands the potential of this model to be used in the study of mechanisms behind melanoma metastatic progression. When vemurafenib, a selective inhibitor of mutant $BRAF^{V600}$ used in patients with advanced melanoma, was tested in this model, it was possible to confirm a decrease in the number of proliferative and apoptotic cancer cells. Moreover, peripheral melanoma cells started to produce melanin, which suggests induction of melanogenesis, as already verified effect with the treatment of vemurafenib, demonstrating the potential future use of this model as a personalized platform to test drugs.

## References


[1] Bray F, et al. Global cancer statistics 2018: GLOBOCAN estimates of incidence and mortality worldwide for 36 cancers in 185 countries. CA Cancer J Clin 2018;68:394–424.

[2] Olszanski AJ. Current and future roles of targeted therapy and immunotherapy in advanced melanoma. J Manag Care Pharm 2014;20:S1–79.

[3] Wada-Ohno M, Ito T, Furue M. Adjuvant therapy for melanoma. Curr Treat Options Oncol 2019;20:63.

[4] Kozar I, Margue C, Rothengatter S, Haan C, Kreis S. Many ways to resistance: how melanoma cells evade targeted therapies. Biochim Biophys Acta - Rev Cancer 2019;1871:313–22.

[5] Fares CM, Van Allen EM, Drake CG, Allison JP, Hu-Lieskovan S. Mechanisms of resistance to immune checkpoint blockade: why does checkpoint inhibitor immunotherapy not work for all patients? Am Soc Clin Oncol Educ B 2019;39:147–64. Available from: https://doi.org/10.1200/edbk_240837.

[6] Khazaei Z, et al. Global incidence and mortality of skin cancer by histological subtype and its relationship with the human development index (HDI); an ecology study in 2018. WCRJ 2019;6:e1265.

[7] Fauci JM, et al. Cancer facts & figures. Atlanta: American Cancer Society 2016. Available from: https://www.cancer.org/research/cancer-facts-statistics/all-cancer-facts-figures/cancer-facts-figures-2016.html.

[8] Schadendorf D, et al. Melanoma. Lancet 2018;392:971–84.

[9] Coricovac D, et al. Cutaneous melanoma—a long road from experimental models to clinical outcome: a review. Int J Mol Sci 2018;19:E1566.

[10] Rabbie R, Ferguson P, Molina-Aguilar C, Adams DJ, Robles-Espinoza CD. Melanoma subtypes: genomic profiles, prognostic molecular markers and therapeutic possibilities. J Pathol 2019;247:539–51.

[11] Hayward NK, et al. Whole-genome landscapes of major melanoma subtypes. Nature 2017;545:175–80.

[12] Clark WH, Elder DE, Van Horn M. The biologic forms of malignant melanoma. Hum Pathol 1986;17:443–50.

[13] Clark WH, From L, Bernardino EA, Mihm MC. The histogenesis and biologic behavior of primary human malignant melanomas of the skin. Cancer Res 1969;29:705–27.

[14] Rozeman EA, Dekker TJA, Haanen JBAG, Blank CU. Advanced melanoma: current treatment options, biomarkers, and future perspectives. Am J Clin Dermatol 2018;19:303–17.

[15] Tucker MA. Melanoma epidemiology. Hematol Oncol Clin North Am 2009;23:383–95.

[16] Craig S, Earnshaw CH, Virós A. Ultraviolet light and melanoma. J Pathol 2018;244:578–85.

[17] Bliss JM, et al. Risk of cutaneous melanoma associated with pigmentation characteristics and freckling: systematic overview of 10 case-control studies. Int J Cancer 1995;62:367–76.

[18] Dębniak T. Familial malignant melanoma - overview. Hered Cancer Clin Pract 2004;2:123.

[19] Le Clair MZ, Cockburn MG. Tanning bed use and melanoma: establishing risk and improving prevention interventions. Prev Med Rep 2016;3:139–44.

[20] Shain AH, Bastian BC. From melanocytes to melanomas. Nat Rev Cancer 2016;16:345–58.

[21] Valverde P, Healy E, Jackson I, Rees JL, Thody AJ. Variants of the melanocyte-stimulating hormone receptor gene are associated with red hair and fair skin in humans. Nat Genet 1995;11:328–30.

[22] Wenczl E, et al. Pheo)melanin photosensitizes UVA-induced DNA damage in cultured human melanocytes. J Invest Dermatol 1998;111:678–82.

[23] Tanaka H, et al. The pro-oxidant activity of pheomelanin is significantly enhanced by UVA irradiation: benzothiazole moieties are more reactive than benzothiazine moieties. Int J Mol Sci 2018;19:E2889.

[24] Zhang J, Bowden GT. Activation of p38 MAP kinase and JNK pathways by UVA irradiation. Photochem Photobiol Sci 2012;11:54–61.

[25] Chadyšiene R, Girgždys A, Chadyšienn R. Assessment of ultraviolet (UV) radiation from technical sources. J Env Eng Landsc Manag 2009;17:164–70.

[26] Epe B. DNA damage spectra induced by photosensitization. Photochem Photobiol Sci 2012;11:98–106.

[27] Cortat B, et al. The relative roles of DNA damage induced by UVA irradiation in human cells. Photochem Photobiol Sci 2013;12:1483–95.

[28] Mukhtar H, Bickers DR. Environmental skin cancer: mechanisms, models and human relevance. Cancer Res 1993;53:3439–42.

[29] Bosch R, et al. Mechanisms of photoaging and cutaneous photocarcinogenesis, and photoprotective strategies with phytochemicals. Antioxidants 2015;4:248–68.

[30] Puntervoll HE, et al. Melanoma prone families with CDK4 germline mutation: phenotypic profile and associations with MC1R variants. J Med Genet 2013;50:264–70.

[31] Greene MH, et al. High risk of malignant melanoma in melanoma-prone families with dysplastic nevi. Ann Intern Med 1985;102:458–65.
[32] Hawkes JE, Truong A, Meyer LJ. Genetic predisposition to melanoma. Semin Oncol 2016;43:591–7.
[33] Lynch HT, Fusaro RM, Lynch JF, Brand R. Pancreatic cancer and the FAMMM syndrome. Familial Cancer 2008;7:103–12.
[34] Soura E, Eliades PJ, Shannon K, Stratigos AJ, Tsao H. Hereditary melanoma: update on syndromes and management emerging melanoma cancer complexes and genetic counseling. J Am Acad Dermatol 2016;74:395–407.
[35] Middlebrooks CD, et al. Analysis of the CDKN2A gene in FAMMM syndrome families reveals early age of onset for additional syndromic cancers. Cancer Res 2019;79:2992–3000.
[36] Jiao Y, Feng Y, Wang X. Regulation of tumor suppressor gene CDKN2A and encoded p16-INK4a protein by covalent modifications. Biochemistry (Moscow) 2018;83:1289–98.
[37] Sharpless NE, Chin L. The INK4α/ARF locus and melanoma. Oncogene 2003;22:3092–8.
[38] Eckerle Mize D, Bishop M, Resse E, Sluzevich J. In: Riegert-Johnson DL, Boardman LA, Hefferon T, Roberts M, editors. Familial atypical multiple mole melanoma syndrome in cancer syndromes. Bethesda, MD: National Center for Biotechnology Information; 2009.
[39] Dobrowolski R, Hein R, Buettner R, Bosserhoff AK. Loss of p14ARF expression in melanoma. Arch Dermatol Res 2002;293:545–51.
[40] Udayakumar D, Mahato B, Gabree M, Tsao H. Genetic determinants of cutaneous melanoma predisposition. SemCutan Med Surg 2010;29:190–5.
[41] Cancer Genome Atlas Network. Genomic classification of cutaneous melanoma. Cell 2015;161:1681–96.
[42] Davies H, et al. Mutations of the BRAF gene in human cancer. Nature 2002;417:949–54.
[43] Shain AH, et al. The genetic evolution of melanoma from precursor lesions. N Engl J Med 2015;373: 1926–36.
[44] Clark GS, et al. Surgical management of melanoma in situ on chronically sun-damaged skin. Cancer Control 2008;15:216–24.
[45] Wilson MA, Schuchter LM. Chemotherapy for melanoma. Cancer Treat Res 2016;167:209–29.
[46] Jilaveanu LB, Aziz SA, Kluger HM. Chemotherapy and biologic therapies for melanoma: do they work? Clin Dermatol 2009;27:614–25.
[47] Kirkwood JM, Ibrahim JG, Sondak VK, Ernstoff MS, Ross M. Interferon alfa-2a for melanoma metastases [5]. Lancet 2002;359:978–9.
[48] Atzpodien J, et al. Combination chemotherapy with or without s.c. IL-2 and IFN-α: results of a prospectively randomized trial of the cooperative advanced malignant melanoma chemoimmunotherapy group (ACIMM). Br J Cancer 2002;86:179–84.
[49] Pasquali S, Hadjinicolaou AV, Chiarion Sileni V, Rossi CR, Mocellin S. Systemic treatments for metastatic cutaneous melanoma (Review). Cochrane Database Syst Rev Syst 2018;2:CD011123.
[50] Liu D, Jenkins RW, Sullivan RJ. Mechanisms of resistance to immune checkpoint blockade. Am J Clin Dermatol 2019;20:41–54.
[51] Payandeh Z, et al. Immune therapy of melanoma: overview of therapeutic vaccines. J Cell Physiol 2019;234:14612–21.
[52] Balch CM, et al. Final version of 2009 AJCC melanoma staging and classification. J Clin Oncol 2009;27:6199–206.
[53] Dimitriou F, Braun RP, Mangana J. Update on adjuvant melanoma therapy. Curr Opin Oncol 2018;30:118–24.
[54] Postow MA, Callahan MK, Wolchok JD. Immune checkpoint blockade in cancer therapy. J Clin Oncol 2015;33:1974–82.
[55] Sullivan RJ, Flaherty KT. Pembrolizumab for treatment of patients with advanced or unresectable melanoma. Clin Cancer Res 2015;21:2892–7.
[56] Raedler LA. Opdivo (nivolumab): second PD-1 inhibitor receives FDA approval for unresectable or metastatic melanoma. Am Health Drug Benefits 2015;8:180–3.
[57] Jin H-T, Ahmed R, Okazaki T. Role of PD-1 in regulating T-cell immunity. Curr Top Microbiol Immunol 2010;350:17–37.
[58] Kleffel S, et al. Melanoma cell-intrinsic PD-1 receptor functions promote tumor growth. Cell 2015;162:1242–56.

[59] Five-year outcomes for opdivo (nivolumab) in combination with yervoy (ipilimumab) demonstrate durable long-term survival benefits in patients with advanced melanoma | BMS newsroom. Available from: <https://news.bms.com/press-release/corporatefinancial-news/five-year-outcomes-opdivo-nivolumab-combination-yervoy-ipilimu> [accessed 16.02.2020].

[60] Padma VV. An overview of targeted cancer therapy. BioMedicine (Neth) 2015;5:1–6.

[61] Sumimoto H, Imabayashi F, Iwata T, Kawakami Y. The BRAF-MAPK signaling pathway is essential for cancer-immune evasion in human melanoma cells. J Exp Med 2006;203:1651–6.

[62] Inamdar GS, Madhunapantula SV, Robertson GP. Targeting the MAPK pathway in melanoma: why some approaches succeed and other fail. Biochem Pharmacol 2010;80:624–37.

[63] Flaherty KT, Yasothan U, Kirkpatrick P. Vemurafenib. Nat Rev Drug Discov 2011;10:811–12.

[64] Chapman PB, et al. Improved survival with vemurafenib in melanoma with BRAF V600E mutation. N Engl J Med 2011;364:2507–16.

[65] Hauschild A, et al. Dabrafenib in BRAF-mutated metastatic melanoma: a multicentre, open-label, phase 3 randomised controlled trial. Lancet 2012;380:358–65.

[66] Rissmann R, Hessel MHM, Cohen AF. Vemurafenib/dabrafenib and trametinib. Br J Clin Pharmacol 2015;80:765–7.

[67] Larkin J, et al. Combined vemurafenib and cobimetinib in *BRAF* -mutated melanoma. N Engl J Med 2014;371:1867–76.

[68] Saji Joseph J, Tebogo Malindisa S, Ntwasa M. Two-dimensional (2D) and three-dimensional (3D) cell culturing in drug discovery. London: IntechOpen; 2019. Available from: https://doi.org/10.1016/j.ygyno.2011.05.023.

[69] Marconi A, Quadri M, Saltari A, Pincelli C. Progress in melanoma modelling in vitro. Exp Dermatol 2018;27:578–86.

[70] Lin R-Z, Chang H-Y. Recent advances in three-dimensional multicellular spheroid culture for biomedical research. Biotechnol J 2008;3:1172–84.

[71] Saltari A, et al. CD271 down-regulation promotes melanoma progression and invasion in three-dimensional models and in zebrafish. J Invest Dermatol 2016;136:2049–58.

[72] Vö Rsmann H, et al. Development of a human three-dimensional organotypic skin-melanoma spheroid model for in vitro drug testing. Cell Death Dis 2013;4:719.

[73] Müller I, Kulms D. A 3D organotypic melanoma spheroid skin model. J Vis Exp 2018;2018.

[74] Zhuang L, et al. Activation of the extracellular signal regulated kinase (ERK) pathway in human melanoma. J Clin Pathol 2005;58:1163–9.

[75] Haass NK, et al. The mitogen-activated protein/extracellular signal-regulated kinase kinase inhibitor AZD6244 (ARRY-142886) induces growth arrest in melanoma cells and tumor regression when combined with docetaxel. Clin Cancer Res 2008;14:230–9.

[76] Organ SL, Tsao MS. An overview of the c-MET signaling pathway. Therap Adv Med Oncol 2011;3:S7–19.

[77] Qin Y, et al. Hypoxia-driven mechanism of vemurafenib resistance in melanoma. Mol Cancer Ther 2016;15:2442–54.

[78] Klicks J, Maßlo C, Kluth A, Rudolf R, Hafner M. A novel spheroid-based co-culture model mimics loss of keratinocyte differentiation, melanoma cell invasion, and drug-induced selection of ABCB5-expressing cells. BMC Cancer 2019;19:402. Available from: https://doi.org/10.1186/s12885-019-5606-4.

[79] Xiao J, Egger ME, Mcmasters KM, Hao H. Differential expression of ABCB5 in BRAF inhibitor-resistant melanoma cell lines. BMC Cancer 18, 2018, 675. Available from: https://doi.org/10.1186/s12885-018-4583-3.

[80] Li L, Fukunaga-Kalabis M, Herlyn M. The three-dimensional human skin reconstruct model: a tool to study normal skin and melanoma progression. J Vis Exp 2011;54:2937. Available from: https://doi.org/10.3791/2937.

[81] Meier F, et al. Human melanoma progression in skin reconstructs biological significance of bFGF. Am J Pathol 2000;156:193–200.

[82] Kiowski G, et al. Engineering melanoma progression in a humanized environment in vivo. J Invest Dermatol 2012;132:144–53.

[83] Lee JT, et al. PLX4032, a potent inhibitor of the B-Raf V600E oncogene, selectively inhibits V600E-positive melanomas. Pigment Cell Melanoma Res 2010;23:820–7.

[84] Yu H, et al. The role of BRAF mutation and p53 inactivation during transformation of a subpopulation of primary human melanocytes. Am J Pathol 2009;174:2367–77.

[85] Gong H, Cheng W, Wang Y. Tumor necrosis factor related apoptosis inducing ligand inhibits the growth and aggressiveness of colon carcinoma via the exogenous apoptosis signaling pathway. Exp Ther Med 2018;17:41.

[86] Meier F, et al. Combined targeting of MAPK and AKT signalling pathways is a promising strategy for melanoma treatment. Br J Dermatol 2007;156:1204–13.

[87] Hill DS, et al. A novel fully humanized 3D skin equivalent to model early melanoma invasion. Mol Cancer Ther 2015;14:2665–73.

[88] Bourland J, Fradette J, Auger FA. Tissue-engineered 3D melanoma model with blood and lymphatic capillaries for drug development. Sci Rep 2018;8:13191.

CHAPTER

# 14

# 3D scaffold materials for skin cancer modeling

*Amir Zarebkohan*[1,2,*], *Roghayeh Sheervalilou*[3,*], *Roya Ghods*[4,5], *Subhas C. Kundu*[6] *and* *Mazaher Gholipourmalekabadi*[5,7,8]

[1]Department of Medical Nanotechnology, Faculty of Advanced Medical Sciences, Tabriz University of Medical Sciences, Tabriz, Iran [2]Drug Applied Research Center, Tabriz University of Medical Sciences, Tabriz, Iran [3]Cellular and Molecular Research Center, Resistant Tuberculosis Institute, Zahedan University of Medical Sciences, Zahedan, Iran [4]Oncopathology Research Center, Iran University of Medical Sciences, Tehran, Iran [5]Department of Molecular Medicine, Faculty of Advanced Technologies in Medicine, Iran University of Medicine Sciences, Tehran, Iran [6]3B's Research Group, I3Bs—Research Institute on Biomaterials, Biodegradables and Biomimetics, University of Minho, Headquarters of the European Institute of Excellence on Tissue Engineering and Regenerative Medicine, Guimarães, Portugal [7]Cellular and Molecular Research Center, Iran University of Medical Sciences, Tehran, Iran [8]Department of Tissue Engineering & Regenerative Medicine, Faculty of Advanced Technologies in Medicine, Iran University of Medical Sciences, Tehran, Iran

## 14.1 Introduction

Today, skin cancer is recognized as the most common malignancy in the population [1–3]. Based on histological features of the malignant cells, skin cancer can be classified into two primary categories: nonmelanoma and melanoma. The most common type of skin malignancy, the nonmelanoma form comprises of basal cell carcinoma (BCC) and squamous cell carcinoma (SCC). The rarely fatal BCC and SCC account for 80% and 20% of all nonmelanoma skin cancer cases. Although melanoma is only observed in 4% of

* These authors contributed equally to this work.

*Biomaterials for 3D Tumor Modeling*
DOI: https://doi.org/10.1016/B978-0-12-818128-7.00014-9

patients with skin cancer, it single-handedly stands responsible for 74% of the death toll caused by skin malignancies [4,5]. The annual incidence rate of skin cancer is estimated to be 5.4 in a million in the United States. Therefore, it is not an overstatement to say that melanomas are responsible for over 10,000 deaths per year [6].

The different diagnostic methods are available for detection of skin cancer, including visual identification, clinical screening followed by dermoscopic analysis, biopsy, and histopathological examinations [6]. Skin cancer can be cured if it is detected in early stages, however, advanced forms of skin cancer almost always have a poor prognosis, with a median survival of about a few months [7] due to the high-metastatic potential and the resistance to chemo- and radiotherapies [8].

Proper models of skin malignancies are required to give us a better understanding of the molecular networks involved in the progression of the disease, and the possible strategies to develop novel therapeutics [9]. According to the European Union Legislation, it is crucial to carefully analyze all the chemical agents to which mankind is exposed. As of recent, it has been suggested to lessen the overwhelming number of experiments conducted on animal models [10]. The main disadvantage of such experiments is that they often poorly predict the human immune response, leading to the development of incompetent drugs. This does not even reach the clinical trials [11–13]. Hence, to accomplish the best imitation of cancer cell behavior outside the human body, development of identical models appears to be a prerequisite. Over the last four decades, it has been shown that tumor microenvironment plays a crucial role in the formation, growth, differentiation, and metastasis of tumors. It has also been demonstrated that intravasation and extravasation of cancer cells strongly depend on the balance between the stimulatory and inhibitory factors of growth in tumor microenvironment. [14]. Thus, more cell-based testing methods are required to pave the way for discovery of processes involved in tumor development, and drug research [15,16]. It has been suggested that the application of cell-based systems demonstrating in vivo behavior might reduce costs, improve the efficacy, and increase the predictive accuracy of the drug discovery [17].

Two-dimensional (2D) cell cultures and animal models have been incorporated in cancer investigations as the routine approach. However, it has widely been suggested that both of them poorly correlate with human disease states [18,19]. The 2D cell cultures cannot thoroughly mimic the microenvironment of tumor cells that naturally occur in the body. In other words, the cells in 2D media exhibit the shape and gene expression patterns of key genes involved in tumor development, that are somehow different from the three-dimensional (3D) models in reality [18,19]. Hence, animal models have been proposed due to their advantages over cell cultures, and similarity to 3D context [18]. Murine models possess intrinsic microenvironment complexity that closely represents the genetic basis of tumor development and metastasis to human molecular networks. This factor has been considered in better prediction of drug pharmacodynamics and efficacy in comparison to 2D cultures. However, there are some limitations in this area, as well as the inability to mimic human-specific features associated with tumors, and the response to therapeutic agents [19]. Overall, because of the limitations of 2D cell cultures in mimicry of the natural tumor microenvironment, and the high costs and issues associated with animal models, both approaches have become less valuable for routine cancer research. In this respect, 3D in vitro cancer models have emerged as an alternative strategy to conventional methods, and have been shown to possess the potential to replicate the natural microenvironment of tumors in a relatively simple and inexpensive manner [20].

In living organisms, the in situ environment of a single cell consists of the cells immediately adjacent to it, along with the extracellular matrix (ECM). The ECM and the receptors found within are important to sustain the epithelial polarity and differentiation [21,22]. These receptors also determine the phenotype of mammary epithelial cells regardless of their genotype [23]. Several studies have indicated that 3D culturing of skin cells might lead to a more physiologically relevant demonstration of morphology seen with fibroblasts and keratinocytes, compared with that of the monolayered cells in 2D cultures [24,25]. Interestingly, this seems to hold true whether the cells are cultured in a physiologically relevant dermis or a simple 3D polystyrene matrix [26]. Thus, the importance of the development of appropriate 3D models is most evident, because 2D models might not be able to simulate these properties.

According to the American Cancer Society statistics in 2019, 104,350 new cases and 11,650 deaths have been estimated for patients with skin cancer [27]. Hence, identification and avoidance of risk factors are required to intercept these increasing trends. Else, early diagnosis and treatment are needed to decrease the number of deaths associated with skin cancer. On the other hands, prognosis is improved when lesions are detected in early stages [28].

Today, the methods commonly used for simulating tumor and cancer cell growth include 2D culture, in vitro spheroids, tumor formation in ordinary 3D gels, bioprinting, and last but not least, using animal models (Fig. 14.1) with various rate of advantages and disadvantages [29,30].

As of recent, developing 3D system has been considered in skin cancer research. The computational and in-silico simulation studies have also been applied following the increase in our knowledge and understanding of the effects of various factors on different cellular and biological stimulatory and inhibitory cascades [31].

To extend our knowledge about human diseases, it is imperative that the engineered tissue models should faithfully mimic to their in vivo counterparts. Inability to simulate the characteristic morphology, phenotype, and behavior of a naturally occurring tissue will result in a limited efficacy for 2D in vitro models. For instance, human keratinocyte monolayer cultures display a limited stratification, partial differentiation, and hyperproliferative growth [32]. Such divergent characteristics may negatively affect the complex cellular responses due to the lack of a 3D tissue architecture. If a signaling pathway should be biologically meaningful, this would only be mediated by the linkage of adhesion and growth. That is, an optimal level of functioning is only reached when the cells are organized in 3D pattern. This is a far-fetched goal in 2D culture systems [33,34]. It is, therefore, pivotal to develop 3D cultures that can exhibit the architectural features seen in vivo. This would allow the engineering of tissue models that could be useful in the study of human diseases in their appropriate tissue context [35]. Obviously, there is a need for immunocompetent and scalable models of human skin disease with a vast selection of biomarkers [9].

In this chapter, we review various approaches that are currently available for generating 3D tumor models such as different types of tumor spheroid formation techniques, bioprinting, microfluidics, assembling strategies, and in vitro organoid models with emphasize on the skin cancers. In addition, the effects of mechanobiological reciprocal

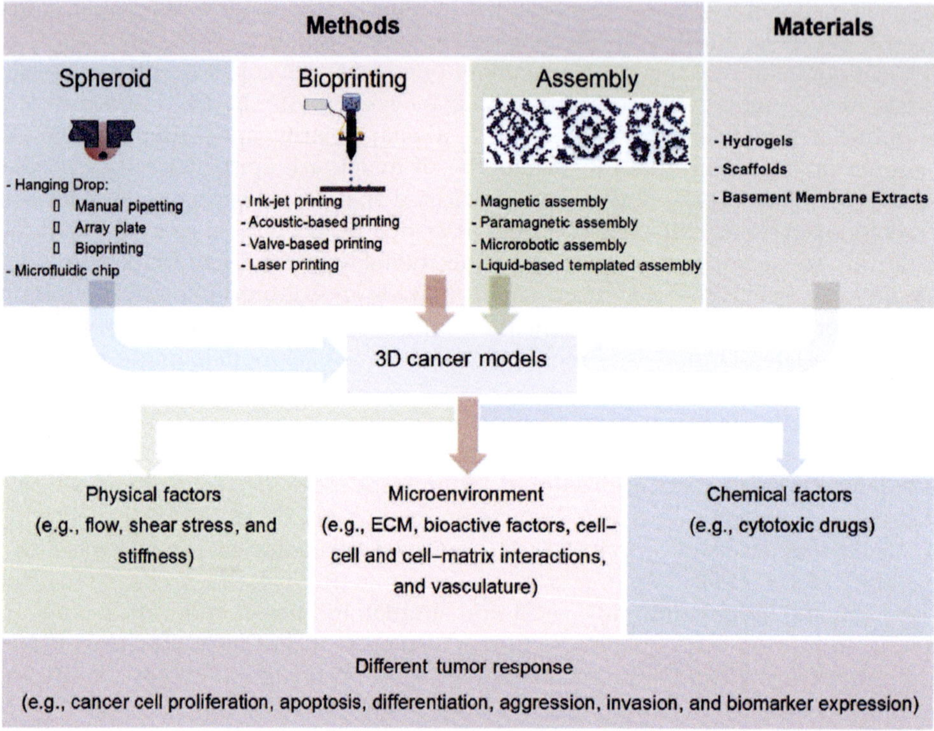


FIGURE 14.1 Different materials and methods for designing of 3D cancer models. Bioprinting, spheroids, and assembly are the main technologies used for engineering of 3D cancer models. Hydrogels and basement membrane extracts are the most common materials used for fabrication of 3D cancer models. Mimicking the physical and chemical factors as well as composition of the native microenvironment make it possible to develop a 3D cancer model with high similarity to cancer in vivo, which can profoundly affect the cancer cell responses such as proliferation, aggression, invasion, and others. Source: *Reprinted by copyright permissions from Asghar W, et al. Engineering cancer microenvironments for in vitro 3-D tumor models. Mater Today 2015;18(10):539–53 [29].*

interactions between cancer cells, such as malignant skin cells, and culture media cues, architect of culture media, biological factors, and chemical factors on the hallmarks of cancers are discussed. There are a few articles using biomaterials for studying skin cancer in 3D in vitro models. Therefore, we highlight some of the future directions for 3D skin cancer models toward applications in drug and gene delivery systems, drug discovery, and screening.

## 14.2 Effective factors in cell culture; 2D and 3D models

Over a century, the conventional 2D cell cultures have been used as in vitro models for studying a wide variety of factors as well as cell behaviors and responses to stimulations from biophysical and biochemical cues, exploring various drug administrations, nanopharmaceutical designing, and modeling of cancer. These approaches have significantly advanced our knowledge about cell behaviors. The growing body of evidence shows that under some circumstances, the 2D models can result in cell bioactivities that might deviate the in vivo response to a larger extent [36]. The cross-talk between different cells have many disadvantages, which can be divided into a few parts. The reason for such division is that the growth, differentiation, motility, and migration of cancer cells in tumor microenvironment can be mediated by using physicochemical cues [37].

In summary, there are some limitations about stimulating some factors in 2D cultures. Here, we intend to present the important differences between 2D and 3D cell culture systems separately and briefly to state why 3D in vitro tumor models are needed. Engineering of ideal 3D tumor models involves cancer and noncancer cells, appropriate biomaterial scaffolds to provide 3D substrate with high mimic to particular tissue matrix, cytokines, vascular perfusion, and biomechanical and biophysical cues (Fig. 14.2).

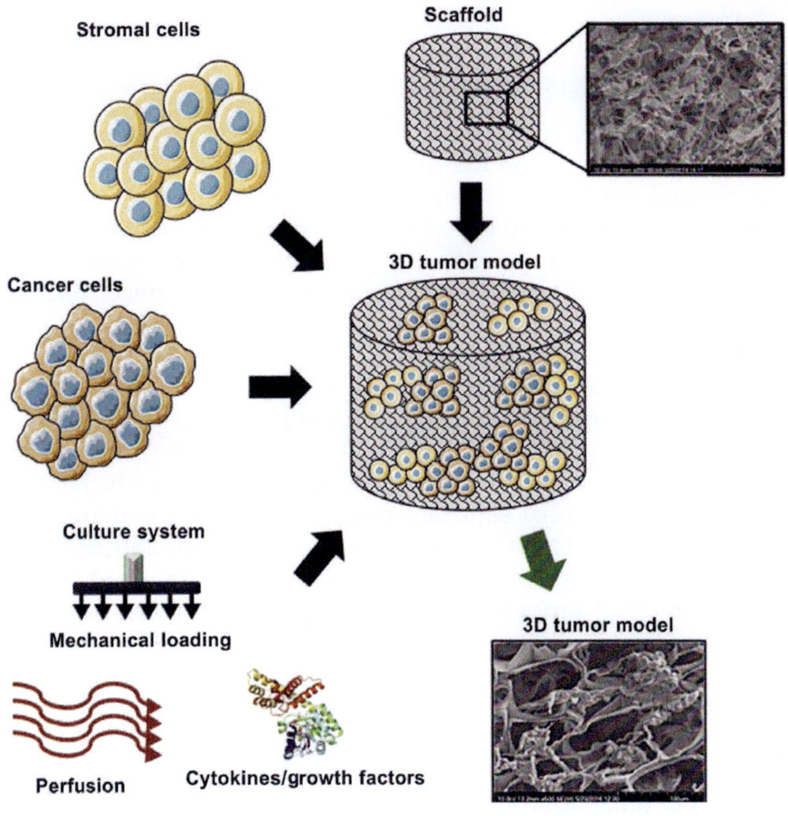


FIGURE 14.2 Engineering of 3D tumor model. Engineering of 3D tumor models involves cancer cells, and non-cancerous stromal supporting cells, biomaterial scaffolds for providing 3D substrate with high similarity to particular tissue matrix, cytokines, vascular perfusion, and biomechanical and biophysical cues. Source: *Reprinted by copyright permissions from Liu Z, Vunjak-Novakovic G. Modeling tumor microenvironments using custom-designed biomaterial scaffolds. Curr Opin Chem Eng 2016;11:94–105 [30].*

### 14.2.1 Ethical and economical parameters

One of the most important hardships regarding the common methods is that working on animals is laced with lot of ethical issues. In more advanced phases of research that require human subjects, ethical and economic issues show up again [38]. Thus, development of 3D in vitro models with the most similarity to tumor microenvironment is expected to revolutionize the screening of malignancies and introduction of different treatment strategies for cancers.

The importance of a 3D tissue structure is only fathomed if we take into account the fact that nonmalignant and malignant epithelial cells are potentially similar under 2D culture conditions; however they behave very differently when organized in 3D cultures [39]. The growth of nontransformed epithelial cells on a 3D basement membrane matrix leads to formation of organized, growth-arrested acini structures analogous to the architecture of normal tissue found in humans. In contrast, the cells derived from carcinomas formed disorganized cellular masses, that do not respond to the growth inhibitory cues presented by the matrix [40].

### 14.2.2 Biological parameters

To explain the effects of cell-to-cell and cell-to-extracellular medium adhesion, and ultimately, tumor tissue architecture on the cell behavior, a wide range of studies have been conducted [41]. Several studies have demonstrated that intracellular signaling is usually subjected to great alterations due to changes. They occur in the extracellular physical elements. Interestingly, resistance to anoikis is a phenomenon that occurs in cancer cells. In this mechanism, following the loss of attachments to ECM, the cancer cells enter into the blood and lymphatic vessels. Through the second step, the cancer cells should survive the hostile environment of the circulation, and therefore, resist anoikis. It has been suggested that loss of contact with the ECM triggers anoikis, a specialized form of cell death, through the activity of c-Poly (ADP-ribose) polymerase (PARP) and c-caspase 3. Metastatic cancer cells might be able to develop resistance to anoikis by activating several signaling pathways that infringe the extrinsic and mitochondrial mediated apoptosis. For example, some stressful conditions trigger special signaling pathways, as well as autophagic dormancy. This phenomena induces notable decrease in cancer cell integrin adhesion to extracellular PAR proteins. Detached cells can evoke anoikis (a form of programmed cell death) [42,43]. According to the mentioned process, there is no way for simulation of such complicated possesses in 2D cultures.

#### ***14.2.2.1 Angiogenesis capabilities***

Another crucial issue, which has widely been explored in cancer biology, is angiogenesis. A wide range of studies have attested to the hypothesis that angiogenesis starts from dysplasia [14]. On the other hand, the role of various stromal cells in this process, like tumor-associated macrophages (TAMs), myeloid dendritic cells (MDCs), fibroblasts in the tumor microenvironment, endothelial cells, and fat cells within this medium is inevitable [44]. Furthermore, the effects of various physical features of culture medium, such as environmental hypoxia, oxygen, and nutrient levels on convoking the mentioned cells for angiogenesis are undeniable [45]. Another noteworthy factor in tumor development is hypoxia that affects the tumor growth through induction of necrosis and releasing preinflammatory material, due to the lack of proper oxygen and nutrients [14]. Obviously, in 2D culture media, it is impossible to effectively activate the cells for angiogenesis in a similar way to tumor microenvironment in the body.

#### 14.2.2.2 *Attachment capabilities to the extracellular matrix*

Another important factor affecting the adhesion behavior of cancer cell growth and invasion is the components of ECM, such as collagen. As a result of the presence of the Arg–Gly–Asp (RGD) tripeptide in its sequence, this protein can enhance the integrins' adhesion to the cancer cell surface, and play an important role in inducing growth, invasion, and infiltration through the epithelial–mesenchymal transition (EMT) pathway [46,47]. Accordingly, there are other proteins in addition to integrins and cadherins [48,49], which have an important role in creating tensile forces within the cell cultures, including platelet endothelial cell adhesion molecule-1 (PECAM-1) and vinculin. These forces could alter internal signaling pathways to induce cellular motility and invasiveness. Unfortunately, because of the absence of these molecules in traditional 2D culture media, the evaluation of cancer cell behavior was not readily possible [50].

### 14.2.3 Physical parameters

Other critical but neglected factors that play important parts in tumor cell behavior are the physical parameters of the medium, as well as the stiffness of the environment, hardness of the medium, and density of the cultured cells [30]. Interestingly, tumorigenesis can induce physical and chemical alterations in ECM surrounding the tumor. Because of these modifications, cellular hallmark behaviors such as recognition of matrix geometry and rigidity, cell polarization and motility, cytoskeletal reorganization, and proliferation can be altered. In other words, tumor microenvironment can manipulate the pathways involved in tumor formation, progression, and metastasis [30,51]. From a physical point of view, this happens due to an increase in tumor area rigidity. As a result of the changes made in ECM, the flexibility and critical traction of cells change, and this leads to promotion in cancer cell invasion [50].

#### 14.2.3.1 *Cell density, proteins, and adhesion molecules*

As mentioned above, physical parameters can affect the cell properties. In this respect, studies have shown that inability of a normal cell in adaption with physical cues of the culture media may trigger malignant cascades. For example, growing of cells in an improper microenvironment can induce tumor development process [48]. Additionally, another effective factor is the cellular density, which strictly depends on the balance between cell-to-cell and cell-to-culture medium forces [49]. Recently, Khetan et al. have reported that E-cadherin might control the myosin-related contractions within the cells. The consequent alteration in the contractility of the cells may dramatically affect the cancer cell behavior. On the other hand, nearly every single strong cell-to-cell and cell-to-substrate connection can directly increase the contractility, motility, and invasiveness of cancer cells [50]. Diescher et al. have demonstrated that the cells cultured on polystyrene substrate have showed an increased level of actin and myosin expression in in vitro. It has been concluded that due to the rigid nature of the medium, the cultured cells exhibited a more invasive behavior [52,53]. The results of this study show how the adhesion of these molecules to each other changed the tissue properties and mechanical forces, and how these properties determine the tissue shape, cell dynamics, dimensions, cell contraction capabilities, based on the rigidity and flexibility of the culture medium [50]. Consequently, we may conclude that the change in the expression profile of actinomyosin complex leads to alterations in phenotype and behavior of the cancer cells [54]. In other words, this is what exactly happens in vivo, where ECM compound, cell-to-cell and cell-to-medium adhesion affect the cell behavior and morphology [51,55].

#### 14.2.3.2 *Surface properties*

Another important physical parameter is the surface smoothness, which can affect the cell behavior. It has been shown that a moderate smoothness of substrate could be effective on cell proliferation [56]. Interestingly, this idea has been inspired by the patterns of natural histology [57]. It seems that the local adhesion and contraction ability arising from cellular contact surface is another important factor. This may affect the cell behavior within the medium [58]. For example, nanopatterns on the substrate largely affect the cell behavior [59]. The effects of nano- and microscaffold patterns on the behavior of cells have been studied. It has been well documented that the behavior of immune cells, such as infiltration and the secretion of proinflammatory agents in the microenvironment of the tumors, is altered by these histopathological patterns. These patterns are effective even on cellular response to drugs, as well as cellular behavior [57]. One of the most important signaling routes involved in this process is the Wingless-related integration site (Wnt)/β-catenin pathway [60]. Clearly, these physical effects cannot be observed in 2D cell culture, hence, the unrealistic cellular responses to stimulators or drugs.

### 14.2.4 Tumor microenvironmental properties

In numerous studies, the behavior of cancer stem cells and normal stem cells within tumors, have been explored and demonstrated. These interactions are largely dependent on the medium in which they are located [61]. The crucial issue about cancer stem cells is dependence of metastasis to these cells [14]. Investigations have also shown that these cells have a very strong cross-talk with stromal cells, and are able to change their behavior. Even the cells of the immune system by secreting preinflammatory agents and growth factors like colony-stimulating factor-1 (CSF-1) have influenced [62]. However, the findings have demonstrated that these cells cannot represent their own normal behavior in terms of growth, differentiation, and expansion due to their inability to form niches in 2D culture, as opposed to 3D culture systems [63,64].

As mentioned before, tumor microenvironment contains a large variety of cells, including TAM, neutrophils, regulatory T cells (Tregs), T-helper cells, B cells, mesenchymal stem cells, and tumor-associated fibroblasts. Each of these cells, depending on the type of cancer and its stage, has various functions, such as immune system paralysis, secretion of the proinflammatory factors and matrix metalloproteinases (MMPs), angiogenesis, and various growth factors in response to the cancer cells [14,65]. It has been shown that cancer stem cells do not have the capability to condense and increase their number in traditional 2D systems. The mechanism involved in this increase is not completely understood yet [63]. However, it has been observed that cancer stem cells affect the formation of tumor microenvironments and their properties. Reciprocally, the tumor environment also affects the behavior of the stem cells [66,67]. Studies have shown that the pluripotent and stemness properties, and the increased expression of P-glycoprotein (P-gp) in stem cells, are largely dependent on tumor microenvironment, and this can affect the drug resistance of the cells [68]. It has also been suggested that increase in the expression of ECM proteins might lead to a diminished drug transfer through these proteins. This is considered as a defense mechanism for cancer stem cells. The feature may undergo alterations based on the medium in which the cells are cultured, and negatively affect the results [69]. Interleukin-32 (IL-32$\gamma$) is a proinflammatory factor, whose tumor suppressor function depends on the expression of matrix metalloproteinase-1 (MMP-1) and integrin $\alpha$V (ITGAV) in ECM. As noted earlier, ECM proteins and their remodeling enzymes depend entirely on the model used in the study [70]. This suggests that numerous studies are needed to examine a variety of 3D cancer models, and their effects on the expression of several proteins and signaling pathways.

Elasticity is another important factor in properties of culture system, that may also affect the fate of cancer stem cells, along with their adhesion, self-renewal, and stemness [71]. For instance, it has been shown that the more malignant the stem cells become, the more elasticity and viscoelasticity they exhibit, that leads to an increase in the invasive behavior and migration of cancer cells [72]. With due attention to the importance of these cells, co-culture of all these cells in 2D culture media is impossible. It seems that the best option for simulation of this architecture will be advanced 3D culture systems like organoids.

### 14.2.5 Hydrophobicity/hydrophilicity effects

Hydrophobic/hydrophilic properties of culture medium are important in that they can lead to variable outcomes in the conformation and absorption of the proteins in different levels [73]. For example, hydrophilic properties of the culture medium have a diminishing effect on the expression of proinflammatory cytokines [74]. Changes in this feature may leave a major impact on shape, adhesion, mobility, and cellular differentiation [75].

To date, a wide range of naturally derived, synthetic, and natural/synthetic biomaterials blends have been examined for development of ideal 3D tumor models with highest similarity to native tissues, with various range of success (Fig. 14.3).

## 14.3 Skin cancers

Through the last few decades, the global incidence rate of skin cancer has significantly heightened due chiefly to the increased exposure to sunlight, ultraviolet (UV), and also

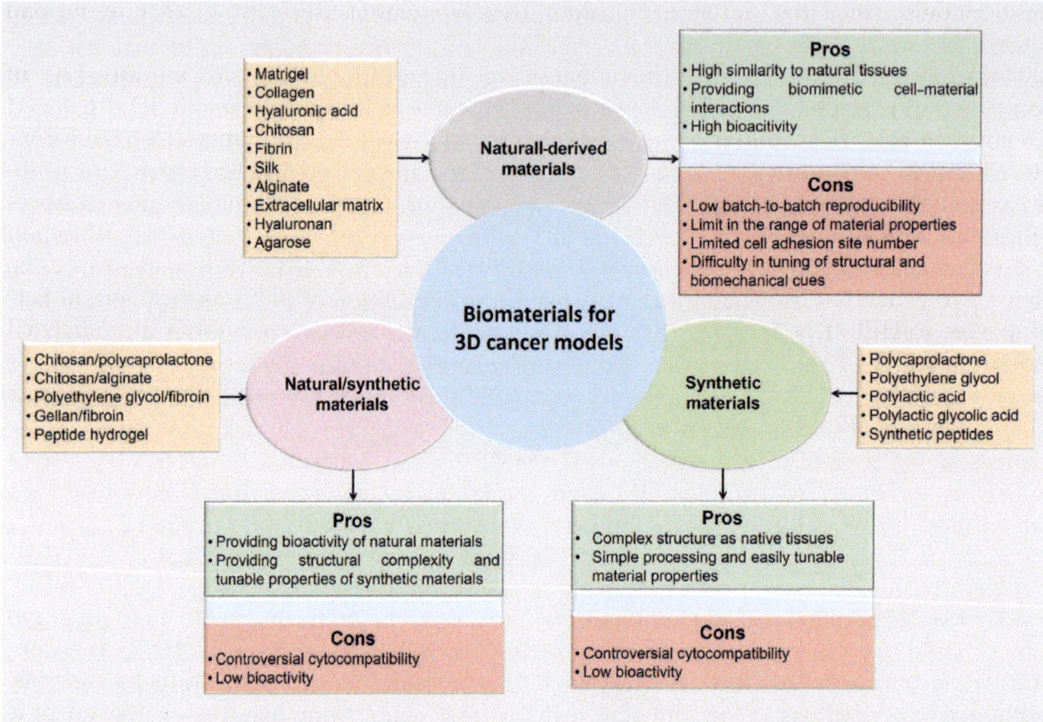


**FIGURE 14.3** Biomaterials usually used for fabrications of 3D tumor models. In general natural-derived, synthetic, and natural/synthetic biomaterials and their blends are the main category of biomaterials used for engineering of 3D tumor models. Some of the main different materials used and their different advantages (Pros) and disadvantages (Cons) are presented here.

aging [9]. According to the latest World Health Organization (WHO) statistics, skin cancer constitutes 4% of the global cancer prevalence. The mortality rate due to skin cancer, depending on the type of cancer, is less than 8% [76]. Generally, these cancers comprise nonmelanomatous tumors and malignant melanoma. BCC and SCC are derived from cutaneous keratinocytes, and often have a good prognosis and respond to treatment. However, malignant melanoma is highly aggressive and has a high-mortality rate. WHO reports 132,000 and 2–3 million new cases globally each year for melanoma and nonmelanoma skin cancers, respectively [9]. The risk factors include prolonged exposure to UV, immunosuppressive drugs, diet, smoking, chronic ulcers in skin, race, and skin factors [76]. According to the published statistics, the incidence of skin cancers is increasing. As a result, there is an urgent need to understand the various aspects of this disease [77].

There are quite a few animal models that can be explored to extract more information regarding the human skin cancer [78,79]. The animal models of melanoma, SCC and BCC are usually prepared through induction of carcinogenesis by exposing the cells to UV-radiation, chemicals, or viruses. Else, there are also genetically engineered models and the xenograft models [78,79]. However, there is a great limitation in the potential of these models, since the timeline of tumor development is different to that in human in vivo [9].

More recently, several studies have sought to develop 3D models for investigation of skin cancer [9]. Sun et al. in 2006, have evaluated the sensitivities of 2D and 3D organized human skin cells to oxidative stress and a potentially toxic heavy metal. They have concluded that 3D co-cultures have relayed a more physiologically accurate response to xenobiotic materials than models. Most of the experiments have relied on 2D monocultures [10]. In 2012, Commandeur et al. have explored the effects of the activation and inhibition of epidermal growth factor receptor (EGFR) on normal and malignant human skin cells. They have generated human skin cancer models with either normal human epidermal keratinocytes (NHEKs) or SCC cell line keratinocytes and have exposed them to epidermal growth factor (EGF) and erlotinib. They have concluded that 3D organotypic SCC models might be suitable for further investigations on the morphological and functional impacts of modifying EGFR signaling in SCC [80].

## 14.4 Modeling of skin cancer

Because of simple handling, affordability, genetic similarity to humans, and a high-reproductive potential, rats are the most suitable animal for in vivo modeling of skin cancer. Interestingly, the architecture and structure of skin are different for humans and rats. Human skin epithelium is the source of most skin cancers, while the same compartment is covered diversely with follicles in rats [81]. Considering the simultaneous growth cycle of follicles and a lot of cancers, this cycle has a great impact on the epithelial level in rats, thus, they cannot be a good model for this kind of cancer [82]. Additionally, severity of skin inhibition in rats is less than humans and absorption is more [81]. Also, epidermis in rats is less than 25-μm thick, and has cellular layers, while human epidermis is more than 100-μm thick and is composed of 6–10 cellular layers [83]. Expressions of some ECM proteins like laminin, collagen, fibronectin, tenascin, hemi-desmosomes, and adhesive fibrils are of crucial importance for real in vivo simulation of human skin in order to study the cancer cell behaviors [84,85].

A primary disadvantages of human skin cancer models are the incapability to precisely imitate tumor microenvironment in animals. There are differences in the activity of immune cells and the preinflammatory secreted by them [86]. Another main issue in development of skin cancer models is their genetic instability compared with other cancer types, which lead to unwanted resistance to drug-delivery systems [87]. Regarding this, a set of mutations occur randomly in various types of cancer. For this a very useful approach for finding new treatments and medications is designing and presenting models in which mutations happen individually with a time interval [83].

Up to now, the different models have been designed to investigate for the development of various treatments for melanoma, including cancer cell transplantation models from human to mice, genetically modified models, chemically induced models, UV-induced models, and xenograft models by melanoma injection in in vivo studies [88]. Keratin 76 (Krt76), a protein expressed in the epithelial, is reported to be associated with poor prognosis in SCC-derived cancers. New evidence has suggested that the repression of Krt76 may lead to formation of tumors instigated by carcinogens. Given the importance of this protein, the role of epithelial cells in the development of cancer in response to carcinogens is more pronounced [89]. In a study conducted in 1983, it has been observed that keratinization behavior and invasion of cancer cells with keratinocytic origins have been different in both in vivo and in vitro conditions [90]. Tissue engineering 3D models in cancer biology has been founded by Janvier et al. To do so, they have used forehead foreskin fibroblasts in a collagen-coated culture system [91,92].

## 14.4.1 In vitro skin cancer modeling

### *14.4.1.1 Spheroid formation*

One of the reasons for invasive behavior of melanoma is that a population of melanoma cells display characteristics of stem cells. The spheroid formation method for generation of 3D models of melanoma has been suggested to be a suitable method for investigating the behavior of melanoma stem cells (Fig. 14.1). Melanoma cells cultured in these 3D environments prominently express markers and invoicing factors for stem cells, as well as invasive and migratory markers [93]. One of the 3D models, useful for studying skin cancer cells is a spheroid. However, this model is not capable of imitating the entire aspects of tumor microenvironment like angiogenesis [90]. This 3D model has been used for simulation of avascular, solid, and melanoma tumors [94].

### *14.4.1.2 Natural-based 3D scaffolds*

Natural materials have been used in order to construct 3D in vitro culture systems for various types of cancer cells (Fig. 14.3). These materials include alginate, fibrin, and proteins derived from silk and synthetic polymers like polyvinyl alcohol and polyethylene glycol (PEG) [95–97]. Moreover, some researchers have used different decellularized tissues to further increase the similarity of the culture medium to ECM [98–100]. Hyaluronic acid is a biologically derived factor, which has widely been used in tissue engineering and drug-delivery researches. This material can bind to the cluster of differentiation 44 (CD44) receptors of the cancer cell surface and change the intracellular signaling routes [30,101].

### *14.4.1.3 Peptide-derived hydrogels*

The rigidity of the ideal substrate is effective in the growth and randomization of spontaneous cells. Particularly in the case of skin cells, the supportive layer should be able to provide biochemical clues for growth and differentiation of the cells (Fig. 14.3). The well-known commercially owned cultivator of Genzyme (MA, United States, Epicel) containing mouse fibroblasts, provides keratinocytes with a good growth rate. Basement membrane matrix (Matrigel), which is made of ECM components such as laminin, nidogen, collagen, and heparan sulfate proteoglycans, is another well-known media suitable for 3D in vitro cell culture [102,103]. In addition, the stimulatory effects of the chemical compounds bound to the surface of the culture medium influence the behavior and extent of the appearance and differentiation of skin cancer cells [104]. For example, it has been shown that keratinocytes can grow on a substrate coated with glutaraldehyde-fixed fibroblasts very well [105]. It should also be noted that some of these chemical-fixated substrates can maintain the nondifferentiated stem cell state [106]. The change in the culture medium hardness can be of great use in order to study the ECM remodeling, the effect of various therapeutic methods on them, and pharmacological properties in a better manner. Some 3D culture system are peptides, which can transform to hydrogel at room temperature [107]. One of the most famous commercially available 3D culture medium is RADARADARADARADA (RADA-16). This culture medium contains cell-bound peptide fragments that can change into nanofiber in a self-assembly manner and become gelatinous [108].

The most important biological agents in ECM, which are responsible for cellular connection and transfer of forces, are created by intracellular skeleton. They include collagen, elastin, laminin, fibronectin, chondroitin sulfate, heparin sulfate, and hyaluronic acid [109,110]. These structures in ECM create areas called "focal adhesion complexes" that are responsible for transferring environmental forces and signals into the intracellular skeleton. These signals have a great effect on the shape, differentiation and metastasis of the cancer cells. For instance, anoikis is one of the most prominent phenomenon, which leads to dormancy of the cancer cells due to the decrease in the connection of cells to ECM. Dormancy is a phase in which the cancer cells enter to escape different treatments, environmental stress, and the metastasis. They do not respond to the growth stimulants; that is, they enter a silent phase and escape the immune system [111]. The proteins present in ECM not only affect the connection and transferring of extracellular forces into the cell, but also ultimately determine the hydrophobic/hydrophilicity properties of the culture media. These properties have a great effect on cell behavior and survival [56,58]. The presence of various amino acid sequences such as Ile–Lys–Val–Ala (IKAV), RGD, and Tyr–Ile–Gly–Ser–Arg (YIGSR) not only are identified by different cellular receptors [112,113], but also affect the ECM hardness (rigidity). The focal adhesion complex proteins can stimulate EMT route through epigenetics [90,94,114] due to modifications being made within the expression.

#### 14.4.1.4 3D *fiber scaffolding in vitro models*

Advantages of using 3D in vitro models such as fiber scaffolding include easy handling, cost-effectiveness, long-term preservation, capability of combining with various chemical and biological materials, capability of connecting to surrounding tissues, ability to control and check the effects of different mechanical properties, and precise control of the environmental conditions [115]. Another widely used technique for making scaffolds based on nanofibers is electrospinning method. These 3D nanofiber mat substrates can mimic the structure of the normal ECM by creating micro-/nanopores [116]. A limited number of medications for the treatment of melanoma have been used with drug-delivery systems as well as in being loaded on various scaffolds with cobimetinib, dabrafenib, and tramitinib. In this respect, nanotechnology has not only revolutionized targeted drug delivery to skin cancer, but also has offered many benefits in the design of ECM mimicking 3D environments, and increased the drug efficacies [117,118].

A major issue in treatment of skin cancer is the presence of various physiological barriers that do not allow the drug to enter the tumor environment. In a study, a hydrogel complex of a carboxyl methylcellulose (CMC) type with doxorubicin (DOX) and citric acid has been developed. It has shown that the design of the gel conformation has a greater effect on killing melanoma than normal cells [119]. As pointed out in previous sections, the release of various cytokines from cancer cells has clearly shown their role in the growth, differentiation and metastasis of tumors [14]. Accordingly, some chemokine such as stromal cell-derived factor 1 (SDF-1α) and erythropoietin, type II interferon (IFN-γ), and chemokine 21 ligands can reduce the metastasis induced by these cells and increase the longevity of the animals [120,121]. Consequently, scaffolds containing these chemokines can act as excellent tumor niche.

In 2017, Velasco et al. have provided a 3D in vitro environment derived from polyglycolic acid (PGA), poly-L-lactide acid (PLLA), and PPy-I (polypyrrole), which had a good ability adhesion and growing of keratinocytes and fibroblasts derived from skin tissue [122]. The 3D scaffolds constructed from poly(lactic-*co*-glycolic acid)−poly(lactic acid)−methoxy poly(ethylene glycol) (PLGA−PLA−mPEG) have been stimulated EMT pathways and have thus been provided a precise simulation of the behavior of skin tumors [123]. Studies have shown that the shear forces imposed on cancer models affect the behavior of cancer cells due to capillary binding, and the subsequent intravasation and extravasation [124].

#### 14.4.1.5 *Chemical additives in 3D culture*

Other important chemical agents affecting the behavior of the cancer cells are metal ions (magnesium, strontium, and others) [125], talin [113], paxilin, inorganic substrates, organic molecules, and pH [113]. The reports have indicated that these chemical agents affect the absorption of various proteins on the substrate of the culture medium and as a result, interacting with the cells [50]. Curan et al. have showed that modifying levels by means of chemical agents such as methyl (CH3), carboxyl (COOH), hydroxyl (OH), and Silan have affected the differentiation and growth of cells [126]. To simulate the effects of the immune system on the growth of skin cancer, a nanofiber scaffold containing molybdenum nanoparticles has been developed. In that study, polycaprolactone nanofibers have been used as a matrix containing molybdenum. A 50% inhibition of melanoma and non-melanoma cells growth have been also demonstrated [127].

#### 14.4.1.6 *Biomaterials based 3D models*

As a result of the minimal side effects of natural-derived materials in the recent years, various types of materials such as alginate, gelatin, chitosan, silk derivatives, and combinations of them with each other or with synthetic polymers have been developed (Fig. 14.3) [128−131]. For example, Kaplan et al. recently have reviewed all the silk-based studies and products have been approved in the area of skin tissue repair [25].

#### 14.4.1.7 *3D cell cultures using microfluidic devices*

Microfluidic systems also allow the control of barriers against delivery of medicine to the body [132−135]. In this regard, systems have been designed that examine the role of different layers of skin [136,137]. In a recent study, a 3D model of human melanoma has been designed to explore the capability of capillary generation, ECM, basement membrane, and epidermis. This 3D system has induced the formation of lymphatic vessels in an out-of-body environment [38].

Microfluidic systems have also been used for separating cells without labeling them. For instance, a system has been designed to separate normal skin cells from melanoma cells [138]. In recent years, the lab-on-a-chip technology has widely been studied, and some tumor related commercial products have also been manufactured using microfluidic devices [139]. One of the interesting tasks innovated using these systems is the study of cumulative and adhesive behavior of cancer cells in a dynamic media as opposed to a 2D static media [140]. Khavari et al. have presented a culturing method for rats suffering from immunodeficiency, which could represent several aspects of cancer such as regenerative purposes. This has been suggested to be used in pharmacological studies and screening for skin cancer [83].

### 14.4.2 In vivo models

The next problem with melanoma modeling is the development and investigation of angiogenesis. Under normal conditions, angiogenesis in melanoma occurs in two stages of radial and vertical growth [141,142]. It seems that the use of human xenograft models is a good way to examine the process of angiogenesis in melanoma. Cutaneous squamous cell carcinoma (cSCC) is another type of malignant cutaneous cancer derived from keratinocytes, responsible for 20% of deaths from skin cancers. In a study, it has been shown that oncostatin M, a proinflammatory factor of the interleukin 6 (IL-6) family, has been increased in the cells of this type of cancer. It has been witnessed that the growth of mouse tumorigenic epidermal cell line (PDV.C57) and malignant cancerous keratinocytic cells growth rapidly increased in the presence of oncostatin M [143].

Similar to other cancer cells, B16 melanoma cells secrete various factors, such as vascular endothelial growth factor-A (VEGFA), placental growth factor (PLGF), transforming growth factor beta (TGFβ), and tumor necrosis factor (TNF), which can stimulate the properties of premetastatic niches [144,145]. The exosomes secreted from B16-F10 melanoma cell class also contain factors that increase the migration and angiogenesis of the cells through the MET pathway [146]. All of the exosomes secreted from the melanoma cells can metastasize to different tissues based on the type of integrin molecules that they express. For example, an exosome secreting melanoma with integrin $\alpha_v$ metastasizes to the lung, β3 to the brain, and β5 to the liver [144]. As a result, the use of different models that propel cancer cells to secrete specific exosomes can also be investigated. In addition, fibronectin sediment has been observed in melanoma tumor models in lung metastatic environments. As stated before, the increase in fibronectin in the tumor environment plays an important role in the accumulation of blood cancer cells. Also, the accumulation of periostin, integrin αvβ3, and integrin αvβ5 has been led to an the increase in melanoma invasion to premetastatic niche [147]. It is worth noting that the identification of different factors of trapping and the design of media containing such signals can be used in the future to collect cancer cells in the blood to prevent metastases to other areas [57]. The receptor for activated protein kinase C (RACK1) protein is known as a marker of melanoma, which is capable of binding to proteins involved in tumorigenic activity. The mutated genetic models of this protein in the melanoma cells have shown a decrease in growth and invasion [148]. With due attention to the histological differences in the differentiation of human and animal cells, attempts have been made to find new animal models that could better represent the behavior of human calls. In recent years, various species of fish such as zebrafish (*Danio rerio*), platyfish (*Xiphophorus* sp.), and medaka (*Oryzias latipes*) have been considered due to the structural similarity of the skin cancer model and their ease of genetic manipulation [149].

During the last few years, a number of studies have suggested that the amount of hyaluronan in melanoma is associated with a poor prognosis [150,151]. Hyaluronan is a major glycosaminoglycan found in ECM with numerous important roles in tumor progression [152]. In this regard, the interaction between hyaluronan, tumor, and stromal cells have been widely investigated with the focus on the process of tumor progression and metastasis. In 2019, Sapudom et al. generated in vivo mouse models of melanoma with low and high expressions of hyaluronan. They noted a meaningful positive correlation between hyaluronan quantity in the skin with tissue stiffness and tumor weight. Furthermore, they reconstituted in vitro models of different biomimetic fibroblasts derived from a mouse model to recapitulate the behavior of melanoma cells at the tumor level. These models included decellularized fibroblast matrix (FbECM), fibroblast embedded into 3D collagen matrices (FbColl), and well-defined hyaluronan-functionalized 3D collagen matrices (HAColl). They observed that HYAL1-treated FbECM and FbColl, as well as HAColl, supplemented with low-molecular weight hyaluronan accelerated the proliferation and invasion of melanoma cells [153].

## 14.4.3 New insights in 3D models of skin cancer

### *14.4.3.1 Microfluidic approach*

Recently, there has been a transition from 2D to 3D cell culture models, which is expected to revolutionize the biomimetic tissue models for better outcomes. However, challenges still remain in mimicking the complex nature of multicellular tissues, as there is no straight-forward way to create an artificial system with an extensive vascular tree, able to continuously monitor gradients and medium exchange with astonishing precision. Microfluidic approach is suggested to be a valuable tool for improving the physiological resemblance of 3D culture system, as it offers control over fluids in micrometer-sized channels. This approach is useful since it lets researchers spatially manipulate co-cultures, perfusion flow, and signaling gradients [154]. Thus, a precise mimicking of biophysical regulation of tumor microenvironment, mechanical forces, microstructure confinement, and ECM can be achieved through application of microfluidic devices [155]. In 2013, Atac et al. have used microfluidic Boyden chamber-like tools and successfully co-cultured hair follicles with slices of skin tissue. Their findings suggested that a dynamically perfused chip-based bioreactor platform might be able to give out variable mechanical shear stress, thus, extending culture periods. This strategy was accompanied with a significant improvement of culture conditions for integrated in vitro skin models, ex vivo skin organ cultures, and biopsies of single hair follicular units [156].

With due attention to the success of tumor-on-chip approach in development of better tumor microenvironment models, it is expected that skin cancer on-chip models based on microfluidic approach will become a reality in near future [155].

#### 14.4.3.2 *Personalized medicine*

Melanoma has been globally recognized as one of the most notable examples of malignancies that can be safely approached with evidence-based personalized medicine methods [157]. As of recent, new insights on 3D models of cancer have been proposed. The next big obstacle to overcome is the full validation of these models and their ultimate implementation in personalized medicine applications [154].

In these models, the cultured tissues and cells isolated directly from patients. As stated above, due to the high-invasive nature of skin cancer, such methods are very suitable for obtaining appropriate and effective treatments [158]. As a result, the growth and characteristics of the cancer cells can be well documented even in commercial 3D tumor models [159]. The importance of the 3D culture environments is also their ability to simulate the behavior of the immune system with cancer cells, especially in immunotherapy-based therapies [93,160]. Immune cells inside and outside the tumor microenvironment can be affected by the growth factors secreted from cancer cells [14].

More recently, studies on skin cancer, as well as melanoma, have focused on 3D models for preclinical studies in order to compromise between standard 2D cultures and xenografts of human tumors [157]. In a study on nonmelanoma skin cancer (SCC subtype), M2 macrophages were found to be highly stimulated in collagen-rich environments, due to the close imitation of the 3D culture medium [160]. It is worth noting that these types of macrophages have an excitatory effect on the growth and progression of cancerous cells. Spheroids made from malignant cells of the skin of melanoma, in addition to stimulating the growth and transplantation of T-cell lymphocytes, cause cancer cells to escape from the immune system [93]. In the recent years, an anticluster of differentiation 40 (CD40) monoclonal antibody (mAb) has been developed and suggested as promising therapeutic agent for melanoma [89]. Based on the available evidences, it is obvious that the effects of these immunotherapy drugs should be tested in 3D cancer environments in order to determine the rate of their effects.

## 14.5 Conclusion and future prospective

Over the past century, skin cancer incidence has been significantly increased probably due to risk sunlight and UV exposure. Regarding with poor prognosis of skin cancer, especially malignant melanoma, there is a need for further investigations in development of ideal 3D modeling in vitro. The importance of 3D in vitro culture system in precise imitation of cancer cell behavior in vivo for drug screening is now well demonstrated. A more accurate mimicry of cancer cell behaviors such as angiogenesis, immunotherapy, and hypoxic circumstances is undeniable. It seems by developing the new microfluidic devices with high accuracy and precision, hydrogels derived from natural or synthetic materials containing various agents such as growth stimulators or inhibitors. The use of cancer organoid models for personalized medicine may have a huge impact on the future of skin cancer medicine.

## Acknowledgment

Our sincere thanks to Professor Mohammad Ali Shokrgozar for providing helpful comments on this write up. SCK holds presently European Research Area Chair and Full Professor Position at 3Bs Research Group, University of Minho, Portugal and supported by the European Union Framework Programme for Research and Innovation HORIZON 2020 under grant agreement nọ 668983 - FoReCaST.

## Conflict of interest

The authors declare no potential conflict of interest.

## References

[1] Rogers HW, et al. Incidence estimate of nonmelanoma skin cancer (keratinocyte carcinomas) in the US population, 2012. JAMA Dermatol 2015;151(10):1081–6.
[2] American Cancer Society. Cancer facts & figures. Atlanta: American Cancer Society; 2016.
[3] Stern RS. Prevalence of a history of skin cancer in 2007: results of an incidence-based model. Arch Dermatol 2010;146(3):279–82.
[4] Cakir BÖ, Adamson P, Cingi C. Epidemiology and economic burden of nonmelanoma skin cancer. Facial Plast Surg Clin North Am 2012;20(4):419–22.
[5] Dubas LE, Ingraffea A. Nonmelanoma skin cancer. Facial Plastic Surg Clin North Am 2013;21(1):43–53.
[6] Esteva A, et al. Dermatologist-level classification of skin cancer with deep neural networks. Nature 2017;542 (7639):115.
[7] Rigel DS, Russak J, Friedman R. The evolution of melanoma diagnosis: 25 years beyond the ABCDs. CA: A Cancer J Clin 2010;60(5):301–16.
[8] Lee N, Barthel SR, Schatton T. Melanoma stem cells and metastasis: mimicking hematopoietic cell trafficking? Lab Invest 2014;94(1):13.
[9] Bergers LI, et al. Immune-competent human skin disease models. Drug Discov Today 2016;21(9):1479–88.
[10] Sun T, et al. Culture of skin cells in 3D rather than 2D improves their ability to survive exposure to cytotoxic agents. J Biotechnol 2006;122(3):372–81.
[11] GlaxoSmithKline. The investigational MAGE-A3 antigen-specific cancer immunotherapeutic does not meet first co-primary endpoint in Phase III melanoma clinical trial; 2013.
[12] Mak IW, Evaniew N, Ghert M. Lost in translation: animal models and clinical trials in cancer treatment. Am J Transl Res 2014;6(2):114.
[13] Seok J, et al. Inflammation and host response to injury, large scale collaborative research program. Genomic responses in mouse models poorly mimic human inflammatory diseases. Proc Natl Acad Sci USA 2013;110 (9):3507–12.
[14] Hanahan D, Weinberg RA. Hallmarks of cancer: the next generation. Cell 2011;144(5):646–74.
[15] Stratowa C, et al. A comparative cell-based high throughput screening strategy for the discovery of selective tyrosine kinase inhibitors with anticancer activity. Anticancer Drug Des 1999;14(5):393–402.
[16] Taylor DL, Woo ES, Giuliano KA. Real-time molecular and cellular analysis: the new frontier of drug discovery. Curr Opin Biotechnol 2001;12(1):75–81.
[17] Bhadriraju K, Chen CS. Engineering cellular microenvironments to improve cell-based drug testing. Drug Discov Today 2002;7(11):612–20.
[18] Carvalho MR, et al. Evaluating biomaterial-and microfluidic-based 3D tumor models. Trends Biotechnol 2015;33(11):667–78.
[19] Carvalho MR, Reis RL, Oliveira JM. Mimicking the 3D biology of osteochondral tissue with microfluidic-based solutions: breakthroughs towards boosting drug testing and discovery. Drug Discov Today 2018;23 (3):711–18.
[20] Yamada KM, Cukierman E. Modeling tissue morphogenesis and cancer in 3D. Cell 2007;130(4):601–10.

[21] Roskelley C, Bissell M. Dynamic reciprocity revisited: a continuous, bidirectional flow of information between cells and the extracellular matrix regulates mammary epithelial cell function. Biochem Cell Biol 1995;73(7-8):391–7.

[22] Ingber DE, et al. Cell shape, cytoskeletal mechanics, and cell cycle control in angiogenesis. J Biomech 1995;28 (12):1471–84.

[23] Weaver VM, et al. Reversion of the malignant phenotype of human breast cells in three-dimensional culture and in vivo by integrin blocking antibodies. J Cell Biol 1997;137(1):231–45.

[24] Birgersdotter A, Sandberg R, Ernberg I. Gene expression perturbation in vitro—a growing case for three-dimensional (3D) culture systems. Semin Cancer Biol 2005;15(5):405–12.

[25] Gholipourmalekabadi M, et al. Silk fibroin for skin injury repair: Where do things stand? Adv Drug Deliv Rev 2019.

[26] Sun T, et al. Self-organization of skin cells in three-dimensional electrospun polystyrene scaffolds. Tissue Eng 2005;11(7-8):1023–33.

[27] Siegel RL, Miller KD, Jemal A. Cancer statistics, 2019. CA: Cancer J Clin 2019;69(1):7–34.

[28] Leiter U, Eigentler T, Garbe C. Epidemiology of skin cancer. In: Reichrath J, editor. Sunlight, vitamin D and skin cancer. New York: Springer; 2014. p. 120–40.

[29] Asghar W, et al. Engineering cancer microenvironments for in vitro 3-D tumor models. Mater Today 2015;18 (10):539–53.

[30] Liu Z, Vunjak-Novakovic G. Modeling tumor microenvironments using custom-designed biomaterial scaffolds. Curr Opin Chem Eng 2016;11:94–105.

[31] Herberg M, Roeder I. Computational modelling of embryonic stem-cell fate control. Development 2015;142 (13):2250–60.

[32] Freeman AE, et al. Growth and characterization of human skin epithelial cell cultures. In Vitro 1976;12 (5):352–62.

[33] O'Brien LE, Zegers MM, Mostov KE. Building epithelial architecture: insights from three-dimensional culture models. Nat Rev Mol Cell Biol 2002;3(7):531.

[34] Simitzi C, et al. Controlling the outgrowth and functions of neural stem cells: the effect of surface topography. ChemPhysChem 2018;19(10):1143–63.

[35] Garlick JA. Engineering skin to study human disease–tissue models for cancer biology and wound repair. In: Garlick JA, editor. Tissue engineering II. New York: Springer; 2006. p. 207–39.

[36] Duval K, et al. Modeling physiological events in 2D vs. 3D cell culture. Physiology 2017;32(4):266–77.

[37] Kim JB, Stein R, O'hare MJ. Three-dimensional in vitro tissue culture models of breast cancer—a review. Breast Cancer Res Treat 2004;85(3):281–91.

[38] Bourland J, Fradette J, Auger FA. Tissue-engineered 3D melanoma model with blood and lymphatic capillaries for drug development. Sci Rep 2018;8(1):13191.

[39] Bissell MJ, et al. The organizing principle: microenvironmental influences in the normal and malignant breast. Differentiation 2002;70(9-10):537–46.

[40] Smalley KS, Lioni M, Herlyn M. Life ins't flat: taking cancer biology to the next dimension. In Vitro Cell Dev Biol Anim 2006;42(8-9):242–7.

[41] Hutmacher DW, et al. Translating tissue engineering technology platforms into cancer research. J Cell Mol Med 2009;13(8a):1417–27.

[42] Gao X-l, et al. Cancer cell dormancy: mechanisms and implications of cancer recurrence and metastasis. Onco Targets Ther 2017;10:5219.

[43] Dey S, et al. ATF4-dependent induction of heme oxygenase 1 prevents anoikis and promotes metastasis. J Clin Invest 2015;125(7):2592–608.

[44] De Palma M, Biziato D, Petrova TV. Microenvironmental regulation of tumour angiogenesis. Nat Rev Cancer 2017;17(8):457.

[45] Zhou J, et al. Tumor hypoxia and cancer progression. Cancer Lett 2006;237(1):10–21.

[46] Saltzman WM. Tissue engineering: engineering principles for the design of replacement organs and tissues. Oxford; New York: Oxford University Press; 2004.

[47] Ala-aho R, Kähäri V-M. Collagenases in cancer. Biochimie 2005;87(3–4):273–86.

[48] Montazeri L, et al. The interaction of plasma proteins with nano-size fluoride-substituted apatite powders. Ceram Int 2013;39(6):6145–52.

[49] Liu X, Wang S. Three-dimensional nano-biointerface as a new platform for guiding cell fate. Chem Soc Rev 2014;43(8):2385–401.
[50] Khetan S, et al. Degradation-mediated cellular traction directs stem cell fate in covalently crosslinked three-dimensional hydrogels. Nat Mater 2013;12(5):458.
[51] Lee G, et al. Cancer mechanobiology: microenvironmental sensing and metastasis. ACS Biomater Sci Eng 2019;5(8):3685.
[52] Douezan S, et al. Spreading dynamics and wetting transition of cellular aggregates. Proc Natl Acad Sci USA 2011;108(18):7315–20.
[53] Bazellières E, et al. Control of cell–cell forces and collective cell dynamics by the intercellular adhesome. Nat Cell Biol 2015;17(4):409.
[54] Friedl P, Gilmour D. Collective cell migration in morphogenesis, regeneration and cancer. Nat Rev Mol Cell Biol 2009;10(7):445.
[55] Discher DE, Mooney DJ, Zandstra PW. Growth factors, matrices, and forces combine and control stem cells. Science 2009;324(5935):1673–7.
[56] Keselowsky BG, Collard DM, García AJ. Integrin binding specificity regulates biomaterial surface chemistry effects on cell differentiation. Proc Natl Acad Sci USA 2005;102(17):5953–7.
[57] Peinado H, et al. Pre-metastatic niches: organ-specific homes for metastases. Nat Rev Cancer 2017;17 (5):302–17.
[58] Brodbeck WG, et al. Biomaterial adherent macrophage apoptosis is increased by hydrophilic and anionic substrates in vivo. Proc Natl Acad Sci USA 2002;99(16):10287–92.
[59] Ayala R, et al. Engineering the cell–material interface for controlling stem cell adhesion, migration, and differentiation. Biomaterials 2011;32(15):3700–11.
[60] Garcıa AJ, Vega MD, Boettiger D. Modulation of cell proliferation and differentiation through substrate-dependent changes in fibronectin conformation. Mol Biol Cell 1999;10(3):785–98.
[61] Rodriguez-Hernandez I, et al. Rho, ROCK and actomyosin contractility in metastasis as drug targets. F1000Research 2016;5:783.
[62] Qian BZ, Pollard JW. Macrophage diversity enhances tumor progression and metastasis. Cell 2010;141 (1):39–51.
[63] Clark AG, Vignjevic DM. Modes of cancer cell invasion and the role of the microenvironment. Curr Opin Cell Biol 2015;36:13–22.
[64] Ouderkirk JL, Krendel M. Non-muscle myosins in tumor progression, cancer cell invasion, and metastasis. Cytoskeleton 2014;71(8):447–63.
[65] Adjei IM, Blanka S. Modulation of the tumor microenvironment for cancer treatment: a biomaterials approach. J Funct Biomater 2015;6(1):81–103.
[66] Lyssiotis CA, et al. Chemical control of stem cell fate and developmental potential. Angew Chem Int Ed 2011;50(1):200–42.
[67] Gossett DR, et al. Hydrodynamic stretching of single cells for large population mechanical phenotyping. Proc Natl Acad Sci USA 2012;109(20):7630–5.
[68] Lobo NA, et al. The biology of cancer stem cells. Annu Rev Cell Dev Biol 2007;23:675–99.
[69] Borovski T, et al. Cancer stem cell niche: the place to be. Cancer Res 2011;71(3):634–9.
[70] Lee YS, et al. Inhibition of skin carcinogenesis by suppression of NF-kappaB dependent ITGAV and TIMP-1 expression in IL-32gamma overexpressed condition. J Exp Clin Cancer Res 2018;37(1):293.
[71] Tanner K. Regulation of the basement membrane by epithelia generated forces. Phys Biol 2012;9(6):065003.
[72] Würth R, et al. CXCL12 modulation of CXCR4 and CXCR7 activity in human glioblastoma stem-like cells and regulation of the tumor microenvironment. Front Cell Neurosci 2014;8:144.
[73] Kievit FM, et al. Proliferation and enrichment of CD133 + glioblastoma cancer stem cells on 3D chitosan-alginate scaffolds. Biomaterials 2014;35(33):9137–43.
[74] Bleau A-M, et al. Metastatic dormancy: a complex network between cancer stem cells and their microenvironment. Histol Histopathol 2014;29(12):1499–510.
[75] Rao W, et al. Enhanced enrichment of prostate cancer stem-like cells with miniaturized 3D culture in liquid core-hydrogel shell microcapsules. Biomaterials 2014;35(27):7762–73.
[76] Simões M, Sousa J, Pais A. Skin cancer and new treatment perspectives: a review. Cancer Lett 2015;357 (1):8–42.

[77] Linos E, Katz KA, Colditz GA. Skin cancer-the importance of prevention. JAMA Intern Med 2016;176 (10):1435–6.
[78] van der Weyden L, et al. Cross-species models of human melanoma. J Pathol 2016;238(2):152–65.
[79] Kasper M, et al. Basal cell carcinoma—molecular biology and potential new therapies. J Clin Investig 2012;122(2):455–63.
[80] Commandeur S, et al. Epidermal growth factor receptor activation and inhibition in 3D in vitro models of normal skin and human cutaneous squamous cell carcinoma. Cancer Sci 2012;103(12):2120–6.
[81] Menon GK. New insights into skin structure: scratching the surface. Adv Drug Deliv Rev 2002;54(Suppl. 1): S3–17.
[82] Stenn KS, Paus R. Controls of hair follicle cycling. Physiol Rev 2001;81(1):449–94.
[83] Khavari PA. Modelling cancer in human skin tissue. Nat Rev Cancer 2006;6(4):270–80.
[84] Ortiz-Urda S, et al. Stable nonviral genetic correction of inherited human skin disease. Nat Med 2002;8:1166.
[85] Robbins PB, et al. In vivo restoration of laminin 5 beta 3 expression and function in junctional epidermolysis bullosa. Proc Natl Acad Sci USA 2001;98(9):5193–8.
[86] Nickoloff BJ, Wrone-Smith T. Injection of pre-psoriatic skin with CD4 + T cells induces psoriasis. Am J Pathol 1999;155(1):145–58.
[87] Lazarov M, et al. CDK4 coexpression with Ras generates malignant human epidermal tumorigenesis. Nat Med 2002;8(10):1105–14.
[88] Saleh J. Murine models of melanoma. Pathol Res Pract 2018;214(9):1235–8.
[89] Sequeira I, et al. Immunomodulatory role of Keratin 76 in oral and gastric cancer. Nat Commun 2018;9 (1):3437.
[90] Lee H-O, et al. FAP-overexpressing fibroblasts produce an extracellular matrix that enhances invasive velocity and directionality of pancreatic cancer cells. BMC Cancer 2011;11(1):245.
[91] Janvier R, et al. Stromal fibroblasts are required for PC-3 human prostate cancer cells to produce capillary-like formation of endothelial cells in a three-dimensional co-culture system. Anticancer Res 1997;17 (3a):1551–7.
[92] Chwalek K, Bray LJ, Werner C. Tissue-engineered 3D tumor angiogenesis models: potential technologies for anti-cancer drug discovery. Adv Drug Deliv Rev 2014;79-80:30–9.
[93] Ramgolam K, et al. Melanoma spheroids grown under neural crest cell conditions are highly plastic migratory/invasive tumor cells endowed with immunomodulator function. PLoS One 2011;6(4):e18784.
[94] Gill BJ, et al. A synthetic matrix with independently tunable biochemistry and mechanical properties to study epithelial morphogenesis and EMT in a lung adenocarcinoma model. Cancer Res 2012;72 (22):6013–23.
[95] Liu J, et al. Soft fibrin gels promote selection and growth of tumorigenic cells. Nat Mater 2012;11(8):734.
[96] Talukdar S, Kundu SC. A non-mulberry silk fibroin protein based 3D in vitro tumor model for evaluation of anticancer drug activity. Adv Funct Mater 2012;22(22):4778–88.
[97] Buwalda SJ, et al. Hydrogels in a historical perspective: from simple networks to smart materials. J Control Release 2014;190:254–73.
[98] Dunne LW, et al. Human decellularized adipose tissue scaffold as a model for breast cancer cell growth and drug treatments. Biomaterials 2014;35(18):4940–9.
[99] Gholipourmalekabadi M, et al. Decellularized human amniotic membrane: how viable is it as a delivery system for human adipose tissue-derived stromal cells? Cell Prolif 2016;49(1):115–21.
[100] Gholipourmalekabadi M, et al. Development of a cost-effective and simple protocol for decellularization and preservation of human amniotic membrane as a soft tissue replacement and delivery system for bone marrow stromal cells. Adv Healthc Mater 2015;4(6):918–26.
[101] Saneja A, et al. CD44 targeted PLGA nanomedicines for cancer chemotherapy. Eur J Pharm Sci 2018;121:47–58.
[102] Hasegawa Y, et al. CCL2 enhances pluripotency of human induced pluripotent stem cells by activating hypoxia related genes. Sci Rep 2014;4:5228.
[103] Sun Y, Chen CS, Fu J. Forcing stem cells to behave: a biophysical perspective of the cellular microenvironment. Annu Rev Biophys 2012;41:519–42.
[104] Perestrelo AR, et al. Locust bean gum as an alternative polymeric coating for embryonic stem cell culture. Mater Sci Eng C 2014;40:336–44.

[105] Verfaillie C, Catanzaro P. Direct contact with stroma inhibits proliferation of human long-term culture initiating cells. Leukemia 1996;10(3):498–504.
[106] Joddar B, Ito Y. Artificial niche substrates for embryonic and induced pluripotent stem cell cultures. J Biotechnol 2013;168(2):218–28.
[107] Huang H, et al. Peptide hydrogelation and cell encapsulation for 3D culture of MCF-7 breast cancer cells. PLoS One 2013;8(3):e59482.
[108] Yang Z, Zhao X. A 3D model of ovarian cancer cell lines on peptide nanofiber scaffold to explore the cell–scaffold interaction and chemotherapeutic resistance of anticancer drugs. Int J Nanomed 2011;6:303.
[109] Reilly GC, Engler AJ. Intrinsic extracellular matrix properties regulate stem cell differentiation. J Biomech 2010;43(1):55–62.
[110] Higuchi A, et al. Biomimetic cell culture proteins as extracellular matrices for stem cell differentiation. Chem Rev 2012;112(8):4507–40.
[111] Vera-Ramirez L, Hunter KW. Tumor cell dormancy as an adaptive cell stress response mechanism. F1000Research 2017;6:2134.
[112] Marklein RA, Burdick JA. Controlling stem cell fate with material design. Adv Mater 2010;22(2):175–89.
[113] Mashinchian O, et al. Regulation of stem cell fate by nanomaterial substrates. Nanomedicine 2015;10 (5):829–47.
[114] Hoshiba T, Tanaka M. Breast cancer cell behaviors on staged tumorigenesis-mimicking matrices derived from tumor cells at various malignant stages. Biochem Biophys Res Commun 2013;439(2):291–6.
[115] Naves LB, Almeida L, Rajamani L. 11 - Nanofiber composites in skin tissue engineering. In: Ramalingam M, Ramakrishna S, editors. Nanofiber composites for biomedical applications. Cambridge, UK: Woodhead Publishing; 2017. p. 275–300.
[116] Cheon YW, et al. Enhanced chondrogenic responses of human articular chondrocytes onto silk fibroin/wool keratose scaffolds treated with microwave-induced argon plasma. Artif Organs 2010;34(5):384–92.
[117] Naves LB, et al. Nanotechnology for the treatment of melanoma skin cancer. Prog Biomater 2017;6 (1–2):13–26.
[118] Gholipourmalekabadi M, et al. Targeted drug delivery based on gold nanoparticle derivatives. Curr Pharm Des 2017;23(20):2918–29.
[119] Carvalho SM, et al. Synthesis and in vitro assessment of anticancer hydrogels composed by carboxymethylcellulose-doxorubicin as potential transdermal delivery systems for treatment of skin cancer. J Mol Liq 2018;266:425–40.
[120] Ko C-Y, et al. The use of chemokine-releasing tissue engineering scaffolds in a model of inflammatory response-mediated melanoma cancer metastasis. Biomaterials 2012;33(3):876–85.
[121] Phan-Lai V, et al. CCL21 and IFN$\gamma$ recruit and activate tumor specific T cells in 3D scaffold model of breast cancer. Anticancer Agents Med Chem 2014;14(2):204–10.
[122] Ruiz-Velasco G, et al. Polymeric scaffolds for skin. Macromol Symp 2017;374:1600133.
[123] Girard YK, et al. A 3D fibrous scaffold inducing tumoroids: a platform for anticancer drug development. PLoS One 2013;8(10):e75345.
[124] Riggi N, Aguet M, Stamenkovic I. Cancer metastasis: a reappraisal of its underlying mechanisms and their relevance to treatment. Annu Rev Pathol Mech Dis 2018;13:117–40.
[125] Yan J, et al. Bone integration capability of a series of strontium-containing hydroxyapatite coatings formed by micro-arc oxidation. J Biomed Mater Res Part A 2013;101(9):2465–80.
[126] Curran JM, Chen R, Hunt JA. The guidance of human mesenchymal stem cell differentiation in vitro by controlled modifications to the cell substrate. Biomaterials 2006;27(27):4783–93.
[127] Janani I, et al. Selectivity and sensitivity of molybdenum oxide-polycaprolactone nanofiber composites on skin cancer: preliminary in-vitro and in-vivo implications. J Trace Elem Med Biol 2018;49:60–71.
[128] Gholipourmalekabadi M, et al. 3D protein-based bilayer artificial skin for the guided scarless healing of third-degree burn wounds in vivo. Biomacromolecules 2018;19(7):2409–22.
[129] Gholipourmalekabadi M, et al. Silk fibroin/amniotic membrane 3D bi-layered artificial skin. Biomed Mater 2018;13(3):035003.
[130] Yang W, et al. Preparation and characterisation of a novel silk fibroin/hyaluronic acid/sodium alginate scaffold for skin repair. Int J Biol Macromol 2019;130:58–67.

[131] Sharma S, Batra S. Recent advances of chitosan composites in artificial skin: the next era for potential biomedical application. In: Grumezescu AM, Grumezescu V, editors. Materials for biomedical engineering. New York: Elsevier; 2019. p. 97–119.
[132] Yu F, et al. Microfluidic platforms for modeling biological barriers in the circulatory system. Drug Discov Today 2018;23(4):815–29.
[133] Abaci HE, et al. Pumpless microfluidic platform for drug testing on human skin equivalents. Lab Chip 2015;15(3):882–8.
[134] Lee S, et al. Construction of 3D multicellular microfluidic chip for an in vitro skin model. Biomed Microdevices 2017;19(2):22.
[135] Wufuer M, et al. Skin-on-a-chip model simulating inflammation, edema and drug-based treatment. Sci Rep 2016;6:37471.
[136] Takeuchi I, Suzuki T, Makino K. Skin permeability and transdermal delivery route of 50-nm indomethacin-loaded PLGA nanoparticles. Colloids Surf B Biointerfaces 2017;159:312–17.
[137] Purohit P, et al. Effect of mixed surfactants on stratum corneum: a drying stress and Raman spectroscopy study. Int J Cosmet Sci 2014;36(4):379–85.
[138] Wlodkowic D, Cooper JM. Tumors on chips: oncology meets microfluidics. Curr Opin Chem Biol 2010;14 (5):556–67.
[139] Maschmeyer I, et al. A four-organ-chip for interconnected long-term co-culture of human intestine, liver, skin and kidney equivalents. Lab Chip 2015;15(12):2688–99.
[140] Kwak B, et al. Simulation of complex transport of nanoparticles around a tumor using tumor-microenvironment-on-chip. J Control Release 2014;194:157–67.
[141] Santagiuliana R, Ferrari M, Schrefler BA. Simulation of angiogenesis in a multiphase tumor growth model. Comput Methods Appl Mech Eng 2016;304:197–216.
[142] Ziemys A, et al. Chapter 7 - Multiscale models for transport and biodistribution of therapeutics in cancer. In: Manca D, editor. Computer aided chemical engineering. New York: Elsevier; 2018. p. 209–37.
[143] Simonneau M, et al. Oncostatin M is overexpressed in skin squamous-cell carcinoma and promotes tumor progression. Oncotarget 2018;9(92):36457–73.
[144] Kaplan RN, et al. VEGFR1-positive haematopoietic bone marrow progenitors initiate the pre-metastatic niche. Nature 2005;438(7069):820–7.
[145] Zhou C-p, et al. Human CD133-positive hematopoietic progenitor cells initiate growth and metastasis of colorectal cancer cells. Carcinogenesis 2014;35(12):2771–7.
[146] Peinado H, et al. Melanoma exosomes educate bone marrow progenitor cells toward a pro-metastatic phenotype through MET. Nat Med 2012;18(6):883–91.
[147] Fukuda K, et al. Periostin is a key niche component for wound metastasis of melanoma. PLoS One 2015;10 (6):e0129704.
[148] Campagne C, et al. RACK1 cooperates with NRAS(Q61K) to promote melanoma in vivo. Cell Signal 2017;36:255–66.
[149] Sarasamma S, et al. The power of fish models to elucidate skin cancer pathogenesis and impact the discovery of new therapeutic opportunities. Int J Mol Sci 2018;19(12):E3929.
[150] Vigetti D, et al. Metabolic control of hyaluronan synthases. Matrix Biol 2014;35:8–13.
[151] Viola M, et al. Molecular control of the hyaluronan biosynthesis. Connect Tissue Res 2008;49(3-4):111–14.
[152] Tammi RH, et al. Hyaluronan in human tumors: pathobiological and prognostic messages from cell-associated and stromal hyaluronan. Semin Cancer Biol 2008;18(4):288–95.
[153] Sapudom J, et al. Biomimetic tissue models reveal the role of hyaluronan in melanoma proliferation and invasion. Biomater Sci 2020;8(5):1405–17.
[154] van Duinen V, et al. Microfluidic 3D cell culture: from tools to tissue models. Curr OpBiotechnol 2015;35:118–26.
[155] Mohammadi MH, et al. Organ-on-chip platforms: skin diseases modeling using combined tissue engineering and microfluidic technologies. Adv Healthc Mater 2016;5(19):2454.
[156] Ataç B, et al. Skin and hair on-a-chip: in vitro skin models versus ex vivo tissue maintenance with dynamic perfusion. Lab a chip 2013;13(18):3555–61.
[157] Levesque MP, et al. Metastatic melanoma moves on: translational science in the era of personalized medicine. Cancer Metastasis Rev 2017;36(1):7–21.

[158] Chin L. The genetics of malignant melanoma: lessons from mouse and man. Nat Rev Cancer 2003;3(8):559.
[159] Li L, et al. The skin cancer chemotherapeutic agent ingenol-3-angelate (PEP005) is a substrate for the epidermal multidrug transporter (ABCB1) and targets tumor vasculature. Cancer Res 2010;70:4509–19.
[160] Linde N, et al. Integrating macrophages into organotypic co-cultures: a 3D in vitro model to study tumor-associated macrophages. PLoS One 2012;7(7):e40058.

SECTION II

# Advanced models for cancer research

CHAPTER

# 15

# Microfluidic systems in cancer research

*David Caballero*[1,2], *Maria Angélica Luque-González*[1,2], *Rui L. Reis*[1,2] *and Subhas C. Kundu*[1,2]

[1]3B's Research Group, I3Bs – Research Institute on Biomaterials, Biodegradables and Biomimetics, University of Minho, Headquarters of the European Institute of Excellence on Tissue Engineering and Regenerative Medicine, Guimarães, Portugal [2]ICVS/3B's – PT Government Associate Laboratory, Braga/Guimarães, Portugal

## 15.1 Introduction

### 15.1.1 Background

In 2018, cancer was responsible for about 10 million deaths, being the second leading cause of death globally [1]. Besides the clinical point of view, the socioeconomic impact of cancer is huge. As an example, the total annual cost of cancer in 2010 in the United States was estimated at 1.16 trillion US dollars [2]. In Europe, the direct health cost of cancer increased from 79 to 86 billion euros during 2005–14 [3]. In order to reduce this healthcare and socioeconomic burden, the development of improved technologies for the early diagnosis and treatment of cancer is of upmost importance. These technologies should be capable to provide new insights about the etiology of the disease, as well as serve as therapeutic platforms to test and screen the effect of compounds and therapies with higher efficiency than traditional approaches. Current technologies focus on the use of regular tissue culture dishes or flasks to culture tumor cells as preclinical in vitro models. These reductionistic models, *albeit* simple to use, fail to recapitulate faithfully the biological and mechanochemical complexities of the native microenvironment in which tumors grow. Under these conditions, in general, cells react to drugs very differently compared with a systemic drug delivery. In this case, the drug may be put aside after a lot of financial investment. Similarly, in vivo experimentation using animal

*Biomaterials for 3D Tumor Modeling*
DOI: https://doi.org/10.1016/B978-0-12-818128-7.00015-0

models, despite providing the needed biological complexity missing in standard in vitro models, are in general poor prognostic models of the human cancer. This is of particular interest for the accurate prediction of drug response in human clinical trials. In addition, this type of model does not allow the identification of how specific components and cells from the tumor microenvironment (TME) influence cancer development and dissemination. Overall, more advanced in vitro cancer models, which better mimic the healthy and cancerous tissue, and their interplay with the surrounding environment, are needed. This includes reproducing tissue–tissue interfaces, fluid flows, and biochemical/mechanical cues that normal and cancer cells experience within living organs. In this regard, microfluidic systems have emerged as a promising alternative to standard technologies in cancer research.

Microfluidics is defined as the "*science of manipulating and processing fluids at the micrometric scale using engineered devices.*" It emerged from the technology developed for the microelectronics industry. This means that microfluidic devices can be mass-produced at a very low cost, which is of upmost importance for the healthcare market. Similarly, the volume of sample and reagents used is very low, including fluids from patients, thus reducing the impact on patients. This also increases the sensitivity of the device and lowers its processing times and energy consumption. Importantly, they are also compatible with existing technologies (e.g., microscopy, screening, and biosensing, among many others), which make microfluidics very attractive for many sectors working in the biomedical sectors, including R&D laboratories, pharmaceutical and biotechnological companies, and hospitals. Importantly, when combined with human cells, microfluidic devices can mimic the fundamental units of human organs and tissues. These models are known as *organ-on-a-chip* (OoC), and have had a dramatic increase during the last decade due to its promising impact in the healthcare sector [4].

OoC mimics with higher fidelity the native human scenario. This means that cells seeded inside the chip show phenotypes and gene expressions similar to those of their native counterparts. This is of upmost importance because, as widely reported, most of the drugs that succeed in preclinical trials, fail when tested in vitro. This is most likely due to the highly artificial cellular environment (flat and static Petri dishes or flasks) where drugs are tested. In contrast, OoC platforms can recreate the biochemical, mechanical, structural, and rheological properties of the native scenario, thus making them ideal models from drug discovery and screening of new drugs. As an example, the presence of biochemical gradients can easily be recreated taking advantage of microfluidics and the properties of fluid flow at the microscale. At this length scale, inertial forces dominate over gravitational ones and liquids display a laminar flow (see Section 15.2). In addition, old drugs that have been discarded, may be revisited to reevaluate their efficiencies in a more realistic environment. Finally, organs-on-chip platforms can also be employed to mimic human diseases and, when combined with cells of a specific patient, to test the efficiency of personalized therapies before moving into the patient. This is of particular interest in cancer research to investigate the effect of chemotherapeutic drugs. Indeed, OoCs have been widely used during the last years in cancer research to investigate the different events occurring in the metastatic cascade, such as cancer cell invasion, intra/extravasation, angiogenesis, or organ-specificity, among others. They have also been employed for the isolation and molecular characterization of circulating tumor cells (CTCs), giving rise to *liquid biopsy-on-*

*a-chip* platforms for the rapid and efficient profiling of tumors. This is of high importance because the number of CTCs circulating in the blood stream is very low (1–10 CTC $mL^{-1}$ of blood). These types of cells can provide fundamental information about the tumor origin and etiology.

Overall, microfluidics shows a huge potential compared with traditional systems to gain increased knowledge not only in cancer research but also in many other areas (biomedicine, pharmacology, biotechnology, and others). In the following sections, we discuss about the contribution of microfluidics in cancer research, ranging from fundamental applications, such as tumor models, to more applied and commercial uses.

### 15.1.2 Traditional systems for tumor diagnosis and modeling

Typically, the appearance of cancer is detected by manual palpation (such as in breast tumors) or through imaging (PET, X-ray, or similar). Then, the virulence of the tumor is analyzed by sampling a biopsy of the tumor and analyzing the expression of specific biomarkers. In general, this procedure is efficient, but it may provide inaccurate results due to the heterogeneous cellular population within a tumor. Multiple biopsies in different regions of the tumor samples may solve this problem, but this procedure is very invasive for patients. In addition, biopsies can damage the tissue and may lead to the dissemination of tumor cells from its original site to other parts of the body. Finally, during the last years, multi-OMICS has allowed the precise identification of cancer biomarkers in body fluids, such as blood and saliva, to provide a precise signature of the molecular profile of the tumor. This "liquid biopsy" approach holds very promising, even though it is very laborious and requires the use of very expensive and sophisticated equipment.

Historically, drug testing and tumor modeling have been performed using simplified approaches, including tissue culture—Petri dishes, flasks, or multiwell plates. These two-dimensional (2D) cell culture approaches are oversimplified and do not recapitulate the complex three-dimensional (3D) architecture of the native scenario as well as the synergistic interaction between all the cells and components of the TME. Consequently, cells display perturbed gene expression and phenotypes, and respond to drugs in a different manner than in vivo. This causes that most of the drugs that progress through preclinical assays, fail when tested in vivo. This has a tremendous economic burden for pharmaceutical and biotechnological companies, and a huge societal and health impact. To overcome this limitation, more realistic 3D tumor models have been developed to investigate tumor progression. This includes the use of *transwell-based assays, hydrogels or scaffolds, microcarriers*, or similar approaches coated or made of ECM proteins [5]. This type of assays allows studying the mechanism of cancer cell migration, invasion, and transendothelial invasion, among others. They also permit the culture of 3D tumor spheroids/organoids, or *tumoroids*, which can recapitulate the interaction of tumor cells with other stromal cells and with the ECM. Tumor spheroids also allow the study of the effect of hypoxia in tumors, since they can form a necrotic core similar to those observed in the native scenario. However, these models are in general static (i.e., lack perfusion), and therefore, they lack the rheological complexities of the in vivo scenario. In addition, they lack gas and nutrient diffusion, vasculature, and shear stress forces, among other limitations. During the last few years, hydrogels/scaffolds encapsulating cells (or tissue biopsies) have been integrated into perfused bioreactors systems, which are capable to inject fresh cell culture media in a well-controlled manner during long-term periods. This approach also permits the injection of cells, compounds, or drugs, mimicking systemic drug delivery. However, 3D tumor models also display serious limitations. In general, they do not reproduce the biological and mechanical properties of the TME. Next, these models are typically not perfused, therefore, they are incapable to mimic the hydrodynamics of the native scenario, such as the fluid shear stress or enhanced interstitial pressure. They are also incapable to decouple fundamental parameters that contribute to cancer progression. Consequently, limited success in the translation of drug candidates in to the clinics has been obtained. Even in vivo models have limited success rate in the discovery and screening of novel therapeutic drugs. These in vivo models range from "simple" models systems, such as the Zebra fish, *Drosophila*, or *Caenorhabditis elegans* to more complex animals, such as mice, pigs, sheep, or primates. These animal models recapitulate with higher fidelity the structural, biochemical, and rheological complexities of humans. This is in particular true for the most advanced in vivo models, such as humanized mouse models. These are immunodeficient mouse strains into which human immune systems are engrafted, and human cancer cells are transplanted. These models are expensive and time-consuming, and the results are not completely predictive of the human scenario.

### 15.1.3 Microfluidics and cancer: main tools and applications

Preclinical experimentation demands for highly reliable and physiologically relevant systems capable of recapitulating the complex human physiology. This includes not only the biological, biomolecular, and structural complexity, but also the hydrodynamic properties of human tissues and organs. This is of upmost importance in cancer research where the rheological properties are also significantly perturbed. The inflammation of the tumor stroma and the presence of leaky blood vessels in the tumor area are the main responsible of this perturbation. In this regard, microfluidics technology permits the development of biomimetic devices, coined *OoC* that replicate in vitro the fundamental functional units of human tissues and organs. Applied to cancer research, OoC platforms result in "*cancer-*, *tumor-*, or *metastasis-on-a-chip*" models. These models allow studying the main events occurring during the dissemination of cancer, including:

1. *Tumor growth*: Proliferation and growth of the tumor on the primary site.
2. *Angiogenesis*: Growth of newly formed blood vessels connecting the tumor and the vasculature.
3. *Lymphangiogenesis*: Growth of newly formed lymphatic vessels connecting the tumor and the lymphatic system.
4. *Tumor cell invasion*: Directional migration of cancer cells out of the tumor mass. Tumor cells perform an epithelial-to-mesenchymal transition (EMT) and migrate directionally along rearranged ECM fibers toward the vasculature.
5. *Intravasation*: Invasion of tumor cells from the stroma into the vasculature becoming CTCs.

6. *CTC transit*: Circulation of tumor cells along the vasculature.
7. *Extravasation*: Invasion of tumor cells into the stroma from the vasculature.
8. *Colonization and organ specificity*: Migration of cancer cells into a secondary metastatic site. The preferred metastatic tissue/organ depends on the origin of the primary tumor.

Other important events involved in cancer progression, such as the formation of cytokine gradients, the increase in interstitial fluid pressure/flow, or the impact of mechanical/dynamic stimuli (e.g., physical constraints or deformations), can also be reproduced using microfluidic-based models. Similarly, microfluidic devices can be employed for (1) drug screening and delivery applications; (2) to monitor, detect, and analyze the presence of a specific compound; or (3) for point-of-care diagnostics. This results in the development of attractive devices for industrial and clinical applications (see Fig. 15.1).

Importantly, microfluidic devices can be combined with existing technologies, such as optical microscopy, screening techniques, or biosensor probes. This provides microfluidic platforms with advanced capabilities that are very attractive in cancer research. All the events in the metastatic cascade can be monitored in real time, the efficiency of a therapeutic compound can be assessed in vitro, and the presence of specific biomarkers can be detected with high precision.

Overall, microfluidic systems pave the way toward improving our knowledge on the mechanism of cancer progression as well as on the development of new therapeutic avenues. This will eventually improve the current treatments against cancer. Herein, we describe the main applications of microfluidics in cancer research. We comment on their main advantages as well as their limitations. We also discuss about their clinical and industrial uses, and the future perspectives. We anticipate that microfluidic devices will provide a revolution in cancer research in the near future unraveling new insights into the mechanisms of cancer growth and dissemination.

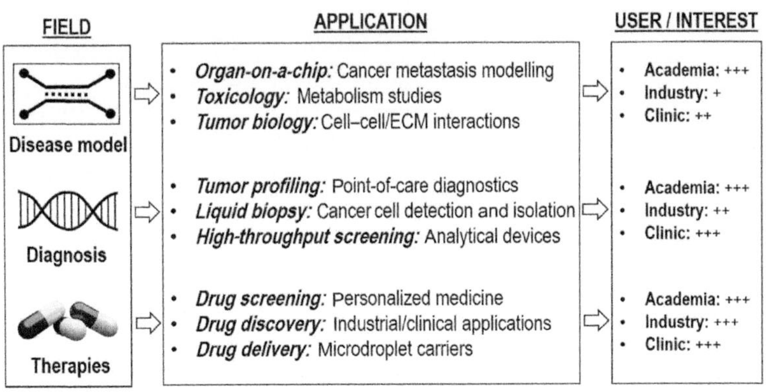


FIGURE 15.1 Main applications of microfluidic devices in cancer research. Microfluidic systems are mainly employed in three research fields (modeling, diagnosis, and therapies). Different applications can be identified on each of the fields; some of them can be indeed shared. Finally, each of the fields/applications is associated to the interest of the different stakeholders involved.

## 15.2 Fundamentals of microfluidics: fluid mechanics in miniaturized devices

The manipulation of small volumes of fluids within confined microchannels with diameters typically in the range of 1 μm to 1 mm is associated with new and unique features related to fluid mechanics at the microscale [6]. This corresponds to fluid volumes in the range of $10^{-6}$–$10^{-9}$ L (in some cases, manipulation of volumes as small as $10^{-12}$–$10^{-15}$ L can be achieved). This is of upmost importance when dealing with cells, which are also in the micrometric scale [7]. The flow of incompressible fluids is described by the *Navier–Stokes* equations, which emerges from the *2nd Newton law* applied to fluids. For a detailed description of their deduction, readers are referred to specialized references in the field [8–10]. In the following, we briefly describe some of these unique features of manipulating liquids at low dimensions.

### 15.2.1 Laminar flow

At the micro- and mesoscopic scale, the fluid displays a laminar flow, where the streamlines of the fluid flow in parallel displaying a Gaussian profile with a zero flow at the solid–fluid interface. In this case, the viscous forces dominate over the inertial ones, whose ratio is quantified by the—dimensionless—Reynolds number (*Re*), defined as:

$$Re = \frac{\rho l v}{\eta},$$

where, $\rho$, $l$, $v$, and $\eta$ correspond to the density of the fluid (in $kg\ m^{-3}$), the most relevant length scale (in $m$), the average flow velocity (in $m\ s^{-1}$), and the viscosity of the fluid (in $kg\ m^{-1}\ s^{-1}$), respectively. If $Re \leq 2000$, then a laminar flow is considered, whereas values of $Re > 2000$ correspond to turbulent flows [11].

### 15.2.2 Diffusion

Diffusion is the spontaneous spreading of particles, molecules, or ions driven by a gradient concentration. Within a microfluidic device, two or more fluids (liquid or gas) flowing along a microchannel at low *Re* (i.e., laminar flow) mix their components mainly by diffusion across their interface only by diffusion. The profile of this interface broadens and dissipates with time along the microchannel. The distance and time of diffusion are governed by the *Einstein–Smoluchowski* equation, as follows:

$$d_{diff} = (2Dt)^{1/2},$$

where, $d_{diff}$ is the diffusion distance (in $m$), $D$ is the diffusion coefficient (in $m^2\ s^{-1}$), and $t$ is the diffusion time (in $s$). Typically, microchannel lengths over distances of several centimeters are needed to increase the interfacial contact between two (or more) different fluids to completely mix by diffusion. For this, serpentine-like channels are employed to accommodate the long channel distances for complete fluid mixture. In contrast, in macroscopic systems with high *Re*, the mixing is instantaneous by turbulence, and macroscopic transport effects usually superpose diffusion.

### 15.2.3 Surface tension

Surface tension, $\gamma$, is a property of a fluid that tends to minimize its surface area (i.e., contract) when in contact with the air because of the cohesive forces between molecules. From a thermodynamic point of view, $\gamma$ is defined as the amount of work $W$ needed to create a unit of surface $A$ of a system, or the force $F$ per unit of length $L$. Therefore

$$\gamma = \frac{dW}{dA} = \frac{F}{L}$$

The surface tension is typically measured in $\mathrm{N\,m^{-1}}$. Table 15.1 shows the $\gamma$ values of some typical fluids. Water has one of the largest $\gamma$ values, threefold higher than typical solvents, oils and soap solutions, which reflect the cohesive energy supplied by hydrogen bonds. Finally, mercury exhibits the highest $\gamma$ value.

Importantly, the surface tension is responsible of the formation of droplets when a fluid stream breaks into individual drops. This effect is known as *Plateau–Rayleigh* instability and is a field of intense research in fluid mechanics [12,13]. In microfluidics, two immiscible fluids within a microchannel can also generate droplets. Competing stresses drive the surface tension to reduce the interfacial area, and viscous stresses act to extend the interface downstream. These stresses destabilize the interface resulting in the formation of microsized droplets.

**TABLE 15.1** Surface tension of some fluids (at 20°C) [6].

| Category | Fluid | Surface tension $\gamma$ ($10^{-3}\ \mathrm{N \cdot m^{-1}}$) |
|---|---|---|
| *Liquids* | Water | 72.8 |
| | Oil (olive) | 32.0 |
| | Oil (mineral) | 29.9 |
| | Soap solution | 25.0 |
| | Glycerin | 63.1 |
| *Solvents* | Toluene | 28.4 |
| | Acetone | 23.7 |
| | Isopropanol | 23.0 |
| | Ethanol | 22.3 |
| *Body fluids* | Plasma blood (at 37°C) | 73.0 |
| | Whole blood (at 37°C) | 58.0 |
| | Tissue fluids (typical) | 50.0 |
| *Others* | Mercury | 465.0 |
| | Hydrogen peroxide | 74.0 |
| | Acetic acid | 27.6 |

### 15.2.4 Capillary forces

Capillary forces, or capillarity, arise at the fluid–air interface in a microfluidic channel. It is the result of the interplay between surface tension of a fluid and its adhesion, resulting from wetting and other cohesive forces, to the microchannel walls with a specific geometry and surface chemistry [14]. Capillary forces drive in passively the fluid forward; no active pumping system, such as a syringe pump, is involved. Indeed, capillary effects have been employed to drive fluids in the absence of active pumping devices [15]. The surface tension maintains intact the surface of the fluid making the fluid front to move forward. The upward force is defined as:

$$F_{cap} = \gamma \cdot 2\pi r,$$

where, $\gamma$ is the surface tension and $r$ the radius of the—capillary—tube or microchannel. The height $h$ to which capillarity will drive the fluid depends on its weight as:

$$\gamma \cdot 2\pi r = \rho g \cdot \left(h\pi r^2\right)$$

Therefore

$$h = \frac{2\gamma}{\rho r g},$$

where, $\rho$ is the density of the fluid and $g$ the gravity. Actually, the effective height depends on the relative strength of the cohesive and adhesive forces. Thus the former equation depends on the contact angle $\theta$ between the fluid and the tube surface.

$$h = \frac{2\gamma cos\theta}{\rho r g}$$

This equation shows that the fluid will rise higher for smaller microchannels.

### 15.2.5 Flow rate and resistance

Flow rate, $Q$, is the amount of fluid that flows per unit of time (in $m^3\ s^{-1}$). Typically, microchannels display a square-shaped profile resulting from the standard microfabrication techniques used (UV-photolithography, replica molding, and rapid prototyping). Under these conditions and assuming a laminar flow, $Q$ can be calculated by solving the *Navier–Stokes* equation assuming that the height of the channels is smaller than its width ($h < w$), resulting in [14]:

$$Q = \frac{h^3 \cdot w \cdot \Delta P}{12\eta \cdot L}\left(1 - 0.63\frac{h}{w}\right),$$

where, $h$ and $w$ are the microchannel height and width, respectively (in $\mu m$), $\Delta P$ is the difference in—capillary—pressure across the channel (in $Pa$), and $L$ is the length of the fluid within the microchannel (in $\mu m$).

The speed and resistance of a fluid flowing in a rectangular-shaped microchannel with width $w$ and height $h$ is therefore:

$$\frac{dL(t)}{dt} = \frac{Q}{w \cdot h} = \frac{h^2 \cdot \Delta P}{12\eta \cdot L}\left(1 - 0.63\frac{h}{w}\right)$$

Finally, the hydrodynamic resistance of the fluid, or flow resistance, within a square-shaped microchannel is defined as:

$$R = \frac{\Delta P}{Q} = \frac{12\eta \cdot L}{h^3 \cdot w}\left(1 - 0.63\frac{h}{w}\right)^{-1},$$

where, $\Delta P$ is described by the *Hagen–Poiseuille* equation. For different microchannel geometries, similar equations can be deduced.

## 15.3 Fabrication principles of microfluidic devices

Microfluidic devices can be applied to a multitude of applications in cancer research, ranging from fundamental studies, drug screening, or cancer disease modeling. This large diversity is associated with the use of different fabrication technologies to recreate the TME and its components, such as the vasculature, the physical tumor stroma, and its biochemical content. In the following, we briefly describe the main microfabrication technologies typically used for the development of microfluidic devices together with the architectural and structural features reminiscent of the TME. We do not aim to provide a detailed description of all the available techniques used for the microfabrication of microfluidic devices, but just a general view of the main methods utilized. For detailed description, readers are referred to specialized review on the topic [16].

### 15.3.1 Molding

#### *15.3.1.1 Replica molding*

The large majority of microfluidic devices are fabricated by replicating the desired features (e.g., microchannels) from a mold (or master). For the fabrication of the mold, a combination of standard UV-photolithography (or laser-assisted photolithography) and soft lithography techniques is mainly employed (see Fig. 15.2). Typically, for UV-photolithography, SU-8 or AZT photoresins are used. Briefly, the fabrication process starts using a clean silicon-based wafer, or alternatively a glass slide, where the UV-photosensitive photoresin is spin-coated at a specific angular velocity to obtain a complete flat resin layer with a well-controlled thickness. The manufacturer of the photoresin usually provides the detailed information about the physicochemical properties of the product, including spin-coating speed, pre-/postbaking times, and exposure dose. After the prebaking steps to stabilize the photoresin, the wafer is deposited in the mask aligner for UV exposure. An acetate or chromium photomask containing the microfluidic chip design and features is deposited upside-down in contact with the photoresin. Next, the sample is

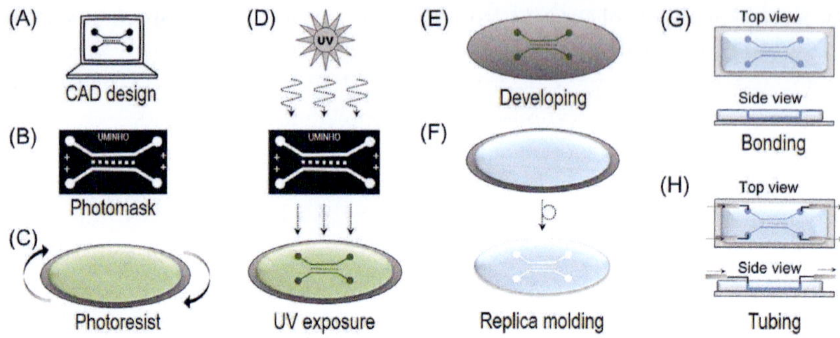


FIGURE 15.2 Scheme showing the typical steps for the fabrication of a microfluidic device by replica molding. (A) Chip design by computer-assisted design (CAD) software. (B) Fabrication of the photomask displaying the same features. (C) Spin-coating of a photosensitive resin (photoresist). (D) UV exposure using the photomask to replicate the features on the photoresist. (E) Replica molding using PDMS. (G) Bonding of the chip replica on a glass slide. (H) Connection of the tubing.

irradiated with the measured dose to cross-link the exposed regions (for negative photoresins; the opposite for positive ones). Then, after the postbaking process, the nonexposed regions are removed by means of a developer, resulting in a negative version of the desired microchannels. Then, the master is replicated using replica molding, typically using polydimethylsiloxane (PDMS) in a 10:1 (*w*/*w* or *v*/*v* prepolymer: cross-linker) ratio. Then, the inlet/outlet are drilled for tube connection, and plasma-bonded to a microscopy glass slide (see Fig. 15.2). Finally, the engineered microchannels are populated with cells (recapitulating e.g., the native organ-level physiology and pathophysiology), and perfused with cell culture medium (or other physiological fluids) by means of external pumping sources, such as a syringe pump. This allows maintaining the viability of the cells over long periods, typically several weeks. Detailed protocols describing the design, fabrication, and characterization of microfluidic devices can be found elsewhere [17–19].

This methodology allows the fabrication of simple "planar" configurations, and structures in the range of about 1–5 μm in lateral resolution and up to 200 μm in thickness can be obtained. Note that more complex 3D configurations can also be obtained by aligning, stacking, and bonding different PDMS layers.

Overall, replica molding offers several advantages, such as its low-infrastructure cost and manufacture simplicity. However, from a processing perspective, the fabrication of microfluidic devices based on PDMS replica molding limits mass production and automation.

#### 15.3.1.2 Hot embossing

Due to its simplicity and high throughput, hot embossing is a very attractive and cost-effective approach for the rapid prototyping of microfluidic devices containing micro- and nanosized features over a large area [20,21]. Briefly, hot embossing uses a solid master (mold) from which a thermoplastic film is replicated. The film is thermally formed under high pressure, and during a specific period, against the mold at a predetermined—glass transition—temperature (which depends on each polymer) that softness the film. Then, the temperature is decreased to withdrawn the mold from the solidified film. For microfluidic applications, the microchannels can be covered and sealed by, for example, a combination of plasma and thermal treatment. Typically, the overall embossing process is relatively fast for simple structures, such as microchannels. However, long cycle times of approximately 30 min may be required to emboss some complex components.

This technique is compatible with a large variety of thermoplastic materials as films, such as polymethylmethacrylate, polysterene, or polycarbonate, which are more attractive for industrial applications due to their low cost and well-controlled surface properties. As a mold, several materials can be employed, including nickel, etched silicon, PDMS, SU-8 resin masters, or epoxy masters.

One of the main advantages of hot embossing (in particular compared with injection molding; see below) is that reduced stress is produced in the thermoplastic, since the viscous film is not transported a long way into the molding tool, reducing the shrinkage during cooling as well as the friction forces acting on the microsized structures during demolding. This makes the resulting molded samples to have good optical properties and to preserve premodifications of the film material.

Finally, it is worth mentioning that hot embossing is very similar to another thermoforming method named nanoimprint lithography (NIL), which uses the same molding approach. Typically, these two techniques are reported indistinctly, even though there are small differences between both methods. Hot embossing is characterized by structuring films that are, in general, much thicker than the height of the mold being embossed. This is in contrast to NIL and therefore, makes it easier to achieve complete filling of the polymer into the mold. Microfluidic systems are usually fabricated by hot embossing rather than NIL.

#### 15.3.1.3 *Microthermoforming*

In microthermoforming, thin polymeric films are heated up to their crystallization temperature to thermoform into thin-walled microstructures [22]. The patterning is accomplished by 3D stretching the film using vacuum or compressed air. After the stretching, the resulting film is thinner that its initial thickness. Sealing of the thermoformed films for the development of microfluidic can be accomplished by heat sealing a thermoformed film with a second one [23]. Alternative sealing methods include the use of adhesive films, lamination, and laser welding [24,25].

Microthermoforming shares obvious similarities with hot embossing technique in terms of the mechanism at work, and therefore, it can be considered as a variant process. However, what occurs to the film inside the molding tool is different. In microthermoforming, the film thickness is typically smaller than the lateral dimensions of the mold, whereas in hot embossing the initial thickness is normally larger than the characteristic depths of the mold [22]. Next, in microthermoforming, the film is formed under high-material coherence, that is, when the film is soft but still in solid state enabling the film to be stretched. In contrast, in hot embossing, the film is formed in low-material coherence, that is, when the film is in a more liquid stated. Finally, in thermoforming, the forming of cavities leads to thin-walled hollow structures whereas in hot embossing, filling up the same cavities results in solid, not hollow structures.

Microthermoforming displays several advantages when compared with hot embossing and injection molding, such as the possibility to fabricate hollow microstructures on thin polymeric films, and in particular, the possibility of forming microstructures with spherical bottoms. However, it displays certain limitations, including the limit of resolution and the possibility to fabricate high-aspect ratio structures. For small structures, the limiting factor of this technique is the thickness of the films to be formed. This is because the film has to bend around and stretch along the mold structures. This limits the resolution of the structures in the low 10-μm range. For a detailed review on the use of microthermoforming, readers are referred to Ref. [22].

#### 15.3.1.4 *Microinjection molding*

Microinjection molding is one of the most common methods to fabricate microfluidic devices using thermoplastic films for industrial applications. One of its main advantages is its simplicity and high reproducibility. Briefly, in microinjection molding, the granular polymer is heated above its glass transition temperature into a rotating forward screw until it is melted when the nozzle is reached. Then, it is injected into the closed microstructured mold by forward motion and is cooled down, prior demolding the polymer. Similar to the abovementioned techniques, sealing of the microstructured features, such as microfluidic channels, can be accomplished by thermal sealing, adhesive films, or solvent bonding, among others [26]. Besides its obvious simplicity, high throughput, low cost, and low-processing times (in the order of some minutes), this method presents some limitations like the possibility of blocking the injection channel due to sudden solidification of the melted polymer within the mold, or the resulting high stress inside the fabricated parts, which can lead to poor postprocessing results (e.g., bonding). However, it is still the favorite methodology for industrial applications, in particular for the fabrication of disposable devices for biomedical applications. In this regard, this methodology has been widely employed for the fabrication of microfluidic devices [27].

### 15.3.2 Sacrificial templating

Sacrificial templating is a simple methodology typically employed to fabricate microfluidic channels within a synthetic or ECM-based hydrogel. It involves the use of 'sacrificial' structures, such as a thin wire, to fabricate hollow channels within a hydrogel matrix. The channels are then coated with the ECM proteins of interest (e.g., fibronectin, collagen, and others) and endothelial cells, mimicking the native vasculature. Other cells, such as tumor and stromal, can be encapsulated within the hydrogel, thus reproducing the TME. Typically, uniaxial and straight microchannel morphologies with diameters of approximately 100 μm can be obtained by this methodology, even though more complex 3D structures with multisized channels ranging from 200 μm to 1 mm have been reported [28].

The advantage of this method compared with UV-photolithography is its simplicity. Sacrificial templating also overcomes the limitations of solid polymeric channels, where the activity is restricted to the channel lumen. In contrast, the hydrogel matrix allows the study of multiple cellular phenomena involved in the progression of cancer, including angiogenesis, invasion, intra/extravasation, or permeability. Importantly, the composition and mechanical properties of the ECM hydrogel can be modulated to mimic the physiological or pathological scenario. Finally, the fabricated devices may be integrated within solid microfluidic devices resulting into hybrid structures.

### 15.3.3 3D (bio)printing

The fabrication of microfluidic devices based on 3D (bio)printing (3DP) has gained significant interest during the last years for its enormous industrial potential [29,30]. Typically, 3DP can be classified into different categories, which differ on their working principle and (bio)material employed. These include (1) *extrusion*, (2) *inkjet deposition*, (3) *stereolithography*, and (4) *two-photon polymerization* methods. Briefly:

1. *Extrusion*: For extrusion, a thermoresistive polymer is dispensed through an extruder. These polymeric templates are typically employed as sacrificial structures to support the adhesion of cells.
2. *Inkjet deposition*: Inkjet deposition dispenses a biomaterial, such as deposited in a dissolvable matrix precursor (e.g., gelatin) or initiator (e.g., calcium chloride), to print layer-by-layer 3D structures.
3. *Stereolithography*: In stereolithography a raster scan laser or projected light with UV wavelength (365 nm) is employed to generate layer-by-layer 3D structures by photopolymerizing a 2D cross-section onto a substrate submerged in a tank containing liquid photosensitive material in a photochemical reaction. The resulting resolution is determined by the laser spot size or by the size of the projector pixel.
4. *Two-photon polymerization*: Two-photon polymerization uses a near-infrared femtosecond laser to polymerize a photocurable epoxy transparent resin into 3D structures. The two photons are absorbed simultaneously by the photosensitive molecule, which induces photochemical reaction between the photoinitiator and the monomers. With this method, resolutions in the order of tens of nanometers can be achieved [31]. However, this is an extremely long time-consuming technique, in particular for the length scales typically needed for microfluidic devices.

The 3DP methods allow the printing of simple microfluidic systems with structures ranging from several micrometer to several centimeter in less than 15 min. However, more complex devices with nm-sized features may take several hours, such as for two-photon polymerization, one of the main drawbacks of this technique. Other limitations of 3DP includes its reproducibility, shape conformity, surface quality, biocompatibility, optical transparency, and material availability [30]. For a detailed review on the use of 3DP for the fabrication of microfluidic devices, readers are referred to references.

## 15.4 Mimicking the tumor microenvironment using microfluidics

### 15.4.1 The tumor microenvironment: an overview

The TME is highly complex and dynamic [32]. It contains a large variety of cell populations (malignant and nonmalignant), including the tumor cells, fibroblasts, immune cells, stem cells, vascular endothelial cells, pericytes, lymphatic cells, and others. These cells interact among them and with the surrounding microenvironment, resulting into a highly complex TME landscape. For instance, cancer-associated fibroblasts (CAFs) are responsible of rearranging the ECM into oriented cross-linked and rigid fibers by lysyl oxidase and transglutaminase present in the TME perpendicularly to the tumor periphery along which tumor cells escape invading the tumor stroma prior colonizing distant organs. Collagen, fibronectin, and elastin fibers, among other ECM components, are the main components of the newly arranged fibers which promote the directed invasion of cancer cells [33]. This rearrangement is mainly mediated by CAFs, which release matrix metalloproteinases (MMPs) that contribute to the degradation of the ECM. There is also a highly dynamic network of chemokines, growth factors, and inflammatory enzymes, among others, which govern the behavior of cells. These biochemical cues are, in part, responsible of the formation of newly formed blood (angiogenesis) and lymphatic vessels (lymphangiogenesis). These two types of vessels, together with the ECM, play a fundamental role in the initial steps of cancer dissemination. The ECM is not just a passive structural support, but it is a dynamic structure, which provides active biochemical and mechanical cues to cells and tissues [34]. Indeed, the mechanochemical properties of the ECM are perturbed during tumor progression, including its structural properties and composition level. The ECM contain angiogenic growth factors and chemokines that interact with the cell surface receptors. This gives to the tissue its mechanical properties. Tumoral ECMs are in general stiffer than their normal counterparts are, and transfer this increased stiffness to the tumor, even though in general cancer cells are softer. The contribution of the TME is so important that some therapies have been addressed targeting the extracellular matrix, and in particular, its mechanical properties. This approach has been denoted as *mechanotherapy*, and focuses on the effect of physical forces on tissues and cells and how they modulate integrin-mediated processes, among others [35]. This influences cell response, such as adhesion, migration, and proliferation, critical events during cancer progression.

Finally, the dynamical properties of the TME are also significantly perturbed. The blood vessels near the TME become more permeable, which facilitates the intravasation of cancer cells, molecules, and compounds into the blood stream. This is known as enhanced permeability and retention (EPR) effect, and is used as a strategy to specifically deliver drugs into the TME [36,37]. The blood vessels around the TME also show a perturbation on their morphology, size, and rheological properties. In particular, the dynamics of the vasculature is very heterogeneous. Some vessels may display flows near to zero, whereas others display significant blood flows. There is also an increase in the interstitial fluid pressure in the TME. The lymphatic vessels are also very leaky. Together with the high pressure, this also facilitates the escape of cells through the lymphatic system.

Overall, during tumor growth and dissemination, there is a major perturbation of all the cellular, biochemical, mechanical, and dynamical properties of the TME, which is challenging to capture and reproduce with standard—static—in vitro tumor models.

### 15.4.2 Microfluidics for reproducing biochemical cues during tumor invasion

Microfluidics has significantly contributed to understand important features in the molecular mechanism of tumor dissemination. This is because microfluidic devices can integrate fundamental features of the native scenario, including—cancer—cells, ECM, and fluid flow, among others. All these features synergize in a complex cascade of events to provide a continuous signaling to—cancer—cells, which modulate their behavior.

As aforementioned, tumor growth induces the recruitment of different cell types, including CAFs or endothelial cells, among others. This activates a cascade of events, including the formation of a gradient of chemokines that promotes angiogenesis and triggers cancer invasion. In the following, we will briefly describe how microfluidics can contribute to copycat in a well-controlled manner the biochemical cues occurring during cancer progression.

### *15.4.2.1 Biochemical gradients*

Gradients of chemokines and growth factors have been shown to regulate a diverse variety of cellular phenomena, such as proliferation or differentiation. They also play a fundamental role during tumor growth and dissemination, such as during the directional migration of tumor cells toward the microvasculature following a gradient of growth factors in a mechanism coined *chemotaxis*. Microfluidics can be employed to recreate these type of gradients in vitro with a tight control over space and time. This is achieved taking advantage of the small length scale of the microfluidic channels, where the flow is laminar (see Section 15.2.1), allowing the formation of gradients in the scale of the cells.

Several approaches have been employed to develop biochemical gradients using microfluidic devices [38]. Examples include (1) *tree-shaped* arrangement, (2) *Y-shaped* channels, (3) *membrane-based*, or (4) *pressure balance* (with and without membrane) methods (see Fig. 15.3). More sophisticated devices can also produce 3D gradients by combining microfluidic devices, stencils, and 3D hydrogels. Among all of them, the most common approach to generate dynamic temporal and spatial concentration gradients is to use an arrangement of *tree-shaped* adjacent microchannels containing two inlets with different concentrations of the molecule of interest. The flow on each channel periodically split into the adjacent channels where they mix with the flow in that channel, resulting in a well-controlled gradient profile [39]. This resulting slope can theoretically be calculated considering the initial compound concentrations and chip characteristics. By adjusting the relative flow rates of solutions, gradients with linear and nonlinear characteristics can be obtained. Typically, the channels display a serpentine-like morphology to guarantee enough diffusive time for complete fluid mixing.

Several works have reported the use of microfluidic devices for the generation of biochemical gradients to study cell behavior. As an example, a Y-shaped platform was

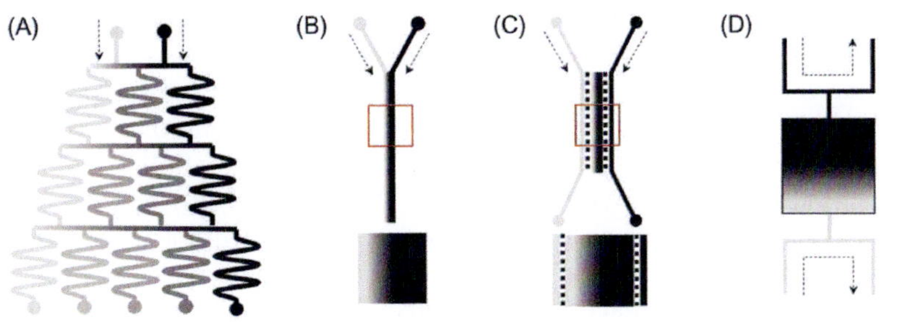


FIGURE 15.3 Most common microfluidic gradient generation methods. (A) Tree-shaped; (B) Y-shaped; (C) membrane-based; and (D) pressure-balance. The dashed arrows show the direction of the fluids.

employed to generate a mesoscopic haptotactic gradient of fibronectin [40]. This gradient was used to investigate the directed migration of cells, a fundamental event during cancer cell invasion. As expected, cell migrated toward the direction of higher fibronectin concentration with large persistence length and directionality. Interestingly, the gradient was combined with a topographical ratchet, which mechanically guided the motion of cells toward one direction. The synergistic interaction of both external cues competed or cooperated (depending on their respective orientation) to guide cell motility. This permitted the identification of the different cellular structures sensitive to the different cues.

One of the advantages of generating gradients using microfluidics is the possibility to perform high-throughput screening studies of the effect of different drugs and/or concentrations on a diverse variety of cells. This saves a significant amount of time and reagents. This approach was recently investigated using a microfluidic device displaying a radial channel configuration [41]. A stable gradient was generated using passive pumping and pressure-balancing strategies. Finally, the effect of the gradient in cell behavior was investigated using highly metastatic MDA-MB-231 and nonmetastatic MCF-7 breast cancer cells.

#### 15.4.2.2 Oxygen gradients and hypoxia

During tumor progression, oxygen gradients develop as they grow beyond their vascular supply. This leads to heterogeneous regions of oxygen depletion, known as hypoxia. Indeed, hypoxia has a critical role in cancer progression and metastasis. Tumor hypoxia causes an advanced and dysfunctional growth of new blood vessels (angiogenesis) toward the tumor. It also triggers EMT to cells, the onset of tumor dissemination and metastasis, as well as the activation of other stromal cells that remodel the ECM to facilitate cancer cell invasion. Hypoxia also perturbs several signaling pathways in cancer cells, alters the expression of growth factors (e.g., VEFG), the metabolism of cancer cells, and contributes to resistance to therapy by inducing cell quiescence.

Microfluidics allows for the highly controlled generation of oxygen gradients with high-spatial control [42,43], allowing to simulate the oxygenation values of tumors and the surrounding normal tissue [44]. However, typical microfluidic tumor models overlook the impact of oxygen levels. Most of the reported microfluidic platform with oxygen gradients are based on the diffusion of oxygen from high to low concentrations, resulting into linear, nonmonotonic, sigmoidal, or exponentially decreasing profiles. On-chip oxygen gradient generators include: (1) direct perfusion of oxygen, (2) addition of oxygen-scavenging agents, (3) polymers and membranes with controlled gas diffusion, (4) off-chip gas mixers, (5) on-chip dilution trees, or (6) spatially confined chemical reactions. The latter provides multiple advantages compared with the other methods, such as a simple fabrication and high-gradient control. These methods were employed to investigate the role of oxygen gradients and hypoxia in tumor progression. As an example, a spatially confined chemical reaction approach has been employed to generate ($NaOCl + H_2O_2 \rightarrow O_2 + NaCl + H_2O$) or scavenge ($C_6H_6O_3 + NaOH$) a nonlinear oxygen gradient within a cell-laden microchannel [45]. The oxygen gradient was generated by confining the areas for chemical reactions, thus without direct chemical contact with the cells. Carcinomic human alveolar basal epithelial A549 cells were monitoring within the device under normoxia, hypoxia, and hyperoxia conditions, as well as with the addition of *Tirapazamine*, an anticancerous drug that is activated to a toxic radical only at very low levels of oxygen. The results showed that indeed the cancerous cells were highly sensitive to the oxygen gradient. More recently, a microfluidic device was employed to generate a linear oxygen gradient with a tunable slope [46]. This slope depended on the dimensions of the diffusion gap between two microchannels. Human brain endothelial cells were seeded on top of a thin PDMS membrane and were exposed to the oxygen gradient. The expression of HIF-1$\alpha$, a hypoxia-inducible transcription factor perturbed during tumor progression, was investigated finding important differences in HIF-activation thresholds when cells were exposed to oxygen gradients or to homogeneous oxygen levels. Finally, a microfluidic device was recently developed simulating the oxygen gradient present in the TME [47]. The device exposed Ishikawa endometrial cancer cells to cross-stream oxygen gradients ranging from hypoxia ($<5\ mg\ L^{-1}$) to hyperoxia ($40\ mg\ L^{-1}$). The oxygen level in the cell culture media was controlled off-chip with gas exchangers and the gradient was generated by diffusion across a microchannel injecting hyperoxia and hypoxic oxygen levels through two inlets in a Y-shaped configuration. Integrated oxygen sensors allowed the monitoring of gradient values. Other studies have addressed the directional migration of cancer cells into microfluidic chips resulting from the presence of oxygen gradients in a mechanism termed *aerotaxis* [48,49].

Overall, microfluidics provides an adequate platform to investigate the behavior of cells in hypoxia and oxygen gradient conditions occurring during tumor growth and progression. This is of upmost importance to investigate the mechanistic determinants of tumor progression as well as to determine the efficiency of therapeutic compounds as a function of the oxygen tension.

#### 15.4.2.3 Microdroplet generation

The use of microdroplets in cell biology originated long time ago with the publication of seminal papers describing the use of droplets to isolate cells [50]. Similar to the formation of biochemical gradients, microfluidic devices can also be employed for the high-throughput generation of microsized droplets (see Section 15.2.3—Surface tension). Microdroplets provide a large number of opportunities in cancer research. Besides all the intrinsic advantages of microfluidics described above, microdroplets also provide a compartmentalized environment in which—individual or multiple—cells, compounds, or even reactions can be isolated and encapsulated. In particular, the encapsulation of cells into a specifically designed microsized droplet provide individualized cell culture models, which can be employed to unravel key insights of cell–cell and cell–ECM interactions. Droplet-based systems have gained much attention into the pharmaceutical industry for the large-scale drug research and screening [51,52]. The use of microfluidics allow rapid and cost-effective diagnostics and screening due to the high-throughput and ultra-fast generation of droplets, typically at a frequency of $10-10^3$ Hz with droplet volumes of $10^{-9}-10^{-12}$ L.

Microdroplets are generated by mixing, within a microchannel, one immiscible fluid (dispersed fluid) into another (continuous fluid), such as water and oil. This allows the high-throughput generation of monodisperse droplets in a fully controlled environment. During this process, biological samples, such as cells, bacteria, DNA, RNA, drugs, and other compounds can be encapsulated and used to perform single-cell analysis [53], generate multicellular spheroids [54,55], or encapsulate drugs for targeted drug delivery [56].

Different methods can be used to generate droplets within microchannels, including squeezing, dripping, jetting, tip-streaming, and tip-multibreaking (for a detailed description about these different methods, readers are referred to specialized reviews on the topic [57,58]) (Fig. 15.4). Depending on the method, different microfluidic *droplet generators* geometries have been designed. In particular:

1. *Cross-flow geometries*: In this type of geometry, the dispersed and continuous phase fluids meet at an angle $0° < \theta < 180°$. This type of configuration includes

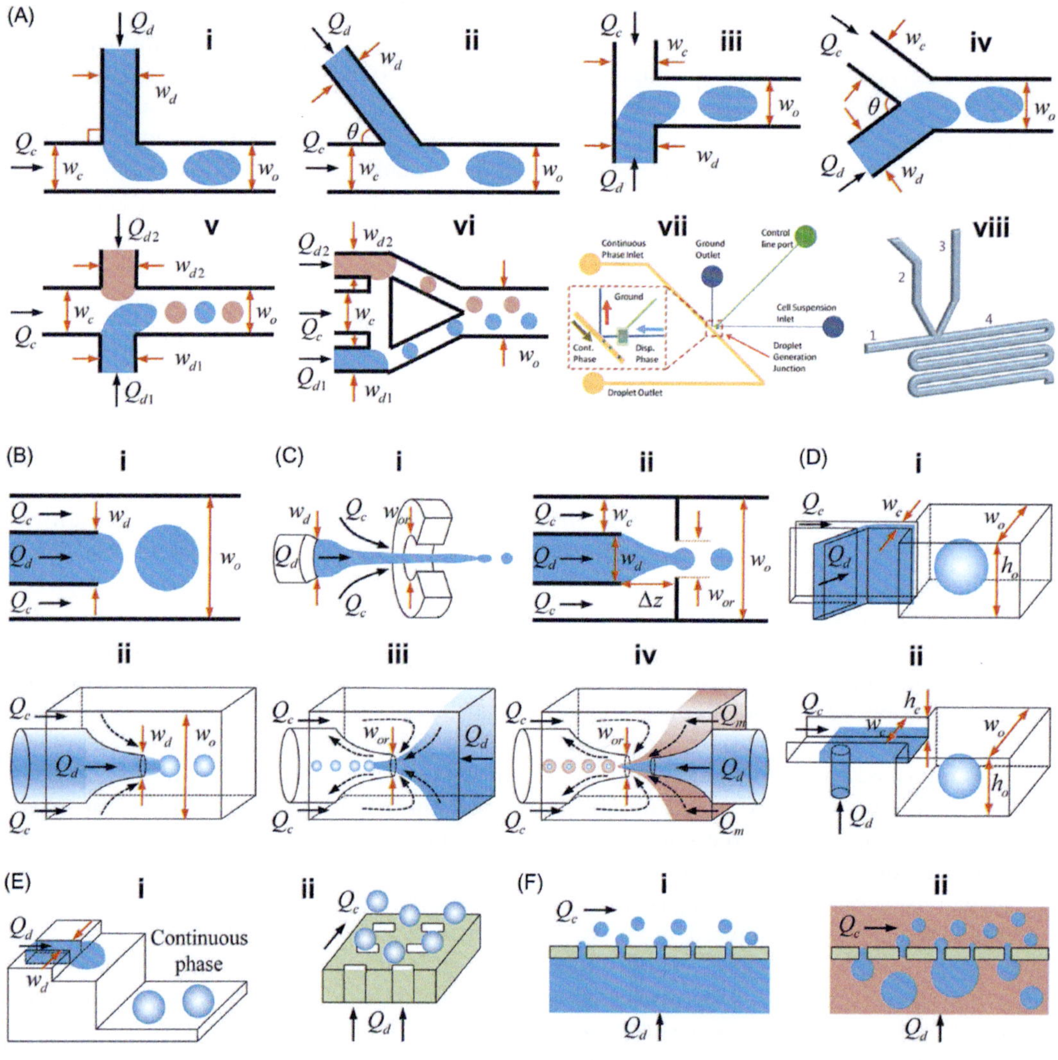


**FIGURE 15.4** Schematic of different methods for the generation of microdroplets using microfluidics. (A) Cross-flow geometry, (B) co-flow geometry, (C) flow-focusing geometry, (D) step emulsification, (E) microchannel emulsification, and (F) membrane emulsification. Source: *Reproduced from reference Zhu P, Wang L. Passive and active droplet generation with microfluidics: a review. Lab Chip 2017;17:34–75 [58] by permission of The Royal Society of Chemistry.*

"T-junctions," where the continuous and dispersed phase fluids intersect (1) orthogonally ($\theta = 90°$), (2) at a certain angle $0° < \theta < 90°$, (3) horizontally ($\theta = 180°$), (4) Y-shaped junctions with intersection angles $0° < \theta < 180°$, (5) double T-junction for the generation of individual (same material) or droplet pairs (two different materials) at the same or separate locations. Modifications of the "T-junction" geometry for droplet generation include the (6) "K-junction" and the (7) "V-junction." This is the most widely used geometry in monodisperse microdroplet generation because of its simplicity and performance.

2. *Co-flow geometries*: In this type of geometry, the dispersed and continuous phase fluids meet in parallel streams. Configurations include (1) quasi-2D planar co-flow, and (2) 3D co-axial. Typically, the size of the generated droplets are larger than the dimension of the dispersed microchannel.
3. *Flow-focusing geometries:* In this type of geometry, two immiscible fluids are hydrodynamically focused. The generated elongated stream flows through a physical constriction to generate of small droplets. This type of configuration includes (1) quasi-2D planar and (2) 3D axisymmetric. A variation of the latter is the microcapillary device, where both fluid phases are injected in opposite direction focusing the flow at the entrance of a small microcapillar. This latter approach can also be combined with co-flow resulting in two-phase droplets.
4. *Step emulsification*: Droplets are generated using variations in microchannel confinement. This cause an abrupt change in capillary pressure and the formation of microdroplets by interfacial tension. Typically, a horizontal or vertical step emulsification geometry is employed. Then, a stabilized co-flow using high-aspect ratio inlet channels reaches an abrupt geometric step toward the release channel resulting into the formation of microdroplets. This approach has several advantages, including less sensitivity to flow rate and pressure fluctuations, improved high throughput, and enhanced monodispersity, among others [58].
5. *Microchannel emulsification*: Microchannel emulsification devices consist of either parallel grooves and terraces or straight-through microsized holes. The generation of microdroplets originates because of the interfacial tension due to the spontaneous transformation of the injected dispersed phase that flows through a multitude of microfabricated microchannel arrays (e.g., microgrooves) toward the continuous phase. This approach involves the absence of external shear forces. Therefore, the droplet size is mainly determined by the microchannel geometry and by the composition of the dispersed and continuous phase. Typically, microchannel emulsification produces monodispersed emulsion droplets with extremely small coefficient of variance [59].
6. *Membrane emulsification*: In this type of geometry, the dispersed fluid phase is forced to pass through a porous membrane toward the continuous phase. Then, due to interfacial tension, microdroplets generate. Therefore, droplet size is determined by the properties of the membrane, including the pore size, distribution, or geometry, among others. Material properties (e.g., fluid viscosities) as well as dynamic parameters (e.g., transmembrane flow/pressure and shear stress) also affect the size of the microdroplets.

Finally, other more sophisticated methods have also been employed for the formation of microdroplets, such as *By-pass* methods [60], which holds very promising for the controlled encapsulation of—cancer—cells.

Microdroplets allow for an extremely large number of simultaneous experiments. This is of upmost importance in anticancer drug research, where thousands of compounds must be screened to evaluate their efficiency prior in vivo experiments. Similarly, microdroplets may also be employed for the targeted delivery of cancer-specific therapeutic compounds, avoiding significant cytotoxicity to neighboring healthy cells. This strategy was recently followed to encapsulate LLL12, a chemical inhibitor of the STAT3 signaling that is overexpressed in several tumors, for the treatment of pancreatic cancer PANC-1/CAPAN-1 cells [61]. The inhibitor was encapsulated together with oxygen in stimuli-responsive microdroplets generated by 3D flow-focusing approach for ultrasound-mediated controlled release at the disease site. The inclusion of oxygen was to deal with hypoxia that reduces the therapeutic efficiency of drugs that rely on reactive oxygen species for cytotoxicity.

Microdroplets also find important applications in cancer diagnosis. This permits the detection of those species that are overexpressed (or perturbed) in cancer cells. As an example, MMPs secreted by tumor cells play a critical role in triggering tumor dissemination since they are involved in the proteolytic cleavage of the ECM. Solid evidences have shown that MMP-9 is involved in tissue remodeling, angiogenesis induction and metastasis [62,63]. The detection of MMP-9 is therefore clinically relevant to evaluate the invasiveness potential of cancer cells. In this regard, the MMP-9 enzymatic activities at the single-cell level were recently investigated using a microdroplet approach [64]. In this work, flow-focusing method was employed to encapsulate within a microsized droplet a single (histiocytic lymphoma U937 or MCF7 breast ductal carcinoma) cancer cells together with peptides and other cofactors. The MMP-9 secreted by the cancer cells cleaved the peptide that was modified with a pair of fluorophore/quencher molecules, which were employed to measure the MMP-9 enzymatic activity and to correlate it with its invasive capability. A similar approach was recently used to profile the MMP activity of CTCs from prostate cancer patients [65]. For this, CTC secretion was analyzed by encapsulating rare cancer cells into microdroplets containing MMPs substrate. CTCs from whole blood were "purified" using a sequence of vortex systems that get rid of undesired cells and plasma proteins, and leave the fluorogenic MMP-cleavable peptide substrate reacting buffer. Finally, the trapped CTCs were released along with the reactive solution of MMP-cleavable substrate toward the microchannel droplet generator. Results showed that CTCs from prostate cancer patients possessed above baseline levels of MMP activity that suggested an active proteolytic process that facilitated invasion or immune evasion.

Overall, microfluidics demonstrates to be a reliable method to copycat the biochemical complexity of the TME, providing important insights into the biochemical mechanisms of tumor growth and progression.

### 15.4.3 Microfluidics for reproducing mechanical cues in tumor invasion

Tumor growth and dissemination also perturb the mechanical properties of the TME. In particular, there is an increase in the stiffness of the ECM and a rearrangement of its morphology, which facilitates the invasion of tumor cells into the stroma and vasculature [66]. Similarly, fluid flow and shear stress are highly perturbed. During migration, cells must physically squeeze deforming the nuclei to migrate through the dense ECM and tight interstitial spaces. In the following, we discuss about the main mechanical cues resulting from the growth and progression of a tumor, and how microfluidics may be employed to reproduce them [67].

#### *15.4.3.1 Physical constrictions*

The deformation of the cell nucleus in cancer cells is associated with reduced levels of A-type lamins (LMNA), resulting in softer nucleus that may facilitate tumor cell deformation and invasion [68,69]. This mechanical interaction was recently investigated using a microfluidic device containing microchannels with precisely defined narrow constrictions that mimicked the native environment [70] (see Fig. 15.5A). The invasive potential of fibroblasts and tumor cells was investigated. The results showed that a larger deformability due to lack of LMNA was responsible of the enhanced invasion of cells. Similar assays were proposed to investigate nuclei deformation [71,72,75] (see Fig. 15.5B and C). In this case, a rupture in the nuclear envelope and DNA damage was observed during constriction, even though tumor cells showed the capability of self-repair and survive. Other compression assays include cell deformation using fine pressure change. Finally, shear stress generated by microfluidic flow may induce mechanical stress on cells, reproducing events occurring on the in vivo scenario. Overall, these works indicate how microfluidics can provide the means to investigate the molecular mechanisms that governs cancer cell migration mimicking the native mechanical environment through narrow constrictions.

#### *15.4.3.2 Anisotropic features*

The structural (architecture and stiffness) properties of the TME greatly influences cell behavior. This mechanical stimulus can also be reproduced using microfluidics to force cells to grow and migrate in channels of a given geometry. These may be further combined with gradients of a chemical compound [76,77]. Among all the possible geometries, anisotropic features can reproduce the rearrangement of the ECM that occurs during the first steps of tumor progression. In this situation, cells migrate along highly oriented and

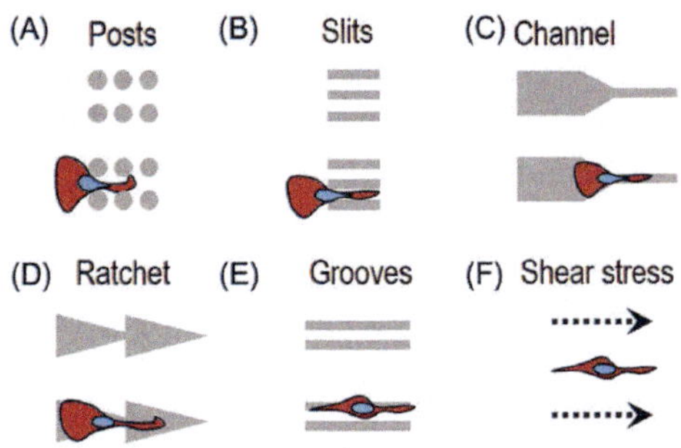


FIGURE 15.5 Typical features integrated in microfluidic systems to reproduce the TME mechanical cues. Deformation of the cell cytoskeleton caused by narrow constrictions (A) posts [70], (B) microslits, (C) channel thinning [71,72], (D) topographical ratchet structures [40], (E) microgrooves [73], and (F) shear stress [74].

aligned fibers. In this case, CAFs mechanically (i.e., force-mediated) remodel the ECM network followed by proteolytic cleavage, resulting into an ECM displaying parallel fiber architecture [78]. Similarly, hollow microsized tracks (i.e., absence of matrix in their center), channels, and holes created via MMPs secreted by the tumor and stromal cells provide contact guidance to cancer cells for fast invasion [79]. Several microfluidic models that reproduce the anisotropic ECM in tumor cell invasion have been developed. Simple approaches include the patterning of ECM protein (e.g., fibronectin, laminin, and others) lines in the range of 1–100 μm in width and several millimeter in length [80,81], induced alignment of ECM proteins on-chip (e.g., template- or shear stress-guided, stretching, or controlled growth of ice crystals, among others) [82–84], topographical grooves [85], ratchet-like structures [40,86,87], or electrospinning of synthetic or natural-origin fibers [88] (see Fig. 15.5D). Regardless of the used methodology, the anisotropy in the TME induced cell alignment, polarization, and directional migration. This strategy was recently employed in a microfluidic device to study the interaction between neurons and cancer cells. The device included two reservoirs interconnected by an array of parallel microsized grooves [73] (see Fig. 15.5E). One reservoir contained—PC-3 prostate, Panc-1 pancreatic, and MCF-7 breast—cancer cells with different rate of perineural invasion. The second reservoir contained primary neurons from rats. The microgrooves allowed the extension of the neurites toward the cancer reservoir, providing a physical support to allow the migration of cancer cells along them and toward the neuronal compartment. The obtained results showed that the highest migration rates correlated with the highest perineural invasion observed in vivo. Interestingly, the blockage of specific signaling pathway allowed investigating the molecular mechanism of perineural invasion. Altogether, this microfluidic device was capable to simulate the directional migration of tumor cells into the neural microenvironment, providing key insights into the mechanism at work of cancer tumor metastasis in the neural microenvironment.

#### 15.4.3.3 Mechanical deformation

Some organs and tissues are exposed to mechanical deformation that can contribute to the progression of cancer. This is the case of the lung, where breathing induces a mechanical stretching of the alveolus. This mechanical stretching was mimicked using a microfluidic model of the lung [89], where human pulmonary microvascular endothelium and human alveolar epithelial cells were seeded on both sides of a porous membrane mimicking the alveolus distribution. The membrane was exposed to cyclic stretching to mimic physiological breathing by means of two adjacent hollow microchannels, and the response of the model to this mechanical stimulation monitored. In a follow-up study, the effect of mechanical stretching on the invasion of human nonsmall-cell lung cancer cells seeded on the primary alveolus was investigated using the same microfluidic mode [90]. Interestingly, results showed that mechanical deformation associated to active breathing decreased by approximately 50% lung cancer cell growth and invasion, whereas larger tumor cell clusters formed in the static alveolus chips.

The same organ (lung) was modeled in a compartmentalized microfluidic chip to investigate the interaction between myofibroblasts and lung cancer cells [91]. The latter were exposed to the culture medium from the fibroblast-containing chamber, which upon cyclic stretching (mimicking breathing), reduced the motility of the lung cancer cells because of a reduction in the expression of α-smooth muscle actin. This result indicated that the mechanical stress influenced fibroblast secretome, and in turn, the behavior of lung cancer cells. Finally, and interestingly, the addition of proinflammatory factor interleukin-1β on the fibroblasts diminished the effects of mechanical stretching, suggesting an interplay between mechanical and chemical stimulations on mediating cancer metastasis by fibroblasts.

#### 15.4.3.4 Modulating matrix stiffness

Tumor growth leads to increased ECM stiffness, a key hallmark in cancer progression [92–94]. As mentioned above, solid evidences have shown that the TME is more rigid than healthy native tissue [94–96]. This enhanced rigidity is mainly a consequence of the dense and cross-linked ECM fiber structures. This malignant transformation of the ECM activates a cascade of events, such as EMT in the tumor. Biomaterials, in particular hydrogels, can be precisely tuned, mimicking the TME, to understand the role of ECM stiffness in tumor dissemination. Most of the microfluidic models employed for cancer research use hydrogels (e.g., collagen, Matrigel, silk, and others) to reproduce the 3D TME. This is due to their simplicity of use and their ability to control their mechanical properties precisely by adjusting the cross-linker amount. As an example, an microfluidic platform was recently used to investigate the invasive potential of HCT-116 colorectal cancer cells into a silk fibroin-based ECM [97]. Different stiffness of silk fibroin matrices (2% and 3%) were employed to analyze cell invasion capacity through the matrix. In particular, the response of cells upon their stimulation with hVCAM-1, a vascular adhesion molecule involved in mediating tumor cell adhesion to vascular endothelial cells and promoting the metastatic process, was investigated. Strikingly, results showed that HCT-116 transmigrated in a larger number through the 2% silk–fibroin hydrogel, in contrast to 3% hydrogel, where almost no cell invasion was observed. The addition of hVCAM-1 enhanced the invasion capability of cells only for softer hydrogels. Altogether, this work showed that in contrast to what it is acknowledged in vivo where stiffer matrices favor the invasion of cells, softer materials may also lead to enhanced invasion of cancer cells. This may provide a new framework to understand cancer cell invasion. Similarly, a microfluidic device was engineered to examine the effect of matrix stiffness in the progression of breast cancer [98]. MCF-7 breast cancer cells were encapsulated within a collagen matrix with specific stiffness values inside the chip. Controlled volumes of nonreactive polyethylene glycol were added during the gelation process of the collagen to modulate—decrease—its mechanical stiffness. The different hydrogels were combined with a pressure gradient that induced an interstitial flow. MCF-7 breast cancer cells displayed a perturbation in the expression levels of ER-α, integrin β1, and E-cadherin depending on the matrix stiffness. In particular, ER-α and integrin β1 displayed a twofold increase in MCF-7 spheroids encapsulated in soft matrices in contrast than those encapsulated in rigid ECMs. Interestingly, the presence of interstitial flow mitigated the effect of matrix stiffness. Overall, this study showed that the stiffness of the ECM modulates the expression of specific—breast—cancer biomarkers.

#### 15.4.3.5 Interstitial fluid pressure and flow

Interstitial fluid pressure and flow has been shown to mechanically act on cells by producing a shear force on the cell surface that leads to changes in their phenotype and migration [99]. The integration of these physical cues into microfluidic devices may allow unraveling their interplay with ECM rigidity to mimic inflammation. Indeed, during early cancer progression, tissue inflammation caused by an increase of the interstitial fluid pressure stimulate the growth and invasion of tumor cells into the microvasculature. Several microfluidic models have been developed to study this important event in cancer progression. Recently, a microfluidic device modeled the hydrodynamical properties of the tumor–lymphatic microenvironment [74] (see Fig. 15.5F). For this, standard transwell Boyden chambers were interconnected with microfluidic channels to simulate the transendothelial and luminal flows. MDA-MB-231 breast cancer cells seeded into the chambers displayed an enhanced intravasation rate into the microchannels because of increased flows, increasing the permeability of lymphatic endothelial cells and, therefore, cancer cells intravasation. Similarly, lymphatic vessels were microengineered together with blood vessels in a microfluidic device to assess the impact of inflammation into vessel permeability [100]. This model contained two microchannels separated by a porous membrane. On both sides, blood and lymphatic endothelial cells were seeded simulating the human microcirculation system. Histamine, an inflammatory promoter, was injected to promote an inflammation inducing an enhancement of vessel permeability. Finally, it is worth noting that many other works report the use of microfluidic systems to investigate in detail the different cellular components involved in tumor dissemination during inflammation processes. For detailed information on the cellular counterparts, readers are referred to recent reviews on this topic [101].

Overall, we are still at the outset of understanding how cells in general, and cancerous ones in particular, respond to the different mechanical cues that occur in vivo. Microfluidics provides unprecedented capabilities to enable a mechanistic understanding of tumor cells–ECM interactions. Importantly, the interplay of this type of cues with biochemical ones can also be investigated in a simple manner, since microfluidics can easily integrate both type of cues and decouple their respective influence in tumor dissemination.

## 15.5 Microfluidic models of cancer

### 15.5.1 Organ-on-a-chip technology

The combination of microfluidic devices with human cells results in the development of OoC models. These are a new generation of organ and tissue in vitro models that truly recapitulate the native physiology and pathophysiology of the native scenario [102]. OoC recreate the functional units of human tissues and organs in vitro and can combine several cell types (epithelial, endothelial, fibroblasts, immune cells, stem cells, and others) in a perfused and well-controlled microenvironment. OoC models have attracted a lot of attention during the last decade in a broad range of sectors, including life sciences, biomedicine, biotechnology, pharmacology, toxicology, cosmetics, agriculture, and food sector. This is because OoC provide a physiologically relevant in vitro environment capable to reproduce the biochemical, mechanical, and the hydrodynamic complexity of the native scenario [4,5]. In particular, for life science-related applications, OoC allow to reproduce the multicellular 3D architecture of tissues and organs, as well as cell–cell, tissue–tissue, and cell/tissue–ECM interactions between the different cells types, including organ-specific epithelial cells, endothelial cells, immune cells, fibroblasts, and others. So far, a long list of OoC models have been reported during the last decade. This includes the lung [89], heart [103], kidney [104], liver [105,106], gut [107,108], spleen [109], bone marrow [110,111], blood–brain barrier [112,113], stomach [114], and many others [115]. These models have been employed for a large plethora of applications, including mechanistic studies of fundamental physiopathological processes [102], pharmacokinetics/pharmacodynamics of drugs [116], or for the evaluation of drug efficacy and toxicity of drugs (absorption, distribution, metabolism, and excretion) [117]. Several individual OoC models can be interconnected on a single chip following a pre-defined physiological order, resulting into multiorgans-on-a-chip (multi-OoC). Multi-OoC models offer an enormous potential for disease modeling as well as comprehensive testing platform for drug discovery and screening applications. In traditional multi-OoC models, the flow of the media is unidirectional and the upstream organs interact with the downstream ones. This offers large possibilities for the prediction of drug efficiency and toxicity to secondary organs in vitro in a reliable manner by mimicking the passage of a drug throughout the human body.

A diverse variety of multi-OoC have been reported to date (see Table 15.2) [133,134]. These models differ in their design, complexity, applications, and in the number and type of tissues/organs modeled. Simplest models include two tissues/organs to study, for example, how the metabolism of a drug or compound in the liver impacts on the secondary organ, such as the heart or the brain. Other multi-OoC include the interconnection of three or four organs for absorption, distribution, metabolism, and excretion (ADME) applications. As an example, a multi-OoC platform comprising the liver, lung, and heart was used to study interorgan response to drug administration [125]. The model interconnected these three independent organ models in a plug and play configuration to investigate the toxicity of *bleomycin*, a drug known to induce lung toxicity. Interestingly, drug response was found to depend on interorgan interactions. In particular, *bleomycin* caused the lung organoid to release cardiotoxic inflammatory cytokines, which was not observed when the drug was tested directly on the heart model, highlighting the utility of the multi-OoC approach. A similar approach interconnecting functional heart, muscle, brain, and liver models was used to assess the long-term systemic toxicity of different drugs with known side effects [127]. The obtained results were in agreement with those reported in the literature, in particular a decrease in cell functionality and viability upon drug treatment, where *doxorubicin*, a well-known chemotherapeutic drug, showed acute cardio- and hepatotoxicity. Taken together, these works demonstrate the potential of multi-OoC platforms as a new generation of in vitro microfluidics-based models for ADME applications. This has attracted the interest of pharmaceutical and biotechnological companies and boosted the development of personalized models of diseases [135]. Multi-OoC have also found applications in cancer research, resulting in the development of *cancer-* or *tumor-on-a-chip (ToC)* systems. These systems are currently being used to explore the mechanistic determinants of tumor growth and progression as well as to investigate the efficacy of therapeutic compounds.

**TABLE 15.2** Most relevant and recent multi-OoC models.

| Organs or tissues | Main applications | Ref. |
|---|---|---|
| *Liver/heart* | Cardiotoxicity; drug screening; long-term monitoring of drug response (heart) and short-term evaluation of acute toxicity (liver). | Zhang et al. [118] |
| *Liver cancer/heart* | Cardiotoxicity of *doxorubicin* caused by the production of toxic metabolites by the liver. | Kamei et al. [119] |
| *Liver/brain neurospheres* | Neurotoxicity assay of *2,5-hexanedione*. This is produced by the metabolization of n-hexane in the liver. Long-term exposure showed a toxic effect on both cell types. | Materne et al. [120] |
| *Gut/liver* | Metastasis; drug screening | Skardal et al. [121] |
| *Gut/liver* | Studies on inflammation due to compound metabolism. | Chen Wen et al. [122] |
| | Pharmacokinetics (permeability, metabolic clearance) of *diclofenac* and *hydrocortisone*. | Tsamandouras et al. [123] |
| *Lung/liver* | Drug toxicity, drug testing, and organs cross-talk studies. | Coppeta et al. [124] |
| *Liver/heart/lung* | Toxicology screening of drugs and disease modeling. The functionality of one organ model influenced the response of others. | Skardal et al. [125] |
| *Intestine/liver/skin/kidney* | ADME of drugs in a long-term culture period; dose systemic toxicity testing of drug candidates. | Maschmeyer et al. [126] |
| *Cardiac/muscle/brain/liver* | Long-term toxicology studies of five drugs (*doxorubicin, atorvastatin, valproic acid, acetaminophen, and N-acetyl-m-aminophenol*) with known side effects after 48 h after administration. | Oleaga et al. [127] |
| *Lung/brain/bone/liver* | Investigation of organ-specificity in the metastasis of lung cancer into bone, brain, and liver tissues. Evaluation of biomarkers characteristic of metastasis into specific tissues. | Xu et al. [128] |
| *Intestine/liver/colorectal cancer/connective tissue* | Drug screening of the prodrug *Capecitabine* and the effect on cancer growth inhibition of its metabolite. Study of the effect of other anticancerous drugs on the intestine and connective tissue. | Satoh et al. [129] |
| *GI tract/lung/liver/bone marrow/kidney* | Interorgan cross-talk; ADME of drugs; screening of cellular biomarkers, gene expression, albumin release, urea, tight junctions. | Miller, Shuler [130] |
| *Intestine/liver/kidney proximal tubule/blood–brain barrier/skeletal muscle* | Studies about the ADME of drugs and their metabolites using an in vivo-like sequential coupling of organs. Drug organ-specificity; pharmacokinetics studies; toxicology studies. | Vernetti et al. [131] |
| *Tumor/lymph node* | Study of T-cells immunosuppression due to interorgan cross-talk secreted factors using a recirculation system. | Shim et al. [132] |

### 15.5.2 Organ-on-a-chip models of cancer metastasis: cancer- or tumor-on-a-chip

Cancer metastasis involves a complex cascade of events, which is first initiated by an uncontrolled growth of the tumor. It is next followed by the invasion of cancer cells into the surrounding stroma. Next, the release of specific growth factors and cytokines promotes the growth of newly formed blood and lymphatic vessels (angiogenesis and lymphangiogenesis) that favor the intravasation of cancer cells into the vasculature becoming CTCs. These CTCs circulate through the vasculature, together with ctDNA/RNA, exosomes, or antibodies, and eventually extravasate colonizing a distant tissue and forming micrometastatic foci, which maturates into a large secondary metastatic tumor (see Figs. 15.6 and 15.7A).

OoC or multi-OoC models of different tissues and organs can be populated with cancer cells. This results in advanced microfluidic models of cancer, which are coined *cancer-* or *ToC*. These models are capable to reproduce one or several of the events occurring in the cascade on cancer [4,5] (see Figs. 15.6 and 15.7). They also allow other key features, namely (1) the culture of other cell types, such as stromal and endothelial cells, (2) the application of shear stress to mimic the hydrodynamic properties of the—cancerous—vasculature, or (3) the injection of anticancerous drugs. Importantly, ToC models allow the local modification of molecular, cellular, biochemical, and mechanical parameters of the TME in a highly controlled manner, such as the presence of cytokine gradients or the enhanced mechanical properties of the tumor stroma. This also allows unraveling their independent or synergistic contribution in the progression of the disease as well as their response to therapy. Importantly, ToC models allow to uncouple the high complexity of all the process of

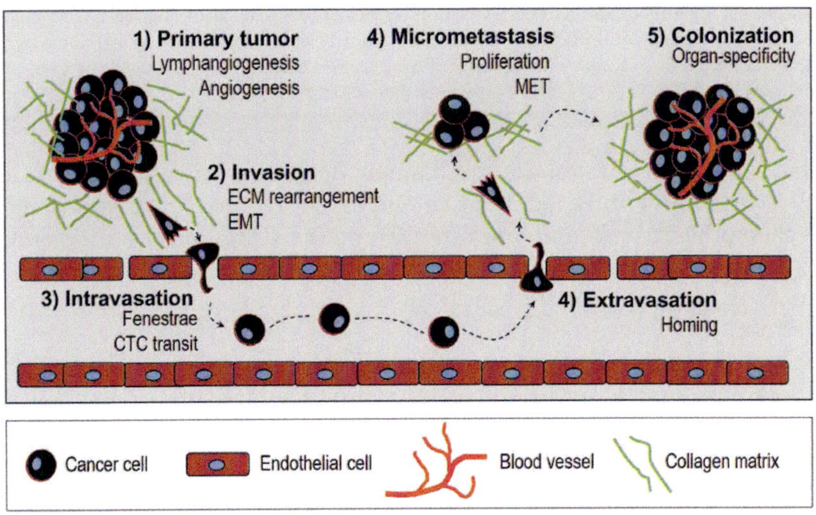


**FIGURE 15.6** Scheme summarizing the main events occurring in the metastatic cascade. Cancer cells escape from the primary tumor and migrate toward the blood/lymphatic vessels where they intravasate becoming CTCs. Next, they circulate along the vasculature and eventually adheres on the vessel wall and extravasate. Cancer cells migrate and form micrometastasis foci, which eventually colonize a secondary organ/tissue.

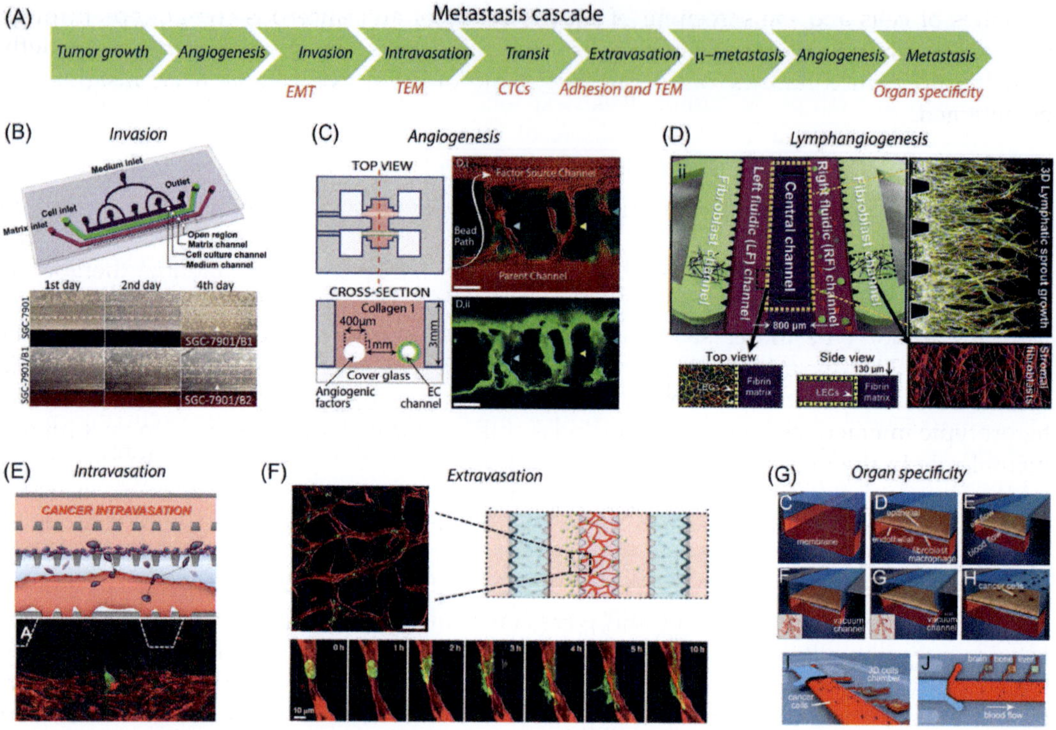


FIGURE 15.7 OoC models of cancer metastasis. (A) Scheme showing the main events in the cascade of metastasis. (B–G) Examples of cancer- or tumor-metastasis-on-a-chip: (B) Cancer cell invasion model. (C) Angiogenesis model. (D) Lymphangiogenesis model. (E) Intravasation model. (F) Extravasation model. (G) Metastasis-on-a-chip device used for the study of organ-specific metastasis. *Reproduced from reference Caballero D, Kaushik S, Correlo VM, Oliveira JM, Reis RL, Kundu SC. Organ-on-chip models of cancer metastasis for future personalized medicine: from chip to the patient. Biomaterials 2017b;149:98–115 [5] by the permission of Elsevier.*

cancer metastasis and to unravel the mechanistic determinants responsible of each single event. Finally, and importantly, ToC can be employed to study the response of the tumor to—novel—chemotherapeutic and immunotherapeutic drugs, or revisit discarded ones, which were previously tested in obsolete preclinical models [136]. Finally, ToC models can also be employed to study the effect on tumors of radiation, leading to a *cancer radiation-on-a-chip* platform.

The design of OoC models of cancer metastasis truly depends on a large list of requirements, such as the organ(s) and tissue(s) to reproduce, and the event in the metastatic cascade to investigate. The design range from simple hydrogel-based microchannels with embedded tumor and endothelial cells to study tumor growth and invasion, to more complex systems based on different microchannels separated by porous membranes (or microslits) to recapitulate tissue–tissue interfaces and multicellular architectures. Some generic commercial chips are also available and have been employed to study cancer progression (see also Section 15.6). Finally, ToC models enable high-resolution and real-time monitoring of tumor progression. Similarly, they also allow the analysis of metabolic activities of cells and the screening of the efficiency of anticancerous compounds through the integration of biosensors and microsized electrodes. In the following, we briefly describe how microfluidics can reproduce some of these events and what insights can be obtained.

#### *15.5.2.1 Tumor growth and invasion models*

The first event during cancer progression is the growth and invasion of the tumor into the surrounding stroma. Several ToC devices have been developed to investigate the mechanistic determinants of this event as well as for the screening of potential therapeutic compounds targeting it (see Fig. 15.7B). These microfluidic chips differ on their architecture, the tumor type under study, or the composition of the cellular microenvironment reproduced. This includes the type and number of cells injected within the microfluidic device, such as endothelial, immune, or stromal cells, and their interactions. Indeed, the heterotypic interactions between tumor cells with stromal cells and the ECM play a fundamental role in the initiation of the tumor. This interaction induces the perturbation of certain genes triggering the progression of the disease. This was investigated using a multiculture ToC device to analyze how the different cells interacted by the analysis of cell type specific gene expression patterns [137]. The different cells types (SKBR3 and BT474 breast cancer cells, HS5 bone marrow stromal cell line, and ThP1 monocytes) were seeded in different compartments and were communicated through paracrine interactions. Interestingly, the obtained results showed that gene expression (production of TGF-β by breast cancer cells and expression of corresponding receptors by stromal cells) was perturbed in co-culture condition, and in particular, in triculture scenario. These results had clinical relevance since stromal-derived gene signature in breast cancer were shown to have a strongest prognostic value. Overall, this work demonstrates how tumor and stromal cells respond to the TME of varying complexities and indicates the importance of heterotypic cultures in tumor modeling.

Among all the phenomena occurring during the first episodes of tumor dissemination, hypoxia is known to play a critical role in tumor growth and invasion, angiogenesis, as well as in metastasis. As previously mentioned, tumor hypoxia causes an advanced and dysfunctional growth of new blood vessels, triggering EMT and directed migration of tumor cells, as well as the activation of other stromal cells that remodel the ECM, facilitating cancer cell invasion. Different ToC-based assays have been reported to investigate the effect of hypoxia and oxygen gradients in tumors due to the advanced capabilities of microfluidic devices to generate well-controlled $O_2$ gradients over long periods. As an example, some works have employed a ToC assay to generate $O_2$ gradients to show that indeed cancer cells migrate toward regions with lower oxygen concentrations in a mechanism termed *aerotaxis* [48,49,138]. Hypoxia has also been related to low efficiency (or even severe rejection reactions) of anticancerous drugs in clinical trials, even though the drugs showed a good pharmacological activity during screening research in vitro. In this regard, the hypoxia-dependent toxicity of anticancerous drugs was recently investigated using a ToC approach [138]. For this, an $O_2$ gradient was generated by using oxygen-scavenging compounds embedded within neighboring microfluidic channels (see Section 15.4.2.2). As a proof-of-concept, adenocarcinomic human alveolar A549 cancer cells were seeded in a microfluidic device and treated with 100 μM *tirapazamine* or *cisplatin* for 12 h under various $O_2$ concentrations ranging from approximately 2% to 11%. The gradient was measured by using an $O_2$-sensitive fluorescence dye, which was quenched by the presence of $O_2$. Results showed an opposite cytotoxic behavior upon the addition of cells under different hypoxic conditions. Upon the addition of *cisplatin*, A549 mortality increased with the increase of $O_2$ tension, whereas the addition of *tirapazamine* caused higher A549 cell death under low-$O_2$ content. Even though the authors proposed a mechanistic explanation for this opposite behavior, further refinement of this work may need to employ of a more realistic TME with the addition of stromal and endothelial cells.

#### 15.5.2.2 Angiogenesis models

Angiogenesis is one of the first events in the cascade of metastasis. It is morphogenetic phenomenon where new blood vessels grow from preexisting vasculature to support tumor growth to provide the tumor with nutrients and oxygen supply. These new vasculature is characterized by being heterogeneous, chaotic and leaky, and therefore, also facilitates the intravasation of cancer cells into the vasculature [36]. Multiple therapeutic strategies focus in inhibiting angiogenesis by using drugs of the pathways involved, in particular the vascular endothelial growth factor (VEGF) [139,140]. VEGF signals to the parental vessel to initiate sprouting as well as the direction toward the new vessels grow. Several anticancerous drugs have been developed to target this event, such *sorafenib* and *sunitinib*. This antiangiogenic capacity was recently demonstrated using a human *microvessel-on-a-chip* model [141]. The addition of VEGF-A triggered the effective endothelial sprouting on the engineered vessel. The obtained results showed that VEGF-A activated Notch signaling through DLL4, and importantly, similar to in vivo scenario. The addition of the—clinically approved—antiangiogenic drugs *sorafenib* or *sunitinib* inhibited the vessel sprouting. However, only the former made the vessels highly permeable, which is a nonphysiological condition. These results indicate that, even though both drugs prevented new vessel growth, their different effects on vessel permeability may explain their distinct effects in cancer treatment.

Other works have reported the use of ToC models to investigate how biochemical cues synergize with mechanical ones in the mechanism of angiogenesis (see Fig. 15.7C). Recently, a ToC platform containing three interconnected channels (one central channel for gel patterning, and two adjacent channels for endothelial cells seeding and gradient generation of angiogenic factors, respectively) was employed to investigate the angiogenic sprouting when cells were exposed to a prolonged gradient of proangiogenic factors (VEGF-165, PMA, and S1P) [142]. This "cocktail" promoted the directed growth of vessels across the collagen gel toward the other adjacent channel, and eventually anastomosed. The obtained results showed that these type of interconnected sprouts matured and remained stable, whereas those who did not bridge both channels retracted and degraded. Interestingly, these results also suggest that the perfusion could act as a stabilization factor for the angiogenic microvessels.

#### 15.5.2.3 Lymphatic system and lymphangiogenesis models

Despite its physiological importance, the role of the lymphatic system in tumor metastasis is less clear and explored than its vascular counterpart. The lymphatic system is responsible, among other functions, of regulating tissue fluid balance by returning the extravasated plasma and proteins back to the blood capillaries. During tumor growth, there is an increase of the intratumoral interstitial fluid pressure that is released through the lymphatic system. Solid evidences suggest that this increased fluid flow has a critical role in the invasion of tumor cells into the lymphatic system, which similar to the vasculature, displays a perturbed—enhanced—permeability [143]. This also triggers the activation of specific genes and promotes the directed chemotaxis of tumor cells toward the lymphatic vessels initiating metastasis.

Several ToC models have been reported to study the role of the lymphatic system in the dissemination of cancer cells (Fig. 15.7D). As an example, a simple biomimetic microfluidic device based on interconnected Boyden chambers together with microfluidic channels was developed to model the TME–lymphatic interaction. The system was employed to investigate the impact of the hydrodynamic flow properties in the invasion of MDA-MB-231 breast cancer cells seeded into the chambers. The obtained results showed that cancer cells displayed an enhanced intravasation rate into the microchannels as a consequence of increased luminal and transendothelial flows, which increased the permeability of the lymphatic endothelial cells and favored the intravasation of cancer cells [74]. This ToC model provided key insights into the mechanism at work of cancer cell invasion into the human circulation. However, it lacked the contribution on this event of the vasculature, that is, the invasion of cancer cells into the blood vessels. In this regard, a human *microcirculation-on-a-chip* was reported where two microchannels separated by a porous membrane were engineered within a microfluidic device. Lymphatic and vascular endothelial cells were seeded on both sides of the membrane mimicking both the native lymphatic and blood vessels [100]. The performance of the ToC model was evaluated by injecting histamine, an inflammatory promoter, to induce an inflammatory response leading to an enhancement of vessel permeability. Unfortunately, the intravasation of cancer cells was not evaluated, but this system provided an interesting platform to investigate the effect of therapeutic drugs, for example, to treat inflammation, which is associated with the initial steps of cancer dissemination. This was investigated using a *TME-on-a-chip* that contained microchannels mimicking both the lymphatic and blood vessels, and a third one containing MCF-7 breast cancer cells. The device was capable to simulate the selective transport of nanoparticles encapsulating drugs and in vivo features of the TME, such as the high-interstitial pressure, enhanced vessel permeability, and poor lymphatic drainage.

Finally, and similar to angiogenesis, the growth of new lymphatic vessels, termed *lymphangiogenesis*, is also promoted by several signaling pathways, such as VEGF. This also facilitates the invasion of tumor cells through the lymphatic system [144]. Lymphangiogenesis is a critical event in the cascade of cancer progression. Despite its critical role, few *lymphangiogenesis-on-a-chip* models have been developed [74,100]. Recently, an ToC of lymphangiogenesis model was developed to mimic the components and dynamics of the lymphatic microenvironment [145]. The obtained results showed a close interplay between mechanical (interstitial flow and endothelial–stromal interaction) and biochemical cues (proangiogenic cues), which was required to start the growth of lymphatic vessels. Interestingly, this data was similar to that obtained in vivo, highlighting the fundamental role of the TME in lymphangiogenesis.

#### 15.5.2.4 Intravasation models

Intravasation is defined as the invasion of metastatic cancer cells into the vascular system, a hallmark feature of cancer metastasis. Typically, during cancer cell intravasation, the endothelium near the tumor region becomes highly permeable. This EPR effect facilitates the invasion of cancer cells into the blood stream, which are then transported through the blood stream toward secondary metastatic sites. All these events are indeed challenging to reproduce using standard in vitro models, in particular because it involves different cell types and dynamic microenvironments. To shed light into the mechanism of intravasation, several ToC models have been reported [146,147] (Fig. 15.7E). Typically, most of these *intravasation-on-a-chip* models integrate two main components: a hollow microchannel coated by endothelial cells mimicking the native vasculature, and a chamber (or reservoir) containing the primary tumor (and stromal cells). Both components are interconnected either by an array of microslits or by a porous membrane, allowing the invasion of cancerous cells into the vascular microchannel. As an example, a 3D ECM hydrogel was employed to develop a simple intravasation-on-a-chip model containing endothelial and tumor cells. This ToC model was used to investigate the mechanism of tumor cells–endothelium interactions [147]. The obtained results showed that cancer cells invaded the 3D ECM upon stimulation and migrated toward the endothelial barrier, revealing a dynamic interplay between vessel permeability and cancer cell intravasation. A more complex ToC platform containing a self-assembled microvascular network and tumor cells in adjacent chambers was recently developed to investigate the mechanism of tumor progression as well as the effect of therapeutic drugs. Interestingly, the delivery of nutrients and drugs to the tumor was performed through the microvasculature. The potential of the platform to investigate tumor invasion and intravasation was tested using both tumor cell lines (MDA-MB-231 breast cancer cells) and patient-derived breast cancer organoids. Tumor invasiveness was stimulated by injecting through the vascular network TGF-β, an EMT promoter. Results showed a twofold increase in the number of cancer cells, which migrated into the microvasculature chamber, even though intravasation was not enhanced, even after the addition of VEGF and thrombin. Despite the mechanism at work of intravasation was not investigated due to the modest results, this work described an interesting method to fabricate native-like vasculature within a microfluidic chip. It also permitted to study multiple events in the metastatic cascade besides intravasation, such as cell proliferation, angiogenesis, or cell migration.

#### 15.5.2.5 Extravasation models

During intravasation, cancer cells invade the blood or lymph vessels and transit along the vasculature becoming CTCs. Depending on the morphological and hydrodynamic properties of the vasculature, the CTCs adhere and arrest in the walls of the endothelium. Next, a large plethora of biochemical factors cells, such as excreted by cells (e.g., cytokine gradients of CXCL-12, EGF, VEGF, HGF, and others) present in the TME stimulate the extravasation of cancer cells across the endothelium (extravasation) into a tissue and the formation of micrometastasis. In order to investigate the impact of biochemical cues in the extravasation of CTCs, several *extravasation-on-a-chip* models have been reported [67,148–151] (Fig. 15.7F). These systems differ in the design of the microfluidic chips, the type of chemokine investigated, or the type of cancer cells employed. As an example, an extravasation-on-a-chip platform was employed to analyze the interaction between CTCs and endothelial cells, transendothelial migration, and the invasion capability of MDA-MB-231 breast cancer cells into a 3D ECM under artificial gradients of SDF-1$\alpha$ [152]. The device included a central microfluidic channel interconnected by transmigration channels, filled with an ECM hydrogel, to two adjacent parallel ones, which were used to inject the chemoattractant. The porous hydrogel compartmentalized the three channels while allowing for the passive diffusion of SDF-1$\alpha$, resulting into a well-controlled gradient. The obtained results showed that MDA-MB-231 cancer cells transmigrated across the endothelium toward the direction with larger concentrations of SDF-1$\alpha$.

Besides biochemical cues, cancer cells are also exposed to mechanical stimuli prior and during extravasation. This includes shear stress forces that CTCs experience due to fluid flow in vivo. These mechanical cues synergy with biochemical ones and influence the invasion capability of cancer cells. This interplay was recently investigated using a simple extravasation-on-a-chip model to assess the role of the mechanical and biochemical microenvironment during the extravasation of HeLa cells [153]. Interestingly, the obtained results revealed a synergistic effect between fluid shear stress, mechanical cyclic stress, and biochemical factors. In particular, it was found that the adhesion of cells in the vasculature, a key step in cancer cell extravasation, was directly correlated with the intensity of fluid shear stress, which was in agreement with the "Seed and Soil" hypothesis [154].

#### 15.5.2.6 Multiorgan and organ specificity models

Metastasis is a multistep process of extreme complexity. Unfortunately, it is yet not well understood the reason why certain type of tumors prefer to invade specific tissues and organs than others. ToC models provide a platform capable to reproduce the structural, biological, and hydrodynamic complexity involved in cancer metastasis and organ specificity. This is of upmost importance since it has been reported that only 0.001%–0.02% of cancer cells that intravasate form a metastatic tumor [155].

Due to its clinical relevance, several *tumor metastasis-on-a-chip* models have been reported during the last decade [4,5,121,150,156,157]. Typically, these models include one chamber with the primary tumor, a microfluidic channel mimicking the vascular network, and a second metastatic chamber. When several secondary metastatic chambers are integrated reproducing different tissues specific to the tumor type, this results in the development of a *tumor organ specificity-on-a-chip* models. Technologically, these are the most complex ToC models since they include all the events occurring during tumor progression. This may be the reason why the amount of this type of multiorgan platforms are scarce. As an example, a *lung metastasis-on-a-chip* model was recently developed to study the organ specificity of lung cancer [128] (Fig. 15.7G). This ToC model contained key elements of metastasis, including the native TME, the microvasculature, immune cells, fibroblasts, and three distant organs to mimic lung metastasis to the brain (astrocytes), bone (osteocytes), and liver (hepatocytes). The obtained results showed that the cells from the different metastatic chambers overexpressed specific proteins characteristic to metastasis, and importantly, similar to in vivo condition. A similar approach was recently reported using a multisite *metastasis-on-a-chip* platform [156]. This platform integrated a channel containing a primary HCT116 colorectal cancer organoid, which was connected to multiple downstream chambers in which liver, lung, and endothelial constructs were seeded. The obtained results showed that CRC cells preferentially invaded the liver and lung constructs, and importantly, similar to the in vivo scenario. Altogether, this validates the applicability of this type of platforms to better understand the mechanism at work of metastasis. Finally, this type of platforms also allows studying the interaction with other cells types, such as stromal and immune cells. In this regard, *cancer-immune system interaction-on-a-chip* models have already been developed [158]. This type of platforms offers a promising approach to investigate the efficiency of immunotherapies against cancer as well as the interaction between cancer and immune cells, and how the latter can intervene in the metastatic cascade.

Overall, the above-described works highlight how microfluidic devices can be employed to mimic all the events occurring in the cascade of cancer metastasis. The resulting ToC models can be then employed to investigate the mechanistic determinants of cancer progression and the effect of drugs in a physiologically relevant environment. This will eventually aid in a better understanding of the disease and in the development of improved therapies.

### 15.5.3 Liquid biopsy-on-a-chip: isolation of CTCs

As abovementioned, the intravasation of tumor cells into the microvasculature results into the generation of CTCs, which is related to the poor outcome of cancer patients. The isolation and analysis of this specific type of cells may provide fundamental information about the etiology of the primary tumor resulting into individualized diagnosis, prognosis, and optimized treatments on cancer patients. OoC technology have emerged as a powerful technology for the specific identification, capture, and analysis of CTCs in peripheral blood. This "liquid biopsy" concept has recently been expanded to include the analysis of other biomarkers, such as ctDNA/RNA, exosomes or antibodies, and body fluids. These devices are denoted *liquid biopsy-on-a-chip* (LBoC) and are employed to capture, detect, and analyze these new type of tumor biomarkers. Most of the LBoC-based methods reported in the literature are related to the capture and isolation of CTCs. This is most likely due to the technical difficulty to detect the other type of biomarkers, as well as their low presence in peripheral blood. Typically, the detection of CTCs is performed based on specific recognition of epithelial surface markers, such as EpCAM, HER2, or cytokeratins. However, these biomarkers are not always present in CTCs as they may gain mesenchymal properties through EMT, which may lead to false negative findings. In contrast, LBoC use different means to isolate the CTCs taking advantage of the particular properties of this type of cells. Isolation methods include physical/chemical entrapments (e.g., channels with microposts), magnetic detection/isolation, or the use of spiral microfluidic channels for hydrodynamic separation, among others.

The importance of this new approach is recently highlighted by a set of articles about this topic, which included the important vision of clinicians [159]. Recently, a simple LBoC device was developed for the isolation of a large amount of CTCs in whole blood. The device was based on leukapheresis but under dynamic conditions (i.e., flow). This method employed a surface coated with an antibody targeting the EpCAM and low-blood flows, which enhanced the amount of trapped CTCs. The method included the combination of multiple antibodies to enhance the number of entrapped CTCs, such as antibodies targeting HER2. However, the combination of antibodies did not increase the capture of CTCs. In this regard, further refinement of this work may include the use of more specific antibodies. Similarly, a LBoC containing an array of microposts functionalized with EpCAM antibodies was employed for the isolation of CTCs [160]. In this seminal paper, CTCs were efficiently identified using peripheral blood from patients with metastatic lung, prostate, pancreatic, breast, and colon cancer with high degree of purity. Finally, a LBoC platform was recently developed for the selective capture of nanosized exosomes from ovarian cancer [161]. In this work, a herringbone-grooved microchannel functionalized with antibodies against cancer exosome membrane biomarkers was employed to isolate exosomes from small volumes of high-grade serous ovarian cancer serum. The obtained result showed that the amount of cancer exosome increased with disease progression.

Overall, the abovementioned works show how LBoC can be employed for the detection, isolation and analysis of tumor biomarkers. Importantly, the use of microfabrication also permits the integration of advanced capabilities into the chips such as biosensors, which may help for the instantaneous identification and enumeration of biomarkers [162]. Finally, it is worth mentioning that other related works report the use of alternative type of microstructures and functionalization methodologies to capture and isolate tumor biomarkers [159].

### 15.5.4 Microfluidics for cancer biomarkers detection

Molecular diagnostics is based on the analysis of biological molecules, such as DNA or proteins. These "biomarkers" act as indicators of biological processes and may provide fundamental information about the etiology and diagnosis of the disease [163]. As discussed above, the current methods for cancer diagnosis typically rely mainly on biopsy staining, which is very invasive and only provides partial information due to tumor heterogeneity. Imaging-based methods, such as CT, MRI, and sonography, are also available, but they have a limited resolution and need expensive equipment. Cancer biomarkers have arisen as an alternative approach. However, the traditional methods for their detections, such as ELISA and PCR, are long, expensive, and with less sensitivity. In this sense, the combination of these diagnostic approaches with microfluidic technology may provide a more sensitive, specific, and cost-effective platform for the detection of cancer biomarkers [164–167]. This was accomplished by combining a droplet-based microfluidic system with a digital PCR that allowed the sensitive and quantitative detection of some mutations of the KRAS oncogene [168]. Importantly, this platform was successfully applied to screen blood plasma samples from patients with metastatic colorectal cancer [169]. Similarly, a microfluidic platform reported the use of a droplet RT-qPCR with magnetic tweezers to quantify RNA for assessing the expression level of HER2, a breast cancer biomarker [170]. Finally, a valveless rotatory microfluidic chip was recently developed that, by using five DNA padlock probes, allowed the specific detection of several single nucleotide polymorphism of TP53 gene, which is known to be present in many cancer types [171].

The levels of expression of specific proteins can also provide valuable information about tumor etiology. However, protein analysis has certain issues, such as their versatility in the number and size, and the potential low abundance of the protein of interest. This could be overcome with the combination of immunoaffinity-based techniques with microfluidic platforms. As an example, a microfluidic mechanical immunoarray was shown to detect two cancer biomarkers, namely prostate-specific antigen (PSA) and interleuking-6 (IL-6) in human serum samples [172]. Next, a nanoporous glass-integrated volumetric bar-chart microfluidic chip was recently employed for the detection of three nonsmall cell lung cancer biomarkers: carcinoembryonic antigen (CEA), CYFRA 21-1, and SCCA in 21 serum samples. Interestingly, the obtained results were comparable with those obtained using ELISA, the typical assays employed in the clinic [173]. Finally, microfluidics has also been successfully implemented on gel electrophoresis assays, such as for the detection and phenotypic characterization of haptoglobin, a biomarker of liver cancer [174].

Overall, these works demonstrate the suitability of microfluidic platforms as screening tools of cancer biomarkers. These platforms display, in general, enhanced performance when compared with their macroscale analogs, and therefore, may find applications in the clinic for advanced cancer theranostics.

## 15.6 Future perspectives

The abovementioned information indicates how microfluidics technology can be employed to investigate the mechanistic determinants of cancer metastasis as well as to screen novel anticancer compounds or even to revisit old ones. They can also be employed to test new types of therapies before moving to the patient, such as immune- or nanotechnology-based therapies (i.e., nanotherapeutics using targeted nanoparticles) or to optimize treatment parameters. However, nowadays the clinical and industrial applicability of OoC technology is threatened by a large plethora of challenges. In the following, we discuss about these limitations and the clinical and industrial future of microfluidics for cancer research.

### 15.6.1 Microfluidic cancer models for clinical applications

Currently, the adoption of microfluidic systems by the clinic is, in general, very limited. This also holds true for cancer research, where these systems may find applications as personalized models of cancer patients for drug screening applications. For this to become a reality, these models may need to fulfill rigid requirements:

1. *Reproduce the complexity of each individual patient to stand for the genetic heterogeneity of tumors*. This means that cells and tissues from patients must be used and kept them alive under dynamic conditions over extended periods to test the efficiency of personalized treatments.
2. *Simple to manipulate and robust*. The abovementioned complexity must be kept low enough to allow the manipulation of the device by the clinical team. This also

includes the development of robust systems. This is in contrast to the devices typically employed by academia, which are in general highly complex, difficult to manipulate, and fragile.

3. *Integrate multiorgan capabilities to assess organ-specificity in metastasis.* This includes the need to encapsulate different cell types from the patient to assess for metastatic-site influence [175,176]. This also involves the technical challenge to isolate the different cell types (tumoral, stromal, endothelial, immune cells, and others) from the same patient, and next, to seed them in the correct proportion and location keeping them alive for extended period of time.
4. *Display high-throughput characteristics.* This parallelization capability is of upmost importance to allow the screening of a large battery of anticancerous compounds efficiently. This may result in the identification of the most convenient treatment and drug concentration.
5. *Mimic native cancer behavior and drug responses.* Very few ToC have clearly demonstrated to mimic cancer phenotypes and the responses to drugs observed in vivo [90]. The validation of ToC models for clinical applications may require studies with a large variety of cancer drugs demonstrating similar effects and behaviors to those observed in humans.
6. *Scaling arguments between organs and tissues.* The dimensions of each organ- and tissue-on-a-chip must be adapted to each individual patient since not all the organs and tissues from patients display the same size. This also involves the number of cell types to reproduce the functional unit of each organ or tissue.
7. *Lack of well-established protocols for the manipulation of patient cells into microfluidic chips in clinical settings.* Clinical-oriented ToC systems must be developed under good manufacturing practices (GMP) conditions using biomaterials displaying rigid requirements in terms of biocompatibility. Strict protocols about the isolation, manipulation, and integration of patient cells into these devices must also be developed and approved.

Currently, the level of development of most of the microfluidic devices limits their clinical applicability, who aim for simple, robust, and multifunctional devices capable to provide a useful readout. In addition, the clinical market may require for a "democratization" of OoC devices. So far, the laboratories at research institutes and universities have developed their own system with custom-made designs, which may lead to different results when compared between the different labs. In this regard, different commercial systems have been proposed like the AIM Biotech 3D cell culture chips, the CNBio's organ-on-chip plates with automated fluid control, or the 384-well OrganoPlate from MIMETAS, among others [5,177]. Clinical applications of such devices may include the molecular diagnosis of the disease, the screening of new drugs and drug combinations [178], drug resistance, optimal drug concentration (thus, shortening Phase I trial duration), or the separation/sorting of cells (e.g., CTCs), among others. They can also provide key insights into the—molecular—mechanism at work of drugs and, importantly, their toxicities before moving to clinical trials. For this, multi-OoC (e.g., liver, kidney and heart, or liver, gut and heart) provide an unprecedented capability to study the direct and indirect toxicity of compounds. In particular, drug metabolism of compounds by the liver may result in the generation of toxic metabolites to vital organs, such as the heart [125].

Finally, the compatibility of OoC, multi-OoC, and ToC systems with existing sensing, imaging, and screening technologies (e.g., ELISA, PCR, LC-MS, multi-OMICS, and others) bring the formers one step closer to real clinical applications in the near future [179]. If the current identified challenges are assessed, it is expected that (multi-) OoC and ToC to become a new paradigm in cancer modeling and drug screening/development in the clinic. Finally, it is worth mentioning that some hospitals are testing OoC systems to evaluate the potential of this technology for clinical applications.

### 15.6.2 Microfluidic cancer models for industrial applications

Pharmaceutical companies are investing a significant amount of resources for the development of novel therapeutic compounds. However, only a few number of these compounds succeed in reaching the clinical stages. This is a consequence of the lack of predictive preclinical models. In this regard, microfluidics is expected to revolutionize the field of drug discovery, developing/screening drugs in a better, safer, and cheaper ways. Interestingly, microfluidic platforms will also permit to revisit old drugs as previously described, which were discarded when tested using standard in vitro models. The reason is that OoC would predict with higher efficiency their outcome in humans than using standard in vitro and in vivo models. Overall, microfludic systems seem to be a powerful platform for biotechnological and pharmaceutical industries to evaluate the efficiency and toxicity of their new drugs and to investigate their pharmacokinetics and pharmacodynamics. However, an end-user perspective on OoC microfluidics technology recently reported that very few systems exhibit the minimum requirements needed to accelerate their transfer to the market [177]. The reasons that were identified were:

1. Lack of automation
2. Difficult to manipulate
3. Noncompatibility with existing technologies
4. Low throughput
5. Too specific or too generic
6. Fabrication technologies not compatible with mass production and/or cost effective

The development of microfluidic systems for cancer research is also accompanied by economic barriers. This restricts their incorporation into the pharmaceutical and biotechnological markets. Indeed, a recent set of articles reported on the challenges and requirements of microfluidic chips in drug development. This study included the important view of all the involved stakeholders, namely researchers, clinicians, industry, and policy makers [180]. In any case, the future looks promising. In 2016, organs-on-a-chip were included by the World Economic Forum in the top 10 list of most promising emerging technologies with the power to improve lives, transform industries and safeguard the planet [181]. In addition, several companies selling OoC-related products for cancer research (and other) applications have emerged during the last years (see Table 15.3) [5]. Importantly, some of these companies have established solid partnerships with important

TABLE 15.3 Most relevant OoC-related companies.

| Company | Area of application | URL |
|---|---|---|
| Emulate | OoC models to evaluate the response of human organs to diseases, drugs, chemicals, cosmetic products, or food additives. | https://emulatebio.com/ |
| Mimetas | OoC platforms for human tissue and disease models. | http://mimetas.com/ |
| TissUse | Dynamic multi-OoC platforms to predict the effect of chemicals and their metabolism. | https://www.tissuse.com/en/ |
| CnBio | OoC devices for human organ and disease modeling, drug discovery, and toxicity. | https://cn-bio.com/ |
| AIM Biotech | Generic microfluidic platforms for biomedical research, drug discovery, and diagnostic applications. | https://www.aimbiotech.com/ |
| Ibidi | Microfluidic-based products for a diverse variety of biomedical applications. | https://ibidi.com/ |
| Nortis | Multiarchitecture microfluidic systems for health research and drug development. | https://www.nortisbio.com/ |
| Micronit | Lab-on-a-chip platform with multiple microwells for organ modeling, sensing and/or drug testing. | https://www.micronit.com |

pharmaceutical companies, such as AstraZeneca, Pfizer, Roche, Merck, and Johnson & Johnson, among others, for using OoC platforms in their laboratories for drug development. In additional, several patents describing methods and apparatus for cancer research applications have been reported, most of them related to the detection and isolation of CTCs [5]. In addition, some of these companies have gathered significant amount of funding to boost the use of OoC technology, not only in biomedicine, but also in other fields, such as cosmetics and food industry, for evaluating the toxicity of additives or supplements. This is the case of Emulate, who has signed an agreement with the U.S. Food and Drug Administration (FDA) to use their OoC systems to assess the toxicity of these type of products on human health. If successful, this will generate a huge demand of OoC in the market. In this regard, the OoC market is expected to be huge in the near future. In 2016, it was valued $5 million, and it is expected by 2023 to reach approximately 170 million [182]. Some reports investigating the OoC market have concluded that heart-, lung-, gut-, kidney-, and liver-on-a-chip devices would have a dominant position in the market, most focused in investigating the toxicity of chemical compounds and their metabolism. This will lead to the rapid development of commercial microfluidic-based systems for cancer applications. Indeed, there are currently commercial microfluidic platforms being used for cancer research, mostly focused on CTC isolation [183]. These type of products will accelerate the transfer of microfluidics from the bench to the bedside, which may result in a more personalized medicine, with the corresponding positive socioeconomic impact.

Overall, despite of all the challenges that OoC sector still needs to address, the future of this technology for cancer research appears very promising and encouraging. However, to become a reality all the involved stakeholders (academia, clinic, and industry) must work together to develop the new generation of OoC, which will be oriented toward more personalized systems, namely *patient-on-a-chip* platforms.

## 15.7 Conclusions

The intrinsic characteristics of microfluidics for the manipulation of fluids make microfluidic-based systems ideal platforms for cancer research. The applications are vast and include drug discovery and screening, tumor profiling, and the development of a diverse variety of ToC models for mechanistic studies. Besides fundamental R&D applications, these types of advanced microfluidic systems may be adopted in the near future by the clinics and the industries to help clinicians for taking decisions (e.g., selection of the best therapeutic approach for a cancer patient) and the relevant pharmaceutical companies in developing more efficient compounds. Some of these systems are currently being used by the industries for a diverse variety of applications. However, there are still multiple challenges to be addressed to make microfluidic devices adopted by the cancer research market. In this regard, all the stakeholders must work together to ensure the utility and validation of microfluidic systems. By doing this, microfluidics will become a fundamental tool in cancer research in the near future.

## Acknowledgments

D.C. acknowledges the financial support from the Portuguese Foundation for Science and Technology (FCT) under the program CEEC Individual 2017 (CEECIND/00352/2017). D.C., M-A L-G and S.C.K also acknowledge the support from FCT under the scope of the project Modeling Cancer Metastasis into the Human Microcirculation System using a Multi-organ-on-a-Chip Approach (2MATCH) (02/SAICT/2017—no 028070) funded by the Programa Operacional Regional do Norte supported by FEDER. Finally, all the authors acknowledge the financial support from the European Union Framework Programme for Research and Innovation Horizon 2020 on Forefront Research in 3D Disease Cancer Models as in vitro Screening Technologies (FoReCaST) under grant agreement no. 668983.

## Conflicts of interest

None.

## References

[1] WHO. Cancer [Online]; 2018. <https://www.who.int/news-room/fact-sheets/detail/cancer>.

[2] WHO. World cancer report 2014 [Stewart BW, Wild CP, editors]. Lyon: International Agency for Research on Cancer; 2014.

[3] Wilking NE, Hofmarcher T, Lindgren P, Jönsson B. The burden and direct cost of cancer in Europe (EU-28). J Clin Oncol 2016;34:6618.

[4] Sontheimer-Phelps A, Hassell BA, Ingber DE. Modelling cancer in microfluidic human organs-on-chips. Nat Rev Cancer 2019;19:65–81.

[5] Caballero D, Kaushik S, Correlo VM, Oliveira JM, Reis RL, Kundu SC. Organ-on-chip models of cancer metastasis for future personalized medicine: from chip to the patient. Biomaterials 2017;149:98–115.

[6] Song, Y., Cheng, D., Zhao, L., Song, Y., Zhao, X., Tian, Q. et al. 2018. Fundamental concepts and physics in microfluidics. In: Song Y, Cheng D, Zhao L, editors. Microfluidics: fundamental, devices and applications. New York: John Wiley & Sons.

[7] Purcell EM. Life at low Reynolds number. Am J Phys 1977;45:3–11.

[8] Rapp BE. Chapter 11 - Conservation of momentum: the Navier-Stokes equation. In: Rapp BE, editor. Microfluidics: modelling, mechanics and mathematics. Oxford: Elsevier; 2017.

[9] Rapp BE. Chapter 15 - Analytical solutions to the Navier-Stokes equation. In: Rapp BE, editor. Microfluidics: modelling, mechanics and mathematics. Oxford: Elsevier; 2017.

[10] Rapp BE. Chapter 28 - Numerical solutions to the Navier-Stokes equation. In: Rapp BE, editor. Microfluidics: modelling, mechanics and mathematics. Oxford: Elsevier; 2017.

[11] Kenis PJA, Ismagilov RF, Whitesides GM. Microfabrication inside capillaries using multiphase laminar flow patterning. Science 1999;285:83–5.

[12] Maxwell JC. Statique expérimentale et théorique des liquides soumis aux seules forces moléculaires. Nature 1874;10:119–21.

[13] Strutt John W. VI. On the capillary phenomena of jets. Proc R Soc Lond 1879;29:71–97.

[14] Olanrewaju A, Beaugrand M, Yafia M, Juncker D. Capillary microfluidics in microchannels: from microfluidic networks to capillaric circuits. Lab Chip 2018;18:2323–47.

[15] Caballero D, Katuri J, Samitier J, Sánchez S. Motion in microfluidic ratchets. Lab Chip 2016;16:4477–81.

[16] Minteer SD. Overview of advances in microfluidics and microfabrication, in microfluidic techniques reviews and protocols. New York: Springer; 2006.

[17] Chung BG, Manbachi A, Khademhosseini A. A microfluidic device with groove patterns for studying cellular behavior. J Vis Exp 2007;7:e270.

[18] Friend J, Yeo L. Fabrication of microfluidic devices using polydimethylsiloxane. Biomicrofluidics 2010;4:026502.

[19] Hoelzle, D., Lake, M., Narciso, C., Cowdrick, K., Storey, T., Zhang, S. et al. Microfluidic device design, fabrication, and testing protocols. Protocol Exchange; 2015.

[20] Jeon JS, Chung S, Kamm RD, Charest JL. Hot embossing for fabrication of a microfluidic 3D cell culture platform. Biomed Microdevices 2011;13:325–33.

[21] Yang S, DeVoe DL. Microfluidic device fabrication by thermoplastic hot-embossing. In: Jenkins G, Mansfield CD, editors. Microfluidic diagnostics: methods and protocols. Totowa, NJ: Humana Press; 2013.

[22] Truckenmüller R, Giselbrecht S, Rivron N, Gottwald E, Saile V, van den Berg A, et al. Thermoforming of film-based biomedical microdevices. Adv Mater 2011;23:1311–29.

[23] Truckenmüller R, Rummler Z, Schaller T, Schomburg WK. Low-cost thermoforming of micro fluidic analysis chips. J Micromech Microeng 2002;12:375–9.

[24] Focke M, Kosse D, Müller C, Reinecke H, Zengerle R, von Stetten F. Lab-on-a-foil: microfluidics on thin and flexible films. Lab Chip 2010;10:1365–86.

[25] Lutz S, Weber P, Focke M, Faltin B, Hoffmann J, Müller C, et al. Microfluidic lab-on-a-foil for nucleic acid analysis based on isothermal recombinase polymerase amplification (RPA). Lab Chip 2010;10:887–93.

[26] Temiz Y, Lovchik RD, Kaigala GV, Delamarche E. Lab-on-a-chip devices: How to close and plug the lab? Microelec Eng 2015;132:156–75.

[27] Chu J-S, Gilchrist MD, Zhang N. Microinjection molding for microfluidics applications. In: Li D, editor. Encyclopedia of microfluidics and nanofluidics. New York, NY: Springer; 2015.

[28] Justin Alexander W, Brooks Roger A, Markaki Athina E. Multi-casting approach for vascular networks in cellularized hydrogels. J R Soc Interface 2016;13:20160768.

[29] Au AK, Huynh W, Horowitz LF, Folch A. 3D-printed microfluidics. Angew Chem Int Ed 2016;55:3862–81.

[30] Waheed S, Cabot JM, Macdonald NP, Lewis T, Guijt RM, Paull B, et al. 3D printed microfluidic devices: enablers and barriers. Lab Chip 2016;16:1993–2013.

[31] Xing J-F, Zheng M-L, Duan X-M. Two-photon polymerization microfabrication of hydrogels: an advanced 3D printing technology for tissue engineering and drug delivery. Chem Soc Rev 2015;44:5031–9.

[32] Balkwill FR, Capasso M, Hagemann T. The tumor microenvironment at a glance. J Cell Sci 2012;125:5591–6.

[33] Levental KR, Yu H, Kass L, Lakins JN, Egeblad M, Erler JT, et al. Matrix crosslinking forces tumor progression by enhancing integrin signaling. Cell 2009;139:891–906.

[34] Lu P, Weaver VM, Werb Z. The extracellular matrix: a dynamic niche in cancer progression. J Cell Biol 2012;196:395–406.

[35] Huang C, Holfeld J, Schaden W, Orgill D, Ogawa R. Mechanotherapy: revisiting physical therapy and recruiting mechanobiology for a new era in medicine. Trends Mol Med 2013;19:555–64.

[36] Caballero D, Blackburn SM, de Pablo M, Samitier J, Albertazzi L. Tumour-vessel-on-a-chip models for drug delivery. Lab Chip 2017;17:3760–71.

[37] Huang YL, Segall JE, Wu M. Microfluidic modeling of the biophysical microenvironment in tumor cell invasion. Lab Chip 2017;17:3221–33.

[38] Wang X, Liu Z, Pang Y. Concentration gradient generation methods based on microfluidic systems. RSC Adv 2017;7:29966–84.

[39] Lin F, Saadi W, Rhee SW, Wang S-J, Mittal S, Jeon NL. Generation of dynamic temporal and spatial concentration gradients using microfluidic devices. Lab Chip 2004;4:164–7.

[40] Comelles J, Caballero D, Voituriez R, Hortigüela V, Wollrab V, Godeau, et al. Cells as active particles in asymmetric potentials: motility under external gradients. Biophys J 2014;107:1513–22.

[41] Wu J, Kumar-Kanojia A, Hombach-Klonisch S, Klonisch T, Lin F. A radial microfluidic platform for higher throughput chemotaxis studies with individual gradient control. Lab Chip 2018;18:3855–64.

[42] Brennan MD, Rexius-Hall ML, Elgass LJ, Eddington DT. Oxygen control with microfluidics. Lab Chip 2014;14:4305–18.

[43] Oomen PE, Skolimowski MD, Verpoorte E. Implementing oxygen control in chip-based cell and tissue culture systems. Lab Chip 2016;16:3394–414.

[44] Brown JM, Wilson WR. Exploiting tumour hypoxia in cancer treatment. Nat Rev Cancer 2004;4:437–47.

[45] Chen Y-A, King AD, Shih H-C, Peng C-C, Wu C-Y, Liao W-H, et al. Generation of oxygen gradients in microfluidic devices for cell culture using spatially confined chemical reactions. Lab Chip 2011;11:3626–33.

[46] Rexius-Hall ML, Rehman J, Eddington DT. A microfluidic oxygen gradient demonstrates differential activation of the hypoxia-regulated transcription factors HIF-1α and HIF-2α. Integr Biol 2017;9:742–50.

[47] Orcheston-Findlay L, Hashemi A, Garrill A, Nock V. A microfluidic gradient generator to simulate the oxygen microenvironment in cancer cell culture. Microelec Eng 2018;195:107–13.

[48] Chang C-W, Cheng Y-J, Tu M, Chen Y-H, Peng C-C, Liao W-H, et al. A polydimethylsiloxane–polycarbonate hybrid microfluidic device capable of generating perpendicular chemical and oxygen gradients for cell culture studies. Lab Chip 2014;14:3762–72.

[49] Yahara D, Yoshida T, Enokida Y, Takahashi E. Directional migration of MDA-MB-231 cells under oxygen concentration gradients. Int J Mol Sci 2016;19:129–34.

[50] Lederberg J. A simple method for isolating individual microbes. J Bacteriol 1954;68:258–9.

[51] Chiu DT, deMello AJ, Di Carlo D, Doyle PS, Hansen C, Maceiczyk RM, et al. Small but perfectly formed? Successes, challenges, and opportunities for microfluidics in the chemical and biological sciences. Chem 2017;2:201–23.

[52] Schneider T, Kreutz J, Chiu DT. The potential impact of droplet microfluidics in biology. Anal Chem 2013;85:3476–82.

[53] Headen DM, García JR, García AJ. Parallel droplet microfluidics for high throughput cell encapsulation and synthetic microgel generation. Microsyst Nanoeng 2018;4:17076.

[54] Chan HF, Zhang Y, Ho Y-P, Chiu Y-L, Jung Y, Leong KW. Rapid formation of multicellular spheroids in double-emulsion droplets with controllable microenvironment. Sci Rep 2013;3:3462.

[55] Moshksayan K, Kashaninejad N, Warkiani ME, Lock JG, Moghadas H, Firoozabadi B, et al. Spheroids-on-a-chip: recent advances and design considerations in microfluidic platforms for spheroid formation and culture. Sens Actuators B: Chem 2018;263:151–76.

[56] Xu J, Zhang S, Machado A, Lecommandoux S, Sandre O, Gu F, et al. Controllable microfluidic production of drug-loaded PLGA nanoparticles using partially water-miscible mixed solvent microdroplets as a precursor. Sci Rep 2017;7:4794.

[57] Shang L, Cheng Y, Zhao Y. Emerging droplet microfluidics. Chem Rev 2017;117:7964–8040.

[58] Zhu P, Wang L. Passive and active droplet generation with microfluidics: a review. Lab Chip 2017;17:34–75.

[59] Khalid N, Kobayashi I, Neves MA, Uemura K, Nakajima M, Nabetani H. Microchannel emulsification study on formulation and stability characterization of monodisperse oil-in-water emulsions encapsulating quercetin. Food Chem 2016;212:27–34.

[60] Sauzade M, Brouzes E. Deterministic trapping, encapsulation and retrieval of single-cells. Lab Chip 2017;17:2186–92.

[61] Xu J, Yuan S, Tian J, Martin KA, Song J, Li C, et al. Ultrasound mediated delivery of oxygen and LLL12 loaded stimuli responsive microdroplets for the treatment of hypoxic cancer cells. Sci Rep 2017;7:44908.
[62] Chakraborti S, Mandal M, Das S, Mandal A, Chakraborti T. Regulation of matrix metalloproteinases: an overview. Mol Cell Biochem 2003;253:269.
[63] Mehner C, Hockla A, Miller E, Ran S, Radisky DC, Radisky ES. Tumor cell-produced matrix metalloproteinase 9 (MMP-9) drives malignant progression and metastasis of basal-like triple negative breast cancer. Oncotarget 2014;5:2736–49.
[64] Yu Z, Zhou L, Zhang T, Shen R, Li C, Fang X, et al. Sensitive detection of MMP9 enzymatic activities in single cell-encapsulated microdroplets as an assay of cancer cell invasiveness. ACS Sensors 2017;2:626–34.
[65] Dhar M, Lam JN, Walser T, Dubinett SM, Rettig MB, Di Carlo D. Functional profiling of circulating tumor cells with an integrated vortex capture and single-cell protease activity assay. Proc Natl Acad Sci USA 2018;115:9986–91.
[66] Sander LM. Modeling contact guidance and invasion by cancer cells. Cancer Res 2014;74:4588–96.
[67] Ma Y-HV, Middleton K, You L, Sun Y. A review of microfluidic approaches for investigating cancer extravasation during metastasis. Microsyst Nanoeng 2018;4:17104.
[68] Denais C, Lammerding J. Nuclear mechanics in cancer. In: Schirmer EC, de las Heras JI, editors. Cancer biology and the nuclear envelope: recent advances may elucidate past paradoxes. New York, NY: Springer; 2014.
[69] Irianto J, Pfeifer CR, Ivanovska IL, Swift J, Discher DE. Nuclear lamins in cancer. Cell Mol Bioeng 2016;9:258–67.
[70] Davidson PM, Sliz J, Isermann P, Denais C, Lammerding J. Design of a microfluidic device to quantify dynamic intra-nuclear deformation during cell migration through confining environments. Integr Biol 2015;7:1534–46.
[71] Hodgson AC, Verstreken CM, Fisher CL, Keyser UF, Pagliara S, Chalut KJ. A microfluidic device for characterizing nuclear deformations. Lab Chip 2017;17:805–13.
[72] Raab M, Gentili M, de Belly H, Thiam H-R, Vargas P, Jimenez AJ, et al. ESCRT III repairs nuclear envelope ruptures during cell migration to limit DNA damage and cell death. Science 2016;352:359–62.
[73] Lei Y, Li J, Wang N, Yang X, Hamada Y, Li Q, et al. An on-chip model for investigating the interaction between neurons and cancer cells. Integr Biol 2016;8:359–67.
[74] Pisano M, Triacca V, Barbee KA, Swartz MA. An in vitro model of the tumor–lymphatic microenvironment with simultaneous transendothelial and luminal flows reveals mechanisms of flow enhanced invasion. Integr Biol 2015;7:525–33.
[75] Le Berre M, Aubertin J, Piel M. Fine control of nuclear confinement identifies a threshold deformation leading to lamina rupture and induction of specific genes. Integr Biol 2012;4:1406–14.
[76] Heuzé ML, Collin O, Terriac E, Lennon-Duménil A-M, Piel M. Cell migration in confinement: a micro-channel-based assay. In: Wells CM, Parsons M, editors. Cell migration: developmental methods and protocols. Totowa, NJ: Humana Press; 2011.
[77] Irimia D, Charras G, Agrawal N, Mitchison T, Toner M. Polar stimulation and constrained cell migration in microfluidic channels. Lab Chip 2007;7:1783–90.
[78] Gritsenko PG, Ilina O, Friedl P. Interstitial guidance of cancer invasion. J Pathol 2012;226:185–99.
[79] Gaggioli C, Hooper S, Hidalgo-Carcedo C, Grosse R, Marshall JF, Harrington K, et al. Fibroblast-led collective invasion of carcinoma cells with differing roles for RhoGTPases in leading and following cells. Nat Cell Biol 2007;9:1392.
[80] Doyle AD, Wang FW, Matsumoto K, Yamada KM. One-dimensional topography underlies three-dimensional fibrillar cell migration. J Cell Biol 2009;184:481–90.
[81] Maiuri P, Terriac E, Paul-Gilloteaux P, Vignaud T, McNally K, Onuffer J, et al. The first World Cell Race. Curr Biol 2012;22:R673–5.
[82] Caballero D, Palacios L, Freitas PP, Samitier J. An interplay between matrix anisotropy and actomyosin contractility regulates 3D-directed cell migration. Adv Funct Mater 2017;27:1702322.
[83] Lanfer B, Freudenberg U, Zimmermann R, Stamov D, Körber V, Werner C. Aligned fibrillar collagen matrices obtained by shear flow deposition. Biomaterials 2008;29:3888–95.
[84] Lee P, Lin R, Moon J, Lee LP. Microfluidic alignment of collagen fibers for in vitro cell culture. Biomed Microdevices 2006;8:35–41.
[85] Chiu LLY, Montgomery M, Liang Y, Liu H, Radisic M. Perfusable branching microvessel bed for vascularization of engineered tissues. Proc Natl Acad Sci U S A 2012;109:E3414–23.

[86] Caballero D, Comelles J, Piel M, Voituriez R, Riveline D. Ratchetaxis: long-range directed cell migration by local cues. Trends Cell Biol 2015;25:815–27.
[87] Caballero D, Voituriez R, Riveline D. The cell ratchet: interplay between efficient protrusions and adhesion determines cell motion. Cell Adh Migr 2015;9:327–34.
[88] Wallin P, Zandén C, Carlberg B, Hellström Erkenstam N, Liu J, Gold J. A method to integrate patterned electrospun fibers with microfluidic systems to generate complex microenvironments for cell culture applications. Biomicrofluidics 2012;6:24131.
[89] Huh D, Matthews BD, Mammoto A, Montoya-Zavala M, Hsin HY, Ingber DE. Reconstituting organ-level lung functions on a chip. Science 2010;328:1662–8.
[90] Hassell BA, Goyal G, Lee E, Sontheimer-Phelps A, Levy O, Chen CS, et al. Human organ chip models recapitulate orthotopic lung cancer growth, therapeutic responses, and tumor dormancy *in vitro*. Cell Rep 2017;21:508–16.
[91] Huang J-W, Pan H-J, Yao W-Y, Tsao Y-W, Liao W-Y, Wu C-W, et al. Interaction between lung cancer cell and myofibroblast influenced by cyclic tensile strain. Lab Chip 2013;13:1114–20.
[92] Acerbi I, Cassereau L, Dean I, Shi Q, Au A, Park C, et al. Human breast cancer invasion and aggression correlates with ECM stiffening and immune cell infiltration. Integr Biol 2015;7:1120–34.
[93] Chaudhuri O, Koshy ST, Branco da Cunha C, Shin J-W, Verbeke CS, Allison KH, et al. Extracellular matrix stiffness and composition jointly regulate the induction of malignant phenotypes in mammary epithelium. Nat Mater 2014;13:970.
[94] Pathak A, Kumar S. Biophysical regulation of tumor cell invasion: moving beyond matrix stiffness. Integr Biol 2011;3:267–78.
[95] Paszek MJ, Zahir N, Johnson KR, Lakins JN, Rozenberg GI, Gefen A, et al. Tensional homeostasis and the malignant phenotype. Cancer Cell 2005;8:241–54.
[96] Plodinec M, Loparic M, Monnier CA, Obermann EC, Zanetti-Dallenbach R, Oertle P, et al. The nanomechanical signature of breast cancer. Nat Nanotechnol 2012;7:757.
[97] Carvalho MR, Maia FR, Vieira S, Reis RL, Oliveira JM. Tuning enzymatically crosslinked silk fibroin hydrogel properties for the development of a colorectal cancer extravasation 3D model on a chip. Glob Chall 2018;2:1700100.
[98] Clay NE, Shin K, Ozcelikkale A, Lee MK, Rich MH, Kim DH, et al. Modulation of matrix softness and interstitial flow for 3D cell culture using a cell-microenvironment-on-a-chip system. ACS Biomater Sci Eng 2016;2:1968–75.
[99] Rutkowski JM, Swartz MA. A driving force for change: interstitial flow as a morphoregulator. Trends Cell Biol 2007;17:44–50.
[100] Sato M, Sasaki N, Ato M, Hirakawa S, Sato K, Sato K. Microcirculation-on-a-chip: a microfluidic platform for assaying blood- and lymphatic-vessel permeability. PLoS One 2015;10:e0137301.
[101] Irimia D, Wang X. Inflammation-on-a-chip: probing the immune system *ex vivo*. Trends Biotechnol 2018;36:923–37.
[102] Ingber DE. Developmentally inspired human 'organs on chips'. Development 2018;145:dev156125.
[103] Nawroth JC, Scudder LL, Halvorson RT, Tresback J, Ferrier JP, Sheehy SP, et al. Automated fabrication of photopatterned gelatin hydrogels for organ-on-chips applications. Biofabrication 2018;10:025004.
[104] Kim S, LesherPerez SC, Kim BCC, Yamanishi C, Labuz JM, Leung B, et al. Pharmacokinetic profile that reduces nephrotoxicity of gentamicin in a perfused kidney-on-a-chip. Biofabrication 2016;8:015021.
[105] Christoffersson J, Aronsson C, Jury M, Selegård R, Aili D, Mandenius C-F. Fabrication of modular hyaluronan-PEG hydrogels to support 3D cultures of hepatocytes in a perfused liver-on-a-chip device. Biofabrication 2018;11:015013.
[106] Lee H, Chae S, Kim JY, Han W, Kim J, Choi Y, et al. Cell-printed 3D liver-on-a-chip possessing a liver microenvironment and biliary system. Biofabrication 2019;11:025001.
[107] Kasendra M, Tovaglieri A, Sontheimer-Phelps A, Jalili-Firoozinezhad S, Bein A, Chalkiadaki A, et al. Development of a primary human small intestine-on-a-chip using biopsy-derived organoids. Sci Rep 2018;8:2871.
[108] Shin W, Kim HJ. Intestinal barrier dysfunction orchestrates the onset of inflammatory host–microbiome cross-talk in a human gut inflammation-on-a-chip. Proc Natl Acad Sci U S A 2018;115:E10539–47.
[109] Rigat-Brugarolas LG, Elizalde-Torrent A, Bernabeu M, De Niz M, Martin-Jaular L, Fernandez-Becerra C, et al. A functional microengineered model of the human splenon-on-a-chip. Lab Chip 2014;14:1715–24.

[110] Sieber S, Wirth L, Cavak N, Koenigsmark M, Marx U, Lauster R, et al. Bone marrow-on-a-chip: long-term culture of human haematopoietic stem cells in a three-dimensional microfluidic environment. J Tissue Eng Regen Med 2018;12:479–89.
[111] Torisawa Y-s, Spina CS, Mammoto T, Mammoto A, Weaver JC, Tat T, et al. Bone marrow–on–a–chip replicates hematopoietic niche physiology in vitro. Nat Methods 2014;11:663.
[112] Chin E, Goh E. Chapter 9 - Blood–brain barrier on a chip. In: Doh J, Fletcher D, Piel M, editors. Methods in cell biology. San Diego: Academic Press; 2018.
[113] Maoz BM, Herland A, FitzGerald EA, Grevesse T, Vidoudez C, Pacheco AR, et al. A linked organ-on-chip model of the human neurovascular unit reveals the metabolic coupling of endothelial and neuronal cells. Nat Biotechnol 2018;36:865.
[114] Lee KK, McCauley HA, Broda TR, Kofron MJ, Wells JM, Hong CI. Human stomach-on-a-chip with luminal flow and peristaltic-like motility. Lab Chip 2018;18:3079–85.
[115] Bhatia SN, Ingber DE. Microfluidic organs-on-chips. Nat Biotechnol 2014;32:760.
[116] Prantil-Baun R, Novak R, Das D, Somayaji MR, Przekwas A, Ingber DE. Physiologically based pharmacokinetic and pharmacodynamic analysis enabled by microfluidically linked organs-on-chips. Annu Rev Pharmacol Toxicol 2018;58:37–64.
[117] Ishida S. Organs-on-a-chip: current applications and consideration points for in vitro ADME-Tox studies. Drug Metab Pharmacokinet 2018;33:49–54.
[118] Zhang YS, Aleman J, Shin SR, Kilic T, Kim D, Mousavi Shaegh SA, et al. Multisensor-integrated organs-on-chips platform for automated and continual in situ monitoring of organoid behaviors. Proc Natl Acad Sci USA 2017;114:E2293–302.
[119] Kamei K-i, Kato Y, Hirai Y, Ito S, Satoh J, Oka A, et al. Integrated heart/cancer on a chip to reproduce the side effects of anti-cancer drugs in vitro. RSC Adv 2017;7:36777–86.
[120] Materne E-M, Ramme AP, Terrasso AP, Serra M, Alves PM, Brito C, et al. A multi-organ chip co-culture of neurospheres and liver equivalents for long-term substance testing. J Biotechnol 2015;205:36–46.
[121] Skardal A, Devarasetty M, Forsythe S, Atala A, Soker S. A reductionist metastasis-on-a-chip platform for in vitro tumor progression modeling and drug screening. Biotechnol Bioeng 2016;113:2020–32.
[122] Chen Wen LK, Edington C, Suter E, Yu J, Velazquez JJ, Velazquez JG, et al. Integrated gut/liver microphysiological systems elucidates inflammatory inter-tissue crosstalk. Biotechnol Bioeng 2017;114:2648–59.
[123] Tsamandouras N, Chen WLK, Edington CD, Stokes CL, Griffith LG, Cirit M. Integrated gut and liver microphysiological systems for quantitative in vitro pharmacokinetic studies. AAPS J 2017;19:1499–512.
[124] Coppeta JR, Mescher MJ, Isenberg BC, Spencer AJ, Kim ES, Lever AR, et al. A portable and reconfigurable multi-organ platform for drug development with onboard microfluidic flow control. Lab Chip 2017;17:134–44.
[125] Skardal A, Murphy SV, Devarasetty M, Mead I, Kang H-W, Seol Y-J, et al. Multi-tissue interactions in an integrated three-tissue organ-on-a-chip platform. Sci Rep 2017;7:8837.
[126] Maschmeyer I, Lorenz AK, Schimek K, Hasenberg T, Ramme AP, Hubner J, et al. A four-organ-chip for interconnected long-term co-culture of human intestine, liver, skin and kidney equivalents. Lab Chip 2015;15:2688–99.
[127] Oleaga C, Bernabini C, Smith AST, Srinivasan B, Jackson M, McLamb W, et al. Multi-organ toxicity demonstration in a functional human in vitro system composed of four organs. Sci Rep 2016;6:20030.
[128] Xu Z, Li E, Guo Z, Yu R, Hao H, Xu Y, et al. Design and construction of a multi-organ microfluidic chip mimicking the in vivo microenvironment of lung cancer metastasis. ACS Appl Mater Interfaces 2016;8:25840–7.
[129] Satoh T, Sugiura S, Shin K, Onuki-Nagasaki R, Ishida S, Kikuchi K, et al. A multi-throughput multi-organ-on-a-chip system on a plate formatted pneumatic pressure-driven medium circulation platform. Lab Chip 2018;18:115–25.
[130] Miller PG, Shuler ML. Design and demonstration of a pumpless 14 compartment microphysiological system. Biotechnol Bioeng 2016;113:2213–27.
[131] Vernetti L, Gough A, Baetz N, Blutt S, Broughman JR, Brown JA, et al. Functional coupling of human microphysiology systems: intestine, liver, kidney proximal tubule, blood-brain barrier and skeletal muscle. Sci Rep 2017;7:42296.
[132] Shim S, Belanger MC, Harris AR, Munson JM, Pompano RR. Two-way communication between ex vivo tissues on a microfluidic chip: application to tumor–lymph node interaction. Lab Chip 2019;19:1013–26.

[133] Lee SH, Sung JH. Organ-on-a-chip technology for reproducing multiorgan. Physiol Adv Healthc Mater 2018;7:1700419.

[134] Wang YI, Carmona C, Hickman JJ, Shuler ML. Multiorgan microphysiological systems for drug development: strategies, advances, and challenges. Adv Healthc Mater 2018;7:1701000.

[135] Rothbauer M, Rosser JM, Zirath H, Ertl P. Tomorrow today: organ-on-a-chip advances towards clinically relevant pharmaceutical and medical in vitro models. Curr Opin Biotech 2019;55:81–6.

[136] Carvalho MR, Barata D, Teixeira LM, Giselbrecht S, Reis RL, Oliveira JM, et al. Colorectal tumor-on-a-chip system: a 3D tool for precision onco-nanomedicine. Sci Adv 2019;5:eaaw1317.

[137] Regier MC, Maccoux LJ, Weinberger EM, Regehr KJ, Berry SM, Beebe DJ, et al. Transitions from mono- to co- to tri-culture uniquely affect gene expression in breast cancer, stromal, and immune compartments. Biomed Microdevices 2016;18:70.

[138] Acosta MA, Jiang X, Huang P-K, Cutler KB, Grant CS, Walker GM, et al. A microfluidic device to study cancer metastasis under chronic and intermittent hypoxia. Biomicrofluidics 2014;8:054117.

[139] Ahmed Z, Bicknell R. Angiogenic signalling pathways. In: Murray C, Martin S, editors. Angiogenesis protocols. *2nd ed.* Totowa, NJ: Humana Press; 2009.

[140] Di Mauro C, Rosa R, D'Amato V, Ciciola P, Servetto A, Marciano R, et al. Hedgehog signalling pathway orchestrates angiogenesis in triple-negative breast cancers. Br J Cancer 2017;116:1425.

[141] Pauty J, Usuba R, Cheng IG, Hespel L, Takahashi H, Kato K, et al. A vascular endothelial growth factor-dependent sprouting angiogenesis assay based on an in vitro human blood vessel model for the study of anti-angiogenic drugs. EBioMedicine 2018;27:225–36.

[142] van Duinen V, Zhu D, Ramakers C, van Zonneveld AJ, Vulto P, Hankemeier T. Perfused 3D angiogenic sprouting in a high-throughput in vitro platform. Angiogenesis 2019;22:157–65.

[143] Swartz MA, Lund AW. Lymphatic and interstitial flow in the tumour microenvironment: linking mechanobiology with immunity. Nat Rev Cancer 2012;12:210.

[144] Stacker SA, Achen MG, Jussila L, Baldwin ME, Alitalo K. Lymphangiogenesis and cancer metastasis. Nat Rev Cancer 2002;2:573–83.

[145] Kim S, Chung M, Jeon NL. Three-dimensional biomimetic model to reconstitute sprouting lymphangiogenesis in vitro. Biomaterials 2016;78:115–28.

[146] Song JW, Cavnar SP, Walker AC, Luker KE, Gupta M, Tung Y-C, et al. Microfluidic endothelium for studying the intravascular adhesion of metastatic breast cancer cells. PLoS One 2009;4:e5756.

[147] Zervantonakis IK, Hughes-Alford SK, Charest JL, Condeelis JS, Gertler FB, Kamm RD. Three-dimensional microfluidic model for tumor cell intravasation and endothelial barrier function. Proc Natl Acad Sci U S A 2012;109:13515–20.

[148] Chen MB, Lamar JM, Li R, Hynes RO, Kamm RD. Elucidation of the roles of tumor integrin β1 in the extravasation stage of the metastasis cascade. Cancer Res 2016;76:2513–24.

[149] Chen MB, Whisler JA, Fröse J, Yu C, Shin Y, Kamm RD. On-chip human microvasculature assay for visualization and quantification of tumor cell extravasation dynamics. Nat Protoc 2017;12:865.

[150] Jeon JS, Bersini S, Gilardi M, Dubini G, Charest JL, Moretti M, et al. Human 3D vascularized organotypic microfluidic assays to study breast cancer cell extravasation. Proc Natl Acad Sci USA 2015;112:214–19.

[151] Jeon JS, Zervantonakis IK, Chung S, Kamm RD, Charest JL. In vitro model of tumor cell extravasation. PLoS One 2013;8:e56910.

[152] Roberts SA, Waziri AE, Agrawal N. Development of a single-cell migration and extravasation platform through selective surface modification. Anal Chem 2016;88:2770–6.

[153] Huang R, Zheng W, Liu W, Zhang W, Long Y, Jiang X. Investigation of tumor cell behaviors on a vascular microenvironment-mimicking microfluidic chip. Sci Rep 2015;5:17768.

[154] Fidler IJ. The pathogenesis of cancer metastasis: the 'seed and soil' hypothesis revisited. Nat Rev Cancer 2003;3:453–8.

[155] Weilbaecher KN, Guise TA, McCauley LK. Cancer to bone: a fatal attraction. Nat Rev Cancer 2011;11:411.

[156] Aleman J, Skardal A. A multi-site metastasis-on-a-chip microphysiological system for assessing metastatic preference of cancer cells. Biotechnol Bioeng 2019;116:936–44.

[157] Bersini S, Jeon JS, Dubini G, Arrigoni C, Chung S, Charest JL, et al. A microfluidic 3D in vitro model for specificity of breast cancer metastasis to bone. Biomaterials 2014;35:2454–61.

[158] Boussommier-Calleja A, Li R, Chen MB, Wong SC, Kamm RD. Microfluidics: a new tool for modeling cancer-immune interactions. Trends Cancer 2016;2:6–19.

[159] Jeffrey SS, Toner M. Liquid biopsy: a perspective for probing blood for cancer. Lab Chip 2019;19:548–9.

[160] Nagrath S, Sequist LV, Maheswaran S, Bell DW, Irimia D, Ulkus L, et al. Isolation of rare circulating tumour cells in cancer patients by microchip technology. Nature 2007;450:1235.

[161] Hisey CL, Dorayappan KDP, Cohn DE, Selvendiran K, Hansford DJ. Microfluidic affinity separation chip for selective capture and release of label-free ovarian cancer exosomes. Lab Chip 2018;18:3144–53.

[162] Chen Y-H, Pulikkathodi AK, Ma Y-D, Wang Y-L, Lee G-B. A microfluidic platform integrated with field-effect transistors for enumeration of circulating tumor cells. Lab Chip 2019;19:618–25.

[163] Kulasinghe A, Wu H, Punyadeera C, Warkiani ME. The use of microfluidic technology for cancer applications and liquid biopsy. Micromachines (Basel) 2018;9:397.

[164] Mughal F, Baldock SJ, Karimiani EG, Telford N, Goddard NJ, Day PJ. Microfluidic channel-assisted screening of hematopoietic malignancies. Genes Chromosomes Cancer 2014;53:255–63.

[165] Roy B, Chattopadhyay G, Mishra D, Das T, Chakraborty S, Maiti TK. On-chip lectin microarray for glycoprofiling of different gastritis types and gastric cancer. Biomicrofluidics 2014;8:034107.

[166] Shadfan BH, Simmons AR, Simmons GW, Ho A, Wong J, Lu KH, et al. A multiplexable, microfluidic platform for the rapid quantitation of a biomarker panel for early ovarian cancer detection at the point-of-care. Cancer Prev Res (Phila) 2015;8:37–48.

[167] Sonnenberg A, Marciniak JY, McCanna J, Krishnan R, Rassenti L, Kipps TJ, et al. Dielectrophoretic isolation and detection of cfc-DNA nanoparticulate biomarkers and virus from blood. Electrophoresis 2013;34:1076–84.

[168] Yeo LY, Chang HC, Chan PPY, Friend JR. Microfluidic devices for bioapplications. Small 2011;7:12–48.

[169] Taly V, Pekin D, Benhaim L, Kotsopoulos SK, Le Corre D, Li XY, et al. Multiplex picodroplet digital PCR to detect KRAS mutations in circulating DNA from the plasma of colorectal cancer patients. Clin Chem 2013;59:1722–31.

[170] Ferraro D, Champ J, Teste B, Serra M, Malaquin L, Viovy JL, et al. Microfluidic platform combining droplets and magnetic tweezers: application to HER2 expression in cancer diagnosis. Sci Rep 2016;6:25540.

[171] Heo HY, Chung S, Kim YT, Kim DH, Seo TS. A valveless rotary microfluidic device for multiplex point mutation identification based on ligation-rolling circle amplification. Biosens Bioelectron 2016;78:140–6.

[172] Chikkaveeraiah BV, Mani V, Patel V, Gutkind JS, Rusling JF. Microfluidic electrochemical immunoarray for ultrasensitive detection of two cancer biomarker proteins in serum. Biosens Bioelectron 2011;26:4477–83.

[173] Li Y, Xuan J, Song Y, Qi W, He B, Wang P, et al. Nanoporous glass integrated in volumetric bar-chart chip for point-of-care diagnostics of non-small cell lung cancer. ACS Nano 2016;10:1640–7.

[174] Huang B, Huang C, Liu P, Wang F, Na N, Ouyang J. Fast haptoglobin phenotyping based on microchip electrophoresis. Talanta 2011;85:333–8.

[175] Holton AB, Sinatra FL, Kreahling J, Conway AJ, Landis DA, Altiok S. Microfluidic biopsy trapping device for the real-time monitoring of tumor microenvironment. PLoS One 2017;12:e0169797.

[176] Lanz HL, Saleh A, Kramer B, Cairns J, Ng CP, Yu J, et al. Therapy response testing of breast cancer in a 3D high-throughput perfused microfluidic platform. BMC Cancer 2017;17:709.

[177] Junaid A, Mashaghi A, Hankemeier T, Vulto P. An end-user perspective on organ-on-a-chip: assays and usability aspects. Curr Opin Biomed Eng 2017;1:15–22.

[178] Eduati F, Utharala R, Madhavan D, Neumann UP, Longerich T, Cramer T, et al. A microfluidics platform for combinatorial drug screening on cancer biopsies. Nat Commun 2018;9:2434.

[179] Li X, Tian T. Recent advances in an organ-on-a-chip: biomarker analysis and applications. Anal Methods 2018;10:3122–30.

[180] Watson DE, Hunziker R, Wikswo JP. Fitting tissue chips and microphysiological systems into the grand scheme of medicine, biology, pharmacology, and toxicology. Exp Biol Med 2017;242:1559–72.

[181] WEF. 2016. *Top 10 Emerging Technologies of 2016* [Online]. World economic forum's meta-council on emerging technologies. <http://www3.weforum.org/docs/GAC16_Top10_Emerging_Technologies_2016_report.pdf>

[182] Bajaj S. Organ-on-chip market by type (heart-on-chip, human-on-chip, intestine-on-chip, kidney-on-chip, liver-on-chip, and lung-on-chip) - global opportunity analysis and industry forecast, 2017-2023. *Emerging and Next Generation Technologies.* Allied Market Research; 2017.

[183] Gwak H, Kim J, Kashefi-Kheyrabadi L, Kwak B, Hyun K-A, Jung H-I. Progress in circulating tumor cell research using microfluidic devices. Micromachines 2018;9:353.

CHAPTER

# 16

# Perfusion-based 3D tumor-on-chip devices for anticancer drug testing

*Nandini Dhiman*[1,2] *and Subha Narayan Rath*[1]

[1]Department of Biomedical Engineering, Indian Institute of Technology Hyderabad, Kandi, India [2]Department of Chemistry and Biotechnology, Faculty of Science, Engineering, and Technology, Swinburne University of Technology, Hawthorn, VIC, Australia

## Abbreviations

| | |
|---|---|
| **5-FU** | fluorouracil |
| **ADME** | absorption, distribution, metabolism, excretion |
| **BioMEMS** | biomedical microelectromechanical systems |
| **CAF** | cancer-associated fibroblasts |
| **CAP** | capecitabine |
| **DFUR** | doxifluridine |
| **ECM** | extracellular matrix |
| **EGF** | epidermal growth factor |
| **EGFR** | epidermal growth factor receptor |
| **GF** | growth factor |
| **GFR** | growth factor receptor |
| **IF** | interstitial flow |
| **IFP** | interstitial fluid pressure |
| **PDMS** | polydimethylsiloxane |
| **PK/PD** | pharmacokinetics and pharmacodynamics |
| **PLA** | polylactic acid |

## 16.1 Introduction

As the global cancer burden increases, we need the discovery of new and efficient drugs to gradually treat with remission free periods while keeping the adverse effects at the lowest possible level. Drug discovery is a very intensive, expensive, and lengthy process with a little success rate [1,2]. The preclinical trials precede the clinical trials; therefore, it becomes imperative that the results obtained during preclinical trials help in

*Biomaterials for 3D Tumor Modeling*
DOI: https://doi.org/10.1016/B978-0-12-818128-7.00016-2

enhancing the drug efficacy and minimizing off-target toxicity in the patients and not do the opposite [3]. Several factors lead to the failure of the drug in the clinical trials or the subsequent therapeutic stages. Preclinical in vitro models have a significant role to play in the initial screening of drug candidates.

The in vitro models that are more physiologically relevant than the currently used two-dimensional (2D) models need to be developed to reduce the failure rate [4]. Physiological relevance can be increased by incorporating the following factors into an in vitro model: co-culture of tumor cells with cells that are known to play a role in cancer progression, the distance between those cells, three-dimensional (3D) extracellular matrix (ECM) in which the cells are housed, dynamic flow surrounding the 3D, and regeneration of vasculature inside the tumor that can carry nutrients and remove waste products. The response of the cancer cells against the anticancer drugs differs significantly depending on which of the above factors is present in the designed cancer model. The microfluidic technology offers a platform, which can ideally include all of these factors to design a new tumor model and then successfully screen new anticancer drugs on it in a more reliable, timely, and high-throughput manner.

In this chapter, we first describe in brief the conventional cancer research models, followed by an introduction to the microfluidic systems, and a brief explanation of the physics behind them. These two sections intend to highlight the growth of 3D culture cancer systems over the past decade and the motivation behind the miniaturization of the tumor models for specific applications, such as disease modeling and drug testing. Next, we discuss in brief the essential tumor components that need to be incorporated in the design to obtain a relevant biomimicking cancer model. We then describe the types of perfusion that can be integrated into the microfluidic devices, including active and passive forms of perfusion. Furthermore, we look into the benefits of perfusion, where we delve deeper into the current applications such as vascularized cancer models, multiorgan chips, modeling complexities of interstitial flow (IF) within tumor model as well as the complicated journey of an anticancer drug through the human system [absorption, distribution, metabolism, excretion (ADME) and toxicity determination]. We end the chapter with a brief description of the challenges faced by perfusion-based microfluidic platforms and some new developments in this field that promise to address those challenges.

## 16.2 Disadvantages of 2D in vitro, 3D in vitro, and animal models

The current 2D models with a monolayer of the cells on a petri dish poorly replicate the tumor present inside the human body (Fig. 16.1). Besides, these models are primarily based on human cell lines, not primary human cancer cells. There are several missing elements from these oversimplified 2D models. For example, the metastatic cancer stage cannot be reproduced well without the presence of multiple other bodily parameters such as the role of other tissues/organs and the flow of blood/fluids between them [6]. The 3D models have overcome some, but not all of the above limitations. The most commonly designed 3D tumor models are spheroid-based models where multiple cells come together to form spherical aggregates called "spheroids" [7]. Several technologies have been established to generate spheroids: hanging drop, suspension culture, and nonadherent surface methods (details can be found in this recently published article [8]). Another form of 3D culture is a scaffold-based

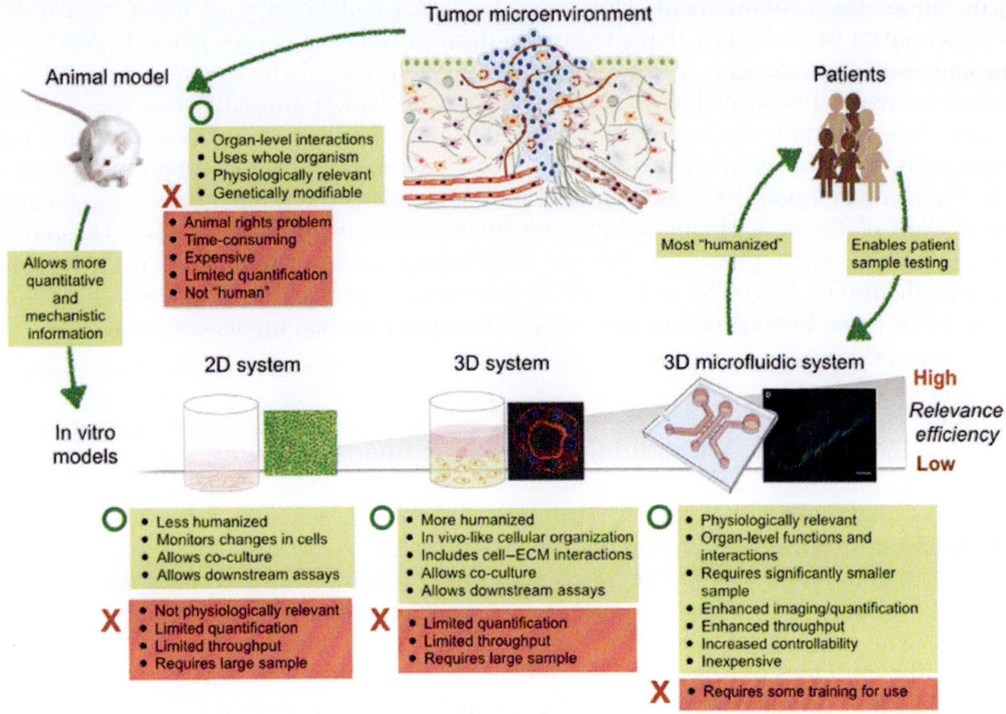


**FIGURE 16.1** Schematics showing a comparison of existing cancer research models. Source: *Reprinted from reference Sung KE, Beebe DJ. Microfluidic 3D models of cancer. Adv Drug Deliv Rev. 2014;79–80:68–78 [5] with permission from Elsevier.*

culture where cells are encapsulated in 3D matrices/hydrogels to form large size tumors [9]. Although the 3D models provide the required ECM, 3D tissue architecture, and facilitate the growth of vasculature, they are primarily developed using human cell lines (Fig. 16.1). They do indeed help in recapitulating some of the complexities of a tumor environment. However, they have several disadvantages of being very large-scale and time-consuming models that fail to accommodate for the spatiotemporal cell arrangements [10].

Animal models are an integral part of the current drug development process (Fig. 16.1). After testing of drugs in the in vitro models, drugs are tested in animal models such as in mice, before they can be subjected to human clinical trials. Despite having several dissimilarities in the biology of a mouse and a human, the mouse models continue to be an essential backbone of the preclinical drug screening. Many types of mouse models have found use in preclinical settings, namely: cell-line mouse xenografts, patient-derived xenografts, immune-competent models (allografts), genetically engineered mouse models, and human organoid cancer mouse models [11,12]. In the cell-line mouse xenografts, human tumor cell lines are transplanted into an immune-deficient mouse. In patient-derived xenograft models, explants from primary human tumors are transplanted into an immune-deficient mouse [13]. In the above two models, tumor modeling is precise, straightforward, and contains many features of the tumor microenvironment. However, the immune-deficiency of these models is a drawback apart from the fact that the human-derived cells are encapsulated in nonhuman (murine) matrix components. In the immune-competent models (or syngeneic models), an immune-competent host, such as a mouse, gets a transplant of an established mouse tumor cell line. The genetically engineered mouse models have a modification done to their genetic makeup that leads to spontaneous formation of a tumor or induced formation if required [14]. The human organoid cancer models are 3D microstructures formed by the self-organization of the cells obtained from the tumor tissue of individuals [12]. The models described above seem promising but are very expensive to develop and to have drugs tested on them (Fig. 16.1). The cost factor is, therefore, a considerable limitation and a motivating factor to explore upcoming technologies to replace animal models in the future.

## 16.3 Microfluidic devices for tumor modeling

Microscale technology holds the promise to overcome the majority, if not all, of the limitations encountered in the conventional in vitro and animal models, as mentioned in the previous section. Based on the integrated circuit (IC) manufacturing technology, the fabrication for microfluidic chips has seen exponential growth over the last two decades [4,15–17]. The immense amount of research undertaken in this field has led to the development of many types of tumor models, which has furthered the attainment of the goal of generating better preclinical in vitro models for anticancer drug research.

A microfluidic device consists of microchannels or microchambers, having a cross-sectional area of 10–1000 μm [18]. The unique physics at this small scale allows the exploration of several parameters to enhance the design and function of the cancer model as per the requirements [19]. The Reynold number (Re) determines the flow in these microchannels and is proportional to flow rate, channel dimensions, and viscosity of the fluid. At the microscale, flow is purely laminar with $Re < 1$, that is, all the fluid components move parallel to the direction of the streamlines, which is very similar to how the fluid flows inside the tissues. Diffusion is the dominant force for particle movement in the microfluidic systems as compared with convection in macroscale systems [19].

Microfluidic-based 2D culture of cells to create tumor-on-chip devices with enhanced functionalities, such as sensors and mechanical movements, for drug screening applications are discussed in another recent review [20]. In this chapter, we mainly describe the devices which house the cancer cells in 3D inside the microchannels. The 3D tumors can be formed by having spheroids formed inside the chip [21,22] or inserting the spheroids into the chip. The latter has not been researched mainly because maintaining the size and integrity of the spheroids during insertion into the microchannels is a significant concern for a successful and reproducible experiment. Another way for accomplishing 3D culture inside microchannels is to mix the cells with the hydrogel (natural or synthetic or composite) and then insert the cell–gel mix into the device for subsequent culture. One more way is to have multiple channels coated with biomaterials/proteins and arranged in such a way that cell monolayers constituted of numerous cell types are very close to each other and the ECM generated by the cells create the 3D environment (organ-on-chip) [23].

Furthermore, to provide the required nutrients, the medium can be applied in two ways: static or dynamic. While the static system implies that the medium is applied once and then be kept undisturbed, it needs to be done intermittently (e.g., once a day) due to the confined nature of the microchannels and quick depletion of the nutrients. The dynamic system can ensure that the media gets perfused into the channels/hydrogel at all times. The active flow is achieved by attaching tubing(s) to the inlet and outlet port(s), which is then connected to a syringe pump for a controlled flow rate or to a peristaltic pump for a more dynamic flow such as oscillatory flow pattern for cardiac tissues.

## 16.4 Tumor components and their inclusion in tumor-on-chip

There is mounting evidence that the response of tumor cells housed inside a 3D gel or as an multicellular tumor spheroids (MTS) to the anticancer drugs is significantly different from the response obtained from a monolayer of the same cells [20]. The tumor microenvironment is a dynamic place, and therefore, to determine near-true drug responses, we need to develop in vitro models, which come very close to replicating that in vivo cellular environment (Fig. 16.2). A high degree of similarity in the in vitro model must be attained with respect to the types of cells present in the vicinity of the tumor cells, the type of ECM around the cells, and the growth factors in the nutrient medium. The following sections explain each of these factors in detail.

### 16.4.1 Cells: monoculture/co-culture

Different cell types reside in a tumor microenvironment that have shown to directly or indirectly affect multiple cancer progression pathways [25,26]. While designing new tumor disease models, incorporating other types of cells with the cancer cells is highly important. For example, one of the cells that are known to be associated with cancer initiation, cancer progression, and cancer metastasis is carcinoma-associated fibroblast (CAF), also known as myofibroblast or activated fibroblast [27]. CAFs are responsible for enhancing anticancer drug resistance via multiple biochemical pathways. For instance, the co-culture of colorectal cancer cells with fibroblasts can lead to increased resistance toward the drug paclitaxel [28]. The co-culture model also has an elevated expression of the ECM proteins

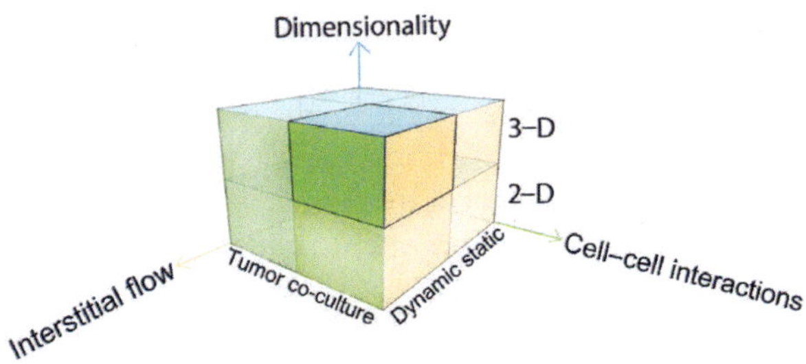


**FIGURE 16.2** Factors to consider while designing 3D tumor models. Source: *Reproduced from Bruce A, Evans R, Mezan R, Shi L, Moses BS, Martin K.H., et al. Three-dimensional microfluidic tri-culture model of the bone marrow microenvironment for study of acute lymphoblastic leukemia. PLoS One 2015;10(10):1–16 [24] under the Creative Commons Attribution 4.0 International Public License (https://creativecommons.org/licenses/by/4.0/)*

(such as fibronectin) led by the fibroblasts, which can further reduce the penetration of the drug into the hydrogel [28].

Other types of cells present in cancer tissue are epithelial cells, endothelial cells, immune cells, and cancer stem cells [26]. Endothelial cells and their incorporation in tumor-on-chip models are discussed in Section 16.6.1. Immune cells present in the circulatory system play an intricate role in cancer modulation which is yet to be fully understood. Chip models must recapitulate tumor-immune interactions. One of the ways to achieve this is by having vascularized tumor spheroids encapsulated inside 3D ECM and subjected to perfusion conditions inside the microfluidic channels [29]. The medium under perfusion flow can have the immune cells flowing along with cancer stem cells. Cancer stem cells are formed mainly via epithelial-to-mesenchymal transition, and therefore their formation can be observed/analyzed in real-time in the growth medium in the vicinity of vascularized tumor spheroids. Different types of immune cells can be introduced, simultaneously or together, to understand the tumor-immune behavior [30,31].

## 16.4.2 ECM: chemical and mechanical cues

The cells produce and manipulate their microenvironment (ECM) and cannot survive without doing so. Therefore, to culture cells, cells must be given the mechanical and biochemical cues that enable their proliferation and other essential activities. While this can be achieved by culturing tumor cells in natural hydrogels like collagen, Matrigel, fibrinogen, and synthetic hydrogels like polylactic acid (PLA), other sources of ECM have also been investigated, which overcome the drawbacks associated with the aforementioned biomaterials.

In a recent study by Romero-López et al., ECM derived from an actual tumor (colon-origin) lead to an enhancement of angiogenesis, tumor cell proliferation, as well as the rate of growth in comparison to the ECM, derived from the healthy colon [32]. The vasculature formed in the tumor ECM can thus be very similar to the one observed in native tumors. The above can be attributed to differential expression of Collagens, IV, VI, and XIV, Fibrillin, Vitronectin, and Laminin in the tumor ECM compared with the healthy ECM. This could also be due to proteins like Fibronectin, Tenascin, Thrombospondin-2, Periostin, and Versican, being expressed only in the tumor ECM and not in the healthy ECM [32]. Besides, the tumor and endothelial cells cultured in tumor ECM can be more glycolytic as the tumor ECM leads to a higher expression of glycolysis-associated genes [32]. This suggests that the ECM plays a very significant role in deciding the fate of cancer cells concerning their dedifferentiation and cell-mediated ECM modifications to enhance metastasis and cancer growth.

Matrix metalloproteinases (MMPs) are responsible for degrading ECM proteins and are the most prominent set of proteinases associated with tumor progression [33]. A microfluidic model incorporating breast carcinoma cells in different 3D configurations analyzed the tumor cell response by inhibiting the MMPs in both mono- and co-culture (with mammary fibroblasts) conditions [34]. While these results are compelling, further research on how the tumor ECM could affect the cell–drug interactions and the response to anticancer drugs is needed.

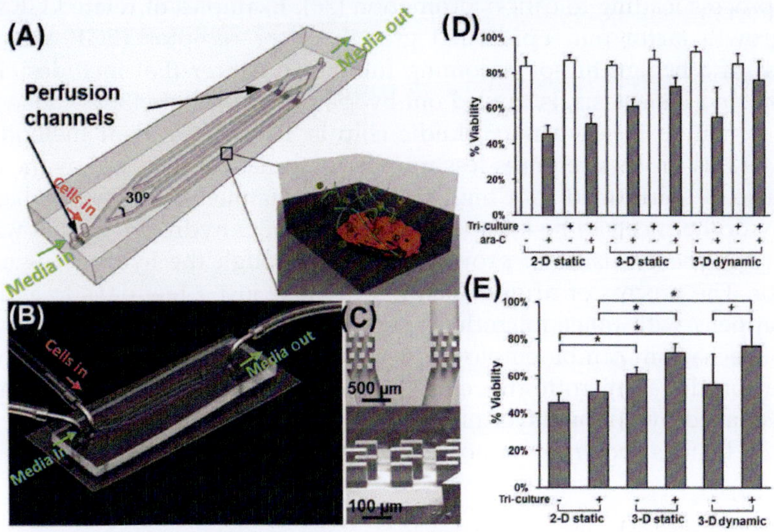


FIGURE 16.3 A 3D microfluidic triculture model and chemoresistance to the drug Ara-C. (A) Schematics of the microfluidic device containing four microchannels (each 2 cm long, 500 μm wide, and 75 μm high). Media channels are near the inlet and outlet. The enlarged boxed area shows an illustration of multiple cell types inside a 3D matrix. (B) Image of the microfluidic device. (C) Scanning electron microscope (SEM) images of top view (upper panel) and side view (lower panel) of the perfusion channels. (D) The effects of the drug Ara-C (Cytarabine) on the viability of tumor cells in different culture conditions. (E) The effects of perfusion-based culture on cell viability in the triculture condition. $^{+}P < .1$, $^{*}P < .05$. Source: *Reproduced from Bruce A, Evans R, Mezan R, Shi L, Moses BS, Martin K.H., et al. Three-dimensional microfluidic tri-culture model of the bone marrow microenvironment for study of acute lymphoblastic leukemia. PLoS One 2015;10(10):1–16 [24] under the Creative Commons Attribution 4.0 International Public License (https://creativecommons.org/licenses/by/4.0/).*

ECM can also be modified by the cells in response to the stress caused by the flow of the fluids around them. A study (Fig. 16.3) showed exactly this, where the drug sensitivity was found to be reduced in a 3D triculture model of leukemic cells, bone marrow stromal cells and human osteoblasts, due to flow-induced remodeling of the matrix and protective shield provided by the ECM [24]. The drug tested in this study was Ara-C, also known as Cytarabine, which is a commonly used anticancer drug for the treatment of acute and chronic leukemia. Fig. 16.3B summarizes the results of chemoresistance where all the culture conditions are shown and Fig. 16.3C specifically highlights only the effects of 3D-culture under triculture condition. As seen in Fig. 16.3C, triculture led to a reduction in leukemic cell death during drug exposure in both 2D and 3D models, and flow did not make any difference in the 3D model in the drug response.

## 16.4.3 Growth factors

Cells receive signals from their surroundings and cells in the vicinity through growth factors (GF) with the help of specific growth factor receptors (GFR) located on the cell membranes or inside the cell [35]. This signaling mechanism led by the polypeptide GF is at the core of the tumor microenvironment and plays a role in tumor initiation and is the heart of the process leading to cancer promotion [36]. Examples of related GF and GFR are epidermal growth factor and epidermal growth factor receptor (EGF and EGFR). The entire process of a benign tumor becoming full-blown cancer that includes clonal expansion, invasion, and metastasis, is carried out by the growth factors [36].

Spheroid formation inside a microfluidic chip is also an excellent method to perform drug testing as this process requires less number of initial cells as well as the drugs. Drug testing can also be performed in a high-throughput manner. Therefore, spheroids generated this way do not need to be encapsulated inside a 3D hydrogel. In this way, they can grow faster since the diffusion of growth medium through the hydrogel would not be a limiting factor. The process of drug evaluation can take just a few days to and not several days as it happens with other microfluidic spheroid-based assays. For instance, the convective flow mechanism can be employed to generate cancer cell spheroids in an accelerated manner inside a microfluidic chip [37]. As shown by other reports, the drug resistance displayed by a monolayer of the same tumor cells in a monolayer culture turns out to be lower than as shown by tumor spheroids [37].

### 16.4.4 Shear stress

Cells in the tumor environment respond radically different to mechanical stimuli as compared with no stimuli. There is a mounting evidence that various tumor-related processes like proliferation, apoptosis, invasion, and metastasis are aggressively affected by the presence of shear stress [38]. In the native tumor environment, cells residing at the periphery of the tumor experience relatively higher mechanical forces while the cells near the center do not and live in hypoxic conditions [39]. For instance, in one study, the shear stress and other mechanical forces acting against the layer of endothelial cells in the vasculature altered cell responses such as the release of chemical factors, inhibition of cell death by apoptosis, differential gene expression, and atherosclerosis [40].

Furthermore, depending on whether the flow was laminar or turbulent, whole new endothelial cell responses, such as proinflammation, were observed [40]. All these factors could help design the in vitro models in such a way that the required physiology for the tumor model can be replicated truthfully for a specific application. For instance, it was recently shown using a 3D microfluidic model, that the laminar shear stress at different magnitudes can inhibit or induce angiogenesis [41]. This study demonstrated that approximately 3 dynes $cm^{-2}$ laminar shear stress repressed angiogenic sprout formation in a bifurcated microchannel design. However, sprout formation was successful when the transvascular flow ($\sim 1\ \mu m\ s^{-1}$) and laminar shear stress were applied simultaneously. This study inferred that the angiogenic sprouting was enhanced by the competitive mechanisms between the flow across the endothelial layer, the bifurcating fluid flow, and laminar shear stress [41]. Another 3D bioreactor system for breast cancer cells showed that cell proliferation, invasiveness, drug resistance were enhanced by an application of pulsatile shear stress [42]. These studies highlight the use of these microphysiological systems to understand the mechanical properties of a tumor microenvironment. Also, these systems could potentially offer patient-specific personalized information on drug transport determining elements.

## 16.5 Types of perfusion methods

A lot of ways have been employed to generate controllable perfusion systems (Table 16.1). They can generally be categorized under two broad categories: pump-driven systems or pump-less systems. Pump-driven systems have an active flow system entering the microfluidic chip where there is a source of pressure like syringe pumps [47,50,59]. Whereas pump-less systems are mainly passive and may sometimes rely on external indirect sources of flow generation like rockers, or by varying channel configurations [49,53].

A detailed review of the types and principles of pumping methods for microfluidic cell culture can be found here [60]. To engage each of these together to achieve a multiorgan microfluidic culture could a technically challenging effort. For example, Edington et al. reported interconnected microfluidic devices (Physiome-on-a-chip; Fig. 16.4) with different flow settings with each device having an internal recirculation system to enable an exchange of molecules [46]. Several pneumatic-based pumps were also fabricated to

**TABLE 16.1** List of different methods to create flow/perfusion in 3D organ/tumor-on-chip microfluidic systems.

| S.no. | Method | Technology | | Organ/disease | Ref. |
|---|---|---|---|---|---|
| 1 | Pump-driven | Syringe pump | | Lung cancer | [43] |
| | | Peristaltic pump | | Liver cancer | [44] |
| | | Pneumatic pump | | Cervical cancer | [45] |
| | | | | Multiorgan (4, 7, and 10 organs) | [46] |
| | | | | Colon cancer and liver organ | [47] |
| | | Electromagnetic pump | | Lung and liver | [48] |
| 2 | Pump-less | Rocking platforms (gravity) | Continuous | Intestine and liver | [49] |
| | | | | Colon cancer, liver cancer, and muscle | [50] |
| | | | Programmable | Multiorgan; lung cancer | [51] |
| | | | | Intestine and liver | [52] |
| | | | | Multiple myeloma (white blood cell cancer) and bone | [53] |
| | | | | Human skin | [54] |
| | | | | Breast cancer, vulva cancer, leukemia, liver, and heart | [55] |
| | | Surface tension | | Uterine cancer, multidrug resistant uterine cancer, liver, and bone marrow | [56] |
| | | Braille actuation | | Liver cancer | [57] |
| | | Osmotic pressure | | Brain | [58] |

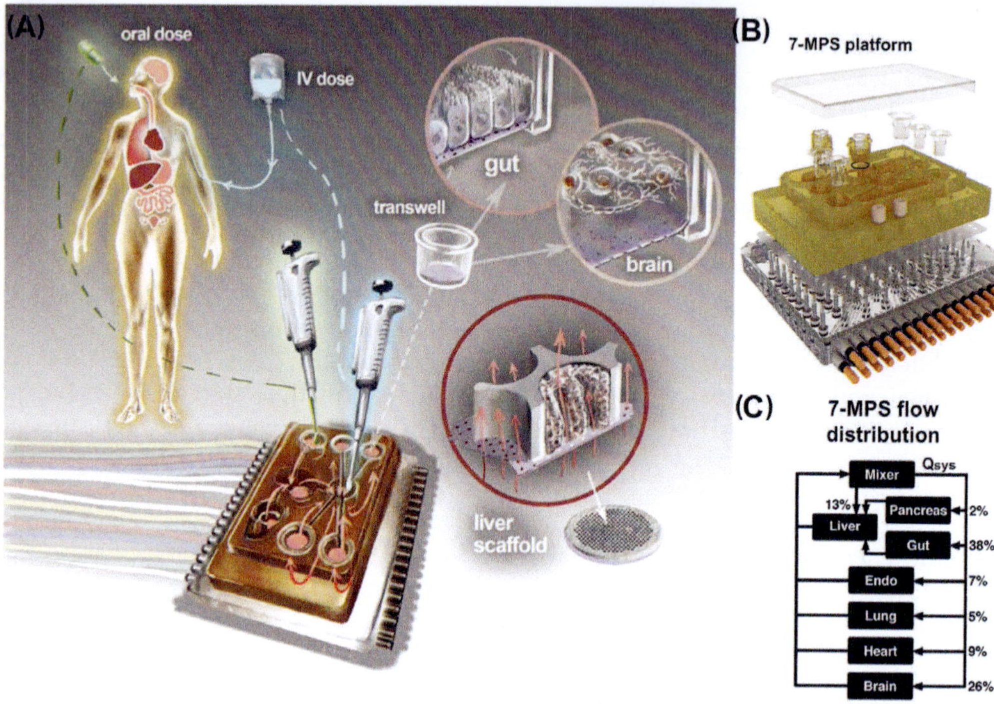


FIGURE 16.4 Physiome-on-a-chip. (A) The Physiome-on-a-chip contains engineered interconnected 3D microphysiological systems that represent several organs of interest. They capture essential features of in vivo organ physiology and are based on corresponding quantitative models designed specifically for applications such as disease modeling or drug screening. (B) MPS platforms and their flow partitioning. Exploded rendering of the 7-MPS platform and (C) corresponding flow partitioning for the 7-way platform that reflects physiological cardiac output. *Reproduced from Edington CD, Chen WLK, Geishecker E, Kassis T, Soenksen LR, Bhushan BM, et al. Interconnected microphysiological systems for quantitative biology and pharmacology studies. Sci Rep [Internet]. 2018;8 (1):4530 [46] under the Creative Commons Attribution 4.0 International Public License (https://creativecommons.org/licenses/by/4.0/)*

achieve precise control over flow rates between each device as well as on the drug distribution [46]. Braille actuation takes advantage of different curing ratio of PDMS and multiple layers in a device with varying hardness [57]. Programmable gravity-based systems make use of the rocking platforms but provide controlled and precise perfusion flow rates in comparison to the continuous gravity-based systems [53]. Depending on the application and number of tumor organs in the design, pump configurations can be accordingly chosen and modified.

## 16.6 Benefits of perfusion and specific applications

Microfluidic technology is exclusively at the forefront to model perfusion in 3D culture since the microscale, and compartmentalized structure enables medium perfusion through a 3D cellularized matrix in a timely and spatially manner [61]. Having a continuous flow of media inside the microfluidic channels regularly leads to media replenishment, helps in waste removal, and maintains proper control over oxygen tension [61]. The flow rate can be varied as desired, and since the flow is laminar, it can be precisely controlled. One critical factor that needs to be considered is that the rate of perfusion is not detrimental to the cells/spheroids cultured inside the gel due to the hydrodynamic shear stress generated by the flow. The following sections entail the benefits of perfusion in an application-specific manner.

### 16.6.1 Vasculature

The inclusion of endothelial cells in a microfluidic culture with cancer cells can be done to induce vasculature in the 3D tumor. Vasculature inside a tumor model is very crucial and can transport nutrients and anticancer agents and can lead to rapid tumor growth. Also, it can ensure that the drug response obtained from the model is pathologically relevant. In microfluidic devices where this is desirable, the dynamic flow of media through the medium channels becomes more significant. Research on microfluidic models is continually showing that the anticancer drug resistance increases when cancer cells are cultured in 3D in the presence of vessel-like structures [62]. For example, in a perfused colorectal cancer-vascular microfluidic model, drug resistance to three drugs, namely, 5-fluorouracil, vincristine, and sorafenib, was found to be significantly higher as compared with in the 2D cancer model [63]. The vascularized microfluidic model was also tested for antiangiogenic drugs (pazopanib and sorafenib), mainly targeting the tumor-derived factors, such as vascular endothelial growth factor receptor (VEGFR) and platelet-derived growth factor receptor (PDGFR) [63].

Two latest studies have modeled perfusion-based vascularized cancers/tumors on the microfluidic platform [29,31]. Nashimoto et al. engineered a vasculature system in a tumor spheroid encapsulated inside a hydrogel-incorporated microfluidic channel. The vasculature led to successful perfusion of the medium inside the spheroid, perfectly mimicking the in vivo tumor microenvironment (Fig. 16.5A). The perfusion is maintained continuously without any obstacle for 24 h of cell culture inside the compartmentalized 3D microfluidic device, and a noticeable increase in tumor proliferation is observed because of the presence of continuous perfusion. However, as compared with the static conditions (Fig. 16.5B), perfusion does not lead to an increase in cell mortality in response to exposure to various concentrations of paclitaxel [29]. This variation could be because of the high-nutrient supply in the vascularized spheroids, where the rate of proliferation is higher than in nonvascularized spheroids. Paclitaxel at a lower dosage ($5\ \mathrm{ng\ mL^{-1}}$) did not suppress the spheroid volumes in vascularized tumors under perfusion condition as compared with spheroids without vasculature (Fig. 16.5B). Whereas, paclitaxel at its highest dose suppressed both the spheroid volume and tumor area (equivalent to E-cadherin$^{+}$ area) in spheroids without and with the vasculature. These results indicate a battle in the tumor spheroid between the antitumor efficacy of paclitaxel and the vasculature-led high-nutrient supply leading to higher cell proliferation [29].

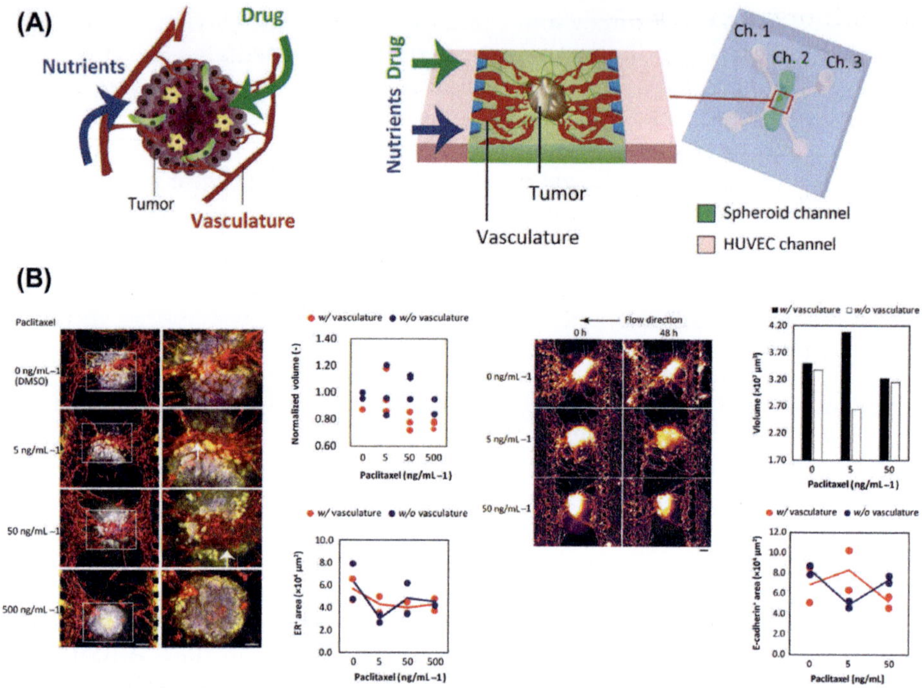


**FIGURE 16.5** Vascularized cancer on a chip: (A) Tumor microenvironment illustration (left) and modeling the same inside a microfluidic device comprising of three channels (right). (B) Testing the drug paclitaxel in a dose-dependent manner on the tumor model with or without the presence of vasculature (left: static condition, right: perfusion condition). HUVECs are red, E-cadherin (MCF-7) are yellow, and nuclei are blue. Scale bars: 200 μm (left column), 100 μm (magnified images). White arrows in the magnified images indicate E-cadherin-negative areas around the large blood vessels. The graphs show normalized spheroid volumes and estrogen receptor-positive (ER + ) area at the center of the spheroids. *Source: Reprinted from Nashimoto Y, Okada R, Hanada S, Arima Y, Nishiyama K, Miura T, et al. Vascularized cancer on a chip: the effect of perfusion on growth and drug delivery of tumor spheroid. Biomaterials [Internet]. 2020;229:119547 [29] with permission from Elsevier.*

In healthy malignant tissues, immune cells such as macrophages play a role in cancer cell invasion and intravasation and monocytes, depending on the subpopulation, enhance or suppress cancer metastases process [64]. In a recent study to understand the behavior of monocytes, a 3D vascularized tumor model comprising of breast cancer cells, and endothelial cells are developed to study the interaction of immune cells (monocytes) and tumor cells in a high-resolution manner [31]. The tumor model is also employed to screen an anticancer drug Blebbistatin and study its effects on the tumor–monocyte interactions and biochemical pathways. Important revelations on monocyte-dependent cancer cell extravasation are reported, including how monocytes reduce extravasation in a noncontact dependent mechanism. Noticeably, this platform successfully shows the monocytes perfusing through the vasculature and hence reaching to the target tumor cells. Such type of tumor modeling is a significant advance and is essential to understand in-depth immune–cancer relationships and pick out potential targets for anticancer therapeutics, especially metastases.

### 16.6.2 Multiorgan systems

The cancer cells during metastasis travel from the organ of origin to numerous sites in other organs. They do so by undergoing a highly complex and intricate process, which involves transitioning from epithelial to mesenchymal cell type, entering the circulatory system, and finally, reaching the destination tissue by reverting to epithelial type [65]. To mimic one or all of these events in an in vitro model, the microfluidic model must contain more than one organ type. The main requirement in designing microfluidic systems containing multiple organ models is for them to be interconnected with channels enabled with the flow. This dynamic, interconnected organ system can better mimic the in vivo human physiological conditions and, therefore, can provide reliable drug responses [66]. Several recent studies have demonstrate the design and fabrication of such models for applications not just limited to anticancer drug screening [46–49,52,55,67,68]. We will discuss some of these in detail in Section 16.6.4.

### 16.6.3 Interstitial flow within 3D hydrogel systems

Convection mechanism drives the fluid transport through the tissue microenvironment, and the fluid carries nutrients and signaling molecules along with it to the cells. This fluid flow is known as IF. In cancer tissues, this flow has been known to affect the morphology of several cell types such as cancer cells, fibroblasts, endothelial cells, and mesenchymal stem cells [69]. The cancer tissues have elevated levels of interstitial pressure owing to leaky and compressed vasculature [70], which results in a reduced delivery rate and low reach of anticancer agents inside the tumor region [39]. All of this has to be mimicked in the tumor models to ascertain a proper drug response.

In 3D microfluidic models containing hydrogels housing the tumor cells or small spheroids, interstitial fluid can be experimented with. Firstly, vasculogenesis has to be achieved, which can be done by integrating endothelial cells in the microchannels alongside the hydrogel–cell channel [71]. The endothelial cells can sprout and form vessel-like structures. This way, both the tumor and vessel structure grow together, and overtime gradually leads to the development of IF and generation of interstitial fluid pressure (IFP). Since high IFP is a significant cause of therapeutic failure in clinical trials, it must be modeled accurately in the in vitro models.

Diffusion of proteins and other factors across the hydrogel in a compartmentalized microfluidic chip can also be modeled for various types of biomolecules and factors like the presence or absence of flow. The concentration gradients can be unstable or stable depending on the perfusion of the medium through the media channels. IF has been shown to impact the direction of tumor cell mechanism with the help of 3D hydrogel-based microfluidic devices [72]. Even the cell seeding concentration in the 3D hydrogel has an impact on the flow-directed migration. While the cells seeded at high density tend to migrate against the flow, the cells seeded at low density tend to migrate along the streamlines [72]. A recently published review article provides an informative summary of the microfluidic models that have tried to understand other properties of tumor ECM transport, such as hydraulic conductivity and interstitial diffusivity [73]. A recent study attempted to quantify the behavior of transmural transport molecules on a chip-based microvascular network [74]. Both the transmural and IF parameters were simulated as seen in the in vivo conditions, and the results obtained can, therefore, determine drug distributions in a better manner.

### 16.6.4 Drug pharmacokinetics and pharmacodynamics

Prediction of drug responses in patients is a great challenge in the current pharmaceutical industry. The microfluidic models need to incorporate pharmacokinetics and pharmacodynamics (PK/PD) models into their design and analysis for a better understanding of the drug behavior and collective organ response. ADME of a drug stands for adsorption, distribution, metabolism, and elimination and is an integral part of PK. A drug after entering the human body is subjected through various intricate and dynamic processes during its ADME. The amount of time the drug stays in the bloodstream and the organ-tissues defines the concentration–time profiles of the drug and is determined by ADME [75]. Depending upon the concentration of the drug at the time of dosage, mathematical pharmacokinetic (PK) models can help estimate the drug concentration–time profiles within each organ [75]. On the other hand, the pharmacological response of a drug, including intensity of therapeutic and adverse effects, as a function of the dose administered is defined by pharmacodynamic (PD) models and combined with PK model, drug pharmacological effects can be determined in a time-dependent manner [75]. For a better prediction of the drug response and an increase in the success rate at the clinical trial stage, we also need the physical counterpart of the mathematical models such as microphysiological tumor models for in vitro screening.

Metabolism dependent toxicity of 5-FU and its oral prodrug Tegafur was tested on a silicon-based microcell culture analog (μCCA), which housed three 3D cell culture chambers containing liver, colon cancer, and myeloblast cells interconnected via microchannels [76]. The pharmacokinetic profile of the drug Tegafur was assessed on the device and was found to be significantly different than the corresponding response in a microtiter 96-well plate. Also, Tegafur was converted to 5-FU in the liver chamber, and some toxicity effects were observed in tumor cells, while this was not seen in the 96 well plates. The hepatotoxicity on the liver cells and hematological toxicity of the drugs were measured in the liver and myeloblasts. The synergetic effects of Tegafur with uracil were also observed on this chip, mainly because the perfusion of the flow was allowed between the 3D cell culture chambers [76]. Real-time cell parameter sensing and measurement were also demonstrated in one of the studies by this group, where they have an oxygen sensor integrated with a μCCA device [77].

Using a similar device design as above, μCCA contained four functional tissues formed using four cell lines, namely, liver, uterine sarcoma, drug-resistant uterine sarcoma, and bone marrow megakaryoblast and was design to perform combinatorial drug screening against multidrug resistant cancer accurately. It was observed that two MDR-suppressor drugs worked well in combination in the μCCA but not in the traditional well-plate assay, mainly because of interconnection between four cell lines and medium flow between each organ, which is primarily missing in static well-plate systems. These results are also easily coupled with PKPD models of the drugs and offer a considerable benefit in terms of measuring correct human drug responses. [56]. In another follow-up study, Shuler's group designed a microscale PK/PD model to assess the toxicity of the drug 5-FU against hepatocarcinoma cells, with or without the modulating effects of uracil, and the results were correlated with a mathematical PK/PD model [50]. All the above studies prove the significant applicability of these devices in preclinical settings. However, they still need to accommodate 3D cell culture into μCCA.

Metabolism dependent toxicity of a prodrug capecitabine (CAP) was tested on a multilayer organ-on-chip containing liver cells on the top layer and lung cancer, breast cancer, and healthy gastric cells in the bottom layer separated by a porous membrane [68]. The cells were cultured as a monolayer of cells on the device surface or permeable membrane. CAP was metabolized by liver cells into DFUR, which was secondarily metabolized into 5-FU in the cancer cells by an intercellular enzyme thymidine phosphorylase, and thereby, exerted cytotoxic effects on the cancer cells. The drug efficacy of 5-FU is dependent on this enzymatic activity and can vary as per the cell type. The data obtained for the drug efficacy using this chip were compared with clinical data collected from the patients by noting the value of effective drug concentration on-chip to the drug plasma concentration in the patients [68].

Three different types of microphysiological systems were developed by Edington et al. [46] designed in such a manner that they could be used for three different periods (2, 3, and 4 weeks) to quantitatively perform PK study of several types of drugs including diclofenac metabolism (Fig. 16.4). This Physiome-on-chip concept can be translated for tumor-on-chip and anticancer drug testing and PK/PD applications [46]. Various cancer types—including multidrug resistance varieties were modeled along with healthy cells in a sophisticated multiorgan-on-a-chip system for evaluating multiple cancer drugs (tamoxifen, imatinib, and diclofenac) efficacy, side-effects (on healthy cells), and ADME properties [55]. The system also had integrated bioMEMS for monitoring the effects of drugs [55]. While the previously mentioned PK/PD studies involved human cell lines, this study included primary human liver cells and induced pluripotent stem cell-derived cardiomyocytes as well. The use of primary cells in the chips can lead the way for patient-specific drug efficacy testing and response predictions.

## 16.7 Specific designs for enhancing perfusion

The continuous efforts toward developing novel designs and devices for perfusion-based microfluidic devices that overcome existing challenges in this technology need to be appreciated as much as the application of the current designs for generating different tumor models. Almost all of the devices fabricated and researched so far have been made out of PDMS. It is an excellent choice as per the ease of fabrication and performing long-term cell culture experiments. PDMS has so many advantages, such as suitability for cell culture, optical imaging, biocompatibility, but it also has a long evolving list of disadvantages [78]. The long-term suitability of tubing material for cell culture and media perfusion are both still a concern. Especially, molecule adsorption on the insides of the tubing can affect the cell-culture parameters. In drug screening applications, some drugs might get adsorbed on the tube and might lead to erroneous drug response results. However, potential solutions have been proposed that have made PDMS-based devices more suitable for microfluidic-based tumor models [79,80]. Bubble generation in a perfusion-based microfluidic system is a big challenge, especially in long-term cell-culture studies, that can lead to cell death and experiment failures. A unique system for bubble elimination was designed for a perfusion-based 3D cell culture system [81]. In this system, two sudden expansion reservoirs were introduced in the flow system to utilize the principle of buoyancy to trap air bubbles. Through this system, they overcame the risk of biological contamination and medium spilling out of the system when filters were used to remove air bubbles. They empirically tested the air-trapping performance and mixing based on diffusion and found the volume of air trapped to be the same as the volume of the expansion reservoir. They also tested the toxic effects of hydrogen peroxide on Ewing's sarcoma cells in a 3D scaffold perfusion culture on this bubble-removal platform [81].

## 16.8 Conclusion

An accurate prediction of systemic drug toxicity and efficacy against established as well as new drug candidates before the drugs reach human clinical trials is the main goal to save millions of dollars in the drug development process. Microfluidic models containing multiple interconnected organs made using primary tumor cells and induced pluripotent stem cell-based other cell types with onboard dynamic flow systems hold great potential toward achieving that goal. Mathematical PK/PD models can also be included in those microfluidic models to understand the behavior and time-based activity of any particular drug or various drug combinations. While the replacement of the animal models is an ultimate aim, in the meantime, the microphysiological tumor models can be employed alongside the data obtained from animal models for better drug response predictions in clinical trials. The microfluidic systems use less number of cells and materials. Therefore, the system can be used in a high-throughput manner for faster and cost-effective drug development procedures. Another decade of research will successfully seal the future application of these devices into highly standard and widely used preclinical models across the healthcare world.

## Acknowledgments

This work was supported by the Ministry of Human Resource Development (MHRD), Government of India.

## References

[1] DiMasi JA. The value of improving the productivity of the drug development process. Pharmacoeconomics [Internet] 2002;20(Suppl. 3):1–10 <http://link.springer.com/10.2165/00019053-200220003-00001> [accessed 30.08.18].

[2] Ocana A, Pandiella A, Siu LL, Tannock IF. Preclinical development of molecular-targeted agents for cancer. Nat Rev Clin Oncol [Internet] 2011;8(4):200–9 <http://www.nature.com/articles/nrclinonc.2010.194> [accessed 18.10.18].

[3] Hay M, Thomas DW, Craighead JL, Economides C, Rosenthal J. Clinical development success rates for investigational drugs. Nat Biotechnol [Internet] 2014;32(1):40–51 <http://www.ncbi.nlm.nih.gov/pubmed/24406927> [accessed 21.08.19].

[4] Bhatia SN, Ingber DE. Microfluidic organs-on-chips. Nat Biotechnol [Internet] 2014;32(8):760–72 <http://www.nature.com/doifinder/10.1038/nbt.2989>.

[5] Sung KE, Beebe DJ. Microfluidic 3D models of cancer. Adv Drug Deliv Rev [Internet] 2014;79–80:68–78 <https://www.sciencedirect.com/science/article/pii/S0169409X14001434#bb0135> [accessed 16.07.19].

[6] Nigam R, Babu KR, Ghosh T, Kumari B, Akula D, Rath SN, et al. Indenone derivatives as inhibitor of human DNA dealkylation repair enzyme AlkBH3. Bioorg Med Chem [Internet] 2018;26(14):4100–12 <https://www.sciencedirect.com/science/article/pii/S0968089618306679> [accessed 29.09.18].

[7] Nath S, Devi GR. Three-dimensional culture systems in cancer research: FOCUS on tumor spheroid model. Pharmacol Ther [Internet] 2016;163:94–108 <https://www.sciencedirect.com/science/article/pii/S0163725816300213?via%3Dihub> [accessed 11.09.18].

[8] Shang M, Soon RH, Lim CT, Khoo BL, Han J. Microfluidic modelling of the tumor microenvironment for anti-cancer drug development. Lab Chip [Internet] 2019;19:369–86 <http://xlink.rsc.org/?DOI = C8LC00970H> [accessed 22.02.19].

[9] Wang D-D, Liu W, Chang J-J, Cheng X, Zhang X-Z, Xu H, et al. Bioengineering three-dimensional culture model of human lung cancer cells: an improved tool for screening EGFR targeted inhibitors. RSC Adv [Internet] 2016;6(29):24083–90 <http://xlink.rsc.org/?DOI = C6RA00229C> [accessed 04.03.19].

[10] Huh D, Hamilton GA, Ingber DE. From 3D cell culture to organs-on-chips. Trends Cell Biol [Internet] 2011;21(12):745–54. Available from: https://doi.org/10.1016/j.tcb.2011.09.005.

[11] Gould SE, Junttila MR, de Sauvage FJ. Translational value of mouse models in oncology drug development. Nat Med [Internet] 2015;21(5):431–9 <http://www.nature.com/articles/nm.3853> [accessed 26.08.19].

[12] Tuveson D, Clevers H. Cancer modeling meets human organoid technology. Science (80-) 2019;364 (6444):952–5.

[13] Xu C, Li X, Liu P, Li M, Luo F. Patient-derived xenograft mouse models: a high fidelity tool for individualized medicine (review). Oncol Lett 2019;17:3–10.

[14] Kersten K, Visser KE, Miltenburg MH, Jonkers J. Genetically engineered mouse models in oncology research and cancer medicine. EMBO Mol Med 2017;9(2):137–53.

[15] Sung JH, Koo J, Shuler ML. Mimicking the human physiology with microphysiological systems (MPS)-organ microphysiological system (MPS). BioChip J [Internet] 2019;13(2):115–26 <https://link.springer.com/content/pdf/10.1007%2Fs13206-019-3201-z.pdf> [accessed 23.07.19].

[16] Sun W, Luo Z, Lee J, Kim H-J, Lee K, Tebon P, et al. Organ-on-a-chip for cancer and immune organs modeling. Adv Healthc Mater [Internet] 2019; <http://doi.wiley.com/10.1002/adhm.201801363> [accessed 06.02.19].

[17] Rothbauer M, Rosser JM, Zirath H, Ertl P. Tomorrow today: organ-on-a-chip advances towards clinically relevant pharmaceutical and medical in vitro models [Internet]. Curr Opin Biotechnol 2019;55:81–6 <https://www.sciencedirect.com/science/article/pii/S0958166918300570> [accessed 05.02.19].

[18] Walker GM, Zeringue HC, Beebe DJ. Microenvironment design considerations for cellular scale studies. Lab Chip 2004;4(2):91–7.

[19] Stone HA, Stroock AD, Ajdari A. Engineering flows in small devices microfluidics toward a lab-on-a-chip. Annu Rev Fluid Mech [Internet] 2004;36(1):381–411 <http://arjournals.annualreviews.org/doi/abs/10.1146/annurev.fluid.36.050802.122124>.

[20] Dhiman N, Kingshott P, Sumer H, Sharma CS, Rath SN. On-chip anticancer drug screening – recent progress in microfluidic platforms to address challenges in chemotherapy. Biosens Bioelectron [Internet] 2019;137 (July):236–54. Available from: https://doi.org/10.1016/j.bios.2019.02.070.

[21] Sarkar S, Cohen N, Sabhachandani P, Konry T. Phenotypic drug profiling in droplet microfluidics for better targeting of drug-resistant tumors. Lab Chip [Internet] 2015;15(23):4441–50 <http://xlink.rsc.org/?DOI = C5LC00923E> [accessed 18.06.18].

[22] Sabhachandani P, Motwani V, Cohen N, Sarkar S, Torchilin V, Konry T. Generation and functional assessment of 3D multicellular spheroids in droplet based microfluidics platform. Lab Chip [Internet] 2016;16 (3):497–505 <http://xlink.rsc.org/?DOI = C5LC01139F> [accessed 20.07.18].

[23] Hassell BA, Goyal G, Lee E, Sontheimer-Phelps A, Levy O, Chen CS, et al. Human organ chip models recapitulate orthotopic lung cancer growth, therapeutic responses, and tumor dormancy in vitro. Cell Rep [Internet] 2017;21(2):508–16. Available from: https://doi.org/10.1016/j.celrep.2017.09.043.

[24] Bruce A, Evans R, Mezan R, Shi L, Moses BS, Martin KH, et al. Three-dimensional microfluidic tri-culture model of the bone marrow microenvironment for study of acute lymphoblastic leukemia. PLoS One 2015;10(10):1–16.

[25] LeBleu VS, Kalluri R. A peek into cancer-associated fibroblasts: origins, functions and translational impact. Dis Model Mech [Internet] 2018;11(4):dmm029447 <http://www.ncbi.nlm.nih.gov/pubmed/29686035> [accessed 05.09.18].

[26] Sontheimer-Phelps A, Hassell BA, Ingber DE, John Paulson HA. Modelling cancer in microfluidic human organs-on-chips. Nat Rev Cancer [Internet] 2019;19:65–81 <https://doi.org/10.1158/0008-5472.CAN-15-1325> [accessed 30.01.19].

[27] Kalluri R, Zeisberg M. Fibroblasts in cancer. Nat Rev Cancer [Internet] 2006;6(5):392–401 <http://www.nature.com/articles/nrc1877> [accessed 05.09.18].

[28] Jeong SY, Lee JH, Shin Y, Chung S, Kuh HJ. Co-culture of tumor spheroids and fibroblasts in a collagen matrix-incorporated microfluidic chip mimics reciprocal activation in solid tumor microenvironment. PLoS One [Internet] 2016;11(7):e0159013. Available from: https://doi.org/10.1371/journal.pone.0159013.

[29] Nashimoto Y, Okada R, Hanada S, Arima Y, Nishiyama K, Miura T, et al. Vascularized cancer on a chip: the effect of perfusion on growth and drug delivery of tumor spheroid. Biomater [Internet] 2020;229:119547. Available from: https://doi.org/10.1016/j.biomaterials.2019.119547.

[30] Mi S, Liu Z, Du Z, Yi X, Sun W. Three-dimensional microfluidic tumor–macrophage system for breast cancer cell invasion. Biotechnol Bioeng [Internet] 2019;116(7):1731–41 <https://onlinelibrary.wiley.com/doi/abs/10.1002/bit.26961> [accessed 11.06.19].

[31] Boussommier-Calleja A, Atiyas Y, Haase K, Headley M, Lewis C, Kamm RD. The effects of monocytes on tumor cell extravasation in a 3D vascularized microfluidic model. Biomaterials 2019;198:180–93.

[32] Romero-López M, Trinh AL, Sobrino A, Hatch MMS, Keating MT, Fimbres C, et al. Recapitulating the human tumor microenvironment: Colon tumor-derived extracellular matrix promotes angiogenesis and tumor cell growth. Biomater [Internet] 2017;116:118–29 <https://www.sciencedirect.com/science/article/pii/S0142961216306585> [accessed 21.08.19].

[33] Kessenbrock K, Plaks V, Werb Z. Matrix metalloproteinases: regulators of the tumor microenvironment. Cell. 2010;141:52–67.

[34] Montanez-Sauri SI, Sung KE, Berthier E, Beebe DJ. Enabling screening in 3D microenvironments: probing matrix and stromal effects on the morphology and proliferation of T47D breast carcinoma cells. Integr Biol (UK) 2013;5(3):631–40.

[35] Zhang X, Nie D, Chakrabarty S. Growth factors in tumor microenvironment. Front Biosci (Landmark Ed [Internet]) 2010;15:151–65 <http://www.ncbi.nlm.nih.gov/pubmed/20036812> [accessed 26.08.19].

[36] Witsch E, Sela M, Yarden Y. Roles for growth factors in cancer progression. Physiology [Internet] 2010;25(2):85–101 <http://www.physiology.org/doi/10.1152/physiol.00045.2009> [accessed 26.08.19].

[37] Uhl CG, Liu Y. Microfluidic device for expedited tumor growth towards drug evaluation. Lab Chip [Internet] 2019;19(8):1458–70 <http://xlink.rsc.org/?DOI = C8LC01250D> [accessed 26.08.19].

[38] Huang Q, Hu X, He W, Zhao Y, Hao S, Wu Q, et al. Fluid shear stress and tumor metastasis. Am J Cancer Res [Internet] 2018;8(5):763–77 <http://www.ncbi.nlm.nih.gov/pubmed/29888101> [accessed 17.06.19].

[39] Jain RK, Martin JD, Stylianopoulos T. The role of mechanical forces in tumor growth and therapy. Annu Rev Biomed Eng [Internet] 2014;16(1):321–46 <http://www.ncbi.nlm.nih.gov/pubmed/25014786> [accessed 21.08.19].

[40] Khan OF, Sefton MV. Perfusion and characterization of an endothelial cell-seeded modular tissue engineered construct formed in a microfluidic remodeling chamber. Biomaterials [Internet] 2010;31(32):8254–61 <https://www.sciencedirect.com/science/article/pii/S0142961210008847?via%3Dihub> [accessed 11.07.19].

[41] Akbari E, Spychalski GB, Rangharajan KK, Prakash S, Song JW. Competing fluid forces control endothelial sprouting in a 3-D microfluidic vessel bifurcation model. Micromachines [Internet] 2019;10(7):451 <https://www.mdpi.com/2072-666X/10/7/451> [accessed 08.11.18].

[42] Novak CM, Horst EN, Taylor CC, Liu CZ, Mehta G. Fluid shear stress stimulates breast cancer cells to display invasive and chemoresistant phenotypes while upregulating *PLAU* in a 3D bioreactor. Biotechnol Bioeng [Internet] 2019;116(11):3084–97 <https://onlinelibrary.wiley.com/doi/abs/10.1002/bit.27119> [accessed 08.11.19].

[43] Wang S, Li E, Gao Y, Wang Y, Guo Z, He J, et al. Study on invadopodia formation for lung carcinoma invasion with a microfluidic 3D culture device. PLoS One 2013;8(2):e56448.

[44] Choucha Snouber L, Bunescu A, Naudot M, Legallais C, Brochot C, Dumas ME, et al. Metabolomics-on-a-chip of hepatotoxicity induced by anticancer drug flutamide and its active metabolite hydroxyflutamide

using HepG2/C3a microfluidic biochips. Toxicol Sci [Internet] 2013;132(1):8–20 <https://academic.oup.com/toxsci/article-lookup/doi/10.1093/toxsci/kfs230> [accessed 22.07.19].

[45] Sugiura S, Edahiro JI, Kikuchi K, Sumaru K, Kanamori T. Pressure-driven perfusion culture microchamber array for a parallel drug cytotoxicity assay. Biotechnol Bioeng 2008;100(6):1156–65.

[46] Edington CD, Chen WLK, Geishecker E, Kassis T, Soenksen LR, Bhushan BM, et al. Interconnected microphysiological systems for quantitative biology and pharmacology studies. Sci Rep [Internet] 2018;8(1):4530 <http://www.nature.com/articles/s41598-018-22749-0> [accessed 23.07.19].

[47] Satoh T, Sugiura S, Shin K, Onuki-Nagasaki R, Ishida S, Kikuchi K, et al. A multi-throughput multi-organ-on-a-chip system on a plate formatted pneumatic pressure-driven medium circulation platform. Lab Chip [Internet] 2018;18(1):115–25 <http://xlink.rsc.org/?DOI = C7LC00952F> [accessed 22.07.19].

[48] Coppeta JR, Mescher MJ, Isenberg BC, Spencer AJ, Kim ES, Lever AR, et al. A portable and reconfigurable multi-organ platform for drug development with onboard microfluidic flow control. Lab Chip [Internet] 2017;17(1):134–44 <http://xlink.rsc.org/?DOI = C6LC01236A> [accessed 22.07.19].

[49] Chen HJ, Miller P, Shuler ML. A pumpless body-on-a-chip model using a primary culture of human intestinal cells and a 3D culture of liver cells. Lab Chip [Internet] 2018;18(14):2036–46 <http://xlink.rsc.org/?DOI = C8LC00111A> [accessed 22.07.19].

[50] Sung JH, Kam C, Shuler ML. A microfluidic device for a pharmacokinetic–pharmacodynamic (PK–PD) model on a chip. Lab Chip [Internet] 2010;10(4):446 <http://xlink.rsc.org/?DOI = b917763a> [accessed 08.07.19].

[51] Miller PG, Shuler ML. Design and demonstration of a pumpless 14 compartment microphysiological system. Biotechnol Bioeng [Internet] 2016;113(10):2213–27 <http://www.ncbi.nlm.nih.gov/pubmed/27070809> [accessed 22.08.19].

[52] Esch MB, Ueno H, Applegate DR, Shuler ML. Modular, pumpless body-on-a-chip platform for the co-culture of GI tract epithelium and 3D primary liver tissue. Lab Chip [Internet] 2016;16(14):2719–29 <http://xlink.rsc.org/?DOI = C6LC00461J> [accessed 22.07.19].

[53] Chen Z, He S, Zilberberg J, Lee W. Pumpless platform for high-throughput dynamic multicellular culture and chemosensitivity evaluation. Lab Chip [Internet] 2019;19(2):254–61 <http://xlink.rsc.org/?DOI = C8LC00872H> [accessed 22.07.19].

[54] Abaci HE, Gledhill K, Guo Z, Christiano AM, Shuler ML. Pumpless microfluidic platform for drug testing on human skin equivalents. Lab Chip [Internet] 2015;15(3):882–8 <http://xlink.rsc.org/?DOI = C4LC00999A> [accessed 22.07.19].

[55] McAleer CW, Long CJ, Elbrecht D, Sasserath T, Bridges LR, Rumsey JW, et al. Multi-organ system for the evaluation of efficacy and off-target toxicity of anticancer therapeutics. Sci Transl Med 2019;11(497): eaav1386.

[56] Tatosian DA, Shuler ML. A novel system for evaluation of drug mixtures for potential efficacy in treating multidrug resistant cancers. Biotechnol Bioeng 2009;103(1):187–98.

[57] Mehta G, Lee J, Cha W, Tung Y-C, Linderman JJ, Takayama S. Hard top soft bottom microfluidic devices for cell culture and chemical analysis. Anal Chem [Internet] 2009;81(10):3714–22 <https://pubs.acs.org/doi/10.1021/ac802178u> [accessed 26.08.19].

[58] Park JY, Kim S-K, Woo D-H, Lee E-J, Kim J-H, Lee S-H. Differentiation of neural progenitor cells in a microfluidic chip-generated cytokine gradient. Stem Cell [Internet] 2009;27(11):2646–54 <http://www.ncbi.nlm.nih.gov/pubmed/19711444> [accessed 22.08.19].

[59] Hattori K, Sugiura S, Kanamori T. Pressure-driven microfluidic perfusion culture device for integrated dose-response assays. J Lab Autom [Internet] 2013;18(6):437–45 <https://journals-sagepub-com.ezproxy.lib.swin.edu.au/doi/pdf/10.1177/2211068213503155> [accessed 15.07.19].

[60] Byun CK, Abi-Samra K, Cho YK, Takayama S. Pumps for microfluidic cell culture. Electrophoresis. 2014;35(2–3):245–57.

[61] Van Duinen V, Trietsch SJ, Joore J, Vulto P, Hankemeier T. Microfluidic 3D cell culture: from tools to tissue models. Curr Opin Biotechnol [Internet] 2015;35:118–26. Available from: https://doi.org/10.1016/j.copbio.2015.05.002 [accessed 01.10.18].

[62] Osaki T, Sivathanu V, Kamm RD. Vascularized microfluidic organ-chips for drug screening, disease models and tissue engineering. Curr Opin Biotechnol [Internet] 2018;52:116–23 <https://www.sciencedirect.com/science/article/pii/S0958166917301969> [accessed 12.09.18].

[63] Sobrino A, Phan DTT, Datta R, Wang X, Hachey SJ, Romero-López M, et al. 3D microtumors in vitro supported by perfused vascular networks. Sci Rep [Internet] 2016;6(1):31589 <http://www.nature.com/articles/srep31589> [accessed 10.09.18].

[64] Wang H, Mooney DJ. Biomaterial-assisted targeted modulation of immune cells in cancer treatment. Nat Mater 2018;17:761–72.

[65] Fidler IJ. The pathogenesis of cancer metastasis: the "seed and soil" hypothesis revisited. Nat Rev Cancer [Internet] 2003;3(6):453–8 <http://www.nature.com/articles/nrc1098> [accessed 24.08.18].

[66] Park D, Lee J, Chung JJ, Jung Y, Kim SH. Integrating organs-on-chips: multiplexing, scaling, vascularization, and innervation. Trends Biotechnol [Internet] 2019; <https://www.sciencedirect.com/science/article/pii/S0167779919301556?via%3Dihub> [accessed 21.07.19].

[67] Kim D, Kim S, Park JY. Floating-on-water fabrication method for thin polydimethylsiloxane membranes. Polymers (Basel) 2019;11(8):1264.

[68] Li Z, Guo Y, Yu Y, Xu C, Xu H, Qin J. Assessment of metabolism-dependent drug efficacy and toxicity on a multilayer organs-on-a-chip. Integr Biol [Internet] 2016;8(10):1022–9 <https://academic.oup.com/ib/article/8/10/1022-1029/5187937> [accessed 22.07.19].

[69] Swartz MA, Fleury ME. Interstitial flow and its effects in soft tissues. Annu Rev Biomed Eng 2007;9(1):229–56.

[70] Baxter LT, Jain RK. Transport of fluid and macromolecules in tumors. I. Role of interstitial pressure and convection. Microvasc Res [Internet] 1989;37(1):77–104 <https://www.sciencedirect.com/science/article/abs/pii/0026286289900745?via%3Dihub> [accessed 21.08.19].

[71] Jeon JS, Bersini S, Gilardi M, Dubini G, Charest JL, Moretti M, et al. Human 3D vascularized organotypic microfluidic assays to study breast cancer cell extravasation. Proc Natl Acad Sci U S A [Internet] 2015;112(1):214–19 <http://www.ncbi.nlm.nih.gov/pubmed/25524628> [accessed 24.08.18].

[72] Polacheck WJ, Charest JL, Kamm RD. Interstitial flow influences direction of tumor cell migration through competing mechanisms. Proc Natl Acad Sci U S A [Internet] 2011;108(27):11115–20 <http://www.ncbi.nlm.nih.gov/pubmed/21690404> [accessed 08.07.19].

[73] Avendano A, Cortes-Medina M, Song JW. Application of 3-D microfluidic models for studying mass transport properties of the tumor interstitial matrix. Front Bioeng Biotechnol 2019;7:6.

[74] Offeddu GS, Possenti L, Loessberg-Zahl JT, Zunino P, Roberts J, Han X, et al. Application of transmural flow across in vitro microvasculature enables direct sampling of interstitial therapeutic molecule distribution. Small. 2019;15:1902393.

[75] Esch MB, King TL, Shuler ML. The role of body-on-a-chip devices in drug and toxicity studies. Annu Rev Biomed Eng [Internet] 2011;13:55–72 <www.annualreviews.org> [accessed 11.07.19].

[76] Sung JH, Shuler ML. A micro cell culture analog (μCCA) with 3-D hydrogel culture of multiple cell lines to assess metabolism-dependent cytotoxicity of anti-cancer drugs. Lab Chip [Internet] 2009;9(10):1385 <http://xlink.rsc.org/?DOI = b901377f> [accessed 19.07.19].

[77] Sin A, Chin KC, Jamil MF, Kostov Y, Rao G, Shuler ML. The design and fabrication of three-chamber microscale cell culture analog devices with integrated dissolved oxygen sensors. Biotechnol Prog [Internet] 2008;20(1):338–45 <http://doi.wiley.com/10.1021/bp034077d> [accessed 22.08.19].

[78] van Meer BJ, de Vries H, Firth KSA, van Weerd J, Tertoolen LGJ, Karperien HBJ, et al. Small molecule absorption by PDMS in the context of drug response bioassays. Biochem Biophys Res Commun [Internet] 2017;482(2):323–8 <https://www.sciencedirect.com/science/article/pii/S0006291X16319179> [accessed 08.07.19].

[79] Gökaltun A, Kang YB, Yarmush ML, Usta OB, Asatekin A. Simple surface modification of poly(dimethylsiloxane) via surface segregating smart polymers for biomicrofluidics. Sci Rep [Internet] 2019;9(1):7377 <http://www.nature.com/articles/s41598-019-43625-5> [accessed 08.07.19].

[80] Toepke MW, Beebe DJ. PDMS absorption of small molecules and consequences in microfluidic applications. Lab Chip [Internet] 2006;6(12):1484 <http://xlink.rsc.org/?DOI = b612140c> [accessed 23.07.19].

[81] Park DH, Jeon HJ, Kim MJ, Nguyen XD, Morten K, Go JS. Development of a microfluidic perfusion 3D cell culture system. J Micromech Microeng [Internet] 2018;28(4):045001 <http://stacks.iop.org/0960-1317/28/i = 4/a = 045001?key = crossref.057ceaa94e61f73eb123768e7383b4eb> [accessed 15.07.19].

CHAPTER

# 17

# Engineering breast cancer models in vitro with 3D bioprinting

*Brad A. Krajina and Daniela F. Duarte Campos*

**Department of Materials Science & Engineering, Stanford University, CA, United States**

## 17.1 Breast cancer microenvironment in vivo

### 17.1.1 Types and stages of breast cancer

Breast cancer commonly begins within the breast lobules, where milk is produced, or inside the ducts [1]. There are at least seventeen distinct histological types of breast cancer recognized by the World Health Organization [2]. The most common type of breast cancer is the invasive ductal carcinoma (IDC), which represents about 80% of the cases, followed by the invasive lobular carcinoma, which has an incidence between 5% and 15%. Stage 0 ductal carcinoma or ductal carcinoma in situ (DCIS) begins with an overgrowth of ductal epithelial cells in the milk ducts. In some cases, these cells may invade neighboring tissues, enter the blood and lymph vessels and settle in other areas of the breast forming new tumors. These tumors are IDCs and they are classified in three stages (Stages 1–3) depending on their size and the regions where they spread (Fig. 17.1). Histomorphological blueprints are consistently associated with each one of the different stages [2,3]. Some tumors may spread to other organs, such as the lungs, bones, or the brain. These are called metastatic tumors or Stage 4.

### 17.1.2 Cancer cell behavior in vivo, microenvironment structure and mechanics

Normal cell behavior is regulated by tissue microenvironments [4–6]. Collagen fibrous matrices play an important role during mammary ductal formation. In breast cancer tissue formation, cancer cells migrate along collagen fibers and remodel their orientation and density [7]. Such structural and mechanical changes directly correlate with the stage of IDC (Fig. 17.2A).


DOI: https://doi.org/10.1016/B978-0-12-818128-7.00017-4

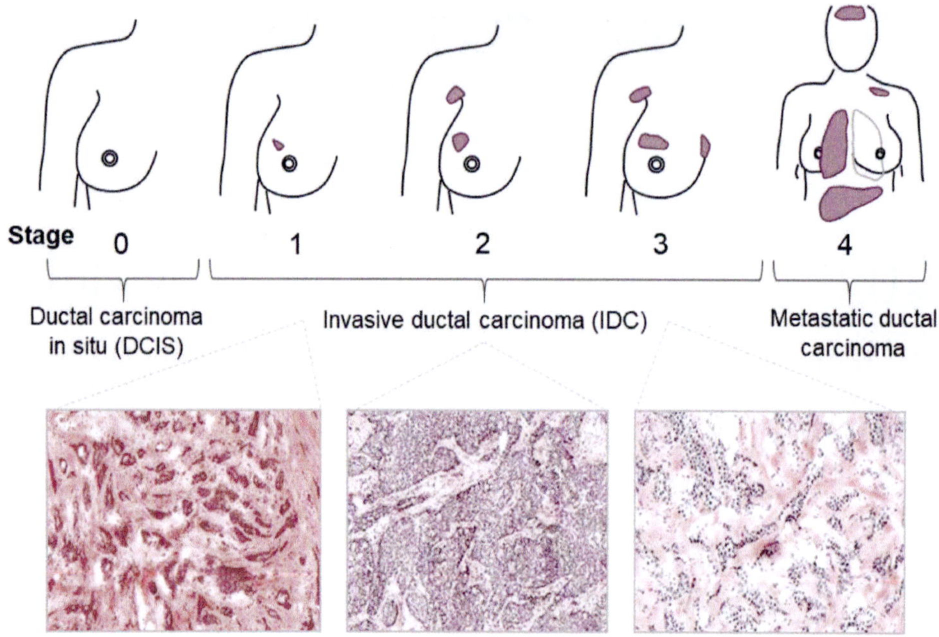


**FIGURE 17.1** Stages of ductal carcinoma: In situ (Stage 0, DCIS), invasive (Stages 1–3, IDC), and metastatic (Stage 4). Significant architectural differences can be identified between the three stages of IDC. Source: *Histological images adapted with permission from Mao X, He J, Li T, Lu Z, Sun J, Meng Y, et al. Application of imaging mass spectrometry for the molecular diagnosis of human breast tumors. Sci Rep 2016;6:21043 [3].*

Tumor-associated collagen structures is a concept that has recently emerged to describe different collagen fiber orientation and density observed in early, advanced, and malignant stages of breast cancer [7,8]. A well-differentiated tumor (Stage 1, Luminal A) demonstrates high homology to the normal breast terminal duct lobular unit, tubule formation, and low mitotic count (Fig. 17.2B). In Stage 1, collagen fibers are dense and wavy bundles, which can be appreciated in benign fibrous lesions, such as fibroma and salivary gland neoplasms [8]. A moderately differentiated tumor (Stage 2, Luminal B) demonstrates intermediate cellular and collagen fiber distribution between stage 1 and 3 tumors. In this stage, type I collagen fibers surrounding the tumor tend to stretch causing linearization of the collagen bundles. A poorly differentiated tumor (Stage 3, HER-2) has a marked degree of cellular pleomorphism, high mitotic activity and no tubule formation. In Stage 3, realignment of collagen bundles is perpendicular to the tumor border, allowing for direct contact with cancer cells and promoting their migration toward the stroma. Malignant tumors like squamous cell and basal cell carcinomas are some examples of malignancies showing collagen signatures as described in Stage 3 [8].

Human breast cancer progression is accompanied by collagen linear reorganization, increased birefringence and stiffening [9]. Imaging techniques, such as second harmonic generation imaging, were used to show changes in collagen fiber density and reorganization [10–13]. The stiffness of breast tissue and tumors may be analyzed by atomic force microscopy (AFM) directly on tissue biopsies (Fig. 17.2C). Adjacent normal breast tissue

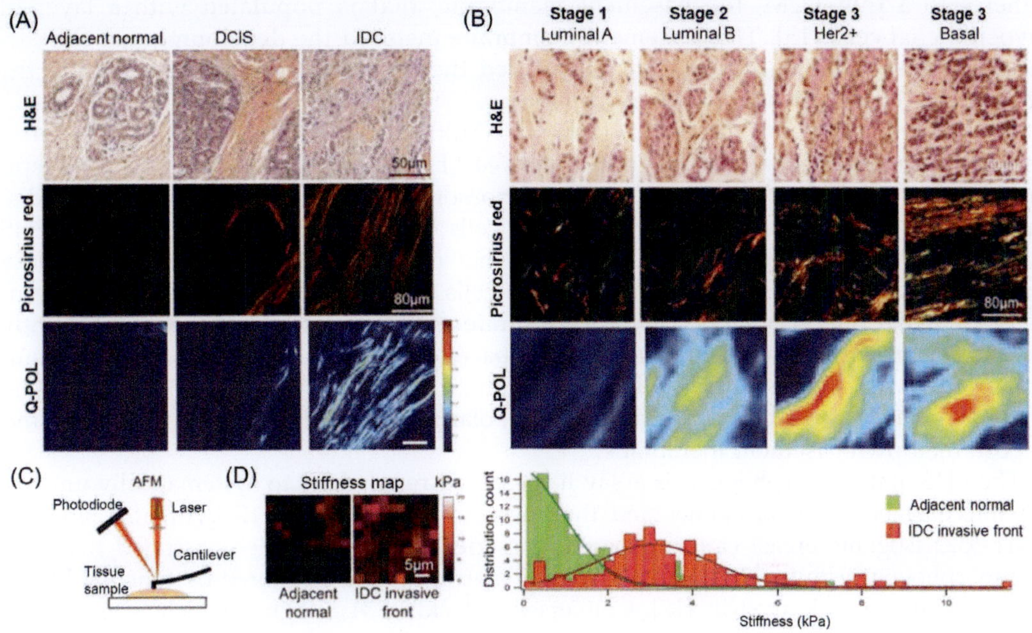


FIGURE 17.2 Breast cancer-associated collagen structures and mechanical properties in the different stages. (A) Collagen fiber linear reorganization in breast cancer progression from normal adjacent tissue and DCIS to IDC. (B) Correlation of collagen fiber orientation and stage of breast cancer (from Stage 1 to 3). (C) AFM is used to analyze the stiffness of breast tissue and tumor biopsies. (D) Heterogeneous and increasing stiffness of IDC at the invasion front in comparison to adjacent and noninvasive stromal tissue. Source: *Images adapted with permission from Acerbi I, Cassereau L, Dean I, Shi Q, Au A, Park C, et al. Human breast cancer invasion and aggression correlates with ECM stiffening and immune cell infiltration. Integr Biol (Camb) 2015;7:1120–34 [9].*

and noninvasive stroma have a ranging stiffness between 0.1 and 2 kPa. As the tumor progresses to IDC its stiffness can increase to up to 10 kPa at the invasive front (Fig. 17.2D). A possible mechanistic explanation for the increase in stiffness may be related with the infiltration of immune cells, mostly protumorigenic macrophages, which stimulate tumor and stromal TGFβ signaling. An increase in TGFβ signaling stimulates the production of collagen, which in turn contributes to the stiffening of the extracellular matrix (ECM).

## 17.2 Biomaterial-based breast cancer in vitro models

### 17.2.1 Mammary morphogenesis in 3D

It has long been appreciated that traditional two-dimensional (2D) monolayer cultures fail to capture hallmarks of mammary gland organization in both healthy tissue and disease [14]. The healthy adult mammary gland is a 3D branching network characterized by a hollow luminal space surrounded by polarized luminal cells that exhibit distinct patterns of protein localization at their apical (luminal) and basal surfaces. The basal surface adheres to a thin ECM, the basement membrane, that is populated with a layer of myoepithelial cells [15]. This basement membrane inspired the development of the earliest 3D organotypic cultures, which employed the basement matrix derived from the Englebreth–Holm–Swarm (EHS) mouse tumor.

In pioneering work by the Bissell group, 3D culture in a basement membrane matrix revealed discrepancies in morphogenesis between benign and malignant mammary epithelial cells that mirrored aspects of the disruption of mammary gland organization in disease [14]. Normal human mammary epithelial cells embedded within an exogenous EHS matrix underwent a morphogenesis program that recapitulated hallmarks of mammary gland organization. After embedding as single cells within an EHS matrix, normal mammary epithelial cells proliferated into nearly uniform spherical clusters that subsequently underwent growth arrest, cleared a luminal space, deposited their own basement membrane, and establish apico-basal polarity. In contrast, malignant breast cancer cells consistently exhibited failure to undergo apico-basal polarization, establish a luminal cavity, and deposit their own basement membrane.

The EHS matrix morphogenesis assay has enabled researchers to systematically unravel the interplay between oncogenes and the microenvironment (Fig. 17.3A) [16]. Using the HMT-3522 isogenic breast cancer progression series, Bissell et al. demonstrated that normal morphogenesis by malignant cells in EHS matrix could be restored by function blocking antibodies for β1 integrin [17]. Conversely, blocking β4 integrin, which is associated with hemidesmisome formation during attachment to the basal lamina, disrupted proper polarization in normal mammary epithelial cells and yielded disorganized colonies reminiscent of the "malignant phenotype." Thus, cell–ECM signaling can override genotype, either by restraining malignant cells or promoting aberrant morphogenesis in genetically benign epithelial cells.

### 17.2.2 Studies on cancer cell migration in 3D (metastasis models)

Migration of cancer cells through complex tissue environments is a critical process in tumor dissemination that is poorly modeled in traditional 2D monolayer culture. Compared with 2D substrates where migration is essentially unimpeded by physical obstacles, 3D environments can present a diverse range of physical and biochemical constraints that restrain or support cell motility. Matrix porosity, fiber topography, mechanics, proteolytic degradability, and fiber orientation represent molecular determinants unique to 3D that cooperatively conspire to control cell migration [18]. In turn, cells migrating in 3D employ a varied arsenal of migration strategies that enable them to adapt to matrix-dependent and cell-intrinsic constraints. These strategies can be adhesion independent or dependent, proteolysis independent or dependent, mesenchymal or ameboid, and individual or multicellular [19].

Spatial confinement by the surrounding matrix can determine modes of 3D migration. Using HT1080 fibrosarcoma cells embedded in collagen matrices, Friedl et al. demonstrated that squeezing of the nucleus through network pores represents a rate-limiting step in 3D migration through dense networks (Fig. 17.3B) [20]. In networks with decreasing pore size, cells increasingly required matrix metalloproteinase (MMP)-mediated

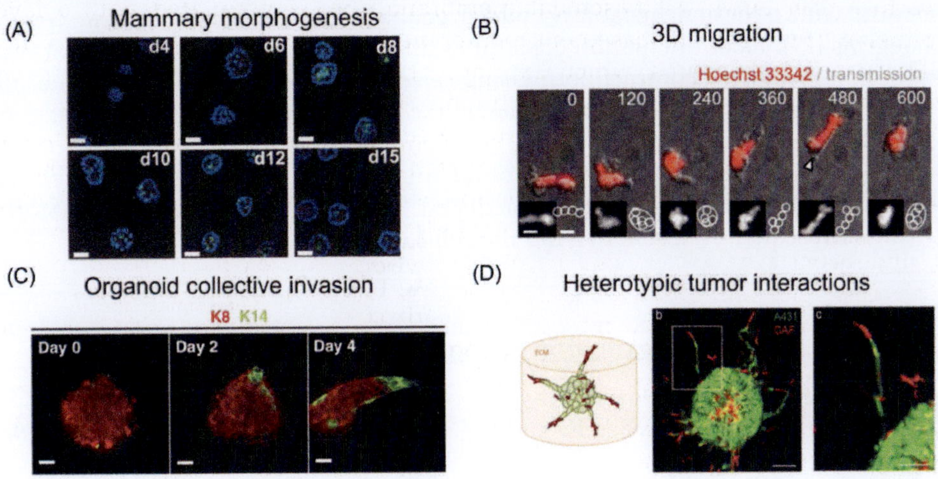


FIGURE 17.3 Examples of nonbioprinted 3D breast cancer studies with biomaterials. (A) Culture of benign mammary epithelial cells in EHS basement membrane matrix drives self-organized cell polarization and lumen formation reminiscent of in vivo mammary glands. (B) Migration studies of cancer cells in 3D biomaterials reveal mechanisms of migration that are unique to 3D. The images shown demonstrate that squeezing of the nucleus through pores is a rate-limiting step in migration through dense collagen matrices. (C) 3D culture of organoids freshly explanted from tumors into biomaterials enable modeling of cell–cell and cell–matrix interactions during collective tumor invasion. In the study shown, keratin-14 positive leader cells were found to initiate collective invasion from organoids in response to fibrous collagen. (D) 3D spheroid models in biomaterials enable the study of heterotypic interactions between cancer cells and other cells that inhabit the tumor microenvironment, such as CAFs, which can lead collective invasion of cancer cells. Source: *(A) Images adapted with permission from Debnath J, Mills KR, Collins NL, Reginato MJ, Muthuswamy SK, Brugge JS. The role of apoptosis in creating and maintaining luminal space within normal and oncogene-expressing mammary acini. Cell 2002;111:29–40 [16]. (B) Images adapted with permission from Wolf K, te Lindert M, Krause M, Alexander S, te Riet J, Willis AL, et al. Physical limits of cell migration: control by ECM space and nuclear deformation and tuning by proteolysis and traction force. J Cell Biol 2013;201:1069–84 [20]. (C) Images adapted with permission from Cheung KJ, Gabrielson E, Werb Z, Ewald AJ. Collective invasion in breast cancer requires a conserved basal epithelial program. Cell 2013;155:1639–51 [26]. (D) Images adapted with permission from Labernadie A, Kato T, Brugués A, Serra-Picamal X, Derzsi S, Arwert E, et al. A mechanically active heterotypic E-cadherin/N-cadherin adhesion enables fibroblasts to drive cancer cell invasion. Nat Cell Biol 2017;19:224–37 [45].*

degradation to generate physical space for migration. In dense matrices, cells undergoing MMP-independent migration underwent nuclear deformation to squeeze through pores, which required integrin-dependent adhesion and actomyosin contractility to propel the nucleus forward. By contrast, in sufficiently porous matrices, cell migration was unimpeded by MMP inhibition.

For cells migrating in a mesenchymal fashion, a 3D fibrous matrix imposes unique dependencies on adhesion dynamics and contractility, which can be controlled through fiber mechanics and organization. In contrast to the contractility-independent motility in 2D, mesenchymal migration in 3D collagen matrices require actomyosin contractility for rapid migration [21,22]. Using collagen matrices with varying fiber architecture and mechanics, the Yamada group demonstrated that fibroblast motility in 3D was optimized by networks with intermediate mechanics that balanced contractility, which propels the cell forward, with adhesion instability, which enables detachment of the rear from activated integrins [21]. Fiber orientation constitutes an additional level of control that couples with adhesion-mediated contractility. Using electrospun Dextran-Methacrylate fibers, Baker et al. developed a family of fibrous networks with independently tunable alignment, fiber diameter, and fiber elasticity [23]. They demonstrated that aligned fiber networks uniquely enable a "sling shot" mode of rapid migration in networks with intermediate fiber mechanics, in which contractility-mediated deformations are elastically stored in fibers, which subsequently recoil, driving the cell forward.

### 17.2.3 3D spheroid and organoid invasion models

Single cell migration may represent an important mode of tumor dissemination, but it does not capture the process whereby individual cells or cell clusters invade and detach from the bulk tumor, which is an inherently collective phenomenon. Collective invasion of strands originating from the primary tumor have been observed in vivo in both mouse models and human breast cancer tissues (Fig. 17.3C) [24–26]. Spheroid invasion models, in which tumor cell aggregates are embedded within a 3D ECM, offer tunable in vitro platforms for dissecting collective aspects of tumor invasion.

Tumor explant organoid cultures arguably most closely mirror the epithelial composition of the primary tumor. Fresh tumor organoids comprise hundreds to thousands of cells, which are obtained by digestion and mechanical disruption of freshly isolated tumor tissues and subsequently embedded in a 3D matrix of interest [7,27,28]. Unlike established cell lines that have been passaged in culture, organoid models can more faithfully recapitulate the genotype and histological features of the origin tissue, such as hormone receptor status and oncogenic mutations, opening opportunities for modeling diverse breast cancer subtypes [29]. Organoid models can also reproduce phenotypic specialization observed within the origin tissue, such as unique expression patterns of invasive leader cells [26]. Compared with in vivo models, fresh organoid explants enjoy the facility of manipulation of in vitro culture, including tunable properties of the embedding ECM, and ability to employ genetic manipulations, such as lentiviral shRNA knockdowns. However, tissue explant cultures require organoids to be generated from fresh animal tissue for each experiment, which can be costly and limits throughput.

As a higher throughput and lower cost solution, spheroids can be generated by self-organization of passaged and resuspended single cells, which can be sourced from passaged tissue organoid explants or immortalized cell lines. Self-organization can occur either through clonal expansion or cell aggregation. For strategies employing clonal expansion, 3D culture in basement membrane (typically EHS matrix) can support spheroid expansion and sequential passaging of a variety of epithelial cells, including immortalized cell lines (e.g., MCF10A) [30] and organoids established from fresh human tumor tissues [29]. Intact spheroids generated in basement membrane cultures can be readily dissociated from the matrix and subsequently transferred to other ECM microenvironments, such as collagen gels [28,31,32]. Alternatively, spheroids can be generated by aggregation of single cells in low-adherent conditions, such as in ultra-low adhesion plates [33], microwells [34], or hanging drop suspension cultures [35]. Aggregation-based methods enable faster processing, superior size control [33], and can accommodate the generation of mosaic spheroids by mixing different cell types [36].

Although 3D basement membrane cultures have proved to be powerful tools for studying disruption of normal mammary gland polarization, basement membrane matrices do not faithfully represent the ECM microenvironment at the leading edge of invasive tumors. Loss of the basement membrane and contact with the surrounding stromal tissue is hypothesized to represent a crucial step in the progression toward invasive breast cancer [37]. Invasive breast cancers are also characterized by abundant deposition of stromal ECM, including fibrous collagen I [7,37,38]. Consequently, collagen gels have formed a natural basis for a variety of spheroid invasion studies.

Collagen spheroid invasion studies have revealed a tight interplay between matrix remodeling and spatially regulated cell specialization in collective invasion. Using breast cancer and fibrosarcoma spheroids embedded in collagen I matrices, the Friedl group demonstrated a multistage proteolysis program that mediates the transition from single to collective strand invasion [39]. They found individual leader cells initiate invasion by proteolytic degradation and traction force, which realigns restraining fibers to form microtracks for migration. Subsequently, single-file follower cells laterally expand the migration tracks proteolytically, eventually forging channels that can accommodate collective invasion. Using several mouse and human organoid models spanning a number of breast cancer subtypes, the Ewald group demonstrated that these invading leader cells represent a basal epithelial subtype that is specifically enriched at the organoid/ECM boundary [26]. They found that collective invasion of basal-like cells is only initiated in fibrillar collagen I-rich environments, and remains latent in basement membrane culture, underscoring the importance of cell–matrix interactions in collective invasion.

### 17.2.4 3D models of heterotypic tumor–stromal interactions

The breast cancer microenvironment is a diverse mosaic of cell types, including heterogeneous epithelial cells, cancer-associated fibroblasts (CAFs) [40], infiltrating immune cells [9], and endothelial cells that vascularize the tumor [41]. The 3D co-culture models enable deconstruction of these heterotypic interactions and their regulation of processes such as invasion, angiogenesis, intravasation, and drug sensitivity. Various strategies have been employed for controlling spatial relationships in co-culture systems, including the use of multilayer gel cultures [42,43], mixed spheroids [36,44,45], microfluidic patterning [46], bottom-up "programmable" assembly [47], and organoid explant cultures [48]. Although significant challenges remain in achieving fidelity to complex in vivo interactions, these models have revealed a variety of heterotypic interactions that reciprocally regulate cellular function.

A defining feature of many solid tumors is the presence of a fibrotic, reactive tumor stroma populated with activated CAFs [49]. The 3D co-culture systems have revealed a potent role for fibroblasts in guiding epithelial invasion. Sahai et al. leveraged co-culture of squamous cell carcinoma (SCC) cells with fibroblasts, in which SCC cells were either plated on top of a fibroblast-populated gel in direct contact or separated by an empty collagen gel layer [42]. When direct contact between fibroblasts and SCC cells was allowed, fibroblast leader cells strikingly initiated collective invasion by SCC follower cells. Invasion by SCC cells was enabled by tracks formed through matrix remodeling of collagen fibers by the leader fibroblasts. Using mixed heterotypic spheroid culture of A431 epidermoid cancer cells and CAFs, the Trepat group demonstrated that initiation of cancer cell invasion by fibroblasts requires a heterophilic adhesion between fibroblast N-cadherin and epithelial E-cadherin, which serves as a mechanical link to guide invasion behind force-generating fibroblasts (Fig. 17.3D) [45]. These studies highlight not only the importance of dimensionality in defining heterotypic interactions, but also the importance of spatial relationships across cell types.

Tumor–vascular interactions are central to myriad processes in the metastatic cascade, including transport of nutrients and drugs [50], intravasation into systemic circulation [51], and conditioning of the perimetastatic niche [52]. Tumors are vascularized with leaky vessels whose compromised barrier integrity may facilitate entry into circulation and alter delivery of macromolecules [50]. The 3D culture models can provide the ECM cues and dimensional constraints to support more physiologically relevant endothelial network assembly.

Both layered and spheroid cultures have demonstrated the capacity of malignant cells to promote endothelial sprouting. Rylander et al. Developed a bilayer hydrogel system consisting of an endothelial monolayer on top of a vacant collagen gel and a breast cancer-populated collagen gel in series [43]. Malignant MDA-MB-231 breast cancer cells drove endothelial proliferation and angiogenic sprouting of capillary-like tube networks into the underlying collagen gel significantly more than noninvasive MCF-7 breast cancer cells. English et al. established "minitumor" spheroids composed of fibroblasts, endothelial cells, and MDA-MB-231 breast cancer cells and performed 3D sprouting assays in collagen matrices [44]. They discovered that the presence of breast cancer cells significantly increased the number and invasion depth of endothelial sprouts and modulated the sensitivity of angiogenesis to growth factors and antiangiogenic drugs.

In vitro co-culture models can also unveil how vasculature modulates cancer cell behavior. In mouse models, the Bissel group showed that dormant tumor cells, which can potentially serve as seeds for high latency metastatic outgrowths, frequently localize to microvasculature at metastatic sites [52]. They engineered an organotypic model of the perimetastatic vascular niche using 2D co-cultures of mesenchymal stem cells and endothelial cells, which spontaneously give rise to capillary networks. By seeding breast cancer cells on these engineered microvascular niches, they demonstrated that endothelial cells induce quiescence in resident breast cancer cells. However, breast cancer cells residing at branch points for neovascularization evaded quiescence, and endothelial interactions promoted their growth. These results suggest that the transition from stable microvasculature to neoangiogenesis may awaken dormant tumor cells and promote metastatic outgrowth.

The lack of mature vessel formation through which fluid can flow in the preceding co-culture systems limits the ability to study tumor intravasation, which is a critical step in the metastatic cascade. Lithography-patterned microfluidic devices enable the incorporation of fluid transport through endothelialized channels and precise spatial control over components of the microenvironment. The Kamm group developed a microfluidic model for tumor intravasation consisting of two flow channels that could be independently seeded with distinct cell types, separated by an intervening collagen matrix [46]. Endothelial cells seeded in one of the channels formed a confluent polarized monolayer that served as a size-selective transendothelial barrier. When a chemoattractive gradient was established in the device, HT1080 fibrosarcoma cells migrated toward the endothelial channel, occasionally breaching the endothelial barrier and intravasating into the luminal space. Using this intravasation model, they demonstrated that macrophage-secreted TNF-α disrupted endothelial barrier function and promoted cancer cell intravasation.

## 17.3 Biomaterials design for in vitro breast cancer models

The daunting complexity of morphogenesis in 3D introduces multifaceted biomaterials design considerations (Fig. 17.4). Matrix mechanics, adhesive ligand presentation, network architecture, and matrix degradability play mutually interdependent roles in instructing

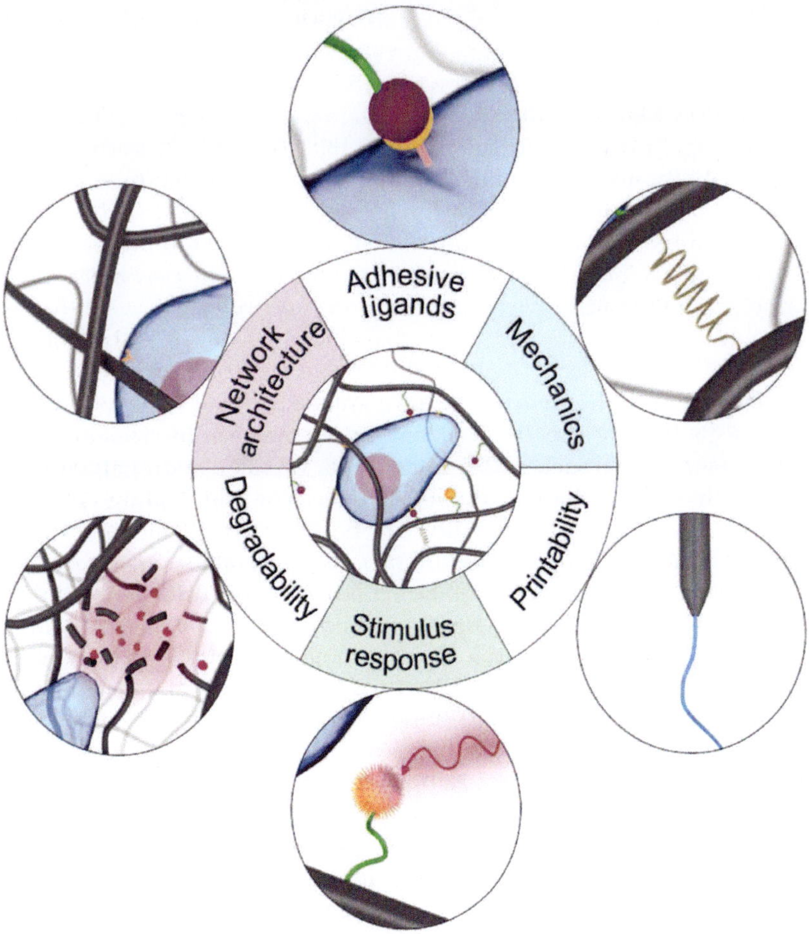


FIGURE 17.4 Illustration of biomaterials design considerations for 3D bioprinted breast cancer models.

cell behavior [53,54]. Stimuli-responsive elements represent another layer of complexity for introducing dynamic and temporally controlled signaling events to model breast cancer progression. To address these challenges, researchers may select from a varied repertoire of materials, including natural ECMs, synthetic hydrogels, and hybrid natural/synthetic materials (Table 17.1).

### 17.3.1 Natural, synthetic, and hybrid biomaterials

Natural ECMs are endogenously equipped with biochemical and biophysical cues that recapitulate features of the tumor microenvironment. The rich milieu of EHS matrix presents complex instructions that are not easily reconstructed using reductionist approaches. Consequently, EHS matrix is often a standard of comparison for many synthetic hydrogels used for epithelial organoids [55,56]. Fibrous collagen I, which is not present in EHS matrix, is abundantly present in the breast cancer stroma [38], and collagen gels more faithfully recapitulate native tissue architecture than many synthetic hydrogels, which are often nanoporous and lack hierarchical structure [57].

However, natural ECMs are challenging to systematically define and tune. The complex composition of matrigel provides intricate signals that are difficult to deconstruct, and its batch-to-batch variability may lead to inconsistent behavior. The stiffness of EHS matrix is typically an order of magnitude less than invasive breast tumors, and its mechanics are not readily tuned in a bioorthogonal fashion [58]. Collagen I gels represent a more defined biochemical environment, but many strategies for tuning their mechanics also alter pore size, fiber diameter, and/or adhesive ligand concentration [59].

In contrast, synthetic polymers and nonnative biopolymers are biologically inert blank canvases that can be systematically decorated with bioactive functionalities and mechanically tuned [53]. Multiarm PEG hydrogels have enjoyed a long history of applications in tissue engineering [60]. They can be prepared through versatile cross-linking strategies, including bioorthogonal approaches. The mechanics of PEG hydrogels are easily tuned across a range of physiological and pathological levels by modulating cross-linking density. Cell-mediated enzymatic degradability can be adjusted by incorporating enzymatically degradable peptide cross-linkers. They also enable systematically defined presentation and composition of integrin-binding adhesive peptides.

Alginate, a polysaccharide derived from seaweed, is also a versatile template for the reductionist ECM approach [54,61]. Like PEG, it is biologically inert from the perspective of mammalian cells, but the carboxylic acid groups on each of its monomers provide chemical handles that can be functionalized with bioactive groups (such as adhesive peptides) or cross-linkable moieties. Unlike PEG, alginate can also be ionically cross-linked with calcium, which produces physical hydrogels whose time-dependent mechanics and malleability can be more systematically tuned than chemical cross-linking [62].

Hybrid materials, consisting of both synthetic and natural ECM components, combine physiologically relevant biochemical signals with the systematic tunability of synthetic materials. For example, interpenetrating networks of EHS matrix and alginate simultaneously provide a complex tumor-derived biochemical environment with the ability to independently tune mechanical properties such as stiffness and plasticity [58,63].

**TABLE 17.1** Examples of biomaterials used for breast cancer models and the tunability of their properties.

| Material class | Material | Material property | Method of tuning | Confounding variables | Selected works (Ref.) |
|---|---|---|---|---|---|
| Natural | EHS matrix | Stiffness | Concentration | Ligand density | [58,111] |
| | | Plasticity | Cross-linking | Cross-linking is not bioorthogonal | [71] |
| | | Ligand density | Concentration | Stiffness | [58,111] |
| | Collagen I | Stiffness | Strain, concentration, gelation temperature, and glycation | Pore size, fiber diameter, and ligand density (unless controlled by strain) | [38,59,66,112] |
| | | Pore size | Concentration and gelation temperature | Stiffness, fiber diameter, and ligand density | [20,59] |
| | | Fiber diameter | Concentration and gelation temperature | Stiffness, pore size, and ligand density | [59,113] |
| | | Fiber alignment | Magnetic and mechanical shear | Can be tuned independently | [114] |
| | | Ligand density | Concentration | Stiffness, pore size, and ligand density | [59,66,112] |
| Synthetic | PEG | Stiffness | Cross-linking density | Pore size and degradability | [84,115] |
| | | Degradability | Degradable linker density | Can be tuned independently | [66,116] |
| | | Ligand type/density | Tethering of adhesive peptides | Can be tuned independently | [66,116] |
| Hybrid | Alginate/EHS | Stiffness | Ionic or covalent cross-linker concentration, and alginate molecular weight | Can be tuned independently | [58,63,71] |
| | | Plasticity | Ionic or covalent cross-linking density and alginate molecular weight, and covalent cross-linking | Can be tuned independently | [58,63,71] |
| | | Ligand type/density | Tethered peptide ligands and EHS concentration | Can be tuned independently | [58,63,71] |
| | Self-assembling peptide (SAP)/ECM protein (laminin, EHS, and collagen I) | Stiffness | SAP concentration | Pore size (modest) | [117] |
| | | Pore size | SAP concentration (weak dependence) | Stiffness | [117] |
| | | Ligand type/density | ECM protein type/concentration | Can be tuned independently | [117] |

Hyaluronic acid is an intrinsically bioactive polymer that is a major constituent of the breast cancer stroma [64], but like alginate it can also be functionalized with adhesive peptides and cross-linked to form tunable hydrogels through a variety of conjugation strategies [65].

### 17.3.2 Matrix stiffness, cross-linking, and network architecture

Breast tumors are fibrotic tissues that are substantially stiffer than healthy mammary glands and are surrounded by a network of cross-linked, fibrillar collagen [38,66]. The 3D in vitro models have proved to be a powerful platform for dissecting the role of mechanotransduction in the malignant phenotype. Seminal work by the Weaver group revealed that simply culturing benign epithelial cells in mechanical environments characteristic of breast tumors drove a malignant phenotype [66]. However, this was demonstrated either on 2D polyacrylamide substrates with an EHS overlay (2.5D), or by tuning collagen concentration to increase stiffness in 3D. Modulating stiffness by collagen density also alters adhesive ligand concentration, pore size, and fiber architecture. 2.5D polyacrylamide overlay cultures independently tune mechanics and stiffness, but do not fully represent a 3D setting.

A variety of biomaterials approaches have revealed a complex interplay between matrix stiffness, network architecture and cross-linking, and biochemistry. The Weaver group developed a collagen tension bioreactor, which enables stiffness to be tuned in collagen gels by application of strain, thereby decoupling stiffness from ligand concentration, pore size, and fiber diameter. This revealed that invasion exhibits a nonmonotonic dependence on collagen density, eventually declining in low porosity/high density matrices. However, at constant collagen concentration, increasing stiffness with strain monotonically promoted invasion [59]. To study the effects of mechanotransduction in the biochemical context of basement membrane proteins, the Mooney group leveraged interpenetrating hydrogels of alginate and EHS matrix [58]. They found that EHS-only matrices robustly supported normal acini formation across a range of concentration-dependent stiffness, but increasing stiffness by alginate cross-linking under constant EHS concentration engendered a malignant phenotype. To interrogate matrix mechanics and remodeling in a fully defined context, the Levental group developed star-PEG hydrogels with independently tunable stiffness, MMP-degradability, and heparin tethering [67]. Stiffer hydrogels with higher cross-linking densities drove invasion, but only with heparin tethering, which supported matrix deposition, and MMP-degradable cross-linkers. In nondegradable hydrogels, increasing stiffness by increasing cross-linking density abolished colony formation, potentially due to confinement effects. Thus, biomaterials approaches to tuning stiffness must carefully address confounding effects of network architecture and biochemistry.

### 17.3.3 Time-dependent and nonlinear mechanics

Mammalian tissues and decellularized natural ECMs often exhibit rich time-scale-dependent and nonlinear mechanical properties that cannot be articulated by elastic stiffness alone [62,63,68–70]. Many covalently cross-linked synthetic hydrogels, such as PEG and polyacrylamide, exhibit linear stress–train responses, and elastically store energy from applied strains in a nearly time-independent manner [62]. By contrast, tissues and fibrous reconstituted ECMs, like collagen I, exhibit nonlinear stress–stiffening [70], dissipate energy from applied strains over time (stress relaxation) [68], and undergo irreversible deformations in response to force (plasticity) [63,69]. The ability of natural ECMs to transform in response to force sets the stage for time-scale-dependent reciprocal dynamics between cancer cells and their matrix that cannot be captured by purely elastic materials.

Alginate-based materials have assumed a leading role in engineered cancer models that move beyond stiffness. To explore the role of matrix plasticity in breast cancer migration, the Chaudhuri group developed nanoporous interpenetrating network hydrogels composed of EHS basement membrane and ionically cross-linked alginate [63]. By varying alginate molecular weight and ionic cross-linker density, they established a family of stiffness-matched hydrogels with tunable plasticity. Independently of broad spectrum MMP-inhibition, matrices with higher plasticity supported significantly higher migration rates of MDA-MB-231 breast cancer cells. Similarly, the Chaudhuri group showed that incorporation of permanent covalent cross-links into either EHS matrix or alginate/EHS interpenetrating networks impeded cancer cell migration independently of stiffness [71].

A number of advancements in engineering nonlinear or time-scale-dependent mechanics remain to be harnessed for cancer modeling. Dynamic covalent cross-links or reversible guest–host supramolecular hydrogels are a versatile means of tuning hydrogel stress-relaxation [72,73]. Nonlinear stress–stiffening has also been achieved using synthetic polyisocyanopeptide matrices, which have been shown to direct mesenchymal stem cell fate through stress–stiffening-mediated mechanotransduction [74]. Interpenetrating hyaluronic acid/collagen I networks containing dynamic covalent cross-links between hyaluronic acid molecules have revealed that focal adhesion assembly in fibrous environments requires stress relaxation for fiber remodeling by fibroblasts [57]. Such materials represent valuable assets in an emerging toolset at the disposal of biomaterials engineers.

### 17.3.4 Stimuli-responsive dynamic materials

The breast cancer microenvironment transforms over time, and current biomaterials are unable to bridge this full spectrum of properties with a single material. Although passively degradable matrices, such as enzymatically or hydrolytically degradable cross-links, have proved instrumental in organoid expansion [55,56], materials whose properties can be spatiotemporally controlled on demand may enable new models of the temporal progression of breast cancer. For example, the Anseth and Deforest groups developed orthogonal photoligation and photolysis strategies to build hydrogels with spatiotemporally controlled degradation and biochemical patterning [75]. Building on the concept of biochemical patterning further, using a "living" photoclick reaction, the Anseth group developed PEG hydrogels in which tethered adhesive peptide ligands were reversibly exchanged in the matrix, presenting the opportunity to present different adhesive ligands within the same matrix in a sequential manner [76]. However, the full potential of dynamically tunable biomaterials in 3D cancer models remains largely unrealized. To this end, the Suggs group designed and a dynamic alginate system using phototriggered release cross-linker

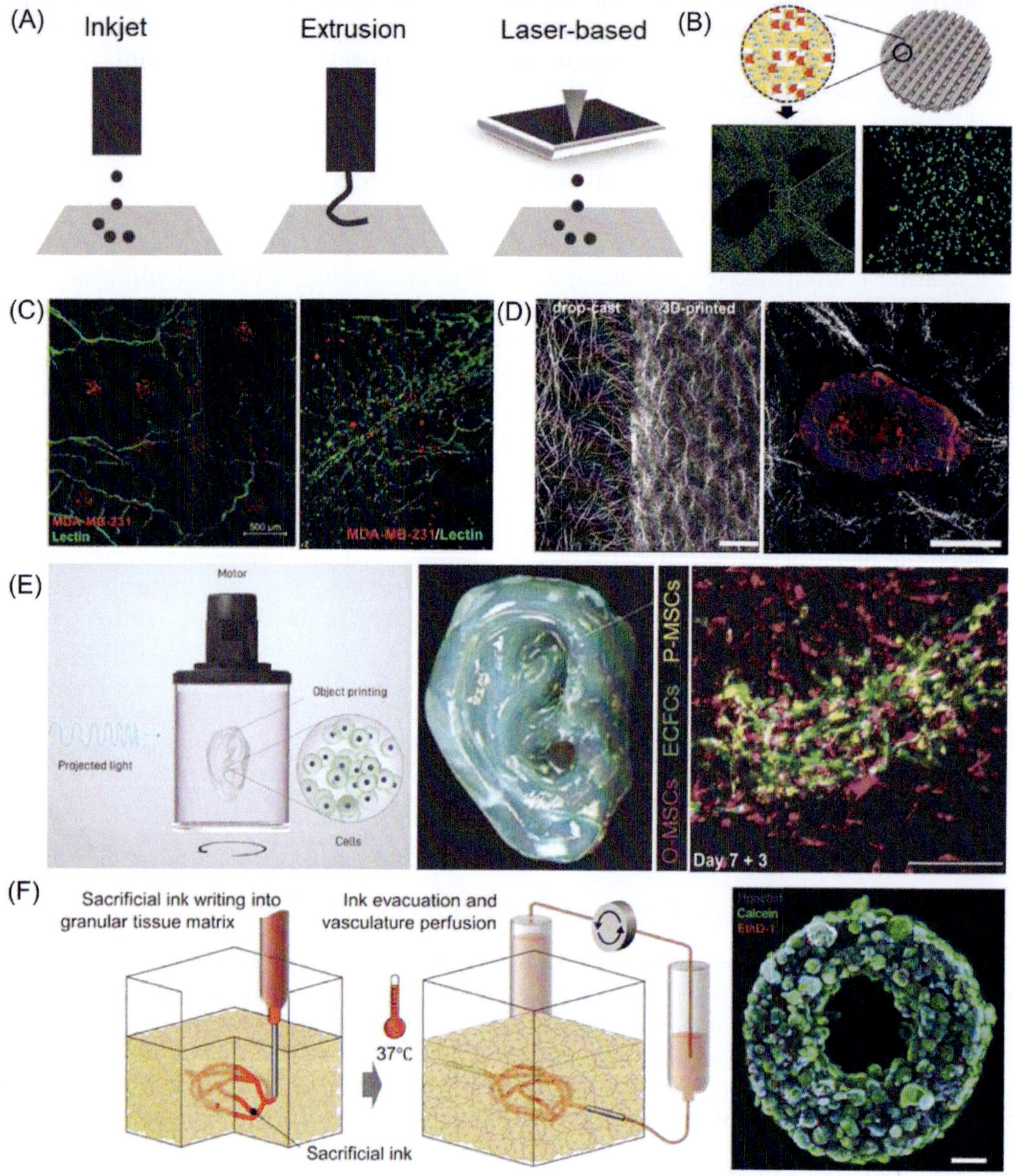


FIGURE 17.5 The 3D bioprinting methods used in breast cancer research and recent alternative biomanufacturing methods. (A) Inkjet-, microextrusion-, and laser-based bioprinting are the most utilized bioprinting methods. (B) Bioprinted breast cancer 3D model by microextrusion. (C) Bioprinted breast cancer cells by laser-based bioprinting showing viable cells. (D) Bioprinted breast cancer cells in collagen-based matrices with aligned fibers. (E) Volumetric and (F) SWIFT biomanufacturing methods may be used in future to develop breast cancer 3D models. Source: *(A) Images adapted with permission from Murphy SV, Atala A. 3D bioprinting of tissues and organs. Nat Biotechnol 2014;32:773–85 [86]. (B) Images adapted with permission from Zhou X, Zhu W, Nowicki M, Miao S, Cui H, Holmes B, et al. 3D bioprinting a cell-laden bone matrix for breast cancer metastasis study. ACS Appl Mater Interfaces*

release from liposomes [77]. Using the MCF10A benign mammary epithelial model, they illustrated that these hydrogels could model the mechanical progression of breast cancer by first allowing polarized acini to form in a soft environment, and subsequently stiffening the matrix to drive invasion [78].

### 17.3.5 Biomaterial inks for 3D bioprinting

The 3D bioprinting using biomaterial inks demands a suite of additional design considerations that must be met in order to balance cellular function with processing constraints. For example, in extrusion-based printing methods, the ink must be both injectable and retain its structure after printing, which requires precise control over gelation in space and time [79,80]. The gelation procedure must also be cell-compatible, which may be a challenge for light-based postprinting cross-linking strategies [81,82]. During long bioprinting times with cell-laden inks, the ink should also prevent cell settling in the cartridge. In inkjet bioprinting, viscosity and surface tension of the ink regulate droplet formation [81,83].

For extrusion-based bioprinting, the use of shear-thinning, rapidly self-healing hydrogels formed from dynamic bonds can reconcile the paradoxical requirements for injectability, maintenance of cell dispersion in the cartridge, and rapid gelation upon printing. By incorporating a secondary cross-linking strategy that is induced after printing, structural fidelity can be further reinforced, such as in dual-cross-linking strategies developed by the Heilshorn and Burdick groups [84]. The use of a solid support during printing can alleviate the need for immediate ink gelation upon printing, which enables printing of materials such as collagen, whose gelation kinetics can be difficult to control [67]. However, the support must also be sufficiently malleable to print into and readily removed after printing. This has been achieved using the "freeform reversible embedding of suspended hydrogels (FRESH)" method pioneered by the Feinberg group, which employs a gelatin microgel support bath that can be melted by heating to physiological temperature [81,82].

## 17.4 3D bioprinting methods and their suitability for breast cancer in vitro engineering

### 17.4.1 Microextrusion- and laser-induced forward transfer used in breast cancer research

The most utilized bioprinting strategies are microextrusion-, inkjet-, and laser-based bioprinting (Fig. 17.5A) [85,86]. Extrusion or microextrusion bioprinting uses pneumatic

*2016;8:30017–30026 [98]. (C) Images adapted with permission from Phamduy TB, Sweat RS, Azimi MS, Burow ME, Murfee WL, Chrisey DB. Printing cancer cells into intact microvascular networks: a model for investigating cancer cell dynamics during angiogenesis. Integr Biol (Camb) 2015;7:1068–1078 [99]. (D) Images adapted with permission from Nerger BA, Brun PT, Nelson CM. Microextrusion printing cell-laden networks of type i collagen with patterned fiber alignment and geometry. Soft Matter 2019;15:5728–5738 [100]. (E) Images adapted with permission from Bernal PN, Delrot P, Loterie D, Li Y, Malda J, Moser C, et al. Volumetric bioprinting of complex living-tissue constructs within seconds. Adv Mater 2019;31:1904209; Skylar-Scott MA, Uzel SGM, Nam LL, Ahrens JH, Truby RL, Damaraju S, et al. Biomanufacturing of organ-specific tissues with high cellular density and embedded vascular channels. Sci Adv 2019;5: eaaw2459[101,102].*

or mechanical dispensing systems to extrude continuous strands of cell-loaded hydrogels. Contrastingly, inkjet bioprinters use thermal or piezoelectric actuators to create drops of cell-loaded hydrogels in the micrometer range, which are dispensed onto the bioprinting stage in a drop-by-drop and layer-by-layer fashion. Laser-assisted bioprinters use a laser pulse to generate drops of cell-loaded hydrogels in the nanometer range, which settle onto the bioprinting stage similarly to inkjet printing.

Before bioprinting was applied to breast cancer research, studies were focused on cell viability after printing [87,88], biomaterial development for bioprinting (bioinks) [89], and cell-biomaterial interactions, such as how bioinks influence cell morphology and differentiation [90–93], as well as if it is feasible to generate vasculogenesis in bioprintable bioinks [94,95].

Tumor microenvironments are highly heterogeneous and complex [85,96]. The microenvironment surrounding breast cancer tissue contains a stromal component or ECM (mostly containing type I collagen), which can express varying stiffness and density depending on the stage of cancer, the presence of different cell types (fibroblasts, adipocytes, endothelial cells, macrophages, and cancer cells) and signaling biomolecules. Breast cancer in vitro models can be assembled by 3D bioprinting with a certain degree of spatial control to replicate physiological cell–ECM interactions and the ECM microarchitecture [97]. Moreover, 3D vascularization can be included in the design of these 3D models. A recent study utilized extrusion bioprinting to model interactions between breast cancer cells and bone stromal cells (Fig. 17.5B) [98]. Gelatin methacrylate hydrogels were bioprinted with mesenchymal stromal cells and co-cultures were initiated with breast cancer cells. The authors showed increased vascular endothelial growth factor (VEGF) secretion and increased cancer cell growth in co-cultures. However, these 3D models exhibited neither resembling tumor microarchitecture nor the formation of vascular tubes. In a further study, laser-assisted bioprinting was used to deposit breast cancer cells and fibroblasts into spatially defined patterns of microvasculature networks (Fig. 17.5C) [99]. Cancer cells were spotted possibly interacting with blood and lymphatic vessels. However, a limitation of this study was that the printed platform did not mimic a tumor–stroma environment. A further study by Nerger et al. showed that it is possible to use extrusion bioprinting to create aligned collagen fiber networks as that observed in breast cancer ECM (Fig. 17.5D) [100]. This type of platform could be used for future drug efficacy studies in breast cancer research. Several other works on bioprinting breast cancer 3D models showing moderate recapitulation of the tumor microenvironments are listed in Table 17.2.

Microextrusion bioprinting of decellularized ECMs can offer control over spatially varying fiber organization that cannot be readily achieved in other in vitro 3D models. Using collagen I/Matrigel bioinks, the Nelson group demonstrated that collagen fiber organization could be patterned along the printing direction. Unlike magnetic alignment of collagen fibers along a single axis, this enabled spatially varying collagen fiber alignment that was dictated by the bioprinting path. They showed that these 3D bioprinted collagen I/Matrigel constructs elicited migration of MDA-MB-231 cells and elongation of normal mammary epithelial clusters along the local ordering of collagen fibers [100].

**TABLE 17.2** List of 3D bioprinting methods used in breast cancer research.

| Bioprinting method | Study | Cells/material | Observations | Selected works (Refs.) |
|---|---|---|---|---|
| Extrusion and microextrusion | Cancer metastasis | BrCa/MSCGelMA | Enhanced VEGF secretion | [98] |
| | | MDA-MB-231/Fib alginate/gelatin | Enhanced proliferation and metabolic activity | [118] |
| | | MCF-7/MCF-12/MDA-MB-468 decellularized mammary ECM | MCF-12 cells invaded surrounding ECM; MCF-7/MDA-MB-468 did not | [119] |
| | Drug efficacy | Epithelial cell linesalginate/Matrigel | Enhanced resistance to paclitaxel in cell spheroids compared with single cell culture | [120] |
| | | MCF-7/HUVECs/Fib alginate/gelatin/Matrigel | Enhanced resistance to doxorubicin in 3D models compared with 2D cultures | [104] |
| | Tumorigenesis | MDA-MB-231/MCF-7 Collagen type I | Decreased 5-hydroxymethylcytosine levels in bioprinted tumoroids | [121] |
| | Mammary duct modeling | Epithelial cell lines GelMA/agarose | Viable cancer cells and COL IV expression | [122] |
| | Cancer microenvironment modeling | MDA-MB-231 Collagen type I/Matrigel | Cancer cells aligned in the direction of collagen fiber networks | [100] |
| Laser-direct writing | Cancer metastasis | MDA-MB-231/adipocytes alginate/collage type I | Enhanced cancer cell proliferation | [123] |
| | | MDA-MB-231/MCF-7/Fib gelatin | Enhanced capillary sprouting compared with nonbioprinted controls | [99] |
| | Cancer microenvironment modeling | MDA-MB-231/ESC alginate/gelatin/chitosan | Viable cancer cells self-assembled to form aggregates in spatially defined cultures | [124] |

## 17.4.2 Volumetric and sacrificial bioprinting as future technologies in cancer research

Current bioprinting methods, such as extrusion and laser-based, have shown already remarkable results, including the generation of biomimetic architectures as that of native tissues and tumor models needed in breast cancer research. Regardless of these outstanding advances, future in vitro models used in cancer research could be improved. Recently, volumetric bioprinting was introduced in the biofabrication and tissue engineering communities as a new, faster, and freeform bioprinting method (Fig. 17.5E) [101]. In contrast to traditional bioprinting methods that are based on the sequential addition of basic building blocks (cell-laden hydrogels or bioinks), volumetric bioprinting is introducing a paradigm shift by creating 3D entire objects at once within seconds. In this new study, the authors have shown that primary human cells (endothelial and mesenchymal stromal cells) survive the bioprinting process. Future works focusing on the biofabrication of breast cancer models could potentially use volumetric bioprinting to create mimicking environments with multimaterial parts, including cancer, endothelial, and stromal cells.

A further bioprinting method that has not yet been utilized in breast cancer research is sacrificial writing into functional tissue (SWIFT, Fig. 17.5F) [102]. SWIFT was introduced recently as an alternative biofabrication process to engineer vascularized 3D tissues and organs at a larger scale than previous bioprinting methods. Skylar-Scott et al. firstly created hundreds of thousands of induced pluripotent stem cell aggregates that were collected in a mold and compacted by centrifugation. Next, a sacrificial ink was patterned in this mold in the shape of a vasculature. Vascular channels were yielded by removal of the sacrificial ink and cell culture media perfused the channels. Despite this method is still in an early stage of development, it could be potentially used in future cancer research. Breast cancer models could be generated by SWIFT using human cancer cell aggregates and stromal supporting cells. The vasculature of the cancer model would be sacrificially patterned by SWIFT.

### 17.4.3 Bioprinting heterotypic cancer models for functional treatment modeling

Bioprinting holds the potential to produce biomimetic microenvironments that reveal functional cancer behaviors that emerge only in the appropriate biophysical, biochemical, and multicellular context. Using glioblastoma-derived decellularized extracellular matrices, Cho et al. bioprinted patient-derived glioblastoma cells and vascular endothelial cells into concentric ring structures on chips that supported radial oxygen gradients. These bioprinted glioblastoma-on-a-chip models reproduced key biochemical and biophysical features of the native glioblastoma microenvironment: a complex brain ECM, the presence of a dysfunctional microvasculature surrounding the tumor, and the existence of oxygen gradients that produce a hypoxic tumor core. These microenvironmental cues synergistically promoted spatially varying gene expression patterns, such as drug resistance SOX2-postive glioblastoma cells near the tumoroid–stromal interface. In turn, this convergence of biomimetic features proved to more faithfully recapitulate drug resistance observed in the original patients compared with 2D culture and 3D spheroid models, as well as bioprinted concentric structures without radial oxygen gradients [103].

Scaffold-free bioprinted structures, in which the extracellular bioink is removed after printing, have also emerged as powerful tools for instructing heterotypic cellular interactions for functional modeling. Using an alginate bioink, the Sears group developed heterotypic solid tumor models consisting of a tumor core surrounded by a stromal compartment. After degradation of the alginate, the bioprinted 3D structured deposited their own ECM. In the case of patient-derived pancreatic cancer cells, this yielded in tumor–stromal morphologies that resembled the origin tissue, exhibited similar levels of Ki67+ proliferative cells, and exhibited chemoresistance to therapeutic doses of gemcitabine. In 3D bioprinted tumor–stromal models of breast cancer cells, 3D bioprinted structures exhibited drug resistance to mTor-targeting therapy that was not observed in 2D monotypic culture, and revealed greater variability among cancer cell proliferation among cell lines derived from distinct breast cancer subtypes [104].

## 17.5 Discussion and outlook

### 17.5.1 Evolution of breast cancer in vitro 3D models: from 2D culture to 3D bioprinting

The breast tumor microenvironment is elaborately structured across hierarchical scales in space and time. 2D monolayer culture methods, which have historically dominated in vitro cell biology, fail to reproduce the spatial relationships that emerge during cancer development. Using 3D culture models, tremendous advancements have been forged in our ability to systematically reconstruct the breast tumor microenvironment in vitro. Biomaterials engineering has occupied a pivotal role in orchestrating these advancements. Naturally derived ECMs have revealed multifaceted roles of the microenvironment in directing breast cancer behavior in 3D, and remain a fertile source for molecular design rules in matrix biology. The marriage of biomaterials, materials science, and synthetic chemistry continues to yield new methods for unprecedented control over 3D microenvironments, offering the opportunity to develop materials that are inspired by nature and supersede the limitations of decellularized matrices.

However, the ECM alone does not fully encapsulate the importance of the third dimension in breast cancer models. A tumor is a community of heterogenous cell types whose communication is governed by spatial organization. This level of organization is not easily reproduced in conventional co-culture systems. Bioprinting offers a powerful means of directing this hierarchical assembly from the top-down. It holds the potential to harness the burgeoning library of engineered biomaterials that mimic the tumor ECM and also deliver precisely controlled assembly of the structural and cellular components that constitute the tumor microenvironment.

### 17.5.2 Advantages and challenges of 3D bioprinting in breast cancer research

Since the last decade there is an increasing trend to improve the biomimicry and complexity of in vitro models used in breast cancer research [104]. Cancer cell spheroids and pioneering 3D cell–hydrogel models placed the first stone toward this goal. Despite major advances of these technologies over simpler 2D monocultures, some aspects remain unanswered and unmet. For example, vascularization is an essential part of an artificial environment used in cancer research [97]. It is through the interaction of cancer cells with the surrounding vasculature that determines the progress and aggressiveness of the cancer [105,106]. Previous 2D cultures and ancient 3D spheroid cultures mostly lacked functional 3D vasculatures. Owing to its spatial control and automation, 3D bioprinting is a fitting technology that may be used to create fully functional vascular networks in in vitro breast cancer models with precise architecture and biological mimicry. A further benefit of using 3D bioprinting in breast cancer research is the possibility to generate in vitro models on-demand and in a high-throughput manner. This will allow these models to be useful not only bedside in personalized medicine, but also benchside in drug development as alternatives to current preclinical testing methods.

Despite the encouraging advantages expected from using 3D bioprinting in breast cancer research, there are also some challenges that may arise. Like in any other method used in a tissue culture laboratory, 3D bioprinting requires high cell densities to be encapsulated in the bioinks in order to generate functional tissue analogues. Thus, there is a need for new cancer cell isolation methods with increased yield. An interesting approach would be to isolate and expand cancer cells from biopsies directly in 3D matrices or dynamic bioreactors with similar mechanical properties as the native tumor and without the need to plate them in 2D culture flasks [107]. This could potentially help increasing cell density after isolation with the added benefit of not interfering with the mechanics surrounding the freshly isolated cells.

### 17.5.3 3D bioprinting in personalized breast cancer research and clinical treatment prognosis

Perhaps the most promising and exciting virtue of 3D bioprinting used in breast cancer research is the option to generate individualized models that fit the genotypic and phenotypic characteristics of individual patients, including health status, age, gender, and cancer stage. Recent advances in biomaterial and bioink development (Fig. 17.6) will certainly contribute to upgrade the mechanics of the tissue models from static to dynamic as that of native tumors in varying stages [108,109]. The orientation of cells and ECM components in bioprinted models may be designed to match the original tumor microstructure and phenotype on a single basis for each patient [110]. The efficacy of multiple drugs for the

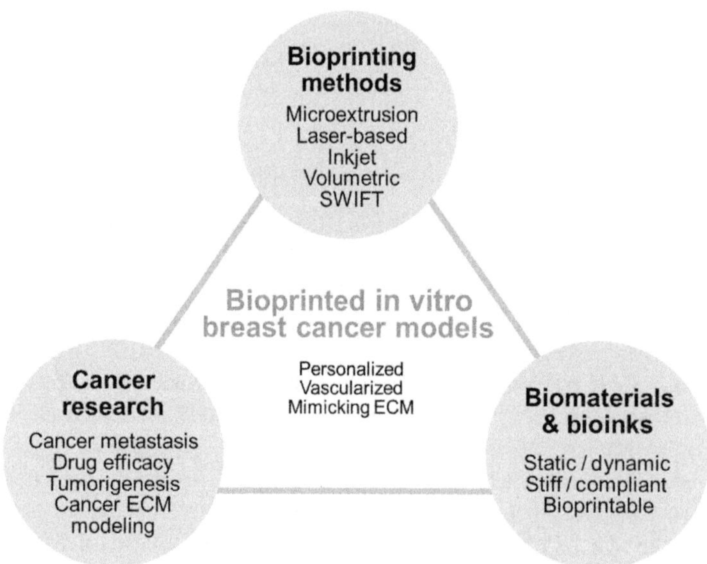


FIGURE 17.6 Interplay between bioprinting methods, biomaterials and bioinks, and cancer research. Enabling technologies, such as bioprinting, combined with biomaterials used in breast cancer research allow for the development of in vitro breast cancer 3D models with patient-specific characteristics (personalized), vasculature, and with mimicking multicellular and multimaterial ECM architectures.

treatment of one patient's cancer may be tested in multiple models in simultaneous, and the most effective therapy may be chosen individually.

## References


[1] van Luijt PA, Heijnsdijk EAM, Fracheboud J, Overbeek LIH, Broeders MJM, Wesseling J, et al. The distribution of ductal carcinoma in situ (DCIS) grade in 4232 women and its impact on overdiagnosis in breast cancer screening. Breast Cancer Res 2016;18:47. Available from: https://doi.org/10.1186/s13058-016-0705-5.

[2] Weigelt B, Reis-Filho JS. Histological and molecular types of breast cancer: is there a unifying taxonomy? Nat Rev Clin Oncol 2009;6:718–30. Available from: https://doi.org/10.1038/nrclinonc.2009.166.

[3] Mao X, He J, Li T, Lu Z, Sun J, Meng Y, et al. Application of imaging mass spectrometry for the molecular diagnosis of human breast tumors. Sci Rep 2016;6:21043. Available from: https://doi.org/10.1038/srep21043.

[4] Clark AG, Vignjevic DM. Modes of cancer cell invasion and the role of the microenvironment. Curr Opin Cell Biol 2015;36:13–22. Available from: https://doi.org/10.1016/J.CEB.2015.06.004.

[5] Soysal SD, Tzankov A, Muenst SE. Role of the tumor microenvironment in breast cancer. Pathobiology 2015;82:142–52. Available from: https://doi.org/10.1159/000430499.

[6] Place AE, Jin Huh S, Polyak K. The microenvironment in breast cancer progression: biology and implications for treatment. Breast Cancer Res 2011;13:227. Available from: https://doi.org/10.1186/bcr2912.

[7] Provenzano PP, Eliceiri KW, Campbell JM, Inman DR, White JG, Keely PJ. Collagen reorganization at the tumor-stromal interface facilitates local invasion. BMC Med 2006;4:38. Available from: https://doi.org/10.1186/1741-7015-4-38.

[8] Pavithra V, Sowmya S, Rao RS, Patil S, Augustine D, Haragannavar VC, et al. Tumor-associated collagen signatures: an insight. World J Dent 2017;8:224–30. Available from: https://doi.org/10.5005/jp-journals-10015-1442.

[9] Acerbi I, Cassereau L, Dean I, Shi Q, Au A, Park C, et al. Human breast cancer invasion and aggression correlates with ECM stiffening and immune cell infiltration. Integr Biol (Camb) 2015;7:1120–34. Available from: https://doi.org/10.1039/c5ib00040h.

[10] Natal RA, Vassallo J, Paiva GR, Pelegati VB, Barbosa GO, Mendonça GR, et al. Collagen analysis by second-harmonic generation microscopy predicts outcome of luminal breast cancer. Tumor Biol 2018;40. Available from: https://doi.org/10.1177/1010428318770953.

[11] Tilbury K, Campagnola PJ. Applications of second-harmonic generation imaging microscopy in ovarian and breast cancer. Perspect Med Chem 2015;7:21–32. Available from: https://doi.org/10.4137/PMC.S13214.

[12] Okoro C, Kelkar V, Sivaguru M, Emmadi R, Toussaint KC. Second-harmonic patterned polarization-analyzed reflection confocal microscopy of stromal collagen in benign and malignant breast tissues. Sci Rep 2018;8. Available from: https://doi.org/10.1038/s41598-018-34693-0.

[13] Gioiella F, Urciuolo F, Imparato G, Brancato V, Netti PA. An engineered breast cancer model on a chip to replicate ECM-activation in vitro during tumor progression. Adv Healthc Mater 2016;5:3074–84. Available from: https://doi.org/10.1002/adhm.201600772.

[14] Petersen OW, Ronnov-Jessen L, Howlett AR, Bissell MJ. Interaction with basement membrane serves to rapidly distinguish growth and differentiation pattern of normal and malignant human breast epithelial cells. Proc Natl Acad Sci USA 1992;89:9064–8. Available from: https://doi.org/10.1073/pnas.89.19.9064.

[15] Weigelt B, Bissell MJ. Unraveling the microenvironmental influences on the normal mammary gland and breast cancer. Semin Cancer Biol 2008;18:311–21. Available from: https://doi.org/10.1016/j.semcancer.2008.03.013.

[16] Debnath J, Mills KR, Collins NL, Reginato MJ, Muthuswamy SK, Brugge JS. The role of apoptosis in creating and maintaining luminal space within normal and oncogene-expressing mammary acini. Cell. 2002;111:29–40. Available from: https://doi.org/10.1016/S0092-8674(02)01001-2.

[17] Weaver VM, Petersen OW, Wang F, Larabell CA, Briand P, Damsky C, et al. Reversion of the malignant phenotype of human breast cells in three-dimensional culture and in vivo by integrin blocking antibodies. J Cell Biol 1997;137:231–45. Available from: https://doi.org/10.1083/jcb.137.1.231.

[18] Doyle AD, Petrie RJ, Kutys ML, Yamada KM. Dimensions in cell migration. Curr Opin Cell Biol 2013;25:642–9. Available from: https://doi.org/10.1016/j.ceb.2013.06.004.

[19] Friedl P, Wolf K. Plasticity of cell migration: a multiscale tuning model. J Cell Biol 2010;188:11–19. Available from: https://doi.org/10.1083/jcb.200909003.

[20] Wolf K, te Lindert M, Krause M, Alexander S, te Riet J, Willis AL, et al. Physical limits of cell migration: control by ECM space and nuclear deformation and tuning by proteolysis and traction force. J Cell Biol 2013;201:1069–84. Available from: https://doi.org/10.1083/jcb.201210152.

[21] Doyle AD, Carvajal N, Jin A, Matsumoto K, Yamada KM. Local 3D matrix microenvironment regulates cell migration through spatiotemporal dynamics of contractility-dependent adhesions. Nat Commun 2015;6:8720. Available from: https://doi.org/10.1038/ncomms9720.

[22] Doyle AD, Wang FW, Matsumoto K, Yamada KM. One-dimensional topography underlies three-dimensional fi brillar cell migration. J Cell Biol 2009;184:481–90. Available from: https://doi.org/10.1083/jcb.200810041.

[23] Wang WY, Davidson CD, Lin D, Baker BM. Actomyosin contractility-dependent matrix stretch and recoil induces rapid cell migration. Nat Commun 2019;10:1186. Available from: https://doi.org/10.1038/s41467-019-09121-0.

[24] Ilina O, Campanello L, Gritsenko PG, Vullings M, Wang C, Bult P, et al. Intravital microscopy of collective invasion plasticity in breast cancer. Dis Model Mech 2018;11:dmm03433. Available from: https://doi.org/10.1242/dmm.034330.

[25] Cheung KJ, Padmanaban V, Silvestri V, Schipper K, Cohen JD, Fairchild AN, et al. Polyclonal breast cancer metastases arise from collective dissemination of keratin 14-expressing tumor cell clusters. Proc Natl Acad Sci USA 2016;113:E854–63. Available from: https://doi.org/10.1073/pnas.1508541113.

[26] Cheung KJ, Gabrielson E, Werb Z, Ewald AJ. Collective invasion in breast cancer requires a conserved basal epithelial program. Cell 2013;155:1639–51. Available from: https://doi.org/10.1016/j.cell.2013.11.029.

[27] Hegerfeldt Y, Tusch M, Bröcker EB, Friedl P. Collective cell movement in primary melanoma explants: plasticity of cell-cell interaction, β1-integrin function, and migration strategies. Cancer Res 2002;62:2125–30.

[28] Nguyen-Ngoc KV, Cheung KJ, Brenot A, Shamir ER, Gray RS, Hines WC, et al. ECM microenvironment regulates collective migration and local dissemination in normal and malignant mammary epithelium. Proc Natl Acad Sci USA 2012;109:E2595–604. Available from: https://doi.org/10.1073/pnas.1212834109.

[29] Sachs N, de Ligt J, Kopper O, Gogola E, Bounova G, Weeber F, et al. A living biobank of breast cancer organoids captures disease heterogeneity. Cell. 2018;172:373–86. Available from: https://doi.org/10.1016/j.cell.2017.11.010.

[30] Debnath J, Muthuswamy SK, Brugge JS. Morphogenesis and oncogenesis of MCF-10A mammary epithelial acini grown in three-dimensional basement membrane cultures. Methods 2003;30:256–68. Available from: https://doi.org/10.1016/S1046-2023(03)00032-X.

[31] Shi Q, Ghosh RP, Engelke H, Rycroft CH, Cassereau L, Sethian JA, et al. Rapid disorganization of mechanically interacting systems of mammary acini. Proc Natl Acad Sci USA 2014;111:658–63. Available from: https://doi.org/10.1073/pnas.1311312110.

[32] Mekhdjian AH, Kai FB, Rubashkin MG, Prahl LS, Przybyla LM, McGregor AL, et al. Integrin-mediated traction force enhances paxillin molecular associations and adhesion dynamics that increase the invasiveness of tumor cells into a three-dimensional extracellular matrix. Mol Biol Cell 2017;28:1467–88. Available from: https://doi.org/10.1091/mbc.E16-09-0654.

[33] Ivascu A, Kubbies M. Rapid generation of single-tumor spheroids for high-throughput cell function and toxicity analysis. J Biomol Screen 2006;11:922–32. Available from: https://doi.org/10.1177/1087057106292763.

[34] Gong X, Lin C, Cheng J, Su J, Zhao H, Liu T, et al. Generation of multicellular tumor spheroids with microwell-based agarose scaffolds for drug testing. PLoS One 2015;10:e0130348. Available from: https://doi.org/10.1371/journal.pone.0130348.

[35] Del Duca D, Werbowetski T, Del Maestro RF. Spheroid preparation from hanging drops: characterization of a model of brain tumor invasion. J Neurooncol 2004;67:295–303. Available from: https://doi.org/10.1023/B:NEON.0000024220.07063.70.

[36] Carey SP, Starchenko A, McGregor AL, Reinhart-King CA. Leading malignant cells initiate collective epithelial cell invasion in a three-dimensional heterotypic tumor spheroid model. Clin Exp Metastasis 2013;30:615–30. Available from: https://doi.org/10.1007/s10585-013-9565-x.

[37] Hu M, Yao J, Carroll DK, Weremowicz S, Chen H, Carrasco D, et al. Regulation of in situ to invasive breast carcinoma transition. Cancer Cell 2008;13:394–406. Available from: https://doi.org/10.1016/j.ccr.2008.03.007.

[38] Levental KR, Yu H, Kass L, Lakins JN, Egeblad M, Erler JT, et al. Matrix crosslinking forces tumor progression by enhancing integrin signaling. Cell. 2009;139:891–906. Available from: https://doi.org/10.1016/j.cell.2009.10.027.

[39] Wolf K, Wu YI, Liu Y, Geiger J, Tam E, Overall C, et al. Multi-step pericellular proteolysis controls the transition from individual to collective cancer cell invasion. Nat Cell Biol 2007;9:893–904. Available from: https://doi.org/10.1038/ncb1616.

[40] Calvo F, Ege N, Grande-Garcia A, Hooper S, Jenkins RP, Chaudhry SI, et al. Mechanotransduction and YAP-dependent matrix remodelling is required for the generation and maintenance of cancer-associated fibroblasts. Nat Cell Biol 2013;15:637–46. Available from: https://doi.org/10.1038/ncb2756.

[41] Weidner N, Folkman J, Pozza F, Bevilacqua P, Allred EN, Moore DH, et al. Tumor angiogenesis: a new significant and independent prognostic indicator in early-stage breast carcinoma. JNCI: J Natl Cancer Inst 1992;84:1875–87. Available from: https://doi.org/10.1093/jnci/84.24.1875.

[42] Gaggioli C, Hooper S, Hidalgo-Carcedo C, Grosse R, Marshall JF, Harrington K, et al. Fibroblast-led collective invasion of carcinoma cells with differing roles for RhoGTPases in leading and following cells. Nat Cell Biol 2007;9:1392–400. Available from: https://doi.org/10.1038/ncb1658.

[43] Szot CS, Buchanan CF, Freeman JW, Rylander MN. In vitro angiogenesis induced by tumor-endothelial cell co-culture in bilayered, collagen I hydrogel bioengineered tumors. Tissue Eng Part C Methods 2013;19:864–74. Available from: https://doi.org/10.1089/ten.TEC.2012.0684.

[44] Correa de Sampaio P, Auslaender D, Krubasik D, Failla AV, Skepper JN, Murphy G, et al. In vitro three dimensional model of tumour-stroma interactions regulating sprouting angiogenesis. PLoS One 2012;7:e30753. Available from: https://doi.org/10.1371/journal.pone.0030753.

[45] Labernadie A, Kato T, Brugués A, Serra-Picamal X, Derzsi S, Arwert E, et al. A mechanically active heterotypic E-cadherin/N-cadherin adhesion enables fibroblasts to drive cancer cell invasion. Nat Cell Biol 2017;19:224–37. Available from: https://doi.org/10.1038/ncb3478.

[46] Zervantonakis IK, Hughes-Alford SK, Charest JL, Condeelis JS, Gertler FB, Kamm RD. Three-dimensional microfluidic model for tumor cell intravasation and endothelial barrier function. Proc Natl Acad Sci USA 2012;109:13515–20. Available from: https://doi.org/10.1073/pnas.1210182109.

[47] Todhunter ME, Jee NY, Hughes AJ, Coyle MC, Cerchiari A, Farlow J, et al. Programmed synthesis of three--dimensional tissues. Nat Methods 2015;12:975–81. Available from: https://doi.org/10.1038/nmeth.3553.

[48] Neal JT, Li X, Zhu J, Giangarra V, Grzeskowiak CL, Ju J, et al. Organoid modeling of the tumor immune microenvironment. Cell 2018;175:1972–88. Available from: https://doi.org/10.1016/j.cell.2018.11.021.

[49] Kalluri R, Zeisberg M. Fibroblasts in cancer. Nat Rev Cancer 2006;6:392–401. Available from: https://doi.org/10.1038/nrc1877.

[50] Hashizume H, Baluk P, Morikawa S, McLean JW, Thurston G, Roberge S, et al. Openings between defective endothelial cells explain tumor vessel leakiness. Am J Pathol 2000;156:1363–80. Available from: https://doi.org/10.1016/S0002-9440(10)65006-7.

[51] Wyckoff JB, Wang Y, Lin EY, Li JF, Goswami S, Stanley ER, et al. Direct visualization of macrophage-assisted tumor cell intravasation in mammary tumors. Cancer Res 2007;67:2649–56. Available from: https://doi.org/10.1158/0008-5472.CAN-06-1823.

[52] Ghajar CM, Peinado H, Mori H, Matei IR, Evason KJ, Brazier H, et al. The perivascular niche regulates breast tumour dormancy. Nat Cell Biol 2013;15:807–17. Available from: https://doi.org/10.1038/ncb2767.

[53] Kratochvil MJ, Seymour AJ, Li TL, Paşca SP, Kuo CJ, Heilshorn SC. Engineered materials for organoid systems. Nat Rev Mater 2019;4:606–22. Available from: https://doi.org/10.1038/s41578-019-0129-9.

[54] Lee JY, Chaudhuri O. Regulation of breast cancer progression by extracellular matrix mechanics: insights from 3D culture models. ACS Biomater Sci Eng 2018;4:302–13. Available from: https://doi.org/10.1021/acsbiomaterials.7b00071.

[55] Gjorevski N, Sachs N, Manfrin A, Giger S, Bragina ME, Ordóñez-Morán P, et al. Designer matrices for intestinal stem cell and organoid culture. Nature. 2016;539:560–4. Available from: https://doi.org/10.1038/nature20168.

[56] Cruz-Acuña R, Quirós M, Farkas AE, Dedhia PH, Huang S, Siuda D, et al. Synthetic hydrogels for human intestinal organoid generation and colonic wound repair. Nat Cell Biol 2017;19:1326–35. Available from: https://doi.org/10.1038/ncb3632.

[57] Lou J, Stowers R, Nam S, Xia Y, Chaudhuri O. Stress relaxing hyaluronic acid-collagen hydrogels promote cell spreading, fiber remodeling, and focal adhesion formation in 3D cell culture. Biomaterials. 2018;154:213–22. Available from: https://doi.org/10.1016/j.biomaterials.2017.11.004.
[58] Chaudhuri O, Koshy ST, Da Cunha C Branco, Shin JW, Verbeke CS, Allison KH, et al. Extracellular matrix stiffness and composition jointly regulate the induction of malignant phenotypes in mammary epithelium. Nat Mater 2014;13:970–8. Available from: https://doi.org/10.1038/nmat4009.
[59] Cassereau L, Miroshnikova YA, Ou G, Lakins J, Weaver VM. A 3D tension bioreactor platform to study the interplay between ECM stiffness and tumor phenotype. J Biotechnol 2015;193:66–9. Available from: https://doi.org/10.1016/j.jbiotec.2014.11.008.
[60] Lutolf MP, Raeber GP, Zisch AH, Tirelli N, Hubbell JA. Cell-responsive synthetic hydrogels. Adv Mater 2003;15:888–92. Available from: https://doi.org/10.1002/adma.200304621.
[61] Rowley JA, Madlambayan G, Mooney DJ. Alginate hydrogels as synthetic extracellular matrix materials. Biomaterials. 1999;20:45–53. Available from: https://doi.org/10.1016/S0142-9612(98)00107-0.
[62] Chaudhuri O, Gu L, Klumpers D, Darnell M, Bencherif SA, Weaver JC, et al. Hydrogels with tunable stress relaxation regulate stem cell fate and activity. Nat Mater 2016;15:326–34. Available from: https://doi.org/10.1038/nmat4489.
[63] Wisdom KM, Adebowale K, Chang J, Lee JY, Nam S, Desai R, et al. Matrix mechanical plasticity regulates cancer cell migration through confining microenvironments. Nat Commun 2018;9. Available from: https://doi.org/10.1038/s41467-018-06641-z.
[64] Auvinen P, Tammi R, Parkkinen J, Tammi M, Ågren U, Johansson R, et al. Hyaluronan in peritumoral stroma and malignant cells associates with breast cancer spreading and predicts survival. Am J Pathol 2000;156:529–36. Available from: https://doi.org/10.1016/S0002-9440(10)64757-8.
[65] Burdick JA, Prestwich GD. Hyaluronic acid hydrogels for biomedical applications. Adv Mater 2011;23:46. Available from: https://doi.org/10.1002/adma.201003963.
[66] Paszek MJ, Zahir N, Johnson KR, Lakins JN, Rozenberg GI, Gefen A, et al. Tensional homeostasis and the malignant phenotype. Cancer Cell 2005;8:241–54. Available from: https://doi.org/10.1016/j.ccr.2005.08.010.
[67] Nowak M, Freudenberg U, Tsurkan MV, Werner C, Levental KR. Modular GAG-matrices to promote mammary epithelial morphogenesis in vitro. Biomaterials. 2017;112:20–30. Available from: https://doi.org/10.1016/j.biomaterials.2016.10.007.
[68] Nam S, Hu KH, Butte MJ, Chaudhuri O. Strain-enhanced stress relaxation impacts nonlinear elasticity in collagen gels. Proc Natl Acad Sci U S A 2016;113:5492–7. Available from: https://doi.org/10.1073/pnas.1523906113.
[69] Nam S, Lee J, Brownfield DG, Chaudhuri O. Viscoplasticity enables mechanical remodeling of matrix by cells. Biophys J 2016;111:2296–308. Available from: https://doi.org/10.1016/j.bpj.2016.10.002.
[70] Licup AJ, Münster S, Sharma A, Sheinman M, Jawerth LM, Fabry B, et al. Stress controls the mechanics of collagen networks. Proc Natl Acad Sci U S A 2015;112:9573–8. Available from: https://doi.org/10.1073/pnas.1504258112.
[71] Wisdom KM, Indana D, Chou PE, Desai R, Kim T, Chaudhuri O. Covalent cross-linking of basement membrane-like matrices physically restricts invasive protrusions in breast cancer cells. Matrix Biol 2019;85–86:94. Available from: https://doi.org/10.1016/j.matbio.2019.05.006.
[72] Rosales AM, Anseth KS. The design of reversible hydrogels to capture extracellular matrix dynamics. Nat Rev Mater 2016;1:15012. Available from: https://doi.org/10.1038/natrevmats.2015.12.
[73] Wang H, Heilshorn SC, Wang H, Heilshorn SC. Adaptable hydrogel networks with reversible linkages for tissue engineering. Adv Mater 2015;27:3710. Available from: https://doi.org/10.1002/adma.201501558.
[74] Das RK, Gocheva V, Hammink R, Zouani OF, Rowan AE. Stress-stiffening-mediated stem-cell commitment switch in soft responsive hydrogels. Nat Mater 2016;15:318–25. Available from: https://doi.org/10.1038/nmat4483.
[75] DeForest CA, Anseth KS. Cytocompatible click-based hydrogels with dynamically tunable properties through orthogonal photoconjugation and photocleavage reactions. Nat Chem 2011;3:925–31. Available from: https://doi.org/10.1038/nchem.1174.
[76] Gandavarapu NR, Azagarsamy MA, Anseth KS. Photo-click living strategy for controlled, reversible exchange of biochemical ligands. Adv Mater 2014;26:2521–6. Available from: https://doi.org/10.1002/adma.201304847.

[77] Stowers RS, Allen SC, Suggs LJ, Anseth KS. Dynamic phototuning of 3D hydrogel stiffness. Proc Natl Acad Sci U S A 2015;112:1953–8. Available from: https://doi.org/10.1073/pnas.1421897112.
[78] Stowers RS, Allen SC, Sanchez K, Davis CL, Ebelt ND, Van Den Berg C, et al. Extracellular matrix stiffening induces a malignant phenotypic transition in breast epithelial cells. Cell Mol Bioeng 2017;10:114–23. Available from: https://doi.org/10.1007/s12195-016-0468-1.
[79] Lee JM, Yeong WY. Design and printing strategies in 3D bioprinting of cell-hydrogels: a review. Adv Healthc Mater 2016;5:2856–65. Available from: https://doi.org/10.1002/adhm.201600435.
[80] Guvendiren M, Molde J, Soares RMD, Kohn J. Designing biomaterials for 3D printing. ACS Biomater Sci Eng 2016;2:1679–93. Available from: https://doi.org/10.1021/acsbiomaterials.6b00121.
[81] Dubbin K, Hori Y, Lewis KK, Heilshorn SC. Dual-stage crosslinking of a gel-phase bioink improves cell viability and homogeneity for 3D bioprinting. Adv. Healthc. Mater 2016;5:2488–92. Available from: https://doi.org/10.1002/adhm.201600636.
[82] Dubbin K, Tabet A, Heilshorn SC. Quantitative criteria to benchmark new and existing bio-inks for cell compatibility. Biofabrication. 2017;9:044102. Available from: https://doi.org/10.1088/1758-5090/aa869f.
[83] Loebel C, Rodell CB, Chen MH, Burdick JA. Shear-thinning and self-healing hydrogels as injectable therapeutics and for 3D-printing. Nat Protoc 2017;12:1521–41. Available from: https://doi.org/10.1038/nprot.2017.053.
[84] Lee A, Hudson AR, Shiwarski DJ, Tashman JW, Hinton TJ, Yerneni S, et al. 3D bioprinting of collagen to rebuild components of the human heart. Science (80-) 2019;365:482–7. Available from: https://doi.org/10.1126/science.aav9051.
[85] Knowlton S, Onal S, Yu CH, Zhao JJ, Tasoglu S. Bioprinting for cancer research. Trends Biotechnol 2015;33:504–13. Available from: https://doi.org/10.1016/J.TIBTECH.2015.06.007.
[86] Murphy SV, Atala A. 3D bioprinting of tissues and organs. Nat Biotechnol 2014;32:773–85. Available from: https://doi.org/10.1038/nbt.2958.
[87] Duarte Campos DF, Blaeser A, Weber M, Jäkel J, Neuss S, Jahnen-Dechent W, et al. Three-dimensional printing of stem cell-laden hydrogels submerged in a hydrophobic high-density fluid. Biofabrication. 2013;5:015003. Available from: https://doi.org/10.1088/1758-5082/5/1/015003.
[88] Blaeser A, Duarte Campos DF, Puster U, Richtering W, Stevens MM, Fischer H. Controlling shear stress in 3D bioprinting is a key factor to balance printing resolution and stem cell integrity. Adv Healthc Mater 2015;5:326. Available from: https://doi.org/10.1002/adhm.201500677.
[89] Forget A, Blaeser A, Miessmer F, Köpf M, Campos DFD, Voelcker NH, et al. Mechanically tunable bioink for 3D bioprinting of human cells. Adv Healthc Mater 2017;6:1700255. Available from: https://doi.org/10.1002/adhm.201700255.
[90] Duarte Campos DF, Blaeser A, Korsten A, Neuss S, Jäkel J, Vogt M, et al. The stiffness and structure of three-dimensional printed hydrogels direct the differentiation of mesenchymal stromal cells toward adipogenic and osteogenic lineages. Tissue Eng Part A 2015;21:740–56. Available from: https://doi.org/10.1089/ten.TEA.2014.0231.
[91] Duarte Campos DF, Blaeser A, Buellesbach K, Sen KS, Xun W, Tillmann W, et al. Bioprinting organotypic hydrogels with improved mesenchymal stem cell remodeling and mineralization properties for bone tissue engineering. Adv Healthc Mater 2016;5:1136. Available from: https://doi.org/10.1002/adhm.201501033.
[92] Kang H-W, Lee SJ, Ko IK, Kengla C, Yoo JJ, Atala A. A 3D bioprinting system to produce human-scale tissue constructs with structural integrity. Nat Biotechnol 2016;34:312–19. Available from: https://doi.org/10.1038/nbt.3413.
[93] Kumar G, Tison CK, Chatterjee K, Pine PS, McDaniel JH, Salit ML, et al. The determination of stem cell fate by 3D scaffold structures through the control of cell shape. Biomaterials. 2011;32:9188–96. Available from: https://doi.org/10.1016/j.biomaterials.2011.08.054.
[94] Kreimendahl F, Köpf M, Thiebes AL, Duarte Campos DF, Blaeser A, Schmitz-Rode T, et al. Three-dimensional printing and angiogenesis: tailored agarose-type I collagen blends comprise three-dimensional printability and angiogenesis potential for tissue-engineered substitutes. Tissue Eng Part C Methods 2017;23:604–15. Available from: https://doi.org/10.1089/ten.tec.2017.0234.
[95] Kolesky DB, Truby RL, Gladman AS, Busbee TA, Homan KA, Lewis JA. 3D bioprinting of vascularized, heterogeneous cell-laden tissue constructs. Adv Mater 2014;26:3124–30. Available from: https://doi.org/10.1002/adma.201305506.

[96] Belgodere JA, King CT, Bursavich JB, Burow ME, Martin EC, Jung JP. Engineering breast cancer microenvironments and 3D bioprinting. Front Bioeng Biotechnol 2018;6:66. Available from: https://doi.org/10.3389/fbioe.2018.00066.

[97] Zhang YS, Duchamp M, Oklu R, Ellisen LW, Langer R, Khademhosseini A. Bioprinting the cancer microenvironment. ACS Biomater Sci Eng 2016;2:1710–21. Available from: https://doi.org/10.1021/acsbiomaterials.6b00246.

[98] Zhou X, Zhu W, Nowicki M, Miao S, Cui H, Holmes B, et al. 3D bioprinting a cell-laden bone matrix for breast cancer metastasis study. ACS Appl Mater Interfaces 2016;8:30017–26. Available from: https://doi.org/10.1021/acsami.6b10673.

[99] Phamduy TB, Sweat RS, Azimi MS, Burow ME, Murfee WL, Chrisey DB. Printing cancer cells into intact microvascular networks: a model for investigating cancer cell dynamics during angiogenesis. Integr Biol (Camb) 2015;7:1068–78. Available from: https://doi.org/10.1039/c5ib00151j.

[100] Nerger BA, Brun PT, Nelson CM. Microextrusion printing cell-laden networks of type i collagen with patterned fiber alignment and geometry. Soft Matter 2019;15:5728–38. Available from: https://doi.org/10.1039/c8sm02605j.

[101] Bernal PN, Delrot P, Loterie D, Li Y, Malda J, Moser C, et al. Volumetric bioprinting of complex living-tissue constructs within seconds. Adv Mater 2019;31:1904209. Available from: https://doi.org/10.1002/adma.201904209.

[102] Skylar-Scott MA, Uzel SGM, Nam LL, Ahrens JH, Truby RL, Damaraju S, et al. Biomanufacturing of organ-specific tissues with high cellular density and embedded vascular channels. Sci Adv 2019;5:eaaw2459. Available from: https://doi.org/10.1126/sciadv.aaw2459.

[103] Yi HG, Jeong YH, Kim Y, Choi YJ, Moon HE, Park SH, et al. A bioprinted human-glioblastoma-on-a-chip for the identification of patient-specific responses to chemoradiotherapy. Nat Biomed Eng 2019;3:509–19. Available from: https://doi.org/10.1038/s41551-019-0363-x.

[104] Langer EM, Allen-Petersen BL, King SM, Kendsersky ND, Turnidge MA, Kuziel GM, et al. Modeling tumor phenotypes in vitro with three-dimensional bioprinting. Cell Rep 2019;26:608–23. Available from: https://doi.org/10.1016/j.celrep.2018.12.090.

[105] Bray LJ, Binner M, Holzheu A, Friedrichs J, Freudenberg U, Hutmacher DW, et al. Multi-parametric hydrogels support 3D in vitro bioengineered microenvironment models of tumour angiogenesis. Biomaterials. 2015;53:609–20. Available from: https://doi.org/10.1016/j.biomaterials.2015.02.124.

[106] Albritton JL, Miller JS. 3D bioprinting: improving in vitro models of metastasis with heterogeneous tumor microenvironments. Dis Model Mech 2017;10:3–14. Available from: https://doi.org/10.1242/dmm.025049.

[107] Smidsrød O, Skjåk-Braek G. Alginate as immobilization matrix for cells. Trends Biotechnol 1990;8:71–8. Available from: http://www.ncbi.nlm.nih.gov/pubmed/1366500 [accessed 15.07.2015].

[108] Bray LJ, Werner C. Evaluation of three-dimensional in vitro models to study tumor angiogenesis. ACS Biomater Sci Eng 2018;4:337–46. Available from: https://doi.org/10.1021/acsbiomaterials.7b00139.

[109] Holen I, Speirs V, Morrissey B, Blyth K. In vivo models in breast cancer research: progress, challenges and future directions. Dis Model Mech 2017;10:359–71. Available from: https://doi.org/10.1242/dmm.028274.

[110] Betsch M, Cristian C, Lin Y-Y, Blaeser A, Schöneberg J, Vogt M, et al. Incorporating 4D into bioprinting: real-time magnetically directed collagen fiber alignment for generating complex multilayered tissues. Adv Healthc Mater 2018;7:1800894. Available from: https://doi.org/10.1002/adhm.201800894.

[111] Carey SP, Martin KE, Reinhart-King CA. Three-dimensional collagen matrix induces a mechanosensitive invasive epithelial phenotype. Sci Rep 2017;7:42088. Available from: https://doi.org/10.1038/srep42088.

[112] Provenzano PP, Inman DR, Eliceiri KW, Keely PJ. Matrix density-induced mechanoregulation of breast cell phenotype, signaling and gene expression through a FAK-ERK linkage. Oncogene. 2009;28:4326–43. Available from: https://doi.org/10.1038/onc.2009.299.

[113] Nguyen-Ngoc KV, Ewald AJ. Mammary ductal elongation and myoepithelial migration are regulated by the composition of the extracellular matrix. J Microsc 2013;251:212–23. Available from: https://doi.org/10.1111/jmi.12017.

[114] Riching KM, Cox BL, Salick MR, Pehlke C, Riching AS, Ponik SM, et al. 3D collagen alignment limits protrusions to enhance breast cancer cell persistence. Biophys J 2015;107:2546–58. Available from: https://doi.org/10.1016/j.bpj.2014.10.035.

[115] Hinton TJ, Jallerat Q, Palchesko RN, Park JH, Grodzicki MS, Shue HJ, et al. Three-dimensional printing of complex biological structures by freeform reversible embedding of suspended hydrogels. Sci Adv 2015;1: e1500758. Available from: https://doi.org/10.1126/sciadv.1500758.

[116] Weiss MS, Bernabé BP, Shikanov A, Bluver DA, Mui MD, Shin S, et al. The impact of adhesion peptides within hydrogels on the phenotype and signaling of normal and cancerous mammary epithelial cells. Biomaterials. 2012;33:3548–59. Available from: https://doi.org/10.1016/j.biomaterials.2012.01.055.

[117] Miroshnikova YA, Jorgens DM, Spirio L, Auer M, Sarang-Sieminski AL, Weaver VM. Engineering strategies to recapitulate epithelial morphogenesis within synthetic three-dimensional extracellular matrix with tunable mechanical properties. Phys Biol 2011;8:026013. Available from: https://doi.org/10.1088/1478-3975/8/2/026013.

[118] Jiang T, Munguia-Lopez JG, Flores-Torres S, Grant J, Vijayakumar S, De Leon-Rodriguez A, et al. Directing the self-assembly of tumour spheroids by bioprinting cellular heterogeneous models within alginate/gelatin hydrogels. Sci Rep 2017;7:4575. Available from: https://doi.org/10.1038/s41598-017-04691-9.

[119] Mollica PA, Booth-Creech EN, Reid JA, Zamponi M, Sullivan SM, Palmer XL, et al. 3D bioprinted mammary organoids and tumoroids in human mammary derived ECM hydrogels. Acta Biomater 2019;95:201–13. Available from: https://doi.org/10.1016/j.actbio.2019.06.017.

[120] Swaminathan S, Hamid Q, Sun W, Clyne AM. Bioprinting of 3D breast epithelial spheroids for human cancer models. Biofabrication. 2019;11:025003. Available from: https://doi.org/10.1088/1758-5090/aafc49.

[121] Reid JA, Palmer XL, Mollica PA, Northam N, Sachs PC, Bruno RD. A 3D bioprinter platform for mechanistic analysis of tumoroids and chimeric mammary organoids. Sci Rep 2019;9:7466. Available from: https://doi.org/10.1038/s41598-019-43922-z.

[122] Duchamp M, Liu T, Genderen AM, Kappings V, Oklu R, Ellisen LW, et al. Sacrificial bioprinting of a mammary ductal carcinoma model. Biotechnol. J. 2019;14:1700703. Available from: https://doi.org/10.1002/biot.201700703.

[123] Vinson BT, Phamduy TB, Shipman J, Riggs B, Strong AL, Sklare SC, et al. Laser direct-write based fabrication of a spatially-defined, biomimetic construct as a potential model for breast cancer cell invasion into adipose tissue. Biofabrication. 2017;9:025013. Available from: https://doi.org/10.1088/1758-5090/aa6bad.

[124] Kingsley DM, Roberge CL, Rudkouskaya A, Faulkner DE, Barroso M, Intes X, et al. Laser-based 3D bioprinting for spatial and size control of tumor spheroids and embryoid bodies. Acta Biomater 2019;95:357–70. Available from: https://doi.org/10.1016/j.actbio.2019.02.014.

CHAPTER

# 18

# A predictive oncology framework—modeling tumor proliferation using a FEM platform

*Gianpaolo Ruocco, Paolo Caccavale and Maria Valeria De Bonis*

**University of Basilicata, College of Engineering, Modeling and Prototyping Laboratory—ModProLab, Potenza, Italy**

## Chapter points

- In order to apply the virtualization of tumor proliferation, the available procedures for transforming the region of interest of a given patient in a virtual volume are discussed.
- The set of governing equations, forming the analytical model for tumor proliferation, is then solved by the finite element method, with the calculations carried out by using a commercial platform.
- After a validation study with literature data on a hepatocellular carcinoma, a sensitivity analysis is conducted on 12 computational cases to demonstrate the flexibility of the proposed model.
- The values for hallmark properties and source terms (diffusivities and growth rates, as well as drug efficiency and local therapy delivery/availability) are assessed and form the knowledge basis for preliminary studies on computational tumor proliferation and oncological biomaterials behavior.

## 18.1 Introduction

### 18.1.1 A vision of feasible virtualized oncological prognoses

Oncology is the branch of medicine dealing with prevention, diagnosis, treatment, and prognosis of cancer. Cancer deeply involves with the physical, social, psychological, emotional, and financial status of patients and our society as a whole. Cancer treatments may

*Biomaterials for 3D Tumor Modeling*
DOI: https://doi.org/10.1016/B978-0-12-818128-7.00018-6

be inherently aleatory at times and costly; therefore personalized medicine protocols are needed in the coming years, to help optimizing the patient's treatment and minimizing its cost for the health administrations. *Development of personalized medicine protocols would help* dealing with this framework, by *optimizing patient's treatment and minimizing its cost for the health administrations*. But, in oncology, the shift from therapeutic paradigm toward personalized care requires new efforts in research and powerful resources in clinics. The fight to cancer will be one of the five missions in "Horizon Europe"—the next research and innovation EU framework program [1].

Quantitative predictions, when driven by the available diagnostic imaging technologies (like computed tomography, magnetic resonance imaging, and positron emission tomography), can play an increasingly important role in the design and execution of oncology practice, addressing personalized protocols in patient prognosis, treatment optimization, response to therapy and control, surveillance after treatment, and research. These predictions can be targeted toward any given patient, when an innovative approach to cancer system biology is exercised by means of mathematical and computational modeling of tumorigenesis and proliferation, starting directly from said diagnostic imaging. *This chapter* proposes a preliminary approach that *addresses this complex framework of personalized and precision cancer prognosis by means of a commercial software platform*.

### 18.1.2 An engineering approach toward predictive oncology

From a materials science perspective, cancer is a complex, multiscaled material, bearing complex (nonlinear) couplings between the nanoscale (genomics), the microscale (cell signals/molecular biology), and the macroscale (tumor mass dynamics). This variety of scales has been attacked so far in a vast body of literature as reviewed by authors in Refs. [2–4], the most promising so far being the ACGT project (summarized by Stamatakos[5]). Among the proposed cancer mechanisms, the continuum mechanics approach exploiting a set of partial differential equations (PDEs) holds the promise to encompass and embody some of the said couplings. PDE formal platforms and related computation, for *reaction-diffusion modeling paradigms*, were initiated by Gatenby and Gawlinski [6] and were more recently employed in a variety of ways, as for example in the works coauthored by Swanson [7], Anderson [8], and Tang [9].

The present work attempts to bring the application potential forward with respect to the available theory. The continuum mechanics approach has been exploited for an occurrence of avascular cancer (no significant transport of chemical/biological species within the tissues by a macroscopic flow) [10], to follow its histology in space and time, and come up with a number of virtual scenarios, including the outcome of the treatment. This approach [named predictive oncology via engineering models (POEM)] can account for both pharmacokinetics (PK) and pharmacodynamics (PD) modeling: (1) it describes the modality and extent of organism action on drug molecules, yielding as outcomes the concentration distribution in time of the drug in the body sites of interest, providing virtual data to evaluate the properties and efficacy, from the histopathological point of view; (2) it also describes the biological responses to drug action, providing virtual topological and quantitative data to evaluate the biological modifications of the body parts of interest.

An essential tool in analysis and design in a variety of industrial sectors, a commercial finite element method (FEM) solver [11] can be exercised to integrate the governing PDEs, to ensure flexible and computationally robust models, allowing for seamless harmonization with diagnostic imaging and desktop hardware. This kind of flexible and fully dimensional (four dimensional in space and time) models will allow, in new sustainable ways:

- biology researchers to study tissue features (with inherent anisotropies) to explain the onset of proliferation and metastasis;
- oncologists and surgeons to realize virtual scenarios of the disease relating directly with actual patienship;
- health care/diagnostic equipment industry to provide augmented reality and an immersive monitoring experience; and
- pharma industry to explore and exercise the PK/PD of perspective drugs of that particular neoplastic occurrence.

### 18.1.3 The cancer liver: a valuable case study

Among the great variety of avascular cancers, the hepatocellular carcinoma (HCC) has been selected in this work because of its general incidence on total diagnosed neoplasias: almost 6%, that is, the sixth one most widespread. Mortality rate is more alarming: liver cancer is the second cause in the world of cancer death (about 9.1%). On the European scale, mortality due to liver cancer is 3%. As for the expected rates, for 2035 the incidence rate will be +71.45%, with a mortality rate of +74.23% [12].

Liver, the largest organ in our body (Fig. 18.1), is one of the most chemically active organs. More than 500 vital functions have been identified that help maintaining the chemical balance of the body. One of the liver's key roles is to metabolize nutrients (carbohydrates, lipids, and proteins) that it receives from the intestine via the bloodstream and process it into energy and nutrients that the body can use. The liver also absorbs toxic substances from the blood, transforming them into harmless waste products. Liver cells also produce about 500 mL of bile each day that is released into the small intestine to aid the digestion of lipids.

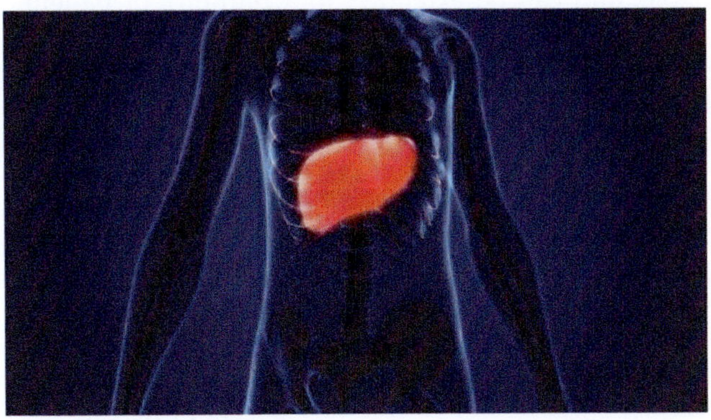

FIGURE 18.1 An artistic rendering of a liver and its position in the human body.

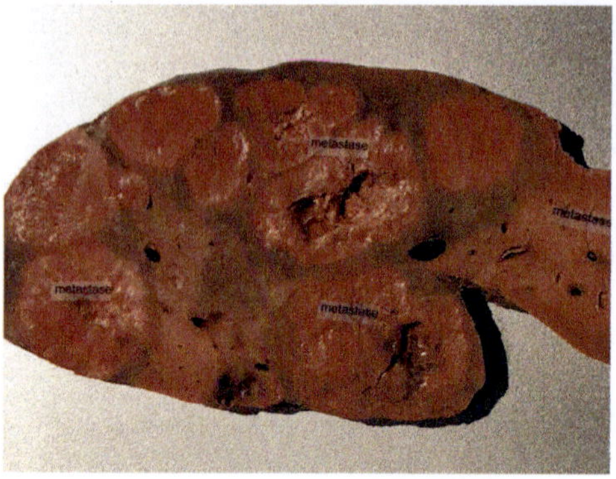

FIGURE 18.2 A realistic reproduction of hepatic metastases and their actual distributions.

Liver is particularly affected by metastases: when a tumor invades the circulatory system, individual cells may break off from the main tumor and enter the bloodstream. They may then stick to the capillary network of other organs as a trapper in a filter. There, they may grow into new tumors (Fig. 18.2). Because of its rich blood supply, the liver is frequently affected by tumors spreading from elsewhere, particularly the bowel or the colon.

## 18.2 A perspective framework of predictive oncology

Stemming from POEM technology, project CAncer Predictions for Therapy and drUg Reposition (CAPTURE) has attracted some interest in the INNOSUP initiative of the EU Horizon 2020 program, in the 2017–19 period [13]. Starting from the available diagnostic images taken from the specific patient, as suggested in a review by Cornelis et al. [14], CAPTURE initiates the histological dynamics bringing forth virtual outcomes of the disease, at any desirable time in the future. CAPTURE can, in principle, test new/repositioned drugs in virtual patients, to reduce the time to clinical deployment. In perspective these models are validable, replicable, and scalable in vertical and horizontal ranges: with finely modified models from a preliminary starting one (as the one reported in the present paper), CAPTURE can deliver for diverse occurrences of cancer having a limited vascularization, when validating data are available, and can be extended to a plurality of patients for the same cancer occurrence.

The workflow of this framework is illustrated in Fig. 18.3, where the various steps and operator concurrences are reported. In the following, the various steps are introduced to some extent, while steps 3 and 4 are explained in detail in Sections 18.3.3 and 18.3.4, respectively.

### 18.2.1 Step 1: Acquisition of diagnostic images

The action thread is initiated by performing a diagnostic imaging technique on the specific patient, in order to create a visual representation of his/her body part, for clinical analysis and intervention. Any current imaging technology, such as radiography, magnetic

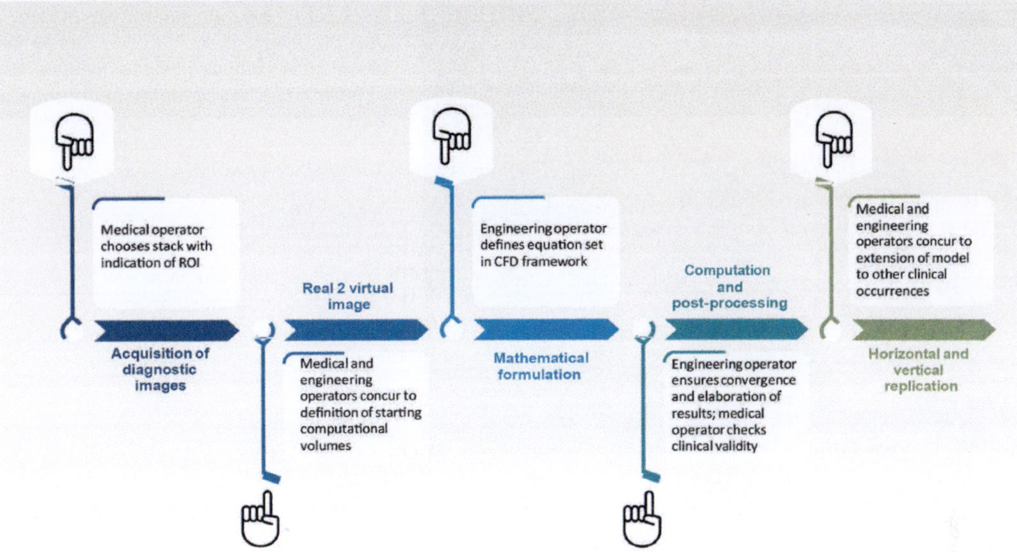


FIGURE 18.3 Five steps for a framework of virtualized tumor proliferation. Each operator's concurrence is also indicated. (1) Acquisition of diagnostic image of any given patient; (2) conversion of diagnostic image to a virtual control volume; (3) mathematical formulation of tumor proliferation and interaction with therapy; (4) computation and postprocessing of results; and (5) exploitation of validated models for more patients, for the same or another cancer occurrences.

resonance, ultrasound, or positron emission tomography yields a stack of images in DICOM format, which is the standard for the communication and management of medical imaging information. One of such images from the clinical stack is reported in Fig. 18.4, for a given Patient *A* affected by HCC. A ROI is highlighted by the white arrow that is where the initial tumor region is found.

Another example is the acquisition of a diagnostic image of a breast, during the scrutiny for adenocarcinoma (Fig. 18.5).

## 18.2.2 Step 2: Real 2 virtual image

In this step, the chosen image is transformed into a dataset which is able to represent, in the computer memory, the real body volume for the specific patient. Two approaches can be employed:

### *18.2.2.1 By using some open-source software*

A possible sequence of operations on software to use is as follows:

- MicroDicom [15] to convert from the *.dcm stack to *.png format

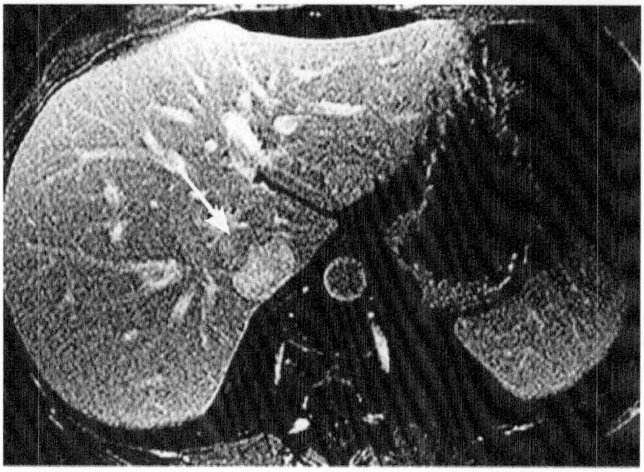

FIGURE 18.4 A sample item from a DICOM image stack. A ROI is highlighted by the arrow initiated by the operator, that is where the initial spreading tumor is found.

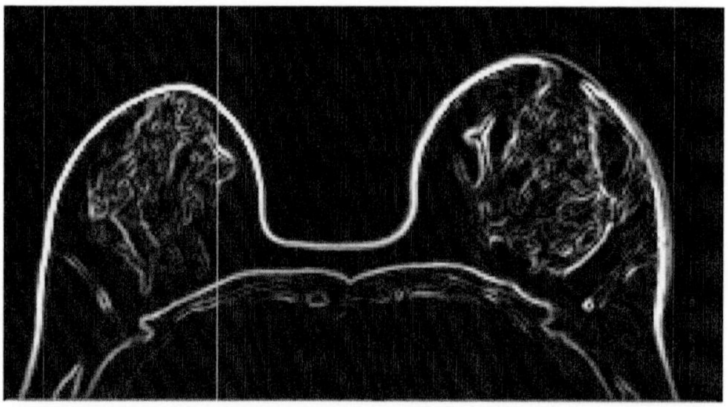

FIGURE 18.5 A sample item from a DICOM image stack.

- ImageJ [16] to convert from *.png to *.txt format
- Rhino3D [17] to convert from *.txt to *.stl format
- Gmsh [18] to convert from *.stl to *.bdf format

Finally, the *.bdf file can be fed into the COMSOL Multiphysics framework [11]. Notably, it is the entire diagnostic stack that is converted in to a detailed representation of the patient body part, ready for the tumor virtualization by the FEM. In need, the geometry can be further manipulated in orientation and detail evidence, such as in Fig. 18.6, in order to increase the augmented reality rendering and advantage the user.

#### *18.2.2.2 By using some proprietary software*

A comprehensive procedure can be performed by using Materialise Mimics [19]. This platform is particularly useful when a detailed tissue segmentation is needed. In Fig. 18.7 the FEM representation of Fig. 18.5 is reported, as an example of this procedure.

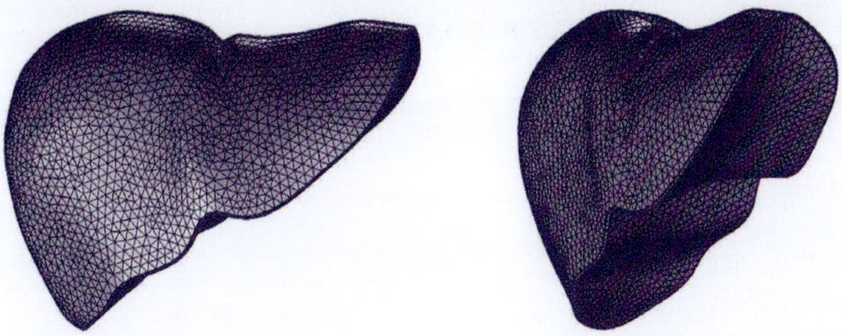

**FIGURE 18.6** Two 3D views of a liver, discretized for a FEM solution.

**FIGURE 18.7** Two 3D views of a breast, discretized for a FEM solution.

## 18.2.3 Step 3: Mathematical formulation

At this point, a mathematical formulation of the biological mass conservation and dynamics can be applied on the virtualized body part, to carry on the task of tumor virtualization. This formulation is directly carried out by the engineer from within the computational framework at hand (as shown in Fig. 18.8 [11]). In particular, a number of governing equations complying the theory of transport phenomena are invoked [10], in particular the transport of dilute species with no macroscopic transport, that is, including Fick's law and every pertinent source terms.

Levels of increasing modeling complexity are generally possible.

### *18.2.3.1 Level 0*

One biological evolving species (the tumor cell population) is enforced, only, following Swanson et al. approach [7].

### *18.2.3.2 Level 1*

An avascular tumor, that is, when no significant mass convection effects are present, is a single-phase biomaterial, featuring sharp boundaries delineating a variety of interacting *biological species* or components [8,20]. Moreover, additional *chemical species* can be administered as a therapy. Up to four competing species are modeled: tumor, necrotic and healthy cells, and therapy drug. Each species is characterized by its own diffusivity and generative/dissipative source terms; moreover, the three biological species enjoy a mass conservation principle.

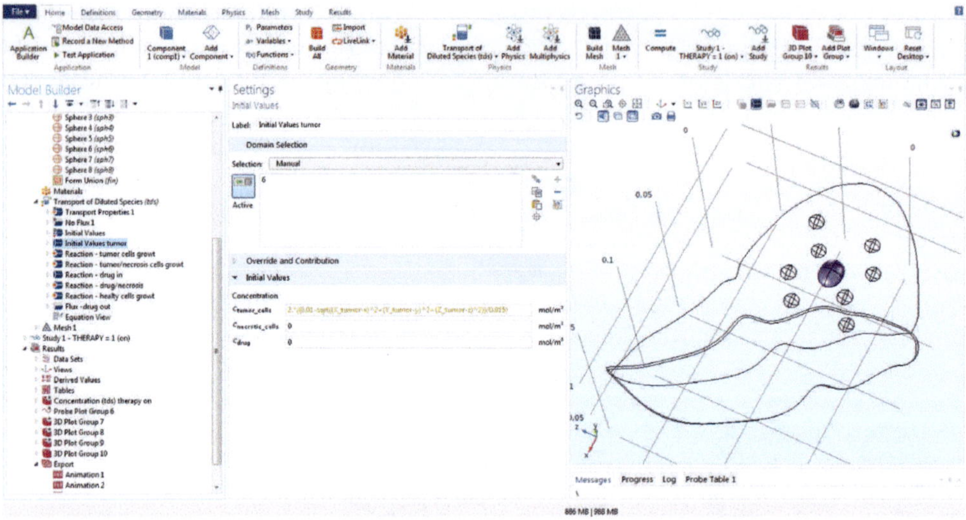

FIGURE 18.8 The COMSOL Multiphysics ®framework in preparation of the initial conditions for Patient A's HCC. In this phase all information must be specified, to allow the pertinent conservation and dynamics formulation as a "Transport of Diluted Species." Here, the ROI of Fig. 18.4 is simplified as an ellipsoid, and up to eight chemoembolization sites of the therapy are indicated (being an option to cure HCC, as explained later).

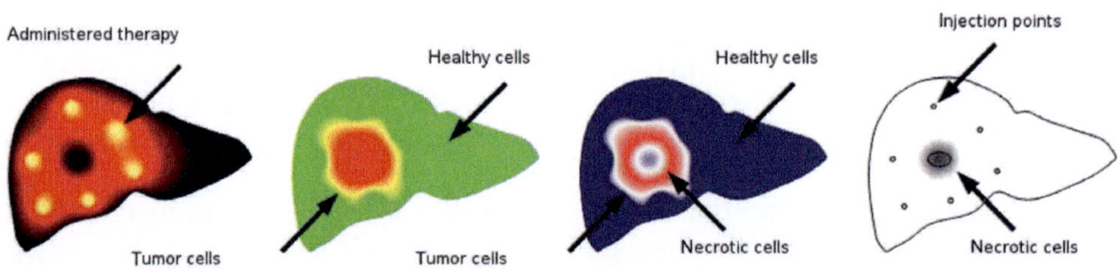


FIGURE 18.9 An illustrative example of four competing biological/chemical species, represented in term of concentration distributions $c$, in Level 1 approach of a HCC modeling. From Left to Right: (1) the distribution of drug (Yellow to Red); (2) the distribution of tumor cells (Red), within an otherwise healthy continuum (Green); (3) the forming necrotic core (Gray); and (4) the positions of six chemoembolization locations.

This approach, illustrated in Fig. 18.9 for a simplified two-dimensional (2D) liver, is adopted in the present work, as explained in detail in Section 18.3.4.

### *18.2.3.3 Level 2*

Following the procedure by Tang et al. [9], tumor cell spreading is a result of a combination of interstitial pressure locally in the given tissue, and the distribution of cell nutrients and metabolites. Some basic cancer hallmark parameter [such as the effect of tumor cells secrete diffusible substances or tumor angiogenesis factor (TAF)] as well as the inherent local therapy effect can also be taken into account. In brief: (1) nutrients distribution is represented by the local concentration of oxygen $C_{O_2}$. As glucose, amino acids, fatty acids, vitamins, and micronutrients are hydrophilic and they do not easily permeate across cell's plasma membrane. For sake of simplicity $C_{O_2}$ could be the representative local nutrient for DNA, RNA, and proteins synthesis, affecting tumor cell metabolism, angiogenesis, growth, and metastasis; (2) metabolites distribution is represented by the local carbon dioxide concentration $C_{CO_2}$. Metabolic wastes accumulation occurs during tumor growth process due to inefficient drainage, causing reduced cell activity and biosynthesis up to necrosis; (3) cancer hallmark parameter distribution is represented by the local TAF concentration $C_{TAF}$. TAF could underlie a variety of biological capabilities, qualified as rational condition for neoplastic diseases. As tumor cells rapidly consume nutrients, hypoxia occurs in the center of avascular tumors once a given threshold is reached, causing local TAF secretion to favor new vessels to sprout from existing ones toward hypoxia regions; and (4) therapy distribution is represented by the local drug concentration $C_d$. The effect of antiangiogenic drugs is to reach a local concentration above a given threshold in order to lower the TAF release rate or induce direct apoptosis; while cytotoxic drugs above a given threshold induce DNA damage to tumor cell in order to prevent or limit cell replication. Similarly than for Level 1 approach, generative/dissipative source term values must be provided, for each of the four competing species.

### 18.2.4 Step 4: Solution and postprocessing

The computational tools offered by the employed framework (discretization of the differential equations and their solution based on complex numerical relaxing methods when multiple and intertwining nonlinearities are inherently present) allow one to carry out the computational solution and postprocessing of the computational results. In this way, Patient A's dynamic histology can be virtualized; but obviously a model validation is mandatory from his/her retrospective data: in principle, a fairly large set of validated patient situations will allow to increase the ability to deploy credible virtual scenarios. As an example, in Fig. 18.10 a time sequence of three-dimensional (3D) distributions is reported. In this case, for about 7 months the tumor has been followed in its free progression in the liver; then at Day 0 the therapy has been administered with chemoembolization around the tumor mass.

This kind of postprocessing will also allow for an immersive experience with oncologists and surgeons when using the available augmented reality gears such as head-mounted displays [21], as during the event whose picture is in Fig. 18.11.

### 18.2.5 Step 5: Replication of the model

Following a clinical evaluation on all of the parameters featuring at the adopted model level, the replication of the model can be tried both horizontally (for a given series of Patients *B*, *C*, *D*, and so on) and vertically, for another patient/cancer occurrence coupling. *This replication lays at the core of the concept of personalized and facilitated medicine.*

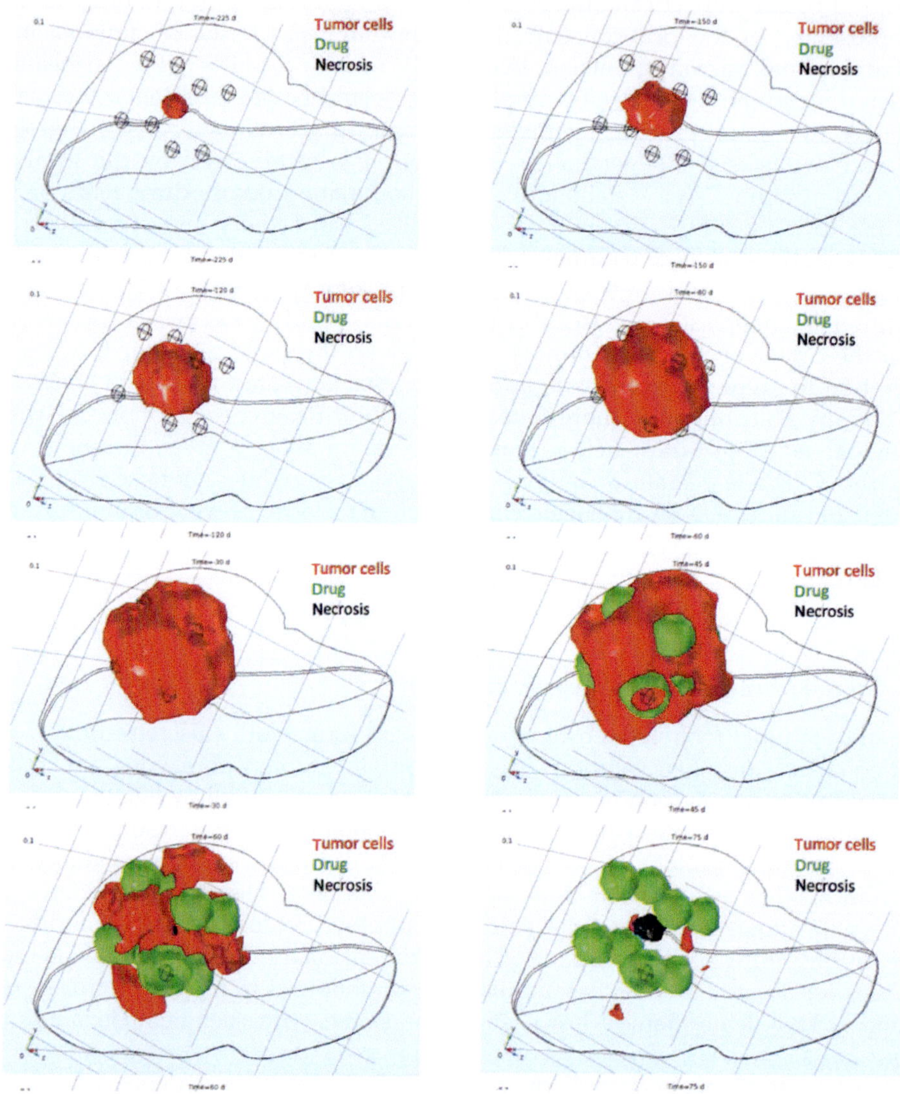


FIGURE 18.10 An illustrative example of tumor (red) and necrotic (black) cells distribution, and drug (green) distribution. Eight times before and after therapy administration have been selected for illustrative purpose: time proceeds from left to right, and from top to bottom.

## 18.3 Detailed model formulation using level 1 modeling

### 18.3.1 The biological conversion logistics-based mechanisms

As implied earlier, in this modeling approach only three complementing species/cell populations are devised, beside the selected drug species: healthy/original cells, cancer/*proliferating*

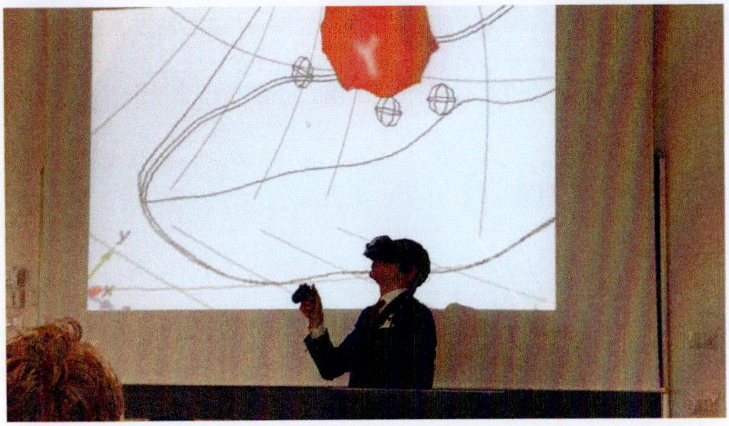

**FIGURE 18.11** A user having an immersive experience into the visualization of Fig. 18.10.

cells, and *necrotic/remnant* cells. One of the simplest approach to devise the evolution (change rate) of two competing/complementing normalized species $0_{1,2}$, stemming from an intuition by T.R. Malthus, is given by the Verhulst logistic growth/decay law $R_1$ of the biological species $0_1$ ([22]):

$$\frac{d\phi_1}{dt} \equiv R_1 = r_1\phi_1\left(1 - \frac{\phi_1}{K}\right), \quad \text{implying} \quad R_1 = -R_2 \tag{18.1}$$

where $R_1$ is generation rate of $0_1$ upon $0_2$, $R_2$ is obvious reverse generation rate of $0_2$ upon $0_1$, $1/r_1$ is a timescale constant, and $K$ is the carrying capacity of the biological matrix [22]. If $0_1$ are meant to represent the species densities or dimensionless volumes in the biological matrix occupied by such biological species, then the following geometric closure applies

$$\phi_1 + \phi_2 = 1 \tag{18.2}$$

and $K - 1$.

This principle can be iterated by inserting a third selectively competing/complementing biological species. In the case at stake, the following conversion mechanisms can be adopted for the healthy h, cancer c, and necrotic species n: h→c and c→n, with a reversal mechanism c→h that applies when a successful drug species d is employed. To this end, any applicable drug $d_j$ can be tried, each having its quantitative PK/PD definitions enforced in the model (as implied later).

From these assumptions, a simple yet general approach results. If required, this scheme can be supplemented with more interacting components, such as a quiescent/hypoxic cell population. Moreover, if desired, a more appropriate logistic law can be provided in Eq. (18.1).

The conversion mechanisms of Eq. (18.1) directly represent the source terms of the following governing multidimensional PDEs.

### 18.3.2 Governing equations

Governing PDEs for reaction-diffusion of tumoral biomass and drug transport, in terms of dimensionless cell densities, are applied to the simplified 3D model of human liver of Fig. 18.6, for i = c, n, or d [23]:

$$\frac{\partial \phi_i}{\partial t} = \nabla \cdot \left[ D_i(\mathbf{x}, t) \nabla \phi_i \right] + R_i \tag{18.3}$$

Similarly than with Eq. (18.2), it is

$$\phi_h + \phi_c + \phi_n = 1 \tag{18.4}$$

In Eq. (18.3), $D_i$ are the applicable diffusion coefficients for each species, which can be cast as space- and time-dependent to account for inherent histological anisotropies and temporal modifications, when such information is available. Using Eq. (18.1), the source terms $R_i$ can be specified for the following two cases:

- *when no therapy is administered*, it is:

$$R_c = r_c \phi_c \phi_h - R_n \tag{18.5}$$

where

$$R_n = r_n \phi_n \phi_c \tag{18.6}$$

- conversely, *when a successful drug is deployed*, it is

$$R_c = (r_c - r_d) \phi_c \phi_h - R_n, \quad \text{with } r_d = e \phi_d \phi_h \tag{18.7}$$

and

$$R_d = f(t) g(\mathbf{x}) \tag{18.8}$$

whereas Eq. (18.6) still holds. The following definitions apply:

- $r_c$ and $r_n$ are biological conversion rates or timescales depending on the specific cancer occurrence (comprising a multiplicity of factors at the aforementioned nanoscale and microscale, such as invasiveness/aggressiveness/malignancy, as deemed appropriate), and must be provided following a clinical analysis;
- $e$ is the drug efficiency, related to the active components of the therapy, and must be provided following a PD evaluation;
- the ($f(t)\, g(\mathbf{x})$) ensemble represents the administration pattern of the therapy, which may vary in time and space, to include the effect of its modulation/inherent decay and its physiological availability/delivery: the former must be provided following a PK evaluation, the latter must be indicated by inherent interventional procedures in HCC treatment.

With the language of transport phenomena, the right hand side of Eq. (18.3) carries two terms: *the diffusive term and the reactive term*, dictating the rate of change of tumor volume. In order for the tumor to be dealt with, two following driving parameters are identified:

- in the diffusive term, the cancer diffusion $D_c$ and the drug diffusion $D_d$;
- in the reactive term, the biological conversion rates $r_c$ and $r_n$, the drug efficiency e and the therapy administration ($f(t)$ $g(\mathbf{x})$) ensemble.

The combination of $D_d$, $e$, and $f$(t) are representative of the aforementioned PK/PD features.

### 18.3.3 Initial conditions, proliferation, and therapy onset

Giving an initial $\Omega_{h0}$ allowable volume, wrapped up in the steady surface $\partial\Omega_e$ which separates the organ at stake from its environment, an initial cancer region of interest (ROI) $\Omega_{c0}$ is detected by a proper diagnostic imaging measurement (as alluded to earlier). Then, with the progress of time, the cancer volume will grow as $\Omega_c(t)$ at the expense of $\Omega_h(t)$. Due to the evolution of the tumor, a necrotic $\Omega_n(t)$ region will eventually coagulate, implying that the geometric closure (stemming from Eq. 18.4)

$$\Omega_h(t) + \Omega_c(t) + \Omega_n(t) = \Omega_{h0} \tag{18.9}$$

will hold at any $t$.

When the therapy is administered, by specifying the functions $f$ and $g$, a region $\Omega_d(t)$ containing the permeating drug $\phi_d$ will superimpose to $\Omega_h(t)$, and the depletion of $\Omega_h(t)$ can be locally stopped and reversed, at the expenses of a receding $\Omega_c(t)$.

### 18.3.4 Boundary conditions

Due to the integration of the PDEs system described by Eq. (18.3), the interface surface between healthy and cancer cells $\partial\Omega_{hc}$ will deform and arrange itself with varying time. The applied drug species $\phi_d$ will perfectly diffuse through this surface, as well as with other interfaces between volumes inherently accounted for, standing the definition of $\phi_{h/c/n}$. The boundary $\partial\Omega_e$ can be subject to selective conditions itself, such as perfect or conditional permeability (pending some physiological particular) to $\phi_d$, which in due time could locally reach this boundary, so that the drug can be naturally disposed of to the organ external environment. In the present model, an outflow of $\phi_d$ through $\partial\Omega_e$ is assumed, based on the value reached by the drug at the equilibrium, when the organ is completely filled in.

## 18.4 A sensitivity analysis of hallmark parameters: results

In this section, we explain how model variables relate to clinical parameters. The main parameter which is measured by the diagnostic image is the experimental tumor volume data.

### 18.4.1 Numerical treatment

To initiate the computational model, the domain of Fig. 18.6 was subjected to mesh yielding for a homogeneous tetrahedral element grid, generated via a prismatic boundary layer grid. After a grid independency iteration, a final mesh of about 42k elements was employed, to optimize result accuracy and computational times.

In order to solve the governing equations system (Eqs. 18.3–18.8), along with its initial and boundary equations, the direct solvers PARDISO was employed used, while the BDF method was applied for the temporal dependence. Execution durations did not exceed 1 h on a Pentium Xeon server (Windows 10 OS, Eightcore-32N at 2.4 GHz, 128 GB RAM) running in serial mode.

### 18.4.2 Model validation: application to a hepatocellular carcinoma—Case 0

In this preliminary study, the data from Ref. [24] have been used for validation. This comparison has been purposeful to focus on the ability of the model platform to identify the hallmark parameters this representation of tumor proliferation depends upon. The case study (or Case 0, in the present context) involved a 72-year-old female patient who developed HCC following adenocarcinoma of sigmoid colon: particularly, the experimental measurements of tumor volume, obtained from patient screening during treatment time, formed the validating data series employed.

These experimental data are plotted against the computed progress of tumor volume ($\phi_c \cdot \Omega_{h0}$) in Fig. 18.12. In this case, the total liver volume $\Omega_{h0}$ was about 1230 $cm^3$ and the nominal drug flow rate delivered was about 6.5 $cm^3\ day^{-1}$; this flow rate was modulated during the administration, by using the function $f$ in Fig. 18.13 at left. It is seen that the adherence to the experimental data is very close: to this end, the driving source terms $R_i$ of Eq. (18.3) had to be optimized, with the values reported in Table 18.1. The value of diffusivity $D_c$ was taken from Ref. [24], while $D_d$ was tentatively set two orders of magnitude larger. Finally, no diffusion was assumed for necrosis, $D_n = 0$. For the validating data to be replicated this close, it is seen from Fig. 18.12 that the tumor had to be initiated at a certain time in the past, that is some 120 days before therapy start ($t = 0$). This lapse was found by an iterative exercise on the model, having adopted few additional assumptions that normally would hinge upon a clinical analysis of few patients' outcomes in retrospective, for the given tumor type. The inflection of the ($\phi_c \cdot \Omega_h$o) curve depends upon said initial time for tumor onset, the initial volume of the ROI $\Omega_{c0}$, its growth rate $r_c$ (standing for the invasiveness/aggressiveness/malignancy of the tumor type at stake) and patient's response to the therapy itself (dictating the promptness of the rate change of tumor volume, right after therapy start).

Relative to the case described so far, featuring the driving parameters in Table 18.1, some 3D representations are reported in Figs. 18.14–18.16:

- evolution of tumor volume, in an untreated scenario, Fig. 18.14 and
- evolution of tumor volume in presence of a successful therapy, Figs. 18.15 and 18.16.

First, in Fig. 18.14 the cancer volume in red is found to fill freely the available liver volume. At the final reported time, $t = 240$ days, a large black lump is visible, representing

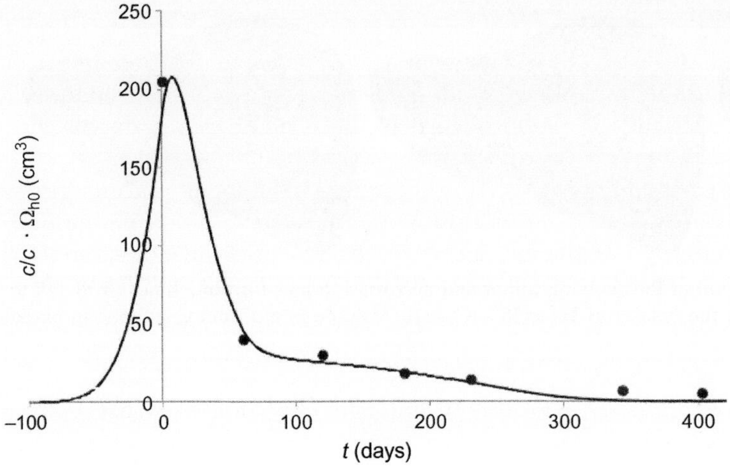


FIGURE 18.12 The experimental data used in this validation (•), and the curve of the simulated tumor volume with time (—).

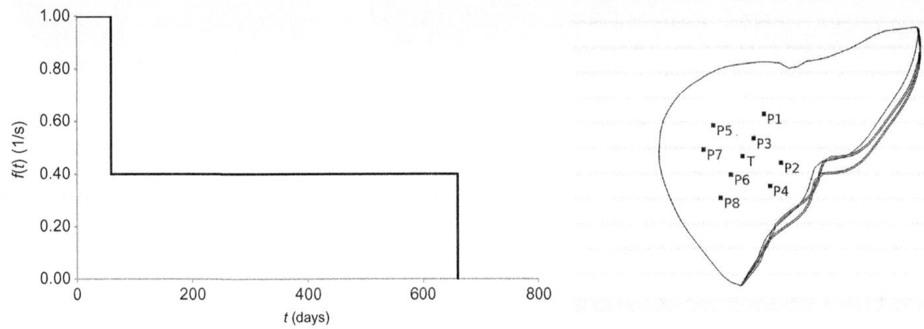


FIGURE 18.13 A segregated description of the $f(t)$ $g(\mathbf{x})$ group for Case 0. Left: the modulation effect $f$ with time $t$; right: the distribution $g$ of the availability/delivery sites. In this case, a cubic stencil of eight points P (having an edge length of 4 cm), centered on the region of interest $T$ has been devised.

TABLE 18.1 Driving parameters of Eq. (18.3) for Case 0 for the various $p$.

| | $D$ (m$^2$ s$^{-1}$) | $r$ (s$^{-1}$) | $e$ (s$^{-1}$) | $f(t)$ $g(\mathbf{x})$ (s$^{-1}$) |
|---|---|---|---|---|
| $\phi_c$ | $5.8 \times 10^{-12}$ | $9.5 \times 10^{-7}$ | – | – |
| $\phi_d$ | $5.8 \times 10^{-10}$ | – | $3.8 \times 10^{-5}$ | see Fig. 18.13 |
| $\phi_n$ | 0.0 | $2.0 \times 10^{-7}$ | – | – |

the forming necrosis. Conversely, when a successful therapy is applied, it is seen in Fig. 18.15 that the green volume representing the drug readily permeates the liver part around the tumor volume in red, after 6 days only, due to the employed pattern $g(\mathbf{x})$. In 60 days the drugs fill almost completely the available liver space. It is seen that around

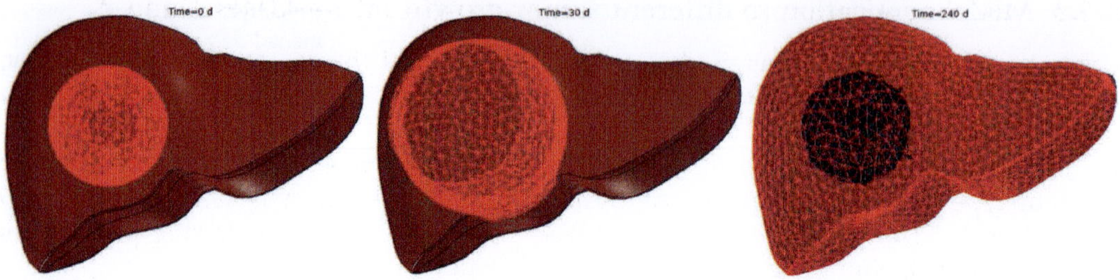


**FIGURE 18.14** A 3D representation of the evolving tumor and necrosis volumes for three times (from left to right: $t=0$, 30, and 240 days), using the dataset in Table 18.1 (Case 0). Volume in red: tumor; volume in black: necrosis.

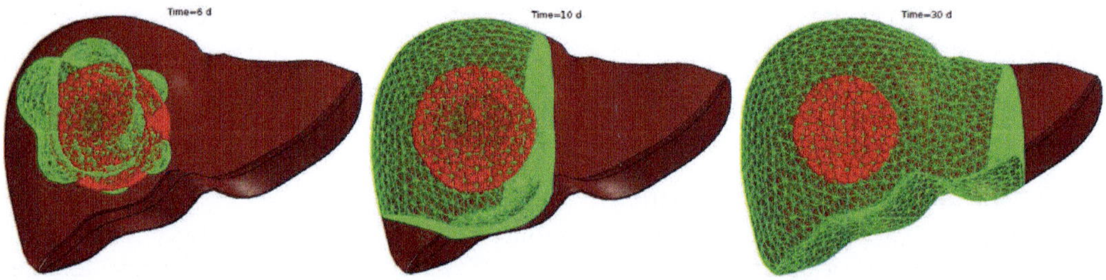


**FIGURE 18.15** A 3D representation of the evolving tumor and drug distribution volumes for three times (from left to right: $t=6$, 10, and 30 days), using the dataset in Table 18.1 (Case 0). Volume in red: tumor; volume in green: drug.

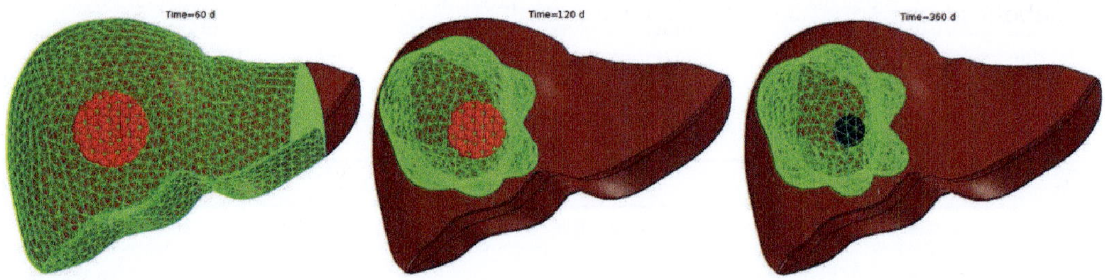


**FIGURE 18.16** A 3D representation of the evolving tumor, drug distribution, and necrosis volumes for three times (from left to right: $t=60$, 120, and 360 days), using the dataset in Table 18.1 (Case 0). Volume in red: tumor; volume in green: drug; volume in black: necrosis.

day 30 the tumor starts to recede from its initial width. At the end of the observed period, after 360 days in Fig. 18.16, the active tumor volume has disappeared, leaving a small necrotic volume in black, only.

### 18.4.3 Model application to different tumor growth rates—Cases 1 and 2

The proposed model is exercised to monitor the role of the main tumor proliferation parameters. In particular, the variation of the tumor growth rate $r_c$ (see Eq. 18.7) has been monitored in Fig. 18.17 to infer upon the *relative importance of the internal reactive term*. This term is paramount to represent the invasiveness/aggressiveness/malignancy behavior of the tumor type at stake. With reference to the value in Table 18.1, $r_c$ is varied from Case 0 to Case 1 ($r_{c1} = 0.90 \cdot r_{c0}$), and to Case 2 ($r_{c2} = 1.1 \cdot r_{c0}$). All other driving parameters were unchanged, including functions $f$ and $g$ of Fig. 18.13.

When less aggressive tumor is devised, as with the … curve (Case 1), for a relatively small variation of $r_c$ (−10%), the tumor is seen to expand, at the time of start therapy, as expected to the lesser extent, in this case some 53% with respect to Case 0. When using this value of $r_{c1}$, the tumor disappeared already after some 60 days from therapy start. When more aggressive tumor is simulated instead, as with the - - curve (Case 2), even for the relatively small increment (+10%), the tumor is seen to respond to the therapy by lagging more than 10 days and expanding about 50% more than with the Case 0, then the loss of control by the therapy is detected around day 60 with the subsequent *run-away* of the disease.

### 18.4.4 Model application to different therapies—Cases 3 to 7

The proposed model is also exercised to monitor the role of main therapy features. In particular, the variation of $e$ and $g(\mathbf{x})$ (see Eqs. 18.7 and 18.8) has been selected to infer upon the *relative importance of the externally-induced reactive terms ensemble*: the effect of the chemical principles of the therapy, and the modalities of its administration. To this end, in Figs. 18.18 and 18.19 the following five additional cases are presented:

- Two cases of variation of drug efficiency $e$: its control on the therapy outcome may represent an opportunity to restrain its toxicity, with positive effects on patient's metabolic alteration while minimizing the treatment cost for the health systems.

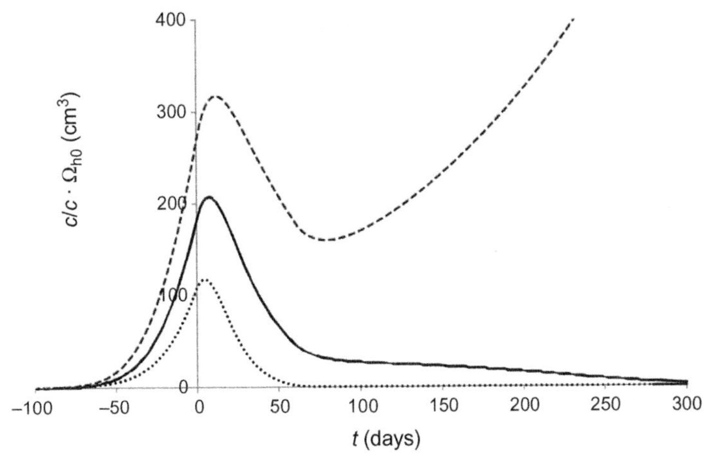


FIGURE 18.17 Comparison of tumor volume ($\phi_c \cdot \Omega_{h0}$) progress with time, for three cases of tumor growth rate $r_c$. $r_{c0} = 9.5 \times 10^{-7}\,s^{-1}$ (Case 0) (—); $r_{c1} = 0.90 \cdot r_{c0}$ (Case 1) (…); $r_{c2} = 1.1 \cdot r_{c0}$ (Case 2) (- -).

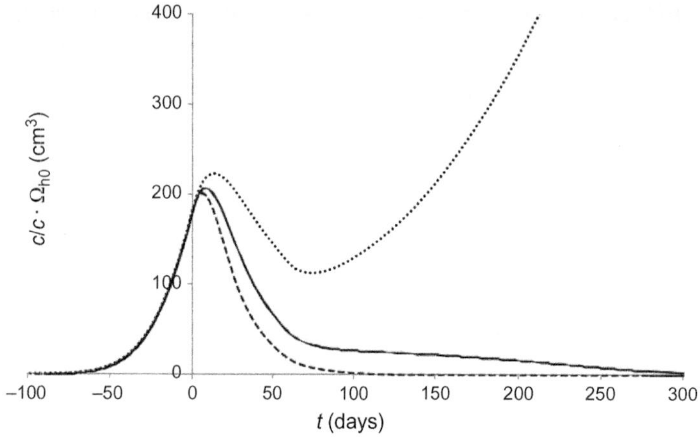


**FIGURE 18.18** Comparison of tumor volume ($\phi_c \cdot \Omega_{h0}$) progress with time, for three cases of drug efficiency $e$. $e_0 = 3.8 \times 10^{-5}\ s^{-1}$ (Case 0) (—); $e_3 = 0.75 \cdot e_0$ (Case 3) (• • •); $e_4 = 1.25 \cdot e_0$ (Case 4) (- -).

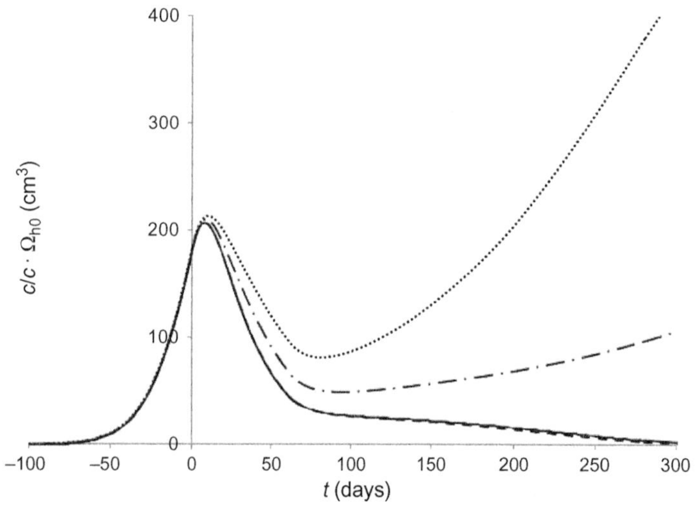


**FIGURE 18.19** Comparison of tumor volume ($\phi_c \cdot \Omega_{h0}$) progress with time, for four cases of therapy administration pattern $g(\mathbf{x})$ (number of points P for drug availability/discharging, see Fig. 18.13 at right). $g_0$ = All P points (Case 0) (—); $g_5$ = Point $P_1$ only (Case 5) (• • •); $g_6$ = Points $P_{1,8}$ only (Case 6) (- • -); $g_7$ = Points $P_{1,2,3,4}$ only (Case 7) (- -).

- Three cases of variation of therapy administration pattern $g(\mathbf{x})$: representing, for HCCs as the one at stake, the pattern for chemoembolization (a minimally invasive procedure performed to restrict blood supply in HCC) for large mass that cannot be ablated or removed by surgery.

For sake of simplicity, in all these cases the modulation function $f(t)$ is left unaltered at the values reported in Fig. 18.13 at left.

First, the effect of changing the drug efficiency $e$ is reported in Fig. 18.18. In particular, $e$ is varied from the value of Case 0 ($e_0 = 3.8 \times 10^{-5}$) to $e_3 = 0.75 \cdot e_0$ of Case 3, and to $e_4 = 1.25 \cdot e_0$ of Case 4. All other driving parameters were unchanged (see Table 18.1).

When a therapy with a lower efficiency $e_3$ is administered, as with the ... curve (Case 3), this has the initial effect to delay the time of curve inflection, that is the patient response being delayed by some 20 days: the tumor volume ($\phi_c \cdot \Omega_{h0}$) keeps in its growth (although affected by the incoming therapy) and starts to recede later. The descent rate of this volume is also slower, and a relative minimum is reached around day 70. Then, the therapy can sustain no longer the tumor activity so that the tumor volume grows again exponentially and uncontrolled (as in the *run-away* seen earlier) until the available space is filled in. Right before the loss of control, when using $e_3$ the tumor volume shrank from about 210 $cm^3$ at therapy start to about 110 $cm^3$ (some 48% of the initial volume) while with $e_0$ (about 25% a larger drug efficiency) the decrement was down to 40 $cm^3$ (or about 80%). A stiffer volume–descent situation is detected instead for a therapy with a higher efficiency $e_4$, as with the - - curve (Case 4). Now, the decrease in tumor volume is slightly faster and after 100 days the disease has disappeared entirely. With this therapy at $e_4$, an increase of 25% in the drug efficiency with respect to $e_0$, a volume smaller by some 30 $cm^3$ (or 15% of the tumor volume at therapy start time) is constantly obtained. It is concluded, therefore, that *the drug efficiency variation scales in a nonlinear fashion with respect to tumor volume reduction*, playing an important role into optimizing a therapy which translate into an economic cost for the community, and a metabolic cost for the patient: with the proposed formulation helping assess this role.

Then, in Fig. 18.19 the effect of changing the spatial administration pattern $g(\mathbf{x})$ is reported. In particular, with reference to Fig. 18.13 at right, $g$ is varied from the pattern of Case 0 ($g_0$ = All of stencil's P points are employed) to Case 5 ($g_5$ = Point $P_1$ is employed, only), Case 6 ($g_6$ = Points $P_{1-8}$ are employed, on opposite corners of the cubic stencil), and Case 7 ($g_7$ = Points $P_{1-2\text{-}3-4}$ are employed, on the same side of the cubic stencil) patterns. All the cases devised here share the same value of drug flow rate deployed. All other driving parameters were unchanged (see Table 18.1).

When a therapy with one-point only of physiological availability/delivery, or $g_5$, is administered, as with the ... curve (Case 5), this has an effect similar to the progress discussed for Case 3 (Fig. 18.18), but with the curve inflection being slightly smoother, and the tumor volume when reaching the beginning of the uncontrolled state much smaller (at about 80 instead of 110 $cm^3$, for Cases 5 and 3, respectively). When the entire drug flow rate is available/delivered is one site, only, the inherent drug diffusion $D_d$ and the resulting delay due to this strong inhomogeneity makes this therapy largely ineffective, leading to sudden tumor run-away. The same poor drug performance is obtained with the progress of the ... curve (Case 6), relative to two opposite-points only of physiological availability/delivery, or $g_6$. Now, the behavior of tumor volume progress is intermediate between Case 0 and Case 5, but the loss of homogeneity in the drug species $\phi_d$ is felt just the same, with the run-away happening at the same time than with Case 5, though with a slowest growing rate. Finally, with the - - curve of Case 7, it is seen that when the therapy is administered on one side only of this cubic stencil, the result can be totally superimposed to the one of Case 0, when all points are involved. This result can be important as a decision support when devising the optimal surgery sites.

### 18.4.5 Model application to different values of tumor and drug diffusivities—Cases 8 to 11

Finally, the proposed model is exercised to monitor the role of tumor and drug diffusivities $D$, see Eq. (18.3). While $D_c$ represents the *ability of the cancer cell species to diffuse through the extra-cellular matrix*, that is by means of chemotaxis, $D_d$ is related to the *PK/PD ensemble of the chemical species/histology couple at stake* (e.g., the concentration dynamics and the related biological responses). To this end, in Figs. 18.20 and 18.21 the following four additional cases are presented:

- Two cases of variation of tumor diffusivity $D_c$: to complete the knowledge of tumor behavior, together with the growth rate $r_c$ monitored above.
- Two cases of variation of drug diffusivity $D_d$: the knowledge of its variation range may represent an opportunity to optimize the therapy outcome.

With reference to the values in Table 18.1, $D_c$ is varied from Case 0 to Case 8 ($D_{c8} = 0.50 \cdot D_{c0}$), and to Case 9 ($D_{c9} = 2.0 \cdot D_{c0}$); while $D_d$ is varied from Case 0 to Case 10 ($D_{d10} = 0.50 \cdot D_{d0}$), and to Case 11 ($D_{d11} = 2.0 \cdot D_{d0}$). All other driving parameters were unchanged, including functions $f$ and $g$ of Fig. 18.13. First, the effect of changing the drug efficiency $D_c$ is reported in Fig. 18.20. A tumor with half a diffusivity $D_8$ with respect to Case 0, as with the ... curve (Case 8), right after the administration of the therapy spreads in the liver causing a maximum cancer volume which is some 40% smaller. However after being attacked by the therapy, the rate of volume reduction is much slower that with Case 0, causing a lagging phase in which the tumor volume ($\phi_c \cdot \Omega_{h0}$) is not fought completely, remaining at an equilibrium value of about 37 cm$^3$. Conversely, a tumor with a double $D_9$ with respect to Case 0, as with the - - curve (Case 9), grows up to a 50% larger volume, but following therapy its steeper descent rate would bring the patient to complete healing after about 120 days from therapy start.

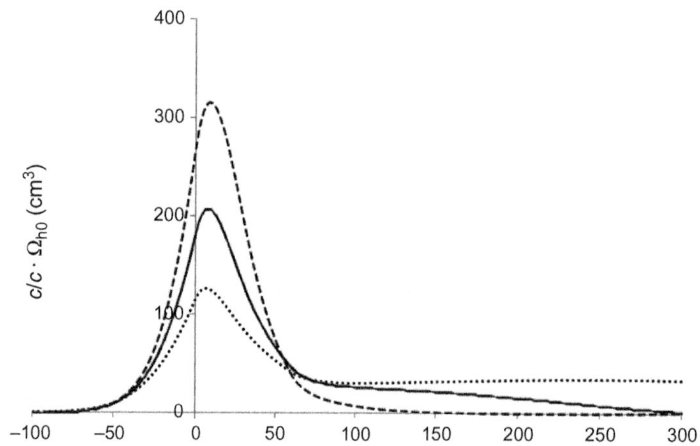


FIGURE 18.20 Comparison of tumor volume ($\phi_c \cdot \Omega_{h0}$) progress with time, for three cases of tumor diffusivity $D_c$. $D_{c0} = 5.8 \times 10^{-12}$ m$^2$ s$^{-1}$ (Case 0) (—); $D_{c8} = 0.50 \cdot D_{c0}$ (Case 8) (• • •); $D_{c9} = 2.0 \cdot D_{C0}$ (Case 9) (- -).

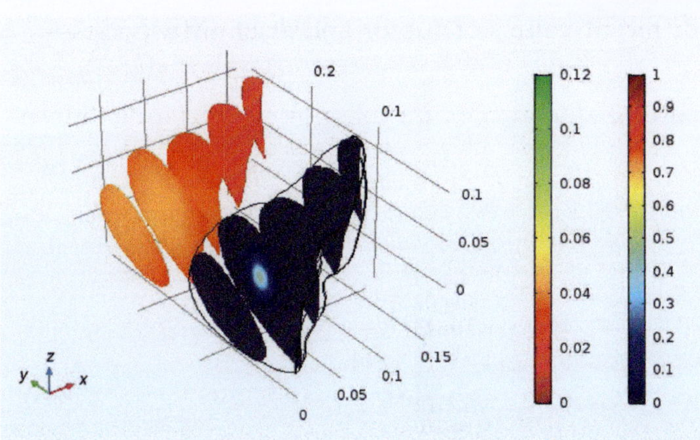


**FIGURE 18.21** Comparison of tumor volume ($\phi_c \cdot \Omega_{h0}$) progress with time, for three cases of drug diffusivity $D_d$. $D_{d0} = 5.8 \times 10^{-10}\,\mathrm{m^2\,s^{-1}}$ (Case 0) (—); $D_{d10} = 0.50 \cdot D_{d0}$ (Case 10) (• • •); $D_{d11} = 2.0 \cdot D_{d0}$ (Case 11) (- -).

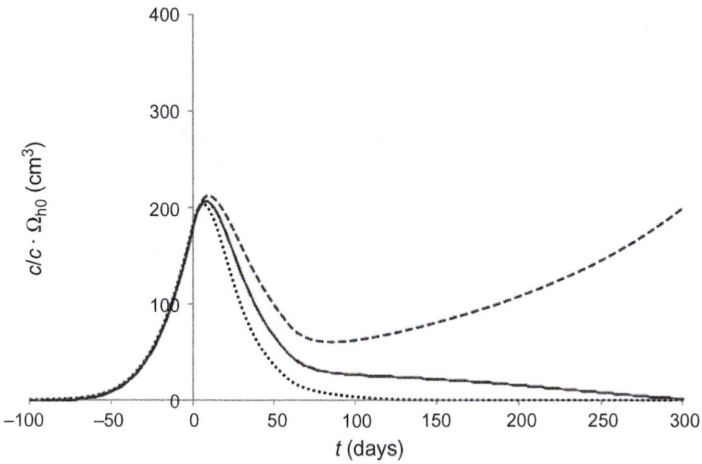


**FIGURE 18.22** Two frame views of the subject liver. In each frame, the upper left slice distribution is for drug, and the lower right distribution is for tumor. Left frame: Case 8; Right frame: Case 9. Color scales are provided at utmost right—left scale: drug concentration $\phi_d$; Right scale: tumor concentration $\phi_c$.

This somewhat surprising behavior is better illustrated in Fig. 18.22: it is seen that a higher tumor diffusivity will allow the proliferated cells to be attacked more profitably by the administered drug, so that with Case 9 (frame at right) after 90 days the patient is almost completely recovered, while with a lower tumor diffusivity (Case 8, frame at left) the cancerous distribution mostly lies still around the initial ROI, with no efficient way to make biochemical contact with the administered drug (provided peripherally, as usually described by Fig. 18.13).

Then, the effect of changing the drug efficiency $D_d$ is reported in Fig. 18.21. A therapy with half a diffusivity $D_{10}$ with respect to Case 0, as with the ... curve (Case 10), accumulates more efficiently around the initial ROI, therefore the rate of tumor volume reduction is higher than with Case 0, and the patient is seen to recover completely right after 100 days from therapy start. Whereas doubling the therapy diffusivity as with the - curve (Case 11), makes the therapy sweep to fast across the organ, causing swift disposal through the liver's boundaries and ultimately a deadly tumor volume run-away.

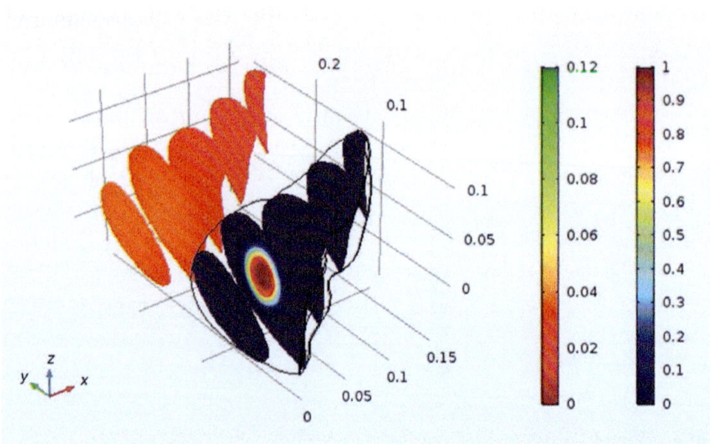


**FIGURE 18.23** Two frame views of the subject liver. In each frame, the upper left slice distribution is for drug, and the lower right distribution is for tumor. Left frame: Case 10; Right frame: Case 11. Color scales are provided at utmost right—Left scale: drug concentration $\phi_d$; Right scale: tumor concentration $\phi_c$.

Once again, this behavior is better illustrated in Fig. 18.23: it is seen that after 90 days the patient is going to recover nicely (Case 10, frame at left) while the decidedly larger tumor lump, at its minimum with this value of drug diffusivity (Case 11, frame at right), will soon expand uncontrolled due to the lack of efficient therapy nearby the initial ROI.

## 18.5 POEM as a tool to empower the clinical decisions

The proposed model is part of a research effort to offer quantitative predictions at oncologists' and researchers' desk, to help identify the optimal therapy: giving users access to future data and histology dynamics (so-called *in silico prognoses*) which will materialize in the patient under observation. In particular, depending on the following values:

- $r_c$, $r_n$, and $D_c$, representative of tumor invasiveness/aggressiveness/malignancy;
- $e$, $D_d$, and $f$(t), representative of drug PK/PD features; and
- $g(\mathbf{x})$, representative of the specific drug delivery mode.

The projected tumor volume is computed, starting from the actual measurement in the specific patient, to help addressing personalized protocols, treatment optimization, response to therapy and control, and surveillance after treatment.

Based on the commercial platform employed, this class of models will be available through an app residing on users' desk, to allow every virtual exercise and what-if scenario at an affordable cost.

## 18.6 Conclusions

In this chapter the potential of systems of reactive/diffusive PDEs describing the onset and proliferation of tumoral cells in tissues and organs, and related therapy framework, is presented. This approach will have an impact onto personalized and precision oncology, to improve quantitative prognosis and to offer a decision tool supporting oncologists and surgeons. Also, this approach may open up new developing perspectives for biology and pharma researchers, as well as for diagnostic device industries.

Up to 12 cases were studied to describe the hallmark features and driving parameters playing a fundamental role in computational tumor prognosis, and compared with one literature dataset of tumoral volume progress. The complex, inherently nonlinear nature of HCC proliferation is highlighted, even for the present simplified representation of a liver. In particular, the effect of the drug efficiency could be monitored to restrain its toxicity, with positive effects on patient while saving the overall cost of therapy and hospitalization; while optimal administration patterns exist, which can be determined for practices such as chemoembolization. Finally, dependence on chemical species diffusivities let the strong nonlinear aspects of the mathematical solution emerge, for the present problem of tumor proliferation and its interaction with administered therapy.

## Acknowledgments

This study has received partial funding from the European Union's Horizon 2020 research and innovation program under agreements No 691556.

## Glossary

**CAPTURE** CAncer Predictions for Therapy and drUg REposition
**FEM** Finite element method
**HCC** Hepatocellular carcinoma
**PD** Pharmacodynamic
**PDE** Partial differential equation
**POEM** Predictive oncology via engineering models
**PK** Pharmacokinetic
**ROI** Region of interest
**TAF** Tumor angiogenesis factor

## References

[1] Science Business®. <http://sciencebusiness.net/framework-programmes/news/eu-council-and-parliament-strike-deal-horizon-research-programme>; 2019 [accessed 01.11.19].

[2] Deisboeck TS, Zhang L, Yoon J, Costa J, Zhang L, Yoon J, et al. In silico cancer modeling: is it ready for prime time? Nat Rev Clin Oncol 2009;6(1):1–34.

[3] Barbolosi D, Ciccolini J, Lacarelle B, Barlesi F, Nicolas A. Computational oncology - mathematical modelling of drug regimens for precision medicine. Nat Rev Clin Oncol 2016;13(4):242–54.

[4] Yankeelov TE, An G, Saut O, Luebeck EG, Popel AS, Ribba B, et al. Multi-scale modeling in clinical oncology: opportunities and barriers to success. Ann Biomed Eng 2016;44(9):2626–41.

[5] Stamatakos G, Dionysiou D, Lunzer A, Belleman R, Kolokotroni E, Georgiadi G, et al. The technologically integrated oncosimulator: combining multiscale cancer modeling with information technology in the in silico oncology context. IEEE J Biomed Health Inform 2014;18(3):840–54.

[6] Gatenby RA, Gawlinski ET. A reaction-diffusion model of cancer invasion. Cancer Res 1996;56(24):5745–53.

[7] Swanson KR, Bridge C, Murray JD, Alvord Jr EC. A reaction-diffusion model of cancer invasion. J Neurol Sci 2003;216(1):1–10.

[8] Anderson ARA, Chaplain MAJ, Newman EL, Steele RJC, Thompson AM. Mathematical modelling of tumour invasion and metastasis. Comput Math Methods Med 2000;2(2):129–54.
[9] Tang L, van de Ven AL, Guo D, Andasari V, Cristini V, Li KC, et al. Computational modeling of 3D tumor growth and angiogenesis for chemotherapy evaluation. PLoS One 2014;9(1):1–12.
[10] Ruocco G. Introduction to transport phenomena modeling. Cham: Springer; 2018. p. 272–9.
[11] COMSOL Multiphsysics® v.5.2. <http://www.comsol.com>; 2019 [accessed 01.04.19].
[12] Cancer fact sheets: Liver cancer. <http://gco.iarc.fr/today/data/pdf/fact-sheet/cancers/cancer-fact-sheets-7.pdf>; 2019 [accessed 01.11.19].
[13] European CLuster Collaboration Platform. <http://clustercollaboration.eu/eu-initiative/innosup-calls>; 2019 [accessed 01.10.19].
[14] Cornelis F, Saut O, Cumsille P, Lombardi D, Iollo A, Palussiere J, et al. In vivo mathematical modeling of tumor growth from imaging data: soon to come in the future? Diagn Interv Imaging 2013;94(6):593–600.
[15] MicroDicom® v.3. <http://microdicom.com>; 2019 [accessed 01.11.19].
[16] ImageJ® v.1.51. <http://imagej.nih.gov>; 2019 [accessed 01.11.19].
[17] Rhino3D® v.5. <http://rhino3d.com>; 2019 [accessed 01.11.19].
[18] Gmsh® v.4.4. <http://gmsh.info>; 2019 [accessed 01.11.19].
[19] Materialise Mimics® v.21. <http://materialise.com/en/medicalsoftware/mimics>; 2019 [accessed 01.11.19].
[20] Cristini V, Lowengrub J. Multiscale modeling of cancer. Cambridge, UK: Cambridge University Press; 2010. p. 24–30.
[21] Oculus®. <http://business.oculus.com>; 2019 [accessed 01.11.19].
[22] Murray JD. Mathematical biology - an introduction. 3rd ed. New York: Springer; 2002. p. 2–3.
[23] Caccavale P, De Bonis MV, Marino G, Ruocco G. Computational modeling of 3D tumor growth for personalized therapy evaluation and prognosis. Inter. Commun. Heat Mass Trans 2020; submitted.
[24] Filipovic N, Djukic T, Saveljic I, Milenkovic P, Jovicic G, Djuric M. Modeling of liver metastatic disease with applied drug therapy. Comput Methods Prog Biomed 2014;115(3):162–70.

# SECTION III

# Tumor models for drug discovery and therapeutics

CHAPTER

# 19

# Tissue-engineered 3D cancer microenvironment for screening therapeutics

*Nancy T. Li*, Ileana L. Co*, Natalie Landon-Brace, Simon Latour and Alison P. McGuigan*

**Department of Chemical Engineering and Applied Chemistry, Institute for Biomaterials and Biomedical Engineering, University of Toronto, Toronto, ON, Canada**

## 19.1 Introduction

A tumor is a complex and evolving biological community comprised of both carcinoma and stromal cells, and heterogeneous biochemical and biophysical signals. The properties of carcinoma cells within a tumor typically depend on both the intrinsic genetic mutations present and the biochemical and biophysical signals the cells experience from their local microenvironment. Understanding how the local tumor microenvironment (TME) impacts the properties of carcinoma cells is important to better understand why tumor cell response to therapy varies in different regions of the tumor, and how to apply this knowledge to create novel therapeutic strategies to target carcinoma cells in therapy-resistant regions. Relating tumor cell properties to specific biochemical and biophysical cues in the microenvironment is often challenging using animal models as precise control over specific and individual signals within the microenvironment is limited. Over the last 15 years, a number of engineered tissue models enabling more precise control and measurement of the TME have emerged, which are providing researchers in the field with the opportunity to explore novel tumor biology in human cells. In this chapter, we provide a brief overview of important features of the TME that engineered culture models aim to capture. We then provide an overview of the currently available tissue engineering strategies for recreating the TME and the typical cell sources used to create these models. We then highlight

* Both authors contributed equally.

*Biomaterials for 3D Tumor Modeling*
DOI: https://doi.org/10.1016/B978-0-12-818128-7.00019-8

examples of how these models have been applied to probe important aspects of the TME biology with a view to identifying novel disease targets or therapeutics. Finally, we provide an outlook for the future and highlight important challenges that must be addressed to ensure in vitro tumor models enhance our understanding and ability to treat tumors.

## 19.2 Tumor microenvironment

The TME is comprised of cellular components including tumor cells, infiltrating cells, resident cells, and also non-cellular components like extracellular matrix (ECM) and small molecule gradients (Fig. 19.1). It is well accepted that the TME plays a pivotal role in tumor progression and influences key features of cancers such as metastasis and drug resistance [1]. In this section, we will provide an overview of the different components of the TME and highlight how it influences tumor progression by focusing on the different cellular and non-cellular components of the TME.

### 19.2.1 Cellular components

Cells comprising the TME can be classified in three broad categories: (1) tumor cells, (2) resident accessory cells such as stromal fibroblasts, and (3) immune cells. Each is considered below.

Tumors cells are heterogeneous and differ between different types of cancer, within the same type of cancer, and even within the same tumor: this feature is known as tumor heterogeneity. This genetic and phenotypic heterogeneity evolves over the course of the disease. Since tumor heterogeneity influences therapy resistance [2], capturing this heterogeneity in tissue-engineered models is a key challenge for drug discovery studies. One strategy to capture this heterogeneity involves using patient-derived xenografts (PDX) or patient-derived organoids (PDO). The use of these approaches for engineered tissue model establishment will be discussed in a following section of this review (Section 19.3).

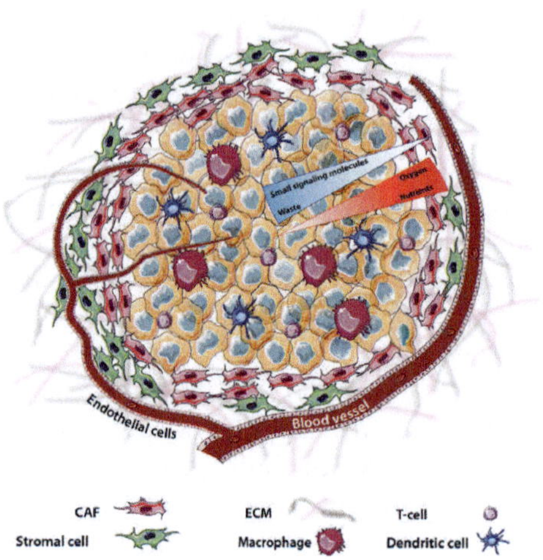


FIGURE 19.1 The tumor microenvironment. Depicted are CAFs, stromal cells, macrophages, T cells, dendritic cells (DCs), and endothelial cells contained in fibrous ECM and irregular blood vessels. Within the tumor islet are gradients of signaling molecules, waste, oxygen, and nutrients.

Fibroblasts and endothelial cells are the most common types of resident accessory cells. Following tumor cells, fibroblasts are the most studied and characterized cells of the TME. During cancer development, resident fibroblasts are transformed into cancer-associated fibroblasts (CAFs) by growth factors such as TGF-β secreted by cancer cells [3]. CAF activation is commonly characterized by its expression of alpha-smooth muscle actin (α-SMA) and fibroblast activation protein (α-FAP) [4]. When activated, populations of CAFs become tumor-promoting by increasing cancer cell survival, promoting cancer cell invasion, angiogenesis, and altering the immune anti-tumor response [3,5,6]. CAFs themselves also secrete tumor-promoting cytokines and growth factors such as TGF-β, and promote cancer cell invasion by modifying ECM via expression of proteases like matrix metalloproteinases (MMPs) and molecules that induce epithelial-to-mesenchymal transition (EMT) of the tumor cells [7,8]. To enhance angiogenesis, CAFs secrete VEGF and PDGF leading to potential dissemination of cancerous cells via blood circulation [5]. Finally, CAFs are central regulators of anti-tumoral immunity through the secretion of both pro-inflammatory cytokines and immunosuppressive factors [3,9]. These factors have both direct and indirect effects on the immunomodulatory properties of immune cells.

The major role of endothelial cells in the TME is to facilitate angiogenesis. Angiogenesis is the development of new blood vessels, which extends the existing vasculature to the growing tumor. The establishment of intra-tumoral vasculature also facilitates metastasis resulting from transendothelial migration of cancer cells [10]. VEGF is an important regulator of angiogenesis as it controls endothelial cell proliferation, elongation and survival [11]. In developing tumors, tumors cells and stromal cells secrete VEGF in response to diverse microenvironmental signals such as hypoxia, inflammation, and growth factors which lead to an abnormal balance of pro- and anti-angiogenic factors. As a result, while angiogenesis is well-regulated in healthy tissue, tumors typically exhibit aberrant vasculature with leaky vessels, excessive branching, and abnormal morphology [12]. The abnormal vasculature leads to irregular mass transport of nutrients from the blood and waste from the tissue, resulting in gradients of small molecules.

The immune compartment is another important part of the TME as there are numerous types of immune cells which contribute in different ways to promote or suppress tumor progression. Immune cells can be resident or be recruited to the tumor. Concerning immune cell infiltration, there is significant infiltration of macrophages and lymphocytes (e.g., T cells, B cells) which result from local inflammation induced by tumor progression.

T cell infiltration in tumors can be described as anti-tumoral by recognition of specific cancer antigens leading to cancer cell destruction. The understanding of these interactions between cancer cells and T cells led to development of immunotherapy, a promising cancer treatment. For example, the study of T cell and cancer cell communication revealed that targeting the programmed death (PD-1/PD-L1) pathway with antibodies such as nivolumab have drastic effects on cancer progression for several types of cancers [13]. On the other hand, a subpopulation of T cells called T regulatory cells have been found to suppress the anti-tumor response, and understanding their role is an important consideration in further development of immunotherapies.

Infiltrating B cells represent a substantial percentage of all infiltrated lymphocytes, however, their role in cancer is still controversial. In fact, B cells exhibit both pro- and anti-tumoral features [14]. On one hand, B lymphocytes produce lymphotoxins which promote angiogenesis in tumors [15] but, on the other hand it was also demonstrated that the infiltration of B cells in melanoma is correlated with better survival [16].

Macrophages that have infiltrated into tumors are called tumor-associated macrophages (TAMs). These TAMs have two distinct origins: monocyte-derived macrophages recruited from systemic blood circulation and macrophages from the resident tissue. Macrophages represent a heterogeneous group that can be broadly categorized in two main categories: pro-inflammatory (M1) or anti-inflammatory (M2) macrophages. M1 macrophages are considered as anti-tumoral due to their ability to recruit other immune cells into the TME and their capacity to present tumoral antigens, while M2 macrophages exhibit tumor-promoting features by enhancing angiogenesis, immunosuppression and tumor growth. TAMs in particular may exhibit a mixed M1/M2 phenotype depending on the balance of chemical cues in the TME, suggesting that there is a spectrum of macrophage phenotypes [17]. Since the number of infiltrated macrophages [18] and the ratio between M1/M2 macrophages [19] have prognostic value in some cancers, they are potentially novel cancer targets, with some molecules that target TAMs already under clinical trial investigation [20]. TAM depletion (e.g., CSF1R inhibition [21]), TAM repolarization to an M1 phenotype [22] and targeting functional molecules of TAM (e.g., Fc receptor inhibition [23]) have been investigated to enhance the effectiveness of immune checkpoint inhibitor therapies that aim to enhance the killing ability of T cells present in the tumor [20].

Natural killer (NK) cells represent another subset of lymphocytes whose main role is cytotoxic activity after activation. NK cell activation and how NK cells recognize cancer cells and eliminate them is well understood [24]. While it has been shown that NK cells are efficient in eliminating circulating tumor cells (CTCs), their efficiency in killing tumor cells is much lower. Accumulating evidence indicates that tumor cells and the surrounding TME are endowed with suppressive properties that affect NK cell reactivity and inhibit their cytotoxic functions, potentially explaining their inefficiency in killing tumors. For example, it has been shown that CAFs in melanoma modulates NK cells anti-tumor cytotoxicity [25]. However, contradictory evidence such as the ability of dendritic cells (DCs) to activate [26] or inhibit [27] NK cell activity suggests that the role of the microenvironment in regulating NK cell activity is still unclear.

DCs are professional antigen-presenting cells (APCs) also found in the TME. In their quiescent state, their major role is to capture antigens. When activated, they can present antigens to T cells by moving previously captured antigens to their surface and expressing co-stimulatory molecules to prime T cells toward the antigen expressing tumor cells. Regarding their role in T cell response priming, DCs were long considered an essential component of anti-tumor immunity. This has led to the development of DC vaccines which are loaded with tumor antigens ex vivo to increase tumor recognition by cytotoxic T cells when delivered to patients [28]. However, some studies have revealed that DCs from the TME exhibit defects in their differentiation, activation, and function, leading to immune escape. These defects could be modulated by small molecule gradients in the TME such as hypoxia or cytokine secretion by macrophages [29,30].

The heterogeneity within the TME and between tumors is a possible source of variation in therapeutic responsiveness between patients. A range of approaches have been suggested to elucidate the complex interactions and cross-talk which occurs in the TME, many of which will be discussed in the following sections. In particular we focus on how researchers have used tissue engineering strategies to understand the roles of these cellular actors. It is apparent that the success of many emerging therapeutic strategies relies on a thorough understanding of the complex tumor-promoting and anti-tumor interactions occurring within the TME.

### 19.2.2 Non-cellular components

As previously mentioned, the abnormal vasculature leads to defects in mass transport and especially in oxygen transport, leading to the development of hypoxic areas which are a feature of most solid tumors [31]. Hypoxia regulates diverse hallmarks of cancer and has been associated with tumor progression, therapeutic resistance, increased invasiveness and local and distant metastasis, abnormal angiogenesis, resistance to apoptosis, and overall worse patient outcomes. The response to hypoxia is regulated through hypoxia-inducible factor (HIF) pathways, mainly through HIF1-α transcription factor activity and its downstream targets genes [31–35]. It has been shown that cells experiencing hypoxia switch to a glycolytic metabolism which increases glucose uptake, decreases mitochondrial oxygen consumption, and alters the pH of the local environment, creating gradients in metabolites such as lactate. Furthermore, severe hypoxia has been found to promote cell survival by activating the unfolded protein response (UPR) [33] due to dysfunctional protein folding in the endoplasmic reticulum (ER) [36]. Hypoxia has also been shown to dysregulate cell behavior. For example, fibroblast activation has been linked to hypoxia and acidification of the TME, leading to altered ECM remodeling which supports tumor invasion [32]. In addition, hypoxia leads to immunosuppression, as shown by the impairment of the antitumor T cell response in the local hypoxic tissue [37]. Further, HIF1-α directly induces PD-L1 expression on tumor cells, which is a T cell immune checkpoint [37]. Downstream, PD-L1 on the tumor cells binds to the PD-1 receptors on T cells causing the T cell to become dysfunctional and resulting in tumor cell protection from T cell killing [37].

Another non-cellular component of the TME that is altered during disease progression is the physical attributes of the microenvironment, such as ECM stiffness, interstitial pressure, and the development of compression forces [38,39]. ECM reorganization and cross-linking are affected by cell types within the microenvironment, such as CAFs. These changes in the physical properties of the ECM can lead to more rapid tumor progression, changes in mechanotransduction, enhanced tissue invasion, and metastasis [40,41]. Further, the highly dense ECM in the TME and increased interstitial pressure can lead to blood vessel collapse within the tumor, further creating localized sites of deoxygenated and nutrient-deficient regions.

The TME integrates this range of complex, heterogeneous, and dynamic cellular and non-cellular components, which results in disease evolution and variable therapy responsiveness between patients. As a result of the spatiotemporal changes in the tissue architecture, small molecule gradients, and ECM remodeling, the relationship between tumor cell properties and local microenvironmental signals are difficult to dissect in vivo. Engineered platforms however, offer the opportunity to modulate specific structural components of the TME to better understand the relationship between tumor cell environment and function, and to generate heterotypic, three-dimensional (3D) culture models that more accurately replicate microenvironmental complexity to better predict the impact of novel therapeutic approaches [42].

## 19.3 Current strategies for creating cell and matrix organization to mimic microenvironment

Many engineered in vitro models have been introduced to recapitulate the TME and facilitate investigation of potential therapeutic approaches. In the following section, we categorize seven major types of models that are used frequently in this field (Fig. 19.2). These models provide an intermediate between conventional two-dimensional (2D) assays which are easy to implement and in vivo studies which capture more biologically complex data. Each model contains unique advantages and limitations which are described in Table 19.1. Examples of the application of these models are discussed further in Section 19.4.

### 19.3.1 Organoid derivation options (patient-derived organoid vs patient-derived xenograft)

Organoids are 3D cellular clusters that have allowed researchers to study disease progression using primary patient material in vitro (Drost and Clevers, 2018). They represent a self-renewing source of patient-specific cells that have been shown to maintain phenotypic heterogeneity, differentiation capacity, patient-specific physiological changes, histological architecture of human tissues, and even response to chemotherapy [43]. PDOs are organoids formed directly from patient tissue which has been digested into single cells and cultured in Matrigel, while PDX organoid models involve growing the patient cells in a mouse before being digested into single cells and cultured in Matrigel domes [33]. Organoid models are typically available from an established tissue bank such as those created by the Human Cancer Models Initiative (HCMI). The HCMI, led by the Wellcome Sanger Institute and Cancer Research UK, generates highly annotated organoid models generated from patient samples for ATCC (https://www.atcc.org/hcmi).

### 19.3.2 Transwell-based assays

Transwell-based assays (i.e., transwell permeable supports or Boyden chambers) have been used extensively in migration studies, in which a migratory cell type is added onto the membrane in the top section of the transwell insert, and a chemokine or second cell type is provided in the well below (Fig. 19.2B). The cells then migrate to the bottom region of the well, allowing for analysis at a defined endpoint. A layer of cells (e.g., HUVEC) or matrix can also be added to the membrane to study invasion instead of migration. This simplistic tool has been used frequently to elucidate cellular interactions which act to promote or inhibit cellular motility. There have been attempts to recreate tools which can replace or extend the current use of the assay, such as the Lockett group who developed a paper-based migration assay significantly cheaper than traditional transwell inserts [44], and Kalli et al. who used transwell inserts in their study of fibroblast activation by the addition of a compressive force to their platform [45]. Further, more complex studies have been performed using novel assays developed based on transwells, such as Sherman et al., in which spheroids were integrated into the system, more closely recapitulating the 3D TME and structure [46].

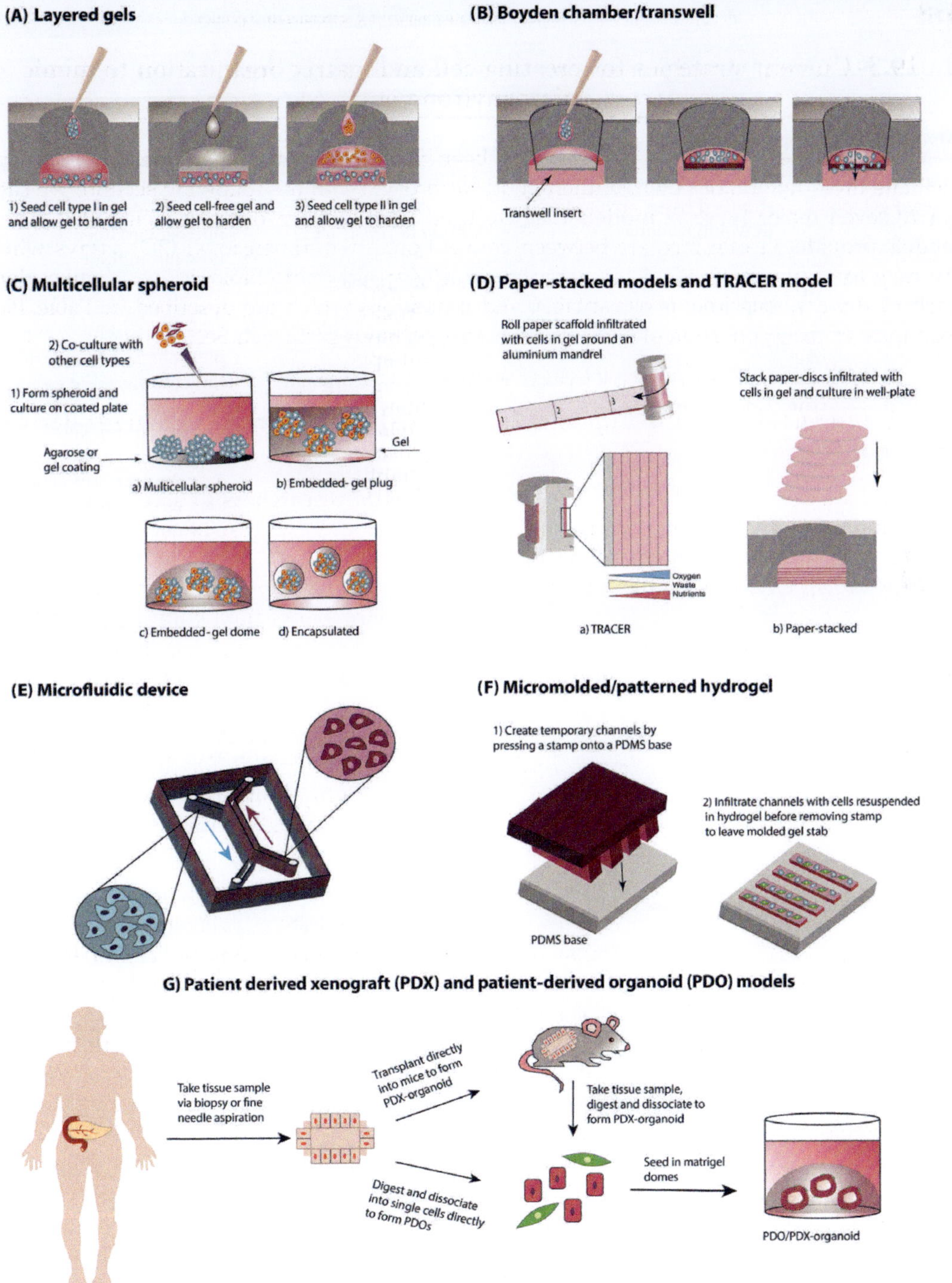


FIGURE 19.2 Heterogeneous models used to recapitulate microenvironmental parameters of the tissue microenvironment. The following models are discussed: (A) the layered gel or organotypic model, (B) Boyden chamber or transwell model, (C) multicellular spheroid models, (D) paper-stacked and TRACER models, (E) microfluidic device, (F) micromolded or patterned hydrogel, (G) PDX and PDO models.

TABLE 19.1 Advantages and limitations of tissue-engineered models.

| Model | Microenvironmental feature mimicked | Advantages | Limitations |
|---|---|---|---|
| Organotypic layered gel model | Heterogeneity of cells, stratified tissue architecture, ECM interactions | • Ability to incorporate different cell types<br>• Simple to construct | • Cannot be used to easily study graded microenvironments<br>• Low throughput |
| Boyden chamber or transwell | Interface between two cell population or cells and ECM | • Study migration/ invasion<br>• Easy implementation | • Cannot be used to study graded microenvironments<br>• Limited capacity to mimic physiological complexity |
| Multicellular tumor spheroid model | Heterogeneity of cell types, TME gradients | • Potential to incorporate different cell types<br>• Easy implementation and high throughput | • Cannot be used to easily study graded microenvironments<br>• Presence of necrotic core over time (i.e., difficulty in long-term cultures) |
| Stacked paper and TRACER model | Hypoxia, small molecule, and lactic acid gradients | • Study graded microenvironments<br>• Incorporate different cell types | • Low throughput<br>• Presence of exogenous material (i.e., cellulose scaffold)<br>• Optimization required for reproducible assembly |
| 2D microfluidic device | Shear stress, fluid flow, interstitial pressure | • Potential for high throughput<br>• Simple and dynamic imaging possible | • Challenging to manufacture and implement<br>• Lacks presence of 3D ECM environment<br>• Low throughput |
| Micromolded or patterned 3D hydrogel model | Heterogeneity of cell types, small molecule gradients | • Flexibility in patterning different cell types<br>• Flat geometry facilitates easier imaging. | • Challenging to manufacture<br>• Limited by hydrogel integrity and remodeling |
| PDX and PDO model | Tumor heterogeneity | • Use of primary cells better captures patient-specific biology<br>• Maintains tumor cell heterogeneity and differentiation capacity | • Expensive to maintain, difficult to access<br>• Currently cannot be used to study graded microenvironments |

## 19.3.3 Organotypic model

Organotypic models are complex culture models which use primary tissue or organ slices as the basis of the experimental model. Through tissue engineering strategies, researchers use hydrogel embedding to generate structures similar to those seen in organ slices such as the layered gel approach which can simulate stratified epidermis [47] (Fig. 19.2A). These cell models are customizable and allow the user to recapitulate or probe the formation of certain tissue organization. However, due to their inherent customizability, these models are low throughput and can be difficult to assemble and analyze.

### 19.3.4 Microfluidic devices

Microfluidic devices have commonly been used to model chemotaxis, but more recently, they have been designed to study the effects of fluid flow [48]. These "organ-on-a-chip" models are typically fabricated with polydimethylsiloxane (PDMS) to generate microchannels and reservoirs in which cells and ECM can be deposited to produce channel regions containing cells encapsulated in a 3D hydrogel phase. These channels containing cell-embedded gels are then perfused with cell culture medium and chemotherapeutic agents.

### 19.3.5 Micromolded 3D gels

Micromolded gel-based models have been used to study tumor cell migration. By generating microtracks in ECM such as collagen hydrogel which mimic the in vivo structures, the migration of cells can be studied [49,50]. These models also allow greater user control over the matrix stiffness and geometry compared with PDMS-based microchannels [51]. Recently, our group (D'Arcangelo et al.) proposed a powerful 3D model with compartmentalized gels, allowing for spatial stratification of multiple cell types and real-time observation of tumor invasion [52]. This 3D gel model allows for observation of the tumor–stromal interface as the cells directly interact with each other and enable the user to probe the mechanisms which drive the microenvironment-dependent cellular behavior.

### 19.3.6 Multicellular spheroid

Multicellular tumor spheroids (MTS) are aggregates typically formed by seeding a high density of cells (cell-line derived) on low adhesive or agarose coated culture plates and is considered to be the first 3D in vitro model to be generated [53]. Current models of the MTS involve embedding these spheroids in a gel [54] to more accurately recapitulate TME architecture in an ECM (Fig. 19.2C-a). Another method to form MTS consists of encapsulating cells in alginate shells [55] (Fig. 19.2C-d). This method allows the formation of MTS with nonadherent cells and is suitable to investigate the role of physical cues on cell behavior. These spheroids can also be mass-produced using high-throughput 3D culture plates [56].

### 19.3.7 Stacked paper models

Stacked paper models, or cells-in-gel-in-paper (CiGiP) models use cellulose-based paper structures as a scaffold for seeding cells and ECM. The paper structure provides mechanical reinforcement of the embedded hydrogel structure and facilitates easy disassembly of the tissue via destacking, for greater spatial analysis. The Whitesides' group pioneered this culture system and has used this system to generate gradients in the microenvironment through the cellular production and consumption of small molecules, allowing for studies of cellular invasion and chemotaxis in the presence of these disease-relevant microenvironmental gradients [42,57]. Similar stacked structures have also been developed by the Lockett group [44], while Boyce et al. developed such as a system which allowed for the study of chemoresistance [58]. Our group has developed a rolled version of the stacked paper model called Tissue Roll for Analysis of Cellular Environments and Response (TRACER), which allows for a more rapid disassembly of the tissue construct (Fig. 19.2D-A). This enabled the analysis of cellular metabolism, which was previously inaccessible due to the slow process of destacking individual layers of these paper–composite structures [59–61].

### 19.3.8 Cell sources used in tissue-engineered models

A variety of cell sources such as mouse and human cell lines, and primary cells from a variety of tissue origins have been used to date in order to establish in vitro 3D tumor models. Different cell sources offer different advantages and disadvantages for establishing and validating 3D platforms, for uncovering novel biological information, and for translation of drug screening applications. Further, as more information is discovered about the growth and maintenance of primary cells in vitro, the possibility of using primary cells instead of cell lines in vitro increases, potentially opening the door to elucidating more complex and physiologically relevant cancer biology. Table 19.2 below summarizes cell sources for modeling a variety of aspects of the TME in vitro.

## 19.4 Modeling important aspects of the tumor microenvironment

### 19.4.1 In vitro models of tumor–fibroblast interactions

The histological structure of the stromal compartment is characterized by the presence of highly organized activated fibroblasts, secreted ECM, and the presence of other cell types. Stromal remodeling is often paralleled to the process of wound healing [106], where some cancer types such as pancreatic cancer develop a large region of remodeled stroma surrounding and within the tumor through what is called the "desmoplastic reaction." This results in the formation of a highly fibrous capsule surrounding the tumor. Tumor–stromal interactions have long been recognized to be tumor promoting and to contribute to tumor invasiveness [3,7,107]. Depending on the type of cancer being studied, the structure and composition of the TME can be drastically different [106]. For example, many tumors show a sharp delineation between the tumor parenchyma and the stromal compartment, in which the basal lamina separates the compartments. Further, the tissue composition can range considerably [106] and the stromal fraction may also include many different cell types. The most abundant cell type is the CAF, but within the population of CAFs there are also subpopulations of CAFs with varying degrees and types of tumor-promoting effects [9]. Ideal models to study CAF–tumor interactions therefore enable control over the relative locations of CAF and tumor cells and a separate analysis of each cell population.

While conventional 2D or in vivo methods have been used extensively to explore aspects of tumor–CAF interactions, the 3D tissue engineering strategies as discussed in Table 19.3 provide the user with control over the organization of CAF and tumor components to enable the recapitulation of microenvironmental features that more closely resemble the

TABLE 19.2 Cell sources currently used in tumor microenvironment models.

| Cell type | Advantages | Disadvantages |
|---|---|---|
| ***Primary human-derived (non-immune, various tissues of origin):*** | | |
| • Cells derived from human tissue samples obtained during diagnostic biopsy, surgical resection, or tissue donation from healthy donors | | |
| • Tumor cells (patient-derived) [62–64]<br>∘ PDO generated directly from tissue sample) [65]<br>∘ Xenograft-derived organoid* (PDX) generated from xenograft of patient tissue sample)<br>• Stromal cells [66–68]<br>∘ Patient-matched CAFs [69]<br>∘ Unmatched (allogenic) CAFs<br>∘ Xenograft-derived CAFs* (human origin)<br>∘ Normal fibroblasts<br>∘ Tissue specific or nontissue specific<br>∘ Donor-matched or unmatched<br>• Endothelial cells<br>∘ Patient-derived<br>∘ Healthy donor-derived<br>∘ Cultured microvascular cells [70,71]<br>∘ Telomerase-immortalized cells [72] | Highly personalized, limited genetic drift (at least at low passage), potential for therapeutic matching or specific mutation identification, greater physiological relevance and potentially improved translational potential | Tissue samples may be rare and/or provide minimal source material, may be difficult to grow/maintain in culture (limited passage number/expansion capacity), may require complex culture conditions, safety risks less well characterized<br>*Xenograft-derived cells may experience contamination with mouse cells due to enrichment of the mouse cell population over time |
| ***Primary human-derived (immune: macrophages, neutrophils, lymphocytes, NK cells, DCs):*** | | |
| • Immune cells isolated from circulating blood, bone marrow, or tumor samples from human subjects. Immunotherapy patient tissue is of particular interest due to its treatment-modified immune parameters | | |
| • Bone marrow derived [73]<br>∘ Peripheral blood/buffy coat-derived<br>∘ Healthy [74–76]<br>∘ Donor-matched [77]<br>∘ Unmatched (allogenic)<br>∘ Immunotherapy patients [78,79]<br>• Endogenous tumor-infiltrating lymphocytes [62,65,80] | Some immune cells cannot be cultured and must be obtained fresh from samples, matched samples will reduce risk of immune cells killing other cell types recognized as foreign, may provide unique insight into tumor–immune interactions in vivo | REB approval with potential restrictions required for use, specific tissue types and patient populations may be more difficult to obtain, potential for donor–donor variability making replicability challenging, may be difficult or impossible to maintain/expand in culture |

(*Continued*)

**TABLE 19.2** (Continued)

| Cell type | Advantages | Disadvantages |
|---|---|---|
| *Human cell lines*: | | |
| • Immortalized cells propagated from a primary tumor or metastatic site, surrounding stromal tissue, or healthy donors | | |
| • Tumor cells [63,78,79,81–85].<br>• Stromal cells<br>◦ Normal fibroblast [74]<br>◦ CAF<br>• Immune cells [86]<br>◦ Macrophages [87]<br>◦ Lymphocytes<br>◦ NK cells [46]<br>• Endothelial cells (most commonly HUVEC) [46,88–97] | Easy to grow, cost effective, commercially available (and quality controlled), typically can be expanded indefinitely, allow direct comparison of results with other studies, minimal variation between batches | Minimal genetic heterogeneity, may have very different genetic profile compared with original cell source, mismatch with tumor cells may impede assessment of tumor–immune interactions recent international concerns about cross-contamination between cell lines |
| *Primary mouse-derived (non-immune)*: | | |
| • Cells derived from mouse tissue including from GEM models, non-GEM models and xenograft models (tissue of mouse origin) | | |
| • Tumor cells (nonxenograft, e.g., GEM) [98,99])<br>• Stromal cells<br>◦ CAFs from GEM models [99]<br>◦ CAFs from xenograft model (mouse origin)<br>◦ Normal fibroblasts | May be easier to obtain compared with primary human tissue, may better recapitulate cell–cell interactions if all mouse cells used, limited genetic drift (at least at low passage), cells can be easily lineage labeled to enable identification of specific cell types, GEM models allow experiments with cells harboring identical genetic driving mutations | Animal models required, non-human source material may not replicate human biology, complex culture conditions may be required, may have limited passage number/expansion capacity in culture |
| *Primary mouse-derived (immune—macrophages, monocytes, neutrophils, lymphocytes, NK cells, DCs)*: | | |
| • Mouse immune cells derived from tissue harboring immune cells, enriched, and propagated in vitro | | |
| • Spleen derived [100–102]<br>• Bone marrow derived [87,103,104]<br>• Peripheral blood derived<br>• Tumor-infiltrating lymphocytes | Tissue is relatively easy to obtain (easier than human), may provide better insight into tumor–immune interactions than immortalized cell lines; variety of immune cell types can be obtained from this tissue, including specific subsets | Access to animal models required, mouse immune system does not accurately represent human immunity; may be challenging to maintain in culture long term |
| *Mouse cell lines*: | | |
| • Immortalized cells propagated from various tissues from healthy animals or mouse models of various diseases | | |
| • Tumor cells [73,100,104]<br>• Stromal cells [76]<br>◦ Normal fibroblasts (commonly 3T3)** [105])<br>• Immune cells [85] | Easy to grow, cost effective, commercially available (and quality controlled), can be expanded indefinitely, allow direct comparison of results with other studies, minimal variation between batches | Mouse origin—may not be representative of human biology; minimal genetic heterogeneity; may have significant genetic drift from tumor source<br>**CAF lines are commercially available but very uncommon |

**xenograft derived cells (top row).*
***stromal cells (bottom row).*

in vivo tumor architecture. While some approaches use mixed co-culture configurations and observe the paracrine signaling in the intermixed system, many aim to recapitulate the initially segregated tissue structure observed in carcinomas. For example, D'Arcangelo et al. produced a 3D model with initial segregation of tumor cells from CAFs in order to observe the formation of invasive strands [52]. These models allow researchers to interrogate disease-relevant questions in vitro and probe either the tumor cell behavior (e.g., invasion) or CAF behavior (e.g., secretion and remodeling of ECM) in the same platform.

### 19.4.2 In vitro models of tumor–immune interactions

Immune cells are important and complex components of the TME that have gained interest in recent years due to their critical role in both preventing and promoting tumor progression. The immune contexture of different tumors can vary [116] and understanding how to modulate the behavior of these immune cells to prevent tumor progression has

**TABLE 19.3** In vitro models of tumor–fibroblast interactions.

| Assay/metric quantified | Models used |
|---|---|
| ***CAF-promoted tumor cell invasion or migration*** | |
| • Migration index (cell count/microscopy) [7,52,60,64,108–112]<br>• Migration velocity (microscopy) [52,108,109]<br>• CAF morphology (microscopy) [52,66] | • Multicellular spheroids [81,108,109]<br>• Microfluidic device/bioreactor [66]<br>• Organotypic model [7,64]<br>• Encapsulated microspheres [110]<br>• Decellularized matrix scaffold [111]<br>• Stacked paper model [60,112]<br>• Patterned hydrogel [52] |
| ***CAF-associated drug sensitivity, chemoresistance, and radiation resistance*** | |
| • Drug penetration and distribution (microscopy) [81,82]<br>• Cell viability in response to treatment [82,110,111]<br>• Tumor volume after treatment [82,113] | • Multicellular spheroids [81,82]<br>• Microfluidic device [113]<br>• Organotypic model [82]<br>• Encapsulated microspheres [110]<br>• Decellularized matrix scaffold [111] |
| ***EMT markers and fibroblast activation marker expression*** | |
| • Histology [60,68,82,111,112,114]<br>• Western blot [108]<br>• ELISA [52,67]<br>• qPCR [67]<br>• Secretome analysis [108] | • Multicellular spheroids [67]<br>• Organotypic model [68,82]<br>• Microfluidic device/bioreactor [114]<br>• Decellularized matrix scaffold [111]<br>• Stacked paper model [60,112]<br>• Patterned hydrogel [52] |
| ***ECM secretion and remodeling*** | |
| • Alignment of collagen and fibronectin fibers (microscopy) [68,114,115]<br>• Collagen expression (Western blot) [52,81] | • Microfluidic device [114,115]<br>• Organotypic model [68,82]<br>• Multicellular spheroids [81]<br>• Patterned hydrogel [52] |

brought about a paradigm shift in the way we treat cancer. While immunotherapy has proved an effective option for melanoma [117] and non small cell lung carcinoma [118] for example, these immunotherapies still remain ineffective in most solid tumors such as pancreatic ductal adenocarcinoma [119] and colorectal carcinoma [120] due to an incomplete understanding of these complex interactions among different immune cells with tumor cells in the TME.

Advances in 3D in vitro models have allowed researchers to more accurately recapitulate the 3D TME architecture compared with 2D in vitro models, and to study human tumor–immune interactions in more detail and throughput than is possible using conventional in vivo mouse models. The questions that most researchers have been trying to address, as well as specific assays performed using 3D in vitro models are summarized in Table 19.4. As early as 1977 [100], most of these questions centered around understanding the tumor-killing capacity of immune cells and their ability to migrate into the tumor, with the most popular in vitro model being the MTS. However, with the promise of immunotherapy as a curative solution for historically hard to treat cancers, other questions such as phenotypic changes, tumor growth, antigen recognition efficacy, and tumor invasiveness have begun being investigated in vitro, specifically in the context of a 3D architecture. More complex models such as microfluidic devices to study immune cell infiltration and tumor killing in real time [122,124], embedded/encapsulated MTS [126,131], layered gels [126,131], and some novel models [123,137] have consequently emerged to assist in probing these questions.

Recently, a few groups have started evaluating the immunotherapeutic potential of engineered CAR-T cells or autologous T cells using these models. This demonstrates the promise of 3D in vitro tools in perhaps discovering novel ways to augment immunotherapies, to predict patient response and to accelerate translation of T cell therapies from bench to clinic [62,122,127]. In addition to this, understanding how the microenvironment affects tumor–immune interactions has gained interest due to an increased recognition of its role in dictating the behavior of immune cells to promote or prevent tumorigenesis [85,137], although outputs on this research as a main focus using 3D in vitro models are scarce.

Overall, these 3D in vitro models provide a valuable opportunity to understand tumor–immune interactions in the context of the human microenvironment which is extremely important in developing and discovering new immunotherapies. Although most research efforts have focused on understanding how T cells can kill tumor cells more efficiently, work on other immune cell types such as how DCs can improve antigen recognition and presentation, how macrophages and myeloid cells might be manipulated to promote an anti-tumorigenic environment, and understanding how small molecule gradients that exist due to a network of cellular cross-talk using innovative 3D in vitro models could be key to augmenting immunotherapies efficiency.

### 19.4.3 In vitro models of hypoxia and small molecular gradients

Gradients of oxygen, nutrients, growth factors, and waste products are typical in tumors and develop due to tumor growth outpacing the available vasculature and/or the formation of dysfunctional vasculature in the TME, resulting in impaired transport of nutrients, oxygen, and waste products [34]. This pathologic environment applies selective pressure on the tumor and influences cell behavior and response—including apoptosis

**TABLE 19.4** In vitro models of tumor–immune interactions.

| Assay/metric assessed | Models used |
|---|---|
| ***Immune cell–mediated tumor killing*** | |
| • $^{15}$Cr-release assay [77,79,80,100,121]<br>• Live/dead viability assay (microscopy) [122],[a] [123–125], [126][a]<br>• Tumor size (microscopy) [121,123], [85],[a] [127][a,b]<br>• Live/dead viability (flow cytometry) [46], [127],[a,b] [125]<br>• Cell death via colorimetric assay (MTT assay) [80], [85][c]<br>*Immune cell types investigated*: T cells, engineered T cells, and NK cells | • MTS [100,121], [127],[a,b] [77,79,80,125]<br>• Embedded MTS [98], [85][c]<br>• Microfluidic [124], [122][a]<br>• Organotypic/layered gels [126][a]<br>• Other/novel [107] |
| ***Immune cell–mediated tumor migration*** | |
| • Migration speed (time-lapse microscopy) [75,87]<br>• Quantification and phenotype (flow cytometry) [128]<br>• Phenotype of invasive tumor cells (IHC/IF) [75,128]<br>• Gene expression of invasive tumor cells (qPCR)<br>• Protein expression of invasive tumor cells (Western blotting) [87]<br>*Immune cells investigated*: Macrophages, T cells, and NK cells | • Microfluidic [75,87]<br>• MTS [128]<br>• Embedded multicellular spheroid [128] |
| ***Immune cell activation and antigen-uptake/recognition related functions*** | |
| • Antigen recognition efficacy (flow cytometry) [62],[a] [78–80,84]<br>• Cytokine secretion of antigen recognition signals (ELISA) [62],[a] [78–80]<br>• Gene expression of tumor/immune antigen expression/recognition signals (qPCR) [78], [62][a]<br>*Immune cells investigated*: T cells, natural killer cells, and DCs | • MTS [62],[a] [78–80,84] |
| ***Phenotypic changes due to tumor–immune cell interactions*** | |
| • Gene expression analysis (qPCR and RNA-seq) [74,129], [122],[a] [63[b],77]<br>• Cytokine secretion (ELISA/multiplex assays) [74,129,130], [122],[a] [63],[b] [131],[b] [126]<br>• Proteomics analysis (LC-MS/MS and western blot) [77,132]<br>• Flow cytometry [133],[a] [77][131][b]<br>• Immunofluorescence/histochemistry [74,104,122,129], [63],[b] [130]<br>*Immune cells investigated*: NK cells, T cells, and macrophages | • MTS [77,129]<br>• Embedded MTS [130], [63][b]<br>• Encapsulated MTS [132]<br>• Scaffold-based [104]<br>• Microfluidic [133],[a] [124], [122][a]<br>• Organotypic/layered gels [131][b]<br>• Transwell/Boyden chamber [130] |
| ***Immune cell migration/infiltration into tumor cells*** | |
| • Migration speed/distance (time-lapse microscopy/microscopy) [124], [133],[a] [101,102,123], [131],[b] [126,134]<br>• Protein expression on immune cells (flow cytometry/western blot) [75,135]<br>• Percentage and phenotype of infiltrating immune cells (flow cytometry) [46], [127][a,b]<br>• Percentage and phenotype of infiltrating immune cells (IHC/IF) [46,62,100,121], [127],[a,b] [78,125,135,136]<br>• Gene expression analysis (qPCR) [133]<br>*Immune cells*: T cells, engineered T cells, NK cells, macrophages, and DCs | • Microfluidic [124], [133],[a] [101,102]<br>• Organotypic [126], [131][b]<br>• MTS [46,62,100,121], [127],[a,b] [78,125,128,134,136]<br>• Transwell/Boyden chambers [135]<br>• Other/novel [123] |

(*Continued*)

TABLE 19.4 (Continued)

| Assay/metric assessed | Models used |
|---|---|
| ***Microenvironmental effects on tumor–immune interactions*** | |
| • Migration of immune cells in hypoxia (microscopy) [124,137]<br>• Tumor killing by immune cells in hypoxia (Live/dead viability via microscopy) [124], [137][a]<br>• Oxygen concentration of tumor–immune co-cultures [85], [137][a]<br>• Cytokine secretion of tumor–immune co-cultures in hypoxia (ELISA) [85]<br>• Immune cell activation in correlation with tumor lactic acid production [78]<br>• Surface marker expression (IF/flow cytometry) [137][a]<br>*Immune cells investigated*: engineered T cells, antigen-reactive T cells, and macrophages | • Microfluidic [124]<br>• Multicellular MTS [78]<br>• Embedded MTS [85]<br>• Other/novel [137][a] |
| ***Tumor growth/formation*** | |
| • H-Thymidine tumor proliferation assay [136]<br>• Tumor size over time (IF/microscopy) [138,139]<br>• Clonogenic assay [100]<br>*Immune cells investigated*: macrophages and leukocytes | • MTS [100,136,138,139] |
| ***Immunotherapeutic potential*** | |
| • Tumor viability (microscopy/flow cytometry) [122],[a] [127],[a] [126]<br>• Tumor size over time (microscopy) [127],[a,b] [85][c]<br>• Gene expression analysis (qPCR) [62], [122][a]<br>• Cytokine secretion (ELISA) [62], [122][a]<br>• Surface marker expression (flow) [133][a]<br>• Antigen recognition efficacy (flow cytometry) [62]<br>• Phenotype of infiltrating immune cells (IF/IHC) [127]<br>• Immune cell migration into tumors (microscopy) [133,137]<br>*Immune cells investigated*: T cells, engineered T cells, NK cells, DCs, and macrophage | • MTS [62,127]<br>• Embedded multicellular spheroid [85][c]<br>• Microfluidic [122],[a] [133])<br>• Organotypic/layered gels [126][a]<br>• Other/novel [137] |

[a]*In response to immunotherapeutic treatment.*
[b]*More than two immune cell types investigated in study.*
[c]*In response/synergy to chemotherapeutic treatment.*

and necrosis of tumor cells—resulting in increased tumor cell heterogeneity and extensive genotypic and phenotypic variance [35]. Due to the significance of these gradients on tumor cell properties a number of in vitro models have emerged and is further discussed in the sections below.

### 19.4.3.1 *Oxygen gradients*

While traditional 2D in vitro models cannot easily recapitulate the hypoxic gradients that arise in tumors, the rise of 3D culture assays and platforms have allowed greater insight into the response of tumor cells to hypoxia [31] (Table 19.5). Thus, the ability to establish hypoxia has increasingly been used as a metric to validate 3D culture platforms which attempt to recapitulate complex features of in vivo biology [145,146] (see examples in Table 19.5). Broadly, these platforms can be divided into two categories: 3D platforms

**TABLE 19.5** In vitro models of hypoxia and small molecular gradients.

| Assay/metric assessed | Models used |
|---|---|
| ***Impact of hypoxia on treatment responsiveness*** | |
| • Cell number [57,137,140]<br>• Cell viability in response to treatment [34,57,137,140]<br>• Mass spectrometry protein analysis for proteins associated with cell cycle/apoptosis/proliferation [140]<br>• Metabolic activity by CellTiter-Glo [57]<br>• Cell proliferation by EdU [57]<br>• Immunofluoresence [137] | • Low-oxygen environment<br>• 3D suspension cell culture [140]<br>• Microfluidic + patterned hydrogel [137]<br>• Cell-generated hypoxia<br>• Stacked paper [34,57] |
| ***Impact of hypoxia on cell behavior (e.g., EMT, extravasation, receptor activity)*** | |
| • Cell viability [34,141–143]<br>• Cell proliferation [34,142]<br>• Spheroid size [141]<br>• qPCR of key genes—e.g., cancer specific, hypoxia-associated [34,141,143]<br>• Protein expression by Western blot—E-cadherin, hypoxia factors, receptors [141,143]<br>• Fluorescent imaging/immunofluorescence [34,141–144]<br>• Migration assay [34,141,144]<br>• ELISA for HIF1-a, cytokine secretion [142–144]<br>• Flow cytometry—ALDH expression [144]<br>• miRNA characterization [144]<br>• Pseudovessel formation [144]<br>• Proteasome activity [143]<br>• Metabolomics analysis [34] | • Low-oxygen environment<br>• Tumor spheroids [141,144]<br>• Hydrogel [141,144]<br>• Paper-based [143]<br>• Cell-generated hypoxia<br>• Microfluidic [142]<br>• Stacked paper [34] |
| ***The establishment and measurement of hypoxia in 3D culture models*** | |
| • Cell number [145]<br>• Cell cycle analysis [145]<br>• Cell viability [145]<br>• Ki67—cell proliferation [145]<br>• qPCR of hypoxia related genes [145,146]<br>• Western blot of hypoxia related proteins [145]<br>• Ki67 [145]<br>• Immunofluoresence [146]<br>• Gel characterization by NMR [146]<br>• Histological analysis [146] | • Tumor cell aggregates in gel + chemical hypoxia-mimetic agent [145]<br>• Gelation reaction-induced hypoxia in a gelatin-based hydrogel [146] |
| ***Impact of small molecules on cell behavior*** | |
| • Western blot [8]<br>• Immunofluorescence [8]<br>• Migration assay [8,133,147]<br>• Weighted genome analysis [8]<br>• Cell proliferation by EdU [147]<br>• Cell aggregate size and density [147]<br>• Live fluorescence imaging—cell migration velocity, quantity [133]<br>• qPCR for expression of chemokines and chemokine receptors [133]<br>• Cell pH by distance from spheroid center [148]<br>• Doxorubicin uptake rate steady state accumulation and effectiveness [149] | • Microfluidic device [133,150]<br>• Tumor spheroids + pH modulating solutions [148]<br>• Transwell [8,147] |

which rely on a low-oxygen environment (e.g., low-oxygen incubator) to generate a hypoxic response, and 3D platforms in which the incorporated cell type(s) generate and define the hypoxic gradient. While platforms using a low-oxygen environment may be easier to develop and control, cell-generated hypoxic gradients likely better recapitulate the pathophysiologic response of tumor cells in vivo.

Thus far, most 3D platforms for modeling hypoxia have focused on the investigation of the influence of hypoxia on treatment resistance [34,57,137,140] and cell invasiveness/migration capacity [141,144]. Hypoxia can also influence response to radiation as demonstrated in two stacked paper models with cell-generated hypoxic gradients [57,136] Furthermore, as immunotherapy has become an increasingly important treatment paradigm, the impact of hypoxia on the performance of immunotherapy has become of great interest. For example, using a microfluidic device in conjunction with a patterned hydrogel, Ando et al. investigated CAR-T cell function in a hypoxic, solid-TME. They demonstrated that in hypoxic conditions, CAR-T cells exhibited both an increased cytotoxicity at the periphery of the micropatterned hydrogels and altered PD-L1 expression, although blocking PD-L1 did not enhance the cytotoxicity of CAR-T cells in hypoxic constructs. Further, they showed that CAR-T cells exhibit a contact-independent killing mechanism in hypoxic tumors, allowing new insight into the mechanism of action of CAR-T cell therapy [137]. While the effect of hypoxia on response to therapy, invasive capacity, and immune cell properties has been of interest in many studies, other aspects of cell behavior in the context of hypoxia such as cancer stem cell selection, promotion of tumor angiogenesis, receptor expression, and activity have also been investigated. The stacked paper–based TRACER platform has also uniquely allowed novel insight into the metabolomic profiles of cancer cells cultured in 3D [34]. Thus, a variety of 3D platforms have helped elucidate the impact of hypoxia on cell behavior within the TME.

#### 19.4.3.2 *Gradients of cytokines and other signaling factors*

While hypoxia has been a focus for many 3D culture platforms, other molecules in the TME exhibit a concentration gradient that can influence cell behavior. Cytokines and other chemotactic factors have been implicated in cancer cell aggregation and extravasation/invasion into surrounding tissues. The most common 3D platforms used to investigate the impact of small molecules on cell behavior are transwell setups [147] and microfluidic devices [150], with a focus on modeling aggregation, invasion and metastasis. The use of microfluidic devices has also provided important insight regarding the paracrine mechanisms driving tumor invasion and metastasis. For example, Jeon et al. used a microvascularized microfluidic device to simulate the bone marrow microenvironment and investigated the extravasation rate of breast cancer cells into this environment [150]. By comparing extravasation into a bone marrow like environment, a primary site of breast cancer metastasis, with a simulated skeletal-muscle environment, they identified adenosine (secreted by myoblasts) as a protective factor against metastasis [150]. While microfluidic devices offer unique ability to monitor cells in real time or reflect particular microenvironmental architecture, they are more challenging to use and fabricate than the transwell model, which is also commonly used to study the effect of small molecule gradients. However, the simplicity of the transwell model has drawbacks as in some cases, transwells have allowed the identification of chemotaxis as an important influence on cell behavior, but have hindered the identification of a particular chemokine [147].

### 19.4.4 In vitro models of tumor vasculature

Endothelial cells in the TME have mainly been investigated for their role in angiogenesis and consequently metastasis. Targeting endothelial cells and, more precisely, targeting tumor angiogenesis represents a promising strategy for cancer treatment. In fact, anti-angiogenic agents have already demonstrated their efficiency, such as bezavicumab, an anti-VEGF drug, which has demonstrated impressive therapeutic effects on diverse pathologies such as breast cancer, colorectal cancer and nonsquamous nonsmall cell lung cancer [151]. One of the major challenges in studying angiogenesis and transendothelial migration in vitro has been a lack of physiologically accurate models. Over the last 10 years, however, a number of tissue-engineered models have been developed to allow assessment of tumor angiogenesis, transendothelial migration of cancer cells and also the impact of the TME on endothelial cell fate and vice versa. For example, using microfluidic models, the Kamm group investigated the process of tumor cells extravasation and demonstrated that extravasation is dependent on soluble factors controlled by endothelial barrier permeability [71]. Miller and colleagues also established a highly tunable, vascularized, flow-directed, 3D, and microfluidic model which revealed that tumor cells from clear cell renal carcinoma induce endothelial cell sprouting driven by biochemical gradient of angiogenic factors [152]. Others models have also been developed to explore the relationship between endothelial cells and tumor cells. These are summarized in Table 19.6.

These models, while powerful, still have limitations. Most importantly, there are very few endothelial cell types that are available. In vivo, endothelial cells represent a heterogeneous population, and it has been shown that tumor endothelial cells are different from normal endothelial cells at the molecular and functional level [156]. While the origin of

**TABLE 19.6** In vitro models of endothelial cells.

| Assay/metric assessed | Models used |
|---|---|
| ***Sprouting induced by presence of endothelial cells*** | |
| • Assessment of vessel-like structure formation [72,91,94,152,153]<br>• Assessment of VEGF secretion [153,154] | • Microfluidic [91,152,153]<br>• Hydrogel plug [72]<br>• Multicellular spheroid [94,154] |
| ***Impact of tumor-endothelial co-culture on cell fate*** | |
| • Endothelial cell proliferation in presence of cancer cells [72,155]<br>• Cancer cell proliferation in presence of EC [95]<br>• Cancer cell migration in presence of EC [95]<br>• Viability of EC in co-culture (AnnV/PI cytometry) [154]<br>• Migration of EC in tumor cells spheroids [154]<br>• Measure of EC cell adhesion molecule and cell junction protein [155] | • Microfluidic [95]<br>• Hydrogel plug [72,155]<br>• Multicellular spheroid [154] |
| ***Transendothelial migration of cancer cells*** | |
| • Live imaging of transendothelial migration to assess cell dynamic [71,96,150] | • Microfluidic [71,96,150] |

these tumor endothelial cells is still unclear, it has been demonstrated that tumor endothelial cells enhance the metastatic process compared with normal endothelial cells [157]. The incorporation of tumor endothelial cells into tissue engineered models represents a challenge for the future that will open new horizons to investigate the potential to target these cells therapeutically.

## 19.5 Future outlook

While 3D culture models are beginning to provide novel biological insights that would not have been possible using mouse models alone, translating these engineered models into industrial-scale practice for novel drug discovery is non trivial and presents some important technical challenges for the future. These include addressing heterogeneity in cell sources selected for drug screening (especially if using primary cells), scaling cell, and device manufacturing, defining standardized culture media formulations (ideally serum-free) in particular for complex models requiring co-culture of multiple cell types, and defining quality control standards to ensure models are recreating known biology and performing as expected.

In addition to industrial translation, expanding the scope of biology that can be modeled is another important future direction. The majority of current culture models have focused on the development of setups containing only tumor cells or co-culture of two cell types. Expanding models to contain three or more cell types [6] to enable higher complexity microenvironmental modeling and the study of cellular function in the presence of multiple cell types [73,75,76,86,103,158–162] is therefore a key opportunity to further expand the utility of engineered tumor models in the future. For example, a model for immune cell infiltration in a triculture system can allow the study of the inhibitory effect of stromal cells on tumor-infiltrating T cells, which is implicated in the success of immunotherapy [76]. Furthermore, these next generation of culture models must be designed to enable measurement of more complex functional and dynamic readouts of cell state and behavior beyond simply viability, proliferation and cell number, which are currently the most commonly used metrics currently. For example, models that capture the dynamic evolution of the TME could potentially provide opportunities to better understand the development of therapy resistance, a phenomenon that results in significant mortality in the clinic.

The role of the microenvironment in promoting tumor growth, resistance to treatment, and disease progression is becoming increasingly clear. Expanding the range of tools available to dissect the complexity of the TME and what factors are most dominant in driving disease progression is essential to identify and assess the effectiveness of novel therapeutic strategies. Indeed, a combination of in vitro culture platforms, in vivo animal models and advanced imaging and molecular analysis tools are all necessary to enable us to understand and treat this complex disease effectively.

## References

[1] Quail DF, Joyce JA. Microenvironmental regulation of tumor progression and metastasis. Nat Med 2013;19:1423–37.

[2] Dagogo-Jack I, Shaw AT. Tumour heterogeneity and resistance to cancer therapies. Nat Rev Clin Oncol 2018;15:81–94.

[3] Bhowmick NA, Neilson EG, Moses HL. Stromal fibroblasts in cancer initiation and progression. Nature 2004;432:332–7.

[4] Pearce OMT, et al. Deconstruction of a metastatic tumor microenvironment reveals a common matrix response in human cancers. Cancer Discov 2018;8:304–19.

[5] Orimo A, et al. Stromal fibroblasts present in invasive human breast carcinomas promote tumor growth and angiogenesis through elevated SDF-1/CXCL12 secretion. Cell 2005;121:335–48.

[6] Seager RJ, Hajal C, Spill F, Kamm RD, Zaman MH. Dynamic interplay between tumour, stroma and immune system can drive or prevent tumour progression. Converg Sci Phys Oncol 2017;3:034002.

[7] Gaggioli C, et al. Fibroblast-led collective invasion of carcinoma cells with differing roles for RhoGTPases in leading and following cells. Nat Cell Biol 2007;9:1392–400.

[8] Underwood TJ, et al. Cancer-associated fibroblasts predict poor outcome and promote periostin-dependent invasion in oesophageal adenocarcinoma. J Pathol 2015;235:466–77.

[9] Ligorio M, et al. Stromal microenvironment shapes the intratumoral architecture of pancreatic. Cancer Cell 2019;178:160–75.

[10] Reymond N, d'Água BB, Ridley AJ. Crossing the endothelial barrier during metastasis. Nat Rev Cancer 2013;13:858–70.

[11] Hicklin DJ, Ellis LM. Role of the vascular endothelial growth factor pathway in tumor growth and angiogenesis. J Clin Oncol 2005;23:1011–27.

[12] De Palma M, Biziato D, Petrova TV. Microenvironmental regulation of tumour angiogenesis. Nat Rev Cancer 2017;17:457–74.

[13] Alsaab HO, et al. PD-1 and PD-L1 checkpoint signaling inhibition for cancer immunotherapy: mechanism, combinations, and clinical outcome. Front Pharmacol 2017;8:561.

[14] Yuen GJ, Demissie E, Pillai S. B lymphocytes and cancer: a love-hate relationship. Trends Cancer 2016;2:747–57.

[15] Hehlgans T, et al. Lymphotoxin-beta receptor immune interaction promotes tumor growth by inducing angiogenesis. Cancer Res 2002;62:4034–40.

[16] Erdag G, et al. Immunotype and immunohistologic characteristics of tumor-infiltrating immune cells are associated with clinical outcome in metastatic melanoma. Cancer Res 2012;72:1070–80.

[17] Mosser DM, Edwards JP. Exploring the full spectrum of macrophage activation. Nat Rev Immunol 2008;8:958–69.

[18] Zhang Q, et al. Prognostic significance of tumor-associated macrophages in solid tumor: a meta-analysis of the literature. PLoS One 2012;7:e50946.

[19] Zhang M, et al. A high M1/M2 ratio of tumor-associated macrophages is associated with extended survival in ovarian cancer patients. J Ovarian Res 2014;7:19.

[20] Cassetta L, Pollard JW. Targeting macrophages: therapeutic approaches in cancer. Nat Rev Drug Discov 2018;17:887–904.

[21] Mantovani A, Marchesi F, Malesci A, Laghi L, Allavena P. Tumour-associated macrophages as treatment targets in oncology. Nat Rev Clin Oncol 2017;14:399–416.

[22] Pyonteck SM, et al. CSF-1R inhibition alters macrophage polarization and blocks glioma progression. Nat Med 2013;19:1264–72.

[23] Arlauckas SP, et al. In vivo imaging reveals a tumor-associated macrophage–mediated resistance pathway in anti–PD-1 therapy. Sci Transl Med 2017;9:eaal3604.

[24] Waldhauer I, Steinle A. NK cells and cancer immunosurveillance. Oncogene 2008;27:5932–43.

[25] Balsamo M, et al. Melanoma-associated fibroblasts modulate NK cell phenotype and antitumor cytotoxicity. Proc Natl Acad Sci U S A 2009;106:20847–52.

[26] Anguille S, et al. Interleukin-15 dendritic cells harness NK cell cytotoxic effector function in a contact- and IL-15-dependent manner. PLoS One 2015;10:e0123340.

[27] Perez-Martinez A, et al. Blood dendritic cells suppress NK cell function and increase the risk of leukemia relapse after hematopoietic cell transplantation. Biol Blood Marrow Transpl 2011;17:598–607.
[28] Perez CR, De Palma M. Engineering dendritic cell vaccines to improve cancer immunotherapy. Nat Commun 2019;10:5408.
[29] Ruffell B, et al. Macrophage IL-10 blocks CD8 + T cell-dependent responses to chemotherapy by suppressing IL-12 expression in intratumoral dendritic cells. Cancer Cell 2014;26:623–37.
[30] Jantsch J, et al. Hypoxia and hypoxia-inducible factor-1 alpha modulate lipopolysaccharide-induced dendritic cell activation and function. J Immunol 2008;180:4697–705.
[31] Nobre AR, Entenberg D, Wang Y, Condeelis J, Aguirre-Ghiso JA. The different routes to metastasis via hypoxia-regulated programs. Trends Cell Biol 2018;28:941–56.
[32] Gilkes DM, Semenza GL, Wirtz D. Hypoxia and the extracellular matrix: drivers of tumour metastasis. Nat Rev Cancer 2014;14:430–9.
[33] Wouters BG, Koritzinsky M. Hypoxia signalling through mTOR and the unfolded protein response in cancer. Nat Rev Cancer 2008;8:851–64.
[34] Rodenhizer D, et al. A three-dimensional engineered tumour for spatial snapshot analysis of cell metabolism and phenotype in hypoxic gradients. Nat Mater 2016;15:227–34.
[35] Rodenhizer D, Cojocari D, Wouters BG, McGuigan AP. Development of TRACER: tissue roll for analysis of cellular environment and response. Biofabrication 2016;8:045008.
[36] Ron D, Walter P. Signal integration in the endoplasmic reticulum unfolded protein response. Nat Rev Mol Cell Biol 2007;8:519–29.
[37] Barsoum IB, Smallwood CA, Siemens DR, Graham CH. A mechanism of hypoxia-mediated escape from adaptive immunity in cancer cells. Cancer Res 2014;74:665–74.
[38] Cox TR, Erler JT. Remodeling and homeostasis of the extracellular matrix: implications for fibrotic diseases and cancer. Dis Model Mech 2011;4:165–78.
[39] Walker C, Mojares E, Del Río Hernández A. Role of extracellular matrix in development and cancer progression. Int J Mol Sci 2018;19:3028.
[40] Tse JM, et al. Mechanical compression drives cancer cells toward invasive phenotype. Proc Natl Acad Sci U S A 2012;109:911–16.
[41] Paszek MJ, et al. Tensional homeostasis and the malignant phenotype. Cancer Cell 2005;8:241–54.
[42] Mosadegh B, et al. A paper-based invasion assay: assessing chemotaxis of cancer cells in gradients of oxygen. Biomaterials 2015;52:262–71.
[43] Longati P, et al. 3D pancreatic carcinoma spheroids induce a matrix-rich, chemoresistant phenotype offering a better model for drug testing. BMC Cancer 2013;13:95.
[44] Kenney RM, Loeser A, Whitman NA, Lockett MR. Paper-based transwell assays: an inexpensive alternative to study cellular invasion. Analyst 2018;144:206–11.
[45] Kalli M, Papageorgis P, Gkretsi V, Stylianopoulos T. Solid stress facilitates fibroblasts activation to promote pancreatic cancer cell migration. Ann Biomed Eng 2018;46:657–69.
[46] Sherman H, Gitschier HJ, Rossi AE. A novel three-dimensional immune oncology model for high-throughput testing of tumoricidal activity. Front Immunol 2018;9:857.
[47] Carlson MW, Alt-Holland A, Egles C, Garlick JA. Three-dimensional tissue models of normal and diseased skin. Curr Protoc Cell Biol 2008; Chapter 19, Unit 19.9.
[48] Polacheck WJ, Charest JL, Kamm RD. Interstitial flow influences direction of tumor cell migration through competing mechanisms. Proc Natl Acad Sci U S A 2011;108:11115–20.
[49] Kraning-Rush CM, Carey SP, Lampi MC, Reinhart-King CA. Microfabricated collagen tracks facilitate single cell metastatic invasion in 3D. Integr Biol (Camb) 2013;5:606–16.
[50] Carey SP, et al. Comparative mechanisms of cancer cell migration through 3D matrix and physiological microtracks. Am J Physiol Cell Physiol 2015;308:C436–47.
[51] Pathak A, Kumar S. Independent regulation of tumor cell migration by matrix stiffness and confinement. Proc Natl Acad Sci U S A 2012;109:10334–9.
[52] D'Arcangelo E, et al. Gels for live analysis of compartmentalized environments (GLAnCE): a tissue model to probe tumour phenotypes at tumour-stroma interfaces. Biomaterials 2019;228:119572. Available from: https://doi.org/10.1016/j.biomaterials.2019.119572.

[53] Mueller-Klieser W. Multicellular spheroids. A review on cellular aggregates in cancer research. J Cancer Res Clin Oncol 1987;113:101–22.
[54] Charoen KM, Fallica B, Colson YL, Zaman MH, Grinstaff MW. Embedded multicellular spheroids as a biomimetic 3D cancer model for evaluating drug and drug-device combinations. Biomaterials 2014;35:2264–71.
[55] Alessandri K, et al. Cellular capsules as a tool for multicellular spheroid production and for investigating the mechanics of tumor progression in vitro. Proc Natl Acad Sci U S A 2013;110:14843–8.
[56] Lee D, Pathak S, Jeong J-H. Design and manufacture of 3D cell culture plate for mass production of cell-spheroids. Sci Rep 2019;9:13976.
[57] Simon KA, et al. Metabolic response of lung cancer cells to radiation in a paper-based 3D cell culture system. Biomaterials 2016;95:47–59.
[58] Boyce MW, LaBonia GJ, Hummon AB, Lockett MR. Assessing chemotherapeutic effectiveness using a paper-based tumor model. Analyst 2017;142:2819–27.
[59] Rodenhizer D, Dean T, Xu B, Cojocari D, McGuigan AP. A three-dimensional engineered heterogeneous tumor model for assessing cellular environment and response. Nat Protoc 2018;13:1917–57.
[60] Young M, et al. A TRACER 3D co-culture tumour model for head and neck cancer. Biomaterials 2018;164:54–69.
[61] Xu B, et al. Patterning cellular compartments within TRACER cultures using sacrificial gelatin printing. Biofabrication 2016;8:035018.
[62] Dangles-Marie V, et al. A three-dimensional tumor cell defect in activating autologous CTLs is associated with inefficient antigen presentation correlated with heat shock protein-70 down-regulation. Cancer Res 2003;63:3682–7.
[63] Augustine TN, Dix-Peek T, Duarte R, Candy GP. Establishment of a heterotypic 3D culture system to evaluate the interaction of TREG lymphocytes and NK cells with breast cancer. J Immunol Methods 2015;426:1–13.
[64] Sawant S, et al. Establishment of 3D co-culture models from different stages of human tongue tumorigenesis: utility in understanding neoplastic progression. PLoS One 2016;11:e0160615.
[65] Neal JT, et al. Organoid modeling of the tumor immune microenvironment. Cell 2018;175:1972–88.
[66] Truong DD, et al. A human organotypic microfluidic tumor model permits investigation of the interplay between patient-derived fibroblasts and breast cancer cells. Cancer Res 2019;79:3139–51.
[67] Dolznig H, et al. Modeling colon adenocarcinomas in vitro a 3D co-culture system induces cancer-relevant pathways upon tumor cell and stromal fibroblast interaction. Am J Pathol 2011;179:487–501.
[68] Colley HE, et al. Development of tissue-engineered models of oral dysplasia and early invasive oral squamous cell carcinoma. Br J Cancer 2011;105:1582–92.
[69] Tsai S, et al. Development of primary human pancreatic cancer organoids, matched stromal and immune cells and 3D tumor microenvironment models. BMC Cancer 2018;18:335.
[70] Gioeli D, et al. Development of a multicellular pancreatic tumor microenvironment system using patient-derived tumor cells. Lab Chip 2019;19:1193–204.
[71] Zervantonakis IK, et al. Three-dimensional microfluidic model for tumor cell intravasation and endothelial barrier function. Proc Natl Acad Sci U S A 2012;109:13515–20.
[72] Szot CS, Buchanan CF, Freeman JW, Rylander MN. In vitro angiogenesis induced by tumor-endothelial cell co-culture in bilayered, collagen I hydrogel bioengineered tumors. Tissue Eng Part C Methods 2013;19:864–74.
[73] Linde N, Gutschalk CM, Hoffmann C, Yilmaz D, Mueller MM. Integrating macrophages into organotypic co-cultures: a 3D in vitro model to study tumor-associated macrophages. PLoS One 2012;7:e40058.
[74] Konur A, Kreutz M, Knüchel R, Krause SW, Andreesen R. Three-dimensional co-culture of human monocytes and macrophages with tumor cells: analysis of macrophage differentiation and activation. Int J Cancer 1996;66:645–52.
[75] Bai J, et al. Contact-dependent carcinoma aggregate dispersion by M2a macrophages via ICAM-1 and β2 integrin interactions. Oncotarget 2015;6:25295–307.
[76] Zboralski D, Hoehlig K, Eulberg D, Frömming A, Vater A. Increasing tumor-infiltrating T cells through inhibition of CXCL12 with NOX-A12 synergizes with PD-1 blockade. Cancer Immunol Res 2017;5:950–6.

[77] He W, et al. Proteomic comparison of 3D and 2D glioma models reveals increased HLA-E expression in 3D models is associated with resistance to NK cell-mediated cytotoxicity. J Proteome Res 2014;13:2272–81.
[78] Feder-Mengus C, et al. Multiple mechanisms underlie defective recognition of melanoma cells cultured in three-dimensional architectures by antigen-specific cytotoxic T lymphocytes. Br J Cancer 2007;96:1072–82.
[79] Ghosh S, et al. Culture of melanoma cells in 3-dimensional architectures results in impaired immunorecognition by cytotoxic T lymphocytes specific for Melan-A/MART-1 tumor-associated antigen. Ann Surg 2005;242:851–7 discussion 858.
[80] Dangles V, et al. Impact of human bladder cancer cell architecture on autologous T-lymphocyte activation. Int J Cancer 2002;98:51–6.
[81] Zoetemelk M, Rausch M, Colin DJ, Dormond O, Nowak-Sliwinska P. Short-term 3D culture systems of various complexity for treatment optimization of colorectal carcinoma. Sci Rep 2019;9:7103.
[82] Jaganathan H, et al. Three-dimensional in vitro co-culture model of breast tumor using magnetic levitation. Sci Rep 2014;4:6468.
[83] Nam S, Khawar IA, Park JK, Chang S, Kuh H-J. Cellular context-dependent interaction between cancer and stellate cells in hetero-type multicellular spheroids of pancreatic tumor. Biochem Biophys Res Commun 2019;515:183–9.
[84] Ramgolam K, et al. Melanoma spheroids grown under neural crest cell conditions are highly plastic migratory/invasive tumor cells endowed with immunomodulator function. PLoS One 2011;6:e18784.
[85] Tevis KM, Cecchi RJ, Colson YL, Grinstaff MW. Mimicking the tumor microenvironment to regulate macrophage phenotype and assessing chemotherapeutic efficacy in embedded cancer cell/macrophage spheroid models. Acta Biomater 2017;50:271–9.
[86] Rebelo SP, et al. 3D-3-culture: a tool to unveil macrophage plasticity in the tumour microenvironment. Biomaterials 2018;163:185–97.
[87] Li R, et al. Macrophage-secreted TNFα and TGFβ1 influence migration speed and persistence of cancer cells in 3D tissue culture via independent pathways. Cancer Res 2017;77:279–90.
[88] Chen MB, et al. Inflamed neutrophils sequestered at entrapped tumor cells via chemotactic confinement promote tumor cell extravasation. Proc Natl Acad Sci U S A 2018;115:7022–7.
[89] Chen MB, et al. On-chip human microvasculature assay for visualization and quantification of tumor cell extravasation dynamics. Nat Protoc 2017;12:865–80.
[90] van der Poel HG, Tillier C, de Blok W, van Muilekom E. Extended nodal dissection reduces sexual function recovery after robot-assisted laparoscopic prostatectomy. J Endourol 2012;26:1192–8.
[91] Nguyen D-HT, et al. Biomimetic model to reconstitute angiogenic sprouting morphogenesis in vitro. Proc Natl Acad Sci U S A 2013;110:6712–17.
[92] Paek J, et al. Microphysiological engineering of self-assembled and perfusable microvascular beds for the production of vascularized three-dimensional human microtissues. ACS Nano 2019;13:7627–43.
[93] Kim A, et al. Malignant peripheral nerve sheath tumors state of the science: leveraging clinical and biological insights into effective therapies. Sarcoma 2017;2017:7429697.
[94] Shoval H, et al. Tumor cells and their crosstalk with endothelial cells in 3D spheroids. Sci Rep 2017;7:10428.
[95] Truong D, et al. A three-dimensional (3D) organotypic microfluidic model for glioma stem cells–vascular interactions. Biomaterials 2019;198:63–77.
[96] Bertulli C, et al. Image-assisted microvessel-on-a-chip platform for studying cancer cell transendothelial migration dynamics. Sci Rep 2018;8:12480.
[97] Bazou D, et al. Vascular beds maintain pancreatic tumour explants for ex vivo drug screening. J Tissue Eng Regen Med 2018;12:e318–22.
[98] Wei WZ, Miller B, Gutierrez RM. Inhibition of tumor growth by peptide specific cytotoxic T lymphocytes in a three-dimensional collagen matrix. J Immunol Methods 1997;200:47–54.
[99] Öhlund D, et al. Distinct populations of inflammatory fibroblasts and myofibroblasts in pancreatic cancer. J Exp Med 2017;214:579–96.
[100] Sutherland RM, MacDonald HR, Howell RL. Multicellular spheroids: a new model target for in vitro studies of immunity to solid tumor allografts. J Natl Cancer Inst 1977;58:1849–53.
[101] Mattei F, et al. A multidisciplinary study using in vivo tumor models and microfluidic cell-on-chip approach to explore the cross-talk between cancer and immune cells. J Immunotoxicol 2014;11:337–46.

[102] Businaro L, et al. Cross talk between cancer and immune cells: exploring complex dynamics in a microfluidic environment. Lab Chip 2013;13:229–39.

[103] Cougoule C, et al. Three-dimensional migration of macrophages requires Hck for podosome organization and extracellular matrix proteolysis. Blood 2010;115:1444–52.

[104] Kim T-E, et al. Three-dimensional culture and interaction of cancer cells and dendritic cells in an electrospun nano-submicron hybrid fibrous scaffold. Int J Nanomed 2016;11:823–35.

[105] Caviggioli F, et al. Evidence-based scar management: how to improve results with technique and technology. Plast Reconstr Surg 2017;139:1371e–3e.

[106] Dvorak HF. Tumors: wounds that do not heal. Similarities between tumor stroma generation and wound healing. N Engl J Med 1986;315:1650–9.

[107] Bhowmick NA, et al. TGF-beta signaling in fibroblasts modulates the oncogenic potential of adjacent epithelia. Science 2004;303:848–51.

[108] Kim S-A, Lee EK, Kuh H-J. Co-culture of 3D tumor spheroids with fibroblasts as a model for epithelial-mesenchymal transition in vitro. Exp Cell Res 2015;335:187–96.

[109] Kaufman LJ, et al. Glioma expansion in collagen I matrices: analyzing collagen concentration-dependent growth and motility patterns. Biophys J 2005;89:635–50.

[110] Fang X, Sittadjody S, Gyabaah K, Opara EC, Balaji KC. Novel 3D co-culture model for epithelial-stromal cells interaction in prostate cancer. PLoS One 2013;8:e75187.

[111] Nietzer S, et al. Mimicking metastases including tumor stroma: a new technique to generate a three-dimensional colorectal cancer model based on a biological decellularized intestinal scaffold. Tissue Eng Part C Methods 2016;22:621–35.

[112] Camci-Unal G, Newsome D, Eustace BK, Whitesides GM. Fibroblasts enhance migration of human lung cancer cells in a paper-based coculture system. Adv Healthc Mater 2016;5:641–7.

[113] Ziółkowska K, et al. Long-term three-dimensional cell culture and anticancer drug activity evaluation in a microfluidic chip. Biosens Bioelectron 2013;40:68–74.

[114] Lee SW, Kwak HS, Kang M-H, Park Y-Y, Jeong GS. Fibroblast-associated tumour microenvironment induces vascular structure-networked tumouroid. Sci Rep 2018;8:2365.

[115] Bischel LL, Beebe DJ, Sung KE. Microfluidic model of ductal carcinoma in situ with 3D, organotypic structure. BMC Cancer 2015;15:12.

[116] Fridman WH, Zitvogel L, Sautès-Fridman C, Kroemer G. The immune contexture in cancer prognosis and treatment. Nat Rev Clin Oncol 2017;14:717–34.

[117] Wolchok JD, et al. Overall survival with combined nivolumab and ipilimumab in advanced melanoma. N Engl J Med 2017;377:1345–56.

[118] Borghaei H, et al. Nivolumab versus docetaxel in advanced nonsquamous non–small-cell lung cancer. N Engl J Med 2015;373:1627–39.

[119] Foucher ED, et al. Pancreatic ductal adenocarcinoma: a strong imbalance of good and bad immunological cops in the tumor microenvironment. Front Immunol 2018;9:1044.

[120] Ganesh K, et al. Immunotherapy in colorectal cancer: rationale, challenges and potential. Nat Rev Gastroenterol Hepatol 2019;16:361–75.

[121] Sacks PG, et al. A multicellular tumor spheroid model of cellular immunity against head and neck cancer. Cancer Immunol Immunother 1990;32:195–200.

[122] Aref AR, et al. 3D microfluidic ex vivo culture of organotypic tumor spheroids to model immune checkpoint blockade. Lab Chip 2018;18:3129–43.

[123] Christakou AE, Ohlin M, Önfelt B, Wiklund M. Ultrasonic three-dimensional on-chip cell culture for dynamic studies of tumor immune surveillance by natural killer cells. Lab Chip 2015;15:3222–31.

[124] Pavesi A, et al. A 3D microfluidic model for preclinical evaluation of TCR-engineered T cells against solid tumors. JCI Insight 2017;2:89762.

[125] Giannattasio A, et al. Cytotoxicity and infiltration of human NK cells in in vivo-like tumor spheroids. BMC Cancer 2015;15:351.

[126] Berencsi K, et al. Colon carcinoma cells induce CXCL11-dependent migration of CXCR3-expressing cytotoxic T lymphocytes in organotypic culture. Cancer Immunol Immunother 2007;56:359–70.

[127] Courau T, et al. Cocultures of human colorectal tumor spheroids with immune cells reveal the therapeutic potential of MICA/B and NKG2A targeting for cancer treatment. J Immunother Cancer 2019;7:74.

[128] Dwyer AR, Ellies LG, Holme AL, Pixley FJ. A three-dimensional co-culture system to investigate macrophage-dependent tumor cell invasion. J Biol Methods 2016;3:49.

[129] Konur A, Kreutz M, Knüchel R, Krause SW, Andreesen R. Cytokine repertoire during maturation of monocytes to macrophages within spheroids of malignant and non-malignant urothelial cell lines. Int J Cancer 1998;78:648–53.

[130] Espinoza-Sánchez NA, Chimal-Ramírez GK, Mantilla A, Fuentes-Pananá EM. IL-1β, IL-8, and matrix metalloproteinases-1, -2, and -10 are enriched upon monocyte-breast cancer cell cocultivation in a matrigel-based three-dimensional system. Front Immunol 2017;8:205.

[131] Tasaki A, et al. Three-dimensional two-layer collagen matrix gel culture model for evaluating complex biological functions of monocyte-derived dendritic cells. J Immunol Methods 2004;287:79–90.

[132] Lu Y, et al. Type conversion of secretomes in a 3D TAM2 and HCC cell co-culture system and functional importance of CXCL2 in HCC. Sci Rep 2016;6:24558.

[133] Parlato S, et al. 3D Microfluidic model for evaluating immunotherapy efficacy by tracking dendritic cell behaviour toward tumor cells. Sci Rep 2017;7:1093.

[134] Martinez-Marin D, et al. PEDF increases the tumoricidal activity of macrophages towards prostate cancer cells in vitro. PLoS One 2017;12:e0174968.

[135] Caruana I, et al. Heparanase promotes tumor infiltration and antitumor activity of CAR-redirected T lymphocytes. Nat Med 2015;21:524–9.

[136] Hauptmann S, Zwadlo-Klarwasser G, Jansen M, Klosterhalfen B, Kirkpatrick CJ. Macrophages and multicellular tumor spheroids in co-culture: a three-dimensional model to study tumor-host interactions. Evidence for macrophage-mediated tumor cell proliferation and migration. Am J Pathol 1993;143:1406–15.

[137] Ando Y, et al. Evaluating CAR-T cell therapy in a hypoxic 3D tumor model. Adv Healthc Mater 2019;8: e1900001.

[138] Long L, Yin M, Min W. 3D co-culture system of tumor-associated macrophages and ovarian cancer cells. Bio Protoc 2018;8:e2815.

[139] Yin M, et al. Tumor-associated macrophages drive spheroid formation during early transcoelomic metastasis of ovarian cancer. J Clin Invest 2016;126:4157–73.

[140] Bielecka ZF, et al. Hypoxic 3D in vitro culture models reveal distinct resistance processes to TKIs in renal cancer cells. Cell Biosci 2017;7:71.

[141] Wang Y, et al. 3D hydrogel breast cancer models for studying the effects of hypoxia on epithelial to mesenchymal transition. Oncotarget 2018;9:32191–203.

[142] Song J, Miermont A, Lim CT, Kamm RD. A 3D microvascular network model to study the impact of hypoxia on the extravasation potential of breast cell lines. Sci Rep 2018;8:17949.

[143] Whitman NA, Lin Z-W, Kenney RM, Albertini L, Lockett MR. Hypoxia differentially regulates estrogen receptor alpha in 2D and 3D culture formats. Arch Biochem Biophys 2019;671:8–17.

[144] Klimkiewicz K, et al. A 3D model of tumour angiogenic microenvironment to monitor hypoxia effects on cell interactions and cancer stem cell selection. Cancer Lett 2017;396:10–20.

[145] Lee HR, Leslie F, Azarin SM. A facile in vitro platform to study cancer cell dormancy under hypoxic microenvironments using CoCl2. J Biol Eng 2018;12:12.

[146] Lewis DM, Blatchley MR, Park KM, Gerecht S. O2-controllable hydrogels for studying cellular responses to hypoxic gradients in three dimensions in vitro and in vivo. Nat Protoc 2017;12:1620–38.

[147] Puliafito A, et al. Three-dimensional chemotaxis-driven aggregation of tumor cells. Sci Rep 2015;5:15205.

[148] Tadmor T. Do lymphocytes count in myeloma? Are we absolutely sure? Leuk Lymphoma 2015;56:1193–4.

[149] Swietach P, Hulikova A, Patiar S, Vaughan-Jones RD, Harris AL. Importance of intracellular pH in determining the uptake and efficacy of the weakly basic chemotherapeutic drug, doxorubicin. PLoS One 2012;7: e35949.

[150] Jeon JS, et al. Human 3D vascularized organotypic microfluidic assays to study breast cancer cell extravasation. Proc Natl Acad Sci U S A 2015;112:214–19.

[151] Shih, T. & Lindley, C. Bevacizumab: an angiogenesis inhibitor for the treatment of solid malignancies. Clin Ther 28, 1779–1802 (2006).

[152] Miller CP, Tsuchida C, Zheng Y, Himmelfarb J, Akilesh S. A 3D human renal cell carcinoma-on-a-chip for the study of tumor angiogenesis. Neoplasia 2018;20:610–20.
[153] Theberge AB, et al. Microfluidic multiculture assay to analyze biomolecular signaling in angiogenesis. Anal Chem 2015;87:3239–46.
[154] Amann A, et al. Development of a 3D angiogenesis model to study tumour - endothelial cell interactions and the effects of anti-angiogenic drugs. Sci Rep 2017;7:2963.
[155] Wang C, et al. Mimicking brain tumor-vasculature microanatomical architecture via co-culture of brain tumor and endothelial cells in 3D hydrogels. Biomaterials 2019;202:35–44.
[156] Hida K, Maishi N, Annan DA, Hida Y. Contribution of tumor endothelial cells in cancer progression. Int J Mol Sci 2018;19:E1272.
[157] Maishi N, et al. Tumour endothelial cells in high metastatic tumours promote metastasis via epigenetic dysregulation of biglycan. Sci Rep 2016;6:28039.
[158] Zhang T, et al. Migration of cytotoxic T lymphocytes toward melanoma cells in three-dimensional organotypic culture is dependent on CCL2 and CCR4. Eur J Immunol 2006;36:457–67.
[159] Berencsi K, et al. In vitro migration of cytotoxic T lymphocyte derived from a colon carcinoma patient is dependent on CCL2 and CCR2. J Transl Med 2011;9:33.
[160] Ehsan SM, Welch-Reardon KM, Waterman ML, Hughes CCW, George SC. A three-dimensional in vitro model of tumor cell intravasation. Integr Biol (Camb) 2014;6:603–10.
[161] Kuen J, Darowski D, Kluge T, Majety M. Pancreatic cancer cell/fibroblast co-culture induces M2 like macrophages that influence therapeutic response in a 3D model. PLoS One 2017;12:e0182039.
[162] Rama-Esendagli D, Esendagli G, Yilmaz G, Guc D. Spheroid formation and invasion capacity are differentially influenced by co-cultures of fibroblast and macrophage cells in breast cancer. Mol Biol Rep 2014;41:2885–92.

CHAPTER

# 20

# Three-dimensional tumor model and their implication in drug screening for tackling chemoresistance

*Manashi Priyadarshini*[1,2], *Sibasish Mohanty*[1,3], *Tanushree Mahapatra*[1], *Pallavi Mohapatra*[1,3] *and Rupesh Dash*[1]

[1]Institute of Life Sciences, Bhubaneswar, India [2]KIIT School of Biotechnology, KIIT University, Bhubaneswar, India [3]Regional Center for Biotechnology, Faridabad, India

## Abbreviations

| | |
|---|---|
| **2D** | Two dimensional |
| **3D** | Three dimensional |
| **API** | Active pharmaceutical ingredient |
| **BME** | Basement membrane extract |
| **CAF** | Cancer-associated fibroblast |
| **Cas9** | CRISPR associated protein 9 |
| **CCL1** | Chemokine C-C motif ligand-1 |
| **CIL** | Contact inhibition of locomotion |
| **CRISPR** | Clustered regularly interspaced short palindromic repeats |
| **CSCs** | Cancer stem cells |
| **CTOS** | Cancer tissue-originated spheroid |
| **DEX** | Dextran |
| **ECM** | Extracellular matrix |
| **EGF** | Epidermal growth factor |
| **EMT** | Epithelial–mesenchymal transition |
| **ESCC** | Esophageal Squamous Cell Carcinoma |
| **IL-11** | Interleukin-11 |
| **IL-17** | Interleukin-17 |
| **IL-6** | Interleukin-6 |
| **iPSC** | Induced pluripotent stem cell |
| **MDR** | Multidrug resistance |
| **PDC** | Patient-derived cell |

*Biomaterials for 3D Tumor Modeling*
DOI: https://doi.org/10.1016/B978-0-12-818128-7.00020-4

| | |
|---|---|
| **PDO** | Patient-derived organoid |
| **PDX** | Patient-derived xenograft |
| **PEG** | Polyethylene glycol |
| **PLA** | Polylactic acid |
| **PLGA** | Poly(lactic-*co*-glycolic) acid |
| **PSCs** | Pluripotent stem cells |
| **SDF-1** | Stromal cell-derived factor 1 |
| **TALEN** | Transcription activator-like effector nuclease |
| **TME** | Tumor microenvironment |
| **VEGF** | Vascular epithelial growth factor |

## 20.1 Chemoresistance in cancer

Across the globe, cancer contributes to the second highest noncommunicable disease-related death with approximately 9 million each year. In the United States alone, approximately 1,762,450 new cancer cases and 606,880 cancer-related deaths are projected in 2019 [1]. Traditionally, surgery, radiotherapy, and chemotherapy are the most common treatment modalities used for treatment of cancer [2]. Some of the recently adopted emerging cancer treatment modalities are hormone-based therapy, stem cell therapy, and immunotherapy [3]. Although, a few successful interventions have been achieved to treat specific types of cancer, many of them have failed to yield the desired result. Chemotherapy remains one of the major treatment regimens for the treatment of cancer.

In spite of having early encouraging response with chemotherapy, many patients experience relapse, with persistent tumor growth and metastatic spread due to chemoresistance. Henceforth, chemoresistance is one of the important factors for treatment failure in cancer. In a tumor mass, chemotherapy targets most of the cells, but a few of them are inherently resistant or acquire resistance to the drug, which ultimately contributes to recurrence. Chemoresistant properties of cancer cells could be attributed to increased drug-efflux, reduced apoptotic-response, genomic instability, and enhanced stem cell-like population. Chemoresistance is broadly classified into two classes, first one is "intrinsic" where the cancer cells have resistance properties against drugs without any chemotherapy and second one is "acquired chemoresistance" where the cancer cells which are initially sensitive to chemotherapy but acquire resistance to a drug after its exposure to chemotherapy for a prolonged duration. Understanding the molecular mechanisms of chemoresistance is a daunting challenge for researchers since last couple of decades. One of the important obstacles for this is lack of appropriate model to study acquired chemoresistance. Three-dimensional culture system (3D culture) recently evolved as an ideal model for studying chemoresistance in cancer, about which, we plan to discuss in this review.

The 3D culture system was first established by Alexs Carrel in 1912, where an explant from a chick embryo was cultured and maintained for 3 months. Further, Leighten improved this system by culturing the tissues on sponge matrix (substrate) [4]. In 1944, Holtfreter established a 3D culture model to observe the changes in shape and arrangement of cells during gastrulation. Since then, the interest in 3D culture has only increased over the years with the discovery of improved imaging techniques [5]. The main advantages of 3D cultures involve maintenance of growth, tissue organization, structure, and interaction with surrounding cells that mimics the exact in vivo condition.

## 20.2 3D tumor culture: an advanced model preferred over 2D culture

The 3D tumor model is an in vitro culture system that mimics in vivo tumor microenvironment (TME) and allows cells to grow and interact with their surrounding environment as they do in the in vivo environment [6]. In 3D tumor model, cadherin interaction among neighboring cells leads to cell aggregation and later compaction, as a result of which, tumor cells form a 3D structure. Sometimes interaction of integrin with extracellular matrix (ECM) also contributes to the formation of 3D tumor structure (Table 20.1). In 3D tumor model, cells form pseudopodia-like extensions and these extensions engulf neighboring cells and finally these cells are attached to each other and form a bulky 3D structure [6]. The size of 3D tumor ranges from 20 μm to 1 mm in diameter [7]. The tumor mass forms various layers, (1) the outer layer of cells that remains exposed to nutrients and proliferates faster, (2) the quiescent layer of viable cells enclosed by outer layer, and (3) the inner layer of narcotic cells, which contributes to the central core of the 3D tumor.

**TABLE 20.1** Main differences between 2D and 3D cultures.

| Cellular characteristics | 2D culture | 3D culture |
|---|---|---|
| Morphology | Cells are flat and sheet-like arranged in monolayer. | Cells retain their natural shape and aggregate in layers to form spheroid-like structure. |
| Arrangement | Cells are not arranged in gradient pattern. | Cells are arranged in gradient pattern. |
| Interaction of cells | Interaction of cells with extracellular matrix (ECM) does not occur. Here cells are attached to the surface. | Interaction of cells with ECM occur. |
| Microenvironment involvement | Tumor microenvironment is not involved in this case. | Tumor microenvironment plays an important role in 3D culture. |
| Proliferation | Cells proliferate faster in 2D culture as compared with 3D culture. | Cells in 3D culture proliferate slower than in 2D monolayer. |
| Cell cycle stage | Here all cells are present in one stage of cell cycle. | 3D model contains quiescent cells, proliferative cells, necrotic cells, and hypoxic cells. |
| Genes and protein expression | Cells in 2D culture show different protein and gene expression than in vivo condition. | Here cells show same protein and gene expression as that of in in vivo environment. |
| Drug sensitivity | It does not show chemotherapeutic drug resistance. | It shows chemotherapeutic drug resistance. |
| Economy | Cost effective. | Costly process. |
| Biological mode | Biologically less relevant model for studying tumor properties. | Biologically more relevant model for studying tumor properties. |

Henceforth, the presence of various layers with different phases of cells in 3D culture mimics the in vivo tumor mass [8].

The 3D tumor model mimics various elements of TMEs like hypoxia, necrosis, migration, cell adhesion, and angiogenesis [9–11]. Henceforth, adaptation to 3D tumor model provides an excellent in vitro platform for studying the mechanism of drugs resistance [12]. The two-dimensional (2D) culture model does not accurately mimic the in vivo TME because 2D monolayer contains only proliferating cells while necrotic cells are removed from the monolayer while changing culture media. Therefore, 2D culture system influences many cellular processes like apoptosis, differentiation, cell proliferation, and expression pattern of genes and proteins [13] (Fig. 20.1). In contrast to 2D monolayer system, in 3D culture model cells are grown as aggregates either within a matrix or in a suspension medium. Unlike the 2D culture model, it comprises of proliferative, necrotic, apoptotic, hypoxic as well as quiescent cells [14]. The rate of cancer cell proliferation significantly differs from 2D to 3D culture, that is the cell proliferation is slower in 3D culture as compared with 2D culture [15,16]. For example, endometrial cancer cell lines like EN-1078D, KLE, and RL95-2 proliferate slower in 3D culture compared with 2D culture, which is due to the reduced expression of the proliferating cell nuclear antigen (PCNA) marker in 3D culture [15]. Similarly, colorectal cancer (CRC) cell lines like CACO-2, DLD-1, HT-29, SW480, and LOVO also proliferate slower in 3D culture as compared with 2D culture [17]. In contrast, JIMT1 breast cancer cells proliferate faster in matrigel culture than in 2D culture. But when the same cell line was grown with synthetic poly(2-hydroxyethyl-methacrylate) 3D scaffold, it showed 7.2-fold slower proliferation rate than in 2D monolayer [18]. This indicates that cell proliferation rate depends upon the type of matrix used in the 3D culture model. It is reported that cell viability does not significantly differ in 3D culture than 2D culture during initial few days. However, cancer cells exhibit reduced cell viability in 3D culture because cell viability depends upon availability of nutrient and oxygen supply, which are less in 3D culture models. In addition, accumulation of waste material further reduces cell viability in 3D models [14]. Earlier, it is reported that the expression pattern of genes, protein, and cell receptor varies significantly in 2D and 3D culture models [19–21]. Genes and proteins, those are responsible for proliferation, angiogenesis, migration, invasion, and chemosensitivity, display diverse expression pattern in 3D tumor model as compared with 2D monolayer system. For example, ovarian and prostate cancer cells grown in 3D culture showed significantly increased mRNA expression of cell surface receptor alpha3/alpha5/beta1 integrin, chemokine-derived factor 1-alpha, and its receptor CXCR4, respectively [22].

The drug resistance properties to cisplatin and paclitaxel are approximately 100–1000 times higher in 3D tumors as compared with traditional 2D monolayer system [4,23]. It is mainly because 3D system provides appropriate TME (interaction between cancer and stromal cells) that contributes to chemoresistance to various chemotherapeutic drugs. Cell adhesion-mediated drug resistance (CAM-DR) to chemotherapeutic drugs (also known as tissue resistance), which can be observed in in vivo tumor mass is clearly mimicked by 3D tumor model [24]. Henceforth, in vitro 3D tumor model is an ideal model for screening various chemotherapeutics for preclinical studies.

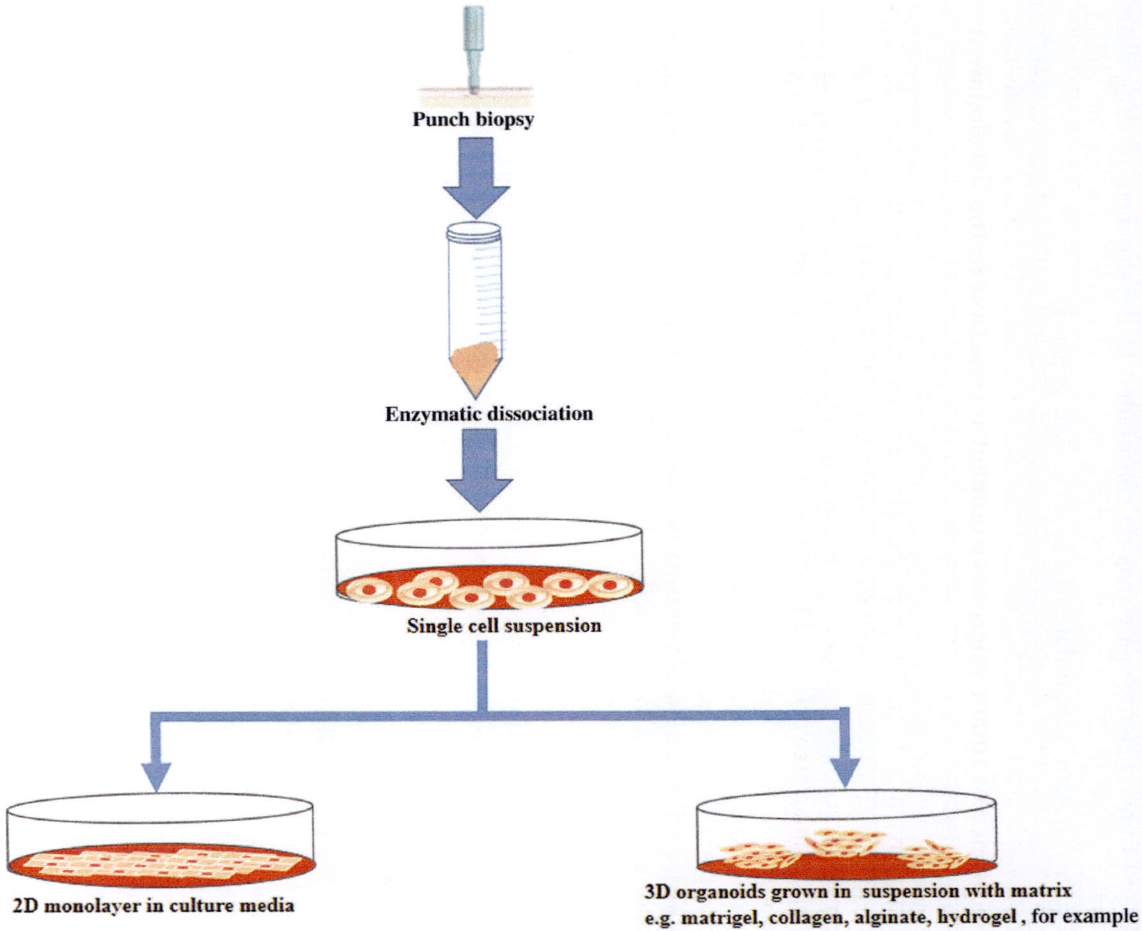


FIGURE 20.1 Difference between 2D and 3B cultures.

## 20.3 3D culture and chemoresistance

### 20.3.1 3D culture acts as a good model to study chemoresistance

The TME and tumor heterogeneity majorly contributed to development of drug resistance in a tumor mass. In 3D tumor model these important components are preserved, thus the 3D tumor system recapitulates the in vivo TME precisely.

### 20.3.2 Importance of tumor microenvironment interaction in the development of chemoresistance

TME plays a pivotal role in tumor progression and in determining the chemotherapy drug efficacy [25]. TME consists of physical, chemical, and biological components that surround the cancer cells. These components include the ECM, interstitial flow, stromal cells, immune cells, vascular networks, and biochemical concentration gradients [25]. Recently it has been observed that the TME plays a significant role in cancer progression including malignancy, invasion, and metastasis. The 3D culture techniques replicate the physiological features of the TME such as cell–cell interactions, fluidic shear stress, and cell–ECM interactions [24]. TME acts as a physical barrier, which limits the access of chemotherapy drugs to tumor cells. Similarly, the interface between faster growing tumor cells and surrounding ECM creates fluidic shear stress, which can induce stemness of tumor cells and ultimately exerts chemoresistance. In addition to this, CAM-DR is the term used for receptor–ligand interaction mediated chemoresistance of the cancer cells along with TME [24,26]. Collagen, laminin, and fibronectin helps in interaction of cancer cells with nearby stromal cells for adherence [27]. During this interaction, integrin-mediating signaling pathways induce the antiapoptotic molecules that helps in acquiring chemoresistance by tumor cells [28]. Previous studies have shown that stromal cells provide necessary factors to tumor cells for acquiring chemoresistance to anticancer drugs via several mechanisms [24]. Stromal cells provide ligand that binds to integrin receptor of cancer cells to activate antiapoptotic signaling [29]. Cytokines mediated chemoresistance is induced in cancer cells when they are being co-cultured with stromal cells. For example, cancer cells acquired chemoresistance when co-cultured with cancer-associated fibroblasts (CAFs) via CAF-secreted chemokines such as interleukin-6 (IL-6), interleukin-11 (IL-11), interleukin-17 (IL-17), chemokine C-C motif ligand-1 (CCL1), and stromal cell-derived factor 1 (SDF-1) [30,31]. Recent studies have also proposed that effector T-cells can abrogate such CAF-induced chemoresistance by modulating the metabolism of CAFs [32].

### 20.3.3 Tumor heterogeneity and chemoresistance

Cancer cells have diverse type of phenotypes ranging from benign to highly aggressive, metastatic cell states. Cancer comprises of heterogeneous cell populations, in which cells having different tumorigenic potential exist within the same single tumor [33]. Intratumor heterogeneity refers to biological differences between malignant cells originated within the tumor mass. Possible explanations for this include the clonal evolution theory and the cancer stem cell (CSC) model. The clonal evolution theory indicates that sequential accumulation of mutation leads to establishment of clones those are phenotypically different among the neighboring cancer cells. These mutations are acquired mostly due to microenvironmental selection pressure [34], that is instability can be inherited or caused by somatic aberrations in genes that maintain genome instability, or can be caused by extrinsic mutagens such as cigarette smoke, ultraviolet light, and chemotherapy [35]. Genomic instability and the dynamic nature of selective pressure can lead to multiple routes of genetic diversification within the adaptive landscapes of tissue ecosystems, which can ultimately result in genetically, epigenetically, and phenotypically diversified cell population within a single tumor [36]. The CSC hypothesis attributes heterogeneity to aberrant differentiation mechanisms and presumes the existence of a hierarchical organization of cancer cells. CSC hypothesis states that only a small fraction of cells can drive tumor progression and are therapy-resistant inherently [36]. Here the CSC may undergo symmetric division to maintain the CSC population or an irreversible asymmetric division to produce non-CSC cancer cells which have less tumorigenic potential. Recently, it was also established that plasticity can be observed between CSC and non-CSC population within the tumor mass, which contributes to the heterogeneity [37].

Both 2D monolayer and 3D culture system demonstrates the capacity of tumor cells to exist in multiple distinct differentiation states (heterogeneity), that is CSCs in self-renewal mode, tumor mass capable of invading the stroma, better differentiated phenotype with a higher level of preservation of epithelial-like cellular organization. Block et al. established the patient-derived 2D monolayer of cells and 3D cultures from tissues of renal cell carcinomas. It was found that both 2D and 3D culture system had clonal subpopulations in culture. However, in 3D organoid culture, unique set of subpopulations can be found which contained sets of mutations those are not present in the corresponding 2D monolayer culture [38]. Similarly, considerable differences in metabolism were observed between the 3D and 2D culture system. In addition, differential drug penetration phenomenon was associated with increased resistance to chemo- and radiotherapy in 3D culture group as comparison to 2D cell cultures [39]. The growing research interest in 3D models of different tumor types is focused on assessing the effects of chemo- and/or radiotherapy, studying tumor metastasis, angiogenesis and cancer cell differentiation, apoptosis as well as epithelial–mesenchymal transition (EMT) and CSCs biology [40,41].

## 20.4 Methods of generating 3D culture system

The 3D models are generated using different sets of methodologies, which offer various degrees of complexity. The most commonly used 3D models are cellular spheroids, tumor spheres, organotypic spheroids, matrix-mediated assembled cellular aggregates, multilayered cultures of cancer cells or tumor slices, organoids, microfluidics-, and microfabricated-based cultures of cancer cells [5]. However, 3D tumor models can be classified broadly into two different categories, that is, spheroids and organoids [15]. Tumor spheroids represent spherical aggregates of tumor cells that are having self-organizing and self-renewing capability. They can originate from single-cell suspensions of tumor cell lines or patient-derived tumor cells or CSCs cultured in nonadherent substrates. On the other hand, organoids are 3D models of primary tumor tissues obtained from fresh biopsies. These are generated by mechanical or enzymatic digestion of the original tumor section into small pieces or slices. These tissue slices are embedded in an ECM, where cell–cell interactions predominate over cell–substrate interactions. Here we will try to emphasize on the methods of generating 3D spheroids and organoids using the natural and synthetic biomaterials. In addition, we will also highlight different aspects of biomaterial-based approaches for generation of 3D tumor models and discuss the potential opportunities that can be addressed with these models in near future.

## 20.4.1 Methods of generating 3D organoids

As mentioned earlier, 3D organoids are generated from fresh biopsy samples from different patients and its formation greatly depends on the self-assembling capacity of cells. Thus, it is essential to create a microenvironment with required niche factors. Unlike spheroid cultures that have employed a variety of materials, organoid cultures are established with animal-derived–ECM matrices, such as matrigel and collagen. These 3D organoids resemble avascular tumor nodules, or micrometastases, and mimic growth kinetics, gradients of nutrient distribution, oxygen concentration, and cell proliferation. Studies have demonstrated that ultra-low–attachment microplates generate organoids of larger size (300–400 μM), one per well, in a few days' time.

## 20.4.2 Methods of generating 3D spheroids

As discussed earlier, spheroids are the aggregation of cells and these spheroids are formed either in suspension or encapsulated or on 3D matrixes. Generation of 3D spheroid culture models can be done through five different models (Fig. 20.2) such as hanging drop

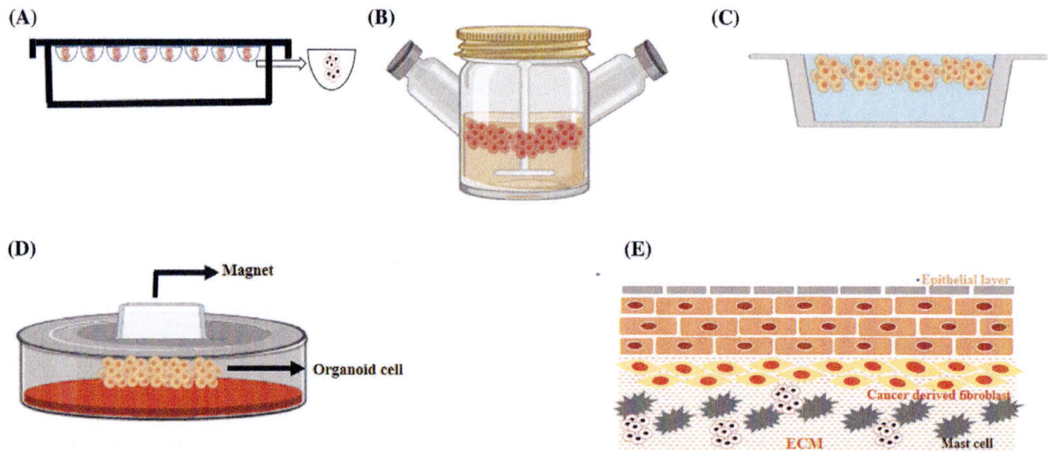


FIGURE 20.2 Conventional methods of 3D spheroid culture: (A) Hanging drop model for 3D culture; (B) suspension 3D cell culture; (C) Nonadherent 3D culture model; (D) magnetic levitation 3D cell culture model; and (E) scaffold-based 3D culture model.

model, nonadherent surface model, suspension culture model, scaffold-based model, and magnetic levitation model.

#### 20.4.2.1 *Hanging drop model*

In this model, drops of cell suspensions are placed in an underlying substratum (polystyrene plastic lid). Then the lid is turned upside down and placed on the top of the plate having phosphate saline buffer, which prevents the dehydration of the drops. Cells aggregate at the tip of the drop at the air–liquid interface and form spheroid-like structures [42]. Currently, customized hanging drop plates are available for generation of spheroids. This method is very simple, yielding one spheroid per drop of cells. The main advantage of hanging drop model in 3D tumor culture is to produce equal-sized spheroids. It is also useful to work with different patient-derived cell line for testing drug screening [43].

#### 20.4.2.2 *Nonadherent surface model*

In nonadherent surface model, low ultra-attachment plates are used to prevent attachment of the cells to the surface. Henceforth, cells will aggregate and form spheroids [44]. In this method the plates are generally coated with poly-2-hydroxyethyl-methacrylate and dried for 3 days before adding the cells to it. The coating is stable, nondegradable, and noncytotoxic.

#### 20.4.2.3 *Suspension culture model*

In this model, the cell suspension is placed inside a container with increasing media viscosity or agitation by adding carboxymethyl cellulose [45]. Agitation prevents attachment of cells to the substratum and promotes cell to cell interaction, as a result cells aggregate with each other to form spheroids. Generally, two apparatus are being used for suspension 3D culture method, one of them is spinner flask, where a stirring bar or rod is used to mix the cell suspension. Here, cell suspension is placed in a container which is present always in a motile state. Oxygen and nutrients are transported easily due to constant motion of the culture fluid. This method allows large yield of spheroids. The second apparatus is bioreactor, where biosynthetic materials are used as a membrane to surround spheroids with sufficient supply of required nutrients, which mimics the TME. In bioreactor, large numbers of spheroids can be grown ($\sim$300) simultaneously. In suspension culture, the size of spheroids is not uniform and it varies from one another [46].

#### 20.4.2.4 *Scaffold-based model*

Scaffold-based spheroids correctly recapitulate the TME with intact cell and ECM interaction. Natural or synthetic hydrogel-based scaffolds are often used for generating tumor spheroids. Most commonly used natural hydrogels are matrigel, collagen, alginate, and fibrin. Similarly, semisynthetic hydrogels are hyaluronan, polypeptides, and most commonly used synthetic hydrogel is polyethylene glycol (PEG) for generation of spheroids [47].

Matrigel and collagen are ideal natural scaffolds for proper reorganization and attachment for 3D culture. Matrigel is a commercially available matrix containing basement membrane proteins like collagen IV, entactin, laminin, metalloproteinase-2, and growth factors [48]. These basement membrane proteins are important for chemotherapeutic resistance, cell adhesion, growth regulation, and polarization [49]. Collagen is also used as a 3D matrix because it regulates cell adhesion and helps in cell migration and chemotaxis.

#### *20.4.2.5 Magnetic levitation model*

Magnetic levitation model utilizes the magnetic field during formation of 3D spheroids. In this process, cells integrated with magnetic nanoparticles (generally in extracellular space) are elevated with magnetic field, enabling cell–cell interaction and assembly of cell aggregates resulting in formation of spheroids [40]. Main advantage of this method is the high growth rate of spheroids without using a matrix or scaffold [34].

## 20.5 3D culture and biomaterials

In 3D culture system, biomaterial is required for providing cell–cell or cell–ECM or cell–surface receptor interaction. Hence biomaterials contribute in assembly of 3D culture for studying cancer cell growth, apoptosis, proteolytic stimuli, drug sensitivity, and polarity [50,51]. Based on the origin of materials, the biomaterials used in 3D culture can be broadly grouped into natural and synthetic biomaterials (Fig. 20.3). Various synthetic and natural biomaterials are used to construct scaffolds that provide physical and mechanical support for adhesion of cells to form a 3D structure [52,53]. Currently, natural biomaterials like collagen, laminin-rich ECM, hyaluronic acid, alginate, and chitosan; and synthetic materials like polyethylene glycol (PEG), poly(lactic-*co*-glycolic) acid, and polycaprolactone are being extensively used as supporting materials for 3D culture of different tissues [54].

### 20.5.1 Cell-derived or natural biomaterials

Natural biomaterials are usually secreted from the tissues. These natural biomaterials are thin and allow high-resolution imaging in 3D culture model [55].

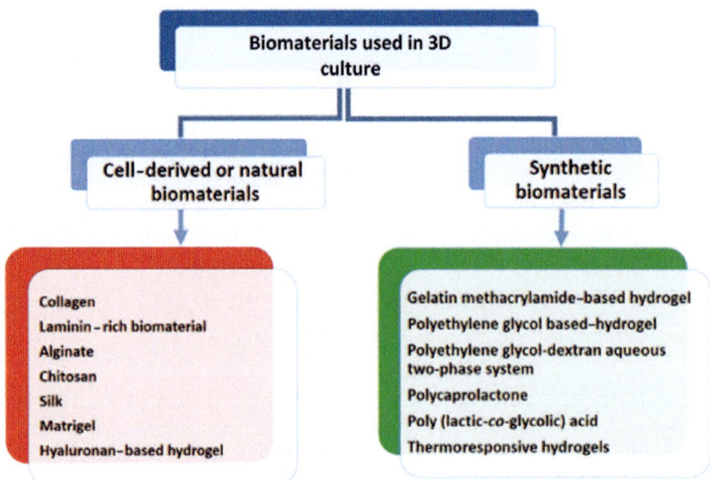


**FIGURE 20.3** Application of 3D organoids in decoding the molecular mechanism responsible for chemoresistance and development of personalized medicine through drug screening.

#### 20.5.1.1 Collagen

Collagen was the first ever matrix to be used for the study of invasion and migration of ovarian and prostate cancer cells in 3D culture system [16,56,57]. As collagen has integrin binding motif Arg–Gly–Asp, it is frequently used for 3D culture of pancreatic, breast, lung, colorectal, and skin cancer cells [58]. Adhesion of cancer cells in collagen matrix allow them for efficient proliferation, survival, migration, and invasion [59,60]. In 3D collagen matrix, tumor cells mimic the morphology and gene expression pattern exactly as in in vivo condition. For example, when MCF7 breast cancer cell line was grown on 3D collagen scaffold, it showed elongated morphology and increased expression of EMT markers like SLUG, SNAIL, ZEB1/2, TWIST1/2, LEF1, and decreased epithelial markers like E-cadherin [61]. In 3D collagen matrix, HT1080 sarcoma cell line showed increased resistance to doxorubicin by altering the activity of focal adhesion kinase (FAK), Rho A, and β1-integrin that leads to the migration of those cells. MDA-MB-231 cells cultured in collagen-3D matrix showed resistance to paclitaxel mainly due to increase in matrix stiffness. When MDA-MB-231 and HT-29 breast cancer cells were cultured on 3D collagen matrix they showed increased expression of HIF1α and VEGF-A that promote angiogenesis [58].

#### 20.5.1.2 Laminin-rich extracellular matrix

In 3D tumor model, laminin matrix is useful for maintenance of biological and biomechanical properties of cancer cells [21]. In 3D culture, laminin-rich ECM induces integrin signaling in tumor cells, which is essential for cancer cell malignancy. Previous studies showed that when breast and prostate nonmalignant cells were cultured in laminin-rich ECM, they underwent growth arrest, and had distinct polarized acinus-structure. On the other hand, when breast and prostate malignant cancer cells were cultured in laminin matrix, they displayed nonpolar and disorganized proliferative growth pattern [15,62,63]. The main advantage of using laminin-rich ECM in 3D culture is that, it differentiates nonmalignant from malignant cell types according to their morphology and gene expression pattern [54].

#### 20.5.1.3 Alginate matrix

Alginate is a biocompatible polymer and it is generally used to study integrin signaling in ECM as it is rich in integrin. It is a natural polysaccharide, extracted from brown sea weeds. It is used as a 3D culture matrix because it provides support for integration of cells and acts as a platform for cellular growth. Ovarian tumor cell lines cultured in alginate hydrogels acquired chemoresistance against paclitaxel [64]. Using alginate in 3D culture system, it was reported that Wnt/β-catenin signaling pathway is essential for maintaining stemness properties of hepatocellular carcinomas [65]. Alginate helps in the interaction and signal transduction between fibronectin and ECM through RGD (Arg–Gly–Asp) peptide sequence [66]. When EpCAM-positive hepatocellular carcinoma cells were cultured in alginate 3D matrix, they mimicked various in vivo phenotypes such as acini, glandular epithelium, apical morphogenesis, enhanced expression of stem cell markers, and activated β-catenin signaling [65].

#### 20.5.1.4 *Chitosan matrix*

Chitosan is structurally similar with glycosaminoglycan of ECM and it is extracted from chitins of crustaceans. It provides platform for attachment of cancerous cells through its surface-charged amino acid groups [67]. When hepatocellular carcinoma and colon cancer cells are cultured on chitosan-based 3D matrix they showed increased expression of stem cell markers like OCT4, NONOG, CD133, CD44, and epithelial markers like EpCAM. Henceforth, the cancer cells displayed drug resistance, increased cell motility, and activated canonical Wnt/B-Catenin-CD 44 signaling [68].

#### 20.5.1.5 *Silk*

Silk is a natural polymer composed of glue-like sericin and filament core protein fibroin [69]. Oxygen and water permeability, slow degradability, cell adhesiveness, relatively low thrombogenicity, and amenability to convenient surface modification make silk an attractive biomaterial for tissue engineering applications including tumor modeling [54]. When osteosarcoma cells were cultured on silk 3D matrix they showed reduced proliferation rate than 2D culture and displayed decreased expression of proliferating markers such as cyclin B, E2F2, PCNA, and Ki67 [42].

#### 20.5.1.6 *Matrigel*

Being a basement membrane extract, matrigel is composed of a complex mixture of over 1000 proteins. Type IV collagen, laminin, and nidogen are some of the major components found in matrigel. It truly augments the self-assembling capacity of cancer cells. Thus matrigel is the most widely used biomaterials in 3D culture as it has several major advantages, such as built-in complex distribution of nutrients and protein gradients, ease in handling and fast gelling kinetics, and the availability of a commercialized product with high quality. Matrigel is being used in 3D culture of ovarian cancer cells to study cell density, growth kinetics, viability, and chemoresistance [70]. In a study involving prostate cancer cells, it mimicked the TME accurately [70,71].

#### 20.5.1.7 *Hyaluronan-based hydrogel*

Hyaluronic acid is an essential component of TME necessary for cell proliferation and migration. One common example of hyaluronan-based hydrogel is bone marrow matrix. It recreates TME, hence it is found to be useful for studying paclitaxel chemoresistance in cancer using 3D culture [72,73]. It also facilitates cancer cell delivery vehicle as it increases vacuolization [74].

### 20.5.2 Synthetic biomaterials

The commonly used synthetic biomaterials for 3D cultures are polyethylene glycol-based hydrogel, polyethylene glycol-dextran aqueous two-phase system, polycaprolactone, poly(lactic-*co*-glycolic) acid, and thermoresponsive hydrogels [63,75].

#### 20.5.2.1 Polyethylene glycol-based hydrogel

A hydrogel is a 3D network of hydrophilic polymers having unique property of holding large amount of water without interruption in structure due to its chemical or physical cross-linking of individual polymer chains. PEG is an excellent biomaterial for hydrogel-based 3D culture system. As it is biologically inert in its native form, PEG conjugates with other natural or synthetic bioactive molecules which can be compatible with cell culture conditions. In addition to simple PEG-based hydrogels for culturing cancer cells, hybrid materials using PEG and synthetic or natural materials are also utilized for specific applications. For example, hyaluronic acid found in certain tumors was used to mimic TME in combination with PEG [54]. Similarly, the use of PEG and dextran in highly aqueous, immiscible polymeric solutions allows a scaffold-free approach to form monoculture as well as coculture spheroids [54]. PEG hydrogels have previously been used to generate 3D culture model in OV-MZ-6 and SKOV-3 ovarian cancer cell lines where these cells form spheroid-like structure [76]. It also recapitulated prostate cancer microenvironment and thus mimics the in vivo tumor mass [77]. PEG hydrogel with RGD peptide allows its interaction with integrin on cancer cell in 3D culture system. On PEG hydrogel, OV-MZ-6 and SKOV-3 ovarian cancer cell lines were found not only to form spheroids and secrete ECM proteins like fibronectin and integrin, but also the cells showed resistance to paclitaxel by enhancing integrin and MMP-9 protein expression. Similarly, a composite hydrogel was formed by interaction of PEG diacrylate with fibrinogen, where the fibrinogen backbone provides cell adhesion-motifs, that facilitate proteolytic degradation by cells [78]. This composite hydrogel is used to generate microsphere for 3D culture of breast cancer cell MCF7, SK-BR-3, and MDA-MB-231 [52].

#### 20.5.2.2 Polyethylene glycol-dextran aqueous two-phase system

The combination of PEG and dextran (DEX) form two immiscible aqueous phases for the formation of aqueous two-phase system in 3D culture [79]. In this system, cancer cell present in DEX aqueous phase remained suspended as a submicroliter drop into nonadherent microwell containing PEG aqueous phase [80,81]. Triple negative breast cancer cell line MDA-MB-231 was grown in aqueous two-phase, which formed spheroids and secretes ECM protein like collagen-1, fibronectin, and laminin. These cells also exhibit enhanced expression of CSC markers like CD24, CD133, and NANOG; and increased expression HIF1-$\alpha$ and carbonic anhydrase 9 [54].

#### 20.5.2.3 Polycaprolactone

Polycaprolactone is formed by condensation of 6-hydroxyhexanoic acid and the ring-opening polymerization of a lactone and epsilon-caprolactone [82]. Previous studies reported that breast cancer cell line MCF7 when cultured in 3D polycaprolactone showed increased expression of stem-like cell markers [83]. MCF7 and T47-D breast cancer cell lines showed enhancement of $CD44^{+}/CD24^{+}$ population in composite electro spun PCL/chitosan nanofiber scaffolds, henceforth these cells acquired resistance to docetaxel and doxorubicin [84]. When prostate cancer cells cultured on polycaprolactone–tricalcium phosphate-based 3D scaffold, it showed increased aggressiveness and increased level of steroidogenic enzymes and PSA which are the biomarkers of prostate cancer progression [85].

#### 20.5.2.4 Poly(lactic-co-glycolic) acid

It is formed by polymerization of polylactic acid and polyglycolic acid (PLGA). Poly (lactic-*co*-glycolic) acid has the ability to promote cell adhesion and proliferation. Highly porous PLGA microspheres were generated by using oil–water emulsion with a coating of collagen-1 which promotes proliferation and expression of epithelial marker E-cadherin [86]. These collagen-coated microspheres were used to establish spheroids of H08910 ovarian cancer cell and U215 glioblastoma cell [86,87]. When U251 cells were cultured in PLGA scaffolds, it showed increased expression of antiapoptotic proteins like Bcl-2 and survivin. These cells also showed elevated expression of angiogenic markers like VEGF and bFGF [88]. Similarly, a complex 3D structure on nanofibrous scaffold which was designed with a mixture of PLGA, PLA, and monopolyethylene glycol using electrospinning method, where MCF7 breast cancer, PC3 prostate cancer, B16 melanoma, BG1 ovarian cancer cells were grown and they showed enhanced expression of EMT marker [89].

#### 20.5.2.5 Thermoresponsive hydrogels

Thermoresponsive hydrogels are temperature-responsive biomaterials facilitating self-assembly of hydrogel in response to temperature change. It provides a supporting platform for spheroid formation and this method requires encapsulation of cancer cell upon incubation at 37°C. Most common example of thermoresponsive hydrogels is poly(*N*-isopropylacrylamide-*co*-acrylic acid) matrigel. It has been reported that liver cancer cell line HepG2 proliferates very well in this type of scaffold performing liver-specific function [65,90].

## 20.6 Drug screening in 3D culture

Currently, the estimated average cost of developing an anticancer drug is approximately $1 billion, which requires a time span of approximately 10 years [91,92]. Furthermore, the requirements of regulatory approval and clinical trials during drug development also contribute to the high pricing of new drugs [93]. Several reports suggest that the success rate of anticancer drugs, from development to approval by the Food and Drug Administration (FDA) is less than 10% [91,92]. Various factors, including the tumor heterogeneity [92,94] and the choice of preclinical models used for evaluating the effectiveness of the cancer drug candidates contribute to the low success rate of discovering new drugs. Preclinical screening for potential drug efficacy and its responsiveness toward cancer cells can be evaluated using specific cell line and defined culture strategies. Till date most of the screening strategies for evaluating drug responses was performed using 2D monolayer culture of cancer cells [95]. However, drug candidates discovered using 2D monolayer culture system showed failure to have success in many clinical trials. The possible explanation for this includes the inability of 2D monolayer system to recapitulate the TME [96]. On the other hand, 3D tumor model especially the 3D organoids have shown better outcomes during trials of candidate drugs [97–100]. Again, the clinical utility of 3D tumor spheroids has been hindered due to its reliance on established cancer cell lines, which often fail to recapitulate the complexity and functionality of primary tumors. To overcome this limitation, organoid provides a more complex and physiologic model than cell line-based spheroids [54]. Organoids have demonstrated excellent potential for disease modeling, drug screening, and tissue engineering for drug testing. Unlike 3D spheroids, organoids recapitulate genetic signatures of their in vivo counterparts apart from resembling them physically. As compared with spheroids, organoids contain several spatially restricted lineages of committed cell types generated from either pluripotent stem cells (PSCs) or organ-specific adult stem cells. These features ultimately make the organoid a better model over spheroids for testing potential drugs on patient-specific tumors and further developing personalized cancer medicines. In addition to this, 3D organoid culture helps in understanding the genetic and epigenetic changes responsible for tumor initiation, progression, metastasis, and drug resistance [101]. The 3D tumor organoid technology has transfigured tumor culture modeling by enabling the ex vivo generation of tumors from individual patient.

### 20.6.1 Importance of organoids for developing personalized medicines

PSCs, such as embryonic stem cells and induced PSCs (iPSCs) or tissue resident adult stem cells are currently preferred in 3D organoid culture system. The 3D organoids originated from PSCs offer several advantages, that is the organoids are highly expandable, can be derived from virtually every tissue, and can recapitulate malignant transformation. On the other hand, organoids derived from adult tissue have a limited success rate and are strongly dependent on a set of highly selective and expensive growth factors [93]. Thus, organoid cultures derived from PSCs tend to have certain advantages over the organoids derived from adult tissues. PSC generated organoids can faithfully mimic the in vivo conditions. Furthermore, human disease models can be generated by using distinct disease-specific iPSCs. However, the generation of a tissue-specific organoids from human PSCs is also time consuming due to the step-wise differentiation to reach respective intermediates and usually does not reach full maturation levels. Nevertheless, organoids can be generated to mimic a number of organs, showcasing astounding similarity to their in vivo counterparts. Organoid generation is possible from virtually every patient and therefore it becomes impactful for biomedical research and the development of novel treatment strategies. In a research study, the sensitivity of the patient-derived organoids to several established drugs were evaluated and compared against the real patient responses. The results indicate that the prediction of a drug that will work was around 88%, while the prediction that a drug will not work was 100% [102]. These data suggests that patient-derived organoids can foretell the response of patients to drugs and thus can be vital for the development of personalized medicines [102] (Fig. 20.4).

### 20.6.2 Organoids in cancer medicine

Since last couple of decades, patient-derived tumor xenografts (PDX) model is very reliable in mimicking the in vivo TME, where freshly resected tumor pieces are transplanted

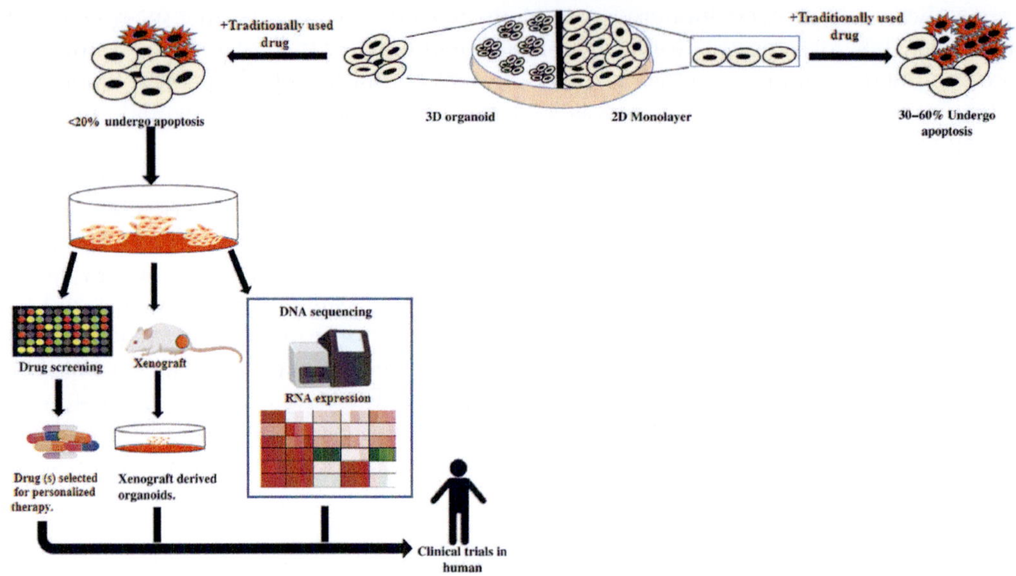


FIGURE 20.4 Application of 3D organoids in decoding the molecular mechanism responsible for chemoresistance and development of personalized medicine through drug screening.

into immunocompromised mice in subcutaneous or heterotopic or orthotropic forms [103]. This model more closely reflects the human in vivo situation in view of tumor heterogeneity and further aspects, but it is labor-, time-, and capacity-intensive. Thus, for a variety of tumor entities, patients might not even survive until a xenograft-based drug screen could reveal an individualized approach for chemotherapy.

Over the last few years a variety of protocols for culturing organoids from different tumors were established. These efforts resulted in generation of patient-derived tumor organoids (PDOs). PDOs can be established either from circulating tumor cells [104] or from solid tumors [105]. PDOs may serve as a simplified approach to predict drug penetration, binding activity, and finally to determine the drug- or radiation-sensitivity in individual patients [106]. Organoids can be established from both healthy tissue and tumor tissue. Hence, the genetic information of both the organoids can be compared and correlated to each other and their pattern of responses to the drug can be evaluated through drug screening assays. The use of both healthy and tumor organoids helps us determining the doses of drugs that are less toxic to normal tissue.

### 20.6.3 Patient-derived organoids used for cancer drug screening

Li et al. [107] established 27 HCC organoid lines, which were derived from 5 different patients and performed the screening for anticancer efficacy of 129 different antineoplastic drugs. They ended up finding a great deal of heterogeneity in terms of response to most of the drugs to different organoid lines. In this study, couple of drugs were found to have pan inhibitory effects on all HCC PDOs [107]. van de Wetering et al. [98] established patient-derived organoids from normal adjacent tissues as well as from CRC tumors. The genetic signature of CRC was found to be preserved in all CRC PDOs. A successful high-throughput drug screening analysis suggests the clinical utility of organoid technology for investigating the molecular basis of drug response [108]. Similarly, PDOs have been successfully established for breast [109], gastric [110], bladder, and ovarian carcinomas [111]; and high-throughput drug screening was performed to reveal the appropriate response regimen. In all these studies involving PDOs, interpatient and intratumor heterogeneity determined the response of different drugs, hence proving the impact of cancer organoids as a useful tool for drug screening. Pauli et al. [112] established 2D culture from various neoplasms and performed a high-throughput screening using 160 different drugs to test their anticancer efficacy. In this study, it was found that only genetic signature of the tumor is not enough to efficiently determine the treatment strategies, but 2D organoid-based high-throughput screening provides additional information to identify the effective combinatorial effective therapeutic strategies. They screened organoids in 2D culture, probably to avoid the time-consuming process of culturing 3D organoids. But the lead drugs or combinatorial strategies were validated using 3D organoid culture [113]. Overall, till now it has been well established that PDOs can be utilized for screening for personalized anticancer drugs. As establishing PDOs are time consuming techniques, researchers have adopted cancer tissue-originated spheroid (CTOS) for developing personalized drugs. The CTOS method has been introduced by many research groups which provides the opportunity for preparation of organoids from both primary tumor as well as xenografts tumor tissues. Kondo et al. in 2018 screened a library of 2427 drugs (which include FDA-approved drugs and small molecules with known targets) with CTOS established from tumor tissues of two CRC patients. Among these 2427 drugs, 15 hit drugs decreased the viability of CRC CTOS line by 40%. Ten of these hit drugs were anticancer reagents and 7 were already approved by the FDA [114].

## 20.7 Future aspects of the 3D tumor organoid model: biobanks for tumor tissues

Biobanking is referred to the storage of 3D tumor organoids after their successful generation. In this case, 3D organoids generated from patients' biopsy samples are stored and thereafter, it can be utilized in future to develop strategies focusing personalized drug treatment in cancer. Hence, these biobanks representing the stored 3D organoids derived from patients will be useful in high-throughput drug screening assays [98,115].

Several organizations have been established, which store 3D organoids generated from patients with variety of neoplasms, so that they can be further analyzed using next generation sequencing and proteomics [116]. Biobanks of 3D organoids for CRC and breast cancer have been generated, consisting of several genetically different tumor organoids with the corresponding normal tissue-derived organoids [99]. These stored breast 3D tumor organoids have now been considered as an emerging tool for studying breast cancer metastasis. Other organoid collections which have been stored in the form of biobanks include liver, prostate, and lung cancers [93,104,115]. Thus, the research community all over the world have started to sense the importance of these tumor organoids in drug screening platforms and personalized medicine.

## 20.8 Limitations of 3D culture technology

Apart from the advantages, several drawbacks are associated with 3D tumor models, which need immediate attention. Lack of uniformity in size and composition are major limitations of tumor spheroid culture. In addition to this, spheroids are established from either cancer cell lines or patient-derived lines but not from the tumor tissues. Establishing PDOs is time-consuming process and requires high resources. To overcome these limitations, optimization of culture conditions are required, which supports appropriate growth and differentiation of stem cells [28]. Microfluidic technology, which is also known as "organ-on-chip," has the potential to be used in screening of several candidate drugs, including novel immunotherapy and chemotherapeutic drugs [117]. The main disadvantage associated with this is the contamination induced by microfluidic chips that affects cellular growth and other processes. Similarly, vascular cells, stroma (CAFs), and immune cells are absent in 3D organoid model. However, addition of these cells and tissue-/organ-specific ECM during organoid culture would help in overcoming these limitations.

## 20.9 Conclusion

The 3D tumor models are excellent tools to study and tackle chemoresistance due to fact that they (1) mimic the in vivo tumor mass representing outer proliferating cell layer, middle quiescent cell layer, and inner nacrotic cell layer, (2) recapitulate accurately the in vivo TME and preserve the tumor heterogeneity, (3) represent an excellent platform for high-throughput screening of chemotherapy drugs/small molecule inhibitors, and (4) have been utilized for establishing personalized therapy for variety of cancer neoplasms. Hence 3D culture model comes out as an indispensable model to replace the tedious preclinical animal models and to discover new therapeutic approaches to overcome chemoresistance in cancer.

## Acknowledgment

This work is supported by the Department of Biotechnology, DST (EMR/2015/000063), India and ICMR (5/13/9/2019-NCD-III), India. SM is a UGC Junior Research Fellow. PM is a CSIR Junior Research Fellow.

## References

[1] Siegel RL, Miller KD, Jemal A. Cancer statistics, 2019. CA Cancer J Clin 2019;69(1):7–34.
[2] Arruebo M, et al. Assessment of the evolution of cancer treatment therapies. Cancers (Basel) 2011;3(3):3279–330.
[3] Schirrmacher V. From chemotherapy to biological therapy: a review of novel concepts to reduce the side effects of systemic cancer treatment (review). Int J Oncol 2019;54(2):407–19.

[4] Hoffman RM. To do tissue culture in two or three dimensions? That is the question. Stem Cell 1993;11 (2):105–11.
[5] Hirschhaeuser F, et al. Multicellular tumor spheroids: an underestimated tool is catching up again. J Biotechnol 2010;148(1):3–15.
[6] Lin RZ, Chang HY. Recent advances in three-dimensional multicellular spheroid culture for biomedical research. Biotechnol J 2008;3(9-10):1172–84.
[7] Burdett E, et al. Engineering tumors: a tissue engineering perspective in cancer biology. Tissue Eng Part B Rev 2010;16(3):351–9.
[8] Shweiki D, et al. Induction of vascular endothelial growth factor expression by hypoxia and by glucose deficiency in multicell spheroids: implications for tumor angiogenesis. Proc Natl Acad Sci U S A 1995;92 (3):768–72.
[9] Justice BA, Badr NA, Felder RA. 3D cell culture opens new dimensions in cell-based assays. Drug Discov Today 2009;14(1–2):102–7.
[10] Reininger-Mack A, Thielecke H, Robitzki AA. 3D-biohybrid systems: applications in drug screening. Trends Biotechnol 2002;20(2):56–61.
[11] Sun T, et al. Culture of skin cells in 3D rather than 2D improves their ability to survive exposure to cytotoxic agents. J Biotechnol 2006;122(3):372–81.
[12] Li H, Fan X, Houghton J. Tumor microenvironment: the role of the tumor stroma in cancer. J Cell Biochem 2007;101(4):805–15.
[13] Huang H, et al. Peptide hydrogelation and cell encapsulation for 3D culture of MCF-7 breast cancer cells. PLoS One 2013;8(3):e59482.
[14] Khaitan D, et al. Establishment and characterization of multicellular spheroids from a human glioma cell line; Implications for tumor therapy. J Transl Med 2006;4:12.
[15] Chitcholtan K, Sykes PH, Evans JJ. The resistance of intracellular mediators to doxorubicin and cisplatin are distinct in 3D and 2D endometrial cancer. J Transl Med 2012;10:38.
[16] Moss NM, et al. Ovarian cancer cell detachment and multicellular aggregate formation are regulated by membrane type 1 matrix metalloproteinase: a potential role in I.p. metastatic dissemination. Cancer Res 2009;69(17):7121–9.
[17] Luca AC, et al. Impact of the 3D microenvironment on phenotype, gene expression, and EGFR inhibition of colorectal cancer cell lines. PLoS One 2013;8(3):e59689.
[18] Hongisto V, et al. High-throughput 3D screening reveals differences in drug sensitivities between culture models of JIMT1 breast cancer cells. PLoS One 2013;8(10):e77232.
[19] Baharvand H, et al. Differentiation of human embryonic stem cells into hepatocytes in 2D and 3D culture systems in vitro. Int J Dev Biol 2006;50(7):645–52.
[20] Benya PD, Shaffer JD. Dedifferentiated chondrocytes reexpress the differentiated collagen phenotype when cultured in agarose gels. Cell 1982;30(1):215–24.
[21] Nelson CM, Bissell MJ. Modeling dynamic reciprocity: engineering three-dimensional culture models of breast architecture, function, and neoplastic transformation. Semin Cancer Biol 2005;15(5):342–52.
[22] Loessner D, et al. Hydrogel microwell arrays allow the assessment of protease-associated enhancement of cancer cell aggregation and survival. Microarrays (Basel) 2013;2(3):208–27.
[23] Miller BE, Miller FR, Heppner GH. Assessing tumor drug sensitivity by a new in vitro assay which preserves tumor heterogeneity and subpopulation interactions. J Cell Physiol Suppl 1984;3:105–16.
[24] Jo Y, et al. Chemoresistance of cancer cells: requirements of tumor microenvironment-mimicking in vitro models in anti-cancer drug development. Theranostics 2018;8(19):5259–75.
[25] Mehta G, et al. Opportunities and challenges for use of tumor spheroids as models to test drug delivery and efficacy. J Control Release 2012;164(2):192–204.
[26] Meads MB, Gatenby RA, Dalton WS. Environment-mediated drug resistance: a major contributor to minimal residual disease. Nat Rev Cancer 2009;9(9):665–74.
[27] Correia AL, Bissell MJ. The tumor microenvironment is a dominant force in multidrug resistance. Drug Resist Updat 2012;15(1–2):39–49.
[28] <Three-Dimensional Organoids in Cancer Research.pdf>.
[29] Lwin T, et al. Cell adhesion induces p27Kip1-associated cell-cycle arrest through down-regulation of the SCFSkp2 ubiquitin ligase pathway in mantle-cell and other non-Hodgkin B-cell lymphomas. Blood 2007;110 (5):1631–8.

[30] Lee J, Shin D, Roh JL. Development of an in vitro cell-sheet cancer model for chemotherapeutic screening. Theranostics 2018;8(14):3964–73.
[31] Pasquier J, et al. Epithelial to mesenchymal transition in a clinical perspective. J Oncol 2015;2015:792182.
[32] Wang W, et al. Effector T cells abrogate stroma-mediated chemoresistance in ovarian cancer. Cell 2016;165(5):1092–105.
[33] Marjanovic ND, Weinberg RA, Chaffer CL. Cell plasticity and heterogeneity in cancer. Clin Chem 2013;59(1):168–79.
[34] Martelotto LG, et al. Breast cancer intra-tumor heterogeneity. Breast Cancer Res 2014;16(3):210.
[35] Hanahan D, Weinberg RA. Hallmarks of cancer: the next generation. Cell 2011;144(5):646–74.
[36] Greaves M, Maley CC. Clonal evolution in cancer. Nature 2012;481(7381):306–13.
[37] Yakisich JS, et al. Cancer cell plasticity: rapid reversal of chemosensitivity and expression of stemness markers in lung and breast cancer tumorspheres. J Cell Physiol 2017;232(9):2280–6.
[38] Fang Y, Eglen RM. Three-dimensional cell cultures in drug discovery and development. SLAS Discov 2017;22(5):456–72.
[39] Weiswald LB, Bellet D, Dangles-Marie V. Spherical cancer models in tumor biology. Neoplasia 2015;17(1):1–15.
[40] Nath S, Devi GR. Three-dimensional culture systems in cancer research: focus on tumor spheroid model. Pharmacol Ther 2016;163:94–108.
[41] Bolck HA, et al. Tracing clonal dynamics reveals that two- and three-dimensional patient-derived cell models capture tumor heterogeneity of clear cell renal cell carcinoma. Eur Urol Focus 2019;.
[42] Nathan SS, et al. Tumor interstitial fluid pressure may regulate angiogenic factors in osteosarcoma. J Orthop Res 2008;26(11):1520–5.
[43] Raghavan S, et al. Formation of stable small cell number three-dimensional ovarian cancer spheroids using hanging drop arrays for preclinical drug sensitivity assays. Gynecol Oncol 2015;138(1):181–9.
[44] Yuhas JM, et al. A simplified method for production and growth of multicellular tumor spheroids. Cancer Res 1977;37(10):3639–43.
[45] Yamada KM, Cukierman E. Modeling tissue morphogenesis and cancer in 3D. Cell 2007;130(4):601–10.
[46] Hirschhaeuser F, et al. Test system for trifunctional antibodies in 3D MCTS culture. J Biomol Screen 2009;14(8):980–90.
[47] Caliari SR, Burdick JA. A practical guide to hydrogels for cell culture. Nat Methods 2016;13(5):405–14.
[48] Kleinman HK, Martin GR. Matrigel: basement membrane matrix with biological activity. Semin Cancer Biol 2005;15(5):378–86.
[49] Souza AG, et al. Comparative assay of 2D and 3D cell culture models: proliferation, gene expression and anticancer drug response. Curr Pharm Des 2018;24(15):1689–94.
[50] Griffith LG, Swartz MA. Capturing complex 3D tissue physiology in vitro. Nat Rev Mol Cell Biol 2006;7(3):211–24.
[51] Hotary KB, et al. Membrane type I matrix metalloproteinase usurps tumor growth control imposed by the three-dimensional extracellular matrix. Cell 2003;114(1):33–45.
[52] Sawicki LA, et al. Tunable synthetic extracellular matrices to investigate breast cancer response to biophysical and biochemical cues. APL Bioeng 2019;3(1):016101.
[53] Zhu W, et al. 3D printing of functional biomaterials for tissue engineering. Curr Opin Biotechnol 2016;40:103–12.
[54] Thakuri PS, et al. Biomaterials-based approaches to tumor spheroid and organoid modeling. Adv Healthc Mater 2018;7(6):e1700980.
[55] Even-Ram S, Yamada KM. Cell migration in 3D matrix. Curr Opin Cell Biol 2005;17(5):524–32.
[56] Barbolina MV, Burkhalter RJ, Stack MS. Diverse mechanisms for activation of Wnt signalling in the ovarian tumour microenvironment. Biochem J 2011;437(1):1–12.
[57] Harjanto D, Maffei JS, Zaman MH. Quantitative analysis of the effect of cancer invasiveness and collagen concentration on 3D matrix remodeling. PLoS One 2011;6(9):e24891.
[58] Szot CS, et al. In vitro angiogenesis induced by tumor-endothelial cell co-culture in bilayered, collagen I hydrogel bioengineered tumors. Tissue Eng Part C Methods 2013;19(11):864–74.
[59] Grzesiak JJ, Bouvet M. The alpha2beta1 integrin mediates the malignant phenotype on type I collagen in pancreatic cancer cell lines. Br J Cancer 2006;94(9):1311–19.

[60] Pickup MW, Mouw JK, Weaver VM. The extracellular matrix modulates the hallmarks of cancer. EMBO Rep 2014;15(12):1243–53.

[61] Chen L, et al. The enhancement of cancer stem cell properties of MCF-7 cells in 3D collagen scaffolds for modeling of cancer and anti-cancer drugs. Biomaterials 2012;33(5):1437–44.

[62] Nielsen HL, et al. Identification of EPSTI1, a novel gene induced by epithelial-stromal interaction in human breast cancer. Genomics 2002;79(5):703–10.

[63] Harma V, et al. A comprehensive panel of three-dimensional models for studies of prostate cancer growth, invasion and drug responses. PLoS One 2010;5(5):e10431.

[64] King SM, et al. Alginate hydrogels for three-dimensional organ culture of ovaries and oviducts. J Vis Exp 2011;(52)).

[65] Takai A, et al. Three-dimensional organotypic culture models of human hepatocellular carcinoma. Sci Rep 2016;6:21174.

[66] Zehnder T, et al. Evaluation of an alginate-gelatine crosslinked hydrogel for bioplotting. Biofabrication 2015;7(2):025001.

[67] Croisier F, Jérôme C. Chitosan-based biomaterials for tissue engineering. Eur Polym J 2013;49(4):780–92.

[68] Chang PH, et al. Chitosan promotes cancer progression and stem cell properties in association with Wnt signaling in colon and hepatocellular carcinoma cells. Sci Rep 2017;8:45751.

[69] Jacobs DJ, et al. Protein flexibility predictions using graph theory. Proteins 2001;44(2):150–65.

[70] Windus LC, Glover TT, Avery VM. Bone-stromal cells up-regulate tumourigenic markers in a tumour-stromal 3D model of prostate cancer. Mol Cancer 2013;12(1):112.

[71] Chu JH, et al. Development of a three-dimensional culture model of prostatic epithelial cells and its use for the study of epithelial-mesenchymal transition and inhibition of PI3K pathway in prostate cancer. Prostate 2009;69(4):428–42.

[72] David L, et al. Hyaluronan hydrogel: an appropriate three-dimensional model for evaluation of anticancer drug sensitivity. Acta Biomater 2008;4(2):256–63.

[73] Burdick JA, Prestwich GD. Hyaluronic acid hydrogels for biomedical applications. Adv Mater 2011;23(12): H41–56.

[74] Serban MA, Scott A, Prestwich GD. Use of hyaluronan-derived hydrogels for three-dimensional cell culture and tumor xenografts. Curr Protoc Cell Biol 2008; Chapter 10: p. Unit 10 14.

[75] Hutmacher DW, et al. Can tissue engineering concepts advance tumor biology research? Trends Biotechnol 2010;28(3):125–33.

[76] Loessner D, Holzapfel BM, Clements JA. Engineered microenvironments provide new insights into ovarian and prostate cancer progression and drug responses. Adv Drug Deliv Rev 2014;79-80:193–213.

[77] Sieh S, et al. Phenotypic characterization of prostate cancer LNCaP cells cultured within a bioengineered microenvironment. PLoS One 2012;7(9):e40217.

[78] Bidault L, et al. Fibrin-based interpenetrating polymer network biomaterials with tunable biodegradability. Polymer 2015;62:19–27.

[79] Sturesson S, Tjerneld F, Johansson G. Partition of macromolecules and cell particles in aqueous two-phase systems based on hydroxypropyl starch and poly(ethylene glycol). Appl Biochem Biotechnol 1990;26 (3):281–95.

[80] Lemmo S, et al. Optimization of aqueous biphasic tumor spheroid microtechnology for anti-cancer drug testing in 3D culture. Cell Mol Bioeng 2014;7(3):344–54.

[81] Atefi E, et al. High throughput, polymeric aqueous two-phase printing of tumor spheroids. Adv Funct Mater 2014;24(41):6509–15.

[82] Labet M, Thielemans W. Synthesis of polycaprolactone: a review. Chem Soc Rev 2009;38(12):3484–504.

[83] Palomeras S, et al. Breast cancer stem cell culture and enrichment using poly(epsilon-caprolactone) scaffolds. Molecules 2016;21(4):537.

[84] Sims-Mourtada J, et al. Enrichment of breast cancer stem-like cells by growth on electrospun polycaprolactone-chitosan nanofiber scaffolds. Int J Nanomed 2014;9:995–1003.

[85] Sieh, Shirly, et al., Paracrine interactions between LNCaP prostate cancer cells and bioengineered bone in 3D in vitro culture reflect molecular changes during bone metastasis. Bone 2014;63:121–31.

[86] Zhang YS, et al. Bioprinting the cancer microenvironment. ACS Biomater Sci Eng 2016;2(10):1710–21.

[87] Lama R, et al. Development, validation and pilot screening of an in vitro multi-cellular three-dimensional cancer spheroid assay for anti-cancer drug testing. Bioorg Med Chem 2013;21(4):922–31.
[88] Gang EH, et al. Highly porous three-dimensional poly(lactide-co-glycolide) (PLGA) microfibrous scaffold prepared by electrospinning method: a comparison study with other PLGA type scaffolds on its biological evaluation. Fibers Polym 2012;13(6):685–91.
[89] Girard YK, et al. A 3D fibrous scaffold inducing tumoroids: a platform for anticancer drug development. PLoS One 2013;8(10):e75345.
[90] Gu J, et al. Effect of particle size in a colloidal hydrogel scaffold for 3D cell culture. Colloids Surf B Biointerfaces 2015;136:1139–47.
[91] Rubin EH, Gilliland DG. Drug development and clinical trials--the path to an approved cancer drug. Nat Rev Clin Oncol 2012;9(4):215–22.
[92] Hay M, et al. Clinical development success rates for investigational drugs. Nat Biotechnol 2014;32:40.
[93] Weeber F, et al. Tumor organoids as a pre-clinical cancer model for drug discovery. Cell Chem Biol 2017;24 (9):1092–100.
[94] Dzobo K, et al. Advances in regenerative medicine and tissue engineering: innovation and transformation of medicine. Stem Cell Int 2018;2018:2495848.
[95] Barretina J, et al. The Cancer Cell Line Encyclopedia enables predictive modelling of anticancer drug sensitivity. Nature 2012;483(7391):603–7.
[96] Caponigro G, Sellers WR. Advances in the preclinical testing of cancer therapeutic hypotheses. Nat Rev Drug Discov 2011;10(3):179–87.
[97] Astashkina A, Grainger DW. Critical analysis of 3-D organoid in vitro cell culture models for high-throughput drug candidate toxicity assessments. Adv Drug Deliv Rev 2014;69-70:1–18.
[98] van de Wetering M, et al. Prospective derivation of a living organoid biobank of colorectal cancer patients. Cell 2015;161(4):933–45.
[99] Sachs N, et al. A living biobank of breast cancer organoids captures disease heterogeneity. Cell 2018;172(1-2):373–86.
[100] Eglen RM, Randle DH. Drug discovery goes three-dimensional: goodbye to flat high-throughput screening? Assay Drug Dev Technol 2015;13(5):262–5.
[101] Verissimo CS, et al. Targeting mutant RAS in patient-derived colorectal cancer organoids by combinatorial drug screening. Elife 2016;5.
[102] Vlachogiannis G, et al. Patient-derived organoids model treatment response of metastatic gastrointestinal cancers. Science 2018;359(6378):920–6.
[103] Morton CL, Houghton PJ. Establishment of human tumor xenografts in immunodeficient mice. Nat Protoc 2007;2(2):247–50.
[104] Gao D, et al. Organoid cultures derived from patients with advanced prostate cancer. Cell 2014;159 (1):176–87.
[105] Mehta G, Sreenivas K. Total synthesis of the novel tricyclic sesquiterpene sulcatine G. Chem Commun (Camb) 2001;(18):1892–3.
[106] Mehta G, et al. Intrauterine exposure to a maternal low protein diet reduces adult bone mass and alters growth plate morphology in rats. Calcif Tissue Int 2002;71(6):493–8.
[107] Li L, et al. Human primary liver cancer organoids reveal intratumor and interpatient drug response heterogeneity. JCI Insight 2019;4(2).
[108] Tsoli M, et al. Integration of genomics, high throughput drug screening, and personalized xenograft models as a novel precision medicine paradigm for high risk pediatric cancer. Cancer Biol Ther 2018;19 (12):1078–87.
[109] Sun Z, Zhang B. High-throughput screening (HTS) of natural products with triple-negative breast cancer (TNBC) organoids. J Clin Oncol 2019;37(15_Suppl.):e12558.
[110] Gao M, et al. Development of patient-derived gastric cancer organoids from endoscopic biopsies and surgical tissues. Ann Surg Oncol 2018;25(9):2767–75.
[111] Tamura H, et al. Evaluation of anticancer agents using patient-derived tumor organoids characteristically similar to source tissues. Oncol Rep 2018;40(2):635–46.
[112] Pauli, C., et al., Personalized in vitro and in vivo cancer models to guide precision medicine. Cancer Discov 2017;7(5):462–77

[113] J. Kondo, M. Inoue, Application of cancer organoid model for drug screening and personalized therapy. Cells 2019;8(5).

[114] Kondo J, et al. High-throughput screening in colorectal cancer tissue-originated spheroids. Cancer Sci 2019;110(1):345–55.

[115] Broutier L, et al. Human primary liver cancer-derived organoid cultures for disease modeling and drug screening. Nat Med 2017;23(12):1424–35.

[116] Fujii M, et al. A colorectal tumor organoid library demonstrates progressive loss of niche factor requirements during tumorigenesis. Cell Stem Cell 2016;18(6):827–38.

[117] Adriani G, Pavesi A, Kamm RD. Studying TCR T cell anti-tumor activity in a microfluidic intrahepatic tumor model. Methods Cell Biol 2018;146:199–214.

CHAPTER

# 21

# Co-culture and 3D tumor models for drug/gene therapy testing

*Lara S. Costard*[1], *Harumi Ramanayake*[1] *and Caroline M. Curtin*[1,2,3]

[1]Tissue Engineering Research Group (TERG), Department of Anatomy and Regenerative Medicine, Royal College of Surgeons in Ireland (RCSI), Dublin, Ireland [2]Trinity Centre of Bioengineering (TCBE), Trinity College Dublin (TCD), Dublin, Ireland [3]Advanced Materials and Bioengineering Research Centre (AMBER), RCSI and TCD, Dublin, Ireland.

## 21.1 Introduction

A very high proportion (around 85%) of new anticancer drugs evaluated in vitro and in animal models fails in clinical trials [1]. Moreover, two-dimensional (2D) models fail to adequately model the microenvironment of the tumor while animal models unfortunately do not always predict the effect in humans, for example, the immune system of mice and rats is very different than that of humans and additionally, gene expression of animal cells differs from that of human cells. Three-dimensional (3D) cancer models have been developed to better model the tumor microenvironment and in doing so, reduce the need for animal models and enable a more precise analysis of effects of anticancer drugs as well as gene therapy as a treatment for cancer. The 3D models have been shown to more closely resemble the in vivo situation with regards cell morphology, gene expression, drug resistance, and cell behavior. Epithelial–mesenchymal transition (EMT), a process that enables primary tumor cells to become invasive and initiate metastasis, is one important aspect that is investigated in 3D as it is not properly represented in 2D. EMT markers, such as vimentin, have been shown to be upregulated in 3D models compared with 2D cultures [2,3]. Furthermore, cancer stem cell (CSC) markers, such as SOX4, have been shown to be upregulated in 3D compared with 2D. This could be a mechanism of the frequently observed higher drug resistance found in 3D models [4,5].

Several 3D models have been generated by utilizing either 3D scaffolds or hydrogels. These can serve either as platforms to model the tumor microenvironment in vitro or can

*Biomaterials for 3D Tumor Modeling*
DOI: https://doi.org/10.1016/B978-0-12-818128-7.00021-6

be used in vivo to achieve a sustained and localized release of anticancer drugs. This chapter reviews the current state-of-the-art in relation to the use of 3D cancer model systems in the form of scaffolds and hydrogels for chemotherapeutic/drug delivery and gene therapy for a selection of the most common cancers: lung, breast, and prostate cancer.

## 21.2 Lung cancer

Lung cancer accounts for most cancer-related deaths worldwide and is one of the most commonly detected forms of cancer. Lung cancers are classified as small cell lung cancer or nonsmall cell lung cancer (NSCLC), whereby NSCLC accounts for approximately 80% of cases. Due to a late occurrence of first symptoms, about 50% of patients detected with lung cancer undergo palliative treatment as they are in an advanced cancer stage or have already developed metastasis. Therefore, the 5-year survival rate is very low, between 4% and 17% [6]. As a result, it is imperative that new treatment options are developed and in order for these to be successful, sufficient testing in an in vitro simulated microenvironment is a necessity. Therefore, this chapter reviews the current 3D biomaterials in the form of scaffolds and hydrogels currently in use for chemotherapeutic and gene therapy treatments.

### 21.2.1 Scaffold chemo/drug treatment

In 2014, Stratmann et al. developed a 3D lung carcinoma model on a decellularized tissue matrix composed of small intestinal decellularized submucosa with preserved mucosa [7]. The group cultured two different types of lung adenocarcinoma, one of which, HCC827, had an activating mutation of the epidermal growth factor receptor (EGFR). As described in other 3D models, cells proliferated slower in 3D compared with 2D. EGFR activation in HCC827 was inhibited by Gefitinib, an anticancer drug acting on EGFR, resulting in a significant reduction in proliferation in 3D but not in 2D. Results from the 3D model correlated with the results of an in silico model that integrates the EGFR signaling network according to clinical data. Other studies have also used in silico models to predict in vivo behavior such as that by Moustakas et al. [8]. Here, activation of tumor growth factor-beta 1 (TGFβ1), known to induce EMT, induced tumor cell invasion and EMT in silico and was confirmed in the 3D model by acquisition of mesenchymal cell morphology and altered expression of EMT markers (fibronectin, E-cadherin, β-catenin, and mucin-1). The group showed that 3D models have the potential of a more accurate prediction of in vivo tumor behavior by combining an in silico model with a 3D model.

Recently, Yan et al. utilized a high-throughput gelatin-based 3D microtumor array for drug screening [9]. The microscaffolds seeded with a lung cancer cell line in 384-well format were shown to predict in vivo drug responses more accurately than 2D monolayer. Doxorubicin (DOX), identified by the array, successfully inhibited tumor growth in a mouse xenograft model, whereas gemcitabine and vinorelbine, which were selected solely through 2D analysis, were not effective in the xenograft model. Furthermore, this array was shown to be suitable for chemosensitivity testing using patient-derived cells.

Li et al. (2018) also described a 3D in vitro tumor model based on silk/fibroin/chitosan scaffolds to culture human lung adenocarcinoma cells (Fig. 21.1) [10]. They showed that

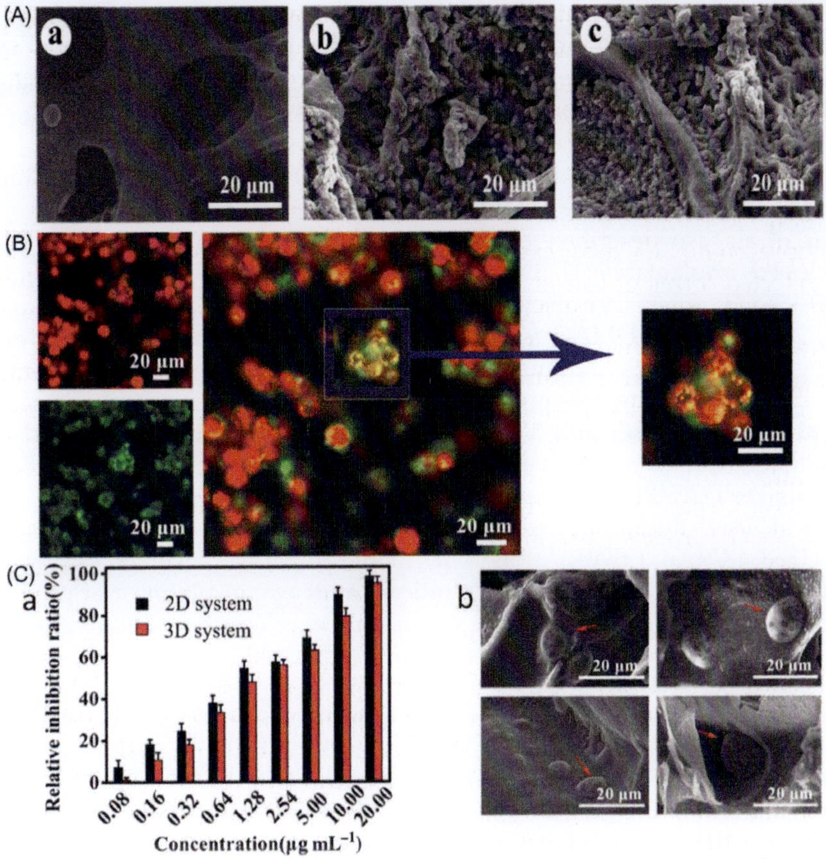


FIGURE 21.1 (A) SEM images of (a) a blank SF/CS scaffold; then, the A549 cells were grown in the SF/CS scaffold after 72 h, (b, c) the cells in the pore. (B) Overall morphology of tumor spheres in 3D scaffolds. Red: LysoTracker and green: EGFP plasmid. (C) (a) Cell cytotoxicity of DOX treatment at various concentrations against A549 cells after 24 h of incubation ($n = 3$). (b) Morphology of the cells treated with DOX on the 3D tumor cell scaffolds. Source: *Reprinted (adapted) with permission from Li J, Zhou Y, Chen W, Yuan Z, You B, Liu Y, et al. A novel 3D in vitro tumor model based on silk fibroin/chitosan scaffolds to mimic the tumor microenvironment. ACS Appl Mater Interfaces 2018;10:36641 – 51. Copyright (2020) American Chemical Society.*

cells cultured in 3D form tumor spheres after 72 h showing morphological as well as biological characteristics much closer to in vivo tumors. The cells proliferated during the spheroid formation at a quicker rate than in the 2D systems, which is in contradiction to previous reports [7,11,12]. Additionally, increased drug resistant to DOX and differences in immune-related protein levels were found. P-glycoprotein, that was overexpressed in various solid tumors, and might play an important role in clinical drug resistance [13] was also upregulated in 3D compared with 2D.

Ardeshirzadeh et al. developed polyethylene oxide/chitosan/graphene oxide electrospun nanofibrous scaffolds to assess controlled release of DOX [14]. The study demonstrated highercytotoxicity of DOX released from the nanofibrous scaffolds compared with free DOX. Another study in 2018 describes a graphene-loaded poly(lactic acid) (PLA) scaffolds for sustained DOX release with similar results [15]. A small burst release and subsequently sustained release was shown. The release rate was significantly decreased by increasing fiber thickness. Enhancement in PLA to chitosan ratio, $TiO_2$ content, and graphene oxide/$TiO_2$/DOX content resulted in a slower release rate. Inhibition of proliferation on target cancer cells was enhanced by increasing DOX content in the nanofibers.

Mellatyar et al. also described the development of nanofibers for drug release [16]. They developed implantable 17-dimethylaminoethylamino-17-demethoxy geldanamycin (17-DMAG)-loaded poly-ε-caprolactone–poly(ethylene glycol) (PCL/PEG) nanofibers to inhibit HSP90 activity. 17-DMAG is a derivative of geldanamycin, that with binding the ATP-binding motif of HSP90 inhibits its activity as a protein chaperone, resulting in degradation of HSP90 [16]. Drug loading enhanced cytotoxicity in the lung cancer cells and showed that free 17-DMAG and 17-DMAG-loaded PCL/PEG nanofiber had dose- and time-dependent cytotoxicity against lung adenocarcinoma cells suggesting that loading 17-DMAG into the PCL/PEG nanofibers improves its inhibitory effects.

Tapias et al. used decellularized lung scaffolds to culture different adenocarcinoma cell lines (H358, PC9, and SW1573) as a 3D model for lung cancer [12]. They found the same relative sensitivity of two different anticancer drugs, cisplatin or erlotinib, on three different cell lines but found a reduced sensitivity to erlotinib in 3D, hypothesizing the cause of this was an increased EMT on the scaffolds as EMT on decellularized lung scaffolds has been described previously [17]. Vishnoio et al. also assessed the effect of cisplatin in their "four-dimensional (4D) model" of lung cancer which consisted of growing nodules from human lung cancer cells on an acellular rat matrix [18]. They showed differential gene expression profiling of tumor cells grown in the 4D model (e.g., upregulation of SOX2 and vimentin, downregulation of E-Cadherin and IL6) which predicts poor survival, whereas the same cells grown in the 3D model or cells grown in a Matrigel predict good survival based on gene expression [17]. Cisplatin treatment led to a significant decrease in the viability of H1299, A549, or H460 lung cancer tumor cells grown under 2D conditions but not in the 4D model. Cisplatin did lead to regression of tumor nodules. Moreover, there was a significantly higher number of circulating tumor cells per tumor area and MMP-2 production per tumor area for all human lung cancer cell lines grown in the 4D model when treated with cisplatin. Furthermore, they could isolate circulatory tumor cells from the 4D lung cancer model that are resistant to cisplatin treatment.

Poly(glycerol monostearate-*co*-epsilon-caprolactone) films were developed by Wolinsky et al. for controlled release [19]. They achieved prolonged and sustained release from low-dose delivery films. The drug-loaded films were applied to a collagen-based scaffold clinically indicated for the mechanical buttressing of lung tissue following surgical resection. They showed sustained release over 7 weeks and thereby prevention of the local growth and establishment of Lewis lung carcinoma tumors in a mouse xenograft model (freedom of local tumor growth of 86%). In comparison, all animals treated with a larger intravenous dose of 10-hydroxycamptothecin or unloaded composites developed rapid local tumors. Table 21.1 fittingly summarizes a selection of the current scaffold and hydrogel models available for chemotherapeutic/drug treatment for lung cancer.

**TABLE 21.1** 3D scaffold models for chemotherapeutic/drug and gene therapy treatment for lung cancer.

| 3D model | Cell type | Therapeutic | Effect | Ref. |
|---|---|---|---|---|
| Silk/fibroin/chitosan scaffolds | A549 (lung adenocarcinoma cells) | DOX | Increased drug resistance compared to 2D | [10] |
| Intestinal decellularized scaffolds | HCC827 and A549 (lung adenocarcinoma cells) | Gefitinib | Significant inhibition of proliferation in 3D, no effect in 2D | [7] |
| Gelatin-based 3D microtumor array | In vitro: 10 different cancer cell lines of various origins (representative results shown of H226, HCC827 and SKBR3)<br>In vivo: Lung cancer cell line (H226) xenografts | DOX, gemcitabine, vinorelbine | DOX effective in 3D (reduces cell viability), the other two only in 2D; results correlated with in vivo results | [9] |
| Decellularized lung scaffolds | H358, PC9, and SW1573 (lung cancer cell lines) | Erlotinib, cisplatin | Reduced sensitivity to erlotinib in 3D | [12] |
| Acellular rat matrix | H1299, A549, or H460 (lung cancer cell lines) | Cisplatin | Significant decrease in the viability of tumor cells grown under 2D conditions but not in the 4D model | [18] |
| Poly(glycerol monostearate-*co*-epsilon-caprolactone) films | Lewis lung carcinoma tumors in vivo | Agent 10-hydroxycamptothecin | Sustained release over 7 weeks and thereby prevention of local growth and establishment of Lewis lung carcinoma tumors | [19] |
| Poly(ethylene glycol) (PCL/PEG) nanofibers | A549 | 17-dimethylaminoethylamino-17-demethoxy geldanamycin (17-DMAG) | Loading 17-DMAG into the PCL/PEG nanofibers improves its inhibitory effects | [20] |
| Graphene-loaded PLA | A549 | DOX | Proliferation inhibition effect on target cancer cells enhanced by increasing DOX content in the nanofibers | [15] |
| Polyethylene oxide/chitosan/graphene oxide electrospun nanofibrous scaffolds | A549 | DOX | Increased and longer lasting cytotoxicity exhibited by DOX[HYPHEN]loaded polyethylene oxide/chitosan/graphene oxide (GO) electrospun nanofibers against A549 cells compared to free DOX and GO/DOX. Higher drug loading capacity and ability for controlled release | [14] |

(*Continued*)

TABLE 21.1 (Continued)

| 3D model | Cell type | Therapeutic | Effect | Ref. |
|---|---|---|---|---|
| Biodegradable polymer scaffolds | Xenograft mouse (A431 and B16F10 cells) | Retrovirally transduced fibroblasts producing thrombospondin-2 | 70% inhibition of malignant tumor growth by cell-based TSP-2 gene-therapy after implantation of Lewis lung carcinomas cells. | [21] |
| Core–shell nanofibrous scaffold | A549 | Suicide gene and 5-fluorocytosine | Significant upregulation of apoptotic genes and consequent downregulation of antiapoptotic genes | [22] |
| Microfluid system | NSCLC cell line (SPCA-1) and stromal cell lines (human lung fibroblast cell line; HFL1) | Gefitinib, paclitaxel (PTX), and gemcitabine | Chemosensitivity of cancer cells to PTX decreased about almost two-fold when co-cultured with fibroblasts as compared with monoculture | [23] |
| Biocompatible poly (lactic-*co*-glycolic acid) electrospinning nanofiber membrane | Co-cultured lung adenocarcinoma cells (A549) and human fetal lung fibroblasts (HFL1) and endothelial cell (HUVECs) | Gefitinib | Increased drug resistance in 3D | [24] |
| 3D matrices (PuraMatrix and Matrigel) | Nonsmall-cell lung adenocarcinoma cells (H460, A549, and H1650) | DOX | Reduced chemoresistance in 3D | [25] |
| AlgiMatrix scaffolds | A549 | DOX | Higher IC50 and reduced uptake in 3D | [26] |
| Polystyrene microwells | Spheroids from lung adenocarcinoma (cancer cell lines NCI-H23 and NCI-H1650) | CI-1033 | Cells grown in 3D were more responsive | [27] |
| Polyacrylamide hydrogel containing an EGFR-phosphorylatable peptide | A549 | Gefitinib and erlotinib | Array was able to detect therapeutic resistance and predict anticancer drugs to overcome this resistance | [28] |
| Thermoresponsive and photocrosslinkable hydrogels | A549 | DOX photocrosslinking to chitosan | Decrease of tumor size at day 30 showing a beneficial effect of sustained release from the chitosan-hydrogels | [29] |
| PEG-hydrogel | Mice inoculated with adenocarcinoma cells (A549) | 5-Fluorouracil (5-FU) | Beneficial effect of sustained release: decreased tumor growth in comparison to the untreated group and a higher tumor inhibition rate compared tothe free 5-FU group | [30] |

| | | | | |
|---|---|---|---|---|
| Photocrosslinkable PTX-incorporated hydrogel | Lewis lung cancer cells were subcutaneously injected into mice | PTX | Reduced tumor growth in mice treated with the developed hydrogel compared with untreated control groups as well as mice injected with thee anticancer drugs or the hydrogel only | [31] |
| Demethoxycurcumin-loaded core–shell hydrogel nanoparticles | NSCLC | Cisplatin/demethoxycurcumin | Efficiently delivered demethoxycurcumin and increased cisplatin-induced cytotoxicity | [32] |
| Interpenetrating network system | Mice bearing human lung cancer xenograft | Gold(III) porphyrin(AuP) | Significantly inhibited tumor growth compared with untreated controls | [33] |
| Injectable hydrogel | Lung adenocarcinoma cells (A549) and HEK-293F for virus production | Metalloproteinase 2 (TIMP2) | Supressed lung cancer cell proliferation in vitro | [34] |
| 3D PEG–fibrin hydrogel | Lung adenocarcinoma cells (A549 or H1299) with fibroblasts (MRC5) and endothelial cells (HUVEC) | Oncolytic human adenoviruses | Degree and kinetics of tumor responses in the PEG-hydrogel resembled the kinetics in a mouse xenograft model for adenovirus treatment better than in 2D | [11] |
| Co-cultured in cell-adhesive glycol-based hydrogels | Lung adenocarcinoma cancer cells (344SQ) and endothelial (HUVECs) and pericyte vascular cells (HVP) | SB431542 (TGF-β1 inhibitor) | Treatment blocked endothelial cell-driven 344SQ migration | [35] |
| Microfluidic system | Adenocarcinoma cells (A549) co-culture with an endothelial monolayer (HUVECs) | EMT blockers (MK-2206, BMS-559626, CI-1033, gefitinib, BMS-536924, PD 0325901, masitinib, AZD 0530, PP1, A83-01, LY 364947, SD-208) | IC50 doses gained by dispersion assays were lower in 3D (in all cases except masitinib) and thereby in a range of plasma doses being effective in clinical trials | [36] |

### 21.2.2 Scaffold gene therapy

As cancers are characterized by a variety of genetic mutations, they are ideally placed for targeted gene therapy treatment. Gene therapy bears several challenges such as efficient and local delivery of negatively charged nucleic acids into cells. The 3D models could help to develop and evaluate systems for successful gene delivery.

In 2002, Streit et al. developed biodegradable polymer scaffolds utilized as an implant system for the continuous in vivo production of thrombospondin-2 (TSP-2), an endogenous inhibitor of tumor growth and angiogenesis, using retroviral transduction [21]. After transplantation into the peritoneal cavity of nude mice, bioimplants maintained high levels of TSP-2 secretion over extended time periods, resulting in increased levels of circulating TSP-2 while mice bearing control grafts and A431 and B16F10 cells formed rapidly growing malignant tumors. In contrast, tumor growth was significantly inhibited by greater than 60% in mice bearing TSP-2-secreting bioimplants. A greater than 70% inhibition of malignant tumor growth by cell-based TSP-2 gene therapy was observed after implantation of Lewis lung carcinomas cells. Using B16F10 cells they also showed an effect on already established tumors by implanting the scaffolds 5 days after tumor implantation and showing a significantly smaller tumor after 19 days compared with control grafts. Furthermore, bioimplant-generated TSP-2 potently inhibited angiogenesis (assessed by vessel volume, size, and total vascular area) of human squamous cell carcinomas, malignant melanomas, and Lewis lung carcinomas that were implanted at a distant site. Sukumar et al. also utilized a core–shell nanofibrous scaffold for therapeutic gene delivery [22]. They delivered cytosine deaminase-uracil phosphoribosyltransferase (CD::UPRT) suicide gene transfection into cancer cells and simultaneously delivered prodrug, 5-fluorocytosine in a controlled and sustained manner to combine anticancer drug with gene therapy. They monitored suicidin release and subsequent drug release resulting in apoptosis and showed significant upregulation of apoptotic genes and consequent downregulation of antiapoptotic genes at 96 h.

### 21.2.3 Scaffold co-culture chemo/drug treatment

In 2013, Xu et al. developed a microfluidic system using continuous media supplementation to co-culture lung cancer cells and stromal cell lines [37]. Cells proliferated faster in 2D while fibroblasts grew in cell clusters within the 3D model. A gradient generator included in the system was utilized to measure sensitivity to different anticancer drugs (gefitinib, PTX, and gemcitabine) assessed by measuring cell survival and compared between 2D and 3D mono- and co-culture. Chemosensitivity of cancer cells to PTX decreased about twofold when co-cultured with fibroblasts as compared with monoculture indicating that stromal cells affect the drug sensitivity of lung cancer cells and highlights the importance of in vitro studies carried out in co-culture systems. By culturing cells from fresh lung cancer tissues, they showed that the system can potentially be applied to guide personalized therapy for cancer treatment. Yang et al. also described the use of patient-derived tumor cells [24]. They developed a "lung-on-a-chip" with a biocompatible poly(lactic-*co*-glycolic acid) electrospinning nanofiber membrane as the chip substrate and cell scaffold to model the alveolar microenvironment. Lung adenocarcinoma cells were co-cultured with human fetal lung fibroblasts and demonstrated increased drug resistance in

3D. The authors evaluated penetration ability through the in vitro tumor and excluded reduced drug penetration in 3D as a factor of reduced effectiveness that has been observed in other studies [25,38]. Prediction of in vivo efficacy (assessed by a xenograft mouse model) of different anticancer drugs was thereby more accurate compared with 2D. Interestingly, by additional co-culturing of endothelial cells they found that lung adenocarcinoma cells can cause endothelial cell apoptosis and tumor invasion.

### 21.2.4 Hydrogel chemo/drug treatment

The 2D cell culture studies are usually carried out in plastic dishes or on glass. Both materials are much stiffer than human tissue and therefore do not adequately mimic an in vivo situation [25]. Prina-Mello et al. compared morphology, gene expression as well as chemoresistance of nonsmall cell lung adenocarcinoma cells cultured in 2D glass dishes and in different soft 3D matrices (PuraMatrix and Matrigel). No upregulation of EMT biomarkers was found in 3D cultures while chemoresistance in two different commercially available hydrogels (PuraMatrix 0.25% and 0.30%) was reduced compared with 2D culture. This contradicts evidence from the literature that the efficacy of anticancer agents is reduced in 3D compared with cells cultured in 2D [4,5]. Earlier, Godugu et al. compared lung adenocarcinoma cells grown on AlgiMatrix scaffolds to 2D cultures [26]. They showed a higher IC50 value for DOX in 3D compared with 2D. They also showed that the uptake of DOX and oil-loaded nanostructured lipid carrier particles in 3D was less efficient than in 2D, which could be one of the reasons for the lower efficacy. Cho et al. also assessed the sustained release of DOX from thermoresponsive and photocrosslinkable hydrogels after conjugation to acrylated chitosan [29]. Photocrosslinking to chitosan did not increase cytotoxicity of DOX in vitro while in vivo tumor volume in athymic nude mice bearing human lung adenocarcinoma xenografts was significantly reduced 1month after injection of photocrosslinking hydrogels.

Tu et al. investigated spheroids from lung adenocarcinoma growing in polystyrene microwells to test anticancer drug efficacy [27]. CI-1033, a phase 2 clinical trial anticancer drug that acts by inhibiting pan-ErbB tyrosine kinase was assessed and the authors showed, by dispersion assay, that cells grown in 3D were more responsive to CI-1033 drug treatment. This contradicts previous studies showing a higher resistance of the 3D model to cytotoxic drugs as mentioned before [4,5,39,40].

As cancers are characterized by a variety of gene mutations, the mutated genes may be potential targets for inhibition by drug or gene therapy. One gene frequently mutated is EGFR. Different anticancer drugs exist to inhibit EGFR but efficacy of these is often hampered by developing resistance. Gosh et al. developed a polyacrylamide hydrogel containing an EGFR-phosphorylatable peptide to evaluate EGFR resistance and efficacy of EGFR targeting drugs (gefitinib and erlotinib) and showed that the array was able to detect therapeutic resistance and predict anticancer drugs to overcome this resistance [28].

Yi et al. developed a PEG-hydrogel system for sustained release of the anticancer drug 5-FU to increase drug retention [30]. Subcutaneous implantation ($100\,mg\,kg^{-1}$) significantly inhibited tumor growth after 14 days in mice inoculated with adenocarcinoma cells in comparison to the untreated controls. Interestingly, comparison between 5-FU-treated control and 5-FU-loaded PEG-hydrogel groups showed no difference in tumor volume until day 21 but a significant difference was observed on day 28 that was attributed to a beneficial effect of sustained release.

Obara et al. developed a photocrosslinkable PTX-incorporated hydrogel to inhibit tumor growth [31]. They demonstrated sustained drug-release from the hydrogel in vitro and when Lewis lung cancer cells were subcutaneously injected into mice, the authors showed reduced tumor growth in mice treated with the developed hydrogel compared with mice injected with thee anticancer drugs. It was postulated that the reduced number of CD34-positive vessels detected indicated inhibition of angiogenesis as a mechanism by which the hydrogel reduced tumor growth.

As mentioned previously, cisplatin is a commonly used chemotherapeutic to treat lung cancer, and acts by impairing DNA repair. However, development of cisplatin resistance decreases its effectiveness. Demethoxycurcumin has been shown to inhibit cisplatin resistance by reducing activity of the PI3K-Akt-snail pathways in lung carcinoma cells [41], but low-water solubility hampered its clinical application. In 2018, Chen et al. showed that a carbomethyl-hexanoyl chitosan with a demethoxycurcumin–polyvinylpyrrolidone core-phase can be used to inhibit cisplatin-induced drug resistance by increasing demethoxycurcumin bioavailability and local delivery. Demethoxycurcumin-loaded core–shell hydrogel nanoparticles that efficiently delivered demethoxycurcumin were developed, and increased cisplatin-induced cytotoxicity [42]. When delivered simultaneously with cisplatin, on-target cisplatin resistance protein, ERCC1, was significantly inhibited via the PI3K-Akt pathway and the posttarget cisplatin resistance pathway was increased including bax and cytochrome c expression. Thymidine phosphorylase, a component of the pyrimidine salvage pathway, was inhibited by the combination treatment suggesting that enhancement of the cytotoxicity to cisplatin via administration of demethoxycurcumin–chitosan nanoparticles was mediated by downregulation of the expression of Thymidine phosphorylase, and ERCC1, regulated via the PI3K-Akt pathway.

Interpenetrating network system consisting of PEG-diacrylate (PEGdA) and modified gelatin, is a biocompatible and biodegradable hydrogel and has been studied for the local delivery of bioactive molecules and drugs. Gold(III) porphyrin (AuP) is a stable metal compound when administered systemically and has been shown to be more effective in killing cells than cisplatin [43]. Lee et al. developed an AuP-loaded interpenetrating network system and showed that this has a higher cytotoxicity against human lung cancer cell lines compared with the interpenetrating network system only [33]. In mice bearing a human lung cancer xenograft, AuP-loaded interpenetrating network significantly inhibited tumor growth and reduced the number of CD-31 positive microvessels compared with untreated controls, whereas interpenetrating network alone did not lead to a significant reduction in tumor volume compared with controls.

Yamada et al. developed an injectable hydrogel for sustained release of tissue inhibitor of metalloproteinase 2 (TIMP2), an endogenous proteinase inhibitor, demonstrating antitumorigenic, antimetastatic, and antiangiogenic activities in vitro and in vivo [34]. They compared TIMP2 release profile from four β-hairpin peptide gels of varying net electrostatic charge. A negatively charged peptide gel enabling encapsulation of 4 mg $mL^{-1}$ of TIMP2, without effects on rheological properties achieved the slow sustained release (0.9%/day) over 28 days. Released TIMP2 was shown to be structurally intact and maintained the ability to inhibit metalloproteinase activity, as well as suppress lung cancer cell proliferation in vitro.

### 21.2.5 Hydrogel co-culture

Besides the development of 3D models, co-culturing of cancer cells with other cell types, such as stromal cells could provide a more accurate environment of a tumor in vivo. Therefore, different co-culture models have been developed to investigate cancer cell behavior under the influence of different other cell types.

Del Bufalo et al. utilized a 3D PEG-fibrin hydrogel to co-culture lung adenocarcinoma cells with fibroblasts (MRC5) and endothelial cells (HUVEC) to model the effect of stroma on tumor cells and validated the effectiveness of oncolytic human adenoviruses as a cancer treatment [11]. Comparing cells in mono- and co-culture they showed an increased proliferation and invasiveness in co-cultured 3D models. Using MSCs to deliver oncolytic adenovirus to fluorescently tagged lung cancer cells in 2D and in 3D hydrogels, they found that the degree and kinetics of tumor responses in the PEG-hydrogel resembled the kinetics in a mouse xenograft model for adenovirus treatment but the response seen in conventional 2D cultures was very different, in which complete oncolysis occurred much faster. They showed furthermore that the tumor cell structures formed in PEG-hydrogels more closely reproduced the tumor's in vivo characteristics, in terms of morphology and phenotype as shown by microscopy and in terms of a slower proliferation rate compared with conventional 2D cultures.

Aref et al. investigated the efficacy of different EMT blockers on adenocarcinoma cells cultured in a microfluidic system that integrates tumor cell spheroids in a 3D hydrogel in co-culture with an endothelial monolayer [36]. Drug candidates inhibiting receptor activation or signal transduction pathways implicated in EMT, such as EGFR or Akt, were tested using dispersion of lung adenocarcinoma cell spheroids to evaluate its effectiveness. They showed significant differences in response to drugs between 2D and 3D, and between mono- and co-culture. IC50 doses gained by dispersion assays were lower in 3D and thereby in a range of plasma doses being effective in clinical trials. This is, as mentioned before, contradictive to previous reports that conclude a higher drug efficacy in 2D compared with 3D [39,40] although higher anticancer drug efficacy on 3D has also been reported previously as well [25,27].

## 21.3 Breast cancer

Breast cancer entails a complex spectrum of disease which has been classified according to gene expression signatures and phenotype. Triple negative breast cancer is one of the aggressive subtypes which is extremely difficult to treat due to its resistance to current treatment modalities. Microscopically all breast cancer has extracellular matrix (ECM). This ECM is made of basement membrane, interstitial ECM, and a network of encapsulating mammary gland allowing for a therapeutically targetable ECM by chemotherapy [44]. Unfortunately, with the current treatment modalities, the tumor microenvironment is completely degraded leading to no clear consensus on their therapeutic value. Thus, it is essential to deepen the understanding of the interaction between breast cancer ECM and delivery system for anticancer drugs against breast cancer cells. As the prevalence of breast cancer has increased over the years, there has been a significant rise in breast cancer research [44]. There is a need to develop more efficient and physiologically relevant models for evaluation of drug interaction with breast cancer cells. As animal testing involves large amount of cost and time, in vitro-based screening models are now in favor. Additionally, in vitro cell culture models allow for the use of human cell lines rather than animal models, enabling researchers to better evaluate human biology and disease models more closely.

This section of the chapter highlights the difference between the cells grown in 2D and on 3D scaffolds utilized for breast cancer research.

### 21.3.1 Scaffolds chemo/drug therapy

Whitman et al. described a wax-patterned paper-based 3D model to assess the anticancer efficacy of oestrogen receptor modulating reagents [17β-estradiol (ER agonist) and ICI 182 780 (ER antagonist)] [45]. The 3D cultures had significantly more oestrogen receptor α than monolayer cultures. Unfortunately, however, they did not increase the accuracy of the 3D model compared with 2D monocultures to predict efficacy of oestrogen receptor modifying agents in vivo.

Recently, Li et al. compared three different biocompatible materials as 3D models for breast cancer research using decellularized lung scaffold, chitosan-gelatin scaffold, and poly-L-lactic acid scaffold. All scaffolds showed increased malignancy of MCF-7 and 4T-1 cells cultured on all three scaffolds compared with cells growing in 2D [46]. In 3D, MFC-7 cells were observed to proliferate slower, build spheroids, and showed an upregulation of breast cancer gene 1 (BRCA1) and human epidermal growth factor receptor 2 (HER-2) [47,48]. Cells grown on chitosan–gelatin and on decellularized lung scaffolds showed a significant increase in tumor size in vivo after 3 weeks. Cells cultured in 3D showed decreased sensitivity to the anticancer drug 5-FU as compared with 2D and showed a reduction in cancer cell viability. This study is in line with earlier work done by Dhiman et al. They cultured MCF-7 cells on a biocompatible chitosan-based scaffold which showed increased lactate production and decreased chemosensitivity to tamoxifen for cells grown in 3D compared with 2D [49]. This presentation is similar to increased lactate production in in vivo tumors. Additionally, Leeper et al. also analyzed tumor responsiveness to tamoxifen using collagen-based scaffolds to culture patient-derived breast cancer tissue MDB-0102-54A [50]. Results regarding expression of apoptotic markers as well as histopathological changes on the scaffolds correlated to inhibition of proliferation, induced apoptosis of luminal epithelium leading to a layer with reduced volume and viability. Unfortunately, tamoxifen had not inhibited the migration of cells to the collagen gel interface and invasion into the surrounding collagen.

Balachander et al. developed a porous PCL scaffold to culture patient-derived triple negative breast cancer cells [2]. They as well showed an increased invasiveness of cells grown on the 3D scaffolds compared with cells grown on 2D monolayer. Cells cultured on 3D in vivo led to an increase in the number of lung metastasis. More recently the research group built macroporous scaffolds from decellularized cancer-associated fibroblasts [51].

They showed that breast cancer cells cultured on the polymer scaffolds form tumors promoting high production of ECM components. They cultured patient-derived breast cancer cells and evaluated chemosensitivity to DOX. No direct comparison to drug response in 2D was done, however, individual dose-dependent responses of the different patient-derived cells indicated a suitability to DOX for the described scaffolds. Treatment with DOX significantly reduced cell viability of cancer cells in the scaffold. A 3D model utilizing human decellularized adipose tissue has also been developed to culture breast cancer cells and evaluate efficacy of chemotherapeutics [52]. Different cell lines (MCF-7, BT474, and SKBR3) were cultured showing slower proliferation on the scaffolds compared with cells grown in 2D and on Matrigels thereby being more similar to cells grown in xenografts. MCF-7 and BT474 cell lines cultured on scaffolds showed decreased chemosensitivity to DOX as compared with 2D cultured cells. They also showed increased chemosensitivity to the EGFR inhibitor and lapatinib on 3D scaffolds compared with 2D and hypothesized that this increased sensitivity is due to the increased EGFR signaling on 3D scaffolds. Horning et al. also evaluated anticancer efficacy of DOX in addition to PTX and tamoxifen using large, porous polymeric microparticles as scaffolds for cancer cell growth and compared with 2D [53]. DOX, PTX, and tamoxifen were shown to be less efficient against cells grown on the microparticles compared with cells growing in 2D. Another study by Yuan et al. developed mesoporous silica nanoparticle composite nanofibers for sustained release of DOX in two phases and demonstrated significant inhibition of tumor growth in vivo [54].

Li et al. described a decellularized porcine lung scaffold and chitosan–gelatin scaffold to culture breast cancer cells growing in spheroids to test the anticancer efficacy of astragalus polysaccharide [55]. They compared the cells (MCF-7) grown on the decellularized lung scaffold to 2D monolayer and chitosan–gelatin scaffolds. They showed anticancer activity of supernatant of astragalus polysaccharide-treated macrophages as well as of 5-FU on the decellularized lung scaffolds compared with untreated controls by measuring effect on proliferation, apoptotic markers and cell viability. After 3 days of treatment with 5-FU, the population of living cells markedly decreased on live dead imaging.

Gupta et al. described regenerative bioscaffolds for postsurgery care to avoid tumor reoccurrence by controlled release of emodin [56]. An in vivo xenograft model using GILM2 breast cancer cells showed no significant effect on tumor size between emodin-loaded and blank scaffold treatment groups but using histological staining they showed that the tumor seemed to invade less when treated with the emodin-loaded scaffolds. Table 21.2 describes the current scaffolds in use for chemo/drug assessment.

### 21.3.2 Scaffolds gene therapy

Gene therapy plays a critical role in medical technologies intended to deliver therapeutic agents in various clinical settings. A personalized approach to cancer treatment using gene silencing by RNA interference is rapidly evolving as a potential therapy. The effector molecules such as a small interfering RNA and microRNA can be used to switch off specific cancer genes. However, major issues to implementing siRNA- and miRNA-based therapies in clinical practice results in the lack of an effective delivery system. Thus, the authors have started research on utilizing collagen-based scaffolds as platforms for gene delivery for breast cancer treatment (Fig. 21.2).

**TABLE 21.2** 3D scaffold models for chemotherapeutic/drug treatment for breast cancer.

| 3D model | Cell type | Therapeutic | Effect | Ref. |
|---|---|---|---|---|
| Wax-patterned paper-based 3D model | MCF-7<br>4T-1 | Tamoxifen | No significant difference between cell survival in 2D versus 3D in both cell lines | [45] |
| Decellularized lung scaffold, chitosan-gelatin scaffold, and poly-L-lactic acid scaffold | MCF-7<br>4T-1 | 5-FU | Chitosan–gelatin and lung scaffolds showed a significant increase in tumor size<br>Cells in 3D showed decreased sensitivity to the anticancer drug 5-FU as compared to 2D | [46] |
| Biocompatible chitosan-based scaffold | MCF-7 | Tamoxifen | Reduced chemosensitivity in 3D | [49] |
| Collagen-based scaffold | MDB-0102-54A | DOX, tamoxifen | Inhibition of proliferation and induced apoptosis of luminal epithelium thus reduced tumor volume<br>No inhibition of cell migration | [50] |
| Macroporous scaffolds from decellularized cancer-associated fibroblasts | Patient-derived breast cancer cells | DOX and mitoxanthrone | DOX group showed up to 60% reduction in cell number in high doses.<br>Mitoxanthrone failed to evoke reduction in cell number even at high doses | [51] |
| Human decellularized adipose tissue Matrigels | MCF-7, BT474, and SKBR3 | DOX, EGFR inhibitor, and lapatinib | Slower proliferation in human decellularized adipose tissue vs 2D monolayer and Matrigels<br>Decreased chemosensitivity to DOX in 3D. Increase chemosensitivity to EGFR inhibitor and lapatinib (MCF-7 and BT474) | [52] |
| Porous polymeric microparticles | MCF-7 | DOX, PTX, and tamoxifen | Decreased chemosensitivity of cells grown on the microparticles | [53] |
| Mesoporous silica nanoparticle composite nanofibers | MCF-7 | DOX | Significant inhibition of tumor growth in vivo | [54] |
| Decellularized porcine lung scaffold chitosan–gelatin | MCF-7 | Astragalus polysaccharide 5-FU | Inhibits cell proliferation in 3D but no significant difference between chitosan–gelatin and decellularized lung scaffold | [55,57] |
| Silk fibroin (SF) and chitosan (CS) emodin-loaded scaffolds | GLIM2 | Emodin | No significant effect on tumor size between emodin-loaded and blank scaffold treatment groups but histological staining showed the tumor seemed to invade less when treated with the emodin-loaded scaffolds | [56] |

A unique approach to increasing DOX's efficiency in treating triple-negative breast cancer MDA-MB-468 is the use of a layer-by-layer nanoparticle scaffold film formation by alternately depositing siRNA targeting multidrug resistance protein 1 and poly-L-arginine in layers atop a DOX-loaded liposome [58]. Results showed significant synergistic effect

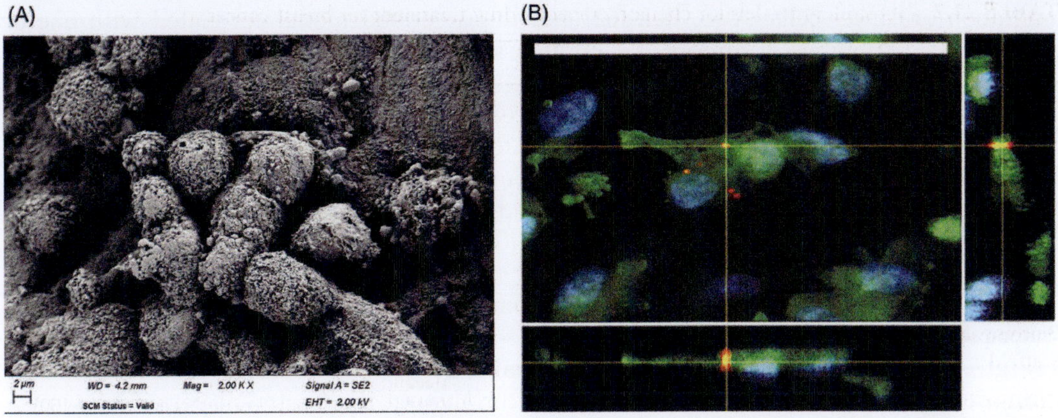


FIGURE 21.2 Breast cancer cells (MDA-MB-231) grown on collagen–glycosaminoglycan scaffolds to model primary breast cancer tumors in 3D. (A) Scanning electron microscopy image of breast cancer cells 7 days after seeding on scaffolds. Scale bar = 2 μm. (B) Confocal image of breast cancer cells transfected with a red fluorescently tagged scrambled microRNA using Lipofectamine RNAiMax. Scale bar = 100 μm.

when DOX and siRNA were used together. The cell viability results showed apoptosis and inhibition of cell proliferation.

Phan-Lai et al. described delivery of IFNγ and CCL21 into breast cancer cells HCC1569 and MDA-MB-231 via plasmids [59]. Initially the cells were seeded on chitosan–alginate scaffolds then they were transfected with IFNγ and CCL21. Seeded cells showed spheroid formation on 3D scaffolds. Histopathology data confirmed increased migration of T cells in the breast cancer microenvironment. Successful release of cytotoxic material was documented however no information on cell viability of the breast cancer cells were documented on this preliminary study.

Finally, Achille et al. showed the first successful demonstration of the delivery of bioactive RNAi-based either Cdk2 (Cdk2i) and EGFP (EGFPi) shRNA plasmid DNA from an electrospun polymer scaffold [60]. The treatment results in disrupting cell cycle regulation and suppressing breast cancer cell proliferation. Cdk2 mRNA expression in cells seeded on the Cdk2i scaffold was decreased by 51% and 30%, in comparison with that of cells plated on control or EGFPi scaffold and a 40% decrease in the proliferation of the MCF-7 breast cancer cells, as well as the presence of increased numbers of dead cells.

### 21.3.3 Scaffolds co-culture chemotherapy

In order to model the periphery of a breast cancer tumor Hume et al. co-cultured tumor fragments with adipocytes in directional collagen scaffolds by creating pores aligned perpendicular to the surface of a nucleation point [61]. The following inhibitors: Rho-associated protein kinase inhibitor, Y-27632, pan-matrix metalloproteinase (MMP) inhibitor, GM6001, and pan-ErbB inhibitor, canertinib were used. Different pathways involved in cancer cell proliferation, migration, and tumor growth were tested regarding their effect to suppress cell migration on the scaffolds. They showed that the model is suitable in assessing effects on cell migration and that the anisotropic arrangement supports migration of cancer cells. Also, Ref. [62] and the described model correlated with one of the identified subtypes of tumors. All inhibitors resulted in suppression of cell migration on the proposed scaffold.

Besides in vitro models for the primary breast cancer tumors, there are several engineered 3D applications to model metastasis sites of breast cancer tumors. Metastasis are the prevalent cause of cancer-related death and therefore it is especially important to develop new treatment strategies to prevent and counteract development of metastasis.

Sublia et al. utilized a silk fibroin scaffold to co-culture breast cancer cells, MDA-MB-231, and osteoblasts, MG63, to investigate their interaction and the potential targeted DOX delivery using folic acid-conjugated silk fibroin nanoparticles for the treatment of bone metastasis [63]. They showed that nanoparticle-mediated drug treatment inhibited cancer cell growth without impacting on bone mineralization. This effect was measured by calcium deposition assay which was maintained throughout the treatment period. However, it also showed a downregulation of osteogenic markers (ALP and osteocalcin) of the MG63 cells. They suggest that their model could be used as a drug screening tool to investigate suitability of drugs to treat bone metastasis.

## 21.3.4 Hydrogels and chemo/drug therapy

Hydrogels are a well-known class of polymer material that have been used for the investigation of intertumoral delivery of anticancer drugs [44]. Hydrogels are able to maintain a protective environment. This polymer possesses unique properties such as interfacial surface tension, pore size, permeability, and controlled release rates of entrapped molecules [1]. The 3D network is maintained in the water-swollen state as a result of either physical or chemical cross-link interactions. In fact, drug release kinetics can be triggered intelligently by dynamic interactions that affect swelling, cross-linking density, and degradation rates of the hydrogels [64].

An increased focus in recent breast cancer research are nanoparticle-loaded hydrogels. However, due to the dense collagen network in the ECM of breast cancer cells, there is a great difficulty in releasing nanoparticles to penetrate deeply into tumors. Shen et al. carried out implantation of a PEG hydrogel loaded with fluorescent polystyrene nanoparticles (FPNPs, size ~100 nm) and losartan (Los) [64]. Results showed enhanced penetration and reduced collagen levels after implantation of FPNP/Los-loaded hydrogel in comparison to standard FPNP-loaded hydrogel. The second part of the study co-delivered DOX to observe tumor growth inhibition. The peritumoral implantation of the DOX-loaded hydrogel in the presence of Los led to considerably better antitumor effects as results showed DOX/Los-loaded hydrogel had the highest inhibition rate of 59%, but the DOX-loaded hydrogel had a rate of only 22%. Similarly, Strong et al. achieved pulsatile delivery of DOX and nucleic acid therapeutic (DNA duplex) by assembling a poly(*N*-isopropylacrylamide-*co*-acrylamide) hydrogel with near-infrared (NIR) absorbing silica–gold nanoshells [65]. This hydrogel has a lower critical solution temperature at approximately 40°C, thus above this temperature the material will have a transition from a hydrated state to a collapsed state. The composite material was loaded with either DOX or nucleic acid therapeutic (DNA duplex) to be released above 40°C. The results showed that there was a 30% decrease in cell proliferation in response to DOX exposure of the cells that were exposed to the irradiated hydrogels in comparison to the nonirradiated cells. Furthermore, delivery of docetaxel and DOX was also demonstrated using biodegradable methoxy PEG–PCL micelles in the study by Wu et al. [66]. PEG–PCL copolymers were cross-linked to docetaxel or DOX using disulfide bonds. In vitro antitumor efficiency was evaluated. It showed that as the concentration of docetaxel and DOX increased within the PEG–PCL micelles, the cellular inhibition ratio in MCF-7 cells also increased. The highest drug concentration group of 1000 nM showed up to 90% cellular inhibition. Likewise, Wang et al. described hydrogels composed of PEGylated [poly(ethylene glycol)] hydrocarbon nanoparticles with PEGylated fluorocarbons for the controlled release of PTX and DOX to MCF-7 breast cancer cells [67]. This hydrogel was injectable and thermosensitive but most importantly it contained two compartment hydrogel networks that allowed PTX and DOX to be integrated in to different compartments. In vitro cell viability study showed less than 15% viability for the group treated with this hydrogel which contained both PTX and DOX whereas all the other groups such as PTX alone, DOX alone, and blank hydrogel group showed significantly higher cell viability comparatively.

In contrast, Liu et al. fabricated a self-assembling peptide hydrogel incorporating the potent chemotherapeutic, PTX, for the treatment of breast cancer [68]. They designed a nanoscale sequencing of alternating hydrophilic and hydrophobic amino acid residues called RADA16 peptide hydrogel. This RADA16 peptide hydrogel was used to encapsulate hydrophobic PTX to form RADA 16-PTX hydrogel. Their data suggested that 1% RADA16-PTX hydrogel maintained relatively high-inhibitory activity even at day 7 as the cell viability level was 40% suggesting that the RADA16 hydrogel could control release of PTX as higher concentration of peptide hydrogel resulted in an increased release time for PTX.

Wang et al. fabricated a biodegradable sequenced amine functionalized polyacrylamide nanoparticle hydrogel polymer [69]. This hydrogel matrix was loaded with sulfosuccinimidyl-4-(*N*-maleimidomethyl) cyclohexane-1-carboxylate which is a noncleavable membrane cross linker which has potential for cytotoxic effects in breast cancer cells. In vitro studies showed that the targeted nanoparticles could be transported efficiently into the human breast cancer cell line, MDA-MB-435. Results showed that as the concentration was increased from 0.1 to 10 μM, the cell survival rate decreased from 100% to 70%.

One of the main issues in drug delivery is that when administered systemically, chemotherapeutic drugs can be taken up by macrophages. Interestingly, Alexander et al. designed a type of hydrogel that is composed of cube-shaped biocompatible elastic microparticles made via layer-by-layer assembly [70]. It contains hydrogen-bonded tannic acid/poly(*N*-vinylpyrrolidone) as hollow polymer shells and their rigid core–shell precursors for their hydrogels. The unique hollow shells are unrecognized by macrophages, however, it allows targeted delivery to breast cancer cells MDA-MB-231 and SUM159. Internalization of cubic shells was 6-fold more efficient in MDA-MB-231 and 2.5-fold more efficient for SUM159 (breast cancer cells), respectively, compared with spherical shells. Alexander et al. suggests that these hydrogels would be ideal for chemotherapeutic delivery to these cells.

### 21.3.5 Hydrogels and gene therapy

Gene therapy is an emerging treatment modality that has been looked at for the treatment of breast cancer in recent years. By targeting the cancer cells using complimentary

RNA or incorporated plasmid DNA, one can achieve isolated cell death and significant reduction in cancer cell proliferation. It is important to look at studies like Vantangoli et al. where they examined specific targets for cancer cell type for gene therapy [71]. There data showed that 3D MCF-7 microtissues increased expression of PDZK1, MYC, CCND1, and TGFA mainly in comparison to MCF-7 cells cultured in a 2D microenvironment. To date, a small number of studies have been performed as summarized in Table 21.3.

Zhao et al. used thermosensitive hydrogels as gene-delivery vectors for sustained and controlled release [72]. This hydrogel was engineered using negatively charged surviving antisense oligonucleotide (Sur-ASON) and positively charged PHB-b-PDMAEMA (PHB-P) co-polymer. This structure was then incorporated with thermosensitive PF127 hydrogel. Results showed it was effective in enhancing the treatment effects of Sur-ASON while reducing the degradation and the possible adverse side effects and allowed for up to 16 days of controlled gene delivery.

Furthermore, in another study hydrogels have also been engineered into pH-responsive hybrid nanoparticles for the intracellular delivery of miRNA in breast cancer cells [73]. Khaled et al. designed these hydrogel particles to be able to respond to fluctuation in temperature and pH changes resulting in the release of a payload in a timely fashion. These nanoparticles, loaded with specific siRNA molecules directed toward the transcript of the membrane receptor CXCR4 significantly decreased the expression of this protein in MDA-MB-231 breast cancer cells. Results further showed less than 0.4 r.u protein levels in the treatment group in comparison to 1.0 r.u in the control group. Additionally, intravenous administration of siRNA-loaded nanoparticles confirmed a preferential deposit at the tumor site that resulted in a reduction of CXCR4 expression.

In a separate study, Gilam et al. showed the downregulation of breast cancer cells by delivering miRNA in an oxidized dextran and PAMAM G5 dendrimer hydrogel-based scaffold [74]. They elaborated that miR-96/miR-182 downregulated palladin protein levels, thereby reducing breast cancer cell migration and invasion. Members of the same group

**TABLE 21.3** 3D hydrogel models for gene therapy treatment for breast cancer.

| 3D model | Cell type | Therapeutic | Effect | Ref. |
|---|---|---|---|---|
| Antisense oligonucleotide (Sur-ASON) and positively charged PHB-b-PDMAEMA (PHB-P) co-polymer | MCF-7 | Surviving antisense oligonucleotide (Sur-ASON) | Hydrogel could achieve the controlled gene release up to maximum 16 days<br>Sur-ASON significantly inhibited drug resistant MCF-7/PDR | [72] |
| Nanostructured inorganic silica core and an organic pH-responsive hydrogel shell | MDA-MB-231 | siRNA molecules targeting membrane receptor CXCR4 | CXCR4 levels in the tumor of the mice models showed high-gene silencing efficacy | [73] |
| Oxidized dextran and PAMAM G5 dendrimer hydrogel | MCF-7<br>Hs578 | miR-96/miR-182 | miR-96/miR-182 downregulated palladin protein level | [74] |
| Sustained siRNA release from siRNA-loaded nanoparticles embedded in a polymeric scaffold | MDA-MB-231 | Oligopeptide-terminated poly (β-amino ester) | Lowest cell viability in siRNA-loaded hydrogel | [75] |

also produced chitosan hydrogels based on a thermosensitive PEG-grafted hydrogel delivery vehicle [75]. Results showed lowest cell viability in the siRNA-loaded hydrogels at 60% in comparison to 96% in the control group.

## 21.4 Prostate cancer

Prostate cancer is one of the leading causes of cancer-related death in men worldwide. A major obstacle to progress prostate cancer survival rates arise from difficulties in treating primary and metastatic disease [76]. Furthermore, inadequacies in the treatment of primary and metastatic prostate cancer arise from the relatively poor comprehension of the advancement of prostate cancer disease, resulting in a significant unmet clinical need [77]. Thus, as described in the previous sections for lung and breast cancer, there exists a distinct requirement to develop physiologically relevant 3D cell culture platforms for two purposes: (1) as possible systems for drug development to enrich our understanding of the disease process and (2) advance the correlation between results attained in the laboratory and in vivo animal and in human subjects. Below describes some of the current scaffold and hydrogel models currently being used for chemotherapeutic/drug delivery and gene therapy for prostate cancer treatment.

### 21.4.1 Scaffold chemo/drug treatment

Fitzgerald et al. [78] described the development of collagen-based scaffolds in the form of collagen–glycosaminoglycan and collagen–nanohydroxyapatite to mimic prostate cancer bone metastasis. Two cell lines, LNCaP and PC3, were assessed across all scaffolds and demonstrated increased resistance to docetaxel chemotherapy. When investigated as systems for the delivery of gene therapeutics, cyclodextrin–siRNA nanoparticles achieved cellular uptake and successfully knocked down the endogenous GAPDH gene in the 3D model (Fig. 21.3).

### 21.4.2 Scaffold gene therapy

Utilizing the same collagen–nanohydroxyapatite scaffold as Fitzgerald et al. [78], anisamide-targeted cyclodextrin nanoparticles were formulated to interact with the sigma receptor which is overexpressed on the surface of prostate cancer cells and deliver siRNA targeting polo-like kinase 1 (PLK1), a cell cycle regulator overexpressed in a variety of human tumors [79]. Targeted formulations showed significantly higher levels of PLK1 mRNA knockdown in two cell lines on the 3D collagen-based scaffolds (46% for PC3 and 37% for DU145).

### 21.4.3 Scaffold co-culture gene therapy

Collagen–nanohydroxyapatite scaffolds as used in the previous studies by members of the authors group were also used to simulate prostate cancer bone metastasis in a co-culture model and to assess the ability of the system to investigate cyclodextrin-mediated

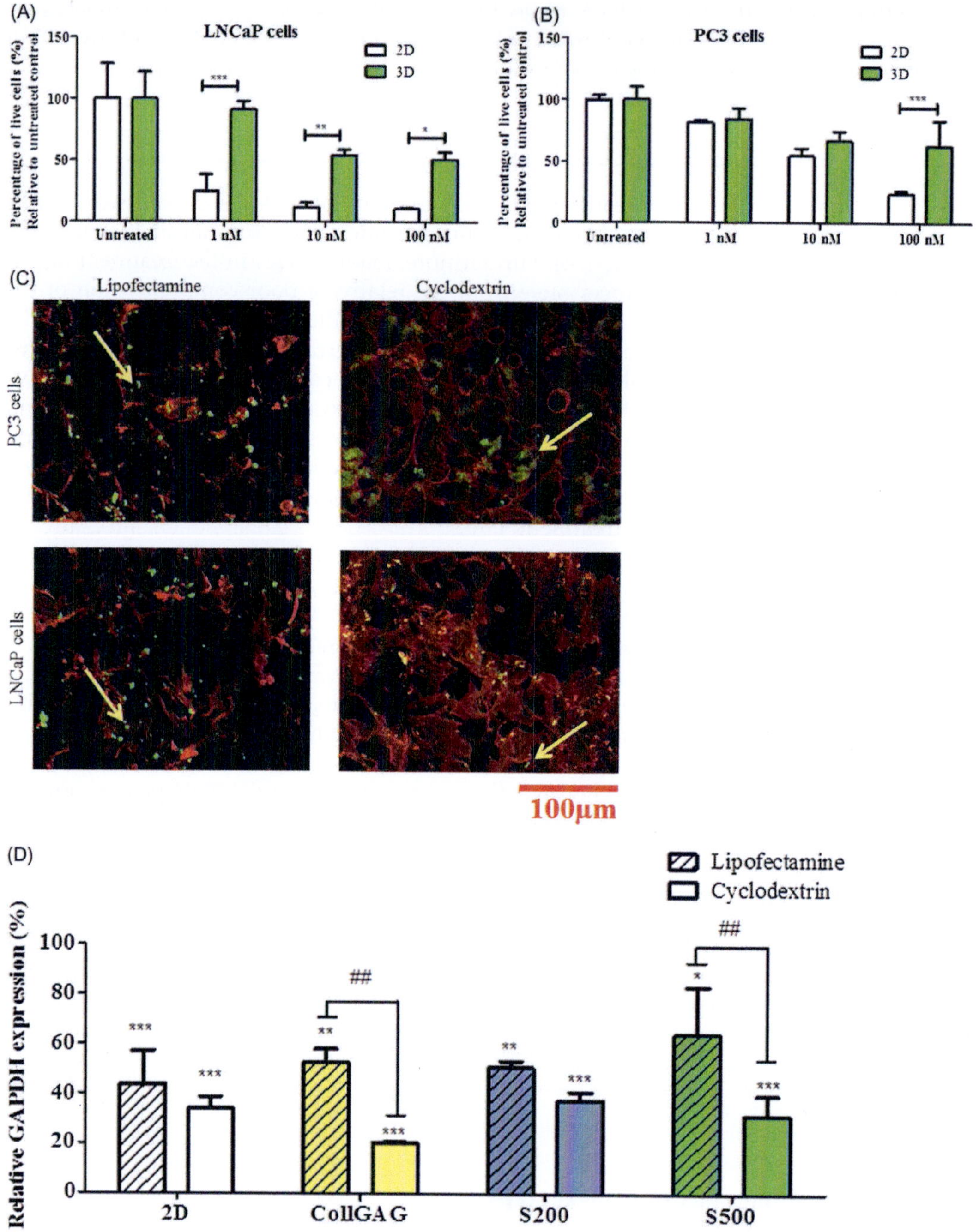


FIGURE 21.3 A 3D model of prostate cancer bone metastasis. (A) The chemosensitivity of LNCaP cell number in 2D and 3D normalized to untreated control after docetaxel treatment. (B) PC3 cell number in 2D and 3D

RelA siRNA (subunit of the NF-κB transcription factor) delivery and gene knockdown [80]. Results showed that the prostate cancer-osteoblast in vitro scaffold co-culture model was more physiologically relevant compared with the monoculture assessment.

### 21.4.4 Hydrogel chemo/drug treatment

In recent years, hydrogels have been utilized within research as efficient drug delivery systems as they can deliver their cargo in a more spatially controlled and temporal manner. A number of studies have focused on the utilization of various hydrogels to act as tunable and efficient drug delivery systems as described below.

Tao et al. [81] recently described the development of an innovative aspartic acid-containing hexapeptide peptide, D3F3, that converts into a hydrogel through crosslinking via physiological levels of zinc ions. The study focused on delivery of DOX. The DOX-hydrogel was capable of good anticancer efficacy when incubated with DU-145 prostate cancer cells, revealed no cytotoxicity against normal fibroblasts did not affect the prostate tissue of rats. The same group also designed alternative hydrogels using histidine-containing hexapeptide derivatives (Nap-1 and ID-1), again crosslinking using zinc ions [81]. This study looked at delivery of bicalutamide (BLT), a nonsteroidal antiandrogen drug used for prostate cancer treatment, which also has many known side effects associated with systemic administration. ID-1-BLT hydrogels showed tunable drug release profiles in a characteristic tumor microenvironment, elevated potency, and selective cytotoxicity against DU145 prostate cancer cells compared with normal fibroblasts and sustained drug delivery at the prostate gland of rats.

Hainline et al. [82] also assessed a self-assembling peptide, bQ13, capable of forming nanofibers and gels more useful for the 3D culture of prostate cancer spheroids when compared with alternatives including Q11, RADA16-I, and Matrigel. Although prostate-specific antigen (PSA) was produced by LNCaP cells in all matrices, PSA production was more responsive to the androgen receptor inhibitor, enzalutamide, in bQ13 gels than in the aforementioned peptide gels thus representing an appealing platform for further adaptation for 3D prostate cancer cell culture.

Schaal et al. [83] reported on development of injectable brachytherapy hydrogels composed of thermally sensitive micelles made of an elastin-like polypeptide (ELP) and labeled with radionuclide $^{131}$Iodine. The authors used these hydrogels to not alone treat a

normalized to untreated control after docetaxel treatment. (C) Uptake into PC3 cells using Lipofectamine and cyclodextrin is shown as well as uptake into LNCaP cells using Lipofectamine and Cyclodextrin. Scale bar 100 μm. Yellow arrows indicate green fluorescent siRNA located within the cell membrane. (D) Endogenous GAPDH gene knockdown obtained in PC3 cells cultured in 2D and on three collagen-based scaffolds (CollGAG, S200, S500) and transfected with 100 nM GAPDH siRNA using Lipofectamine 2000 and SC12-CD-Click-Propylamine cyclodextrin. Results expressed represent the mean ± SD. Each sample was compared with an untreated counterpart normalized to 100% (data not shown) and analyzed for significance using one-way ANOVA (*$P < .05$, **$P < .01$, ***$P < .001$) ($n = 3$). Differences between efficacy of the transfection vector (Lipofectamine 2000® v CD) was analyzed using two-way ANOVA ($^{\#\#}P > .01$). Source: *Reprinted from Fitzgerald KA, Guo J, Tierney EG, Curtin CM, Malhotra M, Darcy R, et al. The use of collagen-based scaffolds to simulate prostate cancer bone metastases with potential for evaluating delivery of nanoparticulate gene therapeutics. Biomaterials 2015;66:53–66, Copyright (2020), with permission from Elsevier.*

human PC-3M-luc-C6 prostate tumor in athymic nude mice but also a human BxPc3-luc2 pancreatic tumor model. Results were impressive demonstrating greater than 95% tumor regression in the prostate tumors with a survival rate of greater than 60 days using the injectable brachytherapy hydrogels compared with 12 days for control mice.

Phage particles, heat-sensitive liposomes, and mesoporous silica nanoparticles (MSNP) capable of self-assembling into hydrogels have also demonstrated a wide spectrum of ligand peptide-based applications for tumor-targeted drug delivery [84]. A strong, significant decrease in tumor growth was detected in a DU145-derived xenograft mouse model of prostate cancer treated with targeted DOX–MSNP-containing hydrogels relative to control groups.

Human cancer angiogenesis for both prostate and breast cancer has also been assessed in a 3D synthetic starPEG maleimide-functionalized heparin hydrogel system [85]. Both LNCaP and PC3 prostate cancer epithelial cell lines were assessed and the PC3 prostate tumor model most closely correlated with reported in vivo results when treated with chemotherapeutics including DOX and PTX. Additionally, this model was used to assess angiogenesis inhibitors to starve tumor cells of oxygen and nutrients. Those assessed included axitinib, a VEGF inhibitor, AMD3100, a potent CXCR4 inhibitor and SU-5402, a specific tyrosine kinase inhibitor for FGFR but it was determined that higher doses would be required to exert more significant and impactful 3D results.

Prostate cancer bone metastatic models have been developed by Fong et al. [86] to aid the culture of patient-derived xenograft (PDX) tumor cells. Thiol-modified hyaluronic acid (HA-SH) and PEG-DA hydrogels were utilized and demonstrated a differential sensitivity of the PDX tumor cells to docetaxel compared with the C4-2B cell line when cultured. Members of this group also published on hydrogels of HA carrying acrylate groups (HA-AC) or reactive thiols (HA-SH) for the in vitro evaluation of DOX-loaded polymer NPs (DOX-NPs) [87]. LNCaP cells grown on 3D hydrogels expressed significantly higher levels of multidrug resistance (MDR) proteins at the mRNA and protein levels compared with cells grown in 2D. LNCaP PCa cells cultured in the HA hydrogel were also more resistant to DOX and DOX-NPs treatments. Another group utilizing HA derivatives carrying complementary aldehyde (HAALD) and hydrazide (HAADH) groups showed LNCaP cells made individual clustered structures that grew and merged, similar of native tumors [88]. The system was used to test the efficacy of anticancer drugs including camptothecin, docetaxel, and rapamycin, either alone and in combination on C4-2B cells and showed different result profiles between 3D and 2D.

### 21.4.5 Hydrogel gene therapy

Chalanqui et al. [89] used thermoresponsive alginate grafted poly(*N*-isopropylacrylamide) hydrogels [Alg-g-P(NIPAAm)] for localized delivery of DNA nanoparticles for treatment of castrate-resistant prostate cancer (CRPC). Nanoparticles were comprised of an amphipathic cell penetrating peptide, RALA, complexed with plasmid (p)DNA. RALA was shown to protect the pDNA from degradation over 30 days and demonstrated a sustained, controlled release profile from the hydrogels compared with noncomplexed pDNA. This study opens up various opportunities for delivery of therapeutic pDNA from thermoresponsive hydrogels with potential for numerous medical applications.

### 21.4.6 Hydrogel co-culture chemo

Antunes et al. [57] rapidly fabricated in-air hyaluronan–methacrylate (HA-MA) and gelatin-methacrylate (GelMA) photocrosslinkable 3D spheroid microgels for development of heterotypic 3D co-culture models of prostate cancer cells (PC-3) and human osteoblasts (hOB) to mimic prostate cancer-to-bone metastasis and the tumor–ECM microenvironment. Drug screening was in the form of cisplatin treatment and showed 3D microgels were more resilient to platin chemotherapeutics than single or co-culture 3D spheroid counterparts.

Engel et al. also described a novel, high-throughput and compatible HA hydrogel system an effective model for high-content analysis of multiple cancer types including prostate cancer and endometrial cancer in both mono- or co-culture [90]. The model was able to recapitulate cytotoxicity of compounds used clinically for prostate cancer and additionally identified drugs that failed clinical trials demonstrating its effectiveness and robustness.

## 21.5 Future outlook

The use of 3D models for cancer research has widely impacted the current landscape of cancer treatments. The fruition of this knowledge now opens up vast and novel therapeutic opportunities, propelling a research area that has grown exponentially in the last 5–8 years. Even though a Nature Review published in 2003 predicted cancer research to move to 3D in vitro models within the next decade it is still only a minority ($\sim$1.4%) of published work carried out using 3D models [91]. Higher costs and effort of 3D platforms compared with 2D in vitro studies might be an aspect hampering a shift to 3D cultures. Furthermore, the versatility and a lack of standardization leads to a low proportion of results being translated to in vivo and clinical studies as of yet [92]. Besides the failure of 2D in vitro models to accurately predict drug efficacy in vivo, rodent models frequently fail to be translated into clinical trials [93] or unfortunately fail when during these [94]. Aspects such as substrate stiffness, absence of ECM molecules and tumor surrounding cells in 2D models lead to a lack in accuracy predicting in vivo outcomes of novel treatments. Further factors that are currently assessed in 3D models and have shown some promise as mentioned in this chapter include culturing with migratory cancer cells, using microfluidic models and analyzing effects of co-culturing with tumor-related cells, such as epithelial or immune cells.

Overall observations of currently developed 3D models conclude a better comparability to in vivo behavior of cells in terms of proliferation, malignancy, and chemoresponsiveness [46]. Increasing complexity of models makes it harder to interpret results appropriately, as more unknown factors can influence study outcomes but once 3D models can be clearly defined and characterized more closely, they offer enormous potential for reducing the need of animal models and provide new insights into mechanisms and treatment options for various cancer types.

## References

[1] Ledford H. Translational research: 4 ways to fix the clinical trial. Nature 2011;477(7366):526–8.

[2] Balachander GM, Balaji SA, Rangarajan A, Chatterjee K. Enhanced metastatic potential in a 3D tissue scaffold toward a comprehensive in vitro model for breast cancer metastasis. ACS Appl Mater Interfaces 2015;7(50):27810–22.

[3] Chen L, Xiao Z, Meng Y, Zhao Y, Han J, Su G, et al. The enhancement of cancer stem cell properties of MCF-7 cells in 3D collagen scaffolds for modeling of cancer and anti-cancer drugs. Biomaterials 2012;33(5):1437–44.

[4] Bielecka ZF, Maliszewska-Olejniczak K, Safir IJ, Szczylik C, Czarnecka AM. Three-dimensional cell culture model utilization in cancer stem cell research. Biol Rev Camb Philos Soc 2017;92(3):1505–20.

[5] Verjans ET, Doijen J, Luyten W, Landuyt B, Schoofs L. Three-dimensional cell culture models for anticancer drug screening: worth the effort? J Cell Physiol 2018;233(4):2993–3003.

[6] Hirsch FR, Scagliotti GV, Mulshine JL, Kwon R, Curran Jr. WJ, Wu YL, et al. Lung cancer: current therapies and new targeted treatments. Lancet (London, Engl) 2017;389(10066):299–311.

[7] Stratmann AT, Fecher D, Wangorsch G, Gottlich C, Walles T, Walles H, et al. Establishment of a human 3D lung cancer model based on a biological tissue matrix combined with a Boolean in silico model. Mol Oncol 2014;8(2):351–65.

[8] Moustakas A, Heldin CH. Signaling networks guiding epithelial-mesenchymal transitions during embryogenesis and cancer progression. Cancer Sci 2007;98(10):1512–20.

[9] Yan X, Zhou L, Wu Z, Wang X, Chen X, Yang F, et al. High throughput scaffold-based 3D micro-tumor array for efficient drug screening and chemosensitivity testing. Biomaterials 2019;198:167–79.

[10] Li J, Zhou Y, Chen W, Yuan Z, You B, Liu Y, et al. 3D in vitro tumor model based on silk fibroin/chitosan scaffolds to mimic the tumor microenvironment. ACS Appl Mater Interfaces 2018;10(43):36641–51.

[11] Del Bufalo F, Manzo T, Hoyos V, Yagyu S, Caruana I, Jacot J, et al. 3D modeling of human cancer: a PEG-fibrin hydrogel system to study the role of tumor microenvironment and recapitulate the in vivo effect of oncolytic adenovirus. Biomaterials 2016;84:76–85.

[12] Tapias LF, Gilpin SE, Ren X, Wei L, Fuchs BC, Tanabe KK, et al. Assessment of proliferation and cytotoxicity in a biomimetic three-dimensional model of lung cancer. Ann Thorac Surg 2015;100(2):414–21.

[13] Jillella AP, Murren JR, Hamid KK, Longley BJ, Edelson RL, Cooper DL. P-glycoprotein expression and multidrug resistance in cutaneous T-cell lymphoma. Cancer Invest 2000;18(7):609–13.

[14] Ardeshirzadeh B, Anaraki NA, Irani M, Rad LR, Shamshiri S. Controlled release of doxorubicin from electrospun PEO/chitosan/graphene oxide nanocomposite nanofibrous scaffolds. Mater Sci Eng C, Mater Biol Appl 2015;48:384–90.

[15] Samadi S, Moradkhani M, Beheshti H, Irani M, Aliabadi M. Fabrication of chitosan/poly(lactic acid)/graphene oxide/TiO2 composite nanofibrous scaffolds for sustained delivery of doxorubicin and treatment of lung cancer. Int J Biol Macromol 2018;110:416–24.

[16] Mellatyar H, Talaei S, Pilehvar-Soltanahmadi Y, Barzegar A, Akbarzadeh A, Shahabi A, et al. Targeted cancer therapy through 17-DMAG as an Hsp90 inhibitor: overview and current state of the art. Biomed Pharmacother 2018;102:608–17.

[17] Mishra DK, Creighton CJ, Zhang Y, Gibbons DL, Kurie JM, Kim MP. Gene expression profile of A549 cells from tissue of 4D model predicts poor prognosis in lung cancer patients. Int J Cancer 2014;134(4):789–98.

[18] Vishnoi M, Mishra DK, Thrall MJ, Kurie JM, Kim MP. Circulating tumor cells from a 4-dimensional lung cancer model are resistant to cisplatin. J Thorac Cardiovasc Surg 2014;148(3):1056–63 discussion 1063-1064.

[19] Wolinsky JB, Liu R, Walpole J, Chirieac LR, Colson YL, Grinstaff MW. Prevention of in vivo lung tumor growth by prolonged local delivery of hydroxycamptothecin using poly(ester-carbonate)-collagen composites. J Control Release 2010;144(3):280–7.

[20] Mellatyar H, Talaei S, Pilehvar-Soltanahmadi Y, Dadashpour M, Barzegar A, Akbarzadeh A, et al. 17-DMAG-loaded nanofibrous scaffold for effective growth inhibition of lung cancer cells through targeting HSP90 gene expression. Biomed Pharmacother 2018;105:1026–32.

[21] Streit M, Stephen AE, Hawighorst T, Matsuda K, Lange-Asschenfeldt B, Brown LF, et al. Systemic inhibition of tumor growth and angiogenesis by thrombospondin-2 using cell-based antiangiogenic gene therapy. Cancer Res 2002;62(7):2004–12.

[22] Sukumar UK, Packirisamy G. Bioactive core-shell nanofiber hybrid scaffold for efficient suicide gene transfection and subsequent time resolved delivery of prodrug for anticancer therapy. ACS Appl Mater Interfaces 2015;7(33):18717–31.

[23] Gao W, Hua J, Jia Z, Ding J, Han Z, Dong Y, et al. Expression of miR-146a-5p in breast cancer and its role in proliferation of breast cancer cells. Oncol Lett 2018;15(6):9884–8.

[24] Yang X, Li K, Zhang X, Liu C, Guo B, Wen W, et al. Nanofiber membrane supported lung-on-a-chip microdevice for anti-cancer drug testing. Lab Chip 2018;18(3):486–95.

[25] Prina-Mello A, Jain N, Liu B, Kilpatrick JI, Tutty MA, Bell AP, et al. Culturing substrates influence the morphological, mechanical and biochemical features of lung adenocarcinoma cells cultured in 2D or 3D. Tissue Cell 2018;50:15–30.

[26] Godugu C, Patel AR, Desai U, Andey T, Sams A, Singh M. AlgiMatrix based 3D cell culture system as an in-vitro tumor model for anticancer studies. PLoS One 2013;8(1):e53708.

[27] Tu TY, Wang Z, Bai J, Sun W, Peng WK, Huang RY, et al. Rapid prototyping of concave microwells for the formation of 3D multicellular cancer aggregates for drug screening. Adv Healthc Mater 2014;3(4):609–16.

[28] Ghosh G, Yan X, Lee AG, Kron SJ, Palecek SP. Quantifying the sensitivities of EGF receptor (EGFR) tyrosine kinase inhibitors in drug resistant non-small cell lung cancer (NSCLC) cells using hydrogel-based peptide array. Biosens Bioelectron 2010;26(2):424–31.

[29] Cho YI, Park S, Jeong SY, Yoo HS. In vivo and in vitro anti-cancer activity of thermo-sensitive and photo-crosslinkable doxorubicin hydrogels composed of chitosan-doxorubicin conjugates. Eur J Pharm Biopharm 2009;73(1):59–65.

[30] Yi H, Cho HJ, Cho SM, Lee DG, Abd El-Aty AM, Yoon SJ, et al. Pharmacokinetic properties and antitumor efficacy of the 5-fluorouracil loaded PEG-hydrogel. BMC Cancer 2010;10(1):211.

[31] Obara K, Ishihara M, Ozeki Y, Ishizuka T, Hayashi T, Nakamura S, et al. Controlled release of paclitaxel from photocrosslinked chitosan hydrogels and its subsequent effect on subcutaneous tumor growth in mice. J Control Release 2005;110(1):79–89.

[32] Chen J, Jiang Q, Jiang XQ, Li DQ, Jiang XC, Wu XB, et al. miR-146a promoted breast cancer proliferation and invasion by regulating NM23-H1. J Biochem 2019;167:41–8.

[33] Lee P, Lok CN, Che CM, Kao WJ. A multifunctional hydrogel delivers gold compound and inhibits human lung cancer xenograft. Pharm Res 2019;36(4):61.

[34] Yamada Y, Chowdhury A, Schneider JP, Stetler-Stevenson WG. Macromolecule-network electrostatics controlling delivery of the biotherapeutic cell modulator TIMP-2. Biomacromolecules 2018;19 (4):1285–93.

[35] Roudsari LC, Jeffs SE, Witt AS, Gill BJ, West JL. A 3D poly(ethylene glycol)-based tumor angiogenesis model to study the influence of vascular cells on lung tumor cell behavior. Sci Rep 2016;6:32726.

[36] Aref AR, Huang RY, Yu W, Chua KN, Sun W, Tu TY, et al. Screening therapeutic EMT blocking agents in a three-dimensional microenvironment. Integr Biol 2013;5(2):381–9.

[37] Xu Z, Gao Y, Hao Y, Li E, Wang Y, Zhang J, et al. Application of a microfluidic chip-based 3D co-culture to test drug sensitivity for individualized treatment of lung cancer. Biomaterials 2013;34(16):4109–17.

[38] Marcucci F, Corti A. Improving drug penetration to curb tumor drug resistance. Drug Discov Today 2012;17 (19–20):1139–46.

[39] Desoize B, Jardillier J. Multicellular resistance: a paradigm for clinical resistance? Crit Rev Oncol/Hematol 2000;36(2–3):193–207.

[40] Liu J, Abate W, Xu J, Corry D, Kaul B, Jackson SK. Three-dimensional spheroid cultures of A549 and HepG2 cells exhibit different lipopolysaccharide (LPS) receptor expression and LPS-induced cytokine response compared with monolayer cultures. Innate Immun 2011;17(3):245–55.

[41] Lin CY, Hung CC, Wang CCN, Lin HY, Huang SH, Sheu MJ. Demethoxycurcumin sensitizes the response of non-small cell lung cancer to cisplatin through downregulation of TP and ERCC1-related pathways. Phytomedicine 2019;53:28–36.

[42] Chen YY, Lin YJ, Huang WT, Hung CC, Lin HY, Tu YC, et al. Demethoxycurcumin-loaded chitosan nanoparticle downregulates DNA repair pathway to improve cisplatin-induced apoptosis in non-small cell lung cancer. Molecules (Basel, Switzerland) 2018;23(12):E3217.

[43] Che CM, Sun RW, Yu WY, Ko CB, Zhu N, Sun H. Gold(III) porphyrins as a new class of anticancer drugs: cytotoxicity, DNA binding and induction of apoptosis in human cervix epitheloid cancer cells. Chem Commun (Cambridge, Engl) 2003;(14):1718–19.

[44] Fleming JM, Yeyeodu ST, McLaughlin A, Schuman D, Taylor DK. In situ drug delivery to breast cancer-associated extracellular matrix. ACS Chem Biol 2018;13(10):2825–40.

[45] Whitman NA, Lin ZW, DiProspero TJ, McIntosh JC, Lockett MR. Screening estrogen receptor modulators in a paper-based breast cancer model. Anal Chem 2018;90(20):11981–8.

[46] Li W, Hu X, Wang S, Xing Y, Wang H, Nie Y, et al. Multiple comparisons of three different sources of biomaterials in the application of tumor tissue engineering in vitro and in vivo. Int J Biol Macromol 2019;130:166–76.
[47] Fiegl H, Millinger S, Goebel G, Muller-Holzner E, Marth C, Laird PW, et al. Breast cancer DNA methylation profiles in cancer cells and tumor stroma: association with HER-2/neu status in primary breast cancer. Cancer Res 2006;66(1):29–33.
[48] Tutt A, Robson M, Garber JE, Domchek SM, Audeh MW, Weitzel JN, et al. Oral poly(ADP-ribose) polymerase inhibitor olaparib in patients with BRCA1 or BRCA2 mutations and advanced breast cancer: a proof-of-concept trial. Lancet (London, Engl) 2010;376(9737):235–44.
[49] Dhiman HK, Ray AR, Panda AK. Three-dimensional chitosan scaffold-based MCF-7 cell culture for the determination of the cytotoxicity of tamoxifen. Biomaterials 2005;26(9):979–86.
[50] Leeper AD, Farrell J, Williams LJ, Thomas JS, Dixon JM, Wedden SE, et al. Determining tamoxifen sensitivity using primary breast cancer tissue in collagen-based three-dimensional culture. Biomaterials 2012;33 (3):907–15.
[51] Nayak B, Balachander GM, Manjunath S, Rangarajan A, Chatterjee K. Tissue mimetic 3D scaffold for breast tumor-derived organoid culture toward personalized chemotherapy. Colloids Surf B Biointerfaces 2019;180:334–43.
[52] Dunne LW, Huang Z, Meng W, Fan X, Zhang N, Zhang Q, et al. Human decellularized adipose tissue scaffold as a model for breast cancer cell growth and drug treatments. Biomaterials 2014;35(18):4940–9.
[53] Horning JL, Sahoo SK, Vijayaraghavalu S, Dimitrijevic S, Vasir JK, Jain TK, et al. 3-D tumor model for in vitro evaluation of anticancer drugs. Mol Pharm 2008;5(5):849–62.
[54] Yuan Z, Pan Y, Cheng R, Sheng L, Wu W, Pan G, et al. Doxorubicin-loaded mesoporous silica nanoparticle composite nanofibers for long-term adjustments of tumor apoptosis. Nanotechnology 2016;27(24):245101.
[55] Li W, Hu X, Wang S, Wang H, Parungao R, Wang Y, et al. Detection and evaluation of anti-cancer efficiency of astragalus polysaccharide via a tissue engineered tumor model. Macromol Biosci 2018;18(11):e1800223.
[56] Gupta V, Mun GH, Choi B, Aseh A, Mildred L, Patel A, et al. Repair and reconstruction of a resected tumor defect using a composite of tissue flap-nanotherapeutic-silk fibroin and chitosan scaffold. Ann Biomed Eng 2011;39(9):2374–87.
[57] Antunes J, Gaspar VM, Ferreira L, Monteiro M, Henrique R, Jeronimo C, et al. In-air production of 3D co-culture tumor spheroid hydrogels for expedited drug screening. Acta Biomater 2019;94:392–409.
[58] Deng ZJ, Morton SW, Ben-Akiva E, Dreaden EC, Shopsowitz KE, Hammond PT. Layer-by-layer nanoparticles for systemic codelivery of an anticancer drug and siRNA for potential triple-negative breast cancer treatment. ACS Nano 2013;7(11):9571–84.
[59] Phan-Lai V, Kievit FM, Florczyk SJ, Wang K, Disis ML, Zhang M. CCL21 and IFN$\gamma$ recruit and activate tumor specific T cells in 3D scaffold model of breast cancer. Anticancer Agents Med Chem 2014;14 (2):204–10.
[60] Achille C, Sundaresh S, Chu B, Hadjiargyrou M. Cdk2 silencing via a DNA/PCL electrospun scaffold suppresses proliferation and increases death of breast cancer cells. PLoS One 2012;7(12):e52356.
[61] Hume RD, Pensa S, Brown EJ, Kreuzaler PA, Hitchcock J, Husmann A, et al. Tumour cell invasiveness and response to chemotherapeutics in adipocyte invested 3D engineered anisotropic collagen scaffolds. Sci Rep 2018;8(1):12658.
[62] Provenzano PP, Eliceiri KW, Campbell JM, Inman DR, White JG, Keely PJ. Collagen reorganization at the tumor-stromal interface facilitates local invasion. BMC Med 2006;4(1):38.
[63] Subia B, Dey T, Sharma S, Kundu SC. Target specific delivery of anticancer drug in silk fibroin based 3D distribution model of bone-breast cancer cells. ACS Appl Mater Interfaces 2015;7(4):2269–79.
[64] Shen H, Gao Q, Ye Q, Yang S, Wu Y, Huang Q, et al. Peritumoral implantation of hydrogel-containing nanoparticles and losartan for enhanced nanoparticle penetration and antitumor effect. Int J Nanomed 2018;13:7409–26.
[65] Strong LE, Dahotre SN, West JL. Hydrogel-nanoparticle composites for optically modulated cancer therapeutic delivery. J Control Release 2014;178:63–8.
[66] Wu J, Zhang H, Hu X, Liu R, Jiang W, Li Z, et al. Reduction-sensitive mixed micelles assembled from amphiphilic prodrugs for self-codelivery of DOX and DTX with synergistic cancer therapy. Colloids Surf B Biointerfaces 2018;161:449–56.

[67] Wang W, Song H, Zhang J, Li P, Li C, Wang C, et al. An injectable, thermosensitive and multicompartment hydrogel for simultaneous encapsulation and independent release of a drug cocktail as an effective combination therapy platform. J Control Release 2015;203:57–66.

[68] Liu J, Zhang L, Yang Z, Zhao X. Controlled release of paclitaxel from a self-assembling peptide hydrogel formed in situ and antitumor study in vitro. Int J Nanomed 2011;6:2143–53.

[69] Wang S, Kim G, Lee YE, Hah HJ, Ethirajan M, Pandey RK, et al. Multifunctional biodegradable polyacrylamide nanocarriers for cancer theranostics--a "see and treat" strategy. ACS Nano 2012;6(8):6843–51.

[70] Alexander JF, Kozlovskaya V, Chen J, Kuncewicz T, Kharlampieva E, Godin B. Cubical shape enhances the interaction of layer-by-layer polymeric particles with breast cancer cells. Adv Healthc Mater 2015;4 (17):2657–66.

[71] Vantangoli MM, Madnick SJ, Wilson S, Boekelheide K. Estradiol exposure differentially alters monolayer versus microtissue MCF-7 human breast carcinoma cultures. PLoS One 2016;11(7):e0157997.

[72] Zhao D, Song H, Zhou X, Chen Y, Liu Q, Gao X, et al. Novel facile thermosensitive hydrogel as sustained and controllable gene release vehicle for breast cancer treatment. Eur J Pharm Sci 2019;134:145–52.

[73] Khaled SZ, Cevenini A, Yazdi IK, Parodi A, Evangelopoulos M, Corbo C, et al. One-pot synthesis of pH-responsive hybrid nanogel particles for the intracellular delivery of small interfering RNA. Biomaterials 2016;87:57–68.

[74] Gilam A, Conde J, Weissglas-Volkov D, Oliva N, Friedman E, Artzi N, et al. Local microRNA delivery targets palladin and prevents metastatic breast cancer. Nat Commun 2016;7:12868.

[75] Segovia N, Pont M, Oliva N, Ramos V, Borros S, Artzi N. Hydrogel doped with nanoparticles for local sustained release of siRNA in breast cancer. Adv Healthc Mater 2015;4(2):271–80.

[76] Higano CS. New treatment options for patients with metastatic castration-resistant prostate cancer. Cancer Treat Rev 2012;38(5):340–5.

[77] Fitzgerald KA, Evans JC, McCarthy J, Guo J, Prencipe M, Kearney M, et al. The role of transcription factors in prostate cancer and potential for future RNA interference therapy. Expert Opin Ther Targets 2014;18 (6):633–49.

[78] Fitzgerald KA, Guo J, Tierney EG, Curtin CM, Malhotra M, Darcy R, et al. The use of collagen-based scaffolds to simulate prostate cancer bone metastases with potential for evaluating delivery of nanoparticulate gene therapeutics. Biomaterials 2015;66:53–66.

[79] Evans JC, Malhotra M, Fitzgerald KA, Guo J, Cronin MF, Curtin CM, et al. Formulation and evaluation of anisamide-targeted amphiphilic cyclodextrin nanoparticles to promote therapeutic gene silencing in a 3D prostate cancer bone metastases model. Mol Pharm 2017;14(1):42–52.

[80] Fitzgerald KA, Guo J, Raftery RM, Castano IM, Curtin CM, Gooding M, et al. Nanoparticle-mediated siRNA delivery assessed in a 3D co-culture model simulating prostate cancer bone metastasis. Int J Pharm 2016;511 (2):1058–69.

[81] Tao M, Liu J, He S, Xu K, Zhong W. In situ hydrogelation of forky peptides in prostate tissue for drug delivery. Soft Matter 2019;15(20):4200–7.

[82] Hainline KM, Gu F, Handley JF, Tian YF, Wu Y, de Wet L, et al. Self-assembling peptide gels for 3D prostate cancer spheroid culture. Macromol Biosci 2019;19(1):e1800249.

[83] Schaal JL, Li X, Mastria E, Bhattacharyya J, Zalutsky MR, Chilkoti A, et al. Injectable polypeptide micelles that form radiation crosslinked hydrogels in situ for intratumoral radiotherapy. J Control Release 2016;228:58–66.

[84] Hosoya H, Dobroff AS, Driessen WH, Cristini V, Brinker LM, Staquicini FI, et al. Integrated nanotechnology platform for tumor-targeted multimodal imaging and therapeutic cargo release. Proc Natl Acad Sci U S A 2016;113(7):1877–82.

[85] Bray LJ, Binner M, Holzheu A, Friedrichs J, Freudenberg U, Hutmacher DW, et al. Multi-parametric hydrogels support 3D in vitro bioengineered microenvironment models of tumour angiogenesis. Biomaterials 2015;53:609–20.

[86] Fong EL, Martinez M, Yang J, Mikos AG, Navone NM, Harrington DA, et al. Hydrogel-based 3D model of patient-derived prostate xenograft tumors suitable for drug screening. Mol Pharm 2014;11(7):2040–50.

[87] Xu X, Sabanayagam CR, Harrington DA, Farach-Carson MC, Jia X. A hydrogel-based tumor model for the evaluation of nanoparticle-based cancer therapeutics. Biomaterials 2014;35(10):3319–30.

[88] Gurski LA, Jha AK, Zhang C, Jia X, Farach-Carson MC. Hyaluronic acid-based hydrogels as 3D matrices for in vitro evaluation of chemotherapeutic drugs using poorly adherent prostate cancer cells. Biomaterials 2009;30(30):6076–85.
[89] Chalanqui MJ, Pentlavalli S, McCrudden C, Chambers P, Ziminska M, Dunne N, et al. Influence of alginate backbone on efficacy of thermo-responsive alginate-g-P(NIPAAm) hydrogel as a vehicle for sustained and controlled gene delivery. Mater Sci Eng C, Mater Biol Appl 2019;95:409–21.
[90] Engel BJ, Constantinou PE, Sablatura LK, Doty NJ, Carson DD, Farach-Carson MC, et al. Multilayered, hyaluronic acid-based hydrogel formulations suitable for automated 3D high throughput drug screening of cancer-stromal cell cocultures. Adv Healthc Mater 2015;4(11):1664–74.
[91] Goodbye, flat biology? Nature 2003;424(6951):861.
[92] Brancato V, Oliveira JM, Correlo VM, Reis RL, Kundu SC. Could 3D models of cancer enhance drug screening? Biomaterials 2019;232:119744.
[93] Olson H, Betton G, Robinson D, Thomas K, Monro A, Kolaja G, et al. Concordance of the toxicity of pharmaceuticals in humans and in animals. Regul Toxicol Pharmacol 2000;32(1):56–67.
[94] Knight A. Systematic reviews of animal experiments demonstrate poor contributions toward human healthcare. Rev Recent Clin Trials 2008;3(2):89–96.

CHAPTER

# 22

# Newly emerged engineering of in vitro 3D tumor models using biomaterials for chemotherapy

*Bo Cai*[1], *Qiying Lv*[1], *Zheng Wang*[1,2] *and Lin Wang*[1,3]

[1]Research Center for Tissue Engineering and Regenerative Medicine, Union Hospital, Tongji Medical College, Huazhong University of Science and Technology, Wuhan, P.R. China [2]Department of Gastrointestinal Surgery, Union Hospital, Tongji Medical College, Huazhong University of Science and Technology, Wuhan, P.R. China [3]Department of Clinical Laboratory, Union Hospital, Tongji Medical College, Huazhong University of Science and Technology, Wuhan, P.R. China

## 22.1 Introduction

Chemotherapy is a commonly used anticancer strategy that employs cytotoxic drugs to kill cancer cells [1]. However, developing a chemotherapeutic drug often fails [only 5% of drug candidates can obtain Food and Drug Administration (FDA, The United States) approval], leading to enormous consumption of time and money [2]. It is estimated that marketing a new anticancer drug needs 14 years and $10 billion on average since designing leading compounds [3,4].

One important reason accounting for such high cost lies in preclinical tumor models [5]. In the process of anticancer drug development, preclinical tumor models are the key tool to simulate practical tumors located in human body. They can be used to assess pharmacokinetics, safety pharmacology, and toxicology of drug candidates, and thus help screen proper candidates and optimize their in vivo performance.

Currently, commonly used tumor models include monolayer cell assays (based on multiwell plates, Petri dishes, and cover slips [6]) and model organisms (such as zebrafish, mice, and pigs [7–10]). However, these tumor models have limitations:

*Biomaterials for 3D Tumor Modeling*
DOI: https://doi.org/10.1016/B978-0-12-818128-7.00022-8

**(1)** For monolayer cell assays, cancer cells grow in two-dimensional (2D) manner. Their morphology, proliferation, differentiation, migration, and communication are different from cells within in vivo physiological environment that has defined three-dimensional (3D) topological structures. The drug performance assessed by monolayer cell assays is undoubtedly biased and cannot reflect what exactly happens in vivo [11].
**(2)** For animal tumor models, despite of their physiological similarity to human body, they are challenged by differences across species, expensive cost, concerns for ethics, and difficulty for real-time monitoring tumor lesions in situ [12,13].

These limitations contribute to the fact that most of drug candidates can successfully pass preclinical tests on cells and animals, but eventually fail in clinical trials on humans. Therefore, tumor models with satisfying biomimetics, low cost, and convenient set-up for preclinical drug evaluation are urgently needed to improve development of chemodrugs.

With technical advances in tissue engineering, in vitro 3D tumor models emerge to be an alternative for monolayer cell assays and model animals in the field of drug assessment [14]. In vitro 3D tumor model is an artificially created environment composed of cancerous cells, normal cells, and biomaterials similar to extracellular matrix (ECM). These models possess 3D tissue-like complexity and heterogeneity mimicking tumors in human body. With advantages of inexpensive fabrication and convenience for tumor monitoring, they are expected to be applied widely in pharmacy industry and cancer research in future [15–20].

Currently, in vitro 3D tumor models include two types:

**(1)** Tumor spheroids: These are aggregates proliferated by a few of cancer cells (from either cell lines or patients' primary tumor tissues) in nonadherent culture conditions. They have been applied in screening drugs and studying mechanisms of tumor invasion, metastasis, matrix remodeling, and angiogenesis [21,22].
**(2)** Artificially engineered tumor models using biomaterials: These models are mainly based on newly emerged micromachining techniques of either microfluidics [23,24] or 3D bioprinting [25–27]. Compared with tumor spheroids, artificially engineered tumor models possess advantages of designable constructions, integrated vasculature, and batch consistency, making them effective platforms for drug development.

This chapter aims to introduce artificially engineered tumor models and their applications in chemotherapy assessment. At the beginning, we present the constitution of these models, including cells and biomaterials. Subsequently, the two fabrication techniques of these models, microfluidics and 3D bioprinting, are demonstrated, respectively. Typical microfluidic and 3D printed tumor models, and their applications in screening, response assessment, toxicity evaluation, and resistance research of chemotherapeutic drugs are reviewed, especially with a focus on the advantages endowed by microfluidics and 3D bioprinting. Finally, the perspectives on limitations and outlook of these models in chemotherapy conclude this chapter.

## 22.2 Constitution of artificially engineered tumor models

Solid tumors are mainly comprised of various cells and ECM [28,29]. As the simulation to mimic tumors in human body, artificially engineered tumor models also contain (1) cells, including cancerous cells (the major population) and other types of cells that infiltrate or surround tumors; and (2) biomaterials with ECM-like functions, which can provide cells with proper physiological environment for proliferation and differentiation. In microfluidic tumor models, biomaterials are used to coat or fill microchannels/chambers inside the microfluidic chips for cell adhesion. In 3D bioprinting technique, biomaterials are usually mixed with cells to form "bio-inks" that are used for specific printers.

## 22.2.1 Cells

Immortalized cancer cell lines are widely used in constructing in vitro 3D tumor models. These cell lines provide the convenience for large-scale subculture and the availability of ample cell-biological assays [30,31]. However, cancer cell lines have drawbacks including potential cross-contamination and genetic/phenotypic instability during long-term culture, which may lead to biased drug assessment [32]. To obtain reliable evaluation results, cells from patient-derived primary tumors have recently been used in tumor model fabrication. These cells maintain physiological and pathological characteristics of in vivo tumors [15,33]. Models constructed by patient-derived tumor cells can effectively facilitate revelation of chemosensitivity of primary tumors and identification of drugs toward personalized cancer medicine [16].

Apart from malignant cells, other constituent cell types, such as stroma cells and immune cells, are also important. They intimately interact with malignant cells and are involved in various aspects of tumor behaviors [34]. For instance, cancer-associated fibroblasts can secrete tumor-promoting factors [such as transforming growth factor-β (TGF-β) [35], stromal cell-derived factor 1 (SDF-1) [36], and platelet-derived growth factor C (PDGF-C) [37]] to promote cancer cells to proliferate or migrate. Nonmalignant cells in tumor environment are also involved in hampering drug access to tumor sites and even contribute to chemotherapy resistance [34]. Therefore, nonmalignant cells should be included into in vitro 3D tumor models as well to precisely replicate in vivo cell–cell interactions and cell–drug interplay [38]. Sources of these cells include homogenous cell lines and cells derived from primary tissues/organs or induced pluripotent stem cells (iPSCs) [31,39].

## 22.2.2 Materials

Selecting appropriate biomaterials as ECM for regulating behaviors of malignant and nonmalignant cells is the key to fabricate in vitro tumor models [40,41]. ECM is a large collection of proteins, glycoproteins, proteoglycans, polysaccharides, and so forth [42]. When engineering in vitro 3D tumor models, natural extracts or synthetic polymers that contain components similar to ECM are utilized.

Commonly used natural materials are extracts from either animal or plant tissues that have similar complexity and structures to native ECM, such as agarose, collagen, and gelatin [43,44]. Among these materials, matrigel is prevalently used for tissue engineering. Matrigel is the basement membrane extract coming from Engelbreth–Holm–Swarm (EHS) mouse sarcoma, and mainly contains collagen type IV, perlecan/heparan sulfate proteoglycan 2 (HSPG2), laminin, and a number of growth factors [45]. Matrigel can improve cell attachment and differentiation. By modifying its concentration and proportion of growth factors, matrigel can provide a range of matrix stiffness and scaffold integrity for different cell engraftment and the formation of various solid tumors [46].

Synthetic polymers, such as poly(ethylene glycol) (PEG), poly(caprolactone) (PCL), poly (lactic-*co*-glycolic) acid (PLGA), poly(*n*-isopropylacrylamide) (pNIPAA), and peptide hydrogels [47], are artificially synthesized materials through controllable chemical engineering. By manipulating the synthesis processes, structures and components of these polymers can be adjusted to provide fine-tuned biochemical and biophysical properties for 3D tumor model fabrication [20,48]. For instance, biomimetic PEG hydrogels can be readily synthesized, and their mechanical properties of stiffness and elasticity as well as their bioactivity can be precisely modified by covalently incorporated with specific functional groups. With these modifications, researchers are able to regulate cellular behaviors of adhesion, migration, growth factor secretion, and epithelial–mesenchymal transition to fulfill requirements of engineering different tumor models [49].

## 22.3 Newly emerged engineering of in vitro 3D tumors for chemotherapy

When proper cells and biomaterials are chosen, they can be orchestrated to construct in vitro 3D tumor models. Current fabrication methods for artificially engineered tumor models are mainly based on the fast development of micromachining techniques, microfluidics and 3D bioprinting.

### 22.3.1 Microfluidic tumor models

Microfluidics (or Lab-on-a-chip technology) is a technique that can manipulate the movement of liquid (with the amount ranging from $10^{-9}$ to $10^{-18}$ L) through specialized micro-/nanostructures and physical/chemical actuators in a microchip with an area of several square centimeters [50]. It allows detection, analysis and manipulation of certain chemical/ biological processes at micro- or nanoscales in an automated and miniaturized manner [51]. This ability of precise fluid manipulation and sensitive signal detection at the cell-level dimension makes microfluidics a useful tool for constructing tumor models in vitro [52,53].

For microfluidic tumor models, the microchips per se can be machined to provide basic scaffolds. After being coated or filled with proper biomaterials and seeded with specified cells, they are able to form various tumor models. This kind of tumor models has several features:

**(1)** At the microscale, microchambers provide confined 3D space to support cell adhesion and proliferation, while microchannels enable continuous transport of soluble chemical/biological factors (such as nutrients, gas, and growth factors) [54]. Other microstructures, such as micropillars, microvalves, and porous membranes, can also be fabricated inside microfluidic chips. They can help to segment compartments for co-culturing different cells [55], to provide actuators for simulating physiological/ pathological mechanical activities [56], or to realize cell–cell communications [57], thus resembling physiologically relevant in vivo tumor environment [52,53].

**(2)** At the macroscale, microfluidic systems can produce several tumor models in parallel within a single chip, significantly improving testing efficiency [58]. Combination of multiple tumor chips and normal tissue chips is also feasible, so that drug efficacy on tumor models and toxicity on normal tissue/organ models can be simultaneously evaluated [59].

All the aforementioned characteristics of microfluidic chips form a tool-kit-like basis to meet various demands from chemotherapy-related testing [23,24,58,60]. In the following sections, some typical microfluidic tumor models and their applications for chemotherapy research and assessment are introduced, focusing on the four specific features provided by microfluidics: fluid transport, easy and efficient set-up, organ-like functions, and multichip integration.

#### 22.3.1.1 *Fluid network for mimicking vasculature*

The elaborate microchannels in a microfluidic chip can form a fluid network, in which various liquids can continuously perfuse throughout the whole chip. This fluid transport capability enables the substance influx and efflux of microchambers, providing continuous support of oxygen/nutrients/drugs for the embedded cells. Combined with vein endothelial cells to mimic blood vessels or vasculature network [12,61,62], versatile tumor models based on microfluidics are applied to reveal mechanisms of cancer angiogenesis, intravasation, and extravasation, which can help to investigate corresponding chemotherapeutic countermeasure and drug responses as a reference to clinical trials [63].

For instance, an in vitro 3D tumor model with a fluid network was developed on a microfluidic platform to screen antitumor drugs against tumor angiogenesis (Fig. 22.1A) [64]. Inside the chip, colorectal and breast cancer cells grew with continuous nutrient delivery through microvessels. After adding anticancer drugs, results revealed that apatinib and vandetanib, which targeted only vascular endothelial growth factor receptors (VEGFRs), could not effectively kill cancer cells, while linifanib and cabozantinib that functioned simultaneously against VEGFRs, platelet-derived growth factor receptors (PDGFRs), and Tunica interna endothelial cell kinase-2 (Tie2) regressed vasculature, thus starving off tumors. Evidently, this vascularized tumor model (VTM) helps to recognize the complexity when targeting tumor angiogenesis for cancer treatment.

With its ability of precise micromachining, microfluidic technique can minimize the footprint of VTMs and array multiple similar models in a single microfluidic chip, thus improving the testing efficiency for vessel-targeted drug screening. For example, 96 tiny VTMs were simultaneously set-up in one microchip for large-scale testing (Fig. 22.1B) [65]. These VTMs could be independently perfused under hydrostatic pressure and manipulated for 96 individual assays. With the help of this highly integrated microfluidic device, two drugs with both antiangiogenic and antitumorous effects were successfully identified from 12 different compounds in a single-blinded manner.

#### 22.3.1.2 *Easy and efficient set-up for massive drug screening*

With the advance of micromachining and soft-lithography technologies, one microfluidic device with tens to even hundreds of similar microstructures can be customized easily

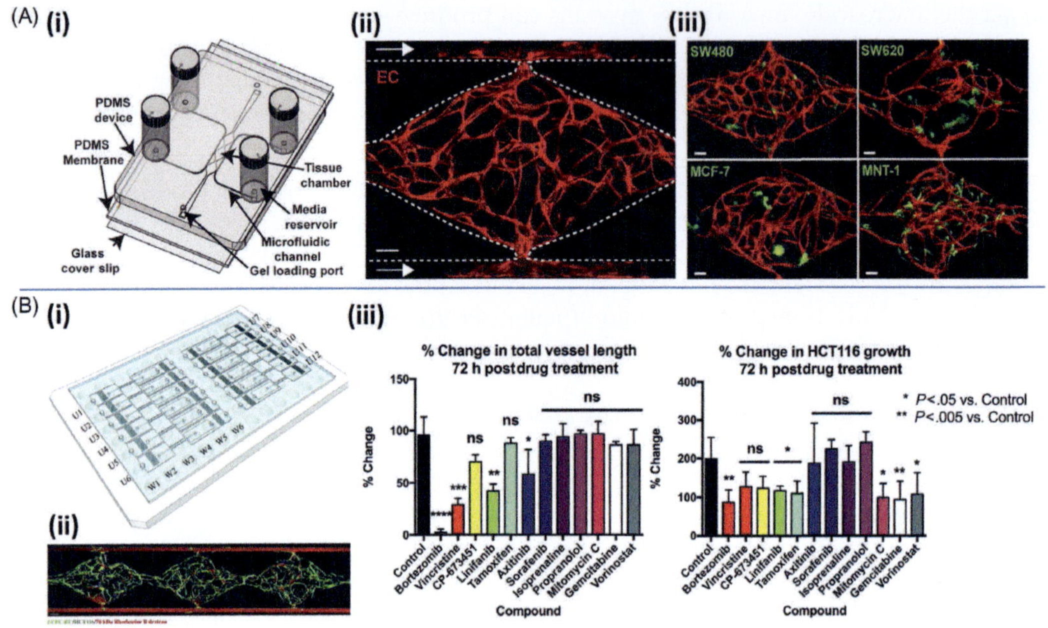


FIGURE 22.1 On-chip vasculature design for assessing chemotherapeutic drugs. (**A**) A vascularized in vitro 3D tumor model to screen antitumor drugs against angiogenesis. (**i**) Schematic of the microfluidic model. (**ii**) A representative tissue chamber with a fully developed vascular network on Day 7. (**iii**) Different tumor cells grown in the vascularized model for assessing vessel-targeting drugs. (**B**) A microfluidic platform with vascularized micro-organs for screening antiangiogenic and antitumorous drugs. (**i**) Schematic of the model. (**ii**) Vascular networks inside the tissue chamber with colorectal cancer cells (HCT116) on Day 7. (**iii**) Drug screening performed on the platform after 7 days of vessel and cancer cell culture. Sources: *(A) Adapted from Sobrino A, Phan DTT, Datta R, Wang X, Hachey SJ, Romero-López M, et al. 3D microtumors in vitro supported by perfused vascular networks. Sci Rep 2016;6:31589 [64], copyright permission Nature Publishing Group 2016. (B) Adapted from Phan DTT, Wang X, Craver BM, Sobrino A, Zhao D, Chen JC, et al. A vascularized and perfused organ-on-a-chip platform for large-scale drug screening applications. Lab Chip 2017;17:511–20 [65], copyright permission Royal Society of Chemistry 2017.*

within 2 days [66]. Therefore, numerous 3D tumor models can be fastly prepared inside a single microchip, thus facilitating efficient screening of chemotherapeutic compounds of different types or dosages at the same time [58].

As an example, by using photopolymerizable poly(ethylene) glycol diacrylate (PEGDA) hydrogel, a microchip was generated through photolithography within a few seconds and was utilized to cultivate 3D brain tumors for parallel drug screening (Fig. 22.2A) [67]. Capable of providing a number of similar culture chambers, this chip enabled high-throughput tumor spheroid formation (uniform 24 glioblastoma multiforme spheroids within 7 days) and massive-parallel testing of drug responses (over 24 assays per chip). It helped to reveal that dual-drug administration of pitavastatin and irinotecan had higher therapeutic efficacy for glioblastoma multiforme compared with individual administration of these two drugs. This study demonstrates that, in combination with tumor spheroid cultivation technology, massive microfluidic channels in one chip acquired the potential to assess combinatorial chemodrug therapy in a high-throughput manner [69].

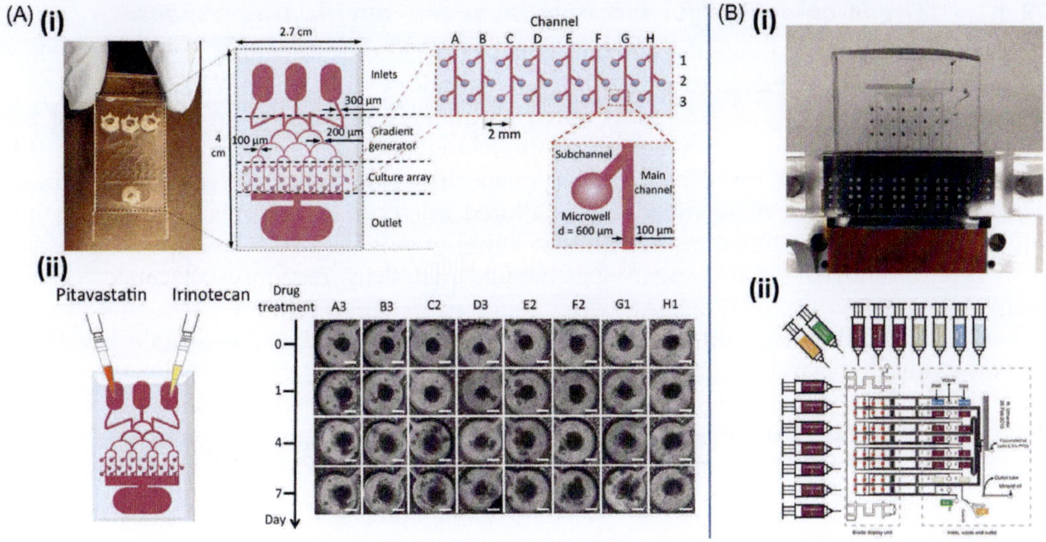


**FIGURE 22.2** On-chip parallel assays for drug screening. (**A**) An in vitro 3D brain tumor model chip for parallel drug screening. (**i**) Photograph and design of the device. (**ii**) Combinatorial drug treatment and screening for glioblastoma multiforme spheroids. (**B**) A droplet-based microfluidic platform for functional screening of drug combinations. (**i**) Photograph of the microfluidic chip mounted on a Braille display aligning the microfluidic channels with braille valves. (**ii**) Schematic of the platform for testing different combinations of 10 individual drug compounds. Sources: *(A) Adapted from Fan Y, Nguyen DT, Akay Y, Xu F, Akay M. Engineering a brain cancer chip for high-throughput drug screening, Sci Rep 2016;6:25062 [67], copyright permission Nature Publishing Group 2016. (B) Adapted from Eduati F, Utharala R, Madhavan D, Neumann UP, Longerich T, Cramer T, et al. A microfluidics platform for combinatorial drug screening on cancer biopsies. Nat Commun 2018;9:2434 [68], copyright permission Nature Publishing Group 2018.*

Another simple and efficient microfluidic regime for drug screening is droplet microfluidics. By fabricating specific microchannels in a microchip, droplet microfluidics can segment biosample contents (such as nucleic acid, proteins, and single cells) into individual droplets [70]. Droplet microfluidics has several advantages, such as small droplet volumes ($10^{-6}$ to $10^{-9}$ L per droplet), fast sample preparation ($10^3$ to $10^6$ droplets with samples inside, generated <1 h), and abundant manipulations for sample treatment (realized by droplet fusion, split, sorting, and injection) [71]. Therefore, it is thought to be an ideal platform for drug testing that requires minimal consumption, high throughput, miniaturization, and automation. As exemplified by a droplet-microfluidic platform for screening drug combinations (Fig. 22.2B) [68], biopsy from pancreatic primary tumors was dispersed into a large number of tiny droplets in a long zigzag tube. By analyzing each single droplet sequentially, over 1200 data points (56 different conditions that were repeated at least 20 times) related to sensitivity and efficiency of drug combinations on the tumor biopsy were collected for analysis. The entire workflow of drug screening could finish within 48 h with low expenditure (<\$150 per patient). For the purpose of accelerating drug screening procedure while reducing the cost, more and more droplet-microfluidic systems have been developed as portable drug screening platforms for laboratories or companies [58].

#### 22.3.1.3 "Organ-on-a-chip" for investigating organ-specific drug response

The development of tissue engineering helps microfluidics to realize "organ-on-a-chip" technology, which can fabricate artificial organs on microfluidic chips [72,73]. By micromachining exact physiological structures (such as glomerulus barriers in kidneys and bile canaliculi in liver), incorporating actuators mimicking biophysical motions (such as breathing and heart-beating), or providing organ-specific environments (such as symbiotic microbes in intestines), these functionally tailored microfluidic systems are able to simulate features as well as functions of different living organs in vitro. The organ-like devices have been employed for investigating behaviors and drug responses of cancer cells in specified organ environment [24,52,60,74].

For example, the "lung-on-a-chip" device [56] was utilized to investigate breathing-motion-induced chemoresistance by inserting lung cancer cells inside the device (Fig. 22.3A) [75]. In this system, two side channels were actuated by oscillating vacuum, and their periodical shrinking/stretching could imitate breathing movement of pulmonary alveoli. Lung epithelial cells and nonsmall-cell lung cancer (NSCLC) cells were co-cultured inside the device, and two endothelial growth factor receptor (EGFR) target drugs

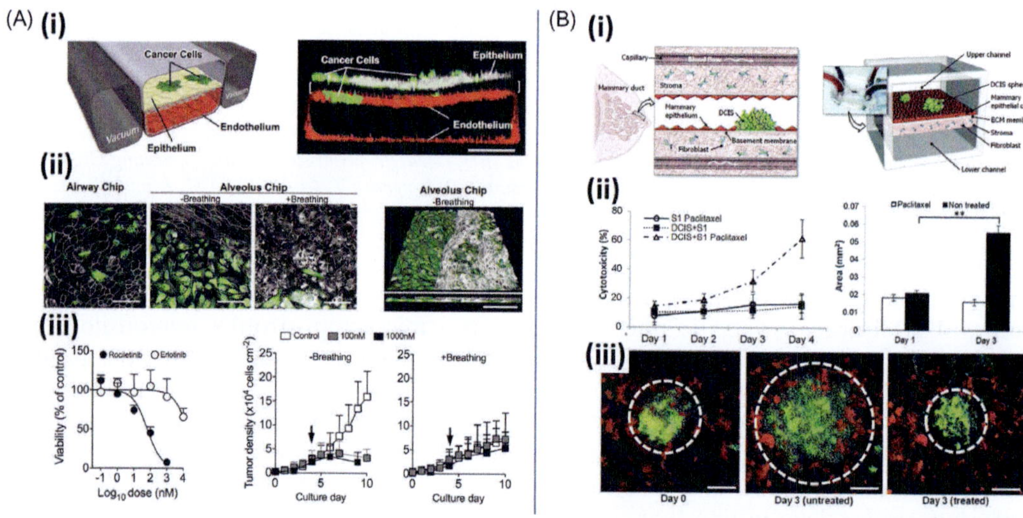

FIGURE 22.3 "Organ-on-a-chip" for investigating organ-specific drug response. (A) A lung cancer model for investigating chemoresistance of tumor cells due to breathing motions. (i) Schematic and confocal micrograph of the "lung-on-a-chip" device embedded with NSCLC tumor cells. (ii) Different growth pattern of lung cancer cells with or without breathing motions. (iii) Different drug response of lung cancer cells to rociletinib with or without breathing motions. (B) A human breast cancer-on-a-chip. (i) The 3D microarchitecture of breast DCIS and the corresponding simulation in a microfluidic device. (ii) Testing of anticancer drugs in the breast cancer-on-a-chip. (iii) Fluorescence micrographs of DCIS spheroids at Day 0 and Day 3 (with or without paclitaxel treatment). Sources: *(A) Adapted from Hassell BA, Goyal G, Lee E, Sontheimer-Phelps A, Levy O, Chen CS, et al. Human organ chip models recapitulate orthotopic lung cancer growth, therapeutic responses, and tumor dormancy in vitro. Cell Rep 2017;21:508–516 [75], copyright permission Elsevier Ltd. 2017. (B) Adapted from Choi Y, Hyun E, Seo J, Blundell C, Kim HC, Lee E, et al. A microengineered pathophysiological model of early-stage breast cancer. Lab Chip 2015;15:3350–7 [76], copyright permission Royal Society of Chemistry 2015.*

(erlotinib and rociletinib) were tested. Drug resistance of NSCLC cells was found to increase significantly when breathing-like motion was introduced, suggesting that high-level chemotherapy resistance might exist in the regions that were functionally aerated and mobile. Another organ-like microchip for modeling breast cancer adopted the similar design for assessing efficacy and toxicity of paclitaxel (Fig. 22.3B) [76]. The 3D microarchitecture of breast ductal carcinoma in situ (DCIS) was replicated in a compartmentalized microfluidic device, through co-culturing human mammary ductal epithelial cells, mammary fibroblasts, and breast tumor spheroids. This luminal-like model partially simulated DCIS lesions that were treated with intravenously administrated paclitaxel, and allowed quantitative assessment of antiproliferative effects of paclitaxel.

With the development of microfluidic "organ-on-a-chip" technology, more and more microfluidic models similar to human organs have been fabricated. After corresponding tumor cells were inserted into these organ models, it is expected to achieve in vitro monitoring of organ-specific chemosensitivity and drug responses that are difficult to be observed in conventional tumor models.

#### 22.3.1.4 *Integration of multimicrochips for systemic drug toxicity evaluation*

Standardization of microfluidic chip fabrication makes it possible to construct a relatively complex biological system by interconnecting multiple microfluidic chips together. The system can co-culture several organ models (such as liver, lung, and kidney) and tumor models simultaneously [60,77], thus allowing drug responses (including therapeutic efficacy on tumors and side effects on normal tissues) to be systemically assessed in a holistic physiologically precise manner [59,72].

Fig. 22.4 shows a typical tumor–organ integrated microfluidic system for monitoring cardiotoxicity of doxorubicin (DOX) during 24 h when it was used to treat hepatocellular carcinoma [78]. This system was constructed by combining a cardiac organoid module (derived from iPSCs), a liver tumor module (fabricated by HepG2/C3A hepatocellular carcinoma cells), a detection microchip containing various physical and electrochemical sensors, and an automated flow control breadboard. A peristaltic pump drove cell culture medium supplemented with DOX to form fluid circulation that flowed around the whole system. When the system kept running, DOX resulted in cancer cell death in the liver tumor model while induced toxicity in cardiac cells in the cardiac organoid model. By monitoring microenvironmental parameters (pH, oxygen, and temperature) and soluble biomarkers [albumin, glutathione S-transferase $\alpha$ (GST-$\alpha$), and creatine kinase MB (CK-MB)], this system was able to continuously monitor the responses of organic and cancerous cells to pharmaceutical compounds in situ.

Currently, efforts are continuously devoted to improving similar integrated platforms. It is possible to fabricate a comprehensive system that can accurately model the human body with a specified tumor, and achieve systemic prediction of chemotherapeutic drugs' efficacy and toxicity.

### 22.3.2 Bioprinted 3D tumor models

The 3D printing is a newly emerged additive manufacturing technology that employs materials to construct a 3D object using an automated "printer" machine. It can pattern

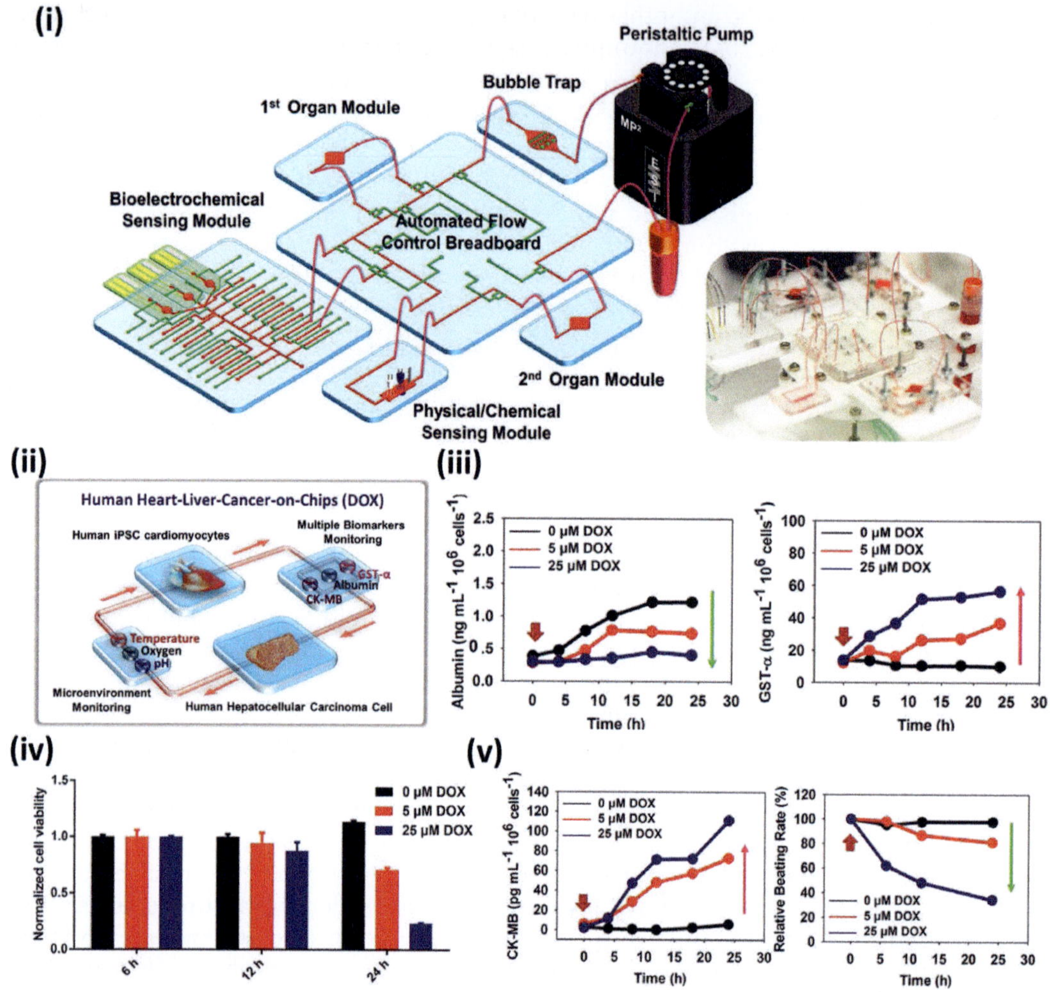


FIGURE 22.4 Integration of tumor and organ microchips for systemic drug toxicity evaluation. **(i)** Schematic of a microfluidic device integrating several modular components as a multiorgan-like system. **(ii)** Schematic of the biomimetic human heart–liver-cancer-on-chip platform. **(iii)** Electrochemical measurements of albumin and GST-α secreted from the liver tumor chip under different DOX concentrations. **(iv)** Normalized tumor cell viability in the presence of DOX from 6 to 24 h. **(v)** Electrochemical measurements of CK-MB and beating analysis of the cardiac organoids affected by different concentrations of DOX. Source: *Adapted from Zhang YS, Aleman J, Shin SR, Kilic T, Kim D, Mousavi Shaegh SA, et al. Multisensor-integrated organs-on-chips platform for automated and continual in situ monitoring of organoid behaviors. Proc Natl Acad Sci U S A 2017;114:E2293–E302 [78], copyright permission National Academy of Society, USA 2017.*

volumetric materials in discrete positions that stack together to form complex 3D constructs, which is thought to be a fast and convenient approach to fabricate models for machining, electronics, and medical service [79,80]. When applied for generating living tissues, 3D printing is redefined as "3D bioprinting" using specific "bioink" that contains

ECM-like biomaterials, cells, and other bioactive factors [81–84]. Compared with other tissue engineering methods, 3D bioprinting has the advantages of automation, precision, reproducibility, and high throughput. It can also manufacture bulk models with microstructures. Therefore, 3D bioprinting is believed to have the ability of producing in vitro models that can exactly recapitulate the features of in vivo tumors [85].

At the early stage of 3D bioprinting development, it was applied to fabricate breast tumor models, due to the distinctly hierarchical and defined structure of breast tissue [86]. These models are used for assessing the efficacy, toxicity, and resistance of chemotherapeutic drugs. For instance, by using a co-extrusion bioprinting technique, a breast tumor model was fabricated for studying the functional interactions among multiple cell types in different spatial domains when anticancer drugs were applied (Fig. 22.5A) [87]. Through printing process using a jet with a co-flow structure, breast cancer cells (MDA-MB-231) were encapsulated in peptide-conjugated alginate fibers, and macrophages (RAW 264.7) were embedded inside the core of the fibers. Chemotherapeutic drug assays on this model revealed that gefitinib and zoledronic acid impaired the migration of macrophages and their co-localization with tumor cells, but the drugs' inhibition efficacy was affected by nonlinear architecture of those fibers. The antitumor effects of these two drugs were restricted by the diffusion distance within the tissue and spatial positioning of multiple cells. Fig. 22.5B demonstrates another 3D bioprinted HER2-positive breast tumor model that was used to examine the anticancer efficacy as well as the chemoresistance of DOX when adipose-derived mesenchymal stem cells (ADMSCs) were presented together with tumor cells [88]. Photocrosslinkable methacrylated hyaluronic acid and gelatin were utilized as the bioink to print 3D constructs with breast cancer cells embedded in the core and ADMSCs in the peripheral region. Cancer cells exhibited higher resistance to DOX-induced apoptosis in the presence of ADMSCs, compared with those cultured alone. By varying the thickness of ADMSC-containing hydrogel layers, this model investigated the ability of fatty tissue to protect tumor cells from DOX-induced apoptosis.

Besides breast cancer, brain tumors are another type of tumors that are usually investigated in the field of 3D bioprinting. Brain tumors (mainly glioblastoma) have the distinct structure surrounded by brain tissue, and they can be simplified and modeled by printing brain tumor cells surrounded by ECM-like bioink doped with normal cells. Fig. 22.5C shows a typical 3D bioprinted glioma model construct to assess drug resistance of temozolomide [89]. The model was printed in a spheroid shape using a bioink formed by gelatin, alginate and fibrinogen. Compared with 2D monolayer models, tumor cells inside this 3D spheroid exhibited higher resistance to temozolomide (with ~20%–60% higher cell viability at the concentrations of 400–1600 $\mu g\ mL^{-1}$). Apart from 3D structure, cellular interactions with normal cells also account for drug resistance of brain tumor cells. By simultaneously bioprinting glioblastoma cells and macrophages to form "minibrains," a glioblastoma model was used as a tool to study the cell–cell interactions and to test therapeutics that targeted these interactions (Fig. 22.5D) [90]. The minibrain was created by printing glioblastoma cells and macrophages in a specific prearranged structure to provide features of a brain tissue with glioblastoma multiforme. Carmustine, a common drug for glioblastoma multiforme, was tested on this minibrain model. This chemotherapeutic drug worked more efficiently in conventional 2D manner [with half maximal inhibitory concentration ($IC_{50}$) of 139 μM] than in 3D manner (with $IC_{50}$ of 581 μM), demonstrating that

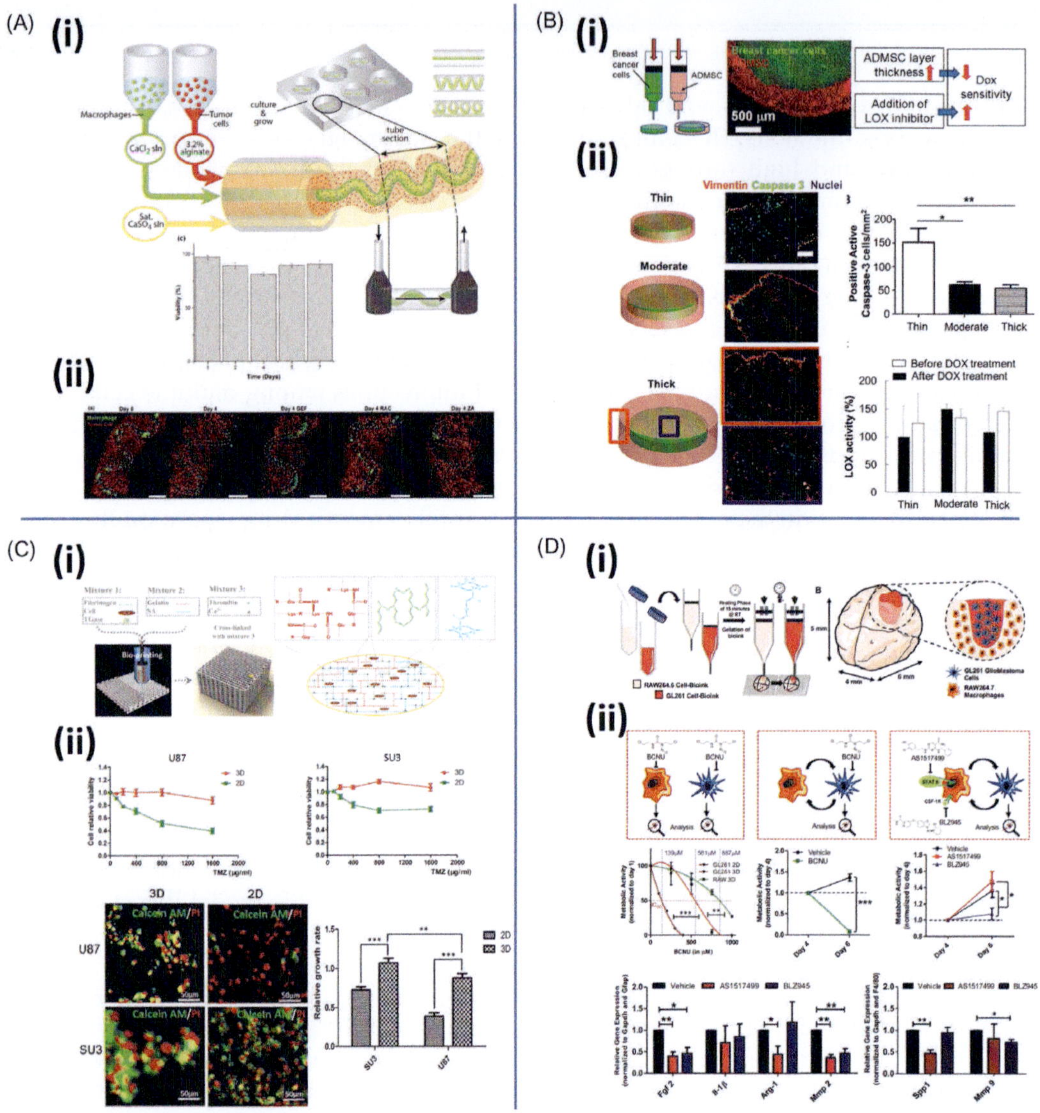


FIGURE 22.5 Bioprinted 3D tumor models for chemotherapy. **(A)** A breast tumor model containing breast cancer cells and macrophages formed by co-extrusion bioprinting. **(i)** Production schematic of the cell-laden hydrogel fibers. **(ii)** The 3D reconstructions of macrophages and tumor cells and the location of the hollow inner channel. **(B)** A 3D bioprinted HER2-positive breast tumor model. **(i)** The co-culture model formed by printing fibers with breast cancer cells embedded in the core and ADMSCs in the peripheral region. **(ii)** The effect of ADMSC-layer thickness on DOX response of breast cancer cells. **(C)** A 3D bioprinted glioma model for evaluating drug resistance of temozolomide. **(i)** Schematic of the printing process and cross-linking of hydrogel components of an in vitro brain tumor model. **(ii)** Resistance of the 3D bioprinted glioma tumor model to high concentration of temozolomide. **(D)** A 3D bioprinted minibrain model consisting of glioblastoma cells and macrophages to study cell–cell interactions and to test therapeutics. **(i)** Schematic of the printing process and the bioprinted minibrains. **(ii)** Comparative performance of chemotherapeutic drug compounds in the bioprinted minibrains. Sources: *(A) Adapted from Grolman JM, Zhang D, Smith AM, Moore JS, Kilian KA. Rapid 3D extrusion of synthetic tumor microenvironments. Adv Mater 2015;27:5512–17 [87]*, copyright permission WILEY-VCH Verlag GmbH & Co. KgaA 2015. *(B) Adapted from Wang Y, Shi W, Kuss M, Mirza S, Qi D, Krasnoslobodtsev A, et al. 3D bioprinting of breast cancer models for drug resistance study. ACS Biomater Sci Eng 2018;4:4401–11 [88], copyright permission American Chemical Society 2018. (C) Adapted from Dai X, Ma C, Lan Q, Xu T. 3D bioprinted glioma stem cells for brain tumor model and applications of drug susceptibility. Biofabrication 2016;8:045005 [89], copyright permission IOP Publishing 2016. Adapted from Heinrich MA, Bansal R, Lammers T, Zhang YS, Michel Schiffelers R, Prakash J. 3D-bioprinted minibrain: a glioblastoma model to study cellular interactions and therapeutics. Adv Mater 2019;31:1806590 [90], copyright permission WILEY-VCH Verlag GmbH & Co. KgaA 2019.*

tumor cells were more resistant to carmustine in 3D environment than they were in 2D conditions. More importantly, this model revealed the significant role of macrophages in anticancer efficacy of chemotherapeutic drugs: co-culturing with macrophages enhanced tumor cell growth and these tumor cells became more sensitive to carmustine due to their faster proliferation when compared with those monocultured ones. This minibrain model emphasized the importance of considering tumors and their microenvironment as a whole when applying chemotherapy.

In summary, 3D bioprinting provides a useful tool for investigating how tissue structures and cellular interactions influence drug performance, especially when dealing with tissues that have heterogeneous structures and are difficult to pattern using other engineering techniques. In recent years, 3D bioprinting has demonstrated its ability in precisely and spatially manipulating various types of cells in a microspace, allowing recapitulation of in vivo cell–cell communication within tumor environment. With the continuously devoted efforts in improving 3D bioprinting techniques, it would be more powerful in creating complex in vitro 3D tumor models for clinically oriented chemotherapeutics testing.

## 22.4 Summary

In vitro 3D tumor models can incorporate cancer cells, matrix molecules, growth factors, and other biochemical cues to simulate in vivo tumors and their microenvironment. In the field of chemotherapy, these models can help predict the efficacy and side effects of chemotherapeutic agents and identify drug candidates for further clinical trials. Owing to the ability of manipulating cells at the microscale, the newly emerged engineering technologies, microfluidics, and 3D bioprinting, have been applied for fabricating in vitro 3D tumor models with higher complexity and functionality. Compared with conventional tumor models of monolayer cell assays, these models are more reliable in generating predictive results with regard to pharmacokinetics, safety pharmacology, toxicology, and resistance of chemotherapeutic agents. Moreover, microfluidics and 3D bioprinting can fabricate tumor models in a facile and efficiency manner, so that they are expected to accelerate the development process and decrease the expenditure of chemotherapeutic drugs, thus well complementing animal models.

However, these artificially engineered in vitro 3D tumor models also face challenges:

**(1)** For microfluidic technology, despite of great efforts devoted to simplifying the manufacture process of microchips, professional equipments, and clean rooms are still necessary. Different types of equipment are also needed to drive microfluidic chips, for example, syringe pumps for liquid manipulation, gas circuit systems for microvalve actuation, and electronic devices for signal detection [51,91]. All these auxiliary instruments require extra professionals, knowledge, and laboratory preparation if microfluidic tumor models are used in laboratories of hospitals or drug R&D (research and design) departments.

**(2)** Compared with microfluidic platforms, 3D bioprinting can produce tumor models in a faster manner and with larger dimensions. However, the current print resolution of 3D bioprinting is at hundreds of microns, which greatly limits tissue fabrication accuracy.

In addition, given that the shear force generated inside bioink during the printing process could damage cells, materials with improved mechanical property and biocompatibility are desired [81,82].

The ultimate goal of applying in vitro 3D tumor models based on either microfluidics or 3D bioprinting is aimed at personalized therapy and precision medicine. At this level, both of the two techniques face several common challenges: firstly, patient-derived tumor cells are difficult to obtain for constructing personalized tumor models. There are logistical complexities for keeping cells viable when transferring them from operating rooms to pathology departments, and finally to laboratories for model construction. Secondly, cellular heterogeneity of primary tumor cells affects reproducibility of tumor model fabrication. The intratumor heterogeneity is thought to partly account for poor reproducibility of drug efficacy assessment conducted in 3D tumor models. Gene expression and phenotype variations during long-term culture in the device may also augment biological differences between tumor models and original tumors. At last, the time-consuming set-up procedure and the difficulty for maintenance limit the practicability of 3D patient-derived tumor models, given that rapid tumor progression requires fast drug screening procedure to identify proper drugs for timely treatment.

To address all these difficulties, the increasing amount of efforts have been devoted to improving and optimizing 3D tumor models. Solutions include developing facile manufacture methods (such as paper-based tumor microchips [92]), and synthesizing materials with enhanced biocompatible performance [93]. Moreover, the burgeoning science such as cancer genomics, system biology, and mathematical modeling would also help researchers to generate in vitro 3D tumor models that could resemble tumors in vivo in a more physiologically and pathologically way. It is reasonable to believe that the artificially engineered in vitro 3D tumor models would undoubtedly contribute more to development of novel chemotherapeutics with improved efficacy and reduced toxicity toward personalized and precise cancer medicine.

## References

[1] Chabner BA, Roberts TG. Chemotherapy and the war on cancer. Nat Rev Cancer 2005;5:65–72.
[2] Kola I, Landis J. Can the pharmaceutical industry reduce attrition rates? Nat Rev Drug Discov 2004;3:711–16.
[3] DiMasi JA, Grabowski HG. Economics of new oncology drug development. J Clin Oncol 2007;25:209–16.
[4] DiMasi JA, Hansen RW, Grabowski HG. The price of innovation: new estimates of drug development costs. J Health Econ 2003;22:151–85.
[5] Santo VE, Rebelo SP, Estrada MF, Alves PM, Boghaert E, Brito C. Drug screening in 3D in vitro tumor models: overcoming current pitfalls of efficacy read-outs. Biotech J 2017;12:1600505.
[6] Caballero D, Kaushik S, Correlo VM, Oliveira JM, Reis RL, Kundu SC. Organ-on-chip models of cancer metastasis for future personalized medicine: from chip to the patient. Biomaterials 2017;149:98–115.
[7] Kirchberger S, Sturtzel C, Pascoal S, Distel M. Quo natas, Danio?—Recent progress in modeling cancer in zebrafish. Front Oncol 2017;7:186.
[8] Kersten K, de Visser KE, van Miltenburg MH, Jonkers J. Genetically engineered mouse models in oncology research and cancer medicine. EMBO Mol Med 2017;9:137–53.
[9] Flisikowska T, Kind A, Schnieke A. The new pig on the block: modelling cancer in pigs. Transgenic Res 2013;22:673–80.
[10] Ruggeri BA, Camp F, Miknyoczki S. Animal models of disease: pre-clinical animal models of cancer and their applications and utility in drug discovery. Biochem Pharmacol 2014;87:150–61.

[11] Unger C, Kramer N, Walzl A, Scherzer M, Hengstschläger M, Dolznig H. Modeling human carcinomas: physiologically relevant 3D models to improve anti-cancer drug development. Adv Drug Deliv Rev 2014;79-80:50–67.

[12] Lin DSY, Guo F, Zhang B. Modeling organ-specific vasculature with organ-on-a-chip devices. Nanotechnology 2018;30:024002.

[13] Fitzgerald KA, Malhotra M, Curtin CM, Brien FJO, Driscoll CMO. Life in 3D is never flat: 3D models to optimise drug delivery. J Control Release 2015;215:39–54.

[14] Kim HN, Habbit NL, Su C-Y, Choi N, Ahn EH, Lipke EA, et al. Microphysiological systems as enabling tools for modeling complexity in the tumor microenvironment and accelerating cancer drug development. Adv Funct Mater 2019;29:1807553.

[15] Meijer TG, Naipal KA, Jager A, Gent DCV. Ex vivo tumor culture systems for functional drug testing and therapy response prediction. Future Sci OA 2017;(3():FSO190.

[16] Morgan MM, Johnson BP, Livingston MK, Schuler LA, Alarid ET, Sung KE, et al. Personalized in vitro cancer models to predict therapeutic response: challenges and a framework for improvement. Pharmacol Ther 2016;165:79–92.

[17] Matsusaki M, Case CP, Akashi M. Three-dimensional cell culture technique and pathophysiology. Adv Drug Deliv Rev 2014;74:95–103.

[18] Breslin S, O'Driscoll L. Three-dimensional cell culture: the missing link in drug discovery. Drug Discov Today 2013;18:240–9.

[19] Cukierman E, Pankov R, Stevens DR, Yamada KM. Taking cell-matrix adhesions to the third dimension. Science 2001;294:1708–12.

[20] Xu X, Farach-Carson MC, Jia X. Three-dimensional in vitro tumor models for cancer research and drug evaluation. Biotech Adv 2014;32:1256–68.

[21] Hamilton G, Rath B. Applicability of tumor spheroids for in vitro chemosensitivity assays. Expert Opin Drug Metab Toxicol 2019;15:15–23.

[22] Weeber F, Ooft SN, Dijkstra KK, Voest EE. Tumor organoids as a pre-clinical cancer model for drug discovery. Cell Chem Biol 2017;24:1092–100.

[23] Shang M, Soon RH, Lim CT, Khoo BL, Han J. Microfluidic modelling of the tumor microenvironment for anti-cancer drug development. Lab Chip 2019;19:369–86.

[24] Kashaninejad N, Nikmaneshi MR, Moghadas H, Kiyoumarsi Oskouei A, Rismanian M, Barisam M, et al. Organ-tumor-on-a-chip for chemosensitivity assay: a critical review. Micromachines-Basel 2016;7:130.

[25] Matai I, Kaur G, Seyedsalehi A, McClinton A, Laurencin CT. Progress in 3D bioprinting technology for tissue/organ regenerative engineering. Biomaterials 2020;226:119536.

[26] Liu T, Delavaux C, Zhang YS. 3D bioprinting for oncology applications. J 3D Print Med 2019;3:55–8.

[27] Ma X, Liu J, Zhu W, Tang M, Lawrence N, Yu C, et al. 3D bioprinting of functional tissue models for personalized drug screening and in vitro disease modeling. Adv Drug Deliv Rev 2018;132:235–51.

[28] Pickup MW, Mouw JK, Weaver VM. The extracellular matrix modulates the hallmarks of cancer. EMBO Rep 2014;15:1243–53.

[29] Quail DF, Joyce JA. Microenvironmental regulation of tumor progression and metastasis. Nat Med 2013;19:1423.

[30] Mirabelli P, Coppola L, Salvatore M. Cancer cell lines are useful model systems for medical research. Cancers 2019;11:1098.

[31] Niu N, Wang L. In vitro human cell line models to predict clinical response to anticancer drugs. Pharmacogenomics 2015;16:273–85.

[32] Lovitt CJ, Shelper TB, Avery VM. Cancer drug discovery: recent innovative approaches to tumor modeling. Expert Opin Drug Discov 2016;11:885–94.

[33] Miserocchi G, Mercatali L, Liverani C, De Vita A, Spadazzi C, Pieri F, et al. Management and potentialities of primary cancer cultures in preclinical and translational studies. J Transl Med 2017;15:229.

[34] Valkenburg KC, de Groot AE, Pienta KJ. Targeting the tumour stroma to improve cancer therapy. Nat Rev Clin Oncol 2018;15:366–81.

[35] Bhowmick NA, Chytil A, Plieth D, Gorska AE, Dumont N, Shappell S, et al. TGF-ß signaling in fibroblasts modulates the oncogenic potential of adjacent epithelia. Science 2004;303:848–51.

[36] Orimo A, Gupta PB, Sgroi DC, Arenzana-Seisdedos F, Delaunay T, Naeem R, et al. Stromal fibroblasts present in invasive human breast carcinomas promote tumor growth and angiogenesis through elevated SDF-1/CXCL12 secretion. Cell 2005;121:335–48.
[37] Crawford Y, Kasman I, Yu L, Zhong C, Wu X, Modrusan Z, et al. PDGF-C mediates the angiogenic and tumorigenic properties of fibroblasts associated with tumors refractory to anti-VEGF treatment. Cancer Cell 2009;15:21–34.
[38] Weigelt B, Ghajar CM, Bissell MJ. The need for complex 3D culture models to unravel novel pathways and identify accurate biomarkers in breast cancer. Adv Drug Deliv Rev 2014;69-70:42–51.
[39] Passier R, Orlova V, Mummery C. Complex tissue and disease modeling using hiPSCs. Cell Stem Cell 2016;18:309–21.
[40] Beri P, Matte BF, Fattet L, Kim D, Yang J, Engler AJ. Biomaterials to model and measure epithelial cancers. Nat Rev Mater 2018;3:418–30.
[41] Gu L, Mooney DJ. Biomaterials and emerging anticancer therapeutics: engineering the microenvironment. Nat Rev Cancer 2015;16:56.
[42] Frantz C, Stewart KM, Weaver VM. The extracellular matrix at a glance. J Cell Sci 2010;123:4195–200.
[43] Sepantafar M, Maheronnaghsh R, Mohammadi H, Radmanesh F, Hasani-sadrabadi MM, Ebrahimi M, et al. Engineered hydrogels in cancer therapy and diagnosis. Trend Biotechnol 2017;35:1074–87.
[44] Tibbitt MW, Anseth KS. Hydrogels as extracellular matrix mimics for 3D cell culture. Biotech Bioeng 2009;103:655–63.
[45] Fridman R, Benton G, Aranoutova I, Kleinman HK, Bonfil RD. Increased initiation and growth of tumor cell lines, cancer stem cells and biopsy material in mice using basement membrane matrix protein (Cultrex or Matrigel) co-injection. Nat Protoc 2012;7:1138–44.
[46] Benton G, Arnaoutova I, George J, Kleinman HK, Koblinski J. Matrigel: from discovery and ECM mimicry to assays and models for cancer research. Adv Drug Deliv Rev 2014;79-80:3–18.
[47] Gill BJ, West JL. Modeling the tumor extracellular matrix: tissue engineering tools repurposed towards new frontiers in cancer biology. J Biomech 2014;47:1969–78.
[48] Hutmacher DW. Biomaterials offer cancer research the third dimension. Nat Mater 2010;9:90–3.
[49] Prestwich GD. Simplifying the extracellular matrix for 3-D cell culture and tissue engineering: a pragmatic approach. J Cell Biochem 2007;101:1370–83.
[50] Squires T, Quake S. Microfluidics: fluid physics at the nanoliter scale. Rev Mod Phys 2005;77:977–1026.
[51] Convery N, Gadegaard N. 30 years of microfluidics. Micro Nano Eng 2019;2:76–91.
[52] Sontheimer-Phelps A, Hassell BA, Ingber DE. Modelling cancer in microfluidic human organs-on-chips. Nat Rev Cancer 2019;19:65–81.
[53] Hachey SJ, Hughes CCW. Applications of tumor chip technology. Lab Chip 2018;18:2893–912.
[54] El-Ali J, Sorger PK, Jensen KF. Cells on chips. Nature 2006;442:403–11.
[55] Liu Y, Kongsuphol P, Chiam SY, Zhang QX, Gourikutty SBN, Saha S, et al. Adipose-on-a-chip: a dynamic microphysiological in vitro model of the human adipose for immune-metabolic analysis in type II diabetes. Lab Chip 2019;19:241–53.
[56] Huh D, Matthews BD, Mammoto A, Montoya-Zavala M, Hsin HY, Ingber DE. Reconstituting organ-level lung functions on a chip. Science 2010;328:1662–8.
[57] Du Y, Li N, Yang H, Luo C, Gong Y, Tong C, et al. Mimicking liver sinusoidal structures and functions using a 3D-configured microfluidic chip. Lab Chip 2017;17:782–94.
[58] Dhiman N, Kingshott P, Sumer H, Sharma CS, Rath SN. On-chip anticancer drug screening – recent progress in microfluidic platforms to address challenges in chemotherapy. Biosens Bioelectron 2019;137:236–54.
[59] Skardal A, Shupe T, Atala A. Organoid-on-a-chip and body-on-a-chip systems for drug screening and disease modeling. Drug Discov Today 2016;21:1399–411.
[60] Low LA, Tagle DA. Tissue chips - innovative tools for drug development and disease modeling. Lab Chip 2017;17:3026–36.
[61] Wang X, Sun Q, Pei J. Microfluidic-based 3D engineered microvascular networks and their applications in vascularized microtumor models. Micromachines-Basel 2018;9:493.
[62] Lee S, Ko J, Park D, Lee S-R, Chung M, Lee Y, et al. Microfluidic-based vascularized microphysiological systems. Lab Chip 2018;18:2686–709.

[63] Caballero D, Blackburn SM, de Pablo M, Samitier J, Albertazzi L. Tumour-vessel-on-a-chip models for drug delivery. Lab Chip 2017;17:3760–71.
[64] Sobrino A, Phan DTT, Datta R, Wang X, Hachey SJ, Romero-López M, et al. 3D microtumors in vitro supported by perfused vascular networks. Sci Rep 2016;6:31589.
[65] Phan DTT, Wang X, Craver BM, Sobrino A, Zhao D, Chen JC, et al. A vascularized and perfused organ-on-a-chip platform for large-scale drug screening applications. Lab Chip 2017;17:511–20.
[66] Lin W-Y, Wang Y, Wang S, Tseng H-R. Integrated microfluidic reactors. Nano Today 2009;4:470–81.
[67] Fan Y, Nguyen DT, Akay Y, Xu F, Akay M. Engineering a brain cancer chip for high-throughput drug screening. Sci Rep 2016;6:25062.
[68] Eduati F, Utharala R, Madhavan D, Neumann UP, Longerich T, Cramer T, et al. A microfluidics platform for combinatorial drug screening on cancer biopsies. Nat Commun 2018;9:2434.
[69] Nath S, Devi GR. Three-dimensional culture systems in cancer research: focus on tumor spheroid model. Pharmacol Ther 2016;163:94–108.
[70] Sánchez Barea J, Lee J, Kang D-K. Recent advances in droplet-based microfluidic technologies for biochemistry and molecular biology. Micromachines-Basel 2019;10:412.
[71] Shang L, Cheng Y, Zhao Y. Emerging droplet microfluidics. Chem Rev 2017;117:7964–8040.
[72] Sung JH, Wang YI, Narasimhan Sriram N, Jackson M, Long C, Hickman JJ, et al. Recent advances in body-on-a-chip systems. Anal Chem 2019;91:330–51.
[73] Zhang B, Korolj A, Lai BFL, Radisic M. Advances in organ-on-a-chip engineering. Nat Rev Mater 2018;3:257–78.
[74] Zhou L, Mao S, Huang Q, He X, Lin J-M. Inhibition of anaerobic probiotics on colorectal cancer cells using intestinal microfluidic systems. Sci China Chem 2018;61:1034–42.
[75] Hassell BA, Goyal G, Lee E, Sontheimer-Phelps A, Levy O, Chen CS, et al. Human organ chip models recapitulate orthotopic lung cancer growth, therapeutic responses, and tumor dormancy in vitro. Cell Rep 2017;21:508–16.
[76] Choi Y, Hyun E, Seo J, Blundell C, Kim HC, Lee E, et al. A microengineered pathophysiological model of early-stage breast cancer. Lab Chip 2015;15:3350–7.
[77] Kanamori T, Sugiura S, Sakai Y. Technical aspects of microphysiological systems (MPS) as a promising wet human-in-vivo simulator. Drug Met Pharmacokinet 2018;33:40–2.
[78] Zhang YS, Aleman J, Shin SR, Kilic T, Kim D, Mousavi Shaegh SA, et al. Multisensor-integrated organs-on-chips platform for automated and continual in situ monitoring of organoid behaviors. Proc Natl Acad Sci U S A 2017;114:E2293–302.
[79] Ngo TD, Kashani A, Imbalzano G, Nguyen KTQ, Hui D. Additive manufacturing (3D printing): a review of materials, methods, applications and challenges. Compos Part B-Eng 2018;143:172–96.
[80] Truby RL, Lewis JA. Printing soft matter in three dimensions. Nature 2016;540:371.
[81] Zhu W, Ma X, Gou M, Mei D, Zhang K, Chen S. 3D printing of functional biomaterials for tissue engineering. Curr Opin Biotechnol 2016;40:103–12.
[82] Chia HN, Wu BM. Recent advances in 3D printing of biomaterials. J Biol Eng 2015;9:4.
[83] Guvendiren M, Molde J, Soares RMD, Kohn J. Designing biomaterials for 3D printing. ACS Biomater Sci Eng 2016;2:1679–93.
[84] Ozbolat IT. Bioprinting scale-up tissue and organ constructs for transplantation. Trend Biotechnol 2015;33:395–400.
[85] Peng W, Datta P, Ayan B, Ozbolat V, Sosnoski D, Ozbolat IT. 3D bioprinting for drug discovery and development in pharmaceutics. Acta Biomater 2017;57:26–46.
[86] Belgodere JA, King CT, Bursavich JB, Burow ME, Martin EC, Jung JP. Engineering breast cancer microenvironments and 3D bioprinting. Front Bioeng Biotech 2018;6:66.
[87] Grolman JM, Zhang D, Smith AM, Moore JS, Kilian KA. Rapid 3D extrusion of synthetic tumor microenvironments. Adv Mater 2015;27:5512–17.
[88] Wang Y, Shi W, Kuss M, Mirza S, Qi D, Krasnoslobodtsev A, et al. 3D bioprinting of breast cancer models for drug resistance study. ACS Biomater Sci Eng 2018;4:4401–11.
[89] Dai X, Ma C, Lan Q, Xu T. 3D bioprinted glioma stem cells for brain tumor model and applications of drug susceptibility. Biofabrication 2016;8:045005.

[90] Heinrich MA, Bansal R, Lammers T, Zhang YS, Michel Schiffelers R, Prakash J. 3D-bioprinted mini-brain: a glioblastoma model to study cellular interactions and therapeutics. Adv Mater 2019;31:1806590.
[91] Sackmann EK, Fulton AL, Beebe DJ. The present and future role of microfluidics in biomedical research. Nature 2014;507:181–9.
[92] Tian C, Tu Q, Liu W, Wang J. Recent advances in microfluidic technologies for organ-on-a-chip. TrAC-Trend Anal Chem 2019;117:146–56.
[93] Aljohani W, Ullah MW, Zhang X, Yang G. Bioprinting and its applications in tissue engineering and regenerative medicine. Int J Biol Macromol 2018;107:261–75.

CHAPTER

# 23

# Marine-derived biomaterials for cancer treatment

*Catarina Oliveira[1,2], Ana C. Carvalho[1,2], Rui L. Reis[1,2], Nuno N. Neves[1,2], Albino Martins[1,2] and Tiago H. Silva[1,2]*

[1]3B's Research Group, I3Bs – Research Institute on Biomaterials, Biodegradables and Biomimetics of University of Minho, Headquarters of the European Institute of Excellence on Tissue Engineering and Regenerative Medicine, Guimarães, Portugal [2]ICVS/3B's, PT Government Associate Laboratory, Braga/Guimarães, Portugal

## 23.1 Introduction

Cancer is one of the leading causes of death worldwide, with an increasing tendency over the last years. Among the different types of cancer, breast, lung, and colon cancer are the most common, lung cancer being the deadliest in both man and woman [1]. The heterogeneity of cancer cells, as well as the different development stages compromise the advances and availability of more effective anticancer therapies. Current treatment approaches of cancer involves surgery, radiation therapy, chemotherapy, targeted, and immunotherapies, which may be used alone or in combination, depending on the cancer types [2]. Radiation therapy uses high doses of radiation to kill cancer cells and shrink tumors, whereas chemotherapy uses anticancer drugs to kill cancer cells. Both treatments may present severe side effects since they affect not only cancer cells but also healthy ones [3,4]. Immunotherapy is a type of biological treatment that helps the immune system to identify and fight cancer and targeted therapies aim to develop more effective treatments that specifically affect cancer cells, aiming to reduce the side effects over noncancerous cells and tissues [5]. To develop effective anticancer therapies, the tumor microenvironment needs to be considered since it comprises not only the cancer cells but also surrounding healthy tissues [6]. In this sense, the use of natural-based approaches with low

*Biomaterials for 3D Tumor Modeling*
DOI: https://doi.org/10.1016/B978-0-12-818128-7.00023-X

toxicities over healthy tissues are interesting possibilities to develop more effective strategies with less side-effects [4].

The oceans represent the largest habitat of the earth, covering more than 70% of its surface. The different oceanic zones and environments offer a wide variety of marine organisms with unique chemical and biological properties not available as products. Each year the shellfish processing industries produce about 75% of waste products from marine-origin organisms that may be further explored and processed to produce new compounds and materials [7–9]. The compounds may present different physicochemical features that result in interesting biological properties to be explored in the biomedical field, namely low toxicity and low-cost production [8,10]. Different biopolymers may be isolated from marine organisms, which can be divided into three different groups: proteins, polysaccharides, and nucleic acids [7,11,12]. The polysaccharides fucoidan, chitosan, carrageenan, and alginate will be the ones described in this book chapter in relation to their anticancer capabilities (Fig. 23.1). Their physicochemical similarities to compounds present in native human tissues and the low cytotoxicity are features of interest. However, they present some limitations like batch-to-batch variability, since the season and local of harvesting may influence the polysaccharide properties, as well as the use of different extraction methods also may influence the final biological performance. In this sense, sustainable and standardized methods are required to obtain controlled and reproducible compounds [13,14].

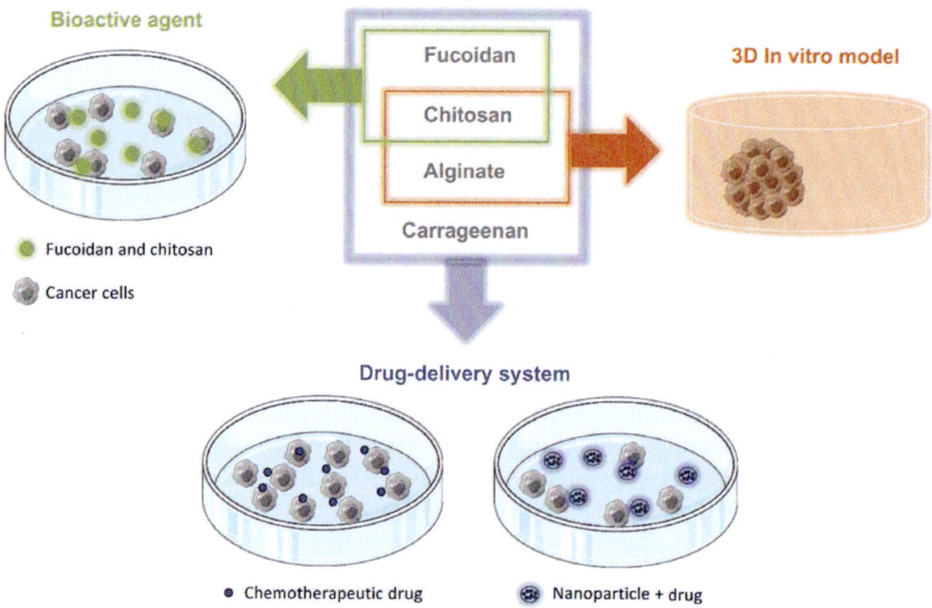


FIGURE 23.1 Main examples on the use of different marine polysaccharides in different anticancer strategies.

## 23.2 Marine biopolymers as bioactive agents

### 23.2.1 Fucoidan

Fucoidan is a sulfated polysaccharide that can be extracted from different species of brown algae. It is mainly composed of fucose and sulfate residues, even if it may also contain other monosaccharides [15]. Fucoidans are currently used in cosmetics and dietary supplements and although its use in biomedical applications has been studied and proposed, there are not yet approved by the regulatory entities. The use of different algae sources and extractions methods, which results in different intrinsic properties like molecular weight and sulfation degrees/patterns, may be the reason why fucoidan has not yet been developed as a regulated therapeutic [16,17]. To be considered as such, fucoidans must be definitely characterized and present consistent properties. Nowadays, this is a concern regarding fucoidan, since not all fucoidans present the desired biological outcomes, leading to inconsistent results [18]. Different biological activities have been reported regarding fucoidan usage as antiinflammatory, antiviral, and both pro- and antiangiogenic [19–21]. The anticancer activity of fucoidan has also been studied for different types of cancer, being breast, colon, and lung the most frequently reported (refer to Table 23.1) [36].

Fucoidan cytotoxic effects over cancer cells has been widely described for different types of cancer, showing dose-dependent behavior particularly for lung and colon cancer [30,33].

**TABLE 23.1** Fucoidan anticancer activity and mechanism of action.

| Cancer | Cell lines | Mechanism of action | References |
|---|---|---|---|
| Breast | MCF-7 | Suppression of cell proliferation and enhancement of apoptosis<br>Cell cycle arrest at G1. Caspase-dependent apoptosis. Induction of apoptosis (mitochondria-mediated) | [22–24] |
| | MDA-MB-231 | Induction of apoptosis (mitochondria-mediated)<br>Induction of apoptosis via PI3K/AKT/GSK3β pathway | [23,25] |
| | T-47D | Induction of apoptosis by activating the initiator and effector caspases | [26] |
| Colon | HT-29 | Suppression of colony formation<br>Inhibition of cell viability. Downregulation of IGF-IR signaling through the main IRS-1/PI3K/AKT pathway<br>Inhibition of cell growth and increased apoptotic cell death | [27–29] |
| | HCT-116<br>DLD-1 | Suppression of colony formation | [28] |
| | HCT-15 | Inhibition of cells proliferation | [30] |
| lung | A549 | Inducing apoptotic responses by upregulating cleaved caspase-3 and poly(ADP ribose) polymerase (PARP) expression.<br>Antiangiogenic activity via blocking VEGF signaling. Inhibited proliferation in a dose-dependent manner | [31–33] |
| | LLCC | Inducing apoptotic responses by upregulating cleaved caspase-3 and poly(ADP ribose) polymerase (PARP) expression.<br>Induces an endoplasmic reticulum stress response by activating the PERK-ATF4-CHOP pathway | [31,34,35] |
| | Mice model | Inhibiting lung carcinoma metastasis through downregulating metastatic factors VEGF and MMPs | [32,35] |

However, complex mechanisms of action may be involved regarding fucoidan anticancer behavior: induction of apoptosis by activating different signaling pathways, inhibition of angiogenesis, and inhibition of colony formation [37,38]. Fucoidan has been tested over different types of cancer and the apoptotic mechanism has been extensively studied and described for different fucoidan extracts and cancers like breast, lung, and colon [26,29,31]. He *et al.* studied the effects of fucoidan over proliferation, migration, and apoptosis of MCF-7 breast cancer cells implanted in female rats. Higher doses of fucoidan significantly increased tumor cell dead when compared with lower dosage [22]. Fucoidan increased the expression of E-cadherin (tumor invasion suppressor gene) while downregulated the expression of MMP-9 (promote tumor invasion and metastasis), suppressing also tumor cell migration. In another study, low-molecular weight (LMW) fucoidan exhibited a decreased cell viability for both MCF-7 and MDA-MB-231 breast cancer cells [23]. It was also associated to an increase in the depolarization of mitochondria membrane leading to early stages of cell apoptosis.

Akt is an important cell signaling pathway that regulates cells growth and survival, and PI3K plays a key role in intracellular signaling. It has been reported that the unusual activation of Akt and PI3K is responsible for decreased cell apoptosis, so inhibiting these pathways may be a possible approach to restore normal apoptosis in cancer cells. Fucoidan has been described to increase apoptosis in breast and colon cancer cells through these signaling pathways [25,27]. Hsu *et al.* studied the toxicity of fucoidan in combination with cisplatin over Lewis lung carcinoma cells (LLCC), showing tumor reduction *in vivo* [31]. This combination presented enhanced apoptotic effects than when administered separately, increasing caspase-3 protein activation and pro-PARP cleavage by caspase-3 in A549 cell line.

Stress is also a possible pathway to modulate endoplasmic reticulum (ER) to induce apoptosis in cancer cells which has also been described when fucoidan was administered in lung cancer model *in vivo* [34]. The inhibition of cell cycle progression in cancer cells has also been studied regarding fucoidan usage [24]. Fucoidan shown increased toxic effects over MCF-7 breast cancer cells than to noncancerous breast cancer cells, arresting cell cycle at G1 with accumulation of cells in the sub-G1 phase. It was also reported the activation of caspases-3, -7, and -9. Fucoidan from different species was also reported to inhibit the formation of colonies for different types of colon cancer cells [28].

Inhibiting angiogenesis may be one possible approach to target tumor cells, since by blocking this process the tumor will not be able to proliferate and metastasize. In this sense, fucoidan antiangiogenic potential is also of interest in the cancer scenario. Fucoidan was reported to block VEGF signaling, a growth factor responsible for endothelial cells proliferation [32]. In another study, fucoidan reduced the expression of VEGF, as well as downregulated matrix metalloproteinases (MMPs—related with tumor invasiveness) expression, showing promising results in a tumor-bearing mice [35].

### 23.2.2 Chitosan

Chitosan is a natural polymer, obtained from chitin, exhibiting biological properties such as biodegradability and biocompatibility for some applications, and presenting similarities to glycosaminoglycans. Chitin is the second most abundant polysaccharide, both in terms of usage and distribution, only surpassed by cellulose [7,39]. Chitin is the major constituent of the cell wall of fungi and can be found in the endoskeletons of cephalopods and in the shells of many crustaceans such as crabs and shrimps, with these last two being the major commercial sources of chitosan [8,10]. Due to the poor solubility of chitin, chitosan has attracted most attention, being obtained by a deacetylation process. This process leads to chitosan with different deacetylation degrees (DD), but also molecular weights (Mw), affecting its performance as structural or bioactive component [40]. The abovementioned properties of chitosan offer interesting bioactive capabilities, such as anticancer, antioxidant, and antiinflammatory [41–43]. It has been also applied in the development of biomaterials for different applications like tissue engineering and regenerative medicine, as well as the development of drug-delivery systems, due to its polyelectrolyte nature, being the only natural cationic polymer [44–46].

Chitosan is commonly solubilized in acidic media, by protonation of free amines in the deacetylated glucosamine units. However, its solubilization in water (neutral pH) is not possible, which have motivated the synthesis of several derivatives, by changing its Mw down to oligosaccharides and by functionalization with different chemical groups, with enhanced biological responses [41,47]. Wimardhani *et al.* studied the cytotoxic effects of two chitosans with distinct molecular weights over oral squamous cell carcinoma (SCC) and a noncancer cell line (HaCaT) [48]. Both chitosans induced the proliferation of HaCaT, although showing cytotoxic effects over two types of SCC cells, while another cell type was resistant to both chitosans. LMW chitosan presented increased cytotoxic effects when compared with high-molecular weight (HMW). In another study, the same authors analyzed the mechanism of action of LMW chitosan and concluded that cell cycle was arrested at G1/S and cells presented early apoptotic behavior [49]. The effects of water soluble LMW chitosan was assessed over HT-29 colon cancer cell line, showing induction of apoptosis by DNA fragmentation, where HMWC did not present cytotoxic effects [50]. High-molecular weight chitosan (HMWC) was hydrolyzed and different DD and Mw fractions of chitosan oligosaccharides (COS—hydrolyzed products of chitosan) were obtained and their toxicity evaluated over different cancer cells, namely A549 [51]. Increased inhibition of cells' proliferation was observed for the fractioned chitosan with higher DD. In a different attempt, the effects of COS were evaluated over tumor growth and the mechanism of action was analyzed [52]. Results point out that COS inhibited hepatocellular carcinoma cells proliferation, arresting cells cycle at S-phase and decreased DNA synthesis *in vitro*. COS also inhibited the activity of MMP in lung cancer cells, showing an antimetastatic potential, confirmed by *in vivo* studies were fewer lung colonies were observed. Other study reported the effects of COS to treat bladder tumors, showing inhibitory and preventive effects, depending on the dose concentration [53]. Authors related the anticancer effects with COS antiinflammatory behavior. Colon cancer cells (HT-29) were also treated with COS to assess its chemopreventive potential which was confirmed by quantifying the expression of different enzymes involved in carcinogenesis [54]. In another study, inhibition of MMP was verified when treated with COS, as well as inhibition of HT-29 cells' invasiveness, showing interesting results regarding the antimetastatic properties of COS [55]. The cytotoxic effects of COS were also evaluated over MDA-MB-231 breast cancer cells, showing a time-dependent toxicity and decrease in cells migration and invasion which was not related with cells' cytotoxicity. This is an important parameter regarding cancer progression and metastasis MMP-9 that was less expressed by cancer cells when treated with COS [56].

Dou *et al.* analyzed the different cellular and molecular mechanisms of oligochitosan over HL-60 (acute leukemia cell line). Results showed that chitosan induced apoptosis by arresting cell cycle at G0/G1 phase in a time and dose-dependent fashion [57]. Chitosan also changed some gene expression molecules involved in intrinsic and extrinsic pathways. For example, galectin-9 is involved in the apoptosis and plays a synergistic effect with oligochitosan. Taken together these results show the potential of chitosan to treat acute leukemia.

## 23.3 Drug-delivery systems

Despite chemotherapy positive results in the treatment of different cancers, there are some associated problems and adverse side effects. Due to drugs' poor bioavailability there is the need to administer them in increased dosages that can compromise the surrounding healthy tissues, having low-therapeutic outcomes and the development of drug resistance [58,59]. In this sense, the development of drug-delivery systems, particularly featuring targeting functionality may overcome some of these current limitations. The carriers will not only protect the encapsulated drugs from the host immune system, but will also be able to transport them specifically to the cancerous tissues, releasing the drug at specific sites, thus reducing side effects as well as the needed dose to exert a therapeutic effect [60]. An increasing knowledge on tumor biology and the use of new materials have led to the development of more efficient systems with improved therapeutic efficacy. Additionally, nanotechnology, by the development of different types of nanoparticles (NPs; being metallic, polymeric, liposomes, micelles, or hybrid), have showed interesting and effective results. They can be used to deliver a variety of therapeutic molecules and tend to accumulate in tumors sites due to a leaky vascularization site and active cellular uptake [61,62]. These systems also increase macromolecules' circulation, protecting drugs from enzymatic degradation and control the drug release, reducing cytotoxicity, and increasing therapeutic index. Nevertheless, these types of system may also present some drawbacks, since nanoparticles may be unstable during circulation and may be associated with uncertain toxicity (also related with accumulation in specific organs), especially for long-term administration [63–65]. The use of natural polymers to develop effective drug-delivery systems have been explored by different groups, investigating several strategies [64]. In this section, drug-delivery systems that have been developed based on marine biopolymers, particularly fucoidan, chitosan, carrageenan, and alginate will be described.

### 23.3.1 Fucoidan-based systems

Fucoidan has been included in nanosystems such as nanoparticles, micelles, and liposomes, for drug-delivery purposes in cancer treatment. Doxorubicin (DOX) have been incorporated into fucoidan/polyethylenimine nanoparticles showing increased toxicity

(MDA-MB-231) and greater cell cycle arrest in G1-S phase in comparison to free DOX [66]. Free DOX induced early and late apoptosis of 11.42% and 0.14% cells, respectively, whereas 66.12% and 0.30% were observed in groups treated with DOX-loaded-NPs. Lu *et al.* used the same chemotherapeutic drug to be encapsulated into protamine/fucoidan nanoparticles and the cytotoxicity was evaluated over the same breast cancer cells, showing increased toxicity when compared with free drug [67]. This higher inhibition of cell proliferation was attributed to different cellular pathways. Protamine/fucoidan NPs showed P-selection-mediated specific internalization along with charge and pH-responsive properties. Another important aspect is that free DOX enters the nuclei rapidly and then diffuses out of the cells quickly, whereas NPs allowed DOX accumulation in cytoplasm being released for longer time periods. Acetylated fucoidan nanoparticles loaded with DOX were produced by dialysis, and cytotoxicity was assessed over colon cancer cells [68]. Over 99% of the DOX released from the nanoparticles were internalized after 2 h, whereas only around 2% of DOX were internalized for fucoidan–DOX mixture and free drug.

In a different attempt to develop more effective breast cancer therapies, fucoidan/chitosan nanoparticles were developed encapsulating a different commercially available drug, gemcitabine [69]. The cytotoxic effects were evaluated not only over breast cancer cells but also over normal endothelial cells. The developed system showed an increased toxicity over breast cancer without further affecting endothelial cells, when compared with the free form of the drug. Two other works studied not only breast cancer but also cervical cancer and melanoma [70]. In the first attempt, fucoidan and methotretaxe nanoparticles were produced by self-assembling and tested over MCF-7 (breast cancer cell line) and HeLa cells (cervical cancer cells), presenting increasing toxicity when compared with the free drug. Cellular internalization assays showed that the nanoparticles are endocytosed by both cell lines, showing higher rate of internalization for HeLa cells. Fucoidan-based nanoparticles aiming on targeting P-selection (higher expressed in tumor cells than normal tissues) were developed by encapsulating different drugs and increased toxic effects over melanoma and breast cancer cells were observed when compared with free drugs and nanoparticles [71]. In another attempt, rutin (Ru), a hydrophobic polyphenolic flavonoid phytochemical with proven chemopreventive activity, has been complexed with fucoidan (Ru-Fu), presenting increased toxicity over HeLa cancer cells other than normal cells [72,73]. Apoptotic evidences were clearer for HeLa cells treated with Ru-Fu than for soluble forms of rutin and fucoidan (alone or in combination). The Ru-Fu anticancer activity was also confirmed by increased apoptotic percentages.

All previously developed systems were based on the release of chemotherapeutic drugs. However, there are reports of other studies were fucoidan was used as the encapsulated bioactive agent or as a coating for nanoparticles. Calcium carbonate microparticles coated with fucoidan and poly-L-ornithine were produced by layer-by-layer technique, followed by incubation with DOX [74]. Cytotoxic effects were increased for the developed system when comparing to free drug, triggering the induction of cell apoptosis. Lower internalization of DOX was observed for pure drug than when delivered by the nanoparticles. DOX was also loaded into gold nanoparticles coated with fucoidan showing an increased efficacy of the drug than its free counterpart [75]. Morphological changes were observed (e.g., membrane integrity loss, cell growth inhibition, cytoplasmic condensation) in

MDA-MB-231 cells treated with the nanoparticles when compared with untreated cells. The developed NPs induced both early and late apoptosis in a dose-dependent manner. DOX was incorporated into gold nanoparticles coated with fucoidan aiming to address eye tumors. The nanoparticles presented toxicity to cancer cells without affecting normal cells [76]. The efficacy of the system was validated *in vitro* and *in vivo*, showing increased toxicity when in combination with laser irradiation, with effectiveness around 90%. In another study, fucoidan was encapsulated into liposomes and the cytotoxic effects over osteosarcoma cells was increased when compared with free fucoidan *in vitro*, inducing apoptosis through activation of the caspase pathway [77]. *In vivo*, fucoidan nanoparticles were able to significantly reduce tumor size and weight. Shin *et al.* developed manganese dioxide nanoparticles coated with fucoidan and the combination with radiation therapy resulted in increased number of apoptotic cells [78]. *In vivo*, the combination of both treatments was also the most effective, resulting in reduced tumor sizes. Copper sulfide nanoparticles were coated with fucoidan and tested over different cancer cell lines like K562 (multidrug-resistant cells), HeLa (colon), and A549 (lung). In all cases, the nanoparticles delivered fucoidan intracellularly, resulting in cells apoptosis [79]. The combination of the system with near-infrared irradiation was effective *in vivo* by eliminating different tumors. A coating with chitosan/fucoidan was performed for silver nanoparticles, showing also toxicity over HeLa cells and increased apoptosis when compared with the untreated cells [73].

### 23.3.2 Chitosan-based systems

The biocompatibility and ability of chitosan to bind organic compounds have attracted tremendous interest over the last years, namely for applications in the pharmaceutical and biomedical fields. Its cationic nature makes it an optimal candidate to develop effective carriers for drug-delivery applications, since they can form different complexes through electrostatic interactions that can facilitate the cellular uptake by the cells [8,39]. Chitosan-based drug-delivery systems for cancer treatment have been extensively studied namely in the form of hydrogels, films and nanoparticles [80–82]. Nanoparticles are the most reported ones and will be further described in this section, for the release of different chemotherapeutic drugs. However, chitosan has also been used to develop nanoparticles that interact with nucleic acids or for gene delivery in cancer [83,84]. Its use as an immune-adjuvant for cancer vaccines has also been recently reported [85].

Chitosan-modified polymeric nanoparticles were developed for the delivery of different anticancer drugs, namely DOX. NPs enhanced cellular accumulation and cellular uptake of DOX in breast and colon cancer cells, thought modulation of specific transporters, when compared with the drug alone [86]. DOX-loaded NPs have been developed and their effects over liver cancer has been studied [87]. The nanoparticles promoted cells apoptosis and arrested cell cycle at G2/M phase and upregulated p53, inhibiting cell survival through this pathway. After *in vitro* production and characterization, the potential use of galactosylated chitosan/graphene oxide/doxorubicin (GC–GO–DOX) for the treatment of liver cancer was evaluated [88]. The galactosylation of chitosan improved cellular uptake and inhibited cells proliferation. *In vivo* results showed a decrease in tumor weight and size. In another recent study, DOX was loaded into chitosan-capped gold nanoparticles (CS-GNPs-DOX) and evaluated over MCF-7 breast cancer cells [89]. The NPs were able to enhance chemo-radiotherapeutic effect affecting cancer cells' viability, inducing necrosis at a very low-radiation dose (0.5 Gy). This synergistic effect between these two systems may be a promising approach to decrease current radiotherapy dosages, reducing the unwanted side effects. It is known that chemotherapy is associated with tumor reinitiating cancer stem-like cells (present $CD44^+$ receptors), being responsible for cancer recurrence. Rao *et al.* developed a DOX-encapsulated polymeric nanoparticle surface-decorated with chitosan that can specifically target this receptor [90]. The cytotoxicity of the DOX-loaded NPs was highly increased when compared with free drug, eliminating $CD44^+$ cancer stem-like cells, with the NPs being also able to reduce tumor size *in vivo*, without side effects over other tissues.

Chitosan nanoparticles loaded with paclitaxel (PTX) were produced [91]. Around 60% of the drug was released in the first 24 h and *in vitro* cytotoxicity was assessed over MDA-MB-231 breast cancer cells, showing increased toxicity when compared with free drug. Apoptosis studies also showed higher percentages of late apoptotic cells than unloaded nanoparticles and naïve drug, confirming the efficacy of the developed system. The effects of chitosan-based micelles loaded with PTX, in an attempt to overcome multidrug resistance (MDR), was validated by the inhibition of p-glycoprotein [92]. The developed micelles showed increased tumor accumulation and therapeutic efficacy *in vivo*. Yang *et al.* [93] proposed the development of estrone-modified glycol chitosan nanoparticles (GCNP-ES) loaded with PTX for the treatment of breast cancer. The developed system exhibited pH-responsiveness and the addition of estrone receptors showed higher internalization index [93]. PTX-loaded GCNP-ES also showed higher cytotoxic effects and cells apoptotic ratio than nonloaded NPs *in vitro* and higher accumulation of NPs at tumor site resulting in tumor shrinkage. A different approach to target lung cancer was developed by using integrin–PLGA–chitosan NPs loaded with PTX [94]. This system showed promising results for cancer treatments, since an increase in the uptake of the NPs was verified and apoptosis was enhanced inducing cell cycle arrest at G2/M phase. Another important aspect was that the cytotoxic effect over normal lung fibroblasts was decreased when compared with cancer cells, confirming its targeting capability. It was also reported that chitosan nanospheres loaded with PTX inhibited lung cancer cells proliferation and enhanced apoptosis [95].

For the last decades, 5-fluorouracil (5-FU) has been the first line chemotherapy in the treatment of various cancers, but there is the need to identify potential delivery systems for 5-FU to achieve better therapeutic efficacy with fewer side effects [96]. 5-FU was loaded into amino-functionalized mesoporous silica nanoparticle (MSN-$NH_2$)-based GCs [97]. Increased drug loading capacity and higher cytotoxicity over human colon cancer cells with 5-FU@MSN-$NH_2$ than with free 5-FU was observed, confirming cells apoptosis and cellular uptake. In a different attempt, 5-FU was loaded into chitosan-PEG-gelatin nanosystems showing its potential use for controlled and targeted therapies [98]. Chitosan nanospheres loaded with 5-Fu were also developed, inhibiting tumor cells' proliferation in a dose and time-dependent manner, as well as the adhesion of cancer to endothelial cells [99]. PTX-loaded chitosan nanoparticles were produced and their cytotoxicity was evaluated over a human ovarian cancer cell lines, showing increased toxicity, cellular uptake efficiency, and apoptotic cells when compared with free PTX [100].

Gemcitabine, a chemotherapeutic drug used to treat different cancers, namely breast and pancreatic cancer, has also been incorporated into chitosan-based systems. Chitosan-capped nanorods containing gemcitabine and further conjugated with a pH-sensitive peptide was evaluated in the treatment of pancreatic cancer, showing increased cytotoxicity when compared with free drug [101]. Chitosan coated iron oxide nanoparticles loading gemcitabine showed a pH related release profile and increased cytotoxicity and cellular uptake than free gemcitabine when tested over two different breast cancer cells [102]. Chitosan-loaded gemcitabine nanoparticles were produced and its effects were evaluated in nude mice inhibiting pancreatic cancer cells invasion and migration, promoting apoptosis and enhancing the efficacy of the chemotherapeutic agent [103].

### 23.3.3 Carrageenan-based systems

Carrageenan is a linear sulfated polysaccharide extracted from the extracellular matrix (ECM) of red marine algae of the Rhodophyceae family. The chemical structure of carrageenan is made of β-D-galactose and 3,6-anhydro-α-D-galactose units combined by glycosidic linkages and containing ester sulfate groups, partially sulfated [7,9,39]. Commercially, carrageenans are divided into three families, according to the number and position of sulfate groups: ι (iota), κ (kappa), and λ (lambda), bearing one, two, and three sulfate groups per disaccharide. The first two are gelling agents and the third is a thickening agent that is most commonly used in the biomedical field [7,9,39]. Recently, the utility of carrageenans in drug-delivery systems for cancer treatment has been demonstrated. Chen *et al.* synthesized a carrageenan oligosaccharide-gold nanoparticle (Car-AuNPs) that was used as a delivery system for pH-triggered delivery of epirubicin, a drug used for chemotherapy [104]. The cytotoxicity of Car-AuNPs, epirubicin (EPI), and epirubicin-loaded Car-AuNPs (EPI-Car-AuNPs) was evaluated using normal (HUVEC) and liver cancer cells (HepG2). Car-AuNPs did not alter the cell viability of normal cells, confirming that these NPs have a good biocompatibility. EPI-Car-AuNPs exhibited lower cytotoxicity over normal cells than the free EPI. Both EPI and EPI-Car-AuNPs exhibited a dose-dependent decrease in cell viability against HepG2 cells, and the inhibition of EPI-Car-AuNPs was more significant than the free EPI at the same dose. Furthermore, the EPI-Car-AuNPs induced cell apoptosis through cell cycle arrest at the G2/M phase in HepG2 cells.

In another study, hybrid nanostructured microparticles (hNPs) based on calcium carbonate and λ-carrageenan (λ-Car) derivatized with folic acid (FA) were designed to deliver DOX [105]. The cytotoxic effect of FA-λ-Car hNPs with or without DOX was evaluated in MG-63 osteoblastic human osteosarcoma-derived cell line. FA-λ-Car-DOX hNPs significantly decreased the cell viability of MG-63 cells, while FA-λ-Car did not. In addition, the cytotoxicity of FA-λ-Car-DOX was similar to the one induced by FA-λ-Car-DOX hNPs supernatants. More recently, a GO-grafted κ-carrageenan nanocarrier conjugated with biotin (GO-κ-Car-biotin) was synthesized to improve DOX delivery [106]. DOX-loaded GO-κ-Car-biotin nanocarrier significantly reduced cell viability of human cervical carcinoma cell line (HeLa) in a concentration-dependent manner, but only a slight decrease in the viability of human dermal fibroblast cell line was observed. Cells treated with DOX-loaded GO-κ-Car-biotin nanocarrier showed irregular-shaped nucleus, shrinkage of chromatin, scattered nuclear granules which suggests nuclear fragmentation. Cell proliferation significantly decreased and LDH leakage significantly increased in a concentration-dependent manner by DOX-loaded GO-κ-car-biotin.

Nanostructured lipid-carrageenan hybrid carriers (NLCCs) were exploited for controlled delivery of the chemotherapeutic agent mitoxantrone hydrochloride (MTO) [107]. The relative cellular uptake of MTO-NLCCs was significantly higher than for MTO solution (MTO-Sol) in MCF-7/MX cells overexpressing the breast cancer resistance protein (BCRP). Accordingly, the cytotoxicity of MTO was significantly enhanced in MTO-NLCCs compared with MTO-Sol. In contrast, no significant growth inhibition was found after incubation with the blank NLCCs, indicating the NLCCs were nontoxic as a vehicle. *In vivo* pharmacokinetics of MTO-Sol and MTO-NLCCs in rats demonstrated that the apparent bioavailability of MTO-NLCCs group significantly increased compared with the MTO-Sol group.

Curcumin has shown antiproliferative effect in multiple cancers and to overcome its poor solubility, Sathuvan *et al.* prepared curcumin loaded κ-carrageenan beads (κ-Car-Cur) for drug-delivery against lung cancer cells (A549) [108]. *In vitro* cell-based assays revealed that κ-Car-Cur composites induced higher cytotoxicity against cancer cells than free curcumin. Additionally, ROS production increased and mitochondrial membrane potential decreased during treatment with κ-Car-Cur, which suggests that this system triggered cellular apoptosis in A549 cancer cells.

### 23.3.4 Alginate-based systems

Alginate is a linear polysaccharide extracted from brown marine algae, composed of α-L-guluronic acid (G) and β-D-mannuronic acid (M) residues linked by 1,4-glycosidic linkages. It has been extensively investigated and used for biomedical applications because of its biocompatibility, low toxicity, relatively low cost, and mild gelation [9,109]. Although a large number of alginate-based delivery systems have been developed in recent years, this section will focus on alginate-based NPs used as delivery vehicles for chemotherapeutic agents, such as DOX and 5-FU. The potent antitumor activity of DOX is accompanied by severe tissue toxicities such as cardiotoxicity and, to try to overcome this issue, several alginate-based NPs have emerged. Prabha and Raj designed sodium alginate–polyvinyl alcohol–bovine serum albumin coated $Fe_3O_4$ NPs as a drug-delivery system for DOX, which were toxic to liver cancer cells (HepG2) and nontoxic to normal liver cells (L02) [110]. Peng *et al.* fabricated alginate modified superparamagnetic iron oxide nanoparticles (SPIONs) with magnetic targeting capability for pH-responsive release of DOX in tumor–cell microenvironments [111]. SPIONs exhibited good cytocompatibility toward liver cancer and normal cells, but DOX-loaded SPIONs showed higher cytotoxicity against HepG2 cells than against L02 cells. *In vivo*, DOX-loaded nanocarriers under an external magnetic field could increase the DOX concentration in tumor tissues, leading to a remarkable inhibition of tumor growth and a significantly reduction of side effects of free DOX. DOX was also covalently conjugated to oxidized sodium alginate to produce an amphiphilic macromolecular prodrug. The prodrug was subsequently self-assembled into nanoparticles (DOX-NPs) in an aqueous solution, which were responsive to the acidic environment in tumor cells [112]. *In vitro* cytotoxicity studies of the DOX-NPs showed an inhibitory effect on the growth of breast cancer cells (MCF-7), but not on normal breast cells (MCF-10A). *In vivo* studies in zebrafish further confirmed an efficient absorption of DOX-NPs. Furthermore, DOX-NPs exhibited an improved cardiotoxicity profile in comparison with free DOX. In another study, natural polyelectrolyte multilayers of chitosan (CHI) and alginate (ALG) were alternately deposited on DOX-loaded poly(lactic-*co*-glycolic acid) (PLGA) NPs [113]. DOX-PLGA (CHI/ALG) NPs exhibited the best antitumor efficiency (83.17%), compared with DOX-PLGA NPs (61.35%) and DOX in solution (23.77%). In a different approach, alginate-based micelles were prepared using alginate-*g*-poly(*N*-isopropylacrylamide) (PNIPAAm) and used to encapsulate DOX [114]. These micelles accumulated at the tumor site of a tumor-bearing mouse model as a result of the enhanced permeability and retention effect. Interestingly, DOX-loaded alginate-*g*-PNIPAAm micelles showed excellent anticancer therapeutic efficacy in a mouse model without any significant side effects.

DOX has been frequently used in combination chemotherapy regimens, namely with PTX. In recent years, a few studies have focused on the development of alginate-based delivery systems of PTX alone or in combination with DOX. One of those studies described that ALG-based NPs containing PTX inhibited cellular proliferation, promoted cell-cycle arrest, and induced apoptosis in human primary breast cancer cells [115]. ALG-coated chitosan hollow nanospheres designed to serve as a carrier for codelivering DOX and PTX, effectively inhibited cell proliferation, and promoted cell apoptosis in A549 lung cancer cells [116]. Other authors compared the cytotoxicity of PTX-loaded ALG NPs, PTX-loaded chitosan–ALG NPs, and PTX-loaded folate-chitosan–ALG NPs against liver cancer cells (HepG2) [117]. It was observed that all tested NPs have effective cellular inhibition properties. Inorganic/organic hybrid ALG/$CaCO_3$ NPs were used to encapsulate DOX and PTX [118]. Both DOX-loaded NPs and PTX-loaded NPs could inhibit cell growth more efficiently than free DOX or PTX, respectively. Moreover, PTX/DOX coloaded NPs exhibited the highest growth inhibition of HeLa cervical cancer cells.

Several alginate-based NPs have been created in an attempt to accomplish site specific and controlled delivery of 5-FU. Polymeric hollow spheres assembled from sodium alginate-*graft*-poly(*N*-isopropylacrylamide) (ALG-*g*-PNIPAM) and β-cyclodextrin (β-CD) were prepared for controlled release of 5-FU [119]. The cell viability of Hela cervical cancer cells declined in a concentration-dependent manner after incubation with 5-FU-loaded particles. Another study reported the development of a GO-based, sodium alginate (ALG) functionalized colon-targeting drug-delivery system, that was loaded with 5-FU [120]. The as-prepared delivery system demonstrated to possess a much lower toxicity and better colon-targeting controlled-release behaviors. GO-ALG/5-FU system significantly inhibited tumor growth and liver metastasis and prolonged the survival time of mice.

## 23.4 Three-dimensional *in vitro* models of cancer

Drug discovery and development in oncology is a very lengthy and costly process with a high-failure rate. A major obstacle for efficient drug discovery is the absence of predictive cell-based assays capable of mimicking physiological conditions, namely tumor initiation, progression, metastasis, and treatment resistance. Typically, cell-based approaches employ two-dimensional (2D) culture systems that do not faithfully represent the complex *in vivo* tumor microenvironment (Fig. 23.2) [121,122]. Conventional cell cultures are grown on flat and rigid surfaces, optimized for growth and cell attachment, and under unrealistic concentration gradients of oxygen, nutrients, growth factors, and treatment agents. More importantly, 2D cell cultures lack the cell–cell and cell–ECM interactions that are present in native tumors and are crucial to the regulation of several cellular functions, including cell proliferation, differentiation, survival, apoptosis, gene and protein expression, and drug metabolism [123,124]. In addition, most conventional and traditional *in vitro* cancer models involve a single cell population (monoculture), disregarding the fact that cancer cells interact with other cells present in the tumor microenvironment such as stromal and immune cells. Thus, it is inappropriate to predict cancer cell behaviors *in vivo* by just considering the response of cancer cells themselves *in vitro* [125,126].

To overcome the limitations associated with 2D cell cultures, novel *in vitro* cancer models, more predictive of *in vivo* drug efficacy, are a current hot-topic of research. Three-dimensional (3D) cell culture systems have shown the potential to recapitulate the natural tumor microenvironment and thus emerged as an alternative method for drug screening [121,127]. The 3D cell culture systems are often classified as scaffold-free/anchorage independent, scaffold-based/anchorage dependent, as well as hybrid 3D culture systems in which formed spheroids are incorporated into scaffolds (Fig. 23.3) [121,127]. Scaffold-free

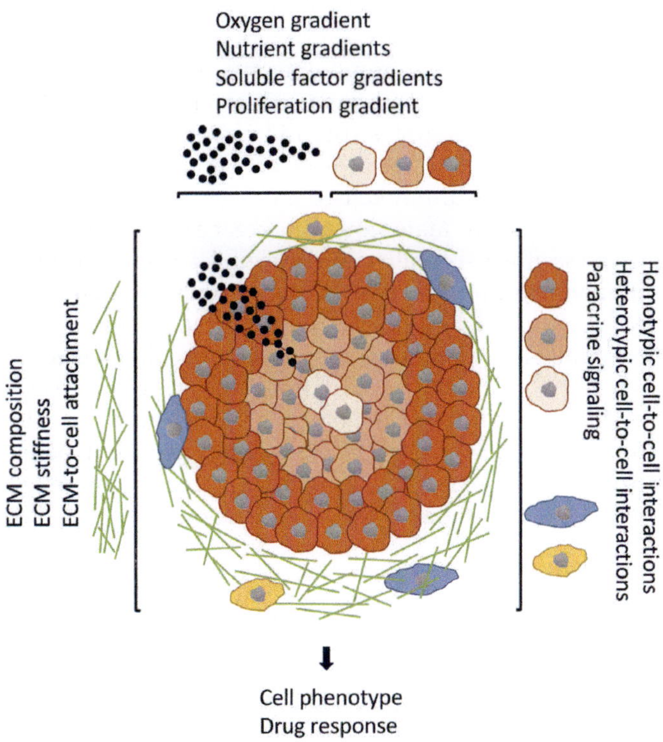


**FIGURE 23.2** Schematic representation of the tumor microenvironment. Cancer cells (orange) encounter a complex microenvironment consisting of extracellular matrix (ECM) proteins and glycoproteins (green), support cells that mediate cell–cell interactions (blue), immune cells (yellow), and soluble factors (black spheres). The tissue microenvironment is further defined by physical factors such as ECM stiffness (indicated by increasing density of ECM proteins), and oxygen (indicated by orange shading of cancer cells) and nutrient and growth factor gradients (indicated by density of black spheres). Source: *Adapted from Langhans SA. Three-dimensional in vitro cell culture models in drug discovery and drug repositioning. Front Pharmacol 2018;9:6 [121].*

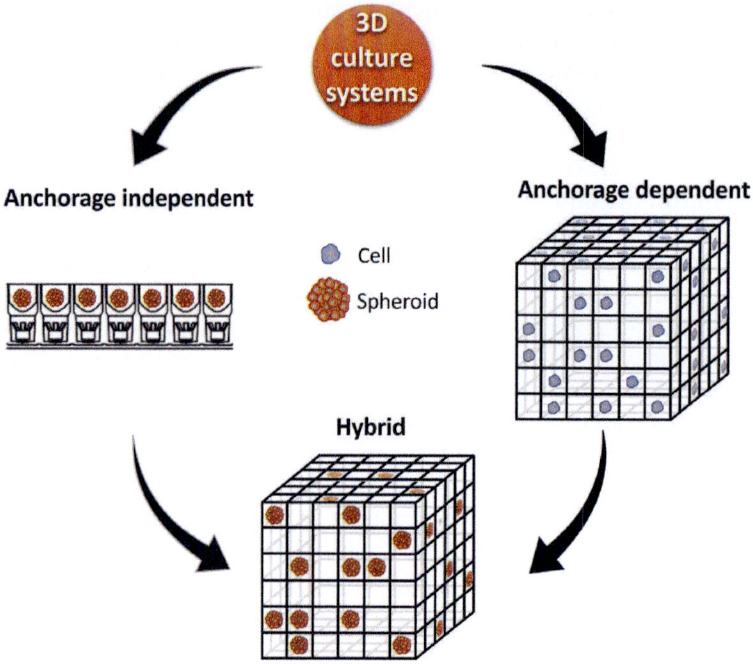


**FIGURE 23.3** Types of 3D cultures. Cells may form spheroids (scaffold-free/anchorage independent systems) or can be encapsulated into scaffolds (scaffold-based/anchorage dependent systems). Spheroids can be embedded in ECM scaffolds, forming hybrid culture systems that combine the advantages of both systems to form a complex microenvironment for 3D cell culture. Source: *Adapted from Langhans SA. Three-dimensional in vitro cell culture models in drug discovery and drug repositioning. Front Pharmacol 2018;9:6 [121].*

3D culture techniques allow the cells to self-assemble forming nonadherent cell aggregates called spheroids. Spheroids, in particular multicellular spheroids, recapitulate physiological characteristics of tumors with regard to cell–cell and cell–ECM interactions [128,129]. Scaffold-based 3D culture systems provide physical support on which cells can aggregate, proliferate, and migrate. In these cultures, cells are embedded into the matrix, and the chemical and physical properties of the scaffold material influence cell characteristics [130,131]. Scaffold materials can either be synthetic or natural in origin. Although synthetic materials typically display better mechanical properties, natural materials have been more extensively used due to their generally observed biocompatibility, biodegradability, and resemblance to the ECM [132,133]. In addition, natural polymers can be engineered to obtain the desirable mechanical and physical properties [134,135].

Several compounds and materials from marine sources have been employed for the *in vitro* development of 3D tumor models [127]. Among marine materials, marine polysaccharides have been receiving increasing attention from researchers and, for the purposes of this book chapter, we will focus on chitosan-based and alginate-based 3D tumor models.

## 23.4.1 Chitosan-based cancer models

Chitosan, as previously stated, is chemically similar to glycosaminoglycans, the major components of tumor ECM [44,136]. Dhiman *et al.* reported the preparation of 3D porous chitosan scaffolds with different degrees of deacetylation for *in vitro* culture of MCF-7 human breast cancer cells [137]. Chitosan scaffold having high degree of deacetylation (80%) favored better adsorption and improved cell growth kinetics. The authors confirmed that MCF-7 cells grown on the chitosan scaffold produce more lactate from glucose in comparison to cells grown on tissue culture plates, suggesting that cells grown on chitosan scaffolds can provide better information on metabolic activities of breast cancer cells than 2D culture. Moreover, higher tamoxifen concentrations were required to achieve comparable cytostatic action in 3D culture, supporting that 3D culture is a better model for the cytotoxic evaluation of anticancer drugs *in vitro* [138].

Chitin-based biomaterials do not provide a friendly interface for the adhesion of specific cell types. [139–141]. Therefore, chitosan-coated surfaces were used to develop 3D multicellular spheroids without additional substrate modification, driving force or 3D scaffold design. Tsai *et al.* demonstrated that colon cancer cells (SW620), mesenchymal stem-like cells (3A6) and foreskin fibroblasts (Hs68) cocultured on chitosan were able to aggregate, forming multicellular spheroids with a core/shell cell distribution [128]. In addition, these 3D coculture multicellular spheroids were able to enhance the resistance of SW620 cells against the cytotoxic effect of chemotherapy drugs.

Chitosan substrates grafted with hyaluronan (HA) also promoted the formation of spheroids made of mesenchymal stem cells (MSCs), which play a crucial role during cancer development [142]. Non-small cell lung cancer cells (A549 and H1299) cultured in chitosan–HA surfaces also formed tumor spheroids, and revealed more cancer stem cell-like properties and epithelial–mesenchymal transition (EMT) markers [143]. Furthermore, cells in tumor spheroids exhibited enhanced invasive activities and MDR. Since the crosstalk between MSCs and the tumor microenvironment could contribute to their tumor-promoting activities, MSCs and A549 lung cancer cells were cocultured on chitosan–HA substrates in order to functionally mimic the tumor microenvironment [144]. The expression levels of the markers related to stemness, EMT, and mobility were upregulated in MSCs-tumor spheroids, indicating that tumorigenicity of cancer cells was enhanced by their aggregation with MSCs.

In another attempt, Florczyk *et al.* developed a chitosan–HA scaffold that promoted tumor spheroid formation and increased stem-like properties of glioblastoma cells (U-118 MG) compared with 2D cultures [130]. Additionally, the invasiveness and drug resistance of glioblastoma cells cultured in this scaffold was significantly enhanced compared with those grown in 2D cultures. Xenograft-derived glioblastoma cells cultured in the same chitosan–HA scaffold had higher expression levels of genes related to EMT and stemness [136]. Scaffold-grown cells also have greater resistance to clinically used alkylating agents when compared with cells cultured on monolayers. More recently, this chitosan–HA scaffold promoted the formation of spheroids of glioma cells (U-87) and increased the expression of stemness and EMT-related genes [145]. Furthermore, U-87 glioma cells derived from the scaffold had a significantly higher tumorigenicity in BALB/c mice than 2D cells.

Other biopolymers were also combined with chitosan for the production of 3D scaffolds with improved properties. For example, Arya *et al.* described a chitosan–gelatin scaffold that promoted spheroid formation of different cancer cells, including non-small cell lung cancer (NCI-H460), cervical cancer (HeLa), breast cancer (MCF-7), and osteosarcoma (MG-63) cells [146]. In another study, a chitosan–gelatin scaffold promoted growth and proliferation of MCF-7 cells, which exhibited characteristics of higher malignancy and resistance to chemotherapy compared with their 2D counterparts [147]. Furthermore, cells grown in the scaffold were found to be more tumorigenic and angiogenic in BALB/c mice xenografts than cells grown in 2D models. Chitosan–fibroin scaffolds may also be an alternative for 3D culture, since A549 lung cancer cells cultured in the scaffold showed *in vivo* tumor-like morphological and biological characteristics and displayed better invasion ability and drug resistance than in 2D models [148]. Additionally, it was observed differences in drug-resistant and immune-related protein levels, which indicated that 3D models might resemble the real-life situation.

Chitosan is only soluble in acidic solutions, which restricts its applications. Recently, there has been a growing interest in the chemical modification of chitosan to improve its solubility in other media, namely in water, thus widening its applications. Wang *et al.* [149] described, for the first time, a method to produce a 3D tumor model using a hydroxyethyl chitosan/glycidyl methacrylate (HECS–GMA) hydrogel produced by a series of chitosan modifications [149]. MCF-7 breast cancer cells cultured in HECS–GMA hydrogels showed a growth status that mimicked the *in vivo* situation better than 2D cultures. In addition, HECS–GMA hydrogels provide an improved platform to assess the efficacy of antiangiogenic drugs such as Endostar and Bevacizumab.

### 23.4.2 Alginate-based cancer models

Despite the absence of cell adhesion sites, most cell lines are able to grow in alginate-based matrices. Several human tumor cell lines have been cultured in alginate-based systems (over a matrix or encapsulated in a gel), representing the *in vivo* environment better than the 2D models (Table 23.2) [121]. For instance, human colorectal cancer (CRC) cells (HCT116) and their corresponding isogenic 5-FU-chemoresistant derived cells (HCT116R) encapsulated in alginate beads were able to proliferate, aggregate and migrate from the alginate environment, and form CRC metastases [150,151]. Indeed, the expression of proliferation- and metastasis-related proteins were significantly increased in alginate-cultured cells compared with the 2D adhered cells. Furthermore, HCT116R cells overexpressed more proteins than HCT116 cells, indicating an increase in cancer malignancy. Other CRC cells (RKO and SW480) were also successfully cultured in the same alginate-beads microenvironment [152]. The experimental results indicate that 3D alginate beads are more representative of the tumor microenvironment than the 2D cell culture. Another report compared the adhesion and other phenotypic characteristics of HCT116 tumor cells cultured in alginate and alginate–gelatin hydrogels [131]. It was concluded that the alginate–gelatin hydrogel is a more suitable 3D system than pure alginate and seems to optimally mimic the physiological behavior of the tumor microenvironment.

Lung tumors, namely the most aggressive solid tumors, have high levels of myeloid cell infiltration, namely tumor-associated macrophages. In an attempt to recapitulate key hallmarks of lung cancer immune microenvironment, a research group developed a 3D culture platform based on alginate microencapsulation and stirred culture systems [153]. This platform enables the coculture of tumor cell spheroids of non-small cell lung cancer cells, cancer-associated fibroblasts and monocytes. It was reported that this 3D model recreated an invasive and immunosuppressive tumor microenvironment, with accumulation of cytokines/chemokines, ECM elements and MMPs, supporting cell migration and

**TABLE 23.2** Alginate-based 3D cell culture models.

| Cancer | Cells | 3D culture systems | References |
|---|---|---|---|
| Colorectal | HCT116 cells<br>RKO cells<br>SW480 cells | Alginate beads | [150–152] |
| | HCT116 | Alginate–gelatin hydrogel | [131] |
| Lung | NCI-H157 cells + lung-derived cancer-associated fibroblasts + THP-1 cells | Alginate beads | [153] |
| | H1650 cells + human dermal fibroblasts<br>H157 cells<br>H460 cells | Alginate hydrogel | [129] |
| Breast | MCF-7 cells + human dermal fibroblasts | Alginate beads | [154] |
| | MDA-MB-231 cells + human mammary fibroblasts | Collagen and alginate hydrogel | [155] |
| | MDA-MB-231 cells + IMR-90 cells | Alginate/gelatin hydrogel | [156] |
| | MCF-7 cells + human dermal fibroblasts<br>HCC1954 cells | Alginate hydrogel | [129] |
| Hepatocellular | MHCC97L cells<br>HCCLM3 cells | Alginate beads | [157] |
| | Hep3B cells | Cell encapsulation in alginate | [158] |
| Brain | U-87 cells<br>SU3 cells | Gelatin/alginate/fibrinogen hydrogel | [159] |
| Non-Hodgkin lymphoma | SUDHL-10 cells +<br>HS-5 cells +<br>Peripheral blood mononuclear cells | Alginate and puramatrix hydrogel | [160] |
| Stomach | MKN74 cells +<br>NST20 cells | RGD-modified alginate hydrogel | [161] |
| Prostate | C4-2 cells + WPMY-1 cells | Double-layered alginate hydrogel microspheres | [162] |
| Cervical | HeLa cells | Gelatin, alginate, and Matrigel hydrogel | [163] |
| Ovarian | Human fallopian fimbriae | Alginate hydrogel | [164] |

promoting cell–cell interactions within the alginate microcapsules. Recently, other researchers applied a low-temperature molding principle of biological manufacturing and 3D bioprinting to fabricate a cell-laden hydrogel-like scaffold, using an alginate–gelatin–lung cancer cells (A549/95-D) suspension as the bioink. The 3D printed cells had enhanced invasion and migration, indicating that the developed *in vitro* model was more biomimetic when compared to 2D models [153].

In order to develop an appropriate *in vitro* model for breast cancer research, Estrada *et al.* combined alginate microencapsulation with agitation-based culture systems, to recapitulate and monitor key aspects of the tumor microenvironment and disease progression [154].

Human breast cancer cells (MCF-7) aggregates were microencapsulated into alginate, either alone or in combination with human fibroblasts. In cocultures occurred a series of phenotypic alterations, typical of advanced stages of cancer, that were not observed in the monocultures of MCF-7 cells. Other researchers constructed a 3D tumor invasion platform, using a hybrid hydrogel composed of collagen and alginate [155]. Multicellular tumor spheroids composed of human breast cancer cells (MDA-MB-231) and mammary fibroblasts were embedded into the collagen/alginate hydrogel, successfully mimicking an early stage of breast tumor invasion, where breast cancer cells invade out of spheroids following mammary fibroblasts into the surrounding gel. In a similar study, MDA-MB-231 cells and IMR-90 stromal fibroblast cells were embedded into a cross-linked alginate/gelatin matrix using 3D bioprinting [156]. The human breast cancer cells formed spheroids and the stromal fibroblast cells migrate through the hydrogel matrix and infiltrate the multicellular tumor spheroids. These mixed MDA-MB-231/IMR-90 multicellular tumor spheroids can be maintained for longer periods of time.

Metastasis resistant to therapy is the major cause of death from hepatocellular carcinoma. Thus, the mechanisms underlying progression and metastasis of this cancer were investigated in a 3D culture system of metastatic hepatocellular carcinoma cells (MHCC97L and HCCLM3) using alginate beads [157]. In this 3D system, cancer cells cultured into alginate beads showed a higher metastatic ability compared with cells cultured in 2D culture systems. The authors advocated that due to the simplicity, large scalability and reproducibility, this metastatic model would be suitable for the molecular analysis of the mechanism of metastasis and the high-throughput screening of antimetastatic drugs. Recently, 3D forming and preformed sphere models of hepatocellular carcinoma cells (Hep3B) encapsulated into alginate and cultured using a micropillar and microwell chip platform were established to test drug efficacy [158]. The authors measured the efficacy of 60 drugs, which were already approved by the Food and Drug Administration (FDA) and whose targets are well known, and sorted the most efficacious drugs by comparing the drug response of the 2D monolayer model with the 3D models. Among the 60 drugs, 45 drugs showed high efficacy in the 2D model, while only 17 drugs showed high efficacy in the 3D preformed sphere models. Thus, a drug efficacy test based on the 3D preformed sphere model could reduce the possible drug candidates to select highly efficacious drugs in drug development or drug repositioning, which would have a tremendous effect on the reduction of the associated time and costs.

Concerning brain cancer, a 3D bioprinted glioma stem cell model was established using a gelatin/alginate/fibrinogen hydrogel [159]. This novel platform mimics the brain tumor microenvironment, with high-cell viability and inherent characteristics for long-term *in vitro* culture, not only expressing typical biomarkers, but also revealing differentiation and vascularization potential. In addition, this 3D bioprinted glioblastoma model showed more resistance to chemotherapeutics compared with 2D model. Hence, this model provides a novel alternative tool for studying gliomagenesis, glioma stem cell biology, drug resistance, and anticancer drug susceptibility *in vitro*.

In addition to the abovementioned, many others 3D *in vitro* culture models were successfully established using alginate alone or combined with other polymers, to mimic the *in vivo* tumor microenvironment of different cancers (Table 23.2), including non-Hodgkin lymphoma [160], stomach [161], prostate [162], cervical [163], and ovarian [164] cancers. Notably, agitation-based technology allowed the generation of 3D alginate-based cell culture models from a list of tumor cell lines representative of various cancer types [129].

### 23.4.3 Chitosan-alginate-based cancer models

The combination of chitosan and alginate to produce structures to be used in the biomedical field has been extensively explored [39]. Chitosan-alginate (ChiAlg) scaffolds have been shown to be a reliable model for replicating the natural ECM and tumor microenvironment of various cancers, including glioblastoma, prostate cancer, and hepatocellular carcinoma [165–169]. Kievit *et al.* investigated the possibility of a ChiAlg scaffold serve as an extracellular matrix that promotes the conversion of cultured glioma cells into a more malignant *in vivo*-like phenotype[165-167]. Initially, this research group showed that human glioma cells (U-87 MG and U-118 MG) were able to proliferate within the ChiAlg scaffolds and secrete factors that promote tumor malignancy [165]. Furthermore, U-87 MG cells cultured into the ChiAlg scaffold and further implanted into nude mice displayed an accelerated tumor growth and improved ability for angiogenesis. In another study, it was reported that ChiAlg scaffolds promoted the proliferation and enrichment of the cancer stem-like cell population (CD133 positive) from glioblastoma cell lines [166]. The constructs further implanted into nude mice formed tumors and displayed increased expression of stem cell related genes. Tumor microenvironment contains a variety of stromal cells that can both affect tumor growth, progression, and stem-like state. Accordingly, Kievit *et al.* reported that U-87 MG cells cocultured with either human astrocytes or endothelial cells (HUVEC) in ChiAlg scaffolds resulted in slower growth of tumor spheroids and increased expression of stem-like cancer cell markers [167]. These works provide strong evidence that it is possible to adjust the *in vitro* tumor microenvironment to better mimic *in vivo* conditions.

The use of ChiAlg scaffolds for culturing human prostate cancer cells (LNCaP, C4-2, and C4-2B) and studying tumor cell interaction with human peripheral blood lymphocytes (PBLs) ex vivo was also proposed [168]. ChiAlg scaffolds supported *in vitro* tumor spheroid formation and the interaction of PBLs with prostate cancer tumors. This study also demonstrated that the use of 3D ChiAlg scaffolds as a culture model supports long-term analysis of interaction of prostate cancer tumor cells with immune cells, providing an *in vitro* platform for rapid immunotherapy development. Leung *et al.* showed that human hepatocellular carcinoma cells [PLC/PRF/5(PLC) and HepG2] grown on ChiAlg scaffolds demonstrated the typical morphology and increased expression of characteristic markers of highly malignant cells [169]. Resistance to DOX was significantly pronounced in ChiAlg scaffold cultured hepatocellular carcinoma cells compared with 2D cultured cells. Furthermore, implantation of ChiAlg scaffolds cultured with hepatocellular carcinoma cells in mice showed accelerated tumor growth. ChiAlg cultured cells induced the formation of large, well rounded blood vessels with well-defined endothelial linings, as opposed to 2D cultured cells.

Altogether, these findings suggest that chitosan and/or alginate-based 3D culture systems can provide a reliable and cost-effective platform to study tumor biology and drug discovery.

## 23.5 Conclusions

The herein described marine biopolymers, besides being extensively studied for many biomedical applications, show promising results as anticancer bioactive agents, particularly the case of fucoidan and chitosan. Nevertheless, there are still some difficulties in establishing a general mechanism of action of such biopolymers, which should be further explored in structure–activity relationship studies. The development of drug-delivery systems, namely the production of nanoparticles, have been extensively reported for the described polysaccharides (fucoidan, chitosan, carrageenan, and alginate), providing increased efficacy to chemotherapeutic drugs. Therefore, effective drug-delivery systems are being proposed, aiming to overcome the limitations and side effects of current cancer treatments. Having this in mind, besides testing the systems over cancer cells, cytotoxic studies over noncancerous cells should be performed, an important aspect that some papers still miss. In addition, the development of 3D systems offers another possibility to study the abovementioned problems, aiming to address the complexity of the tumor microenvironment. Chitosan and/or alginate 3D *in vitro* models have shown promising results. However, substantial progress is still needed to improve the physiological complexity of the model, including cell type, matrix composition, temporal, and spatial presentation of soluble factors.

## 23.6 Acknowledgments

This work was developed under the scope of the Structured projects for R&D&I NORTE-01-0145-FEDER-000021 and NORTE-01-0145-FEDER-000023 supported by the Northern Portugal Regional Operational Programme (NORTE 2020), and the Mobilizing Project ValorMar, POCI-01-0247-FEDER-024517, supported by the Operational Programme for Competitiveness and Internationalization (COMPETE 2020), under the Portugal 2020 Partnership Agreement, cofunded by European Regional Development Fund (ERDF). The authors would like also to thank Norte 2020, for financing the PhD scholarship of C.O. "Norte-08–5369-000037" and FCT for the Investigator grant of A.M. (IF/00376/2014).

## References

[1] Siegel RL, Miller KD, Jemal A. Cancer statistics, 2019. CA Cancer J Clin 2019;69(1):7–34.

[2] Miller KD, et al. Cancer treatment and survivorship statistics, 2016. CA Cancer J Clin 2016;66(4):271–89.

[3] Baskar R, et al. Cancer and radiation therapy: current advances and future directions. Int J Med Sci 2012;9 (3):193–9.

[4] Nurgali K, Jagoe RT, Abalo R. Editorial: adverse effects of cancer chemotherapy: anything new to improve tolerance and reduce sequelae? Front Pharmacol 2018;9:245.

[5] McCune JS. Rapid advances in immunotherapy to treat cancer. Clin Pharmacol Therap 2018;103(4):540–4.

[6] Balkwill FR, Capasso M, Hagemann T. The tumor microenvironment at a glance. J Cell Sci 2012;125(Pt 23):5591–6.

[7] Silva TH, et al. Marine algae sulfated polysaccharides for tissue engineering and drug delivery approaches. Biomatter 2012;2(4):278–89.

[8] Gomathi T, Sudha PN, Venkatesan J, Sukumaran A. Marine biopolymers for anticancer drugs. In: Sudha PN, editor. Industrial Applications of Marine Biopolymers. Boca Raton, FL: CRC Press; 2017.

[9] Silva TH, et al. Materials of marine origin: a review on polymers and ceramics of biomedical interest. Int Mater Rev 2012;57(5):276–306.

[10] Hamed I, Ozogul F, Regenstein J. Industrial applications of crustacean by-products (chitin, chitosan, and chitooligosaccharides): a review. Trends Food Sci Technol 2016;48:40–50.

[11] Huang R-M, et al. Marine nucleosides: structure, bioactivity, synthesis and biosynthesis. Mar drugs 2014;12 (12):5817–38.

[12] Jo C, et al. Marine bioactive peptides: types, structures, and physiological functions. Food Rev Int 2017;33(1):44–61.

[13] Malve H. Exploring the ocean for new drug developments: marine pharmacology. J Pharm Bioallied Sci 2016;8:83–91.

[14] Lindequist U. Marine-derived pharmaceuticals - challenges and opportunities. Biomol Ther (Seoul) 2016;24 (6):561–71.
[15] Atashrazm F, et al. Fucoidan and cancer: a multifunctional molecule with anti-tumor potential. Mar Drugs 2015;13(4):2327.
[16] Fitton JH, Stringer DN, Karpiniec SS. Therapies from fucoidan: an update. Mar Drugs 2015;13(9):5920–46.
[17] Ale MT, Meyer AS. Fucoidans from brown seaweeds: an update on structures, extraction techniques and use of enzymes as tools for structural elucidation. RSC Adv 2013;3(22):8131–41.
[18] Oliveira C, et al. The key role of sulfation and branching on fucoidan antitumor activity. Macromol Biosci 2017;17(5):1600340.
[19] Kyung J, et al. Synergistic anti-inflammatory effects of Laminaria japonica fucoidan and Cistanche tubulosa extract. Lab Anim Res 2012;28(2):91–7.
[20] Thuy TTT, et al. Anti-HIV activity of fucoidans from three brown seaweed species. Carbohydr Polym 2015;115:122–8.
[21] Ustyuzhanina NE, et al. Fucoidans: pro- or antiangiogenic agents? Glycobiology 2014;24(12):1265–74.
[22] He X, et al. Fucoidan promotes apoptosis and inhibits EMT of breast cancer cells. Biol Pharm Bull 2019;42 (3):442–7.
[23] Lu J, et al. Fucoidan extracted from the New Zealand Undaria pinnatifida—physicochemical comparison against five other fucoidans: unique low molecular weight fraction bioactivity in breast cancer cell lines. Mar Drugs 2018;16(12):E461.
[24] Aisha A, et al. Effects of fucoidan and chemotherapeutic agent combinations on malignant and non-malignant breast cell lines. Curr Pharm Biotechnol 2017;18(9):748–57.
[25] Xue M, et al. Caspase-dependent and caspase-independent induction of apoptosis in breast cancer by fucoidan via the PI3K/AKT/GSK3β pathway in vivo and in vitro. Biomed Pharmacother 2017;94:898–908.
[26] Malyarenko OS, et al. Radiosensitizing effect of the fucoidan from brown alga Fucus evanescens and its derivative in human cancer cells. Carbohydr Polym 2019;205:465–71.
[27] Kim I, Nam T. Fucoidan downregulates insulin-like growth factor-I receptor levels in HT-29 human colon cancer cells. Oncol Rep 2018;39(3):1516–22.
[28] Usoltseva RV, et al. Comparison of structure and in vitro anticancer activity of native and modified fucoidans from Sargassum feldmannii and S. duplicatum. Int J Biol Macromol 2019;124:220–8.
[29] Kim I-H, Kwon M-J, Nam T-J. Differences in cell death and cell cycle following fucoidan treatment in high-density HT-29 colon cancer cells. Mol Med Rep 2017;15(6):4116–22.
[30] Somasundaram SN, et al. Cytotoxic effect of fucoidan extracted from Sargassum cinereum on colon cancer cell line HCT-15. Int J Biol Macromol 2016;91:1215–23.
[31] Hsu H-Y, et al. Fucoidan upregulates TLR4/CHOP-mediated caspase-3 and PARP activation to enhance cisplatin-induced cytotoxicity in human lung cancer cells. Cancer Lett 2018;432:112–20.
[32] Chen H, et al. Sulfated fucoidan FP08S2 inhibits lung cancer cell growth in vivo by disrupting angiogenesis via targeting VEGFR2/VEGF and blocking VEGFR2/Erk/VEGF signaling. Cancer Lett 2016;382 (1):44–52.
[33] Alwarsamy M, Gooneratne R, Ravichandran R. Effect of fucoidan from *Turbinaria conoides* on human lung adenocarcinoma epithelial (A549) cells. Carbohydr Polym 2016;152:207–13.
[34] Hsu H-Y, et al. Fucoidan induces Toll-like receptor 4-regulated reactive oxygen species and promotes endoplasmic reticulum stress-mediated apoptosis in lung cancer. Sci Rep 2017;7:44990.
[35] Huang T-H, et al. Prophylactic administration of fucoidan represses cancer metastasis by inhibiting vascular endothelial growth factor (VEGF) and matrix metalloproteinases (MMPs) in Lewis tumor-bearing mice. Mar Drugs 2015;13(4):1882–900.
[36] Kwak J-Y. Fucoidan as a marine anticancer agent in preclinical development. Mar Drugs 2014;12(2):853.
[37] Malyarenko OS, Ermakova SP. Chapter 10 - Fucoidans: anticancer activity and molecular mechanisms of action. In: Venkatesan J, Anil S, Kim S-K, editors. Seaweed polysaccharides. New York: Elsevier; 2017. p. 175–203.
[38] Van Weelden G, et al. Fucoidan structure and activity in relation to anti-cancer mechanisms. Mar Drugs 2019;17(1):E32.
[39] Martins A, da Silva MA, Costa-Pinto AR, Reis RL, Neves NM. Bio-inspired integration of natural materials. In: Brennan AB, Kirschner CM, editors. Bio-inspired materials for biomedical engineering. New York: Wiley Library; 2014.

[40] Younes I, Rinaudo M. Chitin and chitosan preparation from marine sources. Structure, properties and applications. Mar Drugs 2015;13(3):1133–74.
[41] Adhikari HS, Yadav PN. Anticancer activity of chitosan, chitosan derivatives, and their mechanism of action. Int J Biomater 2018;2018:29.
[42] Ngo D-H, Kim S-K. Chapter Two - Antioxidant effects of chitin, chitosan, and their derivatives. In: Kim S-K, editor. Advances in food and nutrition research. New York: Academic Press; 2014. p. 15–31.
[43] Kim S. Competitive biological activities of chitosan and its derivatives: antimicrobial, antioxidant, anticancer, and anti-inflammatory activities. Int J Polym Sci 2018;2018:13.
[44] Croisier F, Jérôme C. Chitosan-based biomaterials for tissue engineering. Eur Polym J 2013;49(4):780–92.
[45] Kim C-H, et al. Chitosan for tissue engineering. In: Chun HJ, et al., editors. Novel biomaterials for regenerative medicine. Singapore: Springer; 2018. p. 475–85.
[46] Ali A, Ahmed S. A review on chitosan and its nanocomposites in drug delivery. Int J Biol Macromol 2018;109:273–86.
[47] Argüelles-Monal MW, et al. Chitosan derivatives: introducing new functionalities with a controlled molecular architecture for innovative materials. Polymers 2018;10(3):E342.
[48] Wimardhani Y, et al. Cytotoxic effects of chitosan against oral cancer cell lines is molecular-weight- dependent and cell-type-specific. Int J Oral Res 2012;3:e1.
[49] Wimardhani YS, et al. Chitosan exerts anticancer activity through induction of apoptosis and cell cycle arrest in oral cancer cells. J Oral Sci 2014;56(2):119–26.
[50] Hossain Z, Takahashi K. Induction of permeability and apoptosis in colon cancer cell line with chitosan. J Food Drug Anal 2008;16:1–8.
[51] Park JK, et al. Effects of the molecular weight and the degree of deacetylation of chitosan oligosaccharides on antitumor activity. Int J Mol Sci 2011;12(1):266–77.
[52] Shen K-T, et al. Inhibitory effects of chitooligosaccharides on tumor growth and metastasis. Food Chem Toxicol 2009;47(8):1864–71.
[53] Fernandes CJ, et al. Inhibition of bladder tumor growth by chitooligosaccharides in an experimental carcinogenesis model. Mar Drugs 2012;10(12):2661.
[54] Nam KS, Kim MK, Shon YH. Chemopreventive effect of chitosan oligosaccharide against colon carcinogenesis. J Microbiol Biotechnol 2007;17(9):1546–9.
[55] Nam KS, Kim MK, Shon YH. Inhibition of proinflammatory cytokine-induced invasiveness of HT-29 cells by chitosan oligosaccharide. J Microbiol Biotechnol 2007;17(12):2042–5.
[56] Nam KS, Shon YH. Suppression of metastasis of human breast cancer cells by chitosan oligosaccharides. J Microbiol Biotechnol 2009;19(6):629–33.
[57] Dou J, et al. Induction of apoptosis in human acute leukemia HL-60 cells by oligochitosan through extrinsic and intrinsic pathway. Carbohydr Polym 2011;86(1):19–24.
[58] Arruebo M, et al. Assessment of the evolution of cancer treatment therapies. Cancers 2011;3(3):3279–330.
[59] Beaver CC, Magnan MA. Managing chemotherapy side effects: achieving reliable and equitable outcomes. Clin J Oncol Nurs 2016;20(6):589–91.
[60] De Jong WH, Borm PJ. Drug delivery and nanoparticles:applications and hazards. Int J Nanomed 2008;3 (2):133–49.
[61] Nichols JW, Bae YH. Odyssey of a cancer nanoparticle: from injection site to site of action. Nano Today 2012;7(6):606–18.
[62] Pérez-Herrero E, Fernández-Medarde A. Advanced targeted therapies in cancer: drug nanocarriers, the future of chemotherapy. Eur J Pharm Biopharm 2015;93:52–79.
[63] Senapati S, et al. Controlled drug delivery vehicles for cancer treatment and their performance. Signal Transduct Target Ther 2018;3(1):7.
[64] Emeje M. Recent applications of natural polymers in nanodrug delivery. J Nanomed Nanotechnol 2011; s4:002.
[65] Agrahari V, Agrahari V, Mitra AK. Nanocarrier fabrication and macromolecule drug delivery: challenges and opportunities. Ther Deliv 2016;7(4):257–78.
[66] Pawar VK, et al. Improved chemotherapy against breast cancer through immunotherapeutic activity of fucoidan decorated electrostatically assembled nanoparticles bearing doxorubicin. Int J Biol Macromol 2018.
[67] Lu K-Y, et al. Development of a new type of multifunctional fucoidan-based nanoparticles for anticancer drug delivery. Carbohydr Polym 2017;165:410–20.

[68] Lee KW, Jeong D, Na K. Doxorubicin loading fucoidan acetate nanoparticles for immune and chemotherapy in cancer treatment. Carbohydr Polym 2013;94(2):850–6.

[69] Oliveira C, et al. Gemcitabine delivered by fucoidan/chitosan nanoparticles presents increased toxicity over human breast cancer cells. Nanomedicine 2018;13(16):2037–50.

[70] Wang P, et al. Poly-allylamine hydrochloride and fucoidan-based self-assembled polyelectrolyte complex nanoparticles for cancer therapeutics. J Biomed Mater Res Part A 2019;107(2):339–47.

[71] Shamay Y, et al. P-selectin is a nanotherapeutic delivery target in the tumor microenvironment. Sci Transl Med 2016;8(345):345ra87.

[72] Deepika MS, et al. A novel rutin-fucoidan complex based phytotherapy for cervical cancer through achieving enhanced bioavailability and cancer cell apoptosis. Biomed Pharmacother 2019;109:1181–95.

[73] Venkatesan J, et al. Preparation, characterization and biological applications of biosynthesized silver nanoparticles with chitosan-fucoidan coating. Molecules (Basel, Switzerland) 2018;23(6):1429.

[74] Wang P, et al. Poly-L-ornithine/fucoidan-coated calcium carbonate microparticles by layer-by-layer self-assembly technique for cancer theranostics. J Mater Sci: Mater Med 2018;29(5):68.

[75] Manivasagan P, et al. Doxorubicin-loaded fucoidan capped gold nanoparticles for drug delivery and photoacoustic imaging. Int J Biol Macromol 2016;91:578–88.

[76] Kim H, et al. Doxorubicin-fucoidan-gold nanoparticles composite for dual-chemo-photothermal treatment on eye tumors. Oncotarget 2017;8(69):113719–33.

[77] Kimura R, et al. Cytotoxic effects of fucoidan nanoparticles against osteosarcoma. Mar Drugs 2013;11 (11):4267.

[78] Shin S-W, et al. Fucoidan-manganese dioxide nanoparticles potentiate radiation therapy by co-targeting tumor hypoxia and angiogenesis. Mar Drugs 2018;16(12):E510.

[79] Jang B, et al. Fucoidan-coated CuS nanoparticles for chemo-and photothermal therapy against cancer. Oncotarget 2018;9(16):12649–61.

[80] Hyun H, et al. Photo-cured glycol chitosan hydrogel for ovarian cancer drug delivery. Mar Drugs 2019;17(1):E41.

[81] Ta HT, Dass CR, Dunstan DE. Injectable chitosan hydrogels for localised cancer therapy. J Control Release 2008;126(3):205–16.

[82] Hong S-C, et al. Chitosan-based multifunctional platforms for local delivery of therapeutics. Mar Drugs 2017;15(3):E60.

[83] Tan ML, Choong PFM, Dass CR. Cancer, chitosan nanoparticles and catalytic nucleic acids. J Pharm Pharmacol 2009;61(1):3–12.

[84] Key J, Park K. Multicomponent, tumor-homing chitosan nanoparticles for cancer imaging and therapy. Int J Mol Sci 2017;18(3):E594.

[85] Babu A, Ramesh R. Multifaceted applications of chitosan in cancer drug delivery and therapy. Mar Drugs 2017;15(4):96.

[86] Khdair A, et al. Modified-chitosan nanoparticles: novel drug delivery systems improve oral bioavailability of doxorubicin. Eur J Pharm Sci 2016;93:38–44.

[87] Ye B-l, et al. Chitosan-coated doxorubicin nano-particles drug delivery system inhibits cell growth of liver cancer via p53/PRC1 pathway. Biochem Biophys Res Commun 2018;495(1):414–20.

[88] Wang C, et al. Design and evaluation of galactosylated chitosan/graphene oxide nanoparticles as a drug delivery system. J Colloid Interface Sci 2018;516:332–41.

[89] Fathy MM, et al. Multifunctional chitosan-capped gold nanoparticles for enhanced cancer chemo-radiotherapy: an invitro study. Phys Medica 2018;48:76–83.

[90] Rao W, et al. Chitosan-decorated doxorubicin-encapsulated nanoparticle targets and eliminates tumor reinitiating cancer stem-like cells. ACS Nano 2015;9(6):5725–40.

[91] Gupta U, et al. Enhanced apoptotic and anticancer potential of paclitaxel loaded biodegradable nanoparticles based on chitosan. Int J Biol Macromol 2017;98:810–19.

[92] Jin X, et al. Paclitaxel-loaded N-octyl-O-sulfate chitosan micelles for superior cancer therapeutic efficacy and overcoming drug resistance. Mol Pharm 2014;11(1):145–57.

[93] Yang H, Tang C, Yin C. Estrone-modified pH-sensitive glycol chitosan nanoparticles for drug delivery in breast cancer. Acta Biomater 2018;73:400–11.

[94] Babu A, et al. Chemodrug delivery using integrin-targeted PLGA-chitosan nanoparticle for lung cancer therapy. Sci Rep 2017;7(1):14674.

[95] Jiang J, et al. Development of drug-loaded chitosan hollow nanoparticles for delivery of paclitaxel to human lung cancer A549 cells. Drug Dev Ind Pharm 2017;43(8):1304–13.
[96] Chandran SP, et al. Nano drug delivery strategy of 5-fluorouracil for the treatment of colorectal cancer. J Cancer Res Pract 2017;4(2):45–8.
[97] Liu W, et al. Galactosylated chitosan-functionalized mesoporous silica nanoparticles for efficient colon cancer cell-targeted drug delivery. R Soc Open Sci 2018;5(12):181027.
[98] Rajan M, et al. Hyaluronidase enzyme core-5-fluorouracil-loaded chitosan-PEG-gelatin polymer nanocomposites as targeted and controlled drug delivery vehicles. Int J Pharm 2013;453(2):514–22.
[99] Cavalli R, et al. New chitosan nanospheres for the delivery of 5-fluorouracil: preparation, characterization and in vitro studies. Curr Drug Deliv 2014;11(2):270–8.
[100] Li F, et al. Anti-tumor activity of paclitaxel-loaded chitosan nanoparticles: an in vitro study. Mater Sci Eng C 2009;29(8):2392–7.
[101] Zeiderman MR, et al. Acidic pH-targeted chitosan capped mesoporous silica coated gold nanorods facilitate detection of pancreatic tumors via multispectral optoacoustic tomography. ACS Biomater Sci Eng 2016;2 (7):1108–20.
[102] Parsian M, et al. Loading of gemcitabine on chitosan magnetic nanoparticles increases the anti-cancer efficacy of the drug. Eur J Pharmacol 2016;784:121–8.
[103] Yu H, et al. The effects of novel chitosan-targeted gemcitabine nanomedicine mediating cisplatin on epithelial mesenchymal transition, invasion and metastasis of pancreatic cancer cells. Biomed Pharmacother 2017;96:650–8.
[104] Chen X, et al. Epirubicin-loaded marine carrageenan oligosaccharide capped gold nanoparticle system for pH-triggered anticancer drug release. Sci Rep 2019;9(1):6754.
[105] Bosio VE, et al. Synthesis and characterization of CaCO3–biopolymer hybrid nanoporous microparticles for controlled release of doxorubicin. Colloids Surf B: Biointerfaces 2014;123:158–69.
[106] Vinothini K, et al. Development of biotin molecule targeted cancer cell drug delivery of doxorubicin loaded κ-carrageenan grafted graphene oxide nanocarrier. Mater Sci Eng C 2019;100:676–87.
[107] Ling G, et al. Nanostructured lipid–carrageenan hybrid carriers (NLCCs) for controlled delivery of mitoxantrone hydrochloride to enhance anticancer activity bypassing the BCRP-mediated efflux. Drug Dev Ind Pharm 2016;42(8):1351–9.
[108] Sathuvan M, et al. κ-Carrageenan: an effective drug carrier to deliver curcumin in cancer cells and to induce apoptosis. Carbohydr Polym 2017;160:184–93.
[109] Lee KY, Mooney DJ. Alginate: properties and biomedical applications. Prog Polym Sci 2012;37(1):106–26.
[110] Prabha G, Raj V. Sodium alginate–polyvinyl alcohol–bovin serum albumin coated Fe3O4 nanoparticles as anticancer drug delivery vehicle: doxorubicin loading and in vitro release study and cytotoxicity to HepG2 and L02 cells. Mater Sci Eng C 2017;79:410–22.
[111] Peng N, et al. High drug loading and pH-responsive targeted nanocarriers from alginate-modified SPIONs for anti-tumor chemotherapy. Biomater Sci 2016;4(12):1802–13.
[112] Gao C, et al. pH-Responsive prodrug nanoparticles based on a sodium alginate derivative for selective corelease of doxorubicin and curcumin into tumor cells. Nanoscale 2017;9(34):12533–42.
[113] Chai F, et al. Doxorubicin-loaded poly (lactic-co-glycolic acid) nanoparticles coated with chitosan/alginate by layer by layer technology for antitumor applications. Int J Nanomed 2017;12:1791–802.
[114] Ahn D-G, et al. Doxorubicin-loaded alginate-g-poly(N-isopropylacrylamide) micelles for cancer imaging and therapy. ACS Appl Mater Interfaces 2014;6(24):22069–77.
[115] Markeb AA, et al. Synthesis, structural characterization, and preclinical efficacy of a novel paclitaxel-loaded alginate nanoparticle for breast cancer treatment. Int J Breast Cancer 2016;2016:7549372.
[116] Tao L, et al. Biodegradable alginate-chitosan hollow nanospheres for codelivery of doxorubicin and paclitaxel for the effect of human lung cancer A549 cells. Biomed Res Int 2018;2018:4607945.
[117] Wang F, et al. Effective method of chitosan-coated alginate nanoparticles for target drug delivery applications. J Biomater Appl 2016;31(1):3–12.
[118] Wu JL, et al. Multi-drug delivery system based on alginate/calcium carbonate hybrid nanoparticles for combination chemotherapy. Colloids Surf B Biointerfaces 2014;123:498–505.
[119] Li G, et al. Polymeric hollow spheres assembled from ALG-g-PNIPAM and β-cyclodextrin for controlled drug release. Int J Biol Macromol 2016;82:381–6.

[120] Zhang B, et al. A colon targeted drug delivery system based on alginate modificated graphene oxide for colorectal liver metastasis. Mater Sci Eng C 2017;79:185–90.
[121] Langhans SA. Three-dimensional in vitro cell culture models in drug discovery and drug repositioning. Front Pharmacol 2018;9:6.
[122] Hu M, Polyak K. Microenvironmental regulation of cancer development. Curr Opin Genet Dev 2008;18(1):27–34.
[123] Tseng H, et al. A spheroid toxicity assay using magnetic 3D bioprinting and real-time mobile device-based imaging. Sci Rep 2015;5:13987.
[124] Satpathy A, et al. Developments with 3D bioprinting for novel drug discovery. Expert Opin Drug Discov 2018;13(12):1115–29.
[125] Salvatore V, et al. Changes in the gene expression of co-cultured human fibroblast cells and osteosarcoma cells: the role of microenvironment. Oncotarget 2015;6(30):28988–98.
[126] Salvatore V, et al. The tumor microenvironment promotes cancer progression and cell migration. Oncotarget 2017;8(6):9608–16.
[127] Xu X, Farach-Carson MC, Jia X. Three-dimensional in vitro tumor models for cancer research and drug evaluation. Biotechnol Adv 2014;32(7):1256–68.
[128] Tsai C-W, Wang J-H, Young T-H. Core/shell multicellular spheroids on chitosan as in vitro 3D coculture tumor models. Artif Cells Nanomed Biotechnol 2018;46(Suppl. 3):S651–60.
[129] Santo VE, et al. Adaptable stirred-tank culture strategies for large scale production of multicellular spheroid-based tumor cell models. J Biotechnol 2016;221:118–29.
[130] Florczyk SJ, et al. Porous chitosan-hyaluronic acid scaffolds as a mimic of glioblastoma microenvironment ECM. Biomaterials 2013;34(38):10143–50.
[131] Ivanovska J, et al. Biofabrication of 3D alginate-based hydrogel for cancer research: comparison of cell spreading, viability, and adhesion characteristics of colorectal HCT116 tumor cells. Tissue Eng Part C Methods 2016;22(7):708–15.
[132] Knight E, Przyborski S. Advances in 3D cell culture technologies enabling tissue-like structures to be created in vitro. J Anat 2015;227(6):746–56.
[133] Singh YP, et al. Injectable hydrogels: a new paradigm for osteochondral tissue engineering. J Mater Chem B 2018;6(35):5499–529.
[134] Peck Y, Wang D-A. Three-dimensionally engineered biomimetic tissue models for in vitro drug evaluation: delivery, efficacy and toxicity. Expert Opin Drug Deliv 2013;10(3):369–83.
[135] Carvalho MR, et al. Evaluating biomaterial- and microfluidic-based 3D tumor models. Trends Biotechnol 2015;33(11):667–78.
[136] Wang K, et al. Culture on 3D chitosan-hyaluronic acid scaffolds enhances stem cell marker expression and drug resistance in human glioblastoma cancer stem cells. Adv Healthc Mater 2016;5(24):3173–81.
[137] Dhiman HK, Ray AR, Panda AK. Characterization and evaluation of chitosan matrix for in vitro growth of MCF-7 breast cancer cell lines. Biomaterials 2004;25(21):5147–54.
[138] Dhiman HK, Ray AR, Panda AK. Three-dimensional chitosan scaffold-based MCF-7 cell culture for the determination of the cytotoxicity of tamoxifen. Biomaterials 2005;26(9):979–86.
[139] Lin S-J, et al. Formation of melanocyte spheroids on the chitosan-coated surface. Biomaterials 2005;26 (12):1413–22.
[140] Huang T-W, et al. Culture of nasal epithelial cells using chitosan-based membranes. Laryngoscope 2009;119 (10):2066–70.
[141] Carvalho CR, et al. Investigation of cell adhesion in chitosan membranes for peripheral nerve regeneration. Mater Sci Eng C 2017;71:1122–34.
[142] Hsu S-h, Huang G-S. Substrate-dependent Wnt signaling in MSC differentiation within biomaterial-derived 3D spheroids. Biomaterials 2013;34(20):4725–38.
[143] Huang Y-J, Hsu S-h. Acquisition of epithelial–mesenchymal transition and cancer stem-like phenotypes within chitosan-hyaluronan membrane-derived 3D tumor spheroids. Biomaterials 2014;35 (38):10070–9.
[144] Han H-W, Hsu S-h. Chitosan-hyaluronan based 3D co-culture platform for studying the crosstalk of lung cancer cells and mesenchymal stem cells. Acta Biomater 2016;42:157–67.
[145] Wang X, et al. Enrichment of glioma stem cell-like cells on 3D porous scaffolds composed of different extracellular matrix. Biochem Biophys Res Commun 2018;498(4):1052–7.

[146] Arya N, Sardana V, Saxena M, Rangarajan A, Katti DS. Recapitulating tumour microenvironment in chitosan–gelatin three-dimensional scaffolds: an improved in vitro tumour model. J R Soc Interface 2012;9(77):3288–302.
[147] Li W, et al. Multiple comparisons of three different sources of biomaterials in the application of tumor tissue engineering in vitro and in vivo. Int J Biol Macromol 2019;130:166–76.
[148] Li J, et al. A novel 3D in vitro tumor model based on silk fibroin/chitosan scaffolds to mimic the tumor microenvironment. ACS Appl Mater Interfaces 2018;10(43):36641–51.
[149] Wang H, et al. Growth of MCF-7 breast cancer cells and efficacy of anti-angiogenic agents in a hydroxyethyl chitosan/glycidyl methacrylate hydrogel. Cancer Cell Int. 2017;17:55.
[150] Shakibaei M, et al. Curcumin potentiates antitumor activity of 5-fluorouracil in a 3D alginate tumor microenvironment of colorectal cancer. BMC Cancer 2015;15:250.
[151] Buhrmann C, et al. Resveratrol chemosensitizes TNF-β-induced survival of 5-FU-treated colorectal cancer cells. Nutrients 2018;10(7):E888.
[152] Buhrmann C, et al. Induction of the epithelial-to-mesenchymal transition of human colorectal cancer by human TNF-β (lymphotoxin) and its reversal by resveratrol. Nutrients 2019;11(3):704.
[153] Rebelo SP, et al. 3D-3-culture: a tool to unveil macrophage plasticity in the tumour microenvironment. Biomaterials 2018;163:185–97.
[154] Estrada MF, et al. Modelling the tumour microenvironment in long-term microencapsulated 3D co-cultures recapitulates phenotypic features of disease progression. Biomaterials 2016;78:50–61.
[155] Liu C, et al. Hybrid collagen alginate hydrogel as a platform for 3D tumor spheroid invasion. Acta Biomater 2018;75:213–25.
[156] Jiang T, et al. Directing the self-assembly of tumour spheroids by bioprinting cellular heterogeneous models within alginate/gelatin hydrogels. Sci Rep 2017;7(1):4575.
[157] Xu X-x, et al. Encapsulated human hepatocellular carcinoma cells by alginate gel beads as an in vitro metastasis model. Exp Cell Res 2013;319(14):2135–44.
[158] Doh I, et al. Drug efficacy comparison of 3D forming and preforming sphere models with a micropillar and microwell chip platform. SLAS Discov 2019;24(4):476–83.
[159] Dai X, et al. 3D bioprinted glioma stem cells for brain tumor model and applications of drug susceptibility. Biofabrication 2016;8(4):045005.
[160] Sabhachandani P, et al. Microfluidic assembly of hydrogel-based immunogenic tumor spheroids for evaluation of anticancer therapies and biomarker release. J Control Release 2019;295:21–30.
[161] Lourenço BN, et al. Bioengineering a novel 3D in vitro model of gastric mucosa for stomach permeability studies. Acta Biomater 2018;82:68–78.
[162] Fang X, et al. Novel 3D co-culture model for epithelial-stromal cells interaction in prostate cancer. PLoS One 2013;8(9):e75187.
[163] Pang Y, et al. TGF-beta induced epithelial-mesenchymal transition in an advanced cervical tumor model by 3D printing. Biofabrication 2018;10(4):044102.
[164] Eddie SL, et al. Three-dimensional modeling of the human fallopian tube fimbriae. Gynecol Oncol 2015;136(2):348–54.
[165] Kievit FM, et al. Chitosan–alginate 3D scaffolds as a mimic of the glioma tumor microenvironment. Biomaterials 2010;31(22):5903–10.
[166] Kievit FM, et al. Proliferation and enrichment of CD133+ glioblastoma cancer stem cells on 3D chitosan-alginate scaffolds. Biomaterials 2014;35(33):9137–43.
[167] Kievit FM, et al. Modeling the tumor microenvironment using chitosan-alginate scaffolds to control the stem-like state of glioblastoma cells. Biomater Sci 2016;4(4):610–13.
[168] Florczyk SJ, et al. 3D porous chitosan–alginate scaffolds: a new matrix for studying prostate cancer cell–lymphocyte interactions in vitro. Adv Healthc Mater 2012;1(5):590–9.
[169] Leung M, et al. Chitosan-alginate scaffold culture system for hepatocellular carcinoma increases malignancy and drug resistance. Pharm Res 2010;27(9):1939–48.

CHAPTER

# 24

# Mesoporous silica nanoparticles for cancer theranostic applications

*Mohammad Eltohamy*

**Glass Research Department, National Research Centre, Giza, Egypt**

## 24.1 Introduction

Despite the advances in cancer treatment in past decades, cancer still remains one of the chief causes of mortality worldwide [1]. Current treatment approaches include chemotherapy, radiation therapy, surgery, laser therapy, and combinational therapy—all of which are associated with severe side effects compromising the quality of patients' life [2,3]. In addition, 90% of failure in chemotherapy is due to multidrug resistance (MDR), which occurs due to exposure of cancer cells to high-drug concentrations [4]. Hence, developing functional materials and/or dexterous methods with enhanced therapeutic efficacy along with minimal side effects are of high concern.

The improvement of drug efficacy and reduction of its side effects can be obtained by target specific delivery of it to the site of need. Nanoparticles (NPs) are extraordinary candidate for this and are commercially available since mid-1990s [5]. However, over the past 30 years only a few nanoformulations have been approved by U.S. Food and Drug Administration (FDA) for clinical trials due to high cytotoxicity of the engineered particle, limited disparity, and cost-effective scalable fabrication methods.

Among the inorganic nanobased formulations, based on the sizes of pores, porous materials are broadly classified into three groups; (1) microporous (<2 nm), (2) mesoporous (between 2 and 50 nm), and (3) macroporous particles ( > 50 nm), respectively. Porous NPs of sizes less than 100 nm are very critical in obtaining long blood circulation (by escaping the hepatic capture or renal clearance) and accumulation at high concentration in cancer tissue [6]. For instance, gold NPs exhibit wide applications in the recognition of small biomolecules via surface adjustment [7]. Moreover, organic NPs (e.g., liposomes and dendrimers), inorganic NPs, and organic/inorganic hybrid NPs, have also been engineered for cancer treatment method [8,9].

*Biomaterials for 3D Tumor Modeling*
DOI: https://doi.org/10.1016/B978-0-12-818128-7.00024-1

In comparison with other NP form, mesoporous silica nanoparticles (MSNs) possess the advantages of ample preparation, remarkable biocompatibility, and wide surface conjugation chemistry. Various strategies including photodynamic therapy, hyperthermia, and drug delivery are typically utilized in acquiring MSNs-based cancer therapy techniques [10,11]. These techniques declare that MSNs are ideal for the development of powerful and versatile cancer therapy models. The capability of these particles is related to the different local surroundings of atoms subjected at surfaces compared with those in the bulk. Consequently, the greater number of surface atoms can easily boost the specific surface area of porous NPs, producing substantial material's reactivity as well as enhanced efficacy in appropriate applications [10]. Based on the literature, porous materials can be classified based on their pore size into three groups including microporous, mesoporous, and macroporous particles which have the pore sizes less than 2 nm, between 2 and 50 nm, and greater than 50 nm, respectively. In addition to relevant surface properties, these mesoporous materials hold good biocompatibility, controllable size, and easy surface modification, which will make them outstanding candidates for various biomedical applications.

Owing to these distinctive worth, research on the MSNs has grown drastically. Large specific surface area and pore volume make MSNs as suitable reservoir for loading therapeutic/diagnostic agents. On the other hand, MSNs are able to safeguard their cargos from premature release and subsequently undesirable degradation in intestines and stomach prior to approaching toward the target place. Several researches have reviewed the internalization of MSNs in a variety of cell lines [12]. Different variables including morphology and size of mesostructures, electrostatic interactions between MSNs as well as the cell membrane and the surface functionalization can impact the uptake mechanisms of MSNs. To attain an appropriate therapeutic effect, MSNs should be able to release their cargo on the cytoplasm by applying different endosomal escape approaches [13–15]. Moreover, physical and chemical surface modifications are used to expand the range of biomedical (diagnostic/therapy = theranostic) applications for MSNs. These modifications result in growth of biocompatibility, inhibition of specific adsorption as well as providing functional groups for additional biomolecule conjugation goals. The widespread physical and chemical ways of modify the surface of MSNs are layer by layer self-assembly and chemical surface functionalization, respectively [16]. In addition to size and shape, surface properties such as charge, functional groups, and the presence of some molecules can change the biocompatibility of MSN.

To the best of our expertise, even though several reviews have described the advances in preparation and biomedical applications of MSNs, all of them only discuss their particular manner in certain applications. No efforts have been conducted to provide a comprehensive evaluation of MSNs in depth in biomedical fields. Therefore, in this chapter, we are going to thoroughly focus on the research concerning about biological applications of MSNs (mainly in the latest several years), based on our own accomplishments and important reports from other research groups in this field, which range from synthetic chemistry and essential parameters of MSNs, to their varied therapeutic and diagnostic applications. On top of that, we will also discuss the biological effects (biodegradation and biocompatibility) of MSNs, as well as their difficulties in clinical translations. It is assumed that such adaptable nanomaterial system will undertake a step forward in growth within the next years.

## 24.2 MSNs chemistry

In general, MSNs have been synthesized by sol–gel chemistry. This method consists of the hydrolysis and condensation of silanes or organosilanes, and the kinetics of both operations are often governed by acid and base. The hydrolysis stage results in creation of reactive silanols species, which are ready to condense with other silanes or organosilanes, by building covalent siloxane bonds and therefore a sol of silicate oligomers [17]. Gradually, such a "hydrothermal" process produces silica frames. To generate mesoporous structure, cationic surfactants, for example, cetyltrimethylammonium bromide (CTAB) are frequently employed as structural-directing agents for directing the reaction passway. The negatively charged reactive silanols species is usually electrostatically linked with positively charged CTAB micelle packing's, hence favoring a template effect [18]. Subsequently, the supportive assembly and aggregate bring about precipitation from a gel, where microphase separation and steady condensation of silicate oligomers develop [19]. The process triggers the development of mesoporous silicate hybrid framework, which only requires a little while in cationic surfactant solutions [18]. Eventually, mesoporous material is attained following the removal of organic templates by extraction or calcination.

It is observed that whenever just a silica precursor, such as tetraethoxysilane, is used in the sol–gel reaction, the final product will be MSN. Rather the sole presence of bridged organosilanes may result in the generation of organic function group NPs, and a mixture of a silanes and an organosilanes will contribute to the creation of hybrid NPs [10]. Thus, correct regulation of sol–gel chemistry by altering the reactants (silica sources and surfactant templates) or experimental conditions (pH, temperature, water content, etc.) is of terrific significance in modulating the structural/compositional characteristics of the resulted nanomaterials. Surface features also play an essential role in identifying the biological doings of MSNs.

The unique mesoporous structure of MSNs gives extremely large specific surface area for them to interact with external substances. Owing to the plentiful silanol groups on the surface of MSNs, these NPs can be covalently bonded by extrinsic organic functional moieties for surface adjustment and modulating there in vivo manners. In the past two decades, several surface functionalization methods are already explored, that pushed ahead the biomedical applications of MSNs and its hybrid derivatives. Ordinarily, these tactics can be categorized into the subsequent two methodologies: The first one is integrating MSNs with biocompatible polymers, such as polyethylene glycol (PEG), for improving in vivo biocompatibility. The second one is modifying MSNs with targeting ligands for specific delivery of therapeutic cargoes. For example, folic acid receptor has been found to be overexpressed on several cancer cells, such as ovarian, endometrial, colorectal, breast, lung, and renal cell carcinoma [20].

In case of theranostic application, which can not only be filled up into the abundant mesopores of MSNs, but also be designed on their extremely large topographical surface via covalent bonding, consequently to endow these NPs with added therapeutic and/or diagnostic functions. In recent work, researchers have built an "all-in-one" theranostic nanosystem by multiple surface functionalization by hybrid derivatives for improved sonodynamic treatment of hepatocellular carcinoma (Fig. 24.1) [22].

This kind of nanoplatforms appeared to be fabricated by first protoporphyrin engineering to trigger sonocatalytic effect for reactive oxygen species (ROS) generation,

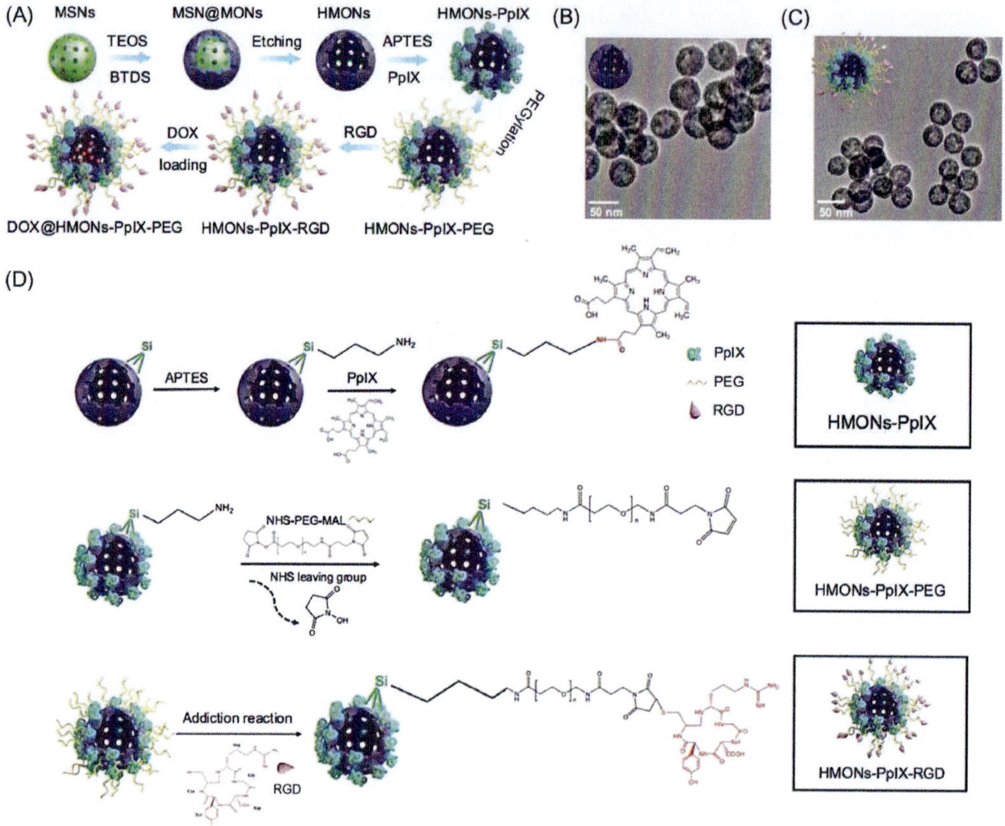


FIGURE 24.1 Surface modification of MSN for improved physiological stability, targeting functioning, and multifunctionality [21]. (A) Schematic illustration for the synthetic process. (B) TEM images MSN and (C) TEM images MSN-RGD. (D) The RGD conjugation on the surface of MSN. Copyright 2018, John Wiley and Son.

accompanied by PEG modification for better physiological balance, then conducted with three amino acid peptide arginine–glycine aspartic acid (RGD) conjugation for targeting integrin αvβ3 on cancer cells, and final nanosystem establishment via doxorubicin (DOX) loading for conferring chemotherapeutic effect. It is noted that such a co-functionalization of various agents leads to more effective utilization of the 3D mesoporous design of NPs, while the accurate power over the dosage and percentage of each functional moieties on the topographical surface is a challenging but important endeavor to ensure that each of them can certainly completely exert their affiliated performs in vivo [23].

## 24.3 Biological effects of MSNs

The expansion of researches on the numerous therapeutic and diagnostic applications of MSNs-based nanosystems guide us to have particular efforts with their biological effect in the complex in vivo center. With all the progress nanosynthetic chemistry, increasingly more MSNs together with diverse structural, compositional, and morphological features are prepared, which not only lead to varied therapeutic and diagnostic routines, but also in the designs in their many biological manners. The underlying nano–bio relationships between these flexible nanosystems and in vivo environment are supposed to be thoroughly inspected for ensuring desired biosafety of MSNs, thus to assist in the ultimate clinical interpretation. The first issue in mastering their biological behaviors is to appraise the biodegradability and approve length of these promising theranostic nanoplatforms (Fig. 24.2).

Generally, the degradation of silica in aqueous media can be summarized as three steps: (1) Hydration: $H_2O$ is adsorbed into the Si–O–Si framework; (2) Hydrolysis: Si–O–Si is hydrolyzed into Si–OH; (3) Ion-exchange: nucleophilic attack of –OH leads to the $Si(OH)_4$ removal. The solubility of the final degradation products of the $Si(OH)_4$ is really important for the powerful biodegradation of MSNs, and its reaction equilibrium is also closely linked to the specific situations of biological microenvironments (pH, ions, etc.). These degradation products of silica are nontoxic and can diffuse into the blood stream or the immune system to be ultimately eliminated out through urine [25].

In a recent review, researchers have provided a systematic elucidation on the in vivo biodegradation dynamics of silica-based materials [26]. They pointed out that the size, morphology, surface area, pore size, and surface functional groups of MSNs are the critical physicochemical variables to affect their biodegradability. Consequently, whenever we design MSNs-based nanosystems, correct regulation of these physicochemical features is highly desired for optimized biodegradation actions. The second point for better knowing the biological behaviors of MSNs is to investigate their biodistribution and biocompatibility for future practical purposes. A systematic study to evaluate the long-term in vivo biodistribution and urinary excretion of spherical MSNs and PEGylated MSNs. The experimental results exhibited that will use both MSNs and PEGylated MSNs with different particle sizes (80, 120, 200, and 360 nm) mainly dispersed in the liver and spleen, and a minority of them were in the lung, kidney, and heart. The PEGylation of MSNs was initially confirmed to be effective in avoiding the capture of MSNs by liver, spleen, and lung, which led to longer blood-circulation lifetime and slower biodegradation. The removal amounts of all the reviewed MSNs and PEGylated MSNs decreased with increase in time duration and particle size, while the latter presented lower excreted amount of degradation products. In addition, histopathological assessment indicated that neither MSNs nor PEGylated MSNs could cause tissue toxicity 1 month after injection, high biocompatibility of MSNs-based nanoplatforms. Comparatively, despite the fact that very recent studies have presented partial evidences within the improved biodegradability and biocompatibility of MSN and its hybrid derivatives [27], more systematic data on their in vivo biodistribution and clearance are still required, to fully reveal the biological characteristics of such an growing organic/inorganic hybrid nanosystem.

It is observed that most of the current researches on the design and fabrication of MSNs-based theranostic nanoplatforms have provided biological information at the end of their reports, such as hematological (aspartate transaminase, alanine transaminase, total protein, total bilirubin, globulin, creatinine, blood urea nitrogen, and albumin) and histological data (i.e., H&E staining of heart, liver, spleen, lung, and kidney), to evidence the

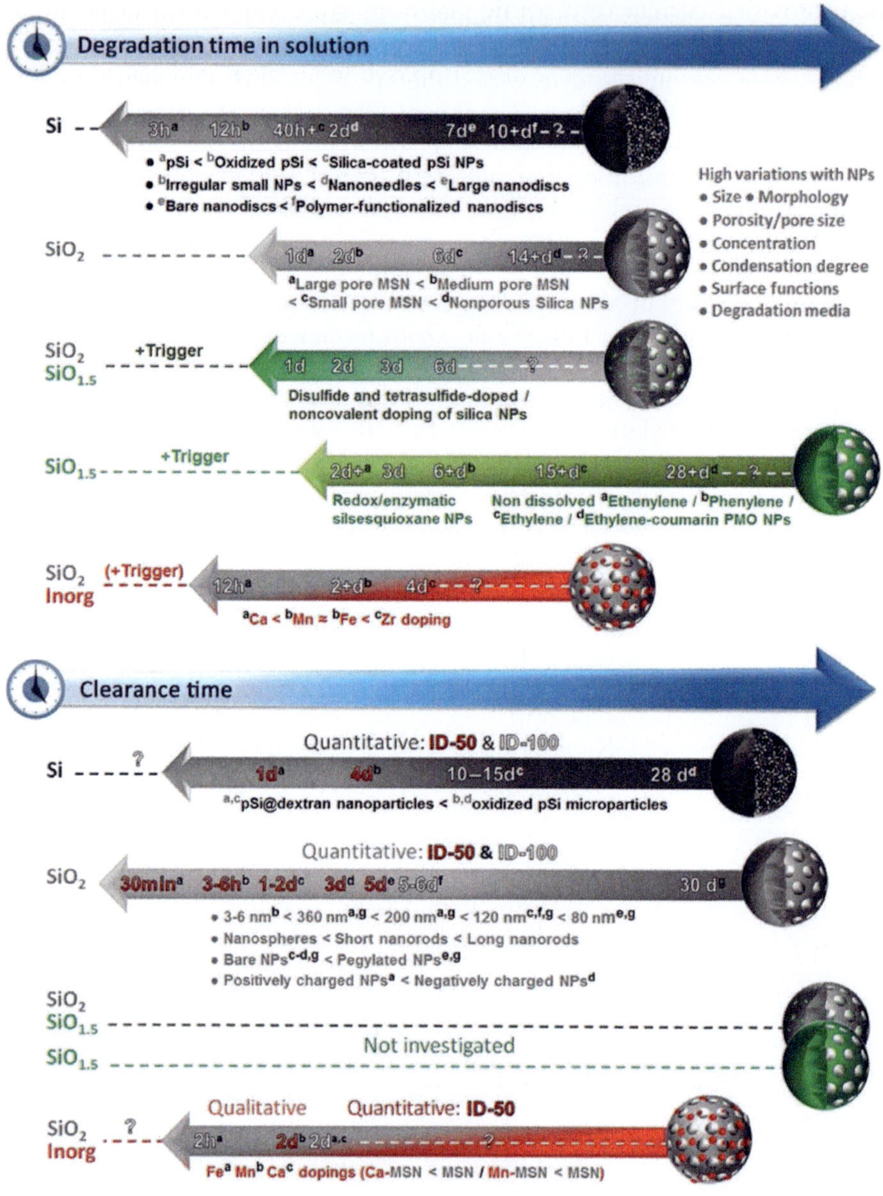


FIGURE 24.2 Explanation of the degradation of MSN NPs in aqueous solutions [24]. Copyright 2017, John Wiley and Sons.

biocompatibility of the set-up nanosystems for even more clinical translation [28]. It is predicted that more detailed in vivo evaluations, such as the examination of genotoxicity, neurotoxicity, reproductive toxicity, embryonic toxicity, etc., will be carried out to further assure the biosafety of MSNs-based nanosystems for future clinical interpretation.

In addition, provided current therapeutic platforms are usually constructed by the incorporation of MSNs along with other functional agents, therefore, the biocompatibilities of such designed functional moieties are also supposed to be taken into consideration.

## 24.4 3D modeling of MSN for cancer therapy

Although the MSN exhibits many outstanding features in tumor theranostics therapy, the lack of cell adhesion sites limits their application as well as the modeling limitations as it exists in 0 dimensions (0D) likewise similar NP. As far the models move from 0D → 3D models, there application toward cancer thereby will be more effective and applicable. In general, 3D modeling have shown great potential in the tumors treatment; owing to the outstanding tumor killing effect and expected to be employed as an adjutant therapy for preventing local return after prosthetics. However, the 0D MSN could hardly remain localized on the sites of the resection-induced cavity, leading to shortage for cancer cell ablation. To address this limitation, and its solution toward development of the MSN into higher diminution (it might be other dimensions are not in the nanoscale range), hydrogels [29], electrospun nanofibers [30], and 3D-printed scaffolds were used [31].

### 24.4.1 Hydrogels

Hydrogels, 3D polymeric networks, consist of cross-linked polymer chains using physical or chemical cross-links [32]. Owing to the mild and easy preparation process, hydrogels have attracted extensive attention in Refs. [29,33,34]. It was found that the hydrogels maintained outstanding photothermal effects and repeated irradiation stability of MSN nameplates, leading to significant tumor killing effects. Owing to the encapsulation function of the hydrogels, the MSN inside the hydrogels can be more stable and safer. It was reported that $Fe_3O_4$@MSN was easily incorporated inside the tumor microenvironment, which led to an inhomogeneous hyperthermia and relatively short-term therapy against cancer [35]. It was demonstrated that MSN entrapped hydrogels not only prevented MSN from rapid degradation and discharge chemotherapy, but also maintained high efficiency, leading to an inhibitory effect on tumor growth. It was believed that such bulks could realize better therapeutic effects and biocompatibility as compared with the intravenous administration of MSN, since the hydrogel was locally implanted and could be repeatedly irradiated by the NIR laser.

### 24.4.2 Electrospun nanofiber scaffolds

Electrospun fibers, 3D interwoven networks, were held together by fibers ranging from micrometers to nanometers [36]. The electrospun fibers have attracted tremendous attention in the biomaterial field, mainly resulting from their physical–structural similarity to the extracellular matrix (ECM), large surface area, large porosity, and easy functionalization by inorganic materials, drugs, and growth factors. MSN-embedded electrospun fibers have been explored for PTT [37]. For example, the incorporation of $Fe_3O_4$@MSN-loaded

DOX nanofibers enabled the temperature of fibers to increase rapidly, leading to the significant death of cervical cancer cells by local hyperthermia treatment combined with DOX [38]. MSN was functionalized onto the surface of electrospun fibers for both capturing and simultaneously killing cancer cells. The surface functionalization of MSN improved the adhesion of cancer cells. Furthermore, MSN-induced photothermal therapy significantly eradicated the captured cancer cells in situ.

### 24.4.3 3D-printed scaffolds

Three-dimensional (3D) printed scaffolds were porous constructs with a concisely defined architecture made from ceramic hydrogels, polymers, and metals [39]. As compared with other 3D porous scaffolds, 3D-printed scaffolds might have enhanced mechanical properties and controlled pore structures. It was reported that the 3D-printed MSN scaffolds exhibited compressive strength higher than corresponding MSN hydrogel scaffolds [40]. With great mechanical properties, the 3D-printed MSN scaffolds with antitumor ability have mainly been used for treating bone tumors. Many MSNs have been reported to be loaded on the surface of 3D-printed scaffolds, including polydopamine [41]. In addition to the surface modification of the scaffolds, the MSN containing photothermal elements were employed directly and incorporated into PTT scaffolds for bone tumor therapy, harnessing the photothermal effect initiated by bioactive elements [39]. On the other hand, MSN-incorporated scaffolds had better biosecurity and bioactivity, as compared with traditional nanosized photothermal agents.

## 24.5 Medical applications of MSNs

Taking advantage of the well-known accomplishments in nanosynthetic chemistry and surface chemistry, a number of nanoplatforms depending on MSNs and its hybrid derivatives were created for various applications in biomedicine, such as stimuli-responsive drug release, targeted drug delivery, gene transportation along with other therapeutic strategies. In this section, we will sum up the cutting edge studies relating to current breakthroughs within this developing area, and provide a deep discussion around the actual nanochemistry of MSNs-based nanoplatforms that underpins there in vivo therapeutic features.

### 24.5.1 Stimuli-responsive drug release

Conventional chemotherapy has problems with acute side effects due to the minimal bioavailability and tumor-specificity of poisonous chemodrugs [42–44]. Hence, great attempts have been intended to assemble smart drug delivery with distinctive respond to one or several physicochemical trigger(s), for release drug at pathological sites [45]. Commonly, MSNs and it is hybrid derivatives are tailored with active moieties for stimuli-responsive cancer chemotherapy, that has designed functional motifs surrounding the opening of mesopores can work as "gatekeepers" to regulate the effluence of inner theranostic agents underneath the incitement of ambient triggers [46]. Furthermore, the platform of MSNs or its hybrid derivatives could also be designed with organic/inorganic dynamic moieties to prejudice tumor-specific biodegradability and consequent controlled drug release. In line with the types and features of ambient triggers, these mesoporous nanosystems can be categorized as (1) endogenous-triggered nanosystems that are addressing endogenous biological stimulus such as pH, redox, and tumor protein; (2) exogenous-triggered nanosystems, which can interact with exterior physical influences such as light, ultrasound (US), etc. The fabrication and application of these nanosystems allow spatio-temporal handle of therapeutic process, thus significantly improving therapeutic efficiencies.

#### 24.5.1.1 pH-responsive

The glycolysis behavior of tumor cell (i.e., Warburg effect) leads to over accumulation of lactate in tumor microenvironment (TME), starting regional low pH values [47]. This specific property of TME has triggered the rigorous design and fabrication of pH-responsive MSNs for better therapeutic need. Polyelectrolyte-coated MSNs for pH-responsive drug release have already been reported [48,49]; noticeable amounts of pH-responsive MSNs have also been reported to manipulate tumor-specific chemotherapy [50]. To overwhelm the MDR of cancer, additionally created pH-responsive DOX-loaded MSNs were used, which could potentiate the cytotoxicity of DOX via chemosensitization of cancer cells [51]. Taking into consideration the 3D heterogeneous structure of tumors, the intrinsic Warburg effect leads to a spatial pH gradient in tumor along with surrounding ordinary tissues, which will impact the in vivo actions of pH-responsive MSNs such as the premature leakage of drug molecules. For this reason, how to benefit from such a pH heterogeneity, to favor accuracy release of drug molecules within tumor tissues after these nanocarriers endure complicated vascular circulation and tissue transmission, are going to be of great relevancy. Surface-to-core diffusion of nanocarriers in entire 3D tumors already developed for cancer targeting, depending on spatial pH variants within/around tumor tissue (Fig. 24.3) [53,54]. Such nanosystem was synthesized of MSNs with large pores, accompanied by the loading of small gold NPs by DNA hybridization. Right after systemic administration, the entire nanosystems accumulate in the tumor via the EPR effect. Then, the mild acidic TME ($pH \approx 6.5$) results in the first changeover of hybrid DNA in the mesopores, resulting in the discharge of inner gold NPs as well as the conjugated DNA, that could pass through the deep tumor region properly because of their small size. Ultimately, the internalization of gold NPs by tumor cells results in the second DNA transition under endosomal pH ($\approx 5.0$) for anticancer drug release. This work tends to make full use of spatial pH variations as specific trigger to enable sequential release of smaller nanocarriers and drug molecules, clearly enhancing the therapeutic effectiveness of cancer.

#### 24.5.1.2 Redox-responsive

Cancer cells possess potent endogenous antioxidant defense mechanisms to eliminate the overexpressed ROS (such as $H_2O_2$ and $O^{2-}$) in TME for establishing a new redox equilibrium [55]. For example, glutathione (GSH) is the significant ROS-detoxifying system in cellular material, and its concentration in tumor cells have been elevated appreciably [56]. Therefore, GSH-responsive nanosystems have also been extensively inquired for higher

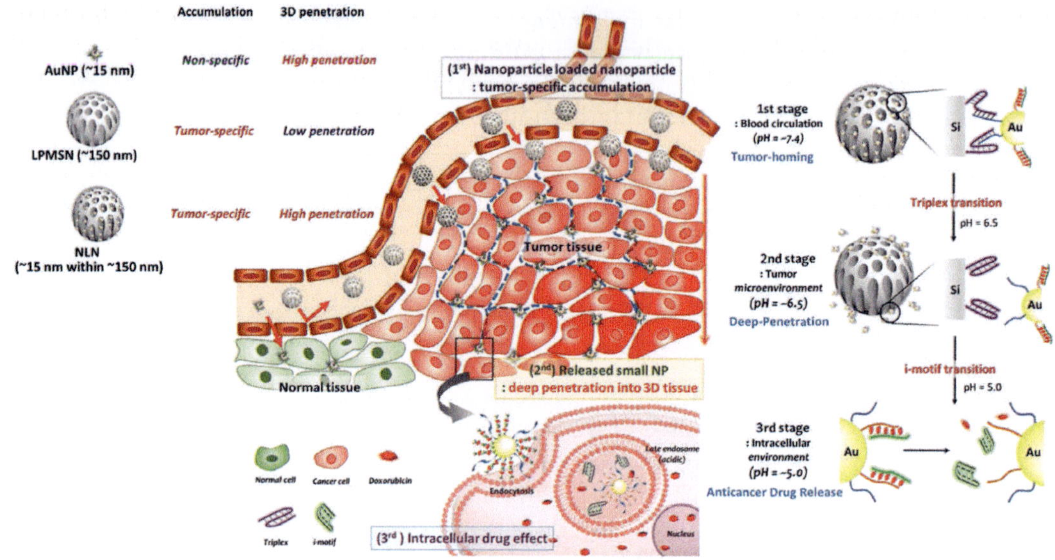


FIGURE 24.3 Schematic illustration for the pH-dependent of gold NP-loaded MSNs [52]. Copyright 2018, John Wiley and Sons.

therapeutic usefulness. Recently, researchers constructed various hybrid-based nanosystems responsive to intratumoral GSH for tumor-specific biodegradability and drug release, by means of engineering redox-active disulfide bonds (–S–S–) into the silica structure. Inside vitro DOX-releasing pattern shows a distinctive reducing-responsive manner that is attributed to the cleavage connected with disulfide bonds in the platform that triggers the disintegration with the hybrid nanocarriers. Glioblastoma multiform is the most common and cancerous brain tumor, which diffusely infiltrates normal brain as well as resides behind blood–brain barrier, resulting in the significant challenge for specific chemotherapy [57]. The overexpressed GSH in TME allows for the biodegradation of hybrid derivatives and the release of as-loaded $Cu_2$-xSe (on the surface) and DOX (in the mesopores), resulting in substantial inhibition of tumor progress (91.1%). Hematoxylin–eosin (H&E) and terminal deoxynucleotidyl transferase-mediated dUTP nick end labelin (TUNEL) staining also existing enhanced cancer cell destruction and apoptosis after the treatment, further confirming the high feasibility of as-fabricated HMSN-based nanosystem for glioblastoma therapy. It is noted that current studies on redox-responsive MSNs just consider to employ endogenous antioxidants as biochemical stimuli for controlled drug release. It is expected that ROS-responsive MSNs will also be developed and fabricated to contribute to the development of precision cancer chemotherapy in the future.

#### 24.5.1.3 *Light-responsive*

As an average remote physical upgrade, light have been broadly applied in the biomedical field, for example, phototherapy (PTT) [58] and photodiagnosis [59,60]. Advances in nanotechnology have catalyzed quantities of photoresponsive nanomedicine for malignant growth restorative applications, for example, controlled medication discharge [61], photothermal treatment (PTT) [62,63], and photodynamic therapy (PDT) [64], in which light gave a high-spatiotemporal controllability to directing remedial procedure. MSNs have likewise been investigated as light-responsive drug delivery systems (DDS) in the previous two decades [65], since first announced photocontrolled arrival of visitor atoms from coumarin-altered MCM-41. For instance, researchers built azobenzene-subordinates improved MSNs for light-responsive load discharge [66]. Under 351 nm light illumination, the isomerization of trans-to-cis azobenzene units brings about the deterioration of β-cyclodextrin rings from the stalks, therefore opening the entryways of the mesopores and discharging the internal load particles in this manner. Be that as it may, these nanosystems may not be capable enough for in vivo applications because of the generally low-tissue infiltration of as-applied UV or noticeable light. Similarly, NIR light lies in the "optical straightforwardness window" of organic tissues, which guarantees further entrance during its tissue proliferation. In this manner, a MSNs-based center shell organized nanosystem for NIR-activated medication discharge conduct [67]. Such a NIR-responsiveness was presented by the development of mesoporous silica, which can change over the illuminated NIR (980 nm in this work) into UV or noticeable light proficiently for encouraging consequent photophysical process [67,68]. From the previous work, the isomerization of trans-to-cis azobenzene to fill in as the "main thrust" to incite the arrival of DOX (zeta potential: +2.2 mV) from the electrostatic adsorption of mesopores (zeta potential: −40.6 mV). The measure of discharged DOX can be finely constrained by shifting the power and time term of NIR light illumination, and its hybrid derivatives treating the plausibility of the MSNs for remote administration of chemotherapeutic procedure.

#### 24.5.1.4 *Magnetic field-responsive*

In contrast to light, magnetic field is free from cells depth/medium restrictions and provides high-approval possibilities in several biomedical areas, such as hyperthermia extirpation, drug release [69], and magnetic resonance imaging (MRI) [70]. MSNs can also be formulated for magnetic field-responsive DDS by integrating supermagnetic iron oxide NPs with MSNs for conferring responsiveness [71]. In previous work, the MCM-41-type MSNs capped with iron oxide NPs for attaining magnetic field-controlled drug release [72]. Under a remote magnetic stimulus, the as-decorated $Fe_3O_4$ NPs could be taken off from the surfaces of MSNs via the breakdown of chemical bonds, resulting in the fast intracellular drug release for cancer therapy. Based on the magnetocaloric effect which could also kick-off local hyperthermia for assisting drug release, reported articles constructed rattle-type $Fe_3O_4$@m$SiO_2$ NPs as a multifunctional platform for simultaneous controlled drug release and hyperthermia ablation [73]. It is anticipated that more magnetic field-responsive MSNs will be fabricated in the future for improved therapeutic efficacy out of the intrinsic superiority of magnetic field.

### 24.5.2 Targeted drug delivery

Even though intrinsic biophysical method of tumor allows the residual accumulation of NPs inside tumor tissue (i.e., EPR effect) to some extent, however, the inert targeting efficiency is often lacking excessive enough intratumoral accumulation and effective cancer therapy. Due to the overexpression of some particular biomacromolecules within the cell membrane/cytoplasm of cancer cells, active-targeting drug delivery has been broadly designed for a long time [74,75], through which amounts of targeting ligands have already been functionalized on as-constructed nanoformulations, for example MSNs-based nanosystems, for accurate placement of cancer cells [76]. Depending on targeted subcellular locations, currently developed targeting approaches consists of cell-membrane targeting and cell-cytoplasm targeting.

#### 24.5.2.1 Cell-membrane targeting

Variety of biomacromolecules, including proteins and nucleic acids, are designed on, or even inlaid into the lipid bilayer of cytomembrane. Some of them are overexpressed particularly on cancer cells and can be designed as target for active intracellular drug delivery. Studies conjugated anti-HER2/neu monoclonal antibody (mAb) on the surface of MSNs to consult targeting properties toward breast cancer cells [77,78]. Such constructed NPs could be internalized into cells through a receptor-mediated endocytosis and eventually, escape from endosome to cytosol, giving a powerful strategy for targeted chemodrugs delivery. It was prepared FA-grafted rattle-type magnetic $Fe_3O_4$/MSNs to favor their active accumulation into FA-receptor-positive HeLa cells [71]. In a successional work, hyaluronic acid (HA) was also functionalized on MSNs to selectively target CD44 protein overexpressed on cancer cells, resulting in superior cytotoxicity of as-delivered hydrophobic camptothecin [79,80]. According to this particular work, further work designed, for example, HA-modified mesoporous carbon/silicon–silica hierarchical framework for targeted delivery of insoluble drugs (Fig. 24.4). Such a composite nanoplatforms combines the benefits of monodispersed size, well-defined morphology, high-hydrophobic drug payload, and exclusive NIR-to-Vis luminescence ability, showing unique feasibility to be applied for cancer theranostics.

#### 24.5.2.2 Cell-cytoplasm targeting

Almost like cytomembrane, cell cytoplasms also existing distinctive biomacromolecules and they are formulated unique target for nanomedicine identification, internalization, and drug release [82]. Commonly, these kinds of nanosystems tend to be required with ever-smaller particle size (<100 nm) as a consequence of decreased dimension of cell cytoplasms. Breakthroughs in nanosynthetic chemistry have preferred the fabrication of MSNs with well-defined structure and lowered particulate dimension, to satisfy the requirement of cell-cytoplasm targeting for more reliable chemodrugs delivery with improved spatial quality. Cell nucleus is the cellular "center" where the genetic data and the transcription machinery reside. It runs an important role in managing cell proliferation, metabolism, gene activation, and cell cycle, which are usually also closely connected to carcinogenesis. Thus, in recent years, cell nucleus has been formulated in the form of talented target for anticancer drug delivery. It is identified the fact that the nuclear membrane is embedded with abundant nuclear pore complexes of 20–70 nm in diameters [83,84], which, fortuitously, supply feasible paths for nanocarriers entry [85]. Various intranuclear drug-delivery approaches have been made by decorating different nuclear localization signal peptides, such as SV40 T antigen, adenoviral, and HIV-1 TAT peptide, on the surface of nanocarriers [81,86,87]. As an example, TAT peptide has the ability to of binding import

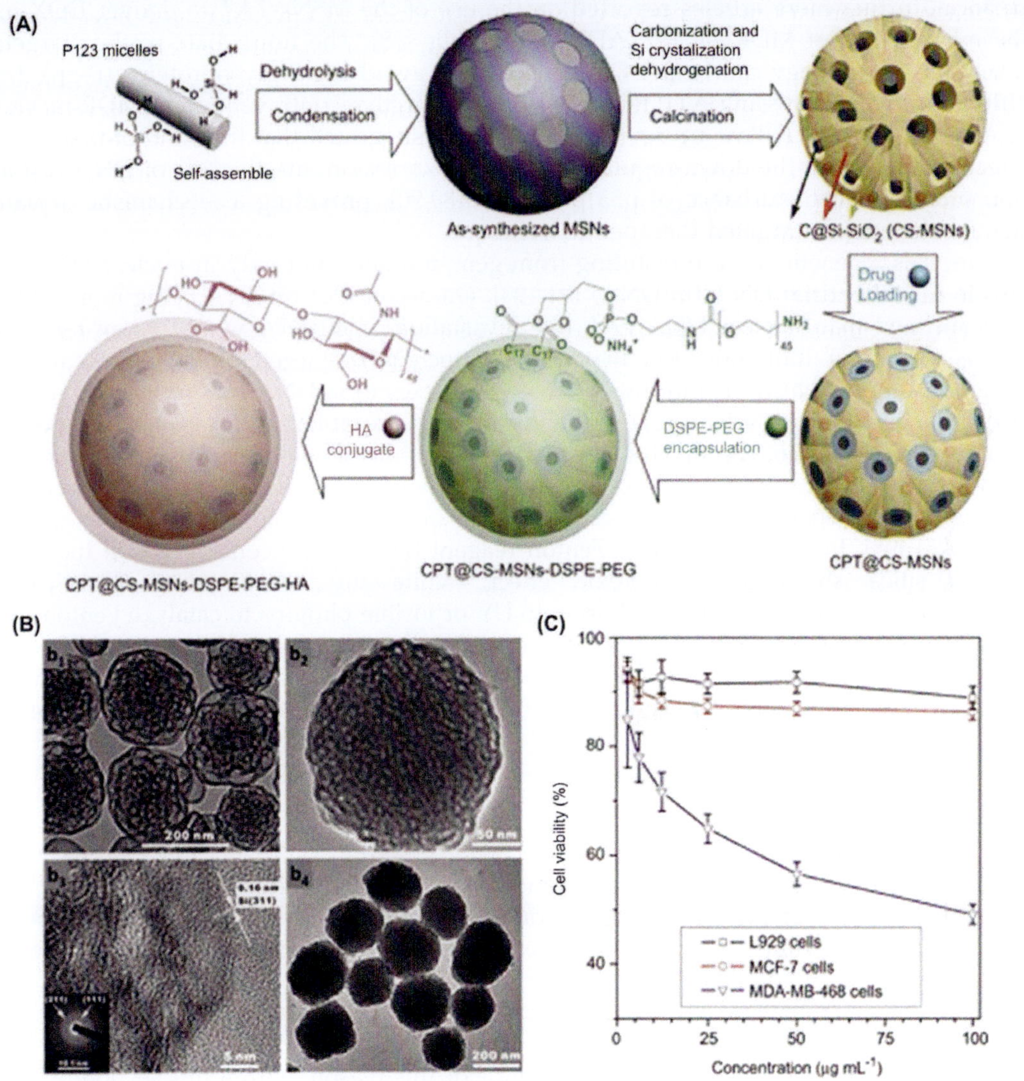


FIGURE 24.4 MSN-HA conjugation for the targeting of CD44 on tumor cells [81]. (A) Synthesis step by step, drug loading, PEG calculation, and HA conjugation. (B) TEM images of as-synthesized MSNs (b1), CS-MSNs (b2, b3), and CPT@CS-MSNs-DSPE-PEG-HA (b4). Inset of (b3) is the corresponding diffraction pattern. (C) Cell-viability assays of L929, MCF-7 and MDA-MB-468 cells after incubation with CPT@CS-MSNs-DSPE-PEG-HA. Copyright 2012, Elsevier Ltd.

receptors importin α and β (karyopherin) and targeting the nuclear pore complexes therefore, assisting NPs to enter the nuclei. Its already conjugated TAT peptide on monodispersed MSNs (MSNs-TAT) for nuclear-targeted drug delivery [81]. In vitro experiments reveal that MSNs-TAT of 50 nm in diameter or smaller can effectively deliver the DOX into the nucleus, resulting in much superior therapeutic efficiencies. According to this perform, further more articles reported on the use of the MSNs-TAT to deliver DOX into the nucleoplasm of MDR MCF-7/ADR cancer cells [88]. The immediate nuclear-targeted drug-delivery strategy can help DOX molecules to avoid the P-glycoproteins (P-gp) drug efflux pump via regressing ATP levels, leading to effective contravention of MDR in MCF-7/ADR cancer cells. Follow-up review further more suggested that this kind of therapeutic effect resulted from the down-regulation of Bcl-2 expression, inactivation of DNA restoration process, and disturbance of p53 pathways [89,90], providing a mechanistic explanation for the nuclear-targeted therapeutic technique.

Cancer is a genetic disease resulting from gene mutation not only in nuclear DNA, but also in mitochondrial DNA (mtDNA) [91–93]. On account of mtDNA being more chemically inclined than nuclear DNA [94], the devastation of mtDNA would be an far better means for tumor inhibition. On that basis, mitochondria-targeting techniques are also designed, among which the binding of $Ru^{2+}$ complexes to mtDNA has been well recorded [95]. The best conjugated aromatic ligands of $Ru^{2+}$ complexes can bind with double-stranded mtDNA helix specifically upon intercalation inside major groove, thus conferring targeting capabilities [96]. Very recently, researchers designed a UCNP@m$SiO_2$ hierarchical design conjugated with $Ru^{2+}$complex for mitochondria-targeting cancer therapy (Fig. 24.5) [98]. In this nanosystem, Fenton reagent ($Fe^{2+}$) had been loaded on the mesoporous silica shell for chemotoxic effect, while the UCNP core ($NaYF_4$:$Yb^{3+}$, $Tm^{3+}$@$NaYF_4$) can nanny NIR irradiation to UV or visible photons to catalyze Fenton reaction for boosting therapeutic final result. However remains less report so far on the construction of mitochondria-targeting nanosystems, it is predicted such a therapeutic approach will be expanded for adding to future oncological researches based on the fast progress nanosynthetic chemistry.

### 24.5.3 Other therapeutic strategies

Besides the immediate delivery of chemodrugs or nucleic acids, MSNs have also been integrated with other organic/inorganic practical materials for croissance of therapeutic functionality. For example, large numbers of organic molecules or inorganic nanomaterials characteristic high photothermal-conversion capability and thus can be used for PTT [97]; some catalytic nanomaterials interact to intratumoral $H_2O_2$ to favor OH generation for chemodynamic therapy (CDT) [99]; 3D hydrogels with excellent mechanical properties can favor osteogenesis for bone tissue engineering. For this reason, within this section, we will elucidate current advances in the incorporation of MSNs together with efficient materials for assorted nanomedical applications, from PTT, US therapy, RT, and CDT to tissue engineering and discuss the structure–property relationships of MSNs that favors the improvement of therapeutic end result.

#### *24.5.3.1 Phototherapy*

PTT have obtained vast concerns nowadays for its affordable, site specificity, and large therapeutic performance [100]. It requires benefit for photoabsorbing agents (PTA) to pass the irradiated NIR optical energy into local hyperthermia, leading to fretting of cancer cells and then tumor elimination. A number of PTA have been investigated previously, such as

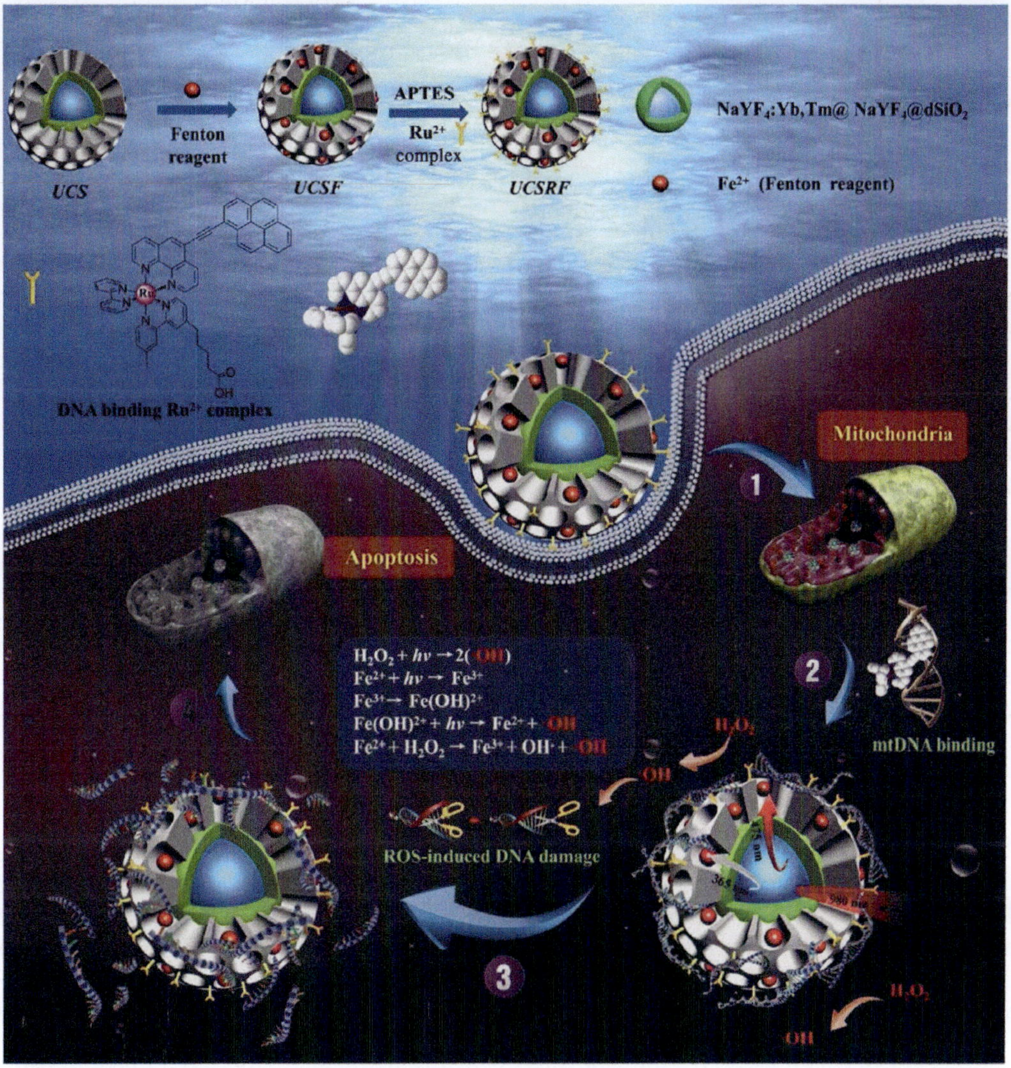


FIGURE 24.5 Conjugation for mtDNA-targeted drug delivery [97]. Copyright 2017, Elsevier Ltd.

gold nanorods [101], copper sulfide NPs [102], carbon-based nanomaterials [103,104], and transition metal dichalcogenide nanosheets [105]. Such a rapid growth has shoved ahead the integration of PTA with MSNs or its hybrid derivatives to combine their individual merits for synergistic enlargement of therapeutic performance, such as great photothermal impact and high-drug loading capability. Au capped $Fe_3O_4@mSiO_2$ designed for combined PTT/chemotherapy [106]. Such a composite nanosystem confirmed high-DOX loading capacity and pH-responsive release actions that is definitely assigned to the electrostatic interaction between DOX and the mesopores of MSNs. Furthermore, in vivo experiment also clarify the synergy between DOX release and photothermal effect for improving therapeutic efficacy. Very recently, Lee et al. also designed rabies virus-mimetic Au@m$SiO_2$ as being a photothermal therapeutic system for the treatment of brain gliomas. The customization of rabies virus glycoprotein 29 (RVG29) on the surface of mesoporous silica shell allows them to avoid the BBB and enter the brain via neuronal paths (possibly over the nicotinic acetylcholine receptor), hence favoring subsequent tumor accumulation and PTT process. The combination of MSNs with two-dimensional (2D) nanosheets have also been investigated very recently [107], for example, the development of a homogeneous coating of mesoporous silica layer onto the surface of 2D $Ti_3C_2$ (MXene) nanosheets for hepatocellular carcinoma therapy [108]. Such a surface-nanopore engineering technique has significantly increased the functionality of 2D $Ti_3C_2$ nanosheets depending on the considerable surface chemistry of MSNs that favors improved dispersity, on-demand drug delivery, and targeting engineering, providing a synergistic therapeutic affect for the treatment of hepatocellular carcinoma.

### 24.5.3.2 *Ultrasound therapy*

As a regular noninvasive diagnostic procedure in clinic, US is also substantially created for several therapeutic applications, such as high-intensity focused US (HIFU) and low-intensity sonodynamic therapy (SDT). Being honored as "bloodless scalpel," HIFU has become successful in the treatment of a range of cancers, including liver cancer, pancreas cancer, breast cancer, kidney cancer, etc. [109]. It can focus the energy of external US radiations on tumor tissues to generate transient hyperthermia and cavitation effects, by which malignant lesions can be ablated. However, the US energy will attenuate within the acoustic propagation channels that will compromise the therapeutic effectiveness of HIFU on deep-seated tumors. Amounts of nanomedicine have been developed and fabricated during the past decade, to affect the acoustic environment of cancer tissues for enhanced therapeutic efficacy. As an evidence of principle, MSNs prepared as "inorganic nanoliposome" to deliver hydrophobic perfluorohexane (PFH) molecules into tumor tissue for the in situ creation of gas bubbles and subsequent improvement of HIFU ablation benefit [110]. This kind of advanced therapeutic end result effects from the biophysical effect of US cavitation that may cause the distinct temperature elevation since the quantity of oscillating bubbles raises [111]. Very recently, we designed a hybrid catalytic nanoreactor for enhanced HIFU cancer surgical treatment, by immobilization of catalase inside the mesopores of dendritic MSN hybrid derivatives [21]. This nanoreactor can effectively catalyze the decomposition of intratumoral $H_2O_2$ along with the generation of O2 bubbles, thus resulting the intensified cavitation effect for higher therapeutic efficiency. SDT will be based on the sonochemistry of US irradiation which could trigger ROS generation during the quick collapse of bubbles (inertial cavitation). Similar to PDT, sonosensitizers can also be requested in SDT to further promote ROS generation under US irradiation for formidable sufficient therapeutic influence. Till now, just like PSs, most of sonosensitizers can also be organic molecules, for example porphyrin derivatives [112], that are hydrophobic and loss of tumor-specificity in biological systems. Therefore, nanocarriers are designed in recent years to deliver these sonosensitive molecules for the advancement of SDT outcome, thus forcing ahead such an emerging ROS-based cancer therapeutic technique for higher medical advances. Recently, a biodegradable nanosystem have been built for highly effective cancer SDT by encapsulating sonosensitizers metalloporphyrin (MnPpIX) within the mesopores of MSN hybrid derivatives. The huge surface area of MSN and its hybrid derivatives provides extremely considerable anchoring sites for covalently immobilization of MnPpIX, thus guaranteeing large loading efficiency while avoiding premature leakage over the circulation process. As a result, the in vivo sonodynamic ROS-generating performance have been considerably increased, leading to specific tumor reductions.

It is observed that SDT is likewise an oxygen-consuming process, in which tumor hypoxia can lead to a low-ROS generation. Hence, nanozyme $MnO_x$ designed and sonosensitizers protoporphyrin (PpIX) into the hollow interior and mesopores of MSN and its hybrid derivatives for hypoxic TME modulation and SDT effectiveness enhancement [113]. The $MnO_x$ component provides catalase-like activity to convert intratumoral $H_2O_2$ into $O_2$, capable of minimize tumor hypoxia for favoring succeeding ROS-generating technique. Within recent paradigm, Chen et al. also prepared fluorocarbon-modified MSN and its hybrid derivatives to load sonosensitizers IR780 for intratumoral oxygen delivery and SDT augmentation. This composite nanosystem provides high-therapeutic efficacy toward hypoxic PANC-1 pancreatic cancer, exhibiting great application prospective in future clinical treatment solution.

#### 24.5.3.3 *Chemodynamic therapy*

CDT is a newly emerged cancer therapeutic technique that uses endogenous biochemical stimuli in TME to activate delivered biomedicines for commencing ROS output, so inducing cancer cell demise and best tumor removal. Several studies already released around the localized Fenton response for tumor-specific therapy by iron nanometallic glasses [114], this kind of chemical-actuated therapeutic technique has acquired robust momentum in the past 2 years, associated with the rise for many TME-responsive ROS-generating nanomedicine for nearing selectivity, specificity, and efficiency of tumor therapy [115], among that Fe-based nanomaterials have achieved broad interest. As an example, Some research already done on MSN and its hybrid derivatives treated the peroxidase-like activity of iron oxide NPs (IONP) [116], which are able to initiating Fenton-like reaction for catalytic $H_2O_2$ decomposition and OH generation [117]. This phenomenal chemical feature has now been extensively developed for cancer CDT [99], by causing oxidative damage of cancer cells for tumor eradication. Due to distinctive structural highlights of mesoporous materials, which benefit the free diffusion of reactants and/or products to assist in catalytic reaction process, IONPs are also incorporated with MSNs for improving CDT efficacy. Previous reports showed introduce of IONPs in the mesopores of MSNs (FeOx-MSNs) by utilizing potassium ferrate as oxidant and iron precursor [118]. The acidic liposomal environment (pH ~ 5.0) stimulate FeOx-MSNs to catalyze the OH generation in TME, leading to successful death of breast cancer ZR75-30 cells. Very recently, engineered FeOx in the framework of MSNs designed to approaching high catalytic overall performance and biodegradability simultaneously, thus further improving the therapeutic benefits.

It can be believed that despite the fact that $H_2O_2$ is overexpressed in tumor tissue and can be catalyzed into OH by delivered Fenton agents, however, its concentration is quite low because of the neutralizing effect of endogenous cellular antioxidant defense mechanisms [119]. To conquer such a restriction for higher $OH^-$ generating efficiency, recently, a constructed a sequential nanocatalyst presented by integrating glucose oxidase (GOD, natural enzyme) and ultrasmall $Fe_3O_4$ NPs (inorganic nanozyme) into dendritic MSNs (DMSNs) to ascertain stepwise catalytic reactions for cancer CDT. The as-loaded GOD could appropriately catalyze the depletion of glucose nutrients in TME for the creation of abounding $H_2O_2$ molecules (biological catalysis), which mementos the subsequent Fenton-like reaction by which $H_2O_2$ can be further decomposed into OH for inducing cancer cell death (chemical catalysis). In this manner, the nontoxic and biocompatible intrinsic substances in TME (i.e., glucose) could be in situ transformed into toxic agents (i.e., ROS) alongside tumors via the established sequential catalytic reactions, resulting in concurrent high-therapeutic outcome and mitigated size effect.

Considering that "catalysis" is needed for most of therapeutic operations, such as PDT, SDT, RT, CDT, and antioxidative therapy, we expect that the as-proposed "nanocatalytic medicine," which bridge "catalysis" and "medicine" for advanced therapeutic strategies, will be further more developed as an crucial subdiscipline in nanomedicine in the future.

## 24.6 Diagnostic application of MSNs

The emergences of MSNs-based therapeutic platforms also have moved frontward the application of MSNs in several diagnostic modalities, such as MRI, fluorescent/luminescent imaging, US imaging, PA imaging, positron emission tomography (PET) imaging, etc., possibly to recognize particular diseases at initial phases, or to guide the therapeutic procedures for adjusted treatment final result. In this section we will sum up latest work in the smart design of MSNs-based diagnostic probes and go over the unification of therapy and diagnosis (i.e., theranostics) that plays a role in ever-greater medical breakthroughs.

### 24.6.1 Magnetic resonance imaging

MRI is a regular noninvasive diagnostic technique that was widely utilized in clinic because of its high-spatial quality, exquisite soft tissue contrast and loss of ionizing radiation [120,121]. To enhance the high quality and specificity of MRI, substantial attempts are already made in developing and fabricating MR contrast agents for giving more extensive visible information [122]. Even though gadolinium(III)-based contrast agents have been popular in clinic, however, they might lead to obvious side effects such as intoxicated kidney function. To resolve this matter, Mn-based contrast agents have been formulated to provide a substitute but useful diagnostic choice. $MnO_x$ developed to engineered species in the mesopores of hollow MSNs for highly efficient T1-weighted MRI [123,124]. Beneath the instruction of such a diagnostic system, the HIFU can be exactly found on the tumor tissue deeply seated in the liver of rabbits, thus clearly bettering HIFU treatment efficacy while mitigating the side effect.

Additional functionalized $MnO_x$ species developed inside the framework of MSNs for TME (pH, GSH)-responsive biodegradability and T1-weighted MRI. The intrinsic breaking-up ability to bond formation in mild reducing or acidic microenvironment favors the "extraction" of paramagnetic $Mn^{2+}$ from the framework, which considerably raises the

T1-weighted MRI outcome for guiding chemotherapeutic method. This work takes full advantage of the chemical characteristics of $MnO_x$ species, to strategy high biodegradability, MRI, and chemotherapy together, ideally contributing to the further progress cancer theranostics. At the same time, according to the metalloporphyrin chemistry, it was chelated Mn ions into the porphyrin ring anchored in the mesopore of MSN and its hybrid derivatives for T1-weighted MRI and cancer SDT. Basically large surface area of MSN-hybrid derivatives boosts the accessibility of $H_2O_2$ to the chelated paramagnetic $Mn^{2+}$, conferring noticeably high MRI overall performance ($r_1 = 9.43\ mM^{-1}\ s^{-2}$, almost twofold that of clinically used Gd-based complex) to guide and monitor SDT process. Furthermore, the TME-responsive biodegradability of MSN-hybrid derivatives also assurances benefit biocompatibility of such a composite theranostic nanosystem. It is actually observed that, within the design of MSNs-based nanoplatforms, the MRI functionality are often designed to guide therapeutic approach (theranostics) instead of to satisfy the only diagnostic purpose.

### 24.6.2 Fluorescent/luminescent imaging

Optical imaging, like fluorescent or luminescent imaging, presents several advantages of diagnostics, for example, high level of sensitivity and simplexes [125]. Even so, the majority of the fluorescent/luminescent probes are organic molecules, that is definitely not qualified enough for in vivo application because of their relatively low biostability which is antagonistic for long-term circulation, plus the short excitation wavelength which are incapable of infiltrating into deep tissue. Due to advances within the production of lanthanide-doped UCNPs, that have offered an achievable tool for converting NIR irradiation into UV or visible light [126,127], these molecular probes have been incorporated with UCNP-based nanosystems for attaining elevated penetration depth and enhanced biostability. Such as, constructed an ultrasensitive nanosensor for in vivo selective hypoxia imaging by encapsulating oxygen indicator tris(4,7-diphenyl-1,10-phenanthroline) ruthenium(II) dichloride ($[Ru(dpp)_3]^{2+}Cl_2$) into the hole of rattle-type UCNP@m$SiO_2$ [125,128]. Below the irradiation of NIR (980 nm), the unconverted visible light (450 and 475 nm) by UCNPs core work extremely well as donor to excite the loaded $[Ru(dpp)^3]^{2+}Cl_2$ via a luminescence resonance energy transfer (LRET) method. Both in vitro and in vivo outcomes indicate that the designed nanosensor can reversibly become quenched or illuminated under normoxic or hypoxic circumstances having a high signal-to-noise proportion, MSN-hybrid derivatives targeting benefit applicability of this type of merged nanosystem for imaging tumor hypoxia.

It really is borne in mind that a lot of chemodrugs, such as DOX, are light-absorbing and is bundled with UCNPs for tracking their release kinetics. This process is dependent on the LRET effect among the donor UCNPs and the acceptor chemodrugs, resulting in the quenching of unconverted luminescence (UCL) emitted from UCNPs. Somewhat, whenever these drugs are released, the LRET influence is going to be weakened and also the quenched UCL will be retrieved. Hence, we can screen chemodrugs release in real-time by detecting UCL signals. Depending on these factors and previous work on NIR-triggered drug release [129], and more completed real-time in vivo quantitative tracking of

NIR-responsive drug release by MR and UCL imaging [129]. The UNCP in the rattle-type UCNP@m$SiO_2$ may up-convert NIR irradiation into visible and UV light, while the DOX molecules can absorb the massive amount of visible light for succeeding UCL monitoring. Most importantly, the rest of the visible light as well as UV light are able to trigger the photoisomerization effect of azobenzene molecules for propelling DOX release. As a consequence, more room in the hollow cavity and mesopores of UCNP@m$SiO_2$ are remaining, favoring the easy access of ambient $H_2O$ molecules to the Gd-UCNP core, therefore maximizing the MR contrast. Researchers deeper fabricated $Ca^{2+}$-doped UCNPs@$SiO_2$@m$SiO_2$ loaded with DOX for drug-delivery monitoring and osteostimulation [130].

### 24.6.3 Positron emission tomography imaging

PET imaging has appeared as a clinical diagnostic technique with ultrasensitiveness and noninvasiveness long time ago [131], that is certainly today playing an essential part in quantitative examination of tumor uptake, along with the analysis of the pharmacokinetic profiles in vivo [132,133]. Frequently, PET imaging demands inner supervision of tracer amounts of a radiolabeled pharmaceutical, which is selective and unique for a goal of curiosity [134].

The previous important advances in the production of radiolabeled nanomaterials [135], among which silica-based NPs are already thoroughly discovered to label with radioisotopes, just like 18F, 64Cu, and 89Zr, for in vivo PET imaging [136]. As a typical paradigm, research workers prepared 89Zr-MSNs via a chelator-free labeling mechanism, to evaluate their in vivo integrity and biological fate [137]. They pointed out that this abundant deprotonated silanol groups in the mesopore or on the surface of MSNs could act as inherent hard oxygen donors to favor firm radiolabeling of 89Zr ($t_{1/2} = 78.4$ h), therefore allowing high-PET imaging efficaciousness. In vivo findings revealed that the detachment rate of 89Zr from the mesopore of MSN was 20-fold sluggish than that of 89Zr through the surface of compressed silica, further more MSN-hybrid derivatives participate in the vital role of mesopores in stabilizing 89Zr. Based on the adaptability of MSNs, it is estimated that more radiolabeled MSNs can be further constructed along with other therapeutic agents for cancer theranostics, thus drive ahead the additional progress this diagnostic technique.

## 24.7 Theranostics application of MSNs

Theranostics is a field of medicine, which brings together precise targeted therapy based on certain targeted diagnostic tests. The theranostics paradigm requires using nanoscience to combine diagnostic and therapeutic applications to create a single agent, permitting diagnosis, drug delivery, and treatment solution response overseeing [138,139]. A particular diagnostic test demonstrates a selected molecular target on a tumor, letting a therapy agent specifically for target that receptor on the tumor, instead of more expand the disease and identify signifies. Featuring theranostic NPs, which merge both therapeutic and diagnostic capabilities in one dosage, has guarantee to direct the biomedical niche in the direction of personalized medicine. Therapeutic approaches such as nucleic acid delivery, chemotherapy, hyperthermia (photothermal ablation), photodynamic, and radiation therapy are combined with one or more imaging functionalities for throughout the vitro and in vivo research. Various imaging probes, such as MRI contrast agents, fluorescent markers, and nuclear imaging agents, can also add onto therapeutic agents or therapeutic delivery gear to accomplish their imaging and, in that way, obtain advice about the trafficking path, kinetics of delivery, and therapeutic efficacy [140]. As opposed to the expansion and utilization of individual materials for both of these goals, theranostics mix these characteristics as one category that has the opportunity to get over unnecessary differences in biodistribution and selectivity between particular imaging and therapeutic agents. Probably the most promising facets of utilizing NPs as therapeutics, diagnostics, and theranostics are their probability of localize within a particular manner toward the site of disease and minimize or get rid of the possible several unfortunate side effects [141,142]. Theranostic nanomedicine can work much better than some other kinds, given that they have superior features in a single system. This unified platform consists of sustained/controlled release, targeted delivery, higher transport efficiency by endocytosis [143], stimulus responsive agent release (i.e., smart delivery), synergetic overall performance (e.g., mix therapy, siRNA co-delivery) [144,145], multimodality diagnosis and/or therapies, and level of quality performances (e.g., oral delivery, escape from MDR protein, autophagy inhibition, etc.) [143,146].

## 24.8 Conclusions and outlook

The unique mesoporous structure, large specific surface area, rich surface chemistry, and tunable system arrangement make MSNs and its hybrid derivatives extremely talented nanoplatforms to become built-in along with other functional element for a lot of downstream biomedical applications. In this chapter, we have displayed the cutting edge of advances within this field and elucidated the essential materials chemistry of those theranostic nanosystems, going to clarify the main therapeutic/diagnostic tips for the development of next-generation innovative nanotheranostic platforms. This chapter commences from the preliminary discussion on the synthetic chemistry of MSNs and its hybrid derivatives, toward the succeeding elucidation on the vital variables of MSNs for the design of biomedicine (i.e., particle size, pore size, shape, chemical composition, and surface customization). Next, current breakthroughs in the architecture of MSNs-based nanosystems for varied therapeutic and diagnostic purposes have already been adequately elucidated, and the fundamental theranostic mechanisms and tactics mentioned in depth. Ultimately, the in vivo biological effects of MSNs-based nanosystems have been outlined, hopefully moving frontward the clinical translation of these emerging systems through the combined attempts among research workers and engineers. It is observed that even though great accomplishments are already done in the realization of MSNs-based biomaterials, there are still several concerns to be resolved, for adding to the further more progress this originating area.

## References

[1] Siegel RL, Miller KD, Jemal A. Cancer statistics, 2019. CA: Cancer J Clin 2019;69(1):7–34.

[2] Lerman C, Rimer B, Blumberg B, Cristinzio S, Engstrom PF, MacElwee N, et al. Effects of coping style and relaxation on cancer chemotherapy side effects and emotional responses. Cancer Nurs 1990;13(5):308–15.

[3] Finger PT. Radiation therapy for orbital tumors: concepts, current use, and ophthalmic radiation side effects. Surv Ophthalmol 2009;54(5):545–68.

[4] Holohan C, Van Schaeybroeck S, Longley DB, Johnston PG. Cancer drug resistance: an evolving paradigm. Nat Rev Cancer 2013;13(10):714–26.

[5] Peer D, Karp JM, Hong S, Farokhzad OC, Margalit R, Langer R. Nanocarriers as an emerging platform for cancer therapy. Nat Nanotechnol 2007;2(12):751.

[6] SalmanOgli A. Nanobio applications of quantum dots in cancer: imaging, sensing, and targeting. Cancer Nanotechnol 2011;2(1–6):1–19.

[7] Sironi L, Borzenkov M, Collini M, D'Alfonso L, Bouzin M, Chirico G. Gold nanostars: synthesis, properties and biomedical application. New York: Springer; 2015. p. 61–74.

[8] Mottaghitalab F, Farokhi M, Fatahi Y, Atyabi F, Dinarvand R. New insights into designing hybrid nanoparticles for lung cancer: diagnosis and treatment. J Control Release 2019;295:250–67.

[9] Li Y, Zhang X, Zhang Z, Wu H, Xu X, Gu Z. Tumor-adapting and tumor-remodeling AuNR@dendrimer-assembly nanohybrids overcome impermeable multidrug-resistant cancer. Mater Horiz 2018;5(6):1047–57.

[10] Tang F, Li L, Chen D. Mesoporous silica nanoparticles: synthesis, biocompatibility and drug delivery. Adv Mater 2012;24(12):1504–34.

[11] Zhang Q, Wang X, Li PZ, Nguyen KT, Wang XJ, Luo Z, et al. Biocompatible, uniform, and redispersible mesoporous silica nanoparticles for cancer-targeted drug delivery in vivo. Adv Funct Mater 2014;24(17):2450–61.

[12] Roggers RA, Lin VS-Y, Trewyn BG. Chemically reducible lipid bilayer coated mesoporous silica nanoparticles demonstrating controlled release and HeLa and normal mouse liver cell biocompatibility and cellular internalization. Mol Pharm 2012;9(9):2770–7.

[13] Sauer AM, Schlossbauer A, Ruthardt N, Cauda V, Bein T, Bräuchle C. Role of endosomal escape for disulfide-based drug delivery from colloidal mesoporous silica evaluated by live-cell imaging. Nano Lett 2010;10(9):3684–91.

[14] Argyo C, Weiss V, Bräuchle C, Bein T. Multifunctional mesoporous silica nanoparticles as a universal platform for drug delivery. Chem Mater 2013;26(1):435–51.

[15] Rosenholm JM, Peuhu E, Eriksson JE, Sahlgren C, Lindén M. Targeted intracellular delivery of hydrophobic agents using mesoporous hybrid silica nanoparticles as carrier systems. Nano Lett 2009;9(9):3308–11.

[16] Cui J, Yan Y, Wang Y, Caruso F. Templated assembly of pH-labile polymer-drug particles for intracellular drug delivery. Adv Funct Mater 2012;22(22):4718–23.

[17] Osterholtz F, Pohl E. Kinetics of the hydrolysis and condensation of organofunctional alkoxysilanes: a review. J Adhes Sci Technol 1992;6(1):127–49.

[18] Wu S-H, Mou C-Y, Lin H-P. Synthesis of mesoporous silica nanoparticles. Chem Soc Rev 2013;42(9):3862–75.

[19] Fedeyko JM, Rimer JD, Lobo RF, Vlachos DG. Spontaneous formation of silica nanoparticles in basic solutions of small tetraalkylammonium cations. J Phys Chem B 2004;108(33):12271–5.

[20] Low PS, Henne WA, Doorneweerd DD. Discovery and development of folic-acid-based receptor targeting for imaging and therapy of cancer and inflammatory diseases. Acc Chem Res 2007;41(1):120–9.

[21] Yang B, Chen Y, Shi J. Mesoporous silica/organosilica nanoparticles: Synthesis, biological effect and biomedical application. Mater. Sci. Eng. R: Reports 2019;137:66–105.

[22] Zhang Z, Wang L, Wang J, Jiang X, Li X, Hu Z, et al. Mesoporous silica-coated gold nanorods as a light-mediated multifunctional theranostic platform for cancer treatment. Adv Mater 2012;24(11):1418–23.

[23] Husseini G, Mohammad A-S, Elsadig A. Systems and methods for targeted breast cancer therapies. Google Patents; 2019.

[24] Croissant JG, Fatieiev Y, Khashab NM. Degradability and clearance of silicon, organosilica, silsesquioxane, silica mixed oxide, and mesoporous silica nanoparticles. Adv Mater 2017;29(9):1604634.

[25] He Q, Zhang Z, Gao F, Li Y, Shi J. In vivo biodistribution and urinary excretion of mesoporous silica nanoparticles: effects of particle size and PEGylation. Small 2011;7(2):271–80.

[26] Lohbauer U, Krämer N, Petschelt A, Frankenberger R. Correlation of in vitro fatigue data and in vivo clinical performance of a glassceramic material. Dent Mater 2008;24(1):39–44.

[27] Omar H, Croissant JG, Alamoudi K, Alsaiari S, Alradwan I, Majrashi MA, et al. Biodegradable magnetic silica@ iron oxide nanovectors with ultra-large mesopores for high protein loading, magnetothermal release, and delivery. J Control Release 2017;259:187–94.
[28] Yu T. In vitro and in vivo impact of silica nanoparticle design on biocompatibility. The University of Utah; 2012.
[29] Chen X, Liu Z, Parker SG, Zhang X, Gooding JJ, Ru Y, et al. Light-induced hydrogel based on tumor-targeting mesoporous silica nanoparticles as a theranostic platform for sustained cancer treatment. ACS Appl Mater Interfaces 2016;8(25):15857–63.
[30] Qiu K, He C, Feng W, Wang W, Zhou X, Yin Z, et al. Doxorubicin-loaded electrospun poly (L-lactic acid)/mesoporous silica nanoparticles composite nanofibers for potential postsurgical cancer treatment. J Mater Chem B 2013;1(36):4601–11.
[31] Chen L, Zhou X, He C. Mesoporous silica nanoparticles for tissue-engineering applications. Wiley Interdiscip Rev Nanomed Nanobiotechnol 2019;11(6):e1573.
[32] Merino S, Martin C, Kostarelos K, Prato M, Vázquez E. Nanocomposite hydrogels: 3D polymer–nanoparticle synergies for on-demand drug delivery. ACS Nano 2015;9(5):4686–97.
[33] Chen X, Liu Z. A pH-responsive hydrogel based on a tumor-targeting mesoporous silica nanocomposite for sustained cancer labeling and therapy. Macromol Rapid Commun 2016;37(18):1533–9.
[34] Chen X, Yuan P, Liu Z, Bai Y, Zhou Y. Dual responsive hydrogels based on functionalized mesoporous silica nanoparticles as an injectable platform for tumor therapy and tissue regeneration. J Mater Chem B 2017;5 (30):5968–73.
[35] Castillo RR, Colilla M, Vallet-Regí M. Advances in mesoporous silica-based nanocarriers for co-delivery and combination therapy against cancer. Expert Opin Drug Deliv 2017;14(2):229–43.
[36] Castaño O, Eltohamy M, Kim HW. Electrospinning technology in tissue regeneration. Methods Mol Biol 2010;811:127–40.
[37] Song G, Wang Q, Wang Y, Lv G, Li C, Zou R, et al. A low-toxic multifunctional nanoplatform based on Cu9S5@ $mSiO_2$ core-shell nanocomposites: combining photothermal-and chemotherapies with infrared thermal imaging for cancer treatment. Adv Funct Mater 2013;23(35):4281–92.
[38] Fatieiev Y, Khashab NM, Croissant JG. Chemical and Biological Engineering, University of New Mexico, 210 University Blvd NE, Albuquerque, New Mexico 87131-0001, United States; Center for Micro-Engineered Materials, Advanced Materials Laboratory.
[39] Ma H, Li T, Huan Z, Zhang M, Yang Z, Wang J, et al. 3D printing of high-strength bioscaffolds for the synergistic treatment of bone cancer. NPG Asia Mater 2018;10(4):31–44.
[40] Rosenholm JM, Zhang J, Linden M, Sahlgren C. Mesoporous silica nanoparticles in tissue engineering–a perspective. Nanomedicine 2016;11(4):391–402.
[41] Ding Y, Li W, Zhang F, Liu Z, Zanjanizadeh Ezazi N, Liu D, et al. Electrospun fibrous architectures for drug delivery, tissue engineering and cancer therapy. Adv Funct Mater 2019;29(2):1802852.
[42] Siegel SE, Moran RG. Problems in the chemotherapy of cancer in the neonate. Am J Pediatr Hematol/Oncol 1981;3(3):287–96.
[43] Murray JM. Some optimal control problems in cancer chemotherapy with a toxicity limit. Math Biosci 1990;100(1):49–67.
[44] Love RR, Leventhal H, Easterling DV, Nerenz DR. Side effects and emotional distress during cancer chemotherapy. Cancer 1989;63(3):604–12.
[45] Zhang J, Yuan Z-F, Wang Y, Chen W-H, Luo G-F, Cheng S-X, et al. Multifunctional envelope-type mesoporous silica nanoparticles for tumor-triggered targeting drug delivery. J Am Chem Soc 2013;135(13):5068–73.
[46] Wen J, Yang K, Liu F, Li H, Xu Y, Sun S. Diverse gatekeepers for mesoporous silica nanoparticle based drug delivery systems. Chem Soc Rev 2017;46(19):6024–45.
[47] Liberti MV, Locasale JW. The Warburg effect: how does it benefit cancer cells? Trends Biochem Sci 2016;41 (3):211–18.
[48] Yang Q, Wang S, Fan P, Wang L, Di Y, Lin K, et al. pH-responsive carrier system based on carboxylic acid modified mesoporous silica and polyelectrolyte for drug delivery. Chem Mater 2005;17 (24):5999–6003.
[49] Tang H, Guo J, Sun Y, Chang B, Ren Q, Yang W. Facile synthesis of pH sensitive polymer-coated mesoporous silica nanoparticles and their application in drug delivery. Int J Pharm 2011;421(2):388–96.

[50] Yang K-N, Zhang C-Q, Wang W, Wang PC, Zhou J-P, Liang X-J. pH-responsive mesoporous silica nanoparticles employed in controlled drug delivery systems for cancer treatment. Cancer Biol Med 2014;11(1):34.
[51] Huang Y, Cole SP, Cai T, Cai Y. Applications of nanoparticle drug delivery systems for the reversal of multidrug resistance in cancer. Oncol Lett 2016;12(1):11–15.
[52] Kim J, Jo C, Lim WG, Jung S, Lee YM, Lim J, et al. Programmed nanoparticle-loaded nanoparticles for deep-penetrating 3D cancer therapy. Adv Mater 2018;30(29):1707557.
[53] Huang K, Ma H, Liu J, Huo S, Kumar A, Wei T, et al. Size-dependent localization and penetration of ultrasmall gold nanoparticles in cancer cells, multicellular spheroids, and tumors in vivo. ACS Nano 2012;6 (5):4483–93.
[54] Jain RK, Stylianopoulos T. Delivering nanomedicine to solid tumors. Nat Rev Clin Oncol 2010;7(11):653.
[55] Nogueira V, Hay N. Molecular pathways: reactive oxygen species homeostasis in cancer cells and implications for cancer therapy. Clin Cancer Res 2013;19(16):4309–14.
[56] Hawk MA, McCallister C, Schafer ZT. Antioxidant activity during tumor progression: a necessity for the survival of cancer cells? Cancers 2016;8(10):92.
[57] Liu H-L, Hua M-Y, Chen P-Y, Chu P-C, Pan C-H, Yang H-W, et al. Blood-brain barrier disruption with focused ultrasound enhances delivery of chemotherapeutic drugs for glioblastoma treatment. Radiology 2010;255(2):415–25.
[58] Cortese D, Kinsey J. Endoscopic management of lung cancer with hematoporphyrin derivative phototherapy. Mayo Clin Proc 1982;543–7.
[59] Moghissi K, Stringer M, Dixon K. Fluorescence photodiagnosis in clinical practice. Photodiagn Photodyn Ther 2008;5(4):235–7.
[60] Jichlinski P, Guillou L, Karlsen SJ, Malmström P-U, Jocham D, Brennhovd B, et al. Hexyl aminolevulinate fluorescence cystoscopy: a new diagnostic tool for photodiagnosis of superficial bladder cancer—a multicenter study. J Urol 2003;170(1):226–9.
[61] Fang W, Tang S, Liu P, Fang X, Gong J, Zheng N. Pd nanosheet-covered hollow mesoporous silica nanoparticles as a platform for the chemo-photothermal treatment of cancer cells. Small 2012;8(24):3816–22.
[62] Huang X, El-Sayed MA. Gold nanoparticles: optical properties and implementations in cancer diagnosis and photothermal therapy. J Adv Res 2010;1(1):13–28.
[63] Manikandan M, Hasan N, Wu H-F. Platinum nanoparticles for the photothermal treatment of Neuro 2A cancer cells. Biomaterials 2013;34(23):5833–42.
[64] Dolmans DE, Fukumura D, Jain RK. Photodynamic therapy for cancer. Nat Rev Cancer 2003;3(5):380.
[65] Knežević NŽ, Trewyn BG, Lin VS-Y. Functionalized mesoporous silica nanoparticle-based visible light responsive controlled release delivery system. Chem Commun 2011;47(10):2817–19.
[66] Yuan Q, Zhang Y, Chen T, Lu D, Zhao Z, Zhang X, et al. Photon-manipulated drug release from a mesoporous nanocontainer controlled by azobenzene-modified nucleic acid. ACS Nano 2012;6(7):6337–44.
[67] Liu J, Bu W, Pan L, Shi J. NIR-triggered anticancer drug delivery by upconverting nanoparticles with integrated azobenzene-modified mesoporous silica. Angew Chem Int Ed 2013;52(16):4375–9.
[68] Yang G, Liu J, Wu Y, Feng L, Liu Z. Near-infrared-light responsive nanoscale drug delivery systems for cancer treatment. Coord Chem Rev 2016;320:100–17.
[69] Strohbehn JW, Douple EB. Hyperthermia and cancer therapy: a review of biomedical engineering contributions and challenges. IEEE Trans Biomed Eng 1984;31(12):779–87.
[70] Jang Jt, Nah H, Lee JH, Moon SH, Kim MG, Cheon J. Critical enhancements of MRI contrast and hyperthermic effects by dopant-controlled magnetic nanoparticles. Angew Chem 2009;121(7):1260–4.
[71] Shin US, Seo J-W, Kundu B, Kim H-W, Eltohamy M. Super-magnetic smart hybrid doxorubicin loaded nanoparticles effectively target breast adenocarcinoma cells. Microporous Mesoporous Mater 2017;243:206–13.
[72] Knežević NŽ, Ruiz-Hernández E, Hennink WE, Vallet-Regí M. Magnetic mesoporous silica-based core/shell nanoparticles for biomedical applications. RSC Adv 2013;3(25):9584–93.
[73] Thomas CR, Ferris DP, Lee J-H, Choi E, Cho MH, Kim ES, et al. Noninvasive remote-controlled release of drug molecules in vitro using magnetic actuation of mechanized nanoparticles. J Am Chem Soc 2010;132 (31):10623–5.
[74] Lee Y, Ishii T, Kim HJ, Nishiyama N, Hayakawa Y, Itaka K, et al. Efficient delivery of bioactive antibodies into the cytoplasm of living cells by charge-conversional polyion complex micelles. Angew Chem Int Ed 2010;49(14):2552–5.

[75] Lee Y, Ishii T, Cabral H, Kim HJ, Seo JH, Nishiyama N, et al. Charge-conversional polyionic complex micelles—efficient nanocarriers for protein delivery into cytoplasm. Angew Chem Int Ed 2009;48(29):5309–12.

[76] Veeranarayanan S, Poulose AC, Mohamed MS, Varghese SH, Nagaoka Y, Yoshida Y, et al. Synergistic targeting of cancer and associated angiogenesis using triple-targeted dual-drug silica nanoformulations for theragnostics. Small 2012;8(22):3476–89.

[77] Sadat SMA, Saeidnia S, Nazarali AJ, Haddadi A. Nano-pharmaceutical formulations for targeted drug delivery against HER2 in breast cancer. Curr Cancer Drug Targets 2015;15(1):71–86.

[78] Tsai C-P, Chen C-Y, Hung Y, Chang F-H, Mou C-Y. Monoclonal antibody-functionalized mesoporous silica nanoparticles (MSN) for selective targeting breast cancer cells. J Mater Chem 2009;19(32):5737–43.

[79] Nairi V, Magnolia S, Piludu M, Nieddu M, Caria CA, Sogos V, et al. Mesoporous silica nanoparticles functionalized with hyaluronic acid. Effect of the biopolymer chain length on cell internalization. Colloids Surf B: Biointerfaces 2018;168:50–9.

[80] Ma M, Chen H, Chen Y, Zhang K, Wang X, Cui X, et al. Hyaluronic acid-conjugated mesoporous silica nanoparticles: excellent colloidal dispersity in physiological fluids and targeting efficacy. J Mater Chem 2012;22(12):5615–21.

[81] He Qianjun, et al. Mesoporous carbon@ silicon-silica nanotheranostics for synchronous delivery of insoluble drugs and luminescence imaging. Biomaterials 2012;33(17):4392–4402.

[82] Sahay G, Alakhova DY, Kabanov AV. Endocytosis of nanomedicines. J Control Release 2010;145(3):182–95.

[83] Davis LI. The nuclear pore complex. Annu Rev Biochem 1995;64(1):865–96.

[84] Wente SR, Rout MP. The nuclear pore complex and nuclear transport. Cold Spring Harb Perspect Biol 2010;2(10):a000562.

[85] Ganta S, Devalapally H, Shahiwala A, Amiji M. A review of stimuli-responsive nanocarriers for drug and gene delivery. J Control Release 2008;126(3):187–204.

[86] Hu M, Chen P, Wang J, Chan C, Scollard DA, Reilly RM. Site-specific conjugation of HIV-1 tat peptides to IgG: a potential route to construct radioimmunoconjugates for targeting intracellular and nuclear epitopes in cancer. Eur J Nucl Med Mol Imaging 2006;33(3):301–10.

[87] Torchilin VP. Tat peptide-mediated intracellular delivery of pharmaceutical nanocarriers. Adv Drug Deliv Rev 2008;60(4–5):548–58.

[88] Pan L, Liu J, He Q, Wang L, Shi J. Overcoming multidrug resistance of cancer cells by direct intranuclear drug delivery using TAT-conjugated mesoporous silica nanoparticles. Biomaterials 2013;34(11):2719–30.

[89] Tjalma WA, Weyler JJ, Bogers JJ, Pollefliet C, Baay M, Goovaerts GC, et al. The importance of biological factors (bcl-2, bax, p53, PCNA, MI, HPV and angiogenesis) in invasive cervical cancer. Eur J Obstet Gynecol Reprod Biol 2001;97(2):223–30.

[90] Burmeister B, Schwerdtle T, Poser I, Hoffmann E, Hartwig A, Müller W-U, et al. Effects of asbestos on initiation of DNA damage, induction of DNA-strand breaks, P53-expression and apoptosis in primary, SV40-transformed and malignant human mesothelial cells. Mutat Res/Genet Toxicol Environ Mutagen 2004;558(1–2):81–92.

[91] Jones JB, Song JJ, Hempen PM, Parmigiani G, Hruban RH, Kern SE. Detection of mitochondrial DNA mutations in pancreatic cancer offers a "mass"-ive advantage over detection of nuclear DNA mutations. Cancer Res 2001;61(4):1299–304.

[92] Kang D, Hamasaki N. Alterations of mitochondrial DNA in common diseases and disease states: aging, neurodegeneration, heart failure, diabetes and cancer. Curr Med Chem 2005;12(4):429–41.

[93] Penta JS, Johnson F, Wachsman JT, Copeland WC. Mitochondrial DNA in human malignancy. Mutat Res/Rev Mutat Res 2001;488(2):119–33.

[94] Kang D, Hamasaki N. Maintenance of mitochondrial DNA integrity: repair and degradation. Curr Genet 2002;41(5):311–22.

[95] Vrabel M, Horakova P, Pivoňková H, Kalachova L, Černocká H, Cahova H, et al. Base-modified DNA labeled by [Ru (bpy) 3] 2+ and [Os (bpy) 3] 2+ complexes: construction by polymerase incorporation of modified nucleoside triphosphates, electrochemical and luminescent properties, and applications. Chem Eur J 2009;15(5):1144–54.

[96] Ossipov D, Gohil S, Chattopadhyaya J. Synthesis of the DNA – [Ru (tpy)(dppz)(CH3CN)] 2+ conjugates and their photo cross-linking studies with the complementary DNA strand. J Am Chem Soc 2002;124 (45):13416–33.
[97] Tang Q, Si W, Huang C, Ding K, Huang W, Chen P, et al. An aza-BODIPY photosensitizer for photoacoustic and photothermal imaging guided dual modal cancer phototherapy. J Mater Chem B 2017;5 (8):1566–73.
[98] Li Z, Zhang Y, Feng N. Mesoporous silica nanoparticles: synthesis, classification, drug loading, pharmacokinetics, biocompatibility, and application in drug delivery. Expert Opin Drug Deliv 2019;16(3):219–37.
[99] Tang Z, Liu Y, He M, Bu W. Chemodynamic therapy: tumour microenvironment-mediated Fenton and Fenton-like reactions. Angew Chem Int Ed 2019;58(4):946–56.
[100] Cheng L, Wang C, Feng L, Yang K, Liu Z. Functional nanomaterials for phototherapies of cancer. Chem Rev 2014;114(21):10869–939.
[101] Huang X, Jain PK, El-Sayed IH, El-Sayed MA. Plasmonic photothermal therapy (PPTT) using gold nanoparticles. Lasers Med Sci 2008;23(3):217.
[102] Wu Y, Wang K, Huang S, Yang C, Wang M. Near-infrared light-responsive semiconductor polymer composite hydrogels: spatial/temporal-controlled release via a photothermal "sponge" effect. ACS Appl Mater Interfaces 2017;9(15):13602–10.
[103] Lin J, Chen X, Huang P. Graphene-based nanomaterials for bioimaging. Adv Drug Deliv Rev 2016;105:242–54.
[104] Liang C, Diao S, Wang C, Gong H, Liu T, Hong G, et al. Tumor metastasis inhibition by imaging-guided photothermal therapy with single-walled carbon nanotubes. Adv Mater 2014;26(32):5646–52.
[105] Li X, Shan J, Zhang W, Su S, Yuwen L, Wang L. Recent advances in synthesis and biomedical applications of two-dimensional transition metal dichalcogenide nanosheets. Small 2017;13(5):1602660.
[106] Cui L, Lin H, Yang C, Han X, Zhang T, Qu F. Synthesis of multifunctional $Fe_3O_4$@ $mSiO_2$@ Au core–shell nanocomposites for pH-responsive drug delivery. Eur J Inorg Chem 2014;2014(36):6156–64.
[107] Liu J, Hu Z-Y, Peng Y, Huang H-W, Li Y, Wu M, et al. 2D ZnO mesoporous single-crystal nanosheets with exposed {0001} polar facets for the depollution of cationic dye molecules by highly selective adsorption and photocatalytic decomposition. Appl Catal B: Environ 2016;181:138–45.
[108] Lin H, Wang X, Yu L, Chen Y, Shi J. Two-dimensional ultrathin MXene ceramic nanosheets for photothermal conversion. Nano Lett 2016;17(1):384–91.
[109] Ibsen S, Schutt CE, Esener S. Microbubble-mediated ultrasound therapy: a review of its potential in cancer treatment. Drug Design Dev Ther 2013;7:375.
[110] Qian X, Zheng Y, Chen Y. Micro/nanoparticle-augmented sonodynamic therapy (SDT): breaking the depth shallow of photoactivation. Adv Mater 2016;28(37):8097–129.
[111] Baker KG, Robertson VJ, Duck FA. A review of therapeutic ultrasound: biophysical effects. Phys Ther 2001;81(7):1351–8.
[112] Miyoshi N, Sostaric JZ, Riesz P. Correlation between sonochemistry of surfactant solutions and human leukemia cell killing by ultrasound and porphyrins. Free Radic Biol Med 2003;34(6):710–19.
[113] Fu J, Li T, Zhu Y, Hao Y. Ultrasound-activated oxygen and ROS generation nanosystem systematically modulates tumor microenvironment and sensitizes sonodynamic therapy for hypoxic solid tumors. Adv Funct Mater 2019;29:1906195.
[114] Zhang C, Bu W, Ni D, Zhang S, Li Q, Yao Z, et al. Synthesis of iron nanometallic glasses and their application in cancer therapy by a localized Fenton reaction. Angew Chem Int Ed 2016;55(6):2101–6.
[115] Yang B, Chen Y, Shi J. Mesoporous silica/organosilica nanoparticles: synthesis, biological effect and biomedical application. Mater Sci Eng: R: Rep 2019;137:66–105.
[116] Su L, Dong W, Wu C, Gong Y, Zhang Y, Li L, et al. The peroxidase and oxidase-like activity of $NiCo_2O_4$ mesoporous spheres: mechanistic understanding and colorimetric biosensing. Anal Chim Acta 2017;951:124–32.
[117] Yang X-J, Xu X-M, Xu J, Han Y-F. Iron oxychloride (FeOCl): an efficient Fenton-like catalyst for producing hydroxyl radicals in degradation of organic contaminants. J Am Chem Soc 2013;135(43):16058–61.
[118] Natour S, Abu-Reziq R. Functionalized magnetic mesoporous silica nanoparticle-supported palladium catalysts for carbonylative sonogashira coupling reactions of aryl iodides. ChemCatChem 2015;7 (14):2230–40.

[119] Chow CK. Interrelationships of cellular antioxidant defense systems. Cellullar antioxidants defense mechanisms, 2. Boca Raton: CRC Press; 1988. p. 217–37.

[120] Ziyeh S, Spreer J, Rössler J, Strecker R, Hochmuth A, Schumacher M, et al. Parkes Weber or Klippel-Trenaunay syndrome? Non-invasive diagnosis with MR projection angiography. Eur Radiol 2004;14 (11):2025–9.

[121] Kupfahl C, Honold M, Meinhardt G, Vogelsberg H, Wagner A, Mahrholdt H, et al. Evaluation of aortic stenosis by cardiovascular magnetic resonance imaging: comparison with established routine clinical techniques. Heart 2004;90(8):893–901.

[122] Stephen ZR, Kievit FM, Zhang M. Magnetite nanoparticles for medical MR imaging. Mater Today 2011;14 (7–8):330–8.

[123] Na HB, Lee JH, An K, Park YI, Park M, Lee IS, et al. Development of a T1 contrast agent for magnetic resonance imaging using MnO nanoparticles. Angew Chem Int Ed 2007;46(28):5397–401.

[124] Chen Y, Yin Q, Ji X, Zhang S, Chen H, Zheng Y, et al. Manganese oxide-based multifunctionalized mesoporous silica nanoparticles for pH-responsive MRI, ultrasonography and circumvention of MDR in cancer cells. Biomaterials 2012;33(29):7126–37.

[125] Okuda K, Okabe Y, Kadonosono T, Ueno T, Youssif BG, Kizaka-Kondoh S, et al. 2-Nitroimidazole-tricarbocyanine conjugate as a near-infrared fluorescent probe for in vivo imaging of tumor hypoxia. Bioconjug Chem 2012;23(3):324–9.

[126] Carling C-J, Boyer J-C, Branda NR. Remote-control photoswitching using NIR light. J Am Chem Soc 2009;131(31):10838–9.

[127] Chen Z, He S, Butt HJ, Wu S. Photon upconversion lithography: patterning of biomaterials using near-infrared light. Adv Mater 2015;27(13):2203–6.

[128] Kiyose K, Hanaoka K, Oushiki D, Nakamura T, Kajimura M, Suematsu M, et al. Hypoxia-sensitive fluorescent probes for in vivo real-time fluorescence imaging of acute ischemia. J Am Chem Soc 2010;132(45):15846–8.

[129] Liu J, Bu J, Bu W, Zhang S, Pan L, Fan W, et al. Real-time in vivo quantitative monitoring of drug release by dual-mode magnetic resonance and upconverted luminescence imaging. Angew Chem Int Ed 2014;53 (18):4551–5.

[130] Wu C, Chang J. Multifunctional mesoporous bioactive glasses for effective delivery of therapeutic ions and drug/growth factors. J Control Release 2014;193:282–95.

[131] Facey K, Bradbury I, Laking G, Payne E. Overview of the clinical effectiveness of positron emission tomography imaging in selected cancers. Health Technol Assess 2007;11(44):iii.

[132] Bahce I, Smit EF, Lubberink M, van der Veldt AA, Yaqub M, Windhorst AD, et al. Development of [11C] erlotinib positron emission tomography for in vivo evaluation of EGF receptor mutational status. Clin Cancer Res 2013;19(1):183–93.

[133] Zhao Y, Sultan D, Detering L, Luehmann H, Liu Y. Facile synthesis, pharmacokinetic and systemic clearance evaluation, and positron emission tomography cancer imaging of 64 Cu–Au alloy nanoclusters. Nanoscale 2014;6(22):13501–9.

[134] Wang J, Maurer L. Positron emission tomography: applications in drug discovery and drug development. Curr Top Med Chem 2005;5(11):1053–75.

[135] Sun X, Cai W, Chen X. Positron emission tomography imaging using radiolabeled inorganic nanomaterials. Acc Chem Res 2015;48(2):286–94.

[136] Anderson CJ, Ferdani R. Copper-64 radiopharmaceuticals for PET imaging of cancer: advances in preclinical and clinical research. Cancer Biother Radiopharm 2009;24(4):379–93.

[137] Wang Q, Hu W, Cai L, Huang Y, Qian Z. Nanomedicines in bone cancer—from diagnostics to therapies. J Biomed Nanotechnol 2017;13(8):911–30.

[138] Menon JU, Jadeja P, Tambe P, Vu K, Yuan B, Nguyen KT. Nanomaterials for photo-based diagnostic and therapeutic applications. Theranostics 2013;3(3):152.

[139] Ahmed N, Fessi H, Elaissari A. Theranostic applications of nanoparticles in cancer. Drug Discov Today 2012;17(17–18):928–34.

[140] Fan W, Yung B, Huang P, Chen X. Nanotechnology for multimodal synergistic cancer therapy. Chem Rev 2017;117(22):13566–638.

[141] Terreno E, Uggeri F, Aime S. Image guided therapy: the advent of theranostic agents. J Control Release 2012;161(2):328–37.

[142] Kumar R, Shin WS, Sunwoo K, Kim WY, Koo S, Bhuniya S, et al. Small conjugate-based theranostic agents: an encouraging approach for cancer therapy. Chem Soc Rev 2015;44(19):6670–83.
[143] Muthu MS, Mei L, Feng S-S. Nanotheranostics: advanced nanomedicine for the integration of diagnosis and therapy. Nanomedicine 2014;9(9):1277–80.
[144] Karimi M, Ghasemi A, Zangabad PS, Rahighi R, Basri SMM, Mirshekari H, et al. Smart micro/nanoparticles in stimulus-responsive drug/gene delivery systems. Chem Soc Rev 2016;45(5):1457–501.
[145] Chen AM, Zhang M, Wei D, Stueber D, Taratula O, Minko T, et al. Co-delivery of doxorubicin and Bcl-2 siRNA by mesoporous silica nanoparticles enhances the efficacy of chemotherapy in multidrug-resistant cancer cells. Small 2009;5(23):2673–7.
[146] Sumer B, Gao J. Theranostic nanomedicine for cancer. Nanomedicine 2008;3(2):137–40.

SECTION IV

# Point-of-care applications

CHAPTER

# 25

# Causes of cancer: physical, chemical, biological carcinogens, and viruses

*Subhayan Das, Moumita Kundu, Bikash Chandra Jena and Mahitosh Mandal*

School of Medical Science and Technology, Indian Institute of Technology Kharagpur, Kharagpur, India

## Abbreviations

| | |
|---|---|
| **AIDS** | acquired immunodeficiency syndrome |
| **ATL** | adult T-cell leukemia/lymphoma |
| **BabA** | blood group antigen binding adhesion |
| **BAP1** | BRCA1 associated protein 1 |
| **BCME** | bis-chloro-methyl-ether |
| **BL** | Burkitt's lymphoma |
| **COX** | cyclooxygenase |
| **CTLs** | cytotoxic T cells |
| **DLBCL** | diffuse large B cell lymphoma |
| **DMBA** | 9,10-dimethyl-1,2-benzanthracene |
| **EBER** | Epstein–Barr virus-encoded small RNA |
| **EBNA** | Epstein–Barr nuclear antigen |
| **EBV** | Epstein–Barr virus |
| **ECM** | extracellular matrix |
| **ESPs** | excretory-secretory products |
| **FGF** | fibroblast growth factor |
| **GLUT** | glucose transporter |
| **HBV** | hepatitis B virus |
| **HCC** | hepatocellular carcinoma |
| **HCV** | hepatitis C virus |
| **HHV** | human herpesvirus |
| **HIV** | human immunodeficiency virus |
| **HL** | Hodgkin's lymphoma |
| **HPRT** | hypoxanthine-guanine phosphoribosyl transferase |
| **HPV** | human papillomavirus |
| **HRS** | Hodgkin Reed–Sternberg |


DOI: https://doi.org/10.1016/B978-0-12-818128-7.00025-3

| | |
|---|---|
| **hTERT** | human telomerase reverse transcriptase |
| **HTLV** | human T-cell lymphotropic virus |
| **IARC** | International Agency for Research on Cancer |
| **IRF** | interferon regulatory factor |
| **JAK/STAT** | Janus kinase/signal transducer and activator of transcription |
| **KS** | Kaposi sarcoma |
| **KSVH** | Kaposi sarcoma herpesvirus |
| **LANA** | latency-associated nuclear antigen |
| **LCR** | long control region |
| **LET** | linear energy transfer |
| **LMP** | latent membrane protein |
| **MALT** | gastric mucosa-associated lymphoid tissue lymphoma |
| **MAPK** | mitogen-activated protein kinase |
| **MCD** | multicentric Castleman disease |
| **MM** | multiple myeloma |
| **MMP** | matrix metalloproteinase |
| **NDMA** | *N*-nitrosodimethylamine |
| **NER** | nucleotide excision repair |
| **NF-κB** | nuclear factor kappa B |
| **NHL** | non-Hodgkin's lymphoma |
| **NMDA** | *N*-nitrosodimethylamine |
| **NOCs** | *N*-nitroso compounds |
| **NOS** | nitrous oxide system |
| **NPC** | nasopharyngeal carcinoma |
| **ORF** | open reading frame |
| **PAHs** | polycyclic aromatic hydrocarbons |
| **PDGF** | platelet-derived growth factor |
| **PEL** | primary effusion lymphoma |
| **PI3K** | phosphoinositide 3-kinase |
| **RB** | retinoblastoma |
| **RNOS** | reactive nitrogen oxide species |
| **ROS** | reactive oxygen species |
| **RTA** | R transactivator |
| **SabA** | sialic acid-binding adhesin |
| **SEA** | *S. haematobium* soluble egg antigen |
| **SMA** | smooth muscle actin |
| **SOD** | superoxide dismutase |
| **TAK** | Tat associated kinase |
| **TGF-β** | transforming growth factor β |
| **TLRs** | Toll-like receptors |
| **TNF** | tumor necrosis factor |
| **TRAF** | TNF receptor-associated factor |
| **UV** | ultraviolet |
| **VEGF** | vascular endothelial growth factor |
| **WHO** | World Health Organization |
| **XAP** | X-associated protein |

## 25.1 Introduction

Cancer is one of the most complex diseases and one of the prominent noncommunicable diseases worldwide. Cancer is primarily a genetic aberration that leads to unregulated replication and cellular growth of the normal cell, which is clearly in excess to the surrounding healthy tissue. Cancerous cells have several distinguishing characteristics, which are documented by Hanahan and Weinberg and denoted as the hallmarks of cancer. The hallmarks are: (1) sustaining proliferative signaling, (2) evading growth suppressors, (3) resisting cell death, (4) enabling replicative immortality, (5) inducing angiogenesis, and (6) activating invasion and metastasis [1]. The authors also suggested four new characteristics in 2011: (1) genome instability and mutation and (2) tumor-promoting inflammation as enabling characteristic as well as (3) reprogramming energy metabolism and (4) evading immune destruction as emerging hallmarks [2]. WHO statistics reported the cancer incidence of every one out of five men and every one out of six women, and death of every one out of eight men and one out of eleven women due to cancer [3]. According to the data, there will be an estimated 18.1 million new cancer cases and 9.6 million cancer deaths by 2018. Lung cancer is presently the most common cancer worldwide and also the leading cause of cancer death. Breast, prostate, and colorectal cancer are the most frequent cancers following lung cancer and colorectal, stomach, and liver cancer follows lung cancer for cancer-associated mortality [3]. Lung cancer most frequently occurs in males, while breast cancer most frequently occurs in females [3]. Much research is going on to find a cure for cancer. However, one of the critical barriers to success is the unavailability of proper cancer models for experimentation. Several in vitro and in vivo models have tried to replicate human cancer. However, most of them have been unable to replicate the uniqueness of human cancer and its microenvironment; one of the reasons being the understanding of the origin and progression of cancer is still unclear.

### 25.1.1 How normal cells become cancerous?

How a normal cell becomes cancerous is still a matter of debate, and many theories exist to explain how the process of carcinogenesis takes place or what are the cells of origin of particular cancer are. Here we would like to discuss only the significant implications in the field of cancer biology and carcinogenesis. Since cancer is a genetic aberration, the underlying mechanisms involved in cancer initiation and progression require detailed understanding. All of the genetic information is encoded in the DNA. Different proteins regulate all of the physiological processes of a living being, and the proteins are made of amino acids. Any variation in the amino acid sequence of the protein can change its structure and function and thus can make it hyperactive or hypoactive [4]. The amino acid sequences of proteins are coded sequentially in specific parts of DNA, which is called a gene moreover, any alteration of the gene sequence can alter the proteins and can cause them to function abnormally. This alteration of gene sequence is called mutation. Since cancers are unrestricted growth of the cells, genes that regulate cellular growth plays a crucial role in cancer development. Genes that enhance cellular proliferation and growth, if are activated by mutation, can cause uncontrolled growth of a particular cell. This type of gene is called as proto-oncogene. When a proto-oncogene becomes mutated and causes oncogenesis, it is referred to as oncogene [5]. Another type of gene which inhibits cell growth and proliferation are supposed to check the uncontrolled growth. If the mutation renders these genes nonfunctional, there is uncontrolled growth and proliferation of the cell [6]. These regulatory genes are called the tumor suppressor gene, and they are equally crucial for carcinogenesis. Cancer is a multistep process with one or more oncogenes activation and one or more tumor suppressor genes inactivation. Although there are some exceptions of this rule, usually, a sequential activation and inactivation of multiple oncogenes and tumor suppressor genes lead to the cancerous changes [7].

## 25.1.2 Stages of carcinogenesis

The progression of cancer is a complex phenomenon and can be subdivided into several stages. Depending upon epidemiological evidence and animal models, carcinogenesis is divided into three stages: (1) initiation, (2) promotion, and (3) progression [8]. The mutation occurs in all three stages of carcinogenesis and gene expression changes [9]. These changes enable the cells to grow uninhibitedly, changing the balance of apoptotic and proliferative signaling in favor of cellular proliferation. In the initial two stages, the apoptotic and proliferative signaling usually try to balance each other and thus restrict the cells from a spurt of growth. However, in the progression stage, this balancing is lost, and cells become cancerous growing uncontrollably [10].

### *25.1.2.1 Initiation*

Initiation is the first stage of carcinogenesis and also the most unappreciated stage. The cells acquire a mutation in this stage but show no sign of occult carcinogenesis. Usually, the cells are not rapidly proliferating or proliferate clonally, which usually cannot be identified with the normal process [11]. A debate exists within the scientific community on which type of cells usually acquires the initiating mutation and how long their latent period exists. Recent evidence suggests that cancer-initiating cells might be a stem cell or progenitor cell which has selective self-renewal abilities [12].

### *25.1.2.2 Promotion*

The stage of promotion is most appreciated in the chemical carcinogenesis, where chemical carcinogens with low-carcinogenic potential were able to induce cancer in experimental models [13]. Usually, this stage is well established in the effects of carcinogens, which do not damage DNA directly [14] instead influence the cellular signaling to either increase the proliferative potential of the cell or inhibit the cellular death mechanisms [15]. The promotion stage is usually slow and reversible. It may take months to years to progress up to occult cancer.

### *25.1.2.3 Progression*

Usually, the first two stages of initiation and promotion histologically lie between preneoplastic lesions to benign neoplasia. However, in the progression stage, the affected cells acquire malignancy, and it is an irreversible stage. All the hallmarks of cancer can be realized in this stage, and there is increased genome instability. This is the final stage of cancer progression [16].

### 25.1.3 Carcinogens

Mutations in proto-oncogene and tumor suppressor gene drive a normal cell to become cancerous. This mutation can be acquired in several ways. Some of the individuals inherit the mutation from the parents, and these types of cancer are called hereditary cancer. This type of cancer accounts for about 5% of all cancers [17]. Sometimes these familial cancers occur as a group of disorders affecting multiple organs or tissues. This type of cancer is called familial cancer syndrome [17]. However, most of the cases, environmental factors are responsible for causing oncogenic mutations and cancer progression. These agents are called carcinogens and are the most critical cause of carcinogenesis. The presence of inherent mutations can augment the effects of carcinogens. According to the origin and mode of action, the carcinogens can be divided into physical, chemical, and biological carcinogens. The mechanisms of different carcinogens and the carcinogenesis process are presented in Fig. 25.1. This chapter deals with each of the subtypes of carcinogens and their mechanisms individually in greater detail.

## 25.2 Physical carcinogens

A carcinogen is any substance which can promote carcinogenesis or cancer formation by physical, chemical, or biological process [18]. A carcinogen can alter the genomic integrity of living beings, can cause metabolic alteration, or interfere with the physiological functions of healthy cells to induce cancer. Physical carcinogens are a selective variety of carcinogen which usually promotes carcinogenesis via their physical attributes avoiding any direct or indirect chemical reaction [18]. Physical carcinogens comprise of several agents such as electromagnetic radiations, ionizing radiations, mechanical traumas, extreme temperatures, solid materials, and colloidal materials. The materials causing carcinogenesis can be further divided into hard materials, soft materials, nonfibrous particles, fibrous particles, and gel materials [19]. Exposure to most of the physical carcinogens is environmental or occupational, and these carcinogens should be considered as an essential public health, economic, and social problem. Because of the nature of exposure of these agents, most of the physical carcinogens are either identified retrospectively from clinical data or are identified during experiments mostly involving animals. The first scientific report of a physical carcinogen was found in experimental models using rats where Bakelite disks implanted in rats, triggered local fibrosarcomas [19].

### 25.2.1 Mechanism of action of physical carcinogens

The molecular mechanisms of physical carcinogens are most diverse amongst all types of carcinogens. The difference in their mechanisms is mainly because of the diversity in their nature and origin. However, one common mechanism is that all of them cause DNA damage by their physical attributes either directly or indirectly. Nonionizing electromagnetic waves can directly damage DNA [20], while ionizing radiations usually generates reactive oxygen species (ROS), which in turn damages the DNA to induce cancer [21]. Hard and soft materials can cause cancer through chronic irritation, ROS generation, or

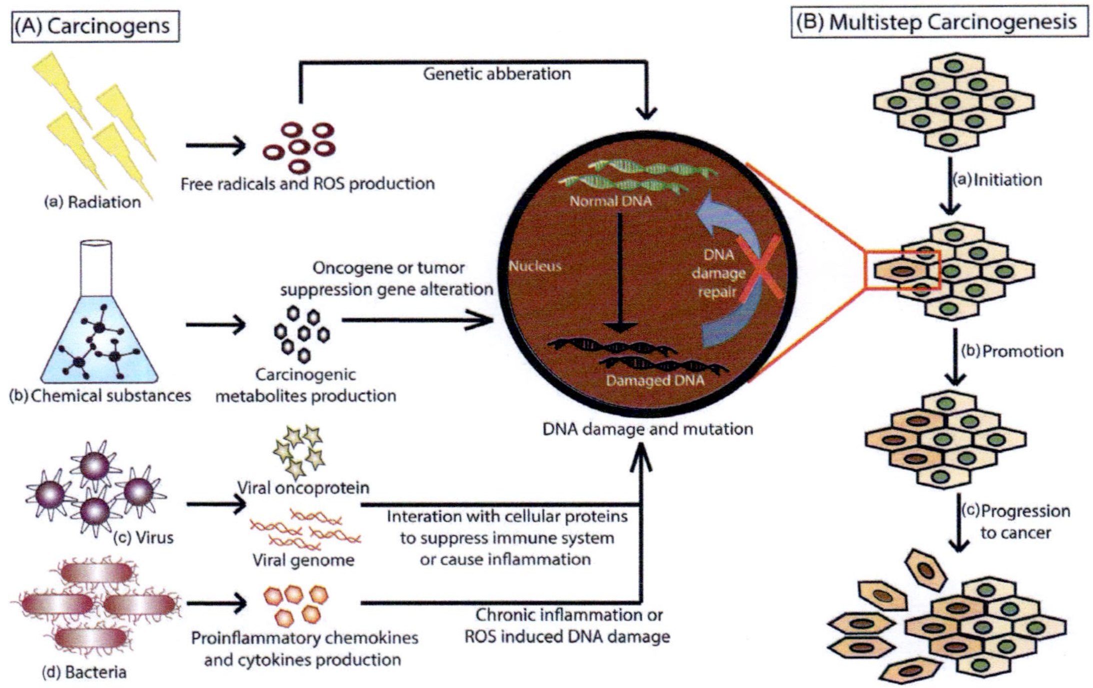


FIGURE 25.1 **Mechanism of carcinogenesis through carcinogens.** (A) Different carcinogens and their mechanism of action. (a) Physical, (b) chemical, (c) viral, and (d) bacterial carcinogenesis. (B) The carcinogenesis process is divided into (a) initiation, (b) promotion, and (c) cancer progression. Most of the carcinogens cause DNA damage by different mechanisms that influence the different stages of carcinogenesis.

damage to chromosomal DNA [19]. Repetitive trauma can cause carcinogenesis by constant irritation that changes the molecular mechanisms in the cell [22]. We have summarized common types of physical carcinogens and their mechanism of actions in Table 25.1.

### 25.2.2 Electromagnetic radiation

Electric and magnetic fields are created by the flow of electron through metal wires and other electronic devices. In modern days the exposure to the electromagnetic field is inevitable. However, whether these exposures measure up to occult carcinogenesis is still a matter of debate. Electromagnetic radiations can be a high frequency, which is also called ionizing radiation or low frequency, which generally originates from magnetic fields from electric power lines and appliances, microwaves, radio waves, visible light, and infrared radiation. The underlying mechanism by which these two radiations induce cancer differs drastically [36]. Various familiar sources of the low-frequency electromagnetic waves are listed in Table 25.2.

**TABLE 25.1** Different physical carcinogens and their mechanism of action.

| Sl no | Physical carcinogen | Mechanism of action | Ref. |
|---|---|---|---|
| 1. | Electromagnetic radiation | Mostly unknown. Possible mechanisms may involve: amplification of innate electric currents in tissues and cells | [23–25] |
| 2. | Ionizing radiation | DNA damage by the production of free radicals | [26–29] |
| 3. | Hard and soft materials | Oxidative stress, chromosomal damage, protein adsorption | [30–32] |
| 4. | Air pollutants | DNA damage by hyper or hypomethylation | [33] |
| 5. | Trauma | Chronic irritation | [34,35] |

**TABLE 25.2** Different common sources of low-energy electromagnetic waves.

| Common household sources | Other sources |
|---|---|
| Electrical appliances such as shavers, hairdryer, etc. | Radio and television signal |
| Cell phone | Radar |
| Laptop, computer | Magnetic resonance imaging (MRI) device |
| Television | Power line |
| Cordless phone | Satellite station |
| Microwave oven | |
| Wireless local area network | |
| Electrical wiring | |

Low-frequency electromagnetic waves cannot damage DNA directly and thus cannot cause cancer by DNA damage. However, there are several possible interactions with human physiology, which could lead to carcinogenesis. Most of the case studies conducted on human subjects exposed to electromagnetic radiation or on animal models have been inconclusive [37–40]. However, most of these studies were short term and may suffer from lack of proper controls. Some long-term retrospective studies have identified a possible association between childhood cancers like leukemia and brain tumors [41–43]. It is possible that these low-frequency radiations have a more adverse effect on developing cells of children than adult cells. A few studies looked into the emissions from household devices [44], cell phones [45], and Wi-Fi devices [46]. However, no association was found with any childhood cancer from these studies. Studies conducted in the 1980s and early-1990s to evaluate workplace exposure of low-frequency electromagnetic waves showed a possible increased risk of leukemia, brain tumors, and male breast cancer [23]. However, methodological limitations make these studies less reliable. Recent studies conducted on the Fischer 344 rats suggested that there is an increased chance of thyroid c cell neoplasm with the exposure to the electromagnetic waves [23]. However, this study also did not explore the possible mechanisms of these carcinogenic changes. Because of the conflicting results, the low-frequency ionizing radiations are categorized as possibly carcinogenic to humans (Group 2B) by the International Agency for Research on Cancer (IARC) [23].

Since the proper mechanism of carcinogenesis by low-frequency electromagnetic waves is not known, several theories have been proposed. Some theories suggest that these radiations may cause amplification of innate electric currents in tissues and cells or modulate those currents via resonance with local field forces [24]. This action may be crucial for altering the voltage-gated channels like $Ca^{+2}$ or $K^{+}$ channels and changes the membrane potential. These phenomena can alter several physiological processes, including increased cellular proliferation and cause carcinogenesis. Low-frequency radiation exposure may suppress the production of melatonin by the pineal gland. This phenomenon can promote the incidence of cancer, especially breast cancer. Low-frequency radiation has been shown to regulate the circadian rhythms in experimental animals and interfere with the normal circadian rhythm. Some of these experiments suggest that the suppression of melatonin has increased the frequency of breast cancer [25].

### 25.2.3 Ionizing radiation

Ionizing radiation causes ionization and generation of free radicals upon exposure to the tissue to induce DNA damage. The radiation dose is measured as the energy absorbed per unit mass. The unit of radiation dose is expressed as Joules per kilogram, which is also known as Gray (Gy). Absorbed energy dose determines the carcinogenic potential. Linear energy transfer or LET ($L$) is another unit of radiation dose which is the deposition rate of energy ($E$) on the absorbing medium per unit distance ($l$) traversed by the radiation ($L = dE/dl$); $L = \text{keV mm}^{-1}$ when the unit of distance traversed is expressed in millimeter [47]. Neutrons, α-particles, heavy ions, and pi mesons are high-LET radiations and γ-rays, X-rays, electrons are low-LET radiations. High-LET radiation deposits higher amounts of energy in the absorbing medium than low-LET radiation. High-LET radiations cause more damage to the cells and tissues than low-LET radiations because, at the similar dose, low-LET radiations cause induction of the same number of free radicals sparingly inside a cell, while high-LET radiations transfer majority of the energy to a confined region within the cell to induce localized damage. This high-LET radiation-induced localized DNA damage is more difficult to repair than the low-LET radiations induced diffused DNA damage [48].

The primary mode of carcinogenesis by the ionizing radiations is DNA damage by the production of free radicals. The absorbed dose of the radiation and the generation of free radicals are interlinked. There is also evidence that this can be correlated to measure the amount of DNA damage caused by ionizing radiation [26]. The DNA damage can be single- or double-stranded damage which can be repaired by the internal DNA repair mechanism present in the cell. Low-LET radiations tend to do single-stranded damage, which could be easily repaired. However, failure to repair the damaged DNA leads to the introduction of mutation, which might ultimately lead to oncogenic progression. UV radiation has a unique property of causing thymine dimers in the DNA [49]. NER mechanism helps to repair these thymine dimers in humans. Sometimes the failure of these repair mechanisms leads to a skin disease known as xeroderma pigmentosum [50].

Cell's oxidative status has a significant role both in the long- and short-term effects of ionizing radiation in the cell. Radiation can produce ROS for a short time to a long time after exposure [51]. ROS can cause DNA damage immediately after radiation exposure or can have a long-term effect to induce repeated DNA damage of failure of the repair system, which causes cancer. Radiations can cause several chromosomal aberrations in addition to DNA damage, which can also contribute to carcinogenesis. Radiation may induce abnormal intrachromosomal crossing over involving individual or both chromatids. It may also result in trisomy, translocations, or deletions by nondisjunction of homologous chromosomes [49]. Delicate sites are more susceptible to chromosomal fragmentation. High-radiation exposure causes such aberrations of the hypoxanthine-guanine phosphoribosyl transferase (HPRT) locus on the X-chromosome [52]. Thus, ionizing radiation can incorporate genome instability, which is one of the critical features of carcinogenesis. The inability to repair genetic aberration can cause pooled mutations. This mutation can occur anywhere in the genome, but when it occurs at the oncogene or tumor suppressor gene, these mutations can activate tumorigenesis [53].

Several studies tried to correlate tumorigenesis with radiation. Some of the significant evidence can be found in the retrospective studies of the Chernobyl accident in 1986 and atomic bomb survivors in Hiroshima and Nagasaki. In the areas affected by the Chernobyl accident, there was an alarmingly increase in the incident of thyroid cancer due to cesium-137 and various isotopes of iodine, including iodine-131 [27]. In the atomic bomb survivors, there was an increased rate of solid tumors in adults and leukemia in children [28]. Occupational exposure to radiation is frequent in nuclear plants, medical industries, mining, and the aviation industry. Some large scale studies found correlations between occupational exposure with solid tumors and leukemia [54,55]. Medical exposure is one of the controversial areas in this field. X-ray exposure is frequent in many patients, and also radiation is a mode of treatment for cancer patients. Retrospective studies conducted on the long-term exposed patients have some correlation with medical exposure and the occurrence of different cancer like sarcoma and leukemia [29].

## 25.2.4 Hard and soft materials

Materials are one of the most common forms of carcinogenesis. Several materials can cause cancer directly or indirectly and include both metals and nonmetals from natural and synthetic origin. These can also take forms of fiber, foam, gel, or solid particulates [56].

### 25.2.4.1 *Asbestos*

Asbestos is the name given to the six naturally occurring silicate fibers. These fibers have unique qualities, which make them very desirable for several industries, including building, furnace, pipe, and railway industries. Therefore, environmental exposure to asbestos is widespread. Several cancers are related to asbestos exposure. Lung cancer and mesothelioma are common among those cancers [30]. However, there are also reports of cancer occurring in the larynx, ovary, stomach, and large intestine after asbestos exposure [30]. Asbestos exposure causes inflammation of the lung tissue, which ultimately causes fibrosis and chronic irritation. This chronic irritation ultimately may lead to cancer [57]. However, there are other mechanisms which might play a crucial role. Several hypotheses have been drawn to explain the mechanism of carcinogenesis by asbestos. There are three dominant theories (1) the "oxidative stress theory" explains that the carcinogenesis occurs due to the ROS generated by phagocytic cells which engulf the asbestos fibers; (2) the "chromosome tangling theory" postulates that chromosome is damaged directly by the asbestos fibers; and (3) the "theory of adsorption of many specific proteins as well as carcinogenic molecules" states that several proteins and chemicals are accumulated by the asbestos fibers which might have the carcinogenic potential. However, none of the theories is proven [31].

### 25.2.4.2 *Erionite*

Erionite is a naturally occurring fibrous form of zeolite. Most of the exposure of erionite occurs in the mining industry. It is most commonly associated with lung cancer and mesothelioma. The mechanism of carcinogenesis is similar to the asbestos. However, a familial predisposition is seen in the people with BAP1 mutation [58].

### 25.2.4.3 *Nonfibrous particulate materials*

Nonfibrous particulates that have a prominent role in carcinogenesis include powdered metallic cobalt, nickel, and crystalline silica. Cobalt can cause rhabdomyosarcoma in experimental animal models [59]. Nickel and silica have also shown similar carcinogenic effects in producing sarcoma and squamous cell carcinomas in rats, respectively [32]. Silica is also shown to induce lung and pleural cancer as well as lymphomas when injected in the pleural cavity [19]. However, the exact mechanism by which these particulate materials work is still mostly unknown.

### 25.2.4.4 *Air pollutants*

Particulate air pollutants possess a severe risk to the health of the population. The air pollutants can enter the respiratory system and cause an inflammatory response. The finer pollutants cause more problems since they can enter the deeper parts of the respiratory system [60]. Long-term exposure to air pollutants increase the risk of lung cancer as well as several other pulmonary disorders [61]. The exact mechanism of carcinogenesis is not known. However, it is suggested that these air pollutants can cause DNA damage by hyper- or hypomethylation, which in turn can cause carcinogenesis [33].

#### *25.2.4.5 Gel materials*

Gel materials causing carcinogenesis are mainly silicone gels used for prosthesis, including artificial breast prosthesis. Animal studies suggest that these gels can cause fibrosarcoma when implanted subcutaneously for a longer duration. However, there is no direct evidence suggesting their correlation with any cancer of human origin [62].

### 25.2.5 Trauma

Different physical trauma is associated with various tumors in human beings and animal models. The role of trauma in the process of carcinogenesis can be explained in several ways. (1) Trauma as the main causative factor: chronic irritation by repeated trauma can cause changes in cellular and molecular levels which might induce carcinogenic changes. A significant example of this is the metaplasia of esophageal mucosa by chronic gastric irritation, which ultimately leads to cancer [34]. (2) Trauma can act as a contributing factor to preexisting neoplastic conditions or lesions. Various animal experiments demonstrate the possibility of trauma-inducing tumors [63–65]. In humans, skin cancer arises secondary to the burns, scars, sinuses, and fistulas, or at sites of chronic infection and inflammation [66]. Some literature suggests a possible correlation between rhabdomyosarcoma and trauma in humans [35].

## 25.3 Chemical carcinogens

The chemical carcinogenesis was first identified by Dr. Katsusaburo Yamagiwa and Koichi Ichikawa in 1915. They observed coal tar application on rabbit ears induced the skin squamous cell papillomas and carcinomas development [67]. Dr. Beremblum and Shubik investigated the carcinogenicity effect of polycyclic, croton oil, and aromatic hydrocarbons on mice and found the development of skin cancer through multiple stages [13]. Another instance of carcinogenesis was reported with low doses of 9,10-dimethyl-1,2-benzanthracene (DMBA) on the skin. Chemical carcinogens are the most diverse group of carcinogenic agents with high-binding efficiency to the genetic material of the living systems. They may bind covalently or noncovalently to the DNA, RNA, and proteins. During the initial phase of cancer, they have tremendous efficacy in gene mutation and alternation of the transcription mechanism of the genomes. They do not possess any common structure, and most of them are activated by the metabolic alternations: alkylation, oxidation, or dealkylation. The chemical compounds usually do not directly stimulate the cancer progression; rather, their metabolites within the organisms are frequently responsible for carcinogenesis. These compounds are gradually transformed into electrophilic reagents and are subjected to calcification or corrosion of the same living cell depending on where the chemical compound is subjected to enzymatic reaction. Then it is converted to a carcinogenic substance.

## 25.3.1 Mechanisms of chemical carcinogenesis

Chemical carcinogens can contribute to carcinogenesis through several possible mechanisms. Chemical carcinogens can damage DNA directly or indirectly causing mutation in vital oncogene or tumor suppressor gene, which causes cancer. The chemical carcinogens can also alter the DNA repair mechanisms, which cause increased genomic instability and cancer progression [68]. Some chemical agents act as cancer promoters and cause mutation in the genome. This type of carcinogen is very active and acts rapidly upon activation [69]. Some chemical carcinogens are promoters of cancer and act rather slowly. These carcinogens usually require long-term exposure and generally influence cellular physiology to promote carcinogenesis [70]. Some of these chemicals act as immunosuppressors and promote cancer progression by permitting cancerous cells to escape immune surveillance [71]. A few chemicals which are derivatives of hormones or can act as hormones upon metabolic activation can induce carcinogenesis without inducing mutation. These agents usually influence cellular growth and often act on the promotion stage of cancer. These carcinogens are called non-DNA reactive carcinogens [72]. Some chemical agents can alter the epigenetic structure of DNA. Demethylation of promoter regions at the CpG islands may cause the upregulation of oncogenes. Similarly, hypermethylation of the promoter region can cause the silencing of the tumor suppressor gene. Several carcinogens also alter histone protein. Histone modification can also activate or deactivate the gene, which can lead to carcinogenesis [73].

## 25.3.2 Types of chemical carcinogens

Chemical carcinogens can be categorized into some chemical classes like polycyclic aromatic hydrocarbons (PAHs), aromatic amines and nitro compounds, alkylating agents, and *N*-nitroso amines and amides. The *N*-nitroso compounds (NOCs) are classified as experimental carcinogens instead of occupational carcinogens. Moreover, occupational exposure of nickel, chromium, arsenic, and asbestos dust are majorly linked to human cancers among the inorganic agents.

### *25.3.2.1 Aromatic amines*

Aromatic amines are the class of chemical compounds that possess an elevated risk of carcinogenesis. This group of organic compound is proven highly carcinogenic, both experimentally in vitro and in vivo. Innumerable studies have demonstrated the carcinogenicity of these compounds when administered to different experimental animals [74]. Bladder cancer is the most prevalent type of cancer caused by exposure to the aromatic amines and mostly seen in the occupationally exposed persons. Also, tobacco smoke is the major source of the aromatic amines, and bladder cancer is predominant in smokers [75]. Many workers develop bladder cancer due to exposure of 2-naphthylamine,4-aminobiphenyl, and benzidine in the industries [76]. Benzidine has major use in analytical chemistry. Therefore, a safer alternative is preferable due to its carcinogenicity. Among the benzidines, *o*-toluidine and 3-amino-9-ethyl carbazole possess a high risk of carcinogenesis. Likewise, the carcinogenicity of 1-naphthylamine becomes high during contamination with the 2-naphthylamine [77].

#### 25.3.2.2 N-*Nitroso compounds*

NOCs cause cancer in a diverse group of animal species, including higher primates. These compounds have a carcinogenic effect on multiple organs in animals [78]. Supporting evidence showed the occurrence of multiple cancer types such as colon, nasopharyngeal, esophageal, and gastric cancer in human species [79,80]. Among the aromatic nitroso compounds, *N*-nitrosodimethylamine (NDMA) is considered to be highly carcinogenic agents exposed to humans [81]. Exposure of humans to NOCs sources such as tobacco products, drugs, diet, and occupational environments possess a high risk of developing cancer [80]. Endogenously nitrite reacts with the degradation products of amino acids in the stomach that give rise to nitrosamines and nitrosamides. This endogenous synthesis contributes to a large portion of cancer-causing NOC agents [82]. Studies have revealed that high consumption of red meat increases the risk of bladder and colon cancer due to the generation of NOCs in the colon [82–85]. Among other nitroso compounds, *N*-nitrosopiperidine, *N*-nitrosomorphiline, and *N*-nitrosopiperizine confer an increased risk of developing bladder, stomach, multiple myeloma, esophagus, and prostate cancer [86].

#### 25.3.2.3 *Dyes*

The ever-increasing use of the dyes in the food sector, the cosmetic sector, etc., holds the high chances of cancer in the blood, bone marrow, and skins. Hair dyes consist of various chemicals and are dependent on dye types. Generally, most of the hair dyes are composed of *p*-phenylenediamine and aminophenyl, which are suggested to be possible carcinogens or mutagens in vitro experiments. Extensive uses of hair care dyes may prompt the development of bladder cancer, breast cancer, non-Hodgkin lymphoma (NHL), and myeloma [87]. Most of the food dyes are synthesized from coal tar, and petroleum has long been controversial because of its adverse effect on human health, and it has also proven in laboratory animals [88,89]. Several azo and other dyes have potential as carcinogenic agents. Methyl or methoxy groups in the dyes sometimes increase the carcinogenic activity of the dyes. 3′-Methyl-4-dimethylaminoazobenzene is a dye used as a dye for coloring polishes, wax products, and soap is a potent inducer of liver tumor, observed in experimental animals [90]. Similarly, *o*-aminoazotoluene and auramine are carcinogenic agents against bladder cancer [91].

#### 25.3.2.4 *Alkylating agents*

The alkylating agents interact directly with the biological materials and do possess the irritation, toxicogenic, mutagenic, and carcinogenic properties. They do not undergo any prior metabolic alternations and act on living systems as such. The alkylating agents are mainly the chemicals of major industrial importance. Among them, sulfur mustard or mustard gas was used during the World war I, which is the potent carcinogen in lungs [92,93]. Bis-chloro-methyl-ether (BCME) is an alkylating agent and a recognized carcinogen. Spontaneously chloridic acid and formaldehyde reaction give rise to BCME. Earlier BCME was used in the organic synthesis. This has been proved to cause occupational lung cancer in the people exposed to the alkylating agents [94]. Very less amount of this BCME is detected in the air. However, BCME is present at a high level in the Friedel crafts mixture containing formaldehyde and commercial chloroalkylation [95]. Methylation of diazomethane is also hazardous to human health. Methylnitrosourea, the precursors of diazomethane is known to alkylate the DNA and induce breast cancer with repeated exposure [96,97]. Both the epidemiological and toxicological data suggest a probable connection between methyl bromide exposure and the risk of developing prostate cancer [98]. Similarly, methyl chloride and methyl iodide, used as the methylating agents are highly mutagenic and possess the risk of developing cancer [99].

#### 25.3.2.5 Natural carcinogens

Naturally occurring nonvolatile metabolites generated by a variety of plants and microorganisms have carcinogenic properties. These are as toxic as the isolated chemicals; if not handled with precautionary measures may cause adverse health issues. Aflatoxin is a toxic compound generated in *Aspergillus* molds that infest staple crops such as rice, peanuts, maize, and wheat. Exposure to aflatoxins possesses the risk of developing liver cancer through the consumption of aflatoxin infested crops [100]. Sterigmatocystin is a mycotoxin and a precursor of aflatoxin B1. This compound is highly carcinogenic that affects several species of crops and experimental animals [101].

#### 25.3.2.6 Inorganic carcinogenic agents

Many inorganic compounds are considered as human carcinogens and pose serious health issues after environmental or occupational exposure. Inorganic arsenic is highly carcinogenic in long-term exposure, affecting the urinary bladder, skin, and lungs. The drinking water contaminated with arsenic is the prime source of arsenic exposure to humans [102]. Plenty of evidence is there in favor of cadmium as the potent human and animal carcinogen. Occupational exposure to cadmium possesses the risk of lung cancer [103]. However, the association between cadmium exposure and tumors at other tissues such as kidney, prostate, and breast cannot be alleviated.

Furthermore, the cadmium toxicity cannot be only restricted to occupational exposure; rather, environmental exposure is also somehow responsible for the enhanced carcinogenicity of cadmium. For example, people in the area's proximity to zinc smelters are highly prone to cadmium toxicity. Cadmium exposure shows the elevated level of ROS in the experimental animals, presumably due to the inactivation of detoxifying enzymes. Also, cadmium interferes with the proteins involved in the DNA damage repairing process and apoptosis inhibition [104]. Similarly, nickel is a metal carcinogen chronic occupational exposure to humans occurs via nickel refining and welding, and electroplating. Chronic exposure to nickel holds a high risk of developing lung fibrosis and respiratory cancers. Nickel generally causes the DNA hypermethylation and histone hypomethylation, which causes silencing of some of the tumor suppressor genes. Some studies are there in favor of the ROS generation in response to nickel exposure that interferes with the DNA damage repair mechanism [105]. Occupational exposure to beryllium is also responsible for lung cancer [106].

#### 25.3.2.7 Solvents and other compounds

Excessive use of many solvents and their continuous exposure due to their volatility contributes to their high-carcinogenic potential, but their carcinogenicity varies. Benzene is highly toxic and can induce acute myeloid leukemia and other hematological malignancies. The workers of the chemical industry, petroleum refineries, and oil pipelines possess the risk of developing leukemia due to benzene exposure [107]. Tetrachloromethane and bromomethane are also a potent carcinogen and possess the risk of developing hepatocellular carcinoma (HCC) [108]. Experiments conducted in vitro described the hexamethylphosphoramide as carcinogenic agents [109]. It causes nasal tumor in the rats those inhaled this toxic substance [110].

## 25.4 Biological carcinogens and viruses

A biological carcinogen is a term used to indicate any biological substance that is directly or indirectly involved in cancer development. The source may include plants, animals, bacteria, fungi, or viruses [111]. According to the IARC Monograph, 11 biological agents are considered to be carcinogenic to humans. These biological agents include seven viruses, three worms, and one bacterium [112]. In 2012, 15.4% of cancers (2.2 million) are caused by viruses (9.97%), bacteria (5.5%), and helminths (0.06%). The major contributing infectious agents are *Helicobacter pylori* (770,000 cases), human papillomavirus (HPV) (640,000), hepatitis B virus (HBV) (420,000), hepatitis C virus (HCV) (170,000), and Epstein–Barr virus (EBV) (120,000) [113]. Other important infectious agents are Kaposi sarcoma herpesvirus (KSHV), human T-cell lymphotropic virus type-1 (HTLV-1), human immunodeficiency virus-1 (HIV-1), *Opisthorchis viverrini*, *Clonorchis sinensis*, and *Schistosoma haematobium*.

### 25.4.1 Mechanisms of biological carcinogenesis

Biological agents may be direct or indirect carcinogens. Some viruses such as HPV, EBV, HTLV-1, and KSHV act as direct carcinogens. They transform normal cells into cancerous cells by incorporating their genome or by interacting with cellular proteins to express oncogenes, which ultimately results in apoptosis inhibition [112]. Some biological agents promote the degradation or metabolic transformation of chemicals in the gastrointestinal tract to convert them into carcinogens. Infectious agents may also modulate the expressions or the activities of different metabolizing enzymes [114]. Inflammation and cancer have a strong correlation. Infectious agents such as *H. pylori*, HBV, and HCV, cause chronic inflammation and ROS-induced DNA damage by producing proinflammatory chemokines and cytokines [112,114]. Viruses like HIV-1 can exhibit their carcinogenicity by suppressing the immune system. Immune suppression increases the risk of cancer [112]. Bacteria and protozoa can also produce or release toxins to affect different cellular processes due to their carcinogenicity [114].

### 25.4.2 Viral carcinogens

There are two types of viral carcinogenesis processes: direct transformation and indirect transformation. The direct transformation process involves different viral proteins with oncogenic potential, while the indirect transformation process involves modulation of the host immune system or induction of inflammation. Viruses that exhibit direct transformation carcinogenesis include HPV, KSHV, HTLV-1, etc. Some oncogenic viral proteins and their role in carcinogenesis are described here. p53 and RB are two vital proteins in the regulation of apoptosis. p53, a tumor suppressor protein, causes cell cycle arrest if any DNA damage is identified. It prevents oncogenic transformations by its DNA repairing activity [115]. E6, an HPV oncogene, promotes ubiquitinylation and subsequent proteasomal degradation of p53 [116]. KSHV can also suppress the transcriptional activity of p53 by expressing LANA1 to inhibit apoptosis [117]. Oncogenic viruses often target pRB, which also regulates apoptosis by governing the E2F transcription factor. During the G1–S phase transition, pRB is phosphorylated and release E2F to express its target genes required for cell proliferation [118]. E7, another HPV protein, stimulates the E2F activity by mimicking the role of pRB to promote cell immortalization [119]. E7 can also directly bind to E2F-1 to activate its transcriptional function [120]. Abnormal telomerase activity can also lead to immortalization. The telomere is TTAGGG nucleotides, which are required for each cell division. Telomere length becomes short with each cell division and finally results in growth arrest. Cells remain in the quiescent or senescent state during this period. Telomerase enzyme helps to add the repeat sequence to telomeres by its reverse transcriptase activity. The hTERT gene is responsible for the expression of the catalytic subunit of telomerase [121,122]. E6 protein of HPV and LMP1 protein of EBV can activate hTERT to maintain the length of the telomere and results in cell immortalization [123,124]. Tax, a protein expressed in HTLV-1, stimulates hTERT transcriptional activity by the NF-κB pathway and induces adult T-cell leukemia [125–127]. These are the major signaling pathways directly regulated by viral proteins. This section also describes the indirect transformation process of different viruses. HBV- and HCV-induced chronic liver infection leads to liver cirrhosis that can induce HCC development. HBV is a DNA virus that does not express any oncoprotein. Its integration with the host genome is believed to be carcinogenic. Moreover, HBx, a protein encoded by this virus, is responsible for HCC development. This virus is also able to modulate the immune system. HCV is an RNA virus that modulates host–protein interaction to induce HCC. NS3 and NS5A are two proteins expressed by the HCV, which affect signal transductions by influencing promoters to alter cell cycle [128,129].

#### 25.4.2.1 *Epstein–Barr virus*

EBV is a member of the family of herpesvirus. It exhibits two phases of infection: latent and lytic phase. In the latent phase, the viral genome is replicated once with each cell cycle, but in the lytic phase, many copies of the viral genome are generated and packaged into viral particles for saliva transmission. The lytic phase involves the infection of normal epithelium, while latent is associated with the infection of epithelial cancers [130]. This DNA virus is responsible for the development of Burkitt's lymphoma (BL), nasopharyngeal carcinoma (NPC), Hodgkin's lymphoma (HL) and NHL. It is transmitted through saliva. It first infects oropharyngeal epithelium, and then B lymphocytes are infected. Acute infection expresses the proteins that induce cell proliferation [131,132]. LMP1 and LMP2a are the major viral proteins involved in the malignant transformation of infected B-cells. In healthy human beings, EBV oncogenic proteins such as LMP1, EBNA2 help cytotoxic T cells (CTLs) to identify infected B-cells to kill them. But in immunosuppressed individuals such as HIV

**TABLE 25.3** EBV latent gene expression profiles and associated cancers.

| Latency pattern | EBV latent gene expression profile | Diseases | Ref. |
|---|---|---|---|
| I | EBNA1, EBER | BL | [134] |
| II | EBNA1, LMP1, LMP2, EBER | NPC, HL | |
| III | EBNA1, EBNA2, EBNA3, LMP1, LMP2, EBER | Posttransplant lymphoproliferative Diseases, infectious mononucleosis | |

infected persons or organ transplant recipients, infected B-cells become immortalized and can easily proliferate due to the inactivity of CTLs [133]. There are three latent gene expression patterns: latency I, latency II, and latency III. Table 25.3 summarizes the genes expressed in different latency patterns. During the initial infection of B-cells latency III, that is, all latent gene expressions are observed. However, chronic infection exhibits EBNA1, LMP2, and EBERs expressions [135]. Latent infection of lymphoblastoid cells expresses EBERs, BARTs, LMP1, LMP2A, LMP2B, EBNA1, EBNA2, EBNA3A, EBNA3B, EBNA3C, and EBNA-LP [133]. Different EBV gene expression patterns are associated with different levels of apoptosis resistance of tumor cells [136,137]. In BL, chromosomal translocation leads to MYC gene activation that regulates the transcription of genes associated with cell cycle, apoptosis, signal transduction, metabolism, and immune functions. In this disease, the latency I gene expression profile is observed, and EBNA1 plays an important role in TP53 mutation. EBER1 and EBER2 regulate apoptosis and IL10 expression through RIG1 and immune response [138]. About 40% of cases of HL are linked with EBV infection. HL tumor cells are known as Hodgkin Reed–Sternberg (HRS) cells, which contain a clonal viral genome [134]. Latent infection expresses LMP1 and LMP2A proteins, which have a major contribution to the oncogenesis process. EBV alters the activity of immune cells surrounding HRS cells. Overexpression of NF-κB in those cells helps in their survival by stimulating antiapoptotic proteins and proinflammatory cytokines. LMP1 helps to activate NF-κB in HRS cells by acting as a CD40 receptor [134,139]. In NPC, LMP1 also has a significant contribution in the same way. Moreover, LMP2, EBNA1, EBER, and BART-miRNAs are also associated with NPC [130,140,141]. Inhibition of several tumor suppressor genes such as RASSF1A, p16/CDKN2A, NF-κB inhibitor alpha, and TNF receptor-associated factor 3 help EBV to promote NPC development by stimulating PI3K-AKT pathway, NF-κB pathway, and cell cycle progression [130]. EBV infection also causes gastric cancer, salivary gland carcinoma, posttransplant lymphoproliferative disorders, AIDS-associated lymphomas, and leiomyosarcomas in immunosuppressed patients [131,142].

#### *25.4.2.2 Hepatitis B virus*

Chronic HBV infection is frequently involved in the development of HCC [143]. HBV can promote HCC by direct viral DNA integration to hepatocytes genome or by modulating apoptosis, DNA repair mechanism, and cell growth through the interaction of viral oncoprotein HBx with different signaling proteins [143–145]. HBV surface antigen is an important marker that has been identified in the serum of about 85% of HCC patients [144,145]. HBV utilizes host cellular enzyme in the integration of its DNA to the host genome, and this integration is not associated with HBV replication. There are four different patterns of HBV DNA insertion depending on the viral sequences present at the integration sites: the pattern I, II, III, and IV. The patterns I and II are more common than patterns III and IV. There are two direct repeat sequences (DR1 and DR2) present in HBV DNA, which are required for replication of HBV DNA minus and plus strand, respectively. Single-stranded linear intermediates of HBV DNA replication takes part in the invasion of host DNA. The 5′ end of minus-strand invasion (pattern II) is the most frequent and 5′ end of plus strand invasion (pattern III) is rarely found. Patterns I and IV result from the invasion of 3′ end of minus and plus strands, respectively [144,146]. Another possible explanation of this HBV integration involves the function of topoisomerase I that cleaves double-stranded HBV DNA to provide linearized viral DNA to be integrated with the host DNA [147]. Viral DNA integration causes mutation in growth-regulating genes [144]. Viral oncoprotein HBx stimulates different gene expressions by promoting their transcription. MAPK, JAK/STAT, and NF-κB pathways are activated by this protein. This protein particularly interacts with CREB, NF-κB, and AP-1 transcription factors [144,145,148]. HBx also inhibits apoptosis of hepatocytes by blocking the nuclear translocation of p53 protein. Moreover, this oncoprotein is also responsible for inhibiting nucleotide excision repair of damaged DNA by interacting with XAP-1 protein [144,148]. HBV caused cirrhosis indirectly increases the risk of HCC development [144,145,148]. The faster rate of cell division does not allow the repairing of mutated DNA. Besides the high activity of topoisomerase I may promote HCC development [144]. This virus also causes cholangiocarcinoma and NHL [149]. The hepatitis B vaccine, developed in 1976, was the first vaccine for the prevention of cancer [150].

#### 25.4.2.3 *Hepatitis C virus*

Hepatitis C virus is also involved in HCC development, with about a 4% annual incidence rate [151]. HCV–HIV coinfection, cirrhosis increases the risk of HCC incidence [152]. Alcoholism and smoking also accelerate the progression of HCC [153]. HCV viral core protein acts on different cell signaling pathways to interact with tumor suppressor proteins such as p53, pRB, and cell cycle proteins [151,154]. This event influences the process of apoptosis, cell growth, and proliferation. Different gene mutations (telomerase reverse transcriptase, tumor protein 53, and β-catenin) promote HCC development by increasing oxidative stress [155,156]. HCV-induced HCC is a multiple-step process initiating from chronic HCV infection through chronic hepatitis and cirrhosis [156,157]. HCV infection upregulates p38, ERK, JNK, and NF-κB pathways [157]. NS5A, an HCV protein, interacts with PI3K/AKT, WNT, and apoptotic pathways. This protein also inhibits TGF-β signaling to prevent Smad nuclear translocation and CDKN1A downregulation [157]. HCV also causes lymphoid malignancy and gallbladder cancer [158].

#### 25.4.2.4 *Kaposi sarcoma herpesvirus*

Kaposi sarcoma herpesvirus (KSHV), also known as human herpesvirus-8 (HHV-8), is a double-stranded DNA gammaherpesvirus [159]. Its infection leads to Kaposi sarcoma (KS), primary effusion lymphoma (PEL), multicentric Castleman disease (MCD), and multiple myeloma (MM) [160,161]. It infects different types of cells, such as endothelial cells, B-cells, and monocytes. It is a sexually transmitted disease, although it can also be transmitted through saliva [160,161]. After infection, it expresses several genes to establish latency. Kaposins, vFLIP, vCyclin, LANA, and various viral microRNAs help the viral genome to be retained as an episome in the host cell nucleus by modulating host immune system, cell growth, and proliferation [162–165]. The lytic phase results in the expression of all viral genes to release progeny virions with the help of ORF50 or replication and transcription activator protein, RTA. Physiologic stress such as hypoxia and chemicals such as sodium butyrate and valproic acid can induce this phase [160]. Both latent and lytic phase proteins are associated with KS malignancy, but the lytic phase plays an indirect role in promoting oncogenesis by inhibiting host cell gene expressions. VEGF, ANGPT2, PDGF, GROa, and IL-6 induced by lytic genes are required for latently infected cell survival and proliferation. They also help to evade the immune system and induce inflammation, angiogenesis, etc. [166]. Latent and lytic proteins inhibit both extrinsic and intrinsic apoptotic pathways. vFLIP, a latent protein, and K1, a lytic protein involve in caspase eight inhibition to block extrinsic apoptosis while Lana-1, vIRF-1, vIRF-3, RTA, K7, and K8 inhibit p53-mediated apoptosis. vBcl2, a viral analogue of Bcl2, inhibits apoptosis by blocking BAX [167]. KSHV also downregulates Th1 response and activates Th2 response to promote KS by chronic inflammation. PI3K/AKT, MAPK, and JNK pathways are activated in KS tissue, which results in NF-κB overexpression. vFLIP protein also upregulates NF-κB [167]. The lytic phase is more linked to MCD as compared with KS and PEL. IL-6 plays a major role in MCD pathogenesis. Moreover, p17, a viral oncoprotein, promotes the malignant transformation of B-cells. JAK/STAT pathway is upregulated in MCD [168,169]. PEL is a very uncommon NHL type. Latent genes such as LANA-1, viral cyclin (vCyc), and vFLIP have an important role in PEL. LANA-1 helps in viral DNA replication and inhibits p53 and pRb. vCyc promotes uncontrolled cell division. vFLIP activates NF-κB pathway [160,164]. This virus is also responsible for KSHV inflammatory cytokine syndrome and multiple myeloma.

#### *25.4.2.5 Human immunodeficiency virus-1*

Human immunodeficiency virus-1 (HIV-1) is an RNA retrovirus that is associated with acquired immunodeficiency syndrome (AIDS). HIV-1 infection depends on several viral and host factors. Symptomatic identification of the primary infection is very difficult, but it slowly damages the immune system to make the individual susceptible to any opportunistic infections. High plasma HIV load and very low $CD4^+$ T-cell count indicates the development of AIDS [170,171]. This virus enters into the host cells with the help of $CD4^+$ protein and chemokine receptors (CCR5 and CXCR4) [171]. KSHV is mainly responsible for KS, but KS is the most frequent malignancy observed in HIV patients [172]. HIV-1 infection helps KSHV in KS progression by suppressing the immune system and generating inflammatory cytokines [173,174]. Moreover, Tat, an HIV protein, plays an important role in promoting KSHV replication by destroying the host immune response. This protein, along with vIL-6, encourages cell proliferation, angiogenesis, and oncogenesis. Tat also activates the PI3K/AKT pathway and overexpresses CD31, CD34, bFGF, VEGF, SMA, and cyclin D1 [174–176]. Inflammatory cytokines and bFGF upregulates $\alpha5\beta1$ and $\alpha V\beta3$ integrins, which in turn promote cell proliferation by modulating signaling pathways. HIV Tat protein increases the expression of MMP-2 and MMP-9, and these MMPs help in the invasion. bFGF and VEGF, overexpressed in HIV-KS, promote angiogenesis [174]. HIV-1 induced immunosuppression also increases the threat of HL, NHL, and cervical cancer [177]. Two most common subtypes of NHL: BL and diffuse large B cell lymphoma (DLBCL) are AIDS-related malignancy, but HL is non-AIDS defining illness [178–180]. EBV infection is associated with almost all cases of HIV-HL and HRS cells of HIV-HL tissue are found to express EBV oncoprotein, LMP1 [179,181]. LMP1 mediated NF-κB activation involves the mechanism of AIDS-related lymphoma [182,183]. Tat-induced IL-6 and IL-10 activation, cell cycle deregulation due to inference with Rb2/p130 protein, MYC translocation, and TP53 mutation are also important factors associated with HIV related lymphoma [184]. HPV infection is primarily related to cervical cancer, but HIV-1 induced immune system disruption increases the risk of HPV infection [185]. Potentiation of HPV protein by HIV–HPV viral protein interaction to deregulate the cell cycle is the most probable cause of the increased risk of cervical cancer [186]. Tat and gp120 loosen tight epithelial junctions to help HPV to get access the basal layer for infection. Tat also increases HPV oncogenic proteins (E6 and E7) expression [187]. HIV patients are also at high threat for developing skin cancer, lung cancer, liver cancer, etc.

#### 25.4.2.6 *Human papillomavirus*

HPV is a DNA virus that is linked to sexually transmitted diseases [188,189]. More than 120 types of HPV are categorized into high- and low-risk groups. Types 6, 11, 42, 43, and 44 are in the low-risk group and types 16, 18, 31, 33, 34, 35, 39, 45, 51, 52, 56, 58, 59, 66, 68, and 70 are in the group of high-risk HPV [189]. Its 7.9 kb long double-stranded genome includes three regions: (1) long control region (LCR), (2) early region consisting of ORFs E1, E2, E3, E4, E5, E6, and E7 involved in replication and tumorigenesis, (3) late region encoding capsid structural proteins L1 and L2 [188,190]. HPV utilizes host cellular proteins for their transcription and replication. E6 and E7 are the major oncoproteins causing cell growth deregulation by interfering with cell cycle proteins. E2 blocks the transcription of E6 and E7 to allow E1 to take part in viral replication. The integration of viral DNA with host cellular genome loss of the E2 region results in the overexpression of E6 and E7. E6 promotes ubiquitination-mediated degradation of p53 by binding to it. E7 disrupts the binding of hypophosphorylated pRB with E2F-1. These phenomena result in apoptosis inhibition, DNA synthesis, and cell growth. E5 activates the MAPK pathway to promote cell growth and differentiation [188,189]. This virus initially infects the cells present near the squamocolumnar junction. This junction forms the stratified epithelial layers of the transformation zone during the maturation of the cervix [191]. This virus is also associated with vaginal cancer, vulvar cancer, penile carcinoma, anal cancer, and oropharyngeal cancer [191]. Types of HPV associated with different cancers are listed in Table 25.4.

#### 25.4.2.7 *Human T-cell lymphotropic virus type-1*

HTLV-1 is an RNA retrovirus that causes adult T-cell leukemia/lymphoma (ATL). It infects mainly T- and B-cells utilizing GLUT1 transporter. Its transmission involves three routes: mother to infant by breastfeeding, sexual transmission, and parenteral transmission [192–194]. The unique 3′ end of its RNA, known as the pX region, encodes regulatory proteins: Tax, HBZ, and Rex [193]. The tax protein has a vital role in the early stage development of ATL, while HBZ helps to maintain the disease [194,195]. During initial HTLV-1 infection, Tax activates the NF-κB pathway by interacting with IKK complexes to induce

**TABLE 25.4** Types of HPV associated with different cancers.

| Type of cancer | Type of associated HPV |
|---|---|
| Vaginal cancer | HPV 16 |
| Vulvar cancer | HPV 16 |
| Head and neck carcinoma | HPV 16 and 33 |
| Lung cancer | HPV 16 and 18 |
| Bladder cancer | HPV 16 and 18 |
| Penile carcinoma | HPV 16 and 18 |
| Anal carcinoma | HPV 16 and 18 |
| Skin cancer | HPV 5 and 8 |

senescence or apoptosis. However, HBZ inhibits this NF-κB hyperactivation to maintain viral latency when Tax protein level is low. This low level of Tax may help in the survival and proliferation of the infected cells by activating different signaling pathways. Moreover, Tax causes genomic instability by increasing the level of ROS and NO, which induce double-strand breaks. It also inactivates DNA Polβ that is involved in base excision repair [192,195]. This protein can modulate p53- and Bax-mediated apoptosis, the transcription activity of CREB in the Wnt pathway, and PI3K pathway [193]. Other malignancies related to this virus are cutaneous T-cell lymphoma, B-cell, and T-cell lymphomas.

### 25.4.3 Bacterial carcinogens

Bacterial carcinogenesis involves chronic inflammation, modulation of hormone activity or its synthesis, oncoprotein activation, or any toxic metabolite generation to stimulate cell proliferation [196]. Bacteria that increase the risk of cancer by chronic inflammation include *H. pylori* (gastric adenocarcinoma), *Salmonella typhimurium* (biliary cancer), *Chlamydia trachomatis* (cervical cancer), etc. [197–199]. *H. pylori* also modulate gastrin hormone to induce gastric adenocarcinoma or colorectal adenocarcinoma [200]. *Mycoplasma* can induce stomach and colon cancer by expressing c-Ras and c-Myc oncoproteins. This bacterium activates NF-κB pathway and inhibits p53-mediated apoptosis [201]. *N*-nitrosamines, a toxic metabolite produced by *Escherichia coli*, may take part in *Schistosoma haematobium* infected bladder cancer [202].

#### *25.4.3.1 Helicobacter pylori*

*H. pylori* is a gram-negative bacterium that causes two most common stomach cancers: adenocarcinoma and gastric mucosa-associated lymphoid tissue (MALT) lymphoma. MALT lymphoma is very rare as compared with adenocarcinoma, but both are linked to chronic gastritis [203,204]. *H. pylori* initially adheres to the surface epithelial cells of the stomach and then colonizes in the proximal duodenum. This leads to gastric metaplasia. The Cag island of the bacterial genome encodes several proteins that help to translocate bacterial proteins into the gastric epithelial cells. Bacterial adhesins (blood group antigen binding adhesion, BabA, and sialic acid-binding adhesin, SabA) help to establish contact between the bacteria and the epithelial cells [205]. *H. pylori* also encode a multimeric vacuolating exotoxin, VacA, which helps in vacuole formation in the epithelial cells [203]. CagA is phosphorylated in the epithelial cells by Src kinases. Phospho-CagA inhibits PAR1/MAPK by activating SHP2 oncoprotein. It also modulates NF-κB, JAK/STAT, TRAF6/TAK1, and PI3K/AKT pathways to inhibit apoptosis and stimulate ROS-induced DNA damage. These events result in actin remodeling, increased cell proliferation, and activation of inflammatory cells and transcription of mitogenic genes [205,206]. Murylpeptides, present in the cell wall of *H. pylori*, interact with NOD1 protein to activate NF-κB through PI3K/AKT pathway and induce cell proliferation. It also promotes the secretion of type-1 interferon and boosts the immune system [205]. VacA protein induces apoptosis to damage gastric mucosa and inhibits T lymphocyte proliferation to disrupt the immune system [203,205]. Gastrin is a peptide hormone secreted from the G cells of the antrum. This hormone promotes gastric cell proliferation, angiogenesis, and invasion. Hypergastrinemia is associated with gastric malignancy [196,200]. This bacterium also causes pancreatic cancer, colorectal adenocarcinoma, and lung cancer.

## 25.4.4 Protozoal carcinogens

### *25.4.4.1* **Opisthorchis viverrini *and* Clonorchis sinensis**

*O. viverrini* and *C. sinensis* are two liver flukes that are associated with cholangiocarcinoma and HCC. These infections can promote carcinogenesis but are not linked to the initiation of the process [207]. Cholangiocarcinoma is a multistep process starting from infection in normal cholangiocytes to malignant cholangiocytes through chronic inflammation, cellular damage, cell proliferation, and mutations [208].

*O. viverrini* causes inflammation and DNA damage by generating ROS and RNOS in cholangiocytes. Oxysterol, produced by this pathogen, also causes DNA damage [209]. Overexpressed IL-6, PDGF, TNF-α, and TGF-β helps to proliferate the damaged cells. Ov-GRN-1, a growth factor released by *O. viverrini*, helps in angiogenesis and wound repairing of the infected bile duct. This protein also promotes cholangiocarcinoma development by mimicking the functions of IL-33. It also activates several cancer pathways, including EMT. Thioredoxin is another protein that inhibits apoptosis [210]. *O. viverrini* proteins upregulate phosphorylation of RB protein and cyclin D1 to induce G1/S phase transition [211].

*C. sinensis* causes mechanical damage to the liver due to its metabolites induced chronic and prolonged inflammation. Its infection also modulates the enzymatic activities of the liver [212,213]. Some metabolites generated by it [excretory-secretory product (ESPs)] cause inflammation and apoptosis inhibition in biliary epithelia due to their high immunogenicity or toxicity [214]. ESPs also modulate histone acetyltransferase and minichromosome maintenance to promote hyperplasia. Granulin, an EPS, promotes migration and invasion through ERK and PI3K/AKT pathway. It overexpresses *N*-cadherin, vimentin, β-catenin, MMP-2, and MMP-9 [215]. Myoglobin, another ESP, diminishes $H_2O_2$ and NO levels. iNOS and SOD expression is also decreased. This protein helps the parasite to survive by reducing oxidative stress [216]. TLRs initiate an inflammatory process by recognizing the infection, but chemokines and cytokines induced prolonged inflammation can cause host tissue damage. High expression of TLR2 and TLR4 stimulates the innate immune system. Th1 mediated response is involved in the elimination of pathogens while Th2 mediated response is beneficial for disease progression. In clonorchiasis, enhanced Th2 cytokines and diminished Th1 cytokines production is observed [213]. Free radicals produced by the pathogen activate NF-κB induced inflammation. iNOS induced by NF-κB, TNF-α, and IL-1β promote NO synthesis. NO causes DNA damage by blocking DDR and stimulating COX-2 production to induce carcinogenesis [213]. *N*-Nitrosodimethylamine (NMDA), an endogenous nitrosation product, also induces a G2/M phase transition, expression of E2F-1, pRb, and cyclin B to modulate cell cycle [208].

#### 25.4.4.2 *Schistosoma haematobium*

*S. haematobium* is a blood fluke that is associated with bladder cancer development. In SH induced carcinogenesis, several protein molecules play various roles. Among these, Bcl2 overexpression and TP53 mutation are very important in tumorigenesis [217,218]. Cox2 also promotes tumorigenesis by adhesion molecules downregulation, ECM degradation, and stimulating angiogenesis process [218,219]. Different chromosomal deletions have been observed in SH induced bladder cancer, but deletion in chromosome 9 is most important. This deletion leads to inhibition of proteins, which are related to the stimulation of Rb and TP53 suppressor pathways. It modulates the PI3K/AKT pathway to downregulate apoptosis [218]. *S. haematobium* soluble egg antigen (SEA) is a key parasite molecule that stimulates VEGF production and also plays a role to modulate apoptosis and cell proliferation. H-IPSE is a parasitic protein that modulates the host immune system to help in egg survival. This protein induces IL-4 production, promotes G1/S phase transition and cell proliferation. These events lead to urothelium damage and ultimately carcinogenesis [220].

### 25.4.5 Other biological carcinogens

There are many more bacteria, viruses, and helminths related to human cancers, which are not mentioned here. Those include *Helicobacter bilis, Helicobacter hepaticus, Campylobacter jejuni, Salmonella typhi, Mycobacterium tuberculosis, Chlamydia pneumoniae, Neisseria gonorrhoeae, Chlamydia trachomatis, Gardnerella vaginalis, Treponema denticola, Bacteroides distasonis, Mycoplasma hominis, Ureaplasma urealyticum, Simian Vacuolating Virus, Human Polyoma Virus, Taenia taeniaeformis, Strongyloides stercoralis, Fasciola hepatica, Trichostrongylus colubriformis,* etc. [114,221]. Protozoa such as *Toxoplasma gondii, Cryptosporidium parvum, Trichomonas vaginalis,* and fungus such as *Paracoccidioides brasiliensis* are also associated with human carcinogenesis [114]. Microbiota present in the human body may influence the susceptibility of the human body to cancer development by inducing chronic inflammation, DNA damage, or producing metabolites with oncogenic potential [222].

## 25.5 Conclusion

Carcinogenesis is a complex process that is not fully understood. The carcinogens play a vital role in either manifesting or promoting the stages of carcinogenesis. We have summarized different cancers associated with different carcinogens in Table 25.5.

**TABLE 25.5** Different types of carcinogens and the associated cancers.

| | Type of carcinogen | Name of carcinogen | Cancer |
|---|---|---|---|
| Physical carcinogen | Radiation | Electromagnetic radiation | Leukemia, brain cancer, breast cancer |
| | | Ionizing radiation | Sarcoma, leukemia |
| | Fibers | Asbestos | Lung cancer, mesothelioma |
| | | Erionite | Lung cancer, mesothelioma |
| | Particulate materials | Cobalt | Rhabdomyosarcoma |
| | | Nickel | Sarcoma and squamous cell carcinomas |
| | | Silica | Sarcoma and squamous cell carcinomas, lung cancer, pleural cancer, lymphomas |
| | | Air pollutants | Lung cancer |
| | Gel materials | Silicone gel | Fibrosarcoma |
| Chemical carcinogen | Aromatic amines | Tobacco smoke | Bladder cancer |
| | | 2-Naphthylamine,4-aminobiphenyl | Bladder cancer |
| | | Benzidine | Bladder cancer |
| | NOCs | *N*-Nitrosodimethylamine | Colon, nasopharyngeal, esophageal, gastric cancer |
| | | *N*-Nitrosopiperidine, *N*-nitrosomorphiline, and *N*-nitrosopiperizine | Bladder, stomach, multiple myeloma, esophagus, prostate cancer |
| | Dyes | Hair dyes (phenylenediamine and aminophenyl) | bladder cancer, breast cancer, NHL, and myeloma |
| | | 3′-Methyl-4-dimethylaminoazobenzene | Liver tumor |
| | | *o*-Aminoazotoluene and auramine | Bladder cancer |
| | Alkylating agents | Bis-chloro-methyl-ether | Lung cancer |
| | | Methyl bromide | Prostate cancer |
| | Natural carcinogens | Aflatoxin | Liver cancer |
| | Inorganic carcinogenic agents | Arsenic | Urinary bladder, skin, and lungs |
| | | Cadmium | Lung cancer |
| | | Nickel | Lung fibrosis and respiratory cancers |
| | | Beryllium | Lung cancer |
| | Solvents | Benzene | Acute myeloid leukemia |

*(Continued)*

TABLE 25.5 (Continued)

| | Type of carcinogen | Name of carcinogen | Cancer |
|---|---|---|---|
| | | Tetrachloromethane and bromomethane | HCC |
| | | Hexamethylphosphoramide | Nasal tumor |
| Biological carcinogen | Viral carcinogens | Epstein–Barr virus | BL, NPC, HL and NHL |
| | | Hepatitis B virus | HCC, cholangiocarcinoma and NHL |
| | | Hepatitis C virus | HCC, lymphoid malignancy and gallbladder cancer |
| | | KSHV | Kaposi sarcoma, primary effusion lymphoma, multicentric Castleman disease, and multiple myeloma |
| | | Human immunodeficiency virus-1 | Kaposi sarcoma, BL, diffuse large B cell lymphoma and cervical cancer |
| | | HPV | Cervical cancer, vaginal cancer, vulvar cancer, bladder cancer, penile carcinoma, anal carcinoma |
| | | Human T-cell lymphotropic virus type-1 | Adult T-cell leukemia/lymphoma |
| | Bacterial carcinogens | *H. pylori* | Adenocarcinoma and gastric mucosa-associated lymphoid tissue (MALT) lymphoma |
| | Protozoal carcinogens | *O. viverrini* and *C. sinensis* | Cholangiocarcinoma and HCC |
| | | *S. haematobium* | Bladder cancer |

Understanding the association of carcinogens with different cancer is vital for not only understanding their molecular mechanisms but also for modeling different cancer through in vivo and in vitro models.

## Acknowledgments

The authors express their gratitude towards the Indian Institute of Technology Kharagpur, Ministry of Human Resource, Govt. of India, Council for Scientific and Industrial Research, Govt. of India and J.C. Bose national fellowship, Science and Engineering Research Board, Department of Science and Technology, Govt. of India for providing the fellowship support to the authors. The authors are also grateful to the Indian Council of Medical Research, Govt. of India, for their support.

## References

[1] Hanahan D, Weinberg RA. The hallmarks of cancer. Cell 2000;100(1):57–70.
[2] Hanahan D, Weinberg RA. Hallmarks of cancer: the next generation. Cell 2011;144(5):646–74.

[3] Bray F, Ferlay J, Soerjomataram I, Siegel RL, Torre LA, Jemal A. Global cancer statistics 2018: GLOBOCAN estimates of incidence and mortality worldwide for 36 cancers in 185 countries. CA Cancer J Clin [Internet] 2018;68(6):394–424. Available from: http://doi.wiley.com/10.3322/caac.21492.

[4] Rivenbark AG, Coleman WB. An introduction to the conspicuous and distinguishing characteristics of neoplasms. Pathobiology of human disease. Elsevier; 2014. p. 349–66.

[5] Weinstein IB, Joe AK. Mechanisms of disease: oncogene addiction—a rationale for molecular targeting in cancer therapy. Nat Clin Pract Oncol 2006;3(8):448–57.

[6] Burkhart DL, Sage J. Cellular mechanisms of tumour suppression by the retinoblastoma gene. Nat Rev Cancer 2008;8(9):671–82.

[7] Schwab M, editor. Encyclopedia of cancer. Berlin, Heidelberg: Springer Berlin Heidelberg; 2017.

[8] Trosko JE. Commentary: is the concept of "tumor promotion" a useful paradigm? Mol Carcinog 2001;30(3):131–7.

[9] Luch A. Nature and nurture – lessons from chemical carcinogenesis. Nat Rev Cancer 2005;5(2):113–25.

[10] Grisham JW, Kaufmann WK, Kaufman DG. The cell cycle and chemical carcinogenesis. Pathol Immunopathol Res 1983;1(1–2):49–66.

[11] Cohen SM, Arnold LL. Chemical carcinogenesis. Toxicol Sci 2011;120(Suppl. 1):S76–92.

[12] Ayob AZ, Ramasamy TS. Cancer stem cells as key drivers of tumour progression. J Biomed Sci 2018;25(1):20.

[13] Berenblum I, Shubik P. The role of croton oil applications, associated with a single painting of a carcinogen, in tumour induction of the mouse's skin. Br J Cancer 1947;1(4):379–82.

[14] Weisburger JH. Worldwide prevention of cancer and other chronic diseases based on knowledge of mechanisms. Mutat Res 1998;402(1–2):331–7.

[15] Mehta R. The potential for the use of cell proliferation and oncogene expression as intermediate markers during liver carcinogenesis. Cancer Lett 1995;93(1):85–102.

[16] Williams GM. Mechanisms of chemical carcinogenesis and application to human cancer risk assessment. Toxicology. 2001;166(1–2):3–10.

[17] Rahner N, Steinke V. Hereditary cancer syndromes. Dtsch Arztebl Int 2008;105(41):706–14.

[18] Moschel RC. Carcinogens. Encyclopedia of genetics. Elsevier; 2001. p. 271–2.

[19] Minardi F, Maltoni C. Chapter 21: Physical carcinogens. Holland-Frei cancer medicine. 6th ed. BC Decker; 2003. p. 313–20.

[20] Havas M. When theory and observation collide: can non-ionizing radiation cause cancer? Env Pollut [Internet] 2017;221:501–5. Available from: <http://www.ncbi.nlm.nih.gov/pubmed/27903411>.

[21] Azzam EI, Jay-Gerin J-P, Pain D. Ionizing radiation-induced metabolic oxidative stress and prolonged cell injury. Cancer Lett [Internet] 2012;327(1–2):48–60. Available from: <http://www.ncbi.nlm.nih.gov/pubmed/22182453>.

[22] Singhvi HR, Malik A, Chaturvedi P. The role of chronic mucosal trauma in oral cancer: a review of literature. Indian J Med Paediatr Oncol [Internet] 2017;38(1):44–50. Available from: <http://www.ncbi.nlm.nih.gov/pubmed/28469336>.

[23] International Agency for Research on Cancer. Non-ionizing radiation, Part 1: static and extremely low-frequency (ELF) electric and magnetic fields. IARC Monogr Eval Carcinog Risks Hum, 80. 2002. p. 1–395.

[24] Sagan LA. Epidemiological and laboratory studies of power frequency electric and magnetic fields. JAMA. 1992;268(5):625–9.

[25] Heath CW. Electromagnetic field exposure and cancer: a review of epidemiologic evidence. CA Cancer J Clin 1996;46(1):29–44.

[26] Little JB. Radiation carcinogenesis. Carcinogenesis 2000;21(3):397–404.

[27] Stsjazhko VA, Tsyb AF, Tronko ND, Souchkevitch G, Baverstock KF. Childhood thyroid cancer since accident at Chernobyl. BMJ. 1995;310(6982):801.

[28] Preston DL, Pierce DA, Shimizu Y, Cullings HM, Fujita S, Funamoto S, et al. Effect of recent changes in atomic bomb survivor dosimetry on cancer mortality risk estimates. Radiat Res 2004;162(4):377–89.

[29] Ron E. Cancer risks from medical radiation. Health Phys 2003;85(1):47–59.

[30] IARC Working Group on the Evaluation of Carcinogenic Risk to Humans. Arsenic, metals, fibres and dusts. IARC Monogr Eval Carcinog Risks Hum 2012;100((C).

[31] Toyokuni S. Mechanisms of asbestos-induced carcinogenesis. Nagoya J Med Sci 2009;71(1–2):1–10.

[32] Hueper WC, Payne WW. Experimental studies in metal carcinogenesis. Chromium, nickel, iron, arsenic. Arch Env Health 1962;5:445–62.

[33] Demetriou CA, Vineis P. Carcinogenicity of ambient air pollution: use of biomarkers, lessons learnt and future directions. J Thorac Dis 2015;7(1):67–95.
[34] Giroux V, Rustgi AK. Metaplasia: tissue injury adaptation and a precursor to the dysplasia-cancer sequence. Nat Rev Cancer 2017;17(10):594–604.
[35] Semerdjian HS, Texter JH, Yawn DH. Rhabdomyosarcoma occurring in repaired exstrophied bladder: a case report. J Urol 1972;108(2):354–6.
[36] International Agency for Research on Cancer. Non-ionizing radiation, Part 2: radiofrequency electromagnetic fields. IARC Monogr Eval Carcinog Risks Hum 2013;102:1.
[37] Kheifets L, Ahlbom A, Crespi CM, Feychting M, Johansen C, Monroe J, et al. A pooled analysis of extremely low-frequency magnetic fields and childhood brain tumors. Am J Epidemiol 2010;172(7):752–61.
[38] Sermage-Faure C, Demoury C, Rudant J, Goujon-Bellec S, Guyot-Goubin A, Deschamps F, et al. Childhood leukaemia close to high-voltage power lines--the Geocap study, 2002-2007. Br J Cancer 2013;108(9):1899–906.
[39] Kroll ME, Swanson J, Vincent TJ, Draper GJ. Childhood cancer and magnetic fields from high-voltage power lines in England and Wales: a case-control study. Br J Cancer 2010;103(7):1122–7.
[40] Kleinerman RA, Kaune WT, Hatch EE, Wacholder S, Linet MS, Robison LL, et al. Are children living near high-voltage power lines at increased risk of acute lymphoblastic leukemia? Am J Epidemiol 2000;151 (5):512–15.
[41] Kheifets L, Ahlbom A, Crespi CM, Draper G, Hagihara J, Lowenthal RM, et al. Pooled analysis of recent studies on magnetic fields and childhood leukaemia. Br J Cancer 2010;103(7):1128–35.
[42] Greenland S, Sheppard AR, Kaune WT, Poole C, Kelsh MA. A pooled analysis of magnetic fields, wire codes, and childhood leukemia. Childhood Leukemia-EMF Study Group. Epidemiology. 2000;11(6):624–34.
[43] Ahlbom A, Day N, Feychting M, Roman E, Skinner J, Dockerty J, et al. A pooled analysis of magnetic fields and childhood leukaemia. Br J Cancer 2000;83(5):692–8.
[44] Hatch EE, Linet MS, Kleinerman RA, Tarone RE, Severson RK, Hartsock CT, et al. Association between childhood acute lymphoblastic leukemia and use of electrical appliances during pregnancy and childhood. Epidemiology. 1998;9(3):234–45.
[45] Elliott P, Toledano MB, Bennett J, Beale L, de Hoogh K, Best N, et al. Mobile phone base stations and early childhood cancers: case-control study. BMJ. 2010;340:c3077.
[46] Peyman A, Khalid M, Calderon C, Addison D, Mee T, Maslanyj M, et al. Assessment of exposure to electromagnetic fields from wireless computer networks (wi-fi) in schools; results of laboratory measurements. Health Phys 2011;100(6):594–612.
[47] Feinendegen LE, Pollycove M, Sondhaus CA. Responses to low doses of ionizing radiation in biological systems. Nonlinearity Biol Toxicol Med 2004;2(3):143–71.
[48] Kim EH, Kim M-S, Lee K-H, Sai S, Jeong YK, Koh J-S, et al. Effect of low- and high-linear energy transfer radiation on in vitro and orthotopic in vivo models of osteosarcoma by activation of caspase-3 and -9. Int J Oncol 2017;51(4):1124–34.
[49] Choudhuri S, Chanderbhan R, Mattia A. Carcinogenesis. Veterinary toxicology. Elsevier; 2018. p. 339–54.
[50] Kraemer KH, Patronas NJ, Schiffmann R, Brooks BP, Tamura D, DiGiovanna JJ. Xeroderma pigmentosum, trichothiodystrophy and Cockayne syndrome: a complex genotype-phenotype relationship. Neuroscience. 2007;145(4):1388–96.
[51] Spitz DR, Azzam EI, Li JJ, Gius D. Metabolic oxidation/reduction reactions and cellular responses to ionizing radiation: a unifying concept in stress response biology. Cancer Metastasis Rev 2004;23(3–4):311–22.
[52] Thacker J. The use of recombinant DNA techniques to study radiation-induced damage, repair and genetic change in mammalian cells. Int J Radiat Biol Relat Stud Phys Chem Med 1986;50(1):1–30.
[53] Bartek J. DNA damage response, genetic instability and cancer: from mechanistic insights to personalized treatment. Mol Oncol 2011;5(4):303–7.
[54] Muirhead CR, O'Hagan JA, Haylock RGE, Phillipson MA, Willcock T, Berridge GLC, et al. Mortality and cancer incidence following occupational radiation exposure: third analysis of the National Registry for Radiation Workers. Br J Cancer 2009;100(1):206–12.
[55] Cardis E, Vrijheid M, Blettner M, Gilbert E, Hakama M, Hill C, et al. Risk of cancer after low doses of ionising radiation: retrospective cohort study in 15 countries. BMJ. 2005;331(7508):77.
[56] Basu AK. DNA damage, mutagenesis and cancer. Int J Mol Sci 2018;19(4).
[57] Kamp DW. Asbestos-induced lung diseases: an update. Transl Res 2009;153(4):143–52.

[58] Carbone M, Yang H. Molecular pathways: targeting mechanisms of asbestos and erionite carcinogenesis in mesothelioma. Clin Cancer Res 2012;18(3):598–604.
[59] Heath JC. The production of malignant tumours by cobalt in the rat. Br J Cancer 1956;10(4):668–73.
[60] Miller FJ, Gardner DE, Graham JA, Lee RE, Wilson WE, Bachmann JD. Size considerations for establishing a standard for inhalable particles. J Air Pollut Control Assoc 1979;29(6):610–15.
[61] Beeson WL, Abbey DE, Knutsen SF. Long-term concentrations of ambient air pollutants and incident lung cancer in California adults: results from the AHSMOG study. Adventist Health Study on Smog. Environ Health Perspect 1998;106(12):813–22.
[62] Brinton LA, Brown SL. Breast implants and cancer. J Natl Cancer Inst 1997;89(18):1341–9.
[63] Haran-Ghera N, Trainin N, Fiore-Donati L, Berenblum I. A possible two-stage mechanism in rhabdomyosarcoma induction in rats. Br J Cancer 1962;16:653–64.
[64] Friedewald WF, Rous P. The initiating and promoting elements in tumor production : an analysis of the effects of tar, benzpyrene, and methylcholanthrene on rabbit skin. J Exp Med 1944;80(2):101–26.
[65] Hollander CF, van Rijssel T. Experimental production of intramandibular carcinoma in mice by mechanical damage. J Natl Cancer Inst 1963;30:337–59.
[66] Kaplan RP. Cancer complicating chronic ulcerative and scarifying mucocutaneous disorders. Adv Dermatol 1987;2:19–46.
[67] Yamagiwa K, Ichikawa K. Experimental study of the pathogenesis of carcinoma. CA Cancer J Clin 1977;27:174–81.
[68] Hoeijmakers JHJ. DNA damage, aging, and cancer. N Engl J Med 2009;361(15):1475–85.
[69] Dragan YP, Sargent L, Xu YD, Xu YH, Pitot HC. The initiation-promotion-progression model of rat hepatocarcinogenesis. Proc Soc Exp Biol Med 1993;202(1):16–24.
[70] Butterworth BE, Bogdanffy MS. A comprehensive approach for integration of toxicity and cancer risk assessments. Regul Toxicol Pharmacol 1999;29(1):23–36.
[71] Schwartz RS. Another look at immunologic surveillance. N Engl J Med 1975;293(4):181–4.
[72] Cohen SM. Human carcinogenic risk evaluation: an alternative approach to the two-year rodent bioassay. Toxicol Sci 2004;80(2):225–9.
[73] Klaunig JE, Kamendulis LM, Xu Y. Epigenetic mechanisms of chemical carcinogenesis. Hum Exp Toxicol 2000;19(10):543–55.
[74] Bartsch H, Hietanen E, Malaveille C. Carcinogenic nitrosamines: free radical aspects of their action. Free Radic Biol Med 1989;7(6):637–44.
[75] Tanaka T, Miyazawa K, Tsukamoto T, Kuno T, Suzuki K. Pathobiology and chemoprevention of bladder cancer. J Oncol 2011;2011:1–23.
[76] Letašiová S, Medveďová A, Šovčíková A, Dušinská M, Volkovová K, Mosoiu C, et al. Bladder cancer, a review of the environmental risk factors. Environ Health 2012;11(Suppl. 1):S11.
[77] Hutchings SJ, Rushton L. Occupational cancer in Britain. Br J Cancer 2012;107(S1):S92–103.
[78] Bogovski P, Bogovski S. Animal species in which N-nitroso compounds induce cancer. Int J Cancer 1981;27 (4):471–4.
[79] Tricker AR, Preussmann R. Carcinogenic N-nitrosamines in the diet: occurrence, formation, mechanisms and carcinogenic potential. Mutat Res 1991;259(3–4):277–89.
[80] Mirvish SS. Role of N-nitroso compounds (NOC) and N-nitrosation in etiology of gastric, esophageal, nasopharyngeal and bladder cancer and contribution to cancer of known exposures to NOC. Cancer Lett 1995;93 (1):17–48.
[81] Grosse Y, Baan R, Straif K, Secretan B, El Ghissassi F, Cogliano V, et al. Carcinogenicity of nitrate, nitrite, and cyanobacterial peptide toxins. Lancet Oncol 2006;7(8):628–9.
[82] Tricker AR. N-Nitroso compounds and man: sources of exposure, endogenous formation and occurrence in body fluids. Eur J Cancer Prev 1997;6(3):226–68.
[83] Jakszyn P, Gonzalez CA, Lujan-Barroso L, Ros MM, Bueno-de-Mesquita HB, Roswall N, et al. Red meat, dietary nitrosamines, and heme iron and risk of bladder cancer in the European prospective investigation into cancer and nutrition (EPIC). Cancer Epidemiol Biomarkers Prev 2011;20(3):555–9.
[84] Hughes R. Dose-dependent effect of dietary meat on endogenous colonic N-nitrosation. Carcinogenesis. 2001;22(1):199–202.
[85] Cross AJ, Pollock JRA, Bingham SA. Haem, not protein or inorganic iron, is responsible for endogenous intestinal N-nitrosation arising from red meat. Cancer Res 2003;63(10):2358–60.

[86] Hidajat M, McElvenny DM, Ritchie P, Darnton A, Mueller W, van Tongeren M, et al. Lifetime exposure to rubber dusts, fumes and N-nitrosamines and cancer mortality in a cohort of British rubber workers with 49 years follow-up. Occup Env Med 2019;76(4):250–8.
[87] Zhang Y, Kim C, Zheng T. Hair dye use and risk of human cancer. Front Biosci (Elite Ed) 2012;4:516–28.
[88] Hallagan JB, Allen DC, Borzelleca JF. The safety and regulatory status of food, drug and cosmetics colour additives exempt from certification. Food Chem Toxicol 1995;33(6):515–28.
[89] Kobylewski S, Jacobson MF. Toxicology of food dyes. Int J Occup Environ Health 2012;18(3):220–46.
[90] Flaks A, Flaks B. Histopathology of the rat liver following a single autoprotective dose of 3′-methyl-4-dimethylaminoazobenzene. J Pathol 1978;126(2):71–8.
[91] Kirsch P, Fleig I, Frentzel-Beyme R, et al. Auramine. Toxicology and occupational health. Arbeitsmed Sozialmed Praventivmed 1978;13:1–28.
[92] Doi M, Hattori N, Yokoyama A, Onari Y, Kanehara M, Masuda K, et al. Effect of mustard gas exposure on incidence of lung cancer: a longitudinal study. Am J Epidemiol 2011;173(6):659–66.
[93] Hosseini-khalili A, Haines DD, Modirian E, Soroush M, Khateri S, Joshi R, et al. Mustard gas exposure and carcinogenesis of lung. Mutat Res 2009;678(1):1–6.
[94] Bernucci I, Turrini D, Landi MT. Bis-chloromethyl ether and carcinogenesis of alkylating agents. Med Lav 1977;88(5):347–55.
[95] Rueping M, Nachtsheim BJ. A review of new developments in the Friedel–Crafts alkylation – From green chemistry to asymmetric catalysis. Beilstein J Org Chem 2010;6:6.
[96] Faustino-Rocha AI, Ferreira R, Oliveira PA, Gama A, Ginja M. N-Methyl-N-nitrosourea as a mammary carcinogenic agent. Tumor Biol 2015;36(12):9095–117.
[97] Perše M, Cerar A, Injac R, Štrukelj B. N-methylnitrosourea induced breast cancer in rat, the histopathology of the resulting tumours and its drawbacks as a model. Pathol Oncol Res 2009;15(1):115–21.
[98] Budnik LT, Kloth S, Velasco-Garrido M, Baur X. Prostate cancer and toxicity from critical use exemptions of methyl bromide: environmental protection helps protect against human health risks. Environ Health 2012;11(1):5.
[99] Bolt HM, Gansewendt B. Mechanisms of carcinogenicity of methyl halides. Crit Rev Toxicol 1993;23 (3):237–53.
[100] Liu Y, Wu F. Global burden of aflatoxin-induced hepatocellular carcinoma: a risk assessment. Environ Health Perspect 2010;118(6):818–24.
[101] Díaz Nieto CH, Granero AM, Zon MA, Fernández H. Sterigmatocystin: a mycotoxin to be seriously considered. Food Chem Toxicol 2018;118:460–70.
[102] Waalkes MP, Liu J, Ward JM, Diwan BA. Mechanisms underlying arsenic carcinogenesis: hypersensitivity of mice exposed to inorganic arsenic during gestation. Toxicology. 2004;198(1–3):31–8.
[103] Chen C, Xun P, Nishijo M, He K. Cadmium exposure and risk of lung cancer: a meta-analysis of cohort and case–control studies among general and occupational populations. J Expo Sci Environ Epidemiol 2016;26 (5):437–44.
[104] Hartwig A. Cadmium and cancer. Met Ions Life Sci 2013;11:491–507.
[105] Kurella V, Obey T, Wakeman TP, Xu B. Nickel induces DNA double strand breaks, mitotic arrest and replication blockage. Cancer Res 2005;65(9 Suppl.):278.
[106] Boffetta P, Fordyce TA, Mandel JS. A mortality study of beryllium workers. Cancer Med 2016;5 (12):3596–605.
[107] McHale CM, Zhang L, Smith MT. Current understanding of the mechanism of benzene-induced leukemia in humans: implications for risk assessment. Carcinogenesis. 2012;33(2):240–52.
[108] Lertprasertsuke N, Shinoda M, Takekoshi S, Tsutsumi Y, Yamamoto Y, Niki E, et al. Different effects of carbon tetrachloride on carcinogen-induced hepatocellular carcinoma and normal liver of the rat: lowered lipid feroxidation and accelerated necrosis in cancer. Jpn J Cancer Res 1991;82(5):503–10.
[109] Ashby J, Styles JA, Paton D. Potentially carcinogenic analogues of the carcinogen hexamethylphosphoramide: evaluation in vitro. Br J Cancer 1978;38(3):418–27.
[110] Lee KP, Trochimowicz HJ. Induction of nasal tumors in rats exposed to hexamethylphosphoramide by inhalation. J Natl Cancer Inst 1982;68(1):157–71.
[111] Baba A, Câtoi C. Chapter 2, Carcinogenesis. Comparative Oncology. The Publishing House of the Romanian Academy; 2007. p. 1–24.

[112] European Agency for Safety and Health at work. Biological agents. In: Safety and health [Internet]; 2003. Available from: <https://osha.europa.eu/en/tools-and-publications/publications/factsheets/41>.

[113] Parkin DM. The global health burden of infection-associated cancers in the year 2002. Int J Cancer [Internet] 2006;118(12):3030–44. Available from: <http://doi.wiley.com/10.1002/ijc.21731>.

[114] Kutikhin AG, Yuzhalin AE, Brusina EB. Infectious agents and cancer. Dordrecht: Springer Netherlands; 2013. p. 1–116. Available from: <http://link.springer.com/10.1007/978-94-007-5955-8>.

[115] Collot-Teixeira S, Bass J, Denis F, Ranger-Rogez S. Human tumor suppressor p53 and DNA viruses. Rev Med Virol [Internet] 2004;14(5):301–19. Available from: <http://doi.wiley.com/10.1002/rmv.431>.

[116] Bernard X, Robinson P, Nominé Y, Masson M, Charbonnier S, Ramirez-Ramos JR, et al. Proteasomal degradation of p53 by human papillomavirus e6 oncoprotein relies on the structural integrity of p53 core domain (Zhang Y, editor). PLoS One [Internet] 2011;6(10):e25981. Available from: <http://dx.plos.org/10.1371/journal.pone.0025981>.

[117] Friborg J, Kong W, Hottiger MO, Nabel GJ. p53 inhibition by the LANA protein of KSHV protects against cell death. Nature [Internet] 1999;402(6764):889–94. Available from: <http://www.nature.com/articles/47266>.

[118] Frolov MV. Molecular mechanisms of E2F-dependent activation and pRB-mediated repression. J Cell Sci [Internet] 2004;117(11):2173–81. Available from: <http://jcs.biologists.org/cgi/doi/10.1242/jcs.01227>.

[119] Gonzalez SL, Stremlau M, He X, Basile JR, Munger K. Degradation of the retinoblastoma tumor suppressor by the human papillomavirus type 16 E7 oncoprotein is important for functional inactivation and is separable from proteasomal degradation of E7. J Virol [Internet] 2001;75(16):7583–91. Available from: <http://jvi.asm.org/cgi/doi/10.1128/JVI.75.16.7583-7591.2001>.

[120] Hwang SG, Lee D, Kim J, Seo T, Choe J. Human papillomavirus type 16 E7 binds to E2F1 and activates E2F1-driven transcription in a retinoblastoma protein-independent manner. J Biol Chem [Internet] 2002;277(4):2923–30 <http://www.jbc.org/lookup/doi/10.1074/jbc.M109113200>.

[121] Jafri MA, Ansari SA, Alqahtani MH, Shay JW. Roles of telomeres and telomerase in cancer, and advances in telomerase-targeted therapies. Genome Med [Internet] 2016;8(1):69. Available from: <http://genomemedicine.biomedcentral.com/articles/10.1186/s13073-016-0324-x>.

[122] Shay JW, Wright WE. Role of telomeres and telomerase in cancer. Semin Cancer Biol [Internet] 2011;21(6):349–53. Available from: <http://genomemedicine.biomedcentral.com/articles/10.1186/s13073-016-0324-x>.

[123] Veldman T, Horikawa I, Barrett JC, Schlegel R. Transcriptional activation of the telomerase hTERT gene by human papillomavirus type 16 E6 oncoprotein. J Virol [Internet] 2001;75(9):4467–72. Available from: <http://www.ncbi.nlm.nih.gov/pubmed/11287602>.

[124] Terrin L, Dal Col J, Rampazzo E, Zancai P, Pedrotti M, Ammirabile G, et al. Latent membrane protein 1 of Epstein-Barr virus activates the hTERT promoter and enhances telomerase activity in B lymphocytes. J Virol [Internet] 2008;82(20):10175–87. Available from: <http://jvi.asm.org/cgi/doi/10.1128/JVI.00321-08>.

[125] Azran I, Schavinsky-Khrapunsky Y, Aboud M. Role of tax protein in human T-cell leukemia virus type-I leukemogenicity. Retrovirology [Internet] 2004;1(6):20. Available from: <http://www.ncbi.nlm.nih.gov/pubmed/15310405>.

[126] Sinha-Datta U. Transcriptional activation of hTERT through the NF-B pathway in HTLV-I-transformed cells. Blood [Internet] 2004;104(8):2523–31. Available from: <http://www.bloodjournal.org/cgi/doi/10.1182/blood-2003-12-4251>.

[127] Nicot CP. Multiple pathways control the reactivation of telomerase in HTLV-I-associated leukemia. Int J Cancer Oncol [Internet] 2015;2(2):1–8. Available from: <http://www.ommegaonline.org/article-details/Multiple-Pathways-Control-the-Reactivation-of-Telomerase-in-HTLV-I-Associated-Leukemia/215>.

[128] Anzola M. Hepatocellular carcinoma: role of hepatitis B and hepatitis C viruses proteins in hepatocarcinogenesis. J Viral Hepat [Internet] 2004;11(5):383–93. Available from: <http://doi.wiley.com/10.1111/j.1365-2893.2004.00521.x>.

[129] Seto W-K, Lai C-L, Yuen M-F. Hepatitis viruses and hepatocellular carcinoma. Cancer and inflammation mechanisms [Internet]. Hoboken: John Wiley & Sons, Inc.; 2014. p. 181–92. Available from: <http://doi.wiley.com/10.1002/9781118826621.ch13>.

[130] Tsao SW, Tsang CM, Lo KW. Epstein-Barr virus infection and nasopharyngeal carcinoma. Philos Trans R Soc Lond B Biol Sci 2017;372(1732).

[131] Thompson MP. Epstein-Barr virus and cancer. Clin Cancer Res [Internet] 2004;10(3):803–21. Available from: <https://www.annualreviews.org/doi/10.1146/annurev-pathmechdis-012418-013023>.

[132] Saha A, Robertson ES. Mechanisms of B-cell oncogenesis induced by Epstein-Barr virus. J Virol [Internet] 2019. Available from: <http://jvi.asm.org/lookup/doi/10.1128/JVI.00238-19>.

[133] Pannone G, Zamparese R, Pace M, Pedicillo M, Cagiano S, Somma P, et al. The role of EBV in the pathogenesis of Burkitt's lymphoma: an Italian hospital based survey. Infect Agent Cancer 2014;9(1):34.

[134] Massini G, Siemer D, Hohaus S. EBV in Hodgkin lymphoma. Mediterr J Hematol Infect Dis [Internet] 2009. Available from: <http://www.mjhid.org/article/view/5114>.

[135] Thorley-Lawson DA. {Epstein-Barr} virus: exploiting the immune system. Nat Rev Immunol [Internet] 2001;1(1):75–82. Available from: https://doi.org/10.1038/35095584. Available from:.

[136] Niedobitek G, Agathanggelou A, Rowe M, Jones EL, Jones DB, Turyaguma P, et al. Heterogeneous expression of Epstein-Barr virus latent proteins in endemic Burkitt's lymphoma. Blood [Internet] 1995;86 (2):659–65. Available from: <http://www.ncbi.nlm.nih.gov/pubmed/7605996>.

[137] Brady G, MacArthur GJ, Farrell PJ. Epstein-Barr virus and Burkitt lymphoma. Postgrad Med J [Internet] 2008;84(993):372–7. Available from: < http://pmj.bmj.com/cgi/doi/10.1136/jcp.2007.047977>.

[138] Grömminger S, Mautner J, Bornkamm GW. Burkitt lymphoma: the role of Epstein-Barr virus revisited. Br J Haematol [Internet] 2012;156(6):719–29. Available from: <http://doi.wiley.com/10.1111/j.1365-2141.2011.09007.x>.

[139] Flavell KJ. Hodgkin's disease and the Epstein-Barr virus. Mol Pathol [Internet] 2000;53(5):262–9. Available from: <http://mp.bmj.com/cgi/doi/10.1136/mp.53.5.262>.

[140] Young LS, Dawson CW. Epstein-Barr virus and nasopharyngeal carcinoma. Chin J Cancer [Internet] 2014;171–86. Available from: <https://www.taylorfrancis.com/books/9780203711279>.

[141] Tsang CM, Tsao SW. The role of Epstein-Barr virus infection in the pathogenesis of nasopharyngeal carcinoma. Virol Sin [Internet] 2015;30(2):107–21. Available from: <http://link.springer.com/10.1007/s12250-015-3592-5>.

[142] Raab-Traub N. Novel mechanisms of EBV-induced oncogenesis. Curr Opin Virol [Internet] 2012;2(4):453–8. Available from: <https://linkinghub.elsevier.com/retrieve/pii/S187962571200106X>.

[143] Xu H-Z, Liu Y-P, Guleng B, Ren J-L. Hepatitis B virus-related hepatocellular carcinoma: pathogenic mechanisms and novel therapeutic interventions. Gastrointest Tumors [Internet] 2014;1(3):135–45. Available from: <https://www.karger.com/Article/FullText/365307>.

[144] Arbuthnot P, Kew M. Hepatitis B virus and hepatocellular carcinoma. Int J Exp Pathol [Internet] 2001;82 (2):77–100. Available from: <http://www.wjgnet.com/1007-9327/full/v20/i20/5951.htm>.

[145] Adrian M, Bisceglie M. Hepatitis B and hepatocellular carcinoma Adrian. Hepatology 2009;49(5).

[146] Wang HP, Rogler CE. Topoisomerase I-mediated integration of hepadnavirus DNA in vitro. J Virol [Internet] 1991;65(5):2381–92. Available from: <http://www.ncbi.nlm.nih.gov/pubmed/1850018>.

[147] Hai H. Role of hepatitis B virus DNA integration in human hepatocarcinogenesis. World J Gastroenterol [Internet] 2014;20(20):6236. Available from: <http://www.wjgnet.com/1007-9327/full/v20/i20/6236.htm>.

[148] Kremsdorf D, Soussan P, Paterlini-Brechot P, Brechot C. Hepatitis B virus-related hepatocellular carcinoma: paradigms for viral-related human carcinogenesis. Oncogene [Internet] 2006;25(27):3823–33. Available from: <http://www.nature.com/articles/1209559>.

[149] Fwu C-W, Chien Y-C, You S-L, Nelson KE, Kirk GD, Kuo H-S, et al. Hepatitis B virus infection and risk of intrahepatic cholangiocarcinoma and non-Hodgkin lymphoma: a cohort study of parous women in Taiwan. Hepatology [Internet] 2011;53(4):1217–25. Available from: <http://doi.wiley.com/10.1002/hep.24150>.

[150] Buynak EB. Vaccine against human hepatitis B. JAMA J Am Med Assoc [Internet] 1976;235(26):2832. Available from: <http://jama.jamanetwork.com/article.aspx?doi = 10.1001/jama.1976.03260520026017>.

[151] Axley P, Ahmed Z, Ravi S, Singal AK. Hepatitis C virus and hepatocellular carcinoma: a narrative review. J Clin Transl Hepatol [Internet] 2018;6(2):1–6. Available from: <http://www.xiahepublishing.com/ArticleFullText.aspx?sid = 2&jid = 1&id = 10.14218%2FJCTH.2017.00067>.

[152] Gjærde LI, Shepherd L, Jablonowska E, Lazzarin A, Rougemont M, Darling K, et al. Trends in incidences and risk factors for hepatocellular carcinoma and other liver events in HIV and hepatitis C virus–coinfected individuals from 2001 to 2014: a multicohort study. Clin Infect Dis [Internet] 2016;63(6):821–9. Available from: <https://academic.oup.com/cid/article-lookup/doi/10.1093/cid/ciw380>.

[153] Chuang SC, Lee YCA, Hashibe M, Dai M, Zheng T, Boffetta P. Interaction between cigarette smoking and hepatitis B and C virus infection on the risk of liver cancer: a meta-analysis. Cancer Epidemiol Biomarkers

Prev [Internet] 2010;19(5):1261–8. Available from: <http://cebp.aacrjournals.org/cgi/doi/10.1158/1055-9965.EPI-09-1297>.

[154] Vescovo T, Refolo G, Vitagliano G, Fimia GM, Piacentini M. Molecular mechanisms of hepatitis C virus–induced hepatocellular carcinoma. Clin Microbiol Infect [Internet] 2016;22(10):853–61. Available from: <https://linkinghub.elsevier.com/retrieve/pii/S1198743X16302439>.

[155] Tomasetti C, Li L, Vogelstein B. Stem cell divisions, somatic mutations, cancer etiology, and cancer prevention. Science (80-) [Internet] 2017;355(6331):1330–4. Available from: <http://www.sciencemag.org/lookup/doi/10.1126/science.aaf9011>.

[156] de Oliveria Andrade L, D'Oliveira A, Melo R, De Souza E, Silva CC, Parana R. Association between hepatitis C and hepatocellular carcinoma. J Glob Infect Dis [Internet]. 2009;1(1):33. Available from: <http://www.jgid.org/text.asp?2009/1/1/33/52979>.

[157] Goossens N, Hoshida Y. Hepatitis C virus-induced hepatocellular carcinoma. Clin Mol Hepatol [Internet] 2015;21(2):105. Available from: <http://e-cmh.org/journal/view.php?doi = 10.3350/cmh.2015.21.2.105>.

[158] Fiorino S, Bacchi-Reggiani L, De Biase D, Fornelli A, Masetti M, Tura A, et al. Possible association between hepatitis C virus and malignancies different from hepatocellular carcinoma: a systematic review. World J Gastroenterol 2015;21(45):12896–953.

[159] Chang Y, Cesarman E, Pessin MS, Lee F, Culpepper J, Knowles DM, et al. Identification of herpesvirus-like DNA sequences in AIDS-associated Kaposi's sarcoma. Science [Internet] 1994;266(5192):1865–9. Available from: <http://www.ncbi.nlm.nih.gov/pubmed/7997879>.

[160] Goncalves PH, Ziegelbauer J, Uldrick TS, Yarchoan R. Kaposi sarcoma herpesvirus-associated cancers and related diseases. Curr Opin HIV AIDS [Internet]. 2017;12(1):47–56. Available from: <http://content.wkhealth.com/linkback/openurl?sid = WKPTLP:landingpage&an = 01222929-201701000-00009>.

[161] Koch S, Schulz TF, Schulz TF. Kaposi sarcoma herpesvirus pathogenesis review. Philos Trans R Soc Lond B Biol Sci 2017;372(1732).

[162] Lee H-R, Amatya R, Jung JU. Multi-step regulation of innate immune signaling by Kaposi's sarcoma-associated herpesvirus. Virus Res [Internet] 2015;209(24):39–44. Available from: <http://doi.wiley.com/10.1002/cncr.27633>.

[163] Valiya Veettil M, Dutta D, Bottero V, Bandyopadhyay C, Gjyshi O, Sharma-Walia N, et al. Glutamate secretion and metabotropic glutamate receptor 1 expression during Kaposi's sarcoma-associated herpesvirus infection promotes cell proliferation. PLoS Pathog 2014;10(10):1–37.

[164] Guasparri I, Keller SA, Cesarman E. KSHV vFLIP is essential for the survival of infected lymphoma cells. J Exp Med 2004;199(7):993–1003.

[165] Cotter MA, Robertson ES. The latency-associated nuclear antigen tethers the Kaposi's sarcoma-associated herpesvirus genome to host chromosomes in body cavity-based lymphoma cells. Virology [Internet] 1999;264(2):254–64. Available from: <http://www.ncbi.nlm.nih.gov/pubmed/10562490>.

[166] Mesri EA, Cesarman E, Boshoff C. Kaposi's sarcoma herpesvirus/human herpesvirus-8 (KSHV/HHV8), and the oncogenesis of Kaposi's sarcoma. Nat Rev Cancer 2010;10(10):707–19.

[167] Douglas JL, Gustin JK, Moses AV, Dezube BJ, Pantanowitz L. Kaposi sarcoma pathogenesis: a triad of viral infection, oncogenesis and chronic inflammation. Transl Biomed [Internet] 2010;1(2):377–88. Available from: <http://www.nature.com/articles/mp2011182>.

[168] Wang H-W, Pittaluga S, Jaffe ES. Multicentric Castleman disease: where are we now? Semin Diagn Pathol [Internet] 2016;33(5):294–306. Available from: <https://linkinghub.elsevier.com/retrieve/pii/S074025701630034X>.

[169] Carbone A, De Paoli P, Gloghini A, Vaccher E. KSHV-associated multicentric Castleman disease: a tangle of different entities requiring multitarget treatment strategies. Int J Cancer 2015;137(2):251–61.

[170] Cohen MS, Shaw GM, McMichael AJ, Haynes BF. Acute HIV-1 infection. N Engl J Med [Internet] 2011;364(20):1943–54. Available from: <http://www.nejm.org/doi/abs/10.1056/NEJMra1011874>.

[171] Naif HM. Pathogenesis of HIV infection. Infect Dis Rep [Internet] 2013;5(1S):6. Available from: <http://www.pagepress.org/journals/index.php/idr/article/view/idr.2013.s1.e6>.

[172] Hoffmann C, Sabranski M, Esser S. HIV-associated Kaposi's sarcoma. Oncol Res Treat 2017;40(3):94–8.

[173] Facciolà A, Rullo EV, Ceccarelli M, D'Aleo F, di Rosa M, Pinzone MR, et al. Kaposi's sarcoma in HIV-infected patients in the era of new antiretrovirals. Eur Rev Med Pharmacol Sci [Internet] 2017;21(24):5869–79. Available from: <http://www.embase.com/search/results?subaction = viewrecord&from = export&id = L619968253>.

[174] Feller L, Lemmer J. Insights into pathogenic events of HIV-associated Kaposi sarcoma and immune reconstitution syndrome related Kaposi sarcoma. Infect Agent Cancer 2008;3(1):1–9.

[175] Zhou F, Xue M, Qin D, Zhu X, Wang C, Zhu J, et al. HIV-1 tat promotes kaposi's sarcoma-associated herpesvirus (KSHV) vIL-6-induced angiogenesis and tumorigenesis by regulating PI3K/PTEN/AKT/GSK-3β signaling pathway. PLoS One 2013;8.

[176] Pyakurel P, Pak F, Mwakigonja AR, Kaaya E, Biberfeld P. KSHV/HHV-8 and HIV infection in Kaposi's sarcoma development. Infect Agent Cancer 2007;2(1):1–8.

[177] Rubinstein PG, Aboulafia DM, Zloza A. Malignancies in HIV/AIDS: from epidemiology to therapeutic challenges. AIDS [Internet] 2014;28(4):453–65. Available from: <https://insights.ovid.com/crossref?an = 00002030-201402200-00001>.

[178] Brunnberg U, Hentrich M, Hoffmann C, Wolf T, Hübel K. HIV-associated malignant lymphoma. Oncol Res Treat 2017;40(3):82–7.

[179] Jacobson CA, Abramson JS. HIV-associated Hodgkin's lymphoma: prognosis and therapy in the Era of cART. Adv Hematol 2012;2012:1–8.

[180] Biggar RJ, Jaffe ES, Goedert JJ, Chaturvedi A, Pfeiffer R, Engels EA. Hodgkin lymphoma and immunodeficiency in persons with HIV/AIDS. Blood. 2006;108(12):3786–91.

[181] Spina M, Carbone A, Gloghini A, Serraino D, Berretta M, Tirelli U. Hodgkin's disease in patients with HIV infection. Adv Hematol 2011;2011.

[182] Cesarman E. Pathology of lymphoma in HIV. Curr Opin Oncol [Internet] 2013;25(5):487–94. Available from: <http://content.wkhealth.com/linkback/openurl?sid = WKPTLP:landingpage&an = 00001622-201309000-00007>.

[183] Carbone A, Gloghini A. AIDS-related lymphomas: from pathogenesis to pathology. Br J Haematol 2005;130 (5):662–70.

[184] Grogg KL, Miller RF, Dogan A. HIV infection and lymphoma. J Clin Pathol 2007;60(12):1365–72.

[185] Holmes RS, Hawes SE, Touré P, Dem A, Feng Q, Weiss NS, et al. HIV infection as a risk factor for cervical cancer and cervical intraepithelial neoplasia in Senegal. Cancer Epidemiol Biomarkers Prev 2009;18(9):2442–6.

[186] Clarke B, Chetty R. Postmodern cancer: the role of human immunodeficiency virus in uterine cervical cancer. J Clin Pathol - Mol Pathol [Internet] 2002;55(1):19–24. Available from: <http://www.embase.com/search/results?subaction = viewrecord&from = export&id = L34123926%5Cnhttps://doi.org/10.1136/mp.55.1.19%5Cnhttp://wt3cf4et2l.search.serialssolutions.com?sid = EMBASE&issn = 13668714&id = doi:10.1136%2Fmp.55.1.19&atitle = Postmodern + cancer%3A + Th>.

[187] Brickman C, Palefsky JM. Human papillomavirus in the HIV-infected host: epidemiology and pathogenesis in the antiretroviral era. Curr HIV/AIDS Rep [Internet] 2015;12(1):6–15. Available from: <http://link.springer.com/10.1007/s11904-014-0254-4>.

[188] Burd E. Human papillomavirus and cervical cancer. Clin Microbiol Rev [Internet] 2003;16(1):1–17. Available from: <http://www.sciencedirect.com/science/article/pii/S0140673607614160>.

[189] Jastreboff AM, Cymet T. Role of the human papilloma virus in the development of cervical intraepithelial neoplasia and malignancy. Postgrad Med J 2002;78(918):225–8. Available from: <http://www.ncbi.nlm.nih.gov/pubmed/11930025>.

[190] Longworth MS, Laimins LA. Pathogenesis of human papillomaviruses in differentiating epithelia. Microbiol Mol Biol Rev 2004;68(2):362–72.

[191] Tulay P, Serakinci N. The role of human papillomaviruses in cancer progression. J Cancer Metastasis Treat 2020;2(6):201.

[192] Giam CZ, Semmes OJ. HTLV-1 infection and adult T-cell leukemia/ lymphoma-A tale of two proteins: tax and HBZ. Viruses. 2016;8(6).

[193] Ghezeldasht SA, Shirdel A, Assarehzadegan MA, Rahimi H, Miri R, Rezaee SAR. Human T lymphotropic virus type I (HTLV-I) oncogenesis : molecular aspects of virus and host interactions in pathogenesis of adult T cell. Iran J Basic Med Sci 2013;16:179–95.

[194] Panfil AR, Martinez MP, Ratner L, Green PL. Human T-cell leukemia virus-associated malignancy. Curr Opin Virol [Internet] 2016;20(24):40–6. Available from: <https://linkinghub.elsevier.com/retrieve/pii/S1879625716301055>.

[195] Bangham CR, Ratner L. How does HTLV-1 cause adult T-cell leukaemia/lymphoma (ATL)? Curr Opin Virol [Internet] 2015;14(24):93–100. Available from: <https://linkinghub.elsevier.com/retrieve/pii/S187962571500139X>.

[196] Chang AH, Parsonnet J. Role of bacteria in oncogenesis. Clin Microbiol Rev [Internet] 2010;23(4):837–57. Available from: <http://cmr.asm.org/cgi/doi/10.1128/CMR.00012-10>.

[197] Zhang XY, Zhang PY, Aboul-Soud MAM. From inflammation to gastric cancer: role of Helicobacter pylori. Oncol Lett 2017;13(2):543–8.

[198] Espinoza JA, Bizama C, García P, Ferreccio C, Javle M, Miquel JF, et al. The inflammatory inception of gallbladder cancer. Biochim Biophys Acta - Rev Cancer [Internet] 2016;1865(2):245–54. Available from: <https://linkinghub.elsevier.com/retrieve/pii/S0304419X16300269>.

[199] Zhu H, Shen Z, Luo H, Zhang W, Zhu X. Chlamydia trachomatis infection-associated risk of cervical cancer: a meta-analysis. Medicine (Baltimore) 2016;95(13):e3077.

[200] Smith JP, Nadella S, Osborne N. Gastrin and gastric cancer. Cell Mol Gastroenterol Hepatol [Internet] 2017;4(1):75–83. Available from: https://doi.org/10.1016/j.jcmgh.2017.03.004.

[201] Logunov DY, Scheblyakov DV, Zubkova OV, Shmarov MM, Rakovskaya IV, Gurova KV, et al. Mycoplasma infection suppresses p53, activates NF-κB and cooperates with oncogenic Ras in rodent fibroblast transformation. Oncogene 2008;27:4521–31.

[202] El-Mosalamy H, Salman TM, Ashmawey AM, Osama N. Role of chronic E. coli infection in the process of bladder cancer- an experimental study. Infect Agent Cancer [Internet] 2012;7(1):19. Available from: <https://infectagentscancer.biomedcentral.com/articles/10.1186/1750-9378-7-19>.

[203] Kim SS, Ruiz VE, Carroll JD, Moss SF. Helicobacter pylori in the pathogenesis of gastric cancer and gastric lymphoma. Cancer Lett [Internet] 2011;305(2):228–38. Available from: <https://linkinghub.elsevier.com/retrieve/pii/S0304383510003629>.

[204] Zhang R-G. Role of Helicobacter pylori infection in pathogenesis of gastric carcinoma. World J Gastrointest Pathophysiol 2016;7(1):97.

[205] Wroblewski LE, Peek RM, Wilson KT. Helicobacter pylori and gastric cancer: factors that modulate disease risk. Clin Microbiol Rev 2010;23(4):713–39.

[206] Hatakeyama M. Helicobacter pylori CagA and gastric cancer: a paradigm for hit-and-run carcinogenesis. Cell Host Microbe [Internet] 2014;15(3):306–16. Available from: <10.1016/j.chom.2014.02.008>.

[207] Abdel-Rahim AY. Parasitic infections and hepatic neoplasia. Dig Dis 2001;19(4):288–91.

[208] Cho BI, Han JK, Hong ST, Lee KH. Clonorchiasis and cholangiocarcinoma: etiologic relationship and imaging diagnosis. Clin Microbiol Rev [Internet] 2004;17(3):540. Available from: <http://europepmc.org/search?query=(DOI:10.1128/CMR.17.3.540-552.2004)%0Ahttp://cmr.asm.org/cgi/doi/10.1128/CMR.17.3.540-552.2004%0Ahttp://cmr.asm.org/content/17/3/540.full.pdf>.

[209] Braconi C, Patel T. Cholangiocarcinoma: new insights into disease pathogenesis and biology. Infect Dis Clin North Am [Internet] 2010;24(4):871–84. Available from: <https://linkinghub.elsevier.com/retrieve/pii/S0891552010000619>.

[210] Sripa B, Tangkawattana S, Brindley PJ. Update on pathogenesis of opisthorchiasis and cholangiocarcinoma. Adv Parasitol 2018;102:97–113.

[211] Mayer DA, Fried B. The role of helminth infections in carcinogenesis. Adv Parasitol 2007;65(07):239–96.

[212] Shi Y, Jiang Z, Yang Y, Zheng P, Wei H, Lin Y, et al. *Clonorchis sinensis* infection and co-infection with the hepatitis B virus are important factors associated with cholangiocarcinoma and hepatocellular carcinoma. Parasitol Res 2017;116(10):2645–9.

[213] Kim TS, Pak JH, Kim JB, Bahk YY. *Clonorchis sinensis*, an oriental liver fluke, as a human biological agent of cholangiocarcinoma: a brief review. BMB Rep 2016;49(11):590–7.

[214] Chen X, Li S, He L, Wang X, Liang P, Chen W, et al. Molecular characterization of severin from *clonorchis sinensis* excretory/secretory products and its potential anti-apoptotic role in hepatocarcinoma PLC cells. PLoS Negl Trop Dis 2013;7(12).

[215] Wang C, Lei H, Tian Y, Shang M, Wu Y, Li Y, et al. *Clonorchis sinensis* granulin: identification, immunolocalization, and function in promoting the metastasis of cholangiocarcinoma and hepatocellular carcinoma. Parasit Vectors 2017;10(1):1–14.

[216] Ren M, He L, Huang Y, Mao Q, Li S, Qu H, et al. Molecular characterization of *Clonorchis sinensis* secretory myoglobin: delineating its role in anti-oxidative survival. Parasit Vectors 2014;7(1):1–11.

[217] Botelho MC, Machado JC, da Costa JMC. Schistosoma haematobium and bladder cancer: what lies beneath? Virulence. 2010;1(2):84–7.

[218] Zaghloul MS. Bladder cancer and schistosomiasis. J Egypt Natl Canc Inst [Internet] 2012;24(4):151–9. Available from: <https://linkinghub.elsevier.com/retrieve/pii/S1110036212000544>.

[219] Mostafa MH, Sheweita SA, O'Connor PJ. Relationship between schistosomiasis and bladder cancer. Clin Microbiol Rev [Internet] 1999;12(1):97–111. Available from: <http://www.ncbi.nlm.nih.gov/pubmed/9880476>.

[220] Ishida K, Hsieh MH. Understanding urogenital schistosomiasis-related bladder cancer: an update. Front Med 2018;5(August):1–7.

[221] Smith AJ, Smith LA. Viral carcinogenesis. Prog Mol Biol Transl Sci [Internet] 2016;121–68. Available from: <http://www.embase.com/search/results?subaction = viewrecord&from = export&id = L613084240%0Ahttps://doi.org/10.1016/bs.pmbts.2016.09.007>.

[222] Bhatt AP, Redinbo MR, Bultman SJ. The role of the microbiome in cancer development and therapy. CA Cancer J Clin [Internet] 2017;67(4):326–44. Available from: <http://doi.wiley.com/10.3322/caac.21398>.

CHAPTER

# 26

# Biodetection and sensing for cancer diagnostics

*Rita Rebelo*[1,2], *Ana I. Barbosa*[1,2], *Subhas C. Kundu*[1,2], *Rui L. Reis*[1,2] and *Vitor M. Correlo*[1,2]

**[1]3B's Research Group, I3Bs – Research Institute on Biomaterials, Biodegradables and Biomimetics, University of Minho, Headquarters of the European Institute of Excellence on Tissue Engineering and Regenerative Medicine, Guimarães, Portugal [2]ICVS/3B's – PT Government Associate Laboratory, Braga/Guimarães, Portugal**

## 26.1 Introduction

Globally, cancer is the second leading reason of death and was responsible for an estimated 9.6 million deaths in 2018 [1]. Cancer can be defined as abnormal and uncontrolled cell growth, which conducts to the formation of a tumor mass. Tumor cells become resistant to apoptosis and other antigrowth defenses, or might metastasize and invade other organs of the body, damaging their function [2]. These malignant cells can affect different organs; however, the most deadly cancers tend to affect lungs, liver, stomach, and bowel [1]. Cancer prognosis depends on tumor type and size, and on the existence of metastases [3]. These tumor features develop with time, which makes early detection extremely important for patient survival, healing, and recovery. In addition, close vigilance after tumor removal to diagnose disease recurrence, also significantly affects cancer prognosis and patient recovery [4]. Currently, in clinical settings, cancer is diagnosed, observing symptoms, by a series of blood tests, followed by imaging analysis and biopsy. Thus, the diagnostic process is long and time consuming, which hinders early detection and treatment, increasing the chances of recurrence. The urgency in early cancer detection has prompted the development of a plethora of biosensing technologies aiming to diagnose rapidly and accurately cancer. In an ideal situation, biosensors would be placed at


DOI: https://doi.org/10.1016/B978-0-12-818128-7.00026-5

healthcare centers or high-risk patient homes, which might be capable of simultaneously detecting a panel of molecules or activities, providing a "disease signature," thereby allowing continuous and real-time monitoring. This technology aims to be specific to the patient's physiology and metabolic responses, contrary to current conventional techniques, providing more personalized treatment [5]. The rapid detection would facilitate decision making by primary healthcare providers.

The molecules or activities which can be measured and evaluated as indicators for normal biological processes, pathogenic processes, or pharmacological response to a therapeutic intervention, are called biomarkers [6]. Cancer biomarkers can be proteins, DNA, mRNAs, enzymes, and metabolites present in tumor tissues or serum released by tumors, or circulating tumor cells (CTC) [7]. Although biomarkers can be a powerful tool in cancer detection, they are only used occasionally for diagnostics and treatment monitoring, since many have not been approved for clinical applications or lack the ideal specificity. Also, their detection is time consuming, and affected by laboratory conditions [8]. Recently cancer detection technologies have been developed by integrating biosensors and microfluidics, or lab-on-chip devices for quantitation of these biomarkers or processes in physiological fluids or tissues. However, these methods still cannot meet current clinical demands, since technological constraints promote lack of accuracy, sensitivity, and specificity for clinical diagnostic applications [9].

This book chapter describes the main cancer biomarkers including the ones currently in clinical use, as well as those that are not yet found in the healthcare centers, presenting high potential to fill gaps in biomarker science. This chapter also discusses the limitations of the current cancer biomarkers, namely how do they influence biosensing technology development? In addition, the section on cancer biosensors presents the advances in technology that are limited for the gaps in cancer biomarkers but can contribute to fill these gaps and meet the current clinical demands (Fig. 26.1). The final section provides information about biosensors already in clinical trials or on the path toward commercialization.

## 26.2 Biomarkers for cancer detection

The application of biomarkers in oncology is wide, including differential diagnosis, risk assessment, screening, determination of prognosis, prediction of response to treatment, and monitoring of disease progression. The diverse uses of cancer biomarkers make them very attractive as they provide a complete picture of a patients' condition. Thus, several new cancer biomarkers have been recently discovered and many are going through analytical validation, clinical validation, and utilization [10]. Nevertheless, many are not approved for applications in clinical settings. Several biological and technical aspects hinder the use of biomarkers in cancer diagnosis. First, there is no ideal biomarker for cancer screening. This means that it is necessary to screen a large number of biomarkers with different clinical thresholds. Second, some biomarkers present low specificity, which results in false positives, and a large number of people subjected to unnecessary diagnostic procedures. Third, although "liquid-biopsy" for cancer detection seems promising, the reliability of liquid base biomarkers has not been determined. Finally, the sensitivity and specificity of cancer biomarker technologies depend on numerous factors that are difficult

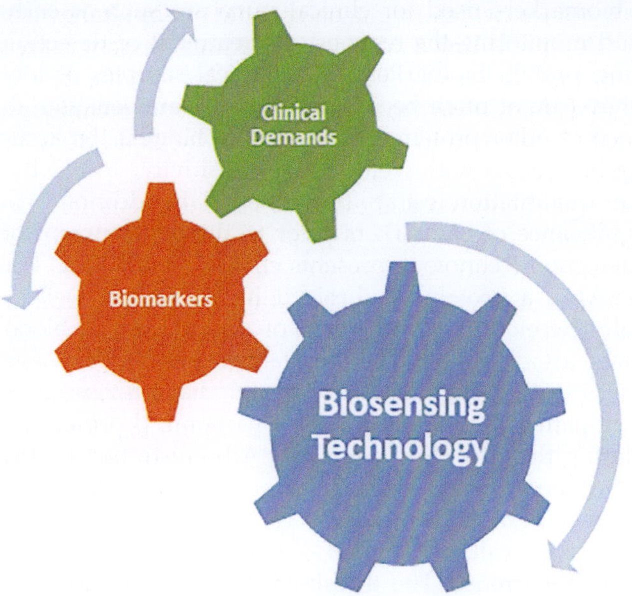


**FIGURE 26.1** Schematic diagram showing the interrelation between clinical demands, existing biomarkers, and biosensing technology.

to control. These include probe-target binding efficiency, specific signal generation, interference for other components in the biological sample, signal amplification, and management of nonspecific binding [11].

Although biosensors can provide the detection of novel cancer biomarker and monitoring platforms with improved clinical performance, and easy use by the patient, the validation of biomarkers is an important factor delaying the clinical application of these technologies. This happens mainly because new biomarkers do not present major clinical significance. Medical doctors, therefore, avoid their usage in order to reduce costs. Interestingly, it has been shown that the combination of more than one biomarker, that have no individual utility can be clinically useful together. This shows the importance of the development of multiplex biosensing technologies [12,13].

Biomarkers are related to the biology of the carcinogenesis process and determine the analytical method and biosensing technology used. The uncontrolled cell growth characteristic of carcinogenesis starts by alterations in DNA, through gene mutations (genetic biomarkers). They are expressed in altered proteins (molecular biomarkers), which affect the metabolism of cells expressed by unrestrained growth, inhibition of apoptosis, or neovascularization (metabolic biomarkers) [10].

### 26.2.1 Protein biomarkers

Protein biomarkers include substances produced by the cancer cells, or by other cells in response to cancer. They can mainly be found in blood and sometimes in the urine [14,15]. Protein biomarkers are the gold standard as cancer biomarkers used in clinical settings.

Currently, they are the only cancer biomarkers used for clinical purposes such as early detection, late disease progression, and monitoring the response to treatment or detection of recurrence [10]. However, detecting protein biomarkers in biological samples at low concentrations is challenging since they cannot often be replicated. They are sensitive to their environment, and the interference of other proteins in the sample hinders the accuracy of the assays. Therefore, creating biosensors with well-defined sensitivity, specificity, and accuracy to a particular biomarker quantitation test are fundamental. In addition, it is important to consider the clinical significance of biomarkers prior to the development of the biosensor. This ensures that the detection technology presents clinical significance. For instance, prostate-specific antigen (PSA) is a protein synthesized by normal as well as malignant cells of the prostate gland. Therefore, the increment of this protein in blood shows a higher probability for men who are diagnosed with prostate cancer. Studies have shown that individuals with PSA levels in the range of 4–10 ng $mL^{-1}$ have a 22%–27% probability of developing cancer, while patients with PSA levels $\geq$10 ng $mL^{-1}$ present an increased risk of 67% [16,17]. Considering this, the Food and Drug Administration (FDA) has approved the determination of PSA serum levels to test asymptomatic men for prostate cancer, in conjunction with digital rectal exam (DRE). This is carried out for men aged 50 years old or older with a cut-off blood PSA value of 4 ng $mL^{-1}$ [18]. PSA is also used as a specific biomarker for prostate cancer recurrence. The threshold after prostatectomy is 0.4 ng $mL^{-1}$ [19,20] and is 2 ng $mL^{-1}$ above the absolute lowest level of PSA after radiotherapy treatment [19]. Consequently, before developing a PSA biosensor it is important to aim for accurate detection within a broad range from 0.4 to $\geq$10 ng $mL^{-1}$. This allows it to be used as a cancer screening and recurrence tool effectively.

Each biomarker presents its own clinical specificities and demands, which must careful be evaluated in the biosensor development process if the goal is achieving successfully commercialization. Table 26.1 presents the most used proteins in cancer monitoring, their current use and clinical demands.

### 26.2.2 Circulating tumor cells

Circulating tumor cells (CTCs) are cancer cells that leave the primary tumor, circulate in the body fluids, and then metastasize to several parts of the body. This process happens even in the primary stages of the disease. CTCs are therefore minimally invasive and multifunctional biomarkers since they can be detected by simple blood analysis, and provide information on tumor progression, target therapy, and response to treatment [28]. The greatest challenge for the detection of CTCs is their low concentration in physiological fluids. CTCs make up a small number of the total cells in circulating blood, that is, 1–10 CTCs per milliliter of whole blood, compared with a few million white blood cells and a billion red blood cells [29,30]. In addition to their low concentration, CTCs may be damaged and be fragmented due to common multistep cell preparation processes. This causes inaccurate detection and misinterpretation. Finally, since CTCs are mainly characterized and are identified by their morphology and immunostaining pattern, their heterogeneity is a major obstacle for their detection. Therefore, the clinical utility of current CTC tests is limited mainly due to technical constraints [31]. Novel biosensing technologies with

TABLE 26.1 Most commonly used protein cancer biomarkers: their use and clinical threshold values.

| Molecule | Use | Cancer type | Clinical thresholds | Ref. |
|---|---|---|---|---|
| PSA | Screening<br>Diagnostics<br>Recurrence | Prostate | 4 ng $mL^{-1}$ biopsy threshold<br>0.4 ng $mL^{-1}$ prostatectomy cut-off<br>2 ng $mL^{-1}$ radiotherapy cut-off | [18–21] |
| CA15-3 | Monitor for response or progression in metastatic disease | Breast | 30 U $mL^{-1}$ | [22] |
| AFP | Diagnostic<br>Evaluate response to chemotherapy<br>Recurrence | Liver | 20 ng $mL^{-1}$ | [23] |
| CEA | Recurrence | Colorectal | 5 ng $mL^{-1}$ | [24] |
| CA 125 | Screening Prognostic<br>Diagnostic<br>Monitor response to therapy | Ovarian | 35 U $mL^{-1}$ | [25] |
| NMP22 | Diagnostics | Bladder | 10 U $mL^{-1}$ | [26] |
| MGMT | Prognostics<br>Diagnostics | Brain | $1.4 \pm 3.3$ fmol $10^6$ $cells^{-1}$ | [27] |

*PSA*, Prostate-specific antigen; *CA 15-5*, carcinoma antigen 15-3; *AFP*, α-fetoprotein; *CEA*, carcinoembryonic antigen; *CA 125*, carcinoma antigen 125; *NMP22*, nuclear mitotic apparatus protein 22; *MGMT*, methyl guanine methyl transferase.

integrated cell isolation, characterization, and quantitation functionalities are urgently needed in order to utilize CTCs as multifunctional biomarkers in current clinical practice.

### 26.2.3 MicroRNAs

MicroRNAs (miRNAs) are endogenous, single-stranded, noncoding small RNA with a length of approximately 22 nucleotides, that operate in RNA silencing and in posttranscription regulation of gene expression. The ability of miRNAs to inhibit the translation of oncogenes and tumor suppressor genes implies that they are involved in carcinogenesis. Many studies have shown that miRNA dysregulation is involved in cancer initiation, invasion and metastasis [32–34]. However, currently, there is no useful miRNA biomarker. There are several reasons for this. First, miRNAs were discovered in 1993, which makes this research field very young. Establishment of a suitable biomarker requires time. Second, not all miRNAs are cell and tissue-specific. Third, miRNA biology is complex, which makes it very difficult to reach conclusions from only one miRNA. Fourth, contradictory results in terms of miRNA quantitation may be due to different detection methods. The sequence homology, low abundance, and common secondary structures of miRNAs complicate efforts to develop accurate, unbiased quantification techniques with both high sensitivity and specificity. Finally, the human miRNA complement is not fully known. Many of the miRNAs that have been investigated are not according to the widely recognized criteria [35]. How miRNAs work and their specific biological function are still not clearly understood to use them as cancer biomarkers in clinics. The importance of creating biosensors for miRNA detection is unquestionable. However, these biosensing technologies may not be easily commercialized and transferred into the clinics in the near future.

### 26.2.4 Circulating tumor DNA

Circulating tumor DNA (ctDNA) is tumor-derived fragmented DNA in the bloodstream that is not associated with cells. Tumor cells may release DNA fragments into the circulation, increasing the tumor DNA quantities the plasma of cancer patients, compared with healthy controls [36]. Because ctDNA may reflect the entire tumor genome, it has potential clinical utility. However, existing ctDNA analysis methods have insufficient sensitivity or patient coverage for broader clinical applicability [11].

### 26.2.5 Biomarker panels

A key aspect of disease biomarkers is that they provide narrow information about a certain health condition. Therefore, discovering panels of biomarkers, which can characterize in detail the patient condition, are needed for early cancer detection and rapid treatment follow-up. Subsequently, when biomarker panels are clinically established, it will be necessary to develop multiplexed biosensing platforms. This is technically challenging since several assays with different parameters would be needed on the same biosensing platform and biological samples [37,38].

## 26.3 Cancer biosensors

As already, discussed, accurate and effective methods for cancer detection are an urgent need for improving the probability of cure. Nowadays, many cancers are only diagnosed after they have metastasized. Thus, suitable biosensors can surpass these limitations and improve the early diagnosis and monitoring of cancer [2,39]. A biosensor is defined as a device that detects a specific analyte (proteins, DNA, and RNA) and converts a biological signal, through a transducer, into a signal which can be detected and analyzed [40]. Normally, for the recognition of cancer biomarkers, antibodies, or other biorecognition molecules are immobilized on the transducer surfaces. After the reaction between the biomarker and recognition molecule, a signal is originated (heat, electron transfer, pH change, etc.). This is then transduced so that the biomarker quantification and monitoring can be accomplished.

Therefore, biosensors can be classified, according to the detection mode and, consequently, signal transduction, for example, electrochemical, optical, or piezoelectric. The most commonly used cancer biosensors are electrochemical and optical [41]. In order to meet clinical demands, and become great tools for cancer detection and monitoring, the biosensors should have the following requirements—biocompatibility, accuracy, high stability, selectivity, ability to be multiplexed, large dynamic range, low limits of detection, and good reproducibility [41,42]. Technologies with all these features capable of quantifying the plethora of biomarkers in a rapid and noninvasive way still do not exist in the market. The development of viable cancer biosensors continues to be an active research area.

### 26.3.1 Electrochemical biosensors

An electrochemical biosensor measures changes in the electrical properties, which are created in the production or consumption of electrons or ions during the biomarker/biosensor interaction. From this interaction, a measurable current (amperometric), potential or charge accumulation (potentiometric), and impedance (impedimetric biosensors) between the electrodes can result. Consequently, electrodes play an essential role in the performance of the electrochemical biosensors, since they are responsible for transducing the signal resulting from the reactions. The material, surface modification, or dimensions greatly influence the sensor performance. Electrochemical biosensing demands a minimum of two electrodes; however, for accuracy commonly three different electrodes are used—a reference, a counter, and a working electrode. The reference electrode presents a stable and well-known electrode potential. This is kept at a certain distance from the reaction site, between the working and counting electrodes. The working electrode is the transduction element and the counter electrode connects with the electrolytic solution, allowing the application of current into the working electrode. All the electrodes should be stable, chemically and conductively, for a reliable measurement [43–57]. However, not all biomarkers are inherently capable of serving as redox partners in electrochemical reactions. It is sometimes necessary to mediate the electrochemical reaction at the working electrode, for instance by the adding of linkers [43].

Kumar et al. developed a Whatman filter paper (WP)-based electrochemical biosensor composed of poly(3,4-ethylenedioxythiophene):poly(styrene sulfonate) and nanostructured iron oxide ($nFe_2O_3$@PEDOT:PSS) nanocomposite for CEA detection [58]. As observed in Fig. 26.2A, $nFe_2O_3$ was dispersed in a PEDOT:PSS solution and WP was immersed in this solution. To enhance the paper's conductivity, it was dipped into a DMSO solution. In the end, an electrochemical paper (EP) was achieved with a conductivity of $2.4 \times 10^2$ S $cm^{-1}$. Finally, a layer of carcinoembryonic antibodies (anti-CEA) was immobilized in the substrate, through physiological adsorption, with BSA used to block nonspecific active sites. The electrochemical response of BSA/anti-CEA/EP electrode was evaluated as a function of CEA concentration (2–100 ng $mL^{-1}$) in a buffer solution. As the CEA concentration increased, the response current decreased, due to the formation of electrically insulating layer of the antigen–antibody complex onto the biosensor's surface, which has perhaps hindered the electron transfer via the redox probe. A linear detection range was obtained for CEA concentration between 4 and 25 ng $mL^{-1}$, and the final sensitivity of the sensor was 10.2 m $Ang^{-1}$ mL $cm^{-2}$. This biosensor shows potential for CEA detection since the clinical threshold of this biomarker is 5 ng $mL^{-1}$ [58].

In a different approach, An et al. developed an alternative strategy to isolate cancer cells from whole blood, using a benzoboric acid-modified gold-plated polymeric substrate, with a regular 3D surface microarray, using a nanoimprinting technique, as demonstrated in Fig. 26.2B. The 3D microarray was coated with gold and was used as a working electrode in the electrochemical detection. Furthermore, the introduction of a 3D structure significantly enhanced the efficiency of cancer cell capture by 3.8 times. This system allowed the direct detection of cancer cells from whole blood samples, with a detection limit is 5 cells $mL^{-1}$. Additionally, isolated CTCs could be recovered by simple exchange reactions without significant damage. The high sensitivity and selectivity of this platform may provide a promising biosensor for an early stage of cancer detection [59].

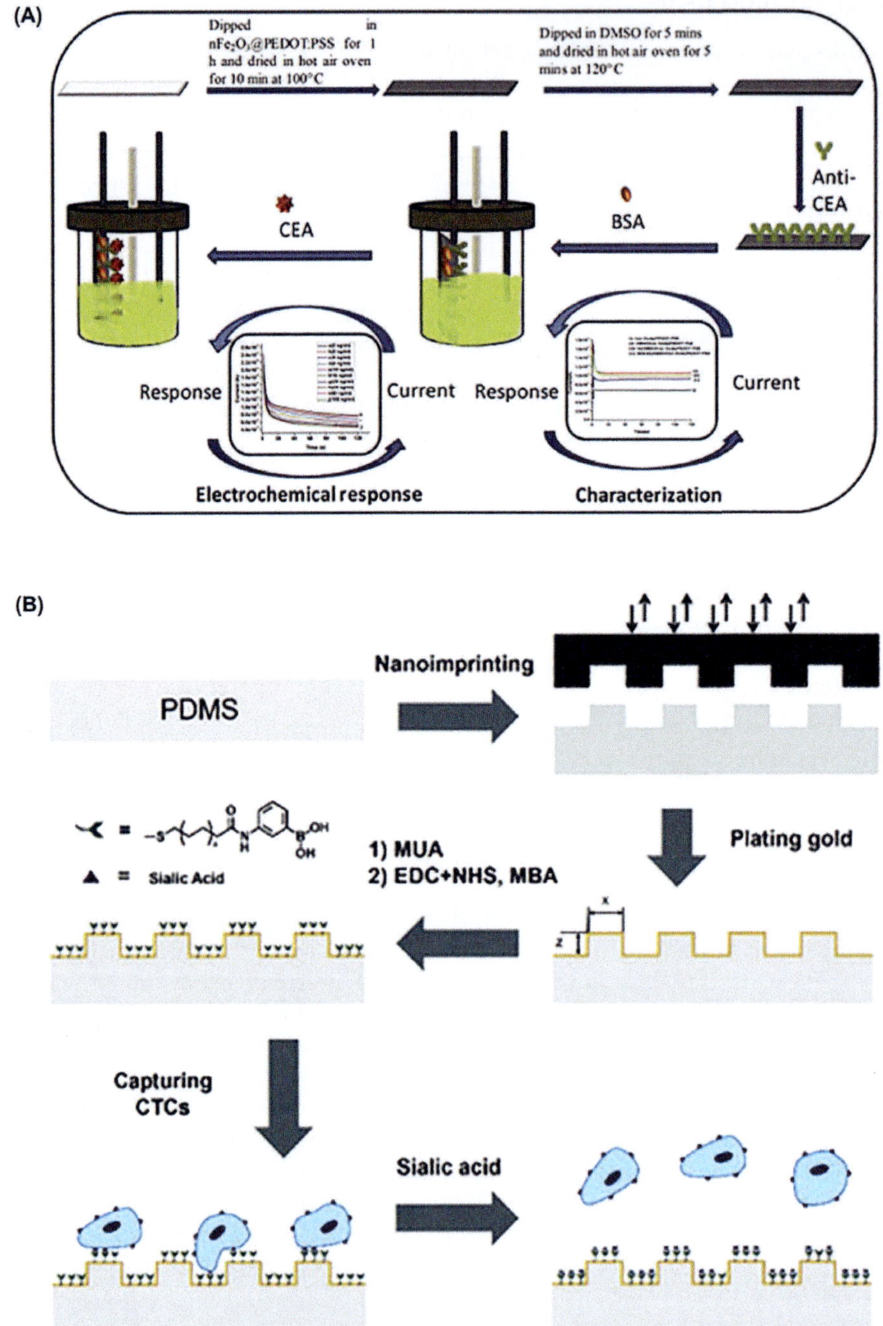


FIGURE 26.2 (A) Schematic diagram of the electrochemical paper-based cancer biosensor fabrication [58] and (B) schematic presentation of the modified gold-plated polymeric substrate with a regular 3D surface array for the capture, detection, and release of cancer cells [59]. Source: *Figures reproduced with permission of Elsevier and Royal Society of Chemistry.*

### 26.3.2 Optical biosensors

Optical biosensors are based on their action mode in the change of optical properties of materials, during the interaction with a biorecognition element [41,60,61]. These types of biosensors present some advantages when compared with conventional systems (e.g., electrochemical), since they can allow real-time and label-free detection. Optical biosensors are immune to electromagnetic interference and present high specificity. They are capable of remote sensing with high sensitivity. They can also provide multiplexed detection within a single device and are cost-effective [60,61].

Overall, there are two main detection types of optical biosensors: (1) fluorescent-based and (2) label-free detection. In the first case, biomarkers are labeled with fluorescent tags, like dyes, and the intensity of the fluorescent signal indicates the presence of the biomarker and its interaction with the biorecognition molecule. Even though this type of detection is very sensitive, since the detection limit is down to a unique molecule, quantitative analyses are very challenging because of fluorescent signal bias. It is not possible to control precisely the number of fluorophores on each molecule [62]. On the other hand, in label-free detection, biomarkers are not labeled or altered as they are detected in their natural form. This allows the quantitative and kinetic measurement of molecular interactions. Some optical label-free detection modes measure refractive index change, which depends on the sample concentration or surface density, instead of sample total mass. This property can be a huge benefit for ultra-small volume detection (femtoliter to nanoliter), providing advantages over fluorescent-based detection, whose signal depends on the number of molecules [60,61,63].

The most commonly used techniques for optical detection are surface plasmon resonance (SPR) and localized surface plasmon resonance (LSPR), fluorescence, chemiluminescence resonance energy transfer, chemiluminescence, and surface-enhanced Raman scattering (SERS).

Kim et al. have developed a LSPR-based biosensor for α-fetoprotein (AFP) detection (Fig. 26.3A). For this purpose, they used a tempered glass functionalized with (3-aminopropyl)triethoxysilane (APTES) to introduce amine groups on the glass surface. After the glass was immersed in a gold nanoparticles (AuNPs) solution which bonded with the amine groups. This is used to enhance the LSPR signal, by changing the local refractive index. The measured absorbance changed, according to the binding of anti-AFP and the AFP in the substrate (Fig. 26.3B). The developed biosensor was able to detect AFP at concentrations between $1\ ng\ mL^{-1}$ and $1\ \mu g\ mL^{-1}$, which includes the clinical threshold of this protein, demonstrating its potential for clinical use [64].

Using the same detection technique of LSPR, Barbosa et al. developed a label-free immunosensor technology based on nanoplasmonic Au-$TiO_2$ thin films (Fig. 26.3C). The authors optimized the immobilization of an antibody monolayer in the thin film's surface, resulting in a redshift in the LSPR band. This work revealed the potential for a cost-effective and label-free biosensor, which can be applied in cancer detection and monitoring [65].

Using a different strategy, Hakimian et al. developed an ultrasensitive optical biosensor for microRNA-155 (miR-155) detection (Fig. 26.3D), aiming for early diagnosis of breast cancer. For that, branched polyethylenimine (PEI) capped AuNP (P-AuNP) was used for target trapping. Since P-AuNPs are highly positively charged, the target miR-155 (negatively charged) electrostatically adsorbs onto the surface, forming the P-AuNPs/miR-155 complex. Another

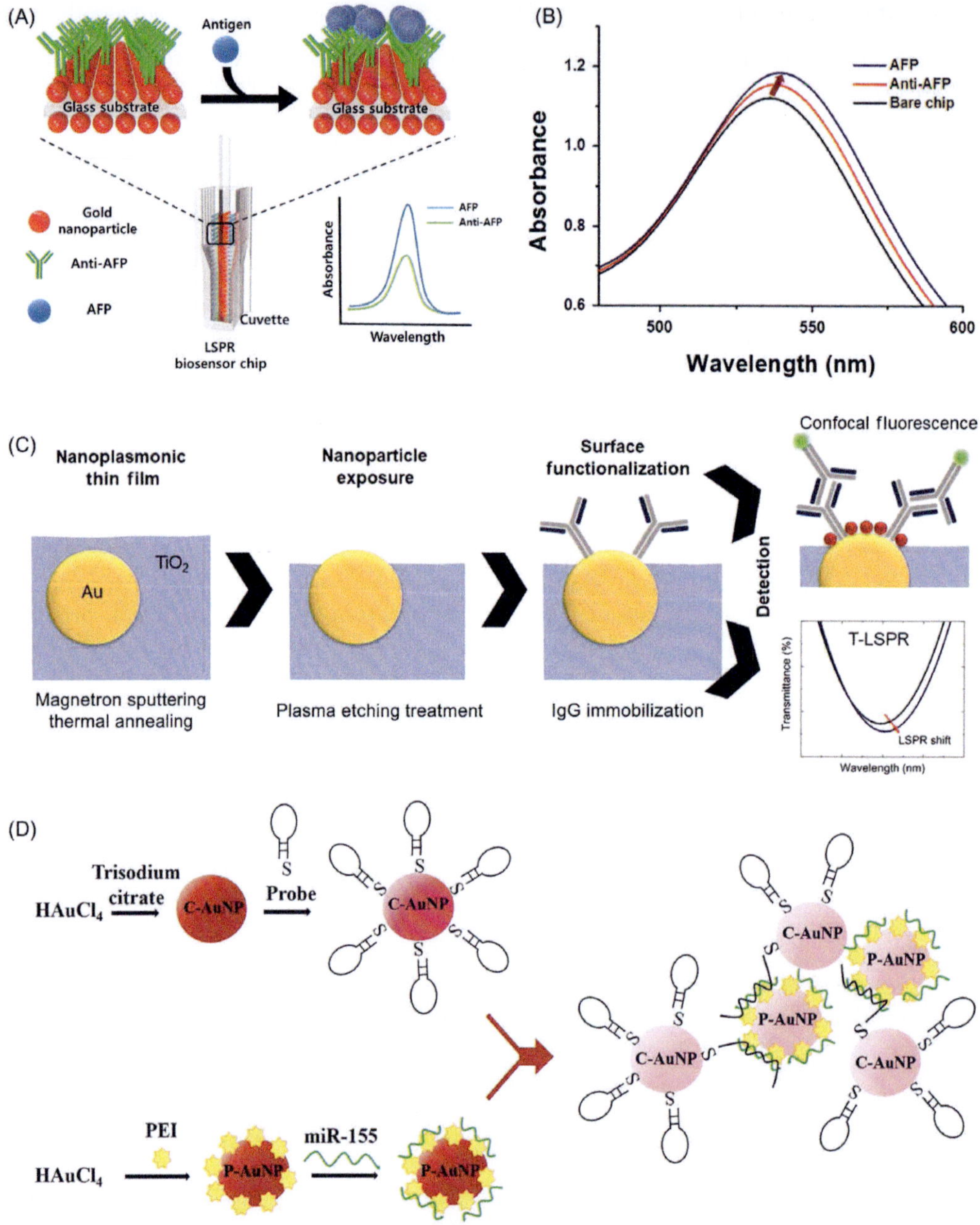


FIGURE 26.3 (A) Schematic diagram of LSPR-based biosensor based on AuNPs nanopatterned biosensor in a glass substrate. (B) Detection of AFP using the LSPR biosensor [64]. (C) Schematic of a LSPR biosensor based on exposed AuNPs in a $TiO_2$ thin-film matrix [65]. (D) Schematic presentation of nanoparticle aggregates, which has been resulted from probe-target hybridization, and enable optical detection of miR-155 [66]. Source: *Figures reproduced with permission from Elsevier and Nature Publishing Group.*

complex is formed by negatively charged citrate-capped AuNPs (C-AuNPs) covalently conjugated with DNA, forming the C-AuNPs specific probe. Finally, by mixing these two complexes, hybridization occurs, due to interparticle crosslinking aggregation, and an optical signal can be measured by UV−Vis absorption and used to quantify the miR-155 content. With this approach, miR-155 detection can be performed at very low concentrations (100 aM) and without the need for expensive and time-consuming signal amplification [66].

### 26.3.3 Piezoelectric biosensors

Piezoelectricity is a physical phenomenon where a material changes its electrical current, and consequently its voltage when it is mechanically stressed. The opposite is also true, that is, when the material is electrically stimulated, a mechanical strain is generated. Usually, anisotropic materials present piezoelectric properties [67]. Aluminum phosphate, aluminum nitride, zinc oxide, quartz, polyvinylidene fluoride, and polylactic acids are examples of piezoelectric materials. Ceramic resonators are an alternative of usually used quartz resonators since it is challenging to produce quartz with suitable piezoelectric properties. Thus, for oscillation frequency stabilization ceramic resonators are used in several biosensors applications [68,69]. The oscillation frequency can be measured since electrodes can be excited on the surface of piezoelectrical materials, causing mechanical oscillations in materials. The bonding of the biomarker on the material surface will change the frequency of oscillation, which is proportional to the amount of bonded analyte [68−70]. Piezoelectric biosensors present some advantages: they do not require external power and do not need to add specific reagents. However, the viscosity of the medium can interfere with the measurement and aspects like fragility should be taken into account. Another drawback is the fact that only some materials present piezoelectrical properties [41,69].

Su et al. used a titanate zirconate (PZT) ceramic resonator to produce a label-free, cost-effective piezoelectric cancer biosensor [70]. Two ceramic resonators were connected in parallel: one as sensing compound and the other as a control, in order to avoid environmental interference, like temperature fluctuations. For the evaluation of biosensor performance, PSA cancer biomarker was used. Protein A and Nafion were used as the linker groups to immobilize antibodies on the resonator surface. PSA antibody or AFP antibody were added onto the modified ceramic resonator surface (Fig. 26.4A) at room temperature. To evaluate the feasibility of this ceramic biosensor for clinical sample analysis, several concentrations of PSA were spiked into the normal human serum, and the resultant frequency changes detected and analyzed (Fig. 26.4B). The biosensor showed a sensitivity of $0.25\ ng\ mL^{-1}$ and a range detection from 0.025 to $2.5 \times 10^2\ ng\ mL^{-1}$ for PSA, compatible with the clinical threshold [70].

Wee et al. reported an electromechanical biosensor for protein detection [PSA and C-reactive protein (CRP)], using self-sensing piezoresistive microcantilevers [71]. The antigen−antibody binding induced stress changes in the biosensor surface and direct nanomechanical response of microfabricated self-sensing microcantilevers. The thin-film piezoresistive microcantilever sensor was fabricated and was also coated with an Au/Cr layer (Fig. 26.4C). The analyte is bound on the Au surface, causing the bending of the cantilever and resistance change of the piezoresistive layer, due to the surface stress induced

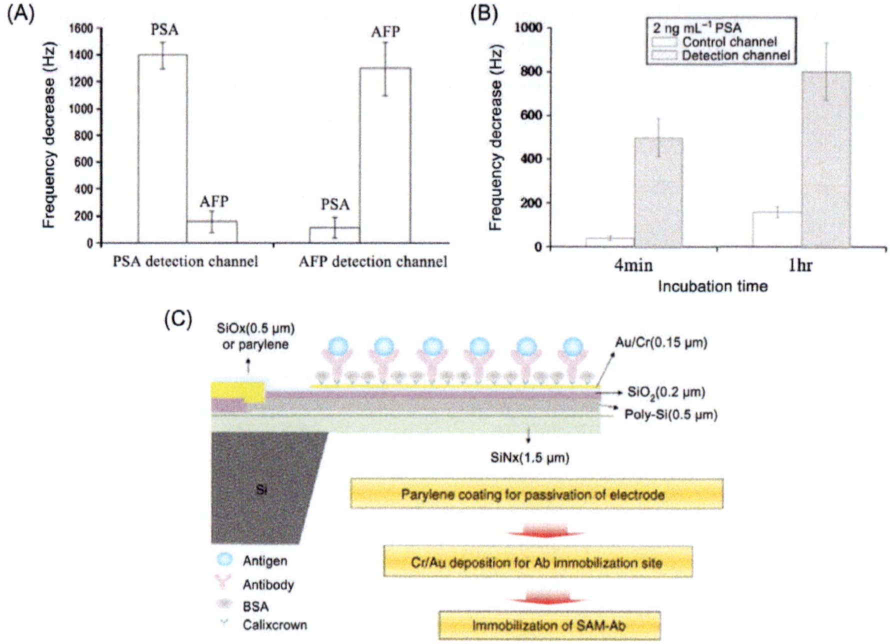


**FIGURE 26.4** Schematic diagrams of (A) Specific antigen-binding evaluation on PSA and AFP detection channels; (B) the frequency changes of the binding of PSA to PSA antibody-immobilized resonator or control channel (without PSA antibody), over time; [70] and (C) monoclonal anti-PSA and CPR immobilization on Au surface of piezoresistive microcantilever sensor [71]. Source: *Figures reproduced with the permissions from Elsevier.*

through the cantilever. The piezoresistive cantilever output voltage was proportional to the PSA and CRP concentration, in the target solution, showing the potential of this biosensor for multiplexed biomarker detection [71]. This detection of different biomarkers shows that the multiplex biosensing platforms can be used for simple, rapid and sensitive measurements of several biomarkers. This helps to fill the gap between the biosensor methods and laboratory techniques used nowadays.

One of the most critical aspects of biosensors is to choose an adequate detection mode, which matches the appropriate biological and electronic components in order to generate a relevant signal during analysis. In order to make sure the analyte is the only analyzed molecule, its isolation from the remaining molecules is essential. Additionally, the interaction between the analyte and the biosensor, as well as its stability is crucial for good detection [41]. The advantages and disadvantages of different detection modes are summarized in Table 26.2.

## 26.4 Commercialization and clinical trials of cancer biosensors

In contrast to the expectations expressed in market surveys conducted during the 1990s, the market for biosensors has not expanded as foreseen. This has essentially been due to

TABLE 26.2 Advantages and disadvantages of the most common detection modes used in biosensors.

| | Subtypes | Advantages | Disadvantages |
|---|---|---|---|
| Electrochemical | Amperometric<br>Potentiometric<br>Impedimetric | High sensitivity<br>Specificity | Not all biomarkers are inherently capable of serving the redox partners in electrochemical reactions<br>May need electrochemical mediators |
| Optical | Label-based detection<br>Label-free detection | Real-time and label-free detection<br>Immunity to electromagnetic interference<br>High specificity and sensitivity<br>Remote sensing<br>Multiplex detection<br>Cost-effective | May be necessary labeled biomarkers with fluorescent tags<br>Signal may not be strong enough |
| Piezoelectrical | – | Do not require external power<br>Specificity<br>Sensitivity<br>Cost-effective | Not all materials have piezoelectrical properties |

*Table reproduced with permission from Elsevier Rebelo R, Barbosa AI, Caballero D, Kwon IK, Oliveira JM, Kundu SC, et al. Biosens Bioelectron. 2019;130, 20–39 [41].*

unpredicted technological problems. Continuous and real-time monitoring of specific biomarkers, as well as in vivo diagnostics still show several limitations. These may probably remain at the experimental stages for some more time. However, it also is necessary to know the market regulations and requirements to guide the technical development and adoption of biosensors.

Biosensors, as for any new medical device, have to go through many challenges to achieve clinical validation and to fulfill the regulatory requirements to demonstrate safety and efficacy. The first challenge is to classify the risk of the device into different categories like Class I, Class II, and Class III for low, medium, and high risk. This classification depends on the device properties such as cycle life/power rechargeability, magnetic resonance imaging (MRI), and other imaging techniques safety, biocompatibility, and power specifications, among others [72–75]. Furthermore, before validation, a detailed toxicity comparison and efficacy between animal and preclinical studies, as well as the device toxicity are required. Finally, the consequences of the biosensor's failure need to be assessed. During the regulatory process it is essential to meet the specifications of regulatory entities [e.g., FDA or Conformité Européenne (CE)], demonstrating the safety and efficiency of the device [76,77]. Usually, this process begins in the laboratory, followed by animal testing and, finally, clinical trials. Table 26.3 reports some ongoing and completed clinical trials using cancer biosensors. During all these processes, several phases of R&D are involved to ensure device safety under the strict supervision of the FDA or CE.

**TABLE 26.3** List of ongoing and terminated/completed clinical trials of biosensors in cancer detection (Information is obtained from http://clinicaltrials.gov).

| NCT number | Date and status | Study | Patients age | Follow-up | Procedure |
|---|---|---|---|---|---|
| NCT02957370 | 2015–21 Recruiting | Electrophage and colorimetric aptamer sensors for clinical staging and monitoring of bladder cancer | ≥ 18 years old | n.d. | Discover urinary biomarker (s) that is distinct to bladder cancer, while developing molecular sensors that can detect the urinary anomalies |
| NCT03173729 | 2017–22 Recruiting | Point of care, real-time urine metabolomics test to diagnose colorectal cancers and polyps in low and middle income countries | ≥ 18 years old | n.d. | Adapt a 3-metabolite biosensor that identifies patients with colorectal cancer and precancerous polyps to Nigerian patients. In phase 2, evaluate the point-of-care biosensor device in Nigeria |
| NCT00813878 | 2001–12 Terminated | Optical biosensor for the early detection of breast cancer | 18–120 years | 5 years | To determine CEA levels in nipple secretions and blood samples from breast cancer patients and normal controls. After, an optical biosensor will be developed to detect fluorescent-labeled antibodies directed against CEA found in serum and breast sections |
| NCT02195076 | 2014–17 Terminated | A clinical study design to evaluate the specificity and sensitivity of our novel system for detection of breast and lung cancer by odor signature | 30–80 years | 3 years | Contribute and detect the patient at the earliest possible stage, in a noninvasive, nonradioactive, and nontoxic way |

## 26.5 Conclusions

The early detection of cancers and continuous follow-up using rapid and reliable technologies will significantly support cancer prognosis and treatment. Biosensing technologies capable of detecting several biomarkers at different clinical thresholds and of different biological natures are essential to meet clinical demands for cancer diagnostics. Currently, protein biomarkers are the most commonly used biomarkers in cancer detection. However, they are only used as secondary diagnostic tools since they cannot provide the desired specificity. Novel biomarkers such as CTCs and miRNA are being studied; however, they are still not used in clinics either due to their lack of approval as cancer biomarkers or due to technological constraints required for their quantitation. Biosensors that can detect these potential biomarkers, and improve the quantitation of the clinically used biomarkers are being developed. These biosensing technologies use different detection methods such as electrochemical, optical, and piezoresistive. Several of the developed technologies present improved performance to current methods, yet only a few are being tested in clinical trials. This difficulty in commercialization is a consequence of their limited sensitivity, specificity, limits of detection, and capacity for multiplexing. Additionally, the challenges and restricted requirements in the medical device commercialization process delays the full clinical validation of biosensors. A cancer diagnosis will significantly be improved by the biodetection of cancer biomarkers. This process is dependent on novel biomarker validation and biosensing technologies capable of meeting clinical demands.

## Acknowledgments

The authors acknowledge the FRONTHERA project (Frontiers of technology for theranostics of cancer, metabolic and neurodegenerative diseases) n° NORTE-01-0145-FEDER-0000232, the European Union Framework Programme for Research and Innovation Horizon 2020 under grant agreement n° 668983—FoReCaST (Forefront Research in 3D Disease Cancer Models as in vitro Screening Technologies), and FCT grants POCI-01-0145-FEDER-031590 and PTDC/BTM-ORG/28168/2017.

## References

[1] Cancer Research UK. Worldwide Cancer Mortality Statistics https://www.cancerresearchuk.org/health-professional/cancer-statistics/worldwide-cancer/mortality (accessed Jun 4, 2020).

[2] Bohunicky B, Mousa SA. Biosensors: The New Wave in Cancer Diagnosis. Nanotechnol. Sci. Appl 2011;4:1–10.

[3] Sopik V, Narod SA. The Relationship between Tumour Size, Nodal Status and Distant Metastases: On the Origins of Breast Cancer. Breast Cancer Res. Treat. 2018;170(3):647–56.

[4] Gospodarowicz M, O'Sullivan B. Prognostic Factors in Cancer. Semin. Surg. Oncol. 2003;21(1):13–18.

[5] Ahmed MU, Saaem I, Wu PC, Brown AS. Personalized Diagnostics and Biosensors: A Review of the Biology and Technology Needed for Personalized Medicine. Crit. Rev. Biotechnol. 2014;34(2):180–96.

[6] Golubnitschaja O, Flammer J. What Are the Biomarkers for Glaucoma? Surv. Ophthalmol. 2007;52(6 SUPPL.): S155–61.

[7] Sawyers CL. The Cancer Biomarker Problem. Nature April 3, 2008;548–52.

[8] Nagpal M, Singh S, Singh P, Chauhan P, Zaidi M. Tumor Markers: A Diagnostic Tool. Natl. J. Maxillofac. Surg. 2016;7(1):17.

[9] Henry NL, Hayes DF. Cancer Biomarkers. Mol. Oncol 2012;6(2):140–6.

[10] Wu L, Qu X. Cancer Biomarker Detection: Recent Achievements and Challenges. Chem. Soc. Rev. 2015;44 (10):2963–97.

[11] Diamandis EP. The Failure of Protein Cancer Biomarkers to Reach the Clinic: Why, and What Can Be Done to Address the Problem? BMC Medicine. 2012;87.

[12] Hanash SM, Pitteri SJ, Faca VM. Mining the Plasma Proteome for Cancer Biomarkers. Nature April 3, 2008;571–9.

[13] Tarro G, Perna A, Esposito C. Early Diagnosis of Lung Cancer by Detection of Tumor Liberated Protein. Journal of Cellular Physiology April 2005;1–5.

[14] Brawer MK, Beatie J, Wener MH, Vessella RL, Preston SD, Lange PH. Screening for Prostatic Carcinoma with Prostate Specific Antigen: Results of the Second Year. J. Urol. 1993;150(1):106–9.

[15] Barry MJ. Prostate-Specific–Antigen Testing for Early Diagnosis of Prostate Cancer. . N. Engl. J. Med 2002;344(18):1373–7.

[16] De Angelis G, Rittenhouse HG, Mikolajczyk SD, Blair Shamel L, Semjonow A. Twenty Years of PSA: From Prostate Antigen to Tumor Marker. Rev. Urol 2007;9(3):113–23.

[17] Parsons JK, Partin AW, Trock B, Bruzek DJ, Cheli C, Sokoll LJ. Complexed Prostate-Specific Antigen for the Diagnosis of Biochemical Recurrence after Radical Prostatectomy. BJU Int. 2007;99(4):758–61.

[18] Amling CL, Bergstralh EJ, Blute ML, Slezak JM, Zincke H. Defining Prostate Specific Antigen Progression after Radical Prostatectomy: What Is the Most Appropriate Cut Point? J. Urol. 2001;165(4):1146–51.

[19] Kuban DA, Thames HD, Levy LB. PSA after Radiation for Prostate Cancer. ONCOLOGY. 2004;595–604.

[20] Wang W, Xu X, Tian B, Wang Y, Du L, Sun T, Shi Y, Zhao X, Jing J. The Diagnostic Value of Serum Tumor Markers CEA, CA19-9, CA125, CA15-3, and TPS in Metastatic Breast Cancer. Clin. Chim. Acta 2017;470 (December 2016):51–5.

[21] Arrieta O, Cacho B, Morales-Espinosa D, Ruelas-Villavicencio A, Flores-Estrada D, Hernández-Pedro N. The Progressive Elevation of Alpha Fetoprotein for the Diagnosis of Hepatocellular Carcinoma in Patients with Liver Cirrhosis. BMC Cancer 2007;7:28.

[22] Locker GY, Hamilton S, Harris J, Jessup JM, Kemeny N, Macdonald JS, Somerfield MR, Hayes DF, Bast RC. ASCO 2006 Update of Recommendations for the Use of Tumor Markers in Gastrointestinal Cancer. Journal of Clinical Oncology 2006;5313–27.

[23] Gupta D, Lis CG. Role of CA125 in Predicting Ovarian Cancer Survival - a Review of the Epidemiological Literature. J. Ovarian Res 2009;2(13):1–20.

[24] Miyake M, Goodison S, Giacoia EG, Rizwani W, Ross S, Rosser CJ. Influencing Factors on the NMP-22 Urine Assay: An Experimental Model. BMC Urol. 2012;12:23.

[25] Christmann M, Verbeek B, Roos WP, Kaina B. O6-Methylguanine-DNA Methyltransferase (MGMT) in Normal Tissues and Tumors: Enzyme Activity, Promoter Methylation and Immunohistochemistry. Biochim. Biophys. Acta - Rev. Cancer 2011;1816(2):179–90.

[26] Yap TA, Lorente D, Omlin A, Olmos D, De Bono JS. Circulating Tumor Cells: A Multifunctional Biomarker. Clin. Cancer Res. 2014;20(10):2553–8.

[27] Joosse SA, Pantel K. Biologic Challenges in the Detection of Circulating Tumor Cells. Cancer Res. 2013;73 (1):8–11.

[28] Balic Matija, Dandachi Nadia, Lin Henry, D.H.R. Advances in the detection and characterization of disseminated tumour cells facilitate clinical translation, Volume 18 Number 5. http://www.ncbi.nlm.nih.gov/pubmed/16433139 (accessed Jul 24, 2019).

[29] Hong B, Zu Y. Detecting Circulating Tumor Cells: Current Challenges and New Trends. Theranostics. Ivyspring International Publisher; 2013. p. 377–94.

[30] Flatmark K, Høye E, Fromm B. MicroRNAs as Cancer Biomarkers. Scand. J. Clin. Lab. Invest. 2016;76 (sup245):S80–3.

[31] Jayanthi VSPKSA, Das AB, Saxena U. Biosensors and Bioelectronics Recent Advances in Biosensor Development for the Detection of Cancer Biomarkers. Biosens. Bioelectron 2017;91(November 2016):15–23.

[32] Rebelo R, Barbosa AI, Caballero D, Kwon IK, Oliveira JM, Kundu SC, Reis RL, Correlo VM. 3D Biosensors in Advanced Medical Diagnostics of High Mortality Diseases. Biosensors and Bioelectronics 2019;20–39 Elsevier January 11.

[33] Jeong-Yeol Yoon. Introduction to Biosensors. In From Electric Circuits to Immunosensors; Springer International Publishing: Cham 2016;257–97.

[34] Grieshaber D, Mackenzie R, Vörös J, Reimhult E. Electrochemical Biosensors -Sensor Principles and Architectures. Sensors 2008;8:1400–58.

[35] Zanghelini F, Frías IAM, Rêgo MJBM, Pitta MGR, Sacilloti M, Oliveira MDL, Andrade CAS. Biosensing Breast Cancer Cells Based on a Three-Dimensional TIO 2 Nanomembrane Transducer. Biosens. Bioelectron 2017;92:313–20.

[36] Sabury S, Kazemi SH, Sharif F. Graphene–Gold Nanoparticle Composite: Application as a Good Scaffold for Construction of Glucose Oxidase Biosensor. Mater. Sci. Eng. C 2015;49:297–304.

[37] Colombo L, Baruzzi AM, Garay FS. Analysis and Optimization of a Hydrogel Matrix for the Development of a Sandwich-Type Glucose Biosensor. Sensors Actuators, B Chem 2015;211:125–30.

[38] Slaughter G, Sunday J. Fabrication of Enzymatic Glucose Hydrogel Biosensor Based on Hydrothermally Grown ZnO Nanoclusters. IEEE Sens. J. 2014;14(5):1573–6.

[39] Ju YM, Yu B, Koob TJ, Moussy Y, Moussy F. A Novel Porous Collagen Scaffold around an Implantable Biosensor for Improving Biocompatibility. I. IIn Vitro/i / Iin Vivo/i Stability of the Scaffold and Iin Vitro/i Sensitivity of the Glucose Sensor with Scaffold. J. Biomed. Mater. Res. Part A 2008;87A(1):136–46.

[40] Guenther M. Hydrogel-Based Piezoresistive Biochemical Microsensors. SPIE Smart 2010;1–5.
[41] Pedrosa VA, Yan J, Simonian AL, Revzin A. Micropatterned Nanocomposite Hydrogels for Biosensing Applications. Electroanalysis 2011;23(5):1142–9.
[42] Kazemi SH, Ghodsi E, Abdollahi S, Nadri S. Porous Graphene Oxide Nanostructure as an Excellent Scaffold for Label-Free Electrochemical Biosensor: Detection of Cardiac Troponin I. Mater. Sci. Eng. C 2016;69:447–52.
[43] Li L, Wang Y, Pan L, Shi Y, Cheng W, Shi Y, Yu G. A Nanostructured Conductive Hydrogels-Based Biosensor Platform for Human Metabolite Detection. Nano Lett. 2015;15(2):1146–51.
[44] Rong Q, Han H, Feng F, Ma Z. Network Nanostructured Polypyrrole Hydrogel/Au Composites as Enhanced Electrochemical Biosensing Platform. Sci. Rep 2015;5(1):11440.
[45] Devillers M, Ahmad L, Korri-Youssoufi H, Salmon L. Carbohydrate-Based Electrochemical Biosensor for Detection of a Cancer Biomarker in Human Plasma. Biosens. Bioelectron 2017;96:178–85.
[46] Elshafey R, Tavares AC, Siaj M, Zourob M. Electrochemical Impedance Immunosensor Based on Gold Nanoparticles–Protein G for the Detection of Cancer Marker Epidermal Growth Factor Receptor in Human Plasma and Brain Tissue. Biosens. Bioelectron 2013;50:143–9.
[47] Kunduru V, Bothara M, Grosch J, Sengupta S, Patra PK, Prasad S. Nanostructured Surfaces for Enhanced Protein Detection toward Clinical Diagnostics. *Nanomedicine*. Nanotechnology, Biol. Med 2010;6(5):642–50.
[48] Chen H, Mei Q, Jia S, Koh K, Wang K, Liu X. High Specific Detection of Osteopontin Using a Three-Dimensional Copolymer Layer Support Based on Electrochemical Impedance Spectroscopy. Analyst 2014;139 (18):4476–81.
[49] Kumar S, Umar M, Sai A, Kumar S, Augustine S, Srivastava S, Malhotra BD. Electrochemical Paper Based Cancer Biosensor Using Iron Oxide Nanoparticles Decorated PEDOT: PSS. Anal. Chim. Acta 2019;1056:135–45.
[50] An L, Wang G, Han Y, Li T, Jin P, Liu S. Electrochemical Biosensor for Cancer Cell Detection Based on a Surface 3D Micro-Array. Lab Chip 2018;18(2):335–42.
[51] Dambors P, Svitel J, Katrlík J. Optical Biosensors. Essays Biochem 2016;60:91–100.
[52] Fan X, White IM, Shopova SI, Zhu H, Suter JD, Sun Y. Sensitive Optical Biosensors for Unlabeled Targets: A Review. Anal. Chim. Acta 2008;620:8–26.
[53] Frascella F, Petri C, Ricciardi S, Napione L, Munzert P, Jonas U, Dostalek J, Bussolino F, Fabrizio Pirri C, Descrovi E. Hydrogel-Terminated Photonic Crystal for Label-Free Detection of Angiopoietin-1. J. Light. Technol 2016;34(15):3641–5.
[54] Kim D, Kim J, Hwan C, Su N, Yeong S, Lee H, Lee G, Vilian ATE, Han Y, Kim W, et al. Rapid and Label-Free Bioanalytical Method of Alpha Fetoprotein Detection Using LSPR Chip. J. Cristal Growth 2017;469:131–5.
[55] Barbosa AI, Borges J, Meira DI, Costa D, Rodrigues MS, Rebelo R, Correlo VM, Vaz F, Reis RL. Development of Label-Free Plasmonic Au-TiO2 Thin Film Immunosensor Devices. Mater. Sci. Eng. C 2019;100:424–32.
[56] Hakimian F, Ghourchian H, Sadat Hashemi A, Reza Arastoo M, Behnam Rad M. Ultrasensitive Optical Biosensor for Detection of MiRNA-155 Using Positively Charged Au Nanoparticles. Sci. Rep 2018;8:2943.
[57] Yang W, Fang F, Fang D-N. Fracture and Fatigue of Ferroelectrics. Compr. Struct. Integr 2003;645–86.
[58] Pohanka M. Overview of Piezoelectric Biosensors, Immunosensors and DNA Sensors and Their Applications. Materials (Basel), 11. 2018. p. 448.
[59] Pohanka M. The Piezoelectric Biosensors: Principles and Applications, a Review. Int. J. Electrochem. Sci 2017;12:496–506.
[60] Su L, Zou L, Fong C, Wong W, Wei F, Wong K-Y, Wu RSS, Yang M. Biosensors and Bioelectronics Detection of Cancer Biomarkers by Piezoelectric Biosensor Using PZT Ceramic Resonator as the Transducer. Biosens. Bioelectron 2013;46:155–61.
[61] Su L, Zou L, Fong C-C, Wong W-L, Wei F, Wong K-Y, Wu RSS, Yang M. Detection of Cancer Biomarkers by Piezoelectric Biosensor Using PZT Ceramic Resonator as the Transducer. Biosens. Bioelectron 2013;46:155–61.
[62] Wee KW, Kang GY, Park J, Kang JY, Yoon DS, Park JH, Kim TS. Novel Electrical Detection of Label-Free Disease Marker Proteins Using Piezoresistive Self-Sensing Micro-Cantilevers. Biosens. Bioelectron 2005;20 (10):1932–8.
[63] Mazzocchi RA. Medical Sensors – Defining a Pathway to Commercialization. ACS Sensors 2016;1:1167–70.
[64] Pina S, Rebelo R, Correlo VM, Oliveira JM, Reis RL. Bioceramics for Osteochondral Tissue Engineering and Regeneration. In: Oliveira JM, Pina S, Reis RL, San Roman J, editors. Osteochondral Tissue Engineering: Nanotechnology, Scaffolding-Related Developments and Translation. Cham: Springer International Publishing; 2018. p. 53–75.

[65] Fiedler BA, Ferguson M. Overview of Medical Device Clinical Trials. Managing Medical Devices Within a Regulatory Framework. Elsevier; 2017. p. 17–32.
[66] Meng E, Sheybani R. Insight: Implantable Medical Devices. Lab Chip 2014;14(17):3233–40.
[67] Zhang B, Radisic M. Organ-on-A-Chip Devices Advance to Market. Lab Chip 2017;17(14):2395–420.
[68] Fadel TR, Farrell DF, Friedersdorf LE, Griep MH, Hoover MD, Meador MA, Meyyappan M. Toward the Responsible Development and Commercialization of Sensor Nanotechnologies. ACS Sensors 2016;1(3):207–16.
[69] Pohanka M. Int J Electrochem Sci 2017;12:496–506.
[70] Su L, Zou L, Fong C, Wong W, Wei F, Wong K-Y, et al. Biosens Bioelectron 2013;46:155–61.
[71] Wee KW, Kang GY, Park J, Kang JY, Yoon DS, Park JH, et al. Biosens Bioelectron 2005;20:1932–8.
[72] Mazzocchi RA. ACS Sens 2016;1:1167–70.
[73] Pina S, Rebelo R, Correlo VM, Oliveira JM, Reis RL. Osteochondral tissue engineering. In. In: Oliveira JM, Pina S, Reis RL, San Roman J, editors. Nanotechnology, scaffolding-related developments and translation. Cham: Springer International Publishing; 2018. p. 53–75.
[74] Fiedler BA, Ferguson M. Managing medical devices within a regulatory framework. New York: Elsevier; 2017. p. 17–32.
[75] Meng E, Sheybani R. Lab Chip 2014;14:3233–40.
[76] Zhang B, Radisic M. Lab Chip 2017;17:2395–420.
[77] Fadel TR, Farrell DF, Friedersdorf LE, Griep MH, Hoover MD, Meador MA, et al. ACS Sens 2016;1:207–16.

CHAPTER

# 27

# Understanding the impact of controlled oxygen delivery to 3D cancer cell culture

*David Grosh[1],*, William J. Wulftange[1,2],*, Robert W. Robey[2], Thomas J. Pohida[3], Nicole Y. Morgan[1] and Michael M. Gottesman[2]*

[1]Trans-NIH Shared Resource on Biomedical Engineering and Physical Science (BEPS), National Institute of Biomedical Imaging and Bioengineering (NIBIB), National Institutes of Health, Bethesda, MD, United States [2]Laboratory of Cell Biology (LCB), Center for Cancer Research, National Cancer Institute (NCI), National Institutes of Health, Bethesda, MD, United States [3]Signal Processing and Instrumentation Section (SPIS), Computational Bioscience and Engineering Laboratory, Center for Information Technology, National Institutes of Health, Bethesda, MD, United States

## 27.1 Introduction

In 1943, the first stable cell culture line was established, forever changing the way researchers model the in vivo physiological environment [1]. Since then, thousands of cell lines have been developed, from both normal and cancerous tissues, and tissue culture has become an indispensable part of biomedical research and drug development. Yet despite their ubiquity, in vitro models are widely understood to be limited in their ability to accurately recapitulate the in vivo environment. Extensive efforts are underway to develop co-cultures, to understand the impact of mechanical and biochemical cues from the extracellular matrix (ECM), and to develop three-dimensional (3D) culture models. Techniques such as culturing 3D spheroids, using "organ-on-a-chip" microfluidic technology, and the

* The authors contributed equally to the chapter.


DOI: https://doi.org/10.1016/B978-0-12-818128-7.00027-7

use of complex bioreactor systems are now widely employed for tissue culture. While the delivery of physiologically relevant oxygen concentrations to tissue culture has always been a challenge, it is particularly an issue for these 3D models, in which oxygen transport through media convection is minimal.

The significance of oxygen concentrations in tissues and the concept of hypoxia were first recognized through the formative work of Gray and colleagues in the 1950s [2,3]; since the development of the first $O_2$-sensitive electrodes [4], this field has rapidly grown, with over 5000 papers in PubMed listing oxygen or hypoxia in the title or as a keyword in the last year alone [5]. Recent scientific research centered around oxygenation has shifted from an initial focus on cellular $O_2$ utilization and the mechanisms of reactive oxygen species (ROS) production to a state of reduced $O_2$ availability, termed hypoxia [6]. The development of microscale oxygen-sensing electrodes enabled direct measurements of oxygen in situ and led to a series of studies that found extremely low levels of oxygenation in solid tumors [7]. This was followed by the discovery that hypoxic conditions can select for cells that are more resistant and adapted to stress, potentially resulting in increased malignancy and metastasis [8,9]. Taken together, the emerging consensus is that variations in oxygenation affect cancer progression and that understanding this is necessary to accurately model in vivo cancer behavior and predict the efficacy of therapeutic interventions.

The aim of this chapter is to provide an overview of current knowledge regarding oxygen distribution in vivo, to address the effects of differing levels of oxygenation on cell culture, and to highlight novel efforts to control oxygen delivery to in vitro cultures. We will begin by providing a background on physiological oxygen concentrations and their normal variations, including tumor hypoxia and the differences between normoxia, physoxia, and hypoxia. The importance of oxygen levels throughout cancer progression and how cells respond to these varying states of oxygenation will then be discussed. State-of-the-art in vitro techniques for controlling oxygen delivery will be compared with traditional cell culture methods. Finally, we conclude with common techniques for measuring oxygenation and characterizing the effects of oxygenation on cell populations.

## 27.2 What is known about physiological oxygen levels?

### 27.2.1 Normoxia versus physoxia

Most tissue culture is performed at atmospheric levels of oxygen, supplemented with 5% carbon dioxide ($CO_2$). Taking into account the water vapor concentration in a humidified incubator at 37 °C, this corresponds to an oxygen concentration of approximately 18.6%, or 140 mmHg in traditional cell culture incubators [10]. The actual pericellular oxygen concentrations experienced by adherent cells will be somewhat lower, depending on diffusion of oxygen through the media and convection. Usage of the term "normoxia" is somewhat imprecise; in some cases, it is used to refer to atmospheric oxygen concentrations, whereas other times it is used to refer to normal physiological concentrations. In this chapter, we will use "normoxia" to mean 18.6% oxygen, equivalent to the atmospheric levels of oxygen exposed to traditional cell culture in normoxic incubators. Physiological concentrations of oxygen vary by location, tissue type, and time, but are always substantially lower than atmospheric levels, even in the alveoli (Table 27.1).

TABLE 27.1 Estimated oxygen levels in normal tissues and tumors.[a]

| $pO_2$ (mmHg) | % Oxygen | Description |
|---|---|---|
| 760 | 100 | Atmospheric pressure |
| 160 | 21 | Amount of oxygen in air at standard atmospheric pressure |
| 100 | 13.5 | Approximate $pO_2$ of inspired oxygen in bronchial alveoli |
| 70 | 9.5 | Approximate arterial blood oxygen level |
| 50 | 6.5 | Approximate venous blood oxygen level |
| *Our definitions* | | |
| 140 | 18.6 | Normoxia: "normal" or atmospheric levels of oxygen in standard in vitro culture incubators/flasks. |
| 30–57 | 4–7.5 | Physoxia: the level of oxygen seen in physiological peripheral tissues. The typical range is from 4% to 7.5% oxygen. As an approximation, an average of around 5%–6% can be utilized. |
| 8–38 | 1–5 | Physiological hypoxia: the level of oxygen at which homeostatic hypoxic responses are triggered. The expected range is between 1% and 5% oxygen, where these are the approximate lower and upper bounds. |
| 8 | 1 | Pathological hypoxia: the level of oxygen at which homeostatic mechanisms fail, resulting in persistent hypoxic conditions. |

[a]*All values are approximated, as it is impossible to assign exact values for tissue levels. These values are merely meant to serve as a guide.*
From McKeown SR. Defining normoxia, physoxia and hypoxia in tumours—implications for treatment response. Br J Radiol 2014;87 [5]; Keeley TP, Mann GE. Defining physiological normoxia for improved translation of cell physiology to animal models and humans. Physiol Rev 2019;99:161–234 [6].

Peripheral tissues typically have oxygen concentrations of around 5% oxygen or 38 mmHg [5,6,11]. This physiological level of oxygenation, often referred to as "physoxia" varies for different tissues but lies in the range of 3%–7% oxygen.

### 27.2.2 Hypoxia (physiological vs pathological)

At low levels of oxygen, tissues experience hypoxia, which can be differentiated into two categories: physiological and pathological hypoxia. Physiological hypoxia is the level of oxygen that triggers tissues in vivo to respond and enact measures to maintain homeostasis, including vasodilation and the upregulation of hypoxia genes [5,12,13]. The oxygen levels that trigger these physiological responses vary between tissues, but any reduction that disrupts normal oxygen homeostasis constitutes physiological hypoxia. Since "physoxia" is between 3% and 7% oxygen, physiological hypoxia can be placed slightly lower, in the range of 2%–6%. Exact measurements for physiological hypoxia are difficult, as normal homeostasis mechanisms can quickly return tissue to physoxic levels of oxygen [13]. When homeostatic mechanisms fail to reverse the lowered levels of oxygenation, tissues experience pathological hypoxia, the long-term variant of hypoxia arising from conditions such as ischemic diseases or cancer [5]. The typical range of oxygenation in pathological hypoxia lies between 0% and 6% oxygen, depending on the tissue.

### 27.2.3 Tumor hypoxia

The rapid growth of tumors often results in deeply hypoxic tissues, with oxygen levels typically less than 2% [5]. This hypoxia can persist despite enhanced angiogenesis, as the newly grown tumor vasculature is known to be of poor quality and chaotic nature, with many porous vessels that are prone to collapse [5]. While it has been established that tumor hypoxia exists at levels of 0%–4% oxygen, individual tumor types exhibit significant heterogeneity in their levels of hypoxia [2,8]. Variations in the median levels of oxygenation observed for thirteen listed cancer types are provided in the review article written by McKeown et al. [2,3,5]; all tumors were hypoxic relative to normal tissue, but the range of median per cent oxygen was seen to vary from 0.3% to 4.2% in different cancer types. Interestingly, several aggressive cancer types, including pancreatic and prostate cancer, were shown to be especially hypoxic, with median oxygen percentages of 0.3%–1.2%, while other cancers such as breast and brain cancer showed median oxygen percentages of 1.2%–1.7% [5].

## 27.3 Importance of oxygen levels in various stages of cancer progression

### 27.3.1 Hypoxia

Hypoxia is a force behind several of the hallmarks of cancer outlined by Robert A. Weinberg. In particular, hypoxia has been shown to be a driver of both metastasis as well as angiogenesis [14]. This critical microenvironmental factor arises when there is an imbalance between the supply and consumption of oxygen. Oxygen is a critical reagent needed by cells to complete oxidative phosphorylation and produce adenosine triphosphate (ATP). As oxygen becomes scarce and cells begin to rely more on other metabolic options such as the lactic acid cycle, other fates for these cancerous cells become more attractive [15]. This was first documented by Warburg et al., who confirmed that tumors switch to using mainly glycolysis for their energy needs when faced with hypoxic microenvironments. Warburg conducted this early hypoxia experiment by exposing cancer cells in vitro to intermittent hypoxia and measuring metabolites in the culture broth [16]. Much of the early work on hypoxia and its role in cancer was generated by exposing in vitro cell cultures to hypoxia or hypoxia-mimetic agents, and the methods for producing hypoxic cell cultures have not drastically changed since then. In the tumor microenvironment, low oxygen tensions can lead to reduced cellular proliferation and apoptosis resulting in necrosis, but they can also give rise to a metastatic phenotype after preferentially selecting for cells that are resilient to hypoxia and stress. Metastatic cancer cells depart their site of origin and migrate to new sites with more plentiful nutrients. In patients, this departure is known as metastatic disease, which carries a poor prognosis and is the leading cause of death among cancer patients.

It is important to understand how metastasis is mediated by hypoxia, as this may lead to pharmaceutical developments that work to mitigate this process. In the tumor microenvironment, vasculature may not be homogenously distributed through the tumor mass, meaning that cells more than 100–200 μm away from any blood vessel are likely hypoxic or even necrotic. Approximately 150 μm is the diffusion limit of oxygen in tissues, as defined by Krogh and further verified through histological analysis of tumor samples by Thomlinson and Gray [17,18]. Further complicating oxygen delivery to tumor masses is a network of irregular and leaky vasculature with sluggish blood flow, in contrast to normal vasculature which is generally ordered, with regulated flow [19]. Broadly speaking, when cells are in this hypoxic region of the tumor mass, typical fates are metastasis or apoptosis. In order to metastasize, cells must adopt an invasive phenotype, dislodge themselves from this primary site of origin, and travel through the stroma to a nearby blood vessel or lymph node. Primarily, the response to low oxygen tension is mediated by the hypoxia-inducible transcription factor (HIF-1), which coordinates cellular response by changing gene expression in an attempt to restore oxygen homeostasis, by adjusting glucose uptake, angiogenesis, proliferation activity, and in some cases triggering apoptosis [20,21].

HIF-1 is an αβ-heterodimeric transcription factor that is activated under conditions of hypoxia. Specifically, the β-subunit is constitutively expressed in the nucleus while the α-subunit posttranslationally exists in the cytosol where its degradation is mediated by the presence of oxygen. When oxygen is present, HIF-1α is hydroxylated by proline hydroxylases and this hydroxylated form can then be marked for ubiquitination by the product of the Von Hippel-Lindau tumor suppressor gene (pVHL). In a hypoxic environment, proline hydroxylases are inhibited, so HIF-1α can accumulate and translocate into the nucleus, where it dimerizes with HIF-1β and is then able to transcriptionally regulate numerous processes by binding to hypoxia response elements in the promoter regions of certain genes [22,23]. Increased HIF-1α signaling has been shown to be indicative of disease progression in a multitude of cancers including those of the prostate, liver, and pancreas [24–26]. A major downstream effect of this HIF-1 signaling is a massive shift in metabolism toward the glycolytic pathway, known as the Warburg effect, which allows cancer cells to maintain redox homeostasis under unfavorable oxygen conditions [16,27]. This glucose-reliant metabolism is triggered by HIF-1α, which causes the upregulation of GLUT1 transporters, thus allowing the uptake of more glucose by cancer cells. The product of the glycolytic pathway is pyruvate, which in normal energy metabolism is converted to acetyl coenzyme A (acetyl-coA) and then shuttled into mitochondria to undergo oxidative phosphorylation. However, HIF-1α also upregulates pyruvate dehydrogenase kinase 1 (PDK1), which inhibits the conversion of pyruvate to acetyl-coA. Instead, pyruvate is converted to lactate via the enzyme lactate dehydrogenase A (LDHA) and effluxed from cells via the monocarboxylate transporter 4 (MCT4); both of these enzymes are upregulated through HIF-1α signaling. Ultimately, HIF-1α allows the progression of the Warburg effect through increased glucose uptake and a shift away from oxidative phosphorylation [27,28]. When tumors are confronted with low oxygen tensions, local pH levels are altered and the consequences of glucose deprivation, such as an imbalance of NADPH and glutathione, exert adverse effects on highly proliferative cancer cells [29].

While the role of glycolysis and the lactic acid cycle in cancer cell metabolism is well documented, in vitro cancer cell lines have been shown to adopt alternative metabolic states when confronted with hypoxia. For example, Valli et al. have shown that cancer cells can take on a lipogenic state when confronted with hypoxia and can adapt metabolically to hypoxia through HIF-independent mechanisms [30,31]. Due to the hypoxic heterogeneity of tumors, it is no surprise that there would be an equal amount of heterogeneity in the metabolic pathways active at any given time.

### 27.3.2 Angiogenesis

During states of cellular hypoxia, HIF signaling can initiate angiogenesis when the HIF-1 $\alpha\beta$ heterodimer binds to hypoxia response elements in the promoter regions of angiogenic genes, such as VEGF. Normal angiogenesis occurs when the existing vasculature begins to dilate in order to sprout new capillaries through the migration, proliferation, and maturation of endothelial and endothelial progenitor cells. Using hypoxia chambers, this process has been shown to be initiated by nitric oxide and hypoxia signaling, both of which lead to an increase in vascular permeability in existing vessels [32]. As intercellular adhesion molecules such as platelet endothelial cell adhesion molecules (PECAM-1) and vascular endothelial (VE)–cadherin are reorganized on the cell surface, plasma proteins seep through into the surrounding ECM, creating an architecture that allows endothelial cell migration. Additionally, the ECM undergoes modification through the enzymatic actions of matrix metallopeptidases (MMPs), which are secreted by the migrating endothelial cells in response to HIF-1$\alpha$ signaling [33]. Angiogenic and anti-angiogenic factors are carefully balanced during this process, so that pathologies do not develop from this temporary endothelial permeability. From mouse and cell culture studies, factors such as angiopoietin-1 have been shown to contribute to this process by the suppression of vascular protective effects [34]. Following the migration of endothelial cells into this newly formed space, proliferation must occur, but since endothelial cells are normally quiescent, a stimulating cytokine such as VEGF, which is a transcriptional target of HIF-1$\alpha$, is required. The direction of vessel growth is dictated by gradients of chemoattractants acting on transmembrane receptors such as VEGF-R2, which may also be regulated by hypoxia through Notch signaling modulation [35–37]. As these newly formed vessels grow through the ECM, their interactions with smooth muscle cells and pericytes complete their maturation. In cancerous tissue, this choreographed generation of new vessels quickly becomes chaotic, as the expression of angiogenic factors is much less organized than in normal tissue. The careful balance of angiogenic and antiangiogenic factors can be disrupted as either of these factors can be aberrantly secreted, resulting in malformed vessels with compromised cell–cell junctions.

Hypoxia in tumors arises when the balance of proliferation and apoptosis becomes uneven, tripping the angiogenic switch [38]. Tumors undergo the same steps outlined above in generating new vasculature, but with poor coordination between the steps. The imbalance between angiogenic and antiangiogenic factors can result in permeable vessels with irregular diameters and random patterning. The morphology of these vessels is highly heterogeneous and although they usually incorporate all of the necessary components, their organization is abnormal [39]. Endothelial cells do not consistently form monolayers, normal pericyte and endothelial cell architecture is disorganized, and abluminal projections result in regions with turbulent flow or stagnation of blood. Despite the leakiness of the vasculature found in a tumor, there is an unusually low amount of convective movement of macromolecules out of the tumor vessels and into the tumor. This is caused by a high interstitial pressure in the tumor which results from stagnant flow in the tumor vasculature and lack of adequate fluid clearance through lymphatics [40]. Due to the reduced blood flow through these vessels, diffusion is the main driver of molecular movement, and this makes for a highly heterogeneous tumor microenvironment. As noted earlier, oxygen has a diffusion limit of approximately 150 μm, and beyond this limit, cancer cells need to adapt to hypoxic or anoxic conditions or metastasize to overcome this inadequacy. The abnormality, and specifically the leakiness, of tumor microvasculature has been extensively studied, as understanding the relative contribution of these factors could lead to insights into the pathogenesis that drives metastatic phenotypes and therapies that target these pathologies.

### 27.3.3 Metastasis

Prior to metastasis, cancer cells undergo a change of phenotype in what resembles an epithelial to mesenchymal transition (EMT), which allows them to escape their niche and migrate toward vessels and new tissues. Elucidating the mechanisms and switches facilitating this transformation is of critical importance in understanding what drives metastasis. Although both hypoxia and the EMT have been extensively researched in terms of cancer and cancer metastasis, only in recent years have signaling commonalities been discovered. One specific example of this is the relationship between HIF-1α and transforming growth factor β (TGF-β), and the downstream effect on E-cadherin, one of the adhesion molecules that connect epithelial cells [41]. Down regulation of E-cadherin in the tumor microenvironment contributes to the EMT, and can be caused by TGF-β, which signals through serine/threonine kinase receptors, triggering cytosolic SMAD proteins to alter activity of transcription factors like SNAIL, ultimately repressing E-cadherin expression [42]. Hypoxia, specifically HIF-1α, has been shown in vitro to induce TGF-β signaling in cancer and noncancer cells, suggesting a link between low oxygen tensions and EMT [43–47]. This HIF-1α/TGF-β axis is an example of how a cancer cell might free itself from its origin and adopt an aggressive phenotype on its way to a secondary location. When an aggressive cancer cell has moved through stroma and found a vessel, the cell must extravasate and become what is known as a circulating tumor cell (CTC), which will travel throughout the body in an anchorage-independent manner. This transition from anchorage-dependent to anchorage-independent phenotype is marked by a cell's colony-forming ability in semisolid media. The driving forces behind this phenotypic change remain largely unknown [48].

As the CTC travels, its destination is already being prepared for its arrival, in what is known as the "pre-metastatic niche." These niches are created, in part, from the accumulation of matrix-modifying proteins and the infiltration of bone marrow-derived cells. While many factors have been identified that drive this process, we will focus here on the role hypoxia has been shown to play. Proteins of particular interest are lysyl oxidase (LOX) and LOX-like proteins, which are regulated by HIF signaling and have recently been implicated as key players in the establishment of the pre-metastatic niche [49]. Recent work using both cell cultures and mouse models has demonstrated that hypoxia-mediated signaling increases the expression and secretion of LOX and LOX-like proteins from primary breast tumors, and that these proteins facilitate cross-linking in matrix proteins such as collagen. This modified matrix enhances the attachment of CD11b + myeloid cells, which can secrete MMPs in order to recruit other bone marrow-derived cells and ultimately, circulating breast cancer cells [50–52].

## 27.4 Techniques for measuring oxygenation

The development of oxygen-sensing electrodes in the early 1950s enabled direct measurement of oxygen levels in fluids and tissues for the first time, greatly advancing study of the mechanisms of oxygen usage and physiology [4]. Since then, a number of additional techniques for measuring oxygen have been developed, each with varying degrees of spatial and temporal resolution. The most commonly used of these techniques in current research include oxygen-sensing wire electrodes, absorptiometric probes, luminescent probes, and finally the use of spectroscopic imaging. Understanding the capabilities and limitations of these techniques and their complementary uses in measuring oxygenation is crucial in hypoxia research and assessing levels of oxygenation in tumor culture.

### 27.4.1 Oxygen-sensing electrodes

Wire electrodes such as the Clark electrode (Fig. 27.1A), invented in 1956, remain one of the most widely used tools to directly measure oxygen levels [54], by using voltammetry to measure reduction of oxygen at the cathode [55]. In a Clark-type electrode, the platinum cathode is covered by an oxygen-permeable membrane, typically Teflon, to prevent biofouling [56]. This membrane must be less than 20 μm in thickness in order for the oxygen diffusion to be sufficiently rapid. As oxygen is reduced, the resulting electrical current is proportional to the available oxygen or oxygen partial pressure present in the sample [54]. The original Clark electrode design was groundbreaking and allowed for accurate in situ oxygen measurement, but with an electrode diameter of approximately 1 μm, spatially precise oxygen measurements were difficult [6]. The Whalen electrode, developed in the early 1970s using a gold cathode with a recessed tapered tip about 1 m in diameter, can measure dissolved oxygen over smaller areas and in vivo [57–59]. More recently, microfabricated Clark electrodes integrated with microstructures have been used to measure oxygen over smaller areas; these devices can be incorporated into tissue culture plates for local measurements down to the single-cell level [57–59]. However, with the more recent development of optical probes, the use of oxygen-sensing electrodes has declined. This is due to four main disadvantages of electrodes: (1) they consume oxygen, which is especially problematic for measurement of very small or poorly oxygenated samples, (2) they necessitate invasive contact and therefore can cause mechanical tissue damage, (3) they require an aqueous solution with dissolved oxygen, and finally (4) they are unable to measure oxygenation over large areas due to the area constraints of each electrode [60].

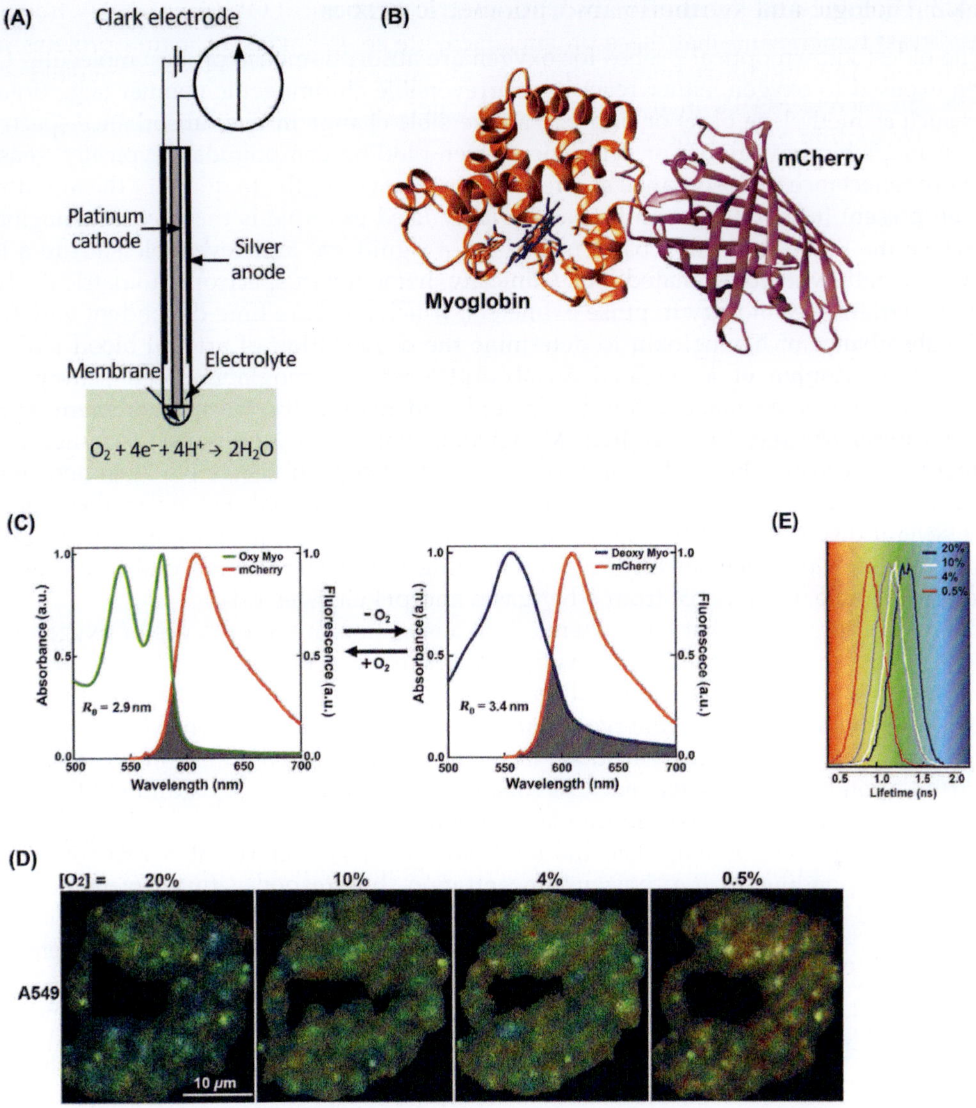


FIGURE 27.1 Techniques for measuring oxygenation in cells and tissues (oxygen-sensing electrodes and luminescent probes). (A) Diagram of a common type of $O_2$-sensitive electrode invented in 1956 by Leland Clark. (B) Myo-mCherry FLIM-FRET probe for use in fluorescent imaging. Diagram depicting the coupling of myoglobin's C-terminus to mCherry to create the myo-mCherry construct used in Ref. [53]. (C) Spectral overlap of mCherry with the absorption of myoglobin in the presence (left) and absence (right) of oxygen. The change in spectral overlap allows for measurement of the oxygenation state of myo-mCherry using FLIM. (D) A549 cells transfected with myo-mCherry and imaged using FLIM in oxygen environments of 20%, 10%, 4%, and 0.5%. (E) Cells are colored based on the fluorescence lifetime distribution histogram of myo-mCherry, where red correlates to lower lifetimes and oxygen concentrations while blue correlates to higher lifetimes and oxygen concentrations. *Source: Panels B–E adapted from Penjweini R, Andreoni A, Rosales T, Kim J, Brenner MD, Sackett DL, et al. Intracellular oxygen mapping using a myoglobin-mCherry probe with fluorescence lifetime imaging. J Biomed Opt 2018;23:1–14 [53]. Used by permission of the authors.*

### 27.4.2 Biologic and synthetic absorptiometric probes

The oldest known optical probes for oxygen are absorptiometric probes: molecules that, when exposed to oxygen, either react in an irreversible chromogenic manner (e.g., organic dyes such as methylene blue) or undergo a reversible change in their absorbance spectrum (e.g., hemoglobin/myoglobin or synthetic oxygen binding compounds). Typically, sensors measure reflectance or absorbance at a few discrete wavelengths to quantify the amount of oxygen present [61,62]. One common and widely used example is the use of hemoglobin, for which the Soret absorption band undergoes a significant and reversible shift to a longer wavelength when oxygenated [61]. Clinically, hemoglobin spectrophotometric analysis is the underlying technology in pulse oximeters, which measure time-dependent variations in the absorbance of hemoglobin to determine the oxygenation of arterial blood [63]. For use in vitro, Zhujun et al. created an absorptiometric hemoglobin-based sensor, with encased deoxygenated hemoglobin in a resin-based matrix, for the measurement of $pO_2$ levels between 20 and 100 Torr [62]. Myoglobin, which also undergoes a change in its absorption spectrum upon binding oxygen, was used in a similar manner, when McCurley and Valentine encased myoglobin in a sol–gel [64,65]. However, these probes have several disadvantages that limit their use for in vitro applications: both hemoglobin and myoglobin degrade relatively rapidly, limiting the lifespan of the probe, and experience considerable interference from other gases and pH changes [61].

Chemical compounds that have altered light absorption in the presence of oxygen were developed to supplement biological absorptiometric oxygen probes. These compounds can be much more stable and generally fall into two categories: molecules that reversibly bind oxygen and molecules that undergo an irreversible chemical reaction when exposed to oxygen [61]. The first category includes molecules that undergo a shift in the absorption spectrum upon binding oxygen, such as cobalt-organic compounds. Baldini et al. used bis (histidinato)cobalt(II) [$Co(His)_2$] in thin layer chromatography plates covered by silicone to create an oxygen sensor [66]. The immobilized $Co(His)_2$ underwent a reversible color change corresponding to the oxygen concentration, but response times were generally long (upward of 30 min) and color change varied with pH [61]. Other synthetic compounds, including chemically reduced leuco dyes such as methylene blue or phenothiazine, undergo irreversible chromogenic redox reactions when exposed to molecular oxygen [67,68]. Due to the irreversible nature of this color change, these compounds can be useful for measuring historical conditions, but they are unable to measure changes in oxygen levels over time [61].

### 27.4.3 Fluorescent and phosphorescent luminescent probes

Although readout of absorptiometric probes can be more straightforward, optical oxygen probes relying on luminescence are generally much more sensitive. These luminescent probes typically rely on molecular oxygen quenching the luminescence (either fluorescence or phosphorescence) of the probe in a concentration-dependent manner [61]. When one of these molecules is excited by an incident photon with sufficient energy, there are several pathways possible for releasing that energy. For a good luminescent probe, the primary pathway is radiative, that is, the molecule emits a photon of light at a slightly longer wavelength, thereby returning to its ground state; an energy-level diagram depicting these processes and competing pathways is shown in Fig. 27.2A [69]. Quenching arises from nonradiative decay pathways from the excited state, typically through static quenching, collisional (dynamic) quenching, or other mechanisms such as Förster resonance energy transfer (FRET) [70]. Static quenching occurs when the luminophore forms a stable chemical complex, for example by binding to oxygen; any bound luminophore is dark, but the luminescence lifetime of the free luminophores is unchanged [71]. Collisional quenching

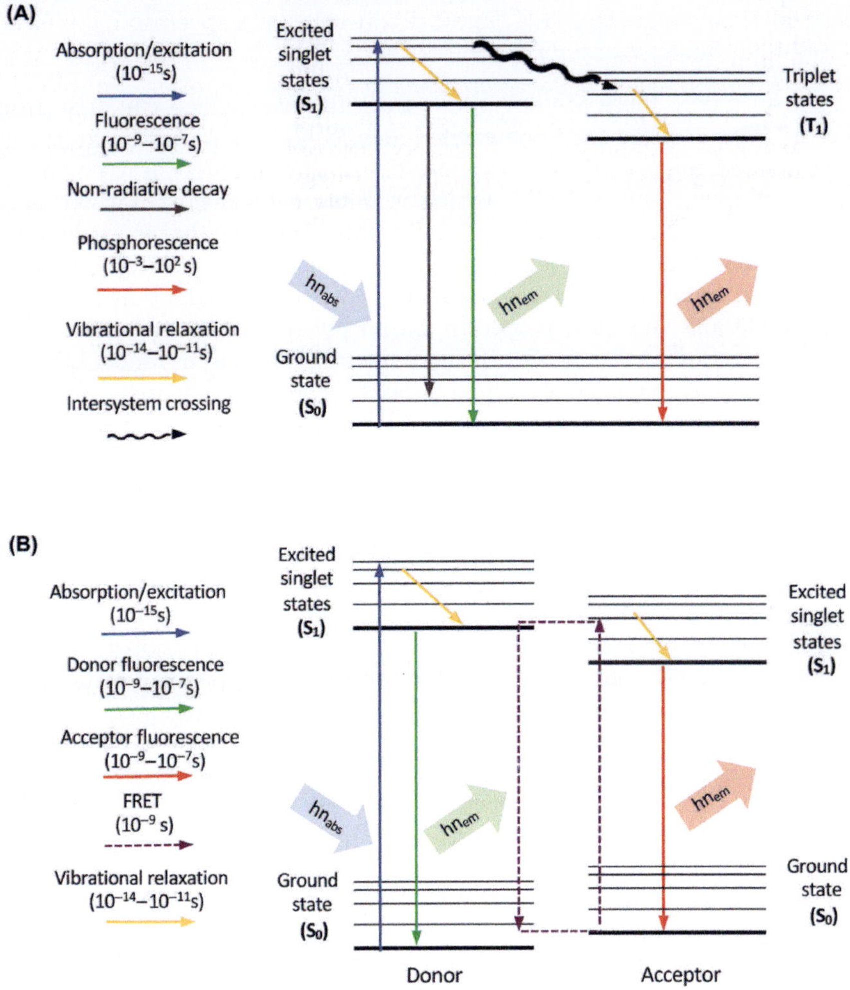


FIGURE 27.2 Typical Jablonski diagrams depicting the different mechanisms (radiative vs nonradiative) for energy transfer in a typical luminophore. (A) Jablonski diagram showing radiative (fluorescent and phosphorescent) and nonradiative (quenching) energy transfer mechanisms used to return the atom back to its ground state after excitation. (B) Jablonski diagram depicting the effect of nonradiative FRET transfer. The energy released from the donor atom's relaxation is absorbed by a nearby acceptor, leading to its excitation, and the emission of a photon by the acceptor. *Source: Adapted from Hochreiter B, Garcia AP, Schmid JA. Fluorescent proteins as genetically encoded FRET biosensors in life sciences. Sensors (Basel) 2015;15:26281–314 [69].*

occurs when molecular triplet oxygen collides with the excited luminophore, exciting oxygen to the singlet state and returning the luminophore to its ground state without the emission of light [71]. This quenching results in a decrease in luminescence intensity as well as an increased decay rate. In an ideal quencher system, the decrease in luminescent intensity is proportional to the concentration of the quencher (oxygen), as expressed in the Stern–Volmer equation (Eq. 27.1), where $F$ and $F_0$ are the luminescent intensities of the oxygen probe with and without oxygen present, $K_{sv}$ is the Stern–Volmer constant, and $[O_2]$ is the oxygen concentration present in the sample [61,72].

$$\frac{F_0}{F} = \frac{\tau}{\tau_0} = 1 + K_{sv}[O_2] \tag{27.1}$$

In most cases though, there are deviations from linearity which can arise from factors such as the presence of more than one excited state of the fluorophore producing fluorescence or the ionic strength of the fluorophore disrupting its quenching constant [73]. An alternate form of quenching is FRET, which is a resonant energy transfer between the excited luminophore and a nearby molecule through resonance that is extremely sensitive to the distance between the donor and acceptor, as depicted in Fig. 27.2B [69,74].

The first use of luminescent probes for imaging oxygen distribution in tissues was in 1988 by Rumsey et al., who used porphyrin-based phosphorescent probes with platinum ($Pt^{II}$) and palladium ($Pd^{II}$) to track the effects of reperfusion in a rat model [75]. Since then, a large number of other molecular oxygen probes have been developed, as cataloged in several recent comprehensive reviews [61,76,77]. Broadly speaking, luminescent oxygen-sensitive probes can be divided into three main categories: (1) organic probes, such as fullerenes and aromatic hydrocarbons, (2) luminescent nanomaterials, and (3) metal–ligand complexes, including porphyrins and transition metal–ligand complexes [61]. Commercial devices incorporating these probes typically embed the oxygen-sensitive molecules in an inert carrier such as a polymer or sol–gel in order to prolong the life of the sensor and, in biological systems, to mitigate any cytotoxic effects. The embedded probes can be used in many physical formats: for example, incorporated into micro- or nanoparticles for oxygen-sensing in vitro or coated onto a film to obtain an image of spatial variations. One especially useful device is a fiber-optic cable with its tip coated by an embedded probe, as is done in the OxyLite system. This design provides the same time-resolved, spatially localized sensing of a traditional wire electrode, allowing for precise oxygen measurements over small areas, but without oxygen consumption [78].

The main detection modalities used for these luminescent oxygen probes are luminescence intensity and lifetime. Especially for imaging applications, measuring the intensity is simpler, requiring only a camera and a steady-state light source. However, interpreting the data becomes much more complicated, as the luminescent intensity can be skewed by the concentration of the luminophore, photobleaching, and spatial variations in light scattering, detector sensitivity, and excitation light intensity [61,79]. Quantitation can be improved by including an oxygen-insensitive luminophore as well and measuring the ratio between the two probes, but the overall accuracy is still somewhat limited. In contrast, luminescence lifetime, measured as the average time for which an excited luminophore remains in its excited state before emitting a photon and returning to its initial ground state, is independent of these parameters. Most commercial optical oxygen sensors measure luminescence lifetime for this reason. For lifetime imaging applications, which require faster data acquisition rates than point measurements, fluorescent probes, which have emission lifetimes in the nanosecond range, are preferable to phosphorescent probes, which have emission lifetimes about three orders of magnitude longer [80]. Fluorescent lifetime imaging microscopy (FLIM) performed with a pulsed laser and a single-photon counter can be used to measure the spatially resolved decay kinetics of a luminophore at frame rates near the millisecond range [79,80]. In general, luminescent optical sensing of oxygen not only can provide excellent spatial resolution, but also allows for flexibility in oxygen sensing, as measurements can be taken over both nanoscale and large areas [60]. In addition, luminescent probes can allow for the specific detection of biological molecules in complex cellular environments with high signal to background ratios, allowing for intracellular measurements and imaging in real-time [81].

One example of a novel luminescent oxygen probe measured using FLIM-FRET is a fusion of myoglobin protein and mCherry fluorescent protein (Myo-mCherry), shown in Fig. 27.1B [53]. This interesting combination incorporates absorptiometric and luminescent probe concepts to create a powerful FLIM measurement tool. As was mentioned before, FRET is a resonant energy transfer between an excited luminophore, the donor, and a nearby molecule, the acceptor [74]. In myo-mCherry, mCherry is the donor, and myoglobin serves as the acceptor. When this novel probe construct is deoxygenated, the intensity and lifetime of fluorescent emission from mCherry fluorescent protein is decreased, thus allowing for the differentiation between the oxygenated and deoxygenated versions of the oxygen-sensing probe using FLIM, as seen in Fig. 27.1C–E [53]. This probe also allows for intracellular oxygen measurement and can be expressed by transfected cells without cytotoxicity, providing a novel manner to determine cellular oxygen consumption and usage on the single-cell level. The ability to effectively measure oxygenation, quantify the extent of oxygen distribution, and determine the existence of oxygen gradients within cancer culture in situ, has typically been limited in 3D culture platforms. Thus, the development of new techniques using probes such as this holds great promise.

### 27.4.4 Spectroscopic imaging: magnetic, paramagnetic, and electron spin resonance

Spectroscopic imaging is occasionally used in vitro, but is especially valuable for making contact between in vivo measurements and in vitro models. In general, magnetic resonance imaging (MRI), positron emission tomography (PET), and electron paramagnetic resonance (EPR) are the main methods for assessing oxygenation in vivo in both animal and clinical models. Several reviews and studies have offered a more in-depth explanation of these spectroscopic methods [82–88]. In general, all three of these methods allow for 3D, noninvasive imaging with sufficient depth of penetration for clinical studies, and provide the ability to report changes in perfusion, visualize oxygen metabolism, and identify hypoxic areas in vivo. Despite the advantages of spectroscopic imaging, the equipment and facilities necessary for these techniques are expensive and specialized, making them inaccessible for many laboratories [6]. In addition, while the depth of penetration of these techniques is well-suited to in vivo usage, their spatial resolution is lacking when compared to some of the methods described above. For these reasons, spectroscopic imaging is rarely used in in vitro research, but is crucial for understanding and properly modeling in vivo physiologically hypoxic conditions.

## 27.5 Traditional/current strategies for controlling oxygen concentration in vitro

### 27.5.1 Hypoxia chambers and two-dimensional models

Traditional cell culture (Fig. 27.3A) is carried out under atmospheric oxygenation, and although this is widely accepted as not physiologically relevant, the requirements for mimicking every cell's in vivo environment are too complex to implement in a high-throughput manner. Cells experience a multitude of oxygenation conditions that go beyond "normoxic" and

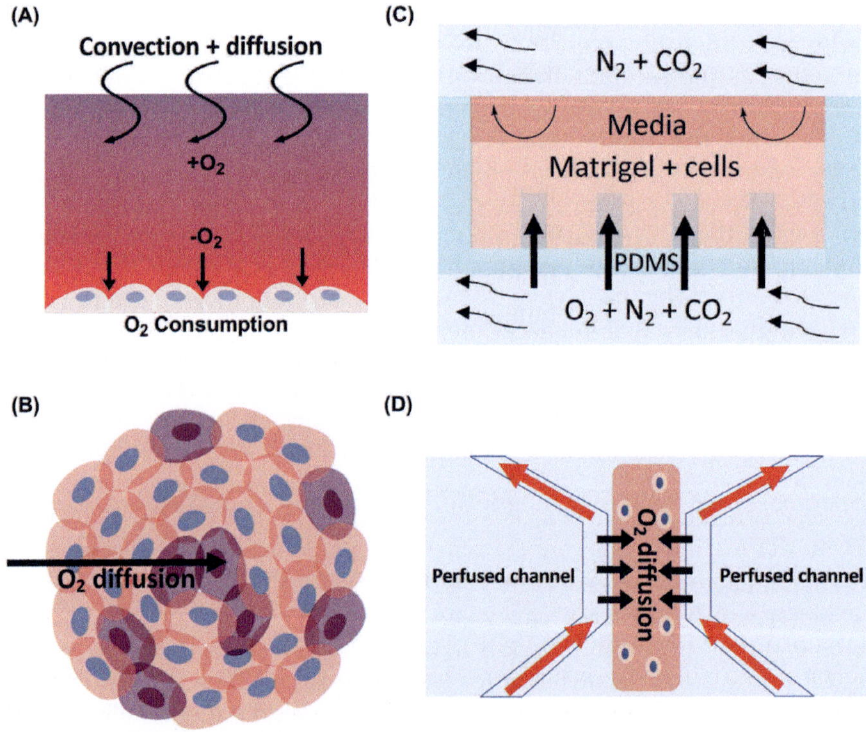


FIGURE 27.3 The kinetics of oxygen delivery in in vitro culture. (A) Schematic of oxygen delivery within 2D static, monolayer cell culture where convection and diffusion deliver oxygen to a monolayer of cells. (B) Oxygen delivery in 3D spheroid culture systems where only diffusion provides oxygen to cells within the spheroid. (C) Oxygen delivery though a microfluidic micropillar system. Oxygen diffuses from below through a PDMS oxygen-permeable substrate, convective mixing occurs in the media above [89]. (D) Oxygen delivery within a generalized microfluidic perfusion system. Oxygen diffuses from perfused media to cell culture.

"hypoxic." Gradients are formed radially around blood vessels and the $pO_2$ decreases as blood travels further away from the alveoli, where gas exchange occurs. In addition to spatial gradients, temporal control is highly relevant in consideration of oxygenation. For example, cardiomyocytes experience cyclical hypoxia with each heartbeat [10]. Spatiotemporal control of oxygenation in tissue culture protocols is still rarely considered in experimental design. In recognition of this shortcoming of traditional culture protocols, several methods have been adopted in order to induce a uniform hypoxic environment to two-dimensional (2D) cell cultures.

Commercially available cell culture incubators are designed to provide high humidity, constant temperature, and an atmosphere that consists of room air supplemented with $CO_2$. When completely humidified using an open water pan and supplemented to 5.0% (38 mmHg) $CO_2$, the atmosphere in a normoxic incubator contains approximately 18.6% (141 mmHg) oxygen, which is substantially higher than physiological values. These incubators rely on diffusion of oxygen through cell culture media, which is further mixed by convective forces resulting from media evaporation and vibrations of the incubator. However, diffusion and convective mixing is not sufficient for many cell lines as the cells' energy requirements exceed the amount of oxygen delivered. As oxygen diffuses through the cell culture media, a gradient from the air–media interface is established along the vertical axis. The pericellular oxygen concentration will depend on the confluency of cells, the depth of the media, and the oxygen consumption rate (OCR) of the cell line, with the maximum delivery of oxygen to the cells being inversely related to the height of the overlaying media. This highlights an important consideration when designing monolayer hypoxia experiments: the atmospheric content of oxygen in an incubator will be higher than the oxygen concentration that is delivered to the cells. This insight is often overlooked when designing experiments and oxygen concentrations are often reported as the atmospheric concentration of the incubator. Measuring pericellular oxygen levels in real time is difficult and requires special microelectrodes that are not common instruments in cell biology laboratories. Pericellular oxygen can be modeled using Fick's law since direct measurement is often not feasible for standard cell culture experiments. Through direct measurements and approximation with Fick's law, Metzen et al. demonstrated that in rat renal and hepatocellular carcinoma cell lines grown in 24-well polystyrene dishes with 1 mL of media per well, the OCRs exceeded the delivery of oxygen, creating a hypoxic pericellular space, despite being in an air–5% $CO_2$ atmosphere [90]. Cellular oxygen levels become even more difficult to predict when cells are cultured in three dimensions. Unintended hypoxia will cause experimental results to deviate even further from physiological relevance, which emphasizes the need for greater control over cell culture oxygenation and the real-time measurement thereof.

In an effort to recapitulate physiological and pathological oxygen levels, several simple adaptations to standard protocols are widely used. Sealed hypoxia chambers that accommodate standard cell culture platforms can be placed inside standard $CO_2$ incubators, a method that most tissue culture labs can implement. A standard incubator hypoxia chamber can be used under sterile conditions by placing tissue culture dishes inside the chamber along with a petri dish of sterile water. Once the desired cultures are inside, the chamber can be purged for several minutes using a hypoxic gas mixture, then sealed using an O-ring and ring clamp. A better alternative is to use tri-gas incubators, which are commercially available, to actively control the oxygen, $CO_2$, and $N_2$ content of the incubator atmosphere. Tri-gas incubators monitor oxygen content in addition to $CO_2$ and, for hypoxic conditions, purge excess atmospheric oxygen using $N_2$ until the desired atmospheric composition is reached. Closure of the hypoxia chamber often marks the beginning of an experiment, yet it may take minutes or hours until the oxygen content of the culture medium reaches the desired concentration. In an informative paper on hypoxia tissue culture, Wenger et al. calculated that a 10 cm petri dish with 10 mL of medium would take 45 min to reach the desired oxygen concentration if a culture dish was put into a 1% $O_2$ atmosphere from room air [10]. The duration of this transition could be reduced by preequilibrating culture media by leaving the medium in a gas permeable vessel or with the cap open in a sterile low oxygen environment. This preequilibration takes a considerable amount of time since only the surface of the medium is able to exchange oxygen, so stirring or bubbling gas through the medium is often preferable. Gas permeable cell culture plates have also become commercially available. They allow oxygen diffusion to occur at the basal side of the tissue culture well and lend themselves nicely to the study of hypoxia, as the delivery of oxygen is no longer dependent on diffusion through media and so is more easily controlled [91]. Another factor often overlooked in hypoxia research is that opening and closing the incubator doors or hypoxia chamber can disrupt the established hypoxic atmosphere, subjecting cell cultures to unwanted fluctuations in oxygen content. Uncontrolled oxygenation disrupts HIF signaling and could obscure pharmacokinetic results that rely on precise oxygen delivery. Many cell types, including endothelial cells and myocardiocytes have diminished capacity to scavenge free radicals after cellular injury that occurs when subjected to prolonged hypoxia. Upon reperfusion with oxygen, cells increase their production of ROS but are not able to effectively neutralize them due to injury sustained during hypoxia, leading to further cell damage [92]. In vivo, this phenomenon is known as "reperfusion injury," but there is an insufficient understanding of the effect of cyclical and intermittent hypoxia in vitro.

After accounting for appropriate hypoxia chamber setup, careful attention must be given to the OCR of the tissue at hand. Wagner et al. compiled an extensive list of OCRs of frequently used cell lines that revealed pronounced disparities in oxygen consumption. Cells of different tissue origin had OCRs that spanned several orders of magnitude, with hepatocytes at the high end of the spectrum. Even within tissue types there existed wide disparities between species and growth durations; rat hepatocytes had an OCR of 430 amol $cell^{-1}$ $s^{-1}$, while porcine hepatocytes had an OCR of 900 amol $cell^{-1}$ $s^{-1}$ 4 days after seeding, which dropped to 300 amol $cell^{-1}$ $s^{-1}$ 15 days after seeding. These large variances were also documented among human cancer cell lines. HepG2 is an immortalized cell line derived from the well-differentiated hepatocellular carcinoma of a 15-year-old male of European descent and was documented to have an OCR of 110 amol $s^{-1}$ ng-$protein^{-1}$. This is an order of magnitude smaller than the MCF-7 breast cancer cell line, derived from a 69-year-old Caucasian female, which has an OCR of 1300 amol $s^{-1}$ ng-$protein^{-1}$. These OCRs were all measured with a Clark electrode [93]. This compilation of data demonstrates the need to consider redox biology and oxygen requirements when designing experiments. It also shows that standard 5% $CO_2$ as well as tri-gas incubators produce vastly different pericellular microenvironments for different cell lines even when the atmospheric oxygen concentration and media depth are held constant.

### 27.5.2 Three-dimensional models: spheroids

The complexity of a tumor, regardless of oxygenation, cannot be accurately replicated in two dimensions. Cells grown in 2D adopt cytoskeletal arrangements to accommodate their planar environment, leading to artificial polarity. Monolayer cultures (Fig. 27.3A) lead to the selection of cell phenotypes that are well-suited to survive on polystyrene, a phenotype likely not observed in vivo. Examining singular aspects of cancer, such as response to a particular drug, may not mimic the response seen clinically, as there is an intricate interplay between cells, chemical gradients, and ECM which is absent in monolayer tissue culture. Three-dimensional tissue culture allows for the construction of a microenvironment that incorporates the complexity of cell–cell and cell–ECM interactions, which are essential for morphogenesis, proliferation, and differentiation (Fig. 27.3C and D) [94]. By spatially distributing growth factors and ECM proteins, cells can polarize, redistribute surface receptors, and structurally organize themselves into functional units. Bissell et al. demonstrated that mammary epithelial cells grown in 3D displayed highly concerted behavior by polarizing, arresting growth, and self-organizing into acini which were capable of producing an endogenous basement membrane and secreting sialomucin into the apical lumen. This behavior is in stark contrast with that of tumorigenic mammary epithelial cells, which do not arrest growth cycles, nor undergo polarization but instead grow into a disorganized mass when grown in 3D [95]. Tissue architecture is also relevant in the context of metastasis. Paul et al. found that mammal mammary tumor cell homing was influenced by vessel architecture in zebrafish [96]. This complements our previous discussion of metastasis concerning LOX- and MMP-mediated ECM remodeling which enhanced metastatic invasion through HIF signaling. Mounting evidence supports the notion that 3D architecture is essential for physiologically relevant experimental design and that controlled oxygenation is imperative to understanding the oxygen levels and hypoxia that occurs in vivo.

The simplest implementation of 3D culture is the growth of spheroids (Fig. 27.3B), clusters of cells varying in size that are typically supported by an ECM. Spheroid culture has been widely implemented in round bottom wells, hanging drop systems, and hydrogels. Round bottom wells have been used to generate homogenous, single spheroids by addition of a defined number of cells to a well, followed by centrifugation in order to pellet cells at the bottom of the low adhesion coated well. This method has been used to generate spheroids of cancer cell lines of consistent size and can even be used on cell lines that otherwise tend to form loose masses of cells if small amounts of basement membrane are added prior to centrifugation [97]. Hanging drop systems (Fig. 27.4A) are used to form spheroids by injecting cells, suspended in a liquid, into a tray and subjecting them to gravity, allowing the cells to cluster at the bottom of a droplet, forming a spheroid. Since the hanging drops generate spheroids of consistent sizes, this platform has been used to investigate and compare anticancer therapies, including hypoxia activated tirapazamine [98]. Another advantage of the hanging drop platform is that heterotypic cell types can be added sequentially, allowing radial control of cell organization. For example, Timmins et al. cultured several cancer cell lines for several days in a hanging drop system, then added a suspension of endothelial cells, which penetrated the cancer spheroids without any exogenous angiogenic factors and established a network of luminal vessels [105].

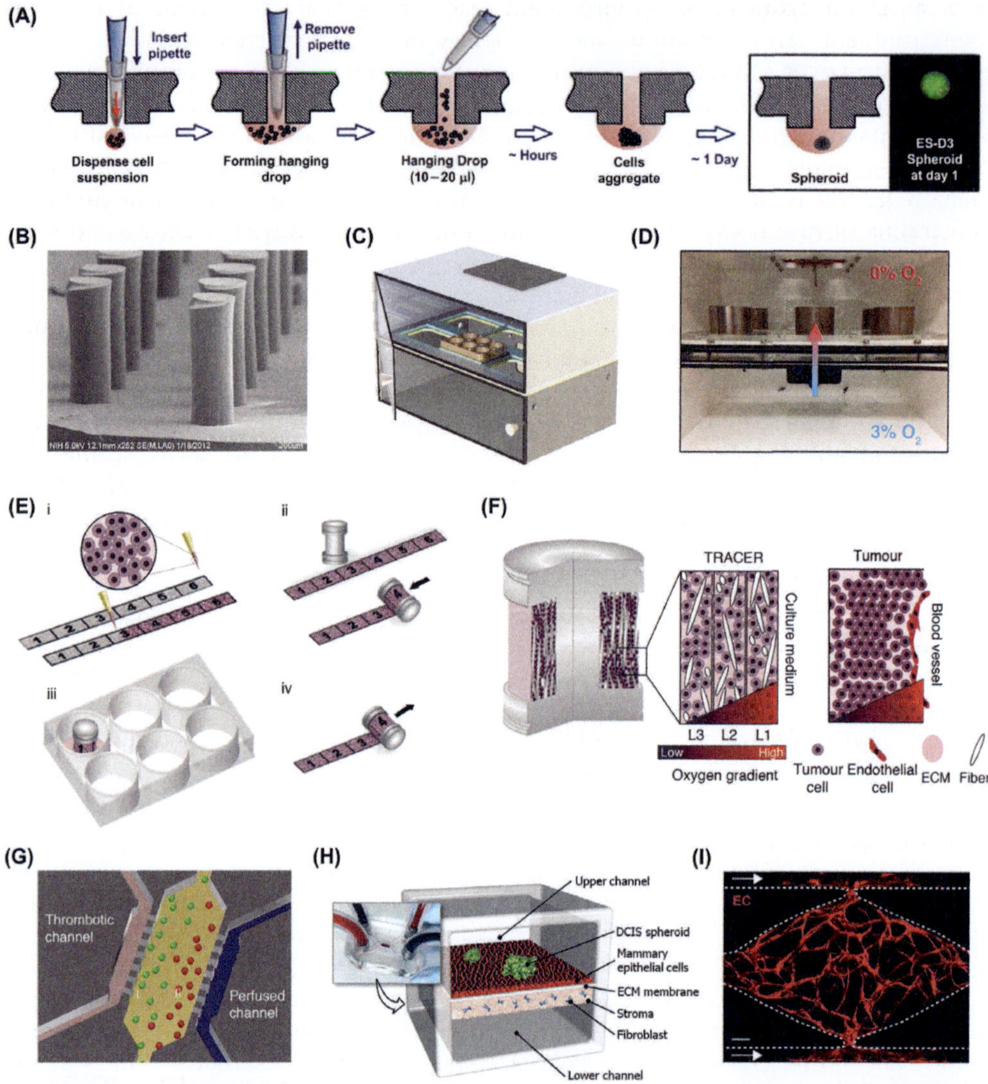


**FIGURE 27.4** Strategies for controlling oxygenation in in vitro culture. (A) Protocol for seeding cells into a hanging drop well to facilitate spheroid formation. (B) Scanning electron microscope image of silicone hydrogel micropillars that function as oxygen reservoirs, allowing diffusion of oxygen into surrounding cell cultures at a much more rapid rate compared with diffusion through media. (C, D) Two-chamber bioreactor that allows oxygen to diffuse from the bottom chamber into the anoxic top chamber, creating a gradient through the PDMS micropillars, allowing spatial control of the oxygenation of cell cultures. (E) Schematic of the TRACER platform. (i) Cells are deposited on the biocomposite strip, (ii) the strip is then rolled onto the core, (iii) the TRACER is then placed in a normal culture dish with media and cultured, and (iv) the biocomposite strip can be unrolled for quick access to specific regions of cell growth [100,101]. (F) The TRACER platform has an oxygen impermeable core that facilitates the formation of a unidirectional oxygenation gradient that models a tumor being oxygenated from a blood vessel. Each layer of TRACER can be isolated to analyze specific positions in the formed oxygen gradient. (G) Schematic of the gliobastoma-on-a-chip in which one channel is artificially obstructed to mimic a

By creating an artificial 3D environment, each individual component of the tumor microenvironment can be modified independently on a reproducible platform. This is a direct advantage over ex vivo tumor culture in which biopsies of tumor samples are plated with collagen and prescreened against chemotherapeutics. While highly patient-specific, tumor explants are heterogeneous, meaning their behavior may not be consistent between patient tissues. Explant tissues are also not conducive to long-term culture, with cellular architecture generally beginning to degrade within 1 month [106]. In order to circumvent the constraints of tumor explants while maintaining their organized structure and spatial distribution of nutrients, hydrogels have become a promising option. Hydrogels are matrices composed of synthetic or natural biocompatible polymers that are not water soluble and can be supplemented with growth factors and cell nutrients to accommodate embedded tissues. Owing to their microporous nature, hydrogels permit the passive transport of nutrients and waste and can accommodate the establishment of chemical gradients around growing 3D tissues. These hydrogel polymer networks often undergo thermally driven polymerization and depolymerization, allowing for the control of matrix stiffness and mechanical cues, an aspect shown to be critical in the tumor microenvironment [107]. Many hydrogels used in tissue culture are intended to recapitulate the ECM and contain polysaccharides such as heparan sulfate, hyaluronic acid, and versican, along with growth factors and other biologically derived materials like chitosan, alginate, or silk fibrils [108]. Matrigel, a gelatinous protein mixture secreted by Engelbreth–Holm–Swarm mouse sarcoma cells, has become the gold-standard hydrogel used for 3D culture. Hydrogel-based scaffolding can be further enhanced by co-culturing heterotypic cancer cell lines or a monotypic cancer cell line with other cell types such as macrophages, endothelial cells, or fibroblasts. These spheroids are known as multicellular tumor spheroids (MCTS). An underlying assumption of most MCTS oxygen distribution modeling is that the spheroids

◀ thrombotic event and cellular response of cells in the (I) starved and (II) enriched regions can be analyzed. (H) Schematic of the breast cancer-on-a-chip in which two channels are separated by an ECM membrane. Human mammary epithelial cells and DCIS spheroids are seeded in the upper channel and human mammary fibroblasts are seeded in the lower channel. (I) Image of red, fluorescent endothelial cells that have formed a vascular network between a pressure gradient in a VMT microdevice. Stromal cells are not fluorescently labeled. Scale bar is 100 μm [104]. *Source: (A) Reprinted with permission of the Royal Society of Chemistry from Tung YC, Hsiao AY, Allen SG, Torisawa YS, Ho M, Takayama S. High-throughput 3D spheroid culture and drug testing using a 384 hanging drop array. Analyst 2011;136:473–8 [98]; (B) Reprinted with permission of Elsevier from Jaeger AA, Das CK, Morgan NY, Pursley RH, McQueen PG, Hall MD, et al. Microfabricated polymeric vessel mimetics for 3-D cancer cell culture. Biomaterials 2013;34:8301–13 [99]; (C, D) Reprinted with permission of Elsevier from Wulftange WJ, Rose MA, Garmendia-Cedillos M, da Silva D, Poprawski JE, Srinivasachar D, et al. Spatial control of oxygen delivery to three-dimensional cultures alters cancer cell growth and gene expression. J Cell Physiol 2019;234:20608–22 [89]; (E, F) Reprinted with permission of Springer Nature from Rodenhizer D, Gaude E, Cojocari D, Mahadevan R, Frezza C, Wouters BG, et al. A three-dimensional engineered tumour for spatial snapshot analysis of cell metabolism and phenotype in hypoxic gradients. Nat Mater 2016;15:227–34 [100]; (G) Reprinted with permission of Oxford University Press from Ayuso JM, Monge R, Martinez-Gonzalez A, Virumbrales-Munoz M, Llamazares GA, Berganzo J, et al. Glioblastoma on a microfluidic chip: generating pseudopalisades and enhancing aggressiveness through blood vessel obstruction events. Neuro Oncol 2017;19:503–13 [102]; (H) Reprinted with permission of the Royal Society of Chemistry from Choi Y, Hyun E, Seo J, Blundell C, Kim HC, Lee E, et al. A microengineered pathophysiological model of early-stage breast cancer. Lab Chip 2015;15:3350–57 [103]; (I) Reprinted from Sobrino A, Phan DT, Datta R, Wang X, Hachey SJ, Romero-Lopez M, et al. 3D microtumors in vitro supported by perfused vascular networks. Sci Rep 2016;6:31589 [104]. All permissions conveyed through Copyright Clearance Center, Inc.*

are perfectly spherical, which is rarely the case [109]. Ellipsoidal modeling of tumor spheroid oxygen diffusion has recently been studied, but the disparity in cell OCRs and packing density make precise modeling of spheroid transport challenging. Innovations in fluorescent imaging such as the FLIM-FRET probe discussed earlier in this chapter could greatly improve this type of modeling.

A major advantage of 3D culture is that spheroids allow for the concurrent culture of cells at different growth stages; physically unconstrained cells at the periphery of the spheroid will be actively proliferating, while cells physically constrained inside the spheroid may be quiescent, hypoxic, apoptotic, or even necrotic. Hypoxic gradients in spheroid cancer culture are established naturally as the OCR and diffusion limitation of oxygen prevent medial cells from being oxygenated (Fig. 27.3B); this usually occurs when a spheroid's diameter exceeds 400 μm, although this value is cell line dependent. The natural tendency of spheroids to develop a hypoxic and necrotic core demonstrates a striking benefit over monolayer culture for the study of HIF-mediated behavior. Unfortunately, reproducibility of these spatiotemporal studies is a defining limitation of hydrogels, as spheroid size is difficult to uniformly control, and batch effects of natural hydrogels may mask more subtle transcriptional signatures. Maintaining a consistent ratio of medial cells in hypoxia and peripheral cells in normoxia requires stringent experimental endpoints and seeding densities. Such observations are facilitated by the nonhomogeneous oxygen distributions of oxygen within MCTS, which has been well documented, correlating to the clinical observations that tumor hypoxia rarely follows a highly geometric distribution [110,111]. While systems such as round bottom wells and hanging drop platforms offer more control over spheroid size and cell distribution, the apparent difficulty of definitively predicting the chemical gradients in tumor spheroids has led researchers to develop platforms which allow for more precise control over cancer tissue oxygenation.

### 27.5.3 Other strategies for controlling oxygen delivery in 3D: lab-on-chip systems, bioreactors

For appropriate oxygenation of 2D cancer tissue cultures, factors such as media depth and OCR of the pertinent cell line(s) must be considered. Controlling oxygenation becomes even more complicated when dealing with 3D models, as gradients form within the cultures and physiological recapitulation of tumor heterogeneity requires heterotypic cell types with disparate OCRs to be co-cultured. Some recent work has begun to address this problem with engineered tissue culture platforms that use microfabrication, microfluidics, and novel materials to enable spatially controlled oxygenation with high reproducibility.

One such example is the recent work from the Gottesman lab that uses soft-lithography protocols to fabricate a gas permeable membrane for cancer cell culture (Figs. 27.3C and 27.4B–D) [89]. Silicone hydrogel and poly(dimethylsiloxane) (PDMS) membranes that permitted oxygenation from the basal side of cell cultures were fabricated to mimic a capillary bed with pillars 25–100 μm in diameter and approximately 250 μm in height. Cells are mixed in Matrigel, seeded around the pillars, and placed in a two-chamber bioreactor, in which the bottom chamber is maintained at 3%–8% oxygen, and the top chamber, where media is placed, is maintained in anoxia. With a grid of pillars delivering oxygen independent from media diffusion limitations, oxygen distribution can be tuned by adjusting pillar size and spacing. Indeed, this bioreactor system is able to achieve larger avascular ovarian cancer spheroids with more consistent hypoxic staining compared with spheroids grown in Matrigel in standard incubators [99]. This bioreactor was subsequently expanded upon to accommodate six 6-well plates, each well containing the same pillared motif. This study reveals that both MCF-7 (breast cancer cell line) and OVCAR-8 (ovarian cancer cell line), when grown in a 3% to 0% oxygen gradient, exhibit growth and transcriptional patterns that are highly similar to controls grown in standard 21% oxygen [89]. Additionally, their findings revealed large transcriptional differences between 3D cultures grown in 3% oxygen and 21% oxygen, which is not surprising, but what was remarkable was that the gradient cultures, given a gradient of oxygen from 3% to 0%, resembled cultures grown in 21% oxygen more than the cultures grown in 3% oxygen [89]. This study shows that the geometry of oxygen delivery is very important when designing hypoxia experiments and that geometric considerations, in addition to concentration, need to be taken into account when designing novel hypoxia platforms and experiments.

Another method of controlling oxygen delivery through geometry uses paper-based scaffolds in which cells in a hydrogel are embedded. This approach was first reported by Whitesides and colleagues, who presented the fabrication of multilaminate structures in which chromatography paper provided structural support for cell-laden hydrogels. Lateral dimensions can be controlled by adjusting the dispensed hydrogel volume, additionally allowing for control over cell number and density. The stacking of the paper provides the vertical arrangement, with the precision determined by the paper thickness ($\sim$100 m). For subsequent analysis, the paper layers are simply separated, eliminating the need to section tissues. These cancer cell impregnated paper stacks exhibit gradient-dependent hypoxic gene expression, and can also be surgically implanted into mice to investigate the effects of in vivo molecular gradients and angiogenic processes [112]. Other adaptations using paper-based scaffolds have since been published [113,114]. One model in particular allows for rapid disassembly, conducive to snapshot analytic methods such as metabolomics. The tissue roll for analysis of cellular environment and response (TRACER) model, developed by McGuigan and colleagues, uses a paper scaffold in which cells are seeded in a hydrogel, then rolled around an oxygen impermeable mandrel, and placed into a standard tissue culture dish filled with media (Fig. 27.4E and F). Nutrients can only enter the cellular architecture from the outermost layer, creating a gradient of oxygen and nutrients for the medial layers, similar to what may be observed in an avascular tumor in vivo. Since cell migration was not observed between layers, each region could be assessed independently as a cell-derived microenvironment, metabolically modified by the implanted cells [100]. Models such as TRACER use novel, defined geometries in order to recapitulate a physiological environment with minimal instrumentation beyond what is already found in a conventional tissue culture laboratory. In contrast, microfluidic lab-on-a-chip devices can provide much more spatial and temporal control, but typically require additional equipment to maintain flow, and careful consideration of mass transport principles in the experimental design.

Microfluidic devices are often composed of poly(dimethylsiloxane) (PDMS), a biocompatible silicone polymer with high gas permeability widely used in conjunction with soft-lithography techniques. Alternatively, thermoplastics which are not highly gas permeable can be used exclusively or in conjunction with PDMS to control gas diffusion from the external atmosphere or gas-equilibrated media into the tissue culture regions. A common approach to investigate hypoxia-mediated cancer response in microfluidic devices is to have one or more perfused channels flanking a tissue culture region, in order to investigate cellular response to lateral stimuli (Fig. 27.3D, Fig. 27.4G). This device format has been used to investigate migration, angiogenesis, and cross-talk between dissimilar microenvironments [115–118]. Ayuso et al. reported on one such device in which human glioblastoma multiforme (GBM) were cultured in order to recapitulate a common histological finding of GBM: pseudopalisades. Pseudopalisades are hypoxic foci, surrounded by high cellularity, and are formed by actively migrating cells [119]. Observation of pseudopalisades is universal in clinical GBM cases, suggesting they are an important part of GBM pathophysiology, but in vitro recapitulation remains challenging. To emulate the occluded brain microvasculature, one channel of the device was sealed, while the other media channel was continuously perfused. The GBM cells in a collagen hydrogel between these two channels were observed and their migratory, metabolic, and morphological response was characterized. As expected, hypoxia was more pronounced near the occluded channel and less so adjacent to the perfused channel. Directional migration was also correlated to hypoxia by imaging fluorescent hypoxia markers when one channel was occluded, compared with bilaterally unrestricted channels [102]. This work demonstrates how a clinical observation, such as the hypoxia-mediated pseudopalisades seen in GBM biopsies, can be recapitulated in vitro using microfluidic devices.

A common approach to validating these novel in vitro cancer models is to analyze the response of cancer cells to a chemotherapeutic agent that is already approved for clinical use to determine if the novel platform can recapitulate clinically observed effects. In the lab of Dongeun Huh, researchers created a PDMS microfluidic device that had two stacked channels that ran parallel to each other with an ECM membrane as a partition (Fig. 27.4H). In the lower channel, mammary fibroblast cells were cultured and in the upper channel, human mammary epithelial cells were cultured. After several days, breast ductal carcinoma in situ (DCIS) spheroids were introduced to the epithelial channel and allowed to integrate. All of the cell lines used had been transduced to express fluorescent proteins, so morphology and spatial interaction of the heterotypic cell culture could be imaged. Additionally, paclitaxel was shown to inhibit growth of the embedded DCIS colonies, when added to the lower channel, simulating intravenous administration [103]. The results of this study demonstrate the potential that these microphysiological systems could have in the drug discovery pipeline. Controlled oxygenation was not incorporated into this breast cancer-on-a-chip, but minor modifications could allow more precise oxygen delivery. For example, a gas-impermeable material could be used to line the external sides of the device, resulting in oxygen being delivered solely through the media. Endothelial cells might also be added to the lower channel in order to create a stromal–vascular interface. The microflow in these devices can serve as an excellent model for vasculature and the associated oxygen and nutrient delivery in vivo, which has motivated much recent work on incorporating endothelial cells into microfluidic co-culture systems.

The Hughes lab introduced an in vitro vascularized microtumor (VMT) that incorporated human tumor and stromal cells growing in a 3D ECM and relied upon nutrient delivery via living, perfused vessels. The PDMS platform was constructed using photolithography and has two parallel channels separated by a diamond-shaped chamber. One channel acts as a high-pressure arteriole, while the other acts as a low-pressure venule, creating a pressure gradient through the diamond-shaped chamber in between, driving anastomoses by the endothelial cells (Fig. 27.4I). A slurry of ECM and endothelial cells was injected into the center chamber and cultured over the course of 5–7 days, during which the cells organized into a network of vasculature that connected to both outer channels. Stromal cells that were NG2 + PDGF-β + were essential to the formation of networks, illustrating the strength of devices that have the capacity for co-culture of heterotypic cell types. Several cancers were investigated including colorectal cancer (HCT116, SW620, and SW480 cell lines), breast cancer (MCF-7 and MDA-MB-231 cell lines), and a melanoma cell line (MNT-1). These cancer cells were added along with the initial ECM and endothelial cell slurry and allowed to colonize alongside the vasculature. Each line exhibited reproducibly different patterns of growth, illustrating the ability of cancer cells to remodel their microenvironment. Additionally, drug resistance was found to be higher in VMTs when compared with dose–response assays performed in 2D monolayer culture, and the hierarchy of drug effectiveness observed in the VMT matches closely with data obtained from zebrafish models. VMTs offer an exciting platform for cancer therapy investigation, as effects of new therapeutic agents that target the tumor directly or indirectly through the vasculature can be observed [104].

Processes such as metastasis can also be researched using vascular microfluidic devices as demonstrated by the device reported on by the Kamm group. This three-channel device had two lateral channels in which media could be flowed and a central channel in which cells embedded in a fibrin gel could be cultured. The central channel was seeded with a triculture of smooth muscle lineage and osteo-differentiated primary bone marrow-derived mesenchymal stem cells and primary human umbilical vein endothelial cells in order to create a bone marrow microenvironment. After 4 days of culture, the endothelial cells formed a microvascular network that anastomosed with the lateral channels, at which point a bone-seeking clone of the breast cancer cell line MDA-MB-231 was introduced into the microvascular network. Once inside the vasculature lumen, cancer cells roll on, and adhere to the endothelium or became physically trapped, then extravasate into the ECM by extending filopodia between the endothelial cells. Under static conditions, cancer cells extravasated at a higher rate and migrated further into the ECM when compared with cells that extravasated under flow [120]. Vascular devices sacrifice the defined organization that is found in many microfluidic devices in order to create physiologically relevant delivery of nutrients through microvasculature. However, neither of these devices has yet been used to control oxygen delivery in a physiological manner. Instead of oxygen gradients being created in a fashion similar to those of physiological gradients along vessels, these PDMS devices are bound to a glass coverslip on the bottom with the top of the PDMS device exposed to atmospheric gases. Oxygen delivery to the cells in the device occurs through a combination of diffusion through this PDMS layer and perfusion of the media through the developing vasculature, likely causing a top–down gradient of oxygen. The geometric delivery of nutrients, chemotherapeutic agents, and shear stress has a higher degree of physiological relevance in these devices, but the development of a functional microvasculature sufficient for oxygen delivery is still a challenge.

Many of these novel in vitro models use immortalized cell lines as they are simple to work with, produce relatively consistent results, and are widely available. Recent work has shifted to some use of patient-derived samples in order to increase the translation efficacy of these novel tissue culture devices. Yi et al. created a platform that attempts to mimic the biochemical and biophysical properties of native GBM using 3D bioprinting technologies and ex vivo patient GBM samples [121]. The team decellularized porcine brain ECM to create an ink that protected cells while being dispensed from the bioprinter and then acted as cellular architecture once seeded. The patient cells were printed in a central region, surrounded by vascularized stroma, a media channel, and then oxygen-permeable silicone walls. When the top surface of this construct was sealed by a gas-impermeable material such as glass, a radial oxygen gradient was formed across the tissue culture region. This GBM-on-a-chip platform was able to replicate patient-specific tumor response trends to concurrent chemoradiotherapy with temozolomide, and outperformed traditional monolayer and spheroid culture systems. There is tremendous interest in combining in vitro models that recapitulate the tumor microenvironment with patient-derived tissue to mimic clinical outcomes and predict personalized treatments, and this study is a very promising step. Ultimately, novel platforms like this could be incorporated into drug development pipelines, patient treatment plans, and basic science research.

## 27.6 Characterizing the effects of oxygenation on cells and tissues

While hypoxia can be detected by staining with pimonidazole or luminescent imaging, these techniques only verify that hypoxia is present, not necessarily how tissues are responding. In order to examine the physiological response to hypoxia, molecular biology techniques must be used to obtain read outs of cell transcription and protein expression.

### 27.6.1 RNA-Seq

Next-generation sequencing (NGS) has revolutionized the way disease mechanisms can be elucidated. High-throughput transcriptome profiling is known as RNA-Seq and can create a depiction of the bulk of RNA transcripts present in a sample. Hypoxia has a complex effect on numerous cellular processes including angiogenesis, metastasis, pH homeostasis, autophagy, cell survival and proliferation, and metabolism. The advantage of RNA-Seq in relation to hypoxia is that it provides a broad scope of what is happening in a population of cells so that subsequent experiments can be more targeted in nature in terms of gene investigation. The technique may provide insights into transcriptional events that are not anticipated and would have gone unnoticed in gene-specific analyses, such as microarrays. RNA-Seq analysis can yield a list of thousands of genes that are differentially expressed for any given experiment, so there is constant effort to catalog genes and gene families into hierarchical repositories. The Gene Set Enrichment Analysis (GSEA) platform contains the Molecular Signature Database which organizes gene collections based on shared characteristics [122,123]. Of particular interest is the "Hallmark Hypoxia" gene set which contains genes upregulated by hypoxia and is frequently used to compare hypoxia gene expression in cancer cell lines [124–127]. These gene sets are constantly evolving as new discoveries are made, but gene ontology pipelines draw lists from multiple databases during analysis. Once target genes have been identified, differential expression is validated by quantitative polymerase chain reaction (qPCR).

There are several special considerations that must be taken when preparing RNA samples from hypoxia experiments. As discussed earlier, HIFs are key transcriptional regulators for hypoxia-mediated gene expression and HIF-1α is rapidly degraded in oxygenated environments, which will begin the process of obfuscating cellular hypoxia processes. Samples should be lysed as soon as possible for hypoxia experiments and preferably in the experimental hypoxic environment. Ideally, tissues should be lysed immediately following an experiment, but for culture systems other than monolayer culture, this can be difficult. When cells are in Matrigel or other hydrogel ECMs, the encapsulated cells or their RNA must be isolated prior to standard RNA work-up protocols. One technique is to digest the ECM at 4°C and then lyse the cells using standard RNA isolation protocols. Alternatively, recent work has detailed a protocol in which cells are lysed first and then RNA is isolated from mechanically degraded ECM [128].

Currently, most RNA-Seq data is generated from a mass of homogenous cells, but due to lowered costs, single-cell RNA-Seq (scRNA-Seq) is becoming more common, which can reveal rare cell populations and complex heterogeneities observed in cancer [129]. More recently, an RNA-Seq technology called "Slide-seq" has been reported in which spatial distribution of a tissue is preserved in the sequencing process, permitting high spatial resolution mapping at the single-cell level [130]. Combined with spatial control of oxygen, these technologies will undoubtedly lead to new insights into the complex interplay of cancer and hypoxia.

### 27.6.2 qPCR of downstream targets

When investigating specific genes, quantitative polymerase chain reaction (qPCR) is an analytical tool that allows for the detection of absolute or relative quantities of a transcript present in a sample, and it is often used to validate RNA-Seq results once particular genes of interest are identified. As discussed earlier, HIF-1 signaling is transduced when the HIF-1 αβ heterodimer recognizes and binds DNA at hypoxia response elements [131], although it has been shown that HIF-1α can interact with transcription factors other than HIF-1β [132]. HIF-1 lies at the center of the transcriptional cascade that controls the cellular response to changes in oxygen tension. There are hundreds of genes that are either directly or indirectly regulated by HIF-1, which orchestrates metabolism, angiogenesis, growth, and metastasis in cancer. Since HIF-1α is not regulated at the transcription level, downstream targets such as VEGF-A and angiopoietin-2, which are expressed in order to sprout new vasculature to accommodate tumor growth, are more suitable qPCR targets. In relation to metastasis, LOX genes have also been associated with poorer prognosis in head, neck, and breast cancer, meaning expression levels of this gene family are also good candidates for qPCR both in vitro and in vivo [133]. Other hypoxia-related genes of interest include BCL-2, which is a pro-apoptotic gene that has altered expression in cancers and has been shown to be modulated by hypoxia, PDK1 and LDHA, which are both involved in metabolism, MMP-2, involved in matrix-remodeling, and VHL, which is integral to HIF-1 signaling [21,28,33,134,135]. These genes and the related collections on GSEA serve as lists to validate the classical effects of hypoxia. However, the expression of these genes may vary between tissues and even within tissues. Intratumor heterogeneity is well documented across a multitude of clinical cancers, which highlights the need for spatiotemporal consideration when performing gene-specific analyses [21,28,33,35,38,49,50,133–138].

### 27.6.3 Pimonidazole staining

Clinical assessment of hypoxia in cancers can also be characterized using exogenous bioreductive compounds such as pimonidazole, a small molecule that is reduced in hypoxic cells, including tumor cells. Reduced pimonidazole binds thiol-containing proteins and molecules such as glutathione, causing accumulation of the molecule, which can be detected using immunohistochemical methods. Pimonidazole represents a translational characterization technique, as it can be applied both in vitro and in vivo. In vitro, tissue cultures can be incubated with pimonidazole, then fixed and analyzed using flow cytometry or immunohistochemical (IHC) staining. In patients, pimonidazole can be administered intravenously, providing a direct route into tumor vasculature. A tumor biopsy can then be taken and IHC staining for piminidazole used to reveal areas of hypoxia within patient samples [139]. Studies have reported prognostic correlations between pimonidazole staining for soft tissue sarcoma, brain tumors, and head and neck cancers [140]. Since pimonidazole requires viable cells for its metabolism, areas of necrosis in tumors are not stained, potentially excluding important findings in histological analyses. Conversely, areas of keratinizing tissues have exhibited nonspecific pimonidazole staining, complicating the clinical interpretation of the results [141].

### 27.6.4 Real-time imaging of growth

In combination with the previously mentioned methods, longitudinal cancer cell growth monitoring and behavioral characterization are important methods for characterizing the effect of oxygenation on cancer cell culture. A well-known characteristic of cancer cells is their amazing proliferative ability, regardless of their hypoxic, nutrient-lacking microenvironment [142]. For this reason, real-time live-cell imaging and growth monitoring is important, as this allows for the assessment of differences in growth, migration, proliferation, and metastasis in cancer cells in and out of hypoxic in vitro environments. There are several systems for real-time hypoxic imaging, most of which involve the combination of inverted automated bright-field microscopy with the control of local atmosphere [143–146]. One such system is the Lionheart FX, which allows for real-time imaging of cancer culture over multiple days and has an environmental control in which humidity, temperature, and most importantly oxygen percentage can be adjusted and controlled. In this way, the real-time assessment of cancer growth in varying states of oxygenation can be easily tracked and measured.

### 27.6.5 Metabolic characterization and imaging

Since the formulation of Warburg's foundational hypothesis (the Warburg effect), it has been confirmed that rapidly proliferating tumor cells deviate from traditional cellular metabolic pathways, relying on glycolysis rather than oxidative phosphorylation for their energy needs [147]. Cancer metabolic processes have been shown to be very heterogeneous, with hypoxic conditions known to be a main driver of cancer metabolic variation [148]. As a result, there is considerable interest in using metabolic imaging to quantify and measure changes in cancer metabolic activity in situ, both in vivo and in hypoxic environments in vitro.

### 27.6.6 In vivo metabolic imaging

In vivo metabolic imaging overlaps significantly with the spectroscopic imaging methods discussed earlier. The most common clinical techniques are PET and MRI imaging. PET imaging is most often used with a radioactive glucose analog, FDG ($^{18}$F, Fluorodeoxyglucose) [147,149]. Uptake of this glucose analog is higher in cancer cells and thus metabolic activity, glucose consumption, and cancer proliferation can be easily tracked by visualizing the $^{18}$F FDG tracer within the body [147]. Hyperpolarized MRI imaging has also been used to track metabolic pathways in vivo in both preclinical and clinical models [147,150]. In recent studies, hyperpolarized [$^{13}$C] pyruvate and [U-$^{2}$H, U-$^{13}$C] glucose were injected and tracked through MRI imaging [151,152] to measure glucose utilization in vivo through alternate metabolic pathways and through the uptake of different metabolites. While these techniques are powerful, as was mentioned before, they require expensive, specialized equipment, and when compared with in vitro techniques, offer less spatial resolution. However, these in vivo methods provide researchers with important data regarding the metabolic behavior of tumors within physiological hypoxic cancer microenvironments, providing a gold standard for evaluating the performance and relevance of in vitro systems.

### 27.6.7 In vitro metabolic imaging

The most prevalent in vitro techniques for metabolic imaging and characterization in current research are Seahorse analysis, luminescent imaging, and metabolomics. Luminescent imaging, using the same principles previously mentioned, can be used to image metabolic activity in cancer culture, specifically with luminescent oxygen probes to track oxygen consumption and the autofluorescence of NADH to track glycolytic activity [53,153–155]. The use of luminescent oxygen probes to track and monitor oxygen was discussed earlier, with oxygen serving as a quencher of probe luminescence. For luminescent metabolic imaging, probes can be tailored to track intracellular oxygen metabolism, using this as a measure of oxygen consumption and therefore an estimate of oxidative phosphorylation [53]. Penjweini et al. developed a FLIM-FRET measurement system using a myo-mCherry probe, discussed earlier, that accomplishes exactly this. The added myoglobin protein component allows for more sensitive oxygen dependent lifetime behavior, which decreases when the probe complex is deoxygenated [53]. In this particular system, two-photon excited fluorescent microscopy is used due to its efficacy in metabolic imaging [53,154]. In combination with metabolic oxygen sensing, the autofluorescence of NADH has been used to track glycolytic metabolism. Specifically, the ratio of free to enzyme-bound NADH is used as a biomarker of glycolysis [155]. These two NADH variants have detectably different lifetimes and separate functions. Free NADH mainly exists in the cytoplasm as a result of glycolysis and enzyme-bound NADH is almost exclusively found in the mitochondria for use in oxidative phosphorylation [155–157]. Therefore, an increase in the free to enzyme-bound NADH ratio is an indicator of increased glycolytic activity [157,158]. Using luminescent metabolic imaging, a full complement of metabolic activity can be measured in cell culture; however, the true power of luminescent metabolic imaging lies in its ability for single-cell measurement and use in situ, allowing for cell heterogeneity to be studied and the continuation of culture without disruption.

The more commonly used method for measuring cellular metabolism in vitro is the Seahorse system, which uses live-cell real-time metabolic assays to measure OCR and extracellular acidification rate (ECAR) in a multiwell plate [159]. The Seahorse XF24 Extracellular Flux Analyzer quantifies oxygen concentration and the continuous flux in $H^+$ proton concentration in culture media through two luminescent sensors, one that responds to changes in local pH and another that is quenched by oxygen [159,160]. These continuously recorded measurements are then converted to OCR and ECAR values, allowing for an accurate estimate of oxidative phosphorylation and glycolysis [159]. When compared with luminescent metabolic imaging using intracellular probes, Seahorse analysis has many of the same benefits, as mitochondrial respiration can be directly correlated with glycolytic activity in the measured cells. However, cells used with the Seahorse system must be cultured in assay plates, and the data recorded is a generalized population average, which forecloses the possibility of measuring cell–cell heterogeneity or co-cultures.

In contrast to these in situ methods, metabolomics is an endpoint technique in which individual metabolites from tissues or culture samples are studied through the coupling of electrophoresis or liquid–gas chromatography to mass spectrometry [161]. In general, this process first involves the extraction of metabolites using organic and aqueous solvents, tailored to the specific metabolites of interest, followed by the separation of these metabolites with chromatography or electrophoresis based on their intrinsic charge and polarity [162–165]. Using the full range of metabolomic techniques, tumor specimens and culture samples can be analyzed to better understand tumor-specific metabolism, observe changes in different metabolite levels, and potentially discover novel tissue-specific biomarkers for cancer [161]. This metabolomic analysis can be used in conjunction with in situ luminescent imaging or Seahorse analysis to connect observed changes in metabolic activity to specific metabolites and metabolic pathways. All of these in vitro methods allow for the characterization of cancer cell metabolic behavior when exposed to hypoxic environments and are crucial in understanding deviations in cellular behavior resulting from differing states of oxygenation.

## 27.7 Conclusions and future prospects

Oxygen is universally understood to be a crucial variable in developing physiologically relevant in vitro models. This is especially true in cancer culture, where oxygen levels affect a wide range of cellular functions including metabolic activity, gene expression, drug resistance, and malignancy. However, traditional in vitro cell culture techniques generally fail to provide active control of oxygenation, which particularly in 3D, can give rise to poorly defined and nonphysiological oxygen concentrations. Oxygen levels in vivo vary with location and tissue type, and in diseases such as cancer, severe hypoxia is often observed. In this chapter, we have outlined the current understanding of oxygen distribution in vivo for both normal and pathological conditions, and discussed the effects arising from oxygen gradients and induced hypoxia observed in vitro. Recent developments in oxygen measurement techniques have greatly enhanced the ability to quantitatively assess oxygen concentrations in real time at high resolution in vitro. In combination with devices and assay platforms that provide controlled oxygen delivery, the ability to measure and quantify oxygen concentrations is especially important in developing models for the impact of oxygenation levels on cell culture. We have provided some examples of recent work in this area, with a few representative devices that provide controlled oxygenation, and discussed the observed biological effects, but this is an area of expanding interest.

Of course, the development of physiologically relevant culture systems also requires 3D co-culture, biomechanics, biochemical cues, and ECM selection; however, in moving to these more complicated tissue geometries it is essential to include controlled oxygenation, as simple diffusion through aqueous media is insufficient over distances of more than a few hundred microns. For anything beyond smaller spheroids grown in a hanging drop geometry, microfabricated structures or cultured vasculature may be needed. Oxygen delivery in these systems can be numerically modeled and validated using the imaging techniques discussed here, and this pairing can then be used to directly assess tumor response to differing states and concentrations of oxygen. Although there has been a considerable amount of recent work to study the interplay of oxygenation and cancer, including metastasis, there has been limited research aimed at the kinetics: understanding the cellular and tissue responses to temporal changes in oxygenation levels such as hypoxic shock. Using the systems and technologies outlined above, researchers should expand this area of research to investigate cancer behavior and further understand cellular response to changing oxygenation conditions.

We hope that the review of research found in this chapter may aid future researchers in their attempts to recapitulate the in vivo cancer microenvironment and provide a steppingstone to further the development of in vitro cancer culture models.

## Acknowledgments

This research was funded by the Intramural Research Program of the National Institutes of Health. We would like to thank G. Leiman in the Laboratory of Cell Biology for editorial assistance.

## Disclaimer

Any mention of particular commercial products in this paper should not be considered as a recommendation or endorsement.

## References

[1] Earle WR, Schilling EL, Stark TH, Straus NP, Brown MF, Shelton E. Production of malignancy in vitro. IV. The mouse fibroblast cultures and changes seen in the living cells. J Natl Cancer Inst 1943;4:165–212.

[2] Gray LH, Conger AD, Ebert M, Hornsey S, Scott OCA. The concentration of oxygen dissolved in tissues at the time of irradiation as a factor in radiotherapy. Br J Radiol 1953;26:638–48.

[3] Gray LH, Thomlinson RH. The histological structure of some human lung cancers and the possible implications for radiotherapy. Br J Cancer 1955;9:539–49.
[4] Clark Jr LC, Wolf R, Granger D, Taylor Z. Continuous recording of blood oxygen tensions by polarography. J Appl Physiol 1953;6:189–93.
[5] McKeown SR. Defining normoxia, physoxia and hypoxia in tumours—implications for treatment response. Br J Radiol 2014;87.
[6] Keeley TP, Mann GE. Defining physiological normoxia for improved translation of cell physiology to animal models and humans. Physiol Rev 2019;99:161–234.
[7] Vaupel P, Höckel M, Mayer A. Detection and characterization of tumor hypoxia using pO2 histography. Antiox Redox Signal 2007;9:1221–36.
[8] Graeber TG, Osmanian C, Jacks T, Housman DE, Koch CJ, Lowe SW, et al. Hypoxia-mediated selection of cells with diminished apoptotic potential in solid tumours. Nature 1996;379:88–91.
[9] Vaupel P. Metabolic microenvironment of tumor cells: a key factor in malignant progression. Exp Oncol 2010;32:125–7.
[10] Wenger RH, Kurtcuoglu V, Scholz CC, Marti HH, Hoogewijs D. Frequently asked questions in hypoxia research. Hypoxia (Auckl) 2015;3:35–43.
[11] Carreau A, Hafny-Rahbi BE, Matejuk A, Grillon C, Kiedaa C. Why is the partial oxygen pressure of human tissues a crucial parameter? Small molecules and hypoxia. J Cell Mol Med 2011;15:1239–53.
[12] Höckel M, Vaupel P. Tumor hypoxia: definitions and current clinical, biologic, and molecular aspects. J Natl Cancer Inst 2001;93:266–76.
[13] Cassavaugh J, Lounsbury KM. Hypoxia-mediated biological control. J Cell Biochem 2012;112:735–44.
[14] Hanahan D, Weinberg RA. Hallmarks of cancer: the next generation. Cell 2011;144:646–74.
[15] Papandreou I, Krishna C, Kaper F, Cai D, Giaccia AJ, Denko NC. Anoxia is necessary for tumor cell toxicity caused by a low-oxygen environment. Cancer Res 2005;65:3171–7.
[16] Warburg O. On the origin of cancer cells. Science 1956;123:309–14.
[17] Krogh A. The anatomy and physiology of capillaries. New Haven, CT: Yale University Press; 1922.
[18] Thomlinson RH, Gray LH. The histological structure of some human lung cancers and the possible implications for radiotherapy. Br J Cancer 1955;9:539–49.
[19] Brown JM, Giaccia AJ. The unique physiology of solid tumors: opportunities (and problems) for cancer therapy. Cancer Res 1998;58:1408–16.
[20] Rankin EB, Giaccia AJ. Hypoxic control of metastasis. Science 2016;352:175–80.
[21] Semenza GL. Hypoxia-inducible factors in physiology and medicine. Cell 2012;148:399–408.
[22] Stewart GD, Ross JA, McLaren DB, Parker CC, Habib FK, Riddick AC. The relevance of a hypoxic tumour microenvironment in prostate cancer. BJU Int 2010;105:8–13.
[23] Carroll VA, Ashcroft M. Targeting the molecular basis for tumour hypoxia. Expert Rev Mol Med 2005;7:1–16.
[24] Ide T, Kitajima Y, Miyoshi A, Ohtsuka T, Mitsuno M, Ohtaka K, et al. Tumor-stromal cell interaction under hypoxia increases the invasiveness of pancreatic cancer cells through the hepatocyte growth factor/c-Met pathway. Int J Cancer 2006;119:2750–9.
[25] Miyoshi A, Kitajima Y, Ide T, Ohtaka K, Nagasawa H, Uto Y, et al. Hypoxia accelerates cancer invasion of hepatoma cells by upregulating MMP expression in an HIF-1alpha-independent manner. Int J Oncol 2006;29:1533–9.
[26] Vergis R, Corbishley CM, Norman AR, Bartlett J, Jhavar S, Borre M, et al. Intrinsic markers of tumour hypoxia and angiogenesis in localised prostate cancer and outcome of radical treatment: a retrospective analysis of two randomised radiotherapy trials and one surgical cohort study. Lancet Oncol 2008;9:342–51.
[27] Kitajima Y, Miyazaki K. The critical impact of HIF-1a on gastric cancer biology. Cancers (Basel), 5. 2013. p. 15–26.
[28] Semenza GL. HIF-1: upstream and downstream of cancer metabolism. Curr Opin Genet Dev 2010;20:51–6.
[29] North S, Moenner M, Bikfalvi A. Recent developments in the regulation of the angiogenic switch by cellular stress factors in tumors. Cancer Lett 2005;218:1–14.
[30] Valli A, Morotti M, Zois CE, Albers PK, Soga T, Feldinger K, et al. Adaptation to HIF1alpha deletion in hypoxic cancer cells by upregulation of GLUT14 and creatine metabolism. Mol Cancer Res 2019;17:1531–44.
[31] Valli A, Rodriguez M, Moutsianas L, Fischer R, Fedele V, Huang HL, et al. Hypoxia induces a lipogenic cancer cell phenotype via HIF1alpha-dependent and -independent pathways. Oncotarget 2015;6:1920–41.

[32] Kimura H, Weisz A, Kurashima Y, Hashimoto K, Ogura T, D'Acquisto F, et al. Hypoxia response element of the human vascular endothelial growth factor gene mediates transcriptional regulation by nitric oxide: control of hypoxia-inducible factor-1 activity by nitric oxide. Blood 2000;95:189–97.
[33] Ben-Yosef Y, Miller A, Shapiro S, Lahat N. Hypoxia of endothelial cells leads to MMP-2-dependent survival and death. Am J Physiol Cell Physiol 2005;289:C1321–31.
[34] Conway EM, Collen D, Carmeliet P. Molecular mechanisms of blood vessel growth. Cardiovasc Res 2001;49:507–21.
[35] Krock BL, Skuli N, Simon MC. Hypoxia-induced angiogenesis: good and evil. Genes Cancer 2011;2:1117–33.
[36] Suchting S, Freitas C, le Noble F, Benedito R, Breant C, Duarte A, et al. The Notch ligand Delta-like 4 negatively regulates endothelial tip cell formation and vessel branching. Proc Natl Acad Sci U S A 2007;104:3225–30.
[37] Hellstrom M, Phng LK, Hofmann JJ, Wallgard E, Coultas L, Lindblom P, et al. Dll4 signalling through Notch1 regulates formation of tip cells during angiogenesis. Nature 2007;445:776–80.
[38] Folkman J. Angiogenesis in cancer, vascular, rheumatoid and other disease. Nat Med 1995;1:27–31.
[39] McDonald DM, Foss AJ. Endothelial cells of tumor vessels: abnormal but not absent. Cancer Metastasis Rev 2000;19:109–20.
[40] McDonald DM, Baluk P. Significance of blood vessel leakiness in cancer. Cancer Res 2002;62:5381–5.
[41] van Roy F, Berx G. The cell-cell adhesion molecule E-cadherin. Cell Mol Life Sci 2008;65:3756–88.
[42] Acloque H, Adams MS, Fishwick K, Bronner-Fraser M, Nieto MA. Epithelial-mesenchymal transitions: the importance of changing cell state in development and disease. J Clin Invest 2009;119:1438–49.
[43] Orphanides C, Fine LG, Norman JT. Hypoxia stimulates proximal tubular cell matrix production via a TGF-beta1-independent mechanism. Kidney Int 1997;52:637–47.
[44] Mingyuan X, Qianqian P, Shengquan X, Chenyi Y, Rui L, Yichen S, et al. Hypoxia-inducible factor-1alpha activates transforming growth factor-beta1/Smad signaling and increases collagen deposition in dermal fibroblasts. Oncotarget 2018;9:3188–97.
[45] Tang YA, Chen YF, Bao Y, Mahara S, Yatim S, Oguz G, et al. Hypoxic tumor microenvironment activates GLI2 via HIF-1alpha and TGF-beta2 to promote chemoresistance in colorectal cancer. Proc Natl Acad Sci U S A 2018;115:E5990–9.
[46] Jiang J, Tang YL, Liang XH. EMT: a new vision of hypoxia promoting cancer progression. Cancer Biol Ther 2011;11:714–23.
[47] Bierie B, Moses HL. TGF-beta and cancer. Cytokine Growth Factor Rev 2006;17:29–40.
[48] Mori S, Chang JT, Andrechek ER, Matsumura N, Baba T, Yao G, et al. Anchorage-independent cell growth signature identifies tumors with metastatic potential. Oncogene 2009;28:2796–805.
[49] Schietke R, Warnecke C, Wacker I, Schodel J, Mole DR, Campean V, et al. The lysyl oxidases LOX and LOXL2 are necessary and sufficient to repress E-cadherin in hypoxia: insights into cellular transformation processes mediated by HIF-1. J Biol Chem 2010;285:6658–69.
[50] Erler JT, Bennewith KL, Cox TR, Lang G, Bird D, Koong A, et al. Hypoxia-induced lysyl oxidase is a critical mediator of bone marrow cell recruitment to form the premetastatic niche. Cancer Cell 2009;15:35–44.
[51] Peinado H, Zhang H, Matei IR, Costa-Silva B, Hoshino A, Rodrigues G, et al. Pre-metastatic niches: organ-specific homes for metastases. Nat Rev Cancer 2017;17:302–17.
[52] Peinado H, Lavotshkin S, Lyden D. The secreted factors responsible for pre-metastatic niche formation: old sayings and new thoughts. Semin Cancer Biol 2011;21:139–46.
[53] Penjweini R, Andreoni A, Rosales T, Kim J, Brenner MD, Sackett DL, et al. Intracellular oxygen mapping using a myoglobin-mCherry probe with fluorescence lifetime imaging. J Biomed Opt 2018;23:1–14.
[54] Allen JR. pH electrodes, ion-selective electrodes, and oxygen sensors: electrochemical sensors used in the medical field. Lab Med 2003;34:544–7.
[55] Skoog DA, West DM, Holler JF. Fundamentals of analytical chemistry. 7th ed. Philadelphia, PA: Saunders College Publishing; 1996.
[56] Toffaletti JG, Rackley CR. Chapter three - monitoring oxygen status. In: Makowski GS, editor. Advances in clinical chemistry. Elsevier; 2016. p. 103–24.
[57] Wu C-C, Yasukawa T, Shiku H, Matsue T. Fabrication of miniature Clark oxygen sensor integrated with microstructure. Sens Actuat B: Chem 2005;110:342–9.
[58] Whalen WJ, Nair P, Ganfield RA. Measurements of oxygen tension in tissues with a micro oxygen electrode. Microvasc Res 1973;5:254–62.

[59] Suzuki H, Hirakawa T, Watanabe I, Kikuchi Y. Determination of blood pO2 using a micromachined Clark-type oxygen electrode. Anal Chim Acta 2001;431:249–59.

[60] Wolfbeis OS. Luminescent sensing and imaging of oxygen: fierce competition to the Clark electrode. BioEssays 2005;37:921–8.

[61] Wanga X-d, Wolfbeis OS. Optical methods for sensing and imaging oxygen: materials, spectroscopies and applications. Chem Soc Rev 2014;43:3666–761.

[62] Zhujun Z, Seitz WR. Optical sensor for oxygen based on immobilized hemoglobin. Anal Chem 1986;58:220–2.

[63] Cohen SH ZJV, Aggoun A. Pulse oximetry optical sensor using oxygenbound haemoglobin. Opt Express 2016;24:10115–31.

[64] Chung KE, Lan EH, Davidson MS, Dunn BS, Valentine JS, Zink JI. Measurement of dissolved oxygen in water using glass-encapsulated myoglobin. Anal Chem 1995;67:1505–9.

[65] McCurley MF, Bayer GJ, Glazier SA. A fluorescence assay for dissolved oxygen using sol-gel encapsulated myoglobin and an analogy to the inner filter effect. Sens Actuat B: Chem 1996;36:491–6.

[66] Baldini F, Bacci M, Cosi F, Del Bianco A. Absorption-based optical-fibre oxygen sensor. Sens Actuat B: Chem 1992;7:752–7.

[67] Goodfellow GI, Webber HM. Absorptiometric determination of low oxygen concentrations in power-station waters. Part I. Manual method. Analyst 1979;104:1105–18.

[68] Hamlin PA, Lambert JL. Determination of dissolved oxygen using photoreduced leuco phenothiazine dyes. Anal Chem 1971;43:618–20.

[69] Hochreiter B, Garcia AP, Schmid JA. Fluorescent proteins as genetically encoded FRET biosensors in life sciences. Sensors (Basel) 2015;15:26281–314.

[70] Lakowicz JR. Chapter 8 - quenching of fluorescence. Principles of fluorescence spectroscopy. 3rd ed. Boston, MA: Springer US; 2006. p. 277–330.

[71] Albani JR. Chapter 4 - fluorescence quenching. In: Albani JR, editor. Structure and dynamics of macromolecules: absorption and fluorescence studies. Amsterdam: Elsevier Science; 2004. p. 141–92.

[72] Shonat RD, Wilson DF, Riva CE, Pawlowski M. Oxygen distribution in the retinal and choroidal vessels of the cat as measured by a new phosphorescence imaging method. Appl Opt 1992;31:3711–18.

[73] Boaz H, Rollefson GK. The quenching of fluorescence. Deviations from the Stern-Volmer law. J Am Chem Soc 1950;72:3435–43.

[74] Albani JR. Chapter 6 - Forster energy transfer. In: Albani JR, editor. Structure and dynamics of macromolecules: absorption and fluorescence studies. Amsterdam: Elsevier Science; 2004. p. 217–36.

[75] Rumsey W, Vanderkooi J, Wilson D. Imaging of phosphorescence: a novel method for measuring oxygen distribution in perfused tissue. Science 1988;241:1649–51.

[76] Amao Y. Probes and polymers for optical sensing of oxygen. Microchim Acta 2003;143:1–12.

[77] Dmitriev RI, Papkovsky DB. Optical probes and techniques for O2 measurement in live cells and tissue. Cell Mol Life Sci 2012;69:2025–39.

[78] Urano M, Chen Y, Humm J, Koutcher JA, Zanzonico P, Ling C. Measurements of tumor tissue oxygen tension using a time-resolved luminescence-based optical oxylite probe: comparison with a paired survival assay. Radiat Res 2002;158:167–73.

[79] Jensen EC. Types of imaging, Part 2: an overview of fluorescence microscopy. Anat Rec (Hoboken) 2012;295:1621–7.

[80] Bastiaens PI, Squire A. Fluorescence lifetime imaging microscopy: spatial resolution of biochemical processes in the cell. Trends Cell Biol 1999;9:48–52.

[81] Sandhu S, Kydd L, Jaworski J. Luminescent probe based techniques for hypoxia imaging. J Nanomed Res 2017;6:00160.

[82] Gaertner C, Souvatzoglou F, Brix G M, Beer A J. Imaging of hypoxia using PET and MRI. Curr Pharm Biotechnol 2012;13:552–70.

[83] Ruiz-Cabello J, Barnett BP, Bottomley PA, Bulte JWM. Fluorine (19F) MRS and MRI in biomedicine. NMR Biomed 2011;24:114–29.

[84] Eberle B, Weiler N, Markstaller K, Kauczor H, Deninger A, Ebert M, et al. Analysis of intrapulmonary O(2) concentration by MR imaging of inhaled hyperpolarized helium-3. J Appl Physiol (1985) 1999;87:2043–52.

[85] Lopci E, Grassi I, Chiti A, Nanni C, Cicoria G, Toschi L, et al. PET radiopharmaceuticals for imaging of tumor hypoxia: a review of the evidence. Am J Nucl Med Mol Imaging 2014;4:365–84.

[86] Ahmad R, Kuppusamy P. Theory, instrumentation, and applications of electron paramagnetic resonance oximetry. Chem Rev 2010;110:3212–36.
[87] Gertsenshteyn I, Epel B, Giurcanu M, Leoni L, Barth E, Fan X, et al. Preliminary investigation of hypoxia within tumor using EPRI, DCE-MRI, and PET-CT with 18F-FMISO to improve radiotherapy. J Nucl Med 2019;60:347.
[88] Rivera BK, Naidu SK, Subramanian K, Joseph M, Hou H, Khan N, et al. Real-time, in vivo determination of dynamic changes in lung and heart tissue oxygenation using EPR oximetry. Adv Exp Med Biol 2014;812:81–6.
[89] Wulftange WJ, Rose MA, Garmendia-Cedillos M, da Silva D, Poprawski JE, Srinivasachar D, et al. Spatial control of oxygen delivery to three-dimensional cultures alters cancer cell growth and gene expression. J Cell Physiol 2019;234:20608–22.
[90] Metzen E, Wolff M, Fandrey J, Jelkmann W. Pericellular PO2 and O2 consumption in monolayer cell cultures. Respir Physiol 1995;100:101–6.
[91] Polak J, Studer-Rabeler K, McHugh H, Hussain MA, Shimoda LA. System for exposing cultured cells to intermittent hypoxia utilizing gas permeable cultureware. Gen Physiol Biophys 2015;34:235–47.
[92] Zweier JL, Talukder MAH. The role of oxidants and free radicals in reperfusion injury. Cardiovasc Res 2006;70:181–90.
[93] Wagner BA, Venkataraman S, Buettner GR. The rate of oxygen utilization by cells. Free Radic Biol Med 2011;51:700–12.
[94] Kleinman HK, Philp D, Hoffman MP. Role of the extracellular matrix in morphogenesis. Curr Opin Biotechnol 2003;14:526–32.
[95] Bissell MJ, Radisky DC, Rizki A, Weaver VM, Petersen OW. The organizing principle: microenvironmental influences in the normal and malignant breast. Differentiation 2002;70:537–46.
[96] Paul CD, Bishop K, Devine A, Paine EL, Staunton JR, Thomas SM, et al. Tissue architectural cues drive organ targeting of tumor cells in zebrafish. Cell Syst 2019;9:187–206.
[97] Ivascu A, Kubbies M. Rapid generation of single-tumor spheroids for high-throughput cell function and toxicity analysis. J Biomol Screen 2006;11:922–32.
[98] Tung YC, Hsiao AY, Allen SG, Torisawa YS, Ho M, Takayama S. High-throughput 3D spheroid culture and drug testing using a 384 hanging drop array. Analyst 2011;136:473–8.
[99] Jaeger AA, Das CK, Morgan NY, Pursley RH, McQueen PG, Hall MD, et al. Microfabricated polymeric vessel mimetics for 3-D cancer cell culture. Biomaterials 2013;34:8301–13.
[100] Rodenhizer D, Gaude E, Cojocari D, Mahadevan R, Frezza C, Wouters BG, et al. A three-dimensional engineered tumour for spatial snapshot analysis of cell metabolism and phenotype in hypoxic gradients. Nat Mater 2016;15:227–34.
[101] Tomaino B, Cappello P, Capello M, Fredolini C, Sperduti I, Migliorini P, et al. Circulating autoantibodies to phosphorylated alpha-enolase are a hallmark of pancreatic cancer. J Proteome Res 2011;10:105–12.
[102] Ayuso JM, Monge R, Martinez-Gonzalez A, Virumbrales-Munoz M, Llamazares GA, Berganzo J, et al. Glioblastoma on a microfluidic chip: generating pseudopalisades and enhancing aggressiveness through blood vessel obstruction events. Neuro Oncol 2017;19:503–13.
[103] Choi Y, Hyun E, Seo J, Blundell C, Kim HC, Lee E, et al. A microengineered pathophysiological model of early-stage breast cancer. Lab Chip 2015;15:3350–7.
[104] Sobrino A, Phan DT, Datta R, Wang X, Hachey SJ, Romero-Lopez M, et al. 3D microtumors in vitro supported by perfused vascular networks. Sci Rep 2016;6:31589.
[105] Timmins NE, Dietmair S, Nielsen LK. Hanging-drop multicellular spheroids as a model of tumour angiogenesis. Angiogenesis 2004;7:97–103.
[106] Nath S, Devi GR. Three-dimensional culture systems in cancer research: focus on tumor spheroid model. Pharmacol Ther 2016;163:94–108.
[107] Swaminathan V, Mythreye K, O'Brien ET, Berchuck A, Blobe GC, Superfine R. Mechanical stiffness grades metastatic potential in patient tumor cells and in cancer cell lines. Cancer Res 2011;71:5075–80.
[108] Tibbitt MW, Anseth KS. Hydrogels as extracellular matrix mimics for 3D cell culture. Biotechnol Bioeng 2009;103:655–63.
[109] Grimes DR, Currell FJ. Oxygen diffusion in ellipsoidal tumour spheroids. J R Soc Interface 2018;15.
[110] Grimes DR, Fletcher AG, Partridge M. Oxygen consumption dynamics in steady-state tumour models. R Soc Open Sci 2014;1:140080.

[111] Mueller-Klieser W. Method for the determination of oxygen consumption rates and diffusion coefficients in multicellular spheroids. Biophys J 1984;46:343–8.

[112] Derda R, Laromaine A, Mammoto A, Tang SK, Mammoto T, Ingber DE, et al. Paper-supported 3D cell culture for tissue-based bioassays. Proc Natl Acad Sci U S A 2009;106:18457–62.

[113] Kim SH, Lee HR, Yu SJ, Han ME, Lee DY, Kim SY, et al. Hydrogel-laden paper scaffold system for origami-based tissue engineering. Proc Natl Acad Sci U S A 2015;112:15426–31.

[114] Kenney RM, Boyce MW, Truong AS, Bagnell CR, Lockett MR. Real-time imaging of cancer cell chemotaxis in paper-based scaffolds. Analyst 2016;141:661–8.

[115] Verbridge SS, Choi NW, Zheng Y, Brooks DJ, Stroock AD, Fischbach C. Oxygen-controlled three-dimensional cultures to analyze tumor angiogenesis. Tissue Eng Part A 2010;16:2133–41.

[116] Xu H, Rahimpour S, Nesvick CL, Zhang X, Ma J, Zhang M, et al. Activation of hypoxia signaling induces phenotypic transformation of glioma cells: implications for bevacizumab antiangiogenic therapy. Oncotarget 2015;6:11882–93.

[117] Businaro L, De Ninno A, Schiavoni G, Lucarini V, Ciasca G, Gerardino A, et al. Cross talk between cancer and immune cells: exploring complex dynamics in a microfluidic environment. Lab Chip 2013;13:229–39.

[118] Yu J, Berthier E, Craig A, de Groot TE, Sparks S, Ingram PN, et al. Reconfigurable open microfluidics for studying the spatiotemporal dynamics of paracrine signalling. Nat Biomed Eng 2019;.

[119] Brat DJ, Castellano-Sanchez AA, Hunter SB, Pecot M, Cohen C, Hammond EH, et al. Pseudopalisades in glioblastoma are hypoxic, express extracellular matrix proteases, and are formed by an actively migrating cell population. Cancer Res 2004;64:920–7.

[120] Jeon JS, Bersini S, Gilardi M, Dubini G, Charest JL, Moretti M, et al. Human 3D vascularized organotypic microfluidic assays to study breast cancer cell extravasation. Proc Natl Acad Sci U S A 2015;112:214–19.

[121] Yi HG, Jeong YH, Kim Y, Choi YJ, Moon HE, Park SH, et al. A bioprinted human-glioblastoma-on-a-chip for the identification of patient-specific responses to chemoradiotherapy. Nat Biomed Eng 2019;3:509–19.

[122] Liberzon A, Subramanian A, Pinchback R, Thorvaldsdottir H, Tamayo P, Mesirov JP. Molecular signatures database (MSigDB) 3.0. Bioinformatics 2011;27:1739–40.

[123] Subramanian A, Tamayo P, Mootha VK, Mukherjee S, Ebert BL, Gillette MA, et al. Gene set enrichment analysis: a knowledge-based approach for interpreting genome-wide expression profiles. Proc Natl Acad Sci U S A 2005;102:15545–50.

[124] Xia X, Kung AL. Preferential binding of HIF-1 to transcriptionally active loci determines cell-type specific response to hypoxia. Genome Biol 2009;10:R113.

[125] Leithner K, Wohlkoenig C, Stacher E, Lindenmann J, Hofmann NA, Galle B, et al. Hypoxia increases membrane metallo-endopeptidase expression in a novel lung cancer ex vivo model - role of tumor stroma cells. BMC Cancer 2014;14:40.

[126] Olbryt M, Jarzab M, Jazowiecka-Rakus J, Simek K, Szala S, Sochanik A. Gene expression profile of B 16(F10) murine melanoma cells exposed to hypoxic conditions in vitro. Gene Expr 2006;13:191–203.

[127] Wang Y, Roche O, Yan MS, Finak G, Evans AJ, Metcalf JL, et al. Regulation of endocytosis via the oxygen-sensing pathway. Nat Med 2009;15:319–24.

[128] Tekin H, Simmons S, Cummings B, Gao L, Adiconis X, Hession CC, et al. Effects of 3D culturing conditions on the transcriptomic profile of stem-cell-derived neurons. Nat Biomed Eng 2018;2:540–54.

[129] Patel AP, Tirosh I, Trombetta JJ, Shalek AK, Gillespie SM, Wakimoto H, et al. Single-cell RNA-seq highlights intratumoral heterogeneity in primary glioblastoma. Science 2014;344:1396–401.

[130] Rodriques SG, Stickels RR, Goeva A, Martin CA, Murray E, Vanderburg CR, et al. Slide-seq: a scalable technology for measuring genome-wide expression at high spatial resolution. Science 2019;363:1463–7.

[131] Goyama S, Shrestha M, Schibler J, Rosenfeldt L, Miller W, O'Brien E, et al. Protease-activated receptor 1 (PAR-1) inhibits proliferation but enhances leukemia stem cell activity in acute myeloid leukemia. Blood 2016;128.

[132] Koshiji M, To KK, Hammer S, Kumamoto K, Harris AL, Modrich P, et al. HIF-1alpha induces genetic instability by transcriptionally downregulating MutSalpha expression. Mol Cell 2005;17:793–803.

[133] Erler JT, Giaccia AJ. Lysyl oxidase mediates hypoxic control of metastasis. Cancer Res 2006;66:10238–41.

[134] Frenzel A, Grespi F, Chmelewskij W, Villunger A. Bcl2 family proteins in carcinogenesis and the treatment of cancer. Apoptosis 2009;14:584–96.

[135] Sermeus A, Genin M, Maincent A, Fransolet M, Notte A, Leclere L, et al. Hypoxia-induced modulation of apoptosis and BCL-2 family proteins in different cancer cell types. PLoS One 2012;7:e47519.

[136] Sottoriva A, Spiteri I, Piccirillo SG, Touloumis A, Collins VP, Marioni JC, et al. Intratumor heterogeneity in human glioblastoma reflects cancer evolutionary dynamics. Proc Natl Acad Sci U S A 2013;110:4009–14.

[137] Bachtiary B, Boutros PC, Pintilie M, Shi W, Bastianutto C, Li JH, et al. Gene expression profiling in cervical cancer: an exploration of intratumor heterogeneity. Clin Cancer Res 2006;12:5632–40.

[138] Ng CK, Martelotto LG, Gauthier A, Wen HC, Piscuoglio S, Lim RS, et al. Intra-tumor genetic heterogeneity and alternative driver genetic alterations in breast cancers with heterogeneous HER2 gene amplification. Genome Biol 2015;16:107.

[139] Varia MA, Calkins-Adams DP, Rinker LH, Kennedy AS, Novotny DB, Fowler Jr. WC, et al. Pimonidazole: a novel hypoxia marker for complementary study of tumor hypoxia and cell proliferation in cervical carcinoma. Gynecol Oncol 1998;71:270–7.

[140] Vaupel P, Mayer A. Hypoxia in cancer: significance and impact on clinical outcome. Cancer Metastasis Rev 2007;26:225–39.

[141] Janssen HL, Hoebers FJ, Sprong D, Goethals L, Williams KJ, Stratford IJ, et al. Differentiation-associated staining with anti-pimonidazole antibodies in head and neck tumors. Radiother Oncol 2004;70:91–7.

[142] Zhou J, Schmid T, Schnitzer S, Brüne B. Tumor hypoxia and cancer progression. Cancer Lett 2006;237:10–21.

[143] Gürkan G, Gürkan K. Incu-stream 1.0: an open-hardware live-cell imaging system based on inverted bright-field microscopy and automated mechanical scanning for real-time and long-term imaging of microplates in incubator. Inst Electr Electron Eng Access 2019;7:58764–79.

[144] Walzik MP, Vollmar V, Lachnit T, Dietz H, Haug S, Bachmann H, et al. A portable low-cost long-term live-cell imaging platform for biomedical research and education. Biosens Bioelectron 2015;64:639–49.

[145] Kim SB, Koo KI, Bae H, Dokmeci MR, Hamilton GA, Bahinski A, et al. A mini-microscope for in situ monitoring of cells. Lab Chip 2012;12:3976–82.

[146] Clayton J. Automated label-free cell counting. Genet Eng Biotechnol N 2017;37:12–13.

[147] Sengupta D, Pratx G. Imaging metabolic heterogeneity in cancer. Mol Cancer 2016;15:4.

[148] Gatenby RA, Gillies RJ. Why do cancers have high aerobic glycolysis? Nat Rev Cancer 2004;4:891–9.

[149] Gambhir SS. Molecular imaging of cancer with positron emission tomography. Nat Rev Cancer 2002;2:683–93.

[150] Golman K, Zandt RI, Lerche M, Pehrson R, Ardenkjaer-Larsen JH. Metabolic imaging by hyperpolarized 13C magnetic resonance imaging for in vivo tumor diagnosis. Cancer Res 2006;66:10855–60.

[151] Nelson SJ, Kurhanewicz J, Vigneron DB, Larson PE, Harzstark AL, Ferrone M, et al. Metabolic imaging of patients with prostate cancer using hyperpolarized [1-(13)C]pyruvate. Sci Transl Med 2013;5:198ra108.

[152] Rodrigues TB, Serrao EM, Kennedy BW, Hu DE, Kettunen MI, Brindle KM. Magnetic resonance imaging of tumor glycolysis using hyperpolarized 13C-labeled glucose. Nat Med 2014;20:93–7.

[153] Sud D, Mycek M-A. Calibration and validation of an optical sensor for intracellular oxygen measurements. J Biomed Opt 2009;14:1–3.

[154] Georgakoudi I, Quinn KP. Optical imaging using endogenous contrast to assess metabolic state. Annu Rev Biomed Eng 2012;14:351–67.

[155] Heikal AA. Intracellular coenzymes as natural biomarkers for metabolic activities and mitochondrial anomalies. Biomarkers Med 2010;4:241–63.

[156] Drozdowicz-Tomsia K, Anwer AG, Cahill MA, Madlum KN, Maki AM, Baker MS, et al. Multiphoton fluorescence lifetime imaging microscopy reveals free-to-bound NADH ratio changes associated with metabolic inhibition. J Biomed Opt 2014;19:086016.

[157] Zheng J. Energy metabolism of cancer: Glycolysis versus oxidative phosphorylation. Oncol Lett 2012;4:1151–7.

[158] Vergen J, Hecht C, Zholudeva LV, Marquardt MM, Hallworth R, Nichols MG. Metabolic imaging using two-photon excited NADH intensity and fluorescence lifetime imaging. Microsc Microanal 2012;18:761–70.

[159] Plitzko B, Loesgen S. Measurement of oxygen consumption rate (OCR) and extracellular acidification rate (ECAR) in culture cells for assessment of the energy metabolism. Bio-protocol 2018;8:e2850.

[160] Wu M, Neilson A, Swift AL, Moran R, Tamagnine J, Parslow D, et al. Multiparameter metabolic analysis reveals a close link between attenuated mitochondrial bioenergetic function and enhanced glycolysis dependency in human tumor cells. Am J Physiol - Cell Physiol 2007;292:125–36.

[161] Hirayama A, Kami K, Sugimoto M, Sugawara M, Toki N, Onozuka H, et al. Quantitative metabolome profiling of colon and stomach cancer microenvironment by capillary electrophoresis time-of-flight mass spectrometry. Cancer Res 2009;69:4918–25.

[162] Lee DY, Bowen BP, Northen TR. Mass spectrometry-based metabolomics, analysis of metabolite-protein interactions, and imaging. BioTechniques 2010;49:557–65.
[163] Prasad B, Garg A, Takwani H, Singh S. Metabolite identification by liquid chromatography-mass spectrometry. TrAC Trends Anal Chem 2011;30:360–87.
[164] Wilson ID. High-performance liquid chromatography-mass spectrometry (HPLC-MS)-based drug metabolite profiling. Methods Mol Biol 2011;708:173–90.
[165] Kang YP, Ward NP, DeNicola GM. Recent advances in cancer metabolism: a technological perspective. Exp Mol Med 2018;50:31.

CHAPTER

# 28

# Tissue engineering strategies for the treatment of skeletal maxillofacial defects resulting from neoplasms resections

*J.P. Ribeiro*[1,2], *Esteban A. Astudillo-Ortiz*[1,2,3], *Pedro S. Babo*[1,2] *and Manuela E. Gomes*[1,2]

[1]Research Group, I3B's—Research Institute on Biomaterials, Biodegradables and Biomimetics, University of Minho, Headquarters of the European Institute of Excellence on Tissue Engineering and Regenerative Medicine, Guimarães, Portugal [2]ICVS/3B's–PT Government Associate Laboratory, Braga/Guimarães, Portugal [3]GIRO Research Group in Oral Rehabilitation, University of Cuenca, Cuenca, Ecuado

## 28.1 Background

### 28.1.1 Oral and maxillofacial neoplasms

Oral and maxillofacial neoplasms are malignancies affecting the oral mucosa or underlying structures in the area between the vermillion border of the lip and the facial pillars superolaterally, and the base of the tongue inferiorly. Other structures typically involved include the maxilla, mandible, other facial skeleton and skin, and salivary glands.

The World Health Organization (WHO) estimates that annually about 657,000 new cases of cancers of the oral cavity and pharynx occur worldwide (Fig. 28.1), causing more than 330,000 deaths.

Skeletal maxillofacial neoplasms (SMNs) are particularly pernicious as both their presence and treatment present significant cosmetic and functional limitations, reducing patients quality of life [1]. These are often asymptomatic for long time, and very rarely cause pain, being difficult to detect in early stages even using radiologic tools [2]. The WHO, in their last

*Biomaterials for 3D Tumor Modeling*
DOI: https://doi.org/10.1016/B978-0-12-818128-7.00028-9

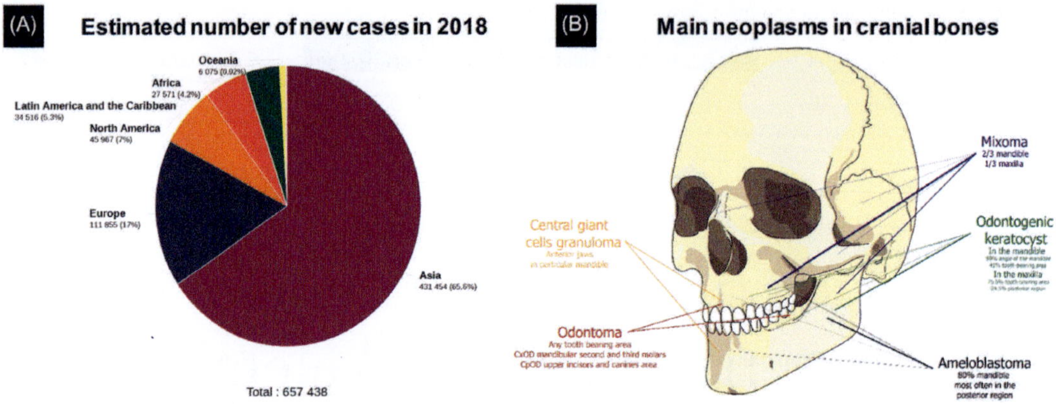


FIGURE 28.1 **Oral and maxillofacial cancer prevalence.** (A) Estimated number of new cases of new oral cavity and pharynx cancer per continent in 2018, for both sexes and all ages. (B) Distribution of the main neoplasms affecting cranial bones. The thickness of the lines indicates its greater or lesser prevalence according to the literature. Source: *(A) Globocan2018; Graph production: Global Cancer Observatory (http://gco.iarc.fr) IARC.*

classification of Head and Neck Tumors in 2017 [3], considered 13 groups of odontogenic and maxillofacial bone tumors involving around 62 diseases. Most of them are considered rare, sporadic, unknown, or only a few cases have been reported worldwide.

The most prevalent SMNs, summarized in Table 28.1, include myxoma, ameloblastoma, odontoma, odontogenic keratocyst, and central giant cells glioblastoma, and can be of dental origin (odontogenic) or arise from bony tissues. The benign or malignant character of bone maxillofacial neoplasms is based on specific histological criteria, including the presence or absence of necrosis, mitotic figures as well as basic understanding of the entity [4].

#### 28.1.1.1 *Myxoma*

Myxoma are a group of rare benign tumors of connective-tissue origin that occur in both hard (central) and soft tissues of the body (Fig. 28.2) [5,6]. The treatment to myxomas consists in wide surgical excision with a 2.0 cm margin of bone, resulting in most of the excisions of mandibular myxomas in segmental mandibular resection (76.3%) or hemimandibulectomy (10.5%) [7].

#### 28.1.1.2 *Ameloblastoma*

Ameloblastoma is a benign intraosseous progressively growing epithelial odontogenic neoplasm characterized by expansion and tendency of local recurrence if not adequately removed [3]. Ameloblastoma can be associated with unerupted third molar teeth (Fig. 28.3) [8].

Treatment can be conservative or radical. Enucleation, curettage, and surgical excision are considered as part of the conventional treatment. Radical treatment consists of bone resection [9].

TABLE 28.1 Summary of the most prevalent maxillary neoplasms.

| Name | Definition | Epidemiology | Localization | Treatment | Prognosis |
|---|---|---|---|---|---|
| Myxoma | Spindle-shaped cells dispersed in abundant myxoid ECM | 2%–5% of all the tumors | 2/3 mandible 1/3 maxilla | Curettage/ complete excision with free margins | RRA 25% |
| Ameloblastoma | Benign intraosseous progressively growing epithelial odontogenic neoplasm | 1% of all the tumors | 80% mandible, most often in the posterior region | Wide surgical excision | RRA 60%–80% 50% within 5 years |
| Odontoma | Tumor-like malformation (hamartoma) composed of dental hard and soft tissues | Most common odontogenic tumors | Any tooth-bearing area | Conservative surgery | Unusual recurrence |
| Odontogenic keratocyst | Odontogenic cyst with a regular lining of parakeratinized stratified squamous epithelium with palisading hyperchromatic basal cells | 10%–20% of the odontogenic cysts. The third most common cyst of the jaws | 80% mandible | Enucleation or surgical resection for large lesions | RRA 25% |
| Central giant cells granuloma | Osteolytic lesion of the jaws characterized by osteoclast-type giant cells in a vascular stroma | 10% of benign gnathic tumors | More frequent in the anterior jaws, in particular, the mandible | Local curettage | Higher recurrence rate |

*ECM*, Extracellular matrix; *RRA*, recurrence rate average.

#### 28.1.1.3 *Odontoma*

Odontoma are calcifying benign tumors of odontogenic origin characterized by their slow growth [10,11]. They consist of enamel, dentine, cementum, and pulpal tissue [3,10].

Odontoma are frequently associated with an unerupted teeth and are usually detected on routine radiographs (Figs. 28.4 and 28.5) [3]. The traditional treatment consists of surgical elimination, and there is a small likelihood of recurrence [11].

#### 28.1.1.4 *Odontogenic keratocyst*

The odontogenic keratocyst is a locally aggressive, cystic jaw lesion with a putatively high-growth potential and a propensity for recurrence (Fig. 28.6) [12]. Treatment is usually the enucleation or surgical resection of large lesions [3,13].

#### 28.1.1.5 *Central giant cells granuloma*

Central giant cell granuloma (CGCG) is benign but sometimes aggressive lesion of the maxillofacial area, characterized by osteoclast-type giant cells in a vascular stroma (Fig. 28.7) [3,14]. The treatment of this lesion depends on the severity of the lesion. Usually the surgical enucleation with curettage or resection is recommended [15]. Conservative therapies can also be used in combination with the surgical treatment, or as alternative,

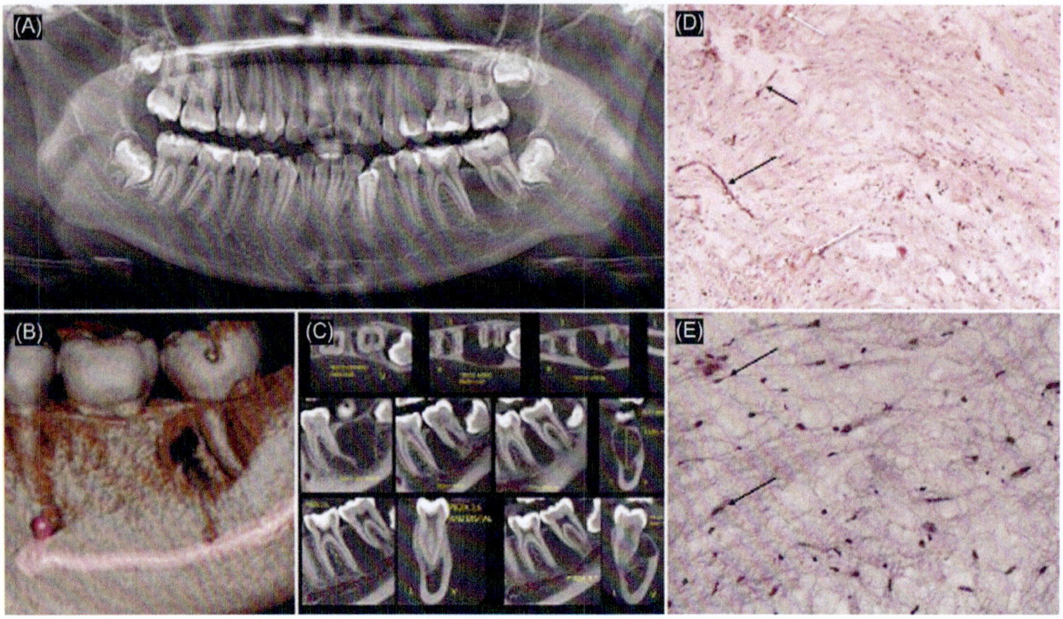


FIGURE 28.2 **Myxoma in a 16 years old patient.** (A) The panoramic radiograph shows a significant lesion between the left mandibular first and second molar. (B) The 3D tomographic reconstruction evidenced a substantial bone loss that involves the mesial root of the mandibular second molar. (C) Different tomographic cuts display the size and configuration of the lesion. (D) Histopathologic analysis at 100 × showing rests of odontogenic epithelium (*black arrows*) with fibrous tissue (*white arrows*). (E) Optical micrograph at 400 ×, showing spindle-shaped cells (*black arrows*) in a myxoid stroma. Source: *(C) Courtesy Dr. Geovanny Vanegas Abad. (E) Courtesy Dr. Fernanda Torres Calle.*

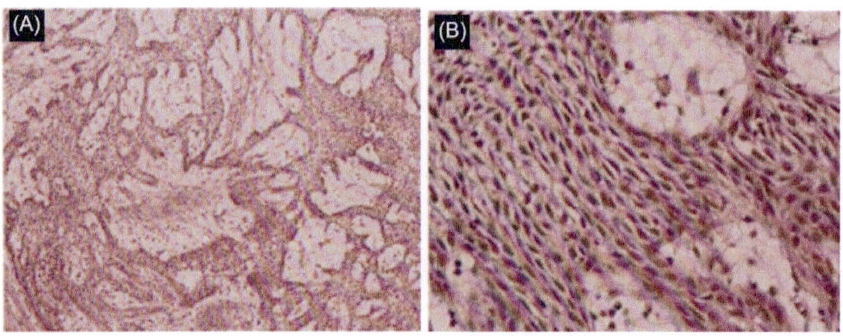


FIGURE 28.3 **Unicystic ameloblastoma. Histopathological analysis at (A) 100 × and (B) 400 ×, respectively, showing epithelial invasion of the ameloblastoma capsule in a plexiform disposition.** Source: *Courtesy Dr. Fernanda Torres Calle.*

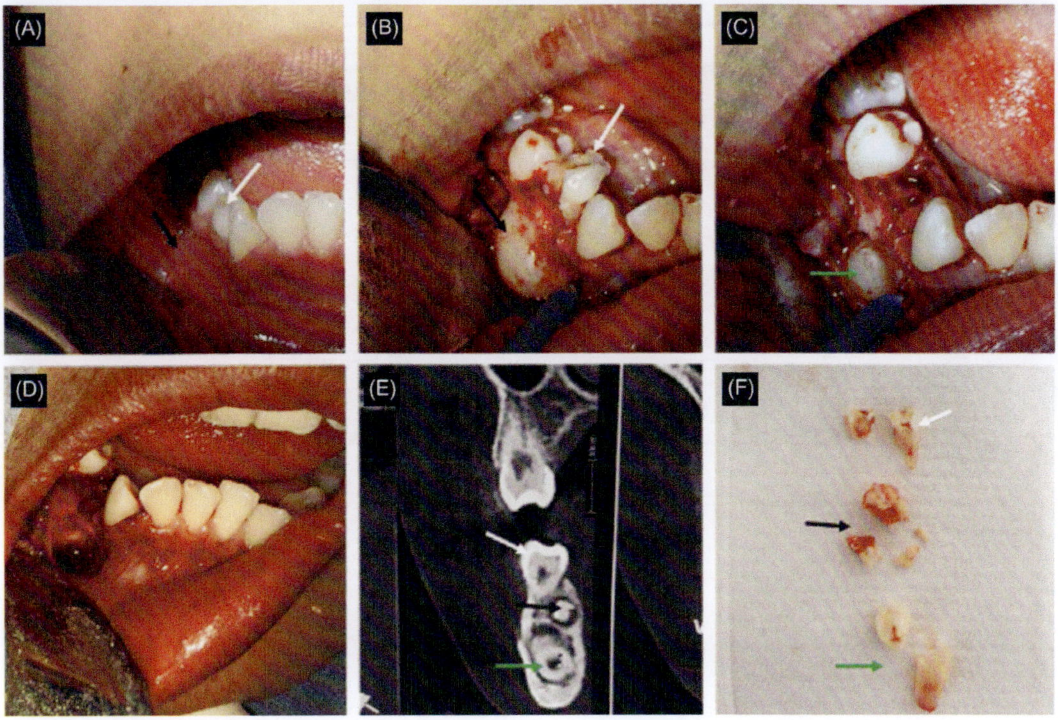


FIGURE 28.4 **Complex type odontoma.** (A) Initial situation: temporary tooth (*white arrow*), odontoma (*black arrow*). (B) Incision and flap lift. Increased vestibular bone table volume. (C) Included premolar (*green arrow*). (D) Surgical bed after lesion enucleation. (E) Tomographic cuts. (F) Odontoma fragments. *Source: Courtesy Dr. Paúl Guillén Guerrero.*

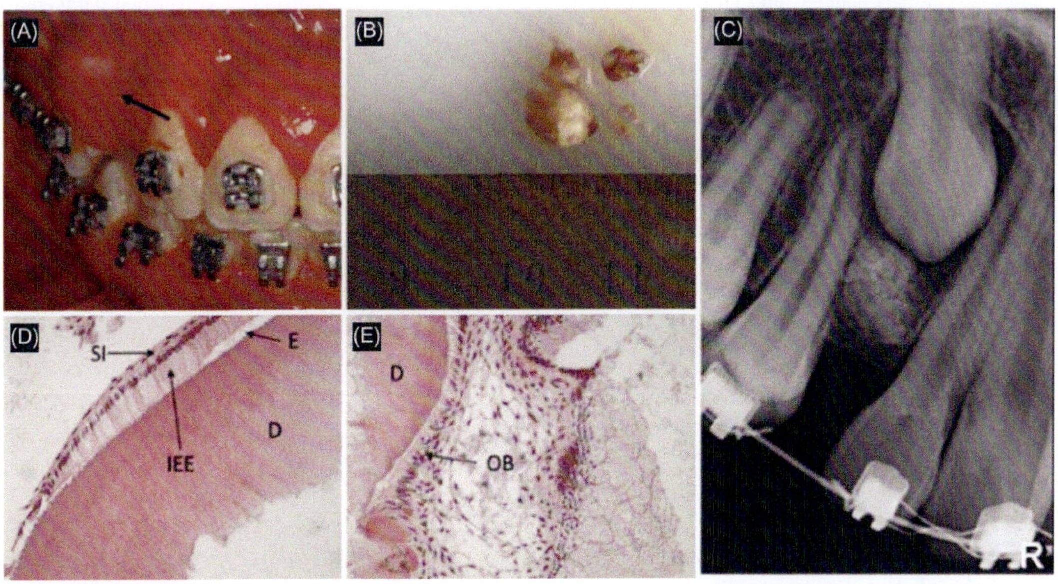


FIGURE 28.5 **Compound type odontoma.** (A) Initial situation, swelling in the gingival mucosa (*arrow*). (B) Enucleated odontoma. (C) Periapical X-ray showing the odontoma blocking the canine eruption. (D and E) Histopathological analysis shows developmental tissues as stratum intermedium (SI), internal enamel epithelium (IEE), odontoblasts (OB), enamel (E), and dentine (D). Source*: Courtesy Dr. Fernanda Torres and Dr. Andrés Vega Yepez.*

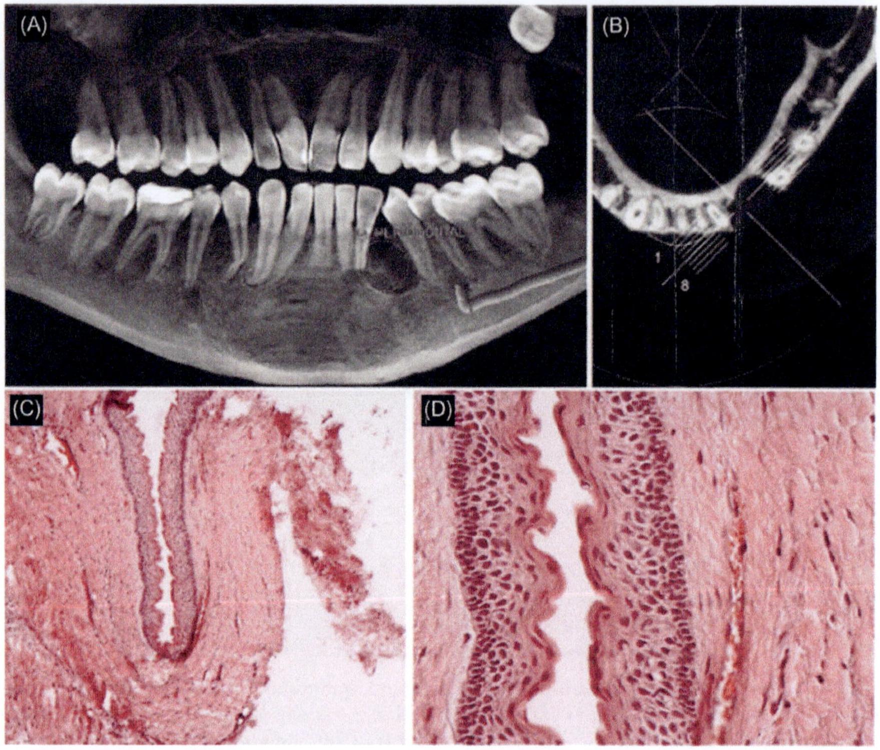


FIGURE 28.6 **Odontogenic keratocyst.** (A) Radiograph of a mandibular OKC, the lesion is placed between the left canine and mandibular first premolar. (B) Tomographic axial cut shows the dimension of the cyst that perforates the compact bone. (C and D) Epithelial lining of OKC shows typical features of a corrugated parakeratinized surface layer and a regular layer of columnar basal cells with nuclear palisading at 10 × and 40 ×. Source: *Courtesy Dr. Fernanda Torres Calle.*

and include the administration of corticosteroids, calcitonin, interferon, or bisphosphonates [15].

## 28.1.2 Currently used therapies

The traditional clinical options to treat these tumors can be more or less conservative, depending on size, location, and malignancy [16]. Considering Hupp [16], the three primary modalities of surgical excision of maxillary tumors are enucleation, segmental, or partial resection, and compound resection.

Smaller bony defects (<4–6 cm) are commonly treated with enucleation followed by filling of the defect with nonvascularized cortico-cancellous grafts harvested from the anterior or posterior iliac crest [17]. On the other hand, when it has been determined that the lesion is aggressive by histopathological examination, excision can be accompanied by resection of the lesion and adjacent bone margins (Fig. 28.8) [16]. Finally, when the

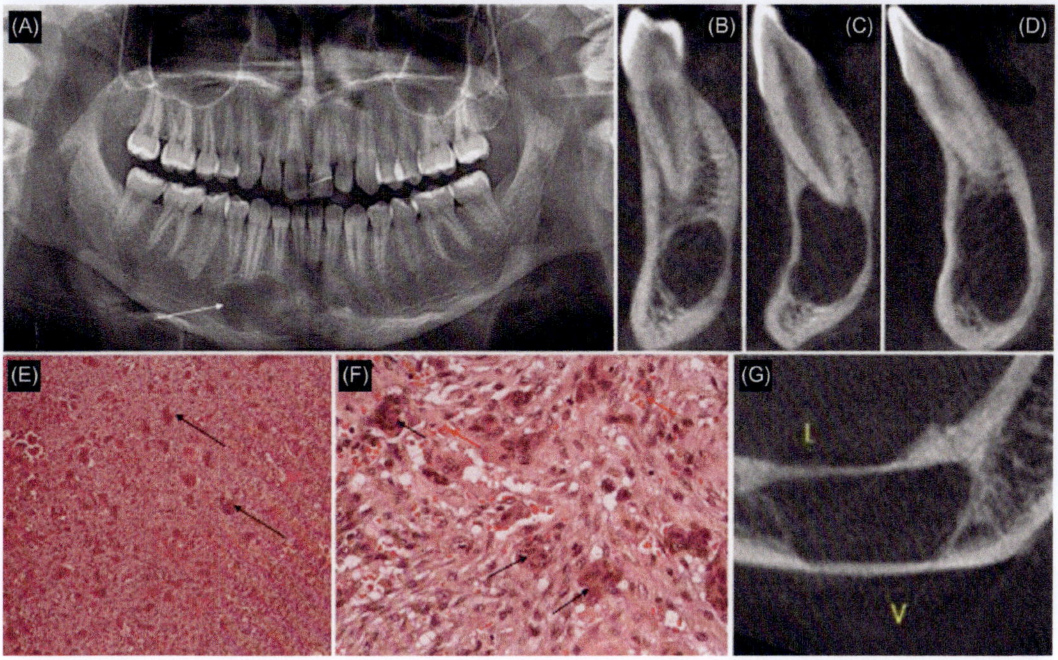


**FIGURE 28.7** **Central giant cells granuloma.** (A) Panoramic radiograph showing a radiolucent area in the canine region. (B–D) Tomographic sagittal cut exhibit a well-circumscribed radiolucency in the anterior mandible. (E and F) Microscopic analysis at 10 × and 40 ×, respectively, of young connective tissue with proliferation of cylindrical and polyhedral cells and foreign body type giant cells (*black arrow*) and erythrocytes infiltrate (*red arrows*). (G) Tomographic axial cut shows a well-circumscribed lesion that occupies the entire width of the spinal bone. Source: *(B–D, G) Courtesy Dr. Geovanny Vanegas Abad. (E and F) Courtesy Dr. Fernanda Torres Calle.*

malignancy or the recurrence rate is very high, the treatment includes total resection of the tumor by removing all of the affected bone [16], sometimes accompanied with an adjuvant chemo- or radiotherapy treatment [3]. The loss of mandibular bone continuity due to neoplasm resections creates "socially mutilated" patients due to the significant loss of function and aesthetics [18]. Likewise, the loss of a part of the jaw creates a communication between the maxillary sinuses or nasal cavity with the oral cavity, which causes great difficulties in speaking or eating. Therefore, after the tumors ablation, patients have to be submitted to subsequent reconstruction surgeries to rehabilitate the function and the aesthetics of the tissues, or, in most extreme cases, to prepare the tissues to receive the prosthetics implants. The "gold standard" for the reconstruction of large bone defects is the autologous bone graft. Usually, it requires the collection of a vascularized bone flap from tibia or fibula [19], iliac crest [20], or ribs [21] which is then adapted to the defect and properly fixed to the preserved bone (Figs. 28.8 and 28.9). Further than the patient discomfort, and risk of donor site morbidity, these procedures have limited aesthetic recovery and represent a heavy burden on national healthcare systems worldwide [20]. Moreover, adverse effects of chemo- and radiotherapies can postpone surgical reconstruction

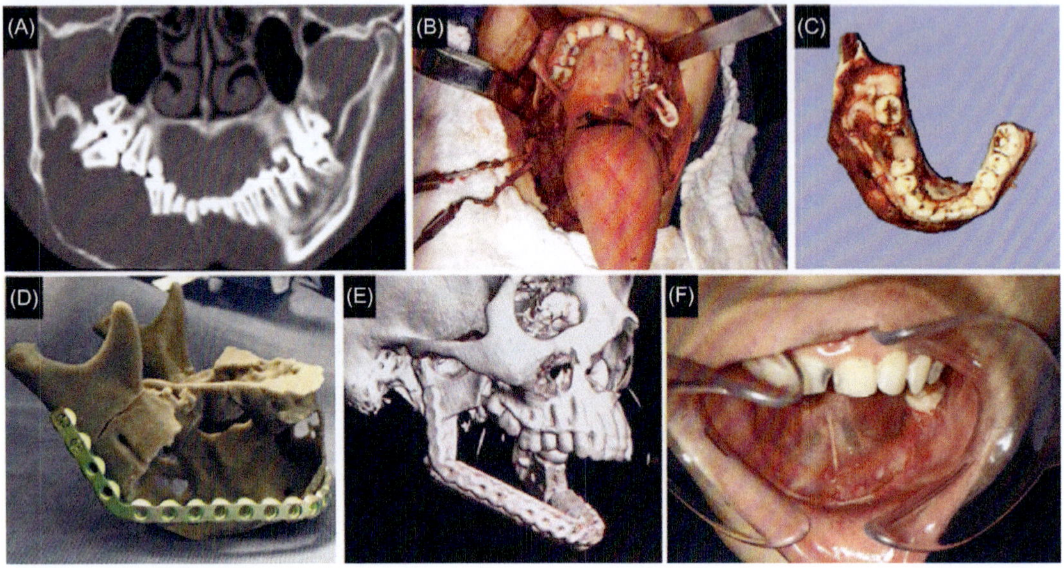

FIGURE 28.8 **Treatment of an ameloblastoma with block resection.** (A) Coronal cut of a tomography showing a significant ameloblastoma lesion. (B) Mandibular block resection with safety margins. (C) Mandibular resected block. (D) 3D-printed surgical guide for mandible reconstruction. (E) Tomography of the reconstruction with vascularized fibula free flap. (F) Control at 21 days. Source: *Courtesy Dr. Sid Brahim Mohamed Salek.*

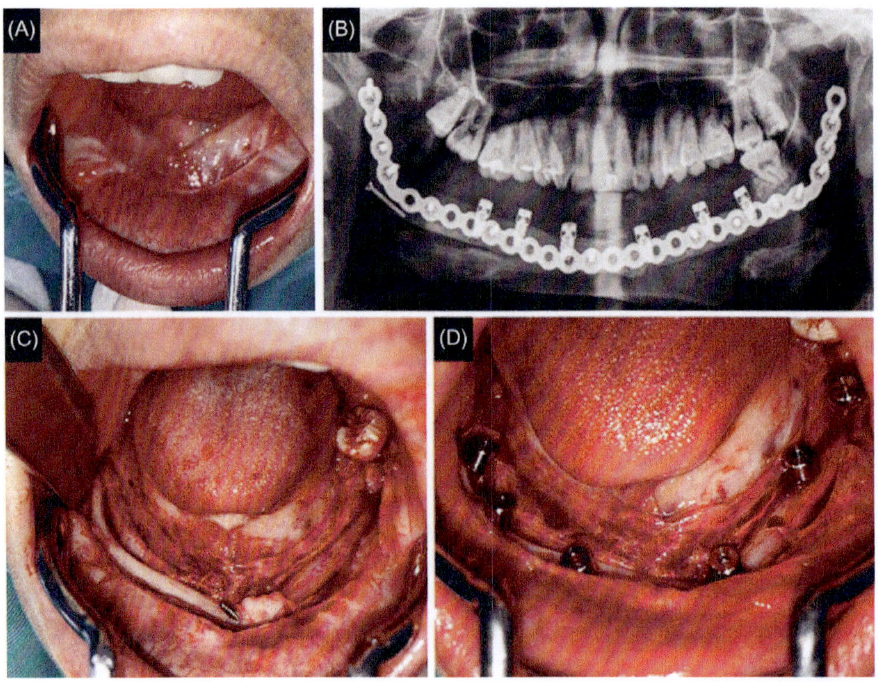

FIGURE 28.9 **Implant-based oral rehabilitation.** (A and C) Control after eleven months. (B and D) Dental implants placed in the bone graft. Source: *Courtesy Dr. Sid Brahim Mohamed Salek.*

procedures due to complications such as thrombocytopenia, which increases patient's risk for hemorrhaging, and granulocytopenia, which increases the susceptibility for infection [25].

Some alternatives have been used for small bone defects such as guided bone regeneration (GBR). It consists on the use of barrier membranes to protect the blood clot to promote bone first intentional repair, while preventing infection and soft tissue collapse into the defect, avoiding the colonization of the defect with epithelial cells or fibroblasts [26,27]. However, oral and maxillofacial surgeons and dentists still face serious difficulties in reconstructing bone maxillofacial defects after tumor resection and in the osteointegration of implantation devices in patients subjected to radiotherapy.

## 28.2 Tissue engineering for reconstruction of ablated skeletal maxillofacial tissues

The tissue engineering (TE) is a multidisciplinary discipline which, during the last three decades, has explored the principles of biology and engineering to the development of functional substitutes for damaged tissues [28]. The most traditional TE concept consists of three main pillars (Fig. 28.10): cells, scaffolds, and growth factors (GFs) [31]. Along the years, it has been shown that there are several other complements to these pillars which can enhance the regeneration of the desired tissue. For example, the use of bioreactors to culture TE constructs under specific stimuli has been explored for better mimic the natural physico-chemical environment of native tissue [32].

The oral and maxillofacial regions are quite complex since they consist of several tissues namely, bone, cartilage and soft tissues, nerves, and blood vessels. The restoration of functional and aesthetics demands of the skeletal maxillofacial tissues after oncologic surgery is one of the major challenges for bone TE [34]. In addition to the surgical treatment, the patients may need to be subject to radiotherapy and/or chemotherapy which compromises patient's tissues self-regenerative potential, jeopardizing the success of oral and maxillofacial rehabilitation [29]. Therefore, clinicians have procured inspiration in the most recent advances in bone TE field to find solutions to the surgical rehabilitation of skeletal maxillofacial tissues.

The TE approaches for oral and maxillofacial regeneration besides biocompatible, should be osteoconductive, osteoinductive, and osteogenic [33,35], that is, should (1) act as a provisional template for new bone growth, while (2) stimulate osteoprogenitor cells to develop into preosteoblasts or bone-forming cell lineage [35,37]; (3) but also provide stable anchorage between the living bone and the implant [37]. This has been endured by the right combination of scaffolds, cells, GFs, and culturing conditions, and, in the past few years, the exploration of the emergent microfabrication and three-dimensional (3D) printing fields (Table 28.2).

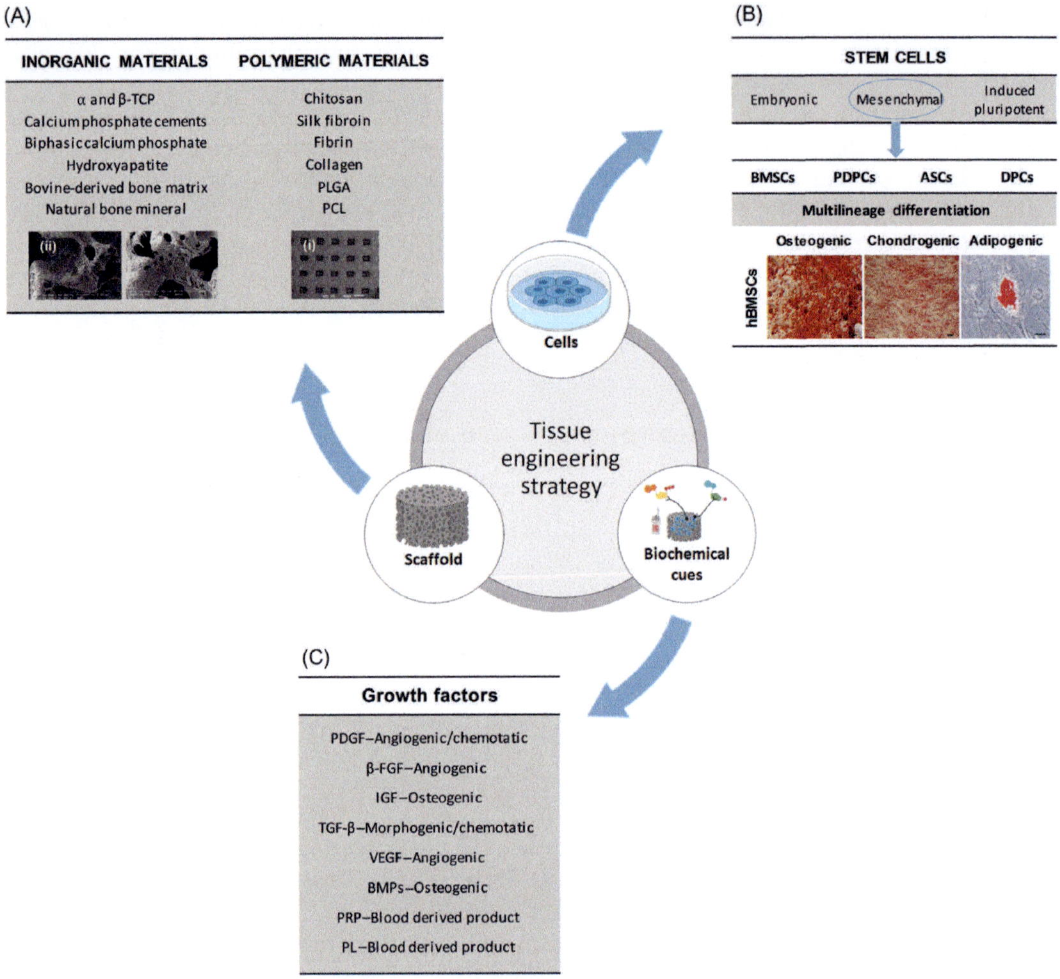


**FIGURE 28.10 Schematic of the three pillars for bone tissue engineering.** (A) Scaffolds from inorganic origin, namely (i) ceramic meshes (β-tricalcium phosphate) [22] or organic origin, such as (ii) 3D printing polymeric scaffold (polycaprolactone) [23]; (B) stem cells, in particular, multipotent mesenchymal stem cells isolated from different tissues—bone marrow derived stem cells (BMSCs), periosteal-derived progenitor cells (PDPCs), ASCs, and dental pulp stem cells (DPCs) (trilineage differentiation example); and (C) biochemical cues including growth factors and blood derivatives. *BMP*, Bone morphogenetic protein; *FGF*, fibroblast growth factor; *IGF*, insulin-like growth factor; *PDGF*, platelet-derived growth factor; *PL*, platelet lysate; *PRP*, platelet-rich plasma; *TCP*, tricalcium phosphate; *TGF*, transforming growth factor; *VEGF*, vascular endothelial growth factor. Source*: (B) Adapted from D. Chen, X. Zhang, Y. He, J. Lu, H. Shen, Y. Jiang, et al., Co-culturing mesenchymal stem cells from bone marrow and periosteum enhances osteogenesis and neovascularization of tissue-engineered bone, J Tissue Eng Regen Med 2012;6 (10):822–32.[24].*

**TABLE 28.2** Tissue engineering strategies for bone maxillofacial regeneration.

| Clinical target | Scaffold | Cell type | Biochemical cues | Type of study | Experimental design | Outcomes | References |
|---|---|---|---|---|---|---|---|
| Large mandibular defects | β-TCP granules | ASCs | BMP-2 | Clinical report | ASCs were combined with β-TCP granules and BMP-2 and cultured in basal medium for 3 weeks. The whole was implanted in a parasymphyseal defect created by ameloblastoma resection. | Bone regeneration 10 months after implantation allowed prosthodontic rehabilitation in the grafted site. | [17] |
| | | BMSCs-hBMP-2 | BMP-2 | In vitro and in vivo | MSCs were genetically modified with a recombinant human adenovirus.<br>Type 5 encoding human BMP-2 and later implanted onto ectopic and mandibular defects in (NOD/SCID) mice. | Large mandibular defects. | [128] |
| | β-TCP granules (Chronos) | ASCs | BMP-2 | Clinical report | ASCs were mixed with the BMP-2 laden β-TCP granules and implanted in the rectus abdominis muscle for maturation. Then after 8 months, the microvascular custom-made ectopic bone flap was transplanted to a hemimaxillectomy defect after keratocysts resection. | 8 months of after implantation the construct produced a well-ossified and vascularized bone graft ectopically. | [48] |
| | PCL/β-TCP (1:1) 3D-printed scaffold | BMSCs | | In vivo | BMSCs were seeded into the scaffolds and incubated in a rotational oxygen-permeable bioreactor system for 14 days.<br><br>Constructs were implanted into mandibular 2 × 2 cm defects in minipigs. | Implanted BMSCs enhanced new bone formation and graft osteointegration 8 weeks after implantation. | [30] |

(*Continued*)

TABLE 28.2 (Continued)

| Clinical target | Scaffold | Cell type | Biochemical cues | Type of study | Experimental design | Outcomes | References |
|---|---|---|---|---|---|---|---|
| Maxillofacial bone regeneration | BBM + PRP | ASCs | PRP | In vitro and in vivo | Preosteogenic differentiated ASCs carried in BBM mixed with PRP were subcutaneously implanted in subcutaneous model in Balb/c mice. | Increased expression of bone-related genes in a callus-like tissue ectopically, 8 weeks after implantation. | [46] |
| | 3D porous PS scaffolds (Alvetex) | ASCs | | In vitro | ACSs spheroids or ASCs seeded PS scaffolds were cultured in OM for 14 days. | Increased osteogenic differentiation, maturation and matrix mineralization by ASCs in PS scaffolds. | [166] |
| | PLGA/PCL fibrous 3D scaffolds | ASCs | BMP-2 | In vitro | Rat ASCs seeded in PLGA/PCL–BMP-2 were cultured in OM over 14 days. | 3D scaffold incorporated with BMP-2 significantly increased proliferation and osteogenic differentiation of rat ASCs. | [127] |
| | Deciduous red deer antler | Mice-BMSCs | BMP-2 TGF-β | In vitro and in vivo | Mice-BMSCs were seeded in slices of deer antler or OsteoSet and preculture in OM or in BMP-2 + TGF-β supplemented OM. Implantation in bone defects surgically induced on the left parietal bone of CD-1 mice. | Tissue-engineered grafts were proven to be more efficient in bone reconstruction than alloplastic commercial biomaterials (OsteoSet). | [40] |
| | CPC/CHT | BMSCs | | In vitro | BMSCs seeded over CPC/CHT constructs and cultured in OM. | CHT enhanced cell the mechanical properties of CPC, cell adhesion, proliferation, and osteogenic differentiation. | [49] |
| | BCP/CHT | BMSCs | Gelatin microspheres laden with FGF | In vitro and in vivo | Cells seeded over porous BCP/CHT scaffolds incorporating FGF-laden gelatin microspheres were cultured in vitro in osteoblastic conditions before implantation in transcortical mandibular defects in skeletally mature female Yorkshire pigs. | FGF release enhanced new vascularization in the defect area and new bone formation. | [50] |

| | | | | | | |
|---|---|---|---|---|---|---|
| | HUVECs BMSCs | | In vitro and in vivo | Spheroids (2% HUVECs and 98% BMSCs) were cultured under osteogenic and endothelial medium before subcutaneous implantation in the dorsal region of two male nude mice. | Enhanced osteogenic differentiation associated to prevascularized spheroids of HUVECs/BMSCs co-cultures. | [122] |
| Hap | Periosteal graft | | In vivo | Four different implants were inserted in premaxillary bone defects in rats; (1) premaxillar mucoperiosteal graft; (2) Hap/premaxillar mucoperiosteal graft; (3) Hap/femoral periosteal graft; (4) and femoral periosteal graft. | The periosteal graft provides satisfactory support to the Hap implant, allowing the growth of new bone. | [91] |
| | Periosteal graft | PRP | Clinical report | Membranous mandibular periosteum cultured in basal medium were applied in severe human alveolar bone defects. | Periosteal explant transplantation increased bone regeneration. | [167] |
| BBM | PDLCs | PDGF-BB hBMP-2 | In vitro | PDLCs seeded over BBM were cultured under PDGF-BB or rhBMP-2 supplemented media. | PDGF-BB improved cell proliferation, while rhBMP-2 did not support the cell mitosis. | [45] |
| Silk fibroin/CHT (SFCS scaffolds) | Periosteal graft | | In vivo | Autologous bone or SFCS were implanted over periosteal grafts placed over the latissimus dorsi muscle of adult sheep. | SFCS combined with periosteal graft produced comparable ectopic vascularized bone grafts as the autologous bone control. | [58] |
| PLGA fleeces | PDPCs | | In vitro and in vivo | Rabbit PDPCs seeded in PLGA fleeces were cultured in vitro in ONM. The tissue constructs were also implanted in critical | PDPCs/PLGA constructs expressed osteogenic markers in vitro and induced new bone grew not only from the margins | [54] |

(*Continued*)

TABLE 28.2 (Continued)

| Clinical target | Scaffold | Cell type | Biochemical cues | Type of study | Experimental design | Outcomes | References |
|---|---|---|---|---|---|---|---|
| | | | | | size calvarial defects in New Zealand White Rabbits. | of the defects but even from islands in the center portion of the defect 4 weeks after implantation, leading to a full regeneration of the injury site. | |
| | β-TCP | PDPCs BMSCs | | In vitro and in vivo | Individual or co-cultures of human PDPCs and BMSCs were cultured in OM. PDPCs + BMSCs/β-TCP implants were inserted in the dorsal surface of CD-1 athymic and immunodeficient nude mice. | Co-culturing MSCs showed synergistic effect on osteogenic differentiation both in vitro and in vivo, enhancing ectopic bone growth and neovascularization. | [24] |
| | PGLA/ β-TCP (cerasorbs) + fibrin | PDPCs | TGF-β1 | In vitro | PDPCs were encapsulated into the PLGA/β-TCP constructs within a fibrin matrix and cultured in OM + TGF-β1 supplementation up to 40 days. | De novo synthesis of human preosseous tissue. The addition of TGF-β1 increased new bone formation with significantly higher concentrations of osteogenic marker proteins. | [55] |
| | Fibrin/ α-TCP | PDPCs | – | In vitro | Cells were embedded in fibrin beads containing 7.5% a-TCP particles. Beads were cultured in Ham's F-12 medium. | DNA content, alkaline phosphatase, and osteocalcin content were significantly higher in the constructs containing α-TCP. | [60] |
| Sinus floor augmentation | β-TCP/Hap (Kasios) | BMSCs | – | Clinical report | BMSCs laden β-TCP/Hap scaffolds were implanted into the subantral sinus cavity in patients with loss | 4 months after surgery, mean new bone formation was 41.34%. 93% of the implants were considered | [22] |

| | | | | | | | |
|---|---|---|---|---|---|---|---|
| | | | | | of bone height in the posterior maxilla, before placement of metallic implants. | clinically successful 6 months postimplantation. | |
| | BBM (Bio-Oss) | BMSCs from iliac crest | – | In vivo | Bilateral augmentations of the sinus floor in six adult sheep. | BMSCs combined with BBM consistently support new bone formation, 8 and 16 weeks after implantation. | [43] |
| | Collagen<br>BBM (Bio-Oss) | PDPCs<br>Autologous osteoblasts | – | Clinical | Removal of the bone, elevating the sinus membrane and implantation of the constructs. | Both strategies could create new bone tissue with sufficient stability for successful implant placement. | [42] |
| | PRP | BMSCs | – | Clinical report | BMSCs were combined with PRP and mixed with thrombin + $CaCl_2$ to produce a cell-laden gel. Mucoperiosteal flap was elevated and the injectable implants were applied in the maxillary sinus in patients' posterior alveolar ridge atrophy. | All 41 dental implants were clinically stable. Mineralized tissue height at 2 years increased $8.8 \pm 1.6$ mm. No adverse effects and remarkable bone absorption was seen in the 2- to 6-year follow-up time. | [168] |
| Maxillary sinus floor elevation | BBM (Osteobone) | BMSCs-hBMP-2 | | In vivo | Cells were seeded in BBM and used in maxillary sinus floor elevation surgeries made bilaterally in Male New Zealand rabbits. | The BMSCs-hBMP-2 combined with BBM promoted new bone formation in rabbit maxillary sinus. | [41] |
| Prosthodontic rehabilitation | PTTM-enhanced Ti implants | – | – | Pilot clinical cases | PTTM-enhanced titanium implants were placed in both maxilla and mandibles of postablative cancer patients. | PTTM-enhanced dental implants were clinically effective in the prosthetic rehabilitation of postoncological patients. | [29] |

(*Continued*)

TABLE 28.2 (Continued)

| Clinical target | Scaffold | Cell type | Biochemical cues | Type of study | Experimental design | Outcomes | References |
|---|---|---|---|---|---|---|---|
| Temporal–mandibular joint replacement | BBM | BMSCs | | In vitro | Culture for 5 weeks in a custom-made perfusion bioreactor. | Formation of confluent layers of lamellar bone and osteoids, homogeneously distributed throughout the scaffold. | [47] |

*α-MEM*, Alpha minimum essential medium; *β-TCP*, beta-tricalcium phosphate; *ASCs*, adipose tissue-derived stem cells; *BCP*, biphasic calcium phosphate; *BBM*, Bovine derived bone matrix; *BMP-2*, bone morphogenetic protein-2; *BMSCs*, bone marrow derived stem cells; *BMSCs-hBMP-2*, genetically modified BMSCs overexpressing recombinant human BMP-2; *CHT*, chitosan; *CPC*, calcium phosphate cement; *DMEM*, Dulbecco's modified medium; *FGF-2*, fibroblast growth factor-2; *Hap*, hydroxyapatite; *HUVECs*, human umbilical vein endothelial cells; *MSCs*, mesenchymal stem cells; *NBM*, natural bone mineral; *OM*, osteogenic medium; *PCL*, polycaprolactone; *PDGF-BB*, platelet-derived growth factor; *PDLCs*, periodontal ligament cells; PDPCs, periosteal-derived progenitor cells; *PLGA*, poly (glycolide-*co*-lactide) acid; *PRP*, platelet-rich plasma; *PS*, polystyrene; *PTTM*, porous tantalum trabecular metal; *TGF-β*, transforming growth factor-beta; *Ti*, titanium.

### 28.2.1 Scaffolds

The use of "scaffolds" or engineered biomaterials, stems from the need for an adequate matrix to drive the regeneration of target tissues [33,35]. Its function lies on providing structural support with adequate mechanical properties, architecture and volume. The scaffolds procure to mimick tissue 3D environment, and biophysical/biochemical cues for progenitor cells to attach, migrate, proliferate, generate their natural extracellular matrix (ECM), and differentiate until the mature tissue is fully regenerated. An ideal scaffold should accomplish several requirements, such as biocompatibility, appropriate mechanical strength and degradation, adequate porosity, and pore interconnectivity [36,33,35,38,39].

The biomaterials that have been used for the regeneration of skeletal maxillofacial tissues can be inorganic, such as ceramics, metals, or metallic alloys, or organic, including both synthetic or natural origin molecules as polymers [33,38].

#### *28.2.1.1 Inorganic materials*

Ceramic materials, such as β-tricalcium phosphate (β-TCP), hydroxyapatite (Hap), biphasic calcium phosphate (BCP), and calcium phosphate cements (CPC), or xenogenic bone mineral matrices, are the most used TE bone graft substitutes for the reconstruction of oral and maxillofacial region [32,35,38,40–46]. The ceramics are highly biocompatible and closely resemble the natural bone apatite in terms of composition and compressive strength [32]. Their intrinsic osteogenic and osteoconductive properties have been widely explored alone [17,41–43,47–49] or in combination with other biomaterials [50,51]. Moreover, they can be produced in a large range of shapes, sizes, and porosity using industrialized manufacturing techniques, as 3D printing technologies [32,35]. Although, their very low-degradation rate [51], brittleness, and limited tensile strength [52], might compromise the quality of the regenerated tissues [53].

Metal alloys have also been used for their biocompatibility and osteointegrative potential. Brauner et al. demonstrated that porous tantalum trabecular metal (PTTM)-enhanced titanium dental implants improved the osteointegration of the implants and, subsequently, boosted the efficacy of the clinical treatment used for the prosthodontic rehabilitation of postablative cancer patients submitted to postsurgery radiotherapy (Fig. 28.11A) [29].

#### *28.2.1.2 Synthetic polymeric materials*

Several synthetic polymers are already approved by the FDA for biomedical applications [56,57]. Among the most used synthetic polymers are poly(glycolic acid) (PGA), poly (lactic acid) (PLA), poly(glycolic-*co*-lactic acid) (PGLA), and PCL [56,57].

In general, these polymers are biodegradable and biocompatible and can be produced in highly controlled conditions, enabling to tailor their average molecular weight, porosity, mechanical properties, etc. TE has explored these materials to produce scaffolds with reproducible and predictable features, such as degradation kinetics or mechanical properties [36,32]. However, synthetic polymeric biomaterials usually do not have adequate biofunctional cues to promote cellular attachment and differentiation [32]. So, synthetic polymeric biomaterials often require either some surface modification, or the blending with natural polymers [e.g., starch, chitosan (CHT)] or ceramics to enhance their

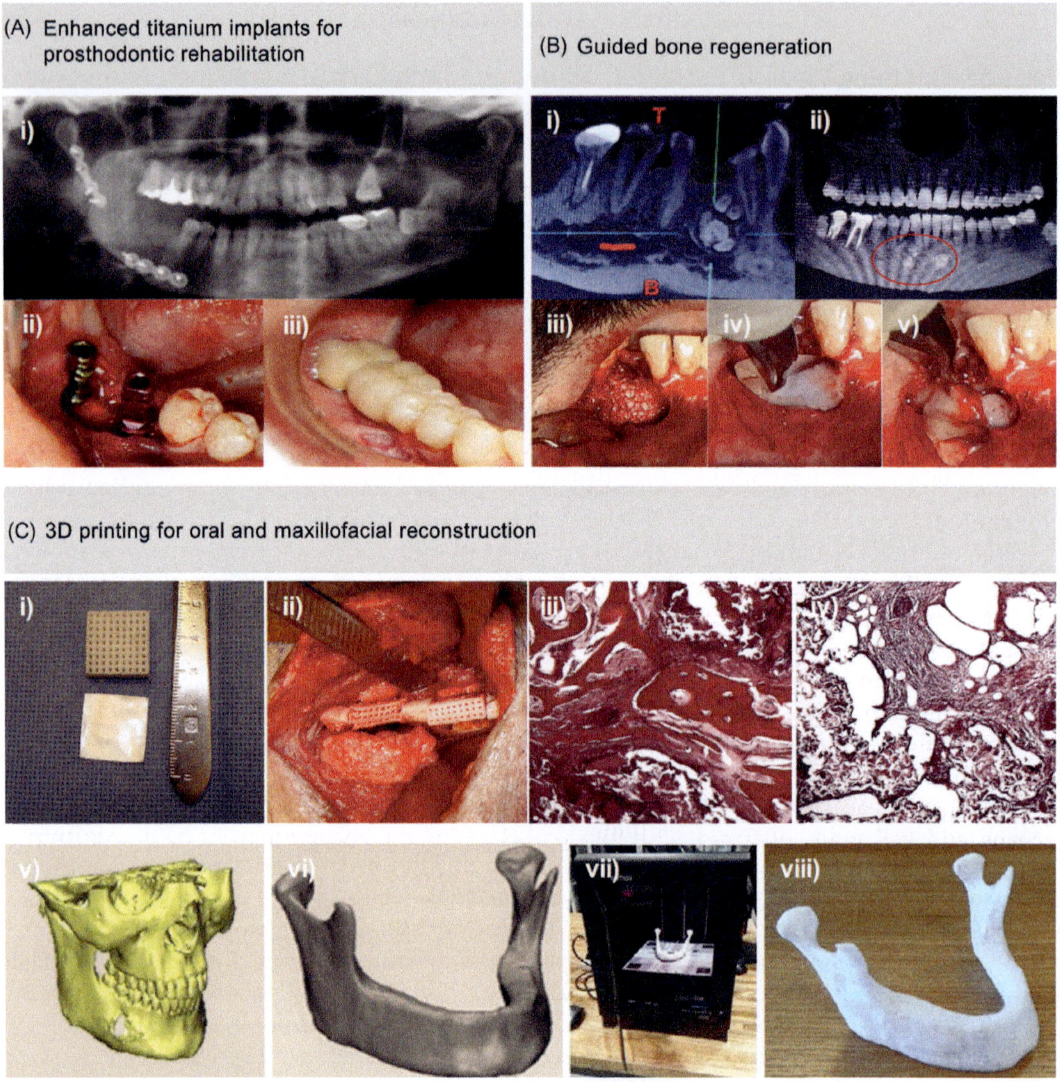


**FIGURE 28.11 Tissue engineering-based therapies for skeletal maxillofacial rehabilitation.** (A) Porous tantalum trabecular metal (PTTM)-enhanced titanium dental implants for prosthodontic rehabilitation in postablative oral and maxillofacial cancer patients: (i) X-ray of the oral cavity after cancer treatments; (ii) implantation of the dental implants; (iii) postimplantation image of the oral cavity. (B) Guided bone regeneration in a patient with an odontoma in 43 zone: odontome identification (i and ii); filling of the mandible defect with (iii) Bioss from Geistlich mixed with autologous liquid fibrin; (iv) covering with EPTFE membrane and (v) PRF. (C) The 3D-printed TE scaffolds for bone resection reconstruction: (i) general appearance of PCL/β-TCP (1:1) 3D-printed scaffolds; (ii) scaffolds seeded with BMSCs (left) or unseeded (right), implanted in mandibular body osteotomies in Yucatan minipigs; (iii) new bone formation in the core of scaffolds seeded with BMSCs and (iv) absence of new bone in plain scaffolds, 8 weeks after implantation. Patient designed whole mandible reconstruction: (v) original model of the cancerous mandible from patient's CT scan data; (vi) final 3D template of the reconstructed mandible; (vii) 3D printing, by fused deposition modeling in ABS polymer; (viii) reconstructed mandible implant. Source: *(A) Adapted from Brauner E, Guarino G, Jamshir S, Papi P, Valentini V, Pompa V, et al. Evaluation of highly porous dental implants in postablative oral and maxillofacial cancer patients. Implant Dent 2015;24(5):631–7 [29]. (B) Courtesy Dr. Antonio Dõlera Ortiz. (C-iv) Adapted from Konopnicki S, Sharaf B, Resnick C, Patenaude A, Pogal-Sussman T, Hwang K-G, et al. Tissue-engineered bone with 3-dimensionally printed β-tricalcium phosphate and polycaprolactone scaffolds and early implantation: an in vivo pilot study in a porcine mandible model. J Oral Maxillofac Surg 2015;73 (5):1016.e1–e11 [30]. (C-viii) Adapted from Ref. [164] A. Aldaadaa, N. Owji and J. Knowles, Three-dimensional printing in maxillofacial surgery: hype versus reality, J Tissue Eng 9; 2018:1–5.*

biocompatibility and bioactivity [30,54,55]. In a preclinical study, Konopnicki and coworkers showed that 3D-printed scaffolds of poly(caprolactone) (PCL)/β-TCP (1:1) blends seeded with pig bone marrow cells were able to regenerate mandibular critical defects in minipigs (Fig. 28.11C) [30].

#### *28.2.1.3 Natural polymers*

Several molecules derived from biologic/natural sources such as collagen [42,44,58], hyaluronic acid (HA) [51,53,59], fibrin [60], CHT [49,50,58], gelatin [50], or silk fibroin [58] are commonly used in TE. Most of them have already FDA approval for clinical use in temporary applications [36,32]. Unlike synthetic polymers, the nature-derived polymers are more challenging to tailor and engineer using industrial processing technologies and present higher batch-to-batch variability. Nevertheless, their similarity with ECM, biocompatibility, and intrinsic biofunctionality, make them ideal to functionalize other biomaterials or even be used as raw materials to produce matrices.

Collagen is the main structural protein present in the ECM of all the animals. Scaffolds based on collagen have been used in oral and maxillofacial applications as membranes [43,22,61], matrices [42,44] or sponges [62]. For GBR, collagen membranes are able to avoid soft tissues collapse into bone defects by while maintaining the ability space stability for new bone growth [22,61]. Collagen matrices have been used for cells [42] or GFs delivery [62] showing promising results in the regeneration of the skeletal maxillofacial defects by improving the cellular attachment and 3D spreading, and promoting faster new bone formation in clinical studies.

CHT is the deacetylated derivative of chitin, commonly found in shells of marine crustaceans and cell walls of fungi. This linear β-(1-4) linked polysaccharide of glucosamine and *N*-acetyl glucosamine, is a polyvalent material used in TE applications as scaffolds, injectable hydrogels, microspheres, composites with ceramics [63], or as carrier agent both for cells and GFs [58,64]. For bone TE, CHT/calcium phosphate composites processed as porous scaffolds were shown to be valuable matrices to support bone and cartilage regeneration [49]. Along with CHT, the incorporation of calcium phosphates promotes osteogenic differentiation, new bone growth, and enhance the ability for osteoblasts to migrate through the pores of the scaffold [49,50,65].

HA is the backbone of vertebrates' ECM. Further than its structural role, HA is implied in several relevant biological processes such as angiogenesis and endothelial cell migration [66]. HA biocompatibility, degradability by means of hyaluronidases, and capacity to be modified through its carboxyl [67] and hydroxyl [68] groups to produce stable hydrogels makes it particularly interesting for TE. HA has been processed into microparticles capable of sustained drug delivery [67] or hydrogels [59,68,69] capable of supporting cell encapsulation. In particular, the HA microspheres, have been explored to be used as porogenic and drug delivery materials to enhance the degradability and osteoconductive/osteointegrative properties of injectable calcium phosphate cements [51,53,70].

Autologous fibrin- or platelet-rich blood derivatives have also been explored (Fig. 28.11B) as matrices for bone regeneration. The similarity of fibrin matrices with blood clot, namely the presence of cell- and GFs-binding motifs [71,72], support their use in regenerative approaches. Despite fibrin's weak mechanical properties, fibrins matrices are excellent cells or GFs carriers, and can be incorporated with inorganic compounds, like ceramics (Fig. 28.11B), in order to enhance their osteogenic properties [60,73].

## 28.2.2 Cells

The inclusion of cells capable of participating in the regeneration of the tissue that is being replaced and integrate in host tissues is a keystone in TE strategies [38]. Stem cells are unspecialized cells with the ability for indefinitely self-renewal and differentiate into multiple cell lineages [74–76]. Commonly, stem cells are classified into three main types: embryonic stem cells (ESCs), usually isolated from the inner wall of the preimplantation blastocysts [38,77,78]; adult stem cells, found in adult tissues [76]; and induced pluripotent stem cells (iPSCs), which are produced artificially by genetic manipulation of somatic cells [78,79].

While ESCs and iPSCs are pluripotent stem cells potentially capable to differentiate into all types of cells from the three germinal layers [38,77,79], their use in TE strategies is impaired by technical, ethical, and safety concerns [38,79]. In contrast, adult stem cells can only differentiate into a restricted number of cell types from certain tissues [38,74]. These cells, also named as postnatal stem cells, or somatic stem cells have been widely explored for TE applications.

### *28.2.2.1 Mesenchymal stem cells*

Since the first isolation of clonogenic cells capable of multilineage differentiation in vitro from bone marrow, termed mesenchymal stem cells (MSCs) [80], different studies have shown that MSCs can be found in several tissues [74,80]. Populations of MSCs-like cells have been found in periosteum, muscle, liver, brain, epithelium, skin, peripheral blood, umbilical cord blood, amniotic membrane, dental, and adipose tissues [35,38,74]. After harvesting, MSCs can be expanded ex vivo, and then be induced to differentiate into the desired cell type.

MSCs from these sources exhibit the ability to undergo osteogenic, chondrogenic, and adipogenic differentiation in vitro [32,81–88], however they have distinct differentiation potential, possibly due to niche effect [89]. Remarkably, these cells have immunomodulatory and chemotactic effects, preventing TE graft rejection and enhancing host regenerative potential [89,90]. For bone TE, MSCs are most commonly obtained from bone marrow aspirates, adipose tissue, and periosteum.

#### 28.2.2.1.1 Bone marrow derived stem cells

Bone marrow derived stem cells (BMSCs) are the most used MSCs in bone TE, and can be isolated from iliac crest aspirates, core biopsies, or surplus bone removed surgically [35]. The preculture in vitro in osteogenic conditions, incorporation into osteogenic biomaterials and/or supplementation with GFs have been explored by TE to engineer implantable grafts for critical bone defects regeneration. Urkmez et al., assessed the combinatory effect of porcine BMSCs, and scaffolds composed of CHT flakes and BCP granules incorporating fibroblast growth factor (FGF) laden gelatin microspheres [50]. The cells precultured under osteoblastic conditions and seeded onto the engineered scaffolds were able to induce new vascularization and new bone formation 8 weeks after implantation in transcortical mandibular defects in skeletally mature female Yorkshire pigs [50]. In another study, aiming to develop a successful sinus floor augmentation strategy for patients that had suffered a loss of height in the posterior maxilla, Shayesteh et al. implanted iliac crest's BMSCs combined with porous HA/TCP ceramic matrices [22]. Numerous areas of osteoid and orthotopic bone formation were seen 4 months after implantation, as confirmed by evaluation of biopsy specimens [22].

#### 28.2.2.1.2 Periosteal-derived progenitor cells

Another interesting source of MSCs-like cells are the periosteal-derived progenitor cells (PDPCs). MSCs isolated from periosteum could be a useful resource for bone regeneration [74]. PDPCs have been shown to differentiate into several cell lineages such as osteogenic, adipogenic, chondrogenic, and myogenic [32,81–83]. For bone TE, the PDPCs have been combined with osteogenic biomaterials, and/or GFs for oral and maxillofacial defects rehabilitation, presenting very good results in terms of new bone growth in critical defects, neovascularization, and higher osteogenic ability than BMSCs [42,54,55,60,24,91]. Redlich et al., with the purpose of repairing critical size calvarial defects in New Zealand White Rabbits, seeded rabbits' tibial periosteum PDPCs, cultured in osteogenic conditions, in PGLA fleeces (Ethisorb) [54]. The expression of osteogenic markers was observed in vitro and, 4 weeks after implantation, histological analysis revealed that the group treated with the biocomplex showed orthotopic new bone growth, not only from the margins of the defects but also from islands in the core of the defect, leading to a full regeneration of the injury site [54]. Arnold et al., using the same polymeric fleeces (Ethisorb) combined with β-TCP and transforming growth factor beta1 (TGF-β1), seeded the developed construct with PDPCs from human anteomedial portion of tibia and cultivated in vitro under osteogenic conditions [55]. After 40 days, histological images demonstrated that the samples that included TGF-β1 presented enhanced new bone formation and displayed significantly higher osteogenic markers expression than the constructs without GFs, whereas the inclusion of the inorganic part enhanced the osteoconductivity of the construct [55]. Remarkably, some studies have suggested that, osteogenic potential of PDPCs is consistent, regardless the bone from which PDPCs were isolated [91].

#### 28.2.2.1.3 Adipose tissue-derived stem cells

Adipose tissue-derived stem cells (ASCs) are MSCs obtained from the adipose tissues [92]. ASCs high abundance in adipose tissue, lower collection morbidity [85,87,88], higher proliferation rate in vitro comparing to BMSCs [38,84], and osteogenic potential [84–88] have been supporting ASCs as a good alternative to BMSCs for bone TE. Despite displaying lower osteogenic potential compared with BMSCs, it is maintained with aging, unlike the rest of the MSCs [93–95]. Clinical studies conducted by Mesimäki et al. or Sandor et al. showed that ASCs hold great potential in oral and maxillofacial rehabilitation [17,48,96]. They both combined autologous ASCs with a tricalcium phosphate scaffold and bone morphogenetic protein-2 (BMP-2) in order to repair critical maxillofacial defects, in patients subject to hemimaxillectomy due to a recurrent keratocyst or patients with anterior mandibular ameloblastoma resection defects, respectively [17,48,96]. The cytokines and chemokines produced by the transplanted ASCs acted as homing signals for endogenous stem cells and progenitor cells migration toward the site of injury [17,48,96]. This ability together with the osteoconductivity of the β-TCP acted synergistically in order to regenerate native-like bone 10 months after implantation in a parasymphyseal defect created by ameloblastoma resection, which allowed prosthodontic rehabilitation in the grafted site [17].

#### 28.2.2.1.4 Dental pulp stem cells

The dental pulp is the soft tissue inside a tooth that contains neural fibers, blood vessels, lymphatics, and MSCs derived from the neural crest [97,98]. Consequently, DPSCs have a strong differentiation potential toward the osteogenic lineage [99]. DPSCs have many advantages for bone TE, including high proliferation rate, excellent osteogenic differentiation potential, low donor site morbidity, and favorable paracrine and immunomodulatory properties [98,100].

There are many different bone TE strategies using DPSCs in combination of scaffolds made of bone ECM [99], alginates [101], bioactive glass [102–104], CHT [102,105], gelatin sponges [106], polycaprolactone (PCL) [107,108], and collagen [109,110]. Other less conventional additives have been explored, to enhance the osteogenic differentiation of the DPSCs for bone regeneration, for example aloe vera [109], platelet-rich plasma (PRP) [111], parathyroid hormone [112], adenosine triphosphate (ATP) [113], tooth powder [114]; drug like aspirin [115], or berberine [116]; or even metal nanoparticles, such as iron [117] or gold [118].

D'Aquino et al. in order to fill the injury site left by the extraction of the mandibular third molar in patients that needed all wisdom teeth extraction seeded DPCs harvested from the extracted teeth onto a collagen sponge scaffold and implanted into the alveolar bone [119]. After 3 months, X-rays showed an optimal vertical repair of the alveolar bone tissue, whereas histological observations from biopsies clearly proved the complete regeneration of bone at the injury site and the enhancement of the expression of osteogenic markers [119]. Also, the combination in vitro of DPCs with HA hydrogels enriched with platelet lysate improved the osteogenic potential of the cells, increasing the cellular metabolism and stimulating the mineralized matrix deposition by DPSCs [120], and promotes implant vascularization [121].

#### 28.2.2.1.5 Co-cultures

Other studies have procured to emulate the organogenesis, combining relevant cell types in co-cultures. Spheroid of human HUVECs and BMSCs (2:98), prevascularized after 10 days in culture, were implanted subcutaneous in the dorsal region of male mice resulting in a significantly increase of the osteogenic marker alkaline phosphatase [122]. This study showed that combination with endothelial cells is likely to have a positive effect on the osteogenic differentiation of BMSCs [122].

In another study, Chen et al. co-culture BMSCs and PDPCs, both harvested from metaphyseal distal part of tibia, and loaded them into porous β-TCP scaffolds to evaluate their osteogenic potential and their capacity for ectopic bone growth [24]. Compared with single cultures, co-culturing MSCs in vitro showed abundant mineralization and a steadily increasing ALP activity and expression of osteogenic specific genes (COL1A1, BMP-2, osteopontin, and osteocalcin) [24]. Eight weeks after implantation in the dorsal surface of

CD-1 nude mice's, similar synergistic effects were confirmed by histological observation of enhanced ectopic new bone formation and neovascularization [24].

### 28.2.3 Biochemical cues

GFs are secreted proteins which bind specific cellular receptors located on target cells to implement a desired function [36,38]. These peptides activate the cellular signaling network, initiating several intracellular cascades that alter and influence many different cellular processes, such as the promotion or prevention of cell adhesion, proliferation, migration, and differentiation, the matrix deposition and the maturation of tissues [36,35,37]. The cellular response depends upon the GF, dose, the cell type, and the bound receptor. Hence, these molecules are therapeutically applied in the regeneration of tissues aiming the enhancement of level of regeneration beyond the limit of spontaneous and natural tissue healing [38,123].

Particularly for oral and maxillofacial reconstruction, the most used GFs are platelet-derived growth factor (PDGF) [45,124], basic FGF (b-FGF) [50], insulin-like growth factor (IGF) [37,40,44,120,125,126], TGF-β [40,55], vascular endothelial growth factor (VEGF) [121,122,125] and bone morphogenetic proteins (BMPs) [17,40,45,48,127,128]. These bioactive molecules can be used as a supplement for cell cultures, impregnated, or coated on the scaffolds [17,40,45,48,50,55,124,126–128].

The morphogenic GFs, such as BMPs and TGF-β, have been widely used in skeletal maxillofacial regeneration for their ability to induce osteogenic differentiation, accelerating the production of bone ECM and, consequently, promote the tissue integration [129]. Specifically, BMPs are chemotactic, mitogenic, and osteogenic, recruiting MSCs from the surrounding tissues toward the bioengineered graft and inducing the differentiation of the recruited cells into the osteogenic lineage [35,38,129,130].

Platelet-rich hemoderivatives (PRHds) are used for maxillofacial bone reconstruction since the 90s. Whitman et al. showed that PRP improve the aggregation and cohesiveness of particulate bone, enhancing the osteogenesis in palatal bone defects [131]. The PRHds are compositions containing supra physiologic concentrations of platelets, usually $1 \times 10^6/\mu L$ [132] which, upon activation, release a cocktail of GFs, including all the above mentioned, and cytokines involved in wound healing orchestration. Among the most clinically explored PRHds are the PRP and the platelet-rich fibrin (PRF), which have been shown to have therapeutic potential for osteointegration of implants [133] or as GTR occlusive barriers [134] (Fig. 28.11B). Recent research on TE field has procured to used PRHds with more consistent batch-to-batch composition (e.g., platelet lysate) which are incorporated into smart materials in order to modulate the spatial-temporal release of the factors, enhancing host tissues healing potential (for more detail see Refs. [135,136]).

### 28.2.4 Bioreactors

Bioreactors have been explored in TE to mimic the human body biophysical, biochemical, and/or hydrodynamic conditions in a sterile, confined, and controlled environment. Four main bioreactor types have been used for bone TE [137]: spinner flasks [138,139], rocker platforms [140,141], rotating wall vessels [142,143], compression [144], and perfusion systems [145,146]. In general, these systems allow a more effective mass transport for feeding in the essential nutrients, cells, and oxygen and remove wastes and carbon dioxide, which yield more homogeneous colonization with cells and matrix deposition within large TE scaffolds [47,137,147]. Bioreactors can also provide complex biophysical stimuli, such as static and dynamic mechanical stresses, and magnetic or electrical signals [137].

The use of bioreactors has been proposed for recellularization of decellularized bone with autologous cells to treat large bone defects. It has been proposed as an alternative for those patients that are not candidates for additional surgery to bone graft harvest either due to radio- or chemotherapies, or to systemic conditions as malnutrition, a common consequence of maxillary neoplasm resection [137]. Some studies [47,148–151] had reached good results regarding the clinical size, shape, and functionality of customized bone grafts using autologous stem cells and a bioreactor system to provide patient-specific bone grafts for maxillary reconstruction. In a study, scaffolds from fully decellularized bovine trabecular bone, anatomically shaped to resemble a human temporal–mandibular junction based on CT scans, were seeded with BMSCs and cultured with interstitial flow of culture medium in a perfusion bioreactor [47]. After 5 weeks of culture in dynamic conditions, tissue growth was evidenced by the formation of confluent layers of lamellar bone and osteoids, homogeneously distributed throughout the scaffold [47]. Conversely, statically cultured constructs exhibited empty pore spaces and loosely packed cells [46]. However, appropriate means to provide construct vascularization are still yet to be achieved [148].

### 28.2.5 Prophylactic tissue engineering constructs

Postoperative proliferation of residual tumor cells may result in the tumor recurrence, which has been strike with the irradiation or chemotherapy after tissue ablation. Given the impaired regenerative potential of irradiated tissues, successful bone regeneration combined with the control of residual cancer cells presents a challenge for bone TE.

Recently, has been explored the delivery of tumor suppressive drugs locally incorporated into the scaffolds to avoid the systemic administration of chemotherapy. Chai and coworkers, demonstrated that polycyclodextrin functionalized porous bioceramics laden with an antibiotic (gentamicin) and an anticancer agent (cisplatin) were cytocompatible and enabled the controlled release of the drugs [152]. Photothermal and magnetic hyperthermia therapies have also been developed, in which specific agents combined with TE scaffolds generate heat locally under near-infrared irradiation or magnetic stimulation, respectively, to ablate tumor cells. Such agents include organic particles [153], gold nanoparticles [154], molybdenum sulfide ($MoS_2$) nanosheets [155] or graphene oxide (GO) [156], and magnetic nanoparticles [157].

## 28.3 Future perspectives and unmet challenges

The field of TE has substantially contributed for the development of therapies aiming the recovery of oral and maxillofacial tissues inflicted by neoplasms. In the particular case of skeletal maxillofacial tissues, two main lines have been approached: (1) the replacement of the "gold standard" bone flap autografts by engineered constructs; and (2) the improvement of impaired regeneration of tissues following surgical resection of tumors and chemo or radiotherapy. Strategies combining biomaterials-based microengineered scaffolds, progenitor cells, and biophysical/biochemical stimuli deeply contributed for the successful outcomes toward the development of successful bioengineered solutions. However, there are still controversial aspects to be clarified, such as the clinical use of GFs. For instance, the unpredictable effects and complications of BMP-supplemented treatments of oncogenic patients, led FDA to restrict the application of this GF to sinus alveolar process augmentation in the United States [158]. Likewise, the proper vascularization of bioengineered bone grafts, and the conjugation of adequate mechanical, processing, and biofunctional properties in a single construct are unmet challenges in bone TE. The bias responses in humans, highlights the complexity behind tissue regeneration after tumor development and defies for new therapeutic solutions with improved efficacy and systematic biological responses.

Overall, the future of TE in the rehabilitation of tissues affected by SMNs is unquestionable. Further than the sole fabrication of tissue substitutes, TE research is evolving to become a holistic research field addressing all the aspects of disease progression and post-implantation issues. The exploitation of imaging technologies and computer modeling to engineer patient-customized therapies [159]; the development of smart drug delivery strategies for spatio-temporal delivery of GF [136]; the incorporation of molecules to prevent opportunistic microbial infections [68,160], or tumor residence [153–157]; or even the development of in vitro models resourcing to artificial tissues to recreate the 3D tumor microenvironment and key physiological interactions [161–163], are driving the development of more efficient and personalized TE-based therapies for patients suffering from SMNs.

Recently, the emerging of microfabrication and 3D printing technologies have empowered the conventional TE approaches, enabling the preparation of hierarchically organized scaffolds emulating the complexity of tissue organization. Mimicking the 3D structural characteristics and functionality of the recessed tissues is one of the most challenging difficulties to overcome [32,35]. The integration of imaging, CAD, and 3D printing technologies can be used to print anatomically identical models of complex bones, such as mandibles [164]. Moreover, the use of 3D printing technology has been explored to produce TE constructs closely mimicking bone microarchitecture [155]. However, in the particular case of bioprinting, which is the 3D printing of cells into constructs, available bioinks are still at an infancy stage and do not have the adequate mechanical properties or biofunctionality to reliably regenerate large bone defects [165].

TE research lines have also been exploring the cumulative experience in the preparation of artificial tissues to produce in vitro tumor models [161–163]. The ideal in vitro cancer models procure to mimic 3D microenvironment and key physiological interactions that are crucial to modulate tumor response to soluble paracrine factors, tumor drug resistance, and invasiveness. These models, further than providing a realistic platform which could potentially minimize the animal-based preclinical cancer models, might be equally essential to pave the way for personalized cancer medicine.

However, the costs for such therapies are still very high and will predictably be affordable just for wealthy populations in the near future, which raises funded ethical concerns. Hence, the future research trends of TE in oral and maxillofacial rehabilitation should focus on three main goals: (1) encourage the clinical translation of the most promising TE approaches for bone regeneration; (2) the development of new printable bioinks and technologies more stable, biofunctional, and cost effective, and its homologation for biomedical application; and, most importantly, (3) to disseminate/communicate this exciting field of knowledge to clinicians and healthcare providers so new solutions for the rehabilitation of SMNs can be successfully implemented, contributing for improved quality of life of patients.

## References

[1] Chaukar DA, Walvekar RR, Das AK, Deshpande MS, Pai PS, Chaturvedi P, et al. Quality of life in head and neck cancer survivors: a cross-sectional survey. Am J Otolaryngol [Internet] 2009;30(3):176–80. Available from: <https://linkinghub.elsevier.com/retrieve/pii/S0196070908000835>.

[2] Prein J. Review of benign tumors of the maxillofacial region and considerations for bone invasion. In: Craniomaxillofacial reconstructive and corrective bone surgery [Internet]. New York: Springer New York. p. 59–64. Available from: <http://link.springer.com/10.1007/978-0-387-22427-5_7>.

[3] El-Naggar A, Chan J, Takata T, Slooweg P, editors. World Health Organization classification of tumours. Lyon: International Agency for Research on Cancer; 2017.

[4] Mintz B, Fleischman RA. Teratocarcinomas and other neoplasms as developmental defects in gene expressioN. Adv Cancer Res. 1981;34:211–78. Available from: https://doi.org/10.1089/107632703322066723.

[5] Saalim M, Sansare K, Karjodkar FR, Farman AG, Goyal SN, Sharma SR. Recurrence rate of odontogenic myxoma after different treatments: a systematic review. Br J Oral Maxillofac Surg 2019;57:7–13.

[6] Agbara R, Fomete B, Athanasius-Chukwudi O, Uchenna-Kevin O, Modupeola-Omotara S. Central myxoma/myxofibroma of the jaws: a clinico-epidemiologic review. Iran J Otorhinolaryngol 2017;29(90):35–41.

[7] Anyanechi CE, Saheeb BD. The clinical presentation of odontogenic myxoma of the jaws: a 22-year retrospective analysis. J Clin Exp Oncol 2017;06(05):193.

[8] McClary AC, West RB, McClary AC, Pollack JR, Fischbein NJ, Holsinger CF, et al. Ameloblastoma: a clinical review and trends in management. Eur Arch Oto-Rhino-Laryngol 2016;273(7):1649–61.

[9] Almeida RDAC, Andrade ESDS, Barbalho JC, Vajgel A, Vasconcelos BCDE. Recurrence rate following treatment for primary multicystic ameloblastoma: systematic review and meta-analysis. Int J Oral Maxillofac Surg 2016;45(3):359–67.

[10] Amado Cuesta S, Gargallo Albiol J, Berini Aytés L, Gay Escoda C. Review of 61 cases of odontoma. Presentation of an erupted complex odontoma. Med Oral 2003;8(5):366–73.

[11] Isola G, Cicciù M, Fiorillo L, Matarese G. Association between odontoma and impacted teeth. J Craniofac Surg 2017;28(3):755–8.

[12] Li T. The odontogenic keratocyst: a cyst, or a cystic neoplasm? J Dent Res 2011;90(2):133–42.

[13] Ahlfors E, Larsson A, Sjögren S. The odontogenic keratocyst: a benign cystic tumor? J Oral Maxillofac Surg 1984;42(1):10–19.

[14] Dolanmaz D, Esen A, Mihmanlı A, Işık K. Management of central giant cell granuloma of the jaws with intralesional steroid injection and review of the literature. Oral Maxillofac Surg 2016;20(2):203–9.

[15] Daroit NB, de Marco RG, Sant'Anna Filho M, Fritscher GG. The challenge in the treatment of central giant cell granuloma – what is the best approach? J Oral Maxillofac. Surgery, Med Pathol 2017;29(2):122–8.

[16] Hupp J, Tucker M, Ellis E. Contemporary oral and maxillofacial surgery. 6th ed St. Louis: Elsevier, Mosby; 2014. p. 448.

[17] Sándor GK, Tuovinen VJ, Wolff J, Patrikoski M, Jokinen J, Nieminen E, et al. Adipose stem cell tissue-engineered construct used to treat large anterior mandibular defect: a case report and review of the clinical application of good manufacturing practice-level adipose stem cells for bone regeneration. J Oral Maxillofac Surg 2013;71(5):938–50.

[18] Moura LB, Carvalho PH, Xavier CB, Post LK, Torriani MA, Santagata M, et al. Autogenous non-vascularized bone graft in segmental mandibular reconstruction: a systematic review. Int J Oral Maxillofac Surg 2016;45 (11):1388–94.

[19] Burk T, Del Valle J, Finn RA, Phillips C. Maximum quantity of bone available for harvest from the anterior iliac crest, posterior iliac crest, and proximal tibia using a standardized surgical approach: a cadaveric study. J Oral Maxillofac Surg 2016;74(12):2532–48.

[20] Osborn TM, Helal D, Mehra P. Iliac crest bone grafting for mandibular reconstruction: 10-year experience outcomes. J Oral Biol Craniofac Res 2018;8(1):25–9.

[21] Hahn HM, Lee YJ, Park DH. Huge radicular cyst of the maxilla treated with complete resection and immediate reconstruction by rib bone graft. J Maxillofac Oral Surg 2019;18(3):378–81.

[22] Shayesteh YS, Khojasteh A, Soleimani M, Alikhasi M, Khoshzaban A, Ahmadbeigi N. Sinus augmentation using human mesenchymal stem cells loaded into a β-tricalcium phosphate/hydroxyapatite scaffold. Oral Surg Oral Med Oral Pathol Oral Radiol Endodontol 2008;106(2):203–9.

[23] Soares CS, Barros LC, Saraiva V, Florit MG, Babo PS, Dias IR , et al. Bioengineered surgical repair of a chronicoronasal fistula in a cat using autologous platelet-rich fibrin and bone marrow with a tailored 3D printed implant. J Feline Med Surg. 2018;20(9):835–43. Available from: https://doi.org/10.1177/1098612X18789549.

[24] Chen D, Zhang X, He Y, Lu J, Shen H, Jiang Y, et al. Co-culturing mesenchymal stem cells from bone marrow and periosteum enhances osteogenesis and neovascularization of tissue-engineered bone. J Tissue Eng Regen Med 2012;6(10):822–32.

[25] Sung EC, Chan SM, Sakurai K, Chung E. Osteonecrosis of the maxilla as a complication to chemotherapy: a case report. Spec Care Dent 2002;22(4):142–6.

[26] Hitti RA. Guided bone regeneration in the oral cavity: a review. Open Pathol J 2011;5(1):33–45.

[27] Ettl T, Gosau M, Sader R, Reichert TE. Jaw cysts - filling or no filling after enucleation? A review. J Cranio-Maxillofac Surg 2012;40(6):485–93.

[28] Langer R, Vacanti J. Tissue engineering. Science (80-) [Internet] 1993;260(5110):920–6. Available from: <http://www.sciencemag.org/cgi/doi/10.1126/science.8493529>.

[29] Brauner E, Guarino G, Jamshir S, Papi P, Valentini V, Pompa V, et al. Evaluation of highly porous dental implants in postablative oral and maxillofacial cancer patients. Implant Dent [Internet] 2015;24(5):631–7. Available from: <http://content.wkhealth.com/linkback/openurl?sid = WKPTLP:landingpage&an = 00008505-201510000-00022>.

[30] Konopnicki S, Sharaf B, Resnick C, Patenaude A, Pogal-Sussman T, Hwang K-G, et al. Tissue-engineered bone with 3-dimensionally printed β-tricalcium phosphate and polycaprolactone scaffolds and early implantation: an in vivo pilot study in a porcine mandible model. J Oral Maxillofac Surg [Internet] 2015;73(5):1016.e1–1016.e11. Available from: <https://linkinghub.elsevier.com/retrieve/pii/S0278239115000762>.

[31] Keyhan SO, Fallahi H, Jahangirnia A, Masoumi SMR, Khosravi MH, Amirzade-Iranaq MH. Tissue engineering applications in maxillofacial surgery. Stem cells in clinical practice and tissue engineering. Croatia: InTech; 2018.

[32] Shaul JL, Davis BK, Burg KJL. Regenerative engineering in maxillofacial reconstruction. Regen Eng Transl Med 2016;2(2):55–68.

[33] Rai R, Raval R, Khandeparker RVS, Chidrawar SK, Khan AA, Ganpat MS. Tissue engineering: step ahead in maxillofacial reconstruction. J Int Oral Health 2015;7(9):138–42.

[34] Cuesta-Gil M, Ochandiano Caicoya S, Riba-García F, Ruiz BD, Navarro Cuéllar C, Navarro Vila C. Oral rehabilitation with osseointegrated implants in oncologic patients. J Oral Maxillofac Surg 2009;67(11):2485–96.

[35] Mobini S, Ayoub A. Bone tissue engineering in the maxillofacial region: the state-of-the-art practice and future prospects. Regen Reconstr Restor 2016;1(1):8.

[36] Smith BT, Shum J, Wong M, Mikos AG, Young S. Bone tissue engineering challenges in oral & maxillofacial surgery. Advances in experimental medicine and biology. Springer New York LLC; 2015. p. 57–78.
[37] Albrektsson T, Johansson C. Osteoinduction, osteoconduction and osseointegration. Eur Spine J 2001;10: S96–101.
[38] Melek LN. Tissue engineering in oral and maxillofacial reconstruction. Tanta Dent J 2015;12(3):211–23.
[39] Mastrogiacomo M, Muraglia A, Komlev V, Peyrin F, Rustichelli F, Crovace A, et al. Tissue engineering of bone: search for a better scaffold. Orthod Craniofac Res 2005;8(4):277–84.
[40] Lucaciu O, Mihaela B, Băciuţ G, Câmpian R, Soriţă O, Bran S, et al. Tissue engineered bone versus alloplastic commercial biomaterials in craniofacial reconstruction. Rom J Morphol Embryol 2010;51(1):129–36.
[41] Sun XJ, Xia LG, Chou LL, Zhong W, Zhang XL, Wang SY, et al. Maxillary sinus floor elevation using a tissue engineered bone complex with BMP-2 gene modified bMSCs and a novel porous ceramic scaffold in rabbits. Arch Oral Biol 2010;55(3):195–202.
[42] Springer IN, Nocini PF, Schlegel KA, De Santis D, Park J, Warnke PH, et al. Two techniques for the preparation of cell-scaffold constructs suitable for sinus augmentation: steps into clinical application. Tissue Eng 2006;12(9):2649–56.
[43] Gutwald R, Haberstroh J, Kuschnierz J, Kister C, Lysek DA, Maglione M, et al. Mesenchymal stem cells and inorganic bovine bone mineral in sinus augmentation: comparison with augmentation by autologous bone in adult sheep. Br J Oral Maxillofac Surg 2010;48(4):285–90.
[44] Eyckmans J. Periosteum derived progenitor cells in bone tissue engineering. Unpublished Thesis; 2007. p. 1–175.
[45] Vavouraki HN, Dereka XE, Vrotsos IA, Markopoulou CE. Ability of a bovine bone graft, alone or enriched with PDGF-BB or rhBMP-2, to promote human periodontal ligament (PDL) cells proliferation. A preliminary study. Cell Tissue Bank 2003;4(1):17–23.
[46] Cvetković VJ, Najdanović JG, Vukelić-Nikolić M, Stojanović S, Najman SJ. Osteogenic potential of in vitro osteo-induced adipose-derived mesenchymal stem cells combined with platelet-rich plasma in an ectopic model. Int Orthop 2015;39(11):2173–80.
[47] Grayson WL, Fröhlich M, Yeager K, Bhumiratana S, Chan ME, Cannizzaro C, et al. Engineering anatomically shaped human bone grafts. Proc Natl Acad Sci U S A 2010;107(8):3299–304.
[48] Mesimäki K, Lindroos B, Törnwall J, Mauno J, Lindqvist C, Kontio R, et al. Novel maxillary reconstruction with ectopic bone formation by GMP adipose stem cells. Int J Oral Maxillofac Surg 2009;38(3):201–9.
[49] Weir MD, Xu HHK. Culture human mesenchymal stem cells with calcium phosphate cement scaffolds for bone repair. J Biomed Mater Res - Part B Appl Biomater 2010;93(1):93–105.
[50] Urkmez AS, Clark SG, Wheeler MB, Goldwasser MS, Jamison RD. Evaluation of chitosan/biphasic calcium phosphate scaffolds for maxillofacial bone tissue engineering. Macromol Symp 2008;269(1):100–5.
[51] Babo PS, Santo VE, Gomes ME, Reis RL. Development of an injectable calcium phosphate/hyaluronic acid microparticles system for platelet lysate sustained delivery aiming bone regeneration. Macromol Biosci [Internet] 2016;16(11):1662–77. Available from: http://doi.wiley.com/10.1002/mabi.201600141.
[52] Nilsson M, Wielanek L, Wang JS, Tanner KE, Lidgren L. Factors influencing the compressive strength of an injectable calcium sulfate-hydroxyapatite cement. J Mater Sci Mater Med 2003;14(5):399–404.
[53] Babo PS, Carvalho PP, Santo VE, Faria S, Gomes ME, Reis RL. Assessment of bone healing ability of calcium phosphate cements loaded with platelet lysate in rat calvarial defects. J Biomater Appl 2016;31(5):637–49.
[54] Redlich A, Perka C, Schultz O, Spitzer R, Häupl T, Burmester GR, et al. Bone engineering on the basis of periosteal cells cultured in polymer fleeces. J Mater Sci Mater Med 1999;10(12):767–72.
[55] Arnold U, Lindenhayn K, Perka C. In vitro-cultivation of human periosteum derived cells in bioresorbable polymer-TCP-composites. Biomaterials. 2002;23(11):2303–10.
[56] Gunatillake PA, Adhikari R, Gadegaard N. Biodegradable synthetic polymers for tissue engineering. Eur Cell Mater 2003;5:1–16.
[57] Marin E, Briceño MI, Caballero-George C. Critical evaluation of biodegradable polymers used in nanodrugs. Int J Nanomed 2013;8:3071–91.
[58] Ríos CN, Skoracki RJ, Miller MJ, Satterfield WC, Mathur AB. In vivo bone formation in silk fibroin and chitosan blend scaffolds via ectopically grafted periosteum as a cell source: a pilot study. Tissue Eng Part A 2009;15(9):2717–25.

[59] Domingues RMA, Silva M, Gershovich P, Betta S, Babo P, Caridade SG, et al. Development of injectable hyaluronic acid/cellulose nanocrystals bionanocomposite hydrogels for tissue engineering applications. Bioconjug Chem [Internet] 2015;26(8):1571–81. Available from: <http://pubs.acs.org/doi/abs/10.1021/acs.bioconjchem.5b00209>.

[60] Spitzer RS, Perka C, Lindenhayn K, Zippel H. Matrix engineering for osteogenic differentiation of rabbit periosteal cells using α-tricalcium phosphate particles in a three-dimensional fibrin culture. J Biomed Mater Res 2002;59(4):690–6.

[61] Ardjomandi N, Duttenhoefer F, Xavier S, Oshima T, Kuenz A, Sauerbier S. In vivo comparison of hard tissue regeneration with ovine mesenchymal stem cells processed with either the FICOLL method or the BMAC method. J Cranio-Maxillofac Surg 2015;43(7):1177–83.

[62] Herford AS, Boyne PJ. Reconstruction of mandibular continuity defects with bone morphogenetic protein-2 (rhBMP-2). J Oral Maxillofac Surg 2008;66(4):616–24.

[63] Oliveira JM, Rodrigues MT, Silva SS, Malafaya PB, Gomes ME, Viegas CA, et al. Novel hydroxyapatite/chitosan bilayered scaffold for osteochondral tissue-engineering applications: scaffold design and its performance when seeded with goat bone marrow stromal cells. Biomaterials [Internet] 2006;27(36):6123–37. Available from: <https://linkinghub.elsevier.com/retrieve/pii/S0142961206006594>.

[64] Di Martino A, Sittinger M, Risbud MV. Chitosan: a versatile biopolymer for orthopaedic tissue-engineering. Biomaterials. 2005;26(30):5983–90.

[65] Xu HHK, Weir MD, Simon CG. Injectable and strong nano-apatite scaffolds for cell/growth factor delivery and bone regeneration. Dent Mater 2008;24(9):1212–22.

[66] Slevin M, Krupinski J, Gaffney J, Matou S, West D, Delisser H, et al. Hyaluronan-mediated angiogenesis in vascular disease: uncovering RHAMM and CD44 receptor signaling pathways. Matrix Biol [Internet] 2007;26(1):58–68. Available from: <https://linkinghub.elsevier.com/retrieve/pii/S0945053X06003519>.

[67] Babo PS, Reis RL, Gomes ME. Production and characterization of hyaluronic acid microparticles for the controlled delivery of growth factors using a spray/dehydration method. J Biomater Appl 2016;31(5):693–707.

[68] Babo PS, Pires RL, Santos L, Franco A, Rodrigues F, Leonor I, et al. Platelet lysate-loaded photocrosslinkable hyaluronic acid hydrogels for periodontal endogenous regenerative technology. ACS Biomater Sci Eng [Internet] 2017;3(7):1359–69. Available from: <http://pubs.acs.org/doi/abs/10.1021/acsbiomaterials.6b00508>.

[69] Neves LS, Babo PS, Gonçalves AI, Costa-Almeida R, Caridade SG, Mano JF, et al. Injectable hyaluronic acid hydrogels enriched with platelet lysate as a cryostable off-the-shelf system for cell-based therapies. Regen Eng Transl Med [Internet] 2017;3(2):53–69. Available from: <http://link.springer.com/10.1007/s40883-017-0029-8>.

[70] Babo PS, Cai X, Plachokova AS, Reis RL, Jansen JA, Gomes ME, et al. The role of a platelet lysate-based compartmentalized system as a carrier of cells and platelet-origin cytokines for periodontal tissue regeneration. Tissue Eng - Part A [Internet] 2016;22(19–20):1164–75. Available from: <http://online.liebertpub.com/doi/10.1089/ten.tea.2016.0226>.

[71] Martino MM, Briquez PS, Ranga A, Lutolf MP, Hubbell JA. Heparin-binding domain of fibrin(ogen) binds growth factors and promotes tissue repair when incorporated within a synthetic matrix. Proc Natl Acad Sci U S A [Internet]. 2013;110(12):4563–8. Available from: <http://www.pnas.org/cgi/doi/10.1073/pnas.1221602110>.

[72] Adamson K, Spain E, Prendergast U, Moran N, Forster RJ, Keyes TE. Fibrinogen motif discriminates platelet and cell capture in peptide-modified gold micropore arrays. Langmuir [Internet] 2018;34(2):715–25. Available from: <https://pubs.acs.org/doi/10.1021/acs.langmuir.7b03279>.

[73] Yamada Y, Boo JS, Ozawa R, Nagasaka T, Okazaki Y, Hata KI, et al. Bone regeneration following injection of mesenchymal stem cells and fibrin glue with a biodegradable scaffold. J Cranio-Maxillofac Surg 2003;31(1):27–33.

[74] Tatullo M, Marrelli M, Paduano F. The regenerative medicine in oral and maxillofacial surgery: the most important innovations in the clinical application of mesenchymal stem cells. Int J Med Sci 2015;12(1):72–7.

[75] Cyranoski D. The cells that sparked a revolution. Nature. 2018;555(7697):429–30.

[76] Gurusamy N, Alsayari A, Rajasingh S, Rajasingh J. Adult stem cells for regenerative therapy. 1st ed. Progress in molecular biology and translational science, vol. 160. New York: Elsevier Inc; 2018. p. 1–22.

[77] Cyranoski D. How human embryonic stem cells sparked a revolution. Nature. 2018;555(7697):428–30.

[78] Csete M. Regenerative medicine. Transfusion medicine and hemostasis: clinical and laboratory aspects. 2nd ed. New York: Elsevier Inc; 2013. p. 559–63.

[79] Yu J, Thomson JA. Induced pluripotent stem cells. 4th ed. Principles of tissue engineering, 581–. New York: Elsevier; 2013. p. 594.

[80] Caplan AI. Mesenchymal stem cells. J Orthop Res 1991;9(5):641–50.

[81] De Bari C, Dell'Accio F, Vanlauwe J, Eyckmans J, Khan IM, Archer CW, et al. Mesenchymal multipotency of adult human periosteal cells demonstrated by single-cell lineage analysis. Arthritis Rheum 2006;54(4):1209–21.

[82] Choi YS, Noh SE, Lim SM, Lee CW, Kim CS, Im MW, et al. Multipotency and growth characteristic of periosteum-derived progenitor cells for chondrogenic, osteogenic, and adipogenic differentiation. Biotechnol Lett 2008;30(4):593–601.

[83] Hutmacher DW, Sittinger M. Periosteal cells in bone tissue engineering. Tissue Eng 2003;9(Suppl. 1):337.

[84] Cowan CM, Shi YY, Aalami OO, Chou YF, Mari C, Thomas R, et al. Adipose-derived adult stromal cells heal critical-size mouse calvarial defects. Nat Biotechnol 2004;22(5):560–7.

[85] Dudas JR, Marra KG, Cooper GM, Penascino VM, Mooney MP, Jiang S, et al. The osteogenic potential of adipose-derived stem cells for the repair of rabbit calvarial defects. Ann Plast Surg 2006;56(5):543–8.

[86] Kern S, Eichler H, Stoeve J, Klüter H, Bieback K. Comparative analysis of mesenchymal stem cells from bone marrow, umbilical cord blood, or adipose tissue. Stem Cell 2006;24(5):1294–301.

[87] Pendleton C, Li Q, Chesler DA, Yuan K, Guerrero-Cazares H, Quinones-Hinojosa A. Mesenchymal stem cells derived from adipose tissue vs bone marrow: in vitro comparison of their tropism towards gliomas (Najbauer J, editor). PLoS One 2013;8(3):e58198.

[88] Zuk PA, Zhu M, Ashjian P, De Ugarte DA, Huang JI, Mizuno H, et al. Human adipose tissue is a source of multipotent stem cells. Mol Biol Cell 2002;13(12):4279–95.

[89] Musina RA, Bekchanova ES, Belyavskii AV, Sukhikh GT. Differentiation potential of mesenchymal stem cells of different origin. Bull Exp Biol Med 2006;141(1):147–51.

[90] Zhang L, Feng G, Wei X, Huang L, Ren A, Dong N, et al. The effects of mesenchymal stem cells in craniofacial tissue engineering. Curr Stem Cell Res Ther 2014;9(3):280–9.

[91] Caria PHF, Kawachi EY, Bertran CA, Camilli JA. Biological assessment of porous-implant hydroxyapatite combined with periosteal grafting in maxillary defects. J Oral Maxillofac Surg 2007;65(5):847–54.

[92] Luby AO, Ranganathan K, Lynn JV, Nelson NS, Donneys A, Buchman SR. Stem cells for bone regeneration: current state and future directions. J Craniofac Surg 2019;30(3):730–5.

[93] Schäffler A, Büchler C. Concise review: adipose tissue-derived stromal cells-basic and clinical implications for novel cell-based therapies. Stem Cell 2007;25(4):818–27.

[94] Davies OG, Cooper PR, Shelton RM, Smith AJ, Scheven BA. A comparison of the in vitro mineralisation and dentinogenic potential of mesenchymal stem cells derived from adipose tissue, bone marrow and dental pulp. J Bone Miner Metab 2015;33(4):371–82.

[95] Kishimoto N, Tran S. Dedifferentiated fat (DFAT) cells: a cell source for oral and maxillofacial tissue engineering. Oral Dis 2018;24(7):42–9.

[96] Sándor GB. Tissue engineering of bone: clinical observations with adipose-derived stem cells, resorbable scaffolds, and growth factors. Ann Maxillofac Surg 2012;2(1):8.

[97] Pettersson LF, Kingham PJ, Wiberg M, Kelk P. In vitro osteogenic differentiation of human mesenchymal stem cells from jawbone compared with dental tissue. Tissue Eng Regen Med 2017;14(6):763–74.

[98] Yasui T, Mabuchi Y, Morikawa S, Onizawa K, Akazawa C, Nakagawa T, et al. Isolation of dental pulp stem cells with high osteogenic potential. Inflamm Regen 2017;37(1):1–10.

[99] Paduano F, Marrelli M, Alom N, Amer M, White LJ, Shakesheff KM, et al. Decellularized bone extracellular matrix and human dental pulp stem cells as a construct for bone regeneration. J Biomater Sci Polym Ed 2017;28(8):730–48.

[100] Lee YC, Chan YH, Hsieh SC, Lew WZ, Feng SW. Comparing the osteogenic potentials and bone regeneration capacities of bone marrow and dental pulp mesenchymal stem cells in a rabbit calvarial bone defect model. Int J Mol Sci 2019;20(20):E5015.

[101] Sancilio S, Gallorini M, Di Nisio C, Marsich E, Di Pietro R, Schweikl H, et al. Alginate/hydroxyapatite-based nanocomposite scaffolds for bone tissue engineering improve dental pulp biomineralization and differentiation. Stem Cell Int 2018;2018:1.

[102] Covarrubias C, Cádiz M, Maureira M, Celhay I, Cuadra F, von Marttens A. Bionanocomposite scaffolds based on chitosan–gelatin and nanodimensional bioactive glass particles: in vitro properties and in vivo bone regeneration. J Biomater Appl 2018;32(9):1155–63.

[103] Sevari SP, Shahnazi S, Chen C, Mitchell JC, Ansari S, Moshaverinia A. Bioactive glass-containing hydrogel delivery system for osteogenic differentiation of human dental pulp stem cells. J Biomed Mater Res Part A 2019;105(11):2957–67.
[104] Tsukamoto J, Naruse K, Nagai Y, Kan S, Nakamura N, Hata M, et al. Efficacy of a self-Assembling peptide hydrogel, spg-178-gel, for bone regeneration and three-dimensional osteogenic induction of dental pulp stem cells. Tissue Eng - Part A 2017;23(23–24):1394–402.
[105] Bakopoulou A, Georgopoulou A, Grivas I, Bekiari C, Prymak O, Loza K, et al. Dental pulp stem cells in chitosan/gelatin scaffolds for enhanced orofacial bone regeneration. Dent Mater 2019;35(2):310–27.
[106] Fu Q, Ren H, Zheng C, Zhuang C, Wu T, Qin J, et al. Improved osteogenic differentiation of human dental pulp stem cells in a layer-by-layer-modified gelatin scaffold. J Biomater Appl 2018;33(4):477–87.
[107] Oliveira NK, Salles THC, Pedroni AC, Miguita L, D'Ávila MA, Marques MM, et al. Osteogenic potential of human dental pulp stem cells cultured onto poly-ε-caprolactone/poly (rotaxane) scaffolds. Dent Mater 2019;35:1740–9.
[108] Wongsupa N, Nuntanaranont T, Kamolmattayakul S, Thuaksuban N. Assessment of bone regeneration of a tissue-engineered bone complex using human dental pulp stem cells/poly(ε-caprolactone)-biphasic calcium phosphate scaffold constructs in rabbit calvarial defects. J Mater Sci Mater Med 2017;28(5):1–13.
[109] Soares IMV, Fernandes GV, de O, Cavalcante LC, Leite YKP, de C, et al. The influence of Aloe vera with mesenchymal stem cells from dental pulp on bone regeneration: characterization and treatment of non-critical defects of the tibia in rats. J Appl Oral Sci 2019;27:1–11.
[110] Chamieh F, Collignon AM, Coyac BR, Lesieur J, Ribes S, Sadoine J, et al. Accelerated craniofacial bone regeneration through dense collagen gel scaffolds seeded with dental pulp stem cells. Sci Rep 2016;6:1–11.
[111] Otero L, Carrillo N, Calvo-Guirado JL, Villamil J, Delgado-Ruíz RA. Osteogenic potential of platelet-rich plasma in dental stem-cell cultures. Br J Oral Maxillofac Surg 2017;55(7):697–702.
[112] Ge X, Li Z, Jing S, Wang Y, Li N, Lu J, et al. Parathyroid hormone enhances the osteo/odontogenic differentiation of dental pulp stem cells via ERK and P38 MAPK pathways. J Cell Physiol 2019;235:1209–21.
[113] Stovall KE, Tran TDN, Suantawee T, Yao S, Gimble JM, Adisakwattana S, et al. Adenosine triphosphate enhances osteoblast differentiation of rat dental pulp stem cells via the PLC–IP 3 pathway and intracellular Ca 2+ signaling. J Cell Physiol 2019;202:1–10.
[114] Wu J, Li N, Fan Y, Wang Y, Gu Y, Li Z, et al. The conditioned medium of calcined tooth powder promotes the osteogenic and odontogenic differentiation of human dental pulp stem cells via MAPK signaling pathways. Stem Cell Int 2019;2019:1–13.
[115] Yuan M, Zhan Y, Hu W, Li Y, Xie X, Miao N, et al. Aspirin promotes osteogenic differentiation of human dental pulp stem cells. Int J Mol Med 2018;42(4):1967–76.
[116] Xin BC, Wu QS, Jin S, Luo AH, Sun DG, Wang F. Berberine promotes osteogenic differentiation of human dental pulp stem cells through activating EGFR-MAPK-Runx2 pathways. Pathol Oncol Res 2019; https://doi.org/10.1007/s12253-019-00746-6.
[117] Xia Y, Chen H, Zhang F, Wang L, Chen B, Reynolds MA, et al. Injectable calcium phosphate scaffold with iron oxide nanoparticles to enhance osteogenesis via dental pulp stem cells. Artif Cells Nanomed Biotechnol 2018;46(Suppl. 1):423–33.
[118] Xia Y, Chen H, Zhang F, Bao C, Weir MD, Reynolds MA, et al. Gold nanoparticles in injectable calcium phosphate cement enhance osteogenic differentiation of human dental pulp stem cells. Nanomed Nanotechnol Biol Med 2018;14(1):35–45.
[119] D'Aquino R, De Rosa A, Lanza V, Tirino V, Laino L, Graziano A, et al. Human mandible bone defect repair by the grafting of dental pulp stem/progenitor cells and collagen sponge biocomplexes. Eur Cell Mater 2009;18:75–83.
[120] Almeida LDF, Babo PS, Silva CR, Rodrigues MT, Hebling J, Reis RL, et al. Hyaluronic acid hydrogels incorporating platelet lysate enhance human pulp cell proliferation and differentiation. J Mater Sci Mater Med 2018;29(6):88.
[121] Silva CR, Babo PS, Gulino M, Costa L, Oliveira JM, Silva-Correia J, et al. Injectable and tunable hyaluronic acid hydrogels releasing chemotactic and angiogenic growth factors for endodontic regeneration. Acta Biomater [Internet] 2018;77:155–71. Available from: <https://linkinghub.elsevier.com/retrieve/pii/S174270611830429X>.
[122] Rouwkema J, De Boer J, Van Blitterswijk CA. Endothelial cells assemble into a 3-dimensional prevascular network in a bone tissue engineering construct. Tissue Eng 2006;12(9):2685–93.

[123] Giannoudis PV, Pountos I. Tissue regeneration: the past, the present and the future. Injury. 2005;36(Suppl. 4):2–5.

[124] Lee Y-M, Park Y-J, Lee S-J, Ku Y, Han S-B, Klokkevold PR, et al. The bone regenerative effect of platelet-derived growth factor-BB delivered with a chitosan/tricalcium phosphate sponge carrier. J Periodontol 2000;71(3):418–24.

[125] Barabaschi GDG, Manoharan V. Engineering mineralized and load bearing tissues. Adv Exp Med Biol 2015;881:79–94.

[126] Yin L, Yang S, He M, Chang Y, Wang K, Zhu Y, et al. Physicochemical and biological characteristics of BMP-2/IGF-1-loaded three-dimensional coaxial electrospun fibrous membranes for bone defect repair. J Mater Sci Mater Med 2017;28(6):1–14.

[127] Hu S, Chen H, Zhou X, Chen G, Hu K, Cheng Y, et al. Thermally induced self-agglomeration 3D scaffolds with BMP-2-loaded core–shell fibers for enhanced osteogenic differentiation of rat adipose-derived stem cells. Int J Nanomed 2018;13:4145–55.

[128] Steinhardt Y, Aslan H, Regev E, Zilberman Y, Kallai I, Gazit D, et al. Maxillofacial-derived stem cells regenerate critical mandibular bone defect. Tissue Eng - Part A 2008;14(11):1763–73.

[129] Chen G, Deng C, Li Y-P. TGF-β and BMP signaling in osteoblast differentiation and bone formation. Int J Biol Sci 2012;8(2):272–88.

[130] Harwood PJ, Giannoudis PV. Application of bone morphogenetic proteins in orthopaedic practice: their efficacy and side effects. Expert Opin Drug Saf 2005;75–89.

[131] Whitman DH, Berry RL, Green DM. Platelet gel: an autologous alternative to fibrin glue with applications in oral and maxillofacial surgery. J Oral Maxillofac Surg 1997;55(11):1294–9.

[132] Marx RE. Platelet-rich plasma: evidence to support its use. J Oral Maxillofac Surg [Internet] 2004;62 (4):489–96. Available from: <http://linkinghub.elsevier.com/retrieve/pii/S0278239103012722>.

[133] Weibrich G, Kleis WKG, Hitzler WE, Hafner G. Comparison of the platelet concentrate collection system with the plasma-rich-in-growth-factors kit to produce platelet-rich plasma: a technical report. Int J Oral Maxillofac Implant 2005;20(1):118–23.

[134] Fu JH, Wang HL. Platelet-rich plasma has no additional benefit during guided tissue regeneration procedure to significantly improve clinical attachment gains in the treatment of periodontal intrabony defects. J Evid Based Dent Pract 2012;12:5–7.

[135] Babo PS, Reis RL, Gomes ME. Periodontal tissue engineering: current strategies and the role of platelet rich hemoderivatives. J Mater Chem B 2017;5(20):3617–28.

[136] Mendes BB, Gómez-Florit M, Babo PS, Domingues RM, Reis RL, Gomes ME. Blood derivatives awaken in regenerative medicine strategies to modulate wound healing. Adv Drug Deliv Rev [Internet] 2017;129:376–93. Available from: <http://linkinghub.elsevier.com/retrieve/pii/S0169409X17303216>.

[137] Amrollahi P, Moghadam F, Tayebi L. Bioreactor design for oral and dental tissue engineering. Biomater Oral Dent Tissue Eng 2017;69:193–204.

[138] Sikavitsas VI, Bancroft GN, Mikos AG. Formation of three-dimensional cell/polymer constructs for bone tissue engineering in a spinner flask and a rotating wall vessel bioreactor. J Biomed Mater Res 2002;62 (1):136–48.

[139] Meinel L, Karageorgiou V, Fajardo R, Snyder B, Shinde-Patil V, Zichner L, et al. Bone tissue engineering using human mesenchymal stem cells: effects of scaffold material and medium flow. Ann Biomed Eng 2004;32(1):112–22.

[140] Edwards LA, Hills F. (12) Patent Application Publication (10) Pub. No.: US 2012/0257061 A1. 2012;1(19).

[141] Poh PSP, Hutmacher DW, Stevens MM, Woodruff MA. Erratum: fabrication and in vitro characterization of bioactive glass composite scaffolds for bone regeneration (Biofabrication (2013) 5 (045005)). Biofabrication 2014;6(2):045005.

[142] Qing-Qing Q, Ducheyne P, Ayyaswamy PS. Fabrication, characterization and evaluation of bioceramic hollow microspheres used as microcarriers for 3-D bone tissue formation in rotating bioreactors. Biomaterials. 1999;20(11):989–1001.

[143] Botchwey EA, Pollack SR, Levine EM, Laurencin CT. Bone tissue engineering in a rotating bioreactor using a microcarrier matrix system. J Biomed Mater Res 2001;55(2):242–53.

[144] Hoffmann W. Novel perfused compression bioreactor system as an in vitro model to investigate fracture healing. Front Bioeng Biotechnol 2015;3:1–6.

[145] Janssen FW, Oostra J, Van Oorschot A, Van Blitterswijk CA. A perfusion bioreactor system capable of producing clinically relevant volumes of tissue-engineered bone: in vivo bone formation showing proof of concept. Biomaterials. 2006;27(3):315–23.

[146] Bancroft GN, Sikavitsas VI, Mikos AG. Design of a flow perfusion bioreactor system for bone tissue-engineering applications. Tissue Eng. 2003;9(3):549–54. Available from: https://doi.org/10.1089/107632703322066723.

[147] Fröhlich M, Grayson WL, Marolt D, Gimble JM, Kregar-Velikonja N, Vunjak-Novakovic G. Bone grafts engineered from human adipose-derived stem cells in perfusion bioreactor culture. Tissue Eng Part A [Internet] 2009;16(1):179–89. Available from: https://doi.org/10.1089/ten.tea.2009.0164.

[148] Frohlich M. Bone grafts engineered from human adipose-derived. Tissue Eng Part A 2010;16(1):179–89.

[149] Costa PF, Vaquette C, Baldwin J, Chhaya M, Gomes ME, Reis RL, et al. Biofabrication of customized bone grafts by combination of additive manufacturing and bioreactor knowhow. Biofabrication. 2014;6(3):035006.

[150] Chase LCG, Rao MSG, Vemuri MS. Mesenchymal stem cell assays and applications. Methods Mol Biol. 2011;698(5):534.

[151] Stevens MM, Marini RP, Schaefer D, Aronson J, Langer R, Shastri VP. In vivo engineering of organs: the bone bioreactor. Proc Natl Acad Sci U S A 2005;102(32):11450–5.

[152] Chai F, Abdelkarim M, Laurent T, Tabary N, Degoutin S, Simon N, et al. Poly-cyclodextrin functionalized porous bioceramics for local chemotherapy and anticancer bone reconstruction. J Biomed Mater Res Part B Appl Biomater [Internet] 2014;102(6):1130–9. Available from: http://doi.wiley.com/10.1002/jbm.b.33094.

[153] Cheng L, He W, Gong H, Wang C, Chen Q, Cheng Z, et al. PEGylated micelle nanoparticles encapsulating a non-fluorescent near-infrared organic dye as a safe and highly-effective photothermal agent for in vivo cancer therapy. Adv Funct Mater [Internet] 2013;23(47):5893–902. Available from: http://doi.wiley.com/10.1002/adfm.201301045.

[154] Li JL, Day D, Gu M. Ultra-low energy threshold for cancer photothermal therapy using transferrin-conjugated gold nanorods. Adv Mater [Internet] 2008;20(20):3866–71. Available from: http://doi.wiley.com/10.1002/adma.200800941.

[155] Wang X, Li T, Ma H, Zhai D, Jiang C, Chang J, et al. A 3D-printed scaffold with MoS2 nanosheets for tumor therapy and tissue regeneration. NPG Asia Mater [Internet] 2017;9(4):e376. Available from: <http://www.nature.com/articles/am201747>.

[156] Ma H, Jiang C, Zhai D, Luo Y, Chen Y, Lv F, et al. A bifunctional biomaterial with photothermal effect for tumor therapy and bone regeneration. Adv Funct Mater [Internet] 2016;26(8):1197–208. Available from: http://doi.wiley.com/10.1002/adfm.201504142.

[157] Zhang J, Zhao S, Zhu M, Zhu Y, Zhang Y, Liu Z, et al. 3D-printed magnetic Fe 3 O 4 /MBG/PCL composite scaffolds with multifunctionality of bone regeneration, local anticancer drug delivery and hyperthermia. J Mater Chem B [Internet] 2014;2(43):7583–95. Available from: <http://xlink.rsc.org/?DOI = C4TB01063A>.

[158] Hustedt JW, Blizzard DJ. The controversy surrounding bone morphogenetic proteins in the spine: a review of current research. Yale J Biol Med [Internet] 2014;87(4):549–61. Available from: <https://www.ncbi.nlm.nih.gov/pubmed/25506287>.

[159] Tao O, Kort-Mascort J, Lin Y, Pham HM, Charbonneau AM, ElKashty OA, et al. The applications of 3D printing for craniofacial tissue engineering. Micromachines. 2019;10(7):E480.

[160] Costa-Almeida R, Franco AR, Pesqueira T, Oliveira MB, Babo PS, Leonor IB, et al. The effects of platelet lysate patches on the activity of tendon-derived cells. Acta Biomater [Internet] 2018;68:29–40. Available from: <http://linkinghub.elsevier.com/retrieve/pii/S1742706118300175>.

[161] Cortini M, Baldini N, Avnet S. New advances in the study of bone tumors: a lesson from the 3D environment. Front Physiol [Internet] 2019;10:915. Available from: <https://www.frontiersin.org/article/10.3389/fphys.2019.00814/full>.

[162] Cavo M, Caria M, Pulsoni I, Beltrame F, Fato M, Scaglione S. A new cell-laden 3D alginate-matrigel hydrogel resembles human breast cancer cell malignant morphology, spread and invasion capability observed "in vivo. Sci Rep [Internet] 2018;8(1):5333. Available from: <http://www.nature.com/articles/s41598-018-23250-4>.

[163] Liu C, Lewin Mejia D, Chiang B, Luker KE, Luker GD. Hybrid collagen alginate hydrogel as a platform for 3D tumor spheroid invasion. Acta Biomater [Internet] 2018;75:213–25. Available from: <https://linkinghub.elsevier.com/retrieve/pii/S1742706118303398>.

[164] Mohammed MI, Fitzpatrick AP, Gibson I. Customised design of a patient specific 3D printed whole mandible implant. KnE Eng 2017;2(2):104.

[165] Aldaadaa A, Owji N, Knowles J. Three-dimensional printing in maxillofacial surgery: hype versus reality. J Tissue Eng 2018;9:1–5.

[166] Rumiński S, Kalaszczyńska I, Długosz A, Lewandowska-Szumieł M. Osteogenic differentiation of human adipose-derived stem cells in 3D conditions - comparison of spheroids and polystyrene scaffolds. Eur Cell Mater 2019;37:382–401.

[167] Mizuno H, Kagami H, Mase J, Mizuno D, Ueda M. Efficacy of membranous cultured periosteum for the treatment of patients with severe periodontitis: a proof-of-concept study. Nagoya J Med Sci 2010;72 (1–2):59–70.

[168] Yamada Y, Nakamura S, Ito K, Kohgo T, Hibi H, Nagasaka T, et al. Injectable tissue-engineered bone using autogenous bone marrow-derived stromal cells for maxillary sinus augmentation: clinical application report from a 2-6-year follow-up. Tissue Eng - Part A 2008;14(10):1699–707.

# Index

*Note*: Page numbers followed by "*f*" and "*t*" refer to figures and tables, respectively.

## B

## D

## E

## H

## I

## Q



## R

## Q



## R

## T

## U



## V

## W



## X



## Y



## Z

CPI Antony Rowe
Eastbourne, UK
September 04, 2020